Anatomy for Surgeons: Volume 3 **THE BACK AND LIMBS**

Anatomy for Surgeons:
Volume 3

THE
BACK
AND
LIMBS *Third Edition*

W. Henry Hollinshead, Ph.D.

Professor Emeritus of Anatomy,
Mayo Graduate School of Medicine
(University of Minnesota),
Rochester, Minnesota;
Visiting Professor Emeritus of Anatomy,
University of North Carolina
School of Medicine,
Chapel Hill, North Carolina

With 636 Illustrations
125 with Color

Harper & Row, Publishers
PHILADELPHIA

Cambridge   London
New York   Mexico City
Hagerstown   São Paulo
San Francisco   Sydney

1817

ANATOMY FOR SURGEONS: VOLUME 3, The Back and Limbs,
third edition.
Copyright © 1982 by Harper & Row, Publishers, Inc.
Copyright © 1969, 1958, by Harper & Row, Publishers, Inc.
All rights reserved. No part of this book
may be used or reproduced in any manner whatsoever
without written permission except in the case of brief quotations
embodied in critical articles and reviews.
Printed in the United States of America.
For information address Harper & Row Publishers, Inc.,
East Washington Square, Philadelphia, Pennsylvania 19105.

3   5   6   4   2

Library of Congress Cataloging in Publication Data

Hollinshead, W. Henry (William Henry), date.
    Anatomy for surgeons.
    Includes bibliographical references and index.
    Contents: —v. 3. The back and limbs.
    1. Anatomy, Surgical and topographical. I. Title.
[DNLM: 1. Anatomy.   2. Surgery.   WO 101 H741a]
QM531.H692   1982   611′.9   81-6909
ISBN 0-06-141266-X (v. 3)   AACR2

Contents

Preface

The third edition of Volume 3 of *Anatomy for Surgeons* attempts to present, as did the first and second editions, anatomic facts and concepts concerning the back and limbs that are of particular interest to the surgeon. It is not intended to be a complete descriptive anatomy of these parts, but is designed to serve both as a ready reference in which the surgeon can find descriptions of the basic anatomy and as a useful review of numerous, sometimes minute, anatomic details described in many other sources. No attempt has been made to describe the indications for, or detailed technique of, specific operations, for these are matters that belong to surgery and not to anatomy. Particular care has been taken throughout this volume, however, to relate anatomic and physiologic details and concepts to underlying surgical procedures.

Most of the anatomy described in the second edition is still both valid and pertinent, for basic anatomy does not change. Our concepts do, however. As a simple example, compare the teaching that the tibia is the sole weight-bearing bone of the leg with the finding that the fibula bears about one fifth of the weight. In sum, our understanding of anatomic details has contributed to, as well as benefited from, new developments in diagnostic and operative procedures on the back and limbs to such an extent as to make a revision of this book necessary. Some of the basic descriptions have been carefully scrutinized, rearranged, or partly rewritten in the interest of greater clarity or accuracy, and numerous minor revisions have been made to incorporate more recent findings and clinical applications.

At the time the second edition was prepared there was a considerable body of evidence on the actions of the muscles as revealed by electromyography. Since that time many additional studies have appeared, and the results of these, some of which are not in accord with previous concepts, appear in almost every chapter. In addition, there is new information on congenital amputations, tennis elbow, various nerve entrapments in both limbs, anatomy and surgery of the flexor tendons and sheaths in the hand, the extensor tendon apparatus of both hand and foot, the knee and its ligaments, and many minor points too numerous to mention.

The Nomina Anatomica (NA), which was adopted in 1955, was used in both preceding editions of this book and is used here. Because many clinicians may be more familiar with the Basel Nomina Anatomica (BNA), however, I have also attempted to include the more commonly used BNA synonyms at least once if they differ appreciably from those of the NA.

I remain deeply indebted to those former colleagues at the Mayo Clinic who were so helpful in the two previous editions. They are listed in each of those editions, and I regret that I could not avail myself of their counsel in the current one. I am indebted also to W. B. Saunders Company for allowing me to continue to use figures from my *Functional Anatomy of the Limbs and Back*. To these, and especially to my publisher, Harper & Row, my thanks.

W. Henry Hollinshead
Chapel Hill, North Carolina

Anatomy for Surgeons: Volume 3 **THE BACK AND LIMBS**

Chapter 1
SOME GENERAL CONSIDERATIONS

Because this volume deals largely with bones, muscles, nerves, and blood vessels, it seems appropriate to consider the general facts and principles that are equally applicable to these structures in whatever region they occur.

Bone, Cartilage, and Joints

BONE

Bone regularly occurs in two forms, compact and spongy (cancellous). Compact bone forms the surface of all bones, and the major part of the body or shaft of long bones; spongy bone occupies the ends of long bones, and permeates the bodies of the short and flat bones. The large medullary cavity in the body of a typical long bone is, in the adult, occupied by yellow marrow. The much subdivided medullary cavity in spongy bone is occupied by red marrow, which produces the granular leukocytes and the red blood cells.

Compact or cortical bone contains the Haversian canals, longitudinally running channels for the accommodation of blood vessels, about which the bone is laid down in concentric lamellae that are now usually called *osteons* or *osteones;* spongy bone also consists of lamellae, but these are in flat branching plates rather than concentric circles.

The articular surfaces at synovial joints are covered, outside the thin plate of cortical bone, by hyaline cartilage, except in a very few joints, such as the sternoclavicular and the temporomandibular, where fibrocartilage is present instead. The remainder of the bone is covered externally by the periosteum, a layer of specialized connective tissue that is bound firmly to the bone by some of its fibers, which enter the bone as Sharpey's fibers. The periosteum contains blood vessels that connect with those in the bone. The outer layer of the periosteum is denser, and contains the periosteal blood vessels, the inner layer is looser and contains, in adults, fibroblasts. The fibroblasts can, under the proper conditions, proliferate and form osteoblasts for the reconstruction of cortical bone. The endosteum is a thinner layer of connective tissue lining the bone where it abuts on the marrow cavity; it also contains cells that are capable of forming bone.

Trabecular Structure and Mechanics

The structure and the mechanical properties of bone have been the subject of numerous investigations. "Wolff's law" states essentially that every bone is constructed in such a fashion as to allow it to resist the forces applied to it; if the direction of the forces changes, there is a corresponding change in the structure of the bone. Koch's mathematical analysis of the compressive and tensile stresses in the femur led him to conclude that this rather complex

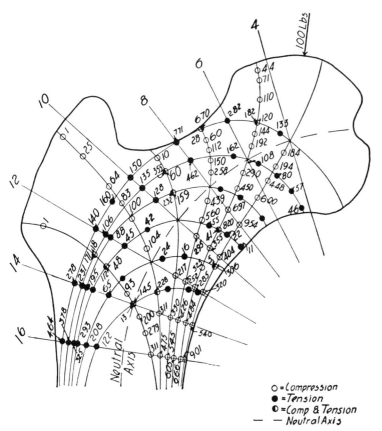

Fig. 1-1. Koch's calculations of the maximal tensile and compressive stresses in the upper end of the femur when there is a load of 100 pounds on the femoral head. The *numbered lines* extending across the femur represent the levels of the cross sections that he studied for the bony architecture. (Koch JC: Am J Anat 21:177, 1917)

bone accords strictly in structure to the best engineering principles, namely, maximal strength with a minimal use of material. Comparing the calculated lines of tension and compression in the head and neck of the femur with the actual arrangement of the trabeculae here, he apparently showed that the trabeculae are arranged in exactly the pattern demanded by his mathematical analysis (Fig. 1-1 and 1-2).

In general, the trabeculae arising from the lateral side of the femur and arching medially correspond to the calculated tensile stresses (*i.e.*, the internal force in the bone that tends to keep two adjacent planes from being pulled apart). The trabeculae arising from the medial side of the femur and arching up-

ward and laterally correspond to the lines of compressive stress. That is, they resist the compressive forces brought about by a load on the head of the femur. Similarly, Koch showed that the absence of trabeculae in most of the body is also according to mathematical and engineering principles; the stresses borne by this part are most economically cared for by a tube such as the cortical bone here. The trabeculae in the lower part of the femur are arranged in accordance with the stresses there.

The principles enunciated by Koch are now generally accepted. They explain the alteration in structure when there are abnormal stresses, such as with a valgus or varus deformity of the femur (*e.g.*, Tobin). They are also in accord with the observed deformation and fracture of bone under conditions of loading, as reported especially by Evans and co-workers in numerous papers (Evans, '57).

The physical properties of bone vary somewhat. Koch quoted the tensile strength (the resistance to being pulled apart) of bone along its long axis is varying from about 13,-200 lb/in^2 to about 17,700 lb/in^2, its compressive strength (resistance to being crumbled) along its long axis as varying from about 18,000 lb/in^2 to 24,700 lb/in^2. These figures may be compared with a tensile strength of 65,000 lb/in^2, and a compressive

one of 60,000 lb/in^2 for medium steel; a tensile strength of 28,000 lb and a compressive one of 42,000 lb for copper; a tensile strength of 1,500 lb and a compressive one of 15,000 lb for granite; and a tensile strength of 12,500 lb and a compressive one of 7,000 lb for white oak when the load parallels the grain, but both tensile and compressive strengths of only about 2,000 lb/in^2 when the load is at right angles to the grain. Koch pointed out that the tensile strength of bone is regularly less than its compressive strength, a finding substantiated by the observation (Evans) that fractures of the femur regularly originate on the

Fig. 1-2. A thin frontal section of the femur to show the trabecular structure. Note the correspondence between the trabeculae here and the lines of tension and compression in Fig. 1-1. (Koch JC: Am J Anat 21:177, 1917)

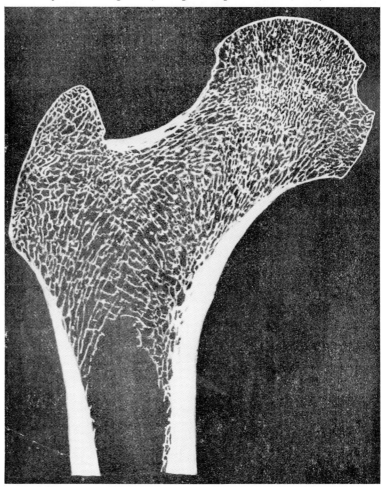

convexity of a femur distorted by a load or a blow and therefore represent a failure under tensile, not compressive, strain.

Evans and Lebow restudied, by engineering technics, the tensile strength of bone as a whole, and found that it varies not only from bone to bone but within different parts of the same bone. Thus, the middle third of the femur was found to have a greater tensile strength than the proximal and distal thirds, and the middle third of the tibia had a greater tensile strength than did the middle third of the femur. The tensile strength of wet specimens, which more nearly approximates living bone, was found to be considerably less than that of dried specimens, concerning which figures are usually quoted. The average tensile strength of the middle third of wet specimens of the femur was only a little more than 12,000 lb/in^2, while that of dried specimens exceeded 16,000 lb. In contrast, however, wet specimens elongated under tension much farther than did dried bone (*i.e.*, were able to absorb a greater amount of energy). Ascenzi and Bonucci found these differences to hold also for individual osteons. They added that the tensile strength varied according to the arrangement of the collagen fibers in successive laminae of an osteon and was greatest when a longitudinal direction of the fibers predominated. Evans and Bang reported that the osteons apparently contribute most to hardness of the bone, and the interstitial lamellae contribute most to tensile strength. Weaver and Chalmers reported that the compressive strength of the trabeculae of cancellous bone bears a constant relationship to the mineral content and that both apparently increase to approximately the age of 30; they also reported suggestive evidence that both may increase in response to stress.

Constituents

Cortical bone contains about 25% to 30% water (Evans and Lebow). Of the dry weight about 60% to 70% is mineral, apparently tiny crystals of $Ca_{10}(PO_4)_6(OH)_2$ (an hydroxyapatite), with carbonate and citrate bound to their surfaces (McLean and Urist). The fibrous connective tissue, collagen or ossein (os-

tein), constitutes about 30% to 40% of the interstitial substance. In contrast to the usual statement that the relative amount of collagen is greatest in young individuals and decreases with age (thus accounting for the occurrence of "greenstick" fractures in the young and increasing brittleness in the elderly), Mueller and co-workers found the organic fraction remaining constant. They said the change is in water content and mineralization, the latter increasing as the former decreases from birth to the sixth or seventh decade. The loss of calcium salts as in rickets and osteomalacia is, of course, of great importance clinically. In osteoporosis the mineral content is normal, but the amount of bony tissue is reduced (Mueller and co-workers), apparently as a result of increased resorption of bone (Jowsey and co-workers). There is also, apparently, a normal diminution of cortical bone with age. Although the sizes of the osteons remain constant, the sizes of the Haversian canals increase in the cortex of the femur, and the cortex of the ribs decreases in thickness (Jowsey).

The collagenous fibers of the osteon surrounding a Haversian canal vary from an almost longitudinal to an almost circular arrangement around the canal, and in many osteons the direction of the fibers alternate from one lamella to the next (see, *e.g.*, Cooper and co-workers). This type of osteon, however, has less tensile strength than the one in which the majority of fibers are longitudinally arranged (Ascenzi and Bonucci). The alternating arrangement, from essentially longitudinal to essentially circular, tends to be true also of the lamellae forming the external surface of cortical bone (Smith, '60).

Blood and Nerve Supply

The blood supply to bones varies according to the shape of the bone. In long bones, however, there are generally three sets of vessels: one or more nutrient arteries, accompanied by paired veins; periosteal vessels; and metaphyseal and epiphyseal vessels that penetrate the ends of the bones. The nutrient artery of most long bones is of somewhat variable origin when there are two or more associated vessels that might give rise to it, and may be multiple instead of single. Thus the concept that "the" nutrient artery is the first vessel to invade the cartilage matrix of the forming bone, and hence, that the line of intersection of the nutrient canal with the center of the bone marrow indicates the original center of the bone, from which the amount of growth at either end can be calculated, is not now generally adhered to (*e.g.,* Hendryson).

The nutrient artery or arteries, with their accompanying veins, usually penetrate the cortex obliquely (their direction, whether distal or proximal, indicating the end at which the lesser growth in length of the bone has occurred) and branch to run toward both ends of the bone; they supply the greater part of the marrow, and they anastomose with the metaphyseal arteries at the ends of the bones. The blood flow to the cortex is then derived from the medullary vessels (*e.g.,* Nelson and co-workers, Brookes and co-workers) through radially directed branches that enter the cortex and quickly divide into small vessels that run in the Haversian canals. Most of the vessels in the Haversian canals are capillaries or venules, although an occasional canal contains an arteriole; venous drainage from the ends of a long bone is largely into veins accompanying the epiphyseal and metaphyseal arteries, but that from the body of the bone is mostly toward the medullary sinusoids (Nelson and co-workers).

It has long been known that the cortical vessels are connected to periosteal vessels, and various experiments have shown that the periosteal vessels can supply cortical bone. Nelson and co-workers found no arteries entering the cortex from the periosteum, however, and Brookes and co-workers, also noting that connections to periosteal vessels are capillary ones, emphasized that blood flow is normally from the cortex to the periosteum (and also into the muscles attached to the bone) and that reverse flow occurs only when there is ischemia of the bone. Even in the fetus the communication between the cortical and periosteal vessels is through capillaries (Brookes).

The possible importance of the periosteal blood vessels is indicated by the reports that interruption of the main nutrient vessels to the femur of rabbits produces primarily necrosis of the central portion of the marrow (Huggins and Wiege), while if the periosteum is also stripped from most of the femur at the same time, infarction of the entire thickness of the femur throughout about its middle third ensues (Foster, Kelly, and Watts). Thus, the vessels entering the ends of the bones are apparently not capable, without the aid of the reversed flow from the periosteum, of supporting the entire cortex.

In growing long bones, where the metaphyseal and epiphyseal arteries are separated by the epiphyseal cartilage, it is apparently the epiphyseal vessels that are responsible for growth of the epiphyseal cartilage, but it is the metaphyseal ones that are responsible for the transformation of the cartilage into bone (Kistler; Trueta and Amato).

Lymphatics are present in the periosteum, and have been said also to accompany blood vessels into the bone, but little seems to be known about them.

Nerve fibers also have been traced into bone, along the blood vessels. Kuntz and Richins traced nerve fibers into the bone marrow, and found that for the most part they remain related to the blood vessels; through degeneration experiments, they showed that some of them were afferent and some were sympathetic. According to Duncan and Shim, adrenergic fibers enter the bone on the surface of the nutrient artery, and end on the outer surface of the media of smaller vessels. Miller and Kasahara found fibers ending in association with arterioles, the endosteum, and the deep surface of the articular cartilage, but could not confirm reports that nerve fibers extend also into the lamellae of cortical bone.

The concept that there may be nerve fibers in the bone that have a trophic function, in some way governing the growth and repair of bone, is apparently negated by the experiments of Corbin and Hinsey; they destroyed the sympathetic and afferent innervation to one hind limb of a number of cats, and compared the denervated and normally innervated bones and joints at periods ranging from 2 weeks to 3½ years. In the animals whose movements were restricted, no changes at all were found; in those that were allowed to run freely in large cages there was trauma of the anesthetic hip joint, but there were no other changes except those attributable to this trauma. Ring later agreed that sympathectomy has no effect, but reported that denervating the muscles of the leg produced at least a temporary increase in the growth of the length of the tibia. This has, of course, nothing to do with the innervation of bone; Ring reported that tenotomy of all the muscles of the leg has the same effect.

GROWTH AND REPAIR

With the possible exception of the clavicle, which has a peculiar ossification, all the bones of the limbs and vertebral column are preformed in cartilage. Erosion of cartilage through the ingrowth of blood vessels tends to occur at about the middle of the cartilaginous

Fig. 1-3. Schema of the ossification and growth of a long bone. *White* represents cartilage; *stipple,* spongy (endochondral) bone; *black,* compact (perichondral) bone. *A* is the cartilaginous stage; in *B* and *C* both endochondral and perichondral bone appear and increase; in *D* the epiphyseal centers have appeared, in *E* the epiphyses have reached their full growth, and in *F* they have joined the body. In the last two stages the marrow cavity (*light stipple*) appears and spreads through resorption of spongy bone. (Arey LB: Developmental Anatomy [ed 6]. Philadelphia, WB Saunders, 1954)

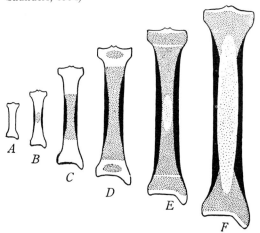

mass, and bone laid down where the cartilage has been eroded establishes the first center of ossification. In the case of long bones the first center of osssification is for the body or shaft (diaphysis) Fig. 1-3; for the various vertebrae, three primary centers appear approximately simultaneously, one (or two uniting quickly into one) for the body and two for the arch. Centers of ossification grow by a continuation of the process through which they arose, erosion of cartilage and replacement by bone; as the bony center for the body of a long bone spreads it soon comes to replace the entire cartilaginous thickness of the body and thereafter growth in diameter of the bone can occur only through the activity of the osteoblasts associated with the periosteum.

Centers of Ossification

Each carpal bone normally ossifies entirely from a single center; so does each tarsal, except for the calcaneus, which normally develops an epiphysis at its posterior end. In the vertebrae, epiphyseal centers appear at the tips of the spinous and transverse processes. The larger long bones develop two epiphyseal centers of ossification (Fig. 1-3), one in each cartilaginous end of the developing bone. The shorter long bones (the metacarpals and metatarsals, and the phalanges of both the hand and foot) typically develop only single epiphyseal centers, which are in the distal ends of the second to fifth metacarpals and metatarsals, in the proximal ends of the first metacarpal and the first metatarsal, and in the proximal ends of all phalanges. Finally, additional epiphyseal centers appear in the cartilage of the ends of some long bones (for instance, centers for the several parts of the distal end of the humerus). As the epiphyseal centers expand, most of the cartilage at the ends of a long bone is also replaced, but a thin layer of cartilage remains over the articular surface of the end of the bone, and a plate of cartilage, the epiphyseal cartilage or epiphyseal plate, persists for a time (it may be years) between the ossified epiphysis and the ossified shaft.

The details of ossification of the various bones are best considered in connection with the regions in which they occur. In general, however, centers for ossification of the long bones appear during the seventh and eighth weeks of fetal life (that of the clavicle, the first to appear, is usually said to be recognizable at 5 weeks), and so do the primary centers for the vertebrae except for the coccygeal ones; by the time of birth the major part of each long bone, the body, is completely ossified. In contrast, the single center of ossification for most of the carpals and tarsals appears only shortly before or, more commonly, after birth, and so do the centers of ossification for the epiphyses, and for the coccygeal segments of the vertebral column; these parts, then, are entirely cartilaginous at birth or have only a small nucleus of ossification in the relatively large cartilaginous mass.

The dates of appearance of the centers of ossification of the carpals and tarsals vary markedly with the bone. For instance, a center is recognizable in the calcaneus at about the sixth month of prenatal life, and centers for all the tarsals appear during approximately the first 4 years of postnatal life, while those for some of the carpals may appear slightly later. That of the pisiform, an extreme example, usually is not visible until the ninth or tenth year in females or the thirteenth to fourteenth year in males (Paterson). Similarly, the dates of first appearance of the epiphyseal centers vary even more markedly according to which one is being considered: centers for the head of the humerus, the lower end of the femur, and the upper end of the tibia are either present at birth or appear shortly thereafter; the epiphyseal center of the olecranon does not appear until about the age of 11, and the epiphyses of the vertebrae typically appear between the ages of 15 and 20.

According to Hill the appearance of ossification centers from about the second month of prenatal life onward is rather accurately correlated with the age of a healthy fetus, sufficiently so that the age can be calculated from the appearance of the centers. Hill, Noback and Robertson, Francis and his co-workers ('39, '40), Flecker ('32, '42), Davies and Parsons, Paterson and others have provided extensive data about the time of appearance of the various centers of ossification, and the time of fusion of epiphyseal centers with the body. Pryor ('36a and b) emphasized the bilateral symmetry of ossification in infants and children, and the fact that identical twins and triplets show similar small anomalies in ossification (of the hands), while there are no more similarities between nonidentical twins than between parents and children.

Pryor ('28) stressed the fact that the carpal bones ossify in the female sooner than they do in the male, and that earlier ossification in general is typical of the female child; Paterson found earlier ossification, not only in regard to the carpals and tarsals but also to the dates of appearance of all the epiphyses and the dates of their union to the body, to be typical of the female. Hill noted that even during the last 3 months of prenatal life the female, as judged by the centers of ossification, matures somewhat more rapidly than does the male. In general, however, the earlier a center of ossification appears, the less likely is there to be any difference between its time of appearance in the male and the female. The later it appears, the greater the gap between the two sexes in the time of appearance. In postnatal life, not only are the dates of appearance of the epiphyses in the female up to 6 months earlier than those of similar epiphyses in the male, but fusion of the epiphyses also typically occurs earlier in the female than in the male; the difference is usually a year or more, and in the case of certain epiphyses may be as much as 5 years (Paterson). As a general rule, the earlier a center of ossification appears in one of the two epiphyses of a long bone, the later this epiphysis is united to the body.

Growth
Once the body of a long bone is ossified, as it generally is long before birth, growth in diameter of the bone necessarily involves an entirely different method, for there is no further peripheral cartilage to be replaced. Therefore, growth in diameter of a long bone is primarily periosteal, with successive layers of bone being laid down on the periphery by the periosteum, while the inner layers of bone, adjacent to the medullary cavity, are being

constantly resorbed in order to enlarge the medullary cavity, and reorganized into Haversian systems or osteons around the blood vessels. According to LeBlond and co-workers, in later stages of growth the endosteum also contributes to bone formation in the shaft near the epiphyses, this contribution accounting for the widening of the cortex and the narrowing of the marrow cavity at these levels. Epker and Frost, taking advantage of the fact that tetracycline is deposited in newly formed bone, reported that periosteal growth may occur at almost any age beyond the growth period.

In contrast, increase in length of the body is endochondral, as was the original development of the bone. Each epiphyseal plate goes through a constant cycle of proliferation, calcification, and absorption of cartilage, with replacement by bone, on the side adjacent to the body. Trueta and Little suggested that one factor in calcification is a too great separation of the deep layers of the plate from the nourishing epiphyseal vessels. It has long been known that one end of any long bone with two epiphyses grows much greater in length than does the other one, and this apparently accounts for the obliquity of the nutrient canal, which almost always points toward the slower growing end (unexplained are occasional cases, quoted by Mysorekar as occurring in about 1% of femurs and found by him in 5% of fibulas, in which the canal points toward the faster growing end); similarly, where only one epiphysis exists, as in the metacarpals, the nutrient canal is slanted away from this epiphysis. Among the large long bones of the limbs, the humerus grows more at its proximal end, and so do the tibia and the fibula. The radius, the ulna, and the femur grow more at their distal ends.

Payton ('32) emphasized that the greater growth at one end of a long bone is not merely a function of time, but depends primarily upon a faster rate of growth at this end. In experiments on pigs, he found that there is, in general, a gradual decrease in the rate of growth of every bone as the animal becomes older, but that, while the rate as a whole decreases, the decrease in rate is less at the faster

growing end than at the slower one. In young pigs, the faster growing end grows about twice as fast as the slower growing one, while in older pigs it grows abut three times as fast.

Growth in diameter of an epiphysis is periosteal, as is that of the body, and growth in length of the epiphysis is endochondral, as is that of the body. (Similarly, growth in diameter of the epiphyseal and articular cartilage seems to be by apposition from the perichondrium—Soloman.) It is now fully accepted, as demonstrated especially by Payton ('32, '33) by feeding madder to pigs, that all new bone formation from the epiphyseal cartilage adds only to the length of the body, and that it is growth and replacement of the articular cartilage only that adds to the length of the epiphysis. (Why the articular cartilage, unlike the epiphyseal one, is not finally replaced entirely by bone is not known. McKibbin and Holdsworth, '67, reported an interesting experiment in which they reversed a piece of articular cartilage, placing its articular surface against endochondral bone; it continued to grow, but there was no ossification at this surface.)

Payton found that the epiphysis at the most rapidly growing end of the bone also grows most rapidly, but there is no correlation between the total length an epiphysis attains and the rapidity of its growth. Further, the epiphysis does not lengthen as rapidly as would be expected, for while bone is being added by the articular cartilage, absorption of bone simultaneously occurs at the junction of the bony epiphysis with the epiphyseal cartilage. Thus, two epiphyses with approximately the same rate of growth may attain different lengths. Some epiphyses (of the pig) may increase in length only 1 mm to 3 mm in more than 500 days, while others, growing at about the same rate, increase 13 mm or more in length.

It has, of course, long been known that hormones, especially those of the anterior lobe of the hypophysis (pituitary gland), of the thyroid gland, and of the gonads strongly influence the growth and maturity of the skeletal system. Without an attempt being made to analyze these effects in any detail, they can be summarized by the reminder that both the growth hormone of the anterior lobe of the hypophysis and thyroxin are necessary for proper skeletal growth, and that lack of either of these hormones inhibits growth.

As studied in the guinea pig, the exact response of cartilage and bone to hormones varies somewhat with the age of the animal (Silberberg and Silberberg, '40). According to Ray and co-workers ('42), growth hormones injected into an otherwise normal animal not only increase the rate of growth of the cartilage, but also increase the rate of osteogenesis; however, pituitary growth hormone injected into animals that have been thyroidectomized and parathyroidectomized in the first few days of life acts differently. Ray and co-workers ('50) reported that under these conditions the growth hormone causes marked increases in the dimensions of the skeleton without evidence of further skeletal maturation. Thyroxin stimulates ossification (Silberberg and Silberberg; Ray and co-workers, '50), and the effects of maturation of the skeleton produced by the anterior lobe of the hypophysis are therefore apparently brought about through the influence of this lobe on the thyroid gland. The optimal balance between dimensional growth and maturation of the skeleton is therefore normally brought about through proper balance between the activity of the anterior lobe of the hypophysis and the thyroid gland.

Sex hormones, in contrast, promote changes in the epiphyseal cartilages. Estrogen, for instance, inhibits the differentiation of epiphyseal cartilage (Silberberg and Silberberg, '41); the earlier union of epiphyses in the female thus seems to be a consequence of the earlier sexual maturity of the female. Similarly, it is well established clinically that delayed epiphyseal union is a common consequence of dysplasia of the gonads, and that early fusion of the epiphyses is associated with sexual precocity.

Histochemical aspects of the development of bone have been studied by, among others, Bevelander and Johnson, and Heller-Steinberg; of significance is the finding of McLean and Bloom and of Bloom and Bloom that under optimal conditions during both early development and later growth, bone matrix is calcified as it is laid down, and thus there is no osteoid (*i.e.*, uncalcified bone matrix) under ideal physiological conditions. They related the appearance of osteoid to a failure in the local supply of the proper minerals.

Attempts to Alter Bone Growth

Disparity in the rate of growth of the limbs, leading to an ultimate discrepancy in lengths, is particularly crippling when it occurs in a lower limb, and its occurrence here, especially, has led to attempts to alter the rate of growth from the epiphyses. Of the disease conditions that may interfere with growth of a limb, a common one is poliomyelitis; White reported that in his series this accounted for 60% of the cases of delayed growth. The exact cause of the slowed rate of growth in such a disease as poliomyelitis is not known, and indeed, there may be multiple causes; however, one of the causes is felt to be the restraining influence of fibrotic muscles, which fail to grow with growth of the bone as do normal muscles. Trueta and Trias reported that disturbance of growth resulting from abnormal pressure on epiphyseal cartilage is probably caused by interference with the blood supply. Regardless of the manner in which compression affects growth, tension has too slight an effect to be clinically significant (Porter).

Because of evidence that decreased vascularity may be a contributing factor to abnormally slow growth of bones, attempts have been made experimentally or clinically to increase the rate of growth. Thus Janes and Elkins, taking advantage of the fact that an arteriovenous fistula produces development of the collateral circulation, reported that the intentional production of such a fistula apparently did result in a decrease in the leg-length discrepancy in one case. Janes has since then created femoral arteriovenous fistulas in a number of children (Jennings and Janes). Kelly and co-workers reported that the procedure in dogs increases both the cortical and the periosteal blood supply, and Vanderhoeft and co-workers analyzed further its effect on the deposition of bone. More recently, Petty and co-workers reported, on the basis of their experience with 28 patients, that while the fistula did on the average reduce the discrepancy, the effect was unpredictable and there was a high rate of complications.

Attempts to increase vascularization and growth of bone by other means apparently have not been notably successful. Sola and co-workers reported that periosteal stripping, especially if done a second time, increased longitudinal growth, but they said the increase in monkeys was slight, and in half of the animals pathological fractures of the femur or tibia occurred. Just-Viera and Yeager reported variable results, with a maximal gain of 2.7% in dogs in which they attempted to produce venous stasis by resection of the veins of the thigh. Keck and Kelly found no increased growth resulting from venous stasis produced by litigation. Barr and co-workers expressed the opinion that vasodilation induced by sympathectomy does, in some cases, have a stimulating effect upon the growth of the shorter extremity of a patient with poliomyelitis. They said, however, that it was not very marked, and can be used, if at all, only in the treatment of minor discrepancies.

Another method of keeping discrepancies to a minimum is by slowing the rate of growth of the normal limb. It is, of course, known that large doses of roentgen rays produce retardation of growth; however, White and others have regarded the margin of safety between a dose that will merely retard the growth of the epiphyseal cartilage and one that will cause permanent damage to the cartilage or to the adjacent joint and the soft tissues as being too little to make clinical use of roentgen rays advisable. Since a resected epiphyseal cartilage is not replaced, the preferred method for permanently arresting growth is to remove one of the two epiphyseal cartilages of a long bone (the Phemister technic). In the case of the lower limb, White destroyed the lower cartilage in the femur, the upper in the tibia. These are the cartilages from which the most growth takes place. Such operations demand careful calculations as to what may be the ultimate expected discrepancy in

length between the two limbs, and of the contribution than can be expected to be made by the normal epiphyses, so that the destruction of cartilage can be done at such a time as to produce, as nearly as possible, the required results. The growth expectancy tables of Green and Anderson ('47) are of great value in timing the operation; Tupman has produced similar tables for British children.

Another method is stapling across the epiphyseal cartilage, thus bringing about increasing pressure upon it as it continues to grow, with speedy cessation of growth because of this pressure. Blount and Clarke regarded stapling of the distal femoral and proximal tibial epiphyses as being the most efficient method of retarding growth, and the easiest and least risky one. A possible advantage of this method is that while stapling stops the growth of the bone at the affected epiphyses almost immediately and almost completely, removal of the staples may allow growth at these cartilages to be resumed. There are pitfalls, however, in that involvement of the periosteum may stimulate the formation of a bridge of bone across the epiphyseal plate, and this bridge may prevent growth after removal of the staples; if the technic is not sound, varus or valgus deformities may be produced. Green and Anderson ('57), on the basis of their experience that growth sometimes did and sometimes did not resume after removal of staples, regarded recurrence of growth as being impossible to predict with any certainty.

D'Aubigué and Duboussent reported that large leg-length discrepancies can often be corrected by shortening the long femur, lengthening the short tibia, or shortening one femur and lengthening the other.

Repair of Bone
Both periosteum and endosteum of young chick embryos are capable of forming bone in vitro (Fell), therefore, repair of a fracture can conceivably be due to either or both. However, Haldeman concluded, from experiments on rabbits and dogs, that the periosteum is much more important than is the endosteum, since union occurred much faster when the

periosteum was present across the fracture than when it was absent, and the endosteum alone often failed to complete the union of the fractured elements. McGaw and Harbin reported, on the other hand, that endosteum and bone marrow can regenerate bone well. They removed measured lengths of both fibulas in dogs, with the periosteum, and on one side filled the defect with fragments of bone marrow and endosteum; on the other side they made no transplant. They found no healing on the latter side, but new bone formed relatively rapidly across the defect on the side where the transplants had been made.

Presumably, many of the factors influencing growth of bone also influence the rapidity of repair. Grauer found that therapeutic doses of viosterol in the guinea pig caused stimulation of the osteogenic layer of the periosteum and therefore quicker healing of fractured bones; however, overdosage stimulated the fibrous layer of the periosteum and cause retardation in repair. Sympathectomy is said to have no apparent effect upon the speed of bone repair, but stasis produced by venous ligation is said to have hastened the healing of the fibula in dogs (McMaster and Roome). Shands ('37) reported that calcium glycerophosphate placed in a defect in the bones of the anterior limb of dogs appeared to increase the rate of healing, but when placed in defects of the vertebral column, it was rapidly removed and appeared rather to inhibit the speed of bone formation.

Vance and Wyatt discussed the process of healing of fractures as visualized by the roentgen ray, and called attention to the fact that fractures may heal with no visible external callus, the inernal callus then being the important one. They said that this is typical of all fractures lying within joint capsules; it is also typical of fractures treated by intramedullary nailing, and of oblique fractures anatomically reduced and held by transfixion screws (Ivins). Lindsay and Howes measured the strength of the healing fibula of the rat and found that strength begins to appear on the sixth day, and increases to the twenty-first day as the callus grows larger; following this,

the strength decreases somewhat, as the callus again becomes smaller and the bone is being reorganized through resorption. Finally, the strength rises once again from the thirty-third to the forty-fifth day, as the definitive Haversian canals and trabeculae of the bone are established.

Bone Grafts

It is now widely accepted that neither bone homografts nor autografts survive, as used clinically, but rather form a substrate into which blood vessels and bone-forming elements grow from the bony bed of the graft.

Hutchinson's observation that bone grafts into muscle, whether from the same or a different animal, promote formation of bone by the fibroblasts in the muscle, and Levander's earlier observation that cortical bone stripped of periosteum promotes formation of new bone from surrounding parenchymal tissue, indicate that bone grafts have a further effect, actually stimulating the development of bone. Levander was able to induce formation of bone in soft tissue in 22% of cases by injecting an alcoholic extract of bone, apparently producing evidence that bone contains a substance that will specifically evoke bone formation from mesenchymal tissue.

It is generally granted that fresh autogenous bone produces the greatest osteogenic effect; Ray and Holloway regarded decalcified bone as being the next best, if mechanical support is not a consideration, while Heiple and co-workers ('63) regarded freeze-dried homogeneous bone as next best. Burger and co-workers reported that decalcified bone treated with chondroitin sulfate produced the most rapid healing of defects in the calvaria of rats.

CARTILAGE

Other than the histological distinction of cartilage as hyaline, fibrocartilage, and elastic, relatively little attention has been paid to its structure. It will be recalled that hyaline cartilage is the usual form, both in the embryo and fetus, where most of the bones are first laid down in hyaline cartilage, and in the adult, where it is hyaline cartilage that forms, for instance, the articular surfaces of the bones. Elastic cartilage, occurring in only a few places such as the external ear and the epiglottis, is distinguished by the fact that it contains elastic fibers. Hyaline cartilage and fibrocartilage both contain collagenous fibers, but in the former the collagenous fibrils are thin and are hidden by the homogeneous matrix deposited about them, while in fibrocartilage the heavy bundles of collagenous tissue are more obvious, and there is less matrix. Fibrocartilage is actually a stage between dense fibrous connective tissue and hyaline cartilage and can be transformed into either; regenerating hyaline cartilage goes through a fibrocartilaginous stage, and many of the so-called fibrocartilages, such as the disk at the temporomandibular joint, become largely or entirely dense fibrous connective tissue.

The physical properties of hyaline cartilage apparently vary enormously. Koch gave the tensile strength of "human cartilage" as 2,250 lb/in^2, its compressive strength as 3,900 lb, and yet quoted the tensile strength of "rib cartilage" as being only 240 lb, thus much less than that of nerve (1,300 lb) and about equal to that of artery (230 lb). Presumably, the higher figures given applied to the hyaline cartilage on joint surfaces.

Except where it forms a surface for a joint, cartilage is covered by a dense perichondrium, a layer of connective tissue firmly attached to the cartilage and the equivalent of periosteum. All cartilage is avascular, but blood vessels can invade and destroy cartilage, as they do normally in the early development of long bones, and in the growth of the long bones from the epiphyseal plates. In the growth of the epiphyseal cartilages the cartilage cells proliferate, then become large and vesicular and arranged in rows, after which the interstitial substance becomes calcified and is then invaded by blood vessels that destroy it and thus leave space for the deposition of bone.

According to Harris ('50), slipping of an epiphysis, as determined experimentally on the upper tibial epiphysis of the rat, regularly occurs in that layer of cartilage in which the cells are hypertrophic but in which mineral substance has not yet been precipitated. He also quoted evidence that after removal of the perichondrium from about the epiphyseal plate, the epiphysis can be detached much more easily, and the line of separation is always through this layer of cartilage cells. Noting that slipping of the upper femoral epiphysis in human beings is nearly always accompanied by abnormalities of growth that appear to be caused by some endocrine disorder, he recalled that hypophyseal growth hormone increases the thickness of the hypertrophic layer of cartilage cells because it produces proliferation of the cells adjacent to this layer, and an increased output of growth hormone therefore makes the epiphyseal cartilages more susceptible to shearing stresses.

Factors influencing the growth of epiphyseal cartilage have already been mentioned in connection with the growth of bone. Normally this growth is orderly, but according to Hume, multiple exostoses arise as a result of unregulated growth from the edges of the epiphyseal cartilages—most frequently from those of the lower end of the femur and the upper end of the tibia. Salter and Harris, and Siffert, have discussed the effect of various types of injuries to the epiphyseal plates.

Growth and Repair of Cartilage

During the formative stages the cartilage of the epiphyseal plates possesses a great ability to grow, and the articular cartilages, by their growth, account for the growth of the bony epiphyses. Otherwise, growth and repair of cartilage are limited even under the best conditions. For instance, Shands ('31) found little evidence of regeneration of the hyaline cartilage of joints in dogs 4 weeks after defects had been made in it, and regeneration occurred more actively only after 12 weeks. He found also that superficial defects that did not involve the underlying bone provoked less complete regeneration than did deeper defects that did involve bone. According to his observations, the regenerative changes consist of, first, the appearance of fibrin; second, the appearance of granulation tissue; then replacement of the granulation tissue by connective tissue; then the appearance of cartilage cells in the connective tissue with the transformation of this gradually into fibrocartilage; and, finally, transformation into new hyaline cartilage.

Bennett and Bauer also agreed that repairs of articular cartilage in dogs is not particularly good; they said that it occurs best near the junction of the cartilage with the perichondrium, and on a weight-bearing surface; they found it resulting both from proliferation of regenerating cartilage cells and from vascular tissue arising from the margins of the perichondrium, or from the subchondral bone marrow when this was involved in the defect. They also found that young puppies repaired articular cartilages no better than older dogs; the growth activity of articular cartilage on its attached surface apparently has no effect upon the ability of the cartilage to repair its superficial surface.

The limited ability of cartilage to regenerate is apparently shown also by the behavior of cartilage grafts. While Dupertuis found that cartilage from young rabbits transplanted into the ears of those same or other young rabbits, and even into adult rabbits, would grow, he found that cartilage from the ears of adult rabbits would not increase in size regardless of the age of the host; thus, to some extent, the growth of cartilage seems to be conditioned by the age of the donor.

Grafts of dead cartilage are gradually resorbed, but autogenous grafts of living cartilage usually survive (Peer), although they show none of the ability of bone to proliferate or to excite proliferation in the surrounding tissue; according to Peer, they usually neither grow nor decrease in size. However, the fact that living cartilage grafts do survive seems to indicate that autogenous rib cartilage is better for a graft in plastic repair than is dead cartilage. Hagerty and co-workers reported that dead cartilage suffers a much more rapid loss of polysaccharide from the ground substance than does living cartilage.

JOINTS

Any attempt to classify joints strictly into categories is probably inaccurate. Of the three main types of joints, fibrous (synarthroses), cartilaginous (amphiarthroses), and synovial (diarthroses), the subclassifications among the last, such as plane, spheroid, condylar, ginglymoid, and the like (Fig. 1-4), are undoubtedly the least satisfactory, since movements at joints rarely correspond exactly to the types of movements implied by these terms. They are in general, however, useful terms in conveying a broad idea of the primary types of movement allowed at a joint.

By far the great majority of joints are, of course, synovial ones, diarthroses; the other types are of so little concern in the present volume that they will be passed over with no further description.

STRUCTURE OF A SYNOVIAL JOINT

The two bones that participate in a synovial joint are united by an articular capsule, usually subdivided into a fibrous portion and the synovial membrane proper, which surrounds the articular cavity (Fig. 1-5). The surfaces between which movement of the bones occurs are covered by a layer of hyaline cartilage, the articular cartilage, and the joint is lubricated by the synovia (synovial fluid). Folds of synovial membrane project into many joints, and some joints, such as that of the knee, contain menisci that partially subdivide the joint cavity, while others, such as the sternoclavicular joint, contain articular disks that normally separate the joint cavity completely into two parts.

Articular Capsule

The articular capsule of most joints consists of united but relatively distinct parts, fibrous and synovial. The fibrous membrane of the capsule is, with a few minor exceptions (such as the capsules of the joints between the bones of the middle ear cavity) composed very largely of collagenous fibers. Because of their collagenous nature, articular capsules are typically inelastic; special thickenings of the fibrous layer, in which the collagenous bundles have predominantly the same direction, form the ligaments of joints.

The synovial layer of an articular capsule is also composed of connective tissue, but the connective tissue is usually less dense than that of the fibrous capsule, and more cellular. The free surface of the synovial membrane is not formed by a continuous layer of mesothelial cells, but (Sigurdson) is simply modified connective tissue which is particularly rich in cells of the same type that are also found deeper in the membrane. The synovial membrane usually rests for the most part directly against the fibrous one, and it may project through a gap in the fibrous membrane to become continuous with a bursa.

Where the attachment of the fibrous capsule of the joint is some distance away from the edges of the articular cartilage, most notably about the neck of the femur, the synovial membrane is reflected around the bone to the level of the articular cartilage, gradually blending with the periosteum. In many joints, the synovial membrane in its deeper portion contains a large amount of fat, which may protrude into the joint as a fat pad. A conspicuous feature of the synovial membrane, except where it is adjacent to large amounts of fat, is the network of small blood vessels only a little beneath the surface. Folds of synovial membrane projecting into joints are common, even in such small ones as the intercarpal and interphalangeal joints (Grant) and should

Fig. 1-4. Major types of synovial joints: *a,* plane; *b* and *c,* spheroid or cotyloid; *d,* condylar; *e,* ellipsoid; *f,* trochoid; *g,* sellar; and *h,* ginglymus or hinge.

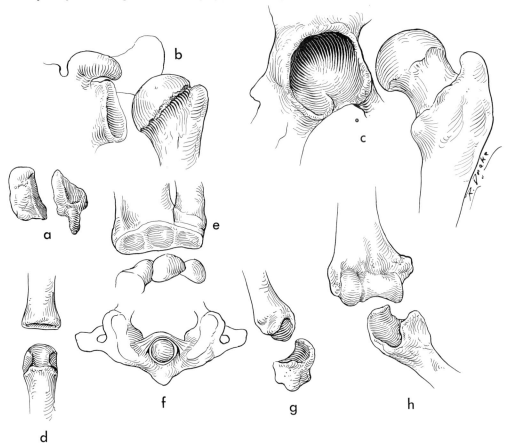

not, therefore, be regarded as pathologic. Bick described pathologic changes in synovial tissue in various affections of joints.

Synovial Fluid

Synovial fluid is thick and sticky, and is either clear or yellowish; its important constituent is mucin, and except for that the components which form normal synovial fluid are said to be derived by dialysis from the blood (*e.g.,* Gardner, '50). The exact origin of the mucin has apparently never been determined; Kling, as well as others, has said that mucin granules can be demonstrated in the lining cells of synovial membranes, but Davies ('43) found that a presumably reliable histochemical test did not indicate that the stainable material in the cells is actually mucin. A common viewpoint has been that the cells do contain mucin, which is liberated either by the degeneration of these cells or by actual secretion. Kling expressed the opinion that pathologic effusions in the joints are derived usually from both cell secretion and circulatory extravasation, and that only a very small percentage are actually pure exudates or transudates. Davies ('44) found a very marked variation in the protein content of the synovial fluid in various joints from cattle. Coggeshall and co-workers obtained an average of 0.45 ml, with a range of from 0.13 ml to 2.0 ml, of synovial fluid from 29 human knee joints which were apparently normal.

It has been said that articular cartilage is nourished mainly by the synovial fluid. Although they could not refute a contribution from the fluid, McKibbin and Holdsworth ('66) found that articular cartilage degenerates when it is separated from its blood supply in the underlying bone but left exposed to synovial fluid. Mitchell and Cruess found changes in staining that they attributed to a loss of polysaccharide from the ground substance, but no degeneration, during a period of about 60 days in which a joint was free of synovial fluid as a result of synovectomy. The question is not settled, apparently, for Maroudas and co-workers, in a study of the

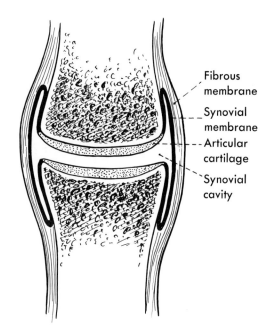

Fibrous membrane

Synovial membrane

Articular cartilage

Synovial cavity

Fig. 1-5. Diagram of a typical synovial joint.

permeability of articular cartilage, concluded that the cartilage is nourished by diffusion from the joint cavity. Diffusion was faster, they found, if the joint was allowed to move (and was faster *in vitro* if the fluid was agitated); the lack of such mixing in an immobilized joint, they suggested, explains why nutrition of the cartilage is partly dependent on joint movement. Honner and Thompson, using rabbits, concluded that in the immature animal the articular cartilage is nourished both by the synovial fluid and by the blood vessels in the adjacent bone, but in the adult by the synovial fluid only.

Linn found that although synovial fluid is largely ineffective in lubricating metal surfaces, it is much more effective than physiological saline in reducing friction between cartilages.

Synovial fluid typically also contains cells, Key finding from 1 to 300 nucleated cells and about an equal number of red cells per cubic millimeter. Coggeshall and his co-workers found an average of 63 nucleated cells per cubic millimeter of fluid, with a range from 13 to 180; about 4.3% of these were synovial cells, and 2.2% could not be identified; of the remainder, 63% were mononuclear phagocytes (monocytes and clasmatocytes), 24.6% were lymphocytes, and 6.5% were polymorphonuclear leukocytes. Thus the cytology of synovial fluid approximates that of tissue spaces.

MacConaill ('50) has discussed in considerable detail the mechanisms by which synovial fluid serves effectively as lubrication in a moving joint. Among these is the circumstance that in a moving joint contact between the surfaces may be at one point only, for the two adjacent surfaces of a joint have been said to be slightly incongruent except in one fixed position for each joint. (Linn, admitting that many joints are incongruent, said this is not true of the loaded tibiotalar joint, where the elasticity of the two cartilaginous articular surfaces produces congruity. Lubrication between these surfaces, he said, then depends upon fluid forced out of the cartilage by the pressure upon it.) Another mechanism is the movement of the fat pads connected with joints: they so move that the fluid is kept in a properly thin layer at the points along which the joint surfaces are in contact, the fat pad in front of the advancing surface receding before it, while that behind this surface follows behind it, and the fluid between the two is drawn out into a film. MacConaill also regarded intra-articular disks, such as that of the sternoclavicular joint, and menisci, such as those at the knee, as being primarily related to lubrication of the joint, rather than being concerned with weight bearing or making the surfaces more nearly congruous.

A number of substances can be absorbed fairly readily from joint cavities (see the review by Gardner, '50); apparently, increased movement of a joint hastens absorption from that joint. Sigurdson reported that immobilization of the ankle joint in rabbits for 53 days produced no marked changes in the morphology of the joint capsule, but that the amount of synovial fluid was considerably reduced. The increased rate of absorption from the articular cavity and the decreased rate of production of synovial fluid with activity and

immobilization, respectively, may be largely a vascular effect.

Innervation

The nerve supply to a joint is usually multiple, for each joint is typically supplied by all the nerves that supply muscles acting across it (Hilton's law), so that in the limbs, for instance, most of the nerves passing each joint give off one or more branches to that joint. The branches to a joint may come directly to it from the nerve, or they may be derived from muscular branches. There is, moreover, considerable overlap between the part supplied by all the branches of one nerve and that supplied by all the branches of another (e.g., Gardner, '50, '53); thus, section even of all the branches from a single nerve, when they are multiple, may fail to relieve pain from a particular part of the joint. Stilwell ('57a and b) emphasized also that joints receive fibers from cutaneous nerves as well as from deeper ones.

Figure 1-6 shows, diagrammatically, the nerve and blood supply to a typical joint. Within the capsule of a joint, some nerve fibers follow blood vessels, but others leave the vessels to spread out in the tissue of the joint, these being more common in the regions in which the capsule is susceptible to compression or deformation during movement (Gardner, '50). Gardner ('44) described in some detail the termination of nerves in the knee joints of the cat, and similar studies of other joints have been reported by, among others, Ralston and co-workers and by Stilwell ('57a and b). Nerve endings in joint capsules and their associated ligaments, apart from vasomotor fibers on blood vessels, are of three general types, the Ruffini type, Pacinian type, and free nerve endings, and are probably responsible for appreciation of movement, pressure, and pain from the joints. Only free nerve endings occur, apparently, in the synovial membrane. Although some of these may be for pain, most of the fibers here are sympathetic ones (Samuel).

In contrast to the apparent lack of effect of denervation on bone, Finterbush and Fried-

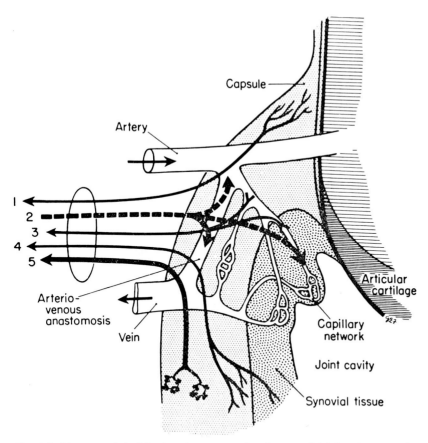

Fig. 1-6. Diagram of the blood and nerve supply of a synovial joint. An artery is shown supplying both the neighboring epiphysis and the joint capsule; it is connected to the vein both through capillaries and an arteriovenous anastomosis. The *numerals* indicate various types of nerve fibers: *1,* a fiber for pain, ending in the capsule close to the periosteum; *2,* a sympathetic postganglionic fiber to blood vessels; *3,* an afferent fiber, probably for pain, from the blood vessels; *4* another afferent fiber for pain; *5,* a proprioceptive (afferent) fiber from the capsule. (Gardner E: Am Acad Orth Surgeons, Instructional Course Lectures 10:251, 1953)

man reported that denervation caused progressive atrophy of all structures of the joint; this was apparently not a result of trauma, since they found it when the joint had been immobilized by a plaster cast.

Kellgren and Samuel investigated clinically the sensitivity of the various components of joints to pain. They noted that the bone and periosteum at the end of the joint certainly play an important part in the appreciation of vibratory and other mechanical stresses, and may contribute to pain because of the sensitivity of the periosteum; they agreed with others that articular cartilage has no nerve supply, and expressed the opinion that cancellous bone may be somewhat less sensitive than the periosteum. In their experiments, they found that either pin prick or injection of 6% saline solution into the fibrous capsule would evoke pain; in the case of the knee, they found that the tibial collateral ligament, the patellar ligament, and both the medial and lateral parts of the capsule all had numerous points that responded to pain or pressure, or both, when stimulated with the point of a needle.

They also found, however, that when the synovial membrane was stimulated by pricking or by scratching with the needle, or by crushing and pulling with artery forceps, sensation was usually not evoked if care was taken to involve only the synovial membrane and not the fibrous one. In only two patients did they evoke pain from the synovial membrane by needle prick, this being in a small area close to the upper margin of the patella. They therefore concluded that both general sensation and pain from a joint arise largely in the fibrous portion of the capsule.

Eyring and Murray reported differences in pressure within the major joints, following injection of saline, according to the position of the joint, and suggested that (1) patients tend to hold an affected joint in the position that causes least pain, and (2) rapid increase in pressure is more important than the absolute pressure in producing pain.

Blood Supply and Lymphatics

The blood supply of joints is likewise multiple and, while it may all come from one major vessel, may consist either of direct articular branches from this vessel or of smaller branches from all the vessels in the neighborhood of the joint. In the case of the larger joints, the vessels consist of both descending ones that originate proximal to the joint, and recurrent ones that originate distal to it.

The vessels to a joint anastomose freely with each other, and supply not only the capsule but also the ends of the bones enclosed within it. The greatest vascularity of the capsule is in the synovial membrane, a particularly dense capillary network being found close to the lining of this membrane (Fig. 1-6) unless there is a considerable quantity of fat here. Davies and Edwards regarded the placement of the terminal capillary network close to the surface of areolar areas of the synovial membrane as indicating that these regions are responsible for the interchange between the blood and the synovial fluid; it also accounts for the ease with which blood is extravasated into the joint. The villi and the fringes of synovial membrane are less well supplied with blood, Davies and Edwards found, and the ligaments are still less vascular.

Venostasis has been supposed to contribute to degenerative changes in joints, but McMaster ('37), in a clinical study, found no evidence that this alone predisposes to damage.

Kuhns, and Davies ('46) studied the lymphatics of joint capsules. Both agreed that the lymphatic network in the synovial membrane is rich, but Kuhns described the lymphatics as being most numerous "just beneath" the lining cell layer of the synovial membrane, while Davies said that the network of capillaries in the synovial membrane lies closer to the larger blood vessels and not as close to the lining cells as do the smaller blood vessels. Kuhns said that inflammation of the synovial tissue decreases the ability of the lymphatics to absorb material, and that if the inflammatory process is sufficiently severe or prolonged the lymphatics seem to disappear, although they reappear again after the inflammatory reaction has subsided. Davies failed to find lymphatics in the villi of the synovial membrane. He said the efferent lymphatics leave the capsule of the joint in groups of two or three, or even more, and accompany the main blood vessels of the joint, but especially run toward the flexor aspect of the limb; they are said to communicate with lymphatics of the periosteum. While it has been said that some synovial lymphatics drain directly into regional veins, Davies found no evidence of it in his study.

Disks and Menisci

A few joints contain fibrocartilaginous disks that completely divide them into two parts, and the knee contains menisci that partially subdivide the joint cavity, but the great majority of joint cavities have neither. The function of disks and menisci, beyond the fact that they participate in sliding and rotatory movements, is not agreed upon, but there are two chief concepts, namely, that they have to do with establishing better congruity between the joint surfaces, and that they have to do with properly spreading the synovial fluid so as to provide for lubrication during movement.

Sullivan rejected the concept, once held, that articular disks primarily play the part of shock absorbers, and pointed out that most of the bones associated with articular disks are also ones that are more frequently fractured. He advocated the concept that while the disks do not increase freedom of movement, and are not concerned particularly with lubrication (which, according to his concept, is most easily effected by the folds of the synovial membrane), they do compensate somewhat for the incongruity of the opposed surfaces at a joint, and thus help to maintain stability.

In contrast to that, MacConaill ('32, '50) argued that the real function is to spread the synovial fluid in a thin film in order to facilitate movement of the joint. He said ('32) that the menisci of the knee, and the cartilage of the inferior radioulnar joint, are congruent with the articular surfaces of the adjacent bones only in the weight-carrying position, and ('50) that the essential mechanical features of an intra-articular cartilage are that it be wedge-shaped, and hinged to the capsule so that its inclination in regard to the moving articular surface can be varied slightly. According to his analysis, intra-articular disks or menisci are found in joints where three conditions occur: a marked rotatory movement about an axis perpendicular to the articular surfaces; a marked flatness of the surfaces; and a force tending to bring the surfaces together during the rotatory movement (*i.e.,* a screwlike movement). This combination of factors apparently demands, from an engineering standpoint, a special mechanism to insure lubrication; MacConaill pointed to the sternoclavicular joint and to the knee joint as being typical examples of joints in which these conditions exist. According to this concept, therefore, menisci and disks are primarily concerned in keeping the film of synovial fluid thin enough for it to be able to lubricate properly.

MacConaill said that it is only the more central part of menisci and disks that are fibrocartilaginous and devoid of synovial membrane, and that these are the only parts of the disk that take weight in an uninjured joint; these parts are also avascular and have no nerve supply. In contrast, the more peripheral parts of all disks or menisci within joints are said to be largely fibrous, covered by synovial membrane, and to have both a vascular and a very rich nerve supply.

Walmsley and Bruce studied the regeneration of the semilunar cartilage of the knee joint of rabbits; they described new menisci as arising from the fibrous tissue beneath the surface layer of the synovial membrane, and spreading rather rapidly; they regarded the rapidity of inward spread as supporting the concept that the menisci are primarily related to distribution of synovial fluid rather than to transmission of weight. They also found no evidence of formation of fibrocartilage within the new fibrous tissue. Menisci so regenerated, they said, differ from the ones removed in that they have no horns of attachment; rather, they attach around their periphery to the articular capsule and to the nonarticular parts of the adjacent femur and tibia (the edges of the articular cartilages).

Movements

Details of the type of movement allowed at the various joints are necessarily given in connection with the individual joints. However, MacConaill ('50) emphasized that all joints have certain features in common. He noted that it is very difficult or impossible to lubricate effectively flat surfaces or those with only slight curves, and that the best lubrication is obtained when there is a continued approach of a moving surface to, combined with a sliding motion upon, a fixed surface; according to engineering principles, he said, there is only one way of securing such an approach, and that is that the surfaces must not be parallel to but rather inclined to each other, and the moving surface must move in the direction of inclination. If both the fixed and moving surfaces are curved, but one surface has a greater curvature than the other, then there can be no strictly parallel motion between the two surfaces, for they are incongruous, and as they move they touch at one point only. He quoted "Walmsley's law" as stating that these conditions hold true in joints, and that the articulating surfaces of each joint are incongruous except in one special position. As already noted, Linn found the weight-bearing tibiotalar joint to be largely congruent in every position and therefore demanding more than a peripheral film of fluid for its lubrication.

Based upon the concept of incongruity of the joint surfaces except in one position, MacConaill defined the position of greatest stability, which he called the synarthrodial or close-packed one, as that in which one part of the convex surface is of the same curvature as the whole concave surface, so that the two surfaces here fit together over a relatively large area, rather than a point.

There is as yet no complete agreement on the role played by the ligaments of a joint in limiting and guiding movements at that joint. However, it is becoming ever more obvious that the primary restraining influence against undue movement of a part in response to gravity is provided by ligaments rather than muscles (see following chapters). For example, rather than muscles crossing it, the capsule of the shoulder joint supports the limb until it is excessively loaded; in the fully flexed position, the muscles of the back relax, and further movement is checked by ligaments; the stability of the extended hip joint depends on tightening of its ligaments by extension, and they then resist further extension; and the joints of the arches of the foot are maintained primarily by ligaments.

A good example of the guiding rather than the restraining effect that ligaments may have is the rotation which regularly accompanies flexion and extension at the carpometacarpal joint of the thumb; it has been shown that this rotation occurs even upon passive movement as long as certain ligaments at the joint are intact, but disappears when these ligaments are cut. Thus, although the direction of pull of the muscles is also very important in rotation, it appears that the ligaments play a very definite part in helping to guide this movement.

The strength of ligaments obviously varies greatly. That of any particular ligament, if the anterior cruciate ligament is typical, is also affected by disuse. Noyes and co-workers found that immobilization of the knee in monkeys led to failure of the ligament at a much lower load than normal, and that the ligament was also more easily stretched. Even 20 weeks of activity did not restore the ligament to its original strength, although its extensibility returned about to normal.

It has been shown that the articular cartilage of the knee joint often presents evidence of damage some time after removal of the menisci, and this has been interpreted as resulting from attrition; the concept of the lubricating function of the menisci fits in well with this interpretation. Similarly, minor lesions of the articular cartilage of otherwise normal joints have been thought to be due primarily to attrition as the result of motion. It may be, however, that the latter lesions are largely degenerative instead of a result of motion itself. At least, Lanier divided inbred male mice into two groups, one of which was subjected to forced exercise for almost a year, while the other was restrained as much as possible from any exercise. He found that the quiescent animals had more minor lesions in the articular cartilage than had the active ones, and that there was no difference in incidence of severe lesions. Fairbank described changes in the shape of a femoral condyle, visible on roentgenographic examination, as sometimes occurring after meniscotomy in man, but found no such changes in 33% following removal of the medial meniscus, and in 50% following removal of the lateral meniscus. He attributed the changes to loss of a weight-bearing, rather than a lubricating, function of the menisci. However, he observed no correlation between the clinical and the roentgenographic findings.

Clinical aspects of immobilization and manipulation of joints have been discussed by Heyman. Evans and co-workers reported that a slight range of movement does not prevent changes in and about a joint resulting from immobilization. The primary changes, they found, are proliferation of intercapsular connective tissue and the formation of adhesions, but after prolonged immobilization, limitation of movement is contributed to both by contraction of muscles and by the capsule, with the muscles contributing most. Immobilization has been reported to lead to degenerative changes in the articular cartilages (Evans and co-workers), but Fingerbush and Friedman found that immobilization of a nondenervated limb caused proliferative changes, first in the synovial membrane and then in the cartilage. Crelin and Southwick reported that immobilization under pressure produced death of the cartilage within 16 days if no movement at all was possible, and they agreed with previous workers that this results from forcing nutrients out of the cartilage. Only a little movement, however, prevented the necrosis, apparently by allowing nutrients to re-enter as the bearing surface shifted.

Bruner quoted evidence that the cause of stiffness that results clinically from immobilization of joints is not so much the immobility itself as the cessation of functional activity around the joint; the resulting venous and lymphatic stasis leads to edema of the periarticular tissues with the formation of exudates that later become organized to form fibrous adhesions and thickened joint capsules. Thus continuous activity of joints distal and proximal to an immobilized one is beneficial: for instance, if the fingers are kept active, he said, a wrist immobilized for many months because of fracture of the scaphoid may be quite supple when it is removed from the cast.

"GANGLIA"

"Ganglia," occurring more commonly in the region of joints and sometimes communicating with them, are one cause of pressure on nerves as they cross joints. Ganglia have often been regarded as actual diverticula of the synovial membrane and cavity, and are, at any rate, more logically discussed in connection with joints than elsewhere. De Orsay and co-workers noted that ganglia have been said to be formed both by the mucinous degeneration of connective tissue, and by secretion rather than degeneration, with the formation

of a pseudo-joint, and that they occur more commonly in persons of the second to fourth decades of life. Aspiration of the tumor is said to be disappointing, although rupture, according to De Orsay and his co-workers, effects a cure in about half the cases; if surgical excision is necessary it effects, under the best operative conditions, a permanent cure in about 85%. Brooks ('52) said that the precise etiology of a simple ganglion has always been a matter of some doubt, but that in most cases a definite connection with the neighboring joint or the joint capsule has been established at operation; however, the histologic appearance of the excised ganglion apparently does not usually resemble that of the synovial membrane, perhaps because there is a metaplastic change here. Warren expressed his opinion that any of the connective tissue elements about a joint, including even the perineurium of a nerve crossing the joint, can form ganglia, and reported a case in which a ganglion had apparently formed in the common peroneal nerve.

Muscle and Tendon

MORPHOLOGY

Muscles vary much in shape and size, and their shapes, as well as their positions and actions, have been used in naming them. They may be thin and ribbonlike, as are the infrahyoid muscles, broad and flat as are the rhomboidei, fan-shaped as is the pectoralis major, or of quite irregular shape, and may have several heads of origin or tendons of insertion. Further, in the more complicated muscles such as the deltoid, there may be intramuscular tendons that serve as the insertion or origin of large groups of muscle fibers. Obviously, the arrangement of the muscle fibers is in part correlated with the shape of a muscle, but not entirely so, and the functional efficiency of a muscle depends more upon the arrangement of the fibers than it does upon the shape of the muscle.

Tendons also vary in shape, although not

as markedly as muscles; in general, they simply range from rounded cords, such as the tendons at the wrist, to very broad flat tendons, such as those of the lateral abdominal muscles, which are usually called aponeuroses. There is no clear line of division other than customary usage between an aponeurosis, as in the abdominal wall, and a flattened tendon such as that of the pectoralis major. Similarly, there is not always a clear distinction between fascia and tendon; the surface of a muscle may be tendinous in character and give rise to muscle fibers, and is then sometimes referred to as tendon, sometimes as fascia.

MUSCLE

A muscle consists of a large number of voluntary muscle fibers bound together by connective tissue, with connective tissue also about the principal blood vessels and nerves. At its ends, the muscle as a whole is continuous with tendon (even in the case of so-called fleshy attachments of muscles), through which it usually attains an attachment into bone. Much of the connective tissue in and about muscle is collagenous, but elastic fibers are present in varying numbers.

The connective tissue sheath immediately surrounding a muscle is the perimysium, also called the epimysium or external perimysium, and often referred to as the fascia of or on the muscle. It may be modified, and aponeurotic or tendinous rather than fascial in character, when it gives rise to muscle fibers. It blends with the covering of the tendons of the muscle, the peritendineum (epitendineum), and may or may not be fused with the deep fascia covering a part; for instance, the perimysium of the triceps brachii is fused to the brachial fascia, that of the biceps brachii is not. From its deep surface the perimysium sends connective tissue septa into the muscle; these septa, a part of the perimysium (but sometimes distinguished from the covering tissue by that being called "epimysium" or "external perimysium," the septa alone being called "perimysium" or "internal perimysium"), both accompany blood vessels and nerves and

subdivide the muscle, by repeated branching, into smaller and smaller bundles of muscle fibers. These muscle bundles or fasciculi are in turn invaded by smaller amounts of connective tissue, the endomysium, continuous with the perimysium, and ultimately surrounding each individual muscle fiber by a delicate layer of reticular tissue.

The muscle cell itself is provided with a sarcolemma, once described as a structureless covering membrane, but now defined as consisting of a very thin inner membrane composed of two layers of protein and an intermediate layer of lipid, and an outer thicker coat of amorphous material.

The individual muscle fibers are paralleled by capillaries, which communicate around the fibers with each other, and form a longitudinally arranged mesh within each fasciculus. Each muscle fiber has a motor end plate (sole plate), a modified portion of sarcoplasm in which the nerve fiber ends after penetrating the sarcolemma. Special groups of muscle fibers are surrounded by a connective tissue sheath within which there is an intricate twisting and winding of nerve fibers; each of these muscle fibers has a motor ending, but the group as a whole, with its afferent nerves, forms a muscle spindle.

The innervation and blood supply of muscle are discussed further in following sections.

Arrangement and Length of Muscle Fasciculi and Fibers

A muscle fasciculus is a more or less obvious bundle of muscle fibers; in fine-textured muscles the fasciculi contain fewer muscle fibers, in coarse-textured ones more. Within different muscles, the fasciculi bear different relations to the line of pull of the muscle (Fig. 1-7): they may be parallel to it, as in strap muscles, and run the whole length of the muscle, or they may be arranged at an angle to it, their length then varying directly with the width of the muscle and how closely they parallel its length. The length of fasciculi is measured from tendon to tendon; those muscles that do not have longitudinally running fasciculi therefore have tendons either along the edges or in the substance of the muscle.

Since an oblique arrangement of the muscle fasciculi presents an appearance somewhat resembling the arrangement of the barbs in relation to the quill of a feather, muscles with this arrangement are usually termed "penniform (pennate)" ones. Depending upon the arrangement, they are subdivided into unipennate, in which the fasciculi, although at an angle to the length of the muscle, parallel each other; bipennate, in which there are two groups of fasciculi that converge upon a centrally placed tendon, and hence are at an angle to each other; and more complicated forms, such as multipennate (see Fig. 1-13) with many groups of fasciculi attaching to many intermuscular tendons, and fusiform, in which the fasciculi are not only at an angle to the line of pull but also describe a curve, as, for example, the rectus femoris.

There is no strict correlation between the length of the muscle fibers composing a fasciculus and the length of the fasciculus, for within a single fasciculus of a rabbit's muscle Huber found fibers that ranged from 0.14 cm to 3.04 cm long. However, even in rather long fasciculi there may be individual fibers that run the whole length of the fasciculus, which may in turn, of course, be the length of the muscle. It is said that fibers as long as 34 cm have been isolated, but probably the majority of fibers in man are no longer than 10 cm to 15 cm, and Elftman said the average length is about 5 cm. Those fibers that run the length of a fasciculus have rounded ends where they are attached to tendons; those ending within a fasciculus, whether at one end or both, have tapered intrafascicular ends (Huber) and are bound to similar ends, which they overlap, by connective tissue (it is this overlap, making it difficult to decide where a fiber ends and another begins, that accounts, presumably, for the many contradictory statements concerning the lengths of muscle fibers).

"Weber's law," or the "Weber-Fick law," states that there is a constant relationship between the length of a muscle fiber and the amount by which it can shorten; the contracted fiber is about 60% of the length of the relaxed and physiologically extended fiber. In muscles with parallel fibers, the total shorten-

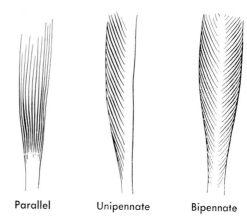

Parallel   Unipennate   Bipennate

Fig. 1-7. Varying arrangements of muscle fibers.

ing of a muscle is obviously exactly equivalent to the total shortening of the fasciculi of which it is composed. In pennate muscles, in contrast, it might be expected that the shortening of the entire muscle would be less than the shortening of its fasciculi, because the fibers are pulling at an angle to the tendon of insertion. However, Pfuhl pointed out that in pennate muscles the thickening of each muscle fiber as it contracts necessarily displaces the fiber above it upward, and this in turn moves the tendon of insertion upward (Fig. 1-8); he calculated that, actually, the tendon of insertion moves upward slightly more than the fibers shorten, the difference between the two being less the more nearly the fibers are parallel.

In practice, this means that the distance over which a muscle can shorten is determined by the length of the fibers or of the fasciculi that compose it. Haines ('32, '34) calculated that the contraction of muscles amounts to about 57% of the length of the physiologically extended fiber, now referred to as the "rest length." Elftman put the amount of contraction as less, to about 60% of the rest length, and pointed out that a fiber develops its greatest tension at its rest length and loses tension as it contracts until, at its shortest, it can develop no tension at all. Thus, he said, if a muscle acting across a single joint is to perform work at the end of the movement, its fibers must be almost three times as long as the distance between which they shorten during complete movement of a joint. Thus the length of the fasciculi determines the range of contraction of a muscle, and muscles that have similar ranges of action have fasciculi of similar lengths. Conversely, the habitual range of contraction of a muscle determines

Fig. 1-8. Action of a bipennate muscle. As the fibers shorten and become more transverse they also become thicker, as indicated by the difference in spacing of the lines. This results in the distal end of the muscle moving upward a greater distance than the fibers have shortened: in this example, the stretched fiber is about 6.5 cm long, the contracted one 3.3 cm, but the whole muscle has shortened by about 3.8 cm. (Pfuhl W: Zeit f Anat u Entw-gesch 106:749, 1937)

proximal

distal

its fascicular length; physiological stretching of a muscle is apparently necessary to proper growth in length of its muscle fibers, and in a fully developed muscle in which the range of action is reduced for a period of time the muscle fibers contract, and thus shorten the fasciculi to a point at which contraction amounting to about 40% of their newly acquired extended length is sufficient to bring about the required action. This phenomenon is of particular importance in orthopedic surgery in regard to tendon lengthening procedures (*e.g.,* Arkin), and the myogenic contractures that may follow paralysis of antagonistic muscles.

Crawford showed that growing muscle does adapt itself to the length over which it has to contract. He transplanted the tendon of the tibialis anterior in front of the extensor retinaculum, so that it would have to contract more in order to produce complete dorsiflexion of the foot, and found that the muscle belly increased in length and the tendon became shorter than on the normal side. Comer reported that both growth of bone and the range of motion influenced the length of the muscle as a whole, but that the amount of contraction of the entire muscle belly varied from one muscle to another.

While the length of the muscle fasciculi apparently determines entirely the range of action of the muscle, the strength of the muscle is directly proportional to the area of cross section of the fasciculi; when these two effects are correlated, therefore, it becomes obvious that long and straplike muscles have a great range of movement, because of the length of their fasciculi, but relatively little strength, because of the small size of their cross section. On the other hand, the unipennate, bipennate, multipennate, and fusiform muscles are particularly strong, because the cross-sectional mass at a right angle to the fasciculi is large; similarly, because their fasciculi are shorter, their range of movement is always less than that of a strap muscle of the same total length, but will vary markedly from muscle to muscle depending upon the obliquity of the fibers to the line of pull, which is of course a determining factor of their length.

TENDON

A tendon consists largely of heavy, parallel bundles of collagenous fibers, so closely packed as to form a very dense tissue. Between fiber bundles are fibroblasts, usually regarded as participating in the new formation of fibers after a tendon is injured. Between larger bundles there is a little loose connective tissue, and in this are the blood vessels and nerves that supply the tendon. This connective tissue is the equivalent of the perimysium of muscle, and like the latter is continuous with a thicker layer of connective tissue that surrounds the structure as a whole. The covering tissue of the tendon is the peritendineum, sometimes also called the epitendineum. It is a dense but thin fibro-elastic membrane, sometimes thicker and looser on one side of the tendon where it may contain major longitudinal vascular channels. Where a tendon runs through a tendon sheath, the inner synovial layer of the sheath is fused to the peritendineum so as to form a smooth gliding surface on the tendon.

The attachment of tendon and muscle fibers to each other is commented upon in the following section. Most tendons at their non-muscular end attach into bone, blending in part with the periosteum but also continuing, as fibers of Sharpey, into the bone to blend with its collagenous tissue. The tendons of those muscles that insert into skin simply blend with the dense connective tissue of the corium (dermis).

As a tendon inserts into bone, it regularly fans out, so that there are always some tendon fibers in the direct line of pull as the angle between tendon and bone is altered by movement of the latter (Fig. 1-9). Movement, or the contraction of a part of a muscle, necessarily throws all of the pull of the muscle on only a part of the tendon; for this reason it is important that the large parallel fiber bundles of a tendon, themselves formed, according to Mollier, by the intertwining of smaller fiber bundles, intertwine with each other close to the attachment of the tendon to bone (Fig. 1-10). Thus the force of contraction of any part of a muscle can be widely spread at its tendinous attachments.

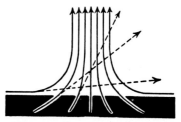

Fig. 1-9. Fanning of tendon fibers as a tendon inserts into bone (*black*), and the manner in which the different angles of insertion of the fibers keep some of them always in the line of pull (*broken arrows*). The fibers almost parallel to the bone are inserting into the periosteum. (Mollier G: Morph Jahrb 79:161, 1937)

ATTACHMENTS BETWEEN MUSCLE AND TENDON FIBERS

Even though individual muscle fibers may have intrafascicular endings through which they are directly attached (by connective tissue) only to other muscle fibers, all muscles as a whole are attached at their origins and insertions by tendon fibers.

The exact way in which muscle fibers are attached to tendon fibers was long disputed, for it was once maintained that myofibrils were continuous with tendon fibrils. However, Goss showed that the tendon fibrils are really attached to the delicate connective tissue sheath associated with the sarcolemma; the fibers of this sheath are circular over the main portion of the muscle fiber, but become abruptly longitudinal near the end of the muscle to become continuous with those of the tendon. Subsequently, Long studied the development of muscle and tendon, and agreed with Goss; he found that the sarcolemma completely encloses the myofibrils, and that the fibers of the tendon develop from reticular fibers continuous with those surrounding the muscle fibers.

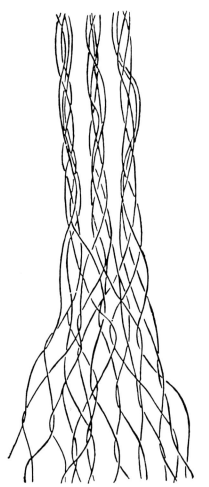

Fig. 1-10. Intertwining of tendon fibers. The large primary tendon bundles, composed of intertwining secondary and tertiary bundles, parallel each other after leaving the muscle, but as they approach their attachment to bone, in the bottom of the figure, lose their identity by intertwining with each other. (Mollier G: Morph Jahrb 79:161, 1937)

STRENGTH OF MUSCLE AND TENDON

The disparity between the structural strength of muscle and tendon is enormous; Koch quoted an ultimate tensile strength for muscle of 77 lb/in^2 and one for human tendon of 9,850 lb/in^2, and tensile strengths much higher than this have been reported for tendons: Cronkite tested a number of different tendons and found their strength to vary from about 8,600 lb/in^2 to almost 18,000 lb/in^2; he found, however, no apparent correlation be-

tween the tensile strength of the tendon and either its size or its function. It is the disparity between the strength of muscle and tendon that allows a very large muscle to act through a small tendon or even a small part of a small tendon.

McMaster ('33) pointed out that neither experimentally nor clinically will undue strain upon a normal tendon cause its rupture; instead, rupture occurs in the belly of the muscle, at the musculotendinous junction, at the junction of the tendon with the periosteum, or is associated with fracture of the particular portion of bone into which the tendon inserts. Tendon itself, therefore, is stronger than its attachments. McMaster found experimentally that rupture of a tendon, rather than of its attachments, will occur only when the tendon is weakened to the degree of having about half of its fibers cut. These findings agree with observations that tendons more commonly ruptured, such as those of the supraspinatus muscle or the long head of the biceps brachii, are also those in which degenerative and attritional changes are most commonly seen.

NERVE SUPPLY

Types of Nerve Fibers

The nerves to muscles (Fig. 1-11), although frequently referred to as motor, contain numerous afferent fibers (some 40% to 60%). Since tendons themselves are noncontractile, the nerve supply to a tendon consists entirely of afferent elements. Stilwell ('57a and b) found proprioceptive endings especially abundant close to the musculotendinous junction, and Ralston and co-workers noted that periosteal endings are especially abundant close to the attachments of tendons. Becton and co-workers found flexor tendons of the fingers to be generally sparsely innervated except in the region of the vincula, and extensor tendons to be more richly innervated over the joints than over the phalanges. Like joints, tendons of the hand and foot derive their innervation from both deep and superficial or "cutaneous" nerves.

Afferent fibers in muscle are for both proprioception and pain; the proprioceptive fibers end peripherally in connection with muscle spindles, groups of two to twelve muscle fibers enclosed within a connective tissue sheath. In addition to their two types of sensory endings, described as primary or "annulospiral" and secondary or "flower spray," the muscle fibers in the spindle receive a motor innervation that is different from the innervation of the extrafusal muscle fibers (which is by alpha or large nerve fibers); through this innervation, gamma fibers, the central nerv-

ous system is able to exert a control over the activity of the muscle spindles (see Mathews). It is the primary or annulospiral ending that, through monosynaptic contact with ventral horn cells, is responsible for the stretch reflex. According to Patton and Mortensen, the primary endings both facilitate contraction of that muscle and inhibit the antagonist, while the secondary endings facilitate flexion and inhibit extension regardless of the muscle in which they are located. It is generally granted, however, that no impulses

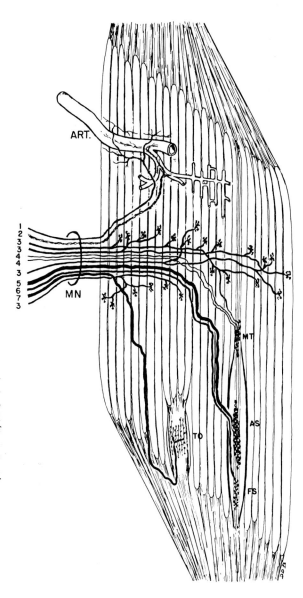

Fig. 1-11. Diagram of the nerve supply to a voluntary muscle. The nerve, *MN*, contains medium-sized motor fibers (*3, 3, 3*) that are distributed to motor end-plates of ordinary muscle fibers, and small ones (*4, 4*) that go to the end-plates, *MT*, of the fibers of the muscle spindles. Small afferent fibers (*1*), largely for pain, are distributed in the connective tissue around blood vessels, *ART.*, and sympathetic fibers (*2*) to the smooth muscle of smaller arteries and arterioles. The largest afferent fibers (*5, 6, 7*) are distributed to the annulospiral and flower spray endings, *AS* and *FS*, of muscle spindles, and to the receptor endings of tendon organs, *TO*. (Adams RD, Denny-Brown D, Pearson CM: Diseases of Muscle [ed 2]. New York, Hoeber, 1962)

from the muscle spindle reach consciousness.

Pain fibers within the muscle may be associated primarily with blood vessels, but their presence is attested by the pain associated with injury or inflammation of muscle; that they may run in the nerve supplying the muscle, rather than accompany the vessels into the muscle, is shown by the fact that pain elicited by injection of saline solution into muscle could not be evoked after the motor supply of the muscle was interrupted (Bakke and Wolff).

The concept that voluntary muscle may have a sympathetic innervation in addition to its voluntary one has been decisively disproved.

General Distribution

Since the nerve supply to a muscle must reach every muscle fiber, the pattern of branching to and within a muscle varies somewhat with the arrangement and length of the muscle fasciculi, and the direction from which the nerve approaches the muscle (Fig. 1-12). Thus, to take two simple examples, if a nerve runs largely at right angles to the muscle fasciculi it will cross many or all of them; if the fasciculi are short, all the muscle fibers can be supplied by small branches of the nerve given off during its course, while if they are long the nerve gives off larger branches at approximately right angles to its course, these running proximally and distally in the muscle (Fig. 1-13) to reach each muscle fiber in these locations. If, on the other hand, the supplying nerve runs in the same direction as a long muscle, its intramuscular branches must not only cross the fasciculi, but often also run as recurrent branches to reach fibers lying above the level of entrance of the nerve, and as descending branches to reach fibers lying below

Fig. 1-12. Types of nerve distribution in muscles of different fascicular arrangement. (Bardeen CR: Am J Anat 6:259, 1907)

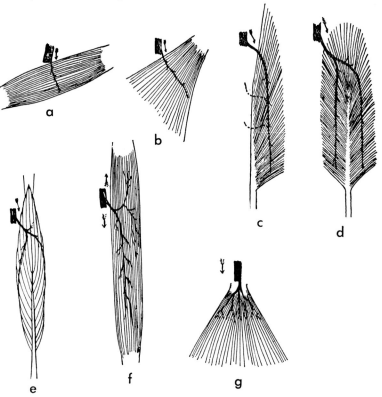

this level (Fig. 1-12f). Frohse studied in some detail the macroscopic branches of nerves in muscles; he noted that in the long muscles the nerves usually enter above the middle of the muscle, more commonly in about the region of junction of the proximal and middle thirds, so that in many muscles there must be fibers distributed in both directions, some of these fibers also continuing to the tendons. Many muscles, of course, receive two or more nerves, which in the case of the longer muscles frequently enter the muscle some distance apart.

"Motor points" of muscles (Chaps. 3 and 7) are not exact anatomical points, but since they are sites from which the contraction of an entire muscle or a specific part of the muscle can be elicited without producing contraction elsewhere, they correspond either to the point of entrance of the nerve into the muscle or part of the muscle, or to some point along the course of the specific nerve after it has left its origin; they are usually assumed to be at the point of entrance of the nerve into the muscle, but obviously this is only an approximation.

Segmental Innervation of Muscles

Most muscles have a plurisegmental innervation, that is, are supplied by nerve fibers originating from two or more spinal nerves; this fact is well established, but the exact segmental innervation of individual muscles is not as well known. Among most texts there is a very considerable variation in the nerve segments that are listed as contributing to the innervation of individual muscles. This is probably inevitable, and presumably depends upon two facts: one is that segmental innervation in man can be determined only occasionally either experimentally or clinically, and in either case the determination has depended largely upon being able to diagnose the nerve lesion accurately and to recognize a degree of contraction or weakness in a muscle. Electromyography now allows, of course, accurate assessment of the amount of degeneration in many muscles and, presumably, will eventually contribute to better knowledge of segmental innervation. The other factor tending to make all listings of segmental innervation

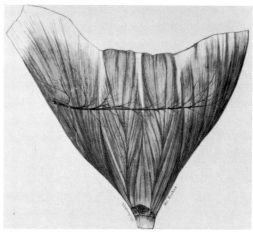

Fig. 1-13. Intramuscular distribution of the nerve to the deltoid. (Frohse F, Fränkel M: Vol 2, Section 2, Part 2 of K. von Bardeleben's Handbuch der Anatomie des Menschen. Jena, Fischer, 1908)

of muscles inaccurate is that there is actual variation in the segmental innervation of the same muscle among different individuals (as in animals; *e.g.,* Van Harrevald) just as there are variations in the segmental nerves that contribute to the brachial and lumbosacral plexuses, and in the contribution of the plexuses to the various peripheral nerves.

The probable segmental innervations of the muscles are discussed in the introductory chapters on the limbs, and noted also in connection with the discussions of the muscles themselves. A general summary of the segmental nerve supply to important functional groups of muscles is given on page 182. Last has enunciated some general rules for segmental innervation of muscles, which at least reduce the necessity for attempting to memorize tables of muscle innervations in the lower limb. His schema points out that, in general, the anterior muscles that move a given joint are supplied by two consecutive segmental nerves, while the posterior muscles that move the same joint in the opposite direction are typically supplied by the two succeeding segmental nerves; thus, for instance, flexion, medial rotation, and adduction of the hip joint are carried out by muscles that are largely supplied by the second and third lumbar nerves, while extension, lateral rotation, and abduction are carried out by muscles that are largely supplied by the fourth and fifth lumbar nerves. Further, the second general rule is that the innervation of the muscles acting across a joint shifts one segment distally with each joint that lies above it; for instance, hip flexors are innervated by L2 and L3, knee extensors by L3 and L4, dorsiflexors at the ankle by L4 and L5.

Obviously, these general rules do not take into account the fact that muscles often are innervated by more than two segmental nerves, or the variation of segmental innervation that may exist. Unfortunately, none of these "rules" holds very well for the upper limb, if the available data are correct, although there is a general tendency for proximal muscles to be innervated through higher segments than distal ones, and for flexor muscles at a given joint to be innervated from a higher level than the extensor ones.

Some muscle fibers have two motor end plates on them, and it has been suggested that perhaps sometimes the endings may represent two separate segmental nerves. In the cases that Feindel and co-workers found, however, both endings were always derived from the same axon. Even if some are not, and an occasional muscle fiber has a double segmental innervation, it is of no practical importance.

Degeneration and Regeneration of the Nerve Supply

Voluntary muscle undergoes atrophy after its nerve supply is interrupted, the muscle fibers being replaced, as they atrophy, by fibrous tissue produced by an increase in the interstitial tissue. However, this same process occurs when the innervation is anatomically intact but is nonfunctional because it comes from an isolated section of the spinal cord (Tower, '37); degeneration of muscle associated with interruption of its nerve supply seems to be largely an atrophy of total disuse. It has been shown both experimentally and clinically that electrical stimulation of denervated muscle does reduce the atrophy of such muscle, and ultimate recovery of function seemed to be better when the atrophy was minimal (Jackson).

Bowden and Gutmann followed in some detail, through biopsies, the changes occur-ring in human muscle following denervation and during the onset of reinnervation; they described the atrophy as consisting of a progressive shrinkage of muscle fibers, with, in the later stages of denervation, a breaking up of the fibers either into longitudinal parts or by fragmentation into round or oval segments, and finally, with deposits of hyaline material in these fragments. They denied that there was any metaplasia of muscle fibers into fibrous tissue, described by some authors, but did describe an increase in the amount of connective tissue and fat in the muscle. This tended gradually to distort the pattern of nerve branching within the muscle. They found that few if any muscle fibers undergo complete disintegration earlier than three years after denervation, although after this time they gradually break up and disappear completely and are replaced by connective tissue and fat. By three years, however, they said, the muscle fibers may be so shrunken and the interstitial connective tissue so increased that the chances of the muscle obtaining a normal innervation and returning to almost normal function are rather poor. Sunderland ('49) said that clinical evidence showed plainly that very good restoration of function can occur in human muscles that have been denervated for at least 12 months; that is, that within 12 months the degenerative changes in muscle are still readily reversible.

Gutmann and Young studied the process of reinnervation of muscle experimentally, and found nerve fibers returning rapidly to the motor end plates of the individual muscle fibers after the nerve had been crushed only; after long degeneration, they found, ingrowing nerve fibers may come in contact with a part of the muscle fiber that does not have an end plate, and develop a new end plate there. With long periods of atrophy, however, the number of endings in the reinnervated muscle never returns to normal. Edds pointed out that in a partially denervated muscle the remaining axons may develop collaterals that grow out to reinnervate the denervated muscle fibers, so that in such cases a single nerve fiber may support from 1.5 to more than 30

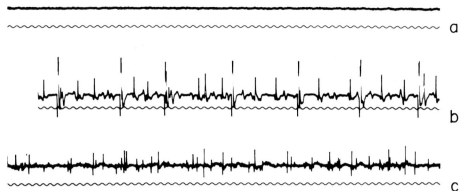

Fig. 1-14. Electromyograms, or tracings of action potentials of muscle. Electrical activity was detected by a needle electrode inserted into the muscle and recorded on a cathode ray oscillograph. An upward deflection indicates a change of voltage in the negative direction at the electrode. (*a*) A recording from a normal resting muscle shows no electrical activity; (*b*) one from a normal muscle during voluntary contraction records the action potentials of motor units close to the electrode; (*c*) a recording from denervated muscle showing action potentials of fibrillating muscle fibers. The time signal (*wavy line* below each recording) is 100 cycles/ sec. The amplitudes shown in *b* cannot be directly compared with those in *c*, for the recording conditions were different: the amplitudes in *b* range up to 1 milivolt, while those in *c* range up to only 0.1 millivolt. (Courtesy of Dr. E. H. Lambert)

times as many muscle fibers as it does under normal conditions. Thus, a paretic as opposed to a paralyzed muscle may recover function even when there is no actual regeneration of nerve fibers from the level of the lesion.

Regenerating nerve fibers to voluntary muscle all grow at about the same rate (in a given nerve and at a given level), hence, denervated muscles tend to recover their function in the order of their distance from the site of the lesion. Sunderland and his co-workers, in a series of papers cited in the chapters dealing with the various nerves, have reported in considerable detail the distances from certain skeletal landmarks to the entrance of branches into the muscles of both upper and lower limbs.

Fibrillation of denervated muscle (Fig. 1-14c) has been said to be a result of heightened excitability of the fibers to traces of acetylcholine (Denny-Brown and Pennybacker) and according to Tower ('39) typically involves only a part of individual muscle fibers, the length of fiber involved varying from a fraction of a millimeter to several millimeters in muscles whose fibers were from several millimeters to several centimeters in length. Tower

observed fibrillation persisting in denervated muscle for as long as 6 months to a year, but showed that if the muscle is tenotomized no fibrillation occurs. While she was unwilling to deny that fibrillation may contribute to the atrophy of denervated muscle (for the atrophy of tenotomized muscle, she said, was somewhat less severe) it is obvious that fibrillation is not the cause of atrophy, as some workers supposed.

VESSELS

Muscles have one or more arteries entering them. They may all be derived from the same chief vessel, or from different stems. An artery that runs between two muscles typically supplies both, but each muscle has its own blood supply otherwise; vessels do not enter and supply one muscle and then proceed into the next one to supply it. Within the muscle the blood vessels lie, as do the nerves, in the interstitial connective tissue; muscles have a particularly abundant blood supply, with capillaries intimately related to the individual muscle fibers. Since activity of a muscle increases the blood flow through it, the de-

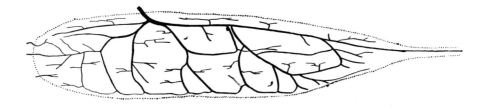

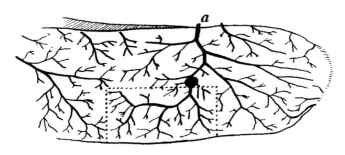

Fig. 1-15. Two types of vascular patterns to and in a voluntary muscle. The upper figure represents the anterior tibial muscle of the rabbit, supplied by two vessels that arise from a single stem and form through their branches a series of loops within the muscle. The lower figure represents a part of the adductor magnus muscle of the rabbit with two separate entering vessels from the profunda femoris and the femoral, respectively, and no large anastomoses between them. The outlined quadrangular area indicates a devascularized zone that was found after interruption of the branch from the femoral artery (*a*) at the level indicated by the black circle. (Clark WELeG, Blomfield LB: J. Anat 79:15, 1945)

creased blood flow resulting from forced inactivity may be a contributing factor to degeneration.

Clark and Blomfield found that the efficiency of anastomosis between intramuscular branches in the rabbit varies with different muscles (Fig.1-15), and that ligation of a chief arterial branch to a muscle may or may not result in localized necrosis. Blomfield's study of the intravascular patterns in a number of human muscles indicates that this is probably true also in man. He described five different types; some of these patterns, he suggested, predispose the muscles to damage as a result of vascular injury or obstruction, while others are such as to provide a good collateral circulation.

In the first group are muscles such as the gastrocnemius, which is supplied by a group of vessels arising from a common stem, entering the upper end, and running distally in the muscle; and muscles such as the biceps bra-chii, in which the intramuscular branches radiate from a common stem that enters close to the middle of the muscle. In the latter group belong muscles such as the soleus and the peroneus longus, in which a succession of vessels entering along the length of the muscle form a longitudinal anastomotic chain, and those such as the tibialis anterior, extensor digitorum longus, and the long flexors in the leg, in which there is a complicated pattern of anastomosing loops formed by a series of entering vessels. In between are muscles such as the extensor hallucis longus, in which a series of entering vessels anastomose somewhat sparsely in what is described as "an open quadrilateral pattern." It is to be noted that adjacent muscles—for instance, the gastrocnemius and the soleus, the tibialis anterior and the extensor hallucis longus—may have quite different patterns.

Further, the sources of the entering vessels also affect the susceptibility of the muscle to damage from vascular causes; if there is a succession of entering vessels all derived from one major vessel, the muscle may be as likely to be damaged by a vascular accident as a muscle in which the entering stem is single, and is more likely to be damaged than is a muscle in which several entering vessels arise from different sources. Here again, Blomfield found, adjoining muscles differ. The tibialis anterior and the extensor digitorum longus, although they adjoin and have a similar intramuscular pattern, differ in that all the vessels entering the former muscle are derived from the anterior tibial artery, while in the latter muscle the vessels are derived both from the anterior tibial artery and from branches of the posterior tibial that penetrate the interosseous membrane. The extensor hallucis longus, with a different intramuscular pattern, is intermediate: it gets most of its supply from the anterior tibial, but some additional supply from the posterior tibial. Obviously, there are no hard-and-fast rules that can be applied to the vascular supply of muscles; each muscle must be studied individually.

The blood supply to a tendon is also multiple, with vessels entering along its length. Edwards ('46) described a coarse network of arteries within tendons, some of the vessels continuing from the junction with the muscle, others entering from the surrounding loose connective tissue where no synovial tendon sheath exists; where such sheaths do exist, several blood vessels usually come in together, typically two veins to each artery. Edwards found no new vessels added at the insertion of a tendon into bone, but said that the vessels at this level do continue to anastomose with the periosteal vessels. Where a tendon has a long course through a tendon sheath, there tends to be one long main channel running on the surface of the tendon.

Lymphatics are found in muscle primarily in connection with the perivascular connective tissue, but may also occur immediately adjacent to the epimysium. Edwards described lymphatics in tendons as being particularly numerous, and forming a network as

do the arteries. The efferent lymphatics tend to run with the blood vessels.

BURSAE AND TENDON SHEATHS

Bursae and tendon sheaths are both formed by synovial membranes, similar in their structure to the synovial membrane of joints. The distinction between a bursa and a tendon sheath is purely that of gross relationship; a bursa lies between a muscle or tendon and some other structure, and allows freer movement than would be provided simply by the usual loose connective tissue. Bursae originate as clefts in the tissue adjacent to tendons and muscles; some of them (*e.g.*, the subscapular) usually undergo secondary fusion with a joint cavity, and others remain independent. Many bursae are well developed at birth and some, such as the subacromial, may be found as early as the third month of fetal life; probably, however, subcutaneous bursae do not usually appear until after birth (Black). Some appear adventitiously, as a result of friction in unusual locations, and many are quite variable in their occurrence.

In contrast to bursae, tendon sheaths surround tendons, and form a tube (Fig. 1-16) consisting of two layers, an outer one that encloses the cavity of the sheath, and an inner one that closely surrounds the tendon. These are usually called the parietal and visceral layers respectively. At each end of the tendon sheath, parietal and visceral layers become continuous; however, the parietal layer is thicker than the visceral one, and the latter tends to be very thin and intimately attached to the tendon, thus forming its smooth gliding surface.

Parietal and visceral layers of a tendon sheath may also be connected by a mesotendon; this consists of two layers of synovial membrane enclosing between them a variable amount of connective tissue, and usually transmitting blood vessels and lymphatics to the tendon. Mesotendons are largely attached to the deep surface of a tendon, therefore to the posterior aspect of flexor tendons and to the anterior aspect of dorsal ones. The same

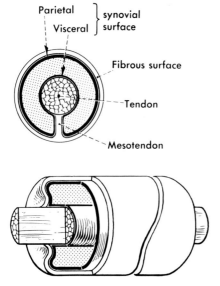

Fig. 1-16. Diagram of a tendon sheath. Cut edges of synovial membrane are represented by the heavier lines. *Above* is a cross section in which the parts are labeled, a mesotendon being included. *Below* is a view from the side with a large segment removed, and parts cut away at various levels on the left side.

tendons in different individuals show great variation in their mesotendons, which may persist throughout the length of the tendon sheath, disappear completely, disappear at one end but persist at the other, be much fenestrated, or reduced to a few threads. The vincula connected with the flexor tendons of the fingers and toes are remains of mesotendons.

Shields described tendon sheaths as developing simply by a separation of mesenchymal cells that surround the tendon, with retraction of their connecting cytoplasmic processes to form a cavity. He expressed the belief that their formation is a specific characteristic of the cells concerned, and not merely a result of a collection of fluid in the tissue.

FUNCTION OF MUSCLE

Although all details in the contraction of a single muscle fiber are not agreed upon and are far too complicated to be considered here, it is agreed that contraction involves inter-

action between the actin and the myosin filaments in the muscle fiber, and that release of phosphate ions from adenosine triphosphate is the source of energy for the contraction. (The three-volume work edited by Bourne is an extensive exposition on the structure and function of muscle.)

It is also true that knowledge of the functions of the muscles as a whole is somewhat incomplete, although the use of needle electrodes in electromyography has much increased our knowledge of the normal action of muscles, and supplemented and refined knowledge obtained by electrical stimulation (Duchenne) and from careful clinical studies such as those of Wright. Basmajian ('67), who has contributed greatly to the field himself, has also brought together the many scattered reports on the results of electromyography.

In general, the "actions" given for the various muscles are their primary actions, that is, the movements which they are believed to bring about rather than merely to modify. It is well known that many muscles also contract synergistically with the prime movers, and one of the problems of interpreting the results of electromyography is sometimes that of determining whether a given muscle is acting as a prime mover or as a synergist. Basmajian has objected to our old classification of muscles as agonist and antagonist, pointing out that contraction of the latter, when it does occur, is not in opposition to the movement, but is synergistic in order to prevent some unwanted effect by the agonist. He stated ('72) that the "antagonist" usually relaxes completely. However, Patton and Mortensen studied the flexors and extensors of the elbow and found co-contraction, which they attributed to stimulation of the muscle spindles and tendon organs. It was more common and in greater degree during extension than during flexion, and increased with the load.

In assessing the action of a muscle, note must be made of the relative cross-sectional area of the muscle, since this determines its strength, and of the length of its fasciculi, since this determines the range of movement it can produce. The fibers of pluriarticular muscles, for instance, are regularly too short to allow them to produce full movements at all the joints they cross; as a well-known example, the fasciculi of the long flexors of the fingers are of such lengths that when they contract to their maximum, which is slightly less than half of their extended length, they cannot flex the fingers and the wrist simultaneously, because this involves too great an approximation of the insertion to the origin.

As pointed out by MacConaill ('49), the spatial relationship of the origin and insertion of a muscle to the joint upon which it acts also determines in part its function: in general, a muscle arising near a joint and inserting far away from it (a "shunt" muscle) tends to exert force primarily along the bone in the direction of the joint, while one arising far from the joint and inserting close to it (a "spurt" muscle) tends to exert its force in such a manner as to produce movement along a curve. Spurt muscles, which provide acceleration, are therefore used in all movements, fast or slow, of a joint. Shunt muscles, which resist centrifugal force, are then used primarily when rapid movements generate that force. In the case of the forearm flexors, Basmajian has identified the brachialis and biceps brachii as spurt muscles, while the brachioradialis, which becomes active particularly during fast movement, is the shunt muscle. Similar identification at other joints has ap-

parently not been made electromyographically.

Another consequence of the varying distance of insertion from a joint is the difference in actions brought about by the length of the lever arm: muscles inserting close to a joint produce a wider excursion of the parts distal to the joint, for a given amount of shortening (Fig. 1-17), but also a less powerful one for a given muscle mass.

Another feature that modifies the actions of muscles is the presence of fibrous sheaths or retinacula, such as those at the wrist and on the fingers; these, by holding the tendons parallel to the bones that they cross, transfer the major part of the effect of muscular contraction to the bones upon which the tendons insert.

It is true that muscles actively contract, and passively elongate as a result of the pull of antagonistic muscles or other forces, and from this standpoint it is often said that muscles can only pull, and never push. This statement is largely true, and yet it is also true that as a muscle contracts it widens, and therefore exerts pressure (pushes) on its surroundings; this expansion of the diameter of a muscle has been shown by Pfuhl to be an important element in the total excursion of the tendons of the more complicated types of muscles (Fig. 1-8), and is, of course, also important clini-

Fig. 1-17. Effect of the place of attachment of a muscle on the range of movement. Both muscles are shown as shortening the same amount, but *a,* attached closer to the joint, moves the lever over a far greater range than does *b.*

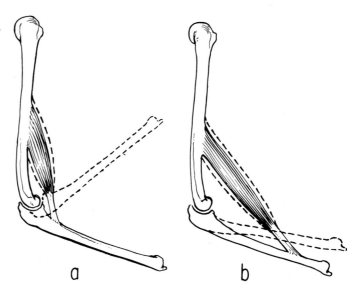

a   b

cally from the standpoint of the massage that it affords the veins and lymphatics.

Biarticular Muscles

There are a number of muscles that pass across two joints and are therefore so situated that they are regularly regarded as being able to act upon either joint or upon both. It is generally acknowledged that these muscles cannot act effectively at both joints simultaneously because of the length of their fasciculi. The idea, once advanced, that such a muscle can act differentially at one joint by contracting only at the end crossing that joint has been disproved by electromyography (Basmajian).

The chief advantage of two-joint muscles, as described by Elftman, is that by being extended at one joint, such as the hip, they can then develop greater tension to be used in flexion of the other joint, in this case the knee.

Motor Unit

It is well known that a single nerve fiber usually innervates a number of muscle fibers through intramuscular branching, and the muscle fibers innervated by a single nerve fiber are known as a "motor unit" since they necessarily contract at the same time. The size of the motor unit varies markedly with the muscle considered, but is probably dependent largely on the fineness of the movement that must be carried out by the muscle. In regard to the muscles moving the eyeball, for instance, the ratio between nerve fibers and muscle fibers indicates that the motor unit here is probably not much larger than three muscle fibers; in the larger muscles of the limb, it probably varies from a hundred to several hundred muscle fibers. Elftman quoted a motor unit of approximately 1,600 fibers for the quadriceps.

It is now recognized (*e.g.,* Feindel and co-workers) that all the muscle fibers constituting a motor unit are not arranged as a group, but rather in a number of groups that are intermingled with groups of other motor units. Nevertheless, there are sufficient fibers in a group to allow it to be identified electromyographically. Harrison and Mortensen, and subsequently Basmajian, showed that persons can learn to contract a specific motor unit.

Contracture

Contracture is a general term used to designate a prolonged or more or less permanent shortening of muscles, and is usually classified into three subtypes. *Hypertonic contracture* is a result of an increased flow of impulses from the central nervous system, such as occurs in spasticity and rigidity, and the essential factor of the contracture is the maintained hypertonicity of one group of muscles at a joint as compared with others. This includes also hysterical contracture and reflex contracture, and is characterized by the fact that while it persists during waking hours it disappears during deep sleep or anesthesia. A second group, *myostatic* or *myogenic contracture,* is characterized by the fact that a relatively permanent shortening of the muscle fibers occurs, as a result of lack of proper stretching of the muscle over a period of time; while the initial shortening originates as a result of nerve impulses, the shortening is eventually maintained even in the entire absence of nerve impulses, for the muscle adjusts to its new length by actually becoming shorter. Myogenic contractures may appear as a result of immobilization over a period of time, as a result of division of the tendon of a muscle, or in the innervated muscles at a joint where there is muscle imbalance consequent to paralysis of an antagonistic muscle group. The third type of contracture, *fibrotic,* differs sharply from both of these, in that it is actually not a result of muscle function, but rather one of degeneration of muscle; this type of contracture is seen, for instance, in Volkmann's contracture, or after any lesion that leads to degeneration of the muscle with its replacement by connective tissue.

Steindler has discussed clinical aspects of contractures in regard to disability of the wrist and fingers; Davenport and Ranson have discussed contractures resulting from tenotomy; and Sherman has discussed those produced by peripheral nerve lesions.

Following tenotomy, a muscle shortens because it is no longer subject to the tension of opposing muscles; in the case of the gastrocnemius of the rat, Davenport and Ranson found a shortening of from 6% to 22% for the entire muscle, a decrease of 32% in length for the individual muscle fibers; the obliquity of the fibers in relation to the muscle accounts for the difference in shortening found. They also found a loss in weight of the muscle, but no marked histological changes indicative of degeneration (their animals lived a maximum of 44 days after tenotomy, and in some of the cases the tendon had healed and the muscle had resumed function at the time of necropsy). The shortened muscle fibers had definitely become wider, and stimulation experiments showed that as a result of this shortening the excursion of the tendon was reduced from an average of 7.5 mm in the normal rat to one of 4.6 mm for the muscles in contracture.

The loss in weight of a muscle that is not functioning properly indicates a partial atrophy of muscle fibers; this in turn invites fibrosis, and if the latter feature becomes marked a myogenic contracture may be gradually converted into one at least partially fibrotic.

Sherman found contractures occurring in the dorsiflexors of the foot of cats in which the sciatic nerve had been sectioned and prevented from regenerating; his findings suggest that these contractures resulted from the formation of fibrous tissue in muscles relatively fixed in a shortened state, but he apparently did not study this group in detail. Contractures also developed in animals in which primary and secondary suture of the sciatic nerve, with consequent regeneration, had taken place. Sherman noted that while at first the contractures were rather elastic, they later became more resistant; however, as function in all the muscles recurred the incidence of contracture dropped off, indicating that the contractures were reversible. He found residual contractures at the end of 150 days to be both more severe and more numerous after secondary suture of the nerve (33%) than after primary suture (10%).

Sherman's evidence appears to indicate that contractures following denervation of two antagonistic groups of muscles in a limb may be of two kinds: one, the result of denervation, is fibrotic, and occurs in the muscles that are more habitually kept in a shortened position; the other, the result of the earlier reinnervation of some muscles than of others, is myogenic, for it tends to develop in the muscles that recover first. The longer the difference in time in the recovery of opposing muscles, the more likely contracture was to develop in the muscle that recovered first.

Sharrard, discussing deformities of the lower limb that may result from paralysis in children, found muscular contraction responsible for none of them. He said contractures occurring in the acute phase of poliomyelitis bear no relation to the muscular paralysis, but are caused by the deposition of collagen in tendons, fascia, and ligaments, and deformity can be prevented by passive movement, as can also the deformity that may develop in a flail limb. In contrast, he said, deformity always develops in children when only some muscles of a limb are paralyzed, but it results from muscle imbalance brought about by the fact that the innervated muscle does not grow properly. It shows no fibrosis, he said, unless it has been injured by overstretching, and its relative shortness increases during the period of growth.

REPAIR OF MUSCLE AND TENDON

Muscle

It is well known that increase in the size of individual muscles, as with exercise, is produced by hypertrophy of the individual muscle cells, and not by formation of new ones. Similarly, it has usually been said that there is no regeneration of muscle in the sense of replacement of fibers that have been destroyed. Thus Clark, and Clark and Wajda studied the regeneration of mammalian voluntary muscle, in which they had injured muscle fibers by crushing them or interrupting a part of the blood supply of the muscle, and found that the "new" muscle fibers all arose as direct and continuous outgrowths from the muscle fibers still present. They described these fibers as growing at the rate of about 1.2 mm to as much as 1.7 mm a day, and (in the rabbit) achieving an average width of about two thirds that of normal in 21 days and about 97% of normal within 4 months. Gay and Hunt described muscle fibers that had been transected as uniting again by a similar process of budding from the cut ends, with eventual union of the buds.

As reviewed by Carlson, there is now no doubt that muscle fibers are regenerated and become anatomically and physiologically normal; the new myoblasts originate either from the nuclei of injured cells or from the satellite cells and fuse to form myotubes. Carlson expressed the opinion that small areas of degenerated muscle may be replaced fairly commonly, but noted that there is no technic for producing the massive regeneration required to restore a sectioned muscle to normal.

Clark and Wajda noted that immobilization of the limb severely impaired the process of repair and regeneration, and that tenotomy of the muscle involved, while not returning the regenerating muscle to the original stage of degeneration, did produce evidence of secondary degeneration and of a greater abundance of fibro-elastic tissue in the muscle. These observations emphasize again the important effect of tension upon the health and growth of muscle fibers. Davenport and Ranson also emphasized this, quoting evidence that if an injured muscle is kept under too great tension it degenerates further rather than regenerating, and yet if a muscle is permanently tenotomized and thus never subject to tension it also degenerates. Saunders and Sissons found no evidence (in rats) that denervation particularly affected the process of regeneration in muscles.

Tendon

Mason and Shearon studied the histology of healing tendon in the dog and found that while union of the cut ends first occurred as a result of proliferation of the tissue of the tendon sheath (peritendineum), within 4 to 5 days the tendon itself began to proliferate and send cells into the callus formed by the sheath. (Dodd and co-workers agreed and said the fibroblasts arise in both instances from or in association with capillaries. Potenzo apparently thought that healing is brought about entirely through tissue extrinsic to the tendon.) Tendon can bridge a small gap in about 2 weeks. Because of the part that the peritendineum plays in the process, Mason and Shearon suggested that this should be carefully approximated in attempting to repair the tendon. They expressed the belief that movements of sutured tendons may be started cautiously by the fifth or sixth day, but that no force should be exerted before the third week.

Mason and Allen measured the tensile strength of healing tendon, and showed that during the first 5 days after suture there is rapid diminution in the strength of union. This is followed by an increase in strength up to about the sixteenth day, followed by a third stage in which the tendon's strength gradually increases until it returns to normal. The stages of diminution and initial increase in strength of the suture line obviously correspond to the period in which there is proliferation of connective tissue and union of the two ends by this, and it is not until after about 14 days that the strength of union returns to the strength of the immediately sutured tendon; thereafter, strength increases as the scar is converted into tendon. They noted that use of the tendon had no effect upon the speed of repair during the first two stages of repair unless it resulted in separation of the ends, but did accelerate the rapidity of healing during the third stage. Allowing the muscle to function during the first two stages led to an increased reaction and to separation at the suture line. Active use of the muscle even after 3 weeks sometimes produced stretching at the suture line. Rothman and Slogoff reported that the blood supply of normal tendon decreased significantly with immobilization for 6 weeks, emphasizing the value of movement in the vascularization of tendon.

These studies emphasize the fact that if motion of a sutured tendon is to be started early to prevent adhesions and improve the blood supply, it must be done very gently and over a limited range. The part must otherwise be kept in such a position that contraction of the muscle concerned cannot exert a pull upon the suture line.

Kernwein found that when tendon is implanted into bone through a drill hole, the process of repair of the bone results in an ossification about the tendon with gradual incorporation of the tendon into the bone; thus, eventually, the usual relationship between tendon and bone is attained. As in the case of sutured tendon, he noted that the strength of the repair falls for a while below that of the initial fixation, and advised that the part should be immobilized for about 21 days until this initial period is past. He also noted that shredding the portion of tendon inserted into bone increases the surface area and accelerates the rise of tensile strength between tendon and bone.

Whiston and Walmsley said that tendon implanted into bone undergoes degeneration and is eventually entirely replaced by new tendon tissue; Flynn and co-workers said this is also true of all tendon grafts. Heiple and co-workers ('67), however, transplanted tendon in which the collagen had been labeled with radioactive carbon and found radioactivity present in the "new" tendon as long as 24 months after transplant. Thus, the collagen either persists as an element of the new tendon or is re-used locally by the fibroblasts that migrate into the graft. The latter seems unlikely.

Blood Vessels and Lymphatics

A majority of the problems connected with surgery of the blood vessels and lymphatics, other than their regional anatomy, are purely physiological and clinical rather than anatomical. Hence, many important considerations are not discussed in this section.

Saunders and co-workers have emphasized that the peripheral vascular bed (of muscle and skin) does not correspond to the usual image invoked by the term "vascular tree."

Instead of arteries branching until they end in capillaries, small arteries anastomose with each other to form a coarse mesh, from which branches are given off to form a fine mesh; the veins are similarly disposed, and the meshes thus form a considerable part of the peripheral vascular bed. The vessels of the two finer meshes are connected to each other both by capillaries and by arteriovenous anastomoses.

GENERAL STRUCTURE

The histological structure of all these vessels is essentially similar, the differences lying in the thickness of their walls and the varying degree in which different tissues are represented here.

Regardless of their caliber, all the vessels are lined with a continuous layer of endothelium (or, in the case of sinusoids of the liver and certain other organs, with this and phagocytic reticuloendothelial cells). In vessels of appreciable size three coats (tunics) are distinguished, an intima, a media, and an adventitia. The intima consists of the endothelium and a variable but always relatively slight adjacent amount of connective tissue; the media, primarily of smooth muscle; and the adventitia, primarily of connective tissue.

Arteries

The *tunica intima* of small arteries and arterioles consists only of endothelium and an internal elastic membrane; that of arteries of medium size (most of the named arteries and their larger branches) also has a well-developed internal elastic membrane, and between this and the endothelium there may be collagenous tissue and other elastic fibers; in the largest arteries there is a thin layer of connective tissue adjacent to the endothelium, but the internal elastic membrane may not be marked because of the presence of similar elastic membranes in the media.

The *tunica media* of small arteries and arterioles consists of smooth muscle cells arranged circularly in the wall of the vessel; some vessels, for instance, the bronchial arteries, also contain longitudinally arranged smooth muscle. In the great majority of arteries, those of so-called medium size, the media is likewise almost exclusively smooth muscle, with only a small amount of reticular and elastic fibers. The tunica media of the largest arteries contains relatively little smooth muscle, its place being largely taken by concentric layers of elastic tissue.

The *tunica adventitia* of all the arteries consists of mixed collagenous and elastic tissue, the fibers passing predominantly in a longitudinal direction. It blends with the surrounding connective tissue and is sufficiently loose to allow movement of the artery and changes in the size of its lumen. In medium-sized arteries the innermost part of the adventitia forms an external elastic membrane.

Veins

Veins (see Fig. 1-22) differ fundamentally in structure from arteries in regard to the media only, which is always thinner than that of an artery of corresponding size. Both the muscular and the elastic tissue are often poorly represented in the media of veins, and there may be no clear distinction between intima, media, and adventitia. The smallest veins have no media at all; veins of medium size typically have an adventitia that is much thicker than the media, and, in the largest veins, the media as a muscular layer may be poorly developed or absent, its place being taken by a thicker connective tissue layer joining the intima and the adventitia. In these large veins, the tunica adventitia may contain longitudinally arranged smooth muscle.

Valves are discussed later in this chapter.

Lymphatics

The small lymphatics have only a little connective tissue and a few smooth muscle cells associated with them, and it is not until they attain the diameter of approximately 0.2 mm or more that the three coats typical of blood vessels can be seen. As in the case of veins, the distinction between the layers may be poor. As the lymphatics unite and become larger, their walls become thicker. The thoracic duct, the largest lymphatic, is said to have a better developed muscular layer, or tunica media, than do the largest veins.

Connections between the lymphatic system and the venous system other than in the neck have been considered rare, although some animals have been said to have such connections in the abdomen normally. However, Pressman and co-workers believe they have demonstrated much finer connections within lymph nodes through which, when the lymphatic pressure is elevated, lymph moves directly into the blood stream.

The endothelium of lymphatic capillaries is said to differ somewhat from that of blood capillaries in its fine structure (Leak and Burke); some of the cells present open spaces between them, suggesting that in these locations fluid and large molecules can move very readily into the lymphatic.

Vasa Vasorum

Blood vessels whose diameters are more than approximately 1 mm receive blood vessels, vasa vasorum, from adjacent small blood vessels; these penetrate the adventitia and form a capillary network in it, but in the arteries do not penetrate farther than the external layer of the media, if that far; in veins, however, they are said to go even as far as the intima. Lymphatics are said to have their own blood supply when they are much smaller than blood vessels large enough to have vasa vasorum (Evans, '07).

The larger veins and arteries are said to have also a network of lymphatics in their adventitia; these connect with the perivascular lymphatics, which regularly accompany most of the blood vessels.

INNERVATION

Apparently, little or nothing is known of the innervation of lymphatics, and the innervation of arteries has been studied more extensively than has that of veins. Woollard described in some detail the nerves to vessels of the lower limb in experimental animals. The nerves contain both afferent and sympathetic fibers, and Woollard found that the distal

part of the vascular tree is especially abun-
dantly supplied with afferent fibers; he described them as ending in part in the adventitia of the blood vessels, but as also sending
collaterals into adjacent connective tissue to
end there. The sympathetic fibers, of course,
end on the media of the blood vessels. Woollard said the innervation of veins is similar to
that of arteries, but that veins are provided
with fewer nerve fibers.

It is obvious that the sympathetic supply to
many blood vessels is primarily vasoconstrictor. The innervation to blood vessels within
voluntary muscle, however, is said to include
both vasoconstrictor and vasodilator fibers
(Dorr and Brody).

Anatomy of Vascular Nerves
The source of nerve supply to the vessels of
the limbs has been of special interest in rela-

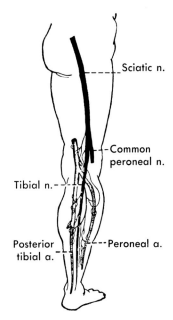

Fig. 1-19. Nerve supply to arteries of the leg. (After
Potts, from White JC, Smithwick RH, Simeone FA:
The Autonomic Nervous System [ed 3]. New York,
Macmillan, 1952)

Fig. 1-18. Nerve supply to some of the arteries of the
forearm and hand. (Kramer JG, Todd TW: Anat Rec
8:243, 1914)

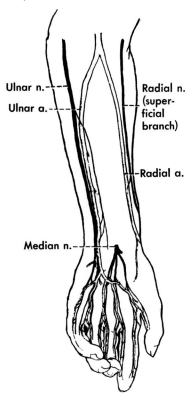

tion to sympathectomy for various vasospastic
conditions. It is now well recognized that,
contrary to previous concepts, periarterial
plexuses in the limbs of man do not extend
from the base of the limb to the periphery,
but are renewed at intervals by the accession
of new fibers from adjacent peripheral nerves;
these new fibers, depending on the location of
the artery in question, may be derived from
one of the major nerve trunks (Figs. 1-18 and
1-19), or from cutaneous nerves (for instance,
Pick; Mustalish and Pick). Pick has reported a
particularly careful study of the dissectable
vascular nerves to arteries of the upper extremity, beginning with the subclavian, and
found them arising from the sympathetic
trunk; trunks and cords of the brachial
plexus; the radial, ulnar, and musculocutaneous nerves in the arm; the median, ulnar,
lateral antebrachial cutaneous, and superficial radial in the forearm; and both the deep
ulnar and digital branches of the median and
ulnar in the hand. The same general pattern
applies to the lower limb; for instance, the

medial and lateral plantar arteries receive branches from the medial and lateral plantar nerves. Woollard's observations indicate that even the nerve supply to vessels within a muscle, although usually entering along the main vessel, may be supplemented by direct branches from the nerves within the muscle.

In agreement with the anatomy of the periarterial plexuses, Blair and co-workers showed that periarterial sympathectomy in man results in the disappearance of nerve fibers along the femoral artery for only a short distance distal to the point of operation, beyond which these vessels receive additional fibers from adjacent nerve trunks (in man the vascular nerves are said to be especially numerous in the popliteal space); Moore and co-authors were unable to demonstrate vasodilation in the hind limb of the cat following periarterial sympathectomy, but did produce it by section of the sciatic or femoral nerves. Thus, early reports that periarterial sympathectomy produced hyperemia lasting for at least some weeks (e.g., Wikle) are difficult to explain. White and co-authors said the rise in temperature of a limb treated by periarterial sympathectomy is a nonspecific effect resulting from the absorption of protein decomposition products released by the destruction of tissue.

While the anatomy of the nerve supply to blood vessels seems, in general, to be thoroughly established, the exact role that the innervation plays in the complicated problem of blood flow is probably not yet clearly understood, and there are varying opinions among clinicians as to the applied physiology of the vasomotor nerves. Burton and Yamada pointed out that the greater the vasoconstriction, the higher will be the "critical closing pressure," that is, the level of blood pressure at which blood will cease to flow through a given vessel (Fig. 1-20): therefore vasoconstriction is an especially important factor in conditions in which the blood pressure is generally lowered, as after extensive hemorrhage, or in which it is lowered locally by partial occlusion of an artery. Lumbar sympathetic block or sympathectomy is certainly indicated when a large vasospastic element is known or suspected to be a contributory factor to ischemia. However, interruption of the sympathetic outflow to a limb will obviously fail to increase the blood flow to it if the small arteries are so involved as to be no longer capable of dilating; similarly, relaxation in the vascular bed of a given artery is useless unless either the blood flow through that artery can be thereby increased, or there are collateral channels capable of increasing their flow to the relaxed vascular bed.

While there have been statements to the effect that the only real result of sympathectomy is to shift the blood flow from the muscles to the skin, much of the disagreement among clinicians as to the effectiveness of sympathectomy in various disease conditions is probably due to variation in the degree to which vasospasm contributes to the ischemia.

Because all available evidence indicates so clearly that periarterial sympathectomy does not denervate, either efferently or afferently, any large proportion of the vascular bed distal to the level of operation, this technic has been generally abandoned in favor of the more rational lumbar sympathectomy (chain ganglionectomy) or block. Unexplained, but perhaps based upon altered pressure relations rather than further sympathectomy, is the observation (Ivins) that a pulseless artery, already presumably denervated by lumbar sympathectomy, may begin to pulsate immediately after it is divided.

Denervation for Ischemia

Treatment of vascular disturbance, especially of the lower limb, by sympathectomy and sympathetic block has been discussed by, among others, Shumacker; Goetz; Olson and Leming; and Palumbo and co-workers. Most writers have emphasized that sympathectomy is not a cure-all in the treatment of ischemia, although it may be a valuable adjunct, and that before sympathectomy for chronic conditions is undertaken the patient should be very carefully studied; obviously, if temporary sympathetic block has no effect, neither will sympathectomy or permanent block. Goetz discussed in detail technics of evaluat-

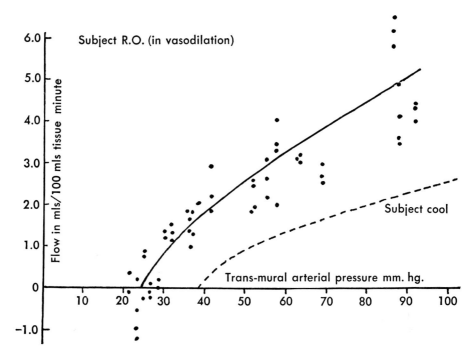

Fig. 1-20. Effects of vasodilation and vasoconstriction on blood flow in the human arm, and the cessation of flow while there is still a positive arterial pressure. Note that with vasoconstriction (*subject cool*) the flow was much lower at any given pressure, and stopped entirely a little below 40 mm of mercury; with vasodilation (*solid line*) the flow stopped only at a much lower arterial pressure, and at any given pressure was very much higher than with vasoconstriction. (Burton AC, Yamada S: J Applied Physiol 4:329, 1951)

ing the possible results in patients with chronic vascular disease.

Olson and Leming reported that in their experience lumbar sympathectomy gave most favorable results in patients with arteriosclerosis and diabetes. Palumbo and his colleagues reported most favorable results in patients with thromboangiitis obliterans (Buerger's disease), varicose or chronic ulcers, and frozen feet; in a whole series of conditions for which they did sympathectomy they considered the results to be good in 65%, fair in 28%, and poor in 7%. Golding and co-workers reported that sympathectomy for frostbite intensifies edema if done too soon, but produces faster healing if done 24 to 48 hours after the part has thawed.

White and co-workers, reviewing the literature on the subject, found evidence that sympathectomy may be useful both in arteriosclerosis and thromboangiitis obliterans, but

warned that the patient must be very carefully studied, and said that the operation should not be done in the presence of rapidly advancing gangrene or infection. Strandness and Bell found little objective evidence that lumbar sympathectomy is useful in the treatment of intermittent claudication.

Sympathectomy often was used to improve the circulation during the period when injuries to major vessels were commonly treated by ligation rather than repair. For instance, Palumbo and his colleagues said that in their experience it apparently prevented or delayed the need for major amputation in many cases, and that when it did not prevent amputation it frequently lowered the level at which amputation could be performed. Similarly, Linton found sympathectomy effective in preventing gangrene following excision of popliteal aneurysms.

Several writers have pointed out that sympathectomy as a last resort, in an attempt to lower the level of amputation in a gangrenous limb, cannot be expected to succeed. Further, with the various methods now available to restore the continuity of vessels, sympathectomy is indicated only when there is evidence of spasm in the peripheral vascular bed.

COLLATERAL CIRCULATION AND VASCULAR LIGATION AND REPAIR

Collateral Circulation as Tested by Ligation

Anatomical studies are useful in indicating the possible collateral pathways available when a given vessel is obstructed, and the size of these anastomotic pathways obviously affords some information as to the possible adequacy of this collateral circulation. However, they cannot be definitive, and the expected incidence and extent of gangrene following ligation of various arteries can only be ascertained through past clinical experience. Learmonth pointed out that if the collateral channels are short there is little loss of pressure in the distal part, but if they are long there is a great loss of pressure and flow; and several authors have pointed out that the best levels of ligation, from the standpoint of developing as direct a collateral circulation as possible, are immediately below a major branch proximal to the injury, and immediately above one distal to the injury.

Statistics on the incidence of gangrene following ligation of various vessels vary widely, depending in part on the level of ligation, the reason for ligation, the length of time elapsing between injury and treatment, and other factors. However, even particularly favorable reports, such as shown in Figure 1-21, indicate that ligation should be a method of last resort. DeBakey and Simeone reported that during World War II, in which almost all arterial wounds of the extremities were treated by ligation, amputation was necessary in 40.3%, ranging from a low of about 26% in injuries to the brachial artery to a high of 72.5% in injuries to the popliteal. If the brachial artery was ligated above the profunda, the percent-

age of amputation rose to 55.7. For injuries to other major arteries, they reported amputation in 28.6% when the subclavian was involved, 39.3% for the radial and ulnar arteries, 43.2% for the axillary, 46.7% for the external iliac, 53.8% for the common iliac, 56.1% for the femoral (and if above the profunda, 81.1%), and 69.2% for the anterior and posterior tibials. A very important factor was the time lag, inevitable in battle, between the time of injury and that of medical and surgical attention; a lag of more than 20 hours raised the total amputation rate to 63%, one of 10 hours or less lowered it to 36.7%. Gage and Ochsner, presumably quoting statistics from civilian practice, said gangrene is reported to have followed ligation of the subclavian artery in 25% of cases, of the axillary in 16.6%, of the femoral in 25.0%, of the popliteal in 41.6% (or, in other cases, varying from 20% to 40% according to the level of ligation), and from the tibial artery never.

Although concomitant ligation of the corresponding vein was once regarded as improving the collateral circulation, clinical experience during World War II indicates that it does not effectively lower the amputation rate (Elkin and DeBakey). The rate was admittedly much lower after quadruple ligation for arteriovenous aneurysm, but this is because of the collateral circulation already developed around the aneurysm (e.g., Holman, '54). Nevertheless, a real functional disability, as a result of either arterial or venous insufficiency, was found in a high percentage of patients treated by this method (Foley, Allen, and Janes), and end-to-end suture has now replaced ligation. Holman emphasized that with the newer methods of vascular repair it is feasible and advisable to treat arteriovenous fistulas early; failure to do so puts an extra burden upon the heart, and may also make repair more difficult because of increasing injury to the vascular wall and enlargement of the arterial segment proximal to the fistula.

REPAIR OF INJURED VESSELS

Lateral arteriorrhaphy, the simplest method of repairing a wound in the wall of an artery, has been found unsatisfactory except for the

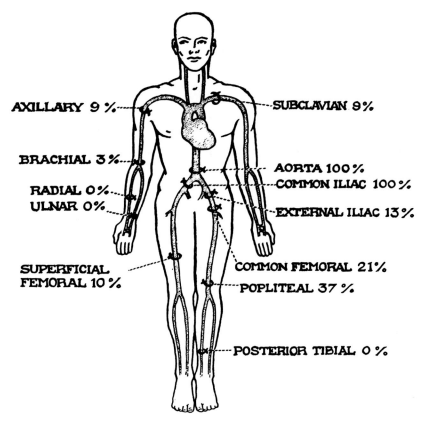

AXILLARY 9 %

SUBCLAVIAN 9%

BRACHIAL 3%

AORTA 100%

COMMON ILIAC 100%

RADIAL 0%

ULNAR 0%

EXTERNAL ILIAC 13%

SUPERFICIAL
FEMORAL 10 %

COMMON FEMORAL 21%

POPLITEAL 37 %

POSTERIOR TIBIAL 0 %

Fig. 1-21. Expectation of gangrene following ligation of certain major vessels; far higher percentages have also been reported—see text. (After Bailey, from Ferguson LK, Holt JH: Am J Surg 79:344, 1950)

very smallest wounds. Because it necessarily produces constriction, it invites thrombosis, reduces the blood pressure and flow distal to the constriction and provides the proper conditions for later development of a poststenotic aneurysm.

In the first place, with any given level of general blood pressure, a constriction of an artery necessarily reduces the blood pressure distal to that constriction, and Burton and Yamada have shown that flow in an artery drops off much more rapidly, with reduction in pressure, than would be expected from the laws governing flow through a rigid tube: as one example, a reduction of the mean blood pressure from 100 mm of mercury to 85 mm, which would be expected according to classic theory to give a reduction in flow of only about 15%, actually reduced arterial flow by 60%. Thus, a decrease in blood pressure may

have very disastrous results insofar as nutrition of tissues is concerned. Further, they also demonstrated that blood vessels will collapse and cease to conduct blood at all while there is still a positive head of arterial pressure; the level at which collapse occurs varies with conditions, but vasoconstriction produces collapse at a higher pressure, and therefore a pressure that produces adequate flow with vasodilation may fail to produce any at all with vasoconstriction (Fig. 1-20). The second important consequence of a local constriction, the production of an aneurysm distal to the constriction, has been demonstrated by Holman ('55); he produced aortic aneurysms at will in puppies by constricting the aorta, and analyzed the factors leading to the production of poststenotic aneurysms.

In view of these possible results, it is obvious that lateral arteriorrhaphy, except for the most minute wounds, is not acceptable. The practice now is to divide completely an injured artery and then suture it end to end.

The superior results of arterial reconstruction through end-to-end suture or the use of grafts are now well known and can be adequately illustrated by two brief references. Thus, in comparison with the over-all amputation rate of 40% reported to follow arterial ligation in World War II, Jahnke and co-workers reported an over-all amputation rate of only about 10% following the reparative operations of the Korean War. In the case of repair of the popliteal artery, the amputation rate dropped to about 18% to 20% from the 75% or more reported to follow ligation. And, in a more recent review, Dillard and co-workers quoted a similar, and slightly more favorable, outcome to treating arterial injuries in the Vietnam War, and reported that in their experience with 67 injuries to vessels of the extremities in civilians, amputation was necessary in only two instances (both among 10 injuries to the popliteal artery).

Jahnke and his colleagues, Holman ('54), and others have described the technic of end-to-end suture. Important points are that about 1 cm of a lacerated artery be removed both proximal and distal to the level of injury, in order to insure an intact arterial wall, and that care be taken that the suture lines are not on tension. Jahnke and his colleagues reported that 75% of the repairs during the Korean War were made by end-to-end suture, while in 25% arterial or autogenous vein grafts (from the saphenous or cephalic) from 1 inch to 3 inches (2.5 cm to 7.6 cm) long were used. Dillard and co-workers likewise used end-to-end suture most commonly, and autogenous vein graft next most commonly.

Grafts are also used to replace sections of arteries involved in aneurysms or by thrombosis. Arterial homografts, when available in the proper size, were once preferred by some workers (*e.g.,* Brown and Hufnagel). More recently, tubes made of one of the newer synthetics have been preferred for larger vessels, while autogenous vein grafts have been favored for bypass of arteries in the extremities (for instance, Dale and DeWeese; Royle).

VEINS

Kampmeier and Birch studied in some detail the development of valves in the venous system. They found the first valves arising at about 3½ months of fetal life, the earliest valves of the lower extremity appearing in the deep veins in the region of the femoral triangle, in the popliteal fossa, and in the upper end of the great saphenous vein. They also reviewed the literature and reported upon the distribution of valves, especially in the veins of the lower limb. In general, they pointed out, both the superficial and deep veins of the limbs are well supplied with valves, but they are much more scarce in other parts of the venous system. Smaller veins are said to have, as a rule, valves with a single cusp, larger ones to have (Fig. 1-22) valves with two or three cusps. They quoted evidence that the emissary veins of the skull may have valves that direct the blood peripherally, that the lingual, superior thyroid, facial, and temporal veins typically have valves at their mouths, where they join each other or the internal jugular, and that the internal jugular usually has a valve at its lower extremity, although it is often imperfect. They also said that a valve typically occurs where the deep cervical vein joins the vertebral, and another where the vertebral joins the brachiocephalic (innominate); they found no valves in the anterior jugular but said the external jugular usually has an incompetent valve at its junction with the subclavian, and another one some 1 to 2 inches above the clavicle.

They also said that the brachiocephalic (innominate) veins, the superior vena cava, and the inferior vena cava have no valves of their own, but that the mouths of the veins which enter into them do have valves; that the intercostal veins which enter the azygos vein are provided with valves at their mouths, but those which enter the hemiazygos are not; and that no valves are found in the venous plexuses of the vertebral column and spinal cord except where the chief channels empty into the intervertebral veins and these in turn empty into the vertebral, intercostal, lumbar, and sacral veins. Valves have also been de-

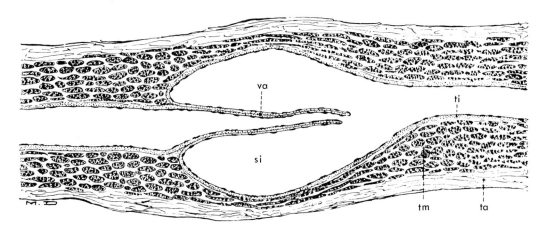

Fig. 1-22. Longitudinal section of the great saphenous vein of a newborn infant: *ti* is the intima, *tm* the media, and *ta* the adventitia; *va* is one of the cusps of a bicuspid valve, *si* the valve sinus. Note the thinness of the venous wall here. (Kampmeier OF, Birch CLaF: Am J Anat 38:451, 1927)

scribed in various parts of the portal system. Thus, while it is generally true that valves are scarce in the veins of the head and neck, the thorax, and the abdomen, it is obvious that they are not completely lacking; experimental and clinical observations show, however, that many of the valves are either incompetent to begin with, in adults, or else are readily induced to become incompetent.

Kelly pointed out that the valves of a vein do not necessarily become incompetent as a result of dilation of the vein, and reported the case of a man with very large veins in one forearm (as a result of an arteriovenous aneurysm) in whom the valves were still competent in spite of the great increase in diameter of the vessels. Edwards ('36) said that in the superficial veins of the extremities, at least, the valves are so oriented that the aperture between the cusps of a bicuspid valve is parallel with the surface of the skin, so that pressure upon the skin pushes the cusps together.

Acute massive venous occlusion obviously interferes markedly with the venous return from an affected limb, and according to Osius may or may not result in gangrene, but does usually result in severe shock. Veal and co-workers stated that the cyanosis and acute pain resulting from the venous congestion can frequently be prevented if the limb is mark-edly elevated and exercised rapidly, and that this may make amputation unnecessary; sympathectomy, they said, does no good. However, Osius said that there is intense vasospasm, and expressed the opinion that immediate thrombectomy should be performed in acute cases, and should be supplemented by paravertebral block to relieve the arterial spasm, and by anticoagulent therapy.

Nerves

Composition of Nerves

All spinal nerves are typically formed by the union of ventral and dorsal (anterior and posterior, motor and sensory) roots, and hence are mixed nerves distal to the point of union of the two roots. The two primary branches, ventral and dorsal (anterior and posterior), of each spinal nerve are typically also mixed nerves; the dorsal branches supply muscles and skin of the back, the ventral branches supply most of the musculature and skin lateral and anterior to the vertebral column, including, of course, the limbs. The dorsal branches of the spinal nerves tend to remain segmental, while the ventral branches regularly join each other to form plexuses (the cervical, brachial, and lumbosacral), except

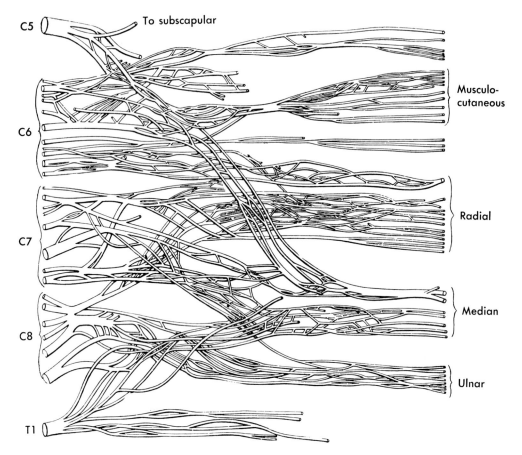

C5

To subscapular

Musculo-
cutaneous

C6

Radial

C7

Median

C8

Ulnar

T1

Fig. 1-23. Interchange of fiber bundles in the brachial plexus as revealed by dissection following maceration of the connective tissue. (Kerr AT: Am J Anat 23:285, 1918)

in the thoracic region, where they are kept apart by the presence of the ribs.

In the plexuses the spinal nerves interchange fibers (Fig. 1-23) so that the major branches originating from the plexuses are no longer segmental, but rather contain fibers derived from no less than two and in some cases from all of the spinal nerves entering the plexus. The branches arising from a plexus are also mixed nerves, and the major branches typically contain all the elements found in a spinal nerve (voluntary motor, autonomic, and afferent fibers). Even the cutaneous branches that arise from a plexus, and the small muscular branches that arise directly from the plexus instead of one of the larger peripheral nerves, are actually mixed nerves, although they are sometimes referred to as being sensory and motor: cutaneous

nerves contain, of course, not only afferent fibers but also the autonomic fibers to the area of their distribution. Muscular branches of nerves regularly contain not only voluntary motor fibers but also afferent ones from the muscle. The larger peripheral nerves are more completely mixed, for they are typically distributed to the skin and to muscles and other deeper structures.

The dorsal roots of spinal nerves are typically larger than the ventral ones and usually contain more nerve fibers, although the ratio between the numbers of fibers in the two roots varies markedly from one spinal nerve to another. The voluntary motor fibers are of course distributed entirely to voluntary muscle, but not with equal density to all the voluntary muscles of the body; the general rule seems to be that the larger and more proximally situated muscles of a limb, which have to do primarily with coarser movements, receive a relatively low number of nerve fibers in comparison with the number of muscle fibers they contain, so that the motor unit (the number of muscle fibers innervated by a single nerve fiber) in these muscles is a large one; on the other hand, more peripherally situated muscles such as those in the hand, which, generally speaking, have to do with finer and more accurate movements, receive a much higher number of motor nerve fibers in relation to their number of muscle fibers, so that their motor unit is a relatively small one.

CUTANEOUS NERVES

The largest number of afferent fibers in the peripheral nerves are distributed to skin, while a smaller number go to the muscles and other deep tissues. Ingbert estimated that about 79% of the myelinated nerve fibers (his technic did not allow him to investigate unmyelinated fibers) in the dorsal roots of spinal nerves go to skin, while about 21% are distributed to the muscles or other deep structures. He also emphasized by his anatomical studies the obvious clinical fact that the distribution of nerve fibers to the skin and the sensitivity of the skin vary from one region of the body to another. His estimates and calculations indicated that, based upon the number of myelinated nerve fibers and the area of skin supplied by the nerve, one nerve fiber in the dorsal root of a spinal nerve must be distributed on the average to about 1.08 mm^2 of skin of the head and neck, 1.3 mm^2 of skin of the upper limb, 2.45 mm^2 of skin of the lower limb, and 3.15 mm^2 of the skin of the trunk.

The calculations of Ingbert were for large areas of skin, and did not take into account the obvious differences within these areas, for instance, between the sensitivity of skin on the forearm and that on tips of the fingers. Therefore, these figures can be taken as only gross approximations. However, Woollard and co-workers found that the minimal distance on the flexor surface of the forearm at which two successive painful stimuli could be recognized as being at different sites is about

1 cm; this is, presumably, evidence that in this location a single fiber for pain is distributed over slightly less than 1 cm^2 of territory, and is overlapped in distribution, for about half this diameter, by fibers ending proximal, distal, medial, and lateral to it.

Dermatomes

It is also known, that there is overlap between the distribution of adjacent cutaneous nerves. Where the peripheral nerves are still largely segmental, as on the front of the thorax and abdomen, this overlap is very marked, for it is a general rule that the area of distribution of any spinal nerve is overlapped completely by the areas of distribution of the spinal nerves immediately above and below it, and that because of this overlap, injury to a single spinal nerve does not produce any marked loss of sensitivity. Dermatomes, which represent the areas of distribution of the various spinal nerves, have usually been described as the total areas of distribution of the various nerves (determined, for instance, on the basis of remaining sensibility following section of a nerve above and below) and Foerster found great overlap. His dermatomes are shown in Figures 1-24 and 1-25. This method of determination of the dermatomes outlines the maximal distribution from a given dorsal root ganglion.

Keegan ('47, and other papers) has protested, however, that the concept that there is no loss of sensation with injury to a single nerve root is a result of lack of careful testing; he reported definite hypalgesia following injury to individual nerve roots, and defined dermatomes on the basis of this hypalgesia. His dermatomes (Fig. 1-26) are quite different from those of previous workers, and, in the case of limbs, are both simpler and more obviously logical, representing as they do longitudinal strips rather than the irregular patches of previous investigators.

Keegan's dermatomes have been criticized (Last) on the basis that they do not indicate the complete overlap assumed by previous workers, and that they do not vary appreciably with the known variations in the brachial and lumbosacral plexuses. However,

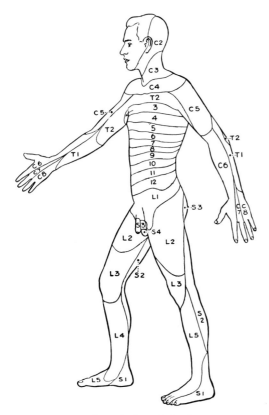

on which charts are more useful. Fletcher and Kitchell, by recording from dorsal root filaments in the dog while stimulating hairs, found that in this animal the dermatomes do apparently extend into the hind limb as continuous strips, as Keegan described for man.

Overlap in Peripheral Nerves

In regard to overlap between two peripheral nerves of a limb both of which contain fibers from adjacent spinal nerves, it is well known clinically that the overlap is much less than on the trunk: section of a single peripheral cutaneous nerve, such as one of the antebrachials, does usually give rise to a total loss of sensation over a certain area (Fig. 1-27),

Fig. 1-24. Anterolateral view of the dermatomes according to Foerster. (After Foerster, from Haymaker W, Woodhall B: Peripheral Nerve Injuries [ed 2]. Philadelphia, WB Saunders, 1953)

Fig. 1-25. Posterior view of the dermatomes according to Foerster. (After Foerster, from Haymaker W, Woodhall B: Peripheral Nerve Injuries [ed 2]. Philadelphia, WB Saunders, 1953)

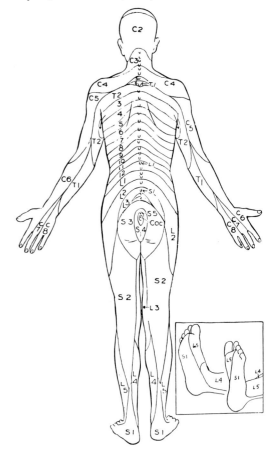

Keegan did not deny an overlap, and indeed his finding of hypalgesia rather than analgesia indicates one; further, he said his dermatomes are correct and constant, regardless of variation in the plexuses, because it is the position of a nerve in the total series, not whether it is classified as a lumbar or a sacral nerve on the basis of variations in the number of vertebrae, that is important. Certainly Keegan's dermatomes are not incompatible with known anatomic facts (except for the distribution of C5 and T1 to the upper part of the thorax, which is really supplied by C3 and C4 through the supraclavicular nerves), and from the developmental standpoint they represent much more nearly the theoretical distribution of nerves to the skin of the limbs than do the dermatomes that have been previously accepted. There is as yet no consensus

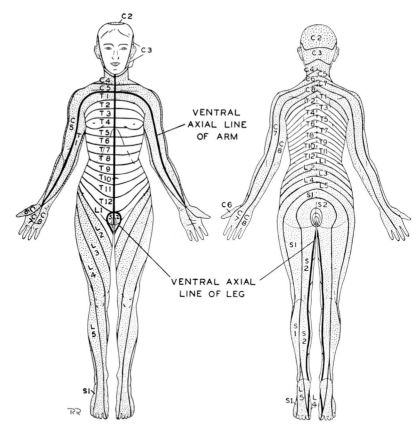

Fig. 1-26. The dermatomes according to Keegan. (Keegan JJ, Garrett FD: Anat Rec 102:409, 1948)

although this area varies in size according to the distribution of the sectioned nerve and of adjacent nerves, and is never as large as the total area supplied by the nerve. Therefore there obviously is not the complete overlap by two adjacent nerves that is found in regard to the distribution of nerve roots. Moreover, it is also known that the amount of overlap by adjacent nerves depends in part at least upon the type of sensation being considered: the area over which the sensation of touch is lost following section of a peripheral nerve is regularly greater than the area of loss of the sensation of pain; thus the fibers for light touch in two adjacent nerves overlap less than do the fibers for pain. This varies, however, with the area of skin considered: Woollard and colleagues found that on the forearm the margin of the area anesthetic to pain lay about a centimeter inside of that anesthetic to touch fol-

lowing infiltration of cutaneous nerves with procaine; on the thenar eminence there was less difference in distribution of touch and pain fibers, and on the fingers there was no detectable difference in the distribution of pain and touch fibers. Woollard and his coworkers found further that, within the area in which touch was lost but pain was still present, the number of points from which pain could be produced was below normal, thus giving further evidence that this area is actually one of overlap between adjacent nerves.

MUSCULAR BRANCHES

Mention has already been made of the segmental innervation of muscles, and there need be repeated here only the simple fact that most muscles have at least two spinal

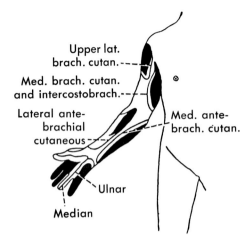

Fig. 1-27. Average sensory loss after separate interruption of the individual cutaneous nerves to the front of the upper limb. The *black* areas are those of total sensory loss; the *outlined* areas adjacent to the black are those in which touch is lost but some pain and temperature are preserved; the remaining areas therefore are those of almost total sensory overlap between adjoining nerves. (After Lewandowsky and Foerster, from Woltman HW, Kernohan JW, Goldstein NP. In Baker AB (ed): Clinical Neurology, Vol 4. New York, Hoeber, 1962)

nerve segments contributing to their innervation. It is usually assumed that the segmental motor and afferent supplies of a muscle are similar, and this is probably usually true. There may, however, be exceptions to it; Harris ('04) quoted Sherrington as having shown that, in regard to the knee jerk, the third lumbar nerve is the motor root but the fourth lumbar is the afferent one, and he himself expressed the opinion that motor and afferent fibers from two adjacent nerves are not necessarily distributed in the same proportion to the muscles they supply. Specifically, he said that the deltoid, the biceps, and the brachialis muscles, usually granted to be supplied from both C5 and C6, receive most of their motor supply from C5, and most of their afferent supply from C6.

As Harris also pointed out, it is entirely possible that the relative number of afferent fibers in comparison with the size of a muscle may vary, just as the size of the motor unit is known to do. Thus he noted that the dorsal root of C5 is only a little larger than the ventral root, and suggested that this was due in part to the fact that this nerve is distributed to the large and proximal muscles which do not need the delicacy of afferent control that more peripherally situated muscles do.

GENERAL STRUCTURE OF NERVES

A typical peripheral nerve consists of nerve fibers (axis cylinders, axons) of various diameters from large myelinated to small unmyelinated, which are surrounded and held together by connective tissue.

Nerve Fibers

Since the voluntary motor fibers arise from cells located in the central nervous system and the sensory fibers arise from cells in the spinal ganglia immediately adjacent to the central nervous system, the fibers composing peripheral nerves may be very long (in the case of those to the foot, for instance, 3 feet or more in a tall person). Further, since a nerve fiber separated from its cell body degenerates, lesions of peripheral nerves that produce degeneration lead to the degeneration of the entire length of the fiber distal to the lesion. This may be a matter of feet over which regeneration must occur.

In comparison with their length, nerve fibers are very slender: the largest ones measure about 18 μm in diameter, and the smallest ones less than one micron. Two features, the speed of conduction and the occurrence of a myelinated sheath, seem to be correlated with the diameter of the fibers. The largest fibers conduct fastest; and the smaller the fiber, the slower its conduction. Thus Heinbecker and co-workers found that fibers from about 18 μm to 6 μm in diameter conduct the nerve impulse at speeds from 120 m/sec to 60 m/sec (these are also designated as "A" fibers), while fibers of 2 μm or less in diameter ("C" fibers) conduct at rates of only 1.5 m/sec to 0.3 m/sec. "B" fibers, although overlapping with "A" fibers in size, are generally intermediate in diameter and conduct at an intermediate rate.

The presence of myelin (Fig. 1-28), a fatty sheath immediately surrounding some nerve fibers, is apparently also correlated with the diameter of the fiber. Duncan found, in a series of animals of varying size, that all fibers of more than about 2 μm in diameter had myelin sheaths, that those of between 2 μm and 1 μm were of the transitional size, and that in those of less than 1 μm no myelin was demonstrable by light microscopy. Unmyelinated fibers are now known to be imbedded, sometimes singly but usually in groups of up to about 14, within the cytoplasm of elongated cells, called Schwann cells. If myelin is present, the cells of Schwann lie outside the myelin and enwrap it and a single axon; the myelin is at intervals, at the point of meeting of two adjacent cells of Schwann, completely interrupted, these interruptions forming the nodes of Ranvier. The axon is constricted at each node of Ranvier, although of course not interrupted. The larger the diameter of the fiber, the thicker the myelin sheath, and the longer the distance between the nodes of Ranvier.

Corbin and Gardner reported a gradual loss of nerve fibers with increasing age, and so did Cottrell; the former workers studied myelinated fibers in the nerve roots, and found that while the number of such fibers increased after birth (presumably as a result of growth in diameter of the fibers) there was a gradual loss of fibers after the third decade, amounting to as much as 32% for an individual 89 years old. Cottrell studied peripheral nerves, and found that with increasing age there were vascular changes in the nerve, a reduction in the number of nerve fibers, and an increase in connective tissue elements, which in some areas replaced the nerve fibers. Corbin and Gardner suggested that the decrease in fibers of the sensory roots, presumably as a result of loss of spinal ganglion cells, may account for the decrease in vibratory sensibility usually found after the third decade of life. Cottrell felt that the changes in the peripheral nerves might explain many of the indefinite motor and sensory complaints of old age.

Connective Tissue

Larger nerves are surrounded by a layer of connective tissue, the epineurium, that ex-

tends into the nerve and divides it into bundles (Fig. 1-29) that are called funiculi or fasciculi. The epineurium consists largely of collagenous tissue but also contains elastic fibers, both types of fibers running for the most part longitudinally. It may contain fat and varies in thickness both from nerve to

Fig. 1-28. Schema of the structure of a myelinated nerve fiber, *A* in longitudinal section and *B* in cross section. Myelin on the right side is shown as if stained with osmic acid, on the left as if dissolved. *Ax* and *Ax*[1] are fibrillar and afibrillar parts of the axis cylinder or nerve fiber, *x* its surface or axolemma. *M* indicates a droplet of myelin, *MS* the space occupied by a droplet, *NR* a node of Ranvier, *NS* the nucleus of a Schwann cell, *Pa* a periaxial shrinkage space, *PN* the protoplasmic net containing the myelin, *S* the neurilemmal sheath, and *SL* Schmidt-Lantermann's incisures. (After Nemiloff, from Peele TL: The Neuroanatomical Basis for Clinical Neurology [ed 2]. New York, McGraw-Hill, 1961)

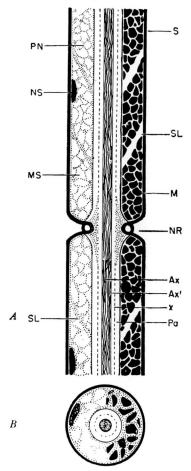

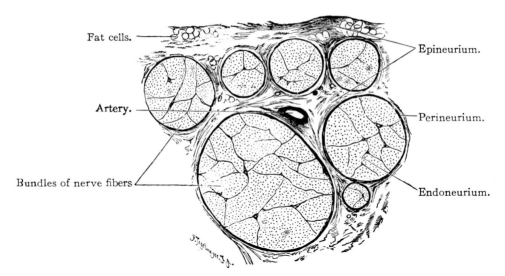

Fat cells.

Epineurium.

Artery.

Perineurium.

Bundles of nerve fibers

Endoneurium.

Fig. 1-29. Cross section of a part of the human median nerve. (Bremer JL, Weatherford HL: A Textbook of Histology. Philadelphia, Blakiston, 1944)

nerve and from one part of a nerve to another, generally being thicker where a nerve crosses a joint and typically composing some 30% to 75% of the substance of the nerve (Sunderland, '66). On small nerve branches it is poorly organized and does not completely surround the funiculi (Burkel).

Each funiculus within a large nerve is surrounded by a layer, the perineurium, that has received much emphasis in recent years. In larger nerves it consists of from 7 to 15 concentric layers of connective tissue, separated by clefts, perineural spaces, that are apparently lined by flattened epithelial cells. The outer layer of the perineurium blends with the epineurium; the inner layer is an epithelium. As a funiculus divides, the perineurium does also, and in small nerves it consists of some 1 to 3 layers that surround from 1 to 10 fibers (Burkel). The perineurium thus forms a system of tubes that eventually come to surround each individual myelinated fiber (and presumably, each unmyelinated fiber or group of fibers embedded in a Schwann cell). Shanthaveerappa and Bourne said that the perineural epithelium is continuous with the sheath covering the motor end plate, but Burkel said it stops just short of the motor end plate, the open end of the tube thus allowing

easy access of material that can follow the tubes to the central nervous system.

Within the perineurium, the endoneurium forms a delicate connective tissue tube that in turn surrounds each individual axon with its myelin and Schwann cell coverings. The endoneurium is usually described as consisting of an outer layer of longitudinally running collagenous fibers and an inner layer of fine reticulum (*e.g.,* Sunderland, '66), but Gamble and Eames could not recognize an inner layer in human nerves, and Thomas ('64) recognized it only around larger myelinated fibers. The outer layer has apparently been called the sheath of Key and Retzius, or of Henle; the inner, the neurolemma or neurilemma, a term that traditionally meant the Schwann cell.

Accounts differ as to how many connective tissue fibers are arranged circularly or spirally in the nerve, but Sunderland regarded the perineurium as contributing most to the elasticity of the nerve. The longitudinal fibers of the epineurium apparently follow a wavy course, and the nerve bundles within the perineurium do also, so that a certain amount of stretch is allowed. According to Sanders, the amount of stretch that can occur without appreciable damage amounts to about 10% of the mobilized length of a nerve. Sunderland ('45b) regarded fat in the epineurium, especially prominent in parts of a nerve subject to pressure and stretch, as contributing to stretch of the nerve as a whole.

Intrinsic Anatomy of Nerve Trunks

It is obviously possible for two adjacent nerve trunks to be bound together with a common epineurium, so that grossly they appear to be a single nerve and yet actually interchange no fibers with each other; the most prominent example of this is the sciatic nerve, in which the tibial and common peroneal components are usually enclosed in a common epineurium, but nevertheless maintain their identities. The sciatic nerve, therefore, is actually two nerves with a common sheath and, as is well known, these two nerves may fail to unite with each other or, if they are united, can be separated with no more damage to nerve fibers than results from the necessary tearing of the epineurium in the proper plane. In the same way, the components of the brachial plexus are typically bound to each other by epineurium both proximal and distal to the levels at which there is interchange of fibers among the nerves. Careful dissection of the connective tissue from around the brachial plexus may allow one to trace many of the larger nerve branches to their segmental origins (Fig. 1-23). It is largely the variation in the extent of common wrapping of epineurium that accounts for the usual variations in the level of origin of branches, whether from the plexuses or from a peripheral nerve.

However, since all the major nerve trunks consist of a mixture of afferent and motor fibers, since these in turn are derived from several different spinal nerves, and since the peripheral branches of the nerves are in turn usually a mixture of sensory and motor fibers, and likewise derived from several spinal nerves, it is obvious that within most larger nerve trunks there has to be a very considerable rearrangement and regrouping of nerve fibers. It is this rearrangement of funiculi and fasciculi, involving branching and anastomosing, that is referred to as the intrinsic anatomy or intraneural topography of nerves. It is of some practical important to the surgeon. Regenerating nerve fibers grow down the remains of the nearest portion of nerve trunk, so that if regeneration that will approximate the previous pattern is expected, it become important to unite the two ends of a severed nerve with as little torsion between them as possible. For instance, if one half of a nerve were shown to be entirely afferent and the other entirely motor, 180° of torsion in uniting the severed ends would be expected to give poor functional regeneration, for in this case the motor fibers would grow out along the afferent branches of the nerve, the afferent fibers along the motor branches.

Sunderland and Ray ('48) also found the intrinsic anatomy of the tibial and common peroneal components of the sciatic nerve to be somewhat variable; in general, they said, communications between the funiculi take place *along the entire course* of the nerve. In successive cross sections through the sciatic nerve they found a constantly changing pattern: at the buttocks, most funiculi seemed to contain fibers destined for most or all of the peripheral branches, while at successively lower levels a regrouping is gradually effected until eventually, at a variable level above the gross origin of a nerve, groups of funiculi destined solely for that specific nerve appear. They found that the fibers representing individual nerve branches remained as discrete trunks within the main nerve for distances of from 1 mm to 187 mm (in the case of the deep peroneal the bundles for it were distinct for 26 mm above the level of the neck of the fibula), but as a rule the more obvious macroscopic plexus formation was found to be in the region adjoining the origin of relatively large branches. They emphasized, however, that anastomosis and division occur repeatedly along most of the length of the nerve, and the maximal length of nerve trunk that they saw with a constant pattern was 6 mm. Thus, according to their findings, there is no abrupt rearrangement of fibers, but rather it is a constant one from the level of the plexus to the levels of origins of the different branches.

Sunderland ('45c) found essentially the

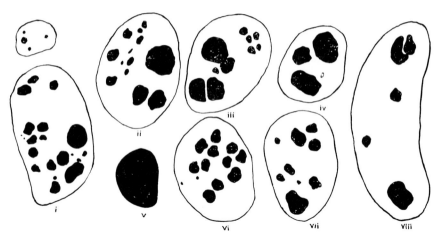

Fig. 1-30. Changes in the number and arrangement of the funiculi (*black*) in the radial nerve, indicating the repeated changes in intraneural morphology, as seen in transverse sections of the nerve at eight different levels. (Sunderland S, Brain 68:243, 1945)

same structure in the radial, median, and ulnar nerves: he said the funicular pattern was continually modified along the entire length of each nerve (Fig. 1-30) through repeated division, anastomosis, and migration of nerve bundles; in the cases of these nerves, the longest section of any nerve which he found with a constant pattern was 15 mm and the average length over which the pattern remained constant varied between 0.25 mm and 5 mm. In contrast to some authors, who believed the pattern of bundles within a nerve to be constant at a given level, Sunderland found the pattern at a given level to vary even on the two sides of a body, and from subject to subject. A variable distance above the external or gross origin of peripheral branches the fibers destined for these branches became grouped together within the nerve; in the case of the radial nerve, Sunderland found, the location of individual bundles was relatively constant from one level to another, but in the median nerve there were constant variations in the location of individual groups of fibers, and the ulnar nerve lay somewhat between the two in this regard.

O'Connell pointed out that even within smaller nerves there is a regrouping of fibers; he said that there is little reassorting of fiber bundles in sensory nerves, but that in motor ones there are two plexuses, a fine one which,

he thought, is probably due to regrouping of sensory and motor fibers, and a coarser one, which he regarded as being a re-sorting of the fibers from the several segments contributing to the innervation of the muscle.

These reports of the intricate and changing anatomy of the larger nerves particularly emphasize the importance of suturing severed nerves with as little rotation of either end as possible, for the more complex the internal anatomy the greater the chance that a slight amount of rotation may approximate fasciculi that normally do not belong together. Sunderland ('45b) pointed out that the amount of fat present among the fasciculi may affect the chances of successful regeneration after anastomosis, since the more fat that is present, the more widely the fasciculi spread, and therefore the more difficult it is to obtain end-to-end apposition of corresponding fasciculi. Bora found it possible to stitch the perineurium rather than the epineurium in cats, and thus obtain better alignment of the fasciculi.

Variations in the levels of external origin of the motor branches of nerves of the upper limb have been recorded by Linnell, and those of the major branches of both the upper and lower limbs by Sunderland and his collaborators in a series of papers; these workers have also recorded the distance from certain bony points to the level at which a branch enters an individual muscle, thus indicating the relative distance over which a severed nerve must regenerate in order to reinnervate the various muscles. Data concerning these features are, in this text, quoted in connection with the descriptions of the individual nerves.

Blood Supply

Within a nerve there are longitudinally running arterial branches that supply the nerve fibers and on the surfaces of the larger nerves there are also one or two superficially located longitudinal pathways that give rise to the smaller vessels within the nerve. These superficially located pathways are in turn fed at irregular intervals by twigs from adjacent blood vessels—in the case of the larger nerves, macroscopically visible ones (Fig. 1-31); in smaller nerves the intraneural blood vessels receive accessions from microscopic twigs in the surrounding loose connective tissue. Ramage pointed out that there is considerable variation in the origin of the blood supply to any given nerve, but that it is typically from several different sources. When the twigs to a nerve reach it they generally divide into proximal and distal branches, which anastomose with each other up and down the length of the nerve and have even been reported as forming a part of the collaterial circulation in such a case as ligation of the popliteal artery.

Adams ('42) agreed that this is the fundamental pattern of the blood supply to a nerve, and reviewed the literature on the effect of local interference with the blood supply; he found that there was evidence that this affects the passage of the nerve impulse and will eventually lead to nerve block. The larger and more rapidly conducting fibers were said to be affected first, and Adams found no evidence that subjective symptoms such as pain or vertigo may result from ischemia of the nerve itself (others have regarded pain as resulting from ischemia of nerve; see the end of this section).

It is generally agreed that the anastomosing channels on and in a nerve are normally fully adequate to support function of the nerve after one or more entering branches have

been interrupted. Adams ('43), noting that ligation of the inferior gluteal artery, only one source of supply to the sciatic nerve, has been said to interfere in experimental animals with the function of the sciatic nerve, found no effect from this in the rabbit; nor, except where the nerve had been so freed from its surroundings that direct damage might have resulted from this manipulation, did he find either clinical or microscopic evidence of degeneration of fibers from ligation of most of the arteries said to supply the sciatic nerve in the rabbit.

Bacsich and Wyburn ('45a) found no essential disturbance in the vascular pattern either after crushing nerves or crushing and extensively mobilizing them, nor did they find any difference in the pattern or evidence of diminished circulation after cutting re-

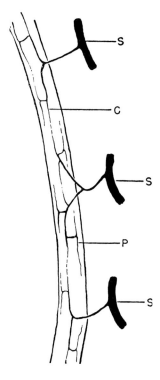

Fig. 1-31. Diagram of the blood supply to a peripheral nerve. *S* identifies neighboring arteries; *C,* the central longitudinal artery of the nerve; *P,* peripheral or secondary longitudinal arteries. (After Adams, from Peele, TL: The Neuroanatomical Basis for Clinical Neurology [ed 2]. New York, McGraw-Hill, 1961)

gional nutrient vessels. They thus agreed that the longitudinal vessels in a nerve trunk can usually supply adequate circulation to a considerable length of nerve. Sunderland ('45d and '45e) also supported this concept; he pointed out that while the blood supply of the nerves of the upper limb in man is derived especially from vessels anatomically associated with the nerves, the number and the site of origin of these contributing vessels vary widely, and said that it is impossible to free a nerve from surrounding tissue for more than a few centimeters without sacrificing some of the vessels supplying it. He also pointed out that nerves have been stripped of all surrounding connective tissue for distances up to 15 cm, and yet found to bleed when divided. More recently Smith ('66) on the basis of injection experiments on human median nerves, contended that the circulation is adequate for no more than 6 cm to 8 cm.

Sunderland did suggest that in freeing a large nerve any superficially running channel should be left on the nerve if possible, and that care should be taken not to injure main channels along the nerve by tearing off the entering vessels too close. He ('45f) described the blood supply to the sciatic nerve as being derived in general from the inferior gluteal, the cruciate anastomosis, and the perforating branches of the profunda; the tibial nerve was found to be supplied by branches from the posterior tibial, the common peroneal by branches from the (circumflex) fibular, and the deep peroneal by branches from the anterior tibial. As evidence of the irregularity of the entering vessels, however, he reported that the branches to the tibial nerve varied in number from 2 to 11, those to the deep peroneal from 2 to 13.

Denny-Brown and Brenner ('44a) cited experiments which, they believed, indicated that pressure on a nerve causes impairment of conduction not by a direct effect on the nerve fibers, but rather by producing ischemia. Richards agreed that ischemic paralysis undoubtedly occurs, but found evidence that pure ischemia is much less injurious to nerve than ischemia produced by or associated with compression. He cited evidence that nerves will recover rapidly after at least 12 hours of pure ischemia, but that persistent paralysis has resulted from as short a period as 20 minutes of ischemia produced by constriction of a limb; he therefore suggested that either pressure itself is directly injurious to nerve fibers or that it so damages the blood vessels locally that after it is removed circulation is not restored to that portion of the nerve.

It is known that some nerves are more susceptible to pressure than others; according to Richards, this is undoubtedly, in part, a consequence of their anatomical position, and the relative ease with which they can therefore be compressed. Sunderland ('45f) expressed the opinion that the common peroneal nerve is more susceptible to pressure than others, not because of a difference in its vascular supply, but because it contains a less amount of supporting adipose tissue.

Richards pointed out that the severe ischemia associated, for instance, with Volkmann's contracture results in degeneration of the nerves in the ischemic region; he also pointed out that the first result of ischemia is an increased irritability of the nerve fibers, and suggested that perhaps the lowered blood supply to the nerves, as well as that to the muscles, may contribute to the pain in "ischemic neuritis," which may occur after occlusion of major arteries of the lower limb.

The adequacy of the anastomotic channels along and within a nerve is of obvious surgical importance, because this adequacy prevents damage to the nerve following necessary mobilization of it during repair. However, it cannot be assumed that the vascular pattern in all nerves is entirely normal, and Richards quoted evidence that pathologic changes in the arteria nervorum are not uncommon. It has been both claimed and denied that the neuropathy associated with diabetes mellitus is a result of peripheral vascular disease; it has apparently been shown beyond doubt (Kernohan and Woltman) that the peripheral neuritis sometimes associated with periarteritis nodosa is due directly to ischemia, occlusion of nutrient arteries leading to focal areas of infarction with subsequent degeneration of the more distal parts of the affected fibers.

In attempting to analyze the relationship between ischemia and neuritic pain, Richards concluded that pain is not a prominent feature when a nerve is suddenly deprived entirely of its blood supply. It does arise when a nerve is gradually or by a series of episodes deprived of a part of its blood supply, but not enough to cause necrosis of the entire nerve. Richards regarded the specific neurologic lesion responsible for ischemic pain in nerve as being unknown, but suggested that the basic mechanism might be similar to that of causalgia, which has been explained as a result of defective insulation in the nerve trunk.

LESIONS OF NERVES

Types

It is obvious that lesions of nerves may vary in severity from one, such as temporary ischemia, which only transiently interferes with the function of nerve fibers and produces no anatomical changes in them, to one in which the entire nerve is divided, with subsequent greater or less retraction of the severed ends (Seddon, '43). Complete anatomical division of the nerve, also referred to as neurotmesis, demands surgical intervention with approximation of the nerve ends or grafting, if necessary, to produce successful regeneration; transient nerve block or neurapraxia obviously demands no surgical intervention, and the same is true of the more severe lesion, axonotmesis, in which degeneration of axons distal to the lesion occurs, but there is no loss of continuity of the nerve trunk as a whole. Obviously, it becomes important to distinguish clinically between those lesions that require surgical treatment and those that do not, but the symptoms may be identical.

Seddon suggested that if the level of the lesion is known the diffential diagnosis can be made by calculating the expected time of beginning recovery of function on the basis that regenerating motor fibers grown about 1.5

mm per day, and allowing some 40 to 50 days beyond this time for the latent period (really two periods: the first, one during which regeneration is initiated; the second, one in which the outgrowing fibers have reached the organ, so that anatomical regeneration is complete, but the fibers are maturing before they can begin to function).

Neurapraxia implies a condition in which there is no degeneration of nerve fibers; obviously, many neural lesions necessarily involve a mixture of this and axonotmesis, that is, a lesion in which some of the nerve fibers have merely stopped conducting, others have actually degenerated. The usual causes of neurapraxia and axonotmesis, separately or combined, are pressure and ischemia. Allen suggested that degeneration of axons results only from direct pressure, and asphyxia produces temporary paralysis only, but, as already noted in the discussion of the blood supply of nerves, other authors have thought much of the effect of pressure is actually due to the ischemia it produces. Denny-Brown and Brenner ('44b) compressed peripheral nerves for only 2 hours under a tension of from 170 g to 430 g and found that while there was no gross defect in sensation and the distal portions of the nerve fibers did not degenerate, this did induce a paralysis that lasted for from 5 to 8 days; less pressure, continued over a longer time, induced similar changes. These effects, they thought, were due entirely to ischemia, and not directly to the effect of pressure on the nerve fibers. Weisl and Osborne described moderate pressure as producing edema both proximal and distal to the constricted area, with the fluid largely between the axons; eventually, they said, fibroblasts invade the fluid and produce a fibrosis that is usually irreversible, so that decompression of a long-standing lesion is not likely to restore function.

Allen's evidence agreed with that of Denny-Brown and Brenner that motor fibers are paralyzed more readily by pressure (or asphyxia) than are afferent fibers, and his evidence that pressure over a narrow segment of nerve is less injurious than that over a wide segment can obviously be interpreted to mean that at least a part of the resulting dysfunction and degeneration, if any, is a result of ischemia.

Sunderland ('45a) followed seven cases of compression injuries of the radial nerve in which recovery was spontaneous. In these, practically all the motor fibers were so damaged that the muscles they supplied were no longer functional, and in the more severe cases sensory defects also appeared; the appearance of sensory defects was correlated with very slow recovery of full motor function, presumably indicating that if pressure is severe enough to cause cessation of conduction in sensory fibers, it still more severely injures the more susceptible motor fibers, perhaps to the point at which they degenerate and have to regenerate. Of his patients, five recovered completely after about 3 to 4 months following injury, and the remaining two recovered much more slowly. Seddon said that in his experience neurapraxia usually lasted clinically for abut 10 weeks, and that paralysis was almost always complete in these cases, but some patients recovered in as little as a week and some required 6 months for recovery; obviously, the latter cases may include axonotmesis, with regeneration of the motor fibers.

DEGENERATION AND REGENERATION

The literature on morphologic aspects of degeneration and regeneration of peripheral nerve fibers is enormous. The subject has been studied extensively and intensively from many angles; such studies as those of Glees and Young and his colleagues (*e.g.,* Holmes and Young) have been supplemented more recently by electronmicroscopic studies.

In brief, degeneration of a peripheral nerve involves the breaking up of both the myelin and the axis cylinder, with subsequent phagocytosis of the remains of both; with their disappearance, the tubular structure surrounding each individual nerve fiber is left empty, and the empty tubes then collapse to about half their original diameter (Holmes and Young; Glees). However, the Schwann

cells proliferate to form bands within the endoneural tubes, and the endoneural and the perineural sheaths, thus filled, then assume almost their former sizes (Glees). According to Thomas ('64) the Schwann cells proliferate within a tube formed by the basement membrane surrounding the Schwann cells, and the collagen fibers of the endoneurium or inner endoneurium increase to form a tube of densely packed longitudinal fibers around that. Finally, Shanthaveerappa and Bourne said there is no change in the perineural epithelium in a degenerating nerve and apparently regarded it as being responsible for maintaining a tubular structure around the degenerating end of the nerve fiber.

Regeneration begins as a result of the sprouting of a number of branches from the central stump of each individual nerve fiber, and if these branches succeed in entering the peripheral stump they grow downward on the surface of the Schwann cells within the endoneural tubes. If Thomas' observations are correct, they grown within the tube formed by the basement membrane of the Schwann cells. A single tube, according to Holmes and Young, may receive as many as 20 regenerating fibers, but usually only one of these will become large, while the remainder subsequently degenerate. Glees described the endoneural tubes as being responsible for guiding the fibers to their ultimate destination, although the Schwann cells have been more often regarded as responsible, and Shanthaveerappa and Bourne said that the perineural epithelium guides a regenerating nerve fiber back to the original motor end plate of a muscle fiber. For practical purposes it perhaps makes no difference, since a fiber growing within an endoneural tube is necessarily in contact with the Schwann cells that fill the tube, and the morphology of the Schwann bands is necessarily that of the endoneural tubes and, in the most distal part of a nerve, of the tube of perineural epithelium around each fiber: when any one of these branches, the others do also; and when a regenerating fiber reaches such a branching it commonly grown down along both branches (Glees).

Thus, since nerve fibers grow down within

an endoneural tube, a regenerating fiber growing into the wrong endoneural tube may attain a connection that it did not previously possess. Ford and Woodhall analyzed a number of disorders of movement and sensation following regeneration of various peripheral nerves, and expressed the belief that they all can be explained by assuming that the regenerating fibers formed connections that they did not originally possess.

During the early stages of regeneration, the regenerating fibers are both much smaller and far more numerous than are the fibers in the central stump that are giving rise to them; because of this multiplicity of regenerating axons, a limited number of fibers in the central stump can, under the proper conditions, furnish a much larger number of mature fibers to the peripheral stump. Thomas ('48), for instance, after cross-anastomosing a proximal portion of the nerve of the lateral head of the gastrocnemius to a distal segment of the common peroneal, found that the latter regularly contained more fibers after regeneration than did the central stump supplying the fibers; on the basis of the weights of the peroneal muscles, he concluded that through this branching the nerve was able to support a great deal more muscle than it did originally.

Normally, however, the number of nerve fibers in a regenerating nerve is reduced, and the sizes of the fibers are increased, through the formation of functional connections by some of the fibers. Thus Aitken and co-workers reported that if a motor nerve is prevented from reaching the voluntary muscle to which it is regenerating it may have four or five times the number of nerve fibers it normally does, and they will be mostly small. On the other hand, if it is allowed to reach the type of tissue which it normally innervates, the fibers simultaneously become reduced in number and increased in size, presumably as a result of some of them making functional connections with the tissue (for instance, motor end plates in muscle) and thus excluding other fibers from making such connections. A connection with the end organ, therefore, seems to be one of the factors necessary for the attainment of maturity by regenerating fibers.

While increase in size of regenerating fibers is a sign of approaching maturity, connection with the periphery is not the only factor that determines the successful attainment of normal size by a regenerated fiber. Hammond and Hinsey, and Simpson and Young, for instance, have shown, by cross anastomosis between nerves containing different sizes of fibers, that the ultimate size of a regenerating fiber depends also upon the source from which it is derived and the size of the tube into which it grows: a regenerated fiber derived from a small fiber, and growing into a large tube, will not exceed its normal size in spite of the fact that it has more room within the tube, but a regenerating fiber derived from a large one and growing into a small tube will be prevented from reaching normal size by the size of the tube in which it lies; it may somewhat inflate the tube, but if the tube is small enough the fiber will never reach a normal diameter. Also, whether or not a regenerating fiber becomes myelinated depends upon the fiber from which it is derived: a myelinated fiber growing into a normally unmyelinated nerve branch will become myelinated, while an unmyelinated fiber growing into the sheath of a myelinated one does not assume a myelin sheath (Simpson and Young).

Factors Affecting Regeneration
Some of these factors have already been discussed in the preceding section, in which it was pointed out that the effectiveness of regeneration is conditioned in part by the peripheral pathway along which the fibers grow, and by their making connections with peripheral organs of the type that they previously had innervated. According to Bueker and Meyers, another factor in successful nerve regeneration is the maturity of the fibers at the time they are injured; they found that the crushed sciatic nerves of young rats regenerated much less well in a period of 100 days than did those of older rats, and attributed this to the immaturity of the fibers at the time of injury in the young animals.

Infection might be supposed to interfere with regeneration of nerves, but Davis and co-workers said that while infection increases the number of adhesions and may cause chronic perineuritis, it does not invade the fiber bundles, and does not interfere with regeneration unless the inflammatory process has resulted in scar tissue between the ends of the nerve. Nevertheless, anything that induces fibrosis in or about the nerve may be expected to interfere with healing, and one of the reasons for delaying suture of nerves in war injuries has been the difficulty in securing a clean wound under emergency conditions.

Bacsich and Wyburn ('45b) reported that local destruction of the vascular supply of a nerve, either by section of a local nutrient artery or by destroying the longitudinal anastomosis in the epineurium, did not interfere with the rate of regeneration of the crushed sciatic nerve of rabbits, thus indicating that injury to the vascular supply during repair is not likely to be severe enough to interfere with this process if the circulation to and in the nerve is otherwise normal.

Holmes and Young reported that as a rule fewer nerve fibers will enter a peripheral nerve stump that has been "long degenerated," but said they will grow as rapidly once they have entered it as they will in freshly degenerated stumps; Glees quoted evidence that the morphology of a peripheral segment of degenerated nerve may persist for as long as 4 to 5 years. It appears probable that the difficulty nerves have in entering a long-degenerated stump may be due largely to scar formation at the end of that stump, and that unsuccessful functional regeneration following long-delayed suture results largely from atrophic changes in the organ to be innervated and from fibrotic changes in the degenerated nerve trunk, rather than in an inability of the nerve fibers themselves to regenerate. However, it is generally believed by clinicians that fibrosis and shrinkage of a degenerated nerve become prominent enough after a few months to interfere with regeneration.

Somewhat surprisingly, Stopford reported that when resuture of a nerve is necessary, the prognosis for recovery is better the nearer the suture is to the spinal cord; he said that

a delay of more than 18 months between the primary and secondary suture decreases the chances of successful regeneration if the suture is in the distal part of the limb, but is not so important if it is in the proximal part.

Sherman and co-workers found that in the cat a delay of 60 days between the time of nerve division and suture was sufficient to retard and to a certain extent prevent successful regeneration of sensory fibers, although apparently the delay in suture did not have a comparable effect upon regeneration of motor fibers.

SURGICAL REPAIR OF INJURED NERVES

Since successful regeneration of nerve fibers demands that they have a scaffolding along which they can grow to reach their previous connections, surgical repair of severed nerves is a necessity. Lacking such repair, or good apposition of the ends, the proximal end forms a neuroma, which is usually described as an unsuccessful attempt of the nerve to regenerate, with the formation of a tangled skein of nerve fibers and connective tissue. Klein and co-workers maintained, however, that in the five human neuromas which they studied there was no tangle of nerve fibers, but rather, while the proximal end of the neuroma differed very little from any peripheral nerve, the distal portions consisted of collagenous connective tissue, reticular fibers, and sheath cells (a connective tissue mass rather than a regenerating nerve).

Suture

For small nerves that can be brought together without tension, the plasma clot "suture" method (Tarlov) or a modification thereof is excellent, and Seddon ('63) reported that for such nerves as the digital ones, human fibrinogen and thrombin is superior to suture. However, if the suture line must resist tension—and severed nerves usually retract, so they may have to be united under some tension—actual suture of the nerve is necessary. Obviously, less disorganization of the nerve is produced if the sutures are taken through the epineurium rather than through the entire nerve; in any case, it is the connective tissue on the nerve that holds the suture, so the suture will be stronger the better this connective tissue is developed. Spurling and Woodhall cited the thinness and lack of strength of the epineurium of a freshly divided nerve as being one factor that led them to delay suture of nerves in war injuries. They said, however, that after 15 to 25 days enough thickening of the epineurium occurs to increase its tensile strength and allow it to hold the sutures much better. They also said that the distal end of a severed nerve maintains its normal size until about the end of the third month, and expressed the opinion that better functional regeneration occurs if the suture is made before this time.

Nerve Grafts

Sanders reported that a stretch of about 10% of the mobilized length of a nerve may be accomplished without damage to the nerve, but that a total of as much as 30%, even if applied during two successive operations, may be harmful. Seddon ('63) said that if the wrist had to be flexed more than 30 degrees, or the elbow or knee more than 90 degrees, to bring the ends together without tension, end-to-end suture should not be attempted. Thus, in those cases in which the ends of nerves cannot be brought together without undue stretch, some method of bridging the gap is necessary. Sanders reviewed the literature, and concluded that autografts appeared to be the best method for bridging such gaps.

Gutmann and Sanders found that in rabbits autografts may allow as rapid and complete recovery as end-to-end suture, but that grafts preserved in alcohol are replaced by the tissues of the host, and poor and late recovery occurs. Homografts also have not proved successful. Barnes and co-workers ('46) reported failure in all of their eight cases in which homografts in man were used to bridge a large gap; and Spurling and co-workers likewise reported total failure of eight homografts. In studying these grafts subsequently, they found evidence that marked fibrosis of the graft, apparently occurring after the nerve

fibers had entered it, was the cause of the failure of the fibers to grow through it.

Predegeneration of the nerve to be used for grafting aparently does not increase the speed of regeneration; according to Barnes and co-workers ('45) the only advantage of predegeneration is a firmer consistency making for easier handling. This advantage, they suggested, is more than counterbalanced by the disadvantage of prolonged predegeneration, during which changes can also occur in the muscles themselves.

Sunderland and Ray ('47) examined a number of cutaneous nerves in regard to their suitability for use as autografts. They noted that best results might be expected when the funicular pattern of the graft approximately fitted the nerve to be repaired, and when the graft contained fibers of about the same size as those expected to regenerate through it. Likewise, the graft should be from a nerve having few branches, if possible. On the basis of these criteria, they considered the superficial branch of the radial, the medial antebrachial cutaneous (which Seddon, '63, generally preferred), the medial brachial cutaneous, the sural nerve, the lateral femoral cutaneous and the posterior femoral cutaneous. Of these, the sural nerve and the superficial branch of the radial were said to be much superior to others: they show little variation and both have considerable length without branching; also, they are easily exposed and the resultant sensory loss does not impair any particular function. The sural nerve, they found, provides greater length than does the superficial radial, but the latter usually has a greater and more constant cross-sectional diameter. Both of the nerves are marked by a minimum of connective tissue between the funiculi, thus providing a maximum of pathways along which fibers can regenerate.

Seddon ('54) reported good results, that is, equivalent to the best seen after suture of the same nerve at the same level, in 42% of 67 cases in which severe injuries were repaired by autogenous nerve grafts, and expressed the opinion that, with further lapse of time, the good results in this series might reach 50%. Partial but useful recovery occurred in other cases, for a total of 68% in which the operation seemed to have yielded worthwhile results. Seddon discussed a number of factors influencing the success of the repair: among them are repair as early as possible (although success, he said, may be obtainable at least up to a year after injury) in order to avoid degenerative changes and fibrosis both in the distal segment of nerve and in the muscle; adequate resection of the nerve stump; and a graft longer than the gap to be bridged, in order to allow for shrinkage.

Rate of Regeneration
The rate of regeneration of peripheral nerves is frequently given as being about 1.5 mm per day, with the additional comment that a latent period must be allowed both for the regenerative process to become organized and the regenerating fibers to enter the distal end of the nerve, and for the regenerated fibers to attain functional maturity after they have reached the end organs. Obviously, the required latent period will tend to vary markedly according to the type of lesion and how it is repaired: if axonotmesis alone has occurred, the first latent period willl be relatively short, while if scar tissue has formed between poorly apposed ends of a sutured nerve, the latent period may be much longer. Seddon ('43) suggested that a latent period of about 40 to 50 days should be added to the calculated time for regeneration before a decision is made to re-explore a sutured nerve; Sunderland ('46) suggested that much more, 11 weeks to 23 weeks, according to the severity of the lesion, should be allowed for the latent period. He found the initial delay in the re-entrance of the regenerating axons into the distal, injured, segment was up to 10 weeks with minimal injuries (axonotmesis), and up to 4 months where there had been extensive injury associated with fracture of the humerus, a gunshot wound, infection with extensive scarring, and the like.

Not only does the latent period vary, but apparently the rate of regeneration does also, perhaps in part with the nerve concerned, and quite certainly according to the level of the lesion and the distance of the regenerating

fibers from the central nervous system. According to Sunderland ('47c), fibers injured close to the central nervous system always show an initially high rate of growth as compared with fibers injured some distance away, and both the actual rate of growth of regenerating fibers and functional maturation with the appearance of myelin and restoration of the diameter of the fiber proceed more slowly the farther peripherally the regenerating fibers grow. Thus, there is no constant rate of growth in regenerating nerve fibers, but rather a steadily diminishing one. Sunderland also said that the actual rate of regeneration is always faster after axonotmesis than after suture of a nerve, whether the rate of growth is measured in the proximal or the distal portion of the nerve.

Marble and co-workers calculated the rate of regeneration in the ulnar nerve as varying from 3.6 mm to 0.58 mm a day, with an estimated average of 1.24 mm, but expressed the belief that the true rate is closer to 1.5 mm per day. They estimated the rate of regeneration of the median nerve at 1.09 mm a day. Sunderland ('46) estimated the rate of regeneration in the radial nerve after axonotmesis as being 1.9 mm per day in the proximal portion of the nerve, and 0.8 mm in distal parts; at the same levels, after suture, he found the rates were 1.2 mm and 0.6 mm per day. He also ('47b) calculated the rate of regeneration of motor fibers in the distal part of the ulnar nerve and in the tibial nerve in the leg, and ('47a) the rate of regeneration of sensory fibers in 12 peripheral nerve injuries. For the ulnar nerve, he found values of from 0.6 mm to 0.4 mm per day, thus at a slower rate than he had found for the radial nerve (but, of course, the terminal branches of the ulnar nerve are generally much distal to the terminal ones of the radial); in the tibial, he found the rate of regeneration to range from 2 mm to 1.2 mm per day. As in his other work, he found that in every case the nerves regenerated faster after axonotmesis than after suture.

In the case of sensory nerves, he apparently established the fact that they, like motor nerves, grow progressively more slowly as they advance peripherally. He calculated the rate of growth as between 2.5 mm and 2 mm per day during the early stages of regeneration, and said that in the earliest stages perhaps they grew as fast as 3 mm per day; after 120 days of regeneration, he found, the rate had dropped to about 2 mm; from 120 to 200 days it averaged about 1.6 mm; and from 200 to 500 it dropped to between 0.8 mm and 0.5 mm per day.

Extent of Functional Regeneration

It is probably true, as Marble and co-workers observed in their series, that perfect regeneration never occurs. However, if a nerve is severed, nerve anastomosis or nerve graft offers the only hope of recovery, and may give rather adequate functional results. It is usually said that regeneration to the small muscles of the hand is less successful than to other muscles; Marble and his colleagues reported that in their series, however, the radial nerve, usually regarded as regenerating better than the other nerves of the arm, regenerated less well than did either the ulnar or the median. Adopting as a criterion of successful regeneration the reattainment of a useful limb, they said the ulnar nerve successfully regenerated in 88% of their cases, the median in 82% and the radial in 78%.

Assessment of the degree of functional regeneration must include an investigation of both the motor and the sensory components, if they are both involved, for Pollock ('34) found no regular correlation between the amount of regeneration of sensory function and that of motor function. As a rule, motor function recovers sooner and better than does sensory, although Pollock had some cases in which there was recovery of sensation without evidence of recovery of motor function. In the experiments of Sherman and co-workers on the cat, sensation usually began to return, after section of a mixed nerve, only after motor recovery had begun. In none of their cases was recovery of sensation complete by the time good motor recovery had occurred. Pollock ('41) discussed sensory and motor loss following nerve lesions, and methods both for testing the function of regenerating motor

nerves and estimating the disability caused by a loss of nerves that did not properly regenerate.

Since cutaneous nerves contain both sensory and autonomic fibers, it might be expected that these will regenerate together; and since loss of autonomic fibers of the skin results in an increase of the skin resistance, the latter could presumably be used as an objective measurement of degeneration and regeneration to skin. Herz and his colleagues reported that the area of increased skin resistance coincided well with the area of sensory loss when there was a complete nerve lesion, and that therefore skin resistance is valuable in mapping anesthetic areas. On the other hand, when there were only partial nerve lesions or some degree of recovery, they were able to find no correlation between the skin resistance and the presence of hypesthesia or hyperesthesia; they concluded, therefore, that skin-resistance tests do not give definite information on partial dysfunction of cutaneous nerves.

Kredal and Phemister reported that sympathetic fibers regenerate into pedicle skin flaps, and that the earliest return of sweating in such grafts, in their experience, was at the end of 11 months. They said vasomotor function usually returns in part, that pilomotor function may also return, and that the amount of restoration of these functions somewhat parallels the amount of return of cutaneous sensation. They reported little evidence of recovery of function in free transplants, and none in Thiersch (split-thickness) grafts. McCarroll reported, in contrast, that sensations of pain and touch may return in split-thickness grafts, free full-thickness grafts, and pedicle flaps. In pedicle flaps, he said, restoration usually takes place first along the proximal border and the edge attached first, but in grafts regeneration occurs over all parts of the graft simultaneously rather than by progression from the borders of the graft. He concluded that complete recovery of sensation in all types of skin grafts is therefore possible unless the cutaneous nerves of the involved region have been destroyed and fail to regenerate. The rate of return, he found, is inversely proportional to the thickness of the graft used.

There is an additional phenomenon that should be considered in assessing the progress of regeneration. This is the fact that totally uninjured nerve fibers which lie adjacent to denervated muscle or skin will "regenerate," that is, give rise to branches that invade the denervated area. Weddell and co-workers demonstrated this in the rabbit, Livingston produced evidence that it happens in the skin of man, and Wedell directly observed, in the rabbit's ear, the ingrowth from uninjured fibers into a denervated area. Sherman and co-workers proved that this phenomenon could result in a rather marked shrinkage of the area of analgesia in the limb of a cat in the complete absence of regeneration; and Hollinshead showed that this may also involve autonomic fibers, and account for the reappearance of sweating in the absence of regeneration (in the paw of the cat).

Insofar as the return of sensation is concerned, the distinction between this process of ingrowth and that of real regeneration is that in the latter the return of sensation typically appears proximally and progresses distally, according to the distance over which the fibers must regenerate, while in the former there is shrinkage of the analgesic area from all sides, since all the intact nerves surrounding the affected area typically participate in its reinnervation.

The same phenomenon also occurs in voluntary muscle (Van Harreveld; Edds). Edds eliminated about 60% of the axons in the long thoracic nerve by sectioning the sixth cervical root of this nerve, and studied the serratus anterior muscle and its nerve supply at variable intervals of time. He found that within 2 weeks after section the intact axons within the muscle developed delicate collateral sprouts that grew out toward the denervated muscle fibers along adjacent empty neurilemmal sheaths. As in regeneration after degeneration of fibers, he found that large numbers of new fibers may arise, but that only those which reached the motor end plates survived and attained almost normal size. Such spreading (in the rat) was found to reach a maximal rate

during the first month but continued at least sporadically through the fourth month. Through this process individual nerve fibers come to support more muscle fibers than they previously did. In his experiments, they supported from 1.5 times to more than 30 times as many.

Obviously, this process may account for the return of function in a partially denervated muscle, and such a return of function may be erroneously attributed to regeneration. Also, this process may account for the recovery of function in a muscle partially paralyzed, for instance, by poliomyelitis, as easily as can the assumption that such recovery depends upon temporary rather than permanent damage to some of the nerve elements supplying it. That it may be an important phenomenon is indicated by Van Harreveld's finding that stimulation of the fifth lumbar nerve in the rabbit, after axonal spread had occurred consequent to destruction of the sixth lumbar nerve, provoked a contraction from the sartorius 4.4 times as strong as that obtained by stimulation of L5 in normal animals.

In recent years, knowledge of the spread of nerve fibers from innervated to denervated muscle has been put to practical use. Nerve–muscle pedicles have been used successfully to reinnervate muscles of the larynx and of the face.

Skin

The skin is of relatively little interest to most surgeons, and will be commented upon only briefly here.

Histologically, the skin consists of two chief layers: the epidermis, composed of epithelium, and tightly attached to the dermis or corium, which is composed of dense connective tissue. The skin is bound down with varying tightness to the deeper tissues by looser connective tissue, the tela subcutanea or superficial fascia, which in places contains an obvious layer of fat and is then referred to as the panniculus adiposus. There is no sharp line of division between the dermis and the tela subcutanea, since connective tissue fibers of the two layers are continuous with each other.

The epidermis, a stratified squamous epithelium, varies much in thickness, especially in regard to its outer layer, the stratum corneum, which on the soles of the feet is far thicker than all other layers combined. This layer is composed of dead and completely cornified (keratinized) cells, and undergoes constant desquamation from its surface. Deep to the stratum corneum, in regions of thicker epidermis where the layers are well differentiated, can be recognized in order the stratum lucidum and the stratum granulosum, containing precursors of keratin, and deep to these is the stratum germinativum or Malpighian layer, usually divided into a thicker portion, the stratum spinosum, and a layer of cylindrical cells adjacent to the dermis, the stratum basale or stratum cylindricum. Melanin pigment occurs in the epidermis, and pigmented melanoblasts that lie at the junction of epidermis and dermis send processes outward among the epithelial cells.

The function of the epidermis is largely protective. As an epithelial layer, it contains no blood vessels, but derives its nourishment from the vessels in the underlying dermis. Similarly, the nerve endings responsible for sensation from the skin lie largely in the dermis.

Nails, hair, and the various glands associated with the skin are all derived from the epidermis, but hairs and glands regularly project down into the dermis, and in the case of larger glands (*e.g.,* sweat glands of the axilla and mammary glands) into the tela subcutanea.

The dermis or corium is composed largely of bundles of dense collagenous tissue, arranged mainly but not entirely parallel to the surface of the skin, interlaced with a thick network of elastic fibers. It is less dense where it borders the epidermis, especially in the papillae, where vascular loops and nerve networks are usually found.

The collagenous bundles and accompanying elastic tissue of the dermis are not arranged in haphazard order, but the main bundles in any region are parallel to each other; this arrangement gives rise to the delicate lines on the skin, tension lines, which become especially obvious with the loss of elastic tissue in old age. Since these tension lines correspond in direction to the predominant direction of the collagenous bundles of the dermis, they also indicate the disposition of the cleavage lines (of Langer): the cleavage lines of the skin are determined by observing the direction in which the skin will split when a sharp round instrument is inserted into it, the split occurring in the direction of the major bundles of the dermis. The pattern of the cleavage lines varies on different parts of the body (Fig. 1-32). Incisions through the skin along the cleavage lines tend to separate bundles in the dermis rather than to cut across them, hence produce a minimum of gaping, while incisions at right angles to them produce maximal disruption of the connective tissue and hence a maximum of gaping and subsequent scarring.

Theoretically, all incisions should be made along rather than across the cleavage lines. For many reasons, surgeons tend to pay little attention to these lines in planning their incisions, except where the lines also correspond to the flexion creases produced by muscular action. They do, in general, so correspond, and it is an accepted principle that in approaches to joints or to structures overlying joints the incision, regardless of its original direction, should almost parallel the chief flexion creases (Fig. 1-33).

BLOOD SUPPLY AND INNERVATION

The chief vascular and nerve trunks supplying the skin, usually referred to as superficial vessels and cutaneous or superficial nerves, run in the tela subcutanea, in which they are dissectible. As they proceed they give off into the skin numerous tiny branches that penetrate the dermis and ramify in it, their terminals in many instances lying close against the epidermis, especially in the dermal papillae. It is certain that no blood vessels penetrate the epidermis; nerve fibers have been described in the epidermis, but they are sparse

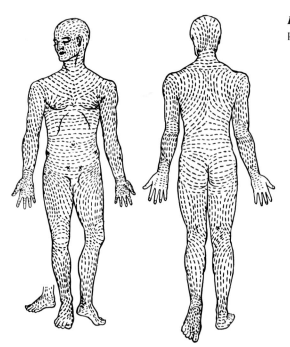

Fig. 1-32. Cleavage lines of the skin, anterior and posterior views. (Cox HT: Brit J Surg 29:234, 1941)

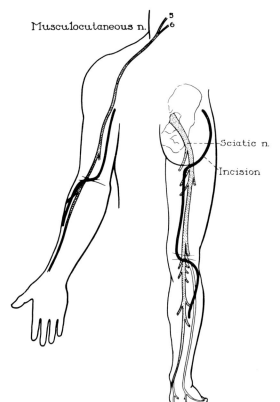

Fig. 1-33. Lines of incision (*heavy black*) for exposure of the musculocutaneous nerve, left, and of the sciatic nerve, right. The change from vertical to horizontal over the front of the elbow and the back of the knee is typical of all vertical incisions that have to cross joints. (Right, from Craig WMcK. In Walters W, (ed): Lewis' Practice of Surgery, Chap 6, Vol 3. Hagerstown, Md, Prior, 1955)

in adult man, and the usual termination of cutaneous nerves is in the form of netlike endings, or in specialized end organs such as lamellated (Pacinian) corpuscles, in the dermis or the tela subcutanea. The gross distribution of nerves to the skin is discussed in the several following chapters on the limbs, and earlier in the present chapter. Except on the trunk, dermatomes must be determined by clinical means.

References

ADAMS WE: The blood supply of nerves: Historical review. J Anat 76:323, 1942

ADAMS WE: Blood supply of nerves: II. The effects of exclusion of its regional sources of supply on the sciatic nerve of the rabbit. J Anat 77:243, 1943

AITKEN JT, SHARMAN M, YOUNG JZ: Maturation of regenerating nerve fibres with various peripheral connexions. J Anat 81:1, 1947

ALLEN FM: Effects of ligations on nerves of the extremities. Ann Surg 108:1088, 1938

ARKIN AM: Absolute muscle power: The internal kinesiology of muscle. Arch Surg 42:395, 1941

ASCENZI A, BONUCCI E: The tensile properties of single osteons. Anat Rec 158:375, 1967

BACSICH P, WYBURN GM: The vascular pattern of peripheral nerve during repair after experimental crush injury. J Anat 79:9, 1945a

BACSICH P, WYBURN GM: The effect of interference with the blood supply on the regeneration of peripheral nerves. J Anat 79:74, 1945b

BAKKE JL, WOLFF HG: Occupational pressure neuritis of the deep palmar branch of the ulnar nerve. Arch Neurol Psychiat 60:549, 1948

BARNES R, BACSICH P, WYBURN GM: A histological study of a predegenerated nerve autograft. Brit J Surg 33:130, 1945

BARNES R, BACSICH P, WYBURN GM, KERR AS: A study of the fate of nerve homografts in man. Brit J Surg 34:34, 1946

BARR JS, STINCHFIELD AJ, REIDY JA: Sympathetic ganglionectomy and limb length in poliomyelitis. J Bone Joint Surg 32-A:793, 1950

BASMAJIAN JV: Muscles Alive. Their Functions Revealed by Electromyography (ed 2). Baltimore, Williams & Wilkins, 1967

BASMAJIAN JV: Electromyography comes of age. The conscious control of individual motor units in man may be used to improve his physical performance. Science 176:603, 1972

BECTON JL, WINKELMANN RK, LIPSCOMB PR: Innerva-tion of human finger tendons as determined by histochemical techniques. J Bone Joint Surg 48-A:1519, 1966

BENNETT GA, BAUER W: Further studies concerning the repair of articular cartilage in dog joints. J Bone Joint Surg 17:141, 1935

BEVELANDER G, JOHNSON PL: A histochemical study of the development of membrane bone. Anat Rec 108:1, 1950

BICK EM: Surgical pathology of synovial tissue. J Bone Joint Surg 12:33, 1930

BLACK BM: The prenatal incidence, structure, and development of some human synovial bursae. Anat Rec 60:333, 1934

BLAIR DM, DUFF D, BINGHAM JA: The anatomical result of peri-arterial sympathectomy. Brit J Surg 18:215, 1930

BLOMFIELD LB: Intramuscular vascular patterns in man. Proc Roy Soc Med 38:617, 1945

BLOOM W, BLOOM MA: Calcification and ossification: Calcification of developing bones in embryonic and newborn rats. Anat Rec 78:497, 1940

BLOUNT WP, CLARKE GR: Control of bone growth by epiphyseal stapling: A preliminary report. J Bone Joint Surg 31-A:464, 1949

BORA FW, JR: Peripheral-nerve repair in cats: The fascicular stitch. J Bone Joint Surg 49-A:659, 1967

BOURNE GH (ed): The Structure and Function of Muscle. New York, Academic Press, 1960

BOWDEN REM, GUTMANN E: Denervation and re-innervation of human voluntary muscle. Brain 67:273, 1944

BROOKES M: Cortical vascularization and growth in foetal tubular bones. J Anat 97:597, 1963

BROOKES M, ELKIN AC, HARRISON RG, HEALD CB: A new concept of capillary circulation in bone cortex: Some clinical applications. The Lancet, May 20, 1961, p. 1078

BROOKS DM: Nerve compression by simple ganglia: A review of thirteen collected cases. J Bone Joint Surg 34-B:391, 1952

BROWN RB, HUFNAGEL CA: Arterial grafting in military surgery. JAMA 157:419, 1955

BRUNER JM: Problems of postoperative position and motion in surgery of the hand. J Bone Joint Surg 35-A:355, 1953

BUEKER ED, MEYERS CE: The maturity of peripheral nerves at the time of injury as a factor in nerve regeneration. Anat Rec 109:723, 1951

BURGER M, SHERMAN BS, SOBEL AE: Observations of the influence of chondroitin sulphate on the rate of bone repair. J Bone Joint Surg 44-B:675, 1962

BURKEL WE: The histological fine structure of perineurium. Anat Rec 158:177, 1967

BURTON AC, YAMADA S: Relation between blood pressure and flow in the human forearm. J Appl Physiol 4:329, 1951

CARLSON BM: The regeneration of skeletal muscle. A review. Am J Anat 137:119, 1973

CLARK WE LE G: An experimental study of the regeneration of mammalian striped muscle. J Anat 80:24, 1946

CLARK WE LE G, BLOMFIELD LB: The efficiency of intramuscular anastomoses, with observations on the regeneration of devascularized muscle. J Anat 79:15, 1945

CLARK WE LE G, WAJDA HS: The growth and maturation of regenerating striated muscle fibres. J Anat 81:56, 1947

COGGESHALL HC, WARREN CF, BAUER W: The cytology of normal human synovial fluid. Anat Rec 77:129, 1940

COMER RD: An experimental study of the "laws" of muscle and tendon growth. Anat Rec 125:665, 1956

COOPER RR, MILGRAM JW, ROBINSON RA: Morphology of the osteon: An electron microscopic study. J Bone Joint Surg 48-A:1239, 1966

CORBIN KB, GARDNER ED: Decrease in number of myelinated fibers in human spinal roots with age. Anat Rec 68:63, 1937

CORBIN KB, HINSEY JC: Influence of the nervous system on bone and joints. Anat Rec 75:307, 1939

COTTRELL L: Histologic variations with age in apparently normal peripheral nerve trunks. Arch Neurol Psychiat 43:1138, 1940

CRAWFORD GNC: An experimental study of muscle growth in the rabbit. J Bone Joint Surg 36-B:294, 1954

CRELIN ES, SOUTHWICK WO: Changes induced by sustained pressure in the knee joint articular cartilage of adult rabbits. Anat Rec 149:113, 1964

CRONKITE AE: The tensile strength of human tendons. Anat Rec 64:173, 1936

DALE WA, DEWEESE JA: Autogenous venous grafts for arterial repair. Surgery 55:870, 1964

D'AUGIGUÉ RM, DUBOUSSENT J: Surgical correction of large length discrepancies in the lower extremities of children and adults: An analysis of twenty consecutive cases. J Bone Joint Surg 53-A:411, 1971

DAVENPORT HK, RANSON SW: Contracture resulting from tenotomy. Arch Surg 21:995, 1930

DAVIES DA, PARSONS FG: The age order of the appearance and union of the normal epiphyses as seen by x-rays. J Anat 62:58, 1927

DAVIES DV: The staining reactions of normal synovial membrane with special reference to the origin of synovial mucin. J Anat 77:160, 1943

DAVIES DV: Observations on the volume, viscosity and nitrogen content of synovial fluid, with a note on the histological appearance of the synovial membrane. J Anat 78:68, 1944

DAVIES DV: The lymphatics of the synovial membrane. J Anat 80:21, 1946

DAVIES DV, EDWARDS DAW: The blood supply of the synovial membrane and intra-articular structures. Ann Roy Col Surgeons England 2:142, 1948

DAVIS L, PERRET G, HILLER F: Experimental studies in peripheral nerve surgery: IV. The effect of infection on regeneration and functional recovery. Surg Gynecol Obstet 81:302, 1945

DEBAKEY ME, SIMEONE FA: Battle injuries of the arteries in World War II: An analysis of 2,471 cases. Ann Surg 123:534, 1946

DENNY-BROWN D, BRENNER C: Paralysis of nerve induced by direct pressure and by tourniquet. Arch Neurol Psychiat 51:1, 1944a

DENNY-BROWN D, BRENNER C: Lesion in peripheral nerve resulting from compression by spring clip. Arch Neurol Psychiat 52:1, 1944b

DENNY-BROWN D, and PENNYPACKER JB: Fibrillation and fasciculation in voluntary muscle. Brain 61:311, 1938

DE ORSAY RH, MECRAY PM, JR, and FERGUSON LK: Pathology and treatment of ganglion. Am J Surg 36:313, 1937

DILLARD BM, NELSON DL, and NORMAN HG, JR: Review of 85 major traumatic arterial injuries. Surgery 63:391, 1968

DODD RM, SIGEL B, DUNN MR: Localization of new cell formation in tendon healing by tritiated thymidine autoradiography. Surg Gynecol Obstet 122:805, 1966

DORR LD, BRODY MJ: Functional separation of adrenergic and cholinergic fibers to skeletal muscle vessels. Am J Physiol 208:417, 1965

DUCHENNE GB: Physiology of Motion: Demonstrated by Means of Electrical Stimulation and Clinical Observation and Applied to the Study of Paralysis and Deformities. Translated and edited by E. B. Kaplan. Philadelphia, JB Lippincott, 1949

DUNCAN CP, SHIM S-S: The autonomic nerve supply of bone: An experimental study of the intraosseous adrenergic nervi vasorum in the rabbit. J Bone Joint Surg 59-B 323, 1977

DUNCAN D: A relation between axone diameter and myelination determined by measurement of myelinated spinal root fibers. J Comp Neurol 60:437, 1934

DUPERTUIS SM: Actual growth of young cartilage transplants in rabbits: Experimental studies. Arch Surg 43:32, 1941

EDDS MV, JR: Cytological evidence for the spreading of intact axons in partially denervated muscles of the adult rat. Anat Rec 103:534, 1949 (abstr.)

EDWARDS DAW: The blood supply and lymphatic drainage of tendons. J Anat 80:147, 1946

EDWARDS EA: The orientation of venous valves in relation to body surfaces. Anat Rec 64:369, 1936

ELFTMAN H: Biomechanics of muscle: With particular application to studies of gait. J Bone Joint Surg 48-A:363, 1966

ELKIN DC, DEBAKEY ME: Surgery in World War II: Vascular Surgery. Washington Office of the Surgeon General, Department of the Army, 1955

EPKER BN, FROST HM: Periosteal appositional bone

growth from age two to age seventy in man. A tetracycline evaluation. Anat Rec 154:573, 1966

EVANS EB, EGGERS GWN, BUTLER JK, BLUMEL J: Experimental immobilization and remobilization of rat knee joints. J Bone Joint Surg 42-A:737, 1960

EVANS FG: Stress and Strain in Bones: Their Relation to Fractures and Osteogenesis. Springfield, Ill., Thomas, 1957

EVANS FG, BANG S: Differences and relationships between the physical properties and the microscopic structure of human femoral, tibial and fibular cortical bone. Am J Anat 120:79, 1967

EVANS FG, LEBOW M: The strength of human compact bone as revealed by engineering technics. Am J Surg 83:326, 1952

EVANS HM: The blood-supply of lymphatic vessels in man. Am J Anat 7:195, 1907

EYRING EJ, MURRAY WR: The effect of joint position on the pressure of intra-articular effusion. J Bone Joint Surg 46-A:1235, 1964

FAIRBANK TJ: Knee joint changes afterr meniscectomy. J Bone Joint Surg 30-B:664, 1948

FEINDEL W, HINSHAW JR, WEDDELL G: The pattern of motor innervation in mammalian striated muscle. J Anat 86:35, 1952

FELL HB: The osteogenic capacity *in vitro* of periosteum and endosteum isolated from the limb skeleton of fowl embryos and young chicks. J Anat 66:157, 1932

FINTERBUSH A, FRIEDMAN B: The effect of sensory denervation on rabbits' knee joints: A light and microscopic study. J Bone Joint Surg 57-A:949, 1975

FLECKER H: Roentgenographic observations of the times of appearance of epiphyses and their fusion with the diaphyses. J Anat 67:118, 1932

FLECKER H: Time of appearance and fusion of ossification centers as observed by roentgenographic methods. Am J Roentgenol 47:97, 1942

FLETCHER TF, KITCHELL RL: The lumbar, sacral and coccygeal tactile dermatomes of the dog. J Comp Neurol 128:171, 1966

FLYNN EJ, WILSON JT, CHILD CG, III, GRAHAM JH: Heterogenous and autogenous tendon transplants. An experimental study of preserved bovine-tendon transplants in dogs and autogenous-tendon transplants in dogs. J Bone Joint Surg 42-A:91, 1960

FOLEY PJ, ALLEN EV, JANES JM: Surgical treatment of acquired arteriovenous fistulas. Am J Surg 91:611, 1956

FORD FR, WOODHALL B: Phenomena due to misdirection of regenerating fibers of cranial, spinal and autonomic nerves: Clinical observations. Arch Surg 36:480, 1938

FOSTER LN, KELLY RP, JR, WATTS WM: Experimental infarction of bone and bone marrow: Sequelae of severance of the nutrient artery and stripping of periosteum. J Bone Joint Surg 33-A:396, 1951

FRANCIS CC: The appearance of centers of ossification from 6 to 15 years. Am J Phys Anthropol 27:127, 1940

FRANCIS CC, WERLE PP, BEHM A: The appearance of centers of ossification from birth to 5 years. Am J Phys Anthropol 24:273, 1939

FROHSE F: Ueber die Verzweigung der Nerven zu und in den menschlichen Muskeln. Anat Anz 14:321, 1898

GAGE M, OCHSNER A: The prevention of ischemic gangrene following surgical operations upon the major peripheral arteries by chemical section of the cervicodorsal and lumbar sympathetics. Ann Surg 112:938, 1940

GAMBLE HJ, EAMES RA: An electron microscope study of the connective tissues of the human peripheral nerve. J Anat 98:655, 1964

GARDNER E: The distribution and determination of nerves in the knee joint of the cat. J Comp Neurol 80:11, 1944

GARDNER E: Physiology of movable joints. Physiol Rev 30:127, 1950

GARDNER E: Physiology of Joints. Am Acad Orth Surgeons Instructional Course Lectures 10:251, 1953

GAY AJ, JR, HUNT TE: Reuniting of skeletal muscle fibers after transection. Anat Rec 120:853, 1954

GLEES P: Observations on the structure of the connective tissue sheaths of cutaneous nerves. J Anat 77:153, 1943

GOETZ RH: The diagnosis and treatment of vascular diseases: With special consideration of clinical plethysmography and the surgical physiology of the autonomic nervous system. Brit J Surg 37:25, 146, 1949

GOLDING MR, MARTINEZ A, DEJONG P, MENDOSA M, FRIES CC, SAWYER PN, HENNIGAR GR, WESOLOWSKI SA: The role of sympathectomy in frostbite, with a review of 68 cases. Surgery 57:774, 1965

GOSS CM: The attachment of skeletal muscle fibers. Am J Anat 74:259, 1944

GRANT JCB: Interarticular synovial folds. Brit J Surg 18:636, 1931

GRAUER RC: The effect of viosterol on the periosteum in experimental fractures. Arch Surg 25:1035, 1932

GREEN WT, ANDERSON M: Experiences with epiphyseal arrest in correcting discrepancies in length of the lower extremities in infantile paralysis: A method of predicting the effect. J Bone Joint Surg 29:659, 1947

GREEN WT, ANDERSON M: Epiphyseal arrest for the correction of discrepancies in length of the lower extremities. J Bone Joint Surg 39-A:853, 1957

GUTMANN E, SANDERS FK: Functional recovery following nerve grafts and other types of nerve bridge. Brain 65:373, 1942

GUTMANN E, YOUNG JZ: The re-innervation of muscle after various periods of atrophy. J Anat 78:15, 1944

HAGERTY RF, BRAID HL, BONNER WM, JR, HENNIGAR

GR, LEE WH, JR: Viable and nonviable human cartilage homografts. Surg Gynecol Obstet 125:485, 1967

HAINES RW: The laws of muscle and tendon growth. J Anat 66:578, 1932

HAINES RW: On muscles of full and of short action. J Anat 69:20, 1934

HALDEMAN KO: The rôle of periosteum in the healing of fractures: An experimental study. Arch Surg 24:440, 1932

HAMMOND WS, HINSEY JC: The diameters of the nerve fibers in normal and regenerating nerves. J Comp Neurol 83:79, 1945

HARRIS W: The true form of the brachial plexus, and its motor distribution. J Anat Physiol 38:399, 1904

HARRIS WR: The endocrine basis for slipping of the upper femoral epiphysis: An experimental study. J Bone Joint Surg 32-B:5, 1950

HARRISON VF, MORTENSEN OA: Identification and voluntary control of single motor unit activity in the tibialis anterior muscle. Anat Rec 144:109, 1962

HEINBECKER P, BISHOP GH, O'LEARY JL: Functional and histologic studies of somatic and autonomic nerves of man. Arch Neurol Psychiat 35:1233, 1936

HEIPLE KG, CHASE SW, HERNDON CH: A comparative study of the healing process following different types of bone transplantation. J Bone Joint Surg 45-A:1593, 1963

HEIPLE KG, NASH CL, JR, KLEIN L: A Study of ^{14}C-labeled collagen of rat homograft tendon. J Bone Joint Surg 49-A:1109, 1967

HELLER-STEINBERG M: Ground substance, bone salts, and cellular activity in bone formation and destruction. Am J Anat 89:347, 1951

HENDRYSON IE: An evaluation of the estimated percentage of growth from the distal epiphyseal line. J Bone Joint Surg 27:208, 1945

HERZ E, GLASER GH, MOLDOVER J, and HOEN TI: Electrical skin resistance test in evaluation of peripheral nerve injuries. Arch Neurol Psychiat 56:365, 1946

HEYMAN CH: Manipulation of joints. J Bone Joint Surg 12:23, 1930

HILL AH: Fetal age assessment by centers of ossification. Am J Phys Anthropol 24:251, 1939

HOLLINSHEAD WH: An attempt to innervate sweat glands through preganglionic fibers. J Comp Neurol 89:193, 1948

HOLMAN E: Fundamental principles governing the care of traumatic arteriovenous aneurysms. Angiology 5:145, 1954

HOLMAN E: The development of arterial aneurysms. Surg Gynecol Obstet 100:599, 1955

HOLMES W, and YOUNG JZ: Nerve regeneration after immediate and delayed suture. J Anat 77:63, 1942

HONNER R, THOMPSON RC: The nutritional pathways of articular cartilage: An autoradiographic study

in rabbits using ^{35}S injected intravenously. J Bone Joint Surg 53-A:742, 1971

HUBER GC: On the form and arrangement in fasciculi of striated voluntary muscle fibers. Anat Rec 11:149, 1916

HUGGINS C, WIEGE E: The effect on the bone marrow of disruption of the nutrient artery and vein. Ann Surg 110:940, 1939

HUME JB: The causation of multiple exostoses. Brit J Surg 17:236, 1929

HUTCHISON J: The fate of experimental bone autografts and homografts. Brit J Surg 39:552, 1952

INGBERT CE: On the density of the cutaneous innervation in man. J Comp Neurol 13:209, 1903

IVINS JC: Personal communication to the author

JACKSON S: The role of galvanism in the treatment of denervated voluntary muscle in man. Brain 68:300, 1945

JAHNKE EJ, JR, HUGHES CW, HOWARD JM: The rationale of arterial repair on the battlefield. Am J Surg 87:396, 1954

JANES JM, ELKINS EC: The effect of a surgically induced arteriovenous fistula on bone growth: Report of case. Proc Staff Meet Mayo Clin 27:335, 1952

JENNINGS WK, and JANES JM: Variations in blood pressure and pulse rate with creation or closure of arteriovenous fistulas in children. Proc Staff Meet Mayo Clin 36:62, 1961

JOWSEY J: Studies of Haversian systems in man and some animals. J Anat 100:857, 1966

JOWSEY J, KELLY PJ, RIGGS BL, BIANCO AJ, JR, SCHOLZ DA, GERSHON-COHEN J: Quantitative microradiographic studies of normal and osteoporotic bone. J Bone Joint Surg 47-A:785, 1965

JUST-VIERA JO, and YEAGER GH: Venous stasis. I. Effects of venous resection on bone growth. Surgery 58:694, 1965

KAMPMEIER OF, BIRCH C LA F: The origin and development of the venous valves, with particular reference to the saphenous district. Am J Anat 38:451, 1927

KECK SW, KELLY PJ: The effect of venous stasis on intraosseous pressure and longitudinal bone growth in the dog. J Bone Joint Surg 47-A:539, 1965

KEEGAN JJ: Relations of nerve roots to abnormalities of lumbar and cervical portions of the spine. Arch Surg 55:246, 1947

KELLGREN JH, SAMUEL EP: The sensitivity and innervation of the articular capsule. J Bone Joint Surg 32-B:84, 1950

KELLY PJ, JANES JM, PETERSON LFA: The effect of arteriovenous fistulae on the vascular pattern of the femora of immature dogs. A microangiographic study. J Bone Joint Surg 41-A:1101, 1959

KELLY RE: Is it true that the valves in a vein necessarily become incompetent when the vein dilates? Brit J Surg 18:53, 1930

KERNOHAN JW, WOLTMAN HW: Periarteritis nodosa: A clinicopathologic study with special reference to the nervous system. Arch Neurol Psychiat 39: 655, 1938

KERNWEIN GA: A study of tendon implantations into bone. Surg Gynecol Obstet 75:794, 1942

KEY JA: Cytology of the synovial fluid of normal joints. Anat Rec 40:193, 1928

KISTLER GH: Effects of circulatory disturbances on the structure and healing of bone: Injuries of the head of the femur in young rabbits. Arch Surg 33:225, 1936

KLEIN M, MAYER G, WEISS AG: Donées morphologiques sur les cicatrices des nerfs chez l'homme. Arch d'anat, d'histol, et d'embryol 30:242, 1947

KLING DH: The nature and origin of synovial fluid. Arch Surg 23:543, 1931

KOCH JC: The laws of bone architecture. Am J Anat 21:177, 1917

KREDEL FE, PHEMISTER DB: Recovery of sympathetic nerve function in skin transplants. Arch Neurol Psychiat 42:403, 1939

KUHNS JG: Lymphatic drainage of joints. Arch Surg 27:345, 1933

KUNTZ A, RICHINS CA: Innervation of the bone marrow. J Comp Neurol 83:213, 1945

LANIER RR: The effects of exercise on the knee-joints of inbred mice. Anat Rec 94:311, 1946

LAST RJ: Innervation of the limbs. J Bone Joint Surg 31-B:452, 1949

LEAK LV, BURKE JF: Fine structure of the lymphatic capillary and the adjoining connective tissue area. Am J Anat 118:785, 1966

LEARMONTH J: Collateral circulation, natural and artificial. Surg Gynecol Obstet 90:385, 1950

LEBLOND CP, WILKINSON GW, BÉLANGER LF, ROBICHON J: Radio-autographic visualization of bone formation in the rat. Am J Anat 86:289, 1950

LEVANDER G: A study of bone regeneration. Surg Gynecol Obstet 67:705, 1938

LINDSAY MK, and HOWES EL: The breaking strength of healing fractures. J Bone Joint Surg 13:491, 1931

LINN FC: Lubrication of animal joints. I. The arthrotripsometer. J Bone Joint Surg 49-A:1079, 1967

LINTON RR: The arteriosclerotic popliteal aneurysm: A report of fourteen patients treated by a preliminary lumbar sympathetic ganglioectomy and aneurysmectomy. Surgery 26:41, 1949

LIVINGSTON WK: Evidence of active invasion of denervated areas by sensory fibers from neighboring nerves in man. J Neurosurg 4:140, 1947

LONG ME: The development of the muscle-tendon attachment in the rat. Am J Anat 81:159, 1947

MACCONAILL MA: The function of intraarticular fibrocartilages, with special reference to the knee and inferior radio-ulnar joints. J Anat 66:210, 1932

MACCONAILL MA: The movements of bones and joints: 2. Function of the musculature. J Bone Joint Surg 31-B:100, 1949

MACCONAILL MA: The movements of bones and joints: 3. The synovial fluid and its assistants. J Bone Joint Surg 32-B:244, 1950

MARBLE HC, HAMLIN E, JR, WATKINS AL: Regeneration in the ulnar, median and radial nerves. Am J Surg 55:274, 1942

MAROUDAS A, BULLOUGH P, SWANSON SAV, and FREEMAN MAR: The permeability of articular cartilage. J Bone Joint Surg 50-B:166, 1968

MASON ML, ALLEN HS: The rate of healing of tendons. Ann Surg 113:424, 1941

MASON ML, SHEARON CG: The process of tendon repair: An experimental study of tendon suture and tendon graft. Arch Surg 25:615, 1932

MATHEWS PBC: Muscle spindles and their motor control. Physiol Rev 44:219, 1964

MCCARROLL HR: The regeneration of sensation in transplanted skin. Ann surg 108:309, 1938

MCGAW WH, and HARBIN M: The rôle of bone marrow and endosteum in bone regeneration: Experimental study of bone marrow and endosteal transplants. J Bone Joint Surg 16:816, 1934

MCKIBBIN B, and HOLDSWORTH FW: The nutrition of immature joint cartilage in the lamb. J Bone Joint Surg 48-B:793, 1966

MCKIBBIN B, and HOLDSWORTH FW: The dual nature of epiphysial cartilage. J Bone Joint Surg 49-B:351, 1967

MCLEAN FC, and BLOOM W: Calcification and ossification: Calcification in normal growing bone. Anat Rec 78:333, 1940

MCLEAN FC, and URIST MR: Bone. An Introduction to the Physiology of Skeletal Tissue (ed. 2). Chicago, University of Chicago Press, 1961

MCMASTER PE: Tendon and muscle ruptures: Clinical and experimental studies on causes and location of subcutaneous ruptures. J Bone Joint Surg 15:705, 1933

MCMASTER PE: Influence of venous stasis on the production of chronic arthritis. Arch Surg 35:833, 1937

MCMASTER PE, ROOME NW: The effect of sympathectomy and of venous stasis on bone repair: Experimental study. J Bone Joint Surg 16:365, 1934

MILLER MR, KASAHARA M: Observations on the innervation of human long bones. Anat Rec 145:13, 1963

MITCHELL NS, CRUESS RL: The effect of synovectomy on articular cartilage. J Bone Joint Surg 49-A:1099, 1967

MOLLIER G: Beziehungen Zwischen Form und Funktion der Sehnen in Muskel-Sehnen-Knochen System. Morph Jahrb 79:161, 1937

MOORE RM, WILLIAMS JH, SINGLETON AO, JR: Vasoconstrictor fibers: Peripheral course as revealed by a roentgenographic method. Arch Surg 26:308, 1933

MUELLER KH, TRIAS A, RAY RD: Bone density and composition. Age-related and pathological changes in water and mineral content. J Bone Joint Surg 48-A:140, 1966

MUSTALISH AC, PICK J: On the innervation of the blood vessels in the human foot. Anat Rec 149:587, 1964

MYSOREKAR VR: Diaphysial nutrient foramina in human long bones. J Anat 101:813, 1967

NELSON GE, JR, KELLY PJ, PETERSON LFA, JANES JM: Blood supply of the human tibia. J Bone Joint Surg 42-A:625, 1960

NOBACK CR, ROBERTSON GG: Sequences of appearance of ossification centers in the human skeleton during the first five prenatal months. Am J Anat 89:1, 1951

NOYES FR, TORVIC PJ, HYDE WH, DELUCAS JL: Biomechanics of ligament failure: II An analysis of immobilization, exercise, and reconditioning effects in primates. J Bone Joint Surg 56-A:1406, 1974

O'CONNELL JEA: The intraneural plexus and its significance. J Anat 70:468, 1936

OLSON KC, LEMING BL: Lumbar sympathectomy in peripheral vascular diseases. Am J Surg 84:202, 1952

OSIUS EA: Acute massive venous occlusion. AMA Arch Surg 65:19, 1952

PALUMBO LT, QUIRIN LF, CONKLING RW: Lumbar sympathectomy in the treatment of peripheral vascular diseases. Surg Gynecol Obstet 96:162, 1953

PATERSON RS: A radiological investigation of the epiphyses of the long bones. J Anat 64:28, 1929

PATTON NJ, MORTENSEN OA: An electromyographic study of reciprocal activity of muscles. Anat Rec 170:255, 1971

PAYTON CG: The growth in length of the long bones in the madder-fed pig. J Anat 66:414, 1932

PAYTON CG: The growth of the epiphyses of the long bones in the madder-fed pig. J Anat 67:371, 1933

PEER LA: The fate of living and dead cartilage transplanted in humans. Surg Gynecol Obstet 68:603, 1939

PETTY W, WINTER RB, FELDER D: Arteriovenous fistula for treatment of discrepancy in leg length. J Bone Joint Surg 56-A:581, 1974

PFUHL W: Die gefiederten Muskeln, ihre Form und ihre Wirkungsweise. Ztschr f Anat 106:749, 1937

PICK J: The innervation of the arteries in the upper limb of man. Anat Rec 130:103, 1958

POLLOCK LJ: The relation of recovery of different sensory branches of peripheral nerves to motor recovery. Surg Gynecol Obstet 59:858, 1934

POLLOCK LJ: Evaluation of incapacity produced by injuries of the peripheral nerves. Surg Gynecol Obstet 73:462, 1941

PORTER RW: The effect of tension across a growing epiphysis. J Bone Joint Surg 60-B:252, 1978

POTENZO AD: Concepts of tendon healing and repair AAOS Symposium on tension surgery in the hand. St. Louis, 1975 The CV Mosby Co, p 18

PRESSMAN JF, DUNN RF, BURTZ M: Lymph node ultrastructure related to direct lymphaticovenous communication. Surg Gynecol Obstet 124:963, 1967

PRYOR JW: Difference in the ossification of the male and female skeleton. J Anat 62:499, 1928

PRYOR JW: Bilateral symmetry as seen in ossification. Am J Anat 58:87, 1936a

PRYOR JW: Ossification as additional evidence in differentiating identicals and fraternals in multiple births: Am J Anat 59:409, 1936b

RALSTON HJ, III, MILLER MR, KASAHARA M: Nerve endings in human fasciae, tendons, ligaments, periosteum, and joint synovial membrane. Anat Rec 136:137, 1960

RAMAGE D: The blood supply to the peripheral nerves of the superior extremity. J Anat 61:198, 1927

RAY RD, EVANS HM, BECKS H: The effect of growth hormone injections on the costochondral junction of the rat rib. Anat Rec 82:67, 1942

RAY RD, HOLLOWAY, JA: Bone implants. Preliminary reports of an experimental study. J Bone Joint Surg 39-A: 1119, 1957

RAY RD, SIMPSON ME, LI CH, ASLING CW, EVANS HM: Effects of the pituitary growth hormone and of thyroxin on growth and differentiation of the skeleton of the rat thyroidectomized at birth. Am J Anat 86:479, 1950

RICHARDS RL: Ischaemic lesions of peripheral nerves: A review. J Neurol Neurosurg & Psychiat 14:76, 1951

RING PA: The influence of the nervous system upon the growth of bones. J Bone Joint Surg 43-B:121, 1961

ROTHMAN RH, SLOGOFF S: The effect of immobilization on the vascular bed of tendon. Surg Gynecol Obstet 124:1064, 1967

ROYLE JP: Autogenous vein bypass: An improved technique. Surgery 60:795, 1966

SALTER RB, HARRIS WR: Injuries involving the epiphyseal plate. J Bone Joint Surg 45-A:587, 1963

SAMUEL EP: The autonomic and somatic innervation of the articular capsule. Anat Rec 113:53, 1952

SANDERS FK: The repair of large gaps in the peripheral nerves. Brain 65:281, 1942

SAUNDERS JH, SISSONS HA: The effect of denervation on the regeneration of skeletal muscle after injury. J Bone Joint Surg 35-B:113, 1953

SAUNDERS RL DE CH, LAWRENCE J, and MACIVER DA: "Microradiographic Studies of the Vascular Patterns in Muscle and Skin," in X-Ray Microscopy and Microradiography. New York, Academic Press, 1957, p. 539

SEDDON HJ: Three types of nerve injury. Brain 66:237, 1943

SEDDON HJ: (ed.). Peripheral Nerve Injuries. London, Her Majesty's Stationery Office, 1954

SEDDON HJ: Nerve grafting. J Bone Joint Surg 45-B:447, 1963

SHANDS AR, JR: The regeneration of hyaline cartilage in joints: An experimental study. Arch Surg 22:137, 1931

SHANDS AR, JR: Studies in bone formation: The effect of the local presence of calcium salts on osteogenesis. J Bone Joint Surg 19:1065, 1937

SHANTHAVEERAPPA TR, BOURNE GH: The effects of transection of the nerve trunk on the perineural epithelium with special reference to its role in nerve degeneration and regeneration. Anat Rec 150:35, 1964

SHARRARD WJW: Paralytic deformity in the lower limb. J Bone Joint Surg 49-B:731, 1967

SHERMAN IC: Contractures following experimentally produced peripheral-nerve lesions. J Bone Joint Surg 30-A:474, 1948

SHERMAN IC, TIGAY EL, ARIEFF AJ, and SCHILLER MA: Return of sensation after experimentally produced lesions in sciatic nerve of cat. J Neurophysiol 12:1, 1949

SHIELDS RT: On the development of tendon sheaths. Contrib Embryol 15:53, 1923

SHUMAKER HB, JR: Sympathectomy in the treatment of peripheral vascular disease. Surgery 13:1, 1943

SIEGLING JA: Growth of the epiphyses. J Bone Joint Surg 23:23, 1941

SIFFERT RS: The growth plate and its affections. J Bone Joint Surg 48-A:546, 1966

SIGURDSON LA: The structure and function of articular synovial membranes. J Bone Joint Surg 28:603, 1930

SILBERBERG M, SILBERBERG R: The response of cartilage and bone of the newborn guinea pig to stimulation by various hormones (anterior hypophyseal extract, estrogen, thyroxin). Anat Rec 78:549, 1940

SILBERBERG M, SILBERBERG R: Further investigations concerning the influence of estrogen on skeletal tissues. Am J Anat 69:295, 1941

SIMPSON SA, YOUNG JZ: Regeneration of fiber diameter after cross-unions of visceral and somatic nerves. J Anat 79:48, 1945

SMITH JW: Collagen fibre patterns in mammalian bone. J Anat 94:329, 1960

SMITH JW: Factors influencing nerve repair. Part II. Collateral circulation of peripheral nerves. Arch Surg 93:433, 1966

SOLA CK, SILBERMAN FS, CABRINI RL: Stimulation of the longitudinal growth of long bones by periosteal stripping. An experimental study on dogs and monkeys. J Bone Joint Surg 45-A:1679, 1963

SOLOMON L: Diametric growth of the epiphysial plate. J Bone Joint Surg 48-B:170, 1966

SPURLING RG, LYONS WR, WHITCOMB BB, WOODHALL B: The failure of whole fresh homogenous nerve grafts in man. J Neurosurg 2:79, 1945

SPURLING RG, WOODHALL B: Experiences with early nerve surgery in peripheral nerve injuries. Ann Surg 123:731, 1946

STEINDLER A: The mechanics of muscular contractures in wrist and fingers. J Bone Joint Surg 14:1, 1932

STILWELL DL, JR: The innervation of deep structures of the foot. Am J Anat 101:59, 1957a

STILWELL DL, JR: The innervation of deep structures of the hand. Am J Anat 101:75, 1957b

STOPFORD JSB: The results of secondary suture of peripheral nerves. Brain 43:1, 1920

STRANDNESS DE, JR, BELL JW: Critical evaluation of the results of lumbar sympathectomy. Ann Surg 160:1021, 1964

SULLIVAN WE: The function of articular discs. Anat Rec 24:49, 1922

SUNDERLAND S: Traumatic injuries of peripheral nerves: Simple compression injuries of the radial nerve. Brain 68:56, 1945a

SUNDERLAND S: The adipose tissue of peripheral nerves. Brain 68:118, 1945b

SUNDERLAND S: The intraneural topography of the radial, median and ulnar nerves. Brain 68:243, 1945c

SUNDERLAND S: Blood supply of the nerves of the upper limb in man. Arch Neurol Psychiat 53:91, 1945d

SUNDERLAND S: Blood supply of peripheral nerves: Practical considerations. Arch Neurol Psychiat 54:280, 1945e

SUNDERLAND S: Blood supply of the sciatic nerve and its popliteal divisions in man. Arch Neurol Psychiat 54:283, 1945f

SUNDERLAND S: Course and rate of regeneration of motor fibers following lesions of the radial nerve. Arch Neurol Psychiat 56:133, 1946

SUNDERLAND S: Rate of regeneration of sensory nerve fibers. Arch Neurol Psychiat 58:1, 1947a

SUNDERLAND S: Rate of regeneration of motor fibers in the ulnar and sciatic nerves. Arch Neurol Psychiat 58:7, 1947b

SUNDERLAND S: Rate of regeneration in human peripheral nerves: Analysis of the interval between injury and onset of recovery. Arch Neurol Psychiat 58:251, 1947c

SUNDERLAND S: The capacity of muscles to function efficiently following reinnervation after long periods of denervation. Anat Rec 103:511, 1949 (abstr.)

SUNDERLAND S: The connective tissues of peripheral nerves. Brain 88:841, 1966

SUNDERLAND S, RAY LJ: The selection and use of autografts for bridging gaps in injured nerves. Brain 70:75, 1947

SUNDERLAND S, RAY LJ: The intraneural topography of the sciatic nerve and its popliteal divisions in man. Brain 71:242, 1948

DE TAKATS G: Revascularization of the arteriosclerotic extremity. AMA Arch Surg 70:5, 1955

TARLOV IM: Plasma clot suture of nerves—illustrated technique. Surgery 15:257, 1944

THOMAS PK: Changes in the endoneurial sheaths of peripheral myelinated nerve fibres during Wallerian degeneration. J Anat 98:175, 1964

THOMAS RW: Motor-nerve fiber branching in regeneration and its relation to trophic support of muscle. Anat Rec 100:718, 1948 (abstr.)

TOBIN WJ: The internal architecture of the femur and its clinical significance. The upper end. J Bone Joint Surg 37-A:57, 1955

TOWER SS: Trophic control of non-nervous tissues by the nervous system: A study of muscle and bone innervated from an isolated and quiescent region of spinal cord. J Comp Neurol 67:241, 1937

TOWER SS: Persistence of fibrillation in denervated skeletal muscle and its nonoccurrence in muscle after tenotomy. Arch Neurol Psychiat 42:219, 1939

TRUETA J, AMATO VP: The vascular contribution to osteogenesis. III. Changes in the growth cartilage caused by experimentally induced ischaemia. J Bone Joint Surg 42-B:571, 1960

TRUETA J, LITTLE K: The vascular contribution to osteogenesis II. Studies with the electron microscope. J Bone Joint Surg 42-B:367, 1960

TRUETA J, TRIAS A: The vascular contribution to osteogenesis. IV. The effect of pressure upon the epiphysial cartilage of the rabbit. J Bone Joint Surg 43-B:800, 1961

TUPMAN GS: A study of bone growth in normal children and its relationship to skeletal maturation. J Bone Joint Surg 44-B:42, 1962

VANCE RV, WYATT GM: Roentgenological manifestations of bone repair: Healing of fractures without external callus. Am J Surg 59:404, 1943

VANDERHOEFT PJ, KELLY PJ, JANES JM, PETERSON LFA: Growth and structure of bone distal to an arteriovenous fistula: Quantitative analysis of tetracycline-induced transverse growth patterns. J Bone Joint Surg 45-B:582, 1963

VAN HARREVELD A: Re-innervation of denervated muscle fibers by adjacent functioning motor units. Am J Physiol 144:477, 1945

VEAL JR, DUGAN TJ, JAMISON WL, BAUERSFELD RS: Acute massive venous occlusion of the lower extremities. Surgery 29:355, 1951

WALMSLEY R, BRUCE J: The early stages of replacement of the semilunar cartilages of the knee joints in rabbits after operative excision. J Anat 72:260, 1938

WARREN R: Ganglion of the common peroneal nerve: Case report. Ann Surg 124:152, 1946

WEAVER JK, CHALMERS J: Cancellous bone: Its strength and changes with aging and an evaluation of some methods for measuring its mineral content. I. Age changes in cancellous bone. J Bone Joint Surg 48-A:289, 1966

WEDDELL G: Axonal regeneration in cutaneous nerve plexuses. J Anat 77:49, 1942

WEDDELL G, GUTTMANN L, GUTMANN E: The local extension of nerve fibres into denervated areas of skin. J Neurol Psychiat 4:206, 1941

WEISL H, OSBORNE GV: The pathological changes in rats' nerves subject to moderate compression. J Bone Joint Surg 46-B:297, 1964

WHISTON TB, WALMSLEY R: Some observations on the reaction of bone and tendon after tunnelling of bone and insertion of tendon. J Bone Joint Surg 42-B:377, 1960

WHITE JC, SMITHWICK RH, SIMEONE FA: The Autonomic Nervous System: Anatomy, Physiology, and Surgical Application (ed. 3). New York, Macmillan, 1952

WHITE JW: Leg-length discrepancies. Am Acad Orth Surgeons, Instructional Course Lectures 6:201, 1949

WIKLE HT: Periarterial sympathectomy. Am J Surg 12:54, 1931

WOOLLARD HH: The innervation of blood vessels. Heart 13:319, 1926

WOOLLARD HH, WEDDELL G, HARPMAN JA: Observations on the neurohistological basis of cutaneous pain. J Anat 74:413, 1940

WRIGHT WG: Muscle Function. New York, Hoeber, 1928; Hafner, 1962

Chapter 2
THE
BACK

Cutaneous Innervation

The skin of the back, separated from the back muscles by musculature of the shoulder (Fig. 2-1), is thicker than that of the anterior surface of the thorax and abdomen, but needs comment only in regard to its innervation.

For a variable distance on each side of the midline, from the scalp to the tip of the coccyx, the skin of the back is supplied by dorsal (posterior) rami of spinal nerves (Fig. 2-2); lateral to this is the distribution of the lateral cutaneous branches of the ventral (anterior) rami. These rami supply most of the skin of the thorax and abdomen, and at the appropriate levels form the plexuses for the limbs. A typical dorsal ramus, a mixed nerve, divides into medial and lateral branches that supply the intrinsic muscles of the back, and usually only one of these two branches emerges from the muscles to supply the overlying skin. It is, generally, the medial branches of the dorsal rami of the cervical nerves and of the upper six or seven thoracic nerves that become cutaneous, but the lateral branches of the lower thoracic, lumbar, and sacral nerves that do this.

There is considerable variation in the distribution of the dorsal rami of the spinal nerves to the skin. However, as a rule the first cervical nerve has no cutaneous branch, and the cutaneous branches of the second and third are distributed primarily to the scalp. The dorsal branches of the third, fourth, and

fifth cervical nerves regularly reach the skin of the neck; those of the sixth, seventh, and eighth typically have no cutaneous branches (Pearson and co-workers).

All the dorsal rami of the thoracic nerves usually have cutaneous branches. The lower ones tend to run somewhat downward to their distribution.

The upper three lumbars also have downwardly distributed cutaneous branches, but the fourth and fifth lumbars have none.

The lateral branches of the dorsal rami of the first three sacral nerves continue, in a similar fashion, to the skin; the dorsal rami of the lower two sacrals and the coccygeal do not divide into medial and lateral branches, but do have cutaneous branches, which unite to form a single nerve that supplies skin in the neighborhood of the coccyx.

The Vertebral Column

The vertebral column (Fig. 2-3) is composed of 33 vertebrae as a rule, commonly arranged as seven cervical, twelve thoracic, five lumbar, five (fused) sacral, and four (or three or five) fused coccygeal vertebrae. Except between the first and second cervical vertebrae, because there is no body to the first or atlas vertebra, and in the sacral and coccygeal regions where the vertebrae fuse, the bodies of the

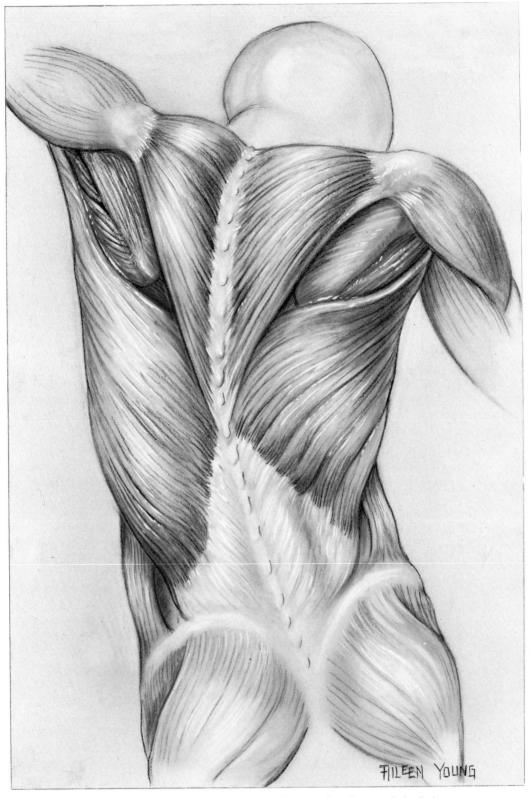

Fig. 2-1. Large superficial muscles of the back. These, the trapezius above and the latissimus dorsi below, are muscles of the upper limb that almost completely cover the true muscles of the back.

loss in height in old age, an additional loss resulting likewise from the usual increase in the curvature of the thoracic region. As discussed in a later section, the disks contribute a varying amount to the total length of each of the three presacral segments of the vertebral column, and also contribute, to a variable degree, to the curves of the vertebral column.

In the early fetus the vertebral column is essentially C-shaped, with its concavity anterior (Fig. 2-4); by the time of birth the cervical curve has begun to reverse itself and the lumbar column is approximately straight, but the thoracic and sacral portions retain some of their original concavity and thus constitute the primary curves of the vertebral column. As the infant begins to hold up his head and sit up, the cervical curvature, with its convexity situated anteriorly, begins to become permanent; this results in a better balance of the head upon the body. Similarly, as the child begins to sit up, stand, and walk, the lumbar

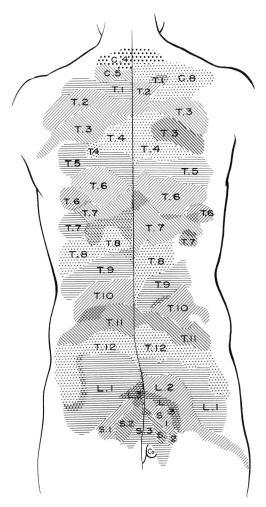

Fig. 2-2. The distribution of the dorsal (posterior) rami to skin of the back; areas supplied by the medial branches of these rami are in *black*, those supplied by the lateral branches are in *red*. The ventral (anterior) rami supply the remaining skin of the back. (Redrawn from Johnston HM: J Anat Physiol 43:80, 1908)

vertebrae are separated from each other by the intervertebral disks. These disks account for a very considerable portion, about one fourth, of the total length of the vertebral column above the sacrum. Shrinkage of the disks through dehydration as a result of pressure may result in the loss of about 0.75 inch in height by a well-developed man during the course of a single day, and of about 0.5 inch by a woman (Inman and Saunders). Shrinkage of the disks also accounts for some of the

Fig. 2-3. The vertebral column from in front and from the side.

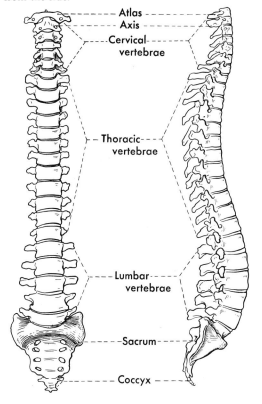

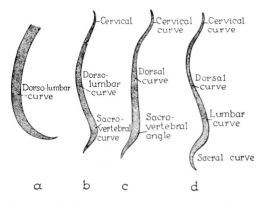

Cervical

Cervical
curve

Cervical
curve

Dorso-lumbar
curve

Dorso-
lumbar
curve

Dorsal
curve

Dorsal
curve

Lumbar
curve

Sacro-
vertebral
curve

Sacro-
vertebral
angle

Sacral curve

a   b   c   d

Fig. 2-4. Development of the curves of the vertebral column: *a* is the single curve present until the third fetal month; *b* shows the appearance of the sacrovertebral curve in the fourth month; *c* the curves at birth; *d* the curves of the adult. (After Keith, from Coventry MB: Am Acad Orth Surgeons, Instructional Course Lectures 6:218, 1949)

curve, also convex anteriorly, develops; this allows a better balance of the body upon the sacrum, through which the weight of the upper part of the body is transmitted to the ilia and the lower limbs. Roughly speaking, therefore, the cervical curvature begins to consolidate at about the second or third month after birth, the lumbar curvature toward or after the end of the first year.

These two curves are secondary or compensatory curves, and so distribute the weight of the body that, according to most reports, a line dropped vertically through the center of gravity of the presacral region of the body passes through the vertebral bodies at about the cervicothoracic, thoracolumbar, and lumbosacral junctions. This would imply that the weight of the body above the sacrum tends to increase the lumbar curve. Asmussen recalculated the vertical line of gravity and put it considerably farther forward. In 150 of 200 boys, he found, it lay in front of the center of the body of the fourth lumbar vertebra, so that the weight should tend to flex the column, and in only 50 did it lie behind this center. He also found, through electromyograms on adults during quiet standing, that it was the erector spinae that usually showed postural activity, to prevent flexion, although in a few cases the abdominal muscles were active

in preventing hyperextension. Variations in the lumbar curve, to adjust the column to the line of gravity, are common; the lumbar curve is typically more pronounced in children and in women than in men, and because of its variation there can be no strict definition of the boundary between a normal lumbar curvature and the abnormal one designated as lordosis.

There is very commonly also a slight lateral curvature or physiological scoliosis in the thoracic region of the vertebral column. This has been held to be either a result of muscular pull with right handedness, or a result of pressure on the vertebral column by the arch and upper part of the descending aorta. Both of these explanations have been supported by observations that in some left-handed individuals, and in those with right-sided aorta, the convexity of the curve is to the left.

The ability of the vertebral column to absorb a very considerable shock, such as is delivered to it by a jump from a height, is contributed to both by its curved shape and by the compressibility of the intervertebral disks, which are fairly efficient shock absorbers.

MORPHOLOGY OF INDIVIDUAL VERTEBRAE

Typical Vertebra

A typical vertebra (Fig. 2-5) consists of a body and an arch, the vertebral arch. The vertebral body and arch enclose the vertebral foramen, in which the spinal cord lies. The arch consists of three chief parts: two limbs or roots, the pedicles, form its sides and unite it to the body, and a roof composed of two laminae that meet in the midline and span the pedicles. The laminae are provided with a number of projections that serve for attachments of muscles or for articulations with other bones. From the midline the spinous process or spine projects dorsally or dorsally and inferiorly; it serves for the attachment of muscles. Projecting laterally from the region of junction of each lamina and pedicle is the transverse process, largely for muscular attachments but in most of the thoracic region

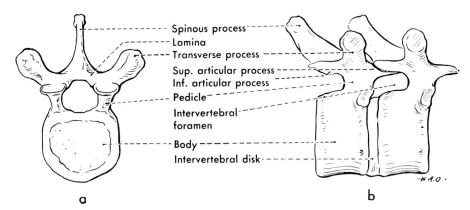

Fig. 2-5. Parts of a typical vertebra.

also articulating with a rib. From the upper border of each lamina, at about its junction with the pedicle, the superior articular process or zygapophysis projects upward, and from the lower border at about the junction of each lamina and pedicle the inferior articular process or inferior zygapophysis projects downward; these form the zygaphophyseal or synovial joints between two adjacent vertebrae. Except for the first two cervical vertebrae, any two adjacent vertebrae down to the sacrum are articulated also at their bodies by an intervening intervertebral disk.

A thin lamina of hyaline cartilage intervenes between the bony end of the body and the fibers of the intervertebral disk but does not, in an adult vertebra, cover the thicker periphery formed by the epiphyseal rings; the articular surfaces of the articular processes are, of course, covered with hyaline cartilage.

The relationship of the synovial joints to the planes of the body varies from one region to another, but in general the superior articular processes tend to face posteriorly, while the inferior articular processes face anteriorly, and thus from a posterior approach overlie the superior processes. This is particularly important in the lower lumbar and lumbosacral region, where the tendency of one vertebra to slide forward on the succeeding one is resisted by the overlapping of its inferior articular processes with the superior articular processes of the next vertebra.

The pedicles of each vertebra have a considerably less vertical diameter than do the bodies; further, each pedicle is notched deeply on its lower surface, rather slightly on its upper surface, to form the inferior and superior vertebral incisures which, when the vertebrae are together, form the superior and inferior boundaries of the intervertebral foramina. The posterior boundaries of these foramina are formed by the articular processes, the anterior by the bodies of the vertebrae, especially the upper of the two, and by the intervertebral disk. In general, the intervertebral foramina are considerably longer vertically than is necessary to allow exit of the spinal nerves; this is especially true in the lumbar region, where each spinal nerve occupies only the upper part of the foramen.

A vertebra contains no central medullary cavity, as do long bones, but is, rather, composed of spongy bone throughout. Except at the ends of the body, the spongy bone is covered by a thin layer of cortical bone; at these ends the plates of hyaline cartilage abut directly against the spongy bone.

Bick and Copel ('52) reported that the characteristic structural change in the vertebral bodies of aged persons is a loss of trabecular substance, apparently as a result of failure to form collagen. Osteophytosis about the periphery of the ends of the bodies ("lipping" and "spurs") is also common. Small areas of avascular necrosis may be found in senescent vertebrae, but Bick and Copel reported that they apparently never coalesce or become large enough to cause collapse of the vertebral body.

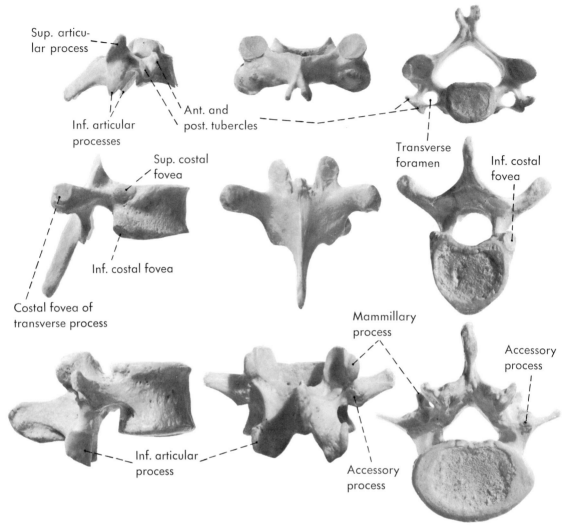

Sup. articu-
lar process

Inf. articular
processes

Ant. and
post. tubercles

Transverse
foramen

Inf. costal
fovea

Sup. costal
fovea

Inf. costal fovea

Costal fovea of
transverse process

Mammillary
process

Accessory
process

Inf. articular
process

Accessory
process

Fig. 2-6. Lateral, posterior, and inferior views of the fourth cervical, seventh thoracic, and fourth lumbar vertebrae.

REGIONAL CHARACTERISTICS

Although they are all built upon the same fundamental plan, typical cervical, thoracic, and lumbar vertebrae are easily distinguished from each other (Fig. 2-6).

Cervical Vertebrae

The cervical vertebrae are all typically small, and while the first, the second, and the seventh are somewhat modified from a typical one, all seven are similar in that they contain not only the vertebral foramen but also a transverse foramen in each transverse process.

The anterior length of the bodies is usually very slightly less than their posterior length (Aeby), so the cervical curve is due to the disks rather than the bodies. In a typical cervical vertebra (Fig. 2-7) the body measures about half again as much from side to side as it does in its anteroposterior dimension. The upper surface of the body is concave from side to side, with upwardly projecting lips that overlap the beveled lateral surfaces of the lower end of the vertebra above; also, it is somewhat convex in an anteroposterior direction, being beveled especially toward the anterior surface. The lower surface is somewhat convex from side to side, concave in an anteroposterior direction. The anterior lip of the concavity is especially prominent and overlaps the vertebra below.

The pedicles are short, and the articular processes are also short and relatively bulky; the articular surfaces of the superior zygapophyses face upward and posteriorly, those of the inferior downward and anteriorly. The posterior root of the transverse process, developmentally a true transverse process, arises near the junction of pedicle and lamina, while the anterior root of the transverse process, developmentally a rib and therefore called a costal process, arises from the side of the body. Between the two is the transverse foramen, limited distally by a bar of bone, the costotransverse lamella, which unites the two ele-

ments of the process. Each element ends in a tubercle, and between the anterior and posterior tubercles is a sulcus for the spinal nerve. The spinous processes of the vertebrae, except the sixth and especially the seventh, are typically short, and those of the third through the sixth are also typically bifid in white persons, but more commonly undivided in Negroes.

The *atlas* or first cervical vertebra (Fig. 2-8) lacks a body, and therefore consists of an anterior and a posterior arch, and thickened lateral masses that bear the transverse processes and the superior and inferior articular foveae. The anterior arch is short. It bears on its posterior or internal surface a facet (fovea) for articulation with the dens, and on its anterior surface an anterior tubercle to which attach the upper ends of the two longus colli muscles. The posterior arch is much longer than

Fig. 2-7. The fifth cervical vertebra from above, *a,* and in front, *b.* (Disse J: Vol 1, Section 1, of K von Bardeleben's Handbuch der Anatomie des Menschen. Jena, Fischer, 1896)

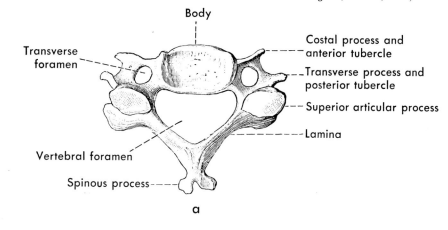

a

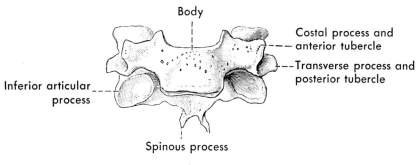

b

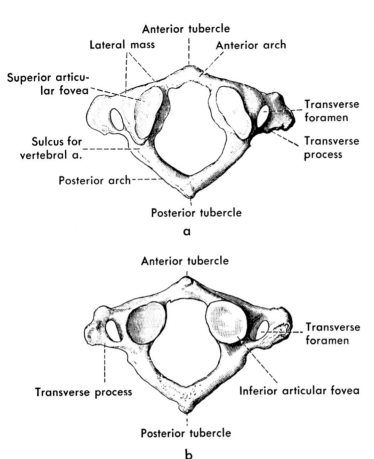

Anterior tubercle

Lateral mass   Anterior arch

Superior articu-
lar fovea

Transverse
foramen

Sulcus for
vertebral a.

Transverse
process

Posterior arch

Posterior tubercle

a

Anterior tubercle

Transverse
foramen

Transverse process

Inferior articular fovea

Posterior tubercle

b

Fig. 2-8. The atlas from above, *a*, and below, *b*. (Disse J: Vol 1, Section 1, of K von Bardeleben's Handbuch der Anatomie des Menschen. Jena, Fischer, 1896)

the anterior one; in place of a spinous process it bears a small posterior tubercle, and laterally, on its upper surface, are the sulci for the verterbral arteries. The superior articular foveae on the lateral masses are concave for articulation with the occipital condyles, kidney-shaped, and face medially, anteriorly, and upward; the inferior articular foveae are almost flat and face inferiorly, anteriorly, and slightly medially, for articulation with the superior zygapophyses of the axis. The transverse processes of the atlas are longer and heavier than those of the other cervical vertebrae, but like the latter contain transverse foramina for transmission of the vertebral arteries.

The *axis* (epistropheus) or second cervical vertebra is distinguished by the projection of the dens (odontoid process) from the upper end of the body (Fig. 2-9). Its lamina is thick,

and its spine heavy and long as compared with the succeeding cervical vertebrae. The inferior vertebral notches (incisures) are especially deep, and there are no superior ones. The superior articular surfaces are large, only slightly convex, and face upward, posteriorly, and a little laterally; they are placed on heavy masses that arise from the body and pedicles rather than at the junction of pedicles and lamina. (The second spinal nerves therefore pass posterior or dorsal to the lateral joints with the atlas, instead of anterior to the synovial joints as do most of the spinal nerves.) The inferior articular processes are essentially similar to the articular processes of other cervical vertebrae. The transverse processes end in a single tubercle each, instead of bifurcating and ending in anterior and posterior tubercles.

The seventh cervical vertebra is the third

atypical cervical vertebra. Its outstanding characteristic is its particularly long and often unbifurcated spine (Lanier, '39a, found it bifurcated in 55% of 100 skeletons of white persons, in none of 100 negro skeletons); this is, typically, the first prominence palpable at the back of the base of the neck, and hence gives to the seventh cervical vertebra its other name, vertebra prominens. The seventh cervical vertebra typically also has small transverse foramina, for they very rarely transmit the vertebral arteries, and the anterior root of the transverse process is usually much smaller than the posterior. Sometimes, however, this anterior root develops separately and then forms a cervical rib; Lanier quoted the incidence of cervical rib as being 2.03% in 1,527 vertebral columns and, in another series of 200, as 2% on the right, 1% on the left, and 0.5% bilaterally.

Thoracic Vertebrae

The thoracic vertebrae are intermediate in size between the cervical and lumbar vertebrae, and increase in size as they are traced downward. Their chief identifying feature is that they present on each side of the body foveae or facets, usually superior and inferior, for articulation with the heads of the ribs (Fig. 2-10). Aeby found the anterior lengths of the bodies to be usually about 1.5 to 2 mm less than their posterior lengths, and the thoracic curve is primarily a result of this difference.

Fig. 2-9. The axis from above, *a*, and from the front, *b*. (Disse J: Vol 1, Section 1, of K von Bardeleben's Handbuch der Anatomie des Menschen. Jena, Fischer, 1896)

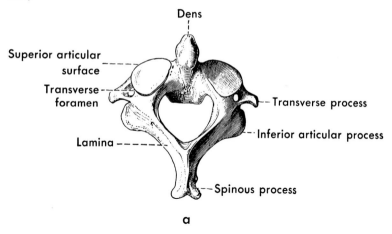

a

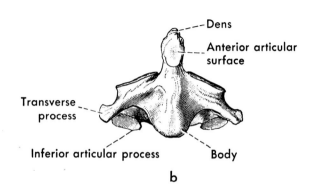

b

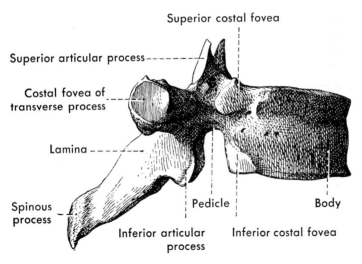

Superior costal fovea

Superior articular process ------

Costal fovea of transverse process

Lamina ------

Spinous process

Pedicle

Inferior articular process

Body

Inferior costal fovea

a

Fig. 2-10. Lateral, *a*, and superior, *b*, views of a thoracic vertebra. (Poirier P, Charpy A: Traité d' Anatomie Humaine [ed 3], Vol 1. Paris, Masson et Cie, 1911)

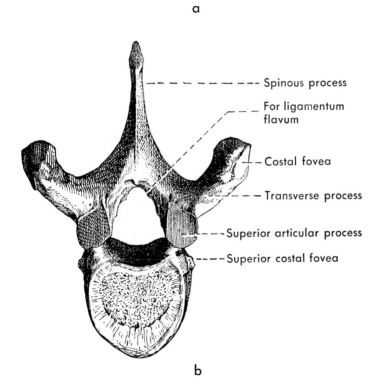

Spinous process

For ligamentum flavum

Costal fovea

Transverse process

Superior articular process

Superior costal fovea

b

The upper and lower surfaces of the body are almost flat; the posterior surface is slightly concave from side to side. The inferior vertebral notches of the pedicles are deep, the superior ones shallow. In relation to the sizes of the bodies, the laminae are long in the superior-inferior direction, and each overlaps somewhat the lamina below it. The spinous processes are typically long and slender, and directed markedly caudad so as to overlap the succeeding process, but in the lower thoracic region they become shorter and broader, and directed more posteriorly, foreshadowing the change into lumbar spinous processes. The articular surfaces of the superior zygapophyses are flat and face posteriorly and slightly inferiorly and medially. The transverse processes are long and heavy and project posteriorly and somewhat upward as well as laterally. Their ends are enlarged and, in the case of the upper 10 vertebrae, approximately, bear foveae on their lateroanterior surfaces for articulation with the tubercles of the ribs; they are elsewhere roughened for the attachments of muscles. The upper and lower thoracic vertebrae resemble each other in that their spines are shorter and slant downward less than do those of the middle thoracic vertebrae, but differ from each other in the size of their bodies, those of the upper ones being much smaller than those of the lower.

The body of the first thoracic vertebra has a superior costal fovea for the head of the first rib and a small inferior costal fovea for the upper portion of the head of the second rib. The second thoracic vertebra, therefore, has a large superior fovea for the major part of the head of the second rib, and an inferior fovea for part of the head of the third rib. Each of the third through the eighth thoracic vertebrae typically articulates with two ribs through superior and inferior foveae of about equal size (demifacets). The bodies of the ninth through the twelfth have only superior foveae for the heads of the similarly numbered ribs.

The transverse processes of the eleventh (usually) and twelfth thoracic vertebrae have no costal foveae for the tubercles of the ribs, and the twelfth thoracic vertebra closely resembles an upper lumbar one except for the facets for the heads of the ribs. It has a mammillary process, as do lumbar vertebrae, and Kaplan ('45) pointed out that this process usually overhangs the synovial joint between the last thoracic and first lumbar vertebrae; he regarded this as a useful feature in identifying the last thoracic vertebra at operation.

Lumbar Vertebrae

The lumbar vertebrae are large and heavy (Figs. 2-11 and 2-12). The large bodies are wider transversely than they are deep anteroposteriorly, and both of these dimensions exceed their lengths. The bodies of at least the lower three lumbar vertebrae tend to be somewhat wedge-shaped as viewed from the side, their length (height) anteriorly being greater than their length posteriorly, and this is marked in the case of the fifth lumbar vertebra. Aeby found the anterior height of the first lumbar vertebra to be about 0.4 mm less than its posterior height, the two heights of the second lumbar to be about equal, while in the remainder the anterior height exceeded the posterior one by from 0.9 mm for L3 to 6.2 mm for L5.

The pedicles are short and heavy, and arise

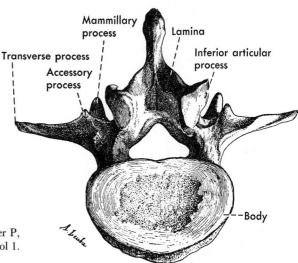

Fig. 2-11. A lumbar vertebra from below. (Poirier P, Charpy A: Traité d'Anatomie Humaine [ed 3], Vol 1. Paris, Masson et Cie, 1911)

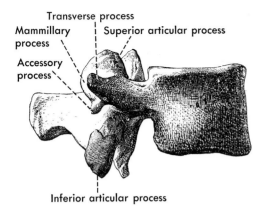

Transverse process

Mammillary
process

Superior articular process

Accessory
process

Inferior articular process

Fig. 2-12. A lumbar vertebra from the side. (Poirier P, Charpy A: Traité d' Anatomie Humaine [ed 3], Vol 1. Paris, Masson et Cie, 1911)

from the upper part of the body, so that they have a shallow superior vertebral notch but a deep inferior one. The transverse processes are long and thin, flattened anteroposteriorly, and project somewhat posteriorly and slightly upward as well as laterally. On the dorsal surface of the base of each transverse process is a small tubercle, the accessory process; Lanier ('39a) found accessory processes seldom missing (3.5%) from the first lumbar vertebra but often (34%) from the fifth lumbar. The laminae are not as long in the vertical direction as are the bodies, and therefore between each two laminae there is an appreciable space, through which lumbar punctures are done. The laminae are directed somewhat caudally from their attachments to the pedicles, therefore the upper border of each lamina tends to be V-shaped.

The spinous processes are broad when viewed laterally, narrow but expanding into an enlarged extremity when viewed from above or below. The lateral thickening of the tip makes it more difficult to strip muscles subperiosteally (desirable to minimize bleeding) from lumbar spinous processes; care must be taken to follow the curve of the enlargement and, as its anterior border is reached, to direct the instrument toward the midline, in order to follow closely the surface of the bone. (Another anatomic factor influencing the ease of reflecting the musculature is that most of the muscles attaching to the spinous processes

are inserting into them, and therefore slant inward as they are followed up. Hence, if the muscles are reflected by working in a caudocranial direction, their direction of slant will tend to guide the instrument to the bone; if stripping in the opposite direction is attempted, the instrument tends to slide laterally into the muscle.)

The superior articular processes are heavy, and on their posterior borders have rounded enlargements, the mammillary processes. Their articular surfaces are somewhat concave in an anteroposterior direction; those of the first lumbar vertebra face decidedly medially and slightly backward, and those of the remaining vertebrae tend successively to face slightly more posteriorly and less medially. The inferior articular processes project markedly downward; in concordance with the shape and direction of the superior articular surfaces, their articular surfaces are slightly convex, and those of the upper lumbar vertebrae face largely laterally and slightly anteriorly. In succeeding vertebrae there is a tendency for them to face more and more anteriorly; the inferior articular processes of the fifth lumbar vertebra have practically flat articular surfaces, and face largely anteriorly rather than laterally.

Sacrum

The sacrum (Fig. 2-13), typically composed of five fused sacral vertebrae, is somewhat triangular in an anterior or posterior view, its wide base articulating both with the fifth lumbar vertebra and with the wings of the ilium. The central portion of the sacrum consists of the fused bodies of the sacral vertebrae, and on its pelvic surface are transverse lines indicating the regions of fusion. Similarly, on its dorsal surface there is typically a series of four spinous tubercles, which are more or less united to form a median sacral crest. On each side of the median crest, just medial to the dorsal sacral foramina, a series of tubercles representing the articular processes forms the intermediate (articular) crest. The roof of the sacral canal in its lower part is typically deficient in regard to the formation of bone, thus presenting a dorsally located sacral hiatus;

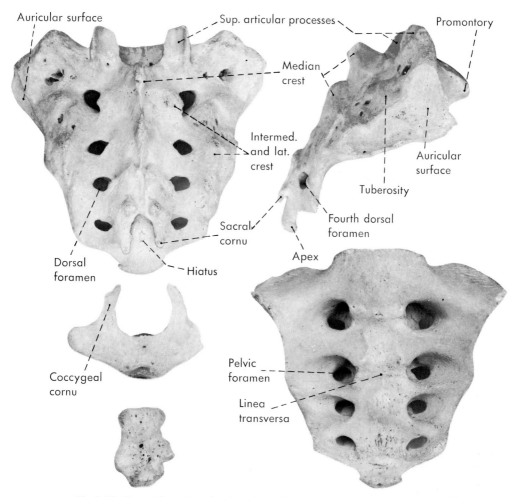

Fig. 2-13. Dorsal, lateral, and pelvic views of the sacrum, and a posterior view of the coccyx

the bone that forms the sides of the hiatus ends, just above the apex of the sacrum, in two sacral cornua. Each large lateral mass of the sacrum (pars lateralis, formerly ala) represents the fused transverse and costal processes of the sacral vertebrae; where these have met lateral to the dorsal sacral foramina they present a series of tubercles, of which the first only is well developed, that together form an indistinct lateral sacral crest. The boundary between the pars lateralis and the central part is the line formed by the four dorsal (posterior) or pelvic (anterior) sacral foramina. These foramina are large, and the pelvic and dorsal ones are typically located almost opposite each other, so that one can look through

both from either surface; where they meet they communicate with the sacral canal by an intervertebral foramen, through which the spinal nerve makes its exit. The dorsal (posterior) branches of the spinal nerves then emerge through the dorsal sacral foramina, and the ventral (anterior) branches emerge through the pelvic sacral foramina; leading laterally from the upper three pelvic sacral foramina are converging grooves that house the sacral roots of the sacral plexus.

Viewed from above, the middle of the base of the sacrum is shaped like the body of a lumbar vertebra; viewed from the side, it has a thickened pelvic edge, the promontory of the sacrum. From the dorsal aspect of the base project superior articular processes, facing primarily posteriorly, for articulation with the inferior articular processes of the last lumbar vertebra.

The upper, wider part of the pars lateralis of the sacrum articulates on each side with the iliac parts of the two hip bones, and is wider posteriorly than anteriorly; thus the weight of the body, tending to force the sacrum downward and forward, tends to wedge the sacrum firmly between the two hip bones. On its dorsal surface, lateral to the lateral crests, it presents roughened areas, the tuberosities of the sacrum, which receive very heavy sacroiliac ligaments; on its lateral surfaces are smooth articular areas that somewhat resemble in shape an external ear seen from the medial side, and are called the auricular surfaces. Each of these represents the part of the sacroiliac articulation that presents a synovial cavity. According to Derry the auricular surface of the female sacrum is typically smaller than that of the male, and usually involves the sacrum only at the level of the first two sacral vertebrae (unless the sacrum has six vertebrae), while in the male the auricular surface involves also a part or all of the lateral portion of the third sacral vertebra.

Coccyx

The coccyx (Fig. 2-13) typically consists of four vertebral bodies, sometimes fused together; they are rudimentary, with no laminae and few processes. The first segment, much the largest, articulates on its upper surface, through a small disk, with the apex of the sacrum, and has projecting upward from its dorsal surface paired cornua, representing pedicles and superior articular processes. The coccygeal cornua are united to the sacral cornua by ligaments. The first segment of the coccyx also presents short transverse processes; rudimentary transverse processes are also often identifiable on the much smaller second segment, while the remaining two (more or less) segments are merely small nodules of bone. There is usually a rudimentary fibrocartilaginous disk between the first and second coccygeal segments, allowing movement at this joint. The remaining segments of the coccyx are regularly fused together, and in advanced age all segments of the coccyx may be fused, and this in turn may be fused to the sacrum.

BLOOD SUPPLY

The blood supply of a typical vertebra is from the segmental vessels most closely associated with it. Because of the method of formation of the vertebrae, each thoracic and lumbar vertebra is typically related at about its middle to paired segmental arteries (intercostal or lumbar) of aortic origin, and their accompanying veins. In the cervical region the segmental arteries are branches of the vertebrals, and in the sacral region they are branches of the lateral sacral arteries. Schift and Parke found the dens to be supplied by two sets of arteries derived from the vertebrals, and by tiny branches arising from the internal carotid arteries just before they enter the carotid canals. Althoff described still other vessels, including branches from the ascending pharyngeal arteries, and said that the vessels in the dens filled as well with injected material after the dens had been fractured as when it was intact.

Although Willis ('49) described the arteries to the anterolateral surfaces of the vertebral bodies of the adult as periosteal vessels only, Harris and Jones demonstrated small vessels from the vertebral arteries entering the anterolateral surfaces of adult cervical vertebrae; Stilwell ('59) demonstrated that in all regions of the vertebral column of the rabbit and monkey; and Wiley and Trueta described them as present in all regions of the adult human vertebral column; they arise from the segmental vessels as these lie close against the vertebral bodies and anastomose freely with the larger vessels that enter the vertebra posteriorly. It is agreed that the latter, which arise from the spinal branches of the segmental vessels (Fig. 2-14), are more important. The spinal branches, in turn, arise from the dorsal or posterior rami of the segmental vessels, as these course backward around the sides of the vertebrae. They arise close to the intervertebral foramina, and as they enter these divide into three terminal branches: a dorsal, an intermediate, and a ventral one

The dorsal branches of the spinal arteries help supply the spinal dura and the tissue of the epidural space. They anastomose with similar branches above and below them to form small channels accompanying the posterior internal vertebral venous plexuses, and in general, lie dorsolateral to the dura. However, the largest part of each dorsal branch enters the vertebral arch and supplies the pedicle, the transverse process, the lamina, and the spinous process, all the dorsal part of the vertebra.

The intermediate or middle branch of a spinal artery is not distributed to the vertebral column. Rather, it supplies the dura of the associated nerve roots, and its radicular branch may pierce the dura, continue along a nerve root intradurally, and help to supply the spinal cord.

The ventral branches of the spinal arteries are the ones that supply the vertebral bodies. They give twigs to the anterolateral part of the spinal dura and to the tissue of the epidural space; and they or their branches anastomose with the similar vessels above and below to form small channels accompanying the anterior internal vertebral venous plexus. Typically, however, the ventral branches divide into two major terminals, an ascending and a descending one, which run obliquely upward or downward toward the centers of the two adjacent vertebral bodies. They pass deep to the posterior longitudianl ligament, between this and the posterior aspect of the vertebral bodies and pierce each body toward its middle, under cover of the ligament. Thus, each vertebral body receives on its posterior aspect blood from four arteries, two from each side, one ascending and one descending.

In the mature adult, the blood vessels within a vertebral body do not reach the disk, but in the fetus and child they pass through the cartilage at the ends of the bodies to supply the adjacent disks.

The vertebral column is intimately asso-

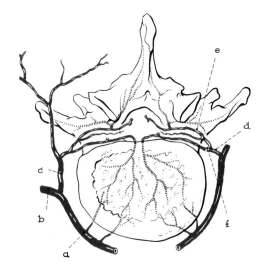

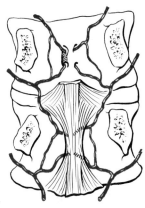

Fig. 2-14. Diagram of the blood supply to a vertebra as seen from below and from behind with the laminae removed. *a* is a segmental (in this case a lumbar) artery, *b* its ventral continuation, *c* its dorsal branch; *d* is the spinal branch, and *e* and *f* are the spinal branch's dorsal and ventral twigs to tissue of the epidural space and to the vertebral column; the unlabeled middle twig is to the nerve roots, and, at some levels, to the spinal cord. In the lower figure, only the branches to the vertebral bodies are seen. The cut surfaces of the pedicles are recognizable and a part of the posterior longitudinal ligament has been removed.

ciated with interconnecting venous plexuses, and the venous drainage of the bones is into these plexuses. A part of the venous drainage of the body is through the middle of the posterior surface, in the same region as the entrance of its chief arterial supply, and thus

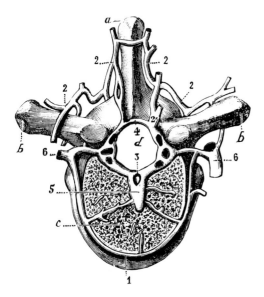

Fig. 2-15. Venous drainage of a vertebra. *a* and *b* are the spinous and transverse processes, *c* the body of the vertebra; *1, 2, 3,* and *4* are the vertebral venous plexuses, *5* the chief vein of the vertebral bodies, and *6* segmental veins and their tributaries. (After Breschet, from Testut L: Traité d' Anatomie Humaine, Vol 2. Paris, Doin, 1891)

into the anterior internal vertebral venous plexus; the rest is anterolaterally (Fig. 2-15). Crock and co-workers described the venous drainage in more detail, noting that in addition to the central or basivertebral vein, large venous channels parallel the cartilage plates at the end of the vertebral body, and connect with both the basivertebral vein and the external plexus; according to these workers, they receive vessels from a hitherto undescribed capillary plexus located in the endplate cartilage.

Nothing is known specifically concerning the nerve supply to a bony vertebra (see the general discussion of the nerve supply of bone in Chapter 1), but Pedersen and his co-workers found twigs from the meningeal branches of the spinal nerves following blood vessels into the vertebrae, and pain may be the first symptom of an intravertebral tumor. It is known that the posterior longitudinal ligament and probably the posterior part of the anulus (annulus) fibrosus, or outer part of the intervertebral disk, has an afferent innervation.

DEVELOPMENT

MESENCHYMAL STAGE

At about the fourth week of embryonic life the vertebral column begins to develop through migration of mesenchyme from the medial part, the sclerotome, of the segmented mesodermal somites. This mesenchyme lies lateral to the notochord and at first corresponds in its segmentation to the somites from which it is derived; each sclerotome is separated from the next one by an intersegmental artery. Soon, however, each sclerotomal derivative divides into cranial and caudal portions, which then separate from each other to unite with the caudal and cranial portions, respectively, of the sclerotomal derivatives above and below it (Fig. 2-16). Thus the segmentation of the skeletal portion of the somite is shifted so that it no longer corresponds to the original segmentation, which is, however, retained by the myotomes. In consequence, as the vertebrae and the segmental muscles develop, the latter no longer correspond to the vertebrae in position; rather, they cross intervertebral joints, since they stretch from about the middle of one developing vertebra to about the middle of the next. It is this method of development, also, that results in the originally intersegmental arteries being related to the middle of each vertebra rather than lying between vertebrae.

Bardeen ('05a and b, and '08) has provided an account of the basic development of the vertebral column in its early stages, and Ehrenhaft has given a particularly good account of its relation to abnormalities. Briefly, the paired primordia of the vertebral column, formed by the fusion of cranial and caudal halves of adjacent sclerotomes, grow medially, about the notochord, to form the vertebral centra, or major part of the bodies; dorsally, about the neural tube, to form the neural arches; and ventrolaterally, to form the transverse processes and the primordia of the ribs. Between the developing vertebrae looser mesenchyme derived from their ends forms the intervertebral disks. As the vertebral body develops, that part of the notochord within the body becomes slimmer and

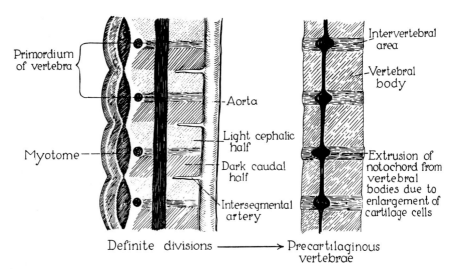

Fig. 2-16. Development of the segmentation of the vertebral column. The original segmentation, marked by the intersegmental arteries and corresponding to that of the myotomes, can still be seen on the *left;* in the definitive stage, *right,* the caudal half of one segment has fused with the cranial half of the next segment below. (After Keyes and Compere, from Coventry, MB: Am Acad Orth Surgeons, Instructional Course Lectures 6:218, 1949)

finally disappears, but a bit of the notochord remains between each two successive vertebral bodies, in the center of each developing intervertebral disk. These pieces of notochord subsequently enlarge with the development of the disks and become the nuclei pulposi, the softer centers of the disks. Mesenchyme adjacent to the developing vertebrae differentiates into ligaments.

CHONDRIFICATION AND OSSIFICATION

While the mesenchymal neural arches are still growing dorsally and the costal processes are still growing ventrolaterally, centers of chondrification (two, right and left, for each centrum, one for each half of the arch, and one for each costal process) appear, in a craniocaudal sequence, and the centers spread until the vertebra is formed largely of cartilage. The arches continue to grow dorsally, but do not meet to complete the arch until some time during the third month; if they fail to meet and develop properly, the resultant defect, spina bifida, may allow herniation of the developing meninges and spinal cord. Before

this growth is complete, centers of ossification appear in the cartilaginous centra of the vertebrae. They are first seen in the lower thoracic and upper lumbar vertebrae, then progressively appear both cranially and caudally. The centers for the centra are usually described as a single one for each vertebra, but are said to be derived from two centers, dorsal and ventral, which fuse together almost as soon as they appear (Ehrenhaft). The centers for each arch are paired, one for each side of the arch; they appear usually in a craniocaudal sequence, although upper thoracic ones may appear first (Noback and Robertson); the centers for the arches of the cervical vertebrae typically appear before those of the centra do (Flecker).

At birth, a typical vertebra consists of three bony pieces, united by cartilage: one piece is the centrum, the others are the two sides of the arch. Centers for the coccygeal vertebrae, one for each centrum but none for an arch, appear between the end of the first year of life and puberty. Subsequent to birth, the two sides of the arch normally unite before they join the body (Fig. 2-17); complete bony laminae are formed by the thoracic and lumbar

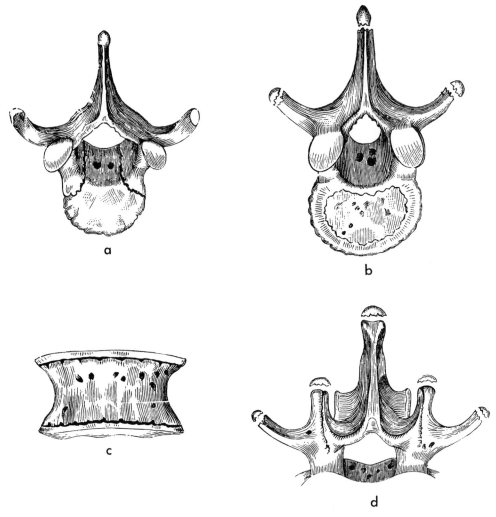

Fig. 2-17. Ossification of typical vertebrae. *a* is a thoracic vertebra of a 3-year-old child; the laminae are fused, but the arch has not yet joined the centrum. *b* shows the epiphyses (separated) of the processes of a thoracic vertebra from a 16-year-old child, and the epiphyseal ring on the body. *c* is the body of a lumbar vertebra showing the two epiphyseal cartilages. *d* is a lumbar vertebra from a child of 16 showing not only epiphyses for the spinous and transverse processes, but also for the mammillary processes. (Kopsch F: Rauber's Lehrbuch der Anatomie des Menschen, Part 2. Leipzig, Thieme, 1914)

vertebrae during the first year of postnatal life, by the cervical during the early part of the second year, and by the sacral much later, between the seventh and tenth years. The completed neural arch joins the centrum of the vertebra during the third year in the cervical region, between the fourth and fifth years in the thoracic region, during the sixth year in the lumbar region, and the seventh or later in the sacral. Each neural arch, besides forming the vertebral arch, contributes a small part to the centrum to complete the vertebral body on each side (Fig. 2-17a).

The costal processes of the cervical and lumbar vertebrae normally fuse with and contribute to the transverse processes; those in the thoracic region develop joint cavities between themselves and the vertebrae; and those in the sacral region fuse together to form the pars lateralis, which fuses with the arches a little before the arches in turn fuse to the centra.

At about the time of puberty secondary centers of ossification, usually referred to as epiphyses, appear on the processes of most of the vertebrae and at each end of the vertebral body. Those for the processes are typically three, one for the tip of the spinous process, one for the tip of each transverse process. However, on those vertebrae whose spinous processes fork there is a secondary center for each tip, and the lumbar vertebrae have an additional center for each mammillary process.

The epiphyses of the vertebral bodies are rings of bone formed in thickened cartilage about the peripheries of the upper and lower surfaces of the body of each vertebra. These develop from a number of separate centers of ossification that soon fuse to form the ring; the two epiphyseal rings unite to the body of the vertebra between the seventeenth and twenty-fifth years. Bick and Copel ('50, '51) objected to calling these rings "epiphyses," saying that they lie outside the true epiphyseal cartilaginous plates and take no part in the longitudinal growth of the vertebrae; but even in the long bones it is the cartilaginous epiphyseal plates, perhaps better called diaphyseal plates, which are primarily responsible for the growth in length of the bone, and the bony epiphyses contribute little to this length. Bick and Copel preferred to call the epiphyseal rings "vertebral rings," and said they correspond more closely to traction apophyses, developing where the fibers of the intervertebral ligaments insert into the ends of the bones. They said that the cartilage from which these portions of the bones are formed begins to calcify at about the age of 6 years, to ossify at about the age of 13, and to fuse with the bony vertebral bodies at about the age of 17. By 18 years, they found, fusion between the ring and the body proper was apparently complete, and by 20 years the ring could no longer be histologically identified.

The occurrence of anomalous ossification of one side of a neural arch from two centers, once advanced as the cause of spondylolysis, has not been confirmed.

Atlas and Axis

The atypical atlas and axis necessarily develop somewhat differently from most of the vertebrae (Fig. 2-18), for the developing body of the atlas separates from this bone and fuses with the axis to form the dens (odontoid process); the vertebral arch of the atlas, then presenting an anterior gap where the body should be, grows together anterior to the dens to form the anterior arch of the atlas. One center of ossification appears for each half of the posterior arch of the atlas, and gives rise also to the corresponding lateral mass that bears the articular foveae for the occipital bone and the axis; the two halves of the arch fuse posteriorly during the third year of life. A center of ossification appears in the cartilaginous anterior arch during the first year of life, and fuses with the lateral masses about the seventh year to complete the atlas.

The centers of ossification in the axis are one for each half of the vertebral arch, appearing, in general, during or shortly after the seventh week of fetal life as do the centers for the arches of the other cervical vertebrae; one for the lower part of the body; and paired centers for the major part of the dens and an upper part of the body. The paired centers for the dens and upper part of the body fuse together about 2 months after their appearance, so that at birth the axis consists of four parts. The two parts of the lamina customarily unite during the second year, and join the center for the lower part of the body during the third year. The now unpaired element for the upper part of the body and for the dens begins to fuse with the rest of the bone at about the age of 3 years (e.g., Macalister, 1894) and is usually not completely united until about the age of 6; the fusion begins superficially and a disk of cartilage may persist for years in the center of the area of fusion. An epiphysis for the tip of the dens appears during the second year and fuses with the remainder of this process between the sixth and twelfth year; at about the age of puberty an epiphysis also appears on the lower surface of the body and fuses between the ages of 17 and 25.

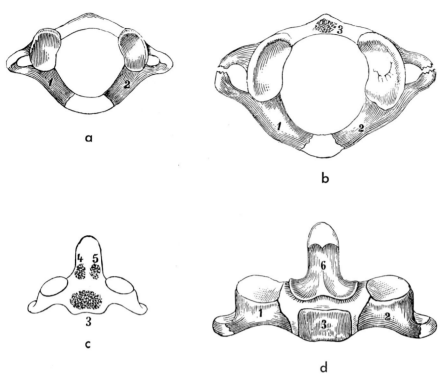

Fig. 2-18. Ossification of the atlas and axis. *a* is the atlas before birth, *1* and *2* being the centers forming the posterior arch and transverse processes. *b* is the atlas of a 1-year-old child, in which a center, *3*, for the anterior arch has also appeared. *c* is an early stage in the ossification of the axis, with a center in the body and paired centers in the dens; *d* is the axis of a newborn child, in which the centers marked *4* and *5* in *c* have fused to form *6; 3* is the center for the major part of the body, *1* and *2* are the centers for the arch. (Kopsch F: Rauber's Lehrbuch der Anatomie des Menschen, Part 2. Leipzig, Thieme, 1914)

GROWTH

The ends of the growing vertebral bodies are covered by cartilaginous plates, which are histologically similar in both the dog (Haas) and man (Bick and Copel, '51) to the epiphyseal plates of long bones. Haas showed experimentally in the dog that growth in length of the bodies occurs only at the ends, apparently from the epiphyseal cartilaginous plates, just as does growth of the body of a long bone. Bick and Copel found a columnar arrangement of cartilage cells (indicative of osteogenesis) on the deep surfaces of the cartilaginous plates at both ends of all vertebrae they studied up to those of a female of 17 years, at which age these formations began to disappear, indicating the completion of growth. Unlike epiphyseal cartilages of long bones, the cartilage plates at the ends of the

bodies do not completely disappear; rather, they persist as thin hyaline plates on the upper and lower surfaces of each body, adjacent to the intervertebral disks. Thus, the deep part of the cartilaginous plates during the growth periods seems to represent a true epiphyseal plate, while the surface persists as the equivalent of an articular cartilage.

Vertebral Column as a Whole

Ligaments

The bodies of the individual vertebrae are bound together not only by the intervertebral disks (see Figs. 2-28 and 2-29), but also by the anterior and posterior longitudinal ligaments, which run superficial to the disks.

The *anterior longitudinal ligament* (Fig. 2-19) is

a broad band placed on the anterior and anterolateral surfaces of the vertebral bodies, from the axis (with a narrower extension to the skull) to the upper part of the pelvic surface of the sacrum. It consists of several sets of fibers, of which the deepest merely extend over one intervertebral disk, between the bodies of adjacent vertebrae; these bind the ligament firmly to the disks and the margins of the vertebrae. Other fibers extend over two or three vertebrae, while the most superficial ones extend over four or five vertebrae. The anterior longitudinal ligament is thickest over the middle of a vertebra, where it helps to fill up the usual concavity on the anterior surface of the body, but is not here firmly attached to the body of the vertebra. Its edges, on the sides of the vertebral bodies, are thinner than the center, and are composed of fibers which pass for only a segment or two; these parts have been termed the lateral longitudinal ligaments.

The *posterior longitudinal ligament* lies on the posterior surfaces of the bodies of the vertebrae, therefore within the vertebral canal, and extends from the body of the axis to the sacrum; at its attachment to the body of the axis it is continuous with the tectorial membrane, which passes from this vertebra to the occipital bone. Like the anterior longitudinal ligament, the posterior longitudinal ligament is attached firmly to the ends of the vertebrae and to the intervening disks, but over the centers of the vertebral bodies it is separated from the bone by intervals through which the arteries and the posterior veins of the bodies course in their entrance into or exit from the bone. Also similarly, it is composed of fibers of varying lengths, the deepest running from one vertebra to the next, the longest passing over several vertebrae.

In the cervical region the posterior longitudinal ligament is broad, but in the thoracic and lumbar regions it takes a different form: it is broad over the ends of two adjacent vertebrae and the intervening disk, then narrows over the middle of the body, to expand again over the next disk and adjacent vertebral ends. It thus resembles a series of hourglass formations with their denticulated ends over the intervertebral disks. The lateral expan-

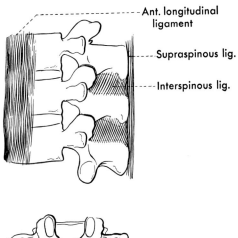

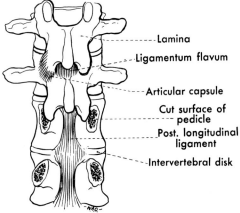

Fig. 2-19. Diagram of the ligaments of the vertebral column

sions over the disks are thin, while the central part of the band is much thicker. This is presumably why most posterior protrusions of intervertebral disks are posterolateral rather than midline ones.

The special posterior ligaments associated with the dens are described in a following section in connection with the median atlantoaxial joint.

The anterior longitudinal ligament is so placed as to tend to limit extension of the vertebral column; it is therefore particularly important to the lumbar region, where it helps to resist the tendency of the weight of the body to increase the lumbar curvature. The posterior longitudinal and other posteriorly placed ligaments have a similar function in the thoracic region, where the weight of the body tends to produce flexion of the vertebral column, increasing the thoracic curvature. In

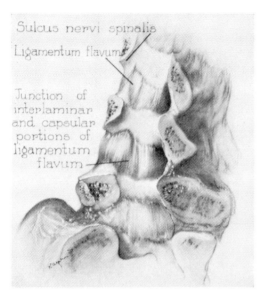

Sulcus nervi spinalis

Ligamentum flavum

Junction of
interlaminar
and capsular
portions of
ligamentum
flavum

Fig. 2-20. Ligamenta flava of the lumbar region, seen from the vertebral canal. The cut surfaces of the pedicles are on the left, those of the laminae on the right. (Naffziger HC, Inman V, Saunders JBdeCM: Surg Gynecol Obstet 66:288, 1938 [by permission of Surgery, Gynecology & Obstetrics])

anterior crush injuries of the vertebral bodies, the anterior longitudinal ligament is ordinarily not injured, so that when the injured region is placed in hyperextension this ligament can act as a splint to hold and fix the fragments in their proper positions.

Various parts of the vertebral arches are also united by ligaments. The zygapophyseal joints (between superior and inferior articular processes of adjacent vertebrae) are provided with thin and lax articular capsules that have a synovial lining. These obviously contribute no particular strength to the articulations of the vertebral column. Intertransverse ligaments, usually lacking in the cervical region and best developed in the thoracic region, also contribute little to the stability of the column.

The important posteriorly placed ligaments are the ligamenta flava, between the laminae, and the associated but much less important supraspinous and interspinous ligaments (Fig. 2-19; see also Fig. 2-28).

The *ligamenta flava* (Figs. 2-19 and 2-20) are rather heavy bands, composed largely of elastic tissue and, consequently, distinctly yellowish (hence the name *ligamentum flavum*, "yellow ligament"). Their elasticity allows flexion and a return to extension without their becoming unduly folded, and in the normal position they are under tension; if the arches are removed from their attachments to the bodies the ligaments so pull the arches together that there is a loss of 14% in the length of that portion of the vertebral column (Steindler). Nachemson and Evans found the ligaments to be almost perfectly elastic up to the point of rupture, but elongating less, and rupturing under less stress, with advancing age.

These paired ligaments almost fill the space between two adjacent laminae, but are separated in the midline from each other by a narrow slit through which pass veins that connect the posterior external and the posterior internal vertebral venous plexuses with each other. Each ligamentum flavum consists of a flattened band that arises from the anterior surface of the lower edge of a lamina and stretches inferiorly to attach to the upper part of the posterior surface of the succeeding lamina. Each band extends laterally from almost the midline to the capsule of the zygapophyseal (synovial) joint. In the lumbar region its thin lateral part is said to blend with this capsule (Naffziger and co-workers), which is composed of collagenous rather than elastic tissue. Here also, the anterolateral edges of the ligamenta flava may bulge markedly forward so as to fill much of the lower part of the intervertebral foramina (see Fig. 2-60).

Hypertrophy (thickening and fibrosis) of a ligamentum flavum has been said to be one cause of compression of a nerve root at the intervertebral foramen in the lumbar region, where the ligaments are thicker than elsewhere. Dockerty and Love found some fibrosis in 90% of ligamenta flava removed at the time of operation on a disk, and reported the average thickness of such ligaments as 5.1 mm, compared with an average thickness of 2.8 mm in apparently normal ligaments. Naffziger and his colleagues regarded enlargement of the ligament as being a relatively uncommon cause of injury to a nerve as

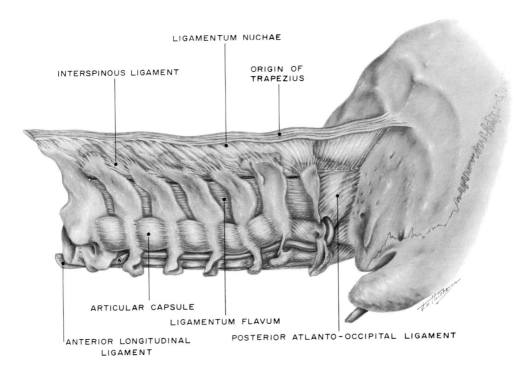

LIGAMENTUM NUCHAE

INTERSPINOUS LIGAMENT

ORIGIN OF
TRAPEZIUS

ARTICULAR CAPSULE

LIGAMENTUM FLAVUM

ANTERIOR LONGITUDINAL
LIGAMENT

POSTERIOR ATLANTO-OCCIPITAL LIGAMENT

Fig. 2-21. The ligamentum nuchae and other ligaments of the cervical column. (Janes JM, Hooshmand H: Mayo Clin Proc 40:353, 1965)

compared with protrusion of an intervertebral disk.

In the cervical region hyperextension of the vertebral column may produce a decrease in the interlaminal space beyond the normal elasticity of the ligamenta flava, and result in their becoming folded and bulging forward, with the possibility of compressing the spinal cord.

The *supraspinous ligament* (NA, supraspinal) is a thin bundle of fibrous tissue which runs over the tips of the spinous processes, shorter fibers connecting adjacent ones, longer ones connecting several in a row; on its deep surface it blends with the interspinous ligaments, and in the cervical region it is continuous with the ligamentum nuchae. Jonck denied that it is present in the lumbar region, saying that tendons of origin of the erector spinae, and longitudinal fibers connecting them, have apparently been mistaken for the ligament. Rissanen came to a similar conclusion regarding the lower end, usually described as attaching to the sacrum, but he did describe the ligament as present in the lumbar region;

he said that it ended at the L4 spinous process in 73% of his specimens, reached L5 in 5% and ended at L3 in 22%. Although it was histologically tendinous at younger ages, he found that with advancing age fibrocartilage appeared in it, portions became infiltrated with fat, and after age 40 these areas degenerated into cystic pouches.

The *ligamentum nuchae*, the equivalent of the supraspinous ligament in the cervical region, is thick on its posterior edge between the external occipital protuberance and the spinous process of the seventh cervical vertebra, where the trapezius and splenius capitis muscles arise; as it extends deeply to the posterior tubercles of the atlas and the spinous processes of C2 to C6 it is no more than a thin, translucent, midline intermuscular septum (Fig. 2-21). Scapinelli found fibrocartilage and bone sometimes lying in the thicker posterior border of the ligament.

The *interspinous ligaments* (NA, interspinal) in the cervical region are also thin and translucent; one or more are frequently absent, represented by a few shreds, or perforated, apparently usually as a result of wear and tear (Halliday and co-workers). They have not been carefully studied in the thoracic region. In the lumbar region, Jonck described them as containing a few elastic fibers among the dense collagenous ones, originating both from the upper surface of the base of the spinous process and the adjacent upper border of the lamina to run to the spinous process above. Rissanen found them intact in children, although with areas of fat, and after age 20 found cavities in them, especially at the L4–L5 and L5–S1 levels, and with further age degenerated ligaments were frequently ruptured, especially in the lower three interspaces.

The absence of a supraspinous ligament in at least the lower lumbar region, where it might be considered particularly important in checking movement, and the degeneration of the interspinous ligaments in the lumbar and cervical regions, cast doubt upon the importance of these posterior ligaments in checking movements and emphasize the difficulty of assigning rupture of them to some specific trauma. Traumatic rupture in the lumbar region has long been considered a cause of low back pain, although to what extent pain is caused by adjacent structures is not clear, nor is it agreed whether pain from close to the midline is referred to the limb. Sinclair, Feindel, and their co-workers produced experimental evidence that stimulation confined to the midline leads to local pain only, and that the more laterally lying dorsal rami of the spinal nerves must be involved before pain is referred to the distribution of the sciatic nerve. Pedersen and co-workers regarded the evidence as indicating that referral of pain from posterior structures of the back is largely a question of sufficient intensity of the stimulus.

These workers traced branches of the dorsal rami to the interspinous ligaments and pointed out that as a result of the obliquely descending course of these rami, the interspinous ligaments are for the most part supplied by the nerve one segment above the level at which they lie.

Numerical Variations in Vertebrae

The vertebral column may consist of 32 or 34 vertebrae instead of the usual 33, but such variations are not common, nor are they of any importance if they involve only changes from the usual number of sacral or coccygeal vertebrae. They may, however, be associated with a reduction or an increase in the number of thoracolumbar vertebrae (Bardeen, '00). Much more common than variations in total number is a shift from the normal vertebral formula, whereby a vertebra is subtracted from the region to which it belongs and added to an adjacent region; according to Mitchell ('36), this occurs in about 20% of human skeletons, and the most common change from normal is the addition of a coccygeal vertebra to the sacrum; the next most common, he said, is elongation or shortening of the thoracolumbar column.

The fusion of a coccygeal vertebra to the sacrum is of no practical importance, nor does a shift in the relative number of rib-bearing and non-rib-bearing vertebrae in the thoracolumbar column affect the strength of that column. (Bardeen, '00, described a tendency for the last thoracic vertebra to lose its ribs, as indicated by the finding of ones less than 2 inches long in 28% of 59 bodies, and evidence of lumbar ribs was said by Le Cocq to have been found in 1.22% of 500 roentgenograms.) However, fusion of the fifth lumbar vertebra with the sacrum (sacralization or assimilation, Fig. 2-22) decreases the length and adds to the strength of the thoracolumbar column, while transformation of the first sacral vertebra into a lumbar one (lumbarization) increases the length and mobility and thereby weakens this portion of the column. Apparently, unilateral sacralization or lumbarization both tend to make the lumbar region more susceptible to damage because of the asymmetry of the articulations; they are commonly recognized as causes of backache.

Variations in the length of the vertebral column are usually reported in terms of the number of vertebrae above the sacrum, often called presacral vertebrae, or, since the cervical vertebrae are constant in number except in truly anomalous conditions, in terms of the number of thoracolumbar vertebrae only. Bardeen ('04) reported, from a survey of the literature, that more than 90% of vertebral columns have 24 presacral vertebrae, from 2 to 4% have 23, and from 3 to 7% have 25. Among 1,000 skeletons the tendency to reduction and the tendency to increase in length were said to be equal, 4% each. Willis ('23b), reporting upon 850 thoracolumbar columns, found that 88.5% had 17 vertebrae, the normal number; 4.4% had 17 minus, that is, presented partial sacralization of the fifth lumbar vertebra; 1.1% had 16, complete sacralization of the fifth lumbar; 2.1% had 17 plus, partial lumbarization of the first sacral; and 3.9% had 18, complete lumbarization of the first sacral vertebra. Lanier ('39a) found lumbarization of the first sacral vertebra or partial to complete sacralization of the fifth lumbar in 11% of 100 skeletons of white persons, 20% of 100 skeletons of Negroes.

Changes in the number of vertebrae do not necessarily add to or detract from the total length of the vertebral column (Dwight), but commonly do; even if they do not, they increase or decrease the number of joints, and thus change both the mobility of the column and the number of points that are a potential source of weakness. When the number of presacral vertebrae is decreased, those bearing ribs are still typically 12, but when there is an additional vertebra it may be, on the basis of presence or absence of a rib, described as a thirteenth thoracic or a sixth lumbar one (or, rarely, there may be one of each; Fig. 2-23), and the sacrum may consist of either four or five segments (Gladstone; Decker).

Most statistical studies of the presacral region of the vertebral column have been made upon skeletal collections, and only in shorter series and isolated cases have variations in the vertebral column and the lumbosacral plexus been studied simultaneously. In such studies, the nerve must of course be designated by its numerical order rather than named according to the vertebra with which it is associated. It is known that a normal plexus may be associated with an abnormal column, and that the extremes of variation in the plexus found

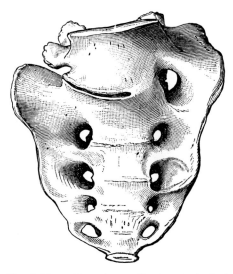

Fig. 2-22. Unilateral sacralization (assimilation) of the last lumbar vertebra. (Disse J: Vol 1, Section 1, of K von Bardeleben's Handbuch der Anatomie des Menschen. Jena, Fischer, 1896)

with abnormal columns may also occur with normal columns. However, plexuses that deviate from the normal (Chap. 7) are more frequently associated with abnormal vertebral segmentation; the plexus tends to shift slightly cranially with shortening of the column, slightly caudally with its lengthening (Bardeen and Elting; Horwitz, '39a).

Keegan ('47) maintained that the cutaneous distribution of the segmental spinal nerves to the limbs is constant, regardless of variation in the vertebral column. One would then expect that the distribution of the peripheral cutaneous nerves would vary according to their segmental origin, but Bardeen ('07) was unable to find such a correlation for the major cutaneous nerves of the lower limb.

SYNOVIAL JOINTS

Except for the special joints between the skull and the atlas and the latter and the axis, the synovial joints of the vertebral column are zygapophyseal, formed by the articular processes. The facets of the superior articular processes tend to face posteriorly, those of the inferior articular processes tend to face an-

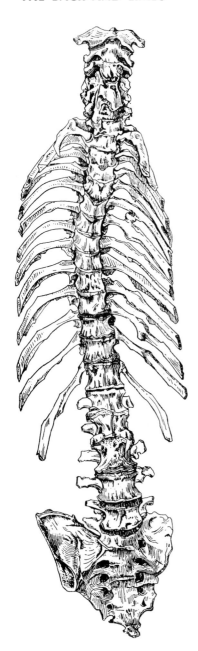

the various portions of the vertebral column. These relationships have been mentioned in part in connection with the description of the various vertebrae, and are commented upon again in a following section, in connection with the movements of the vertebral column. Otherwise, the zygapophyseal joints are all essentially similar, and can be very briefly described.

The joints are usually described as of the plane variety, although they vary from being almost flat to being decidedly concave-convex; each joint is surrounded by a thin articular capsule, necessarily somewhat lax in order to permit movements between the articular processes. The joints are innervated by twigs from the medial branches of the posterior rami (Fig. 2-24); Stilwell ('56) emphasized

Fig. 2-24. Nerve supply to the intervertebral synovial joints. Only the medial branch of the dorsal (posterior) ramus is shown here, and its terminal branches are omitted. (Pedersen HE, Blunck CFJ, Gardner E: J Bone Joint Surg 38-A:377, 1956)

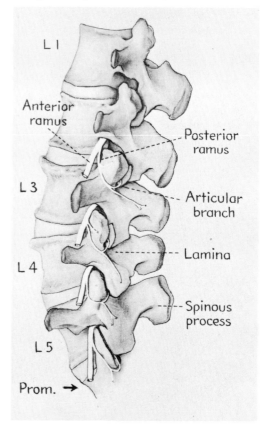

Fig. 2-23. A vertebral column with supernumerary presacral vertebrae. There are 13 thoracic and six lumbar vertebrae; the last lumbar is partly sacralized. (Dwight, T: Anat Anz 19:321, 1901)

teriorly. However, since the vertebral bodies are united by fibrocartilages which allow limited movement in all directions, the varying relationship of the synovial joints to the frontal and sagittal planes is of importance in determining the types of movement allowed in

that in the monkey each joint receives fibers from two adjacent nerves.

Kraft and Levinthal regarded pinching of the synovial membrane of a lumbar synovial joint, especially likely to occur as one straightens from a twisting or rotatory movement of flexion, as being a not uncommon cause of sudden severe low back pain. They said this is most likely to occur between the fourth and fifth lumbar vertebrae, and that the pain is usually in the lumbosacral region. They regarded narrowing of the disk space as predisposing to this, and said that many patients with true disk syndrome report having had periodic attacks of "low back pain," but with no extension of pain down the leg, before there were any signs of true herniation of the disk.

Atlantooccipital and Atlantoaxial Joints
The atlas is joined to the skull by the anterior and posterior atlantooccipital membranes and by two synovial atlantooccipital joints,

formed by the occipital condyles and the superior articular facets of the atlas. The *anterior atlantooccipital membrane* is attached to the anterior arch of the atlas, and blends on its edges with the capsules of the synovial joints; in its middle, it receives some accession from fibers of the anterior longitudinal ligament. The *posterior atlantooccipital membrane* is attached below to the posterior arch of the atlas, and above to the posterior margin of the foramen magnum (Fig. 2-25); it, like the anterior one, blends laterally with the capsules of the synovial joints. On either side it is penetrated, just above the posterior arch of the atlas, by the vertebral artery as this courses toward the foramen magnum, and by the dorsal (suboccipital nerve) and ventral branches of the first cervical nerve.

The occipital condyles, with their convex articular surfaces, are set obliquely in regard to the frontal plane of the body, their anterior ends being closer together than their posterior ones. The concave superior articular surfaces

Fig. 2-25. Articulations of the axis, atlas, and skull in median longitudinal section. (Redrawn from Fick R: Vol 2, Section 1, Part 1 of K von Bardeleben's Handbuch der Anatomie des Menschen. Jena, Fischer, 1904)

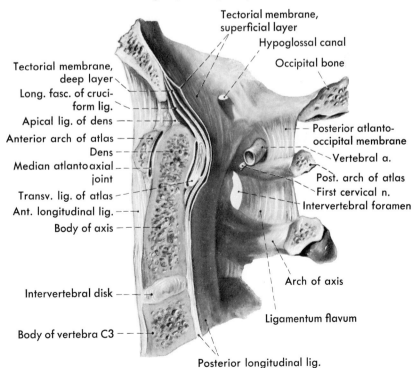

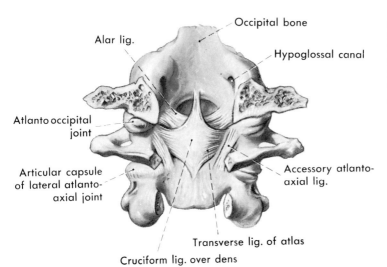

Occipital bone

Alar lig.

Hypoglossal canal

Atlanto occipital joint

Articular capsule of lateral atlanto-axial joint

Accessory atlanto-axial lig.

Transverse lig. of atlas

Cruciform lig. over dens

Fig. 2-26. Articulations of the axis, atlas, and skull, viewed from the dorsum after removal of the arches and of the tectorial membrane and posterior longitudinal ligament. (Redrawn from Fick, R: Vol 2, Section 1, Part 1 of K von Bardeleben's Handbuch der Anatomie des Menschen, Jena, Fischer, 1904)

of the atlas so fit the occipital condyles that, for practical purposes, the two joints act as one. Although a slight forward slipping of one condyle and backward slipping of the other, has been said to allow a very small amount of lateral bending and rotation, Hohl said that movement at the joint is limited to flexion and extension and totals only about 15°. The closely adjacent fibrous and synovial layers of the capsules of the atlantooccipital joints are attached at the margins of the articular surfaces.

The atlas and the axis are united by three synovial joints: paired lateral ones and a midline one between the dens (odontoid process) and the arch of the atlas. The *lateral atlantoaxial joints* betweens the superior articular facets of the axis and the inferior ones of the atlas are approximately flat, and inclined laterally and inferiorly rather than being in the transverse plane. The capsule of each joint is attached close to the articular surface.

The *middle atlantoaxial joint* (between dens and atlas) is slightly more complex. The dens projects upward from the axis to lie on the posterior surface of the anterior arch of the atlas, from which it is separated by a synovial cavity; it is held in place on the arch by ligaments that attach both to the atlas and to the occipital bone of the skull, and on its dorsal aspect it has a second synovial cavity that intervenes between it and the transverse ligament of the atlas.

The ligaments associated with the dens are the tectorial membrane, the cruciform ligament of the atlas, the alar ligaments, and the apical ligament of the dens. The tectorial membrane (Fig. 2-25) is a broad band of fibers that stretches from the lower part of the body of the axis, where it is continuous with the posterior longitudinal ligament, to an upper attachment on the base of the occipital bone; it covers the other ligaments connected with the dens. Deep or anterior to the tectorial membrane is the cruciform (cruciate) ligament of the atlas. The heavier part of this, the transverse ligament of the atlas (Figs. 2-25 and 2-26), passes between tubercles situated on the medial sides of the lateral masses of the atlas, and holds the dens forward against the anterior arch of the atlas; the synovial cavities between it and the dens and between the dens and the arch allow rotational and sliding movements of the atlas on the dens. The cruciate form is completed by the longitudinal fasciculi, small bundles of fibers that run downward from the transverse ligament to attach to the body of the axis, and upward to attach to the base of the occipital bone just within the foramen magnum.

The transverse ligament of the atlas is sometimes ruptured by trauma, allowing the atlas to be displaced forward on the axis. Fielding and co-workers found experimentally much difference in the strength of this ligament, but in every case the force applied caused its rupture rather than fracture of the dens. In only five of 20 cases did the attachment of the ligament to bone fail; usually there was a rupture in the middle.

From the apex of the dens there stretches upward to the base of the occipital bone, under cover of the superior part of the cruciform ligament, a slender band of fibers that is the apical ligament of the dens. The alar ligaments (Fig. 2-27), much broader and much stronger than the apical ligament, run upward and laterally from each side of the dens to attach to the medial side of each occipital condyle. The alar ligaments tend to restrict rotation upon the dens.

INTERVERTEBRAL DISKS

The intervertebral (once called "fibrocartilaginous") disks form the chief union between the vertebral bodies, from the lower surface of the body of the second cervical vertebra through the junction between the last lumbar vertebra and the sacrum. They are assisted in their function of holding the vertebrae together by the ligaments of the vertebral column, especially those connected with the bodies of the vertebrae. The disks (or discs) play, however, a very much more important part than merely holding the vertebral bodies together: through their elasticity they allow a limited amount of movement between each two adjacent vertebral bodies, movement that would be largely impossible if the vertebral

bodies were in direct apposition; further, their soft centers distribute the weight over the surfaces of the vertebral bodies during movement, preventing it from being concentrated exclusively on the edge toward which the column is bent.

It is generally stated that the intervertebral disks together form approximately 25% of the length of the vertebral column, by which is meant the length of the vertebral column above the sacrum. This varies somewhat with the age of the individual. Also, the disks do not contribute equal proportions of the lengths of the cervical, thoracic, and lumbar regions. Thus Aeby reported that in an adult the intervertebral disks in the cervical region contribute from 20% to 24% of the length of the cervical column, a percentage slightly below that of the newborn; that the adult thoracic disks contribute from 18% to 24% of the length of the thoracic column, while in the newborn the disks contribute 31%; and that the lumbar disks contribute from 30% to 36% of the length of the lumbar column in the adult, only slightly less than they contribute in the newborn. These are, of course, only average measurements, for even one disk is not necessarily of the same height (thickness) in its anterior and posterior aspects.

The disks in the cervical region are thicker anteriorly than they are posteriorly; they are entirely responsible for the normal cervical curvature. The disks here are also different in that they do not conform entirely to the sur-

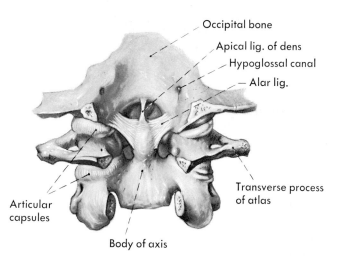

Fig. 2-27. Articulations of the axis, atlas, and the skull, viewed from the dorsum after removal of the cruciform ligament. (Redrawn from Fick R: Vol 2, Section 1, Part 1 of K von Bardeleben's Handbuch der Anatomie des Menschen, Jena, Fischer, 1904)

Occipital bone

Apical lig. of dens

Hypoglossal canal

Alar lig.

Transverse process of atlas

Articular capsules

Body of axis

face of the vertebral bodies with which they are connected; they measure slightly less from side to side than do the vertebral bodies, so that the lower rounded edge of one vertebra comes very close to the overlapping upper edge of the next, especially posterolaterally (this prominent projection is sometimes known as the uncus or uncinate process). It is here that "Luschka's joints," once described as true synovial joints, are found. Orofino and co-workers found no cavities here in fetal columns, and clefts but no synovial linings in adults. Payne and Spillane found clefts here in all adults, but again no synovial linings, and both groups of workers agreed that they are degenerative fissures in the tissue, not synovial joints.

The disks in the thoracic region are about the same height whether measured anteriorly or posteriorly, for the thoracic curve is due largely to the shapes of the bodies of the vertebrae. Also, although the percentage of the length of the thoracic column formed by the disks is, according to Aeby, only a little less than that which the cervical disks contribute to their region, it is not equally distributed in the thoracic column; the disks of the upper part are much thinner than those of the lower. For this reason, if for no other, mobility of the upper part of the thoracic vertebral column would be expected to be restricted as compared with the lower, and is indeed very much less.

In the lumbar vertebral column, the disks all tend to be higher anteriorly than posteriorly; this reaches its maximum in the fifth lumbar disk (the disks are properly named according to the vertebra below which they lie). Through its decided wedge shape, the fifth disk contributes markedly to the lumbosacral angle. In the upper part of the lumbar column the normal curve is due entirely to the disks, but in the lower part the shape of the vertebral bodies also contributes, the last lumbar vertebra being the most wedge-shaped.

Each disk consists of two parts (or three, according to some authors); the two parts are an outer portion composed of fibrocartilage, the anulus (annulus) fibrosus, and a central, soft and mucoid portion, the nucleus pulposus

(Fig. 2-28). Except at the periphery of the vertebral body, the intervertebral disk is bothered by thin plates of hyaline cartilage interposed between it and the spongy bone of the vertebral body; some authors (e.g., Coventry and co-workers '45a) prefer to regard this cartilaginous plate as a part of the disk, but most anatomists regard it as a part of the vertebral body. It is generally granted that the cartilaginous plates are responsible for growth in length of the vertebrae, and therefore are epiphyseal cartilages during the growth period. It is also true that in the adult the plates function with the anulus fibrosus as seals for the soft nucleus pulposus: as long as they are intact, they prevent the nucleus pulposus from herniating into the spongy body of the vertebra.

Anulus Fibrosus

The anulus (annulus) fibrosus (Figs. 2-28 to 2-31), forming the periphery of the disk, is composed of fibrocartilage in which the fibrous element predominates. The fibers of the anulus run obliquely between the vertebrae and are arranged primarily in concentric laminae; the fibers composing each successive lamina have a different slant from that of the preceding lamina, and the direction of fibers in successive laminae tends to alternate so that the fibers of each lamina cross those of the two adjacent ones at an angle. (Peacock objected to this description, saying the arrangement of fibers is much more complicated than this, but Horton confirmed it and said it is a change in the angle between two adjacent layers of fibers that gives elasticity to the anulus. Jonck said the angle varies between $30°$ and $60°$, depending on the state of the disk and the pressure to which it is subjected.) The most superficial posterior fibers blend with the anterior longitudinal ligament, and the most superficial posterior fibers blend with the posterior longitudinal ligament. Peripheral fibers also pass over the edge of the hyaline cartilage to sink into the bone of the vertebral body as Sharpey's fibers; and the majority of fibers, situated more deeply, insert into the hyaline cartilage at each end of the disk.

The anulus fibrosus blends deeply, toward

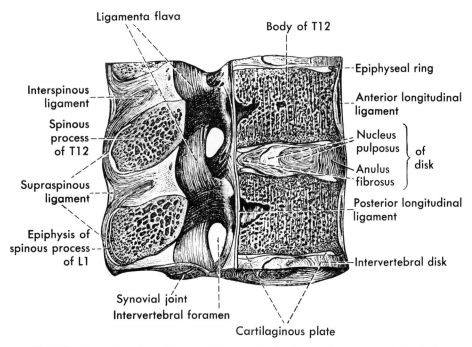

Ligamenta flava

Body of T12

Interspinous
ligament

Spinous
process
of T12

Supraspinous
ligament

Epiphysis of
spinous process
of L1

Epiphyseal ring

Anterior longitudinal
ligament

Nucleus
pulposus

Anulus
fibrosus

Posterior longitudinal
ligament

Intervertebral disk

of
disk

Synovial joint
Intervertebral foramen

Cartilaginous plate

Fig. 2-28. Sagittal section of a part of the vertebral column, showing especially the intervertebral disks. (Toldt C: An Atlas of Human Anatomy [ed 2], Vol 1. New York, Macmillan, 1928)

Fig. 2-29. Surface view of the anulus fibrosus of two intervertebral disks. (Toldt C: An Atlas of Human Anatomy [ed 2], Vol 1. New York, Macmillan, 1928)

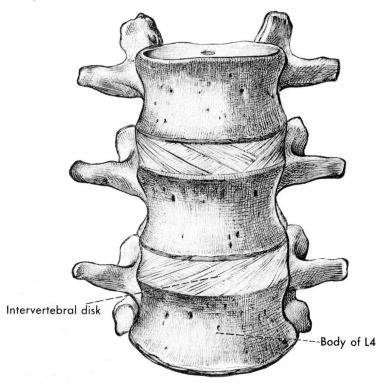

Intervertebral disk

Body of L4

Fig. 2-30. Histological section through the denser lamellae of the anulus fibrosus. Note the alternating arrangements of the layers of fiber bundles. (Fick R: Vol 2, Section 1, Part 1 of K. von Bardeleben's Handbuch der Anatomie des Menschen. Jena, Fischer, 1904)

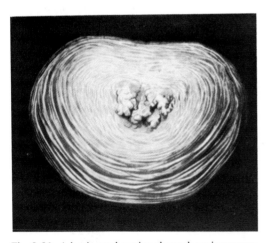

Fig. 2-31. A horizontal section through an intervertebral disk; the nucleus pulposus is here easily distinguishable from the surrounding, lamellated, anulus fibrosus. (Fick R: Vol 2, Section 1, Part 1 of K von Bardeleben's Handbuch der Anatomie des Menschen. Jena, Fischer, 1904)

the center of the disk, with the nucleus pulposus; according to Coventry and his coworkers, fibers pass from the nucleus pulposus into the anulus to join the lamellae of the anulus and insert obliquely into the cartilaginous plates. Although there is thus no absolute boundary between the nucleus pulposus and the anulus, the two are relatively distinct in younger people, and until degenerative phenomena, dehydration, and the formation

of fibrocartilage in the nucleus (Peacock) become prominent there is a great contrast between the soft and gelatinous center of the healthy nucleus and the very firm fibrous consistency of the outer portion of the anulus.

The anulus is somewhat thicker anteriorly than posteriorly (that is, the nucleus is eccentrically located), and this is presumably one of the factors responsible for the more common protrusion of the nucleus pulposus posteriorly. Further, the anulus is obviously strengthened anteriorly by the broad and rather strong anterior longitudinal ligament, while the posterior longitudinal ligament is both less strong and, in the particularly important lumbar region, less broad; it reinforces the anulus primarily in the midline rather than posterolaterally.

Nucleus Pulposus

The nucleus pulposus is said to consist of a delicate network of collagen fibers surrounded by a mucoprotein gel with a high polysaccharide content (Sylvén and co-workers), and it is apparently this gel that is responsible for imbibition of water by the disk (Mitchell and co-workers). It is generally acknowledged that the nucleus typically has a very high water content, usually said to be over 80%. However, the exact percentage of water varies with the age of the individual (Key, '49, quoted a water content of 88% in the full-term fetus, one of about 70% in late adult life). It varies even with the pressure that the disk has borne in preceding hours; loss in water content of the disk as a result of pressure while standing and sitting apparently accounts for the loss of height that individuals undergo during a day's time (for instance, Saunders, '40). Thus the disk normally undergoes a constant cycle of loss of water and therefore a diminution in volume with weight-bearing, and a reabsorption of water and increase in volume when the pressure on it is sufficiently lessened. With increasing age and the appearance of degenerative changes in the nucleus, absorption of water does not quite equal its loss, so the nucleus gradually becomes dehydrated.

Changes In The Disk With Age

According to Coventry and his colleagues ('45b) the nucleus pulposus develops rapidly during the first 10 years of life, and the interlacing fibrils in the gel ground substance seem to begin to grow coarser even toward the end of the first decade. They found that development continues to take place in the nucleus pulposus during the second decade, for its fluid content seems to increase and it begins to separate more clearly from the anulus. This development continues in the third decade; however, during the latter part of the third decade the nucleus again becomes less distinct as the fibrils become coarser, and begins to blend again with the anulus. The nucleus usually reaches its height of development in the third decade; thereafter there is loss of fluid, with a gradual fibrous replacement of the nucleus, and in the sixth and seventh decades, if not before, much of the nucleus blends with the fibrocartilaginous elements of the anulus and becomes fibrocartilage (Coventry and co-workers; Peacock; and others). Mitchell and co-workers attributed dehydration to the disappearance of the protein polysaccharide gel of the nucleus; they said this apparently must fall to a certain level before prolapse of the disk can occur.

Paralleling these changes in the nucleus pulposus, Coventry and his colleagues also found changes in the anulus fibrosus. This appeared to continue its differentiation through much of the second decade, but in the latter part of this decade there was evidence of degeneration: the fibers became less distinct, the number of nuclei seemed to decrease, and early hyalinization was often found. During the third decade, the anulus tends to show considerable evidence of wear and tear, for the fibers become coarse and hyaline in appearance, and fissuring of the lamellae occurs. During the fourth, fifth, and subsequent decades the degenerative changes in the anulus become more marked, and include the appearance of nests of cartilage cells among thick and swollen fibers; pigmentation, often referred to as "brown degeneration," occurs.

Physical Properties of the Disk

It is customary to speak of compression of the disk, yet it is obvious that, because of its high water content, the normal nucleus pulposus is essentially incompressible. The compressibility of the disk therefore lies in the possibility of a change in shape of the nucleus pulposus, a flattening of it under pressure, with a consequent peripheral bulging of the anulus fibrosus; similarly, with movements of the vertebral column, the nucleus and the anulus are differentially compressed: with flexion of the vertebral column, for isntance, the anterior fibers of the anulus and the anterior portion of the nucleus pulposus are compressed so that the anulus tends to bulge anteriorly and the nucleus pulposus becomes wedge-shaped, narrow anteriorly and wide posteriorly; at the same time, the posterior portion of the anulus is put upon a stretch, and, if it is intact, resists posterior movement of the nucleus. The elasticity of the disk therefore lies in the ability of the disk as a whole to be deformed and to regain its shape, not in a true compressibility of the nucleus pulposus.

The intervertebral disks are normally under pressure even when they are not bearing weight, because of the tension exerted by the posterior ligaments. Petter showed that they expand an average of 1.08 mm when they are removed from the body of a fresh cadaver, and that an average of 30.2 lb of pressure was necessary to reduce a disk to its former size, thus indicating that in the supine position the disks are under a pressure of about 30 lb. Nachemson and Evans reported comparable intradiscal pressures, and showed that those are entirely attributable to the ligamenta flava (although *in vivo* they may be increased by muscular tension). While the pressure increased somewhat in forward bending, this amounted to only a small part of the increased pressure during flexion found in living persons.

The isolated disk has no pressure of its own, or at least too little to measure (Nachemson, '60). Obviously, the pressure is greater in the erect position, and increases in successively lower disks with the added weight of the body at each level; further, if a disk is loaded on a tilt rather than vertically, there is, with the same load, an increase in intradiscal pressure (Nachemson, '63), indicating that movement of the vertebral column increases the pressure upon the disks still more. Thus, Nachemson ('65) measured the intradiscal pressure of the third and fourth lumbar disks in living persons and found the total load on the disk to be two or more times the calculated weight of the body in the standing position (loads of from 86.4 kg to 150.4 kg or about 190 lb to 330 lb), and considerably more in the same persons (140.6 kg to 179.2 kg, or about 309 lb to 394 lb) when they were sitting. When the subjects leaned forward approximately 20° the increase in pressure on the disk averaged 30%; with a weight of 10 kg (22 lb) in each hand, a still greater increase occurred; with one subject, for instance, the added weight of 20 kg increased the load on the disk when leaning forward from 231 kg to 336 kg (about 508 lb to 739 lb, or 230 lb more pressure on the disk from an added weight of only 44 lb). In lifting heavy objects, however, the pressure of the thoracic and abdominal contents transmits some of the weight directly to the pelvis, perhaps relieving the lumbrosacral disk of as much as 30% of the load (*e.g.,* Morris and co-workers). Nachemson and Morris showed a similar reduction from the sitting to the standing position, and almost as much more when the standing subject wore an inflated corset.

Virgin, among the earlier workers, investigated the mechanical properties of the disk by subjecting one to five disks from each of 51 cadavers to loads of from 50 lb to 500 lb. He noted that if the disks were compressed while dry, beads of moisture appeared upon them when the load level reached about 300 lb, giving evidence that the deformation was accompanied by gross loss of fluid; also, disks recovered their elasticity much better when they were immersed in physiologic fluid (Ringer's solution) than when they were tested dry, thus confirming deductions made by others that loss and reabsorption of fluid are important factors in the physiology of the disk. He found that the disk was unable to maintain its height under a constant load of even 50 lb, but lost very little after 48 hours of continuous load and recovered completely after the load was removed. When loads on disks were gradually increased from 50 lb to 500 lb and then decreased again to 50 lb, the recovery of the disk to its previous height varied considerably: lower thoracic and upper lumbar disks recovered best, while the lowest lumbar recovered most poorly, and maintained, under the conditions of the experiment, a certain amount of "set" at a diminished height. There was also a variation according to age: disks from young subjects showed a particularly poor recovery from overloading, and those of aged people with actual degeneration of the disk recovered only moderately well. The best recovery, in his experiments, was in disks from persons of the middle decades, and in subjects of the seventh decade in whom the disks showed no marked degenerative changes. Recovery was particularly poor, he found, in disks from persons who had suffered from chronic passive congestion, and in cases in which Schmorl's bodies (nodes) were particularly marked.

Nachemson ('60), through measuring intradiscal pressures, calculated that a load is not borne equally by the nucleus and the anulus, for the vertical stress on the nucleus is about 50% more than the applied load and that on the anulus only about 50% of the applied load. (In contrast, Markolf and Morris reported that the anulus alone, without nucleus or cartilaginous end plates, reacts almost normally to compression, and concluded that it is the primary factor in weight bearing, the nucleus serving primarily to regulate the height of the disk.)

The tangential stress on the anulus, however, is three or more times the load; Nachemson ('63) correlated this with posterior protrusions in the lumbar region, since on tilting the tangential stress on the anulus is much increased at the same time that the vertical stress is similarly increased in the side of the anulus toward which tilting (bending) occurs. (In this later paper, Nachemson retracted an earlier statement that the articular processes bear approximately 15% of the total load in the intact vertebral column; recalculation, he said, indicates that they bear no appreciable part of it.)

Blood and Nerve Supply

A blood supply to the disk reaches it in the fetus both from its periphery and from vessels in the bodies of adjacent vertebrae that grow through the cartilaginous plates and run toward but do not reach the nucleus pulposus (Ehrenhaft). These vessels are commonly described as degenerating "after birth"; Coventry and his colleagues ('45a) quoted evidence that the vessels from the vertebrae begin to become scarred at the age of 8 months, and are completely occluded between the ages of 20 and 30 years. Ehrenhaft described the disks as being avascular by the age of 25. Persistence of the channels in the cartilage through which the blood vessels entered the disk facilitates herniation of disk substance into the body of a vertebra, to form Schmorl's bodies. While there is agreement that the normal adult disk is avascular, Crock and colleagues described a capillary plexus in the cartilage plate and suggested that this is related to nutrition of the disk.

The periphery of the anulus fibrosus is commonly avascular except in the fetus, but Coventry and his colleagues ('45b) found evidence of vascularization of the posterior part of the anulus in specimens from the fourth decade onward. The secondary vascularization is probably usually regarded as an attempt to repair degenerative changes in the anulus, but Lindblom and Hultquist described it as being concerned with absorption of nectrotic portions of the disk, not with repair.

Pedersen and co-workers investigated the meningeal (sinuvertebral or recurrent) nerves, which supply the structures in the vertebral canal, both by dissection and in serial sections of fetal columns, and Stilwell ('56) studied the innervation of all the tissues of the vertebral column in the monkey with essentially similar findings in regard to the meningeal nerves. (Stilwell also reported finding nerve endings of the type believed associated with pain in both longitudinal ligaments, in the periosteum of the vertebral bodies, in loose tissue on the surface of the anulus, in the thoracolumbar fascia, and in the capsules of the synovial joints. Of these structures, the innervation to those on the posterior surfaces of the vertebral bodies, to the dura, and to the inner surface of the ligamentum flavum is supplied by the meningeal nerves, the remainder from the outer surface of the vertebral column, therefore primarily by the dorsal rami of the spinal nerves.)

The meningeal nerves arise from the spinal nerves close to or in common with the rami communicantes and consist of both sympathetic and spinal (afferent) fibers. After a meningeal branch enters the vertebral canal through the intervertebral foramen it branches as it proceeds toward the posterior longitudinal ligament, and both ascends and descends (Figs. 2-32 and 2-33); thus the nerve endings at any level are apparently derived from at least two segmental spinal nerves. Fibers and endings lie both in and on the surface of the posterior longitudinal ligament, but have been found only on the surface of the anulus.

MOBILITY AND STABILITY OF THE VERTEBRAL COLUMN

Stability of the vertebral column is of course supplied by the vertebral musculature, the ligaments, including the intervertebral disks, and the curvatures. To a certain extent, mobility and stability are incompatible, yet it is the elasticity of the ligamentum flavum, the compressibility and elasticity of the disks, and the positions of the synovial joints that permit and at the same time limit the mobility of the vertebral column. The size of the intervertebral disks in relation to the bony segments are important in determining how much movement is allowed, and the direction in which the articular facets face are similarly important in determining in what directions movement is allowed. Movements are apparently checked primarily by the ligaments, including the anulus fibrosus, and not by the muscles. In the completely flexed condition, for instance, there is only a little activity in the erector spinae, which is very active during

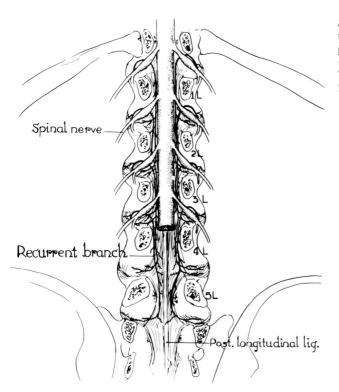

Spinal nerve

Recurrent branch

Post. longitudinal lig.

Fig. 2-32. Diagram of the innervation of the posterior longitudinal ligament and the posterior surface of the anulus fibrosus. (After Spurling and Roofe, from Craig WMcK, Walsh MN: J Bone Joint Surg 23:417, 1941)

flexion, and the ligaments primarily support the weight (Morris and co-workers; Pauly).

Movements

Careful roentgenographic studies of movements of the first two cervical vertebrae have much altered usual concepts of the movements here. Thus, Hohl has been unable to confirm the usually described slight movement of lateral flexion and rotation at the atlantooccipital joint, reporting that movement here is limited to about 15 ° of flexion-extension. Similarly, the common statements that the atlantoaxial joints allow only rotation (and, as quoted by Payne and Spillane, that rotation here accounts for about 90% of the total rotation in the cervical region) have not been borne out by further study. Fielding has shown that rotation here amounts to about 45 ° in either direction, only about half of the total rotation in the cervical region (the remaining rotation is distributed about equally among the lower five interspaces); and Hohl has shown that the atlantoaxial joint normally allows approximately 15 ° of flexion-ex-

tension (as much as at the atlantooccipital joint), with the atlas gliding downward and upward on the dens. Further, when it is combined with rotation, there is enough lateral flexion to allow the articular surface of the atlas to protrude approximately 2 mm beyond that of the axis—a relatively small amount, but enough to be mistaken for subluxation. The alar ligaments presumably limit rotation, but the transverse ligament of the dens obviously does not limit other movements as much as had been supposed.

In the succeeding portions of the cervical region the intervertebral disks are relatively thick, commonly almost a fourth of the height of the associated vertebra (or, if measured through their middles, from one third to one half, according to Payne and Spillane). In addition, the anteroposterior cupping of the lower surface of one vertebra over the beveled upper surface of the vertebra below and the lateral beveling to fit into the lateral cupping of the vertebra below allow a certain amount of sliding of one vertebra upon the other. Further, the planes of the interarticular joints

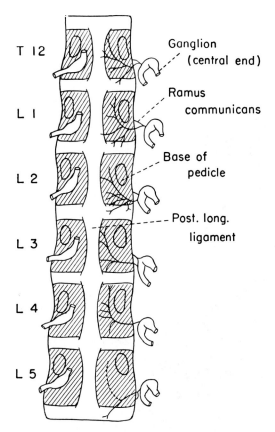

Fig. 2-33. Diagram of the meningeal branches of the spinal nerves as shown by dissection. The arches have been removed to the bases of the pedicles; spinal nerves on the right side are reflected to show the origins of the meningeal branches. Meningeal nerves of the left side are not shown. (Pedersen HE, Blunck CFJ, and Gardner E: J Bone Joint Surg 38-A:377, 1956)

tended position, flexion usually carries the cervical column little beyond the straight position. Extension, also free in the cervical region, is accompanied by a downward and backward slipping of the inferior upon the superior articular processes, and a compression of the posterior parts of the disks.

Because the surfaces of the articular processes are are neither in the horizontal plane nor exactly parts of the arc of a circle, lateral flexion and rotation in the cervical region always accompany each other: because of the slant of the articular processes, downward movement of the inferior articular process on the side toward which flexion is occurring is necessarily accompanied by a backward movement (and an upward and forward movement of the inferior articular process on the other side), thus producing rotation to the side of flexion. In the same way, attempted rotation to one side produces not only forward movement on one side, backward movement on the other, but also upward movement on one side, downward on the other, producing flexion to the same side as the rotation. In this combined movement, the disk is compressed on the concave side, and also twisted.

In the thoracic region, and especially in the upper thoracic region, movements are limited by the relatively small size of the intervertebral disks; further, the ribs and sternum also tend to limit movements here. The synovial joints lie somewhat in the frontal plane (Fig.

Fig. 2-34. Horizontal section through the synovial joints between the fourth and fifth cervical vertebrae. (Poirer P, and Charpy A: Traité d' Anatomie Humaine [ed 3], Vol 1. Paris, Masson et Cie, 1911)

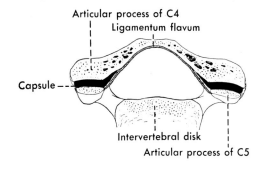

(Fig. 2-34), on an oblique slant between the horizontal and the vertical, permit rather free movements in all directions.

In flexion of the cervical region the inferior articular processes slide upward and forward on the superior ones, so that the upper vertebra moves forward on the lower one, and the anterior part of the disk is compressed; the movement is checked primarily by the posteriorly lying ligaments and by the intervertebral disks. Flexion is the freest of all movements of the vertebral column as a whole and is especially free in the cervical region; however, since the neck starts from a hyperex-

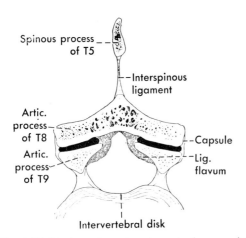

Fig. 2-35. Horizontal section through the synovial joints between the eighth and ninth thoracic vertebrae. (Poirier P, and Charpy A: Traité d' Anatomie Humaine [ed 3], Vol 1. Paris, Masson et Cie, 1911)

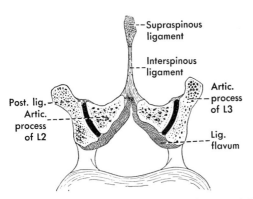

Fig. 2-36. Horizontal section through the synovial joints between the second and third lumbar vertebrae. (Poirer P, Charpy A: Traité d' Anatomie Humaine [ed 3], Vol 1. Paris, Masson et Cie, 1911)

2-35), approximately on the arc of a circle whose center is usually in the vertebral bodies, but in front of them for the upper one or two and the last one (Davis). The almost frontal direction of the joints means that movements of flexion and extension are brought about primarily by an upward and downward, rather than a forward and backward, sliding of the inferior articular processes upon the superior processes below, and are therefore more quickly checked by the longitudinally arranged ligaments. Lateral

bending is limited primarily by the ribs with their attachments to the sternum, for the direction of the articular facets would not tend to limit this appreciably. Rotation, as would be expected from the positions of the facets, is relatively free but limited by the sternum and ribs. Gregersen and Lucas found an average 6° maximal rotation of the thoracic vertebrae at each level except the thoracolumbar one, which varied much (one individual had 11° of rotation here, another a total of only 3° between T12 and L4); this presumably reflects the fact that facets at the thoracolumbar junction vary in their planes, sometimes resembling lumbar rather than thoracic joints in this respect.

In the lumbar region the disks are relatively thick, especially anteriorly, but the planes of the synovial joints gradually change between the upper and lower levels: The superior articular processes of the first lumbar vertebra face largely medially and only a little posteriorly, its inferior articular processes largely laterally and only a little anteriorly. Thus the articular surfaces of the first lumbar vertebra are more nearly in the sagittal than in the frontal plane, and this is also true of the second (Fig. 2-36). While there is a great deal of variation among different vertebral columns, there tends to be a gradual transition in the relation of the articular surfaces to the planes of the body as the joints are followed from the thoracolumbar to the lumbosacral ones: the superior articular processes gradually face more posteriorly and less medially, while the inferior articular processes therefore gradually face more anteriorly and less laterally (Fig. 2-37). As a result, rotation in the upper lumbar region is possible at all only to the extent that the joints allow a little forward sliding of one inferior articular process while the other slides backward. Total rotation in the lumbar column was quoted by Rissanen as amounting to about 5°, but Gregersen and Lucas measured it as 18° (from one side to the other) in one individual, and it is usually assumed that much of this takes place in the lower joints, which are farther from the sagittal plane. The lumbosacral joint typically departs farthest from the sagittal, but Willis ('59) found much variation in this. Although the angle between the joints and the sagittal plane varied from 20° to 90°, the average was 49°, thus only a little closer to the frontal than to the sagittal plane. Variations in the planes of this joint are reflected in the amount of rotation allowed; Gregersen and Lucas reported it as 17°, 13°, and 9° in three persons; in the last instance, this accounted for half of the total rotation of the lumbar column.

Flexion and extension are fairly free in the lumbar region; Rissanen quote 90° of extension from the neutral position, 25° of flexion, as typical, with vertebra L4 or L5 moving the most. Lateral flexion, dependent on the ability of one inferior articular process to slip downward as the other slips upward, is usually accompanied by a variable amount of torsion to the other side (Miles and Sullivan), indicating that as the process slips downward it also slips forward. Maximal movement was most often between vertebrae L2 and L3, but fairly frequently at any of the other joints except the lumbosacral one, and sometimes one or more joints flexed to the side opposite that of major bending.

As reviewed here, it is obvious that the several regions of the vertebral column contribute variably to the total possible movement, and some movements, such as flexion in the lumbar part of the column, are more important than others. Davis presented estimates as follows of the amount of permanent disability produced by a number of spinal fusions: Occipitocervical fusion, extremely disabling, and of the first two cervical vertebrae, negligible (both assessments possibly incorrect, since according to more recent information either should interfere about equally with flexion; Hamblen reported a number of cases in which one of these was necessary to provide stability of the upper cervical column); fusion of the last four cervical vertebrae or of the last three cervical and the first thoracic, 15%; upper and midthoracic fusions, negligible; lumbosacral fusion, fusion of the thoracolumbar junction, and midlumbar fusion, each 15%.

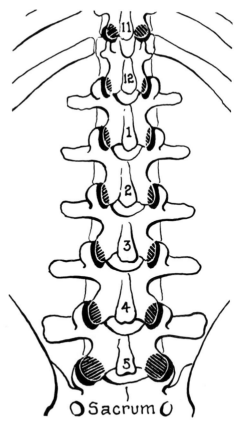

Fig. 2-37. Changes in the position of the synovial joints of the lumbar portion of the vertebral column; the obliquely lined areas indicate the contiguous joint surfaces. (Dandy WE: Ann Surg 118:639, 1943)

Movers of the Vertebral Column

The actions of the posterior muscles of the vertebral column on that structure are discussed more fully in the Musculature of the Back section in this chapter. It need only be pointed out here that these muscles are primarily extensor ones for the vertebral column and are therefore most active in controlling flexion against gravity and in restoring the erect position. Some of them show some activity in rotation and in lateral flexion, but much of both of these movements is brought about by such muscles as the anterolateral abdominal ones, the intercostals, and the quadratus lumborum.

Lumbosacral Junction

The cervical and lumbar curves are gradual and therefore relatively strong; at the levels of change from one curve to another, the vertebral column is at somewhat of a disadvantage, both because of the change in direction of the curve and the additional leverage afforded by the length of vertebral column above the level of the change (hence fracture is more common at the junction of one region with another). This is especially true at the lumbosacral junction, with the tremendous leverage exerted upon it by the entire length of the column above the sacrum, and with the very sharp change of angle, not approximated at the cervicothoracic or the thoracolumbar junction. Further, with forward shifts of the line of gravity the lumbar curve is increased to shift the weight backward, and this increases the tendency toward anterior displacement at the lumbosacral junction. This junction is therefore subject to particular strain, and is of special clinical importance.

The lumbosacral synovial joints allow flexion, extension, and lateral flexion, as do the interlumbar joints. In addition, the joints between the articular processes of the fifth lumbar and the first sacral vertebrae usually approach the frontal plane more than do the other joints, and probably allow more rota-

Fig. 2-38. The buttressing action of the articular processes of the lumbosacral joint; they interlock, and hence resist anterior dislocation of the fifth lumbar vertebra. (Mitchell GAG: J Bone Joint Surg 16:233, 1934)

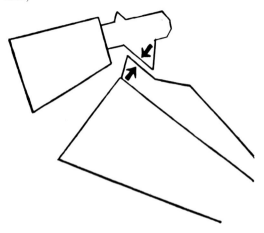

tion. This relationship of the articular processes actually strengthens the lumbosacral joint, however, for the whole weight of the body tends to force the fifth lumbar vertebra forward, so as to produce a subluxation, and the locking of the inferior articular processes of the fifth lumbar vertebra behind the superior processes of the first sacral obviously strongly resists such subluxation (Fig. 2-38). The tendency to subluxation on the part of the fifth vertebra is, of course, much greater than is that on the part of the other lumbar vertebrae, because of the relatively sharp sacrovertebral angle: According to Mitchell ('34) the average inclination of the upper surface of the first sacral vertebra in 28 European skeletons was 41° to the horizontal plane, and any figures within 5° or 6° on either side of the usually given 42.5° are to be considered within the limits of normality. On the other hand, the average inclination of the upper surface of L5 in Mitchell's specimens was only 19°; thus the fourth lumbar vertebra has much less tendency to slide forward and downward than does the fifth lumbar.

Mitchell called attention to the fact that the sacrovertebral angle varies inversely with the inclination of the upper surface of the base of the sacrum, since the more this is inclined from the horizontal, the smaller is the sacrovertebral angle; further, since the upper surface of the base of the sacrum is at about a right angle to the pelvic surface, the slant of the upper surface depends primarily upon the obliquity of the sacrum. The sacrum is usually directed slightly more obliquely posteriorly in the female than in the male, on the average, but much wider variations exist among specimens of the same sex than between males and females as a whole. According to Mitchell, the sacrovertebral angle becomes somewhat more prominent with age, because of the weight of the body bearing upon it; the sacrum may, he said, sometimes become almost horizontal, so that the angle becomes very prominent. In such cases a compensatory lordosis normally develops. He said that he had seen extremes of 32° and 68° in the inclination of the upper surface of the first sacral segment, and that extremes of 28°

and 80° have been reported. Other things being equal, it is obvious that stability at the lumbosacral junction is favored by a more vertical position of the sacrum.

Rather than sacral obliquity alone, which can be compensated for in part by lordosis, a better measure of the strength of the lumbosacral joint is the sacrovertebral angle: the smaller this angle becomes, the greater the strain on the inferior articular processes in preventing the fifth lumbar vertebra from subluxating anteriorly. According to Mitchell, the sacrovertebral angle has been defined and measured in two ways: as the angle between a line drawn through the approximate centers of the first two sacral segments and a vertical line approximating the center of gravity; and, in a hemisected column, as the angle between the midline of the fifth lumbar vertebra and the midline of the first sacral (Fig. 2-39). The first measurement, according to Mitchell, has been said to give an average of about 102° in the female, 117° in the male; the second, he said, has given an average of 137° in the female, of 140° in the male, but with a range of from 128° to 160°. The measurements on the hemisected vertebral column might more appropriately be called the lumbosacral angle, Mitchell suggested; the larger the lumbosacral angle the more stable the vertebral column. The lumosacral angle has also been defined as that angle which the upper end of the sacrum makes with the horizontal, but this does not alter the truth of the preceding statement.

Mitchell pointed out that for best stability the lumbosacral angle should be such that a line through the middle of the fifth lumbar vertebra makes a right angle with a horizontal line through the middle of the disk below it; however, the latter angle normally exceeds a right angle by about 15° to 18°, averaging about 16°, and the lumbar convexity compensates for the excess sacral obliquity. According to Mitchell, if the lumbosacral angle fails by 25° or more to compensate for the sacral obliquity, definite lordosis is usually present.

Obviously, spondylolysis (interarticular or isthmal defect, discussed in the following sec-

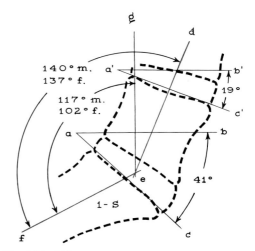

Fig. 2-39. Difference in the slant from the horizontal of the lumbosacral and last interlumbar joints (angles b, a, c, and b', a', c'), and two definitions and measurements of the lumbosacral angle (*red lines*, forming angles d, e, f, and g, e, f). *1-S* is the first sacral segment. Measurements are those given by Mitchell.

tions) particularly predisposes to instability of the lumbosacral junction; relaxation of ligaments and muscles might be expected to have a lesser effect; asymmetrical articular processes, lumbarization of the first sacral vertebra, and sacralization, partial or complete, of the fifth lumbar vertebra have also been thought to be responsible for derangements in the mechanics of this joint. Of the latter, bilateral sacralization would seem to have an opposite effect, for, since this tends to shorten the column above the lumosacral joint, it should strengthen rather than weaken the joint. Mitchell said that the fifth lumbar nerve may be particularly susceptible, because of its relations: its dorsal branch runs close to the outer side of the lumbosacral synovial joint and may therefore be vulnerable to swelling or arthritic change about the joint. Further, this nerve, typically the largest of all the lumbar nerves, emerges through a foramen that is usually the smallest of all the lumbar foramina; Mitchell noted also that the nerve is sometimes crossed anteriorly by a ligamentous band that stretches from the front of the wing of the sacrum to the body of the last lumbar vertebra.

The rather frequent occurrence of congenital abnormalities of the last lumbar vertebra, and of narrowing of the lumbrosacral disk (found in 26.4% of 500 roentgenograms by Brav and co-workers) combined with the rather limited anteroposterior extent of the lower lumbar intervertebral foramina (Larmon), probably all contribute toward making this region a common source of low back pain.

CONGENITAL ABNORMALITIES

Except for spina bifida, commonly in the lumbar or sacral regions, real anomalies of the vertebral column are rather rare. Disturbances of ossification may of course affect any part of the vertebral column, and lead, for instance, to congenital kyphosis or scoliosis; similarly, fusion may occur between any two adjacent vertebrae, thus restricting the mobility of that part of the vertebral column (although Butler noted that fusions found in adults are not necessarily congenital, but may develop at an early age as a consequence of osteochondrosis), or a hemivertebra may develop and produce congenital scoliosis. Brailsford reported that 0.3% of more than 3,000 vertebral columns of clinical patients contained hemivertebrae. Ehrenhaft said hemivertebrae probably arise as a result of lack of vascularization of the defective side, but that when two adjacent ones are present the defect is probably a result of the failure of one side of the vertebra to shift its metamerism by the usual division and recombination of cranial and caudal parts. The very rare sagittal splitting of vertebral bodies, he said, may result from splitting of them by the notochord; the also very rare division of the bodies into anterior and posterior halves may result from failure of union of the two centers of ossification for each body that he described as normally present in the earliest stages of ossification. Congenital weakness of the cartilaginous plates at the ends of the vertebrae, as a result of improper healing of the defects left by the blood vessels, which pass from the bodies to the intervertebral disks, may lead to multiple herniations of the disk substance

into the spongy bone of the vertebral body, with resulting juvenile kyphosis (Ehrenhaft).

Congenital absence of lumbar articular processes is very rare; Keim and Keagy reported three cases of unilateral absence at the lumbosacral joint.

SPINA BIFIDA

Failure of the two parts of a lamina to fuse in the midline results in a bony defect here, spina bifida or rachischisis, which is relatively common in the lumbar region; usually the defect is slight, and therefore termed "spina bifida occulta" (Fig. 2-40). In larger defects, the meninges may herniate and give rise to a meningocele; in a still more severe condition the spinal cord is also present in the herniated sac, and the defect is associated with a meningomyelocele. In the greatest defect, all the vertebrae fail to close dorsally, and this is known as holorachischisis.

In general, if the bony defect is in the upper lumbar region or above and is large, defects of the spinal cord are frequently associated with it; in the extreme condition, the spinal cord itself, instead of forming a tube, retains its original condition of a flattened plate. Even in the occult type there may be involvement of the nervous system: spina bifida occulta may be associated with a fistulous tract leading from the skin to the dura of the spinal cord, or with a fibrous stalk that leads from a dimple or port-wine discoloration to the dura. Adhesions of the cord and dura at the site of spina bifida have been thought to play a part in the Arnold-Chiari deformity (caudal displacement of the lower part of the brain stem through the foramen magnum, and of the upper part of the spinal cord so that the upper cervical nerves run upward toward their exits), often associated with spina bifida.

Incidence

Spina bifida that involves herniation of the meninges or of those and the spinal cord is said to occur in only about 1 in 1,000 births, but Willis ('23a) found spina bifida, in the anatomic sense of a posterior defect of the vertebral arch, in nine (1.2%) of 748 adult lumbar vertebral columns that he examined. In all of these it was the last lumbar vertebra that presented the spina bifida, although in one the next to the last also was involved; in seven of these cases the last lumbar vertebra was the fifth lumbar, in the remaining two it was the sixth lumbar.

Some degree of spina bifida in the sacral region is still more common, although usually not of clinical importance from the standpoint of involvement of the spinal cord. Trotter found the sacral canal entirely open posteriorly in 2% of male sacra, and in 0.3% of female. In about 25% of sacra there was a posterior defect in the sacral canal above the level of the sacral hiatus, commonly a failure of the halves of the lamina of the first sacral segment to meet in their superior parts, or at all, or smaller defects in the bony walls. She also found that in 47% of more than 1,200 sacra the sacral hiatus was longer than is considered to be normal; the apex of the hiatus most commonly (34%) lies at the level of the lower third of the body of the fourth sacral segment. While such sacral defects may, very rarely, be associated with either a dural sac or a spinal cord that extends abnormally low, they commonly are not; their usual clinical significance, as Trotter pointed out, lies in the field of caudal analgesia, where a particularly long hiatus may permit a needle of the usual length to penetrate the dural sac instead of remaining in the epidural space, and defects in the sacral wall above the hiatus may allow the needle to escape from the sacral canal with the result that the anesthetic is delivered subcutaneously rather than epidurally. Wheeler reported that the first sacral segment is defective posteriorly in about 12% to 13% of columns, the entire sacrum in about 2.9%.

The incidence of spina bifida as observed clinically is definitely higher than that observed in studies of collections of skeletons, apparently because spina bifida occulta in the lumbar (and perhaps in the upper sacral) region may be associated in adult life with symptoms of low back pain. Keiller said, however, that spina bifida occulta associated with symptoms composes only about 5% of all cases of spina bifida. The exact cause of the

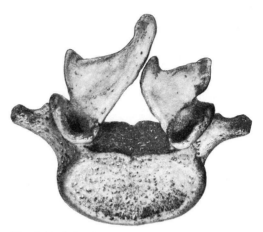

Fig. 2-40. Spina bifida occulta of a lumbar vertebra. (Retouched from Willis TA: Am J Anat 32:95, 1923)

symptoms is not always apparent but may rest in part on weakness of musculature resulting from the fact that failure of fusion of the vertebral arch decreases or abolishes the formation of the spinous process and thus decreases the bony area available for the attachment of muscles. Dittrich pointed out that there may be fibroadipose masses that extend through the defect and so overlie the dural sac as to produce pressure upon nerve roots in the canal; when this occurs the knee and ankle jerks are usually absent or diminished, he said, and limitation of motion is due primarily to muscle spasm.

Wheeler reported spina bifida of the last lumbar vertebra as being manifested in 2.3% of 1,000 roentgenograms of the lumbar columns in adults (almost twice the incidence found by Willis in skeletal material). Brailsford reported a still higher incidence of spina bifida occulta of the fifth lumbar vertebra, namely, 6% of about 3,000 roentgenograms of patients; he also reported this as involving the first and second sacral vertebrae in 11%, a figure much higher than Trotter found in her extensive series.

James and Lassman reported that the first evidence of a neurologic lesion in spina bifida is often in the gait, resulting from overaction of the evertors of the foot; disturbances of bladder and bowel control are, of course, common symptoms. They found traction of some sort to be a more common cause of symptoms than pressure.

Meningocele and Meningomyelocele

Of the cases of spina bifida recognizable at birth because of herniation of the contents of the vertebral canal (this is the type usually reported to occur in about 1 per 1,000 births) meningoceles constitute a relatively small percent (Keiller said about 10%) and the more severe meningomyeloceles (Fig. 2-41) constitute the great majority. Owens reported that in his series 87% were meningomyeloceles, and only 13% were meningoceles.

The most common associated congenital abnormality in Owens' series was clubfoot, this having occurred in about 14%. Fisher and co-workers reported a study of 471 cases of spina bifida (and 59 cases of cranium bifidum) and found hydrocephalus to be the most marked associated anomaly, present in 150 cases; clubfoot was encountered in 25 cases; other congenital anomalies of bones, either of the vertebral column, the ribs, or the limbs were present in 28 cases; congenital dislocation of the hip was present in 6 cases; and anomalies of the urinary tract were found in 21 cases. Miscellaneous congenital anomalies

Fig. 2-41. Diagram of meningomyelocele. (Rasmussen, AT In Baker, AB (ed): Clinical Neurology. Vol 3, New York, Hoeber, 1955)

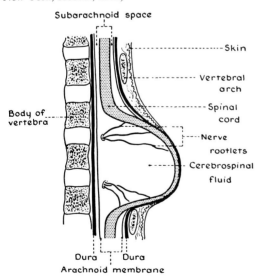

(of the ear, harelip, imperforate anus, and the like) were encountered in 26 cases.

Keiller cited a series of 385 cases of spina bifida in which an operation had been performed and said that 2 cases involved the occipital region of the skull, 9.5% were in the cervical region, 4.5% in the thoracic region, 34% in the lumbar region, 24% in the lumbosacral region, and 23% were sacral. She pointed out that a large proportion of untreated patients with severe defects die within a few months after birth, and that the nonviable patients born with this defect very commonly have associated anomalies such as harelip, ectopia of the bladder, or clubfoot. The series of Fisher and his co-workers involved herniations through the skull other than in the occipital region, but 34 were in this region and 11 in the occipitocervical region. Of 468 hernial sacs limited to the vertebral level, 34 were in the cervical, four in the cervicothoracic, 29 in the thoracic, 14 in the thoracolumbar, 233 in the lumbar, 94 in the lumbosacral, and 60 in the sacral and pelvic regions. Thus about three fourths of the sacs were at least partly in the lumbar region.

Anterior sacral meningocele or anterior spina bifida, resulting from a developmental defect in the vertebral body, may occur either alone or in connection with the common type of spina bifida, although it is rare. Amacher and co-workers, reporting 5 cases of their own, found less than 80 recorded in the literature. Saunders reported a case of combined anterior and posterior spina bifida in which a cleft extended through both the anterior and posterior surfaces of the vertebrae from the first lumbar to the second sacral, and the spinal cord was similarly split, so that an anal fistula opened dorsally through the cleft. He was able to find reports of 37 cases of combined anterior and posterior spina bifida.

CERVICAL ABNORMALITIES

Congenital abnormalities of the cervical region are not particularly common and often are not severe. According to Schrock, fusion of two or three cervical vertebral bodies is not really rare although such fusions seldom give rise to symptoms; he said the most common fusion is between the second and third cervical vertebrae, but multiple vertebral fusions or absence of one or more may produce an abnormally short neck. He also noted that most cases of undescended scapula or Sprengel's deformity (Chap. 4) are associated with abnormalities of the cervical or thoracic vertebral column or of the thoracic cage. Spillane and co-workers found that in patients with symptoms, cervical fusion was usually associated with atlantoaxial fusion.

Malformations of the *atlas* in particular include defects of the anterior arch or of both the anterior and posterior arches (Macalister, 1893; Childress and Wilson); absence of the posterior arch (Brown, '41); partial or complete fusion of the atlas with the occipital bone (Lanier, '39a and b); an accessory suboccipital bone (Lanier, '39b); and fusion of the atlas and axis (Smith).

Defects in the anterior arch are apparently rare, and defects in both arches still more rare. Childress and Wilson found only 24 cases of the latter reported, but said a posterior defect had been reported in about 3%.

Incomplete closure of the posterior arch (Fig. 2-42) was found by Wheeler in 1.4% of 745 atlases, and in about 60% of the cases in which it occurs it is associated with atlantooccipital fusion. Fusion of the atlas to the skull may be more or less complete, but involves at least one lateral mass. The fusion may involve both lateral masses and a midpart of the anterior arch or, more rarely, the entire anterior arch; sometimes, the posterior arch is largely fused in the midline, leaving only lateral foramina for the vertebral arteries. Lanier ('39a) found one case of atlantooccipital fusion among 200 American white and Negro skeletons, and said that some degree of fusion has been reported in 0.5% of 4,000 European skeletons, 1% of 800 Japanese; however, in a larger series of 1,246 American skulls he ('39b) found only two cases (0.16%) of atlantooccipital fusion. Fusion of the atlas and skull is often referred to as "assimilation of the atlas."

Fig. 2-42. An atlas with incomplete posterior arch. (Disse J: Vol 1, Section 1, of K von Bardeleben's Handbuch der Anatomie des Menschen. Jena, Fischer, 1896)

An accessory suboccipital element, either as a free ossicle or as a mass fused to the upper part of the atlas, may intervene between the atlas proper and the skull, and has been interpreted as a remains of an occipital vertebra. Among the 1,246 skulls that Lanier ('39b) studied, 13, or 1.04%, showed some evidence of an "occipital vertebra"; thus it seems to be much more common than is fusion between the atlas and the skull. Fusion of the dens (developmentally the body of the atlas) to the atlas rather than the axis is apparently rare; fusion of it to both bones has usually been regarded as the result of disease, but Smith ('07) reported an apparently congenital case of such fusion.

An anomaly peculiar to the *axis* is failure of the dens to be fused with that bone; this was reported by Todd and D'Errico as having an incidence of about 3 per 1,000 skeletons. The dens may then be entirely separate, or fused to either the atlas or the occipital bone. A separate dens is called an os odontoideum, and has usually been regarded as a congenital anomaly. However, Fielding and Griffin, and Hawkins and co-workers, reported cases in which a normal dens had been fractured in infancy, and appeared later as an os odontoideum. (Sherk and co-workers said the dens is fractured three times as often in young children as in adults, but that the fracture usually heals quickly if the head is kept in hyperextension.)

As with destruction of the dens by disease,

atlanto-axial union is seriously weakened by nonfusion of the dens with the axis, for the articular facets are such as not to resist either anterior or posterior dislocation of the head and atlas upon the axis. Wollin has presented a number of clinical cases of separate dens, and many of Hamblen's cases of occipitocervical fusion were to overcome the instability resulting from this condition. The dens is sometimes hypoplastic or congenitally absent, although Giannestras and colleagues found less than 20 reported instances of congenital absence associated with symptoms. Freiberger and colleagues reported a case in which the major part of the dens of a child disappeared following a fall; if the earlier presence of the dens had not been established by roentgenogram, this obviously might have been regarded as a case of congenital absence. Another anomaly of the dens is failure of its apical part, developed from a separate center of ossification, to fuse with the rest of the structure, forming the so-called *ossiculum terminale* (Sherk and Nicholson).

Fig. 2-43. A last lumbar vertebra, in this case L6, with spina bifida and marked asymmetry. *a* identifies spina bifida occulta; *b* the articular facets between L4 and L5, and the transverse process of L6 articulated to, not fused with, the sacrum; *c,* the horizontal position of the lumbosacral facet on the right side; and *d,* the left transverse process of L6, sacralized and articulating with the ilium. (Courtesy of Dr. H. W. Meyerding)

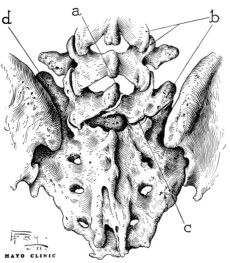

THORACOLUMBAR ABNORMALITIES

Anomalies of the thoracic region are largely limited to such things as hemivertebrae, thus congenital scoliosis; defect or absence of one or more vertebral bodies, producing kyphosis, occasional spina bifida; or partial or complete transformation of the twelfth thoracic vertebra to one of lumbar type.

Willis ('23a), noting that abnormal shortness of a spinous process in the lower thoracic region had been reported as a common variation, said that the eleventh thoracic vertebra fairly frequently, and the twelfth more rarely, shows a defective or absent spinous process even though the laminae have fused. He regarded it as an imperfect form of spina bifida.

The common congenital abnormalities in the lumbar region are spina bifida and variation in the length of the lumbar column, both of these conditions having already been discussed. Spondylolysis (separate or separated neural arch, interarticular or isthmal defect), once regarded as probably congenital in origin, has not been shown to be this, and is therefore discussed in a following section. It has already been noted that sacralization of the last lumbar vertebra may be complete, or may involve only one side, and that the last lumbar vertebra is also the one most often involved in spina bifida occulta. Willis ('23a) noted that it is otherwise also somewhat variable, for its transverse processes may be much enlarged, and occasionally (he found this in 5 of 748 columns) may make actual contact with the sacrum, even though there is no fusion between the adjoining elements. Later ('59) he emphasized the variations in the articular processes and facets. The latter are not necessarily symmetrical, and a number of variations and anomalies may exist simultaneously (Fig. 2-43).

An unusual effect of a lumbar anomaly was reported by Marr and Uihlein; in this, a bony spicule projected posteriorly from the body of the third lumbar vertebra and involved a low-lying spinal cord. The cord was bifid, completely divided in this region by the spicule (see Fig. 2-93).

SACRAL AND COCCYGEAL ABNORMALITIES

The common anomaly of the sacrum is spina bifida, involving the upper part of the sacrum, the lower part so as to elongate the sacral hiatus, or the entire sacrum.

Very rarely, a lower portion of the vertebral column (the sacrum and coccyx, or these and more or less of the lumbar vertebral column) may be completely absent. Sinclair and co-workers reported a case of absence of both the lumbar and sacral portions of the column, and found 41 previous cases (in which the patients lived) reported in the literature. The condition is typically associated with other skeletal anomalies, especially clubfoot; with the loss of control of voluntary muscle below the level of the anomaly and therefore with general atrophy or underdevelopment of muscles in the lower part of the body and the limbs; and often with incontinence of urine and feces. Clubfoot probably and the other defects certainly indicate concomitant anomalies of the nervous system. The extent of visceral involvement may be incompatible with life.

Blumel and co-workers reviewed 50 cases of absence, partial absence, or malformation of the sacrum, and discussed the frequency of associated anomalies; in 22% of these patients congenital anomalies of some sort were said to have been found also in their kinspeople.

An anatomical and phylogenetic curiosity, of no clinical importance, is the very rare existence of ventrally placed arches in connection with the coccyx. These represent the hemal arches or chevron bones of lower vertebrates; Schultz, reporting a case in adult man, found only 4 previously recorded cases.

ABNORMAL SPINAL CURVATURES

Abnormal spinal curvatures are particularly complicated, and often involve two or more curves and an associated rotation of the vertebral column. For purposes of simplicity, however, they are usually described as ky-phosis (humpback), an abnormal forward curvature of the vertebral column with a dorsal prominence; lordosis, an abnormal posterior curvature of the vertebral column, typically an exaggeration of the lumbar curve; and scoliosis, originally any abnormal curvature of the vertebral column, but now used primarily to describe a lateral one. Any of these three types may be congenital; may be of developmental origin, but make its appearance during a period of active growth of the column after birth; may result from trauma, such as the collapse of a vertebra; or, in the case of lordosis and scoliosis, may be a habitually maintained posture to restore an otherwise abnormal relationship to the center of gravity, as scoliosis may compensate for shortness of one leg. In addition, scoliosis may result from muscle imbalance or from the thoracic distortion produced by thoracoplasty.

KYPHOSIS

Kyphosis typically occurs in the thoracic region, where the center of gravity lies in front of the vertebral column, and the tendency of the weight of the body is therefore to produce an anterior collapse of the thoracic curve. Thus a mild degree of kyphosis is often encountered in elderly persons, here typically involving the whole length of the thoracic vertebral column, and a marked kyphosis is produced when a pathologic process destroys or seriously weakens the bodies of one or more vertebrae; kyphosis was formerly seen often as a result of destruction of vertebral bodies in tuberculous patients. Ferguson attributed preadolescent kyphosis to anterior collapse of vertebrae consequent to failure of ossification in the region of exit of large venous channels from the vertebrae.

Congenital kyphosis, according to Bingold, is of two types: one, seen in chondro-osteodystrophy, cretinism, gargoylism, and achondroplasia, is associated with a widespread disturbance of ossification; the other is the result of a localized malformation of the vertebral column while the rest of the skeleton is normal. Bingold reported that the latter type of deformity is apparently more common in fe-

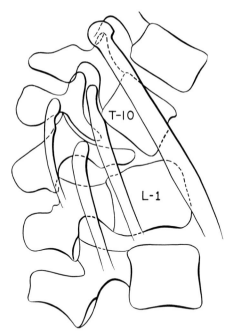

Fig. 2-44. Tracing from a roentgenogram of congenital kyphosis resulting from defective bodies of the last two thoracic vertebrae. (Redrawn from Bingold AC: J Bone Joint Surg 35-B:579, 1953)

males, that no cause is known, and that there is no real evidence that it has a familial basis. Of the three cases he presented, the deformity in one was caused by the absence of the body of the first lumbar vertebra and the presence of only a rudimentary body for the twelfth thoracic vertebra; in the second both the eleventh and twelfth thoracic vertebral bodies were rudimentary (Fig. 2-44); and in the third the body of the second lumbar vertebra was rudimentary. Bingold said that operative correction of the deformity had apparently not been reported; if the angulation is severe enough, however, there may be evidence of damage to the spinal cord. He said that minor degrees of paralysis do not require treatment, but that if severe symptoms develop, such as inability to walk or retention of urine, some effort should be made to decompress the spinal cord. He suggested laminectomy with decompression of the cord by longitudinal and transverse incisions of the dura that do not damage the arachnoid; Love ('56) has in some such cases "transplanted" the spinal cord (*i.e.,*

removed appropriate portions of the vertebral column at the angle of both kyphotic and scoliotic curves) so as to shorten the course of the spinal cord and relieve the compression of it at the angle.

LORDOSIS

Lordosis is commonly purely a compensatory curve, as is the normal lumbar curve of which it is an exaggeration. Thus young children, in whom the center of gravity is apparently farther forward than in the adult, typically show more lordosis than do adults; similarly, with kyphosis in the thoracic region and therefore a forward shifting of the center of gravity, a compensatory lordosis develops. In the same way, the lumbar curvature varies with the obliquity of the sacrum, as already mentioned; thus if the sacrum is more nearly horizontal than usual, whether developmentally or as a result of gradual downward rotation of the pelvis, the lumbar curvature increases; as already noted, it has been said that if the intersection of a line through the middle of the body of the fifth lumbar vertebra with a horizontal line through the lumbosacral disk fails by as much as 25° to approximate a right angle, the lumbar curvature is always abnormal enough to be designated as lordosis. Developmentally, lordosis results from an accentuated wedging of the lumbar vertebrae.

SCOLIOSIS

Scoliosis is the most complicated of the abnormal curvatures and, as emphasized by a great number of writers, is characterized by the fact that there is not only a lateral bending but also a rotation of the vertebral column around its longitudinal axis. Further, scoliosis is often associated also with kyphosis and lordosis (*e.g.,* Hauser). It may be divided into two general types, functional and structural.

Functional scoliosis is often temporary and may result, for instance, from unilateral spasm of the muscles of the back. It may also be a compensatory curve to establish a new line of gravity over a laterally tilted pelvis, such as results from shortness of one limb.

However, if it persists for a long time, especially in a growing individual, a functional scoliosis will be converted into a structural one. In structural scoliosis, as the term implies, a structural change—wedging—of the vertebral column establishes a permanent deformity.

A single scoliotic curve may be present, but usually additional compensatory curves, in the opposite direction, also develop. The secondary or compensatory curves result from an effort to center the head over the sacrum and maintain the shoulders and pelvis level as much as possible. Cobb pointed out that in a proper definition of a primary curve the important factor is not that it appeared first or is more severe, but that it shows a structural change, while a secondary curve is physiologic or functional. Thus, there may be more than one primary curve.

Von Lackum and Miller pointed out that since practically all patients with untreated scoliosis attain an erect posture through the development of two compensatory reversed curves to balance the single primary curve, correction of the primary curve should not be carried beyond the point to which the secondary curves can reverse themselves spontaneously. They said that most difficulty following operation arises from too much correction, not too little, and that in the average fully matured vertebral column the secondary curves can reverse themselves by only about 20°. Evarts and co-workers have warned that any sudden correction of the curvature may so increase the acuteness of the angle of origin of the superior mesenteric artery as to cause serious duodenal compression.

Scoliosis With Paraplegia
Scoliosis may, rarely, be associated with paraplegia as a result of compression of the cord; McKenzie and Dewar reported 5 such cases in 1949, and found 41 previously recorded ones. In all these the deformity was in the thoracic region, and as a result was associated also with marked kyphosis. They noted that in the patients operated upon the cause of the compression was apparently almost always a combination of a tightly stretched dura and a

sharply angulated vertebral canal, and suggested that the cause of the paralysis may be a rapid growth of the vertebral column with an inability of the dura to accommodate itself to such growth (a rapid increase in the scoliosis would produce the same result). The problem is obviously similar to that produced by kyphosis. They reported that striking results have been obtained by laminectomy with section of the dura and, sometimes, division of the denticulate ligaments and of nerve roots that seemed to be abnormally tight.

Kerr in 1953 reported an additional case of paraplegia associated with scoliosis, and found 47 reported cases up to that time. He noted that in the majority of cases the paraplegia occurred during the period of most rapid growth, between the ages of 11 and 19 years. In his case the scoliosis was not idiopathic, but rather resulted apparently from neurofibromatosis that had affected the bone; spinal fusion checked the progress of scoliosis, and also alleviated the symptoms of the paralysis. Spinal fusion, he said, has generally not been regarded as necessary to relieve the paraplegia of scoliosis.

Origin of Scoliosis
Scoliosis may be congenital, resulting, for instance, from the suppression of half of a vertebra (Fig. 2-45) or from other defective development of one or more vertebrae. Most commonly, however, it develops during the growth period, and this type, because of uncertainty of its cause, has been known as idiopathic scoliosis.

It is generally recognized that scoliosis results from unequal growth of vertebrae, producing wedging, and that once this process has started it will tend to worsen because the initial wedging allows more weight to be transmitted to the concave side and less to the convex side, thus retarding epiphyseal growth on the concave side, favoring that on the convex side. Nachlas and Borden, for instance, induced scoliosis in dogs by stapling adjacent vertebral bodies together on one side to produce pressure on the growing epiphyses. It is this also that accounts for the conversion of what was at first a functional scoliosis, result-

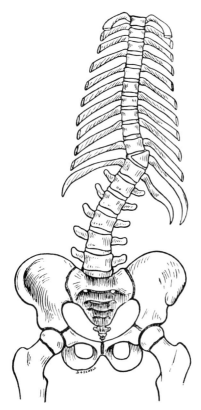

Fig. 2-45. Congenital scoliosis resulting from a hemivertebra. (Wakeley CPG: J Anat 57:147, 1923)

ing, for instance, from dysplasia of the hip, into a structural scoliosis.

It has long been known that poliomyelitis can cause scoliosis as a result of the muscle imbalance it may produce, but there has been much difference of opinion as to how often it, or any muscular imbalance, is a cause. Gruca regarded the primary cause of "idiopathic" scoliosis as weakening of the muscles on the convex side of the primary curve and advocated early restoration of muscle balance as the operation of choice. Zuk also thought he demonstrated electromyographically that the muscles on the convex side are weak. The role of the musculature in the production of scoliosis is apparently now more widely accepted than it once was; Risser, who once suggested metabolic abnormalities resulting from dietary deficiencies as the cause of "idiopathic" scoliosis, has adopted the view that muscle weakness (perhaps sometimes spasm), most

often caused by poliomyelitis, is the common cause. Wynne-Davies emphasized a genetic factor, particularly strong for the scoliosis developing in girls during adolescence.

Risser and Ferguson emphasized that uncorrected scoliotic curves, whether congenital, paralytic, or idiopathic, gradually become worse during the period of active growth in length of the vertebral column, but that this increase in curvature always ceases with the cessation of growth in length. Further, the rapidity of the increase in curvature was found to be closely correlated with the rapidity of growth of the column.

Risser ('64) has again emphasized that cessation of growth of the vertebral column in any one individual coincides rather precisely with the completion of ossification of the epiphysis (apophysis) of the iliac crest, from the anterolateral border of the ilium to its junction with the sacrum; this occurs at close to 14 years of age in girls, 15½ years in boys. Thus the full development of the epiphysis of the iliac crest (Risser's sign) is of great prognostic value, for it indicates that the scoliotic curve has reached a static condition. If the progress of the curve is to be halted, spinal fusions must be done before this time. Indeed, Risser and Norquist suggested that since preventing increase in the curve is easier than correcting it later, mild curves should be considered for treatment.

Congenital Scoliosis

Kuhns and Hormell said that there is a type of congenital scoliosis that is caused by muscular and ligamentous contracture, but that this is rare, and the common type is caused by congenital abnormalities of the vertebrae. Apparently, numerical variation alone is not often a cause of scoliosis. The vertebral deformity may vary from one involving a single segment to one involving the entire column, and in their series was most frequently in the thoracic region. Bony deformities in their cases were rarely confined to the vertebral column only, however, and the greater the amount of vertebral deformity, the more frequently were other congenital deformities also found. Forty-six of 170 patients had neurological disturbances, the most common being shortening and atrophy of one leg.

Winter and his colleagues have more recently reviewed the morphology and treatment of a still larger group of children with scoliosis caused by vertebral defects. In their cases also the apex of the curve was most commonly in the thoracic region and was frequently accompanied by cardiac anomalies. When it was at the cervicothoracic junction there were often congenital abnormalities of the upper limbs and the heart, and deformities in the lumbar region tended to be associated with anomalies of the lower limbs or of the genitourinary tract. In their experience, an "unsegmented bar," or unilateral failure of segmentation between two or more vertebrae, is most likely to cause significant progressive scoliosis.

Idiopathic Scoliosis

James regarded a combination of the site of the primary curve and the age of onset as giving the best guide to prognosis in idiopathic scoliosis, but the location of the primary curve as the more important. In regard to age, he pointed out that there are three peak periods in the establishment of abnormal primary curves: the infantile, below the age of 3 years, the juvenile, between the ages of 5 and 8, and the adolescent, from age 10 until the end of growth.

James reviewed the cases of 79 persons with lumbar scoliosis followed to maturity, and found that except for one in whom the curve appeared at the age of 9, the primary curvature invariably developed during the adolescent years of 10 to the end of growth. Among these, 70 were females and 9 males. The frequency of curves to the left and to the right (scoliotic curves are described according to their convex side) was almost equal; the apex was most commonly at L1, but fairly frequently at L2 or L3, and the curve was usually less than 70°. Lumbar curves, he found, produce relatively little deformity, because the degree of curvature remains small, no ribs are involved in the rotation, and the shoulders remain level. He did note, however, that backache is a common sequel of lumbar

scoliosis; this presumably arises in the synovial joints as a result of osteo-arthritis consequent to the rotation forced upon the vertebral bodies, in spite of the fact that the articular surfaces are not adapted for much rotation. Somerville also noted that deformities resulting from curves in the lumbar region are generally less severe than those resulting from higher curves, but that pain is a noticeable feature.

James studied 26 cases of idiopathic thoracolumbar scoliosis, noting that this is not a common curve; in one of these the onset was at the age of 3 years, in another at 4 years, but in the remainder the curve developed during adolescence, as do most lumbar curves. Twenty of the 26 curves were in girls and most of them were convex to the right (the apex involved the eleventh or the twelfth thoracic vertebra). In general, James noted, the prognosis as to the extent of deformity becomes worse as the apex of the curve ascends; thus in thoracolumbar curves some ribs are always involved in the rotation, the shoulder sometimes drops, and sometimes the hip becomes extremely prominent so that a slight curve may be particularly disfiguring. Among his cases, he classed 31% as being severe, between 70° and 99°; only 2, however, were believed to require correction and fusion. Of 9 adults, 8 complained of pain.

Thoracic idiopathic scoliosis was said by James to be common and also most important. He studied 134 such curves in adults, and found that 66 began during adolescence, 16 during the juvenile period, and 52 dated from the infantile period. The earlier thoracic scoliosis commenced, the worse the prognosis. Widely varying figures have been reported in regard to the incidence of thoracic scoliosis to the right and to the left; James said that thoracic scoliosis in infants is more common in boys than in girls, and is convex to the left in 92% while such scoliosis beginning in adolescence nearly always affects girls and is convex to the right in 90%. He noted that since the deformity is accentuated by distortion of the thorax (Fig. 2-46), dropping of the shoulder, and prominence of the hips, the radiological measuring of the curve does not provide an

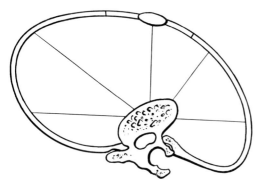

Fig. 2-46. Diagrammatic cross section showing the distortion of the thorax at the level of a thoracic scoliosis. (Redrawn from Taylor RS: Surgery of the Spine and Extremities. Philadelphia, Blakiston, 1923)

Fig. 2-47. Thoracolumbar scoliosis. Note the marked rotation of the vertebral column as indicated by the positions of the spinous processes. The plumb lines indicate the differences in the center of gravity of the head and that over the sacrum. (Huc G, Brisard P: *La Scoliose*. In Ombrédanne L, Mathieu P (eds): Traité de Chirurgie Orthopédique, Vol 2. Paris, Masson et Cie, 1937)

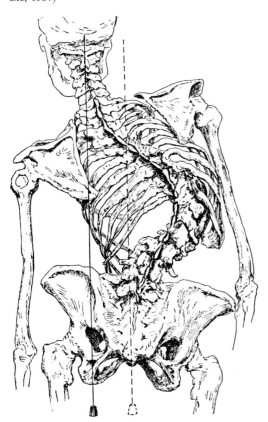

actual picture of the extent of the deformity. Somerville pointed out that since thoracic scoliosis involves stress on the ribs and vertebral bodies rather than on the articular processes, pain is not a prominent feature, as it is likely to be in lumbar scoliosis, in spite of the greater deformity.

Rotatory Component of Scoliosis

Much has been written concerning the exact mechanisms that produce rotation in connection with lateral deviation of the vertebral column.

Arkin ('49, '50) regarded two factors as contributing to the convex-side rotation of scoliosis: one is the wedged shape of the vertebral bodies, which tend to be squeezed toward the convex border, the greater side of the wedge. The other is the force exerted upon the vertebrae, especially the spinous and transverse processes, by the muscles and ligaments; these, of course, tend to follow a straight line rather than a curve, and therefore tend to pull the spinous processes toward the concave side. If the muscles on the convex side are often weak, as is now claimed, this pull would appear to be also a factor initiating rotation. (It will be remembered that the convex side of the scoliotic curve is occupied by the vertebral bodies, therefore weakness of the muscles here would allow the spinous processes to be pulled to the concave side.)

Somerville found the primary curve not so simple as a lateral flexion with rotation, and noted that scoliosis has also been defined as rotation plus lordosis. His analysis followed this line, in general: in extreme scoliosis a posterior view of the vertebral column shows the vertebrae at the apex of the curve almost in a lateral view, with their bodies on the convexity of the curve and their arches in the concavity (Fig. 2-47). In these cases it is obvious that the curve is a lordosis with a great deal of rotation also present. However, he said, the curve is still more complicated. Although early stages in the development of scoliosis show in lateral view a small but definite midthoracic lordosis, the rotation later superimposed upon this produces an apparent kyphosis as viewed from the side. By the time the rotation has been developed to 70° or more, what is seen from a lateral view as kyphosis is obviously really a lateral flexion of the vertebral column, in regard to the vertebrae involved; thus the deformity actually consists, according to his analysis, of lordosis, rotation, and lateral flexion, and he preferred to speak of it as rotational lordosis.

Somerville pointed out that the easiest way for a localized structural lordosis to develop would be through failure of growth of the posterior elements of the vertebrae. To compensate for such a slowly developing structural lordosis the vertebral column above and below could be flexed forward, at first only a little bit. Later, however, further forward flexion would be impossible without changes in the shapes of the bones; then rotation of the affected part of the column allows lateral flexion of the normal parts above and below to center the head once more above the pelvis. Somerville simulated those movements with a model and produced it in some animals by cauterizing the vertebral arches and wiring them together.

Risser and Roaf have both agreed with the basic concept that rotation occurs because anterior parts of the vertebrae are growing faster than posterior parts. Thus, at least in the thoracic region where the articular processes resist dorsiflexion, the vertebral bodies have to rotate in order to accommodate their greater length. As Roaf noted, this does not apply as much to the lumbar region, where part of the growth can be accommodated by increased lordosis. Thus, a lumbar scoliotic curve would be expected to be less severe, as James reported it to be, than a thoracic one.

CHORDOMA

The vertebral column may, of course, be invaded by a number of tumors, but chordoma is unique in that it is believed to arise from remains of the notochord (Fig. 2-48). Sensenig pointed out that adhesions may occur between the notochord and the neural tube (or notochord and pharynx) in early stages of development, and that during further develop-

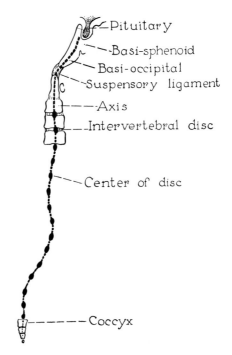

Fig. 2-48. Possible remains of the notochord in the adult, along its normal course. (After Keith, from Coventry MB: Am Acad Orth Surgeons, Instructional Course Lectures 6:218, 1949)

ment, as the site of adhesion is drawn out into a band between the notochord and the neural tube and subsequently separates from both, bits of it may be left within the vertebral centra and the occipital bone, within the vertebral canal, or the cranial cavity. It is presumably from such abnormal remains, which may of course be absorbed, that chordomas develop.

Richards and King noted that chordomas usually arise within or adjacent to vertebral bodies, and not from the normal remains of the notochord in the intervertebral disk. The greatest number of chordomas are found either in the sphenooccipital or the sacrococcygeal region. Sensenig noted that chordomas occur most often in the regions in which the developing notochord shows most variation; he quoted a series of 317 cases in which 37% of the chordomas were cranial ones, only 13% were in the segmented portion of the vertebral column, and 50% were caudal. Zollinger, reporting a chordoma of the third lumbar vertebra that had compressed the spinal cord, said pain is one of the early symptoms of this tumor when it is in a vertebral body.

Hass reviewed the literature on chordomas of the cranium and of the cervical portion of the vertebral column, and pointed out that the majority have been first identified during the third, fourth, or fifth decade, and are more frequent in males than in females in about the ratio of 3 to 2. The majority of cranial chordomas, he said, arise in the region of the sphenooccipital synchondrosis, and are usually located in the midline beneath the dura; if they extend ventrally, it is usually to the nasopharynx. Tumors arising from the cervical vertebrae usually grow anteriorly to involve retropharyngeal tissues.

FRACTURE AND DISLOCATION

Fracture

Fracture of the vertebral column most often results from an automobile accident or a fall, and the most common injury is an anterior compression fracture (Fig. 2-49) of one or more vertebral bodies resulting from forced hyperflexion. (*e.g.,* Davis). If the injuring force exerts a strong rotatory component, there may be a lateral wedge fracture of the verte-

Fig. 2-49. Simple compression fracture (*arrow*) of a lumbar vertebra

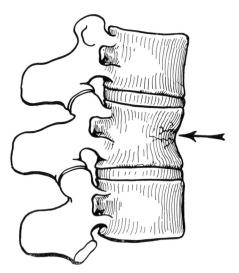

bral body and perhaps a fracture of the vertebral arch (Nicoll). A lower thoracic or an upper lumbar vertebra is most likely to be involved (usually only a single vertebra is), because of the greater leverage here and the change in mobility between the semirigid thoracic and the more mobile lumbar portion; the lower part of the cervical column is next most frequently injured (Davis).

Any fracture of the vertebral column may of course result in sufficient dislocation to injure the spinal cord or spinal nerves; thus while fractures of the cervical region are less common than are fractures of the thoracic and lumbar regions, they are also much more dangerous. Dunlop reported there was little or no damage to the spinal cord or nerves, as reflected by paralysis, in more than half of all fractures of the vertebral column at the time of his writing (1937). This figure may be a reflection of the slower automobile speeds at that time. In such fractures, maintenance in hyperextension is extremely important (Thomson) since flexion tends to induce an anterior dislocation with consequent damage to the spinal cord, while hyperextension splints the injured vertebra between its articular processes on the one hand and the strong anterior longitudinal ligament on the other.

Flexion fractures of the vertebral column necessarily throw a strain upon posteriorly located structures. According to Davis, the capsules of the posterior (interarticular or synovial) joints are most frequently injured in association with crush fractures; other associated injuries in descending order were said to be tears of the interspinous ligaments, of the posterior longitudinal ligament, and of the posterior fibers of the anulus fibrosus of the disk. Davis listed complete dislocation with severance of the spinal cord as being next most frequent to rupture of the posterior part of the anulus, followed in descending order by unilateral dislocation, bilateral dislocation, fractures of the vertebral arch, and fracture-dislocation (Fig. 2-50) with or without paralysis.

Of 166 fractures among 152 patients studied by Nicoll, 66% were in the twelfth thoracic or the first two lumbar vertebrae. Of these, anterior wedge fracture constituted 58%, lateral wedge fracture 14%, fracture-dislocation 19%, and fracture of the vertebral arch without wedging, 9%. Associated fracture of transverse processes occurs in about half the cases of fracture of the arch. Nicoll pointed out that anterior wedging cannot be severe unless posterior ligaments are ruptured and there is at least a subluxation at the zygapophyseal joints. It is this type of fracture, of course, that can be easily converted accidentally into a fracture-dislocation with possible damage to the cord. In Nicoll's series, 62% of the fractures were associated with damage to the spinal cord or the cauda equina.

Jefferson reported four cases of fracture of the atlas, and reviewed the literature on this subject. He found that while injury to the cord was not produced in about half the cases, in only slightly less than half there was an injury severe enough to be fatal. Even when the cord escaped damage, he found, the greater occipital nerve was very frequently injured. According to Dunlop, if the fracturing force is delivered to the head while the head and neck are hyperextended, the fracture is likely to occur in the atlas or the axis; if they are flexed at the time, it is likely to be in or below the third cervical vertebra.

Fractures through the arch of the axis, separating the superior and inferior articular processes and thus allowing the body of the axis to be displaced on the body of C3 are the typical result of judicial hanging (Wood-Jones), but may also result from other causes. Brashear and co-workers found automobile accidents to be the most common cause in their 29 cases.

Dislocation

Except in the cervical region, dislocation without associated fracture is practically impossible, because of the interlocking of the articular processes. Dislocation between the atlas and the axis may occur as a result of congenital absence or fracture (Fig. 2-51) of the dens (*e.g.,* Hamblen, and Blockey and Purser), or as a result of injury to the transverse ligament of the atlas (Davis); the lateral

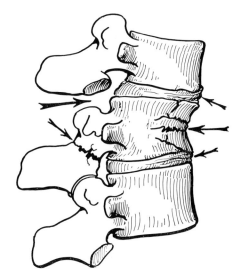

Fig. 2-50. Comminuted fracture of the vertebral body with injury to the disks, fracture of the arch, and dislocation (*arrows*).

synovial joints cannot by themselves prevent even a gradual dislocation between these two vertebrae, but according to Blockey and Purser fractures in children are really epiphyseal separation and usually heal nicely, and even in adults there may be bony healing or enough firm fibrous union to prevent excessive movement or subluxation. Echols and Kleinsasser said that dislocation of the atlas upon the axis may also occur with intact dens and transverse ligament, as the result of such excessive rotation of the head that the articular facets of the atlas slip off the corresponding facets of the axis. They reported such a case, and said that about 27 similar cases had been reported. Obviously, atlantoaxial displacement may cause immediate death or very serious damage to the spinal cord, but moderate displacements here are better tolerated because of the particularly large size of the vertebral foramen of the atlas. Wadia has reported a number of cases, with varying degrees of neurologic involvement.

Dislocation without fracture is possible between the remaining cervical vertebrae because of the slope of the articular processes. Injuries to the cervical portion of the spinal cord, even severe enough to produce instantaneous death, have been reported to occur in

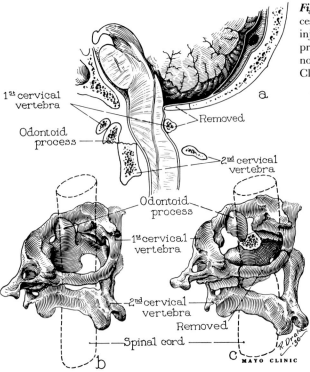

Fig. 2-51. Fracture of the dens (odontoid process) with dislocation of the atlas and skull and injury to the spinal cord. The method of decompression is indicated in *a* and *c*; *b* is a view of the normal course of the cord. (Craig, WMcK: S Clin North America [Aug] 1931, p 841)

the absence of fracture of the vertebral column.

Damage to the Cervical Cord in Flexion and Extension Injuries

It is generally recognized that dislocation or fracture-dislocation of the vertebral column is the most common cause of injury to the cervical portion of the spinal cord resulting from violence. Among 22 paraplegic patients Barnes found 15 flexion injuries and 7 hyperextension ones. Of the injuries produced by hyperflexion, 8 were said to have been produced by dislocation, 4 by crush fracture of a vertebral body, and 3 by acute retropulsion of an intervertebral disk. Of the 7 injuries from hyperextension, 6 occurred in arthritic vertebral columns, the damage to the cord was usually not so severe as in flexion injuries, and the damage to the vertebral column was said to be a rupture of the anterior longitudinal ligament or avulsion of it with a fragment of bone.

It has, however, been generally recognized that there is not necessarily a correlation between the extent of damage to the spinal cord and radiological evidence of damage to the vertebral column in many injuries once regarded as flexion ones. The absence of fracture or dislocation in severe injuries to the spinal cord has been explained as being due to a dislocation during flexion, with spontaneous reduction; but Barnes apparently showed on the cadaver that once anterior dislocation occurs, the articular processes so lock that spontaneous reduction is impossible. It has also been blamed upon acute retropulsion of an intervertebral disk with immediate return of the disk to its normal position, but whether such rupture occurs or not retropulsion cannot be a factor in those cases in which necropsies have shown that both the posterior longitudinal ligament and the anulus were intact and that the lesion of the cord was dorsal rather than ventral. It is now generally accepted that many severe injuries to the cord in the absence of apparent damage to the vertebral column are actually flexion-extension ("whiplash") injuries.

Rupture of the anterior longitudinal ligament may occur in hyperextension injuries, but leave no roentgenological evidence of dislocation. As reported by Berkin and Hirson, this may allow the vertebra above the tear to move backward and downward on the vertebra below during hyperextension sufficiently to force the spinal cord against the lamina and ligamentum flavum of the lower vertebra and yet assume a normal position during flexion. However, Taylor regarded rupture of the anterior longitudinal ligament as not the usual cause of injury in hyperextension, and he described a case in which the cord was nevertheless damaged in its dorsal half. He showed that in cadavers forced hyperextension could cause a forward bulging of the ligamentum flavum so great as to occupy about 30% of the anteroposterior space in the vertebral canal and suggested that injury by the ligamentum flavum is a common cause of damage to the cord without demonstrable dislocation. He also cautioned that since there is no way of telling whether or not the anterior longitudinal ligament is ruptured, these injuries should be treated by immobilization in slight flexion rather than in the extended position used for most injuries to the vertebral column. Payne and Spillane agreed that the ligamentum flavum does indeed bulge forward in hyperextension, and they said that with degeneration of the disks there may also be folds on the anterior wall of the vertebral canal.

SPONDYLOLYSIS AND SPONDYLOLISTHESIS

The anatomical abnormality called "spondylolysis" is difficult to classify. It is exceedingly rare except in the lumbar region, and usually involves the fifth lumbar vertebra only. However, the bony defect of spondylolysis, which disposes to the anterior displacement of the affected vertebra, termed "spondylolisthesis," has not been shown to be congenital; at the same time, it hardly fits into the classification

of fracture, for it shows none of the characteristics of a true fracture. The present discussion, unless otherwise specified, deals only with the lumbar region, since only here is the defect frequent enough to have been studied carefully.

SPONDYLOLYSIS

Spondylolysis is a defect in the bony continuity of the lamina of a vertebra, either on one or both sides. The defect passes between the superior articular process and the remainder of the lamina; it is caudal to the pedicles. As a result, in a bilateral defect the superior articular processes are attached to the vertebral body but the major portion of the lamina, including the spinous and inferior articular processes, are detached (Fig. 2-52). The abnormality has therefore been known, when it is bilateral, as separate neural (vertebral) arch; or, since the defect is between the articular processes, it has also been called interarticular defect or isthmal defect, the pars interarticularis or isthmus being defined as the region between the superior articular processes and the rest of the lamina. This is the narrowest part of the lamina, but according to Krenz and Troup its cortical bone is

Fig. 2-52. Bilateral spondylolysis, seen from above. The separated arch bears the inferior articular processes. (Stewart TD: J Bone Joint Surg 35-A:937, 1953)

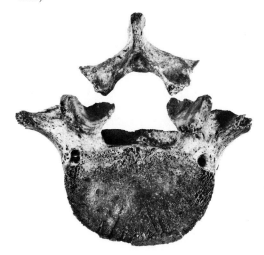

thicker anterolaterally and posteromedially than that of other parts of the lamina, and its trabeculae appear to be stronger.

As will become apparent, the incidence of spondylolysis varies widely according to age, sex, and especially race; Stewart reported an incidence as high as 50% in skeletons of some groups of Eskimos.

Spondylolysis was found by Willis ('23a) in about 4.13% of 748 vertebral columns of American whites and Negroes; in 23 of the 31 cases the defect was bilateral, and in 8 it involved only the right side. In all but 2 of the cases the defect was in the last lumbar segment (which in 3 cases was a 25th segmental—sixth lumbar—vertebra); in the 2 exceptions the defect was in the third lumbar vertebra.

Roche and Rowe ('51a, '52) reported a study of 4,200 skeletons in which they found an isthmal defect in 4.2%. However, when they divided the skeletons on the basis of race and sex, they found that the white male skeletons showed the defect in 6.4%, Negro male skeletons in 2.8%, white female skeletons in 2.3%, and Negro female skeletons in 1.1%. More than one sixth of the defects were unilateral, and when so they were twice as frequent on the right side as on the left. Two of the separations involved the lamina of a sixth lumbar vertebra, 156 involved the fifth, 17 the fourth, 5 the third, and 3 the second; there were no such defects in the first lumbar vertebrae of their series. There seemed to be no correlation between the incidence of defective arch and the length of the vertebral column above the sacrum, for of those vertebral columns that did show the interarticular defect 92.5% had 24 presacral vertebrae, 2.5% had 23, and 5% had 25 (approximately the finding in unselected series of vertebral columns). However, spina bifida occurred much more frequently in vertebral columns showing separate arch than in those that did not.

Origin

The early clinical literature on spondylolysis and spondylolisthesis has been reviewed by Chandler, that on the defect of the arch by Rowe and Roche and by Roche and Rowe;

based upon early reports that sometimes the arch of the fifth lumbar vertebra, and occasionally that of the fourth, presented two centers of ossification on one side, the origin of spondylolysis was once generally believed to be a failure of fusion of the two centers. If it were an embryological abnormality, spondylolysis would have to arise during an early stage of ossification and should be detectable thereafter.

Roche and Rowe ('51b) and Rowe and Roche ('53) studied the ossification of the vertebrae, and reviewed the previous literature. Among 509 fetal vertebral columns that they or others studied, there was no report of double centers of ossification that could give rise, through failure to fuse, to a separate vertebral arch. Thus, even if one assumes that the incidence of defective neural arch is no more than 2% (with a variation from 6.4% in the white adult male to 1.1% in the Negro female) there seems to be little chance that separate arch is of embryological origin.

Roche and Rowe did find one report of a separate center of ossification for the right inferior articular process of both the second and third lumbar vertebrae. However, they said, the centers were so placed that if fusion between them had failed to occur the inferior articular processes would have been separated from the entire remainder of the arch, rather than these and most of the lamina being separated from the superior articular processes. Obviously, therefore, the accessory centers that have been reported could not be responsible for the usual defect.

They also (Rowe and Roche, '50) pointed out that in oblique roentgenographic views of the lumbar portion of the vertebral column of children between the ages of 3 and 6 years there is often visible a line that suggests a possible separation of the arch; they identified it in about 60% of the vertebral columns that they examined in their study. They were able to show, however, that the line was actually the synchondrosis between the pedicle and the body, normally not fused in this age group; when viewed from the proper angle, this may be so superimposed upon the lamina as to appear to be a laminal defect.

Embryological evidence that spondylolysis has its origin in abnormal centers of ossification seems therefore to be completely lacking, yet sufficient specimens have been carefully examined to reveal a number of such cases of double centers of ossification if they actually exist. There is other evidence, indirect but strong, that separation of the arch is not congenital. Some indirect evidence was introduced by Roche through dissection in a case of interarticular defect; it would be expected that if the defect was caused by a failure of fusion of centers of ossification, the gap should be bridged by cartilage or fibrocartilage, representing the unossified portion, and this has often been assumed to be the condition; however, Roche found complete absence of any union on one side, and on the other side the defect was bridged by some strong fibrous ligaments that were apparently derived from the ligamentum flavum; no evidence of any cartilaginous union was seen on either side. Further, Rowe and Roche ('53), having failed to find any instance of separation of the arch at birth, found that its incidence in children is about half that of adults, this also indicating that the lesion is acquired rather than congenital. In their specimens, however, they found no evidence of increased incidence between the years of 20 and 80.

Stewart found even stronger statistical evidence that the defect is an acquired one, for he studied skeletons of Eskimos, in whom the incidence in adults is known to be much higher than it is in Americans. In his study, he found that skeletons from different areas in Alaska varied considerably in the incidence of the defect, and that adult skeletons from some locations showed an incidence of almost 50%. In analyzing his skeletal material from the standpoint of the approximate age of the individual, he found that the defect occurred in young individuals usually only in the fifth lumbar vertebra, where it might be unilateral or bilateral; involvement of more than one vertebra began to appear between the ages of 19 and 24, and multiple defects were most frequent beyond the age of 31. In addition to unilateral or bilateral spondylolysis, he also found unilateral defects through the pedicles

after 13 years of age, but he found only one case in which the pedicles were bilaterally interrupted. In his series, the incidence of defects in the neural arch ranged from about 5% at the age of 6, to 17% at the age of 30, to about 34% between the ages of 30 and 40, skeletons of the last age period also showing hypertrophic arthritis.

Stewart presented evidence that both the interarticular defects and those of the pedicle probably arose as a result of unusual positions assumed by the Eskimos. He, therefore, supported the general opinion that the defect of the arch is not a true fracture, since, as has previously been noted by other workers, there is usually no evidence of an attempt to repair the bone; he classified it rather as a de-ossification of the arch in response to stress and strain, therefore as a fracture of fatigue (pseudofracture).

The combination of failure to find a developmental basis (double centers of ossification) for the defect and demonstration of increasing incidence with age seems to be overwhelmingly indicative that the typical isthmal defect of the arch is acquired and not congenital. However, evidence that other defects of the vertebral arch occur more commonly in individuals with separation of the arch than they do in the average population implies that there may be a genetic factor here. Roche and Rowe ('51a) found, specifically, that lumbar spina bifida in vertebral columns that showed separate arches was about four and one-half times as frequent as it is in unselected vertebral columns, and that spina bifida of the sacrum was about twice as frequent; however, cervical ribs, spina bifida of individual vertebrae other than the fifth lumbar, and atypical numbers of presacral vertebrae occurred in normal proportions. Stewart also found spina bifida sometimes associated with isthmal defect in the fifth lumbar vertebra, especially in skeletons of the younger age groups, but said the incidence of spina bifida decreased rather than increased with age.

Obviously, the specific factors producing spondylolysis are not completely understood. Even though the defect is not congenital,

Rowe and Roche's data indicate that in Americans, at least, it arises during the period of growth; otherwise there should be an increasing incidence with age beyond 20 years. Wiltse and co-workers regarded spondylolysis as a fatigue fracture with a strong hereditary predisposition; and Kettlekamp and Wright, finding an incidence of 28.1% in contemporary Eskimos, agreed that there is probably an inherited factor affecting remodeling of the isthmus. Amuso and Mankin reported one family in which it was apparently inherited as a dominant gene.

Cervical spondylolysis and spondylolisthesis are apparently very rare. Prioleau and Wilson, in 1975, found only seven reported cases, in none of which had there been neurological symptoms; in the case they reported, such symptoms were present.

SPONDYLOLISTHESIS

Spondylolisthesis, an anterior displacement of a vertebra, usually the last lumbar (Fig. 2-53), is a predictable result of spondylolysis. Once there has been dissolution of the laminae

below the pedicles, the weight upon the body of the vertebra involved tends to force this body forward, since it is no longer anchored by the posterior overlap of its inferior articular processes with the vertebra below (Fig. 2-54). Hence the entire vertebral column above the vertebra involved, with the body, pedicles, and superior articular processes of the involved vertebra, gradually slips forward on the disk, which, even with the associated ligaments, is not able to withstand this forward thrust. Since the displacement is usually that of the fifth lumbar vertebra, and the spinal cord practically never reaches this level, it is the nerve roots of the cauda equina that are subjected to tension and pressure. Gill and co-workers reported that most patients with symptomatic spondylolisthesis show signs of compression of the fifth lumbar roots and sometimes the first sacral, not as a result of angulation but of pressure from fibrocartilaginous tissue filling the laminar defect; they said removal of this mass and of the loose laminae has relieved both radicular and low back pain. There may also be protrusion or extrusion of intervertebral disks associated

Fig. 2-53. Diagrams of a normal and a spondylolisthetic vertebral column. *b* and *b'* indicate how vertebral displacement can compress the dural sac and the spinal nerves contained therein against the upper border of the sacrum (*broken line* in *b'*). (Meyerling HW: Collect Papers Mayo Clin 34:557, 1944)

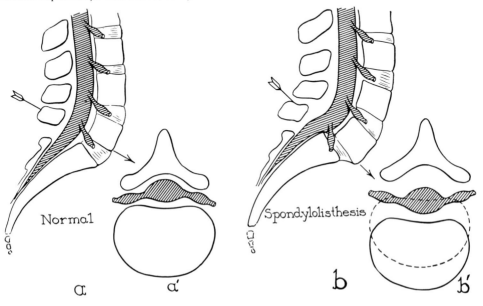

with spondylolysis (*e.g.*, Henderson); or, if the fifth lumbar vertebra is involved, the first sacral nerve may be so compressed by the upper edge of the sacral body as to necessitate removal of this edge before fusion (Lance).

Spondylolisthesis without spondylolysis does occur, but is rare (*e.g.*, Macnab, '50); it is sometimes called pseudospondylolisthesis. Macnab reported its occurrence in 22 patients, and said that the symptoms associated with it may be backache with or without sciatica, but with no signs of direct compression of nerve roots; sciatica, with or without backache, with evidence of compression of the roots; or evidence of compression of the cauda equina as a whole. The last, he found, is rare, and it occurred in none of his series, but compression of the roots of the nerve below the displaced vertebra, apparently between the inferior articular process of the displaced vertebra and the body and disk of the vertebra below, occurred in 11 patients.

Obviously, anterior displacement of a normal fifth lumbar vertebra is practically impossible. Degenerative changes in the lumbosacral synovial joints may, however, allow slight anterior slipping of the fifth lumbar vertebra; or if the changes are unilateral they may produce pseudospondylolisthesis in which the vertebral body seems in a lateral roentgenogram to be displaced forward, but has actually only rotated around the intact joint (Young, '68). In marked spondylolisthesis with an intact vertebral arch the inferior articular processes are said to be deformed: according to Macnab, instead of lying at about right angles to the pedicles and therefore firmly overlapping the sacral superior articular processes, the inferior articular processes form an angle that is much broader and may approach 180°. Such an alteration throws the weight of the body on the vertebral disk and the soft tissues associated with the vertebral column, and as in the usual case of spondylolisthesis they yield to the pressure. This type of spondylolisthesis, the dysplastic type, apparently has a stronger hereditary basis than does the isthmic type: Wynn-Davis and Scott studied first-degree relatives of 35 persons with the isthmic type, and found 15%

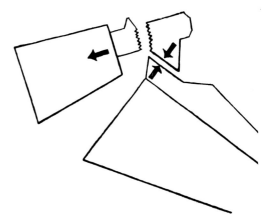

Fig. 2-54. Diagram indicating how spondylolysis nullifies the buttressing action of the articular processes, allowing spondylolisthesis. (Mitchell, GAG: J Bone Joint Surg 16:233, 1934)

to have the same condition, but among the relatives of persons with the dysplastic type 33% were similarly afflicted. (Wynn-Davis and Scott quoted estimates of 5% and 1%, respectively, in the general population.)

ABNORMALITIES OF INTERVERTEBRAL DISKS

Narrowing

Narrowing of an intervertebral space as a result of normal dehydration of the disk is minute, but marked loss of thickness of the disk as a result of degenerative changes and abnormal dehydration, or as a result of protrusion or rupture, is not uncommon. Horwitz ('39b) found narrowing of the intervertebral disk between the fifth lumbar and first sacral vertebrae in 50 of 75 cadavers, narrowing of the disk of L4 in 36, that of L3 in 27, and that of L2 in 19; among these 75, a true herniation of the disk (unilaterally) occurred in only 9, while posterior protrusion of the disk but without herniation occurred at the disk of L5 in 21, at that of L4 in 19, that of L3 in 9 and that of L2 in 1. Many of the cases of narrowing were therefore not associated with protrusion or rupture. Of the third of these cases in which clinical histories were obtainable, 24 presented narrowed disks, and among these were 4 in which the narrowing was associated with Schmorl's bodies (herniation of disk material into the vertebral bodies) and 9 in which it was associated with posterior protrusion of the disk; however, in none was there a history of sciatic pain, and backache was reported as having been a minor symptom, indicating that a narrowed disk is not necessarily associated with irritation of the nerve root and sciatic pain.

Lindblom and Hultquist said that a lumbar disk may decrease in volume from about 15 ml to about 1 ml. However, in studying stages in this process of loss of volume and therefore narrowing they were unable to find sufficient extruded material to account for the loss of volume. They concluded that much of this loss can be accounted for only by the gradual escape of disk tissue through radial ruptures and its absorption as it reaches the surface of the anulus. They regarded the vascularity at the surface of a disk showing degenerative changes as being concerned with the absorption of the extruded material, not with repair of the disk.

Extreme narrowing of a disk, by allowing the adjacent vertebral bodies to come in contact with each other, may bring about morphological changes in the vertebrae (Cloward, '52). While the most common cause of encroachment upon an intervertebral foramen is herniation or bulging of the disk, slipping of the articular facets and arthritic spur formation may also be causes; changes in the soft tissues, secondary to protrusion of the disk (venous engorgement, fibrosis of the ligamentum flavum) probably also contribute to narrowing of the foramen (Love, '47, and others). It is anteroposterior narrowing of the intervertebral foramina that may be a source of pain from impingement upon the nerve root. The lumbar nerves emerge close to the pedicle and across the body of the vertebra, so that they cannot be encroached upon even when the vertebral bodies are in contact. In the cervical region, also, Semmes pointed out, narrowing of an intervertebral interspace cannot be sufficient to compress the nerve root, since the foramen is so much longer than is the disk itself.

Apparently, narrowing of a disk is in itself not a serious source of pain, and probably never directly of root pain. This is probably the result of herniation or of secondary changes in the vertebrae. The low back pain that may be associated with narrowing of a disk is probably primarily a result of the undue strain thrown upon the posterior, synovial, intervertebral joints (if the narrowing is associated with protrusion, the sensitivity of the ligaments connected with the vertebral bodies also contributes).

Spondylosis

It is in the cervical region in particular that degeneration of the disk produces other degenerative changes that are likelsy to produce symptoms, and cervical spondylosis is now recognized as a common cause of nerve root compression. In severe cases the abnormal mobility may allow damage to the spinal cord by infolding of the ligamentum flavum and even the anterior wall of the canal during extension, by pinching the cord between the body of one vertebra and the upper edge of the lamina below (Payne and Spillane), or by posteriorly placed osteophytes that press upon the spinal cord (e.g., Friedenberg and coworkers). Stoltmann and Blackwood ('64) regarded anterior pressure from the vertebral column and posterolateral pressure from protruding ligamenta flava during hyperextension as accounting for most lesions of the cord.

Osteophytes, bony spurs or ridges, have been regarded as developing at the ends of the vertebrae as a result of traction from the longitudinal ligaments, especially the anterior one, but have also been said to develop as a result of pressure (e.g., Nathan). Perhaps they may be caused by either traction or pressure. MacNab ('71) reported that anterior ones in the lumbar region are indeed traction spurs, developing several millimeters from the edge of the vertebral body at the strongest attachment of the anterior longitudinal ligament, and denoting instability of the segments involved. However, Gloobe and Nathan found that osteophytes were much more numerous in bipedal rats (their forelimbs were removed 48 hours after birth) than in normal ones, and concluded that these were a response to pressure.

Nathan found anterior osteophytes in all regions and in all columns from persons more than 40 years old. Posterior osteophytes were more common in the cervical and lumbar regions than the thoracic, but were generally fewer and smaller than anterior ones. Holt and Yates found disk degeneration frequently associated with osteophyte formation in the cervical region, as did Friedenberg and co-workers. Both groups agreed that lower cervical disks are most affected; Friedenberg and co-workers found degeneration in the disk between C5 and C6, and C6 and C7, in 49% of 41 columns they investigated, but found no increase in incidence beyond the fifth decade. Holt and Yates, on the other hand, found the mean age at the time of mild disk degeneration among 120 specimens to be 60 years, and of severe degeneration, 72 years. Rosomoff and Rossmann later quoted an estimate that 75% or more of persons over 50 have some narrowing of the cervical vertebral canal or of intervertebral foramina, and that in half these cases the narrowing is symptomatic.

Osteophyte formation in the cervical region is particularly likely to occur in the region of "Luschka's joint," where the upward projecting lateral borders of the superior surface of a vertebra ("uncus" or "uncinate process") are easily brought into contact with the vertebra above by only a little degeneration of the disk. This process may also, with degeneration of the disk, project into the intervertebral foramen (Payne and Spillane; Friedenberg and co-workers). Arthritis may be an additional cause of nerve root irritation, particularly if it is in the lower cervical region where disease of the disk is more common.

Symptomatic cervical spondylosis has been treated by anterior interbody fusion, as first developed by Cloward ('58) for protruded disks; Robinson and co-workers found this satisfactory when posterior decompression was not necessary and said that following fusion the spurs typically diminish and may disappear entirely. Rosomoff and Rossmann have combined anterior fusion with removal of posterolateral osteophytes and of the poste-

rior lips of adjacent vertebrae to enlarge the foramina. For relief from posterior protrusions that necessitate decompression of the spinal cord, Rogers has done a complete cervical laminectomy, opened the dura, and sectioned all the attachments of the denticulate ligament. (However, Stoltman and Blackwood, '66, found these ligaments in the cervical region to be so lax that they do not prevent anteroposterior displacement of the cord.) Scoville performed a partial or complete bilateral facetectomy to free the dura-clad nerve roots here and followed this with an extensive laminectomy. Both of these operations are designed to mobilize the cord and therefore remove it from anterior pressure.

PROTRUSION AND RUPTURE

The clinical literature upon this subject is enormous, and only aspects pertaining particularly to anatomy can be summarized here.

Types

Protrusions of the intervertebral disks (Figs. 2-55 and 2-56) have been reported under a great variety of names, including "rupture of the intervertebral disk," "herniation of the nucleus pulposus," "rupture of the nucleus pulposus," "disk protrusion," "slipped disk," "herniated intervertebral disk," and others. Once regarded primarily as resulting from herniation of the nucleus pulposus through a rupture of the anulus fibrosus, as implied by several of the terms just mentioned, it is now recognized that the syndrome may be a result of varying degrees of pathologic change in the disk, from simple bulging of the anulus to massive protrusion of much of the interior of the disk. For the most part, therefore, the terms "protrusion" and "rupture" are used interchangeably.

In spite of the frequent allusion to the condition as a herniation of the nucleus pulposus through a ruptured anulus fibrosus, operative evidence indicates that many intervertebral disks that cause symptoms through stretching or compression of nerve roots show no actual rupture of the anulus, but rather simply protrude posterolaterally. Haley and Perry said

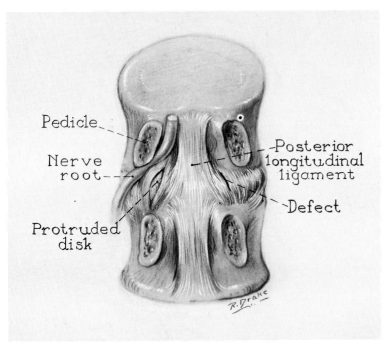

Fig. 2-55. Bulging of an intervertebral disk, *right,* and protrusion with compression of a nerve, *left.* (Love JG. In Baker AB (ed): Clinical Neurology [ed 2], Vol 3. New York, Hoeber, 1962)

that this was true of 32% of all the lesions of the disk that they found, including cervical, thoracic, and lumbar ones.

Further, study of material removed at operation from ruptured disks has plainly shown that the protruding portion is not simply nucleus pulposus that had escaped through a rent in the anulus, but regularly consists at least in part of tissue of the anulus itself (*e.g.,* Deucher and Love). Lewey regarded herniation of the nucleus pulposus as being the exception, and Key ('59) described the extruded material in older individuals as consisting often of necrotic and disintegrating fibrous tissue and cartilage.

According to Lewey, symptoms of compression of nerve roots by abnormal disks may apparently be caused by any one of three mechanisms: a simple bulging of the disk; an actual defect in the anulus fibrosus with material herniating through it into the vertebral canal; and a posterior slipping of the disk from between the two adjacent vertebral bodies.

Protrusions of the intervertebral disks may

occur in children (Key, '50), and Key quoted a 3-year series of patients operated upon for "ruptured" disk in which 2.1% were said to have been between 10 years and 19 years old. However, Key ('49) said that he had seen no completely ruptured disks in young persons, the lesion here being rather a domelike swelling.

Semmes divided rupture of cervical intervertebral disks into two categories: an acute variety in which a nodule of fibrocartilage bulges beneath a stretched and thin posterior longitudinal ligament and produces flattening and compression of the nerve root, with resulting pain in the shoulder and arm, stiffness and pain in the neck, and sometimes even marked weakness of muscles; and a second variety in which the condition is chronic, and the nerve root is found at operation to be not only flattened but also adherent to a protruded portion of the disk.

Location

Herniations of disk material may obviously occur in any direction—vertically (into adjacent vertebral bodies) or anywhere around the periphery of the disk—but it is the posterior or posterolateral ones, which may traumatize an adjacent nerve root or even the spinal cord, that are of prime clinical importance. While disks may simply protrude anteriorly or laterally, such ruptures and protrusions are relatively rare (Coventry and co-workers '45c). Moreover, since such herniations and protrusions do not come in contact with spinal nerves, with the spinal cord, or with other important structures, they have been of little clinical interest. Because the posterior protrusions are commonly lateral rather than midline ones, myelography may not reveal them; however,

roentgenograms may be valuable in eliminating other lesions.

Protrusions of part of the nucleus pulposus into an adjacent vertebral body (Fig. 2-57) are probably relatively common, although only the larger ones can be detected by roentgenogram. Known as Schmorl's bodies or nodes, many of them are thought to be the result of herniation through developmental defects in the cartilage between the disk and the cancellous tissue of the vertebral body, these defects often representing the pathways of the vessels that once supplied the disk. These protrusions, like anterolateral protrusions of the disks, are typically asymptomatic; however, if they are large, as they may be

Fig. 2-56. Normal relations, *a* and *b*, and the distortion produced by a protruded disk, *c* and *d*. (Craig WMcK, Walsh MN: Minnesota Med 22:511, 1939)

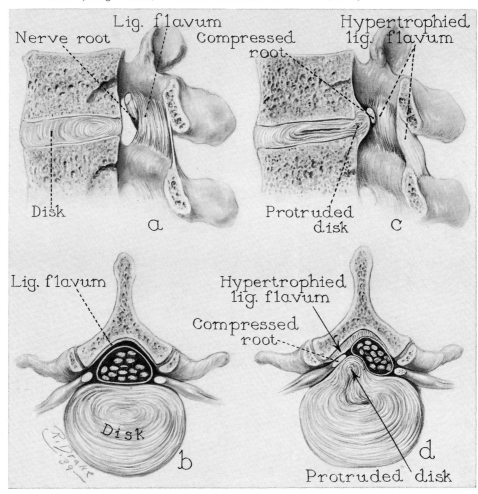

Fig. 2-57. A Schmorl's body (protrusion of the nucleus pulposus into the spongy vertebral body) seen in a longitudinal section of two adjacent vertebrae. (Courtesy of Dr. G. P. Sayre)

especially in adolescents who have done unusually hard labor, they may result in such narrowing of the disks that the growth of the vertebral bodies is interfered with, and kyphosis ("juvenile kyphosis") may result (Ehrenhaft).

Most ruptured lumbar disks are either the fifth or the fourth lumbar; Ross and Jelsma found that among 366 patients with protruded or herniated lumbar disks the disk affected was that of (that below) the fifth lumbar vertebra in 227 cases, the disk of L4 in 104 cases, and that of L3 in only one. They, as have others, noted that multiple protrusions of disks may also occur, and in their series these were found in 18 cases, in 17 of which the affected disks were those of both L5 and L4, while in the other they were those of L5, L4, and L3. Echlin and co-workers reported multiple ruptured disks in 16.6% of 60 patients, and emphasized that there may be a rupture that does not give symptoms accompanying one that does. Haley and Perry reported that among 27 patients with lumbar protrusions, only 1 disk was involved in 12, 2 were involved in 11, and 3 in 4; in their series, as in others, the disk of L5 (lumbosacral disk) was the most frequently involved, that of L4 next most frequently.

Although protrusions or ruptures of lumbar intervertebral disks were described before those of cervical disks, and are far more frequently a cause of symptoms, Haley and Perry found that protrusions of cervical disks may be more common than those of lumbar ones: in 99 spinal columns that they studied anatomically, they found one or more protrusions of disks in 63; 53 of these specimens had protrusions of cervical disks, 7 had protruded thoracic disks, and 27 had lumbar ones; in 36% of the affected columns there were both cervical and lumbar disks involved. They found multiple protrusions of cervical disks to be common, for among their 53 cases a single disk was involved in only 9, while 2 were involved in 15, 3 in 23, and 4 in 6. The cervical disks most frequently involved were the fourth, fifth, and sixth. Protruded thoracic disks, however, were usually found to be single. Thirty-two percent of the protruded disks that they found showed no rupture of the anulus. Semmes found roentgenological evidence of protrusion or rupture of cervical intervertebral disks in 1.2% of 5,557 consecutive patients.

Kristoff and Odom, and others, have pointed out that while early reports of the effect of rupture of cervical disks emphasized the incidence of compression of the spinal cord itself, either unilaterally or bilaterally, posterolateral protrusion or rupture is actually more common, as it is in the lumbar region; in a series of 20 cases they reviewed, in only 4 was there evidence of compression of the spinal cord, while in the remaining 16 there was evidence of compression of nerve roots only. Spurling and Scoville regarded protrusion or rupture of a cervical disk as a fairly common cause of radicular pain affecting the upper limb. They said that the disk involved is usually either the fifth or the sixth cervical, agreeing with later accounts that it is these two disks that are most likely to show degenerative changes. Degeneration of these disks does not necessarily lead to protrusion, however; in the series of Friedenberg and his co-workers only a little more than half of the disks with degenerative changes were protruded, and none of these sufficiently to affect the cord or nerves.

Origin

Earlier discussions of protruded lumbar disks tended to associate them with some specific and sudden severe strain, such as a twisting motion during severe lifting, or a fall from a height. It is now recognized that protrusion of disks, wherever located, is typically preceded by degenerative changes that weaken the disk. Virgin and othes have shown that it is very difficult to rupture a normal disk through pressure applied to it (and if the vertebral body is included, this will be crushed before the anulus ruptures). It is also recognized (*e.g.*, Shutkin) that most patients with ruptured disks give a history of low back pain without sciatica (implying, presumably, some bulging of a disk) which lasted over a period of many months or years before sciatica occurred, and that well-preserved disks are probably the exception rather than the rule during the middle decades of life, when herniation of the disk is more common. It thus appears that degeneration is the most important factor in protrusion, which may then occur either spontaneously (as a result of repetitive normal movements) or be precipitated by unusual strain.

Associated Pain

Pain is the predominating and often the only symptom of protrusion or rupture of an intervertebral disk. In the lumbar region the pain may be purely local and is then usually attributed to stretching or rupture of the sensitive anulus fibrosus and posterior longitudinal ligament, or to abnormal stresses placed upon adjoining muscles, ligaments, and synovial joints as a result of injury to the disk (*e.g.*, O'Connell, '51). Falconer and co-workers said that they had produced low back pain by pressing upon a protruded disk; Smyth and Wright produced backache or pain by pulling on various structures, including the anulus; and Shutkin noted that stimulation of the fibrous capsules of the synovial joints at operation has shown them to be very sensitive and painful. Commonly, low back pain is present long before the appearance of sciatica produced by involvement of one or more nerve roots.

Low back pain with or without sciatica has been attributed to a great variety of causes, only some of which can be mentioned here: among them are fibrosis of the ligamentum flavum (*e.g.*, Bradford and Spurling); tumors of the cauda equina (Bradford and Spurling, and Toumey and co-workers); tension of the iliotibial tract on the iliac crest (Freiberg, and other observers); compression of the sciatic nerve by the piriformis muscle (Freiberg); herniation of muscle through the thoracolumbar (lumbodorsal) fascia (Sargent), or herniation of fat pads; fibrositis of the back (Long and Lamphier); strains or sprains of the sacroiliac joint (Chap. 8); and tear of interspinous ligaments. In all of these conditions muscular spasm may contribute to the pain (Schlesinger and Stinchfield). Regardless of this, it is now generally recognized that protruding disks are a very common, perhaps the most common, cause of low back pain. Keegan ('53) expressed the opinion that postural low back pain, common in and after middle age, is usually a result of bulging of disks with stretching of the posterior longitudinal ligament.

Radicular pain may also result from causes other than protruded disks, but these are now recognized as the most common cause, and sciatic pain is always considered indicative of the possibility of a protruded or ruptured disk. Even when sciatica is associated with a protruded disk, however, the mechanism by which the nerve is involved is not always clear. O'Connell ('51), from a study of 350 cases, suggested that sciatica was a result of compression alone of the nerve root in only 1%, of tension on the root across the protrusion in 31%, of a combination of stretching and compression in 66%, and of sudden injury to the cauda equina in 2%. Breig and Marions implicated an increase in tension as the usual factor, and Smyth and Wright found that a nerve became extremely sensitive so that touch or slight pull insufficient to stretch it could produce sciatic pain.

Cervical Disks

As already noted, clinically important protrusions of cervical disks are most often those

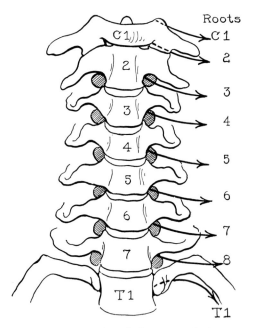

Roots

Fig. 2-58. Relations of cervical nerves to the intervertebral disks. The nerves emerge above the correspondingly numbered vertebrae, the disks are numbered according to the vertebrae below which they lie, therefore protrusion of a disk affects the nerve numbered one greater than the disk. (Love JG. In Baker AB (ed): Clinical Neurology [ed 2], Vol 3. New York, Hoeber, 1962)

of the fifth and sixth (those between vertebrae C5 and C6 and between C6 and C7), so that the nerve affected is most commonly the sixth or the seventh cervical nerve: because the first cervical nerve emerges between the skull and the first cervical vertebra, the succeeding cervical nerves emerge above the correspondingly numbered vertebrae; therefore the nerve affected by protrusion of a cervical disk, although it is the one that emerges at the level of the disk, bears a number that is one greater than the disk (Fig. 2-58).

Yoss and co-workers investigated the sensory and reflex changes and muscle weakness in 100 cases in which the nerve involved by the protrusion was definitely established and found variations similar to those described by others. The most reliable localizing signs, they found, are sensory changes in the digits and localized muscle weakness, and they felt that in 87 of these cases the particular root affected

(C5, C6, C7, or C8, produced by protrusions in the interspaces below vertebrae C4, C5, C6, and C7, respectively) could have been accurately predicted. With compression of nerve C5, they found, pain typically does not extend below the elbow, there is no paresthesia of the digits, and muscle weakness is confined to the shoulder, but the biceps and brachioradialis stretch reflexes are depressed. With C6, pain is on the radial aspect of the forearm, paresthesia most commonly involves the thumb and index finger (but may involve either alone), the biceps, brachioradialis, and extensors of the wrist show weakness, and the biceps and brachioradialis stretch reflexes are depressed. With C7, pain is on the anterior or posterior surfaces of the forearm; paresthesia of the digits varies, but is most commonly confined to the index and middle fingers; the triceps shows weakness, and its reflex is depressed. With C8 compression, pain is on the ulnar side of the forearm, paresthesia tends to affect the little and ring fingers, muscle weakness is in the hand, and the triceps reflex is depressed. Figure 2-59 shows another, very slightly different, presentation of sensory and reflex changes.

Cervical disks may protrude posteriorly and damage the cord directly or interfere with its blood supply. Such protrusions are rare, however; O'Connell ('55) found among 950 protrusions only 14 (1.5%) that involved the spinal cord, of which 8 were cervical ones.

Discography (injection of radiopaque material into a disk) has been used to diagnose rupture of a disk, but Holt found it of no value since in only 10 of 148 asymptomatic and presumably normal disks did the material remain within the disk.

Lumbar Disks

Massive extrusion of tissue from a lumbar intervertebral disk may cause acute compression of the cauda equina (ver Brugghen), as may the very rare presence of extruded disk material intradurally (Slater and coworkers), but the usual protruded or ruptured disk affects only one or two nerves. In the lumbar region the intervertebral foramina are much

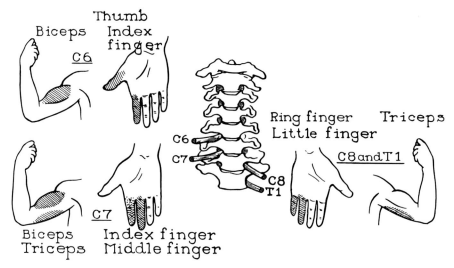

Fig. 2-59. Usual distribution of sensory and reflex changes with compression of the lower three cervical and first thoracic nerves. (Courtesy of Dr. A. Uihlein)

larger than is the diameter of the nerve, and the nerve makes its exit through the foramen close against the pedicle of the vertebra above, and therefore across the posterior aspect of the body of that vertebra (Fig. 2-60); in consequence, protrusion of a lumbar disk does not usually affect the nerve corresponding in number to the protruded disk; rather, the nerve escapes above the disk, and is not subject to pressure from protrusion of it. Thus protrusion or rupture of the fifth lumbar (lumbosacral) disk commonly affects primarily the first sacral nerve, but may affect also the second, very rarely affects the fifth lumbar; similarly, protrusion of the fourth lumbar disk usually does not affect the fourth lumbar nerve, but rather the fifth lumbar and perhaps the first sacral.

Thus, although for quite different reasons, the rule that a protruded disk does not affect the correspondingly numbered nerve, but rather the nerve numbered one greater than it, holds for both the lumbar and the cervical regions. In the cervical region the sixth cervical nerve, for instance, makes its exit across the fifth cervical disk, which lies at the fifth intervertebral foramen. In the lumbar region the fifth lumbar nerve leaves the vertebral canal above the fifth lumbar disk, but the first

Fig. 2-60. View of the third and fourth lumbar intervertebral foramina of the right side, from the vertebral canal. Note that the spinal nerves leave the uppermost part of the intervertebral foramina; thus the third lumbar nerve makes its exit above the third disk, while the fourth nerve (*dotted lines*) crosses the third disk, and can be compressed by it, shortly before it reaches its level of exit. (Naffziger HC, Inman V, Saunders JBdeCM: Surg Gynecol Obstet 66:288, 1938 [by permission of Surgery, Gynecology & Obstetrics].

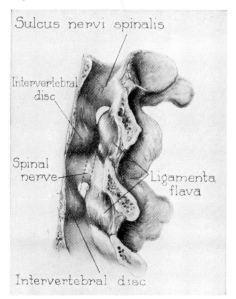

sacral nerve crosses this disk in its course to the first sacral foramen; further, at the level of the fifth lumbar disk the first sacral nerve is the most laterally lying nerve trunk, hence more likely to be involved by a lateral protrusion of the disk. Still further, of all the nerves crossing this disk the first sacral is the least mobile, since at the level of the fifth disk it lies in or is about to enter the special dural sheath that accompanies it as far as the ganglion, and this sheath, attached to the tube of spinal dura at one end and to the margins of the intervertebral foramen at the other, is relatively taut and unyielding. Similarly, the second sacral nerve is next most likely to be affected by protrusion of the fifth disk, both because it lies next most laterally and because it is the next nerve to acquire its own dural sheath, thereby losing the mobility that it has in the lumbar cistern. The greater the length of nerve between the level of the protruded disk and the level of exit of the nerve, the less likely the nerve is to be stretched or compressed across the bulging or herniated disk.

As in the case of protruded cervical disks, actual paralysis of muscles is relatively rare with lumbar disks, for most muscles are supplied from two or more segments. Therefore, the neurological findings with protruded lumbar disks are primarily those of diminished or absent reflexes and hypesthesia or paresthesia, and tend to be minimal; accurate localization by neurological methods is often impossible (Love, '47). Many workers have emphasized that there are no infallible symptoms or findings (Woodhall reported that of 100 patients shown at operation to have had protruded disks, 40 were found preoperatively to have no sensory changes, and in 18 others the changes were equivocal or even falsely localizing); however, the following tend to be typical:

When the fifth lumbar (lumbosacral) disk is involved and the first sacral nerve is affected, there are often pain and numbness or hypalgesia on the posterolateral side of the leg and the outer part of the foot and toes (Ford, and others). As a rule, the patient is unaware of any weakness in muscles on the affected side, unless two or more nerves are simulta-

neously involved, but Falconer and co-workers reported one patient with involvement of the first sacral nerve by a herniated lumbosacral disk who had paralysis of the peronei and the extensor digitorum longus, and weakness of the extensor hallucis and the tibialis anterior (more common with involvement of the fifth lumbar nerve). The distribution of pain, or the sensory findings (Fig. 2-61) plus the altered tendon reflexes are apparently of much more value in diagnosis than is muscular weakness. However, Spurling and Grantham said that in lesions of the first sacral nerve the muscles of the posterior aspect of the calf, especially the gastrocnemius and the soleus, are more commonly affected than others, and that weakness in these may be demonstrated by having the patient walk on his toes. The most common alteration in reflexes, when the first sacral root alone is involved, is diminution or absence of ankle jerk (Falconer and his colleagues; Spurling and Grantham). Ross and Jelsma said that protrusion of the lumbosacral disk affects the ankle jerk in about 50% of cases. O'Connell ('51) reported that of 248 patients with a lesion of the lumbosacral disk the tendon reflexes at both the knee and ankle were normal in 14.5% the ankle jerk was reduced or absent in 85%, and the knee jerk was reduced or absent in only 2.8%.

Involvement of the fifth lumbar nerve, typically through protrusion of the disk of L4, tends to produce extension of pain to the dorsum of the foot and the big toe (O'Connell), or hypesthesia on the anterolateral side of the leg, on the big toe, and on the inner side of the heel (Ford). In Keegan's dermatomal charts (see Fig. 1-26), constructed from his findings of the distribution of hypalgesia in involvement of single nerves, it is the fourth lumbar, not the fifth, that is shown distributed to the dorsum of the big toe. Muscular weakness, if any, is in the anterior muscles of the leg, and is especially likely to affect the extensor of the big toe (Spurling and Grantham). Falconer and his colleagues reported three patients who had footdrop resulting from almost complete paralysis of the anterior tibial and extensor hallucis longus muscles, with the extensor digitorum longus and the peronei

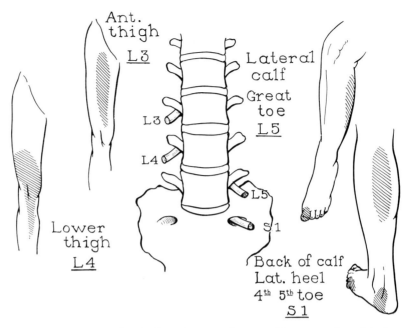

Fig. 2-61. Common distribution of sensory changes with compression of the last three lumbar and the first sacral nerves. (Courtesy of Dr. A. Uihlein)

weakened, as a result of involvement of the fifth lumbar nerve. Love ('47) also noted that complete footdrop may occasionally result from a lesion of a single nerve.

Spurling and Grantham said that there are no very reliable reflex changes associated with herniation of the fourth lumbar disk, and Falconer and his colleagues agreed, as did Ford, that although the ankle jerk may be depressed it is usually normal. O'Connell said that of 198 cases of protrusion of disk L4, ankle and knee reflexes were both normal in only 34%, while the ankle jerk was reduced or absent in 65.1% and the knee jerk reduced or absent in 5%. In 46 cases in which both the fifth lumbar and first sacral nerves were involved by protrusions of the fourth and fifth disks the reflexes were said to be normal in 13%, the ankle jerk was reduced or absent in 82.6%, and the knee jerk reduced or absent in 6.5%, findings similar to those with lesions of the first sacral nerve alone, except for the slightly greater percentage in which the knee jerk was involved. O'Connell's findings apparently tend to substantiate the opinion that there is no diminution or loss of reflexes that is characteristic of a lesion of the fifth lumbar

nerve, but also indicate too much dependence should not be placed in the localizing value of the ankle jerk; see the next paragraph.

Involvement of the fourth lumbar nerve by protrusion of the third disk, or involvement of the third through protrusion of the second disk, may be associated with pain on the anterior surface of the thigh that extends across the patella and along the anteromedial surface of the leg (O'Connell); the most common finding in involvement of the fourth lumbar nerve, or of the third and fourth, is diminution or absence of the knee jerk (Spurling and Grantham); however, O'Connell reported that in eight cases in which both the second and third lumbar disks were protruded the knee jerk was reduced or absent in only 50%, while the ankle jerk was reduced or absent in 75%. Ross and Jelsma reported that a decrease in the knee jerk is found with protrusion of the disk L4 or higher in from 18% to 20% of cases.

Clinical study of the symptoms and myelography have been the traditional methods for localizing a protruded disk. To these there has been added ascending lumbar or lumbosacral venography. O'Dell and co-workers

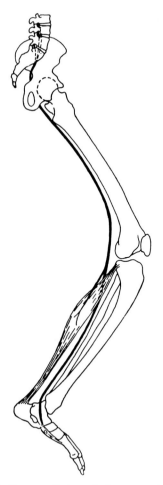

produce the sciatic pain. Sometimes the pain on the affected side can also be reproduced by raising the "well" leg, that is, on the unaffected side. The principle of the leg-raising test (Fig. 2-62) is that sciatic pain is a result of a root of the sciatic nerve's being pressed against or stretched across the protruded disk, and flexion of the hip with the knee extended exerts tension on the sciatic nerve (just as it does on the hamstring muscles); thus the sciatic nerve draws its duraclad roots of origin against the anterior wall of the vertebral canal.

Woodhall and Hayes, pointing out that under normal circumstances the tension on the spinal dura varies (since the change in the length of the vertebral canal between full extension of the vertebral column and full flexion has been said to be about 7 cm), measured on the cadaver the traction produced by straight-leg raising. Flexion of both hips with the knees extended produced an obvious increase in tension on the dura-clad nerve roots of both sides, and drew the spinal dura both downward and forward against the anterior wall of the canal; flexion of one hip only with the knee extended produced a similar increase of tension in the roots of the corresponding side, but flexion of the knee after the hip was flexed immediately relaxed the tension on the nerve roots. (The increased tension is borne by the inelastic dura; the nerve roots within the dural sac are rather loose, and the dura does not move enough really to tighten them.)

Woodhall and Hayes also found that when one thigh alone was flexed with the leg extended the nerve roots on the other side tended to withdraw somewhat from the foramina and to shift toward the side of movement, and to approximate the anterior wall of the vertebral canal, although less strongly so than on the side of flexion. This shift toward the side of flexion is the basis, they surmised, for production of sciatic pain upon raising the leg on the sound side; they said that about one third of patients with protrusion of a disk do give a positive result with the well-leg-raising test, and that in such cases the protrusion is usually large and lies medial to the affected nerve root.

Fig. 2-62. Course of the sciatic nerve: its course between its anchoring points, the hip joint above and the knee joint below, is lengthened by flexion at the hip, and therefore it can be put on tension, when the hip is flexed, by straightening the leg or pressing the nerve forward at the knee. (Cram, RH: J Bone Joint Surg 35-B:192, 1953)

and MacNab and co-workers found this as accurate as myelography (when previous operations had not affected the segmental veins), but advised that it should be used only as an additional procedure to help localize the lesion when the localization was not clear.

Not localizing in regard to the segment involved, but considered one of the better tests for confirming the tentative diagnosis of protruding lumbar disk, is the leg-raising test: with the patient recumbent, the lower limb on the affected side is flexed at the hip while the knee is kept straight, and this should re-

Cram suggested a different method of producing tension on the roots of the sciatic nerve: he flexed the lower limb, with the leg straight, to the point at which pain was produced, then flexed the knee about 20° and further raised the thigh to a point just short of that which caused pain; he then applied pressure over the middle of the popliteal space so as to produce additional pressure on the tibial nerve, thus causing pain in the lower lumbar region or the affected buttock. He showed on the cadaver that both this test and the straight-leg-raising test increased the mechanical tension on the nerve root at the level of a ruptured disk.

Although carefully applied traction may relieve symptoms of nerve irritation in cervical spondylosis, this is apparently not true of protruded lumbar disks. Rothenberg and colleagues reported that traction of as much as 25 lb to 30 lb on each leg, which was all the patients could tolerate, produced no widening of lumbar disks in carefully controlled experiments.

There is as yet no agreement as to when spinal fusion should be done in connection with lumbar disk operations. Sacks described anterior fusions as being useful in selected cases as did Freebody and co-workers. Stauffer and Coventry ('72a) thought they should be utilized only as a "salvage procedure" when the traditional posterior or posterolateral grafting is inadvisable; they also ('72b) found that better results were obtained with a posterolateral fusion than with a strictly posterior one. MacNab and Dall, reporting a similar experience, said a major source of bleeding in posterolateral fusion is the arteries around the joints; they described these in detail, and advised coagulating them by cautery.

Barr and co-workers were not convinced that the combined operation significantly improved their results and said they were tending to employ fusion less frequently than previously. On the other hand, Young ('62) compared the results of disk removal and the combined operation in two large series of patients and reported decidedly better long-term results following fusion. Although they were better in regard to relief of both sciatic and back pain, Young advocated the combined operation only in those instances in which careful evaluation indicates that fusion will probably relieve or prevent the subsequent development of the latter.

Associated Scoliosis

In the cervical region flexion toward the side of the lesion may produce pain by narrowing the foramen about the involved nerve root; however, flexion to the opposite side may also produce or aggravate the pain by causing traction on the nerve root (Love, '68).

A lumbar tilt to one side has been interpreted as evidence that the protruded part of the disk lay on the other side, but subsequent experience has shown that the problem is not so simple. Thus Falconer and co-workers reported that while in about half of their cases of ruptured lumbar disk with sciatica there was scoliosis, and the tilt (concavity of the scoliosis) was away from the side of rupture in 25, it was toward the side of rupture in 17. Similarly, O'Connell ('51) said that in about 50% of cases of protrusion or herniation of a lumbar disk there is a lateral flexion away from the side of protrusion, but that sometimes the bending is toward rather than away from the painful side. An interesting condition, sometimes encountered, is an alternating scoliosis associated with alternating sciatic pain (Love, '68).

Cloward ('52) quoted evidence that scoliosis with its concave side away from the side of the herniation may result from a protective unilateral spasm of the erector spinae muscle, to relieve pressure on the disk on the side of rupture; he expressed the belief that in some cases, the scoliosis results from a large mass of anulus fibrosus, which is completely detached from the rest of the disk and lies in the intervertebral space in such a fashion that it acts as a wedge and hence mechanically maintains flexion toward the opposite side.

A concept supported by O'Connell ('43, '51) as probably explaining the direction of a lumbar list in cases of protruded lumbar disk is that the direction may depend primarily on whether the nerve root that is involved lies

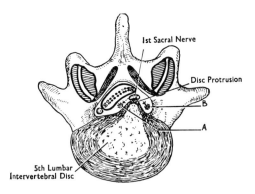

Fig. 2-63. Possible mechanism of alternating scoliosis with a protruded disk: the nerve lies on the apex of the protrusion, and flexion to either side, moving it to *A* or *B*, will relieve the tension upon it. (O'Connell JEA: Brit J Surg 30:315, 1943)

medial or lateral to the protrusion. He suggested that flexion is toward the side of the lesion if the nerve root lies lateral to the protrusion, while it is away from that side if the nerve root lies medial to the protrusion. However, Falconer and his colleagues said that among their cases, which included scoliosis both toward and away from the side of the lesion, there were in both groups cases in which the nerve root lay lateral to the protrusion of the disk, and cases in which it lay medial to it. O'Connell said that in patients with alternating scoliosis it can be shown that the pain is greatest when the scoliosis is corrected, but is relieved by a tilt to either side. At operation on such patients the involved nerve was usually found to be freely mobile and the protrusion to have a trough on each side of its summit, so that the nerve had presumably been slipping back and forth across the protrusion (Fig. 2-63). Love ('47) suggested that a shift in a small protrusion from side to side beneath the posterior longitudinal ligament may be a cause of alternating scoliosis without sciatica.

Whatever the explanation of a tilt toward the side of the lesion, there is general agreement that it may occur. It is therefore obvious that in the lumbar region lateral flexion is not necessarily an attempt to minimize protrusion of the disk. In consequence, it cannot be considered as necessarily indicative of the laterality of the lesion.

Thoracic Disks

Protruded thoracic disks are rare, but are more commonly in or close to the midline than lateral, and therefore often affect the spinal cord. Love and Schorn found them accounting for only 0.5% of all protruded disks, present at every level but most frequent in the lower four thoracic interspaces, and lateral in position in 23%. They found no syndrome typical of protruded thoracic disks. Damage to sensory and motor pathways of the cord varied greatly, and radicular pain, although common, was apparently often produced by stretching of the nerve roots through displacement of the cord rather than by impingement of the protrusion on a nerve root.

Musculature of the Back

Covering Muscles

The true or intrinsic muscles of the back, sometimes referred to as the deep muscles of the back, are almost entirely covered superficially by the trapezius and the latissimus dorsi, which together extend as a rule from the sacrum and crest of the ilium to the occipital bone (Fig. 2-1). Only in the upper cervical region, as the trapezius narrows, do some of the deep muscles appear as a part of the floor of the posterior triangle of the neck. Immediately deep to the trapezius in the thoracic region are the two rhomboidei, also muscles of the shoulder, and described in Chapter 4. Deep to the shoulder muscles, and also lying superficial to the true back muscles, are two muscles connected with the ribs, the serratus posterior superior and the serratus posterior inferior.

The *serratus posterior superior* (Fig. 2-64) is a flat muscle that arises by a thin aponeurosis from the lower part of the ligamentum nuchae and the spinous processes of the seventh cervical and first two or three thoracic vertebrae and the intervening supraspinous ligament. It runs laterally and downward and inserts by four or sometimes three muscular slips into the upper borders of the second, third, fourth, and fifth ribs, or the second, third, fourth, a little lateral to their angles. The *serratus posterior inferior* arises by a thin aponeurosis from the spinous processes of about the lower two thoracic and upper two lumbar vertebrae, and the intervening supraspinous ligament, and is therefore widely separated from the superior serratus; it runs laterally and upward, divides into four or three fleshy slips, and inserts into the lower borders of the last four (or three) ribs, a little beyond their angles. The aponeurosis of the serratus posterior inferior is fused superficially to the aponeurosis of origin of the latissimus dorsi, and on its deep surface to that part of the thoracolumbar (lumbodorsal) fascia superficial to the intrinsic musculature of the back.

The two serratus posterior muscles are inspiratory muscles, the superior one elevating upper ribs, the lower one depressing lower ribs and hence preventing their being pulled upward by the action of the diaphragm. They are innervated by ventral (anterior) rami of spinal nerves, not by dorsal ones as are most of the muscles of the back; the serratus posterior superior is usually innervated by twigs from the first four intercostal nerves, the serratus posterior inferior by twigs from the eighth through eleventh intercostal nerves, each fleshy slip receiving its own innervation.

Associated Fascia

The muscles of the back are separated from the overlying muscles by fascia. In the cervical region the fascial layer immediately adjacent to the back musculature is a posterior continuation of the prevertebral or deep layer of cervical fascia; this is continuous around the sides and front of the vertebral column with the fascia of the opposite side, and thus encircles all the muscles attached to the vertebral column, including the scalenes and the longus muscles.

In the thoracic region the fascia over the muscles is thin and transparent, attached in the midline to the spinous processes of the vertebrae, and laterally, at the edge of the back muscles, to the ribs and to the fascia covering the intercostal muscles.

In the lumbar region the fascia is much

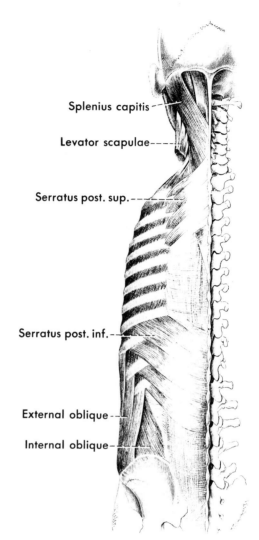

Fig. 2-64. Serratus posterior superior and inferior muscles. (Eisler P: Vol 2, Section 2, Part 1 of K von Bardeleben's Handbuch der Anatomie des Menschen. Jena, Fischer, 1912)

(labels in figure: Splenius capitis, Levator scapulae, Serratus post. sup., Serratus post. inf., External oblique, Internal oblique)

thicker, and represents not only fascia but also the aponeuroses of several muscles. The fascio-aponeurotic layer superficial (posterior) to the musculature of the back is usually called the posterior layer of the thoracolumbar (lumbodorsal) fascia, but at the lateral border of the erector spinae muscle it becomes continuous with another fascio-aponeurotic sheet, the anterior layer of the thoracolumbar fascia, which extends in front of the muscles to attach to the transverse processes of the lumbar vertebrae (Fig. 2-65). The posterior

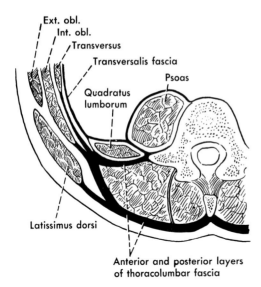

Ext. obl.
Int. obl.
Transversus
Transversalis fascia
Psoas
Quadratus
lumborum
Latissimus dorsi
Anterior and posterior layers
of thoracolumbar fascia

Fig. 2-65. Diagram of the relation of the thoracolumbar fascia to the muscles of the back, in a cross section at the lumbar level.

layer of the thoracolumbar fascia, attached to the spinous processes of the lumbar and sacral vertebrae, to the iliac crest, and to the twelfth rib, is contributed to by the aponeuroses of origin of the latissimus dorsi and the posterior inferior serratus muscles, and by a part of the fused aponeuroses of origin of the internal oblique and transversus muscles; the anterior layer of the thoracolumbar fascia is formed by the remaining part of the aponeuroses of origin of the internal oblique and transversus, so that, essentially, the fused origins of these two muscles simply split to pass on either side of the back muscles to attain an attachment to the vertebral column. MacNab and Dall described a "middle layer" of the thoracolumbar fascia, extending laterally from an attachment to the posterior border of the superior articular process and the pars interarticularis (isthmus) of the lamina in the lower lumbar region.

The ligamentum nuchae, a prominent elastic ligament in the neck of large-head quadrupeds, is in man, except for its thickened posterior edge, a thin intermuscular septum in the posterior midline of the neck (Fig. 2-21).

General Grouping and Structure
The muscles of the back (Fig. 2-66) have been variably classified into two to five or more general groups, according to whether they arise from and insert into spinous processes, arise from spinous processes but insert into transverse processes, or arise from transverse processes and insert into the spinous processes; sometimes distinctions are made on the basis of the length of the muscles and whether or not they insert into the skull. Possibly, the easiest subdivision is one based upon the prevailing direction of the major portion of each muscle, regardless of whether or not it has an attachment onto the skull; on this basis, the longer muscles of the back can be divided into three main groups. (1) The first consists of those that in general arise from the midline, therefore from ligamentum nuchae and spinous processes, and run laterally and upward to insert into transverse processes of the skull, sometimes referred to as the spinotransversalis system (the two splenius muscles form this group). (2) This consists of those that run almost longitudinally, or with only a slight outward slant as they are traced upward, and thus run primarily either from transverse process to rib, from rib to rib or to cervical transverse process, or from one spinous process to another, hence once referred to as the transversocostalis and spinospinalis systems, but now grouped together as the erector spinae (sacrospinalis) muscle. (3) The last consists of those that arise from transverse processes but insert primarily into spinous processes, hence have an upward and inward direction; these form a transversospinalis system, and include the semispinalis, multifidi, and the rotators.

In addition to these, there are strictly segmental muscles, stretching only from one vertebra to the next, the interspinales and the intertransversarii, which could be allocated to subgroups of the erector spinae, but are never considered as a part of this muscle; and the short suboccipital muscles, which also demand special description. The short rotators are just as strictly segmental as the interspinales or intertransversarii, but are most easily classified, with the long rotators, as a part of the transversospinalis group.

The long muscles of the back, being derivatives of originally shorter muscles that attached to every segment, have multiple ori-

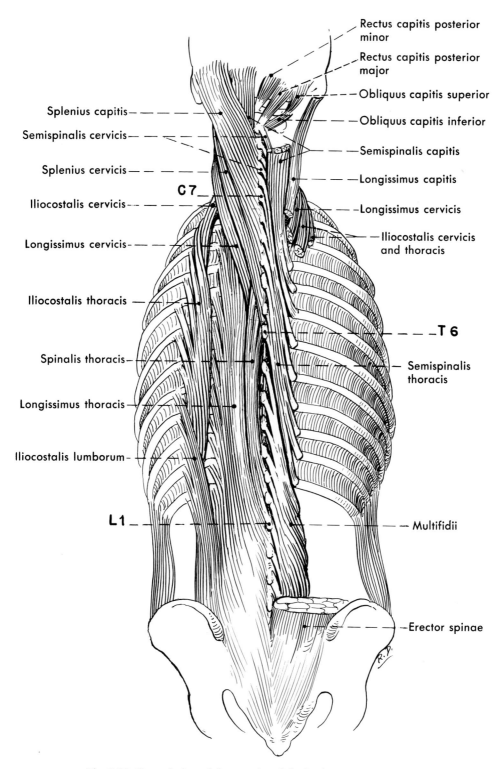

Rectus capitis posterior
minor

Rectus capitis posterior
major

Obliquus capitis superior

Obliquus capitis inferior

Splenius capitis

Semispinalis cervicis

Semispinalis capitis

Splenius cervicis

Longissimus capitis

Iliocostalis cervicis

C 7

Longissimus cervicis

Longissimus cervicis

Iliocostalis cervicis
and thoracis

Iliocostalis thoracis

T 6

Spinalis thoracis

Semispinalis
thoracis

Longissimus thoracis

Iliocostalis lumborum

L 1

Multifidii

Erector spinae

Fig. 2-66. General view of the muscles of the back.

gins, and also multiple insertions unless they insert into the skull. Thus any one of the longer muscles typically arises from several successive vertebrae or ribs, and also inserts on several successive vertebrae or ribs. Moreover, their structure is complex, for a slip of origin from any one vertebra, whether from spinous process, transverse process, or rib, typically blends with slips of origin from neighboring segments; the blended fasciculi then divide into new groupings that go to insertions on several adjacent segments. Because of this structure, the muscles cannot be separated from each other as can, say, muscles of the limbs: most of the named muscles could be subdivided, if it seemed advisable, into a number of blending muscles each of which had one attachment peculiar to it, but others shared with adjacent muscles; but different fascicles would be included depending upon whether the origin or the insertion was used as a basis of classification (see, for instance, the diagram of the semispinalis in Fig. 2-68). In practice, this would make for a great deal of confusion, and the named muscles are the larger and fairly obvious subdivisions. The lines of demarcation between the various subdivisions may be somewhat arbitrary in some cases, but do allow a more understandable description of these intricately interwoven, complex muscles.

SPLENIUS MUSCLES

There are only two major muscles of the back that arise from the midline (the ligamentum nuchae and spinous processes) and run markedly upward and laterally to their insertions; these are the two splenius muscles, the splenius capitis and the splenius cervics (Figs. 2-66 and 2-67).

The splenius capitis is a relatively simple, flat muscle, which arises from about the lower half of the ligamentum nuchae and from the spinous processes of the seventh cervical and the upper three or four thoracic vertebrae. It runs upward and laterally to insert into the occipital bone just below the lateral part of the superior nuchal line, and into the mastoid

process of the temporal bone under cover of the sternocleidomastoid muscle. The splenius capitis is under cover of the trapezius only at its origin, and as it extends laterally it forms a part of the floor of the posterior triangle of the neck before it disappears deep to the sternocleidomastoid muscle.

The splenius cervicis arises from spinous processes below the origin of the splenius capitis, but is a narrower muscle, arising usually from the spinous processes of about the third through the sixth thoracic vertebrae. As it runs upward and laterally it divides into from two to four tendons of insertion that attach into the posterior tubercles of the transverse processes of the upper two to four cervical vertebrae, these insertions being overlapped by the origin of the levator scapulae.

ERECTOR SPINAE (SACROSPINALIS)

This system forms the largest muscular mass of the back, being especially well developed in the lumbar region. Its several divisions have a common origin below, largely tendinous, from the posterior surface of the sacrum including its median crest (remains of spinous processes), from the sacrotuberous ligament, and from a posterior part of the crest of ilium; the tendon of origin between the midline and the ilium tapers to a point inferiorly. Also, the group has a muscular origin from the anterior (deep) surface of the tendon of origin and, laterally, from a part of the crest of the ilium; and a mixed tendinous and muscular origin from the lumbar spinous processes and the spinous processes of the last two thoracic vertebrae.

In the upper lumbar region the mass is largely muscular; it divides here into three vertical columns, a laterally lying iliocostalis, an intermediate longissimus, and a medial spinalis. In general, these columns become more slender as they are traced upward, but the longissimus, thin in the midthoracic region, thickens again in the cervical region.

The longitudinal columns into which the erector spinae divides are in turn divisible

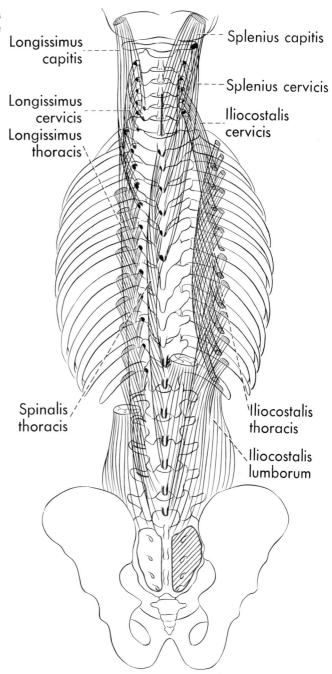

Fig. 2-67. Origins (*red*) and insertions (*black*) of the splenius muscles and of the several components of the erector spinae

Longissimus capitis

Longissimus cervicis

Longissimus thoracis

Spinalis thoracis

Splenius capitis

Splenius cervicis

Iliocostalis cervicis

Iliocostalis thoracis

Iliocostalis lumborum

into subsidiary and overlapping portions (Fig. 2-67) which receive additional names: thus the iliocostalis is divided into the iliocostalis lumborum, iliocostalis thoracis (dorsi), and iliocostalis cervicis; the longissimus is divided into longissimus thoracis (dorsi), cervicis, and capitis; the spinalis is usually also subdivided, but the spinalis thoracis (dorsi) is rather poorly developed, the spinalis cervicis is inconstant, and the spinalis capitis is sometimes regarded as being really a part of another muscle (the semispinalis capitis).

ILIOCOSTALIS

The iliocostalis lumborum originates as the lateral part of the common mass of the erector spinae, and divides into six or seven slips that insert by slender flattened tendons into the lower borders of the lower six or seven ribs, at about their angles. This muscle, more than the other parts of the erector spinae, runs slightly laterally as it is traced upward.

The lower part of the iliocostalis thoracis (iliocostalis dorsi) lies medial to the upper part of the iliocostalis lumborum, arising by six slips from the upper borders of the lower six ribs just medial to the slips of insertion of the iliocostalis lumborum. It forms a rather slender, vertically running, muscle mass which divides into six slips of insertion that attach into the upper six ribs; there may be an additional slip continuing to the transverse process of the seventh cervical vertebra.

The iliocostalis cervicis arises medial to the upper part of the iliocostalis thoracis, by slips from the angles of the upper six ribs, or of the lower four or five of these ribs, and divides usually into three slips of insertion that attach to the posterior tubercles of the transverse processes of the fourth, fifth, and sixth cervical vertebrae.

LONGISSIMUS

The longissimus is the largest as well as the longest of the three divisions of the sacrospinalis.

The longissimus thoracis (dorsi) arises as the intermediate part of the erector spinae, and runs almost straight upward to have a double insertion on each of the lower nine or ten ribs, or more, and the tips of the transverse processes of the corresponding vertebrae. The insertions into the ribs are fleshy, attaching between the tubercles and the angles; the insertions into the transverse processes are by narrow tendons.

The longissimus cervicis arises medial to the upper end of the longissimus thoracis, by long thin tendons from the dorsal aspect of the tips of the transverse processes of about the upper four to six thoracic vertebrae. It forms a rather slender muscle mass, which extends into the cervical region where it is partly covered by the splenius cervicis and iliocostalis cervicis muscles. It divides into slips that are inserted into the posterior tubercles of the transverse processes of the second through the sixth cervical vertebrae; the slips of insertion are mostly tendinous like the slips of origin.

The longissimus capitis is also a fairly slender muscle and originates, medial to the longissimus cervicis, partly from the tendons of origin of this muscle and partly from the posterior aspect of the articular processes of the lower four cervical vertebrae. This muscle runs somewhat laterally as well as upward and, passing under cover of the splenius capitis, inserts into the posterior surface of the mastoid process.

SPINALIS

The spinalis is the smallest division of the erector spinae. It is usually poorly defined, and often separable with difficulty from other muscles.

The spinalis thoracis (dorsi) lies on the medial side of the longissimus thoracis, and in the lower part of its origin is intimately blended with that muscle. Its tendons of origin arise from the spinous processes of about the lower two thoracic and the first two lumbar vertebrae, and form a very slender muscle that runs vertically to divide into a variable number of slender tendons that insert into the spinous processes of some four to eight of the upper thoracic vertebrae. Winckler said that in spite of the variations in origin and insertion usually found in this muscle there is never an attachment to the tenth thoracic vertebra.

The spinalis cervicis is inconstant; in its best-developed form it arises usually from the lower part of the ligamentum nuchae and the spinous process of the seventh cervical vertebra, and sometimes also from the spinous processes of the first or first and second thoracic vertebrae, and inserts above into the spinous process of the axis, and sometimes into the spinous processes of the third and fourth cervical vertebrae. Even when well developed, the spinalis cervicis is rarely much larger in diameter than a pencil, and it may consist of a few bundles of muscle fibers attached at both ends, in the midcervical region, to the ligamentum nuchae.

The spinalis capitis is rarely a separate muscle, but is instead a medial part of the semispinalis capitis. It arises largely with the semispinalis but may have some attachment to upper thoracic spinous processes, and it inserts with the semispinalis capitis into the skull.

TRANSVERSOSPINALIS MUSCLES

These muscles (Figs. 2-66 and 2-68) largely fill the concavity between the sides of the spinous processes and the backs of the laminae and transverse processes, and form a series of fascicles directed in general from the transverse processes upward and medially to the spinous processes. There are three muscles belonging to this group: the semispinalis, the multifidi, and the rotators. In the thoracic region, where all three muscles are represented, it is obvious that the semispinalis consists of the longest muscle fascicles, and therefore makes the sharpest angle with the midline; the underlying multifidi run more obliquely, and therefore consist in general of shorter muscle fascicles; the deepest lying rotators are short muscles, passing over no more than one vertebra, and therefore making a broad angle with the midline. Further, because any one bundle of origin divides, as usual, into several bundles of insertion, each muscle has fascicles of varying lengths. The shortest bundles of the semispinalis are about the same length as the longest bundles of the multifidi, and the shortest ones of the multifidi are about the same length as the longest ones of the rotators; in consequence, these three muscle layers may be difficult to separate accurately from each other.

SEMISPINALIS

The semispinalis is not represented in the lumbar region, so the lowest division of this muscle is the semispinalis thoracis (dorsi), which lies under cover of the spinalis thoracis and the more laterally lying longissimus muscles.

The semispinalis thoracis arises by a series of rather long slender tendons from the transverse processes of about the lower six thoracic vertebrae. These tendons give rise to narrow flattened muscle fascicles, which fuse and then subdivide into other fascicles that are inserted by slender tendons into the spinous processes of the upper four thoracic and the lower two cervical vertebrae. The longest of the fascicles of the semispinalis thoracis pass over as many as six vertebrae between their origin and insertion, while the shortest pass over four vertebrae.

The semispinalis cervicis is not clearly separated from the semispinalis thoracis at its origin from the transverse processes of the upper five or six thoracic vertebrae, but as it extends into the lower part of the neck it becomes thicker and muscular. It is largely or entirely covered by the semispinalis capitis, and its tendons both of origin and of insertion are usually shorter than those of the semispinalis thoracis. Its insertion is into the spinous processes of about the second through the fifth cervical vertebrae, the slip extending to the axis being the largest and almost entirely muscular, the other slips inserting by tendons.

The semispinalis capitis is a large muscle, lying deep to the splenius muscles and medial to the longissimus cervicis and capitis. It arises by tendons that are attached to the tips of the transverse processes of about the upper six thoracic vertebrae, and to the articular processes of the lower three or four cervical vertebrae. These origins give rise to a broad muscle that extends upward almost vertically to insert between the superior and inferior nuchal lines of the occipital bone.

The most medial fibers associated with this muscle may arise from spinous processes of lower cervical and upper thoracic vertebrae, and it is this part of the muscle that is termed the spinalis capitis. The entire muscle is commonly traversed in the upper part of the neck by a tendinous inscription; the medial part of the muscle, the spinalis capitis, is, when well developed, usually interrupted by a tendon

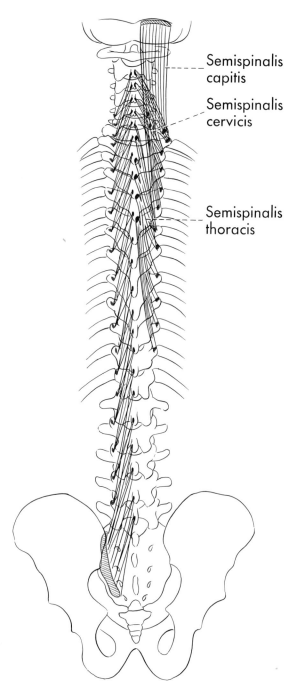

Fig. 2-68. Origins (*red*) and insertions (*black*) of the semispinalis muscles (*right*) and the multifidi (*left*).

Semispinalis capitis

Semispinalis cervicis

Semispinalis thoracis

placed at about the level of the spinous process of the seventh cervical vertebra.

MULTIFIDI

Unlike the semispinalis, the multifidi extend from the sacrum to the second cervical verte-

bra, but do not reach the skull (Fig. 2-68). They also differ in the length of their fasciculi, for while the longest and most superficial fasciculi pass over four vertebrae, as do the shortest fasciculi of the semispinalis, other fasciculi are shorter and pass across three vertebrae, while the shortest and deepest pass

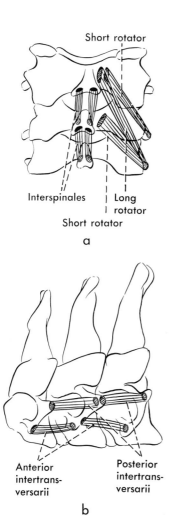

Short rotator

Interspinales | Long
| rotator
Short rotator

a

Anterior
intertrans-
versarii

Posterior
intertrans-
versarii

b

Fig. 2-69. Origins (*red*) and insertions (*black*) of the short muscles of the back in the cervical region.

from the mammillary processes of the lumbar vertebrae. In the thoracic region the thinner fasciculi arise from all the transverse processes. The still thinner parts in the cervical region arise from the articular processes of the lower four cervical vertebrae. The muscles extend upward and medially, the fascicles passing over from four to two vertebrae to attain an attachment on the sides of the spinous processes of all the vertebrae from the fifth lumbar to the axis.

ROTATORS

The rotators (Figs. 2-69, 2-70; see also Fig. 2-72), the deepest members of the transversospinalis system, are very small muscles that lie deep to the multifidi, and may not be clearly separable from them. Like the multifidi, they extend from the sacrum to the second cervical vertebra, but they are distinctly divided into two sets of muscles, the long rotators and the short rotators.

Each long rotator arises from the transverse process of one vertebra, passes beyond the vertebra immediately above, and inserts into the base of the spinous process of the second vertebra above, thus skipping one vertebra. The short rotators also arise from the trans-

across two only. In the sacral region the multifidi are covered by the tendon of origin of the erector spinae, from which some of its fibers take origin, and in the lumbar region they are covered by the muscular mass of the erector; in the thoracic and cervical regions they are covered by the semispinalis.

The multifidi are not easily divisible into several portions, as are most of the muscles of the back. The lumbar portion, the heaviest, arises from the dorsal surface of the sacrum as low as the third or fourth dorsal sacral foramen; from the medial surface of the posterior superior iliac spine; from the deep surface of the tendon of origin of the erector spinae; from the posterior sacroiliac ligament; and

Fig. 2-70. Origins (*red*) and insertions (*black*) of the short muscles of the back in the lumbar region.

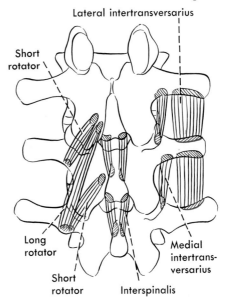

Lateral intertransversarius

Short
rotator

Long
rotator

Short
rotator

Medial
intertrans-
versarius

Interspinalis

verse process of each vertebra, but they insert into the base of the spinous process of the vertebra immediately above, skipping no vertebrae. The long rotators thus have a more oblique course than do the short ones; the latter, in the thoracic region, run almost horizontally to reach their insertions.

SEGMENTAL MUSCLES

Under this term are included those muscles that pass only from one vertebra to the next; the short rotators can also be put in this group, but are more conveniently considered with the long rotators and the other members of the transversospinalis system. The strictly segmental muscles, other than the short rotators, are the interspinales and the intertransversarii.

The *interspinales* (Figs. 2-69 and 2-70) are best developed in the cervical region, where they consist of six pairs of muscles of which the uppermost pass between the spinous processes of the second and third cervical vertebrae, the lowest between those of the seventh cervical and first thoracic. The interspinales of the two sides are situated immediately on either side of the midline, being separated by the interspinous ligament. The interspinales are largely lacking in the thoracic region, but usually occur between the first and second thoracic vertebrae, and between the eleventh and twelfth; additional ones may be present. The lumbar interspinales are usually well developed, and consist of at least four pairs stretching between the spinous processes of the five lumbar vertebrae; there may be a pair between the last thoracic and first lumbar vertebrae, or one between the fifth lumbar and the median crest of the sacrum.

The *intertransversarii*, as their name implies, pass between adjacent transverse processes. They are best developed in the cervical and lumbar regions, largely lacking in the thoracic region.

In the cervical region (Fig. 2-69) there are two intertransversarii on each side: anterior intertransversarii, which pass between the anterior tubercles of two adjacent transverse

processes; and posterior intertransversarii, which pass between the posterior tubercles of the transverse processes. There are seven sets of cervical intertransversarii, the uppermost set connecting the transverse processes of the atlas and axis, the lowest the processes of the seventh cervical and first thoracic vertebrae.

In the thoracic region the intertransversarii are not well developed, but may usually be found in the lower three or four intervertebral spaces, where they pass between the transverse processes of two adjacent thoracic vertebrae. They tend to be subdivided, as are the lumbar intertransversarii, into medial and lateral ones.

The lumbar intertransversarii (Fig. 2-70) are distinctly divided into two sets, as are the cervical ones, but they are medial and lateral rather than anterior and posterior. The lateral intertransversarii pass between the transverse processes of two adjacent vertebrae, while the medial intertransversarii arise from the mammillary process of one vertebra and insert on the accessory process of the vertebra above.

SUBOCCIPITAL MUSCLES

Four muscles comprise this group (Fig. 2-71), which lies deep to the semispinalis in the upper part of the neck, between the level of the spinous process of the axis and the skull.

The *obliquus capitis inferior* arises from the spinous process of the axis and extends laterally and upward to an insertion upon the transverse process of the atlas. The *obliquus capitis superior* arises from the transverse process of the atlas and passes upward, backward, and somewhat medially to be inserted into the occipital bone above the inferior nuchal line, largely concealed by the insertion of the splenius capitis.

The *rectus capitis posterior major* arises from the spinous process of the axis and extends upward and slightly laterally to insert into the occipital bone below the inferior nuchal line, and under cover of the insertion of the semispinalis capitis and obliquus capitis superior. The *rectus capitis posterior minor* arises from

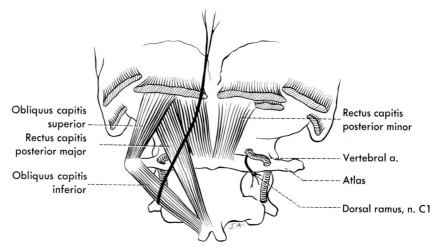

Fig. 2-71. Suboccipital muscles and the suboccipital triangle. The nerve on the left crossing the triangle is the greater occipital nerve.

the posterior tubercle of the atlas and extends upward and a little laterally, largely under cover of the rectus capitis posterior major, to insert into the occipital bone below the inferior nuchal line medial to and somewhat under cover of the insertion of the rectus capitis posterior major.

Suboccipital Triangle

The obliquus capitis inferior, the obliquus capitis superior, and the rectus capitis posterior major together bound a small triangle, the suboccipital triangle. In the floor of this triangle a portion of the posterior arch of the atlas is visible, and on the upper border of the posterior arch the vertebral artery runs transversely between its exit from the transverse foramen of the atlas and its disappearance through the posterior atlantooccipital membrane.

The dorsal or posterior primary branch (suboccipital nerve) of the first cervical nerve emerges above the posterior arch of the atlas, usually between the vertebral artery and the arch, but sometimes above the artery; it supplies all four suboccipital muscles and then typically emerges through the suboccipital triangle to help supply the overlying semispinalis capitis. (The smaller ventral ramus of the first cervical nerve lies deep to the vertebral artery, where it pierces the posterior at-

lantooccipital membrane and then turns forward around the superior articular fovea of the atlas.) Through the suboccipital triangle, also, run anastomotic branches between the deep cervical and occipital arteries, on the one hand, and the vertebral on the other. The major portion of the dorsal branch of the second cervical nerve crosses the triangle; it appears at the lower border of the obliquus capitis inferior, gives branches into the overlying semispinalis capitis and other muscles, and the main part then turns upward superficial to the triangle to reach the scalp as the greater occipital nerve. Before it does so, it may receive a twig from the first cervical nerve and usually receives a communication from the third, the cutaneous part of which is otherwise distributed to the scalp as the third occipital nerve.

RELATED MUSCLES

Muscles other than those usually described as deep back muscles also attain an attachment to the vertebral column, and should be mentioned briefly. These include the levatores costarum, attached posteriorly to the transverse processes in the thoracic region, and lateral and anterior vertebral muscles, largely confined to the cervical region.

LEVATORES COSTARUM

The levatores costarum (levators of the ribs) are small muscle slips that arise from the transverse processes of the seventh cervical and the upper eleven thoracic vertebrae, spread fanlike as they descend, and attach to the posterior surfaces of the ribs immediately medial to the angles (Fig. 2-72). There are 12 levatores costarum breves, and each of these attaches to the rib immediately below the transverse process from which it arises. Long levators are fewer in number, but usually arise from the transverse processes of about the seventh through the tenth thoracic vertebrae, in common with the short levators; like the short levators, they spread out in a fanlike manner, but each passes across the rib associated with the vertebra immediately below the origin of the muscle, to pass to the next rib below.

The levatores costarum probably act exclusively on the ribs, helping to raise them and thus increase the size of the thoracic cavity; theoretically, they might help to extend the vertebral column, to flex it laterally, and to rotate it toward the opposite side. These muscles are usually described as being innervated

Fig. 2-72. Origins (*red*) and insertions (*black*) of the levatores costarum, *left,* and of the rotatores, *right,* in the thoracic region

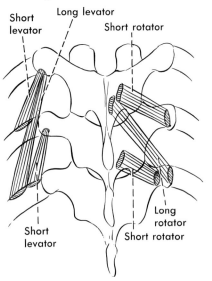

Short levator
Long levator
Short rotator
Short levator
Long rotator
Short rotator

by branches of the intercostal nerves, but Morrison found them to be supplied by dorsal rami.

ANTEROLATERAL MUSCLES

The anterior and lateral muscles associated with the cervical vertebral column are the rectus capitis lateralis, the rectus capitis anterior, the longus colli and the longus capitis, and the three scalenes. In the lumbar region the psoas major, and the minor if there is one, take origin from the front and sides of the vertebral column, and the quadratus lumborum arises and inserts on transverse processes. Of these muscles, the scalenes, arising from transverse processes of the cervical vertebrae and inserting into the first and second ribs, have already been described in connection with the muscles of the neck (Vol. 1); the psoas major and minor, muscles of the lower limb, are considered in a following chapter of the present volume.

The *longus colli* (longus cervicis) lies on the front of the cervical and of the upper thoracic parts of the vertebral column, for it extends from about the third or fourth thoracic vertebra to the atlas (Fig. 2-73). The muscle is broad in the midcervical region where the paired muscles almost completely cover the front of the vertebral column, but tapers at each end. It is usually described as being divided into three portions, an inferior oblique part, a vertical portion, and a superior oblique part. The inferior oblique part typically arises from the bodies of the first three thoracic vertebrae and extends upward and somewhat laterally to insert upon the anterior tubercles of the transverse processes of the fifth and sixth cervical vertebrae. The vertical part arises from the bodies of the first three thoracic and the last three cervical vertebrae and passes almost straight upward to insert into the bodies of the second, third, and fourth cervical vertebrae. The superior oblique portion arises from the anterior tubercles of the third, fourth, and fifth cervical vertebrae and is directed upward and medially to insert into the anterior tubercle of the atlas.

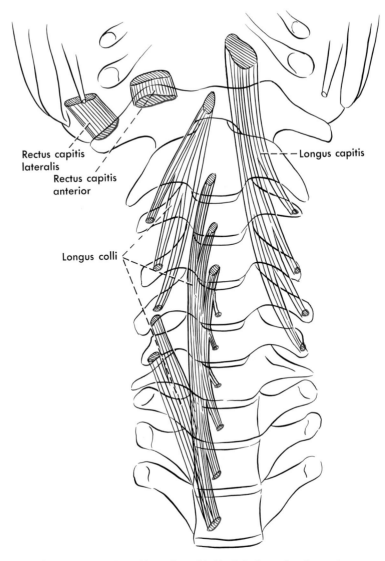

Rectus capitis
lateralis

Rectus capitis
anterior

Longus capitis

Longus colli

Fig. 2-73. Origins (*red*) and insertions (*black*) of the lateral and anterior vertebral muscles in the cervical region.

The *longus capitis* also covers part of the anterior surface of the vertebral column, lying in front of the superior oblique part of the longus colli. The longus capitis arises from the anterior tubercles of the transverse processes of the third, fourth, fifth, and sixth cervical vertebrae; the slips unite and the muscle becomes broad as it is traced upward. It inserts upon the inferior surface of the basilar part of the occipital bone.

The *rectus capitis lateralis* (Fig. 2-73) apparently represents the continuation upward to the skull of the posterior intertransversarii of the cervical region; it arises from the transverse process of the atlas, and inserts into the inferior surface of the jugular process of the occipital bone. The *rectus capitis anterior* arises from the upper surface of the lateral mass of the atlas, mostly under cover of the longus capitis, and inserts into the basilar part of the occipital bone between the insertion of the longus capitis and the occipital condyle.

The prevertebral muscles are all inner-

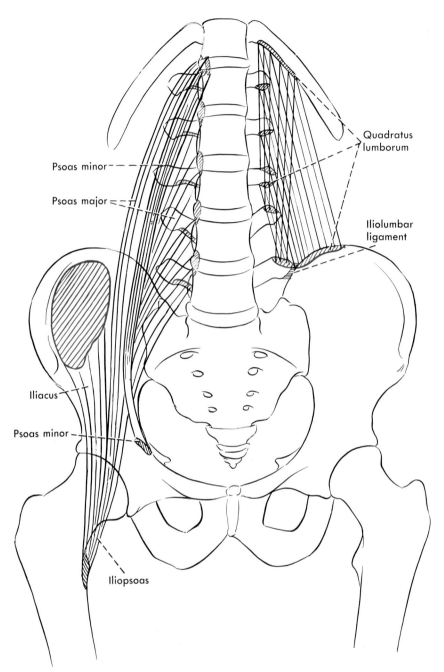

Fig. 2-74. Origins (*red*) and insertions (*black*) of the quadratus lumborum, psoas minor, and iliopsoas muscles.

vated by twigs from the anterior primary branches of cervical nerves: the two recti usually by fibers arising from the loop between the first and second cervical nerves; the longus capitis by twigs from the first three cervical nerves; and the longus colli by branches from the second to the seventh or eighth cervicals.

The *quadratus lumborum* (Fig. 2-74), a muscle of the posterior abdominal wall, arises from the iliac crest, the transverse processes of approximately the lower three lumbar vertebrae, and the layer of thoracolumbar (lumbo-dorsal) fascia immediately dorsal to it. It inserts into the transverse processes of the upper three or four lumbar vertebrae, and into a proximal part of the twelfth rib.

This muscle lies just lateral to the psoas major, in the interval between the iliac crest and the last rib. It is a lateral flexor of the vertebral column and, acting from above, can help tilt the pelvis toward the opposite side. Bilateral action produces extension of the vertebral column, and helps fix the lower ribs, as for inspiration. The muscle is innervated by twigs from the upper three or four lumbar nerves.

INNERVATION AND ACTION OF THE BACK MUSCLES

Innervation

With very few exceptions, the deep muscles of the back are supplied by the dorsal (posterior) rami of the spinal nerves. The exceptions apparently are the lateral intertransversarii of the lower thoracic and lumbar regions, which are said to be supplied by ventral (anterior) rami even though the medial are supplied by the dorsal; and, according to most texts, both the anterior and posterior intertransverse muscles of the cervical region. According to Cave, however, while the anterior intertransversarii are supplied by ventral rami as usually described, each posterior one receives a double innervation, its more lateral fibers being supplied from the ventral ramus but its more medial ones from the dorsal ramus; medial and lateral parts of this muscle are recognized in current anatomic terminology.

The dorsal rami of the spinal nerves contain fibers derived from both the ventral (anterior) and dorsal (posterior) roots, and therefore furnish the muscles of the back with both their motor and afferent fibers; many of the nerves, after supplying the muscles of the back, also continue to supply skin of the back.

The four suboccipital muscles are supplied by the dorsal ramus of the first cervical nerve, which usually continues also into the semispinalis capitis; the upper parts of the back muscles in the cervical region are also supplied by direct twigs from the dorsal rami of the second and third cervical nerves, and by branches that arise from the so-called posterior cervical plexus, a very simple series of loops between the first and second and the second and third dorsal rami, sometimes also with a loop to the fourth.

The dorsal rami of the first two cervical nerves do not divide into medial and lateral branches, but those of the remaining cervical nerves, of the thoracic and lumbar nerves, and of the first three sacral nerves all divide; in general, both medial and lateral branches of the dorsal rami run oliquely downward to supply muscles at and a segment or two below their level of emergence; Stilwell ('56) found the back muscles, like the intervertebral ligaments, receiving fibers at every level from two adjacent nerves. The medial branches supply the more medially lying musculature, the lateral branches the more lateral muscles; one of the two branches, after supplying muscle, commonly continues to supply skin. For the cervical and upper thoracic nerves, the medial branch usually continues to the skin, while for the lower thoracic and lumbar ones it is the lateral branch that so continues. The lower part of the multifidi is supplied both by medial branches from the first three sacral nerves and by twigs from the posterior sacral plexus, a series of simple loops connecting the dorsal rami of the first three sacrals, sometimes also the nerve above or below these. The fourth and fifth sacrals do not usually participate in the innervation of any muscle, but go to skin and tissue about the coccyx.

Function

It is obvious that, for the most part, the deep muscles of the back tend to work together as a whole, all of them extending the vertebral column, some perhaps flexing, but with varying efficiency, toward their own side; some rotating the vertebral column or the head, most frequently toward the opposite side but sometimes toward the same side. Electromyographic studies of the muscles of the back, although not in complete agreement, have added considerably to our knowledge of the function of the various muscle groups. The action of the splenius muscles seems obvious from their anatomy: The splenius capitis and the splenius cervicis, acting unilaterally, should extend the head and neck and laterally flex them toward the same side, at the same time rotating the face toward the same side; both sets of muscles acting together should extend or hyperextend the head and neck. Electromyography has confirmed these actions for the splenius capitis (Takebe and co-workers).

Morris and co-workers recorded from the iliocostalis dorsi, longissimus dorsi, and iliocostalis lumborum parts of the erector spinae, from the multifidi, and from the rotators. They found during flexion that all the muscles tested were active; in complete flexion, the multifidi and rotators became inactive, but some slight activity usually persisted in some part of the erector spinae, commonly the iliocostalis dorsi; all muscles again became active when extension was started, although not necessarily at the same time and with the same force. Pauly's findings were much the same, but he said that in complete flexion the erector spinae was "largely inactive," while the semispinalis capitis and cervicis showed some activity. Pauly also, surprisingly, concluded that the spinalis, the most poorly developed of the group, is the major muscle involved, showing more activity than the longissimus, which in turn is more active than the iliocostalis lumborum. Thus, although all the muscles tested proved to be extensors, in the fully flexed position most of the weight is borne by ligaments. Morris and co-workers found that the rotators usually showed continuous activity while standing at rest and the longissimus showed slight to moderate activity unless the body was in unforced extension. Asmussen also reported that the lower part of the erector spinae was usually active in quiet standing and attributed this to the fact that the line of gravity is usually anterior to the center of the body of the fourth lumbar vertebra; in some people, however, the back muscles were relaxed and the abdominal muscles were the antigravity ones.

The rotators and the multifidi were found to be active in rotation to the opposite side (Morris and co-workers), and the obliquely running semispinalis, not tested, would be ex-

pected to be active also. (The semispinalis capitis, according to Takebe and co-workers, acts primarily as an extensor of the head.) Donisch and Basmajian had a different interpretation of the action of these obliquely running muscles. They studied the transversospinalis system, probably specifically the multifidi, electromyographically and regarded these muscles as acting primarily to adjust adjacent vertebrae rather than to move the vertebral column as a whole.

In Morris's investigation, the ipsilateral erector spinae, especially the longissimus, was also active during rotation, but this was interpreted as stabilizing the column against the pull of the abdominal muscles rather than initiating rotation. The ipsilateral erector spinae also became active on lateral flexion (and some activity was present in all the contralateral muscles after flexion started); Morris and co-workers regarded the ipsilateral activity as initiating lateral flexion, but Pauly said it occurs only when there is also ventral flexion and regarded this also as extensor rather than lateral flexor activity. There thus seems to be doubt whether the erector spinae contributes appreciably to either rotation or lateral flexion. In returning to the erect position from both rotation and lateral flexion there was relatively little muscle activity, suggesting that this return is brought about mostly by the elasticity of the ligaments and muscles.

Davis and co-workers reported that in lifting heavy weights with the knees bent, the hips at first rise faster than the shoulders, so that the trunk becomes more flexed; thus the erector spinae has to bear more of the weight than would be expected.

Of the short muscles of the back that have not been investigated, the interspinales presumably help only in extending the vertebral column, the intertransversarii in flexing it laterally. The obliquus capitis inferior rotates the atlas, thus turning the face toward the same side; the obliquus capitis superior extends the head and bends it laterally toward the same side; the rectus capitis posterior major both extends the head and helps to rotate it toward the same side; and the rectus capitis posterior minor probably only extends the head.

Among the prevertebral muscles, the longus colli and capitis primarily flex the neck and head, or acting unilaterally rotate to the same side, the rectus capitis lateralis aids in flexing the head laterally, the rectus capitis anterior in flexing it anteriorly. The lateral muscles in the neck, the scalenes, act primarily upon the first and second ribs, but may also bend the neck toward the same side and slightly rotate the face toward the opposite side. The psoas major, arising as it does from the lumbar vertebral column, usually acts upon the free limb as a flexor of this but becomes active if one leans back or toward the opposite side; or, if opposed by the weight of most of the trunk as in attempting to sit up from a recumbent position, it may first pull forward primarily upon the lumbar portion of the column rather than flex at the hip joint, thus, with gravity, producing hyperextension of the lumbar column. The quadratus lumborum laterally flexes the column or, acting bilaterally, helps to extend it.

Finally, the obvious contribution to flexion, lateral flexion, and rotation that such muscles as those of the anterolateral wall make must be borne in mind: thus the rectus abdominis is an important flexor of the vertebral column, and the external and internal obliques of the two sides, acting together, also aid in this flexion. Further, the external oblique abdominal muscle of one side and the internal oblique of the other side, acting together, impose a twisting, rotatory, flexion upon the vertebral column, flexing the column anteriorly and to the side of the external oblique, and at the same time rotating the column toward the opposite side.

Although other muscles thus move the vertebral column, the deep muscles of the back are alone capable of maintaining an erect posture and providing stability of the vertebral column during other movements. Paralysis of the musculature of the back makes it impossible to maintain an erect posture; spasm of the musculature on one side produces scoliosis; and soreness of the muscles of the back makes any movement difficult, be-

cause of the importance that these muscles have in providing stability to the trunk while movements of a limb are being executed.

Spinal Cord and Nerve Roots

The spinal cord and its surrounding meninges lie in the vertebral canal, formed by the successive vertebral foramina. The anterior wall of the vertebral canal is therefore formed by the vertebral bodies and the intervening disks of the vertebral column; the lateral walls by the pedicles, between which appear the intervertebral foramina for the exit of the spinal nerves; and the posterior wall, or roof, by the overlapping articular processes laterally, and medially by the remainder of the laminae with their surmounting spinous processes and by the ligamenta flava connecting adjacent laminae. Surgical access to the vertebral canal is therefore usually obtained by laminectomy, in which an appropriate amount of one or more laminae medial to the articular processes, and the intervening ligamentum flavum, are removed. In the lower lumbar region, however, where the laminae are some distance apart—specifically, between the fourth and fifth lumbar vertebrae, and the fifth lumbar and the first sacral—sufficient exposure for examination and operation upon the limited region of an intervertebral disk can often be obtained by removal of ligamentum flavum alone, without removal of adjoining bone (Love, '39).

Epidural Space
The anteroposterior and transverse diameters of the spinal cord and its meninges, while varying at different levels, are everywhere considerably less than the diameters, also varying, of the vertebral canal, thus allowing for movement of the vertebral column without compression of the spinal cord. The space between the walls of the vertebral canal and the outer meninx of the cord, the dura mater, is the epidural space, and is loosely filled with

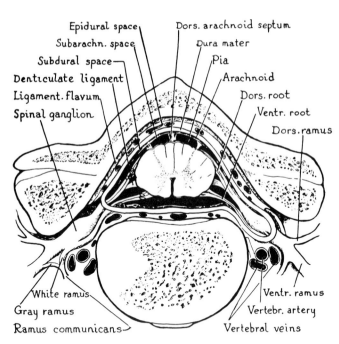

Epidural space
Subarachn. space
Subdural space
Denticulate ligament
Ligament. flavum
Spinal ganglion
Dors. arachnoid septum
Dura mater
Pia
Arachnoid
Dors. root
Ventr. root
Dors. ramus
White ramus
Gray ramus
Ramus communicans
Ventr. ramus
Vertebr. artery
Vertebral veins

Fig. 2-75. Cross section of the spinal cord *in situ*. Note the epidural space. (After Rauber-Kopsch, from Strong OS, Elwyn A: Human Neuroanatomy [ed 3]. Baltimore, Williams & Wilkins, 1953)

fat and connective tissue that serves as padding about the cord; it contains parts of the *internal vertebral venous plexuses* (Fig. 2-75), and the branches of the spinal arteries that supply the vertebral column, the dura, and the tissue of the epidural space.

The posterior internal vertebral venous plexus lies posterior and posterolateral to the spinal cord in the fat of the epidural space. It consists of more or less obvious paired longitudinal channels that communicate freely across the midline with each other, between the laminae, with the external vertebral venous plexus (Fig. 2-76), and anterolaterally with the anterior internal vertebral venous plexus. The anterior internal vertebral venous plexus, consisting also of two main longitudinal channels, lies on the posterior aspect of the vertebral bodies, in part in the epidural fat; the paired channels communicate with each other across the midline largely through veins that pass between the posterior longitudinal ligament and the vertebral bodies, and form a large part of the drainage of the bodies.

The anterior internal vertebral venous plexus connects freely with the posterior internal vertebral venous plexus. It receives the veins that emerge from the dorsal aspect of the vertebral bodies and connects laterally, by segmental branches that emerge through the intervertebral foramina, with the external vertebral venous plexuses. Amsler and Wilber have reported demonstrating the anterior internal plexus and its segmental drainage through the foramina by injection of radiopaque material into the spinous process of a lumbar vertebra and found that failure of one of the segmental veins to fill allows localization of the level of a protruded disk with apparently as much accuracy as does myelography. (Ascending lumbar venography achieves the same result.)

Both internal vertebral venous plexuses connect above, by piercing the spinal dura at its attachment to the skull, with the marginal sinuses and the basilar plexus in the dura of the posterior cranial fossa. They form a part of the system, sometimes referred to as "Batson's veins," which, because of the absence or incompetence of valves, can receive blood from veins of the pelvis and abdomen, instead of sending blood to these vessels, and hence conduct infection or metastasis to the cranial cavity without involvement of the pulmonary circulation. They are also an important part of the venous outflow from the brain, taking over increasing amounts of this function from the internal jugulars as intrathoracic pressure rises, and they are the best-developed collateral channel for drainage of the inferior vena cava (Batson; Norgore; Eckenhoff). In compression of the internal jugulars or inferior vena cava their dilation reduces the epidural space, compressing the dura, and thus elevates the pressure of the cerebrospinal fluid.

The *spinal branches* of the vertebral arteries in the neck, of the dorsal (posterior) branches of the intercostal and lumbar arteries in the thorax and abdomen, and of the lateral sacral arteries in the pelvis, typically divide into three branches as they enter the intervertebral foramina: one passes dorsally toward the laminae, dividing into ascending and descending branches; one passes ventrally toward the vertebral bodies, dividing into ascending and descending branches that disappear deep to the posterior longitudinal ligament; and the middle one runs along the nerve root. The dorsal branches primarily supply the vertebral arches, the ventral branches the vertebral bodies (Fig. 2-14), but the vessels of each set also tend to anastomose with their fellows to form paired longitudinally running channels in the epidural tissue dorsal and ventral to the spinal cord; these are closely associated with the venous plexuses, and supply the epidural tissue and the dura. The middle (radicular) branches of the spinal arteries supply the meninges about the nerve roots, and the larger ones continue along dorsal or ventral roots to help supply the spinal cord.

It is into the connective tissue of the epidural space that injections of anesthetics are made in the operation referred to as caudal or epidural analgesia or anesthesia. Since the aim here is to avoid penetrating the dural-arachnoid sac (which would produce spinal anesthesia) the variations in the caudal limit of this sac (see a following section) should be borne in mind, as should also variations in the sacrum (Lanier and co-workers; Trotter and

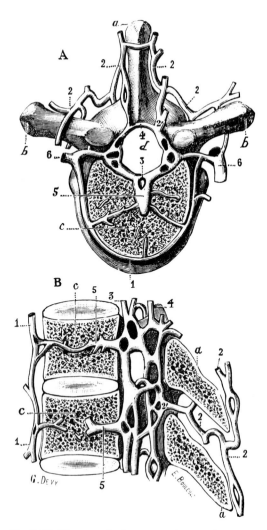

Fig. 2-76. The vertebral venous plexuses in a cross section (*A*) and a longitudinal section (*B*) of the vertebral column. *a, b,* and *c* identify the spinous and transverse processes and the body of the vertebra. *1* is the anterior external venous plexus, *2* the posterior external, *2′* connections from the posterior external to the posterior internal (*4*); *3* is the anterior internal plexus, *5* the major drainage of the vertebral body, *6* segmental veins and their tributaries. (After Breschet, from Testut L: Traité d'Anatomie Humaine, Vol 2. Paris, Doin, 1891)

Letterman; Letterman and Trotter; and Trotter), especially the sacral hiatus through which the needle is introduced in caudal analgesia. Because connective tissue binds the spinal nerves to the edges of the intervertebral foramina as they leave these foramina, mate-

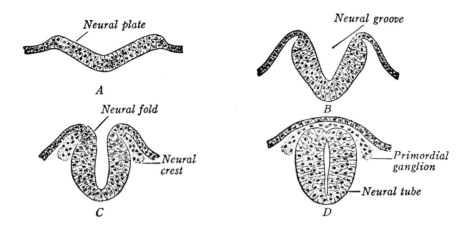

Fig. 2-77. Early development of the spinal cord, in cross section. (Arey LB: Developmental Anatomy [ed 6]. Philadelphia, WB Saunders, 1954)

rial introduced into the epidural space tends to spread within the loose tissue of this space rather than to make an exit through the foramina. Lanier and co-workers found a good deal of variation in the level reached by material injected into the epidural space of cadavers, and Rosenbaum reported that among 46 patients receiving the same (20 ml) dose of anesthetic agent the level of anesthesia varied from involvement of the lower sacral nerves only to involvement of the sixth thoracic. Lins reported that with the proper precautions injections into the epidural space can be made anywhere between the levels of the seventh cervical and fifth lumbar vertebrae.

Development of Cord and Meninges

The central nervous system makes its appearance in about the 19-day-old embryo, as a thickening of the ectoderm in the dorsal midline. The neural plate thus formed undergoes rapid thickening, this being more pronounced at its edges than at its midline, and the plate is converted into a neural groove (Fig. 2-77). With further growth, the lips of the neural groove meet and fuse dorsally to form the definitive neural tube. Even before the neural tube is completed, the cranial end has shown considerable enlargement over the caudal part and is destined to form the brain, while the narrower caudal portion represents the spinal cord. The caudal portion is at first relatively short, but is lengthened by the growth of the caudal end of the embryo.

As the neural tube closes, cells lying at the junction of the neural ectoderm and the undifferentiated, more laterally lying ectoderm destined to help form skin separate from both to form a column of cells on either side, the neural crest. The neural crest becomes segmented, and its cells form the neurons of the dorsal or posterior root ganglia. (Traditionally, these cells have also been described as giving rise to the ganglia of the sympathetic system and to the nonneuronal neurilemma or sheath cells of the spinal nerves, but there is conflicting experimental evidence on both these questions. The weight of evidence seems to indicate that the neural crest definitely does participate in the formation of sympathetic and sheath elements, but that migratory cells from the neural tube itself may also help to form both of them.) Brown and Podosin described, as the syndrome of the neural crest, the condition of congenital insensitivity to pain and of anhydrosis, suggesting failure of the neural crest to contribute properly to the dorsal root and the sympathetic ganglia.

After the closure of the neural tube its walls thicken rapidly and become differentiated into three layers, a central ependymal, an intermediate mantle, and a peripheral marginal layer. The ependymal layer persists as the ependymal lining of the central canal of the spinal cord in the adult, and in the embryo contributes cells to the mantle layer; the mantel layer differentiates into the gray columns (horns) of the spinal cord; the marginal layer is composed largely of the processes of ependymal and neuroglial cells, and into this grow axons from the mantle layer and from the dorsal (posterior) root ganglia to form the white matter of the spinal cord.

At first, the developing nervous system forms a major part of the embryo, and during the first few months of prenatal life the spinal cord grows more rapidly than it does during later prenatal life; however, other parts of the body are developing even more rapidly, so that even as early as the second prenatal month the size of the spinal cord in relation to the body as a whole begins to decline rapidly; after the third month, when all the organ systems are well established, the decline becomes considerably slower. At the end of the third month of fetal life, the developing spinal cord and the vertebral column are still of approximately equal length, but thereafter the spinal cord begins to lag behind in its longitudinal growth. This results in the caudal end of the cord being shifted progressively higher in relation to the vertebral segments (Fig. 2-78); however, the lower end of the cord is attached to the surrounding dura and through this to the more caudal vertebrae, and the spinal nerves are already formed and thus have fixed relations to the vertebral segments. The difference in rate of growth between the vertebral column and the spinal cord results, therefore, in the terminal portion of the cord being drawn out into a functionless thread of tissue, the filum terminale, and in the more caudal nerve roots having to run longitudinally for some distance, where they form the cauda equina, before they reach their levels of exit.

The dura mater, the outer of the meninges, is apparently developed as a condensation from surrounding mesoderm, as is the vertebral column itself. The arachnoid and the pia, the leptomeninges, probably develop from the ectodermal neural crest cells (Harvey and co-workers).

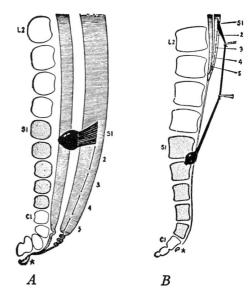

Fig. 2-78. Formation of the filum terminale and cauda equina. In *A*, the condition at 9 weeks, the spinal cord extends to the coccygeal level, and the spinal nerves, represented here by the first sacral, arise and leave the vertebral canal at almost the same level; in *B*, at 6 months, the faster growth of the vertebral column has stretched the caudal attachment of the cord into a long filum terminale and, with the relative upward displacement of their origins, the course of the sacral nerves in the vertebral canal has become a vertical one. (Arey LB: Developmental Anatomy [ed 6]. Philadelphia, WB Saunders, 1954)

MENINGES

Dura Mater and Subdural Space

The spinal dura mater (Fig. 2-79; see also Fig. 2-81) is a tube of dense fibrous connective tissue that extends from the foramen magnum to about the middle of the second sacral vertebra. Superiorly, the dura is continuous at the foramen magnum with the intracranial dura; here, because the inner periosteum of the skull and the cranial dura are fused together, the epidural space terminates.

The rough outer surface of the dura, composed of fibrous connective tissue, blends with connective tissue in the epidural fat, especially in the anterior midline and these trabeculae (sometimes regarded as ligaments of the dura; *e.g.,* Hofmann) partially anchor the dura to the walls of the vertebral canal. The firmer attachments of the spinal dura, in ad-

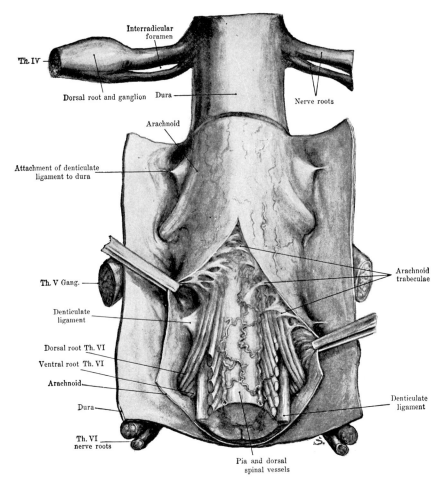

Th. IV

Interradicular foramen

Dorsal root and ganglion   Dura

Nerve roots

Arachnoid

Attachment of denticulate ligament to dura

Arachnoid trabeculae

Th. V Gang.

Denticulate ligament

Dorsal root Th. VI

Ventral root Th. VI

Arachnoid

Dura

Denticulate ligament

Th. VI nerve roots

Pia and dorsal spinal vessels

Fig. 2-79. Meninges about the cord. The dural sac is intact above, reflected below, and the arachnoid has also been opened below to show the arachnoid trabeculae in the subarachnoid space. (Mettler, FA: Neuroanatomy [ed. 2]. St. Louis, Mosby, 1948)

dition to that at the foramen magnum are, however, at the intervertebral foramina, where the dural sleeves surrounding each set of spinal nerve roots and dorsal root ganglia blend with surrounding connective tissue to anchor the dura rather firmly at the foramina; further, the slender fibrous cord of dura continuing downward from the tapering lower end of the dural sac aids the lower spinal nerves in anchoring the caudal end of the dura, for it (the filum of the dura mater) blends with the connective tissue on the dorsal surface of the coccyx. The dural sleeves about the nerve roots protect these roots, so that a pull upon the nerve at the foramen is transmitted through the dural sheath about

the roots to the dural sac itself, rather than transmitted upward along the nerve roots to the spinal cord.

As already noted, the filum of the dura mater (also called external filum terminale) begins at the tapered lower end of the dural sac, and the dural sac itself ends at about the level of the second sacral vertebra. The exact level of ending varies somewhat, however, for Lanier and co-workers found the sac ending somewhat caudal to the middle of the second sacral vertebra (the mean in their series) in 46% of more than 50 cadavers investigated, and cranial to the mean in 37%, the extent of variation being from the middle of the first to the middle of the third sacral vertebra. A lower ending is apparently usual in infants: Wagner (1890) found the dural sac in 20 newborn and young children ending at the middle of the second sacral vertebra in only 10%, at the lower border of this vertebra in another 10%, and at the level of the third sacral vertebra in 80% but at the level of the lower part of this vertebra in only 5%.

In contrast to the outer surface of the dura, the inner surface has a smooth fibroblastic lining separated from the outer surface of the arachnoid, the second of the three meninges, by a small potential space, the subdural space. This space normally contains only enough fluid to moisten its walls and is crossed only occasionally by small veins and fine strands of connective tissue. Arachnoid and dura are continuous with each other through the occasional trabeculae that cross the subdural space, and also fuse together on the nerve roots at or distal to the ganglia, thus obliterating the epidural space here.

Congenital defects in the dura could obviously allow the arachnoid to herniate through it, and such defects, or perhaps diverticula of the dura and arachnoid, are thought to be the origin of the rather rare "extradural cysts" (du Toit and Fainsinger; Tarlov, '53).

Arachnoid, Pia, and Subarachnoid Space

The outer surface of the *arachnoid* is largely smooth, and forms the inner wall of the subdural space. However, from its inner surface arise numerous trabeculae (Fig. 2-79) that pass across the underlying space, the subarachnoid space, to become continuous with the pia mater, the innermost meninx, which is closely applied to the outer surface of the spinal cord and nerve roots. The flattened cells lining the inner surface of the arachnoid continue around the connective tissue cords of the arachnoid trabeculae to become continuous with the exactly similar flattened cells forming the outer surface of the pia mater; thus the subarachnoid space is one between the cellular surfaces of the arachnoid and the pia, and the trabeculae and blood vessels crossing the subarachnoid space are covered by flattened cells continuous with both layers.

Developmentally and phylogenetically, the arachnoid and pia represent a layer, the leptomeninx, which has partially split into two layers, the intervening space being the subarachnoid cavity. Put another way, the arachnoid and the pia are related to each other in very much the same way as are visceral and parietal peritoneum.

The *pia mater* is a thin layer closely applied to the spinal cord and to the nerve roots as they cross the subarachnoid space. As already noted, its outer layer helps to line the subarachnoid cavity, and is composed of flattened cells that are continuous with those of the arachnoid. A very small amount of connective tissue, attaching the pia closely to the nervous system, forms its deep layer.

On the sides of the cord, pia and arachnoid and dura are united by paired bands of dense connective tissue, the *denticulate ligaments* (Fig. 2-80), which pass from the spinal cord to the dura. The denticulate ligaments are firm bands of tissue continuous with the pia, but very much thicker than it. They are attached to the spinal cord along almost its entire length, between the origins of the dorsal and the ventral roots. Hyndman and Van Epps said that the denticulate ligaments attach posterolaterally so as to separate a posterior third from an anterior two thirds, but this varies, and the ligament is often attached very close to the middle of the cord. Laterally, as it extends toward the arachnoid and dura, each denticulate ligament gives rise to a series of pointed processes (from which the ligament derives its name) that pass completely across the subarachnoid space to penetrate the arachnoid and attach to the dura, thus suspending the cord in the subarachnoid space and preventing its undue rotation. Between the denticulations the lateral edge of the denticulate ligament is free and projects only slightly from the spinal cord.

The denticulations or pointed processes vary somewhat in number; the first one typically attaches above the level of the first spinal nerve to the lowermost part of the cranial dura mater, while the lowermost one passes laterally between the roots of the last thoracic and the first lumbar nerves, or between the

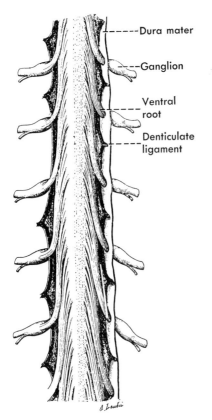

Dura mater

Ganglion

Ventral root

Denticulate ligament

Fig. 2-80. Anterior view of the spinal cord to show a part of the denticulate ligament. On the *right* side the reflected edge of the dura is seen, on the *left* the dura has been cut away. (After Hirschfeld, from Poirer P, Charpy A: Traité d'Anatomie Humaine [ed 2], Vol 3, Fasc 1. Paris, Masson et Cie, 1901)

last two thoracic nerves, to attach to the dural sac between the entrances of these nerves into their special dural sleeves. Thus there are usually 20 or 21 denticulations; since the number varies, and there may be even a different number on the two sides of the body, the position of the last denticulation of the denticulate ligament cannot be used to identify absolutely the last thoracic nerve; the lowermost denticulation may be below this nerve or above it. Presumably, when the spinal cord is unusually long there are also more denticulations than usual.

Since the arachnoid is closely apposed to the dura, the arachnoid sac ends below at essentially the same level as does the dural sac; the shape of its lower end varies, according to the variation in the dural sac, and may be either somewhat long and tapering or rather rounded and blunt. The arachnoid presumably contributes to the filum of the dura, for the filum of the cord joins this.

The subarachnoid cavity (Fig. 2-81), between the arachnoid and pia, varies in size. It is usually particularly small in the thoracic region, where the spinal cord is also small. Throughout the lumbar region both the dura and arachnoid are large tubes, and therefore in the lower lumbar region, below the end of the spinal cord, the subarachnoid cavity is large. This part of the subarachnoid space, the lumbar cistern, is occupied only by the nerve roots of the cauda equina (and by the filum terminale), and therefore contains a rather large amount of cerebrospinal fluid in which the nerve roots can be moved freely about. Thus the introduction of a needle into the subarachnoid space below the level of the spinal cord taps a relatively large accumulation of cerebrospinal fluid, and avoids injury to the nervous system; the nerve roots, floating in the fluid, tend to be pushed aside rather than penetrated by the needle.

The subarachnoid space around the spinal cord communicates freely through the foramen magnum with the cranial subarachnoid space. Most of the cerebrospinal fluid is, of course, formed in the ventricular system of the brain and is absorbed into the superior sagittal cranial venous sinus through the arachnoid villi that project into the sinus; hence the cerebrospinal fluid around the cord moves downward from the cranium, and then upward again. Because the cranial subarachnoid space below the cerebellum and posterior to the medulla, the cerebellomedullary cistern or cisterna magna, can be reached by a needle inserted between the atlas and the occipital bone, it is possible to measure simultaneously the pressure of the cerebrospinal fluid in this cistern and the lumbar cistern, and thereby determine whether or not a suspected block of the spinal subarachnoid space, as by a tumor, does indeed exist.

The subarachnoid space is continued, within the dural-arachnoid sleeve around each dorsal and ventral root, to about the level of junction of these roots, where the

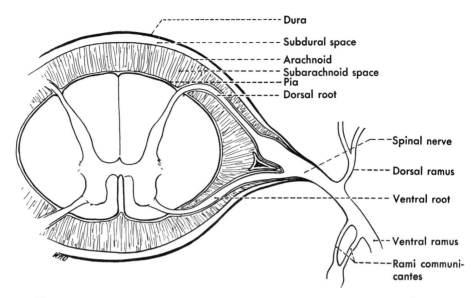

Fig. 2-81. Diagram of the spinal cord, nerve roots, and meninges in cross section

meninges become continuous with the connective tissue of the peripheral nerve. The subarachnoid spaces around the nerve roots are usually thought to be completely closed distally, and continuous with neither the perineural spaces nor the lymphatics in the peripheral nerves; extensions of the radiopaque material along the nerve roots may frequently be seen in roentgenograms, but are usually confined to the subarachnoid space. However, sometimes extensions apparently beyond this are seen (French and Strain reported four such instances among 200 Pantopaque myelograms). It has also been shown that certain substances introduced into the subarachnoid space may reach lymph nodes fairly quickly, and Brierley and Field interpreted this as being due to a special permeability in the region of ending of the meninges about the nerve roots, or even possibly to the presence of stomata here, although these have not been demonstrated.

French and Strain expressed the belief that the apparent continuity between the subarachnoid space and the perineural spaces, which they observed in roentgenograms, is not normal, and that occasional continuity may be due either to local anatomical abnormalities or to the conditions of injection. They said that they were able to establish

such connections in experimental animals through retrograde injections along the nerves.

SPINAL CORD

Length

The level of continuity between medulla and spinal cord is arbitrarily said to be the lower limit of the decussation of the pyramids, the level of origin of the uppermost filaments of the roots of the first cervical nerve, or the foramen magnum; for the surgeon, either of the two latter is obviously preferable, but it sometimes happens that the first cervical nerve lacks a dorsal root, and in this case the level of the foramen magnum, which is usually slightly above that of the uppermost rootlets of the first cervical nerve, is the only readily available landmark. Because of the much greater growth in length of the vertebral column, and even of the dural and arachnoidal sacs, than of the spinal cord, the cord ends considerably above the lower end of the vertebral column and several vertebral segments above the tapered ends of the dural and arachnoidal sacs.

McCotter found the average length of the spinal cord in males to be almost 45 cm

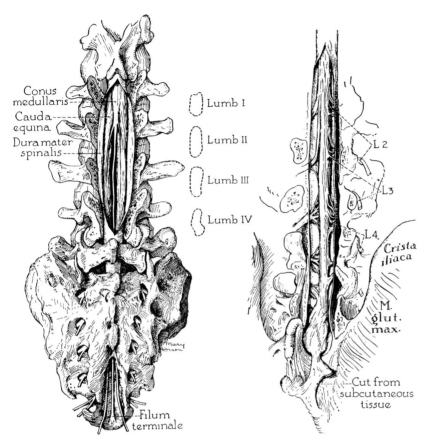

Conus
medullaris----
Cauda----
equina
Dura mater----
spinalis

Lumb I

Lumb II

Lumb III

Lumb IV

L 2

L3

L4

Crista
iliaca

M.
glut.
max.

--Cut from
subcutaneous
tissue

--Filum
terminale

Fig. 2-82. Normal level of ending of the spinal cord, *left,* and a rare example of sacral cord, *right,* exposed by laminectomy. (Reimann AF, Anson BJ: Anat Rec 88:127, 1944)

(about 18 inches), that in females about 42 cm; while the length of the spinal cord tends to vary with the length of the trunk, McCotter found no definite ratio between the length of the vertebral column and that of the spinal cord. In his series of 234 cases the position of the lower end of the cord varied from as high as the middle of the twelfth thoracic vertebra to as low as the lower border of the second lumbar; however, in 77% of his white subjects the termination was between the upper border of the first and the upper border of the second lumbar vertebra. Needles, in a series of comparable size, found a slightly wider variation, from the middle of the twelfth thoracic to the lower third of the third lumbar vertebra; however, the majority ended below the middle of the intervertebral disk between L1 and L2.

Reimann and Anson, in a series of 129 cords, found that 94% terminated within the levels of the first two lumbar vertebrae. In reviewing published data on 801 spinal cords, including the series already cited, they found terminations within the limits reported by Needles except for a case of their own which showed a persistent sacral cord (Fig. 2-82). In this series of 801 spinal cords, the greatest frequency of ending lay opposite the lower third of the first lumbar vertebra, and 51% ended at the level of the lower third of the first lumbar and the upper third of the second (Fig. 2-83).

Cross-Sectional Diameters

The spinal cord varies considerably in its diameters from one region to another, and Elliot found among 98 adult cords a very large range of variation in diameters at identical

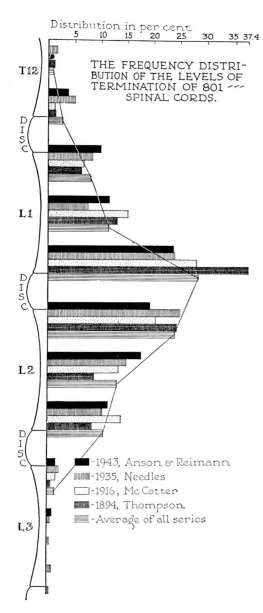

THE FREQUENCY DISTRI-
BUTION OF THE LEVELS OF
TERMINATION OF 801 ---
SPINAL CORDS.

Distribution in per cent

■ -1943, Anson & Reimann
▥ -1935, Needles
▢ -1916, McCotter
▨ -1894, Thompson
☰ -Average of all series

Fig. 2-83. Reported levels of ending, in terms of the relation to upper, middle, and lower thirds of the vertebral bodies, of 801 spinal cords. (Reimann, AF, Anson BJ: Anat Rec 88:127, 1944)

levels. In general, the cord is wider laterally than it is deep in the anteroposterior direction; thus Elliot found average measurements of the cord at the largest part of the cervical enlargement to be 13.2 mm in width, 7.7 mm in anteroposterior depth; at the smallest part of the thoracic region, the corresponding figures were 8 mm and 6.5 mm, and

in the largest part of the lumbosacral swelling, 9.6 mm and 8.0 mm. If these measurements are compared with the size of the vertebral canal, as given by Aeby, it is obvious that the cord has considerable room about it for the subarachnoid and epidural spaces; Aeby reported that in an adult female the vertebral canal in the cervical region was 24.5 mm wide, 14.7 mm in anteroposterior dimension; in the thoracic region, these measurements were 17.2 mm and 16.8 mm; in the lumbar, 23.4 mm and 17.4 mm. Roughly, therefore, the cord occupies perhaps about half of the space, in each direction, in the vertebral canal.

Conus and Filum

The tapered lower end of the spinal cord is the conus medullaris (Fig. 2-84) best defined, according to Tarlov ('38, '53) and others, as

Fig. 2-84. Conus medullaris, filum terminale, and the lumbar cistern, after removal of the nerve roots of the cauda equina, in a lateral view. (Poirer P, Charpy A: Traité d'Anatomie Humaine [ed 2], Vol 3, Fasc 1. Paris, Masson et Cie, 1901)

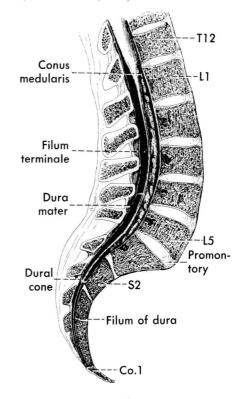

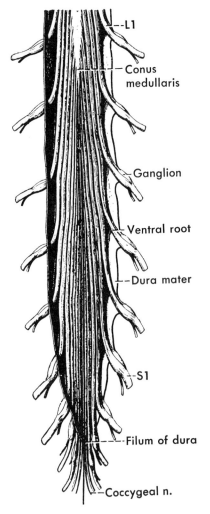

L1

Conus medullaris

Ganglion

Ventral root

Dura mater

S1

Filum of dura

Coccygeal n.

Fig. 2-85. The cauda equina after removal of the anterior part of the dura and arachnoid. (Poirier P, Charpy A: Traité d'Anatomie Humaine [ed 2], Vol 3, Fasc 1. Paris, Masson et Cie, 1901)

beginning at the level of origin of the fourth sacral nerve, thus below the level of the nerve supply to the limbs, and as ending below the origin of the coccygeal nerve. From its tapered point the conus medullaris is continued as the filum terminale (filum of the spinal cord, internal filum terminale); the filum terminale, representing a portion of the spinal cord drawn into a filament because of the unequal growth of cord and vertebral column, consists largely of glia surrounded by pia, but may contain (Harmeier; Tarlov, '53) occasional normal-appearing nerve cells and

nerve fibers. The conus and the filum are surrounded by the numerous nerve roots of the lower lumbar, sacral, and coccygeal nerves, the whole constituting the cauda equina (Fig. 2-85); according to Tarlov, the filum is distinguished from the nerve roots by a very prominent anterior vein. At the lower end of the arachnoidal and dural sacs the filum terminale penetrates the arachnoid sac and continues caudally, surrounded by a sheath derived from the dura (filum of the dura mater), to its attachment on the dorsum of the coccyx. The nerve cells and nerve fibers in the filum may extend for varying distances into the "extradural" segment of the filum (that is, beyond the end of the dural sac).

Apparently there is little growth in the conus medullaris, but a great deal in the filum terminale, between birth and adulthood: Tarlov ('53) found the length of the conus in a newborn infant to measure 1.2 cm while the average measurement for adults was 1.7 cm. The intradural segment of the filum (filum of the cord) was 4.2 cm long in the infant, the extradural segment (filum of the dura) 2.2 cm, while the average measurements for the adult were 15 cm for the intradural segment and 27 cm for the extradural one.

In its upper part, at least, the filum contains a continuation of the central canal of the spinal cord; the junction of conus and filum regularly presents a dilation of the central canal, the ventriculus terminalis (Kernohan). The ependymal cells surrounding the central canal continue into the intradural segment of the filum; the relatively large number of ependymal cells, especially around the terminal ventricle, presumably accounts for the frequency of ependymomas in the upper part of the filum.

External Markings

The gross markings of the spinal cord (Fig. 2-86) are few: along the posterior midline is the posterior (dorsal) median sulcus, and along the anterior midline the larger anterior or ventral fissure. The dorsal (posterior) roots attach to the spinal cord along the posterior lateral (posterolateral) sulcus, not as well

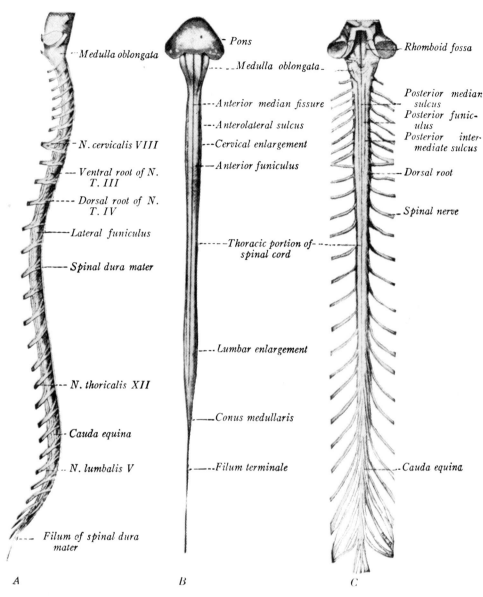

Medulla oblongata

N. cervicalis VIII

Ventral root of N. T. III

Dorsal root of N. T. IV

Lateral funiculus

Spinal dura mater

N. thoricalis XII

Cauda equina

N. lumbalis V

Filum of spinal dura mater

A

Pons

Medulla oblongata

Anterior median fissure

Anterolateral sulcus

Cervical enlargement

Anterior funiculus

Thoracic portion of spinal cord

Lumbar enlargement

Conus medullaris

Filum terminale

B

Rhomboid fossa

Posterior median sulcus

Posterior funiculus

Posterior intermediate sulcus

Dorsal root

Spinal nerve

Cauda equina

C

Fig. 2-86. Lateral, *A,* anterior, *B,* and posterior, *C,* views of the spinal cord. The nerve roots have all been removed in *B.* (After Spalteholz, from Ranson SW: The Anatomy of the Nervous System [ed 10, revised by Clark SL]. Philadelphia, WB Saunders, 1959)

marked in the cervical portion of the cord as elsewhere; however, the cervical cord does present, between the posterior median and the posterolateral sulci, the well-marked posterior intermediate sulci, which indicate the division of each of the paired posterior columns into a medial fasciculus gracilis and a lateral fasciculus cuneatus. While the ventral (anterior) roots emerge in a line from the an-

terolateral surface of the cord, they are not associated, as are the dorsal roots, with a definite sulcus.

The portion of the circumference of the cord between the posterior median and the posterolateral sulci constitutes the posterior funiculi (white columns) of the spinal cord, and at the level of the posterolateral sulcus the gray matter of the posterior horn of the spinal cord almost reaches the surface; the lateral funiculus begins at the posterolateral sulcus and extends anteriorly, but is not clearly demarcated from the anterior funiculus; the boundary between lateral and anterior funiculi is usually defined as the line of emergence of the anterior roots from the spinal cord. The anterior funiculi therefore extend between the lines of emergence of the roots and the anterior median fissure.

Segmentation

Although the spinal cord shows neither gross nor microscopic evidence of true segmentation, the spinal nerves are segmental and it is convenient to refer to spinal cord segments; a spinal segment is defined as the region of spinal cord from which a given spinal nerve arises. Since many of the incoming afferent fibers in the dorsal roots run upward for varying and sometimes very great distances, and it is impossible to tell the exact location of the lower motor neurons that give rise to the fibers of the ventral roots, a spinal segment is arbitrarily described as extending from the level of the upper filaments of one root to the level of the upper filaments of the next root. It does not, therefore, necessarily include all the cell bodies that give rise to the motor fibers, for these may course caudally before making their exits from the surface of the cord, and it never includes a major proportion of the secondary sensory cells on which the incoming afferent fibers of a given nerve root end; the great majority of incoming afferent fibers and their collaterals end in cord segments above the level of entrance of the nerve, or even, in the case of most of the fibers of the posterior funiculi, in the medulla rather than the spinal cord.

Because the cord is so much shorter than the vertebral column, spinal cord segments do not at all correspond in position to the vertebral segments; for instance, while the first cervical segment lies at the level of the body of the first cervical vertebra, the lumbar, sacral, and coccygeal segments of the cord lie in relation to about the last two thoracic and the first two lumbar vertebrae. In regard to length, the cervical segments are intermediate

between those of the thoracic and lumbosacral portions of the cord, but the eight cervical segments of the cord are slightly shorter than the seven cervical segments of the vertebral column; the mid-thoracic segments of the cord are the longest of all, the upper thoracic segments increasing in length, the last two or three decreasing; the first lumbar segment is about the same length as a cervical segment, but thereafter they decrease rapidly in length, and the sacral and coccygeal segments are very short.

The relations of the spinal cord segments to the vertebral bodies and spinous processes are shown in Figure 2-87. This is necessarily no more than a representation of the average condition, since the length of the spinal cord in relation to the vertebral column varies, as already noted. It is probably more nearly accurate in regard to the relation to the vertebral bodies than it is in regard to the relation to the spinous processes, since varying slants and lengths of these processes introduce an additional factor of variation.

Lesions of the Cord

Lesions of the spinal cord may be diffuse, as are generally degenerative and inflammatory ones, or relatively discrete, as may be those commonly resulting from trauma or from a tumor.

The localization of the level of discrete lesions of the cord is often difficult, and involves a careful neurologic examination with detailed testing of sensation, muscular strength, and reflexes. A thorough knowledge of the dermatomes and of the segmental innervations of muscle groups is necessary if one is to localize the uppermost level of the lesion. The question is beyond the confines of this text; however, dermatomal charts are shown in Chapters 1, 3, and 7, and the approximate levels of innervation of the chief muscle groups, plus the segments involved in some of the more commonly tested reflexes, are shown in Table 2-1.

Essentially, transverse lesions of the cord tend to affect all sensation and all motor activity below the level of the lesion, since they interrupt more or less completely all ascending tracts, thus interfering with all sensation below the level of the lesion, and interrupt also the motor tracts, thus producing paralysis below the level of the lesion. Signs and symptoms of lesions of the spinal cord at various levels have been summarized by Rooke and Corbin; Long and Lawton have reviewed the disability that may result from lesions of the cord at different levels, from the standpoint of the ability of a person with a permanent lesion to care for himself and earn a living.

BLOOD SUPPLY

Arteries

The arterial blood supply of the spinal cord is partly from a longitudinally running anterior spinal artery, situated approximately along the anterior median fissure, and two posterior spinal arteries that run approximately along the posterolateral sulci. The anterior spinal artery originates typically by the fusion of paired branches arising from the two vertebral arteries, while the posterior spinal arteries originate from the posterior inferior cerebellar arteries or the vertebral arteries. These three arteries are, however, rather small even at their origins, and as they run down they are therefore reinforced, at irregular intervals, by branches of appreciable size that reach the spinal cord by way of the nerve roots. These *radicular arteries* are the middle branches of the spinal arteries, and vary much in size; they may supply only the nerve roots and their meningeal sleeves. According to Suh and Alexander, there are usually from six to eight anterior radicular arteries that contribute appreciably to the anterior spinal artery, and from five to eight posterior radicular arteries (Fig. 2-88).

The contributions of the radicular arteries to the anterior and posterior spinal arteries are not constant, but Suh and Alexander found that the largest contributions are typically in the lumbar region. Hassler, and Turnbull and co-workers have also stressed the great variability in number, size, and placement of the radicular arteries. Garland

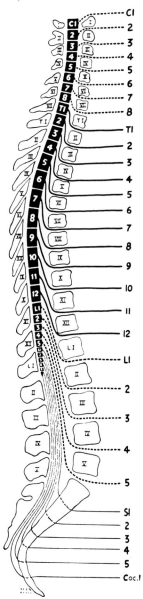

Fig. 2-87. Relation of the spinal cord segments and nerve roots to the vertebrae in a sagittal section of the vertebral column. Spinous processes and bodies of the vertebrae bear Roman numerals, spinal segments and nerves arabic ones. The relative lengths of the cord segments are clearly shown. (Haymaker W, Woodhall B: Peripheral Nerve Injuries [ed 2]. Philadelphia, WB Saunders, 1953)

Table 2-1
The Approximate Segmental Nerve Supply Involved in Certain Movements and Reflexes

NERVES	MUSCLE GROUPS	REFLEXES
Accessory, C1-C4	Flexors, extensors, rotators of head and neck	
Accessory, C3-C5	Elevators of scapula	
Accessory, C5-C7	Upward rotators of scapula	
Accessory, C5, C7, C8	Retractors of scapula	
Accessory, (C5, C6), C7-T1	Depressors of scapula	
C3-C5	Diaphragm	
(C3-C5), C7-T1	Downward rotators of scapula	
(C3, C4), C5-T1	Protractors of scapula	
C5, C6	Abductors at shoulder joint	Biceps
	External (lateral) rotators at shoulder	Brachioradialis
	Flexors at elbow	
C5, C6, (C7)	Flexors (protractors) of arm at shoulder joint	
	Supinators of forearm	
C5-T1	Extensors (retractors) of arm at shoulder joint	
	Internal (medial) rotators at shoulder	
	Adductors at shoulder	
C6, C7, (C8)	Extensors at wrist	
	Pronators of forearm	
C6-T1	Flexors at wrist	
(C6), C7, (C8)	Long extensors of fingers	
(C6), C7, C8, (T1)	Extensor at elbow	Triceps
(C6), C7-T1	Radial abductors at wrist	
C7-T1	Ulnar abductors at wrist	Wrist tendon
	Long flexors of fingers	
C8, T1	Intrinsic muscles of hand	
T1-T12	Musculature of trunk	
T6, T7		Superficial epigastric
T8, T9, (T10)		Superficial abdominal (upper quadrants)
(T10), T11-L1		Superficial abdominal (lower quadrants)
		Cremasteric
L1, L2		
L2-L4, (L5, S1)	Flexors at hip	
L2-S2	Adductors at hip	
	External rotators at hip	
	Internal rotators at knee	
(L2), L3, L4	Extensor at knee	Patellar
(L2, L3), L4-S2, (S3)	Flexors at knee	
L4-S1	Abductors at hip	
	Internal rotators at hip	
	Invertors of foot	
(L4), L5, S1	Dorsiflexors at ankle	
	Extensors of toes	
(L4), L5, S1 (S2)	Evertors of foot	Internal hamstring
(L4), L5-S2	Extensors at hip	External hamstring
	External rotators at knee	
(L4, L5), S1, S2	Plantar flexors at ankle	Achilles (triceps surae)
L5-S2		Plantar
(L5), S1, S2, (S3)	Flexors of toes	
	Intrinsic muscles of foot	
(S2), S3, (S4)	Perineal muscles	Bulbocavernosus
S3-S5, Co 1		Anal

* Numbers in parentheses indicate a variable or a probably minor participation in the reflex or the innervation of the muscle group. (Hollinshead WH. In Baker AB (ed): Clinical Neurology, ed 2, Vol 3. New York, Hoeber, 1962)

and co-workers reported cases of infarction of the cord thought to result from occlusion of a large radicular artery, and Cox and co-workers warned of the danger of spinal cord damage, if, during aortography, the contrast material is delivered close to the mouth of a lumbar artery supplying a large radicular artery. The largest is usually described as running along the ventral root of one of the nerves between T7 and L4, sometimes on the left side, sometimes on the right, and is called the arteria radicularis anterior magnus, or artery of Adamkiewicz. Dommise reported that there is frequently an artery of equal size in the cervical region.

The first large radicular artery to the cervical part of the cord is usually along one of the lower three cervical nerve roots, and because of this it has usually been stated that the anterior spinal artery, derived from the vertebrals, is the chief supply to the cord to about the C6 level. However, Fried and co-workers ('70) found that in the monkey, blood from the major radicular artery in the cervical region flows cephalad, toward the vertebrals, in the anterior spinal artery. According to their findings, the vertebrals contribute little to the spinal cord either through their spinal or radicular branches; the major radicular artery is from either the costocervical trunk or the deep cervical artery.

In contrast to the apparent importance of the major radicular artery to the cervical cord, the arteria radicularis anterior magnus may be of less importance than usually supposed. Fried and co-workers ('69) ligated it in 10 monkeys; 5 of these had no deficit resulting from the ligation, the other 5 had minimal deficits. When, however, they ligated the anterior spinal artery below the arteria radicularis, 10 of 11 monkeys suffered severe to moderate motor deficits. Ligation of the anterior spinal artery above the arteria radicularis

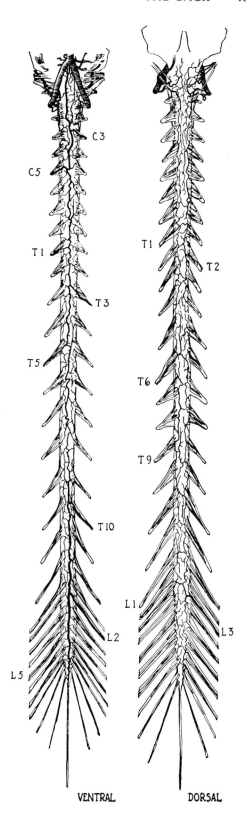

VENTRAL   DORSAL

Fig. 2-88. Ventral and dorsal views of the arterial supply to the spinal cord. Note the great variation in the sizes of the radicular arteries. (After Suh and Alexander, from Steegmann AT. In Baker AB (ed): Clinical Neurology [ed 2], Vol 3. New York, Hoeber, 1962)

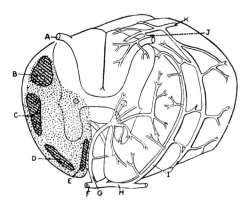

Fig. 2-89. Diagram of the distribution of the arteries to the spinal cord. The shaded area on the left represents the distribution of the anterior spinal artery. *A* is the dorsal root; *B, C, D,* and *E* are the lateral corticospinal, lateral spinothalamic, ventral spinothalamic, and ventral corticospinal tracts, respectively; *F* is the anterior spinal artery, *G* a sulcal artery, *H* an anterior radicular, *I* the superficial arterial plexus, *J* a posterior radicular artery, and *K* the posterior spinal. (Steegmann AT. In Baker AB (ed): Clinical Neurology [ed 2], Vol 3. New York, Hoeber, 1962)

caused no or minor deficits. These results seem to indicate that downward flow from the region of the artery is particularly important, but does not necessarily depend on that artery.

Branches of the anterior spinal artery and of the entering anterior radicular arteries (Fig. 2-89) supply the anterior and intermediate portions of the gray matter and the base of the posterior gray column; they also supply the white matter of the anterior funiculus and the deeper portion of the lateral funiculus (Herren and Alexander). The posterior spinal arteries and posterior radicular arteries supply the posterior funiculus, the major portion of the posterior gray columns, and a superficial part of the lateral funiculus. According to Zeitlin and Lichtenstein, anastomoses between branches of the anterior and posterior spinal arteries are larger in the cervical and lumbar regions of the cord than they are in the thoracic, and occlusion of the anterior spinal artery in the thoracic region produces greater damage than it does at other levels of the cord.

Veins

The venous drainage of the spinal cord (Figs. 2-90 and 2-91) resembles the arterial supply, except that both anterior and posterior spinal veins tend to be single. In addition to their communications with the veins of the medulla, the spinal veins likewise drain laterally, at irregular levels, through radicular veins. As in the case of the radicular arteries, the largest radicular veins are said to be associated with roots of lumbar nerves.

MICROSCOPIC ANATOMY

Even a general consideration of the arrangement of nerve cells and of nerve fibers in the spinal cord and of their functional significance is far beyond the confines of this book. Only a few generalities can be pointed out here.

Gray Matter

In cross section, the gray matter of the spinal cord, centered primarily around the central canal, takes somewhat the shape of a butterfly or an H. The posterior gray columns then appear as the posterior horns of the gray matter, the anterior columns as the anterior horns. In general, the posterior gray columns are composed of cells that receive and forward impulses entering the spinal cord by way of the dorsal (posterior) nerve roots, while the anterior gray columns receive impulses from higher centers as well as from the dorsal roots, and through the large multipolar neurons that they contain send impulses to voluntary muscle. These large neurons are referred to as lower motor neurons, and constitute the final common pathway by which nerve impulses, however instigated originally, reach the motor unit of voluntary muscle.

The preganglionic neurons of the sympathetic system are situated laterally in the gray matter of all the thoracic and of the upper two lumbar segments, at about the junction of anterior and posterior gray columns; the preganglionic neurons of the sacral portion of the parasympathetic system occupy a similar position in about the second, third, and

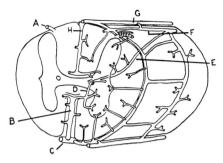

Fig. 2-90. Diagram of the distribution of the veins in the spinal cord. *A* is the dorsal root, *B* an anterior sulcal vein, *C* the anterior venous trunk, *D* a paracentral vein, *E* an anastomotic channel, *F* a posterolateral venous trunk, *G* the posterior venous trunk, and *H* a posterior septal vein. (After Herren and Alexander, from Steegmann AT. In Baker AB, (ed): Clinical Neurology [ed 2], Vol 3. New York, Hoeber, 1962)

fourth sacral segments of the cord. All these preganglionic neurons send their axons through the ventral roots with the voluntary motor fibers that form the bulk of these roots.

A great deal of work has been done upon the functional significance of groupings of nerve cells within the gray matter of the spinal cord, but there are many contradictory statements in regard to these groupings.

White Matter

The gray matter of the spinal cord comes almost to the surface along the posterolateral sulci, since the posterior horns are long; the shorter anterior horns, however, do not reach close to the surface of the cord. Thus the white matter of the spinal cord, largely surrounding the gray matter, is very clearly divided into posterior funiculi or columns, lying posterior and medial to the posterior gray columns and to the entering dorsal roots of the spinal nerves; however, the lateral funiculi (white columns), lying in general lateral to the gray matter and between the posterior and anterior gray columns, are not clearly distinguished from the anterior funiculi,

Fig. 2-91. Ventral and dorsal views of the veins of the spinal cord. Note the variation in the sizes of the radicular veins. (After Suh and Alexander, from Steegmann AT. In Baker AB (ed): Clinical Neurology [ed 2], Vol 3. New York, Hoeber, 1962)

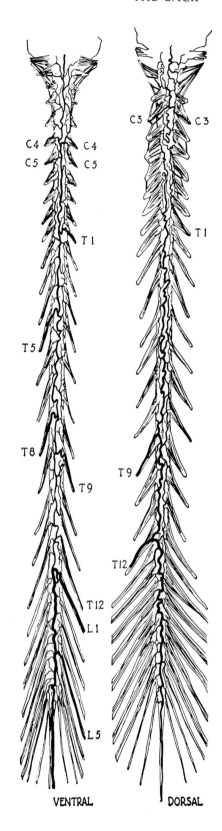

VENTRAL   DORSAL

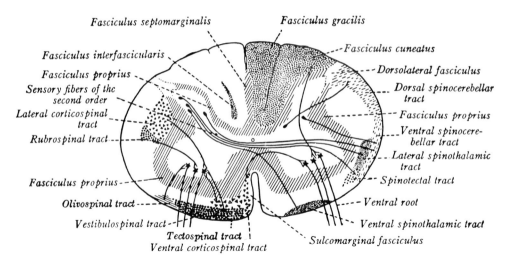

Fasciculus septomarginalis   *Fasciculus gracilis*

Fasciculus interfascicularis   ---*Fasciculus cuneatus*

Fasciculus proprius   ---*Dorsolateral fasciculus*

Sensory fibers of the   ---*Dorsal spinocerebellar*
second order   *tract*

Lateral corticospinal   --*Fasciculus proprius*
tract

Rubrospinal tract   ---*Ventral spinocere-*
bellar tract

---*Lateral spinothalamic*
tract

Fasciculus proprius --   -- *Spinotectal tract*

Olivospinal tract ----   ---*Ventral root*

Vestibulospinal tract --   --*Ventral spinothalamic tract*

Tectospinal tract   *Sulcomarginal fasciculus*
Ventral corticospinal tract

Fig. 2-92. Chief tracts of the spinal cord. The gray matter is surrounded by the fasciculus proprius (*diagonal lines*), consisting of a mixture of relatively short ascending and descending fibers. Otherwise, ascending tracts are shown on the right, descending ones on the left. (Ranson SW: The Anatomy of the Nervous System [ed 10, revised by Clark SL]. Philadelphia, WB Saunders, 1959)

usually defined as lying medial and in part anterior to the anterior gray columns, and separated from the lateral funiculi only by the lines of emergence of the anterior spinal rootlets.

The two posterior funiculi, separated from each other by the posterior median septum, consist almost entirely of primary afferent ascending fibers, that is, the direct continuation of fibers arising from the cells of the dorsal root ganglia. Other afferent fibers enter the posterior gray columns and make synapse there, secondary afferent fibers then passing into the lateral or anterior funiculi of the same or the opposite side. Descending fibers that influence the lower motor neurons and the autonomic nervous system are likewise located in the lateral and anterior funiculi, so that these two white columns consist of a mixture of ascending and descending fibers of many different functions.

There is a very considerable overlap between the various fiber tracts of the lateral and anterior funiculi, and it is therefore possible to give only the approximate location of the specific tracts. One diagram indicating the approximate location of the more important tracts of the cord is reproduced in Figure 2-92.

Briefly, it might be noted that the posterior funiculi are concerned primarily with deep sensation other than pain: proprioception (muscle and joint sense), pressure, and the conduction of impulses that are interpreted and built up at higher levels into an appreciation of vibratory sensibility, two-point tactile discrimination, shape and texture (stereognosis), and tactile localization. The posterior columns also conduct some impulses having to do with light touch (but other fibers concerned with light touch ascend in the lateral funiculi); they are composed of primary afferent fibers, their cells of origin being in the spinal ganglia, and conduct impulses from the same side of the body.

The lateral funiculi contain a mixture of ascending and descending fiber tracts. The ascending tracts are composed of secondary afferent neurons, and are largely crossed: their cells of origin therefore lie, for the most part, in the opposite posterior gray column. Of the ascending tracts, the lateral spinothalamic is probably of the most clinical interest, since it is composed of fibers mediating pain and temperature sensations. As indicated in the diagram, these fibers cross the midline anterior to the central canal, and turn upward to form a tract in the anterolateral part of the lateral columns. Interruption over several segments of the fibers of the lateral spinothalamic tract as they cross the midline to reach the position in which they ascend gives rise to a localized segmental loss of pain and temperature sensation, such as occurs in syringomyelia; it is usually the lateral spinothalamic tract that is interrupted, unilaterally or bilaterally, for the alleviation of intractable pain in the operation generally referred to as anterolateral chordotomy. (Mayer and co-workers reported that proper placement of the electrode in percutaneous chordotomy can be improved by using the electrode first as a stimulator to determine the threshold for pain.) The two spinocerebellar tracts, of less importance clinically than the lateral spinothalamic, conduct impulses to the cerebellum, not to consciousness.

The largest descending tract in the lateral funiculus is the lateral corticospinal (pyramidal) tract, composed largely of fibers that have crossed at the level of the medulla and thereafter end on the same side of the spinal cord in which the tract is located. Associated with this corticospinal tract are numerous extrapyramidal fibers that have to do also with the contraction of voluntary muscle.

The chief ascending tract in the anterior funiculus is the anterior (ventral) spinothalamic tract, composed of crossed fibers concerned with light touch. Injury to it is of no great importance, since impulses for light touch are also conducted ipsilaterally in the posterior funiculus; hence an approximate hemisection of the cord does not appreciably affect light touch. Descending tracts in the anterior funiculus are more numerous than ascending ones, and except for the anterior or ventral corticospinal (pyramidal) tract are classified as extrapyramidal tracts; of these, the vestibulospinal tract is usually thought to be among the more important. As is true of the extrapyramidal tracts of the lateral funiculus, those of the anterior funiculus are concerned with the co-ordination of contraction of muscle and with the maintenance of a proper balance in muscle tone. Reticulospinal tracts, not shown in Figure 2-92, include many fibers with this function and are located in both the anterior and lateral funiculi.

ANOMALIES OF SPINAL CORD

The common anomaly of the spinal cord, spina bifida, has already been discussed in connection with the vertebral column, although whether the vertebral defect or the defect of the cord is the primary abnormality is not agreed upon. The term "spina bifida," although originally describing the vertebral defect alone, has come to include also associated defects of the spinal cord and meninges, particularly meningocele and myelomeningocele. Spina bifida anterior, in which the vertebral bodies are defective and the contents of the vertebral canal protrude into the abdomen or pelvis, has also been mentioned.

Diplomyelia is a doubling of the spinal cord, often associated with spina bifida. The doubling is usually for only a short distance, commonly in the lumbosacral enlargement; there may be a bony septum protruding in part between the two portions (Fig. 2-93). As a rule, a common arachnoid and dura surround both parts of the cord, each of which has its own central canal; in this respect, diplomyelia differs from pseudo-duplication or diastematomyelia, in which neither of the two parts represents a complete cord.

Other rare anomalies of the spinal cord are congenital myelocystocele, in which the cord is dilated by a cavity within it filled with fluid; incomplete development of the cord (atelomyelia), in which some important tract, most commonly the pyramidal, is lacking; and marked smallness of the cord (micromyelia), usually in the form of an unusually short one or one which is abnormally constricted in diameter at some level.

In the Arnold-Chiari syndrome, sometimes associated with spina bifida, abnormal fixation of the spinal cord toward its lower end, associated with slower growth of the cord than of the vertebral column, results in the cord being pulled downward in relation to the vertebral column so that the upper cervical roots run upward to their foramina rather than slightly downward. The traction on the cord tends, in turn, to pull the medulla and the cerebellum into the foramen magnum, with consequent compression of the hindbrain.

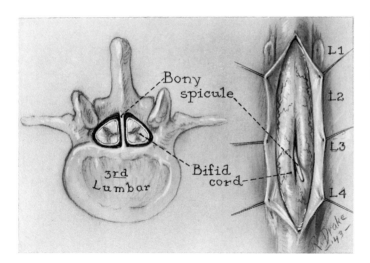

Fig. 2-93. Diplomyelia, the two portions of cord being subdivided by a bony spur. (Marr GF, Uihlein A: S Clin North America [Aug] 1944, p 963)

Garceau reported three cases in which a tight filum terminale was thought to be a cause of traction upon the cord, producing symptoms similar to the Arnold-Chiari syndrome; following section of the filum terminale, there was said to be some neurological recovery, although whether decompression of the foramen magnum might have to be done later was not apparent at the time of his writing. Jones and Love reviewed cases of tight filum (associated with spina bifida occulta) that apparently caused degenerative changes in the lower part of the spinal cord, and they reported that the children among their own cases improved markedly when the filum was sectioned. Campbell, among others, has discussed surgical aspects of a number of congenital deformities of the spinal cord.

SPINAL NERVE ROOTS

The dorsal (posterior) roots of the spinal nerves arise each by a series of filaments that attach to the cord along a fairly well marked posterolateral sulcus, while the ventral (anterior) roots arise in an anterolateral position, but associated with no sulcus.

The dorsal and ventral roots of each spinal nerve run separately through the subarachnoid space surrounding the spinal cord, with a caudal inclination that varies according to the difference in level between their origins from the cord and their levels of exit (Fig. 2-87). At the levels of their exit both roots, still separate, turn laterally, enter special dural-arachnoidal sheaths that project from the dural and arachnoidal sacs of the spinal cord, and unite with each other very close to the ganglion borne by the dorsal root. Because the ganglia regularly lie at the level of the intervertebral foramina, the roots of the spinal nerves vary tremendously in total length, almost all of this being in the length that lies within the subarachnoid cavity about the spinal cord. Average measurements of the latter in eight cadavers, as reported by Poirier and Charpy ('04), range from only 3 mm for the first cervical to 29 mm for the first thoracic, 91 mm for the first lumbar, 185 mm for the first sacral, and 266 mm for the coccygeal. The caudally running lumbar and sacral rootlets form the cauda equina as they lie in the lumbar cistern of the subarachnoid space (Fig. 2-85; see also Fig. 2-95).

Dorsal Root Ganglia

The fibers composing the dorsal roots are derived from cells located in the dorsal root ganglia; while there have been conflicting claims as to whether the number of fibers in a given dorsal root equals the number of cells in the corresponding dorsal root ganglion (e.g., Duncan and Keyser; Barnes and Davenport) it is now rather generally agreed that there are actually no motor fibers emerging from the spinal cord by way of the dorsal roots (Westbrook and Tower). All the sensory nerve cells are not necessarily collected in the ganglia: Peters examined histologically the dorsal roots of selected cervical, thoracic, lumbar, and sacral nerves, and found aberrant ganglion cells in each case; however, they ranged in number only from 3 to 58 for any one root. They were unevenly distributed on the root between the ganglion and the spinal cord. Peters suggested that the presence of such aberrant nerve cells may account for the recurrence of pain following rhizotomy.

The first cervical dorsal root is known to be sometimes absent; however, it may be represented more frequently than has been supposed. Among the nerves investigated by Ouaknine and Nathan, only 18 had a posterior root arising from the spinal cord; in 12 of these there was also an anastomosis between the accessory nerve and the root. In the 22 in which there was no posterior root arising from the cord, an anastomosis between the accessory nerve and the first spinal nerve, which usually had a ganglion upon it, seemed to represent the distal part of the root. In only 10 instances was there neither a root arising from the cord nor an anastomosis with the accessory nerve.

Number of Fibers in the Roots

In general, the dorsal root of any one nerve is usually larger than is the ventral or motor root. However, the relative sizes of the two roots varies with the particular nerve investigated. Thus Ingbert ('03, '04) attempted to enumerate the myelinated nerve fibers in the dorsal and ventral roots of man, and arrived at a figure of about 653,000 myelinated fibers in the dorsal roots of one side, and about 203,-700 myelinated fibers in the ventral roots of one side. These figures can be used for comparison only, since they are obviously incomplete in regard to the total number of fibers in the nerve roots. Davenport and Bothe pointed out that the proportion of unmyelinated fibers in both the dorsal and ventral roots is higher in the thoracic and sacral nerves (which give rise to autonomic and associated afferent fibers) than in the other nerves of

man; they found the greatest proportion of unmyelinated fibers among the nerves they examined (C2, C6, T4, T9, L3, S2, and S5) in the ninth thoracic, where there were 1.76 unmyelinated fibers to 1 myelinated one, and the least proportion of unmyelinated fibers in the second cervical, where there was only 0.52 unmyelinated fiber to 1 myelinated one.

Some concept of the variations in the sizes of the roots of different nerves may be obtained from Ingbert's results. In each of the dorsal roots of the second through the fourth cervical nerves, Ingbert found about 27,000 to 28,000 myelinated fibers, while in the larger nerves contributing to the brachial plexus the number of fibers rose to 50,000 and then dropped to about 17,000 for the first thoracic. There were as few as 7000 dorsal root fibers in some of the thoracic nerves; the number increased to totals of 31,000 to 47,000 in the nerves entering the lumbosacral plexus, and dropped rapidly in the fourth and fifth sacral and the tiny coccygeal. Among ventral roots, the fifth and sixth cervical nerves had the largest number of myelinated fibers, with 13,-500 and 11,700 respectively; the third and fifth lumbars contained 11,000 and 10,000 respectively, while the fourth lumbar had only about 7300; the thoracic nerves had about 6000 to 7000 apiece; the lower sacrals from about 2300 to 1700, and the coccygeal the least of all, 519.

RELATIONS OF NERVE ROOTS

To the Subarachnoid Space

As they leave the spinal cord, the nerve roots run through the subarachnoid cavity, with varying degrees of obliquity, until each dorsal and ventral root reaches approximately the level of the intervertebral foramen through which the nerve to which each belongs makes its exit. At this level the two roots composing a spinal nerve turn somewhat more laterally, if they have been coursing caudally, and penetrate the lateral wall of the main dural-arachnoid sac. As they do so, however, each root (for dorsal and ventral roots are still separate at this level) carries out with it a diverticulum of dura and arachnoid (Fig. 2-94).

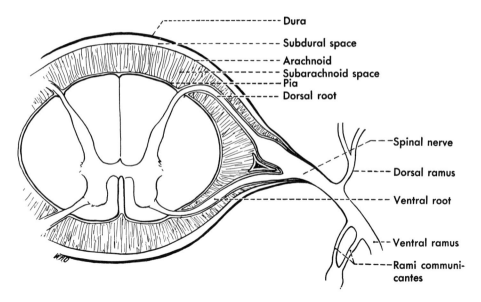

Fig. 2-94. Extension of the meninges and the subarachnoid cavity about the nerve roots in a diagrammatic section of the spinal cord.

Thus each root, as it leaves the main dural sac and extends laterally, or laterally and caudally, to attain the intervertebral foramen, is provided with a dural sleeve lined by an arachnoid sleeve, and therefore containing a continuation of the subarachnoid cavity. (This part of the nerve is nevertheless often referred to as the extradural segment, since it no longer lies in the main dural sac.) The contiguous surfaces of the dural sleeves fuse together to form a septum as or before the proximal end of the spinal ganglion is reached, and may fuse as they leave the main dural sac, but the two extensions of the subarachnoid cavity remain separate.

It is usually said that the dura continues out as a sleeve around the nerve roots and becomes continuous with the epineurium of the spinal nerve just before the two roots unite (distal to the dorsal root ganglion) that the arachnoid here becomes continuous with the perineurium, and the pia with the endoneurium. According to Tarlov ('53), the distance outward along the roots over which the subarachnoid space extends varies somewhat; he said that the transition from arachnoid to perineurium usually takes place just central to the proximal pole of the dorsal root ganglion (rather than on the distal part of the ganglion), and at an approximately corresponding point on the anterior root. This, of course, marks the distal closure of the subarachnoid space, but according to Tarlov there is no continuity between the cellular surfaces of the arachnoid and pia; he said both of these continue into the connective tissue of the nerve trunk. The dura then becomes converted into epineurium immediately thereafter. In some cases, however, he found that all three meninges extend as far outward as the distal pole of the ganglion or even onto a proximal portion of the peripheral nerve, so that the entire ganglion and a part of the peripheral nerve actually lie in the subarachnoid space. Extensions of the subarachnoid space beyond the ganglion are of potential clinical importance, since in paravertebral injections for these nerves the injected material can on occasion be delivered into the subarachnoid space (Tarlov).

Since the roots of the cervical nerves run almost transversely to reach the intervertebral foramina through which they make their exits, their courses through the general subarachnoid cavity are short. With increasing distance between the levels of origin and exit of the nerves, more and more of the total length of the roots comes to lie in the general subarachnoid cavity (see above). Thus the major portions of the nerve roots of the lumbar nerves, and even greater lengths of the sacral and coccygeal nerves, lie in the lumbar cistern as the cauda equina. Their caudally and laterally directed sleeves of dura and arachnoid are somewhat longer than the similar sleeves on the cervical and thoracic nerves, and diverge only slightly from the main dural sac before the nerves turn more sharply laterally around the inferior surfaces of the pedicles.

Within the lumbar cistern the nerve roots of the cauda equina are somewhat loose and have a slightly wavy course (Fig. 2-95). This allows for the increase in length of the dural sac accompanying flexion of the neck and trunk, and for the downward movement of the sac with flexion of the hip, without exerting undue tension on the otherwise unprotected nerve roots; they become straighter, but not actually taut. In contrast, after they enter their special dural sheaths they are practically straight, and since they are also short and covered by inelastic dura they have little "give" to them. It is for this reason that even a small protrusion of an intervertebral disk against the dura-clad nerve may cause root pain.

As already noted, it is usually thought that there is normally no continuity between the subarachnoid spaces around the nerve roots and the perineural spaces in the nerve trunks. However, there is evidence that some substances can pass across the barrier between them, and under certain probably abnormal conditions these may include radiopaque material that apparently spreads distally in the nerves from the subarachnoid space.

Tarlov ('53) and others have described cysts on the nerve roots (Fig. 2-96) at about the point at which dorsal and ventral roots come together, and Tarlov apparently traced their origin to the perineural spaces. They are said to be still another cause of the sciatic syndrome (although they may produce no symptoms) and may largely destroy the nerve root and the dorsal root ganglion. They are to

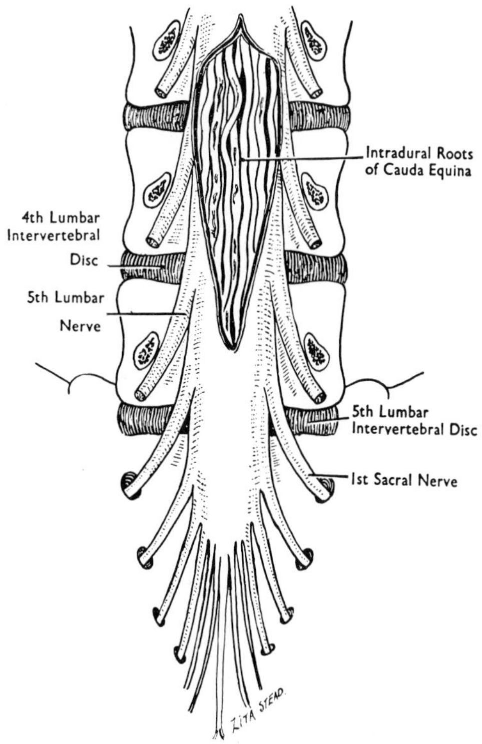

Fig. 2-95. Nerve roots of the cauda equina exposed in the lumbar cistern. This also shows the dura-clad nerves making their exits close to the pedicles. (O'Connell JEA: Brit J Surg 30:315, 1943)

be distinguished, according to Tarlov, from extradural cysts, which have a different origin (discussed earlier). Holt and Yates found arachnoid cysts on at least one cervical nerve root or ganglion in 36 of 120 cervical columns that they studied, with 5% of all the roots examined being so affected.

To the Pia

As the filaments of the spinal nerves leave the spinal cord they come together to form the definitive dorsal and ventral roots of the spinal nerves, each surrounded by its layer of pia and lying within the subarachnoid cavity. As described by Tarlov ('53), the pia along the nerve rootlets and roots consists of two layers, an inner and an outer; the outer one, with its covering layer of low cuboidal or flattened cells, has its connective tissue continued into the endoneurium of the peripheral nerve trunk, but the inner layer turns into the nerve

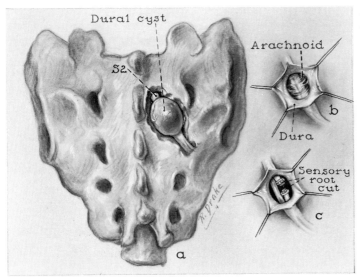

Fig. 2-96. A cyst on a sacral nerve
root. (Courtesy of Dr. H. V. Svien)

root to form the pial ring, an area of connective tissue through which the nerve fibers pass
as they approach the spinal cord.

The level of the pial ring corresponds approximately to the transition between the
glial segment of the nerve (the glia extends
into the roots of the spinal nerves for a short
distance) and the more peripheral segment of
the nerve that has no glia in it. The glial segment of a spinal nerve usually extends (Tarlov, '37) for less than a millimeter into the
proximal portion of the nerve. Therefore, by
far the greater length of nerve in the subarachnoid space lies distal to the pial ring. At
the pial ring the myelin sheaths of the nerve
fibers are narrowed, and this part of the nerve
fiber is sometimes thought to be particularly
susceptible to damage—for instance, from the
lesion of tabes dorsalis.

To the Intervertebral Foramina and Disks

The dorsal and ventral roots come together
just at the distal edge of the dorsal root ganglion to form the mixed nerve trunk, which
then makes its exit from the vertebral canal.
The dorsal root ganglia lie for the most part
in the intervertebral foramina, but the ganglia of the sacral nerves lie in the sacral canal,
and these nerves divide into their ventral and
dorsal branches within the canal, rather than
outside the foramina as in the thoracolumbar

and cervical regions. Further, since there are
no bony intervertebral foramina between the
atlas and the skull, or between the axis and
the atlas, the ganglia of the first two cervical
nerves (if the first cervical nerve has one) lie
simply upon the posterior arches of the atlas
and axis.

Outside the intervertebral foramina each
spinal nerve typically divides into a dorsal
(posterior) branch, which turns posteriorly to
reach the musculature and skin of the back,
and a ventral (anterior) branch which
usually, except in the thoracic region, enters
into the formation of a plexus. The sacral
nerves divide, as mentioned, into their ventral
and dorsal branches while they lie within the
sacral canal, and their ventral branches
emerge through the pelvic sacral foramina,
their dorsal ones through the dorsal sacral
foramina. The ventral branches of the nerves,
or the nerves at about their points of division,
are connected to the sympathetic trunk by
one or more rami communicantes. At this
level, also, tiny meningeal branches (Fig. 2-
33) arise and run a recurrent course through
the intervertebral foramina to be distributed
within the vertebral canal.

Except in the case of protruding lumbar
disks, it is at the level of the foramina by
which they leave the vertebral canal that
nerve roots are most commonly compressed.

It has already been noted that the intervertebral foramina are of such length that even total disappearance of an intervertebral disk will not usually bring the pedicles of the adjacent vertebrae close enough together to pinch the nerve. However, the transverse (horizontal) dimensions of the intervertebral foramina are more limited: for instance, Larmon found that the average diameter of the fourth lumbar nerve was only slightly less than 7 mm, and the vertical diameter of its foramen from pedicle to pedicle 19 mm, but the transverse diameter, from ligamentum flavum to vertebral body, only 7 mm; his measurements of the fifth lumbar nerve and its foramen gave similar results, except that the vertical diameter was less, 12 mm. In the cervical region, likewise, while the foramina are considerably longer than they are wide, the cervical nerves largely fill the horizontal space between the articular processes and the vertebral bodies. Thus any change that tends to narrow the horizontal diameter of an intervertebral foramen may be expected to produce pressure upon the nerve making its exit through that foramen. Causes of such narrowing have been said to include swelling of the synovial membranes of the joints that border the foramina (Larmon reported that injection of material into a lumbar synovial joint reduced the anteroposterior diameter of the foramen from the usual 7 mm to 5 mm, less than the average diameter of the nerve); bony proliferations from the margins of the foramen; subluxation of one vertebra upon another; and, in the cervical and thoracic regions, posterolateral protrusion or rupture of an intervertebral disk. Golub and Silverman reported another possible cause of compression of a nerve at a foramen; they found, specifically in the lumbar region, that one or more foramina are usually crossed by a ligamentous band that limits the nerve to a part of the foramen.

As already pointed out the cervical nerves, in making their exits through the foramina, pass directly across the posterolateral surfaces of the disk at the level of exit. It is only because there are eight cervical nerves, the first passing above the first cervical vertebra, that a protruded cervical disk compresses a nerve numbered one greater than it; here it is purely a question of terminology. In the lumbar region, however, the foramina are so long that the nerve makes its exit across the vertebral body above the disk rather than across the disks. Therefore a protrusion of a lumbar disk does not narrow the foramen at the level of the nerve and rarely affects the nerve that makes its exit there. Rather, a posterolateral protrusion of a lumbar disk produces its effects by narrowing the epidural space and thus compressing, through the dura, the most laterally lying nerve or nerves of the cauda equina at that level—commonly the nerve that makes its exit at the foramen below the level of the disk. Thus in the lumbar region no change in nomenclature could alter the fact that it is not the corresponding nerve, but the one below, that is affected by a protruded disk.

ANOMALIES

Anomalies of nerve roots are rare, but Ethelberg and Riishede reported four anomalies found in 4612 operations. All of these were said to have involved the fifth lumbar and first sacral nerves, and consisted of both nerves sharing a common dural sleeve but then separating, of anastomosis between the roots of the two nerves, and (two cases) of both nerves making their exit through a single intervertebral foramen. Cysts on the nerve roots have already been mentioned, as have also extradural cysts on nerves.

References

AEBY C: Die Altersverschiedenheiten der menschlichen Wirbelsäule. Arch f Anat u Physiol (Anat Abt) p 77 1879

ALTHOFF B: Fracture of the odontoid process: An experimental and clinical study. Acta orthop scandinav suppl 177, 1979

AMACHER AL, DRAKE CG, MCLACHLIN AD: Anterior sacral meningocele. Surg Gynecol Obstet 126:986, 1968

AMSLER FR, JR, and WILBER MC: Intraosseous vertebral venography as a diagnostic aid in evaluating intervertebral-disc disease of the lumbar spine. J Bone Joint Surg 49-A:703, 1967

AMUSO SJ, MANKIN HJ: Hereditary spondylolisthesis and spina bifida: Report of a family in which the lesion is transmitted as an autosomal dominant through three generations. J Bone Joint Surg 49-A:507, 1967

ARKIN AM: The mechanism of the structural changes in scoliosis. J Bone Joint Surg 31-A:519, 1949

ARKIN AM: The mechanism of rotation in combination with lateral deviation in the normal spine. J Bone Joint Surg 32-A:180, 1950

ASMUSSEN E: The weight-carrying function of the human spine. Acta orthop scandinav 29:276, 1960

BARDEEN CR: Costo-vertebral variation in man. Anat Anz 18:377, 1900

BARDEEN CR: Numerical vertebral variation in the human adult and embryo. Anat Anz 25:497, 1904

BARDEEN CR: The development of the thoracic vertebrae in man. Am J Anat 4:163, 1905a

BARDEEN CR: Studies of the development of the human skeleton: (A) The development of the lumbar, sacral and coccygeal vertebrae. (B) The curves and the proportionate regional lengths of the spinal column during the first three months of embryonic development. (C) The development of the skeleton of the posterior limb. Am J Anat 4:265, 1905b

BARDEEN CR: Development and variation of the nerves and the musculature of the inferior extremity and of the neighboring regions of the trunk in man. Am J Anat 6:259, 1907

BARDEEN CR: Early development of the cervical vertebrae and the base of the occipital bone in man. Am J Anat 8:181, 1908

BARDEEN CR, ELTING AW: A statistical study of the variations in the formation and position of the lumbo-sacral plexus in man. Anat Anz 19:124, 209, 1901

BARNES JF, DAVENPORT HA: Cells and fibers in spinal nerves: III. Is a 1:1 ratio in the dorsal root the rule? J Comp Neurol 66:459, 1937

BARNES R: Paraplegia in cervical spine injuries. J Bone Joint Surg 30-B:234, 1948

BARR JS, KUBIK CS, MOLLOY MK, MCNEILL JM, RISEBOROUGH EJ, WHITE JC: Evaluation of end results in treatment of ruptured lumbar intervertebral discs with protrusion of nucleus pulposus. Surg Gynecol Obstet 125:250, 1967

BATSON OV: The function of the vertebral veins and their rôle in the spread of metastases. Ann Surg 112:138, 1940

BERKIN CR, HIRSON C: Hyperextension injury of the neck with paraplegia. J Bone Joint Surg 36-B:57, 1954

BICK EM, COPEL JW: Longitudinal growth of the human vertebra: Contribution to human osteogeny. J Bone Joint Surg 32-A:803, 1950

BICK EM, COPEL JW: The ring apophysis of the human vertebra: Contribution to human osteogeny. II. J Bone Joint Surg 33-A:783, 1951

BICK EM, COPEL JW: The senescent human vertebra: Contribution to human osteogeny. III. J Bone Joint Surg 34-A:110, 1952

BINGOLD AC: Congenital kyphosis. J Bone Joint Surg 35-B:579, 1953

BLOCKEY NJ, and PURSER DW: Fractures of the odontoid process of the axis. J Bone Joint Surg 38-B:794, 1956

BLUMEL J, EVANS EB, EGGERS GWN: Partial and complete agenesis or malformation of the sacrum with associated anomalies: Etiologic and clinical study with special reference to heredity. A preliminary report. J Bone Joint Surg 41-A:497, 1959

BRADFORD FK, SPURLING RG: Intraspinal causes of low back and sciatic pain: Results in 60 consecutive low lumbar laminectomies. Surg Gynecol Obstet 69:446, 1939

BRAILSFORD JF: Deformities of the lumbosacral region of the spine. Brit J Surg 16:562, 1929

BRASHEAR HR, VENTERS GC, PRESTON ET: Fractures of the neural arch of the axis: A report of thirty-nine cases. J Bone Joint Surg 57-A:879, 1975

BRAV EA, MOLTER HA, NEWCOMB WJ: The lumbosacral articulation: A roentgenologic and clinical study with special reference to narrow disc and lower lumbar displacement. Surg Gynecol Obstet 87:549, 1948

BREIG A, MARIONS O: Biomechanics of the lumbosacral nerve roots. Acta Radiol [Diagn] (Stockh) 1:1141, 1963

BRIERLEY JB, FIELD EJ: Fate of an intraneural injection as demonstrated by the use of radio-active phosphorus. J Neurol Neuro-surg Psychiat 12:86, 1949

BROWN CE: Complete absence of the posterior arch of the atlas. Anat Rec 81:499, 1941

BROWN JW, PODOSIN R: A syndrome of the neural crest. Arch Neurol 15:294, 1966

VER BRUGGHEN A: Massive extrusions of the lumbar intervertebral discs. Surg Gynecol Obstet 81:269, 1945

BUTLER RW: Spontaneous anterior fusion of vertebral bodies. J Bone Joint Surg 53-B:230, 1971

CAMPBELL JB: Congenital anomalies of the neural axis: Surgical management based on embryologic considerations. Am J Surg 75:231, 1948

CAPENER N: Spondylolisthesis. Brit J Surg 19:374, 1932

CAVE AJE: The innervation and morphology of the cervical intertransverse muscles. J Anat 71:497, 1937

CHANDLER FA: Lesions of the "isthmus" (pars interarticularis) of the laminae of the lower lumbar vertebrae and their relation to spondylolisthesis. Surg Gynecol Obstet 53:273, 1931

CHILDRESS JC, JR, WILSON FC: Bipartite atlas: Review of the literature and report of a case. J Bone Joint Surg 53-A:578, 1971

CLOWARD RB: Changes in the vertebra caused by ruptured intervertebral discs: Observations on their formation and treatment. Am J Surg 84:151, 1952

CLOWARD RB: The anterior approach for removal of ruptured cervical disks. J Neurosurg 15:602, 1958

COBB JR: The problem of the primary curve. J Bone Joint Surg 42-A:1413, 1960

COLSEN K: Atlanto-axial fracture-dislocation. J Bone Joint Surg 31-B:395, 1949

COVENTRY MB, GHORMLEY RK, KERNOHAN JW: The intervertebral disc: Its microscopy anatomy and pathology. Part I. Anatomy, development, and physiology. J Bone Joint Surg 27:105, 1945a

COVENTRY MB, GHORMLEY RK, KERNOHAN JW: The intervertebral disc: Its microscopic anatomy and pathology. Par II. Changes in the intervertebral disc concomitant with age. J Bone Joint Surg 27:233, 1945b

COVENTRY MB, GHORMLEY RK, KERNOHAN JW: The intervertebral disc: Its microscopic anatomy and pathology. Part III. Pathological changes in the intervertebral disc. J Bone Joint Surg 27:460, 1945c

COX EF, WOLFEL DA, INGLESBY TV: Translumbar aortography: A mechanism of injury—a method of protection. Surgery 60:1146, 1966

CRAM RH: A sign of sciatic nerve root pressure. J Bone Joint Surg 35-B:192, 1953

CROCK HV, YOSHIZAWA H, KAME SK: Observations on the venous drainage of the human vertebral body. J Bone Joint Surg 55-B:528, 1973

DAVENPORT HA, BOTHER RT: Cells and fibers in spinal nerves: II. A study of C2, C6, T4, T9, L3, S2 and S5 in man. J Comp Neurol 59:167, 1934

DAVIS AG: Injuries of the spinal column. Am Acad Orth Surgeons Instructional Course Lectures 6:73, 1949

DAVIS PR: The medial inclination of the human thoracic articular facets. J Anat 93:68, 1959

DAVIS PR, TROUP JDG, BURNARD JH: Movements of the thoracic and lumbar spine when lifting: A chrono-cyclophotographic study. J Anat 99:13, 1965

DECKER HR: Report of the anomalies in a subject with a supernumerary lumbar vertebra. Anat Rec 9:181, 1915

DERRY DE: The influence of sex on the position and composition of the human sacrum. J Anat Physiol 46:184, 1912

DEUCHER WG, LOVE JG: Pathologic aspects of posterior protrusions of the intervertebral disks. Arch Path 27:201, 1939

DITTRICH RJ: Low back pain and spina bifida occulta. Am J Surg 43:739, 1939

DOCKERTY MB, LOVE JG: Thickening and fibrosis (so-called hypertrophy) of the ligamentum flavum: A pathologic study of fifty cases. Proc Staff Meet Mayo Clin 15:161, 1940

DOMMISE GF: The blood supply of the spinal cord. J Bone Joint Surg 56-B:225, 1974

DONISCH EW, BASMAJIAN JV: Electromyography of deep back muscles in man. Am J Anat 133:25, 1972

DUNCAN D, KEYSER LL: Some determinations of the ratio of nerve fibers to nerve cells in the thoracic dorsal roots and ganglia of the cat. J Comp Neurol 64:303, 1936

DUNLOP J: Fractures of the spine. Am J Surg 38:568, 1937

DWIGHT T: Numerical variation in the human spine, with a statement concerning priority. Anat Anz 28:33, 96, 1906

ECHLIN FA, SELVERSTONE B, SCRIBNER WE: Bilateral and multiple ruptured discs as one cause of persistent symptoms following operation for a herniated disc. Surg Gynecol Obstet 83:485, 1946

ECHOLS DH, KLEINSASSER LRJ: Atlanto-axial dislocation: Reduction by skeletal traction. Report of a case. Am J Surg 52:70, 1941

ECKENHOFF JE: The physiologic significance of the vertebral venous plexus. Surg Gynecol Obstet 131:72, 1970

EHRENHAFT JL: Development of the vertebral column as related to certain congenital and pathological changes. Surg Gynecol Obstet 76:282, 1943

ELLIOT HC: Cross-sectional diameters and areas of the human spinal cord. Anat Rec 93:287, 1945

ETHELBERGER S, RIISHEDE J: Malformation of lumbar spinal roots and sheaths in the causation of low backache and sciatica. J Bone Joint Surg 34-B:442, 1952

EVARTS CM, WINTER RB, HALL JE: Vascular compression of the duodenum associated with the treatment of scoliosis: Review of the literature and report of eighteen cases. J Bone Joint Surg 53-A:431, 1971

FALCONER MA, MCGEORGE M, BEGG AC: Observations on the cause and mechanism of symptom-production in sciatica and low-back pain. J Neurol Neurosurg Psychiat 11:13, 1948

FERGUSON AB, JR: The etiology of pre-adolescent kyphosis. J Bone Joint Surg 38-A:149, 1956

FIELDING JW: Normal and selected abnormal motion of the cervical spine from the second cervical vertebra to the seventh cervical vertebra based on cineroentgenography. J Bone Joint Surg 46-A:1779, 1964

FIELDING JW, COCHRAN GVB, LAWSING JF III, HOHL M: Tears of the transverse ligament of the atlas: A clinical and biomechanical study. J Bone Joint Surg 56-A:1638, 1974

FIELDING WF, GRIFFIN PG: Os odontoideum: An acquired lesion. J Bone Joint Surg 56-A:187, 1974

FISHER RG, UIHLEIN A, KEITH HM: Spina bifida and cranium bifidum: Study of 530 cases. Proc Staff Meet Mayo Clin 27:33, 1952

FLECKER H: Time of appearance and fusion of ossification centers as observed by roentgenographic methods. Am J Roentgenol 47:97, 1942

FORD LT: The diagnosis of lesions of the intervertebral

disc. Am Acad Orth Surgeons Instructional Course Lectures 6:33, 1949

FREEBODY D, BENDALL R, TAYLOR RD: Anterior transperitoneal lumbar fusion. J Bone Joint Surg 53-B:617, 1971

FREIBERG AH: Sciatic pain and its relief by operations on muscle and fascia. Arch Surg 34:337, 1937

FREIBERGER RH, WILSON PD, JR, NICHOLAS JA: Acquired absence of the odontoid process: A case report. J Bone Joint Surg 47-A:1231, 1965

FRIED LC, DICHIRO G, DOPPMAN JL: Ligation of major thoracolumbar spinal cord arteries in monkeys. J Neurosurg 31:608, 1969

FRIED LC, DOPPMAN JL, DICHIRO G: Direction of blood flow in the primate cervical cord. J Neurosurg 33:325, 1970

FRENCH JD, STRAIN WH: Peripheral extension of radiopaque media from the subarachnoid space. Surgery 22:380, 1947

FRIEDENBERG ZB, EDEIKEN J, SPENCER N, TOLENTINO SC: Degenerative changes in the cervical spine. J Bone Joint Surg 41-A:61, 1959

GARCEAU GJ: The filum terminale syndrome: The cord-traction syndrome. J Bone Joint Surg 35-A:711, 1953

GARLAND H, GREENBERG J, HARRIMAN DGF: Infarction of the spinal cord. Brain 89:645, 1966

GIANNESTRAS NJ, MAYFIELD FH, PROVENCIO FP, MAUER J: Congenital absence of the odontoid process: Case report. J Bone Joint Surg 46-A:839, 1964

GILL GG, MANNING JG, WHITE HL: Surgical treatment of spondylolisthesis without spine fusion. Excision of the loose lamina with decompression of the nerve roots. J Bone Joint Surg 37-A:493, 1955

GLADSTONE RJ: A case of an additional presacral vertebra. J Anat Physiol 31:530, 1897

GLOOBE H, NATHAN H: Osteophyte formation in experimental bipedal rats. J Comp Path 83:133, 1973

GOLUB BS, SILVERMAN B: Transforaminal ligaments of the lumbar spine. J Bone Joint Surg 51-A:947, 1969

GREGERSEN GG, LUCAS DB: An in vivo study of the axial rotation of the human thoracolumbar spine. J Bone Joint Surg 49-A:247, 1967

GRUCA A: The pathogenesis and treatment of idiopathic scoliosis: A preliminary report. J Bone Joint Surg 40-A:570, 1958

HAAS SL: Growth in length of the vertebrae. Arch Surg 38:245, 1939

HALEY JC, PERRY JH: Protrusions of intervertebral discs: Study of their distribution, characteristics and effects on the nervous system. Am J Surg 80:394, 1950

HALLIDAY DR, SULLIVAN CR, HOLLINSHEAD WH, BAHN RC: Torn cervical ligaments: Necropsy examination of the normal cervical region of the spinal column. J Trauma 4:219, 1964

HAMBLEN DL: Occipito-cervical fusion: Indications, technique and results. J Bone Joint Surg 49-B:33, 1967

HARMEIER JW: The normal histology of the intradural filum terminale. Arch Neurol Psychiat 29:308, 1933

HARRIS RS, and JONES DM: The arterial supply to the adult cervical vertebral bodies. J Bone Joint Surg 38-B:922, 1956

HARVEY SC, BURR HS, VANCAMPENHOUT E: Development of the meninges: Further experiments. Arch Neurol Psychiat 29:683, 1933

HASS GM: Chordomas of the cranium and cervical portion of the spine: Review of the literature with report of a case. Arch Neurol Psychiat 32:300, 1934

HASSLER O: Blood supply to human spinal cord: A microangiographic study. Arch Neurol 15:302, 1966

HAUSER E: Scoliosis: A functional decompensation. Arch Surg 34:1159, 1937

HAWKINS RJ, FIELDING JW, THOMPSON WJ: Os odontoideum: Congenital or acquired. A case report. J Bone Joint Surg 58-A:413, 1976

HENDERSON ED: Results of the surgical treatment of spondylolisthesis. J Bone Joint Surg 48-A:619, 1966

HERREN RY, ALEXANDER L: Sulcal and intrinsic blood vessels of human spinal cord. Arch Neurol Psychiat 41:678, 1939

HOFMANN M: Die Befestigung der Dura mater im Wirbelcanal. Arch f Anat u. Physiol (Anat Abt) p 403, 1898

HOHL M: Normal motions in the upper portion of the cervical spine. J Bone Joint Surg 46-A:1777, 1964

HOLT EP, JR: Fallacy of cervical discography. Report of 50 cases in normal subjects. JAMA 188:799, 1964

HOLT S, YATES PO: Cervical spondylosis and nerve root lesions: Incidence at routine necropsy. J Bone Joint Surg 48-B:407, 1966

HORTON WG: Further observations on the elastic mechanism of the intervertebral disc. J Bone Joint Surg 40-B:552, 1958

HORWITZ MT: The anatomy of (A) the lumbo-sacral nerve plexus—its relation to variations of vertebral segmentation, and (B), the posterior sacral nerve plexus. Anat Rec 74:91, 1939a

HORWITZ T: Lesions of the intervertebral disk and ligamentum flavum of the lumbar vertebrae: An anatomic study of 75 human cadavers. Surgery 6:410, 1939b

HYNDMAN OR, VAN EPPS C: Possibility of differential section of the spinothalamic tract: A clinical and histologic study. Arch Surg 38:1036, 1939

INGBERT C: An enumeration of the medullated nerve fibers in the dorsal roots of the spinal nerves of man. J Comp Neurol 13:53, 1903

INGERT CE: An enumeration of the medullated nerve fibers in the ventral roots of the spinal nerves of man. J Comp Neurol 14:209, 1904

INMAN VT, SAUNDERS JB DE CM: Anatomicophysiological aspects of injuries to the intervertebral discs. J Bone Joint Surg 29:461, 1947

JAMES CCM, LASSMAN LP: Spinal dysraphism: The diagnosis and treatment of progressive lesions in spina bifida occulta. J Bone Joint Surg 44-B:828, 1962

JAMES JIP: Idiopathic scoliosis: The prognosis, diagnosis, and operative indications related to curve patterns and the age at onset. J Bone Joint Surg 36-B:36, 1954

JEFFERSON G: Fracture of the atlas vertebra: Report of four cases, and a review of those previously recorded. Brit J Surg 7:407, 1919

JONCK LM: The mechanical disturbances resulting from lumbar disc space narrowing. J Bone Joint Surg 43-B:362, 1961

JONES PH, LOVE JG: Tight filum terminale. AMA Arch Surg 73:556, 1956

KAPLAN EB: The surgical and anatomic significance of the mammillary tubercle of the last thoracic vertebra. Surgery 17:78, 1945

KEEGAN JJ: Relations of nerve roots to abnormalities of lumbar and cervical portions of the spine. Arch Surg 55:246, 1947

KEEGAN JJ: Alterations of the lumbar curve related to posture and seating. J Bone Joint Surg 35-A:589, 1953

KEILLER VH: A contribution to the anatomy of spina bifida. Brain 45:31, 1922

KEIM HA, KEAGY RD: Congenital absence of lumbar articular facets: A report of three cases. J Bone Joint Surg 49-A:523, 1967

KERNOHAN JW: The ventriculus terminalis: Its growth and development. J Comp Neurol 38:107, 1924

KERR JG: Scoliosis with paraplegia. J Bone Joint Surg 35-A:769, 1953

KETTELKAMP DB, WRIGHT DG: Spondylolysis in Alaskan Eskimos. J Bone Joint Surg 53-A:563, 1979

KEY JA: The intervertebral disc: Anatomy, physiology, and pathology. Am Acad Orth Surgeons Instructional Course Lectures 6:27, 1949

KEY JA: Intervertebral-disc lesions in children and adolescents. J Bone Joint Surg 32-A:97, 1950

KRAFT GL, LEVINTHAL DH: Facet synovial impingement: A new concept in the etiology of lumbar vertebral derangement. Surg Gynecol Obstet 93:439, 1951

KRENZ J, TROUP JDG: The structure of the pars interarticularis of the lower lumbar vertebrae and its relation to the etiology of spondylolysis: With a report of a healing fracture of the fourth lumbar vertebra. J Bone Joint Surg 55-B:735, 1973

KRISTOFF FV, ODOM GL: Ruptured intervertebral disk in the cervical region: A report of twenty cases. Arch Surg 54:287, 1947

KUHNS JG, HORMELL RS: Management of congenital scoliosis: Review of one hundred seventy cases. AMA Arch Surg 65:250, 1952

LANCE EM: Treatment of severe spondylolisthesis with neural involvement. A report of two cases. J Bone Joint Surg 48-A:883, 1966

LANIER RR, JR: The presacral vertebrae of American white and Negro males. Am J Phys Anthropol 25:341, 1939a

LANIER RR, JR: An anomalous cervico-occipital skeleton in man. Anat Rec 73:189, 1939b

LANIER VS, MCKNIGHT HF, TROTTER M: Caudal analgesia: An experimental and anatomical study. Am J Obst Gynec 47:633, 1944

LARMON WA: An anatomic study of the lumbo-sacral region in relation to low back pain and sciatica. Ann Surg 119:892, 1944

LE COCQ E: Anomalies of the lumbosacral spine. Am J Surg 22:118, 1933

LETTERMAN GS, TROTTER M: Variations of the male sacrum: Their significance in caudal anesthesia. Surg Gynecol Obstet 78:551, 1944

LEWEY FH: The mechanism of the intervertebral disc protrusion. Surg Gynecol Obstet 88:592, 1949

LINDBLOM K, HULTQUIST GT: Absorption of protruded disc tissue. J Bone Joint Surg 32-A:557, 1950

LINS HR: Extradural anesthesia. Surgery 18:502, 1945

LONG C II, LAWTON EB: Functional significance of spinal cord lesion level. Arch Phys Med Rehab 36:249, 1955

LONG NGN, LAMPHIER TA: Fibrositis and the disc syndrome. Am J Surg 86:414, 1953

LOVE JG: Removal of protruded intervertebral disks without laminectomy. Proc Staff Meet Mayo Clin 14:800, 1939; 15:4, 1940

LOVE JG: The disc factor in low-back pain with or without sciatica. J Bone Joint Surg 29:438, 1947

LOVE JG: Transplantation of the spinal cord for the relief of paraplegia. Arch Surg 73:757, 1956

LOVE JG: Personal communication to the author. 1968

LOVE JG, SCHORN VG: Thoracic-disk protrusions. JAMA 191:627, 1965

MACALISTER A: Notes on the development and variations of the atlas. J Anat Physiol 27:519, 1893

MACALISTER A: The development and varieties of the second cervical vertebra. J Anat Physiol 28:257, 1894

MACNAB I: Spondylolisthesis with an intact neural arch—the so-called pseudo-spondylolisthesis. J Bone Joint Surg 32-B:325, 1950

MCNAB I: The traction spur: An indication of segmental instability. J Bone Joint Surg 53-A:663, 1971

MACNAB I, DALL D: The blood supply of the lumbar spine and its application to the technique of intertransverse lumbar fusion. J Bone Joint Surg 53-B:628, 1971

MACNAB I, ST LOUIS EL, GRABIAS SL, JACOB R: Selective ascending lumbosacral venography in the assessment of lumbar-disc herniation. An anatomical study and clinical experience. J Bone Joint Surg 58-A:1093, 1976

MARKOLF KL, MORRIS JM: The structural components of the intervertebral disc. A study of their contributions to the ability of the disc to withstand compression forces. J Bone Joint Surg 56-A:675, 1974

MARR GE, UIHLEIN A: Diplomyelia and compression of the spinal cord and not of the cauda equina, by a congenital anomaly of the third lumbar vertebra. S Clin North America (Aug) 1944, p. 963

MAYER DJ, PRICE DP, BECKER DP, YOUNG HF: Threshold for pain from anterolateral quadrant stimulation as a predictor of success of percutaneous cordotomy for relief of pain. J Neurosurg 43:445, 1975

MCCOTTER RE: Regarding the length and extent of the human medulla spinalis. Anat Rec 10:559, 1916

MCKENZIE KG, DEWAR FP: Scoliosis with paraplegia. J Bone Joint Surg 31-B:162, 1949

MILES M, SULLIVAN WE: Lateral bending at the lumbar and lumbosacral joints. Anat Rec 139:387, 1961

MITCHELL GAG: The lumbosacral junction. J Bone Joint Surg 16:233, 1934.

MITCHELL GAG: The significance of lumbo-sacral transitional vertebrae. Brit J Surg 24:147, 1936

MITCHELL PEG, HENDRY NGC, BILLEWICZ WZ: The chemical background of intervertebral disc prolapse. J Bone Joint Surg 43-B:141, 1961

MORRIS JM, BENNER G, LUCAS DB: An electromyographic study of the intrinsic muscles of the back in man. J Anat 96:509, 1962

MORRIS JM, LUCAS DB, BRESLER B: Role of the trunk in stability of the spine. J Bone Joint Surg 43-A:327, 1961

MORRISON AB: The levatores costarum and their nerve supply. J Anat 88:19, 1954

NACHEMSON A: Lumbar intradiscal pressure. Acta orthop scandinav suppl 43:1, 1960

NACHEMSON A: The influence of spinal movements on the lumbar intradiscal pressure and on the tensile stresses in the annulus fibrosus. Acta orthop scandinav 33:183, 1963

NACHEMSON A: The effect of forward leaning on lumbar intradiscal pressure. Acta orthop scandinav 35:314, 1965

NACHEMSON AL, EVANS JH: Some mechanical properties of the third human lumbar interlaminar ligament (ligamentum flavum). J Biomechanics 1:211, 1968

NACHEMSON A, MORRIS JM: In vivo measurements of intradiscal pressure: Discometry, a method for the determination of pressure in the lower lumbar discs. J Bone Joint Surg 46-A:1077, 1964

NACHLAS IW, BORDEN JN: Experimental scoliosis—the role of the epiphysis. Surg Gynecol Obstet 90:672, 1950

NAFFZIGER HC, INMAN V, SAUNDERS JB DE CM: Lesions of the intervertebral disc and ligamenta flava. Surg Gynecol Obstet 66:288, 1938

NATHAN H: Osteophytes of the vertebral column. An anatomical study of their development according to age, race and sex with considerations as to their etiology and significance. J Bone Joint Surg 44-A:243, 1962

NEEDLES JH: The caudal level of termination of the spinal cord in American whites and American Negroes. Anat Rec 63:417, 1935

NEWMAN PH: Sprung back. J Bone Joint Surg 34-B:30, 1952

NICOLL EA: Fractures of the dorso-lumbar spine. J Bone Joint Surg 31-B:376, 1949

NOBACK CR, ROBERTSON GG: Sequences of appearance of ossification centers in the human skeleton during the first five prenatal months. Am J Anat 89:1, 1951

NORGORE M: Clinical anatomy of the vertebral veins. Surgery 17:606, 1945

O'CONNELL JEA: Sciatica and the mechanism of the production of the clinical syndrome in protrusions of the lumbar intervertebral discs. Brit J Surg 30:315, 1943

O'CONNELL JEA: Protrusions of the lumbar intervertebral discs: A clinical review based on five hundred cases treated by excision of the protrusion. J Bone Joint Surg 33-B:8, 1951

O'CONNELL JEA: Involvement of the spinal cord by intervertebral disk protrusions. Brit J Surg 43:225, 1955

O'DELL CW JR, COEL MN, IGNELZI RJ: Ascending lumbar venography in lumbar-disc disease. J Bone Joint Surg 56-A:159, 1977

OROFINO C, SHERMAN MS, SCHECHTER D: Luschka's joint—a degenerative phenomenon. J Bone Joint Surg 42-A:853, 1960

OUAKNINE G, NATHAN H: Anastomotic connections between the eleventh nerve and the posterior root of the first cervical nerve in humans. J Neurosurg 83:189, 1973

OWENS G: Review of spina bifida and cranium bifidum: With follow-up studies of eighty-one cases. Am J Surg 86:410, 1953

PAULY JE: An electromyographic analysis of certain movements and exercises. I. Some deep muscles of the back. Anat Rec 155:223, 1966

PAYNE EE, SPILLANE JD: The cervical spine. An anatomico-pathological study of 70 specimens (using a special technique) with particular reference to the problem of cervical spondylosis. Brain 80:571, 1957

PEACOCK A: Observations on the postnatal structure of the intervertebral disc in man. J Anat 86:162, 1952

PEARSON AA, SAUTER RW, BUCKLEY TF: Further observations on the cutaneous branches of the dorsal primary rami of the spinal nerves. Am J Anat 118:891, 1966

PEDERSEN HE, BLUNCK CFJ, GARDNER E: The anatomy of lumbosacral posterior rami and meningeal branches of spinal nerves (sinu-vertebral nerves):

With an experimental study of their functions. J Bone Joint Surg 38-A:377, 1956

PETERS GA: Sensory nerve cells in the dorsal roots of human spinal nerves. Anat Rec 78:113, 1940

PETTER CK: Methods of measuring the pressure of the intervertebral disc. J Bone Joint Surg 15:365, 1933

POIRIER P, CHARPY A: Traité d'Anatomie Humaine vol. 3, Fasc. 3 (ed. 2). Paris, Masson et Cie., 1904

PRIOLEAU GR, WILSON CB: Cervical spondylolysis with spondylolisthesis: Case report. J Neurosurg 43:751, 1975

REIMANN AF, ANSON BJ: Vertebral level of termination of the spinal cord with report of a case of sacral cord. Anat Rec 88:127, 1944

RICHARDS V, KING D: Chordoma. Surgery 8:409, 1940

RISSER JC: Scoliosis: Past and present. J Bone Joint Surg 46-A:167, 1964

RISSER JC, FERGUSON AB: Scoliosis: Its prognosis. J Bone Joint Surg 18:667, 1936

RISSER JC, NORQUIST DM: A follow-up study of the treatment of scoliosis. J Bone Joint Surg 40-A:555, 1958

RISSANEN PM: The surgical anatomy and pathology of the supraspinous and interspinous ligaments of the lumbar spine with special reference to ligament ruptures. Acta orthop scandinav suppl 46, 1960

ROAF R: The basic anatomy of scoliosis. J Bone Joint Surg 48-B:786, 1966

ROBINSON RA, WALKER AE, FERLIC DC, WIECKING DK: The results of anterior interbody fusion of the cervical spine. J Bone Joint Surg 44-A:1569, 1962

ROCHE MB: The pathology of neural-arch defects: A dissection study. J Bone Joint Surg 31-A:529, 1949

ROCHE MB, ROWE GG: The incidence of separate neural arch and coincident bone variations: A survey of 4,200 skeletons. Anat Rec 109:253, 1951a

ROCHE MB, ROWE GG: Anomalous centers of ossification for inferior articular processes of the lumbar vertebrae. Anat Rec 109:253, 1951b

ROCHE MB, ROWE GG: The incidence of separate neural arch and coincident bone variations: A summary. J Bone Joint Surg 34-A:491, 1952

ROGERS L: The treatment of cervical spondylitic myelopathy by mobilisation of the cervical cord into an enlarged spinal canal. J Neurosurg 18:490, 1961

ROOFE PG: Innervation of annulus fibrosus and posterior longitudinal ligament: Fourth and fifth lumbar level. Arch Neurol Psychiat 44:100, 1940

ROOKE ED, CORBIN KB: Clinical aspects of spinal cord localization. In: Baker AB ed: Clinical Neurology [ed. 2], Vol. 3: New York, Hoeber, 1962

ROSENBAUM MM: Caudal anesthesia for anorectal surgery. Am J Surg 86:636, 1953

ROSOMOFF HL, ROSSMAN· F: Treatment of cervical spondylosis by anterior cervical diskectomy and fusion. Arch Neurol 14:392, 1966

ROSS P, JELSMA F: Postoperative analysis of 366 consecutive cases of herniated lumbar discs. Am J Surg 84:657, 1952

ROTHENBERG SF, MENDELSOHN HA, PUTNAM TJ: The effect of leg traction on ruptured intervertebral discs. Surg Gynecol Obstet 96:564, 1953

ROWE GG, ROCHE MB: The lumbar neural arch: Roentgenographic study of ossification. J Bone Joint Surg 32-A:554, 1950

ROWE GG, ROCHE MB: The etiology of separate neural arch. J Bone Joint Surg 35-A:102, 1953

SACKS S: Anterior interbody fusion of the lumbar spine. J Bone Joint Surg 47-B:211, 1965

SARGENT M: Localized back pain: Separation of fibers of posterior layer of lumbodorsal fascia with herniation of the sacrospinalis muscle as a cause. Am J Surg 71:338, 1946

SAUNDERS JB DE CM, INMAN VT: Pathology of the intervertebral disk. Arch Surg 40:389, 1940

SAUNDERS RL DE CH: Combined anterior and posterior spina bifida in a living neonatal human female. Anat Rec 87:255, 1943

SCAPINELLI R: Sesamoid bones in the ligamentum nuchae of man. J Anat 97:417, 1963

SCHIFF DM, PARKE WW: The arterial supply of the odontoid process. J Bone Joint Surg 55-A:1450, 1973

SCHLESINGER EB, STINCHFIELD FE: The use of muscle relaxants as an aid in the diagnosis and therapy of acute low-back disorders. J Bone Joint Surg 33-A:480, 1951

SCHROCK RD: Congenital abnormalities at the cervicothoracic level. Am Acad Orth Surgeons Instructional Course Lectures 6:228, 1949

SCHULTZ AH: Chevron bones in adult man. Am J Phys Anthropol 28:91, 1941

SCOVILLE WB: Cervical spondylosis treated by bilateral facetectomy and laminectomy. J Neurosurg 18:423, 1961

SEMMES RE: Lateral rupture of cervical intervertebral discs: Incidence and clinical varieties. Am J Surg 75:137, 1948

SENSENIG EC: Adhesions of notochord and neural tube in the formation of chordomas. Am J Anat 98:357, 1956

SHERK HH, NICHOLSON JT: Rotatory atlanto-axial dislocation associated with ossiculum terminale and mongolism. A case report. J Bone Joint Surg 51-A:957, 1969

SHERK HH, NICHOLSON JT, CHUNG SMK: Fractures of the odontoid in young children. J Bone Joint Surg 60-A:921, 1978

SHUTKIN NM: Syndrome of the degenerated intervertebral disc. Am J Surg 84:162, 1952

SINCLAIR DC, FEINDEL WH, WEDDELL G, FALCONER MA: The intervertebral ligaments as a source of segmental pain. J Bone Joint Surg 30-B:515, 1948

SINCLAIR JG, DUREN N, RUDE JC: Congenital lumbosacral defect. Arch Surg 43:473, 1941

SLATER RA, PINEDA A, PORTER RW: Intradural herniation of lumbar intervertebral discs. Arch Surg 90:266, 1965

SMITH EG: On a case of fusion of the atlas and axis. Anat Anz 31:166, 1907

SMYTH MJ, WRIGHT V: Sciatica and the intervertebral disc: An experimental study. J Bone Joint Surg 40-A:1401, 1958

SOMERVILLE EW: Rotational lordosis: The development of the single curve. J Bone Joint Surg 34-B:421, 1952

SPILLANE JD, PALLIS C, JONES AM: Developmental abnormalities in the region of the foramen magnum. Brain 80:11, 1957

SPURLING RG, GRANTHAM EG: Ruptured intervertebral discs in the lower lumbar regions. Am J Surg 75:140, 1948

SPURLING RG, SCOVILLE WB: Lateral rupture of the cervical intervertebral discs: A common cause of shoulder and arm pain. Surg Gynecol Obstet 78:350, 1944

STAUFTER RN, COVENTRY MB: Anterior interbody lumbar spine fusion: Analysis of Mayo Clinic series. J Bone Joint Surg 54-A:756, 1972a

STAUFFER RN, COVENTRY MB: Posterolateral lumbar-spine fusion: Analysis of Mayo Clinic series. J Bone Joint Surg 54-A:1195, 1972b

STEINDLER A: Mechanics of Normal and Pathological Locomotion in Man. Springfield, Ill, Thomas, 1935

STEWART TD: The age incidence of neural-arch defects in Alaskan natives, considered from the standpoint of etiology. J Bone Joint Surg 35-A:937, 1953

STILWELL DL, JR: The nerve supply of the vertebral column and its associated structures in the monkey. Anat Rec 125:139, 1956

STILWELL DL, JR: The vascular supply of vertebral structures. Gross anatomy: rabbit and monkey. Anat Rec 135:169, 1959

STOLTMANN HF, BLACKWOOD W: The role of the ligamenta flava in the pathogenesis of myelopathy in cervical spondylosis. Brain 87:45, 1964

STOLTMANN HF, BLACKWOOD W: An anatomical study of the role of the dentate ligaments in the cervical spinal canal. J Neurosurg 24:43, 1966

SUH TH, ALEXANDER L: Vascular system of the human spinal cord. Arch Neurol Psychiat 41:659, 1939

SYLVÉN B, PAULSON S, HIRSCH C, and SNELLMAN O: Biophysical and physiological investigations on cartilage and other mesenchymal tissues: II. The ultrastructure of bovine and human nuclei pulposi. J Bone Joint Surg 33-A:333, 1951

TAKEBE K, VITTI M, BASMAJIAN JV: The function of semispinalis capitis and splenius capitis muscles. An electromyographic study. Anat Rec 179:477, 1974

TARLOV IM: Structure of the nerve root: II. Differentiation of sensory from motor roots; observations on identification of function in roots of mixed cranial nerves. Arch Neurol Psychiat 37:1338, 1937

TARLOV IM: Structure of the filum terminale. Arch Neurol Psychiat 40:1, 1938

TARLOV IM: Sacral Nerve-root Cysts: Another Cause of the Sciatic or Cauda Equina Syndrome. Springfield, Ill, Thomas, 1953

TAYLOR AR: The mechanism of injury to the spinal cord in the neck without damage to the vertebral column. J Bone Joint Surg 33-B:543, 1951

THOMSON JEM: First aid and transportation of suspected spine injuries. Am J Surg 45:42, 1939

TODD TW, D'ERRICO J, JR: The odontoid ossicle of the second cervical vertebra. Ann Surg 83:20, 1926

DU TOIT JG, FAINSINGER MH: Spinal extradural cysts. J Bone Joint Surg 30-B:613, 1948

TOUMEY JW, POPPEN JL, HURLEY MT: Cauda equina tumors as a cause of the low-back syndrome. J Bone Joint Surg 32-A:249, 1950

TROTTER M: Variations of the sacral canal: Their significance in the administration of caudal analgesia. Anesth Analg 26:192, 1947

TROTTER M, LETTERMAN GS: Variations of the female sacrum: Their significance in continuous caudal anesthesia. Surg Gynecol Obstet 78:419, 1944

TURNBULL IM, BRIEG A, HASSLER O: Blood supply of cervical spinal cord in man: A microangiographic cadaver study. J Neurosurg 24:951, 1966

VIRGIN WJ: Experimental investigations into the physical properties of the intervertebral disc. J Bone Joint Surg 33-B:607, 1951

VON LACKUM WH, MILLER JP: Critical observations of the results in the operative treatment of scoliosis. J Bone Joint Surg 31-A:102, 1949

WADIA NH: Myelopathy complicating congenital atlanto-axial dislocation. (A study of 28 cases). Brain 90:449, 1967

WAGNER CJ: Low back pain and sciatica. Am J Surg 78:203, 1949

WAGNER R: Die Endigung des Duralsackes im Wirbelcanal des Menschen. Arch f Anat u Physiol (Anat. Abt.) p 64, 1890

WESTBROOK WHL, JR, TOWER SS: An analysis of the problem of emergent fibers in posterior spinal roots, dealing with the rate of growth of extraneous fibers into the roots after ganglionectomy. J Comp Neurol 72:383, 1940

WHEELER T: Variability in the spinal column as regards defective neural arches (rudimentary spina bifida). Contrib Embryol 9:95, 1920

WILEY AM, TRUETA J: The vascular anatomy of the spine and its relationship to pyogenic vertebral osteomyelitis. J Bone Joint Surg 41-B:796, 1959

WILLIS TA: The lumbo-sacral vertebral column in man, its stability of form and function. Am J Anat 32:95, 1923a

WILLIS TA : The thoracicolumbar column in white and Negro stocks. Anat Rec 26:31, 1923b

WILLIS TA: Nutrient arteries of the vertebral bodies. J Bone Joint Surg 31-A:538, 1949

WILLIS TA: Lumbosacral anomalies. J Bone Joint Surg 41-A:935, 1959

WILTSE LE, WIDELL EH JR, JACKSON DW: Fatigue fracture: The basic lesion in isthmic spondylolisthesis. J Bone Joint Surg 57-A:17, 1975

WINCKLER G: La structure du muscle spinalis dorsi chez l'homme. Arch d'anat d'histol et d'embryol 23:183, 1937

WINTER RB, MOE JH, EILERS VE: Congenital scoliosis. A study of 234 patients treated and untreated. Part I: Natural history. Part II: Treatment. J Bone Joint Surg 50-A:1, 15, 1968

WOLLIN DG: The os odontoideum: Separate odontoid process. J Bone Joint Surg 45-A:1459, 1963

WOODHALL B: Sensory patterns in the localization of disc lesions. J Bone Joint Surg 29:470, 1947

WOODHALL B, HAYES GJ: The well-leg-raising test of Fajersztajn in the diagnosis of ruptured lumbar intervertebral disc. J Bone Joint Surg 32-A:786, 1950

WOOD-JONES F: The ideal lesion produced by hanging. Lancet 1:53, 1913

WYNNE-DAVIES R: Familial (idiopathic) scoliosis. A family survey. J Bone Joint Surg 50-B:24, 1968

WYNN-DAVIES R, SCOTT JHS: Inheritance and spondylolisthesis: A radiographic family survey. J Bone Joint Surg 61-B:301, 1979

YOSS RE, CORBIN KB, MACCARTY CS, LOVE JG: Significance of symptoms and signs in localization of involved root in cervical disk protrusion. Neurology 7:673, 1957

YOUNG HH: Posterior fusion of vertebrae in treatment for protruded intervertebral disk. J Neurosurg 19:314, 1962

YOUNG HH: Personal communication to the author. 1968

ZEITLIN H, LICHTENSTEIN BW: Occlusion of the anterior spinal artery: Clinicopathologic report of a case and a review of the literature. Arch Neurol Psychiat 36:96, 1936

ZOLLINGER R: Chordoma of the third lumbar vertebra: Report of a case. Am J Surg 19:137, 1933

ZUK T: The role of spinal and abdominal muscles in the pathogenesis of scoliosis. J Bone Joint Surg 44-B:102, 1962

Chapter 3

GENERAL SURVEY OF THE UPPER LIMB

Development

The development of the upper limb has been described especially by Bardeen and Lewis, and by Lewis.

The upper limb first appears in the embryo toward the end of the fourth week as a swelling, the arm bud, on the lateral body wall; the mesenchyme producing the swelling is apparently derived by proliferation from the lateral body wall, and not by migration from the somites. The swelling grows rapidly; its distal end flattens to a plate for the hand, and is partially separated from the remaining proximal portion by a constriction; the plate develops ridges that are separated at first by grooves, and then project freely as the digits (Fig. 3-1). A second constriction, representing the elbow, divides the proximal cylindrical segment into forearm and arm.

Soon after it arises the limb bud points caudad, but very quickly it grows laterally, at approximately a right angle to the body. In this position its surfaces are dorsal and ventral, its borders cranial (preaxial) and caudal (postaxial); the thumb develops on the cranial or preaxial border, which is thus the radial side of the limb. A subsequent ventral bending at the level of the constriction marking the elbow causes the palms of the hands to be turned toward the thorax instead of facing ventrally.

It is usually said that the upper limb during its development undergoes a rotation of 90° about its long axis, in a lateral direction (thus in the opposite direction from the rotation of the lower limb); however, unlike the lower limb, the adult upper limb gives no indication of having undergone such a rotation, and the effect of the so-called rotation, a change from a dorsally or laterally projecting elbow to a caudally projecting one, can actually be brought about by simple adduction of the limb through 90°, rather than by a rotation around its axis. In fact, if the limb in the anatomical position, palm facing forward, is abducted to the horizontal, its surfaces and borders then correspond, with no rotation, to its original surfaces and borders: the radial side of the limb is directed toward the head, the ulnar or postaxial side is still directed caudally; likewise, the original ventral or flexor surface is directed ventrally, and the original dorsal or extensor surface is directed dorsally.

Skeleton and Musculature

At an early stage, while the hand plate is forming, a denser core appears in the center of the limb bud as the forerunner of the skeleton of the free limb. At this same stage, about 5 weeks, similar condensations of mesenchyme at the base of the limb foreshadow the girdle. Within about 2 weeks, the cores of mesenchyme have been transformed into cartilage (with the exception of the clavicle, which develops a center of ossification in its blastemal

5th week   6th week   7th week   8th week

Fig. 3-1. Outline of the development of the upper limb. (After Keith, from Coventry, MB: Am Acad Orth Surgeons, Instructional Course Lectures 6:218, 1949)

forerunner toward the end of the fifth week, and is the first of all bones to ossify) and the first centers of ossification in the bones of the free limb appear toward the end of the eighth week.

At the same time that the condensations of mesenchyme for the skeleton are appearing, similar but less dense condensations, the premuscle masses, appear in the limb. At the base of the limb, a more ventrally placed premuscle mass gives rise to the pectoralis major and minor muscles, while more dorsally placed ones give rise to, respectively, the latissimus dorsi and teres major; the serratus anterior and levator scapulae; the two rhomboids; the sternocleidomastoid and trapezius; the supraspinatus, infraspinatus, teres minor, and deltoid; and the subscapularis. In the free limb the premuscle mass at first forms a simple sheath around the mesenchymal condensation for the skeleton, but in the arm it divides into two masses, a dorsal one from which the triceps differentiates and a flexor one from which the biceps, brachialis, and coracobrachialis arise. Similarly, the premuscle mass in the forearm divides into a flexor and an extensor mass, and the individual muscles differentiate by subdivision of these two masses.

Splitting of the premuscle masses into individual masses for the various muscles and differentiation of muscle fibers from mesenchyme begin in general in the most proximal muscles and extend distally. In general, also, the dorsal or extensor musculature at any level differentiates a little earlier than does the ventral or flexor musculature at that level. By the end of about 7 weeks the differentiation of muscles, progressing distally, has occurred in the hand.

Nerves

The base of the early limb bud is relatively broad in comparison with the length of the body, and lies at the level of a greater number of somites than it does after further growth occurs. Since the segmentation of the spinal nerves is dependent upon that of the somites, the limb bud develops at the level of a number of spinal nerves, typically those from the fifth cervical through the first thoracic. The ventral branches of these spinal nerves join each other to form a plexus (the dorsal branches supply the muscle derived directly from the somites), and the plexus divides into dorsal and ventral (posterior and anterior or extensor and flexor) parts, corresponding to the division of the premuscle mass into dorsal and ventral, or extensor and flexor, parts. The nerves from these divisions grow out with the developing limb, the dorsal, posterior, or extensor portion of the plexus (largely the posterior cord and its branches) supplying the corresponding part of the premuscle mass, the ventral, anterior, or flexor portion (medial and lateral cords) supplying the developing anterior or flexor muscles.

It is, of course, impossible to follow individual nerve fibers through the plexus to their distribution by the definitive nerves of the plexus, but it is usually assumed that, in spite of the plexus, the segmental nerves are distributed to the skin of the limb in very much the same manner that they would be if there were no plexus present. Thus, as shown in Figure 3-2, the fifth cervical nerve should develop a distribution along the radial or preaxial border of the limb, the first thoracic should develop along the caudal or postaxial border of the limb. Further, it is also usually assumed that the most centrally placed nerves, cen-

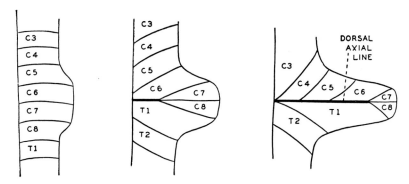

Fig. 3-2. Bolk's commonly accepted view of the development of the dermatomes of the upper limb. (Keegan, JJ, Garrett FD: Anat Rec 102:409, 1948)

tering around C7, are carried farther out on the limb than are the more peripherally placed nerves, C5 and T1, especially on the flexor surface; this allows contiguity between the distributions of nerves originally separated from each other by intervening nerves, the line of meeting being known as the axial line. The ventral axial line extends distally to about the wrist; on the dorsum, however, the centrally lying nerves are not drawn as far from the base of the limb, and the dorsal axial line is limited to the back of the shoulder and a part of the arm (if it exists at all; according to Keegan's concept of the development of the dermatones, Figure 3-3, there is no dorsal axial line).

This usually accepted schema of the development of the segmental contribution to the upper limb explains the dermatomes of the upper limb very well. All dermatomal schemes acknowledge that if one starts on the radial or preaxial border of the limb in the lower part of the arm or upper part of the forearm and proceeds around the dorsum to the postaxial border, one encounters the distribution of the segmental nerves in approximately their regular numerical order, while if one proceeds around the ventral or flexor surface one passes abruptly from the distributions of the more cephalic to those of the more caudal nerves (crossing the ventral axial line, and thus skipping some centrally placed nerves). Foerster's dermatomal scheme (see Figs. 3-36 to 3-38) implies that the cutaneous distributions of the more centrally placed

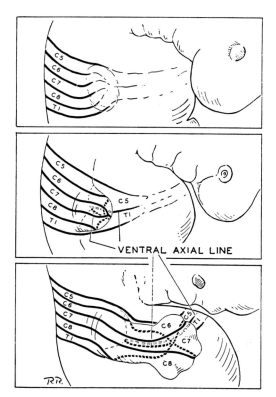

Fig. 3-3. Keegan and Garrett's concept of the development of the dermatomes, leading to the formation of a ventral axial line, but no dorsal axial line. (Keegan JJ, Garrett FD: Anat Rec 102:409, 1948)

nerves are drawn completely into the peripheral part of the limb; Keegan's dermatomal scheme (see Fig. 3-39) implies, in contrast, that all the nerves retain a distribution at the base of the limb.

The segmental innervation of the muscles of the upper limb is far more complicated

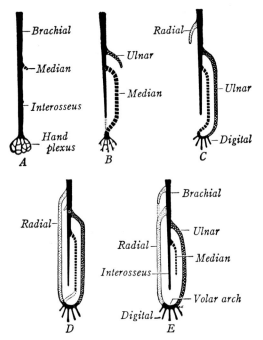

Fig. 3-4. Schema of the development of the arteries of the upper limb. (Arey LB: Developmental Anatomy [ed 6]. Philadelphia, WB Saunders, 1954)

than is the dermatomal distribution to the limb, and knowledge both of the development of the nerves to the muscles and of the adult segmental relationship is certainly incomplete and possibly faulty. It cannot, unfortunately, be assumed that there has been no shifting in relationships between the muscles and the overlying skin, so that muscles covered by certain dermatomes are innervated through the nerves supplying the dermatomes; rather, in spite of Last's schema (Chap. 1) for the innervation of the muscles of the limbs, about all the order that can be made out of the usually accepted statements concerning the segmental innervations of the muscles of the upper limb is that as one proceeds distally in the limb on the flexor surface one tends to encounter first muscles innervated through the upper part of the plexus, later muscles innervated through the lower part, and that the muscles on the extensor side of the free limb tend to be innervated through the intermediate rather than the bordering nerves of the limb.

Arteries

The blood supply to the limb bud is at first in the form of a capillary plexus derived from the several segmental (intersegmental) arteries to which it is adjacent (Woollard). Normally, the seventh segmental artery soon becomes the main stem, and as the aortic arches are resolved into their definitive condition a part of the aortic arch system plus the seventh segmental artery becomes the subclavian-axillary stem on the right side, and the seventh segmental alone becomes the subclavian-axillary stem on the left. The stem extends as a single main channel in the flexor side of the arm and forearm, and breaks up into a plexus that supplies the hand. The portion of the stem in the arm is the future brachial artery, that in the forearm, supplying the hand, the future anterior interosseus artery. The development from this single stem to the condition found in the adult is illustrated in Figure 3-4, and involves simply a succession of branches that in turn take over the chief circulation to the distal part of the limb, the median replacing the interosseus, the ulnar replacing the median, and the radial then joining the ulnar. It might be noted (Fig. 3-4C) that the radial artery first arises from the brachial well above the elbow (marked by the origin of the ulnar artery), and only subsequently (Fig. 3-4D and E) is this high origin replaced by a lower one. Persistence of the more proximal origin of the radial artery leaves this origin in the arm, from the axillary or the upper part of the brachial, and is the obvious explanation for the occurrence of superficial radial arteries (Chaps. 4 and 5), the most common type of superficial accessory artery found in the arm.

Veins and Lymphatics

The chief capillary plexus draining the limb bud lies superficially in it, and the larger venous channels develop along the cranial (radial or preaxial) or caudal (ulnar or postaxial) borders of the bud. The primitive channel along the caudal border of the bud persists as the basilic, axillary, and subclavian veins; at first it opens caudal to the heart into the postcardinal vein (a precursor of the azygos system), but as the heart descends the venous opening shifts cranial to the heart, into the precardinal vein (precursor of the internal jugular), thereby forming the brachiocephalic (innominate) vein of the adult. Much of the primitive venous channel along the cephalic border of the limb bud disappears, but before it does so it develops a secondary proximal extension that becomes the cephalic vein. The cephalic vein originally joins the external jugular vein (a connection that may persist in the adult), but normally develops a termination in the axillary. The deep veins of the limb develop later than the superficial ones, along the arterial stems distal to the axillary artery.

The lymphatic vessels of the upper limb develop from primitive jugular lymph sacs, which also give rise to the lymphatics in the thorax and, persisting as the terminations of the thoracic and right lymphatic ducts, acquire permanent openings into the internal jugular veins. The early development of the lymphatics is primarily along the venous channels.

OSSIFICATION OF THE SKELETON

With the exception of the clavicle, which begins to ossify from a mesenchymal or a "precartilaginous" stage and is sometimes interpreted as a membrane or dermal bone, sometimes as an atypical cartilage one, all the bones of the upper limb are at first formed in cartilage. Ossification appears in the bodies of all the long bones during the early weeks of prenatal development.

Centers of ossification for the clavicle appear at about the fifth or sixth week, this bone being the first of any in the body to show ossification. Unlike most long bones, the clavicle apparently has two centers of ossification for its body. Centers of ossification for the body of the scapula and for the body or shaft of the humerus, radius, ulna, metacarpals, and phalanges all appear in about the eighth to ninth weeks in the approximate centers of the cartilaginous precursors of these bones (Mall, '06). At birth, all the bones of the upper limb

except the carpals and sesamoids are represented by bone. However, the acromion, the coracoid process, and the medial border and inferior angle of the scapula are still cartilaginous. So are both the proximal and distal ends of the humerus (a bony center may be present in the cartilaginous head); both ends of the radius and ulna; all the carpals; the proximal ends of the first metacarpal and of all the phalanges; and the distal ends of the second through the fifth metacarpals. Thus the epiphyseal or secondary centers of ossification for the clavicle, the scapula, and all the long bones of the limb (with the possible exception of the head of the humerus) typically appear after birth; the centers of ossification of the eight carpals, each of which is usually developed from a single center of ossification, also appear after birth.

In the discussion of ossification in general (Chap. 1) it has already been noted that the skeleton of the female develops somewhat faster than that of the male, and that this is especially noticeable in the secondary centers of ossification, which both appear earlier in the female than in the male and fuse to the rest of the bone earlier. Also noted was the general rule, holding for both female and male, that the earlier a secondary center of ossification appears, the later this center tends to be united to the rest of the bone. Because of both sexual and individual differences in the exact times of appearance and fusion of secondary centers, the dates given in the following account must be considered only approximate. More detail and further references to the literature can be found especially in the accounts of the roentgenographic studies of Paterson and of Flecker. The chart of Camp and Cilley is a convenient and practical summary for clinical use.

Shoulder

A secondary or epiphyseal center of ossification at the sternal end of the clavicle appears usually between the eighteenth and twentieth years, and unites with the rest of the bone by or before the twenty-fifth year (Fig. 3-5); Flecker found it as early as age 11 years, 7 months, but failed to find it constantly (by

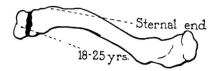

Fig. 3-5. Ossification of the clavicle. The first number, 18, and the second, 25, indicate the usual age at the time of appearance and fusion, respectively, of the epiphysis. (Camp JD, Cilley EIL: Am J Roentgenol 26:905, 1931)

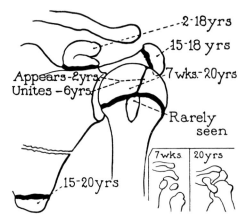

Fig. 3-6. Ossification about the shoulder joint. The first of two numbers indicates the usual age at the time of appearance, the second that at the time of fusion, of the epiphyseal center. (Camp JD, Cilley EIL: Am J Roentgenol 26:905, 1931)

roentgenogram) even in the age group of 19 to 21. Todd and D'Errico described also a small epiphysis at the acromial end, both appearing and uniting to the shaft during the twentieth year.

The scapula has at least six or seven secondary centers of ossification; the more important ones in roentgenographic examination are shown in Figure 3-6. One, for the major part of the coracoid process, appears usually during or after the first year, but may be present at birth; because the coracoid process is phylogenetically a separate bone, this is sometimes considered a primary center. (Two additional centers for the coracoid, apparently inconstant, may appear at about 13 to 16 years, a scalelike one for the angle, another for the tip, and fuse with the main center.) Two centers for the acromion, one for the major part of its free end, the other for its posterolateral tip, appear at about 14 and 16 years, respectively, or several years earlier (its base is developed by extension from the body and spine). A center develops for the inferior angle of the scapula at about 16 years, and one for the medial border at about 17. Finally, one or two centers, a constant one for its upper third, contributing also to the root of the coracoid, and an inconstant one for the remaining margin, complete the ossification of the glenoid cavity. These numerous centers join each other and the body in rapid succession between the fourteenth and twentieth years, those for the glenoid, which do not appear until about the tenth or eleventh years, joining between the sixteenth and eighteenth year.

The upper end of the humerus (Fig. 3-6) has been reported to ossify from one, two, or three secondary centers, in the last case representing centers for the head, for the greater tubercle, and for the lesser tubercle, respectively. Paterson, in his investigations, never found a center for the lesser tubercle; this, he said, always develops by extension from the center for the head. Flecker also failed to find it, but noted that it has been regularly described by anatomists, and that demonstrating it by roentgenogram would be difficult. The center for the head of the humerus is usually said to appear approximately at the time of birth; Flecker found it in 6 out of 10 fetuses of 33 weeks and older. The center for the greater tubercle, which Paterson found in only about 50% of cases (in the others, the greater tubercle apparently developed as an extension from the center for the head) arose during the latter part of the second year or the early part of the third year, according to Paterson; Flecker found it as early as 6 months after birth. If there is a center of ossification for the greater tubercle, it joins the center for the head about the fifth year.

Whether derived from one, two, or three centers, the upper bony end of the humerus, including the head and both tubercles, had, in Paterson's experience, always joined the body (shaft) by the age of 18 years in the female; it had joined the body in 66% of cases by the age of 20 years in the male, and always by the age of 21 years. Flecker reported the majority fused at 17 years in females, at 18 years in males.

Elbow

The lower end of the humerus (Figs. 3-7 and 3-8) is typically ossified from four centers, one for the capitulum, one for the trochlea, and one for each epicondyle. The center for the capitulum is typically the first center about the elbow joint to appear, doing so between the ages of 12 and 18 months; the earliest appearance seen by Paterson was 9 months. The center for the trochlea usually appears during the tenth year in both sexes, but Paterson saw it as early as the eighth year, and Flecker found it in a girl of 7 years. The centers for the capitulum and the trochlea fuse together at about the age of 13 years in the female, 15 in the male, the lateral epicondyle also fusing to the capitulum at about the same time; some 2 years or more later this combined center then fuses with the body—as early as 14 in the female, according to Paterson, but rather regularly during the nineteenth year in the male. Flecker found fusion of the three epiphyseal centers with each other as early as 10 years in a girl, and said fusion with the body had occurred in the majority of boys by the age of 16.

The time of appearance of the center for the lateral epicondyle is said to be variable: sometimes as early as the age of 10, and usually by the age of 14, according to Paterson, and as early as 8.5 years according to Flecker. In only about 70% of cases did Paterson find a separate center; he expressed the opinion that in the remainder the lateral epicondyle was ossified by extension from the capitulum, but Flecker maintained that pictures so interpreted actually are of fusion of the separate center to the capitulum.

Paterson found the center of ossification for the medial epicondyle to be very irregular both in the time of its appearance and in the time of its union with the body. He said he had seen it once before the fifth year, but usually found it toward the end of the fifth year or in the early part of the sixth year in

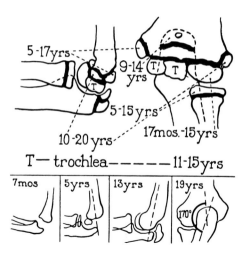

Fig. 3-7. Ossification at the elbow. The first of two numbers indicates the usual age at the time of appearance, the second that at the time of fusion, of the epiphyseal center. (Camp JD, Cilley EIL: Am J Roentgenol 26:905, 1931)

the female, late in the eighth or early in the ninth year in the male; it is, he said, always the last of the four epiphyses of the distal part of the humerus to fuse (it fuses directly to the body, in contrast to the lateral epicondyle), usually doing so at about the age of 14 in the female, and between 18 and 21 in the male.

The head of the radius (Fig. 3-7) develops from a single center of ossification, appearing in the female by the fifth or sixth year, in the male during the sixth or seventh year, or, rarely, the eighth year. Fusion with the body has usually occurred by the fourteenth year in females, by the sixteenth (Flecker) or eighteenth year (Paterson) in the male.

The center for the upper end of the ulna appears at the tip of the olecranon and extends downward, a lower portion of the olecranon, including the coronoid process, developing by extension from the body. Paterson found two centers for the olecranon, one for the main part and one for the tip, in 10% of his series, and noted that sometimes there may be even more; he found a center for the olecranon always present by the age of 11 years, sometimes as early as 9, and Flecker found one as early as 6. The olecranon joins the body at about 14 in the female, about 16 in the male (Flecker) or 19 or later (Paterson).

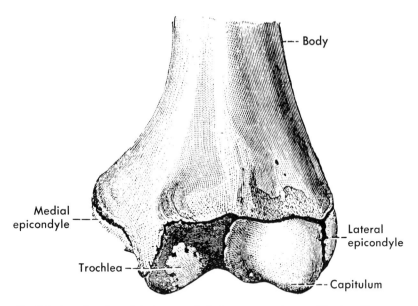

Fig. 3-8. Anterior view of the lower end of a humerus to show ossification. The center for the medial epicondyle has already extended to give origin to the medial border of the trochlea, and is largely united to the body; the center for the trochlea is still small; the centers for the capitulum and the lateral epicondyle are well developed but not yet united. (After Sappey, from Poirier P, Charpy A: Traité d'Anatomie Humaine [ed 3], Vol 1. Paris, Masson et Cie, 1911)

Wrist and Hand

The appearance of centers of ossification in the wrist and hand after birth (at which time the bodies of the long bones are well represented) is spread over a number of years: centers for the capitate and hamate, for instance, apparently always appear during the first 6 months (Fig. 3-9), and usually by 3 months, while the center for the pisiform sometimes does not appear in the male until the age of 13 years, and sesamoids of the digits may appear even later. Greulich and Pyle have, however, found the skeletal maturity of the wrist and hand to be a rather reliable guide to the skeletal and sexual growth and maturity of the individual as a whole.

The center of ossification for the lower epiphysis of the radius (not the first center at the wrist) appears at about 1 year in the female, between 1 and 1.5 years in the male, and unites with the body at about the nineteenth and twenty-first years, respectively (Paterson), 1 to 2 years earlier than this (Flecker), or

Fig. 3-9. Ossification of the wrist and hand. For the epiphyses, the two numbers indicate the usual age at the time of appearance and the time of fusion, respectively; for the carpals, they indicate the range in time of appearance of these bones. (Camp JD, Cilley EIL: Am J Roentgenol 26:905, 1931)

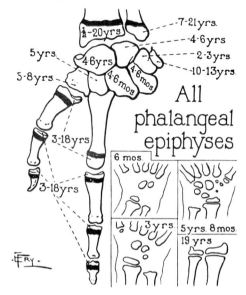

during the fifteenth and sixteenth years in the female, after the eighteenth year in the male (Greulich and Pyle). The lower epiphysis of the ulna, except for the pisiform and the sesamoids the last center to appear in the wrist and hand, becomes evident between the ages of 5 and 7 years, and fuses to the body at about the same time as does the lower radial epiphysis. Paterson found no separate center of ossification for the styloid process, Flecker found one only once; neither found a separate center for the tubercle of the radius, although this has been described as sometimes occurring.

The carpals typically develop from single centers of ossification, most of which first become visible after birth; occasional double centers of ossification have, however, been reported for at least the scaphoid, lunate, triquetral, trapezoid, and hamate, and explain the occurrence of bipartite carpals and perhaps of accessory ones. Paterson found the capitate typically appearing during the first 6 months, and both capitate and hamate always present by 6 months in the female. The appearance of the capitate usually but not always precedes that of the hamate; both have been reported as early as the thirty-sixth week of fetal life but commonly appear about 3 months postpartum.

Following the capitate and hamate there typically appear the triquetral (triquetrum), the lunate, the trapezium and trapezoid (greater and lesser multangulars) together, the scaphoid (navicular), and the pisiform, in about that order: The earliest appearance of the triquetral is said to range in the female from before the end of the first year to the age of 7 or later, and in the male similarly, but the bone is usually first visible in the majority of females during the second or third year, in most males during the third or fourth year. The lunate usually appears in the female during the latter part of the third year, and is almost always present by the age of 4; it tends to appear a year or more later in the male. The trapezium and the trapezoid may be present by the age of 3 in either sex, or still be lacking by the age of 6 in the female and 8 in

the male, but usually are present between 4 and 5 in the female and 5 and 7 in the male. The scaphoid typically appears in the early part of the fifth year in the female, about a year later in the male. The pisiform—somewhat variable—is usually recognizable in the female at the age of 9 to 10, apparently always by age 11, but not recognizable very often in males before 10 or 11, and not constantly present in males until age 14 or later.

The epiphyses of the metacarpals and phalanges, on the proximal ends of the first metacarpal and of all the phalanges, but the distal ends of the second through fifth metacarpals, all usually appear after the first year; some have been found earlier. Paterson said they appear practically simultaneously, except that for the first metacarpal, which tends to lag about 6 months behind. Flecker found a great deal of variation in their times of appearance, and no regular order even in regard to similarly named phalanges of the four fingers: for instance, a center for the middle phalanx of the little finger only was present in a female of 1 year, 2 months, while in another female of 1 year, 11 months, centers were present for the middle phalanges of the index, middle, and ring digits, but not for the little one.

In general, however, the epiphyses for the metacarpals and phalanges appear particularly during the second year in females, and during the second to fourth years in males. Further, the epiphyses of the metacarpal and the three phalanges of the little finger tend to appear later than those of the other three fingers, and for each of the four fingers the centers for the epiphyses of the proximal phalanges, metacarpals, middle phalanges, and distal phalanges tend to appear in that order (Greulich and Pyle). The epiphyses for the first metacarpal and the proximal phalanx of the thumb tend to appear during the time that the epiphyses for the distal phalanges of the other four digits are appearing, but the epiphysis for the distal phalanx of the thumb appears earlier, during the time that the epiphyses for the second to fourth metacarpals are appearing. Sometimes there is an additional epiphysis at the distal end of the first metacarpal, and, more rarely, there may be epiphyses at the proximal ends of other metacarpals.

Sesamoids may include others than the two at the metacarpophalangeal joint of the thumb; these may appear in the tenth year in girls (Flecker), but the majority of children do not show them until about the age of 13. Other bony sesamoids, if there are any, typically appear 2 or 3 years later.

ANOMALIES

Variations and anomalous conditions of the upper limb are for the most part described in the several succeeding chapters, in connection with the specific parts of the limb in which they are found.

Malformations of the upper limb as a whole take the form of absence of all or a part of the limb, and of duplications, abnormal clefts, and fusions in the fingers and hand. Cleft or "lobster-claw" hand, polydactyly, syndactyly, and other abnormalities of the digits are discussed in the chapter on the hand.

Amelia is the absence of a limb; phocomelia is the absence of an unspecified proximal part of a limb; hemimelia is the absence of a part, usually not really half of a limb, that may be located either distally or on the preaxial or postaxial side. Because, as just noted, hemimelia and phocomelia, as well as other terms, need qualification if they are to provide a reasonably accurate description of an existing condition, Frantz and O'Rahilly have suggested a more detailed classification that does provide a more nearly complete description of the skeletal defect. Thus, they classified defects involving one or more segments of a limb, such as the hand, as transverse defects, those involving one side of the limb as longitudinal ones. They further divided transverse defects into terminal ones, involving a variable distal part of the limb, and intercalary ones, involving a part (arm, forearm, leg, thigh) usually intercalated between the hand or foot and the trunk. Their

classification of terminal transverse defects includes amelia, absence of the entire limb; hemimelia, absence of the forearm and hand or leg and foot; partial hemimelia, with a proximal part of the forearm or leg represented; acheiria or apodia, absence of the hand or foot; complete adactylia, absence of all metacarpals or metatarsals and their associated phalanges; and complete aphalangia, or absence of all phalanges. As intercalary transverse lesions, they listed complete phocomelia with the hand or foot attached directly to the trunk; proximal phocomelia, with forearm and hand, or leg and foot, attached directly to the trunk; and distal phocomelia, with hand or foot attached directly to arm or thigh. Their classification of longitudinal defects includes complete paraxial hemimelia (radial or ulnar, tibial or fibular), in which one of the bones of the forearm or leg is missing, and so is a corresponding part of the hand or foot; incomplete paraxial hemimelia (again radial, ulnar, and so forth) with one of the bones represented only in part, and the hand or foot more or less complete; partial adactylia, or absence of all or parts of the first or fifth digit, including the metacarpal or metatarsal; and partial aphalangia, absence of one or more phalanges of any of the digits.

The factors leading to such congenital abnormalities are in general not at all well understood, but many of the malformations of the digits tend to have a strongly hereditary basis. Congenital or intrauterine amputation and congenital bands, which may involve one or more digits or the entire limb at or above the wrist, were said by Streeter to result from a focal defect in the tissues at the level of "amputation." Present evidence, however, implicates external factors. Field and Krag, among others, have presented evidence that amnionic strands left by rupture of the amnion without chorionic rupture cause the condition; and Kino has produced it by amniocentesis in rats. This procedure, he found, causes excessive uterine contractions which interfere with development by producing hemorrhage from the marginal sinuses of the digital rays.

General Anatomy

Since the anatomy of the upper limb is discussed in detail in the following three chapters, only a few very general observations are necessary here.

SKELETON

The skeleton of the upper limb (Fig. 3-10) is for the sake of convenience divided into the pectoral girdle, which consists of the clavicle and the scapula, and the skeleton of the free limb. The development of these bones has already been briefly recounted.

The upper girdle of man, in contrast to the lower girdle, is largely suspended by muscles in order to allow mobility; its only direct connection to the axial skeleton is through the clavicle, which articulates at its medial and more movable end with the sternum, and at its lateral end, through a slightly movable sliding joint, with the scapula.

In the free limb, the humerus is the skeleton of the arm segment, and through the articulation of its head with the glenoid cavity of the scapula forms the shoulder joint; distally it articulates with both the radius and the ulna, the two bones of the forearm.

The articulation of the ulna with the trochlea of the humerus at the elbow joint is almost entirely a hinge joint; the radius, however, so articulates with the capitulum of the humerus and with the lateral surface of the ulna that it can not only hinge on the humerus but also rotate around its long axis on both the humerus and the ulna, the latter movement allowing pronation and supination. The ulna, small at its distal end, is typically excluded from actual participation in the formation of the main part of the wrist joint by the presence of a disk, and the expanded distal end of the radius therefore furnishes most of the articular surface upon which the proximal row of carpal bones moves.

The proximal and distal rows of carpals are each composed of four bones, with movement between the proximal row and the radius

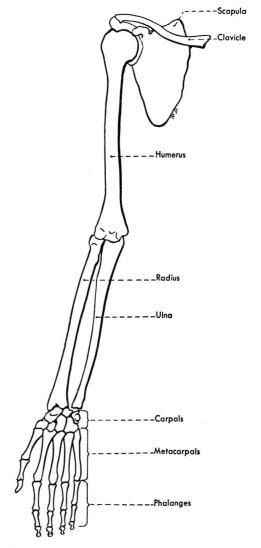

Fig. 3-10. A general view of the skeleton of the upper limb

being relatively free, and supplemented in certain movements by movement of the two rows upon each other at the mid-carpal joint; movements of individual bones are severely limited.

The distal row of carpals is succeeded by the five metacarpals, of which only the first metacarpal, that of the thumb, has a relatively freely movable carpometacarpal joint. The metacarpals are in turn succeeded by the phalanges, two for the thumb and three for

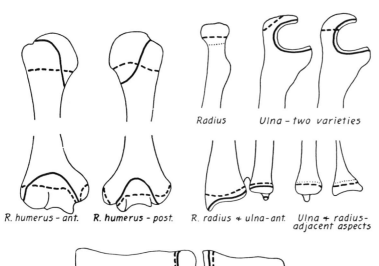

Radius   Ulna - two varieties

R. humerus - ant.   R. humerus - post.   R. radius + ulna-ant.   Ulna + radius-adjacent aspects

Metacarpal   Phalanx

Fig. 3-11. Varying relationships between the epiphyseal lines (*heavy broken lines*) and the attachment of capsules of joints (*solid lines*). The *dotted lines* indicate reflections of the synovial membrane where the attachment of the fibrous capsule is weak or absent. (Mainland D: Anatomy. New York, Hoeber, 1945)

the other digits. These phalanges have been designated by number only, but one could argue as to which end they should be numbered from, so they are usually now specifically designated proximal, middle, and distal. Since the distal phalanx bears the nail it is sometimes also called the ungual phalanx.

It might be noted that there is no regular relation between the attachment of a joint capsule to a bone and the epiphyseal line at that end of the bone. The epiphyseal line may be either entirely inside or outside the joint capsule, or be crossed by it (Fig. 3-11).

Miller ('56) has discussed and illustrated anatomical aspects of paracentesis of the joints of the upper limb. Noting that paracentesis has been increasingly adopted in order to introduce hydrocortisone into affected joints, he described approaches that tend to avoid major nerves and vessels, and pointed out the most useful bony landmarks and the desirability of avoiding injury to the articular cartilage. He stressed both positioning and, where possible, distraction of the joint as a means both of increasing the reliability of bony landmarks and of fascilitating the placement of the needle by enlarging the joint cavity and putting the capsule on a stretch.

MUSCLES

There are approximately 58 muscles concerned directly with movements of the upper limb, of which only 9 have attachment to any part of the skeleton other than that of the limb. These 9 muscles (the pectoralis major and minor, the subclavius, and the serratus anterior on the anterior and anterolateral thoracic wall, and the trapezius, the latissimus dorsi, the two rhomboids, and the levator scapulae on the dorsum) are frequently classed as extrinsic muscles of the upper limb, while the remaining muscles are of course intrinsic ones. Because of their positions, the extrinsic muscles are typically classified as the muscles of the thorax (the anterolateral muscles) or of the back (the dorsally situated muscles). Such a classification according to position, however, should not obscure the facts that these are functionally muscles of the limb, and that phylogenetically, and for the most part also embryologically, they have simply spread out from the base of the limb over the thorax and the back to obtain a firm attachment to the axial skeleton.

The superficially lying muscles of the upper limb and their contribution to the contour of

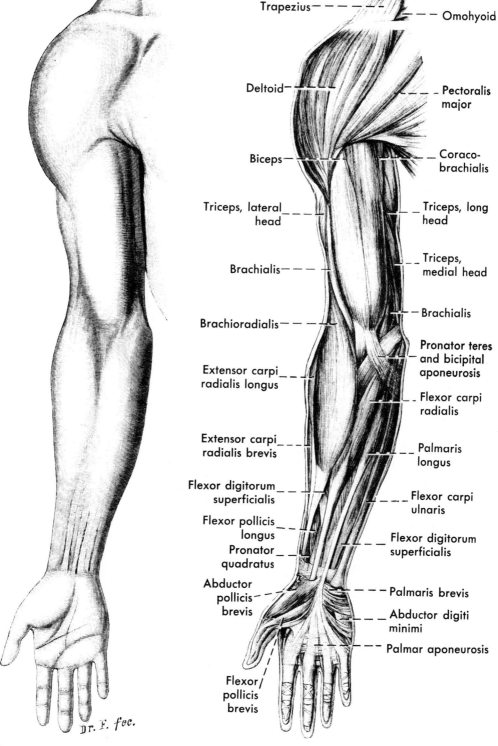

Fig. 3-12. Contour of the upper limb and the superficial muscles of the limb from the front. (Frohse F, Fränkel M: Vol 2, Section 2, Part 2 of K von Bardeleben's Handbuch der Anatomie des Menschen. Jena, Fischer, 1908)

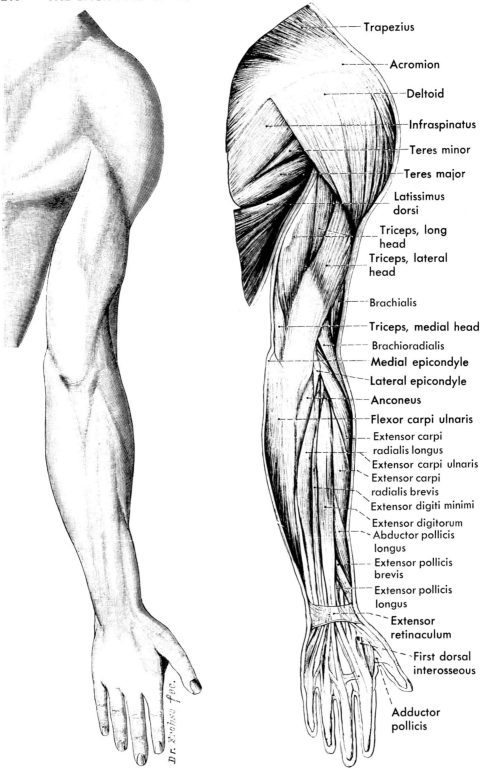

Fig. 3-13. Contour of the upper limb, and the musculature, from behind. (Frohse F, Fränkel, M: Vol 2, Section 2, Part 2 of K von Bardeleben's Handbuch der Anatomie des Menschen. Jena, Fischer, 1908)

the limb are seen in Figures 3-12 and 3-13. The muscles of the upper limb are innervated through branches of the ventral (anterior) rami of the spinal nerves at the original level of the limb bud; the dorsal (posterior) rami of these nerves go to the back, the ventral rami form the brachial plexus. The trapezius, however, is an exception, for it is supplied primarily by the eleventh cranial, or accessory, nerve.

Since the muscles of the limb are in general divisible into dorsally situated and ventrally situated muscles, their innervation is also according to this fundamental pattern. Thus dorsal or posterior branches of the brachial plexus go to most of the muscles of the shoulder, while ventral or anterior branches of the plexus supply the anteriorly situated muscles, the pectorals and the subclavius. In the same way, the muscles on the front of the arm, derived from the flexor premuscle mass, are innervated through the musculocutaneous nerve, a derivative of the anterior portion of the brachial plexus, and the muscles of the flexor aspect of the forearm and the muscles in the hand, also derived from the flexor premuscle mass, are innervated through the median and ulnar nerves, also branches of the anterior part of the plexus. In contrast, the muscles on the posterior or dorsal surface of the arm and forearm are innervated through the radial nerve, the continuation of the posterior part of the brachial plexus into the free limb. Details of the nerve supply of the various muscles are given in the several following chapters in which these muscles are discussed; Table 3-1 summarizes their innervation, and Figures 3-14 and 3-15 show their motor points.

Because the nerves to the upper limb enter first into the brachial plexus, the definitive peripheral nerves leaving the plexus are typically composed of fibers from several segments, and the branches into the individual muscles usually also contain fibers from at least two segments. As already pointed out, the segmental innervation of muscles is probably not accurately known; there is evidence that a fair amount of variation, one segment or more either up or down, exists among individuals.

Bursae and Tendon Sheaths

Bursae are somewhat inconstant; in both limbs the great majority are associated with muscles and tendons, and only a few are subcutaneous. Bursae may communicate with joints, but apparently such communications are always secondary. Whittaker listed the bursae he found in full-term fetuses; there are some, especially subcutaneous ones, usually present in the adult but not present at birth, and most of those that Whittaker found were not constantly present at birth.

The apparently constant subcutaneous bursa in the upper limb of the adult is the subcutaneous olecranon bursa, lying on the dorsal surface of the olecranon deep to the tela subcutanea (superficial fascia); often, also, there is a subcutaneous acromial bursa, over the acromion. Occasional subcutaneous bursae of the upper limb include one over the lateral epicondyle and one over the medial epicondyle of the humerus, the latter being the more frequent; dorsal subcutaneous metacarpophalangeal bursae, on the dorsal aspects of the metacarpophalangeal joints, especially the fifth one; and, apparently frequent, dorsal subcutaneous bursae of the digits over the proximal interphalangeal joints and, rarely, over the distal joints.

The named and rather constant bursae connected with muscles about the shoulder are the subacromial and the subdeltoid, frequently fused into one, and lying beneath the acromion and the coracoacromial ligament, therefore especially above the tendon of the supraspinatus muscle (Chap. 4), and the subtendinous bursae of the following muscles: trapezius, between the muscle and the base of the scapular spine; infraspinatus, between the tendon of the muscle and the capsule of the shoulder joint or the greater tubercle; teres major, between the insertion of this tendon and the humerus; latissimus dorsi, between the tendons of the latissimus dorsi and the teres major; and subscapularis, between the lateral angle of the scapula and the subscapularis muscle, and commonly communicating with the shoulder joint. Gardner and Gray described the development of many of these bursae.

Table 3-1
Innervation of the Muscles of the Upper Limb

MUSCLE GROUP	INDIVIDUAL MUSCLE	PERIPHERAL NERVE SUPPLY	APPROXIMATE SEGMENTAL NERVE SUPPLY*
Extrinsic of shoulder	Pectoralis major	Med. and lat. pectorals	C5-T1
	Pectoralis minor	Medial pectoral	C8-T1
	Subclavius	N. to subclavius	C5, C6
	Serratus anterior	Long thoracic	C5-C7
	Trapezius	Accessory (N. XI)	Also C2-C4?
	Latissimus dorsi	Thoracodorsal	C6-C8
	Rhomboideus major	Dorsal scapular	C5
	Rhomboideus minor	" "	"
	Levator scapulae	Nn. to levator scapulae	C3, C4
Intrinsic of shoulder	Deltoid	Axillary	C5, C6
	Supraspinatus	Suprascapular	"
	Infraspinatus	"	"
	Teres major	Lower subscapular	"
	Teres minor	Axillary	"
	Subscapularis	Both subscapulars	"
Flexor in arm	Biceps brachii	Musculocutaneous	C5, C6
	Coracobrachialis	"	C5-C7
	Brachialis	"	C5, C6
Extensor in arm	Triceps brachii	Radial	C6-C8
	Anconeus	"	C7, C8
Flexor in forearm	Pronator teres	Median	C6, C7
	Flexor carpi radialis	"	"
	Palmaris longus	"	C7, C8
	Flexor carpi ulnaris	Ulnar	C7?, C8, T1
	Flexor digitorum superficialis (sublimis)	Median	C7-T1
	Flexor digitorum profundus	Median and ulnar	"
	Flexor pollicis longus	Median	"
	Pronator quadratus	"	"
Extensor in forearm	Brachioradialis	Radial	C5, C6
	Supinator	"	"
	Ext. carpi radialis longus	"	C6, C7
	Ext. carpi radialis brevis	"	"
	Abductor pollicis longus	"	"
	Extensor pollicis brevis	"	"
	Extensor digitorum	"	C6-C8
	Extensor digiti minimi	"	"
	Extensor carpi ulnaris	"	"
	Extensor pollicis longus	"	C7, C8
	Extensor indicis	"	"
Thenar	Abductor pollicis brevis	Median	C8, T1 (or C6, C7?)
	Flexor pollicis brevis	Median and ulnar	" "
	Opponens pollicis	Median	" "
	Adductor pollicis	Ulnar	C8, T1
Hypothenar	Palmaris brevis	Ulnar	C7?, C8, T1
	Abductor digiti minimi	"	"
	Flexor digiti minimi brevis	"	"
	Opponens digiti minimi	"	"
Other muscles of palm	1st and 2nd lumbricals	Median	C8, T1 (or C6, C7?)
	3rd and 4th lumbricals	Ulnar	C7?, C8, T1
	Dorsal interossei (4)	"	"
	Palmar interossei (3)	"	"

* Question marks in association with the listings of segmental innervations indicate major points of disagreement.

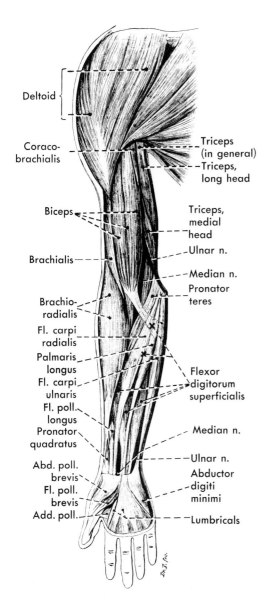

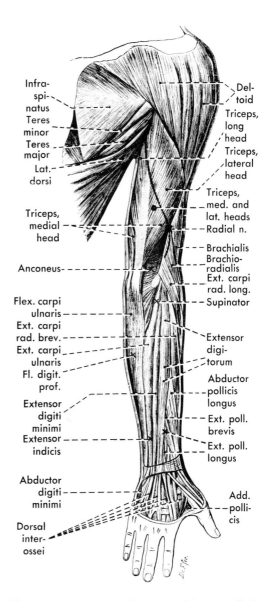

Fig. 3-14. Motor points of muscles of the upper limb, anterior view. (After Cohn, from Frohse F, Fränkel M: Vol 2, Section 2, Part 2, of K von Bardeleben's Handbuch der Anatomie des Menschen. Jena, Fischer, 1908)

Fig. 3-15. Motor points of muscles of the upper limb, posterior view. (After Cohn, from Frohse F, Fränkel M: Vol 2, Section 2, Part 2 of K von Bardeleben's Handbuch der Anatomie des Menschen. Jena, Fischer, 1908)

Usual deep bursae about the elbow include the subtendinous bursa of the triceps brachii (subtendinous olecranon bursa) deep to the tendon of the triceps over the olecranon, and the intratendinous olecranon bursa, within the tendon of the triceps near its insertion, accounts varying as to which is more frequent; the bicipitoradial bursa, between the anterior part of the radial tuberosity and the tendon of the biceps; and, less constant but frequent, the cubital interosseous bursa, medial to the tendon of the biceps, between it and the ulna and its covering muscles. At the wrist, bursae of the flexor carpi radialis, between the tendon of this muscle and the tubercle of the scaphoid bone, and of the extensor carpi radialis brevis, lying between the tendon and the base of the third metacarpal, are rather constant.

Other subtendinous bursae that are not so constant, some of them recognized in the N.A., some of them not mentioned, include the following: About the shoulder, a bursa between the tendon of insertion of the pectoralis major and the long head of the biceps (bursa of the pectoralis major), which is fairly frequent; and the bursa of the coracobrachialis muscle, frequently present between the subscapularis muscle and the overlying tendon of origin of the coracobrachialis from the coracoid process. About the elbow, a bursa of the anconeus muscle, frequently present between the tendon of origin of the muscle and the head of the radius; a bursa of the supinator muscle, between this muscle and the overlying common tendon of origin of the extensor muscles from the lateral epicondyle; and a bursa over the dorsal aspect of the medial epicondyle, rarely present, tending to separate the ulnar nerve from the epicondyle and the edge of the triceps.

In the wrist and hand, there may be a bursa of the flexor carpi ulnaris, lying between the tendon of this muscle and the upper part of the pisiform bone, and a number of inconstant bursae on the dorsal surface. These include a bursa often found between the tendon of insertion of the extensor carpi ulnaris and the base of the fifth metacarpal; a bursa of the extensor carpi radialis longus, between the tendon of this muscle and the base of the second metacarpal; bursae deep to the tendon of insertion of the abductor pollicis longus, and perhaps other bursae between the closely adjacent tendons of this muscle and the extensor pollicis brevis, on the one hand, and the tendons of the long and short radial extensors on the other; intermetacarpophalangeal bursae, between the heads of the metacarpal bones dorsal to the deep transverse metacarpal ligaments; and small bursae sometimes found deep to the tendons of the extensor digitorum to the index and little fingers, and to that of the extensor pollicis longus as it lies on the first metacarpal.

The tendon sheaths in the upper limb are all confined to the wrist and hand, and are described in Chapter 6. It must suffice to point out here that as the flexors and extensors of the wrist and digits cross the wrist joint they are provided, either individually or in groups, with synovial tendon sheaths; these sheaths are more numerous on the dorsum of the wrist than they are on the flexor surface. On the dorsum the sheaths are primarily limited to the region of the extensor retinaculum (dorsal carpal ligament) but in the palm the tendon sheaths continue, or are resumed after interruption, along the palmar surfaces of the digits.

NERVES

Branches of the cervical plexus supply the levator scapulae, one of the muscles of the upper limb, and the accessory (eleventh cranial) nerve (with at least afferent fibers, and possibly some motor ones, from the cervical plexus) supplies the trapezius. Otherwise, all the muscles of the upper limb are supplied by branches of the brachial plexus. Similarly, while skin over the muscles of the limb that are spread over the trunk is innervated by the nerves of the trunk, branches of the brachial plexus supply almost all the skin of the free limb; the exceptions are skin over the upper part of the deltoid muscle, innervated by the supraclavicular nerves (from the cervical plexus), and a small amount of skin on the medial side of the arm, innervated by the intercostobrachial nerve (from the second or second and third intercostals).

Since the brachial plexus is formed in the neck, its formation has been briefly described in Volume 1; there also its relations, especially in regard to neurocirculatory compression, have been discussed in some detail. The following account is concerned particularly with the details of the formation of the plexus and its component parts, and the variations that occur therein.

FORMATION OF BRACHIAL PLEXUS

The brachial plexus (Figs. 3-16 to 3-28) arises by union of the ventral (anterior primary) rami of the fifth, sixth, seventh, and eighth cervical nerves and the first thoracic nerve; to these are added, in many instances, some fibers from C4. Kerr found that among 175 plexuses 62% received a fiber bundle, varying from a tiny twig to a communication the size of the average suprascapular nerve, from C4; 30% received no fibers from C4 but received all the anterior ramus of C5; and 7% received only part of the anterior ramus of C5. There is often, also, a small communication from the second thoracic nerve to the brachial plexus, but this has been described as consisting mostly of small nerve fibers and being probably a sympathetic connection to the nerves of the limb. In rudimentary first thoracic rib, in contrast, there is almost invariably a large contribution from the second thoracic nerve to the brachial plexus; that is, the plexus is definitely "postfixed."

Except in the anomalous condition of rudimentary first thoracic rib, however, the brachial plexus shows little significant variation in its origin. The minor shifts in its origin are probably, according to Kerr, expansions and contractions of the plexus rather than true "prefixation" and "postfixation." Kerr was not able to study the caudal limit of the plexus in many of his cases, but from his own observations and a survey of the literature concluded that there was no evidence that there is a definite shift of the entire plexus cranially or caudally to produce real prefixation or postfixation. Thus he could not determine that when the fourth cervical nerve joins the plexus, the second thoracic does not, and that when the second thoracic joins, the fourth cervical does not.

Trunks, Divisions, Cords, and Chief Branches

In the usual brachial plexus (Figs. 3-16 and 3-17) the fifth and sixth cervical nerves, the fifth carrying also commonly a small contribution from the fourth, unite to form the upper

Fig. 3-16. Schema of the formation of the brachial plexus.

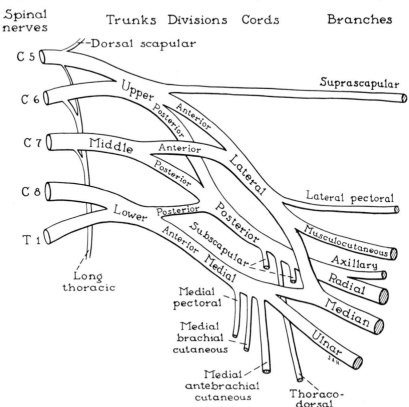

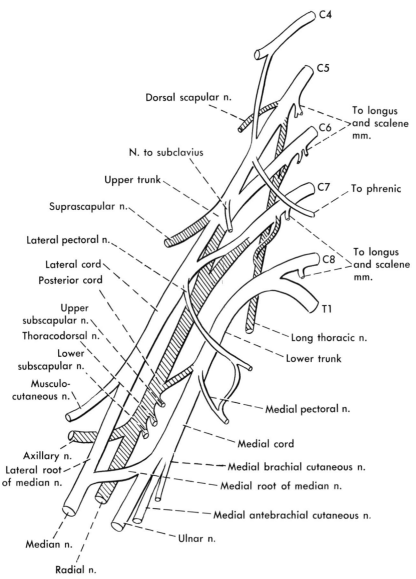

Fig. 3-17. A common form of the brachial plexus and its branches; the posterior derivatives of the plexus are shaded.

trunk of the brachial plexus, the seventh cervical nerve alone forms the middle trunk, and the eighth cervical and first thoracic unite to form the lower trunk. Each trunk then divides to form an anterior and a posterior division: the anterior divisions of upper and middle trunks unite to form the lateral cord (lateral fasciculus), and their posterior divisions unite to form the major part of the posterior cord (fasciculus); the anterior division of the lower trunk continues as the medial cord (fasciculus) of the plexus while its posterior divi-

sion joins the posterior divisions of upper and middle trunks to complete the posterior cord. The division of the lower trunk occurs low in the neck or after the trunk has crossed the first rib, so that the division of the trunk and the formation of the definitive medial and posterior cords often occur in the axilla rather than in the neck.

Each cord gives off one or more branches in the axilla, and ends by dividing into two terminal branches: musculocutaneous and lateral root of the median nerve for the lateral cord, ulnar and medial root of the median nerve for the medial cord, and axillary and radial nerves for the posterior cord.

Before its division, the upper trunk gives off the suprascapular nerve, to the supraspinatus and infraspinatus muscles, and the small nerve to the subclavius, which supplies this muscle and may supply an additional root (accessory phrenic nerve) to the phrenic nerve. Higher, at about the lateral border of the anterior scalene muscle, the fifth cervical nerve gives rise to a root of the phrenic nerve, and close to its emergence from the intervertebral foramen, behind the anterior scalene muscle, the fifth nerve gives off the dorsal scapular nerve for the innervation of the rhomboids; in the same location, the fifth, sixth, and seventh, approximately, give rise to the long thoracic nerve. The courses of these nerves to the muscles are described in connection with the nerves of the shoulder.

The splitting of the brachial plexus into anterior and posterior divisions is fundamental in the innervation of the limb, and a similar division occurs in the lumbosacral plexus. By this division there is a regrouping of fibers, those destined for the original anterior (ventral, flexor) aspect of the limb separating from those destined for the original posterior (dorsal, extensor) aspect. This is particularly easily seen in the distribution of the major nerves of the upper limb, where surfaces are less distorted from their original positions than they are in the lower limb. For instance, the musculocutaneous, median, and ulnar nerves are all derived, through the medial and lateral cords, from the anterior divisions of the plexus, and are distributed to the muscles of the flexor or anterior surface of the arm, forearm, and hand; similarly, the radial nerve, the largest derivative of the posterior divisions, is distributed to the muscles of the extensor or dorsal aspect of the arm and forearm, there being normally no true dorsal muscles of the hand.

General Composition of Cords and Branches

It should be pointed out that certain broad conclusions in regard to the distribution of fibers in various parts of the plexus can be drawn purely from knowledge of its general plan. Thus the suprascapular nerve, the upper trunk, and the small nerve to the subclavius (Fig. 3-17) must usually consist of no more than fourth, maybe, and fifth and sixth cervical nerve fibers. Likewise, the lateral cord contains fourth, maybe, and fifth, sixth, and seventh nerve fibers, while the medial cord, normally derived entirely from the lower trunk, contains only eighth cervical and first thoracic ones.

Although the posterior cord is formed by the posterior divisions of all three trunks, the contributions from these trunks obviously are not equal. In the usual plexus, the anterior and posterior divisions of the upper trunk are of approximately equal size, or the anterior division is greater, but not markedly so; thus the posterior cord receives a goodly number of fibers from both the fifth and sixth cervical nerves. When the axillary nerve arises separately from the plexus, as it may, instead of by way of the posterior cord, it can be seen that this nerve derives most or all of its fibers from the upper trunk, but that fibers from the upper trunk also contribute to the radial nerve.

The middle trunk, derived from the particularly large seventh cervical nerve, as a rule divides unequally: commonly, somewhat less than half of its bulk is continued into the anterior division to join the lateral cord, while more than half of its bulk is continued into the posterior division; thus the posterior cord and the radial nerve contain a particularly large number of seventh cervical fibers. The lower trunk of the brachial plexus divides still more unequally, but here the great bulk of nerve fibers continues into the anterior division to form the medial cord, and the posterior division is typically small; thus the radial nerve can normally contain relatively few fibers originating from the eighth cervical or first thoracic nerves.

In contrast to the other terminal branches of the plexus, the median nerve is complex. Since it takes its origin by two large roots, one from the lateral cord (hence presumably from C5, 6, and 7), and one from the medial cord (hence presumably from C8 and T1), it can be expected to contain fibers from each

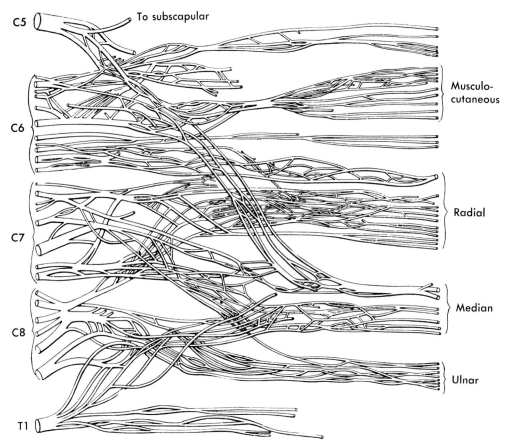

Fig. 3-18. Dissection of the intertwining elements of the brachial plexus after maceration of the surrounding fibrous tissue. (Kerr AT: Am J Anat 23:285, 1918)

of the five spinal nerves regularly entering the plexus, but the proportion of fibers from the various nerves cannot be estimated from the usual dissection.

Beyond such generalities as those just recited, it is impossible to go on the basis of routine dissections. Special dissections of the plexus following destruction of its connective tissue reveal more detail (Fig. 3-18), but this method is limited both by its tediousness and by the numerous intertwinings of the fiber bundles.

Relative Contributions of Spinal Nerves

It is obvious from an inspection of any brachial plexus that some of the nerves contributing to the plexus are larger than others, and thus presumably contribute more fibers to the plexus than do others. And while comparison of sizes of nerves is certainly not an exact criterion of the numbers of fibers contained in them, because of the varying thickness of connective tissue and the varying amount of fat they contain, it is, within limits, a reasonable way of assessing the contributions made by each nerve. Kerr measured the sizes of the ventral rami contributing to 27 brachial plexuses, and found that, disregarding the fourth cervical contribution, the fifth cervical apparently made the smallest contribution to the plexus in 11 cases, and the first thoracic did in 7, while in 9 the contributions from the fifth cervical and the first thoracic were equal but smaller than those of the other nerves.

At the other extreme, the seventh cervical ventral ramus was the largest nerve to the plexus in 7 cases, the eighth was the largest in 6, and the two were equal, but larger than the others, in 6 cases. Thus, either the seventh or the eighth cervical, or both, were the largest contributors in more than 70% of these plexuses. In the remainder, the sixth, seventh, and eighth cervical ventral rami were equal, and largest, in 2 cases; the first thoracic and the last three cervicals were equal in one; the sixth and seventh cervicals in 2; and the fifth, sixth, and seventh cervicals in one.

Harris agreed in general with these findings; he listed the usual sizes of the ventral rami contributing to the brachial plexus as being, in descending order, C8, C7, C6, and C5, with T1 contributing about the same as C5. However, he also said that the relative motor and afferent contributions of the nerves to the plexus are quite different: the fifth cervical nerve, he said, has the smallest dorsal root of any of the nerves entering the plexus, while its ventral or motor root is about as large as the motor root of the sixth, otherwise the largest motor root contributing to the plexus. Thus Harris found that while C5 and C6 contribute fewer total fibers to the plexus than do C7 and C8, they actually contribute more motor ones; in terms of the motor fibers that the nerves contribute, he listed them as being, in descending order, C6 and C5 about equally, then C8, C7, and T1, with T1 contributing about as many as C7. On the other hand, the contribution of afferent fibers, judged by the sizes of the dorsal roots is, again in descending order, C7, C8, C6, T1, and C5.

Sympathetic Contribution to the Brachial Plexus

According to the findings of Ray, Hinsey, and Geohegan the segmental (preganglionic) sympathetic contribution to the brachial plexus is somewhat variable, but extends more caudally than has usually been assumed; they explored the caudal limit of the sympathetic contributions to the upper limb on 9 sides of 7 persons, and found it to vary from T7 to T10. In the 2 patients in whom they explored both sides, they found a variation of one segment between sides, the caudal limit of the sympathetic contribution in one patient being T8 and T9, in the other T9 and T10. They also

confirmed the generally accepted view that the first thoracic nerve does not usually contribute preganglionic sympathetic fibers to the limb; among 11 individuals in whom they tested the upper limit of the sympathetic contribution to the limb on 13 sides, they obtained a response to stimulation of the first thoracic nerve on only one side.

Woollard and Weddell said that the greatest number of postganglionic vasoconstrictor fibers are brought into the plexus by the eighth cervical nerve; Sunderland regarded the eighth cervical and first thoracic as contributing about equally, with an additional contribution of good size through the seventh cervical. In contradistinction to some previous reports, however, Sunderland found that the sympathetic fibers in the lower trunk were not concentrated on the lower surface of this trunk, but instead passed into its substance either centrally or anteriorly, and by the time the trunk crossed the first rib they were widely but irregularly scattered in it; thus he found no anatomical evidence to support the belief that the vascular complications of neurocirculatory compression are a result of irritation of superficially placed sympathetic fibers in the lower trunk.

GROSS VARIATIONS OF BRACHIAL PLEXUS

The common and more obvious variations in the brachial plexus are in its gross form (in the level of junction or separation of its component parts). Kerr found many variations in the level of union and division of the ventral rami and cords, and in the apparent origin of the branches of the plexus. The latter are described in more detail in following sections; the most striking departures from normal in the division of the plexus were those in which the upper and middle trunks failed to divide as usual, but united to form a lateral cord that then divided into lateral cord proper and the major contribution to the posterior cord. Kerr also found three plexuses in which the medial and lateral cords united to form a single cord anterior to the axillary artery, so that only anterior and posterior cords were pres-

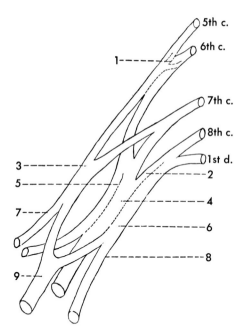

Fig. 3-19. Apparent abnormality of the brachial plexus produced by its connective tissue wrapping (*solid lines*); the medial and posterior cords appeared to form a common trunk. The *broken lines* indicate that proper dissection of the connective tissue, without interrupting nerve fibers, resolved the plexus into a normal one. *1* is the upper trunk, dissectible into anterior and posterior elements; *2*, the posterior division of the lower trunk; *3* is the lateral cord; *4* the radial nerve, dissectible from the axillary nerve, *5; 6* is the medial cord, *7* the musculocutaneous nerve, *8* the ulnar nerve, and *9* the median nerve. (Walsh JF: Am J Med Sci 74:387, 1877)

ent, the anterior one giving off the branches proper to both medial and lateral cords; and there were two others essentially similar except that the musculocutaneous nerve arose above the junction.

Many of the variations may be more apparent than real, since they can often be produced by dissecting more or less connective tissue from around the nerve trunks and their branches; Walsh argued that almost all so-called variations in the plexus, except in regard to the segmental contributions to various nerves, are actually a result of improper or incomplete dissection of connective tissue from around the plexus (Fig. 3-19). An extreme example of improper natural separation is that reported by Singer, in which all the roots of the plexus united to form a single

cord. He found two similar cases reported in the literature, and explained the occurrence of this as being due to an abnormal formation for the axillary artery: the embryo, he said, has both deep and superficial axillary arteries, and the normal persistence of the deep vessel separates the cords; an artery developed from the superficial one, however, fails to do so. Hasan and Narayan reported an additional example of this rare anomaly.

Some observed variations obviously introduce into the peripheral nerves fibers that they do not normally contain, some obviously do not, and in some the true status cannot be decided without detailed dissection. As examples, a contribution from the lateral cord to the ulnar nerve, very common, necessarily introduces into that nerve fibers from at least as high as the seventh cervical nerve; a nerve arising more proximally than usual may often be seen to have its usual segmental origin; but when a nerve arises more distally than usual, after other elements have been added, it may or may not receive fibers from the additional elements.

Apparent Anomalous Plexuses

Aside from apparent improper separation of cords (described in the preceding section), which often introduces no change in the segmental origin of the branches, anomalies of the brachial plexus are rare. In only 11 of 175 plexuses (6.28%) did Kerr find a real anomaly in the formation of the cords; in five cases there was a contribution from the inferior trunk or the eighth cervical nerve to the lateral cord or one of its roots, while in six there was a contribution from the middle trunk (C7) to the inferior trunk or medial cord. Since Kerr and others have shown that fibers of the seventh cervical nerve frequently pass into the ulnar nerve, the second group of cases cited may be merely examples of such a contribution given off higher than usual, and therefore possibly should not be regarded as anomalous. Walsh reported an anomalous plexus in only two of 350 dissections.

A different type of variation in the brachial plexus in the neck and upper part of the axilla is in relation to the subclavian and axillary arteries. This arterial stem is normally derived from the seventh segmental branch of the dorsal aorta, and therefore normally passes between the lateral and the medial cords, representing fifth, sixth, and seventh cervical nerves on the one hand, and eighth cervical and first thoracic on the other. Sometimes, however, the subclavian-axillary stem is derived from the sixth, the eighth, or the ninth segmental artery, and it then has abnormal relations to the brachial plexus (Fig. 3-20), and the brachial plexus is in turn modified by the presence of the abnormally placed artery. In the single case (among 960 brachial plexuses) reported by Miller ('39) in which the sixth segmental artery had formed the subclavian-axillary stem, the artery separated the sixth and seventh cervical nerves, so that the upper trunk consisted of fibers of C5, and C6 only, and the lower of C7, C8, and T1. Miller found 8 cases in which the eighth segmental artery had persisted, and when this had occurred the artery passed between the eighth cervical and first thoracic nerves, thus splitting the medial cord into two parts. In two cases the ninth segmental artery had persisted, the subclavian-axillary stem lay caudal to the first thoracic root and inferomedial to the inferior trunk and medial cord, and hence did not penetrate the plexus at all; in both there was said to be a communication between the pectoral and ulnar nerves across the front of the artery, and in one (Fig. 3-20d) the stem of origin of the humeral circumflex and deep brachial arteries represented a remnant of the seventh segmental artery and passed between the roots of the median nerve. Abnormal vascular relations of the branches of the plexus in particular are discussed in a following section.

COMPOSITION AND VARIATIONS OF BRANCHES

The branches of the brachial plexus vary considerably in their levels of origin, probably much less frequently in regard to the fibers they contain. Nevertheless, they do vary in the latter respect also, and these variations may be clinically more important, since they account for the variations in segmental innervations of muscles that are generally accepted

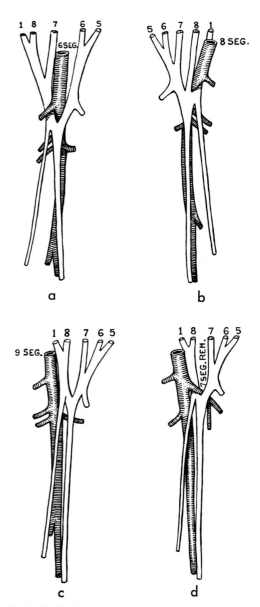

Fig. 3-20. Variations in arterial relations of the brachial plexus brought about by the development of the subclavian-axillary stem from different segmental arteries. In d the main vessel developed from the ninth segmental, but a part of the seventh persisted to give rise to the deep brachial and humeral circumflex arteries. (Miller RA: Am J Anat 64:143, 1939)

as existing. Data on such variations are neither particularly extensive nor precise: because in clinical practice it is often impossible to assess accurately either the amount of damage to a given nerve or the extent of functional loss in a muscle or muscle group, knowledge of the variations comes largely from purely anatomical studies (in part routine dissections, in part special ones in which attempts were made to follow various nerve bundles into the peripheral nerves).

Upper Branches

Four nerves, the dorsal scapular, the long thoracic, the suprascapular, and the nerve to the subclavius, arise from the ventral rami contributing to the plexus or from the upper part of the superior trunk.

The *dorsal scapular nerve* usually arises directly from the ventral ramus of C5, and contains fibers from this nerve only (Table 3-2). Variations of it have apparently not been specifically studied. However, it may arise with the nerve to the subclavius (3 among 83 sides reported by Kerr), or, more commonly, with the contribution of C5 to the long thoracic nerve (44% of 100 sides studied by Horwitz and Tocantins). In one of the 3 cases reported by Kerr, the common stem of the nerve arose from the trunk formed by the union of C4 and C5 instead of directly from C5, and therefore the dorsal scapular nerve may have contained fibers from the fourth cervical nerve. Horwitz and Tocantins did not record the relation of the fourth cervical contribution to the plexus in their series, but implied that they found no fourth cervical fibers entering a common root of the long thoracic and dorsal scapular nerves.

Horwitz and Tocantins found the *long thoracic nerve* formed, as usual, by junction of roots from the fifth, sixth, and seventh cervical nerves in 84% of 100 cases. In 8%, they reported, C7 failed to contribute to the nerve, in another 8% fibers from the eighth cervical joined those from C5, 6, and 7. They apparently found no example in which C5 failed to contribute at all, although this is sometimes seen in the dissecting laboratory; they did report that in 5% of cases the rootlet from C5 went independently to the upper digitations of the serratus, instead of joining the rest of the nerve. They found the roots from C6 and C7 were approximately equal in size, and usually larger than that from C5. In about 50 cases once surveyed by the present writer, there was a decided tendency for the root from C6 to be the largest; Harris also regarded this root as being the most important, and said that atrophy of the serratus is often characteristic of a lesion involving C6.

Horwitz and Tocantins also recorded the varying relations of the long thoracic nerve to the middle scalene muscle: in 84% of their cases, the root from C5 ran through the middle scalene, in 2% it passed posterior to this muscle, and in 14% anterior to it; in 74% the root from C6 ran through the muscle, in 2% posterior to it, and in 24% anterior to it; in the 92% of cases in which the root from C7 was present it ran through the middle scalene in 3% of bodies, posterior to it in 1%, and anterior to it in 88%; and in the 8 bodies in which there was a contribution from C8, this passed anterior to the middle scalene in all cases.

The *suprascapular nerve* commonly arises from the upper trunk of the brachial plexus (Figs. 3-21 and 3-22) very close to its separation into anterior and posterior divisions. Kerr found an origin from the upper trunk in 108 of 172 cases; there was a more proximal origin (Fig. 3-23) in 30 cases, among which it arose from the trunk formed by C4 and C5 in 13, from the anterior and posterior divisions of that trunk in one case each, from C5 alone

Table 3-2
Segmental Composition of Upper Branches of Brachial Plexus

NERVE	USUAL COMPOSITION	APPARENT VARIATIONS
Dorsal scapular	C5	C4, 5; C5,6
Long thoracic	C5, 6, 7	C5, 6; C6, 7; C5-8
Suprascapular	C5 or C5,6	C4, 5
N. to subclavius	C4, 5	C3; C4; C5,6; C6; C6,7; C7

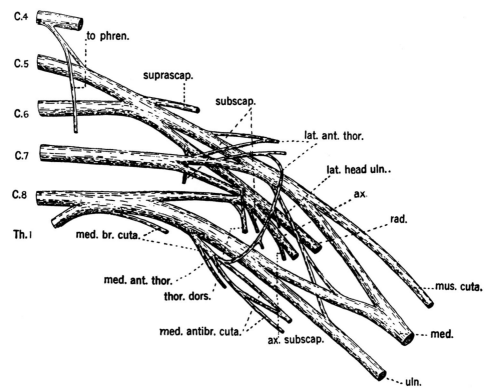

Fig. 3-21. A brachial plexus with a contribution from C4. Note the lateral root (*head*) of the ulnar nerve. In this and the succeeding several figures *ax. subscap.* designates the lower (*axillary*) subscapular nerve, *subscap.*, the upper subscapular, and *cor. br.* a nerve to the coracobrachialis muscle; the pectoral nerves, *lateral* and *medial,* are here designated by their older names of anterior thoracic (*ant. thor.*). (Kerr AT: Am J Anat 23:285, 1918)

in 14, and from the anterior division of C5 in one; there was a more distal origin (Fig. 3-24) in 34 cases, in which it arose from the anterior division of the upper trunk in 12, and from the posterior division of this trunk in 22. Kerr quoted a survey of the literature which indicated that of 34 suprascapular nerves half of them contained fibers from both C5 and C6, half of them fibers from C5 only. In only 30 of the plexuses he studied could he be sure that C6 did not contribute to the nerve, and in half of them C4 could have done so.

The *nerve to the subclavius* was reported by Kerr to arise alone in 54 of 83 cases, in common with a root of the phrenic nerve in 24, in common with the dorsal scapular in 3, and in common with a root of the lateral pectoral (anterior thoracic) nerve in 2. Either alone or combined with another nerve, the nerve to the subclavius took its origin, in Kerr's series,

from the upper trunk or its anterior division in 49.39%, from C5 alone in 26.5%, from the common trunk of C4 and C5 in 21.68%, from C6 once, and from the lateral cord once. The nerve has apparently been reported, by various authors, to consist of fibers derived from C3 alone; C4 alone; C5 and 6; C5, 6, and 7; C6 alone; and C7 and C8. In 48.3% of Kerr's cases it could have contained fibers from no more than C4 and C5, in most of the remainder its origin was such that it could have contained fibers from C6, and in one case only could it have obtained fibers from C7.

Lateral Cord and Branches
In Kerr's series, the lateral cord apparently contained fibers from the fourth, fifth, sixth, and seventh cervical nerves (Fig. 3-21 and Table 3-3) or the fifth, sixth, and seventh (Fig. 3-22) in all the cases (168 out of 175) in which

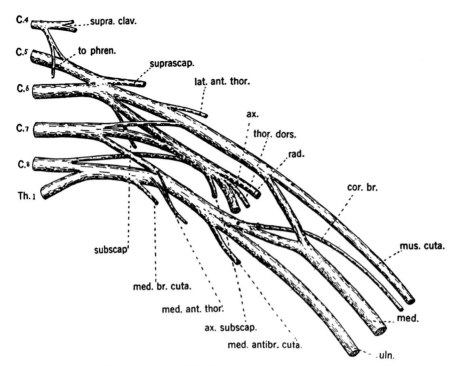

Fig. 3-22. A brachial plexus in which C7 contributes to the lower trunk; C8, but apparently not T1, contributes to the posterior cord; the thoracodorsal nerve arises at the point of division of the axillary and radial, and distal to the lower subscapular; and a nerve to the coracobrachialis muscle arises from the medial root of the median nerve rather than from the musculocutaneous. (Kerr AT: Am J Anat 23:285, 1918)

a lateral cord was formed; in 3 plexuses the lateral cord received also a contribution from lower elements (in 1 case from the lower trunk, in 2 from the eighth cervical nerve), and in 2 more plexuses the middle trunk, before its division, received a contribution from the lower trunk which may or may not have contributed to the lateral cord.

A common variation of the lateral cord is for it to give a contribution to the ulnar nerve. Kerr found this in 42.85% of his series; it is discussed further in the following section. Another common variation is in the size of the lateral root of the median nerve; when this is particularly small, it is common to find a communication from the musculocutaneous to the median nerve in the arm, the assumption being that median nerve fibers have taken the wrong course for a short distance.

The *musculocutaneous nerve* typically arises as one of the terminal branches of the lateral cord (88.57% of Kerr's series), but may be the only continuation of this cord, the lateral root of the median being derived from the seventh cervical only (one case of Kerr's, and in this the musculocutaneous sent a branch to the median nerve in the arm); or it may arise from a single anterior cord, formed by the union of lateral and medial cords (3 of Kerr's cases); it may arise from the anterior division of the upper trunk (4.75% in Kerr's series); or it may arise late, from the median nerve or a combined median and ulnar trunk (5.14% of Kerr's series, but in only 9 of 350 in Walsh's).

Kerr calculated, on the basis of its level of origin, that the lowest possible contribution to the musculocutaneous nerves in his series was from the seventh cervical nerve in 56%, from the eighth in 30.85% and from the first thoracic in 7.42%. In contrast to the possibility that the musculocutaneous nerve could have received fibers from C7 in almost 95% of cases, he concluded from his own records and a review of the literature that it actually does not in almost a third of cases (and a branch from the median nerve to the musculocutaneous in the arm, which could transmit such fibers, is rare). Walsh found a much lower incidence than this, for he traced C7 fibers to the musculocutaneous in only 23 of 73 special dissections. Similarly, Kerr estimated that the musculocutaneous actually receives eighth cervical or eighth cervical and first thoracic fibers in about 6 to 7%. Since the musculocutaneous nerve sends a branch to the median nerve in the arm more frequently than this, it may be that these lower fibers regularly go to the median rather than being distributed through the musculocutaneous. There seems to have been no study of a possible correlation between these two variations, although it is known that the branch to the median may be present when the musculocutaneous receives fibers from no lower than C7.

The *lateral root of the median nerve*, normally derived from the lateral cord, was so derived in the great majority of Kerr's series. Of the remainder, it was derived from the anterior division of the upper trunk, and then joined by seventh cervical fibers in 7; derived only from the upper trunk, the seventh cervical fibers joining the medial root of the median in 1; derived entirely from the seventh cervical in 1 (but in this case the musculocutaneous nerve, from C5 and C6, divided into two major branches, of which one joined the median at about the mid-brachial level); and derived from a stem common to it and the ulnar in 5 cases. There is no reason to suppose that any of these reported variations necessarily produce alterations in the composition of the median nerve. The probable composition of this nerve is discussed in the following section in connection with its medial root.

The *lateral pectoral* (anterior thoracic) *nerve* may arise from the lateral cord, from the anterior division of the upper trunk (Fig. 3-22), from the anterior divisions of the upper and

Fig. 3-23. A brachial plexus with no contribution from C4, and with a very small one from the lower trunk to the posterior cord. In this case also the ulnar nerve has a lateral root. The two upper subscapular nerves arise from the posterior division of the upper trunk instead of from the posterior cord, and the lower trunk gives rise to branches usually arising from the medial cord. (Kerr AT: Am J Anat 23:285, 1918)

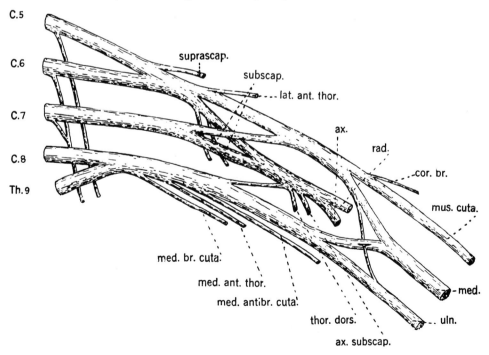

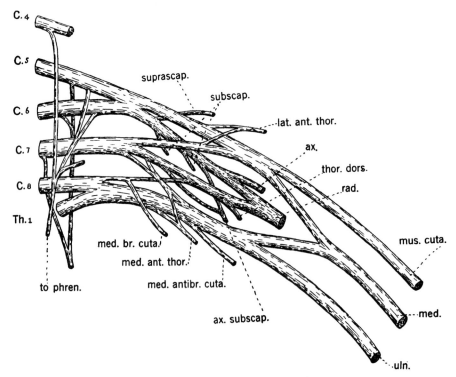

Fig. 3-24. A brachial plexus with most of the smaller branches arising from the anterior and posterior divisions. (Kerr AT: Am J Anat 23:285, 1918)

Table 3-3
Segmental Composition of Lateral Cord and Branches

NERVE	USUAL COMPOSITION	APPARENT VARIATIONS
Lateral cord	C4-7 or C5-7	C4-8
Lateral pectoral	C5-7	C5,6; C6,7; C7
Musculocutaneous	C5-7	C5,6; C5-8; C5-T1
Lateral root of median	C5-7	C5,6; C6,7
Median, both roots	C5-T1	C5-8; C6-T1

middle trunks (Fig. 3-24), or in various ways from the upper part of the plexus. Among 166 cases, Kerr found a single root of origin for this nerve in about 24%, 2 roots (Figs. 3-24 and 3-25) in about 55%, and 3 roots (Fig. 3-21) in about 20%, but these were not always from separate sources; both roots, for instance, may arise from the lateral cord, or one may arise from the anterior division of the upper trunk, the other from the anterior division of the lower trunk or from the lateral cord. Kerr calculated that the lateral pectoral nerve could, in his series, have contained fibers from the fourth or fifth through the seventh cervical nerves, as it is usually said to do, in about 83% and might even have had eighth cervical fibers in one case. Among the remainder there probably was not a contribution from the seventh in 15 cases, C6 and C7 alone contributed in 3 cases, and C7 alone formed the nerve in 9 cases.

Medial Cord and Branches

The medial cord typically represents only the continuation of the anterior division of the lower trunk, and therefore contains only eighth cervical and first thoracic fibers (Table 3-4). Kerr found this to be the usual origin in 94.58% of his specimens, but in 5 the medial cord received a contribution from the seventh cervical nerve also (Fig. 3-24). In 6 others the posterior division of the lower trunk arose from the eighth cervical nerve, and the medial cord therefore obviously received all the fibers contributed to the plexus by the first thoracic nerve (Fig. 3-22), and in one of these there was a contribution from the seventh cervical; in two the medial cord was formed directly by the union of the anterior divisions of the eighth cervical and first thoracic nerves, and in one it represented the entire continuation of the lower trunk, the medial cord itself giving off a branch to the posterior cord. Except for the 6 cases among 175 in which the medial cord received seventh cervical fibers, no change in the composition of the medial cord is necessarily introduced by any of these variations. There seems to be no adequate study as to the incidence of a contribution from the second thoracic nerve to the plexus, and the distribution of fibers of this origin in the medial cord and its branches; possibly such fibers go entirely into the medial brachial cutaneous, to be distributed with second thoracic fibers in the intercostobrachial nerve to skin of the arm, but Harris expressed the view that they help to supply intrinsic muscles of the hand.

The *medial pectoral* (anterior thoracic) *nerve* is commonly the first branch of the medial cord, and did so arise in 69.59% of Kerr's series; in 24.5% it arose from the lower trunk (Fig. 3-23) or the anterior divisions forming this trunk (Fig. 3-24), and in the remainder it arose from the eighth cervical, the medial root of the median, the ulnar, or the medial cord and the seventh cervical nerve. It arose in common with the medial antebrachial cutaneous nerve in 8 instances, with the medial brachial cutaneous in 6, and with both in 8. It presumably receives both eighth cervical and first thoracic nerve fibers in the great majority of cases, but in 3 of Kerr's cases (from 151 satisfactory records) it definitely did not receive first thoracic fibers, and in 3 others it did receive seventh cervical fibers.

Fig. 3-25. A brachial plexus in which the lower trunk contributes to the lateral cord. The ulnar nerve again has a lateral root. (Kerr AT: Am J Anat 23:285, 1918)

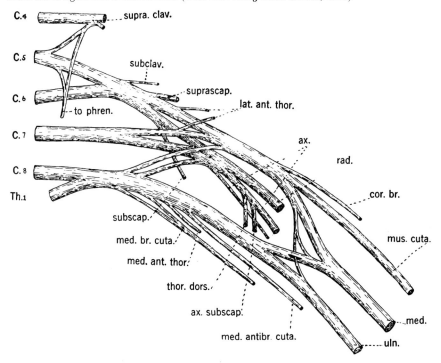

Table 3-4
Segmental Composition of Medial Cord and Branches

NERVE	USUAL COMPOSITION	APPARENT VARIATIONS
Medial cord	C8, T1	C7-T1; C8-T2
Medial pectoral	C8, T1	C7-T1; C8
Medial brachial cutaneous	T1	C8, T1
Medial antebrachial cutaneous	C8, T1	T1
Medial root of median	C8, T1	C7-T1
Median, both roots	C5-T1	C5-8; C6-T1
Ulnar	C7-T1	C5-T1; C6-T1; CT-T2; C8, T1

Kerr found the *medial brachial cutaneous nerve* arising by one or more roots from the medial cord, arising from the eighth cervical and first thoracic nerves or the first thoracic alone, from the lower trunk or its posterior division (Fig. 3-24), or from the ulnar nerve. In some cases there were two medial brachial cutaneous nerves (Fig. 3-21); in others it arose with the medial antebrachial cutaneous, or with the medial pectoral, or with both. While in his series it could have received fibers from both the eighth cervical and first thoracic nerves in 95%, he quoted evidence that it commonly consists of first thoracic fibers only. Walsh traced fibers of the eighth cervical nerve into the medial brachial cutaneous in only 11 or 73 plexuses.

The *medial antebrachial cutaneous nerve* may arise from the lower trunk of the brachial plexus (Fig. 3-23), from the first thoracic nerve only, or from the ulnar nerve, instead of from the medial cord, but Kerr found an origin from the medial cord in 145 of 174 plexuses, or about 83%. It may be doubled (6 cases), or may arise with the medial brachial cutaneous or with the medial anterior thoracic, or both. Walsh found it to consist of fibers derived from C8 and T1 in 53 cases, of fibers from T1 only in 20, and to receive fibers from C7 only once (this in an anomalous plexus).

The *medial root of the median nerve* apparently arose from the medial cord in its usual fashion in all of Kerr's 175 specimens except the five in which there was no medial cord proper, but in more that a third of cases it received an additional branch that came from the lateral cord (Fig. 3-21) or the seventh cervical nerve. Since the median nerve normally derives fibers from the seventh cervical nerve and the lateral cord through its lateral root, these connections presumably do not change its composition.

Kerr found the *median nerve* formed in the usual fashion by roots derived from the lateral and medial cords in 86.85% of plexuses, most of the remaining median nerves having lateral roots that were derived in an abnormal manner. However, he had no cases in which it was possible to show by dissection that all the nerves contributing to the plexus did not enter into the formation of the median nerve; he did show, by maceration, that the fifth cervical failed to contribute to the median nerve in 1 of 27 cases, and quoted evidence that in about 7.5% either the fifth cervical or the first thoracic nerve may fail to send fibers into the median. Walsh found the median to receive fibers from C5 through T1 in 65 cases, from C6 through T1 in 8.

The *ulnar nerve,* commonly a terminal branch of the medial cord, is usually described as receiving only eighth cervical and first thoracic fibers; however, it very commonly receives also fibers from the seventh cervical nerve (or perhaps even higher nerves) through a lateral root, derived from the lateral cord or one of its branches (Figs. 3-21 and 3-25), or the seventh cervical nerve directly. Kerr found a lateral root of the ulnar nerve in only 42.85% of his cases, but pointed out that this root is easily overlooked or broken at dissection, and that his records were derived from students' dissections; he estimated 60% as being close to its true incidence. Linell found it "obvious" in 57% of his short series. Walsh found a lateral root in 265 of 290 plexuses in which he sought it, but said that in 10 of these the fibers returned to the median nerve; thus this higher contribution to the ulnar occurred in about 87% of cases.

Whatever may be the exact percentage, it is obvious that the ulnar nerve receives fibers from at least as high as the seventh cervical nerve more commonly than it does not, and Kerr quoted evidence that it sometimes receives, through its lateral root, fibers from the fifth and sixth cervical nerves. The distribution of the fibers reaching the ulnar nerve through its lateral root is unknown; Harris suggested that they go to the flexor carpi ulnaris but, in at least a few cases, they apparently are destined for muscles of the thumb, and their presence in the ulnar nerve may explain certain cases of anomalous innervation of thumb muscles (Chap. 6).

In 170 of 175 plexuses, Kerr found the main root of the ulnar nerve to be derived as a terminal branch from the medial cord; in 3 cases it arose from an anterior cord that gave rise also to the musculocutaneous and median nerves, and in 2 from a similar trunk formed after the musculocutaneous nerve had been given off.

Posterior Cord and Branches

As already noted, the posterior cord is most commonly formed by the union first of the posterior divisions of the upper and middle trunks, then joined at a lower, often an axillary, level by the small posterior division of the lower trunk (Fig. 3-23). Variations from this exist, however, for the posterior divisions of the three trunks may join simultaneously (Fig. 3-25), those from the middle and lower trunks may join first (Fig. 3-24), or nerves that have split into anterior and posterior divisions before forming the trunks may contribute their posterior divisions to the posterior cord. In 36 of 173 cases in Kerr's series there was no posterior cord proper, for the radial and axillary nerves arose independently from the plexus (Fig. 3-24); Walsh regarded this as the usual condition, and said that a "posterior cord" is always simply the two nerves bound together by connective tissue. None of these variations necessarily introduces any change in the composition of the posterior cord. Whether fibers of the fourth cervical nerve ever enter the posterior cord when that nerve contributes to the plexus is not known (Table 3-5); Kerr quoted evidence that the fifth cervical nerve does not send fibers to the posterior cord in about 20% of cases, and that the first thoracic does not in about 80%.

Kerr found the *upper subscapular nerve* or nerves (there were 2—Fig. 3-24—in 40.76%, and 3 in 5.73%) arising most commonly from the posterior division of the upper trunk (Figs. 3-24 and 3-25), rather than from the posterior cord, and therefore containing fibers from no more than C4 to C6; the next most common origin was from the posterior cord (28.02%) and in about 20% it or one of them arose from the combined posterior divisions of the upper and middle trunks. In 1 case it arose exclusively from the fifth cervical nerve, and in 8 cases from the seventh only; Kerr quoted evi-

Table 3-5
Segmental Composition of Posterior Cord and Branches

NERVE	USUAL COMPOSITION	APPARENT VARIATIONS
Posterior cord	(C4?), C5-8	(C4?), C5-T1; C6-T1; C6-8
Upper subscapular	C5	C5,6; C7
Thoracodorsal	C7,8	C5,6; C5-7; C6-8; C8, T1
Lower subscapular	C5,6	C5; C5-7; C6; C6,7
Axillary	C5,6	C5; C5-7
Radial	C5-8	C5-7; C5-T1; C6-C8; C6-T1; C7,8

dence that the fibers in this nerve are usually derived from the fifth, less commonly from the fifth and sixth.

The *thoracodorsal nerve* in Kerr's series arose most commonly directly from the posterior cord (Fig. 3-23), or indirectly from it through the radial or axillary nerves (Fig. 3-25), having one of these origins in 70.19% of cases. In the others it arose proximal to the formation of the posterior cord in a variable manner. In some instances it arose with the lower subscapular nerve, in one case with both subscapulars, in one it gave off a branch to the teres minor (which was also supplied by the lower subscapular), and in another it gave a branch to the lower subscapular nerve. Kerr quoted evidence that this nerve is formed by fibers of C7 alone in about 50% of cases, and in most others by fibers of the seventh and eighth, sixth to eighth, or fifth to seventh. In most of his own cases no elements entering the plexus could be excluded, but in three the nerve consisted of fifth and sixth cervical fibers only, and in one of eighth cervical and first thoracic fibers only.

The *lower subscapular nerve* arose directly from the posterior cord (Fig. 3-22) in only 48 of Kerr's 157 cases; in 68 others it arose from the axillary nerve (Fig. 3-25), in one from a radial that in turn arose from the posterior cord, and in the remaining 25.48% it arose wholly or in part from a more proximal part of the plexus (Fig. 3-24). In 1 case it received a branch from the upper subscapular nerve and gave off one to the latissimus dorsi; in 5 it arose with the thoracodorsal nerve. Kerr quoted evidence that this nerve usually receives neither eighth cervical nor first thoracic fibers, but rather contains fifth and sixth fibers (most commonly), seventh only, sixth and seventh, sixth only, or either fifth alone or fifth to seventh, in that order. He noted that this nerve may send two or more branches instead of one into the subscapularis muscle, or supply only the teres major, its usual branch to the subscapularis then being replaced by a branch from the axillary nerve.

The *axillary nerve* arose from the posterior cord in 79.76% of Kerr's series, and in the remainder from the upper or upper and middle posterior divisions directly (Fig. 3-24), there being no posterior cord. In the latter cases it did not receive fibers from the lower trunk of the plexus, and Kerr expressed doubt that it ever receives eighth cervical or first thoracic fibers; however, he quoted evidence that while it consists of fifth and sixth fibers only in almost 80% of cases, it may also contain fibers of the seventh, or consist of fifth cervical fibers only. Walsh, in his special dissections, found it to contain fibers from C5 and C6 only in 63 of 73 plexuses, and from C7 also in the remaining 10. In 55.44% of Kerr's series the lower subscapular nerve arose in whole or in part from the axillary nerve.

The *radial nerve* arose from the posterior cord (Fig. 3-25) in 79.7% of Kerr's series, and in the remainder it arose from posterior divisions of the trunks (Fig. 3-24). Kerr quoted evidence that it consisted of fifth through eighth cervical nerve fibers in 98 of 156 cases collected from the literature, of sixth through eighth in 27, of fifth cervical through first thoracic in 25, of sixth through first thoracic in 3, and, in one instance each, of fifth through seventh, of sixth and seventh, and of seventh and eighth. In Walsh's cases the radial received fibers from C5 through C8 in 67, and from T1 also in only 6. Frohse and Fränkel expressed the opinion that the contribution of C5 to the radial nerve is purely afferent, but many writers list C5 as contributing to the innervation of posterior muscles of the forearm.

Anomalous Vascular Relations

Occasionally, as described in the following chapter, a large superficial branch of the axillary artery emerges between the two roots of the median nerve to course into the arm as a superficial radial, ulnar, brachial, or median artery. In addition to variations in arterial-nerve relations in the upper part of the plexus, already recounted, Miller ('39) reported a number of anomalous vascular relations among the branches of the plexus. There were 15 cases, occurring in 8 of 480 bodies, in which the median nerve below the union of its two roots was penetrated and divided by a branch of the axillary artery (Fig. 3-26a)—the subscapular in 10 instances and a stem common to both humeral circumflex arteries in 4, the deep brachial in 1. She also found 3 cases in which a common stem for the humeral circumflex arteries divided the lateral root of the median nerve (Fig. 3-26b), 1 in which it divided the lateral cord (Fig. 3-26c), and 4 (in 2 bodies) in which the lateral thoracic artery divided the medial cord. In 2 cases there was a communication between the musculocutaneous and median nerves around the common stem of the humeral circumflex arteries (Fig. 3-26d), in 1 a communication between the lateral thoracic and ulnar nerves around the axillary artery, and in 1 a communication between the median and ulnar nerves around the subscapular artery. In Miller's series, 8% of bodies had some type of anomalous plexus-artery relationship; she said that 4% showed anomalies of the plexus associated with veins, but did not describe these.

INJURIES OF BRACHIAL PLEXUS

Because of the changing arrangement of the brachial plexus as it is followed distally, injuries of it may result in many diverse paralyses and anesthesias or parasthesias, depending upon the exact level at which it is injured and the extent of injury to the various elements at that level. Thus DeJong described a case in which damage to the plexus had been produced by an anterior dislocation of the shoulder, and the symptoms were referable only to injury of the posterior cord; paralysis was limited to the deltoid, teres minor, subscapularis, latissimus dorsi, teres major, and the muscles of the arm and forearm innervated through the radial nerve.

Injuries to the brachial plexus of the infant may occur during birth, as a result of the strain placed upon the plexus by wide separation of the head and shoulder during a difficult delivery, and may result during adult life from a similar violent separation of head and shoulder, such as may occur in a fall from a motorcycle. Such injuries are typically in the supraclavicular part of the brachial plexus,

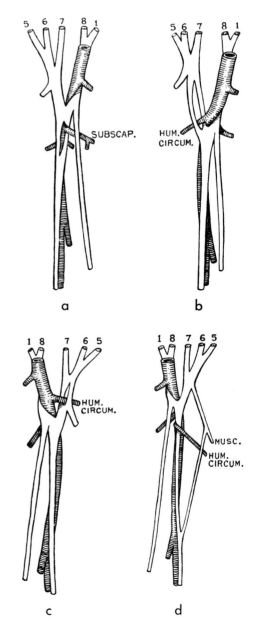

Fig. 3-26. Abnormal arterial relations of the distal part of the brachial plexus; in *a* the median nerve, in *b* its lateral root, and in *c* the lateral cord is penetrated by an arterial trunk, while in *d* a trunk passes in front of the median nerve and through a loop formed by a communication of the musculocutaneous with this nerve. (Miller RA: Am J Anat 64:143, 1939)

but Davis and co-worker said strain or blunt compression of the plexus may damage it anywhere between the level of the roots and that of the most distal point of injury. Barnes, recording his experience with injuries of the plexus of the adult produced by violent strain, found that in these cases the lesion was more commonly in the spinal nerves between the intervertebral foramina and the point at which they joined to form trunks (that is, was in the roots of the brachial plexus). Kolodny expressed the opinion that traction injuries of adults more commonly affect the nerve trunks, although those of infants may affect the nerve roots or the emerging nerves, and explained this apparent difference on the basis of both the difference in the injuring force and the direction of the transverse processes on which the nerves lie. They are somewhat turned upward in the infant, so that the nerves are angulated across their tips, but slanting downward in the adult. Avulsion of the roots of the nerves from the spinal cord is apparently rare (Kolodny; Davis and co-workers), in both infant and adult, but when it does occur is usually accompanied by signs of injury to the spinal cord itself; it can be diagnosed by a cervical myelogram (Wickstrom and colleagues; Tracy and Brannon), which will reveal the resulting distortion of the subarachnoid space.

Types of Involvement of Nerve Fibers

It is generally agreed that traction injuries of the brachial plexus are more commonly not actual avulsion of nerves, but rather a stretching of the nerves that result in tearing of connective tissue and hemorrhage in and about them (Kolodny; Barnes; and others). A plexus so injured usually becomes embedded in scar tissue and the formation of the scar tissue is usually said to produce, in turn, further injury to the nerves (but Barnes disagreed, saying that the paralysis is almost always most extensive immediately after the injury); however, there is in such a case no profound disturbance of intraneural morphology. On the other hand, if nerves are actually ruptured, as they may be, their ends become so embedded in scar tissue that no re-

generation is possible unless the lesion is repaired surgically. Moderate stretching of a nerve apparently causes temporary inhibition of the activity of the motor fibers without seriously affecting the sensory ones (Barnes), and complete recovery can be expected within a period of 2 months. With degenerative lesions recovery is slower and often incomplete, especially in the more distal muscles (Barnes).

In open wounds of the brachial plexus, the extent of damage to the several nerve trunks affected obviously may, and usually does, vary considerably. Brooks said that the damage is seldom uniform, but rather consists of a mixture of neurapraxia (temporary cessation of function of the fibers), interruption of axons without interruption of the nerve as a whole (axonotmesis), and complete severance of nerves (neurotmesis), a variable number of fibers typically escaping damage altogether.

Upper and Lower Lesions

Lesions of the brachial plexus are sometimes classified as of the Erb or Erb-Duchenne type, involving the upper part (C5 and C6) of the plexus and affecting especially the shoulder and arm (Fig. 3-27), or as of the Klumpke or Klumpke-Dejerine type, involving C8 and T1 and affecting especially the distal part of the limb (Fig. 3-28). In a general way such classification may be useful, but it is widely recognized that these categories are so broad that they aid little in diagnosis of the lesion; rather, the extent of damage must be determined from observation of individual muscles, and is usually found complete for some nerves, incomplete for others. Moreover, because of the variation in the segmental supply to the muscles of the limb, accurate diagnosis of the exact site of the lesion may be impossible. Sensory loss following lesions of the brachial plexus is seldom coextensive with the motor loss, and is said to be not as reliable an index to the extent or severity of the lesion.

However, the more common traction injury of the plexus is to an upper part of the plexus, both in birth injuries and in injuries to the plexus of the adult (Wickstrom and colleagues; Kolodny; Davis and his co-workers;

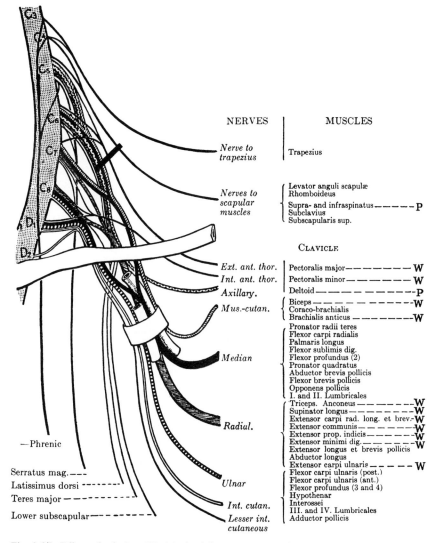

NERVES | MUSCLES

Nerve to trapezius | Trapezius

Nerves to scapular muscles | Levator anguli scapulæ
Rhomboideus
Supra- and infraspinatus ————— P
Subclavius
Subscapularis sup.

CLAVICLE

Ext. ant. thor. | Pectoralis major ———————— W
Int. ant. thor. | Pectoralis minor ———————— W
Axillary. | Deltoid ——————————— P
Mus.-cutan. | Biceps ————— ————— W
Coraco-brachialis
Brachialis anticus —————— W

Median | Pronator radii teres
Flexor carpi radialis
Palmaris longus
Flexor sublimis dig.
Flexor profundus (2)
Pronator quadratus
Abductor brevis pollicis
Flexor brevis pollicis
Opponens pollicis
I. and II. Lumbricales

Radial. | Triceps. Anconeus ———————— W
Supinator longus ————————— W
Extensor carpi rad. long. et brev.-W
Extensor communis —— ————— W
Extensor prop. indicis————— W
Extensor minimi dig.— ———— W
Extensor longus et brevis pollicis W
Abductor longus
Extensor carpi ulnaris — — —— W

Ulnar | Flexor carpi ulnaris (post.)
Flexor carpi ulnaris (ant.)
Flexor profundus (3 and 4)
Hypothenar
Interossei
III. and IV. Lumbricales
Adductor pollicis

Int. cutan.
Lesser int. cutaneous

—Phrenic

Serratus mag.———
Latissimus dorsi ————————
Teres major ——————
Lower subscapular———————

Fig. 3-27. Effect of a lesion (*black bar*) of the upper part of the brachial plexus. *P* indicates paralysis, *W* weakness. (Pollock LJ, Davis L: Peripheral Nerve Injuries. New York, Hoeber, 1933)

Barnes). According to Barnes, this is because the more common injury is produced while the arm is at the side, in which position traction is exerted only on the upper nerves of the plexus; he said that violence must be severe enough in adults to tear the fascia over the brachial plexus and to rupture or avulse the scalene muscles before injurious stretching of the nerves occurs. He also said that if the arm is abducted at the time of injury, the traction falls primarily on the lower roots, but that if it is both abducted and forced behind the trunk, while the head is forced to the opposite side, tension is placed on the entire plexus. Most of the injuries in the 63 cases he reported were of an incomplete or of a mixed type, rather than involving purely upper or lower parts of the brachial plexus; in these, 14 apparently were lesions of either the fifth or the fifth and sixth cervical nerves, 19 were lesions involving, apparently, at least parts of the fifth, sixth, and seventh cervical nerves, 2 were lesions apparently involving C7, C8, and T1, and 28 were lesions of the entire brachial plexus.

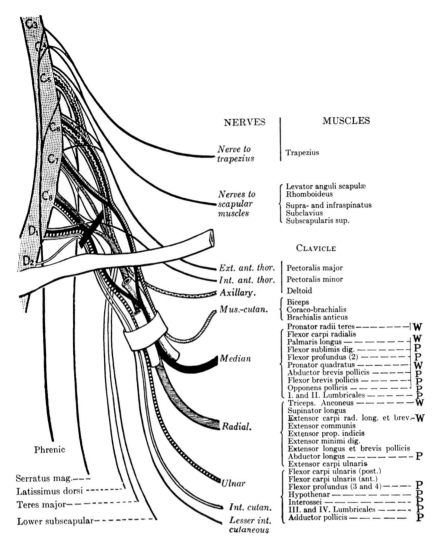

NERVES | MUSCLES

Nerve to trapezius | Trapezius

Nerves to scapular muscles | { Levator anguli scapulæ
Rhomboideus
Supra- and infraspinatus
Subclavius
Subscapularis sup. }

CLAVICLE

Ext. ant. thor. | Pectoralis major
Int. ant. thor. | Pectoralis minor
Axillary. | Deltoid
Mus.-cutan. | { Biceps
Coraco-brachialis
Brachialis anticus }

Median | { Pronator radii teres ------- **W**
Flexor carpi radialis ------ **W**
Palmaris longus — — — — — **W**
Flexor sublimis dig. — — — — — **P**
Flexor profundus (2) ----- **P**
Pronator quadratus ----- **W**
Abductor brevis pollicis ----- **P**
Flexor brevis pollicis — — — — **P**
Opponens pollicis — — — — **P**
I. and II. Lumbricales — — — — **P** }

Radial. | { Triceps. Anconeus — — — — — **W**
Supinator longus
Extensor carpi rad. long. et brev. **W**
Extensor communis
Extensor prop. indicis
Extensor minimi dig.
Extensor longus et brevis pollicis
Abductor longus — — — — — **P**
Extensor carpi ulnaris }

Ulnar | { Flexor carpi ulnaris (post.)
Flexor carpi ulnaris (ant.)
Flexor profundus (3 and 4) ---- **P**
Hypothenar — — — — — — — **P**
Interossei — — — — — — — **P**
III. and IV. Lumbricales — — —, **P**
Adductor pollicis — — — — — **P** }

Int. cutan.
Lesser int. cutaneous

Phrenic
Serratus mag.- - -
Latissimus dorsi - - - - - - -
Teres major- - - - - -
Lower subscapular- - - - - - - -

Fig. 3-28. Effect of a lesion (*black bar*) of the lower part of the brachial plexus. *P* indicates paralysis, *W* weakness (Pollock LJ, Davis L: Peripheral Nerve Injuries. New York, Hoeber, 1933)

Recovery

Barnes reported in detail the extent of recovery in his series of traction injuries of the plexus, and Brooks discussed functional recovery following open wounds and surgical repair of the plexus. They agreed that following injuries of the upper part of the plexus there was generally good recovery of power in the affected muscles of the shoulder and arm. Among Barnes' 14 patients with injury confined to the fifth or fifth and sixth cervical nerves, 11 regained the movements of flexion at the elbow and abduction and external rotation at the shoulder, against both gravity and resistance; 1 had incomplete recovery; and 2 had failed to recover at the time they were last seen. In Barnes' series, those patients who had had greater trauma, with involvement of the fifth, sixth, and seventh nerves, had less satisfactory recovery under conservative treatment: 11 of 19 regained extension of the wrist and fingers, flexion of the elbow, and abduction of the shoulder against gravity and some resistance, 3 had incomplete recovery of

Table 3-6
Composition and Distribution of the Nerves to the Upper Limb

NERVE	USUAL SEGMENTAL COMPOSITION*	MUSCULAR DISTRIBUTION	CUTANEOUS DISTRIBUTION
Accessory	A cranial nerve Also C2-C4?	Trapezius and sternocleido-mastoid	None
Supraclaviculars	C3, C4	None	Base of neck, upper thorax and shoulder
Dorsal scapular	(C4), C5	Both rhomboids	None
Long thoracic	C5-C7	Serratus anterior	,,
Subclavian	(C4), C5, (C6)	Subclavius	,,
Suprascapular	C5, (C6)	Supraspinatus and infraspin-atus	,,
Subscapular, upper	C5	Subscapularis	,,
Thoracodorsal	C7, C8	Latissimus dorsi	,,
Subscapular, lower	C5, C6	Subscapularis and teres major	,,
Lateral pectoral	C5-C7	Pectoralis major	,,
Medial pectoral	C-8, T-1	Pectoralis major and minor	,,
Musculocutaneous	(C4), C5, C6, (C7)	Coracobrachialis, biceps, and brachialis	Forearm
Median	C5-T1	All muscles on flexor side of forearm except flexor carpi ulnaris and an ulnar part of flexor digitorum profundus; short abductor, opponens, and most of short flexor of thumb; first two lumbricals	Palm of hand, digits
Medial brachial cutaneous	(C8), T1	None	Arm
Medial antebrachial cutaneous	C8, T1	None	Forearm
Ulnar	C7-T1	Flexor carpi ulnaris, ulnar part of flexor digitorum profundus, all hypothenar muscles, all interossei, 3rd and 4th lumbricals, adductor pollicis, part of flexor pollicis brevis	Hand
Axillary	C5, C6, (C7)	Deltoid and teres minor	Arm
Radial	(C5), C6-C8, (T1)	All muscles on posterior or extensor side of arm and forearm	Arm, forearm, and hand
Intercostobrachial	T2, (T3)	None	Arm

* The numbers enclosed in parentheses indicate a variable but not rare contribution; the segmental contributions listed here are based largely upon the anatomical findings of Walsh (1877).

the abductors and external rotators (the injury apparently being greatest in the fifth and sixth nerves), 3 had residual paralysis of the extensors of the wrist and fingers (the injury apparently being more severe in the region of C7), and 2 showed no recovery at all. Brooks found that regeneration of fibers following injury to the posterior cord usually resulted in "fair" recovery, most patients being able at least to extent the thumb against gravity

within a period of 2 years; if, however, there was no evidence of recovery in the triceps by the end of 9 months, the prognosis was bad, for none of those patients showed any significant improvement within the 2 years they were followed.

Both workers agreed that when the lesion involved the lower part of the plexus recovery was apt to be disappointing. Brooks reported that when there was a degenerative lesion of

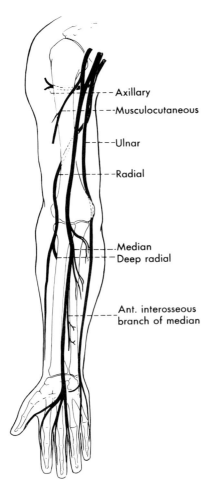

Fig. 3-29. Major nerves of the upper limb.

C8 and T1 or of the medial cord, the more distal muscles never regained useful function, even though the proximal ones might recover well; in any case, he said, persistence of complete paralysis of the musculature innervated by either the ulnar or the median nerve beyond 6 months gave a very poor prognosis. Barnes had no patients with a lesion of only C8 and T1, but had 2 with lesions of C7, C8, and T1; 1 of these, with Horner's syndrome, showed no recovery, the other had a transient lesion of C7, but a degenerative lesion of C8 and T1 from which there was no recovery.

Barnes had 28 patients in whom the entire plexus was affected, but in 4 the lesion was nondegenerative, and these had complete recovery within 6 months; the remaining 24 had incomplete recovery of muscles inner-

vated either by the upper or lower roots of the plexus, or permanent paralysis of the entire limb. The presence of Horner's syndrome worsened the prognosis: of 13 patients in this group who also had Horner's syndrome, 7 had permanent and total paralysis of the upper limb, while 6 regained useful power in the abductors of the shoulder and in the flexors of the elbow, but had no recovery of the muscles of either the forearm or the hand.

Muscle and tendon transplantations and arthrodeses are sometimes feasible in helping to overcome the effects of residual paralysis from injuries of the brachial plexus, just as they are when the lesion is more peripherally situated. Wickstrom and his co-workers reported that the internal rotation that is often the major residual defect in birth paralyses results in flattening of the humeral head and deformity of the glenoid cavity if it is not corrected.

DISTRIBUTION OF PERIPHERAL NERVES

The details of the courses and distributions of the nerves to the upper limb are described in the following three chapters. The major features of their distribution can, however, be most conveniently summarized here.

Table 3-6 lists the nerves of the upper limb, and indicates their general distribution. A diagram of the courses of the major nerves is presented in Figure 3-29. It will be seen from this table that the larger nerves of the limb—the musculocutaneous, median, ulnar, axillary, and radial—all have both muscular and cutaneous distributions, while the other nerves contributing to the innervation of the limb have either a muscular or a cutaneous distribution. In futher summary of the larger nerves, it can be pointed out that the *musculocutaneous nerve* (Fig. 3-30) supplies all the muscles on the flexor surface of the arm, and continues, as the lateral antebrachial cutaneous nerve, to supply skin on the lateral aspect, both anterior and posterior surfaces, of the forearm.

The *ulnar nerve* (Fig. 3-31), a derivative of the anterior part of the brachial plexus as is the musculocutaneous, is limited in its muscular distribution to anteriorly placed muscles,

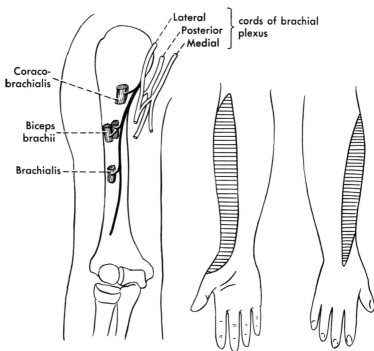

Fig. 3-30. Distribution of the musculocutaneous nerve.

but supplies skin of both the palmar and dorsal surfaces of the hand. The ulnar nerve supplies only two muscles in the forearm, the flexor carpi ulnaris and an ulnar portion of the flexor digitorum profundus, but in the hand it supplies a majority of the muscles, including all the muscles of the hypothenar eminence; typically all the interossei; the two ulnar lumbricals; and the deeply placed muscles of the thumb namely, both heads of the adductor pollicis and a deep portion of the flexor pollicis brevis. Its cutaneous distribution is on the ulnar border of the hand, thus anatomically, disregarding overlap, to about 1.5 fingers on their palmar surfaces, 1.5 or more on their dorsal surfaces, and to corresponding portions of the ulnar side of the palm and dorsum.

The *median nerve* (Fig. 3-32) supplies the flexor muscles of the forearm and hand not supplied by the ulnar—all the flexor muscles of the forearm except the flexor carpi ulnaris and an ulnar part of the flexor digitorum profundus, the superficially placed muscles of the thumb (abductor pollicis brevis, opponens pollicis, and the larger or superficial head of the flexor pollicis brevis), and the first two lumbricals. The median nerve, therefore, is the chief nerve to the muscles of the forearm and to the muscles of the thenar eminence, while the ulnar nerve participates relatively little in the innervation of forearm muscles, but is the important nerve to muscles of the hand with the exception of those of the thenar eminence. The cutaneous distribution of the median nerve is largely to the radial side of the palmar surface of the hand, but also onto the dorsal surface of the distal parts of those fingers that it supplies.

The *axillary nerve* (Fig. 3-33), a derivative of the posterior cord of the brachial plexus, typically supplies only two muscles, the teres minor and the deltoid; its cutaneous distribution is over the region of the deltoid. The *radial nerve* (also Fig. 3-33), the only derivative of the posterior cord in the arm and forearm, therefore necessarily supplies all the muscles on the extensor surface of the arm and also all the extensor muscles of the forearm; its cutaneous distribution is to a limited amount of skin on the posterior aspect of the arm, to a strip down the posterior aspect of the forearm, and to skin on the radial side of the dorsal aspect of the hand.

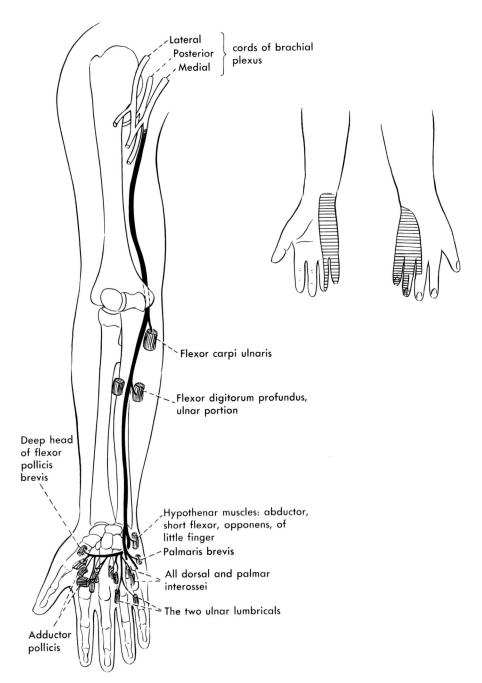

Fig. 3-31. Distribution of the ulnar nerve.

Cutaneous Distribution

The skin over the upper part of the shoulder is supplied by the supraclavicular nerves, derivatives of the cervical plexus and of the third and fourth cervical nerves. A limited area of skin on the medial side of the arm is supplied by the intercostobrachial nerve, derived from the second intercostal nerve, and sometimes also from the third. Otherwise, the skin of the upper limb is innervated entirely by branches of the brachial plexus (Fig. 3-34).

The cutaneous nerves of the arm in addition to those just mentioned are the upper lat-

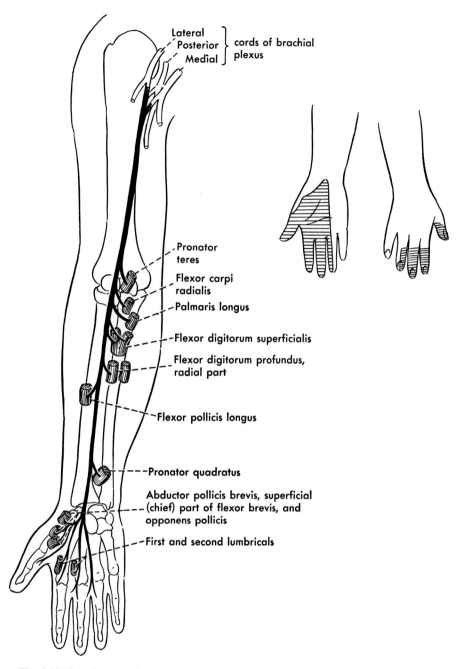

Lateral
Posterior } cords of brachial
Medial   plexus

Pronator
teres

Flexor carpi
radialis

Palmaris longus

Flexor digitorum superficialis

Flexor digitorum profundus,
radial part

Flexor pollicis longus

Pronator quadratus

Abductor pollicis brevis, superficial
(chief) part of flexor brevis, and
opponens pollicis

First and second lumbricals

Fig. 3-32. Distribution of the median nerve.

eral brachial cutaneous nerve, from the axillary; the posterior brachial cutaneous and the lower lateral brachial cutaneous nerve from the radial; the medial brachial cutaneous nerve, a direct derivative of the medial cord of the brachial plexus; and twigs from the medial antebrachial cutaneous nerve, also a direct derivative of the medial cord of the plexus.

The cutaneous nerves of the forearm are simpler: the medial antebrachial cutaneous nerve, from the brachial plexus; the lateral antebrachial cutaneous nerve, the cutaneous continuation of the musculocutaneous; and

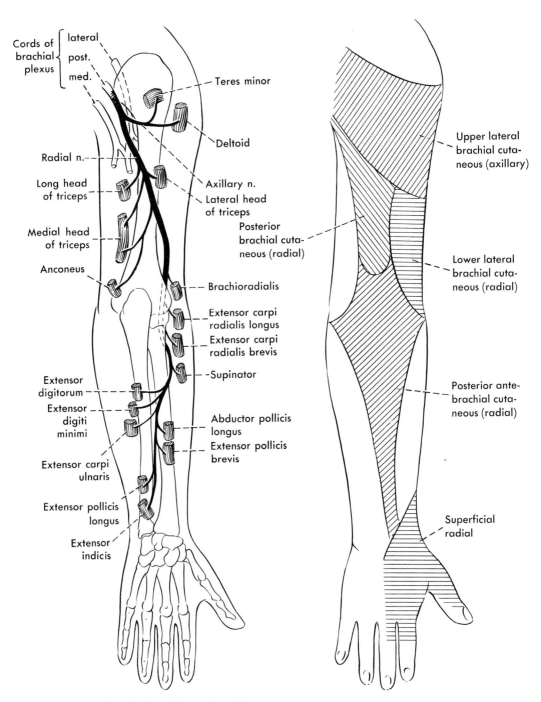

Fig. 3-33. Distribution of the axillary and radial nerves.

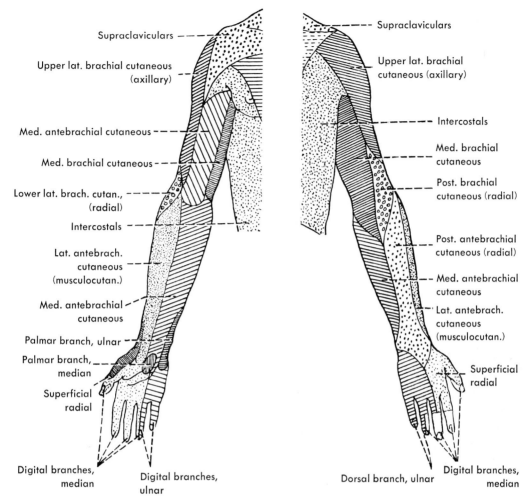

Supraclaviculars

Upper lat. brachial cutaneous
(axillary)

Med. antebrachial cutaneous

Med. brachial cutaneous

Lower lat. brach. cutan.,
(radial)

Intercostals

Lat. antebrach.
cutaneous
(musculocutan.)

Med. antebrachial
cutaneous

Palmar branch, ulnar

Palmar branch,
median

Superficial
radial

Digital branches,
median

Digital branches,
ulnar

Supraclaviculars

Upper lat. brachial
cutaneous (axillary)

Intercostals

Med. brachial
cutaneous

Post. brachial
cutaneous (radial)

Post. antebrachial
cutaneous (radial)

Med. antebrachial
cutaneous

Lat. antebrach.
cutaneous
(musculocutan.)

Superficial
radial

Dorsal branch, ulnar

Digital branches,
median

Fig. 3-34. Cutaneous innervation of the upper limb. The distribution of the intercosto-brachial, not shown, is largely identical with that of the medial brachial cutaneous. (Flatau E: Neurologische Schemata für die ärztliche Praxis. Berlin, Springer, 1915)

the posterior antebrachial cutaneous nerve, a branch of the radial. The skin of the ulnar side of the hand, and of approximately one and one-half digits, but especially variable on the dorsum, is supplied by the ulnar nerve. Skin on the dorsal aspect of the radial side of the hand is supplied by the superficial branch of the radial; and skin on the radial side of the palmar surface of the hand, as also the distal portions on the dorsal aspect of the more radial digits, is supplied by the median nerve.

These nerves are futher mentioned in connection with the discussions of the cutaneous distribution to the different parts of the limb;

Figure 3-35 shows their clinical distribution—areas of overlap and areas of sensory loss commonly found when they are interrupted.

Cutaneous innervation must always be considered from two aspects, the distribution of the definitive peripheral nerves, already discussed, and the dermatomal or segmental innervation of the skin. It has been noted in Chapter 1 that the determination of dermatomes is a clinical and not an anatomical matter, and that the various dermatomal paterns given by different investigators do not necessarily coincide; Figures 3-36 through

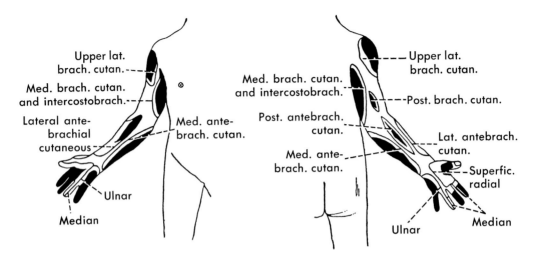

Fig. 3-35. Average areas of complete sensory loss (*black*) after interruption of the various cutaneous nerves of the upper limb. The surrounding *lines* enclose areas in which there is loss of touch but no more than impairment of pain and temperature; *intervening areas* are those of complete overlap between adjacent nerves. (After Lewandowsky M, Foerster O, from Woltman HW, Kernohan JW, Goldstein NP. In Baker AB (ed): Clinical Neurology [ed 2], Vol 4. New York, Hoeber, 1962)

Fig. 3-36. Foerster's dermatomes viewed from the front. (After Foerster, from Haymaker W, Woodhall B: Peripheral Nerve Injuries [ed 2]. Philadelphia, WB Saunders, 1953)

Fig. 3-37. Foerster's dermatomes viewed from behind. (After Foerster, from Haymaker W, Woodhall B: Peripheral Nerve Injuries [ed 2]. Philadelphia, WB Saunders, 1953)

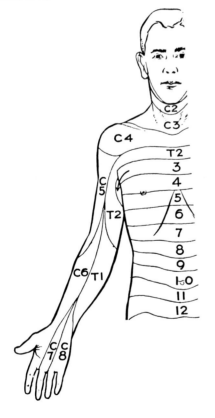

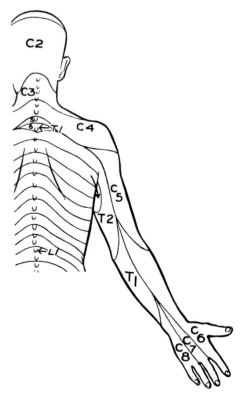

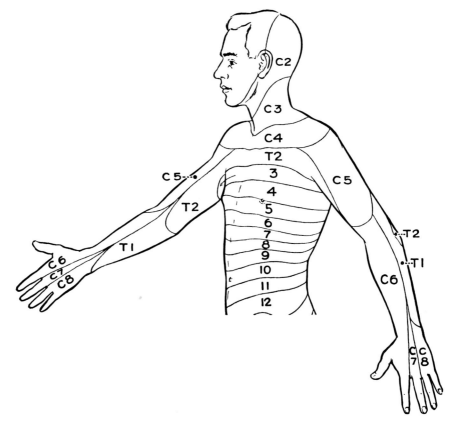

Fig. 3-38. Foerster's dermatomes viewed from the side. (After Foerster, from Haymaker W, Woodhall B: Peripheral Nerve Injuries [ed 2]. Philadelphia, WB Saunders, 1953)

3-39 show the two more commonly used dermatomal charts, those of Foerster and of Keegan. They were determined by different methods, Foerster's method being that of mapping the sensibility remaining after section of nerves above and below the one under investigation, Keegan's being that of determining an area of hypalgesia when it was thought that conduction along only one nerve was interfered with; this variation in method may be the explanation for the differences. The chief differences, it will be noted, are that Keegan's dermatomes constitute long strips extending distally from the base of the limb, while Foerster's are somewhat more patchy and do not all reach the base of the limb; further, Foerster emphasized that there is almost complete overlap between adjacent derma-

tomes, while Keegan's dermatomes, being areas of hypalgesia, are areas in which there appeared to be no functionally adequate overlap by adjacent segmental nerves.

Muscular Distribution

The distribution of nerves to muscles, like that to skin, needs to be considered both from the aspect of the distribution of peripheral nerves and from the aspect of segmental innervation. The former is shown in Table 3-6; the distributions of the larger nerves to the limb have also been briefly summarized at the beginning of this section on peripheral nerves, and are illustrated in Figures 3-30 to 3-33. Lines along which the chief nerves and their motor branches can be stimulated are shown in Figures 3-40 to 3-43. The approximate seg-

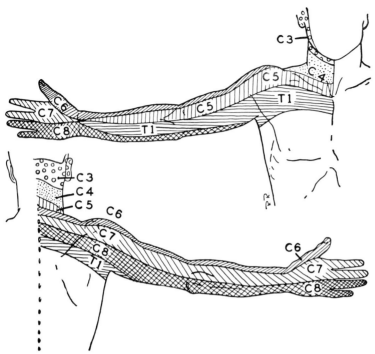

Fig. 3-39. Dermatomes of the upper limb as determined by Keegan. T1 is not distributed to skin of the thorax. (Keegan JJ, Garrett FD: Anat Rec 102:409, 1948)

mental innervations of the muscles of the upper limb are shown in Table 3-1, and discussed in more detail, with accompanying tables, in appropriate places in the following three chapters.

The determination of the segmental innervation of muscles is largely a clinical matter, and is beset with many difficulties. In view of the very considerable diversity in segmental innervation reported by different authors, it seems obvious that no one table can be assumed to be entirely accurate; whether this variability results from a real variability in the distribution of the spinal nerves or from lack of knowledge, it is evidence that such listings should be taken as useful approximations rather than as indubitable facts. The most marked discrepancy in the upper limb is in regard to the innervation of short muscles of the thumb by the median nerve; these have traditionally been listed as being supplied through the upper and middle trunks of the brachial plexus, but there is now considerable

evidence that they are supplied through the lower trunk (C8 and T1).

VESSELS

Arteries

The arterial stem to the upper limb is the subclavian artery, derived on the left side from the arch of the aorta, on the right side from the brachiocephalic (innominate) artery. The subclavian artery supplies two vessels to the shoulder, the transverse cervical (a. transversa colli) and the suprascapular (transverse scapular), while its continuation into the axilla and the arm as the axillary and brachial arteries supplies the rest of the upper limb (Fig. 3-44).

The subclavian artery changes its name to axillary as it crosses the upper border of the first rib, and the axillary in turn becomes the brachial at the lower border of the tendon of the teres major muscle. The brachial artery

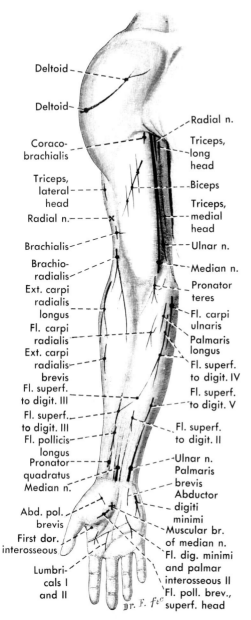

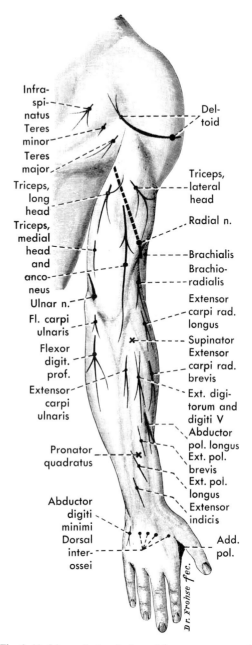

Fig. 3-40. Lines of stimulation of the motor nerves of the upper limb, and motor points, front view. (Frohse F, Fränkel M: Vol 2, Section 2, Part 2 of K von Bardeleben's Handbuch der Anatomie des Menschen. Jena, Fischer, 1908)

Fig. 3-41. Lines of stimulation of the motor nerves of the upper limb, and motor points, posterior view. (Frohse F, Fränkel M: Vol 2, Section 2, Part 2 of K von Bardeleben's Handbuch der Anatomie des Menschen. Jena, Fischer, 1908)

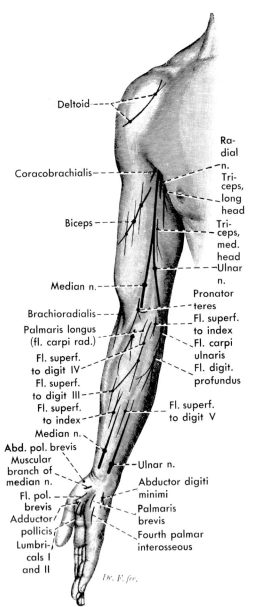

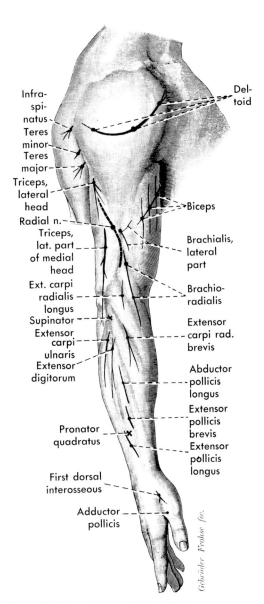

Fig. 3-42. Lines of stimulation of the motor nerves of the upper limb, and motor points, medial view. The leader for the palmaris longus crosses the motor point of the flexor carpi radialis. (Frohse F, Fränkel M: Vol 2, Section 2, Part 2 of K von Bardeleben's Handbuch der Anatomie des Menschen. Jena, Fischer, 1908)

Fig. 3-43. Lines of stimulation of the motor nerves, and motor points, lateral view. (Frohse F, Fränkel M: Vol 2, Section 2, Part 2 of K von Bardeleben's Handbuch der Anatomie des Menschen. Jena, Fischer, 1908)

runs down the medial side of the arm, and in the cubital fossa divides into the radial and ulnar arteries. These course down the anterior or flexor aspect of the forearm, and are primarily distributed to structures of this surface, continuing into the palm of the hand to form the two palmar arches; however, one branch derived from the ulnar (the posterior or dorsal interosseous) passes above the upper border of the interosseous membrane to reach the posteriorly located structures of the forearm, and the radial artery goes to the dorsum of the hand before reaching the palm. Details of the courses and branches of these vessels are supplied in appropriate places in the following chapters.

Veins

Most of the named veins of the upper limb accompany the named arteries, and are typically paired, two for each artery; they have been named "the accompanying veins" (for instance, "venae comitantes of the brachial artery"), but the larger ones are now named like the arteries (e.g., "the brachial veins"). The superficial veins of the upper limb do not accompany arteries.

There are two chief superficial veins of the upper limb (Figs. 3-45 and 3-46), the cephalic and the basilic. Most of the superficial veins of the palmar surface of the hand drain between the digits and around the borders of the palm to join the prominent dorsal venous network, and both the cephalic and basilic veins originate on the dorsum of the hand. The cephalic vein arises primarily from the radial side of the dorsal network, and passes around the radial border of the forearm to occupy an anterolateral position at the elbow; the basilic vein, arising from the ulnar side of the dorsal network, passes around the ulnar border of the forearm to occupy an anteromedial position at the elbow. The two veins here communicate with each other and with the deep veins, and the basilic vein, or the connection between the cephalic and basilic, usually receives a small but somewhat inconstant median forearm vein from the palmar

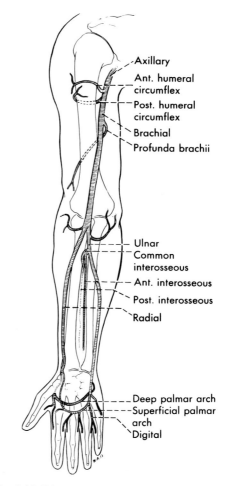

Fig. 3-44. Diagram of the major arteries of the upper limb

surface of the hand. The basilic vein leaves its subcutaneous position at about the junction of middle and lower thirds of the arm by penetrating the deep fascia on the medial side of the arm, but the cephalic vein runs subcutaneously in an anterolateral position on the arm almost to the level of the clavicle, where it passes deeply between the deltoid and pectoralis major muscles.

Lymphatics

The lymphatics of the upper limb are, like the veins, divisible into superficial and deep groups. The deep lymphatics are very limited in number, and accompany the arteries. The

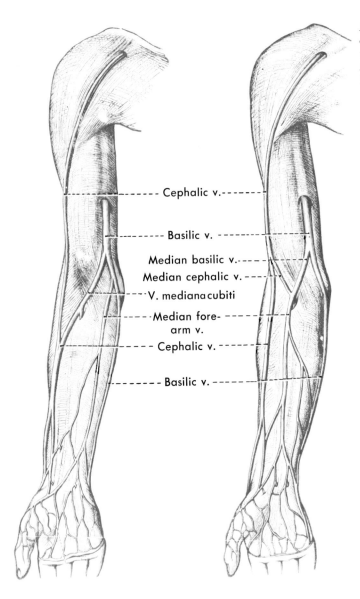

Fig. 3-45. Two patterns of the superficial veins of the upper limb, anterior view. (Toldt C: An Atlas of Human Anatomy [ed 2], Vol 2. New York, Macmillan, 1928)

Cephalic v.

Basilic v.

Median basilic v.

Median cephalic v.

V. mediana cubiti

Median fore-arm v.

Cephalic v.

Basilic v.

superficial lymphatics are numerous (Figs. 3-47 and 3-48), and tend to accompany the superficial veins. Those on the palmar surfaces of the digits and most of those in the palm of the hand drain around the borders of the fingers or of the hand to join dorsally situated vessels, so that most of the drainage from the hand is along the posterior surface of the forearm, although a few vessels drain upward along the anterior surface. As they approach the region of the elbow the vessels running posteriorly pass around the radial and ulnar borders of the forearm to run upward in front of the elbow and along the medial side of the arm, gradually converging toward the axillary lymph nodes. In this course they are joined by vessels from the back of the arm, the more lateral of which drain around the lateral aspect of the arm, the more medial around its medial aspect. A pathway along the cephalic vein that bypasses most of the axillary nodes can also often be demonstrated by lymphangiography and helps to account for the fact that removal of axillary lymph nodes

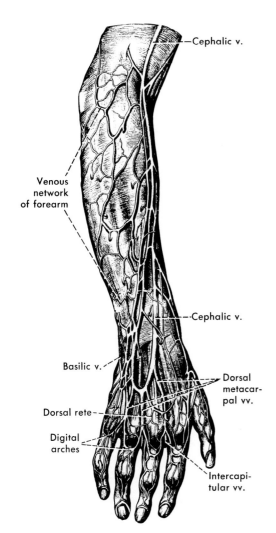

Cephalic v.

Venous
network
of forearm

Cephalic v.

Basilic v.

Dorsal
metacar-
pal vv.

Dorsal rete

Digital
arches

Intercapi-
tular vv.

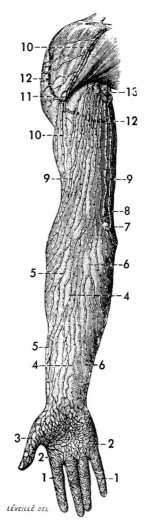

LÉVEILLÉ DEL

Fig. 3-46. Superficial veins of the upper limb, posterior view. (Toldt C: An Atlas of Human Anatomy [ed
2], Vol 2. New York, Macmillan, 1928)

Fig. 3-47. Superficial lymphatics of the upper limb,
anterior view. The numbers largely identify the obvious: for instance, *1, 1* are lymphatics of the fingers, *2,
2* those of the palm. *7* and *8* are supratrochlear nodes,
13 axillary ones. (After Sappey, from Poirier P,
Charpy A: Traité d'Anatomie Humaine [ed 2], Vol 2,
Fasc 4. Paris, Masson et Cie, 1909)

is not necessarily followed by edema of
the limb. The superficial lymphatics from
the region of the shoulder drain for the
most part downward toward the axilla.
Frieden has discussed the lymphatic drainage
from the standpoint of dissection of
lymph nodes in malignant neoplasms of
the limb.

Almost all the lymph nodes of the upper
limb are concentrated in the axilla (Chap. 4).
One or more superficially located lymph
nodes (supratrochlear nodes) are however reg-

ularly found a little above the medial epicondyle of the humerus, lying on the basilic vein;
deep lymph nodes are occasionally found in
the forearm and at the elbow (cubital nodes)
along the courses of the radial, ulnar, and interosseous arteries, and the lower end of the
brachial artery. Most of the drainage from
the forearm passes around rather than
through these inconstant nodes.

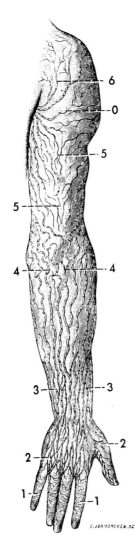

E. JEB MONCKEN, SC

Fig. 3-48. Superficial lymphatics of the upper limb, posterior view. As in the preceding figure, the numbers largely identify the obvious: for instance, *1, 1* are lymphatics of the dorsum of the fingers, *4, 4,* and *5, 5* call attention to the divergence of the lymphatics of the back of the forearm and arm; some of these pass laterally, some medially, around the borders of the limb. (After Sappey, from Poirier P, Charpy A: Traité d'Anatomie Humaine [ed 2], Vol 2, Fasc 4. Paris, Masson et Cie, 1909)

References

BARDEEN CR, LEWIS WH: Development of the limbs, body-wall and back in man. Am J Anat 1:1, 1901

BARNES R: Traction injuries of the brachial plexus in adults. J Bone Joint Surg 31-B:10, 1949

BROOKS DM: Open wounds of the brachial plexus. J Bone Joint Surg 31-B:17, 1949

CAMP JD, CILLEY EIL: Diagrammatic chart showing time of appearance of the various centers of ossification and period of union. Am J Roentgenol 26:905, 1931

DAVIS L, MARTIN J, PERRET G: The treatment of injuries of the brachial plexus. Ann Surg 125:647, 1947

DEJONG RN: Syndrome of involvement of the posterior cord of the brachial plexus. Arch Neurol Psychiat 49:860, 1943

FIELD JH, KRAG DO: Congenital constricting bands and congenital amputation of the fingers: Placental studies. J Bone Joint Surg 55-A:1035, 1973

FLECKER H: Time of appearance and fusion of ossification centers as observed by roentgenographic methods. Am J Roentgenol 47:97, 1942

FRANTZ CH, O'RAHILLY R: Congenital skeletal limb deficiencies. J Bone Joint Surg 43-A:1202, 1961

FRIEDEN JH: The regional lymph node dissection in cancer of the extremities. Surg Gynecol Obstet 89:591, 1949

FROHSE F, FRÄNKEL M: Die Muskeln des menschlichen Armes, Vol 2, Section 2, Part 2 in von Bardeleben K (ed): Handbuch der Anatomie des Menschen Jena, Fischer, 1908

GARDNER E, GRAY DJ: Prenatal development of the human shoulder and acromioclavicular joints. Am J Anat 92:219, 1953

GREULICH WW, PYLE SI: Radiographic Atlas of Skeletal Development of the Hand and Wrist. Stanford, California, Stanford University Press, 1950

HARRIS W: The true form of the brachial plexus, and its motor distribution. J Anat Physiol 38:399, 1904

HASAN M, NARAYAN D: A single cord human brachial plexus. J Anat Soc India 13:103, 1964

HORWITZ MT, TOCANTINS LM: An anatomical study of the role of the long thoracic nerve and the related scapular bursae in the pathogenesis of local paralysis of the serratus anterior muscle. Anat Rec 71:375, 1938

KERR AT: The brachial plexus of nerves in man, the variations in its formation and branches. Am J Anat 23:285, 1918

KINO Y: Clinical and experimental studies of the congenital constriction band syndrome, with an emphasis on its etiology. J Bone Joint Surg 57-A:636, 1975

KOLODNY A: Traction paralysis of the brachial plexus. Am J Surg 51:620, 1941

LEWIS WH: The development of the arm in man. Am J Anat 1:145, 1902

LINELL EA: The distribution of nerves in the upper limb, with reference to variabilities and their clinical significance. J Anat 55:79, 1921

MALL FP: On centers of ossification in human embryos less than one hundred days old. Am J Anat 5:433, 1906

MILLER JA, JR: Joint paracentesis from an anatomic point of view: I. Shoulder, elbow, wrist and hand. Surgery 40:993, 1956

MILLER RA: Observations upon the arrangement of the axillary artery and brachial plexus. Am J Anat 64:143, 1939

PATERSON RS: A radiological investigation of the epiphyses of the long bones. J Anat 64:28, 1929

RAY BS, HINSEY JC, GEOHEGAN WA: Observations on the distribution of the sympathetic nerves to the pupil and upper extremity as determined by stimulation of the anterior roots in man. Ann Surg 118:647, 1943

SINGER E: Human brachial plexus united into a single cord: Description and interpretation. Anat Rec 55:411, 1933

STREETER GL: Focal deficiencies in fetal tissues and their relation to intra-uterine amputation. Contrib Embryol 22:1, 1930

SUNDERLAND S: The distribution of sympathetic fibres in the brachial plexus in man. Brain 71:88, 1948

TODD TW, D'ERRICO J, JR: The clavicular epiphyses. Am J Anat 41:25, 1928

TRACY JF, BRANNON EW: Management of brachial-plexus injuries (traction type). J Bone Joint Surg 40-A:1031, 1958

WALSH JF: The anatomy of the brachial plexus. Am J M Sc 74:387, 1877

WHITTAKER CR: The arrangement of the bursae in the superior extremities of the fulltime foetus. J Anat Physiol 44:133, 1910

WICKSTROM J, HASLAM ET, HUTCHINSON RH: The surgical management of residual deformities of the shoulder following birth injuries of the brachial plexus. J Bone Joint Surg 37-A:27, 1955

WOOLLARD HH: The development of the principal arterial stems in the forelimb of the pig. Contrib Embryol 14:139, 1922

WOOLLARD HH, WEDDELL G: The composition and distribution of vascular nerves in the extremities. J Anat 69:165, 1935

Chapter 4
PECTORAL REGION, AXILLA, AND SHOULDER

Pectoral or Shoulder Girdle

The pectoral or shoulder girdle is formed by two pairs of bones, the clavicles, which are subcutaneous (Fig. 4-1), and the scapulas, more nearly covered by muscle but each with a spine and acromion that are subcutaneous.

Most of the muscles inserting on the scapula or across it into the humerus tend to retract it, thus adducting the arm; the clavicle is therefore particularly important as a brace that keeps the shoulder joint far enough laterally to allow movements of the arm. However, movements of the scapula are important in movements of the arm, for the limited excursion of the free limb at the scapulohumeral joint is under normal circumstances added to by movements of the scapula itself; forward, upward, and downward movements of the arm at the shoulder are typically accompanied by a turning of the glenoid cavity in the corresponding direction. The mobility of the scapula is in turn dependent in part upon the mobility of its one bony brace, the clavicle.

THE CLAVICLE AND ITS JOINTS

Viewed from above (Fig. 4-2) or below (Fig. 4-3), the clavicle presents somewhat the appearance of the italic letter *f*, with the concavity of the medial curve being directed pos-
teriorly, that of the lateral curve anteriorly. Its lateral or acromial end is somewhat flattened and, in addition to rather obvious upper and lower surfaces, presents anteriorly a small articular surface for the acromion of the scapula, and on its lower surface a conoid tubercle and a trapezoid line (both once "coracoid tubercle") for the attachment of the two parts of the coracoclavicular ligament. The medial two thirds, extending to its sternal end, is more rounded, and its expanded sternal end presents a sternal articular surface for the manubrium sterni, and on its lower surface the roughened impression for the costoclavicular ligament; the middle third of the bone is more rounded than the proximal third.

Abnormalities and Fracture

As noted in the preceding chapter, the body of the clavicle arises from two centers of ossification; Taylor described from roentgenograms nonunion of medial and lateral parts of the clavicle (clavicular dysostosis), presumably a result of nonunion of the two centers of ossification. Very rare is cleidocranial dysostosis, a partial or complete absence of the clavicle, associated with defective ossification of bones of the skull.

Fractures of the clavicle are relatively common, accounting, according to McCally and Kelly, for some 8 to 10% of all fractures of bones. Nonunion is rare, but occurred in the

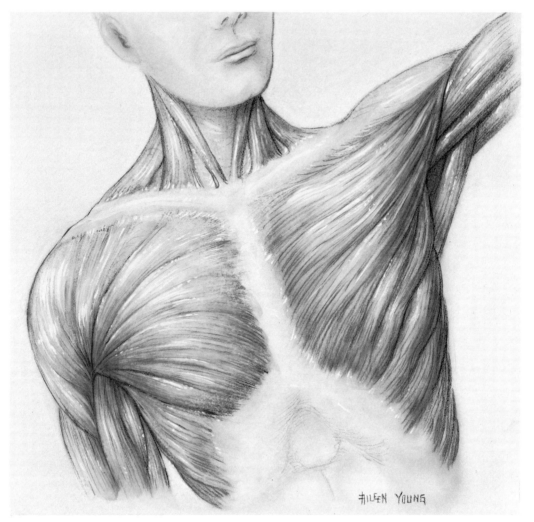

Fig. 4-1. Anterior view of musculature about the shoulder. The sternocleidomastoid with its two heads of origin and the anterior border of the trapezius appear in the neck. The pectoralis major covers most of the thoracic wall, and is bordered by the deltoid over the prominence of the shoulder and by the serratus anterior below. In the lifted arm, the posterior wall of the axilla appears.

middle third of the bone in more than half of the 69 cases Johnson and Collins reviewed. Howard and Shafer reported a few cases of neurovascular compression following injuries to the clavicle.

Sternoclavicular Joint

The sternoclavicular joint (Fig. 4-4) is formed between the clavicular notch in the upper corner of the manubrium sterni and the sternal end of the clavicle. The articular surface of the latter is a distorted oval, for its inferior border is flattened, its anterior border somewhat asymmetrically rounded, and its posterosuperior and inferior borders meet in a rounded postero-inferior angle. Since the anteroposterior and the vertical dimensions of the sternal end of the clavicle are considerably greater than the dimensions of the clavicular notch, the clavicle projects posteriorly well beyond the posterior border of the sternum, and superiorly well above the upper border. The latter projection helps to form, with the jugular (suprasternal) notch of the sternum and the clavicle of the other side, the jugular or suprasternal fossa.

The anterior part of the articular surface of the clavicle tends to be somewhat convex, the posterior part concave; this surface is directed medially, downward, and forward. The approximately oval articular surface of the clavicular notch of the sternum is directed somewhat upward, posteriorly, and laterally. The sternal head of the sternocleidomastoid muscle lies in front of the joint, the lower ends of the sternohyoid and sternothyroid muscles behind it.

The sternoclavicular joint usually contains two separate articular cavities, completely divided by an articular disk; this disk is sometimes perforated in its center. It is strongly attached above to the sternal end of the clavicle and to the interclavicular ligament, while below it is likewise strongly attached to the cartilage of the first rib as this attaches to the sternum. Around the remainder of its periphery it is attached to the capsule of the joint. Upward and downward movements of the clavicle take place primarily between the clavicle and the disk; rotatory ones that accompany other movements, and backward and forward movements, involve movement of both the clavicle and disk on the sternum.

Up-and-down movement of the clavicle is much freer than is anteroposterior movement. The former has a total excursion of about 60°, according to Steindler (somewhat less according to Inman and colleagues, and almost all of it occurring during the first 90° of elevation of the arm); the latter amounts to only 20° to 30° when the movement is a circling one, and still less if it is strictly in the anteroposterior plane. Rotation of the clavicle about its longitudinal axis can amount to 30° or so.

The capsule of the sternoclavicular joint is strengthened by anterior and posterior ster-

Fig. 4-2. The right clavicle from above. Muscular attachments, origins in *red* and insertions in *black*, are shown in *b*.

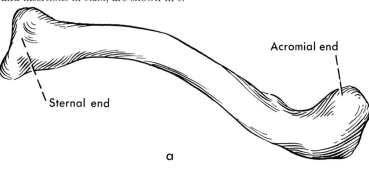

Acromial end

Sternal end

a

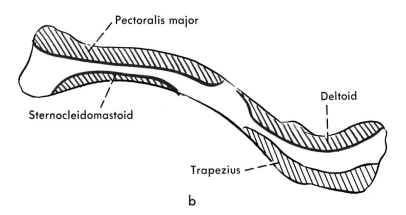

Pectoralis major

Sternocleidomastoid

Deltoid

Trapezius

b

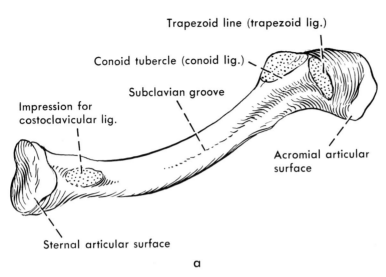

Trapezoid line (trapezoid lig.)

Conoid tubercle (conoid lig.)

Subclavian groove

Impression for
costoclavicular lig.

Acromial articular
surface

Sternal articular surface

a

Fig. 4-3. The right clavicle from below. Attachments of ligaments are shown in *a;* those of muscles, as in Fig. 4-2, in *b.*

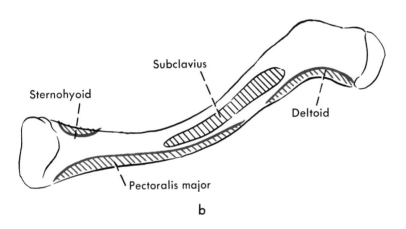

Subclavius

Sternohyoid

Deltoid

Pectoralis major

b

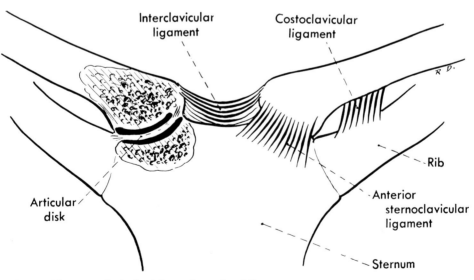

Interclavicular
ligament

Costoclavicular
ligament

Rib

Articular
disk

Anterior
sternoclavicular
ligament

Sternum

Fig. 4-4. The sternoclavicular joint and associated ligaments.

noclavicular ligaments; the posterior one is thicker than the anterior, this accounting for the greater frequency of anterior dislocations at this joint. The fibers of these ligaments slant predominantly downward and inward, and therefore tend to oppose retraction of the clavicle from its articulation with the sternum, or levering upward of its sternal end when the bone is brought against the cartilage of the first rib in depression of the shoulder. Bearn found, however, that when the shoulder is depressed it is the posterosuperior fibers of the capsule of the joint, presumably including the upper fibers of these ligaments, that support the weight and maintain the clavicle in position. Cutting other ligaments or the articular disk in skeletal preparations had no effect on the resistance to dislocation from a downwardly directed force, but cutting the capsule led to tearing the disk from the first costal cartilage. Lateral displacement of the clavicle from the sternum is also opposed by an interclavicular ligament, which passes between the two clavicles and the capsules of the two joints; upward and lateral dislocation of the clavicle is likewise resisted by the costoclavicular ligament, which arises from the cartilage and distal end of the first rib and passes upward, laterally, and slightly backward to insert into the inferior surface of the clavicle lateral to the attachment of the capsule of the sternoclavicular joint. According to Bearn's observations, none of these are as important as the upper part of the capsule itself. Another factor tending to prevent upward dislocation of the sternal end of the clavicle through forcible contact with the first rib is the elasticity of this rib's costal cartilage.

Denham and Dingley reported four cases in which dislocation of the medial end of the clavicle involved epiphyseal separation; in the three cases that required open reduction, the sternoclavicular joint was found to be intact.

The blood supply to the sternoclavicular joint is usually through twigs from the clavicular branch of the thoracoacromial artery, from the internal thoracic (internal mammary), and sometimes from the suprascapular or other vessels close to the joint. Its nerve supply is said to be from the nerve to the sub-clavius and from the medial (anterior) supraclavicular nerve.

Acromioclavicular Joint

The acromioclavicular joint (Fig. 4-5) affords a relatively small surface of contact between the clavicle and acromion. The synovial cavity between these two surfaces is partly or, rarely, completely subdivided by an articular disk. The articular capsule is relatively lax, thus permitting the necessary movement between scapula and clavicle, especially important in rotation of the scapula. The fibers run medially, backward, and upward from the acromion to the clavicle and thus tend to resist lateral rather than medial displacement of the scapula upon the clavicle; on the anterior surface of the joint they are attached in the immediate vicinity of the borders of the joint, but posteriorly they extend approximately 0.75 inch medially along the clavicle. The upper part of the capsule is reinforced by the acromioclavicular ligament.

Movements at the acromioclavicular joint are all gliding ones, and are limited to adjustments between the scapula and clavicle in movements involving them both. The greatest movement is in upward and downward rotation of the scapula; during abduction or flexion of the arm (accompanied by upward scapular rotation) it apparently amounts to only about 20°, and most of this takes place during the early part and again at the last of the movement (Inman and colleagues). Very slight movements forward and backward at the acromioclavicular joint allow the medial margin (vertebral border) to follow the curvature of the thoracic cage while the lateral angle follows the different curve imposed by the clavicle; even more slight up-and-down movements allow the inferior angle of the scapula to retain its contact with the thoracic wall as the clavicle and scapula are elevated and depressed.

The blood supply to the acromioclavicular joint is from the network of small vessels (acromial rete) over the acromion; usual tributaries to this network are the suprascapular (transverse scapular) artery, the posterior humeral circumflex, and the acromial branch of the thoracoacromial artery. The nerve supply

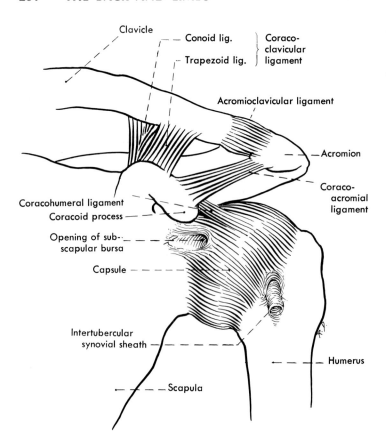

Clavicle

Conoid lig.

Coraco-
clavicular
ligament

Trapezoid lig.

Acromioclavicular ligament

Acromion

Coraco-
acromial
ligament

Coracohumeral ligament

Coracoid process

Opening of sub-
scapular bursa

Capsule

Intertubercular
synovial sheath

Humerus

Scapula

Fig. 4-5. Ligaments of the acromial end of the clavicle and the anterior aspect of the shoulder joint.

is said to be from the pectoral, suprascapular, and axillary nerves.

Coracoclavicular Ligament

The acromioclavicular joint is not the strongest union between the clavicle and scapula; its function is to provide mobility between these bones. The stronger union between the two bones is provided by the coracoclavicular ligament, which consists of two parts, the trapezoid and the conoid ligaments (Fig. 4-5).

The trapezoid ligament gains its name from its shape, and is the more anterior and lateral part of the coracoclavicular ligament. It is almost an inch wide and is attached below to the anterior part of the upper surface of the coracoid process from its angle well out toward its tip. From this attachment the fibers run upward, anteriorly, and somewhat laterally to attach to the trapezoid line on the inferior surface of the clavicle. Its lateral free edge is also most anterior, so that the anterior

surface of the ligament looks not only anteriorly but also downward and medially.

The conoid ligament is the posteromedial component of the coracoclavicular ligament. It arises from a small area on the upper surface of the angle of the coracoid process, and passes upward and slightly backward, expanding in circumference as it does so, to attach to the conoid tubercle on the under surface of the clavicle at the posteromedial end of the trapezoid line.

Lewis found inconstant bursae between layers of the ligaments, but said the major one here usually lies between the two and may extend over the coracoid process to furnish a synovial lining for areas of contact ("coracoclavicular joint") between that process and the clavicle.

The two ligaments are powerful ones, and together insure that forward, backward, and downward movements of the scapula, not adequately resisted by the acromioclavicular joint, are accompanied by movements of the clavicle. Both ligaments, but especially the trapezoid because of the direction of its fibers, tend to prevent medial displacement of the acromion beneath the clavicle.

The coracoclavicular ligament as a whole, therefore, markedly strengthens the union between scapula and clavicle. The importance of the limited movement allowed by this ligamentous union is, however, emphasized by the observation of Inman and co-workers, who deduced that the necessary relaxation of the ligament is brought about by rotation of the clavicle, and found that if such rotation is prevented abduction of the arm is limited to 110°.

Dislocation of the Clavicle

Dislocation at the acromioclavicular joint ("shoulder separation") producing an abnormal overriding of the acromion by the clavicle, is a not infrequent accident. When there is a complete dislocation (Fig. 4-6) the acromioclavicular ligament is torn, as are the deltoid and trapezius muscles; the coracoclavicular ligament is usually torn, but may not be (Urist, Horn). Treatment has varied widely. Imatani and co-workers obtained similar results in patients treated nonoperatively and those treated by operation; they advised minimal immobilization and early rehabilitation. Urist presented arguments against open reduction, including the possibility of producing degenerative changes by transfixing the joint. Horn noted that closed reduction may not be possible, since the joint may have to be cleared of a partly extruded meniscus or of a part of the trapezius, and he said that if power is to be restored to shoulder movements the coracoclavicular ligament and the torn deltoid and trapezius muscles must be repaired.

Dislocation of the clavicle at the sternoclavicular joint is less common than dislocation at the acromioclavicular joint (both being less common than fractures of the clavicle), and is most often an anterior dislocation. It may however be a superior one or, rarely, a posterior one (Kennedy). Kennedy pointed out the dangers of posterior dislocation, because of

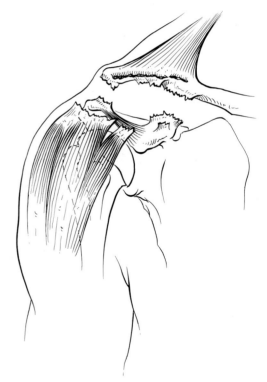

Fig. 4-6. Dislocation at the acromioclavicular joint, with disruption of the acromioclavicular and coracoclavicular ligaments, and of the clavicular origin of the deltoid. (Courtesy of Dr. P. J. Kelly)

the proximity of the trachea and of the great vessels of the upper mediastinum. He said that death had been reported as resulting from tracheal laceration consequent to such a dislocation.

Cyriax reported as "floating clavicle" cases in which the sternal end of the clavicle was abnormally mobile. He thought the condition to be acquired rather than congenital, as the articular surfaces of both clavicle and sternum seemed to be normal. In an extreme case the sternal ends of both clavicles separated about 1.25 inches when the shoulders were thrown back but came almost together when the shoulders were brought forward. He found the condition to be more common unilaterally. In this condition, the medial displacement of the end of the clavicle is halted not by impingement of the clavicle, disk, and sternum, as usual, but by the joint capsule.

Duggan reported a case of recurrent dislo-

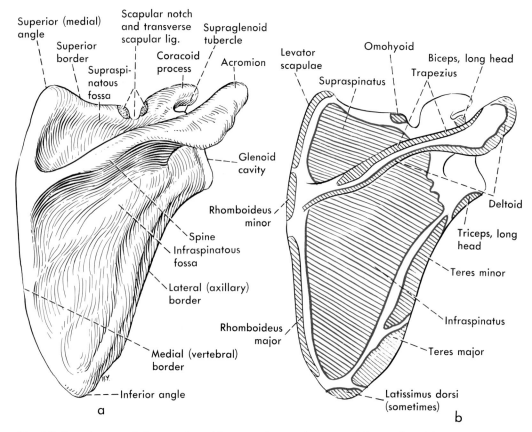

Fig. 4-7. Dorsal aspect of the scapula and its muscular attachments; origins are in *red*, insertions *black*.

cation of the sternoclavicular cartilage, which had to be removed; he said the condition is rare.

SCAPULA

The scapula (Figs. 4-7 and 4-8) has a thin superior (medial) angle and thin superior and medial (vertebral) borders, but its inferior angle and its lateral (axillary) border are thickened for the attachment of muscles. The curved coracoid process projects upward, forward, and laterally from the superior border of the scapula, and just medial to the base of the coracoid process is the scapular notch (scapular incisure, superior scapular notch) which is, in the fresh condition, bridged by the superior transverse scapular ligament (transverse scapular or suprascapular liga-

ment). This ligament converts the notch into a foramen that transmits the suprascapular nerve and sometimes, although rarely, the suprascapular (transverse scapular) vessels.

The costal surface of the scapula, forming the subscapular fossa, is flat or slightly concave; its dorsal surface is divided by the scapular spine into the small supraspinatous (supraspinous) and the larger infraspinatous (infraspinous) fossae. The acromion continues the spine, but has no direct attachment to the body of the scapula; the gap between it and the neck of the scapula, through which the suprascapular nerve and vessels pass from the supraspinatous fossa to the infraspinatous one, is sometimes called the greater scapular or spinoglenoid notch. (There may be here an inferior transverse scapular ligament that produces a foramen for the nerve and vessels.) The flattened free end of the acromion bears on its anteromedial border the facet for artic- ulation with the clavicle, and the triangular coracoacromial ligament is attached to its tip. This ligament arises from most of the length of the horizontal portion of the coracoid pro- cess and completes the coracoacromial arch over the shoulder, separating the overlying deltoid muscle from the underlying supra- spinatus muscle and capsule of the shoulder joint.

The major part of the scapula receives its blood supply through numerous twigs from the vessels supplying the muscles covering it: the subscapular, suprascapular, and circum- flex scapular arteries. The acromion obtains its blood supply largely from the acromial branch of the thoracoacromial artery.

Fractures and Anomalies

In spite of its thinness, the body of the scap- ula, protected as it is both by muscle and by its apposition to the thoracic wall, is rarely fractured. McCally and Kelly noted that most fractures of the scapula are of the acro-

mion, although sometimes they involve the spine or even the body; they estimated the in- cidence of fracture of the scapula as being probably less than 1% of all fractures.

Three interesting anomalies of the scapula are nonunion of the acromion, nonunion of the coracoid, and undescended (congenitally elevated) scapula.

The major part of the acromion develops from a separate center of ossification (Chap. 3), and may fail to fuse with the rest of the bone. The epiphyseal line may then simulate a fracture on roentgenographic examination. Liberson ('37) reported that 62% of the people who have this condition show it bilaterally, and that it occurs in from 1% to 2.7% of per- sons. Other workers, apparently, have re- ported its occurrence in from 3% to 15% of persons.

The major portion of the coracoid process develops from a single center of ossification (Chap. 3) and typically unites with the scap- ula at about the fifteenth year; nonunion of the coracoid to the scapula is much less fre-

Fig. 4-8. Costal surface of the scapula.

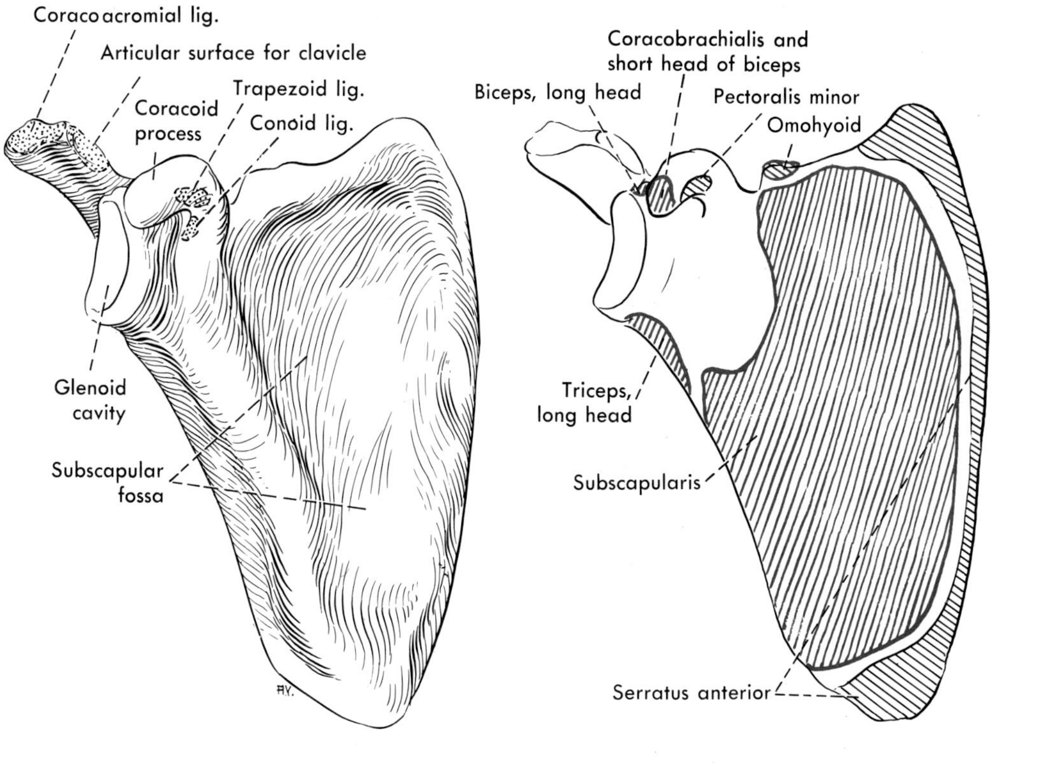

quent than is nonunion of the acromion, but does occur occasionally.

The condition described as undescended scapula or congenital elevation of the scapula, also sometimes known as Sprengel's deformity, is brought about by the attachment of the scapula to the cervical vertebral column through bone, cartilage, or fibrous tissue. This attachment apparently occurs early in development, and prevents the normal descent of the scapula to its usual thoracic level. Jeannopoulos has reported 35 cases; he found that there are always, apparently, associated abnormalities in the cervical or thoracic vertebrae or the thoracic cage. In 11 of his 35 patients the connection between the scapula and the cervical vertebral column was largely bony (such a connection is usually called the omovertebral bone; Woodward said it occurs in about a third of cases). Apparently the "omovertebral bone" is regularly fused to the cervical vertebrae, but the attachment to the scapula varies: Jeannopoulos found bony continuity, cartilaginous union, union by means of a fibrous band, and union by a synovial joint. In discussing operative procedures, he noted that the scapula is not only elevated by the upper and middle fibers of the trapezius, by the levator scapulae, and by the rhomboids, but also rotated downward by the levator and rhomboids; after the connection to the vertebral column has been severed, the usual procedure has been to remove the scapular insertions of these muscles in order to allow the scapula to be placed in an approximately normal position. Woodward reported moving the origins of these muscles downward, instead of resecting them from the scapula. Jeannopoulos noted also that since the scapula is frequently hypoplastic, lowering it to a level at which its inferior angle corresponds to the inferior angle of the opposite side may constitute an overcorrection. In two patients of the series he reviewed, paralysis of the upper extremity followed the repositioning of the scapula, thus emphasizing the danger of severe damage to the brachial plexus if the shoulder is lowered too much. For this reason, some workers regard the more satisfactory procedure as often being simple resection of the omovertebral bone.

SHOULDER JOINT

The shallowly convex glenoid cavity of the scapula is somewhat increased in area by the glenoidal labrum (glenoid lip), a rim of fibrocartilage attached to the periphery of the cavity. Even with this addition, only a small part of the head of the humerus can be in the glenoid cavity at any instant. The glenoid labrum is approximately 0.5 cm wide above and below the glenoid cavity, somewhat less anterior and posterior to it. Recurrent dislocation of the shoulder rather regularly has associated with it partial detachment of the labrum from the bony scapula, but there is no complete agreement as to whether this detachment is or is not a principal cause of recurrence.

Surgical approaches to the shoulder joint usually involve reflection of the deltoid muscle, since this muscle lies anterior, lateral, and posterior to the joint. In an anterior approach, a small anterior part of the deltoid may be split off and retracted medially with the cephalic vein and pectoralis major; the remaining anterior part of the muscle is then separated from the clavicle as necessary. Leslie and Ryan obtained more satisfactory exposure through an axillary approach in which they retracted the deltoid laterally, and partially or totally detached the pectoralis major at its insertion. Section of the coracoid process and reflection of this process downward with the muscles attaching to it allow better exposure of the anterior aspect of the shoulder. A limited lateral approach can be obtained by splitting the lateral part of the deltoid for about 1.5 to 1.75 inches below the acromion (no more, because of the position of the axillary nerve); a better one is obtained by an inverted U incision over the shoulder, sectioning the origin of the acromion process from the scapular spine and disarticulating the acromioclavicular joint, splitting the deltoid at the levels of bone detachment, and reflecting bone and upper deltoid downward. Rather good exposure can also be obtained without sectioning bone, by reflecting the deltoid from clavicle and acromion through an inverted U incision. In a posterior ap-

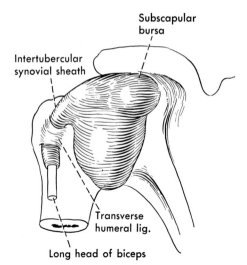

Fig. 4-9. Anterior view of the synovial membrane of the shoulder joint.

proach, the deltoid can be detached from its origin on the spine of the scapula, or a more extensive exposure can be obtained by detaching both the trapezius and the deltoid from the spine, the former from the acromion, sectioning the acromion, and reflecting the deltoid and acromion.

Synovial Membrane

The synovial membrane of the shoulder joint (Figs. 4-9 and 4-10) arises from the edge of the glenoidal labrum and extends around the head of the humerus. Anteriorly and posteriorly it attaches to the humerus at about the epiphyseal line (anatomical neck), and thus to the margin of the articular cartilage. On the medial side of the humerus, however, it attaches approximately a centimeter below the epiphyseal line, and is then reflected upward along the bone to the articular cartilage; likewise it attaches about a centimeter lateral to the epiphyseal line over the upper part of the greater tubercle, and is then reflected back along the bone to the articular cartilage. Further, the synovial membrane passes downward as a diverticulum between the greater and lesser tubercles, here forming a tubular sheath, the *intertubercular synovial sheath*, that surrounds the tendon of the long head of the biceps. The intertubercular synovial sheath is attached posteriorly to the floor of the intertubercular groove, and on its sides

to the greater and lesser tubercles and their crests.

Through a gap in the anterior wall of the articular capsule the synovial membrane usually protrudes between the neck of the scapula and the subscapularis muscle to become continuous with the synovial membrane of the subscapular bursa, which is then some-

Fig. 4-10. Anterior, *a,* and posterior, *b,* views of the attachment of the capsule of the shoulder joint. That of the fibrous capsule is indicated by *broken lines,* that of the synovial by *red lines; red shaded areas* indicate reflection of the latter over the surfaces of the bones.

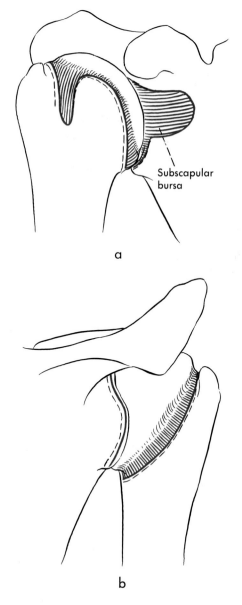

times called the subscapular recess. The position of the opening of this bursa into the shoulder joint varies; DePalma and co-workers found it sometimes above, sometimes below, the middle glenohumeral ligament, sometimes double, and sometimes missing (see Fig. 4-12). Horwitz and Tocantins ('38a) found a subscapular bursa in only 66 of 75 shoulders, and in only 49 of these did it communicate with the shoulder joint. It is said that sometimes the synovial membrane is continuous posteriorly with a bursa deep to the infraspinatus tendon.

Fibrous Capsule and Ligaments

The fibrous articular capsule of the shoulder joint is loose, thus allowing free mobility of the scapulohumeral joint; the strength of this joint is provided not by the capsule and its associated ligaments, but rather by the closely adjacent musculotendinous cuff (rotator cuff) formed by the tendons of the subscapularis, supraspinatus, infraspinatus, and teres minor muscles.

The capsule is generally rather thin, and on the scapula is attached (Fig. 4-10) to the roughened area immediately around the glenoid cavity except at the lower lip where it

Fig. 4-11. The interior of the shoulder joint viewed from behind and below to show the glenohumeral ligaments. The tendon of the long head of the biceps traverses the upper part of the joint cavity.

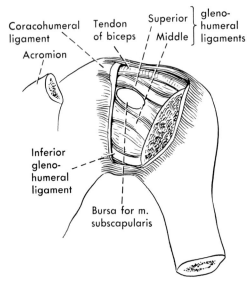

may extend medially several millimeters before attaching to the bone. As it leaves the bone it is closely attached to the glenoidal labrum, and therafter becomes intimately associated with the synovial layer of the shoulder joint. Its lines of attachment to the humerus are similar to those already described for the synovial layer, except that it attaches to the bone where it first meets it, having no upward reflection to the edge of the articular cartilage. Its prolongation downward over the intertubercular synovial sheath is thin, and attaches to the lesser and greater tubercles. As already mentioned in connection with the description of the synovial membrane, the fibrous capsule usually presents a gap anteriorly for the communication between the joint cavity proper and the subscapular bursa, and may occasionally be perforated posteriorly.

The most important and constant thickening of the fibrous capsule of the shoulder joint is the *coracohumeral ligament* (Fig. 4-5), a rather strong band that arises proximally from most of the lateral edge of the coracoid process and extends over the top of the shoulder joint to attach to the greater tubercle. The anterior edge of the ligament forms a prominent thickening of the capsule of the shoulder joint, but the posterior edge blends insensibly with the capsule. It is said that there is sometimes a bursa between it and the capsule close to its attachment to the coracoid process. With the arm hanging by the side, the upper part of the capsule and the coracohumeral ligament become taut enough to maintain the humeral head in position with no muscular action, according to Basmajian and Bazant; with abduction the capsule and ligament necessarily become lax and thus lose their ability to support the humerus.

The other ligaments described in connection with the articular capsule are the *glenohumeral ligaments* or bands, which are variable thickenings on the interior surface of the anterior wall of the articular capsule (Figs. 4-11 and 4-12). These ligaments are therefore visible only when the capsule is opened and viewed from behind. The thickenings may be very slight, or may be fairly well marked; however, while they tend to check lateral rotation, and the superior one especially should help to suspend the humerus, it is doubtful that they add much strength to the shoulder joint.

The inferior glenohumeral ligament is usually said to be the best developed of the three, but DePalma and his co-workers described it as being often indistinct, and no more than a diffuse thickening of the capsule; they said it was well defined in 54, poorly defined in 18, and absent in 24 shoulders. It passes from about the middle of the anterior margin of the glenoid labrum to the lowest point (on the medial side) of the neck of the humerus.

The middle glenohumeral ligament is attached to the upper border of the margin of the glenoid cavity and labrum and to the root of the coracoid process, close to the supraglenoid tubercle, and extends across the joint to attach to the front of the lesser tubercle. DePalma and his co-workers found this ligament well defined in 68 specimens, poorly developed in 16, and absent in 12; they said that it sometimes had no attachment to the labrum, sometimes attached only to this. Above or below the middle glenohumeral ligament, or above and below, the subscapular bursa usually opens into the cavity of the shoulder joint (Fig. 4-12).

The superior glenohumeral ligament is attached above at about the same level as is the middle one, just anterior to the supraglenoid tubercle (origin of the long tendon of the biceps); it approximately parallels this tendon, and attaches to the upper surface of the lesser tubercle of the humerus. DePalma and his co-workers found it to be the most distinct of the three, in their series; it was present in 94 of 96 specimens, absent in only 2. They noted that it varies considerably in its upper attachment, which may be to the biceps tendon, the middle ligament, and the labrum, or to either the biceps tendon or the middle ligament with no attachment to the labrum.

The *transverse humeral ligament* (Fig. 4-9) is composed of a few transverse fibers of the fibrous capsule, extending between the two tubercles across the front of the upper end of the intertubercular synovial sheath. It is often poorly developed, but must be ruptured be-

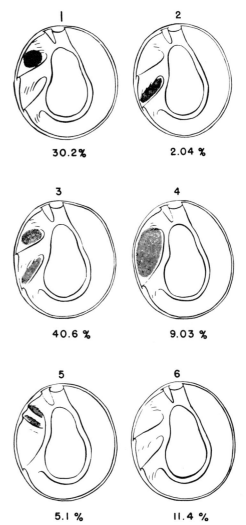

Fig. 4-12. Variations in the opening of the subscapular bursa (subscapular recess). In *1* the opening is between the superior and middle glenohumeral ligaments, in *2* between the middle and the inferior, and in *3* there are openings in both locations; in *4* a very large recess lies between the superior and inferior glenohumeral ligaments, and the middle ligament is missing; in *5* there are two slight dimples associated with a thin synovial fold representing the middle glenohumeral ligament, and in *6* there is no recess or opening into a bursa. (DePalma AF, Callery G, and Bennett GA: Am Acad Orth Surgeons, Instructional Course Lectures 6:255, 1949)

fore the long tendon of the biceps can be displaced. It is usually regarded as being the chief retinaculum for this tendon, but Abbott and Saunders regarded the chief retinaculum below the summits of the tubercles as being a tendinous expansion, attached to both lips of the groove, from the insertion of the sterno-costal portion of the pectoralis major.

Blood and Nerve Supply

The blood supply of the shoulder joint is from the closely associated arteries, the chief ones on the scapular side of the joint being the suprascapular (transverse scapular) and subscapular arteries (and the latter's circumflex branch), those on the humeral side being the anterior and posterior humeral circumflex arteries. The nerve supply, somewhat variable, is commonly from several sources: Gardner found that contributions from the axillary, suprascapular, and lateral pectoral nerves are apparently constant, and that there may be a twig from the posterior cord or one from the radial nerve. The distribution of nerves to the shoulder joint, and some of the variations in these nerves, is shown in Figure 4-13. Gardner noted also that the stellate ganglion and perhaps other sympathetic ganglia send direct twigs to the shoulder joint; these are presumably vasomotor rather than afferent.

Developmental Defects

Abnormalities of the shoulder joint may be developmental, but far more commonly result from degenerative processes and trauma, or a combination of the two. Developmental defects of the skeletal elements of the joint are apparently rare: Owen reported the cases of 5 patients with bilateral hypoplasia of the glenoid articular surface, which was accompanied by flattening of the humeral head; he regarded this hypoplasia as resulting from improper development of the inferior center of ossification that normally contributes to the margins of the glenoid cavity, appearing first at about the age of 13 years. In some of the individuals with this abnormality there were other associated skeletal abnormalities.

Andreasen reported 2 cases in which the head of the humerus was absent, apparently as a result of lack of development of the epiphysis of this head; the associated glenoid cavity was also rudimentary. He found only 6 similar cases previously described; while other skeletal defects were said to be usually associated with this condition, the maldevelop-ment of the humeral head and of the glenoid cavity were the only ones present in his cases.

Dislocation

Acquired abnormalities of the shoulder joint include ruptures of tendons closely associated with it, and calcification in the subacromial bursa; these are discussed in a following section, in connection with the musculature of the shoulder. Among other acquired conditions, dislocation of the shoulder is relatively common and often serious. The most common dislocation is an anterior one, and posterior or retroglenoid dislocation is rare: Wilson and McKeever found only 4 posterior dislocations among 260 consecutive dislocations of the head of the humerus, an incidence of 1.5%.

Wilson and McKeever reported that posterior dislocation of the shoulder may be associated with tearing of the posterior part of the capsule, separation of the capsule from the neck of the scapula, or detachment of the labrum from the rim of the glenoid margin; associated fracture is apparently common, for among 11 patients reviewed by Wilson and McKeever, only 4 had dislocation without fracture, while the remaining 7 had 8 dislocations accompanied by fractures of varying extent. Wilson and McKeever regarded the probability of recurrent dislocation as being very great, unless operative repair is carried out. They advocated attaching the reduced head of the humerus to the acromion by Kirschner wires; repair of the soft tissues, as commonly done for anterior recurrent dislocation, may be effective, but in a case reported by Jones it was not and had to be followed by grafting bone along the posterior edge of the glenoid cavity, a procedure sometimes preferred as the primary repair. Boyd and Sisk repaired recurrent posterior dislocation by rerouting the long tendon of the biceps around the lateral and posterior aspects of the neck of the humerus, and stapling it to the posterior lip of the glenoid labrum. Wilson and McKeever said that posterior dislocation can be produced by forcible internal rotation of the adducted arm, and that extreme internal rotation is typical of posterior

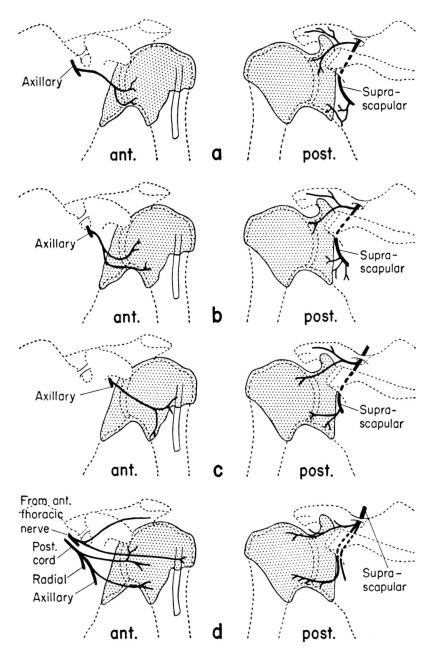

Fig. 4-13. Anterior and posterior views of the distribution of nerves to the shoulder in four cases. The articular capsule is represented by stippling. In *a*, *b*, and *c* the same nerves contribute, but with slightly different patterns, while in *d* there are branches also from the medial pectoral (anterior thoracic) nerve and the posterior cord. (Gardner E: Anat Rec 102:1, 1948)

dislocation. They pointed out that, in contrast to anterior dislocation, there is no danger of damage to the nerve trunks in the axilla, since the displacement is away from these trunks.

In anterior dislocation of the shoulder, the head of the humerus is displaced downward and forward, usually being palpable beneath the anterior axillary fold. Such a displacement necessarily brings the head of the humerus against the great nerves and vessels of the axilla, and pressure upon the brachial plexus may result; according to Milch, about one out of seven patients with so-called simple dislocation has some damage to the brachial plexus. Damage to the great vessels is rarer, although Johnston and Lowry reported a case of damage to the axillary artery; in the past, they said, such damage apparently often resulted from forced reduction rather than from the dislocation itself. Anterior dislocations are fairly frequently accompanied by fracture of the greater tubercle; Milch said that about 10% are.

Recurrent anterior dislocation of the shoulder has been widely discussed, but there is as yet no complete agreement as to its predisposing cause or causes. It seems to be generally agreed, following Bankart, that the most common operative findings in recurrent dislocation of the shoulder are detachment of the labrum from the anterior rim of the glenoid cavity and, usually, stripping of the capsule and periosteum for a variable distance from the anterior surface of the neck of the scapula (Osmond-Clarke; Adams; Gallie and Le Mesurier; Dickson and O'Dell; and others). However, detachment of the labrum is not invariable, for Adams reported that the labrum was still attached in 10 of 79 recurrently dislocated shoulders, and DePalma found it attached in 19.4% of 36 such shoulders.

Detachment of the labrum and stripping of the fibrous capsule from the neck of the scapula, or tearing of it, have been widely regarded as the chief cause of recurrent dislocation. However, this allocates to the thin and loose fibrous capsule and the variable glenohumeral ligaments a restraining and protective role for which they are ill adapted and which, indeed, is generally denied them in all discussions of the shoulder joint other than those of recurrent dislocation—for it is commonly stated that the strength of the shoulder joint lies primarily in the musculotendinous cuff. The detachment of the labrum, if related at all, and certainly the increased capacity of the synovial sac are a result rather than a cause of the initial displacement, just as is a third common finding in long-standing dislocations, a posterior erosion defect in the head or neck of the humerus where it was in contact with the rim of the glenoid. Although once these conditions are established they should add to the ease of recurrence, this does not necessarily indicate that other factors that led to the initial dislocation may not also be equally important in recurrence.

In regard to labral detachment, DePalma and co-workers found it in their careful study of the anatomy of presumably normal shoulders as early as the second decade of life, but more frequently in shoulders of individuals dying in the fifth to ninth decades, with a rate of occurrence of 50% in the fifth decade, 72.2% in the sixth, 96.4% in the seventh, and 100% in the eighth and ninth (Fig. 4-14). They therefore found detachment to be more common in those decades in which recurrent dislocation is rare (saying the latter seldom occurs after the fourth decade), and regarded detachment as being a common result of "wear and tear" at the shoulder joint. Even though a high incidence of detachment is associated with recurrent dislocation, the disparity between the age incidence of dislocation and of detachment of the labrum, as well as the finding that the labrum is not necessarily detached, seems to indicate that the latter cannot be the chief cause of the former.

DePalma and his associates said, and DePalma later reiterated, that the displaced head of the humerus usually lies in an enlarged subscapular bursa (recess) which is normally, of course, simply an outpouching of the shoulder joint. He and Dickson and O'Dell blamed recurrent dislocation on laxity of the musculotendinous cuff, or on muscular imbalance (in favor of the external rotators)

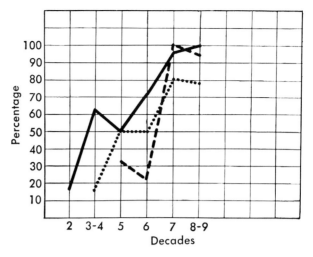

Fig. 4-14. Incidence of detachment of the glenoidal labrum correlated with age. (DePalma AF, Callery G, Bennett GA: Am Acad Orth Surgeons, Instructional Course Lectures 6:255, 1949)

━━━ Labral detachments in each decade

••••••• Biceps tendons in each decade showing degenerative changes

━ ━ ━ Tears in fibrotendinous cuff in each decade

between the external and internal rotators at the shoulder. This concept seems to place the blame for anterior weakness at the shoulder joint where it belongs, on the tendon of the subscapularis muscle, the one normally strong element of the front of the shoulder. DePalma found partial detachment of the subscapularis tendon in 3 of 36 recurrently dislocated shoulders, and reported that after the tendon had been divided, anterior dislocation could easily be produced by external rotation. Moseley and Overgaard have also stressed the loss of resistance in the anterior capsule, including the subscapularis. More recently, Symeonides also implicated the subscapularis muscle as the cause of recurrent anterior dislocation, while Rowe and co-workers doubted that abnormalities of it are a usual cause.

The view that anterior dislocation should be impossible unless resistance of the subscapularis is somehow overcome is anatomically sound. Nevertheless, there is as yet no complete agreement as to the cause of recurrent dislocation. Nicola expressed the opinion that all recurrent dislocations are not alike, and that the underlying pathology varies; Adams, describing a flattening of the head of the humerus in 82% of 68 recurrently dislocated shoulders (and in all shoulders which at

operation showed no detachment of the labrum) suggested that such flattening, because it would allow the head of the humerus to slide more easily over the glenoidal rim, should be regarded as a cause of recurrent dislocation.

With the varying opinions as to the cause of recurrent dislocation, it might be expected that operative repair has taken several forms. Nicola reiterated his belief that passing the long head of the biceps and a part of the coracohumeral ligament through the head of the humerus, thus affording better suspension to the bone, may be all that is required (although, if the labrum is detached, he said, he repairs the capsule also); however, DePalma and his co-workers said that detachment of the glenoidal labrum commonly begins in the region of attachment of the biceps tendon to the labrum, and that adding to the tendon the duty of suspending the arm would necessarily increase the tendency to detachment. It is said by most workers that operations of the Bankart or Putti-Platt type, directed at the anterior part of the shoulder joint (Fig. 4-15), give a far higher rate of cure than does the Nicola operation.

Reattaching the labrum and plicating the redundant fibrous capsule has been the pri-

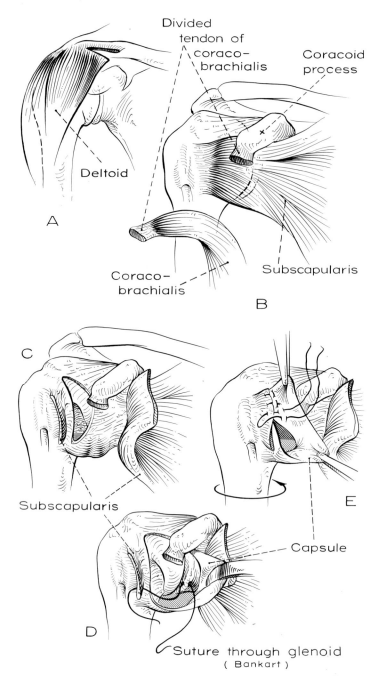

Fig. 4-15. Repair of recurrent dislocation of the shoulder by the Bankart method. In *A* an anterior part of the deltoid has been sectioned at its origin from the clavicle to allow better retraction, and in *B* the coracobrachialis and short head of the biceps are sectioned and retracted. In *C* the subscapularis has been sectioned and separated from the capsule of the joint, which is then incised; in *D* the capsule is retracted and a suture placed for reattachment of the glenoidal labrum (the Bankart procedure), and in *E* the proximal part of the capsule is

(*continued on next page*)

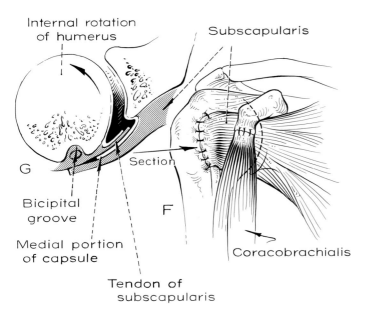

Internal rotation of humerus

Subscapularis

G

Section

F

Bicipital groove

Medial portion of capsule

Tendon of subscapularis

Coracobrachialis

Fig. 4-15. (continued) being sutured to the distal part of the subscapularis. *F* and *G* show completion of the repair, with plication of the capsule and of the tendon of the subscapularis. (Sandow TL, Jr., Janes JM: Proc Staff Meet, Mayo Clin 38:1, 1963)

mary aim of the anterior approach, although usually the tendon of the subscapularis muscle has been shortened by overlap in repairing it, and the superior results obtained have been attributed to the limitation of external rotation thus produced. Emphasizing the latter factor, DePalma, among others, has cut the subscapularis tendon and reattached it to the greater tubercle, fastening it so that it gave support to the head both anteriorly and inferiorly.

Other methods to limit external rotation have also been used: Gallie and Le Mersurier reported success in using fascial slips passed through the humerus and scapula and through the acromion; Dickson and O'Dell felt it was sufficient to attach the pectoralis minor to the humerus. In contrast, Boyd and Hunt reported that they obtained satisfactory repair by stapling the capsule after slitting the subscapularis along its fibers, thus without strengthening by overlap of its tendon. Finally, Bailey quoted evidence that a combination of elimination of the enlarged pouch, the Nicola suspension (in some instances), and

transplantation of the subscapularis has apparently given excellent results in athletes.

Persistent dislocation of the humerus may occur as the result of widespread paralysis of shoulder muscles, as in the "flail shoulder" of victims of poliomyelitis. In this condition muscle transfers are usually not effective, and the treatment of choice is usually arthrodesis if the scapular muscles are strong enough.

The Pectoral Region

The present discussion is concerned primarily with muscles, nerves, and vessels of the pectoral region.

The paired pectoralis major muscles (Figs. 4-1 and 4-16) cover most of the front of the thorax, and each usually completely overlaps the corresponding pectoralis minor, which lies between it and a portion of the thoracic wall. Completely hidden behind the pectoralis major and the clavicle, in the living condition, is the subclavius muscle. These three

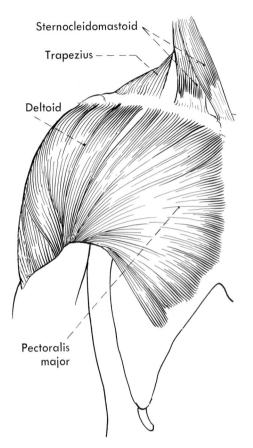

Sternocleidomastoid

Trapezius

Deltoid

Pectoralis
major

Fig. 4-16. Pectoralis major and some neighboring shoulder muscles.

muscles constitute the group to be discussed here. Lateral and inferior to the pectoralis major the serratus anterior hugs the anterolateral thoracic wall, but this is really a muscle of the shoulder, and is discussed with that group of muscles.

Fascia

The superficial fascia (tela subcutanea) of the pectoral region contains a variable amount of fat, but is not noteworthy except for the fact that the breast extends into it; hence, especially clearly in the female, it splits to surround the breast, thus forming a plane of cleavage between this organ and the pectoralis major. The deep fascia about the pectoralis major muscle becomes thickened after it leaves the muscle to extend across to the posterior axillary fold and to the fascia of the latissimus dorsi and the muscles of the arm,

and thus forms the floor of the axillary space.

The *clavipectoral fascia* (Fig. 4-17) is usually defined as the fascia surrounding the pectoralis minor and its continuation upward from the muscle to the clavicle. The latter part has also been called the coracoclavicular or costocoracoid fascia. It is thin and adherent to the pectoralis minor where it surrounds this muscle, and it unites at both edges of the muscle to form a compartment about it. From the lateral edge of the muscle the fascia is continued below to the fascia over the serratus anterior, but above it fuses to the deep surface of the axillary fascia, the floor of the axilla here contributed to largely by the fascia of the pectoralis major. Above the superomedial border of the muscle it blends in part

Fig. 4-17. Diagram of the clavipectoral and related fascia (*broken lines*)

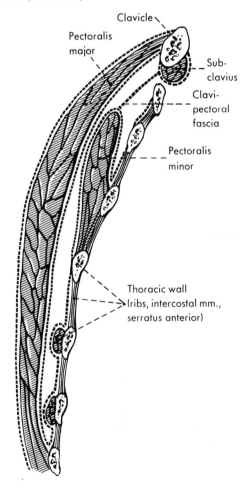

Clavicle

Pectoralis
major

Sub-
clavius

Clavi-
pectoral
fascia

Pectoralis
minor

Thoracic wall
(ribs, intercostal mm.,
serratus anterior)

with fascia over the intercostals, but laterally it passes upward to the clavicle and splits to go around the subclavius muscle, one layer attaching in front of the muscle and the other behind it. The part in front of the subclavius is particularly tough laterally, where it is attached not only to the clavicle but continues beyond the subclavius to the base of the coracoid process and the first costal cartilage. This part is usually called the costocoracoid ligament.

The cephalic vein pierces the clavipectoral fascia to join the axillary vein, and the thoracoacromial artery and the pectoral nerves pierce it after they leave the axillary sheath.

Superficial Vessels and Nerves

These are merely cutaneous twigs, except in the female where the vessels are especially developed to supply the breast.

Perforating (anterior cutaneous) branches of the internal thoracic (internal mammary) vessels pass through the pectoralis major muscle over each intercostal space close to the lateral border of the sternum, accompanied, except at the first space, by anterior cutaneous branches of the intercostal nerves. These vessels and nerves supply skin of the pectoral region on each side of the midline. Skin farther laterally is supplied especially by branches of the lateral thoracic vessels and by the lateral cutaneous branches of the intercostal nerves; these branches run medially, over the lateral edge of the pectoralis major. An uppermost part of the pectoral region gets its cutaneous nerve supply from the supraclavicular nerves of the cervical plexus (See Fig. 3-34).

The superficial lymphatics of the pectoral region include those from the breast; those connected particularly with the limb are shown in Figure 3-47.

MUSCULATURE

Pectoralis Major

The pectoralis major is a large, fan-shaped muscle which, with the muscle of the other side, covers most of the upper part of the thorax, and extends across the front of the axilla

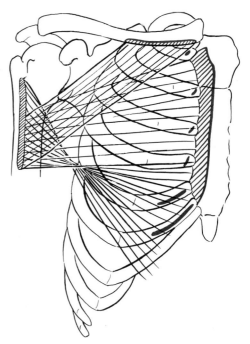

Fig. 4-18. Origin (*red*) and insertion (*black*) of the pectoralis major.

to an insertion on the humerus, forming as it does so the anterior axillary fold.

The pectoralis major has an extensive origin on the thorax (Fig. 4-18): the clavicular portion of the muscle arises from approximately the medial third of the clavicle, and is directed downward and laterally; the sternocostal portion arises from the entire length of the anterior surface of the manubrium and body of the sternum, each muscle occupying about half of this surface of the sternum; a few lower fibers of the sternocostal part usually arise also from the aponeurosis of the external oblique muscle, and if a slip of such origin is pronounced it is referred to as the abdominal part of the muscle. Deep to the sternal origin the sternocostal part arises also from the cartilages of about the first six ribs. There is usually a fairly obvious line of separation between the clavicular and sternocostal portions of the muscle.

The fibers from these various origins converge to form a broad tendon that passes in front of the coracobrachialis muscle and both heads of the biceps brachii, and disappears behind the anterior edge of the deltoid muscle to reach an insertion on the crest of the greater tubercle (lateral lip of the intertubercular or bicipital groove). A peculiarity in the insertion of the muscle lies in the manner in which fiber bundles of different origin are attached to the humerus: The clavicular fibers, running laterally and downward, attach in exactly the manner that might be expected: the most lateral fibers insert highest on the crest, the most medially arising fibers insert lowest. The upper sternocostal fibers pass almost straight laterally and disappear behind the clavicular fibers, and their tendon blends with the tendon derived from the clavicular portion of the muscle. However, the lower sternocostal fibers plus such abdominal ones as exist pass upward behind the upper sternocostal fibers, and as they do so they become so twisted that the fibers of lowest origin insert highest on the crest. Thus the tendon of insertion of the pectoralis major consists of two laminae that are continuous below, the anterior lamina consisting of fibers of clavicular and upper sternocostal origin, the posterior of fibers of lower sternocostal and abdominal origin.

The chief *action* of the pectoralis major is that of an adductor; in consequence of its attachment on the lateral lip of the intertubercular groove it is also an internal rotator, but contracts for internal rotation only against resistance (Scheving and Pauly). The upper fibers alone help to flex the arm (draw it forward and upward) to about the horizontal; the lower fibers alone help to extend the arm until it is by the side, but cannot hyperextend.

The pectoralis major is *innervated* by two nerves, the lateral and medial pectoral (anterior thoracic) nerves. The lateral pectoral nerve, derived from the lateral cord of the brachial plexus and usually containing fibers from C5 through C7, supplies the clavicular and the upper portion of the sternocostal parts of the muscle, while the medial pectoral nerve (from the medial cord, and usually containing fibers from C8 and T1) supplies the lower sternocostal and abdominal parts of the pectoralis major. Through these nerves the pectoralis major receives fibers from all the spinal nerves entering into the brachial plexus.

The chief *blood supply* to the pectoralis major is the pectoral branch of the thoracoacromial artery, which runs downward on the deep surface of the muscle in company with the lateral pectoral nerve; twigs from the perforating branches of the internal thoracic (internal mammary) artery supply the muscle close to its costal origin. The lateral thoracic artery sends branches into its lateral and inferior part, while the deltoid branch of the thoracoacromial gives branches both to the deltoid muscle and to the pectoralis major as it runs between the two.

Anomalous Pectoral Muscles

Occasionally an anomalous muscle, the *sternalis*, is found superficial to the pectoralis major (Fig. 4-19). It more or less parallels the sternum, and has variable attachments, to the sternum, the sheath of the rectus, the sternocleidomastoid, or the pectoralis major. Its incidence is usually reported to be in the neighborhood of 4% of bodies, but it is more common in darkly pigmented races than it is in white persons. Barlow found 38 sternalis muscles among 535 cadavers, these 38 occurring in 33 bodies. Apparently the muscle occurs as often on one side as on the other, and as often unilaterally as bilaterally. The nerve supply is usually reported as being from one or both pectoral nerves, and Barlow said that this has been reported to be three times as frequent as a supply from an intercostal nerve. In his own series, however, a nerve supply from an intercostal was more common. There may be a nerve supply from both sources (Fick).

A number of accessory muscle slips connected with the pectoralis muscle or lying in the axillary region have been described under a variety of names. These anomalous bundles may arise from the lateral border of the pectoralis major, from ribs or costal cartilages, or from the latissimus dorsi, and have a variable insertion. They may pass between pectoralis major and latissimus, thus forming a complete arch across the axilla or, arising from either, they may attach into the fascia of the arm or into the flexor muscles of the arm, or extend down as a "chondroepitrochlearis

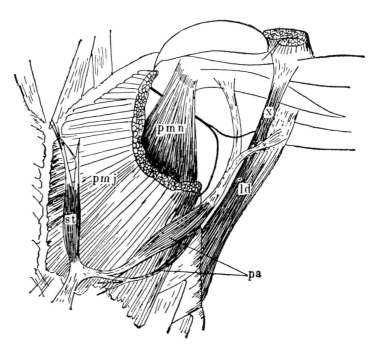

Fig. 4-19. Some anomalous pectoral muscles: *pmj* and *pmn* are the pectoralis major and minor, respectively, and *ld* is the latissimus dorsi; *st* is a sternalis muscle, *pa* an axillary arch muscle associated with the pectoralis major and attaching across the axilla into both the latissimus dorsi and the humerus; *x* is an axillary arch muscle arising from the latissimus and attaching to the pectoralis major at the latter's insertion. (After Gehry, from Wilder HH: The History of the Human Body. New York, Holt, 1923)

muscle" (Landry) as far as the medial epicondyle. Vare and Indurkar reported a case in which fibers arising between the pectoralis major and minor bilaterally inserted into the axillary fascia, and another in which fibers of the latissimus dorsi inserted into the coracoid process.

Many of these muscle bundles are grouped together as "axillary arch" muscles, regardless of whether they arise from the pectoral or the latissimus muscle. Other accessory slips that form less definite arches, arising from the thoracic fascia or the external oblique aponeurosis, or from ribs or costal cartilages, and inserting into the coracoid process, or arising from the sternum or the coracoid process and inserting into the clavicle, have also been described and named. Generally, they are very small and are of purely academic interest; as Huntington pointed out, many of them are better represented in apes than in man. Their innervation varies; apparently the innervation of an axillary arch muscle is more frequently from the medial pectoral nerve (Wilson), thus indicating the derivation of the muscle from the pectoral muscle mass, but some muscles are supplied by the intercostobrachial or by the medial brachial cutaneous

nerves, and those closely connected to the latissimus dorsi may be supplied by its nerve, the thoracodorsal.

Variations and Rupture of the Pectoralis Major

Partial or complete absence of the pectoralis major muscle is not common, but neither is it excessively rare. Clark found that about 200 cases of grave deficiency or absence of the pectoral muscles had been reported by 1915. Apparently the most frequent defect of the muscle is absence of the sternocostal portion, with the clavicular part being present; less frequently the entire muscle is missing. It is said that absence of the clavicular part alone has never been reported. Movements are usually reported to be normal, but Marmor and coworkers reported weakness particularly in adduction and internal rotation in absence of the sternocostal portion. Absence or grave defect of the pectoralis major may or may not be accompanied by absence of the minor.

Rupture of the pectoralis major muscle is apparently very rare; Hayes reviewed the literature and found only 21 cases reported, to which he added 2 of his own. The muscle is usually ruptured near the axillary fold at the musculotendinous junction, but Marmor and co-workers reported that in 1 of their 2 cases the rupture was at the insertion of the tendon into bone. They regarded rupture as resulting typically from excess force applied while the muscle is holding at its maximum power.

Pectoralis Minor

The pectoralis minor muscle (Fig. 4-20) lies deep to the pectoralis major, with no more than a small lateral part being visible before the major is reflected. It is enclosed by a fascial envelope continuous with the clavipectoral fascia, this envelope splitting at one border of the muscle and uniting again at the other border.

The pectoralis minor arises from about the second to the fifth ribs and inserts on the coracoid process, but both its origin and insertion are subject to some variation. Seib found that 34% of 570 muscles arose from the

Fig. 4-20. Pectoralis minor and subclavius muscles; some of the origin of the serratus anterior is also shown.

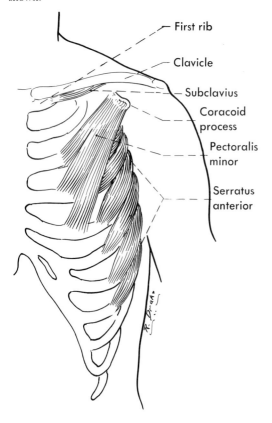

First rib

Clavicle

Subclavius

Coracoid process

Pectoralis minor

Serratus anterior

second to fifth ribs inclusive, 33% arose from the third to fifth ribs, and 17% from the second to fourth ribs (in the remaining 16% the origin was variable, but ranged from the first through the sixth ribs). He also found that among 1,000 muscles examined for their insertion, 15% had an insertion in addition to the usual one on the coracoid process: a variable number of the superficial fibers gave rise to a tendon that crossed the coracoid process to attach to the greater tubercle of the humerus, to the articular capsule or the labrum, or to both (Fig. 4-21). The tendons that attached to the greater tubercle did so indirectly, first joining the coracohumeral ligament or the tendon of the supraspinatus or the adjacent articular capsule; the abnormal insertion in 62.5% of the anomalous muscles was into the humerus. Attachment into the articular capsule or the glenoidal labrum was found in 25% of the abnormal attachments, and the remaining 12.5% of abnormal tendons had both attachments.

Lambert reported a rare case of pectoralis minor that inserted into the clavicle instead of into the coracoid; the muscle of the other side likewise failed to reach the coracoid, attaching instead into the tendon of origin of the biceps and coracobrachialis muscles.

The chief *action* of the pectoralis minor is to depress the lateral angle of the scapula; it may thus assist also in downward rotation of the scapula. If the shoulder is retracted, the pectoralis minor will help to draw the scapula forward. The *nerve supply* is through the medial pectoral nerve, which arises from the medial cord of the brachial plexus (hence contains fibers from C8 and T1), supplies the pectoralis minor, and then continues through or around the lateral edge of the minor to end in the lower part of the pectoralis major. The *blood supply* of the muscle comes largely from pectoral branches of the thoracoacromial artery and from branches of the lateral thoracic artery, this vessel approximately paralleling the inferolateral border of the muscle.

The pectoralis minor muscle may be absent when the pectoralis major is, or it may be present in absence of the major; it has apparently been said that the minor is never absent when the major is present, but Williams reported a case in which the pectoralis major was removed during the course of radical operation on the breast, and although it was well developed there was no pectoralis minor present.

Subclavius

The subclavius (Fig. 4-20) is a small muscle that arises by a tendon from about the junction of the first rib and its cartilage, and passes upward and laterally to insert on the lower surface of the clavicle. It is covered by the pectoralis major and enclosed, as already described, by the clavipectoral fascia. It is innervated by the small subclavian nerve, derived from the upper trunk of the brachial plexus.

Cave and Brown suggested that the shorter medial part of the tendon of the subclavius can be used to stabilize a sternoclavicular joint in which the costoclavicular ligament has been ruptured; they described the larger part (pars libera) as being about 0.5 inch long, lying against the lateral aspect of the costoclavicular ligament in 85% of specimens, and in front of it in 15%. They said a bursa has been described as occurring occasionally between the tendon and the ligament, but they found none. They also said that the muscle has very rarely been reported as being absent (Crerar reported a case in which the muscle was replaced by a strong fibrous band), and that duplication of it seems to be only slightly more frequent than is absence. Reiss and co-workers found electromyographic evidence that the muscle acts primarily to help stabilize the sternoclavicular joint.

NERVES AND VESSELS

The contents of the axilla are discussed in the succeeding section; its larger neural and vascular components pass down the arm, but certain ones appear on the thorax and need mention here (see Figs. 4-22 and 4-23).

Somewhat under cover of the pectoralis major, running approximately along the inferolateral border of the pectoralis minor, is the lateral thoracic artery; this is a vessel that in the female participates prominently in the

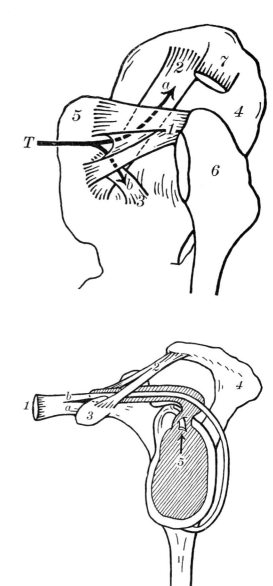

Fig. 4-21. Abnormal attachments of the pectoralis minor. In the upper figure the shoulder is seen from above: *T* represents the tendon of a pectoralis minor that passed through the coracoacromial ligament, *1,* and had double insertions, *a* and *b;* the first followed a glenohumeral ligament, *2,* to the humerus, the second followed a coracoglenoid ligament, *3,* to the glenoid labrum. *4* is the articular capsule of the shoulder joint, *5* the coracoid process, *6* the acromion, and *7* the insertion of the supraspinatus muscle. In the lower figure, a lateral view, *1* is the pectoralis minor tendon, with a normal insertion, *a,* into the coracoid process, *3,* and an abnormal tendon *b,* that passes through the coracoacromial ligament, *2,* to attach to the lower rim of the glenoid cavity. *4* is the acromion, *5* is the tendon of the long head of the biceps. In this unusual case, a diverticulum of the synovial membrane of the shoulder joint (*shaded*) formed a sheath about a part of the abnormal tendon. (Seib GA: Am J Phys Anthropol 23:389, 1938)

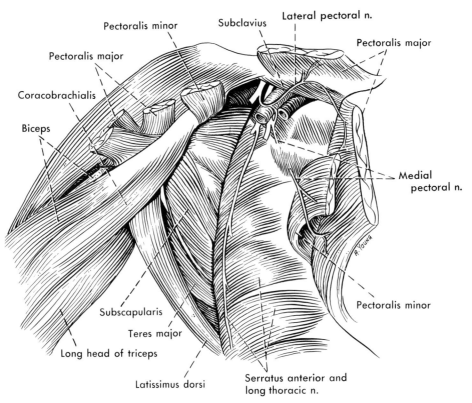

Pectoralis minor
Subclavius
Lateral pectoral n.
Pectoralis major
Pectoralis major
Coracobrachialis
Biceps
Medial pectoral n.
Pectoralis minor
Subscapularis
Teres major
Long head of triceps
Latissimus dorsi
Serratus anterior and long thoracic n.

Fig. 4-22. The axilla from the front, after reflection of its anterior wall and removal of most of its contents.

blood supply of the breast. On the surface of the serratus anterior, situated posterolateral to the lateral thoracic artery, is the long thoracic nerve, the nerve supply to the serratus. Toward the level of the inferior angle of the scapula this nerve is joined in its superficial position on the serratus by a branch of the subscapular artery, other branches of this artery continuing downward along the scapula and into the latissimus dorsi muscle; the lower branches are derived from the thoracodorsal artery, the continuation of the subscapular stem.

Nerves and vessels associated with the superficial surface of the pectoralis major have already been described.

The two pectoral (anterior thoracic) nerves, which supply the pectoral muscles, are not visible until the pectoralis major is reflected, for after leaving the brachial plexus they penetrate the deep surfaces of the pectoral muscles. Likewise, the thoracoacromial artery, the

chief blood supply to the pectoralis major, is largely hidden by this muscle, although its deltoid branch may appear in the groove between the deltoid and pectoralis major, where it accompanies the cephalic vein, and its acromial branch becomes subcutaneous as it approaches and ramifies over the acromion process.

The *lateral pectoral nerve* arises from the lateral cord or from the anterior divisions of the fifth, sixth, and seventh cervical nerves, from which it derives its fibers. It thus often arises in the neck rather than in the axilla, but if so it accompanies the lateral cord into the axilla. Here it crosses in front of the axillary vessels to pierce the clavipectoral fascia and end in the upper part of the pectoralis major, including the entire clavicular and an upper portion of the sternocostal part of the muscle (Fig. 4-22). It accompanies part of the pectoral branch of the thoracoacromial artery.

The *medial pectoral nerve* arises below the

level of the lateral one, and from the medial cord of the brachial plexus. It emerges between the axillary artery and axillary vein, and commonly receives from the lateral anterior thoracic nerve a contribution that forms a loop in front of the artery. It enters the deep surface of the pectoralis minor to supply this muscle, and a part of it continues through or around the lower border of the minor to end in the lower part of the major.

The pectoralis minor probably receives only eighth cervical and first thoracic fibers, for any fibers in the medial pectoral nerve that are derived from the lateral cord through the lateral pectoral nerve usually continue into the major. The pectoralis major receives nerve fibers from all the chief nerves entering into the plexus, since its nerves are derived from both the lateral and medial cords. The fibers are distributed in the muscle rather segmentally, so that the uppermost part is innervated through the fifth cervical nerve, the lowest through the first thoracic.

The *nerve to the subclavius muscle* does not appear in the pectoral region; it arises from the upper trunk of the brachial plexus and runs downward in front of the plexus to reach the upper posterior surface of the muscle.

The Axilla

The axilla (Figs. 4-22, 4-23; see also Figs. 4-29 and 4-32) is the relatively limited pyramidal compartment between the thoracic wall and the arm, bounded anteriorly by the structures forming the anterior axillary fold, and posteriorly by those forming the posterior fold. The floor or base of the axilla is tough connective tissue, the axillary fascia, derived from the fascia of the pectoralis major, latissimus dorsi, serratus anterior, and the brachial fascia around the muscles of the arm, and extending from the anterior to the posterior axillary fold. The apex opens upward into the base of the neck between the clavicle and the first rib, and transmits the great vessels and nerves between the neck and the arm and shoulder.

The anterior wall of the axilla is the pectoralis major and, to a lesser extent, the minor, therefore the axilla can be conveniently explored by reflecting these structures; the posterior wall, from above downward, is the subscapularis muscle (which forms by far the major part), and the teres major and latissimus dorsi muscles. The medial wall is the lateral surface of the thorax, here covered by the serratus anterior; the lateral wall is that portion of the humerus situated between the converging anterior and posterior axillary walls, therefore the intertubercular groove and the long head of the biceps; the pectoralis major inserts upon the lateral lip of the intertubercular groove (crest of the greater tubercle) while the subscapularis inserts on the lesser tubercle and the crest below it, and the teres major and latissimus dorsi insert below the subscapularis. Primarily because the axilla contains the diverging elements derived from the brachial plexus, relationships are particularly close and complicated, and can only be displayed by very careful dissection.

AXILLARY ARTERY

The axillary artery (Figs. 4-23 to 4-28) can conveniently be considered as the central axis of the axilla, since the veins of the axilla tend to follow the artery and its branches, and the brachial plexus is arranged around the artery. This artery is a direct continuation of the subclavian, the change in name occurring as the vessel crosses the first rib. It becomes the brachial as it crosses the lower border of the tendon of the teres major. As the artery passes through the axilla it is at first enclosed, with the vein and the cords of the brachial plexus, in the axillary sheath, a prolongation downward from the deep fascia of the neck. The axillary sheath gradually thins out and disappears.

The axillary artery is described as having three parts, the division being based on the relationship to the pectoralis minor: the first part lies above the muscle, the second part behind it, and the third below it.

The first part of the artery lies upon the

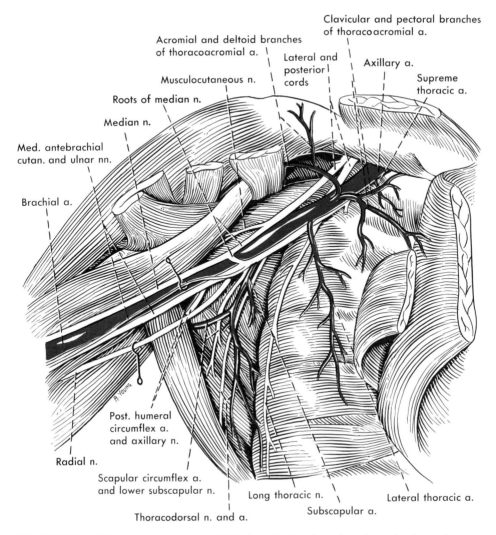

Clavicular and pectoral branches
of thoracoacromial a.

Acromial and deltoid branches
of thoracoacromial a.

Lateral and
posterior
cords

Axillary a.

Supreme
thoracic a.

Musculocutaneous n.

Roots of median n.

Median n.

Med. antebrachial
cutan. and ulnar nn.

Brachial a.

Post. humeral
circumflex a.
and axillary n.

Radial n.

Scapular circumflex a.
and lower subscapular n.

Long thoracic n.

Lateral thoracic a.

Subscapular a.

Thoracodorsal n. and a.

Fig. 4-23. The axillary artery and the brachial plexus in situ. Lymph nodes and veins and
the pectoral nerves have been removed.

first slip of origin of the serratus anterior, and
in this position the medial cord of the brachial plexus and the medial pectoral nerve lie
behind the artery, enclosed in the axillary
sheath. The long thoracic nerve, taking origin
higher in the neck, descends behind the axillary sheath on the surface of the serratus anterior muscle. The axillary vein lies medial to
(and hence below) the artery, and the medial
pectoral nerve passes between these two vessels in its course to the pectoralis minor. Lateral to the artery, and therefore also above it,
are the lateral and posterior cords of the brachial plexus; anterior to it is the clavipectoral

fascia, and the structures—branches of the
thoracoacromial artery and accompanying
veins, cephalic vein, and branches of the lateral anterior thoracic nerve—that penetrate
this fascia. In front of the clavipectoral fascia
is the clavicular head of the pectoralis major;
behind it an anastomotic loop between the
lateral and medial pectoral nerves is usually
present, immediately in front of the artery.

The second part of the axillary artery rests
on connective tissue separating it from the
subscapularis muscle; the posterior cord is
here behind the artery, the lateral cord is anterolateral, lateral, or posterolateral to it, and

the medial cord is medial or posteromedial to it; the axillary vein lies below and medial to the artery, separated from it by the medial cord.

The third part of the artery lies on the subscapularis and teres major muscles, with the axillary and radial nerves behind it until the former nerve enters the quadrangular space. The medial root of the medial nerve here usually passes in front of the artery to join the lateral root, the median nerve so formed then lying lateral or anterolateral to the artery; the musculocutaneous nerve lies lateral to the median nerve until it enters the coracobrachialis muscle. The medial brachial cutaneous and medial antebrachial cutaneous nerves and the ulnar nerve, from before backward, lie between the axillary artery and vein, and are the direct medial relations of the artery; the medial brachial cutaneous nerve escapes between the artery and vein and runs downward in front of the lower end of the axillary vein and the upper part of the basilic.

The relationship of the medial and lateral cords and the roots of the median nerve to the axillary artery is responsible, according to Telford and Mottershead, for at least certain cases of obliteration of the radial pulse with depression of the shoulder, often described as being diagnostic of neurocirculatory compression by the scalene muscles (Vol. 1) or between the clavicle and the first rib. According to the observations of these workers, in 63% of arms the axillary artery, when the arm is at the side, turns laterally for about 1.5 cm. at about the level of the second rib before turning downward again toward the arm, while in 37% the artery makes its lateral curve near the outer border of the first rib; in the latter case, they found, the lateral, posterior, and medial cords of the brachial plexus lay anterior, posteromedial, and medial to the artery, respectively, and the roots of the median nerve crossed it loosely; in the former, however, the cords lay posterolateral, posterior, and posteromedial to the artery, and the two roots of the median nerve embraced the artery tightly. They found diminution or arrest of the radial pulse, but no change in the

upper axillary one, in 64% of apparently normal people when the shoulder was depressed; this change in the radial pulse they attributed to constriction of the axillary artery by the roots of the median nerve.

Telford and Mottershead also denied that the axillary artery can ever be pinched between the clavicle and the first rib, pointing out that downward movement of the clavicle is accompanied by a forward movement, so that the space between clavicle and rib is increased rather than decreased. They did adduce evidence that the artery may be somewhat constricted between forcibly contracting muscles, perhaps especially between the pectoralis minor and subscapularis.

Rupture of the axillary artery has been reported as a result of traumatic dislocation of the shoulder, and Gibson reported 2 cases in which rupture was produced by a fall on the shoulder that did not produce dislocation. He suggested that sudden hyperabduction of the arm pulls on the axillary artery, and if the artery is already weakened it tears at the origin of the subscapular artery. He warned that because the collateral circulation is poor after ligation at this level, primary repair should be done.

VARIATIONS

The variation in the position of the lateral curve of the axillary artery, described by Telford and Mottershead, has already been mentioned. Aside from this, the common variation in the artery is in the manner in which its branches arise (Figs. 4-25 to 4-28). Details of this are given in following sections. On the basis of the origins of the branches, De Garis and Swartley described 23 different types of axillary artery. They said that there is a greater tendency in the Negro than there is in white persons toward clumping of the branches, with two or more arising in common, and that there is also a greater variety of patterns in the Negro than in white persons. In contrast, Trotter and her associates found a sex difference (common origin of two or more vessels being more frequent in the female), but no significant difference between

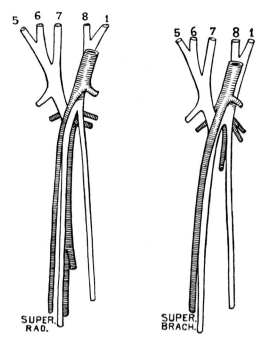

Fig. 4-24. Superficial radial and superficial brachial arteries of axillary origin. They pass in front of rather than behind the loop formed by the union of the two roots of the median nerve. Lower cervical nerves and the first thoracic are numbered; lateral and medial cords and the musculocutaneous, median, and ulnar nerves are shown. (Miller RA: Am J Anat 64:143, 1939)

races in males; the numbers of female limbs, especially of white ones, were perhaps too few to be statistically significant, but Trotter's results indicated that abnormal origin of axillary branches is probably considerably more common in females, as a whole, than in males, and more common in the Negro female than the white female.

A relatively rare variation is the division of the axillary artery into two major stems that are continued down the arm (Fig. 4-24). When this occurs, one of the stems is usually called the "superficial brachial artery," the other the "brachial" or "deep brachial" artery (not to be confused with the profunda brachii, also deep brachial, an artery normally present).

Superficial arteries of the arm are discussed in more detail in the following chapter, since they are more commonly of brachial rather than axillary origin. As pointed out there, the term "superficial brachial" has been used to

designate the more superficial of any two large arterial stems in the arm, although in fact the superficial stem may or may not be a brachial in the sense of giving rise to radial and ulnar arteries. Description is facilitated if appropriate terms such as "superficial brachial," "superficial radial," and "superficial ulnar," are used, but it is not always possible to determine from the literature which particular category is meant; if the forearm is not explored, the term "superficial brachial" often has to suffice.

De Garis and Swartley found among 512 axillas 45 cases in which the axillary artery divided to form superficial and deep stems for the arm, and estimated the incidence of this as being about 13.4% in Negroes, 4.6% in white persons. Miller found high division of the axillary artery in 17 of 480 bodies, the condition being bilateral in 15, unilateral in 2. McCormack and co-workers found 24 cases among 750 limbs. In this high division of the artery, the deeper continuation of the axillary is surrounded in the usual fashion by the brachial plexus, but the superficial artery emerges between the roots of the median nerve and enters the arm superficial to this nerve.

Among De Garis and Swartley's 45 cases, 20 of the superficial arteries proved to be superficial radial ones, and 4 were superficial ulnars. In 3 other cases there was present a superficial artery that emerged between the roots of the median nerve and in the forearm joined the ulnar artery (2 cases) or the radial artery (1 case). These represent, of course, intermediate developmental stages between the normal adult origin of the ulnar and radial arteries and the persistence of superficial ulnar and radial arteries. In Miller's series, 30 of the 32 superficial branches of the axillary were superficial radial arteries (these all occurring bilaterally, in 15 bodies), while one was a superficial ulnar. In the series of McCormack and his co-workers, there were 16 superficial radial and 7 superficial ulnar arteries of axillary origin. Seventeen of De Garis and Swartley's cases represented superficial brachial arteries, giving rise to the radial and ulnar in the forearm, while the deep stem gave rise to the subscapular and humeral circumflex vessels and to the profunda brachii and superior ulnar collateral arteries; only one of Miller's cases represented a superficial brachial artery, and only one of McCormack's was of this type. De Garis and Swartley had one case of a superficial median artery (continuing into the forearm as a sizable median artery), and so did Miller; McCormack and his co-workers apparently had none.

A superficial radial, ulnar, brachial, or median artery obviously presents an abnormal relationship to the brachial plexus, and other branches of the axillary artery, whether normally arranged or not, may do likewise: for instance, a branch of the axillary artery may pass through the median nerve or a cord of the plexus. These relationships are discussed in more detail in connection with variations of the brachial plexus (Chap. 3). It may be noted, however, that of the 8% of bodies in Miller's series that showed abnormal relationships between the artery and the plexus, only 45% of these had variations of both the artery and the plexus; either a normal artery or a normal plexus may have abnormal relations to the other.

BRANCHES

The axillary artery is conveniently described as giving off 6 main branches (Fig. 4-25). These are the supreme thoracic, the thoracoacromial, the lateral thoracic, the subscapular, and the anterior and posterior humeral circumflex arteries. A rule of thumb, to which there are many exceptions, is that the first arises from the first part of the axillary artery, the next two from the second part of the artery, and the last three from the third part. A fairly constant additional branch, found by Huelke in 86% of 178 sides, enters the subscapular muscle directly (*12* in Fig. 4-27) and is now called the subscapular ramus (to be distinguished, of course, from the subscapular artery).

As indicated by the studies of De Garis and Swartley, Trotter and her colleagues, and Huelke, there is no fixed pattern for the branches of the axillary artery. Any vessel may arise proximal or distal to its usual place of origin; two or more of the named vessels

may arise by a common stem (Figs. 4-26 and 4-27), or some of their branches may arise independently from the axillary (Fig. 4-28), and there may be other, unnamed, branches. Even with a usual pattern of origin, the distribution where the vessels overlap, thus particularly on the lateral thoracic wall, varies much.

Variable branches described by De Garis and Swartley include one or more twigs, commonly two, to the shoulder joint (*10* in Figs. 4-26 to 4-28), and one or several, called "alar thoracic," "accessory external mammary," or unnamed, and with variable distribution, to the lateral thoracic wall. In a few instances the suprascapular artery has been seen to arise from the first part of the axillary. Taking these and other variations into account, De Garis and Swartley reported finding from 5 to 11 branches, with 8 being most common. So far as the named branches are concerned, there may be as few as two (Huelke), and both Trotter and colleagues and Huelke agreed that they all arise separately in well under 50% of sides.

Because many of the variations in the branches of the axillary artery are largely of academic interest, they will be mentioned only briefly. The papers of De Garis and Swartley, Trotter and her colleagues, and Huelke should be consulted for details.

Supreme Thoracic Artery
The supreme (highest, superior) thoracic artery is almost always the first branch of the axillary artery, although it may arise from the thoracoacromial, the subclavian, or the lateral thoracic. It is distributed over the medial ends of the first and second, or sometimes only the first, intercostal spaces. De Garis and Swartley described a rare branch that descended behind the lateral thoracic artery, usually supplying two or three intercostal spaces but in one instance reaching the seventh space.

Thoracoacromial Artery
The thoracoacromial artery is usually the second branch to arise from the axillary, and the first large one; it may give rise to the lateral thoracic (Fig. 4-26), reportedly in anywhere from about 7% to 70%; and it may arise from

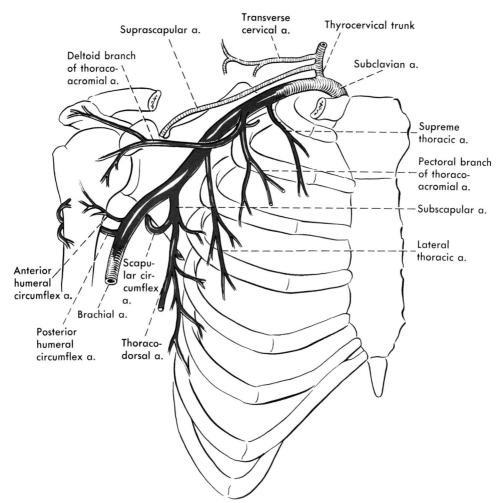

Fig. 4-25. Diagram of the axillary artery and its branches (*red*); arteries to the shoulder that arise in the neck are also shown (arising here from the thyrocervical trunk, although either very frequently comes directly from the subclavian).

the subscapular, or one or more of its branches may arise separately from the axillary (Fig. 4-28). Usually, it has a short trunk that pierces the clavipectoral fascia and divides into two major branches, pectoral and deltoid; the two smaller branches of the artery, the acromial and the clavicular branches, also may arise from the common trunk, but the acromial often arises from the deltoid branch and the clavicular from either that or the pectoral branch.

The pectoral branch or branches of the artery, commonly the largest, descend between the pectoralis major and minor to supply both muscles, giving off to the thoracic wall a branch that anastomoses with intercostal arteries and with the lateral thoracic artery. The deltoid branch descends in the groove between the pectoralis major and the deltoid, giving off branches to both muscles; it lies beside the cephalic vein in this location and ends at about the level of insertion of the deltoid. The acromial branch runs upward and laterally to branch superficially over the acromion, anastomosing here with twigs from the deltoid branch and from the suprascapular and posterior humeral circumflex vessels. The clavicular branch, a particularly small one, runs upward and medially to the sternoclavicular joint to supply this joint and the muscles adjacent to the vessel in its course. It anastomoses with the supreme thoracic artery, with the first perforating branch of the internal thoracic (internal mammary), and with twigs of the suprascapular.

Lateral Thoracic Artery

The lateral thoracic artery is usually described as arising independently from the second part of the axillary artery, but, as already noted, it may arise from the thoracoacromial, and Trotter and her colleagues reported an origin from the subscapular in approximately 25% of cases. It normally passes downward behind the pectoralis minor to run approximately along the inferolateral border of the muscle, giving branches to the pectoral muscles, the serratus anterior, and the outer surfaces of about the third to fifth intercostal spaces. It anastomoses with the intercostal arteries at its level, with pectoral branches of the thoracoacromial, and with branches of the subscapular (thoracodorsal) that also reach the lateral thoracic wall; and it may be largely replaced by the thoracodorsal, or it may spread to take over in part the usual distribution of both this vessel and the thoracoacromial.

Subscapular Artery

The subscapular artery is commonly the largest branch of the axillary and usually arises from the third part of the artery, but may arise from the second or even the first part (Huelke). Miller found 10 instances (all bilateral, in 5 bodies) in which the subscapular artery passed through the median nerve, dividing it into two parts.

The subscapular artery normally has a fairly short course downward on the costal surface of the subscapularis muscle, to which it gives branches, before it divides into two main trunks, scapular circumflex and thoracodorsal. Frequently, the scapular circumflex

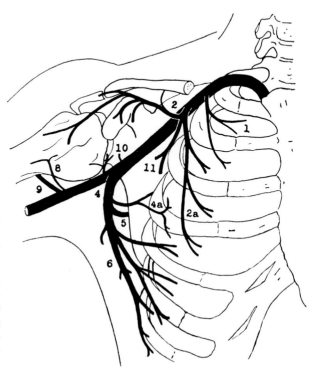

Fig. 4-26. A common type of branching of the axillary artery. *1* is the supreme thoracic artery, *2* the deltoid branch of the thoracoacromial, *2a* the lateral thoracic arising from the thoracoacromial; *4* is the subscapular, *4a* a branch that it gives to the thoracic wall, *5* the scapular circumflex, *6* the thoracodorsal; *8* and *9* are the anterior and posterior humeral circumflex arteries, *10* branches to the subscapularis muscle and the shoulder joint, and *11* the pectoral branch of the thoracoacromial. (De Garis CF, Swartley WB: Am J Anat 41:353, 1928)

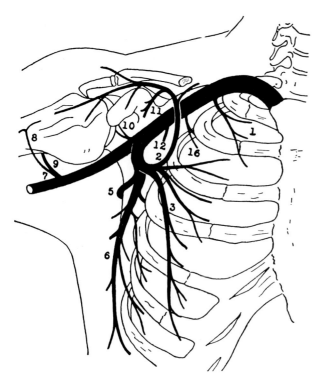

Fig. 4-27. Marked clumping of the branches of the axillary artery, with the thoracoacromial, *2,* and the lateral thoracic, *3,* both arising from the subscapular. *1* is the supreme thoracic, *5* and *6* the scapular circumflex and thoracodorsal arteries, *7* a common stem for the anterior, *8,* and posterior, *9,* humeral circumflex arteries. *10* is a branch to the shoulder joint and muscles of the arm, *11* the pectoral branch of the thoracoacromial, *12* a branch to the subscapularis muscle, and *16* an accessory branch to the thoracic wall. (De Garis CF, Swartley WB: Am J Anat 41:353, 1928)

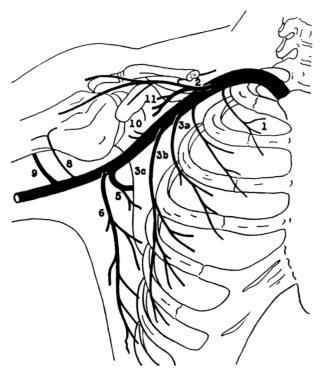

Fig. 4-28. Multiple branching of the axillary artery. *1* is the supreme thoracic, *2* the thoracoacromial and *11* its pectoral branch; *3a, b,* and *c* represent separately arising branches of the lateral thoracic artery, *5* and *6* are the scapular circumflex and thoracodorsal arteries arising separately, *8* and *9* the two humeral circumflex arteries, and *10* branches to the shoulder joint and muscles of the arm. (De Garis CF, Swartley, WB: Am J Anat 41:353, 1928)

is larger than the thoracodorsal, although the latter is the continuation of the subscapular. The *scapular circumflex artery* passes backward around the lower border of the subscapularis muscle into the triangular space (bordered by the subscapularis, teres major, and long head of the triceps); here it usually gives off a branch that runs toward the inferior angle of the scapula between the teres major and the teres minor, and may give off one or more small branches that pass into the subscapular fossa. The larger part of the artery continues into the infraspinatous fossa, passing around the lateral border of the scapula under cover of the teres minor and branching extensively in this fossa between the infraspinatus muscle and the bone. Here it anastomoses with branches of the suprascapular and transverse cervical arteries.

After the scapular circumflex artery has been given off, the *thoracodorsal artery* continues the course of the subscapular, downward along the lateral axillary wall, in company with the thoracodorsal nerve. It gives branches to the posterior wall of the axilla, that is, to the subscapularis, teres major, and latissimus dorsi muscles, and also branches to the lateral thoracic wall; it often divides into two distinct branches, one continuing on the posterior axillary wall, one on the serratus. The former branch or branches anastomose with the deep branch of the transverse cervical artery, the latter with the lateral thoracic artery and branches of the intercostal arteries of about the fourth to sixth interspaces.

Humeral Circumflex Arteries

The anterior and posterior humeral circumflex arteries commonly arise from the third part of the axillary, and are typically the last branches of this artery. The posterior arises more frequently from the subscapular than does the anterior.

Of these two arteries the anterior humeral circumflex is typically small and may be absent. It passes laterally around the front of the surgical neck of the humerus, deep to the coracobrachialis muscle and the two heads of the biceps, and anastomoses deep to the del-

toid with the posterior humeral circumflex artery. It gives branches to these muscles and to the pectoralis major, and a branch which ascends along the long tendon of the biceps to reach the shoulder joint.

The posterior humeral circumflex artery passes backward around the surgical neck of the humerus with the axillary nerve; it reaches the back of the humerus by passing through the quadrangular space (bounded by the subscapularis, teres major, long head of the triceps, and the humerus) and runs with the axillary nerve around the humerus under cover of the deltoid muscle. It gives branches to adjacent muscles, especially the deltoid, and branches to the shoulder joint and to the upper end of the humerus. Through a branch that penetrates the deltoid muscle to reach the acromion, the posterior humeral circumflex artery anastomoses here with the acromial branches of the suprascapular and thoracoacromial arteries; deep to the deltoid it anastomoses with the anterior humeral circumflex; and it communicates with the deltoid branch of the profunda brachii artery, which ascends along the long head of the triceps. This branch may be enlarged and form the stem of origin either of the posterior circumflex from the profunda or of the profunda from the posterior circumflex.

COLLATERAL CIRCULATION

Arterial anastomoses around the shoulder and on the lateral thoracic wall provide channels for collateral circulation to the arm when the third part of the subclavian artery or the first or second part of the axillary artery is ligated, the specific channels available depending both upon the level of ligation and upon the relationship of this to the somewhat variable points of origin of the vessels concerned.

The chief anastomoses around the scapula are between the transverse cervical and suprascapular arteries on the one hand, these being vessels that arise in the neck, and the scapular circumflex branch of the subscapular on the other. Normally, therefore, these are a route from the first part of the subclavian artery to the third part of the axillary. Over the acromion there are small anastomotic connections between the suprascapular, the posterior humeral circumflex, and the thoracoacromial arteries; posteriorly on the arm, the posterior humeral circumflex and the profunda brachii ordinarily anastomose; and, on the lateral thoracic wall, not only do the various branches of the axillary artery that reach the wall anastomose with each other, thus affording circulation, for instance, between the thoracoacromial artery with its high origin from the axillary and the subscapular with its low origin from this same vessel, but these vessels also anastomose with lateral perforating branches of the intercostal arteries, thus affording a source of blood independent of most of the subclavian and of the upper part of the axillary vessels.

AXILLARY VEIN

The axillary vein (Fig. 4-29) is a direct continuation upward of the basilic, the latter vein changing its name at the lower border of the teres major, where it may or may not be joined by a brachial vein; the axillary becomes the subclavian as the vessel crosses the first rib. The vein lies medial to the artery and therefore both medial and inferior to the first part of the artery; the medial brachial and antebrachial cutaneous, and the ulnar, nerves lie between the lower parts of the vein and artery, and the medial brachial cutaneous and

Fig. 4-29. Veins and lymph nodes of the axilla *in situ.*

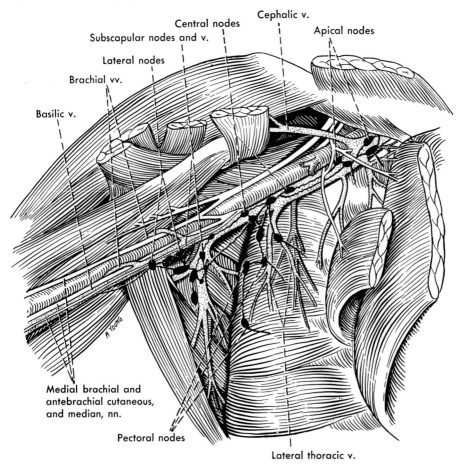

the medial root of the median nerve emerge between artery and vein. Higher, the medial cord of the brachial plexus intervenes between the artery and vein, and still higher the medial pectoral nerve emerges between the two vessels. The lateral pectoral nerve crosses anterior to the axillary artery and vein to reach the pectoralis major, and the medial brachial cutaneous nerve crosses anterior to the vein to continue along its anterior or medial side. Behind the vein, between it and the posterior axillary wall, are the long thoracic nerve, the subscapular nerves, the nerve (thoracodorsal) to the latissimus dorsi, and the subscapular artery. The lateral (brachial), central, and apical groups of axillary lymph nodes lie along the medial side of the vein.

The axillary vein typically contains a single bicuspid valve located at about the level of the lower border of the subscapularis muscle (Weathersby). However, 17% of 52 veins had 2 valves, 9% had none.

Tributaries
The tributaries of the axillary vein are the venae comitantes (brachial veins) of the brachial artery, one or both of which usually join it at about the lower border of the subscapularis muscle, but may do so lower or higher; anterior and posterior humeral circumflex veins; the subscapular vein; the lateral thoracic vein; and, in its upper part, the cephalic vein. The deep axillary tributaries thus usually correspond to the branches of the axillary artery, except the thoracoacromial; the veins accompanying the branches of the thoracoacromial artery do not ordinarily form a common stem, but empty into both the axillary and the cephalic veins.

Instead of piercing the clavipectoral fascia to join the upper end of the axillary vein, the cephalic vein sometimes passes superficially across the clavicle, deep to the platysma, to join the lower end of the external jugular vein (as it does when it first develops).

Occlusion of the Axillary Vein
According to Kaplan and Katz, thrombosis of the axillary vein is relatively rare but is more commonly found in muscular individuals and

therefore in males, and the right arm is more often involved. It has been said that forced expiration accompanying effort may force the vein against the costocoracoid ligament and the subclavius muscle; the fundamental lesion has been said to be a stretching of the wall of the vein with rupture of a valve said to be often present at the level of the subclavius muscle (Weathersby apparently found no valve here in most of the specimens he investigated anatomically). The duration of disability from this lesion is said to be variable—sometimes less than a month or two, sometimes prolonged, with edema, weakness, and stiffness of the arm.

While some authors have expressed the opinion that postoperative edema of the limb following radical mastectomy may be due more to obstruction of the axillary vein than to destruction of the lymphatic drainage, Neuhof reported that he had removed the axillary vein purposely during dissection of the axilla in connection with radical removal of the breast, in order to facilitate more complete removal of the lymph nodes. Of the 11 patients upon whom he reported carrying out this procedure, two were said to have suffered some edema of the arm, and two others fluctuating edema of the hand; he felt that this incidence of edema is about the same as that following radical removal of the breast when the axillary vein is spared. Later, Danese and Howard reported from lymphangiographic studies that axillary lymphatics apparently do not regenerate after radical mastectomy, and that absence of edema appeared to be caused by the persistence of lymph channels, usually along the cephalic vein, that had not been interrupted at operation.

LYMPH NODES OF THE AXILLA

The lymph nodes of the axilla (Figs. 4-29 and 4-30) are usually subdivided for convenience of description into several groups. They vary in number from less than a dozen to about three dozen, and tend to be larger when their number is restricted. The larger groups of nodes—the apical, central, and brachial—lie

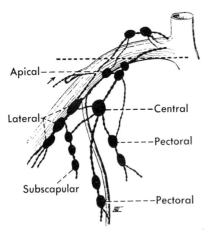

Fig. 4-30. Schema of the axillary lymph nodes. The heavy broken line represents the clavicle, and the nodes above it are therefore lower deep cervical ones. The different groups of axillary nodes are labeled. (Poirier P, Charpy A: Traité d'Anatomie Humaine [ed 2], Vol 2, Fasc 4. Paris, Masson et Cie, 1909)

mostly on the medial side of the axillary artery and vein, but other groups lie on the medial axillary wall. Subdivision of the axillary nodes is largely arbitrary, for the various subgroups are in fact continuous with each other, and they have been variously named.

The apical or subclavian axillary nodes (once also known as the infraclavicular, although that term is now usually reserved for a few more superficially lying nodes of the infraclavicular fossa) lie on the upper part of the axillary vein, especially above the pectoralis minor muscle; they receive the lymphatic drainage from all other groups of axillary nodes, and their efferents form the subclavian lymphatic trunk, which on the left side either joins the thoracic duct or enters independently into the region of junction of internal jugular and subclavian veins, and on the right side either enters independently into the venous system or joins the jugular or the mediastinal lymphatic trunks or both. In addition, however, some of the efferents drain into the lower deep cervical nodes, and hence by way of the jugular trunk into the venous system; consequently, the lower deep cervical nodes may be involved by metastatic growths originating in the field of drainage of the axillary nodes; metastasis from carcinoma of

the breast, for instance, often involves these nodes. The apical nodes also receive the drainage from infraclavicular (deltopectoral) lymph nodes, which lie in the infraclavicular fossa about the upper end of the cephalic vein, and in turn drain some of the skin of the shoulder and an upper part of the mammary gland; the efferents of the infraclavicular nodes pierced the clavipectoral fascia to reach the apical axillary nodes.

The central group of axillary nodes lies on the axillary vein distal to the apical ones, and receives efferents of the three groups still to be described; this group sends its own efferents to the apical group. The central group represents the region of confluence of the lateral, pectoral, and subscapular groups.

The lateral (brachial) group of nodes lies along the lower part of the axillary vein; these nodes receive practically all the drainage, superficial and deep, from the upper limb except for the shoulder. As already noted, the infraclavicular nodes drain some of the skin of the shoulder, and the remainder of the shoulder drains primarily into the subscapular group of nodes.

The subscapular (posterior) nodes lie mostly about the subscapular artery at the lateral (axillary) border of the scapula, distal nodes of this group being said to lie between the teres major and minor muscles on the dorsal surface of the scapula. These nodes receive the drainage from most of the shoulder and from the posterior surface of the thorax, and their efferents join nodes of the upper brachial and central groups.

The pectoral (anterior or thoracic) lymph nodes lie mostly along the inferolateral border of the pectoralis minor, that is, along the lateral thoracic artery; some may lie between the pectoral muscles. They receive the lateral drainage from the mammary gland, as well as other drainage from the anterior thoracic wall; they send their efferents to upper members of the apical group and to the central group.

According to Buschmakin, the abundant blood supply to the various groups of axillary lymph nodes is fairly constant in its origin. He described the apical nodes as receiving their blood supply from the thoracoacromial artery and only occasionally from the supreme thoracic, and said that branches of this artery almost always supply also the lateral and the pectoral nodes. The central nodes were said to receive a blood supply from both the lateral thoracic and thoracodorsal arteries in 75% of cases, while the subscapular nodes are usually supplied by branches of the subscapular or thoracodorsal arteries, and only occasionally by the lateral thoracic.

BRACHIAL PLEXUS

The formation of the brachial plexus, the origins of its branches, and variations in both of these, have been described in the previous

chapter. Here, therefore, only the relations of the plexus and its branches in the axilla need to be discussed.

At the base of the neck much of the brachial plexus lies above the subclavian artery, but the inferior trunk of the plexus lies behind it. As the artery and the plexus cross the first rib and enter the axilla, the lateral cord is still above (superolateral to) the axillary artery (Fig. 4-31), but so is the major part of the posterior cord, namely, that part formed by the union of the posterior divisions of the upper and middle trunks, which may have been joined at this level by the posterior division of the lower trunk to form the complete posterior cord. On the other hand, the lower trunk of the brachial plexus, or the medial cord if the lower trunk is already divided into poste-

Fig. 4-31. Brachial plexus exposed by excision of a part of the clavicle. (Craig, WMcK, MacCarty CS. In Walters W (ed): Lewis' Practice of Surgery, Vol 3, Chap 7. Hagerstown, Prior, 1955)

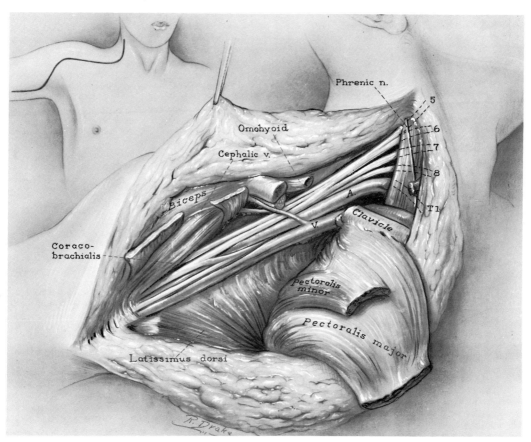

rior and anterior divisions, still lies behind the artery, and therefore intervenes between the artery and the first rib. The latter relationship is presumably one of the difficulties involved in obtaining good anesthesia over the area of distribution of the ulnar nerve in blocks of the brachial plexus at the first rib. Moore and coworkers, agreeing that axillary blocks of the brachial plexus are technically easier than supraclavicular blocks, noted that the former are not free of possible dangerous sequelae.

In the upper part of the axilla the parts of the brachial plexus are closely grouped around the axillary artery, with which they and the axillary vein are enclosed in the axillary sheath. The lateral cord and the major part of the posterior cord lie lateral to and above the artery, the lower trunk or the medial cord behind the artery, and the major continuations of these cords are also grouped closely about the artery.

Lateral Cord

In the lower part of the neck or the upper part of the axilla the anterior rami that form the lateral cord, or this cord itself, gives off the lateral pectoral nerve, which pierces the clavipectoral fascia to go to its distribution to the pectoralis major. Behind or below the pectoralis minor the lateral cord divides into its two terminal branches: the musculocutaneous nerve, which diverges laterally to give off branches to the coracobrachialis muscle and disappear more deeply into the arm by penetrating this muscle; and the lateral root of the median nerve, which continues the course of the lateral cord, thus lying lateral or anterolateral to the axillary artery. The lateral cord frequently gives a contribution, usually concealed behind the roots of the median nerve, to the ulnar nerve (Chap. 3).

Medial Cord

The medial cord of the brachial plexus (Fig. 4-32) gives off, while it lies behind the axillary artery, the medial pectoral nerve to the pectoralis minor and major; this nerve emerges between the axillary artery and the axillary

vein. Lower down, the medial cord passes from behind the axillary artery to attain the position medial to the artery that gives it its name. At about the lower border of the pectoralis minor the medial cord gives off in close succession, or sometimes by a common stem, the small medial brachial cutaneous nerve, which emerges between the axillary artery and vein to lie in front of the vein, and the medial antebrachial cutaneous nerve, which runs downward between the axillary artery and vein (Fig. 4-29). (The medial brachial cutaneous nerve may not be identifiable. Kerr found it arising independently in about 73% of cases, arising with the medial antebrachial in most of the remainder.) This cord then divides into its terminal branches: the ulnar nerve, which passes downward on the posteromedial side of the artery, between it and the vein but behind the medial antebrachial cutaneous nerve; and the medial root of the median nerve, which passes across the front of the axillary artery to join the lateral root of this nerve and form the median nerve lateral or anterolateral to the axillary artery.

Posterior Cord

The posterior cord gives off three branches to the muscles of the shoulder before it divides into its terminal branches. Lying at first above and lateral to the axillary artery, it soon disappears behind the vessel, and in this position commonly gives off at rather close intervals the upper subscapular nerve (or two or more) to an upper part of the subscapularis muscle (the cord lies between the axillary artery and the subscapularis muscle, being separated from the latter only by loose connective tissue); the thoracodorsal nerve, which descends behind the axillary vein and, toward the lower border of the subscapularis muscle, joins a part of the subscapular artery in a course downward to the latissimus dorsi; and, finally, the lower subscapular nerve (often derived from the axillary nerve), which also passes behind the axillary vein toward the axillary or lower border of the subscapularis muscle, and divides into two main branches, of which one supplies a lower part of the sub-

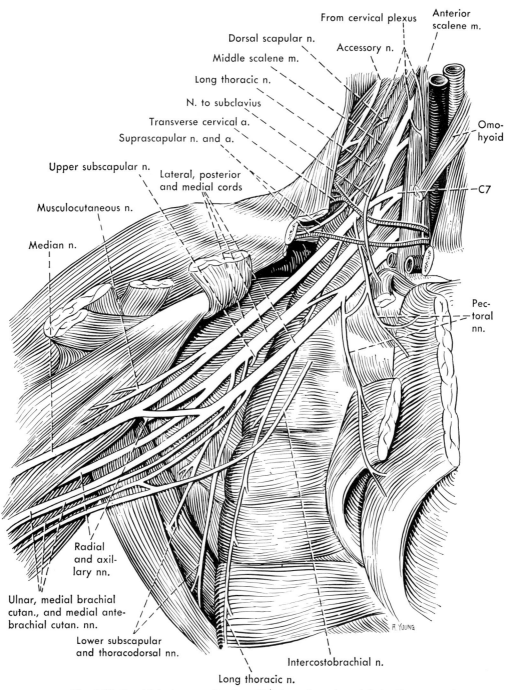

Fig. 4-32. Brachial plexus and its branches in a dissection of the axilla and neck.

scapularis while the other continues into the teres major.

The terminal branches of the posterior cord are the axillary and radial nerves. The axillary nerve passes around the lower border of the subscapularis muscle through the quadrangular space, with the posterior humeral circumflex artery, to be distributed to the teres minor and deltoid muscles and to skin on the lateral surface of the upper part of the arm. The radial nerve passes downward, still behind the axillary artery where it arose and, after it has crossed the tendons of the teres major and latissimus dorsi and thus left the axilla, disappears between the long head of the triceps and the humerus.

In the axilla, or in the upper part of the arm while it is still visible anteriorly, the radial nerve ordinarily gives off three branches: these are the posterior brachial cutaneous nerve and muscular branches to the triceps. The posterior brachial cutaneous nerve may arise independently or in common with one of the muscular branches and is distributed to a limited area of skin on the posteromedial surface of the arm. The muscular branches in the axilla are usually two in number, the upper one passing into the long head of the triceps and the lower one passing downward in the arm, parallel to the ulnar nerve (hence sometimes called the ulnar collateral nerve) to help supply the medial head of the triceps.

The Shoulder

The scapula and the shoulder joint have been discussed in preceding sections of this chapter.

MUSCLES

The muscles of the shoulder are, as noted in the preceding chapter, conveniently divided into two groups, the extrinsic muscles, which arise from the vertebral column or the costal cage and attach to the scapula or humerus, and the intrinsic or shorter muscles that arise from the pectoral girdle and insert upon the humerus. The pectoral and subclavius muscles, already described, are really muscles of the shoulder in their attachments and actions, although they are conveniently considered as a separate group.

EXTRINSIC SHOULDER MUSCLES

Two extrinsic shoulder muscles, the trapezius and latissimus dorsi (Fig. 4-33), completely cover the deep musculature of the back except in the upper cervical region where the splenius muscles appear lateral to the trapezius. The trapezius, the latissimus, the rhomboids, and the levator scapulae are topographically muscles of the back, but are obviously best discussed with the other muscles of the shoulder, with which they really belong.

Trapezius

The anterior and upper border of the trapezius forms the posterior border of the posterior triangle of the neck; the muscle completely covers the supraspinatus muscle and most of the rhomboids, but lateral to its lower border a portion of the infraspinatus muscle can be seen and, commonly, a part of the rhomboideus major as it inserts into the lower part of the vertebral border of the scapula close to the inferior angle is visible. Lower fibers of the trapezius ordinarily overlap upper fibers of the latissimus dorsi; between the trapezius, the latissimus, and the rhomboideus major there is, at least when the scapula is well protracted, a triangle of variable size known as the triangle of auscultation.

The trapezius muscle commonly arises (Fig. 4-34) from a medial part of the superior nuchal line on the occipital bone and from the external occipital protuberance, from the ligamentum nuchae, and from the seventh cervical and all the thoracic vertebral spinous processes and their connecting supraspinous ligaments. That part of the muscle arising in the region of junction of neck and thorax is the thickest and arises from longer tendon fibers than does the remainder of the muscle. The uppermost part is especially thin and

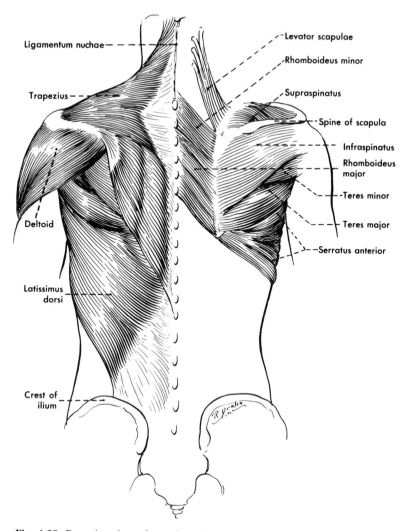

Fig. 4-33. Posterior view of muscles of the shoulder; the trapezius, latissimus, and deltoid have all been removed from the right side.

may fail to reach the skull at all, arising rather from the ligamentum nuchae; likewise, the lower part may fail to attach to all the lower thoracic vertebrae. Beaton and Anson found a dissimilarity in the trapezius muscles of the two sides in 61% of 80 bodies, although usually the variation was upward or downward for only one vertebral segment; the highest point of origin of the muscles they examined varied from the skull to the fifth cervical vertebrae, the lowest point from the eighth thoracic to the second lumbar.

From this extensive origin the fibers of the trapezius converge as they pass inferolat-erally, laterally, and superolaterally toward the shoulder, where they have a more limited but nevertheless relatively broad insertion. The upper fibers insert on the back part of the superior surface of approximately the distal third of the clavicle (see Fig. 4-37); the lower cervical and upper thoracic fibers, the most numerous, insert into the medial border of the acromion and the upper border of the spine of the scapula; and the lower fibers form a flat triangular tendon that inserts into the base of the scapular spine. Between this tendon and the smooth root of the spine there is ordinarily a small bursa.

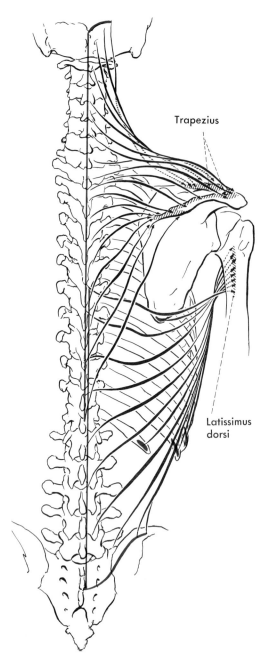

Trapezius

Latissimus
dorsi

Fig. 4-34. Origins (*red*) of the trapezius and latissimus dorsi, and insertion (*black*) of the former. The site of insertion of the latissimus, anteriorly on the humerus, is shown in Fig. 4-37.

The *function* of the trapezius as a whole is to retract the scapula, but its most important action is that of elevation of the lateral angle of the scapula; since it is the only muscle that passes downward to an insertion on the lateral angle, it is the sole muscle that can perform this function (an important one from the standpoint of upward rotation of the scapula, especially if abduction of the arm is to be carried out with considerable force). In the unusual case of bilateral absence of the trapezius reported by Horan and Bonafede the patient could not abduct his arms beyond 90°, and could not maintain this posture against strong counterpressure. The upper and lower fibers together actually help in upward rotation of the scapula, the upper fibers (especially the stronger part near the base of the neck) through their upward pull on the tip of the shoulder, the lower ones through their downward pull on the root of the spine of the scapula. Mortensen and Wiedenbauer have apparently shown, through recording action potentials over the muscle, that the upper third of the muscle is most active during elevation of the shoulder, the lower two thirds during retraction, and that, during both flexion and abduction of the arm, movements normally accompanied by an upward rotation of the scapula, there is a progressive increase in activity from the upper to the lower part of the muscle. Bearn reported that it is the sternoclavicular joint, not the upper trapezius, that supports the shoulder when the subject is relaxed.

There is some disagreement as to the exact *innervation* of the trapezius: both the accessory (spinal accessory) and several cervical nerves send fibers into this muscle, and the functional importance of these two groups of fibers is not altogether agreed upon. The accessory nerve reaches the trapezius by passing through the posterior triangle of the neck, having first passed through or deep to, and supplied, the sternocleidomastoid muscle (Fig. 4-35a); the nerve usually brings with it some fibers derived from the second cervical nerve, and before it enters the anterior or deep surface of the trapezius it is joined by further cervical nerve fibers from the third

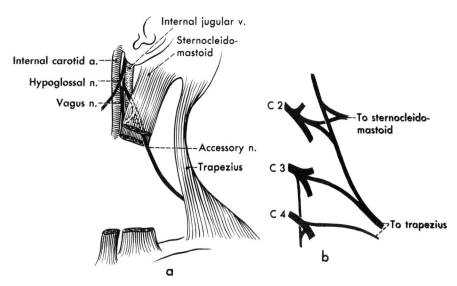

Fig. 4-35. Diagram of the course of the accessory nerve in the neck, *a*, and of the contribution of cervical nerves to it, *b*.

and fourth cervical nerves, or these may enter the muscle directly. The cervical branches to the trapezius (Fig. 4-35*b*) typically arise with the supraclavicular branches of the third and fourth cervical nerves and proceed laterally and posteriorly across the posterior triangle of the neck to the deep surface of the muscle. As the accessory nerve runs downward on the deep surface of the muscle, it is accompanied by the superficial (ascending) branch of the transverse cervical artery (called the superficial cervical artery when it arises separately from the other branch).

Some clinicians insist, apparently on the basis of failure to find obvious abnormalities of movement following sacrifice of the accessory nerve, that both this nerve and the branches of the cervical plexus are motor to the muscle. If one also believes, however, that only the lower part of the muscle is supplied by cervical nerve fibers and that the upper part is supplied exclusively through the accessory nerve, or that only a few scattered fibers receive a cervical innervation, a statement sometimes made, the question becomes purely academic rather than of clinical importance, for the only function of the trapezius not shared by other muscles is that of lifting the lateral angle of the shoulder during

upward rotation of the scapula; there are other muscles that can elevate and retract the scapula. Lockhart and others have observed that in paralysis of the trapezius through injury to the accessory nerve, the superior (medial) angle of the scapula is higher than normal, although the point of the shoulder is lower than normal, thus indicating that under this circumstance the levator scapulae and the rhomboids elevate the medial border unduly, while the weight of the arm pulls down the unsupported lateral angle of the scapula. These combined pulls produce downward rotation.

It is true that section of the accessory nerve does not always produce a pronounced disability of abduction of the arm, a movement produced in part by upward rotation of the scapula; but this, rather than being an indication that cervical nerve fibers are motor to the upper or any part of the trapezius, is probably simply another example of the well-known ability of other muscles to compensate for one that is paralyzed or missing. The serratus anterior muscle is the most active upward rotator of the scapula and can undoubtedly produce this action without the aid of the trapezius if the scapula is sufficiently fixed by other muscles, such as the levator scapulae and the rhomboids. Dewar and Harris explained a case of severe shoulder drop at rest and inability to abduct the arm more than 90° or flex it forward beyond this point, on the basis of incompetence of other muscles; in their case they transplanted the levator scapulae onto the acromion, so that it could aid in upward rotation. They said that while atrophy of the trapezius follows sacrifice of the accessory nerve, and there is a varying degree of drop of the shoulder, the majority of individuals in whom this occurs are able to abduct the arm fully, although not with normal power. Saunders and Johnson described exercises for strengthening the scapular elevators and retractors after paralysis of the trapezius; they also reported that after the accessory nerve had been severed during the course of a radical neck dissection, electromyograms showed paralysis of all three parts of the muscle.

It might be recalled that the accessory nerve itself contains no afferent fibers, yet that the nerve supply to most muscles is by no means exclusively motor, the average "motor nerve" containing in the neighborhood of 40% to 50% of afferent fibers. In the case of the trapezius, these afferent fibers are obviously contributed through the cervical nerves that reach it; Corbin and Harrison apparently showed that in the monkey the cervical nerve contribution to the trapezius does supply such fibers, and, moreover, that it contains afferent fibers only; they were unable to produce contraction of the muscle through stimulation of the peripheral ends of the cervical nerves reaching it. The prevailing evidence thus seems to be that the trapezius receives all its motor fibers from the accessory nerve, all its sensory ones from the cervical plexus.

Major defects of the trapezius are rare; Beaton and Anson found none among 925 subjects examined, and there are relatively few reports of complete absence of the muscle. Sheehan reported an instance of bilateral absence (as did Horan and Bonafede later), and Selden reported a similar case in which the rhomboideus major muscles were likewise absent and the pectoralis major was represented only by slips arising from the clavicle. Selden noted that defects of the shoulder muscles are much rarer than those of the pectoralis major.

Latissimus Dorsi

The latissimus dorsi (Figs. 4-33, 4-34; and see Fig. 4-37), the other broad muscle of the shoulder that covers a major portion of the back, has a particularly wide origin: this is usually from the spinous processes of the lower six thoracic vertebrae, this part being overlapped by the trapezius; from the spinous processes of the lumbar and sacral vertebrae by an aponeurosis usually described as being a part of the posterior layer of the thoracolumbar (lumbodorsal) fascia; similarly, by an aponeurosis from the iliac crest; and, in its more lateral and anterior part, by muscular slips that arise from the lower three or four ribs and interdigitate here with the slips of origin of the external oblique muscle of the abdomen. Finally, as the muscle passes over the inferior angle of the scapula, an additional slip of muscle may arise here and join the deep surface of the latissimus.

The fibers of the latissimus converge as they run laterally. As they reach the lateral part of the lower border of the teres major they spiral around this muscle, passing below and then medial to it; in this process the original anterior surface of the latissimus, applied to the teres major, retains this relationship, and consequently is directed posteriorly as the spiral is completed (Figs. 4-34 and 4-37). In the axilla the muscle fibers end in a flattened tendon that passes, with the tendon of the teres major, around the medial surface of the humerus to insert into the medial wall and the floor of the intertubercular (bicipital) groove of the humerus. Although a bursa between the tendons of the latissimus and the teres major is said to be constant, the insertions of the two muscles may be so closely bound together elsewhere as to make their separation difficult. Close to its insertion the muscle is crossed anteriorly by the axillary vessels and nerves and by the coracobrachialis and short head of the biceps; posterior to it and the teres major is the long head of the triceps.

The chief *actions* of the latissimus dorsi are those of adduction, inward rotation, and extension of the arm, as in a swimming stroke. Through the pull of the humerus on the scapula it can aid in depression or downward rotation of the scapula, a function particularly important in enabling the shoulder to support the weight of the body upon a crutch placed in the armpit. In this action the latissimus is aided by the lower part of the pectoralis major; however, it is said that the latissimus dorsi can adequately perform this function alone, while the pectoralis major cannot.

The *nerve* to the latissimus is the thoracodorsal nerve (Fig. 4-32), derived variably from the posterior cord as this lies on the costal surface of the subscapularis muscle, from either the axillary or radial branches of this cord, or from posterior divisions contributing to the cord; it usually brings into the muscle fibers derived mostly from the seventh cervical nerve, probably, but often with a contribution from both the sixth and eighth cervicals and perhaps sometimes even one from the fifth.

Important variations of the latissimus are apparently rare, although it may vary considerably in its origin. Its association with "axillary arch muscles" has already been noted.

Levator Scapulae

The levator scapulae (Figs. 4-33 and 4-36) arises from the posterior tubercles of the transverse processes of the first three or four cervical vertebrae and passes downward, laterally, and slightly backward to insert into the superior angle of the scapula and along the posterior surface of the medial (vertebral) border of this bone almost as far down as the base of the spine.

At its origin the muscle lies behind the scalenus medius and in front of the splenius cervicis; as it passes downward, the scalenus posterior replaces the medius as the muscle immediately anterior to it. In its upper part the levator is covered by the sternocleidomastoid muscle, in the middle part of its course it forms a portion of the floor of the posterior triangle of the neck, the accessory nerve lying

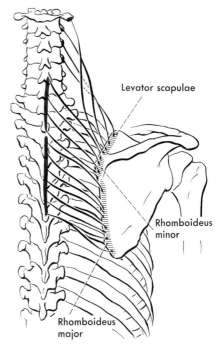

Fig. 4-36. Origins and insertions of the levator scapulae and of the two rhomboids. Origins are *red*, insertions *black*.

on its outer surface here, and in its lower part it is covered by the trapezius. The dorsal scapular nerve (to the rhomboids) courses deep to or through it, and the transverse cervical artery ordinarily divides anterior to its lower end, the deep (descending) branch of this artery coursing downward in front of the muscle to reach the rhomboids with the dorsal scapular nerve, the superficial (ascending) branch passing around the lateral border of the muscle to reach the trapezius.

The chief *function* of the levator scapulae is to elevate the superior angle of the scapula or to fix this during the movement of upward rotation of the scapula. As a result of its action on the medial border of the scapula, it may aid in downward rotation.

The levator scapulae is *innervated* by deep branches of the cervical plexus derived from the third and fourth cervical nerves and entering the anterolateral surface of the muscle close to its origin; it sometimes also receives a supply from the fifth cervical nerve by way of the dorsal scapular nerve.

Rhomboidei

The rhomboideus minor and the rhomboideus major (Figs. 4-33 and 4-36) may be entirely separate, but often separation of the two muscles is arbitrary. Except for a small portion of the lower part of the rhomboideus major, the rhomboids are entirely covered by the trapezius. In turn, they cover the serratus posterior superior and parts of the true muscles of the back.

The fibers of the rhomboideus minor arise from the lower part of the ligmanetum nuchae, from the spinous processes of the seventh cervical and first thoracic vertebrae, and from the supraspinous ligament between these processes. The muscle passes laterally and downward to insert into the posterior surface of the medial border of the scapula at the base of the scapular spine.

The rhomboideus major arises from the spinous processes of the second to fifth thoracic vertebrae and from the intervening supraspinous ligaments; it likewise passes downward and laterally, paralleling the minor, to insert into the posterior surface of the medial border of the scapula from just below the base of the spine to the inferior angle.

The chief *action* of the two rhomboidei is to retract the scapula, at the same time elevating the medial (vertebral) border; they therefore function also in elevation of the shoulder and in downward rotation of the scapula.

The *nerve supply* to both of the rhomboids is through the dorsal scapular nerve, usually derived exclusively from the fifth cervical nerve shortly after this has left the intervertebral foramen. This nerve may arise in common with the nerve to the subclavius or with the fifth cervical root of the long thoracic nerve, but if so it separates to pass posteriorly through the middle scalene; it passes deep to or through the levator scapulae, which it may also supply, and then runs downward on the deep (anterior) surface of the rhomboids, close to their scapular attachment. In the latter position the nerve is accompanied by the deep or descending branch of the transverse cervical artery (called the dorsal or descending scapular artery when it arises separately from the other branch).

Major defects of the rhomboids are apparently rare; as already noted, bilateral absence of them associated with similar absence of the trapezius has been reported at least once, but variations in the development of the minor and in the exact origin of the two muscles, and fusion of the two muscles, are the variations more commonly seen. There may be muscle slips attaching the rhomboids to other scapular muscles.

Serratus Anterior

The serratus anterior is a large muscle that covers much of the lateral aspect of the thorax, and follows the curvature of the thorax from its origin to its insertion. It arises (Fig. 4-37) by fleshy slips from the outer surfaces of the upper eight or nine ribs, the uppermost slip arising from both the first and second ribs and from the fascia between these. This upper slip runs backward and often slightly upward to reach the superior angle of the scapula and insert on its costal surface; the succeeding three slips, from the second, third, and fourth ribs respectively, form a broad mass that inserts along the costal surface of most of the remaining medial border of the scapula; the lower four (or five) slips converge, the lowest ones running markedly upward, to insert on the costal surface of the inferior angle of the scapula.

The upper part of the muscle is covered anteriorly, at its origin, by the pectoralis major, and as it goes toward its insertion forms the medial wall of the axilla. The lower slips of origin are subcutaneous and can be seen easily in muscular individuals; these slips interdigitate with the origins of the external oblique muscle of the abdomen from the same ribs. As the serratus goes to its insertion it becomes closely associated with the costal surface of the subscapularis muscle. The axillary vessels and the accompanying trunks and cords of the brachial plexus lie against the upper part of the muscle as they enter the axilla.

The chief *function* of the serratus anterior is to draw forward (protract) the scapula, and especially its inferior angle; in doing the latter, it produces the upward rotation of the scapula necessary for full abduction of the arm. Because of its course around the curve of the thorax the muscle also holds the medial border of the scapula tightly against the thoracic cage; hence, "winging" of the scapula ordinarily ensues following paralysis of the muscle. In its important function of upward rotation of the scapula the serratus is normally aided by upper fibers of the trapezius; it can, however, often perform this function adequately without the trapezius. Conversely, upward rotation, and therefore full abduction or flexion of the arm, cannot take place when the serratus anterior is paralyzed.

The *nerve supply* to the serratus (see Fig. 4-22) is through the long thoracic nerve, which arises typically from the anterior rami of the fifth, sixth, and seventh cervical nerves close to the intervertebral foramina. It passes down behind the rest of the brachial plexus, and emerges from behind the axillary vessels and the plexus in the upper part of the axilla; it then continues down on the outer (superficial) surface of the serratus anterior to supply this muscle. Horwitz and Tocantins ('38a) found the fifth cervical root of this nerve coursing independently to the upper digitations of the serratus in 5% of their specimens. In the lower part of its course the long thoracic nerve is more or less closely accompanied by thoracodorsal branches of the subscapular artery, which also descend on the superficial aspect of the muscle. Above, the lateral thoracic artery runs on the outer surface of the muscle.

The long thoracic nerve is relatively well protected in the neck, for after it emerges from the scalenes it lies behind the brachial plexus and the subclavian vessels. It is, however, superficially located in the thorax; injury to it here can paralyze the major part of the muscle. It is in particular danger in radical mastectomy, since, unless it is identified, it may be stripped with the fascia from the surface of the serratus. Isolated paralysis of the serratus following injury to the shoulder or as a consequence of infections has been recorded: Horwitz and Tocantins ('38b) found more than 150 cases reported. The exact cause of such an injury is not known, but

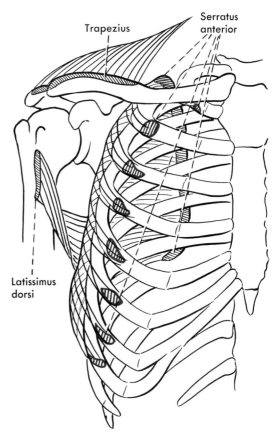

Fig. 4-37. Origin and insertion of the serratus anterior, and insertions of the trapezius and latissimus dorsi, seen from the front. Origins are *red*, insertions *black*.

Horwitz and Tocantins suggested that the nerve can be stretched over the second rib or compressed by contact between the coracoid process and the thoracic wall; they also suggested that it may be compressed by, or involved in, inflammation from adjacent bursae.

Injury to the nerve from an abnormal position of the shoulder has also been reported a few times. Prescott and Zollinger reported two instances of winging of the scapula in patients who lay prone, with an arm abducted, following operation. Lorhan reported an instance of paralysis of the serratus that apparently resulted from the position of the patient during an abdominal operation; he found reports of only eight similar cases.

INTRINSIC SHOULDER MUSCLES

Deltoid

The deltoid muscle (Fig. 4-38; see also Figs. 4-41 and 4-42) has a wide origin, corresponding approximately in its extent to the insertion of the trapezius. Thus it arises from the lateral third of the clavicle, from the lateral border of the acromion, and from almost the whole of the spine of the scapula. The anterior fibers run approximately parallel to each other, but in the powerful middle part of the muscle there are tendinous septa which in the upper part of the muscle give rise to muscle fibers and in the lower part receive the insertions of fibers. This multipennate arrangement markedly increases the strength of this part of the deltoid. All fibers of the deltoid converge to insert into the deltoid tuberosity on the lateral surface of the humerus.

The deltoid is superficially located and gives shape to the prominence of the shoulder; atrophy of it is therefore easily observed. Anteriorly, it covers the coracoid process and therefore the origins of the coracobrachialis muscle and of the short head of the biceps; the long head of the biceps also emerges from deep to the muscle. Laterally and posteriorly, the muscle covers the surgical neck of the humerus and the insertions of the teres minor, infraspinatus, and supraspinatus muscles on the greater tubercle. Deep to the middle fibers of the deltoid, and also deep to the acromion and the coracoacromial ligament, is the subacromial or subdeltoid bursa, which separates the structures just mentioned from the upper surfaces of the supraspinatus, infraspinatus, and subscapularis muscles and tendons. There may be two entirely separate bursae here, a subacromial and a subdeltoid, but more commonly the two communicate and "subacromial" and "subdeltoid" are usually used as synonyms.

The *innervation* of the deltoid muscle is through the axillary nerve, typically therefore by fibers derived mostly from the fifth and sixth cervical nerves. The axillary nerve arises as one of the terminal branches of the posterior cord of the plexus, or directly from posterior divisions of the plexus, and reaches the back of the arm by passing through the quadrangular space with the posterior humeral circumflex vessels. It gives a branch to the teres minor (and often gives off the lower subscapular nerve) and also gives rise to cutaneous branches (the upper lateral brachial cutaneous nerve) that pass either around or through the posterior border of the deltoid muscle to be distributed to skin over this muscle. The remainder of the nerve continues around the surgical neck of the humerus with the posterior humeral circumflex vessels (Fig.

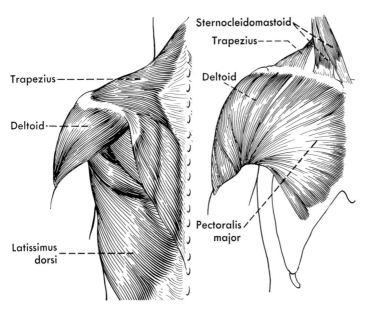

Fig. 4-38. The left deltoid muscle from behind and the right one from the front.

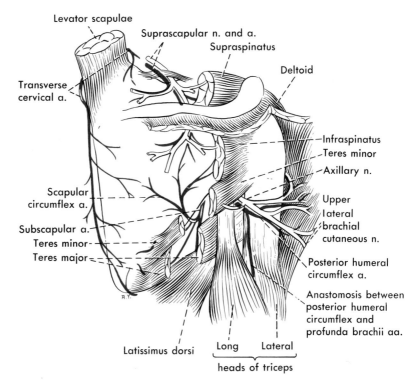

Levator scapulae

Suprascapular n. and a.

Supraspinatus

Deltoid

Transverse cervical a.

Infraspinatus
Teres minor
Axillary n.

Scapular circumflex a.

Upper lateral brachial cutaneous n.

Subscapular a.
Teres minor
Teres major

Posterior humeral circumflex a.

Anastomosis between posterior humeral circumflex and profunda brachii aa.

Latissimus dorsi

Long   Lateral

heads of triceps

Fig. 4-39. Nerves and vessels of the posterior aspect of the shoulder.

4-39) to supply the entire deltoid muscle. As the nerve passes forward around the humerus on the deep surface of the muscle it lies only about 1.5 to 2 inches below the acromion, a position that limits the extent to which the deltoid muscle can be split downward through its lateral tendinous origin if the anterior part of the muscle is not to be denervated.

In its *action* the deltoid is a powerful abductor, being assisted in this by the supraspinatus. Although it was once taught that the supraspinatus was responsible primarily for initiating abduction, both muscles work together from the start (Inman and co-workers). Usually, with paralysis of the deltoid, abduction is limited to about 45°; however, Staples and Watkins reported the case of a patient who could carry out full active abduction of the arm in spite of the fact that the deltoid was paralyzed.

In addition to its chief function of abduction, brought about especially by its middle portion, the anterior fibers flex the arm (and have been reported as hypertrophied in ab-

sence of the pectoralis major) and, probably incidentally, internally rotate it, while its posterior fibers extend and, probably incidentally, outwardly rotate it. The lower posterior fibers, inserting below the axis of movement of the scapulohumeral joint, contract during resisted adduction as the arm nears the side (and the anterior fibers may during the last part of the movement); the muscle is therefore often regarded as being also an adductor. However, Scheving and Pauly, reporting activity in all parts of the muscle whenever movement was resisted, interpreted the greater activity in the posterior deltoid during adduction as at least in part a resistance to the internal rotation that the latissimus and pectoralis major tend to bring about.

Supraspinatus

The supraspinatus muscle (Figs. 4-40 to 4-42) arises from most of the wall of the supraspinatous fossa, which it fills, and from an aponeurotic fascia that covers the muscle. It is largely covered by the trapezius muscle, and passes laterally and somewhat forward be-

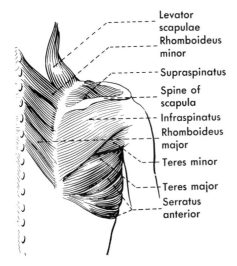

Fig. 4-40. Intrinsic muscles of the back of the shoulder, and related muscles, after removal of the trapezius and deltoid.

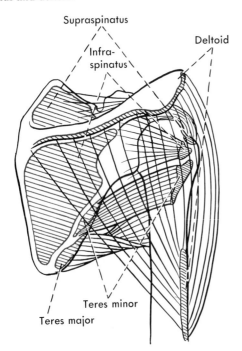

Fig. 4-41. Attachments of intrinsic muscles of the shoulder, from behind. Origins are *red,* insertions *black.*

neath the acromion and the coracoacromial ligament to insert on the upper of three facets of the greater tubercle of the humerus by a broad tendon that covers the top of the shoulder joint and is attached in part to the capsule of this joint.

The supraspinatus muscle is *supplied* by the suprascapular nerve (Fig. 4-39), containing fibers usually derived from the fifth and sixth cervical nerves or the fifth cervical alone. This nerve enters the supraspinatous fossa by passing through the scapular notch, beneath the transverse scapular ligament, while the suprascapular (transverse scapular) vessels, which in general accompany the nerve, pass across rather than under the ligament. Nerve and vessels give branches that ramify between the supraspinatus and the bone, the vessels supplying both muscle and bone, and then continue downward under the root of the acromion to ramify deep to and supply the infraspinatus muscle. In the infraspinatous fossa the artery has anastomoses with the subscapular and transverse cervical arteries.

The supraspinatus *functions* primarily as an abductor of the humerus, assisting the deltoid in this action; as it abducts, it tends slightly to rotate the humerus internally (Duchenne). Like the other muscles closely applied to the capsule of the joint, it has a particularly important role in keeping the head of the humerus in position in the glenoid cavity while the longer muscles are carrying out movements that tend to displace this head. It is also active in resisting a downward pull on the arm that adds unduly to the load on the coracohumeral ligament (Basmajian and Bazant).

The muscle is separated from the overlying coracoacromial arch (acromion and coracoacromial ligament) and deltoid muscle by the subacromial or subdeltoid bursa (see Figs. 4-47 and 4-48); since some slight abduction accompanies almost all movements of the arm, in order to free it from the side, inflammation of this bursa may be crippling. The supraspinatus muscle forms the upper part of the musculotendinous (rotator) cuff and has usually been reported as being most frequently involved when parts of this cuff are torn.

Infraspinatus

The infraspinatus muscles (Figs. 4-40 and 4-41) arises from most of the infraspinatous fossa, except for an area along the medial border that gives insertion to the rhomboids, and the thickened area along the lateral border that gives rise to the teres muscles; also, it does not arise from the neck of the scapula, but is here separated from the bone by a bursa which, it is said, sometimes communicates with the cavity of the shoulder joint. It also arises from the deep fascia overlying it. The fibers converge to form a tendon that attaches to the middle facet of the greater tubercle, some of the deeper fibers attaching also into the capsule of the shoulder joint.

The infraspinatus is partly covered by the deltoid and trapezius muscles; it is largely separated from the supraspinatus by the spine of the scapula, and along its lower border it is closely related to the teres minor and major muscles. Its tendon of insertion forms the upper posterior portion of the musculotendinous or rotator cuff.

The infraspinatus muscle is *innervated* by the suprascapular nerve (Fig. 4-39), which after supplying the supraspinatus passes to the infraspinatous fossa to end in the infraspinatus. The nerve is here accompanied by the suprascapular (transverse scapular) artery; the major branch of the scapular circumflex artery turns around the lateral border of the scapula to run between the muscle and the bone and helps supply both.

The infraspinatus is a lateral rotator of the humerus, and with the other muscles of the musculotendinous cuff plays an especially important role in keeping the head of the humerus in position in the glenoid cavity during movement of the humerus.

Teres Minor

The teres minor (Figs. 4-40 and 4-41) arises from about the middle half of the lateral border of the scapula, and from a septum common to it and the infraspinatus. It inserts both by a flat tendon into the lowest of the three facets of the greater tubercle (and into the capsule of the shoulder joint), and by a muscular insertion into the posterior surface of the body of the humerus below the tubercle.

At its origin the teres minor is immediately adjacent to the teres major, but as the two muscles pass to their insertions they diverge, and the long head of the triceps passes between them. Waterston described three cases in which the teres minor was larger than

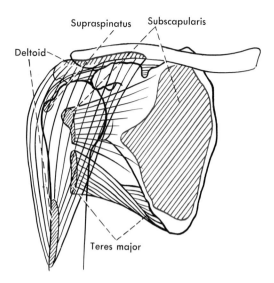

Fig. 4-42. Attachments of intrinsic muscles of the shoulder, from the front. Origins are *red*, insertions *black*.

usual, partly covering the infraspinatus muscle and arising from the fascia over that muscle, but variations of it are rare.

The teres minor gets its nerve supply from a branch of the axillary nerve, probably containing as a rule fifth and sixth cervical fibers, which enters the muscle as the nerve passes beneath its lower border. This nerve usually has a fibrous swelling or pseudoganglion upon it. Like the infraspinatus, the teres minor is primarily an external rotator of the humerus, and it acts with that muscle in steadying the head of the humerus during abduction and flexion.

Teres Major

The teres major (Figs. 4-40 to 4-42) arises on the dorsal surface of the scapula from about the medial (lower) third of the lateral border, and from the intermuscular septa that separate it from and at the same time unite it to the teres minor, the infraspinatus, and the subscapularis. It is a much larger muscle than the teres minor and has a different course; as it passes laterally it is separated from the minor by the long head of the triceps and comes in intimate association with the latissimus dorsi, which at first partly covers its posterior surface but which then spirals below and medial to the teres to lie anterior to it; the

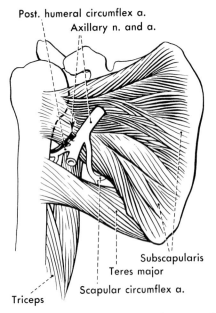

Fig. 4-43. Triangular and quadrangular spaces from the front. The axillary nerve and the posterior humeral circumflex artery are shown traversing the latter, while the scapular circumflex artery is seen entering the former.

teres major also spirals, so that its lowest-arising fibers attach highest on the humerus. It and the latissimus continue close together around the medial side of the humerus to its front, where the tendon of the teres major is attached to the medial lip of the intertubercular groove and that of the latissimus slightly higher and into the depths of the groove.

By virtue of its course around the medial side of the humerus, the teres major is an internal rotator of the arm, with the latissimus dorsi, as well as an extensor and an adductor; Broome and Basmajian found it participating in these movements only when there was resistance to them. Unlike the muscles of the musculotendinous cuff, which are active during the movements of abduction and flexion (to retain the head of the humerus in position), the teres major, which would resist these movements, is inactive then but does contract when the arm is held in an abducted or flexed position (Inman and co-workers). The muscle receives its *innervation* on its anterior or costal surface through the lower sub-

scapular nerve, which supplies a lower lateral portion of the subscapularis muscle and then continues to end in the teres major. The lower subscapular nerve arises most commonly from either the posterior cord or the axillary nerve, and probably consists usually of fibers derived from the fifth and sixth cervical nerves.

Triangular and Quadrangular Spaces

Between the upper border of the teres major muscle and the lower border of the subscapularis, from the front (Fig. 4-43), or between this border of the teres major and the lower border of the teres minor, from behind (Fig. 4-44), there is a triangular gap, with apex located medially, produced by the diverging courses of these muscles. This gap is divided into two spaces by the long head of the triceps, these spaces being commonly called the triangular and quadrangular spaces.

The borders of the triangular space are the teres major below, the subscapularis and the teres minor above, and the long head of the triceps laterally; the anterior part of this space transmits the scapular circumflex artery

Fig. 4-44. Triangular and quadrangular spaces from behind.

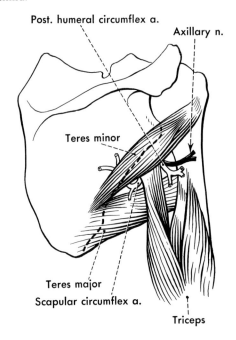

which, however, does not emerge through the posterior part of the space (between teres minor and major), but instead turns around the lateral border of the scapula deep to the teres minor.

The borders of the quadrangular space are the teres major below, the subscapularis and teres minor above, the long head of the triceps medially, and the surgical neck of the humerus laterally. The quadrangular space transmits the axillary nerve and the posterior humeral circumflex vessels in the first part of their course around the surgical neck of the humerus. The nerve to the teres minor is given off while the axillary nerve lies in the quadrangular space.

Subscapularis

The subscapularis muscle (Figs. 4-42 and 4-45) covers most of the costal surface of the scapula, and arises from most of the subscapular fossa; it is usually separated by the subscapular bursa from the neck of the scapula. It may also arise from heavy tendinous septa that produce ridges on the costal surface of the scapula. The fibers converge to form a broad insertion, tendinous above but somewhat more fleshy below, which passes behind the coracobrachialis and short head of the biceps to reach the lesser tubercle of the humerus and extend downward onto the crest below it. This muscle forms the anterior part of the musculotendinous or rotator cuff, and some of the deeper fibers of the muscle insert into the capsule of the shoulder joint.

The subscapularis is the major part of the posterior wall of the axilla, and the axillary vessels and the cords of the brachial plexus pass across it; its medial part is in contact with the serratus anterior as this muscle approaches its insertion. Since it forms the upper border of the quadrangular and triangular spaces, the axillary nerve, posterior humeral circumflex vessels, and scapular circumflex vessels turn around its lower border; this border is also crossed by the great nerves and vessels that continue from the axilla into the arm and, more medially and lower, by the downward continuation (thoracodorsal artery) of the subscapular artery and by the

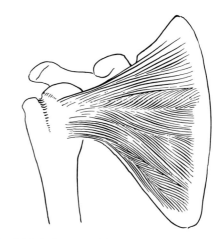

Fig. 4-45. Subscapularis muscle.

thoracodorsal nerve as these pass to the latissimus dorsi and the thoracic wall.

The subscapular bursa, sometimes called subscapular recess, between the muscle and the neck of the scapula is developmentally a separate bursa but commonly becomes continuous with the cavity of the shoulder joint; it may be doubled or missing. Horwitz and Tocantins ('38a) found it in 66 of 75 shoulders; in 49 it communicated with the shoulder joint, and there was an accessory bursa in 3 instances; DePalma and co-workers, from an examination of the synovial membrane of more than 100 shoulder joints, described the subscapular bursa as having double openings into the joint in 40.6% of cases, and as being rudimentary or missing in 16.5%. The subscapular bursa has been described as extending over the upper edge of the muscle to intervene between this and the coracoid process and form a subcoracoid bursa; however, Horwitz and Tocantins found a separate subcoracoid bursa in 89 of 100 shoulders, and said that in the other 11 it communicated with the subacromial, not the subscapular, bursa.

The subscapularis is an important internal rotator of the humerus, and acts with the other muscles of the musculotendinous cuff in preventing displacement of the head of the humerus; it is particularly important in preventing anterior displacement and there is increasing evidence that stretching or tearing of this muscle may be a common, and possibly the usual, cause of recurrent anterior dislocation of the shoulder. The *nerves* to it (Fig. 4-32) enter its costal surface, arising from elements of the posterior cord while they lie on this surface; there are usually two nerves, the upper and lower subscapular nerves, although there may be more. The upper subscapular nerve or nerves, from the posterior cord or from posterior divisions of the trunks of the brachial plexus, supply the upper and larger portion of the muscle; the lower subscapular nerve, usually from the posterior cord or the axillary nerve, is separated from the upper subscapular by the intervening origin of the thoracodorsal nerve and supplies a lower lateral part of the subscapularis muscle before continuing into the teres major. Kerr found one upper subscapular nerve in 53.5% of 157 axillas, 2 in 40.76, and 3 in 5.73; he found the lower subscapular sometimes sending more than one branch into the muscle and in 1 case failing to send any; sometimes there were also one or more branches into the muscle from the axillary nerve or the posterior cord. The upper subscapular nerve probably consists usually of fibers of the fifth and sixth cervical nerves, but Kerr found some that were derived exclusively from the seventh cervical; the lower subscapular apparently usually contains also fifth and sixth fibers, but may contain seventh or consist of either fifth or sixth alone.

MUSCULOTENDINOUS CUFF AND LONG HEAD OF BICEPS

The musculotendinous cuff (Figs. 4-46 and 4-47), often also called the rotator cuff, is formed by those short muscles of the shoulder that are intimately applied to the fibrous capsule of the shoulder joint. As the tendons of these muscles pass across the joint, they tend to fuse with each other and with the fibrous capsule lying deep to them; together they form a musculotendinous mass that is applied to all sides of the joint except its lower surface. The broad tendon of the subscapularis, with its insertion into the lesser tubercle, represents the front of this cuff; it is succeeded above by the tendon of the supraspinatus, inserting into

the upper part of the greater tubercle and thus forming the top of the cuff; the supraspinatus tendon is succeeded below and behind by the tendons of the infraspinatus and teres minor muscles, which thus form the posterior part of the cuff. The upper part of the cuff is surmounted by the subacromial (subdeltoid) bursa, which intervenes between it and the overlying coracoacromial arch.

The fact that these muscles are involved in practically all movements of the shoulder results in painful lesions of them being particularly crippling to the shoulder, as apparently first emphasized by Codman in his well-known book. Since then a great deal has been written concerning degenerative changes, calcification, and tears of this cuff.

Tears of the Cuff

Studies on cadavers and on necropsy material, such as those of Meyer ('28, '37), Wilson and Duff, Grant and Smith, DePalma and co-workers, and others, have shown that pathologic changes in the musculotendinous cuff and also in the long tendon of the biceps, closely related to the joint, are not unusual. They have also shown that while the supraspinatus tendon is commonly more severely involved than other tendons, the lesions are by no means confined to this tendon. Further, while traumatic rupture of the supraspinatus tendon may presumably occur from forcing the tendon and its overlying bursa against the sharp edge of the acromion, it is known clinically that many people with lesions of the supraspinatus have no history of sudden trauma and, further, that lesions of the tendons about the shoulder are seldom observed before the age of 30 or 35, and that the incidence of such lesions increases with advancing age. (Neer noted that it is only the anterior edge of the acromion, or that and spurs on its under surface, that impinge on the supraspinatus tendon, and to prevent weakening the deltoid removed only these parts and the coracoacromial ligament to alleviate chronic impingement.) Meyer described fraying and partial destruction of the walls of the subdeltoid bursa, fraying and partial or complete division of the tendon of the supraspinatus, infra-

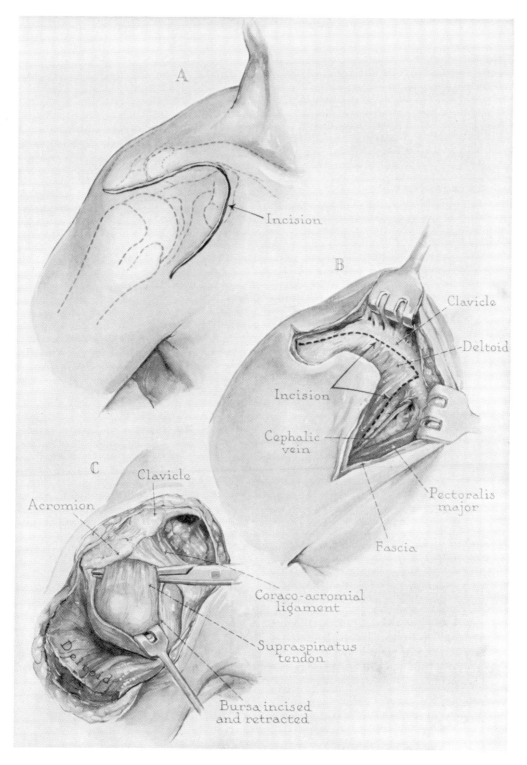

Fig. 4-46. Exposure of the subdeltoid bursa and supraspinatus tendon, the uppermost part of the musculotendinous cuff. (Banks SW, Laufman H: An Atlas of Surgical Exposures of the Extremities. Philadelphia, WB Saunders, 1953)

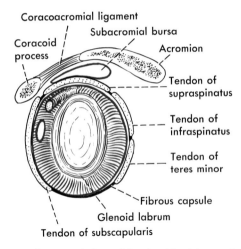

Fig. 4-47. Lateral view of the shoulder joint to show the composition of the musculotendinous cuff. Synovial membrane is *red.*

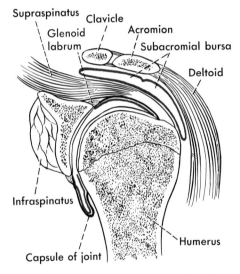

Fig. 4-48. Frontal section through the shoulder, emphasizing the interposition of the supraspinatus tendon between the shoulder joint and the subacromial bursa. Synovial membrane is *red.*

spinatus, and subscapularis muscles, or of the long tendon of the biceps, and dislocation of the long head of the biceps as being observable in many shoulders, but seldom before the age of 30.

Wilson and Duff found a 20% incidence of complete rupture of the supraspinatus tendon in a necropsy series of persons dying after the age of 32, and an incidence of 26.5% among

cadavers; the average age of subjects with complete rupture was 65 years. In their series they also found partial rupture of the supraspinatus in 20% and rupture of the long head of the biceps in 8 of 108 individuals more than 30 years old. As already noted in the discussion of dislocation of the shoulder, De-Palma and co-workers found degenerative changes in the shoulder joint beginning very early; the earliest changes were in the articular cartilage, but they found detachment of the glenoid labrum as early as the second decade. Not until the fifth decade, however, did they observe gross lesions in the synovial side of the musculotendinous cuff; these consisted at first of shredding, fraying, and tearing of the innermost fibers of the supraspinatus and subscapularis tendons but in the later decades, especially, involved the entire thickness of the tendons. In their series, there were tears of the cuff in 4 of 12 shoulders representing the fifth decade, in 4 of 18 of the sixth decade, in 30 of 30 of the seventh decade, and in 23 of 24 of the eighth and ninth decades.

Grant and Smith, defining rupture of the supraspinatus tendon as the existence of an actual communication between the shoulder joint and the subacromial bursa, thus necessarily (Figs. 4-47 and 4-48) involving a complete tear of the tendon, found 11 individuals with bilaterally ruptured tendons and 14 with unilateral rupture. They found no instance of complete rupture up to age 46; in succeeding decades up to age 86 they found an incidence of 25%, 18%, 39%, and 50%, respectively.

The available evidence, therefore, all seems to indicate that lesions of the cuff are usually degenerative rather than traumatic, actual tear of a tendon resulting from ordinary wear and tear imposed upon a degenerative process. Wilson and Duff, for instance, found that unilateral complete rupture of the supraspinatus tendon occurred regularly on the side of the stronger arm, as evaluated by the circumference of the arm at the level of the belly of the biceps muscle.

Ellis noted that the diagnosis of the degree of injury to the rotator cuff may be difficult; he said that if injection of procaine restores the power of voluntary abduction up to about

150 degrees or more it is unlikely that a major lesion of the supraspinatus tendon is present and therefore treatment should be conservative, but that if it fails to increase the voluntary range, arthrography should. be carried out. Prompt passage of contrast medium from the shoulder joint into the subacromial bursa indicates a large tear of the supraspinatus tendon.

The higher incidence of rupture of the supraspinatus as compared with other muscles of the cuff may well be due to two factors: attrition against the overlying acromion, and the greater functional importance of the supraspinatus muscle; it ordinarily acts in any movement of the humerus, slight abduction of the humerus being necessary to allow free play of the arm. DePalma and his co-workers found degenerative lesions in the subscapularis tendon more frequently than in the supraspinatus, although they were usually less severe and actual tears were less common.

It may also be, because tears of the supraspinatus tendon are easier to recognize during operation directed toward this tendon than are tears of other tendons, that some of the latter are overlooked. Hauser noted that he had observed minor tears of the subscapularis muscle during the process of repairing severe tears of the supraspinatus but that usually few major injuries to the muscles around the shoulder other than to the supraspinatus and the biceps are recognized. However, in their dissections of shoulders, DePalma and his co-workers found a high incidence of tears of other elements also; 34.4% of specimens with supraspinatus tears had tears also of the subscapularis, and there was some tearing of the infraspinatus in 54.2% of the supraspinatus tears; they never found involvement of the infraspinatus tendon alone. Hauser reported 2 cases in which the fascia over the subscapularis tendon was intact at operation, and the fact that it had ruptured and avulsed was recognized only after the muscle itself had been exposed. Incidentally, both patients complained primarily of recurrent dislocation of the shoulder, thus emphasizing again the role of the subscapularis muscle in preventing anterior dislocation of the shoulder.

Debeyre and colleagues described osteotomy of the acromion, leaving it attached at both its ends and splitting the trapezius and deltoid, as the most satisfactory approach to the cuff as a whole.

Calcification and Bursitis

Lesions of the musculotendinous cuff are not confined to rupture, although these are the most easily observed ones anatomically. Codman; Bishop; McLaughlin; Caldwell and Unkauf; and many others have pointed out that calcification in the cuff, which may involve particularly the supraspinatus tendon, is also a not uncommon lesion. It may apparently be symptomless or give rise to acute pain. It may, of course, be associated with tears of the tendon.

Bishop reported that from a third to a fourth of individuals with calcification of the cuff have it bilaterally, although only one shoulder may give rise to symptoms, and suggested that the calcareous deposits are the result of interruption of the blood supply to the tendon. It is generally recognized that the deposits are actually in the tendon, presumably as a result of previous degeneration of tendon fibers, and that the subacromial (subdeltoid) bursitis that usually accompanies calcification is a secondary condition. According to Caldwell and Unkauf, for instance, calcium deposits and surrounding fluids cause tension within the tendon, this being the source of the pain, and hence puncture of the tendon relieves the severe pain, even though the fluid and the calcified matter escape into the bursa. McLaughlin described the degree of inflammation of the subacromial bursa as varying not with the total amount of calcium observed in the tendon, but rather with the amount in contact with the floor of the bursa, and said that acute attacks of such bursitis may be self-limiting as the result of rupture of the swelling into the bursa, with subsequent absorption of the calcium.

In the experience of Caldwell and Unkauf, conservative methods of treatment were effective in from 70% to 85% of individuals with calcification and bursitis. In more stubborn cases, Armstrong reported excision of the acromion as valuable in restoring normal movement, apparently through relieving the pressure that this bone may exert on an abnormal tendon or bursa.

Biceps Tendon and The Shoulder Joint

That the long tendon of the biceps is subject to dislocation, and that such dislocation may result in a painful shoulder is now generally accepted. Meyer ('28) reported 33 cases of partial dislocation and 6 of complete dislocation in shoulders of cadavers. The dislocation is a medial one, the biceps slipping out of the intertubercular groove onto (incomplete dislocation) or over (complete dislocation) the lesser tubercle. Meyer estimated that about 4% of bodies and 8% of limbs show some degree of dislocation.

Clinical aspects of dislocation of the biceps tendon have been discussed by Abbott and Saunders and by Hitchcock and Bechtol. The former workers reported 6 cases of sudden traumatic dislocation of this tendon confirmed at operation, and discussed the anatomy involved. Hitchcock and Bechtol likewise reported cases, and analyzed a group of 100 humeri in regard to the conformation of the intertubercular groove; they found a partial or complete supratubercular ridge (see Fig. 5-3), which has the effect both of making the groove more shallow and of affording a

roughened area that may traumatize the tendon, in 67% of their specimens; they also found marked differences in the depth of the groove and in the angle that its medial wall makes with its floor (Fig. 4-49), with 8% having a medial wall that made an angle of less than 45°. Thus the anatomy of some shoulders favors dislocation of the long tendon of the biceps when it is forced against the lesser tubercle either by sudden violent external rotation of the arm or (because of its angulation over the front of the head of the humerus) by forceful flexion of the internally rotated arm. Recommended treatment of painful dislocation of the biceps tendon is attachment of the tendon to the floor of the groove, with removal of the portion above the groove.

Fraying and rupture of the biceps tendon are discussed in the following chapter. Rupture of this tendon removes a steadying influence from shoulder movements, for the humerus ordinarily glides up and down on the biceps tendon during flexion and extension.

MOVEMENTS OF THE SCAPULA AND HUMERUS

The chief actions of the individual muscles of the shoulder have already been described in connection with each muscle; in the present

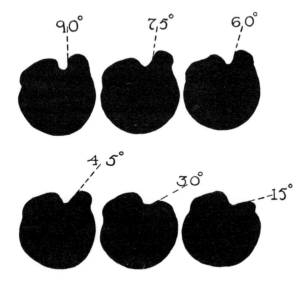

Fig. 4-49. Views of the upper end of the left humerus, to illustrate variations in the depth of the intertubercular groove and in the angle of its medial wall with its floor. (Hitchcock HH, Bechtol CO: J Bone Joint Surg 30-A:263, 1948)

section these actions are summarized in terms of movement of the shoulder and arm.

MOVEMENTS OF THE SCAPULA

Movements of the scapula are conveniently referred to as elevation, depression, protraction (moving the scapula laterally and forward around the thorax), retraction, and upward and downward rotation. In upward rotation, the inferior angle of the scapula swings laterally and forward more than does the superior (medial) angle, so that the glenoid cavity is turned upward; in downward rotation the opposite occurs and the glenoid cavity is turned downward. The scapula and the humerus normally move together, scapular movements enhancing and reinforcing humeral ones. As examples, it has already been pointed out that in abduction of the arm the scapula must be rotated upward if abduction above about 90° is to be obtained; in a similar fashion the glenoid cavity is normally directed forward if the arm is flexed, and the scapula is rotated upward as flexion continues, while the bone is retracted if the arm is extended.

It is recognized that most movements at the shoulder involve humeral and scapular movements simultaneously and not successively: for instance, the older view that abduction of the arm was a movement at the scapulohumeral joint during its first 90° and a rotation of the scapula, with the humerus fixed in its relation to the scapula, for the remainder of abduction, has long been disproved. Instead, there is normally movement of both simultaneously but in different degree, and this is referred to as the scapulohumeral rhythm. Inman and co-workers, measuring movement of the two bones during abduction of the arm in the coronal plane, found variations for the first 30° of abduction, but thereafter, up to 170°, there was 1° of upward rotation of the scapula for every 2° of abduction of the humerus. Freedman and Munro obtained slightly different figures, either because of variation among subjects or because they observed abduction in the plane of the scapula rather than in the coronal one. They found 2°

of scapular rotation for every 3° of humeral movement, and said that although this was almost constant throughout the entire range of movement there was in the final phase a relative increase in glenohumeral movement.

Muscle Groups Moving the Scapula

The muscles responsible for scapular movement are the extrinsic shoulder muscles, largely those inserting directly into the scapula, although the latissimus dorsi and the pectoralis major have an indirect action on the scapula through their action on the humerus. In addition to moving the scapula, the muscles attaching to it also serve to stabilize it and provide a firm base upon which the head of the humerus can move.

The *elevators* of the scapula (Fig. 4-50) include part of the upper fibers of the trapezius (particularly those arising close to the base of the neck and inserting into the acromion, frequently known as the "middle" part of the muscle), especially important since they are the only ones acting on the lateral angle of the scapula; the levator scapulae; and the two rhomboideus muscles. Since all these muscles except the levator scapulae arise from vertebral spinous processes and the ligamentum nuchae, they are also retractors of the scapula; the levator scapulae, however, runs pos-

Fig. 4-50. Elevators of the scapula.

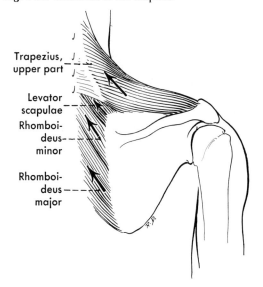

Trapezius, upper part

Levator scapulae

Rhomboideus minor

Rhomboideus major

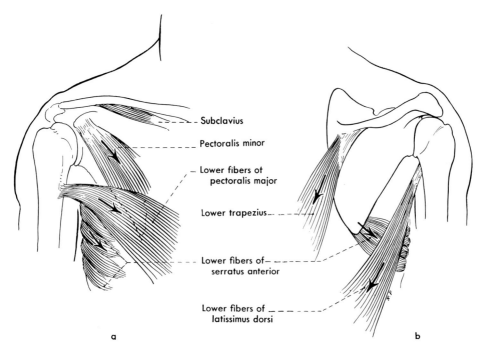

Subclavius

Pectoralis minor

Lower fibers of
pectoralis major

Lower trapezius

Lower fibers of
serratus anterior

Lower fibers of
latissimus dorsi

a

b

Fig. 4-51. Depressors of the scapula from the front, *a*, and behind, *b*.

teriorly from its origin on cervical transverse processes, and therefore cannot retract the scapula.

The weight of the limb itself tends to *depress* the scapula, especially its lateral angle, and this is normally resisted first by the sternoclavicular joint (Bern), then by the fibers of the upper part of the trapezius and by the other elevators; active depression of the scapula is brought about in part by muscles attaching to it and the clavicle, in part by muscles attaching to the humerus (Fig. 4-51). Of the latter, both the lower fibers of the pectoralis major and the entire latissimus dorsi (but particularly its lower part) run upward to their attachment on the humerus and by their pull on this bone can depress the scapula also or, important if the weight of the body is suspended by the shoulders as in chinning oneself or using crutches, can prevent the scapula from being forced upward. The lower fibers of the serratus anterior run upward to their attachment on the inferior angle of the scapula and the lower fibers of the trapezius pass upward to their attachment on the spine of the scapula, so these muscles also are depressors.

Finally, the pectoralis minor, with its insertion on the coracoid process, and presumably the subclavius, with its insertion on the under surface of the clavicle, must also be included among the depressors (although Reiss and coworkers regarded the latter muscle's chief function to be stabilization of the sternoclavicular joint). Many of these, it will be noted, depress especially the lateral angle of the scapula and hence serve also as downward rotators; the serratus anterior, however, is the primary upward rotator of the scapula.

As indicated in Table 4-1, the elevators of the scapula are supplied by the accessory nerve and the upper part of the brachial plexus; the most important depressors are supplied by the middle and lower parts of the plexus. Each muscle has its own peripheral nerve.

Only two muscles can produce *upward rotation* of the scapula, these being the serratus anterior, the primary muscle, and the trapezius (Fig. 4-52). The serratus anterior produces upward rotation by pulling forward the inferior angle more rapidly than the rest of the medial border, while the upper fibers of the

Table 4-1
Innervation of the Muscles That Elevate and Depress the Scapula

| | MUSCLE | NERVE | SEGMENTAL INNERVATION | |
			Reported Range	*Probable Most Important Segment(s)*
Chief elevators	Trapezius, "middle" part	Accessory		
	Levator scapulae	C3 and 4, perhaps dorsal scapular	C3-C5	C3, 4
	Rhomboideus major and minor	Dorsal scapular	C4-C6	C5
Accessory elevators	None			
Chief depressors	Latissimus dorsi	Thoracodorsal	C6-C8	C7
	Pectoralis major, lower part	Pectorals	C6-T1	C7, 8
Accessory depressors	Serratus anterior, lower part	Long thoracic	C5-C8	C6
	Pectoralis minor	Medial pectoral	C8-T1	C8, T1
	Subclavius	N. to subclavius	C3-C8	C5
	Trapezius, lower part	Accessory		

trapezius assist in this action by lifting the lateral angle of the scapula, and the lower fibers of the trapezius assist in this action by pulling downward on the base of the spine. The trapezius starts the movement and entirely supports the lateral angle until the limb is abducted to about 45°, only after which does the serratus contract (Beevor; Wright); hence, with a paralyzed trapezius, the scapula first sags and rotates downward with abduction, rises and rotates upward only after the serratus begins to act (the rhomboids, and presumably the levator scapulae, also contract during scapular rotation and therefore help to steady the scapula).

Downward rotators of the scapula are more numerous; they include both those elevators that work purely on the medial border and the depressors attaching laterally to the scapula or to the humerus—that is, the levator scapulae and the rhomboideus minor and major, medially, and the pectoralis minor, lower fibers of the pectoralis major, and the latissimus dorsi laterally (Fig. 4-53).

The rotators of the scapula are supplied by a number of peripheral nerves, with a wide variety of origins (Table 4-2).

Protraction or lateral and forward movement of the scapula (also called abduction) is

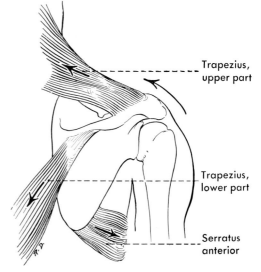

Fig. 4-52. Upward rotators of the scapula.

brought about primarily by the serratus anterior, which at the same time keeps the medial border closely applied to the thoracic wall; the serratus can be assisted in protraction but not in preventing winging of the scapula by the actions of the pectoralis minor and pectoralis major (Fig. 4-54). By acting together as protractors the serratus and the pectoralis minor, which have opposite rotatory effects

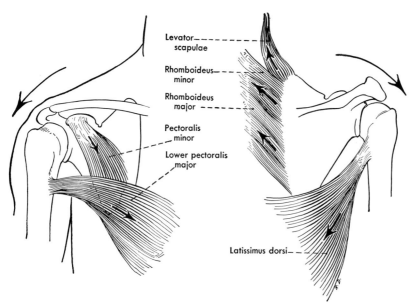

Levator scapulae

Rhomboideus minor

Rhomboideus major

Pectoralis minor

Lower pectoralis major

Latissimus dorsi

Fig. 4-53. Downward rotators of the scapula from the front and behind.

Table 4-2
Innervation of the Muscles That Rotate the Scapula

			SEGMENTAL INNERVATION	
			Reported Range	*Probable Most Important Segment(s)*
	MUSCLE	NERVE		
Chief upward rotators	Trapezius, especially "middle" part	Accessory		
	Serratus anterior	Long thoracic	C5-C8	C6
Accessory upward rotators	None			
Chief downward rotators	Levator scapulae	C3 and C4, perhaps dorsal scapular	C3-C5	C3, 4
	Rhomboideus major and minor	Dorsal scapular	C4-C6	C5
	Pectoralis minor	Medial pectoral	C7-T1	C8, T1
Accessory downward rotators	Pectoralis major, lower part	Pectorals	C6-T1	C7, 8
	Latissimus dorsi	Thoracodorsal	C5-C8	C7

on the scapula, can produce pure protraction.

Retractors (or adductors) of the scapula include the trapezius, especially its middle portion, the rhomboideus major and minor, and the latissimus dorsi, especially its upper fibers (Fig. 4-55).

The protractors and retractors of the scapula, like the other muscle groups, receive their innervation through a number of individual peripheral nerves, collectively of wide origin (Table 4-3).

Paralysis of Scapular Muscles

Harmon ('50) regarded the minimum necessary for effective control of scapular movements to be fair to good function in both the trapezius and serratus anterior muscles; for forward and backward movements of the scapula, a good pectoralis major muscle and the latissimus dorsi are also needed. In the absence of sufficient fixation of the scapula by approximately normal muscles, he pointed out, fixation may result from contracture of

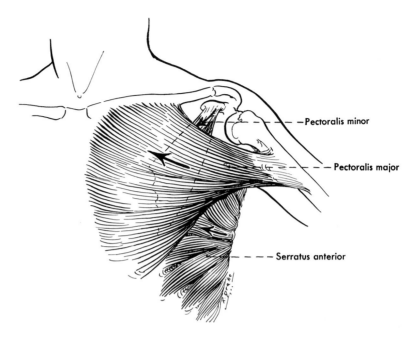

Fig. 4-54. Protractors (abductors) of the scapula.

affected muscles or be brought about operatively by fascial transplants attached to the scapula; in either of the latter cases, apparently, the loss of the last 30° to 60° of abduction always occurs. Durman reported attaching a lower part of the pectoralis major muscle to the scapula by fascia passed through a hole in the bone, to substitute for a paralyzed serratus anterior muscle.

MOVEMENTS AT THE SCAPULOHUMERAL JOINT

Movements at the scapulohumeral or glenohumeral joint are brought about in large part by the muscles of the shoulder, but these are assisted in certain movements by muscles of the arm that pass across the joint.

Movements at the shoulder joint are usually referred to the planes of the body rather than to the planes of the scapula, and Johnston pointed out that, as a result of this, abduction in the frontal (coronal) plane of the body is actually a combination of abduction and extension in terms of the scapular planes (for the glenoid cavity typically points somewhat forward, the scapula in the resting position making an angle of about 30° with the frontal plan of the body); similarly, flex-

ion in the sagittal plane of the body is actually a combination of flexion and abduction. As usually defined, however, abduction implies a lateral movement of the arm, away from the side, approximately in the coronal plane; adduction implies returning the arm to the side; flexion is a forward movement of the arm; extension is a backward one; internal or medial rotation is a turning of the anterior surface of the humerus medially, external or lateral rotation a turning laterally. Of these movements, external rotation from the "anatomic" position is limited, for the so-called anatomic position involves about 45° of external rotation from the natural position of the humerus, and the total range between extreme internal and extreme external rotation is little more than 90°. Extension is also more limited than flexion and abduction, even when these are pure scapulohumeral actions.

Flexion, adduction, and internal rotation, as in bringing the arm across the front of the thorax, frequently accompany each other. A terminal rotation also usually accompanies both complete flexion and complete abduction (that is, extreme elevation of the arm). Although internal rotation has been said to be essential to full flexion, this is so only if the forearm is at the same time extended; if it is

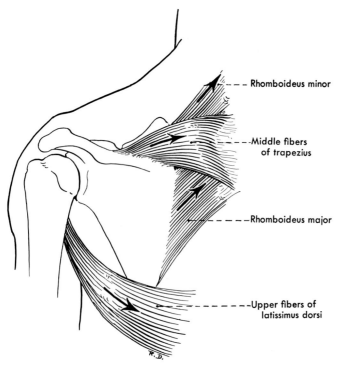

Fig. 4-55. Retractors (adductors) of the scapula.

- - - Rhomboideus minor

- - -Middle fibers of trapezius

- - - - -Rhomboideus major

- - - - -Upper fibers of latissimus dorsi

Table 4-3
Innervation of the Muscles That Protract (Abduct) and Retract (Adduct) the Scapula

	MUSCLE	NERVE	SEGMENTAL INNERVATION	
			Reported Range	Probable Most Important Segment(s)
Chief protractors	Serratus anterior	Long thoracic	C5-C8	C6
	Pectoralis minor	Medial pectoral	C7-T1	C8, T1
Accessory protractor	Pectoralis major	Pectorals	C5-T1	C5, 6, 7
Chief retractors	Trapezius	Accessory		
	Rhomboideus major and minor	Dorsal scapular	C4-C6	C5
Accessory retractor	Latissimus dorsi	Thoracodorsal	C6-C8	C7

flexed, full flexion at the shoulder can be obtained without inward rotation; the rotation observed with extended forearm seems to be a result primarily, therefore, of passive pull by the biceps. External rotation has been said to be essential to full abduction, but Johnston pointed out that this is not true of the more natural abduction in the plane of the scapula. In abduction in the coronal plane, however,

there is external rotation; Martin found that at least part of it is passive, for abduction in a cadaver caused external rotation after the greater tubercle had disappeared beneath the coracoacromial ligament: if this rotation, apparently imposed by the eccentricity of the coracoacromial arch in relation to the shoulder joint, was forcibly prevented, the insertion of the supraspinatus muscle came in contact with the lateral edge of the acromion; further abduction without rotation would obviously injure the supraspinatus, shear off the greater tubercle, lever the head of the humerus out of the socket, fracture the acromion, or result in some combination of these injuries.

Muscle Groups Acting Across Shoulder Joint

Proper action of the muscles moving the arm depends not only upon the prime movers themselves but also on two other factors: the shoulder must be sufficiently stabilized to afford a firm base upon which the arm can move, and the humerus must be held in firm contact with the glenoid cavity as the arm is moved. The first condition requires adequate scapular muscles: weakness of abduction of the arm may be the result, for instance, not of weakness of the deltoid and supraspinatus muscles, but rather of the trapezius or the ser-

ratus anterior. The second condition requires that muscles other than the prime movers often must contract simultaneously with them in order to prevent subluxation of the head of the humerus. As pointed out by Shevlin and co-workers, the contributions of the various muscles to any one movement, such as elevation of the arm, may vary according to the plane in which the movement is carried out. For instance, the activities of different parts of the deltoid and pectoralis major were found to vary with whether the movement was in the frontal or the sagittal plane.

The *flexors* at the shoulder joint (Fig. 4-56) are usually listed as the anterior part of the deltoid, the clavicular part of the pectoralis major, the coracobrachialis, and the biceps. The anterior part of the deltoid is particularly important, according to Duchenne, in flexing and adducting during the movement of

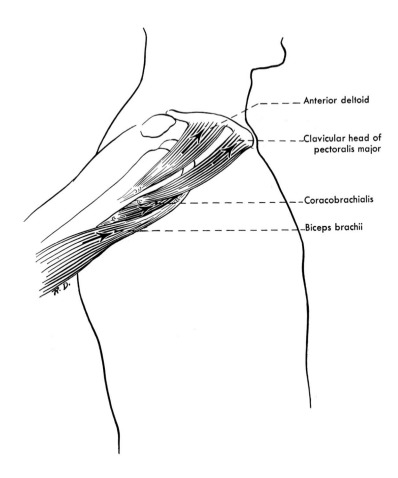

Anterior deltoid

Clavicular head of pectoralis major

Coracobrachialis

Biceps brachii

Fig. 4-56. Flexors of the arm.

Table 4-4
Innervation of the Muscles Flexing the Arm

	MUSCLE	NERVE	SEGMENTAL INNERVATION Reported Range	Probable Most Important Segment(s)
Chief flexors	Deltoid, anterior part	Axillary	C4-C7	C5, 6
	Pectoralis major, clavicular part	Lateral pectoral	C5-C6	C5, 6
	Coracobrachialis*	Musculocutaneous	C5-C8	C6, 7
From hyperextension only	Pectoralis major, sternal part	Both pectorals	C6-T1	C6, 7, 8
Accessory flexor	Biceps (?)	Musculocutaneous	C5-C6	C5, 6

* There is, perhaps, room for argument as to whether the coracobrachialis should be regarded as a primary or an accessory flexor.

bringing the arm across the front of the thorax; he said the clavicular part of the pectoralis major cannot do this. This part only of the pectoralis major can flex from the neutral position, starting with the arm by the side, and it assists flexion only to approximately the horizontal position (Duchenne said that with the arm above the head the clavicular portion is then an extensor; Beevor admitted it is anatomically, but said it does not contract when extension is carried out from this position). The lower part of the pectoralis major can flex only from a hyperextended position, bringing the arm forward until it is by the side. The coracobrachialis is usually considered an important flexor at the shoulder, but Duchenne said it flexes weakly, and Wright said it acts much less strongly in flexion than in adduction. Finally, while both heads of the biceps contract during flexion of the arm, Wright surmised that this is largely to protect the elbow joint and said that the biceps alone cannot produce much flexion at the shoulder. Electromyographic evidence (Chap. 5) has been interpreted as indicating that the biceps does usually participate in flexion.

The flexors of the arm are innervated through several different nerves, but primarily from the upper part of the brachial plexus (Table 4-4).

The chief *extensors* of the arm (Fig. 4-57) are the latissimus dorsi and the posterior part of the deltoid, which are aided by the sternocostal part of the pectoralis major between the

position of flexion and the resting position with the arm at the side. In addition, both the teres major and the long head of the triceps extend the arm a little beyond the resting position, although they apparently contract more in adduction than they do in extension. Of these several muscles, the posterior part of the deltoid is the only one, according to Duchenne, that can extend the arm far enough posteriorly to allow the action of placing the hand in the hip pocket.

Each of these muscles is supplied by a different peripheral nerve, and collectively they are supplied from both upper and lower parts of the plexus (Table 4-5).

The effective *abductors* of the humerus are only two, the supraspinatus and the deltoid (Fig. 4-58). As already noted, the two muscles normally work together throughout the whole range of movement; however, the deltoid is much stronger and also has greater leverage, and it alone can carry out abduction to about the horizontal plane (Duchenne—but, he said, there is then a tendency for the head of the humerus to subluxate downward). Often the supraspinatus cannot abduct that far, but some can (Duchenne; Staples and Watkins; Dehne and Hall). Normally, of course, abduction at the shoulder is accompanied by upward rotation of the scapula so as to allow movement above the horizontal. In the usual movement of abduction it is the middle part of the deltoid that is particularly active. With external rotation of the arm a more anterior part becomes active and with internal rota-

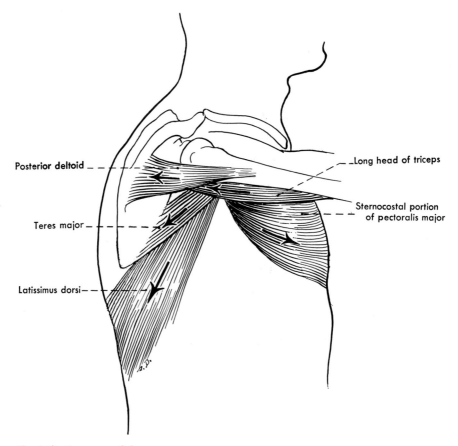

Fig. 4-57. Extensors of the arm.

Table 4-5
Innervation of the Muscles Extending the Arm

| | MUSCLE | NERVE | SEGMENTAL INNERVATION | |
			Reported Range	Probable Most Important Segment(s)
Chief extensors	Latissimus dorsi	Thoracodorsal	C5-C8	C7
	Deltoid, posterior	Axillary	C4-C7	C5, 6
From flexion only	Pectoralis major, sternal part	Both pectorals	C6-T1	C6, 7, 8
Accessory extensors	Teres major	Lower subscapular	C5-C7	C5, 6
	Triceps, long head	Radial	C6-C8	C7, 8

tion a more posterior part; abduction in external rotation is much more powerful than abduction in internal rotation (Duchenne, Wright), and Wright used the latter movement to bring out weakness in the deltoid. With the arm externally rotated, the long head of the biceps is in a position to aid in abduction, and does contract if the movement is resisted (and also, according to Dehne and Hall, when the deltoid is paralyzed). Wright suggested, as she did in considering flexion at the shoulder, that this contraction is largely a synergistic one to protect the elbow joint, and electromyography has apparently indicated the biceps probably does not normally participate in abduction.

While the abductors are each innervated by a different nerve, those nerves are derived exclusively from the upper part of the brachial plexus (Table 4-6).

The potential *adductors* are relatively numerous, and besides the powerful pectoralis major (chiefly the sternocostal part) and latissimus dorsi include the teres major, probably at least the posterior part of the deltoid, and the coracobrachialis and long head of the triceps (Fig. 4-59). Duchenne said the posterior part of the deltoid adducts from about 45°; Scheving and Pauly interpreted the contraction as a resistance to the internal rotation that would be produced by the latissimus and pectoralis major, as did Shevlin and co-workers. Adductor action of the anterior part of the deltoid is important only when it is combined with flexion. According to Duchenne, both the coracobrachialis and the long head of the triceps adduct only weakly, yet he and

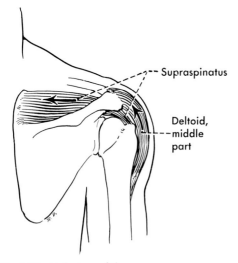

Fig. 4-58. Abductors of the arm.

Wright reported that they both contract strongly when adduction is carried out against resistance; much of this contraction, they said, is to resist the subluxation of the head of the humerus that the downward pull of the primary adductors, pectoralis and latissimus, would tend to produce.

Each of the adductors of the arm is supplied through a different peripheral nerve; collectively, they are supplied from both the upper and lower parts of the plexus but the strongest ones receive their nerve supply from the lower part (Table 4-7).

The *external or lateral rotators* (Fig. 4-60) include the infraspinatus, the teres minor, and the posterior part of the deltoid, according to most observers; Wright said the posterior part of the deltoid does not contract during external rotation. The supraspinatus is often listed as an external rotator, but according to Duchenne it internally rotates slightly as it abducts.

The *internal or medial rotators* (Fig. 4-60) are the subscapularis primarily, and to a less extent the teres major, anterior part of the deltoid, pectoralis major, latissimus dorsi, and perhaps the supraspinatus. According to Wright, the deltoid, pectoralis, and latissimus do not assist in internal rotation unless this is combined with another movement; Scheving and Pauly regarded the latissimus as a more important rotator than the pectoralis major, saying the latter becomes active only against resistance. Duchenne described the internal rotatory action of the teres major as being weak, and said that while isolated action of the supraspinatus both abducts and slightly internally rotates the arm, the muscle does not resist external rotation during the time

Table 4-6
Innervation of the Muscles Abducting the Arm

	MUSCLE	NERVE	SEGMENTAL INNERVATION Reported Range	Probable Most Important Segment(s)
Chief abductors	Deltoid	Axillary	C4-C7	C5, 6
	Supraspinatus	Suprascapular	C4-C6	C5
Accessory abductor	Biceps, long head (?)	Musculocutaneous	C5-C6	C5, 6

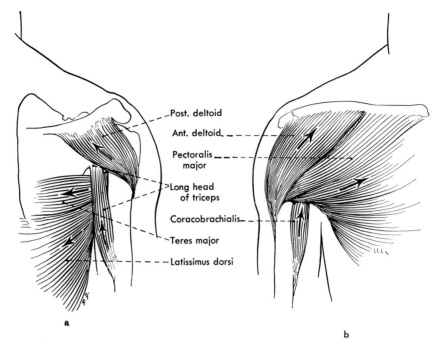

Fig. 4-59. Adductors of the arm from behind, *a*, and in front, *b*.

Table 4-7
Innervation of the Muscles Adducting the Arm

| | MUSCLE | NERVE | SEGMENTAL INNERVATION | |
			Reported Range	Probable Most Important Segment(s)
Chief adductors	Pectoralis major	Pectorals	C5-T1	C6, 7, 8
	Latissimus dorsi	Thoracodorsal	C5-C8	C7
	Teres major	Lower subscapular	C5-C7	C5, 6
Accessory adductors	Deltoid, posterior and anterior parts	Axillary	C4-C7	C5, 6
	Coracobrachialis	Musculocutaneous	C5-C8	C6, 7
	Triceps, long head	Radial	C6-C8	C7, 8

that it is maintaining the arm in abduction.

The rotators are supplied by a number of different peripheral nerves. However, the external rotators are supplied entirely from the upper part of the brachial plexus, while the internal rotators have a wider innervation, including both upper and lower parts of the plexus (Table 4-8).

Paralysis of Intrinsic Shoulder Muscles

The innervation of the various muscle groups acting on the humerus explains the typical position of the limb resulting from lesions more or less limited to the upper nerve trunks contributing to the brachial plexus, the so-called Erb-Duchenne type of paralysis. Since the flexor muscles, including the clavicular portion of the pectoralis major, are supplied primarily with nerve fibers from the fifth and sixth cervical nerves, while the extensors receive their innervation from a lower part of the plexus, the flexors are typically more severely injured and therefore the arm may be in slight extension; since the only abductors,

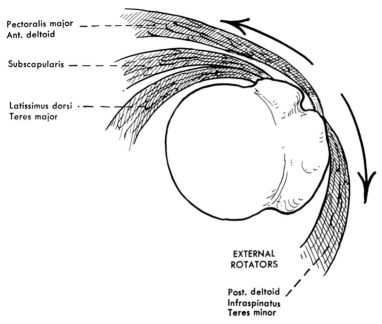

**INTERNAL
ROTATORS**

Pectoralis major — — — —
Ant. deltoid

Subscapularis — — — —

Latissimus dorsi — — —
Teres major

**EXTERNAL
ROTATORS**

Post. deltoid
Infraspinatus
Teres minor

Fig. 4-60. Rotators of the arm.

Table 4-8
Innervation of the Muscles Externally and Internally Rotating the Arm.

	MUSCLE	NERVE	SEGMENTED INNERVATION	
			Reported Range	*Probable Most Important Segment(s)*
Chief external	Infraspinatus	Suprascapular	C4-C6	C5, 6
rotators	Teres minor	Axillary	C4-C7	C5, 6
Accessory external	Deltoid, posterior	Axillary	C4-C7	C5, 6
rotator	part			
Chief internal	Subscapularis	Subscapulars	C5-C7	C5, 6
rotator				
Accessory internal	Teres major	Lower subscapular	C5-C7	C5, 6
rotators	Deltoid, anterior	Axillary	C4-C6	C5, 6
	part			
	Pectoralis major	Pectorals	C5-T1	C5-C8
	Latissimus dorsi	Thoracodorsal	C5-C8	C7
	Supraspinatus (?)	Suprascapular	C4-C6	C5

the supraspinatus and the deltoid, are likewise supplied by the fifth and sixth cervicals, while adductors such as the latissimus and the lower part of the pectoralis major receive fibers from lower parts of the plexus, the arm is adducted; and since the internal rotators, notably the pectoralis major and the latissimus dorsi, receive much of their innervation from the lower part of the brachial plexus, while the external rotators receive theirs from the upper part of the plexus only, the arm may also be somewhat internally rotated. Associated with this partial paralysis of the shoulder muscles there is a loss of flexor power at the elbow, since the muscles of the flexor surface of the arm are also innervated through the upper part of the plexus.

In general, muscle transplants about the shoulder are not considered particularly satisfactory, and arthrodesis is usually preferred; transplants are often reserved for patients with involvement of both shoulders and an arthrodesis on one side.

Harmon, however, regarded the results of transplantation of muscles around the shoulder in limited lesions as being superior to arthrodesis. He said that in order to have a functionally useful scapulohumeral joint there must be fair to good function in the muscles of the musculotendinous cuff, especially the supraspinatus and the external rotators, or effective movements of abduction and external rotation must be obtainable through muscle transplants. In order to obtain abduction and external rotation, he reported shifting the origin of the posterior part of the deltoid to an anterior site and so fastening the pectoralis major, latissimus dorsi, and long head of the triceps that they functioned as abductors and external rotators.

Steindler said·that substitution of the trapezius for a paralyzed deltoid is efficient in stabilizing the shoulder, but that the muscle does not provide sufficient contractile length, and acts rather as a tenodesis; he regarded attachment of the long head of the triceps and short head of the biceps to the acromion as stabilizing the shoulder only enough to prevent subluxation and, while recognizing the shortcomings of arthrodesis, noted that it does produce perfect stabilization. According to him, the most useful ranges of movement at the shoulder are 0° to 90° for abduction and adduction, 0° to 60° for flexion, and 0° to 45° for internal rotation; arthrodesis of the scapulohumeral·joint reduces abduction to less than 90°, flexion to about 50° to 60°, inward rotation to about 30°, and external rotation to less than 20°. However, a limb arthrodesed in slight abduction and flexion will have much of the most useful parts of the normal ranges of arm movements.

Antigravity and Protective Muscles
The muscles of the shoulder are sometimes divided also, according to their actions, into three groups: those that serve especially as antigravity muscles in addition to their action in moving the limb; those that are important in preventing displacement of the head of the humerus during the movements that they and other muscles carry out; and other muscles that participate in movement of the limb, but may play no particular part in the protection of the scapulohumeral joint. There are, of course, no clear distinctions among these groups.

Basmajian and Bazant found that with the arm at the side, loads beyond the supportive ability of the coracohumeral ligament provoke contraction of the supraspinatus and of the posterior, not the middle, part of the deltoid. They attributed the antigravity action of these muscles to the fact that the glenoid cavity is normally turned slightly upward, and thus an outward sliding of the head of the humerus, which these muscles resist, would have to occur to allow downward subluxation (but Scheving and Pauly said that in their study the glenoid cavity was more commonly turned slightly downward). They found no activity in the biceps or triceps, which like the middle deltoid are in a particularly favorable position to resist downward movement.

During abduction and flexion, when the supraspinatus and the deltoid are providing most of the upward movement, the subscapularis, infraspinatus, and teres minor contract to overcome the potential upward displacement and are the chief stabilizing muscles here (Inman and co-workers). However, Inman and co-workers said that the teres major contracts when a position of abduction or flexion is maintained; and Scheving and Pauly, reporting activity in all parts of the deltoid muscle during all movements of the humerus, regarded this as an important stabilizer. Finally, if the observations quoted concerning the contraction of the coracobrachialis and of the long head of the triceps during adduction are correct, these muscles, too, should be regarded as having a protective action at the shoulder.

It thus appears that of the muscles crossing the shoulder joint only the biceps, pectoralis major, and latissimus dorsi can be excluded from the antigravity and protective groups.

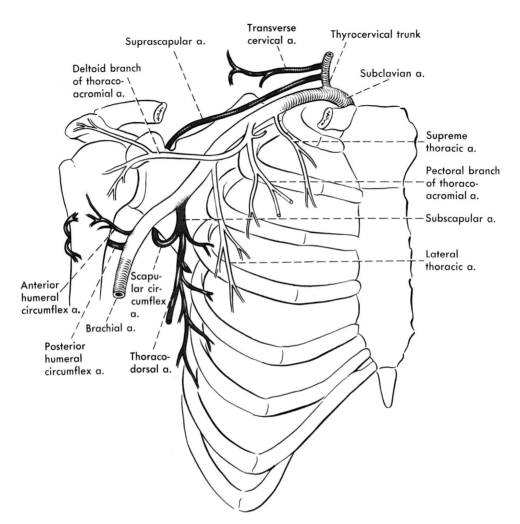

Deltoid branch
of thoraco-
acromial a.

Suprascapular a.

Transverse
cervical a.

Thyrocervical trunk

Subclavian a.

Supreme
thoracic a.

Pectoral branch
of thoraco-
acromial a.

Subscapular a.

Lateral
thoracic a.

Anterior
humeral
circumflex a.

Scapu-
lar cir-
cumflex
a.

Brachial a.

Posterior
humeral
circumflex a.

Thoraco-
dorsal a.

Fig. 4-61. Diagram of the arteries to the shoulder (*red*).

Most of the prime movers are therefore also important synergists in other movements.

VESSELS AND NERVES OF THE SHOULDER

The origins of most of the arteries to the shoulder and the details of their variations have already been described with the axilla. Similarly, the origins of the nerves to the shoulder and their variations have been discussed in Chapter 3. Further, the distributions of the vessels have already been largely discussed in connection with the axillary ar-

tery and its branches, and the distribution of the nerves in the discussions of the various muscles. The present discussion is therefore only a summary, with special reference to their courses. Their usual origins are shown in Figures 4-61 and 4-62.

SUPRASCAPULAR AND TRANSVERSE CERVICAL ARTERIES, SUPRASCAPULAR AND DORSAL SCAPULAR NERVES

These all arise in the neck and are distributed primarily to the back of the shoulder. The suprascapular (transverse scapular) artery and the transverse cervical artery (a. transversa colli) may arise close together from the thyrocervical trunk. If they do, they pass in front of the brachial plexus, and have a parallel course across the base of the neck, the transverse cervical lying above the suprascapular and typically above the level of the clavicle, the suprascapular typically lying behind the clavicle (Fig. 4-32). If, as frequently occurs, either of them or one branch of the transverse cervical artery arises from the second or third part of the subclavian, this vessel runs through or behind the brachial plexus.

Both vessels are paralleled in general by their veins, and the suprascapular artery is also paralleled by the suprascapular nerve. They leave the posterior triangle of the neck by disappearing under the anterior edge of the trapezius muscle. The suprascapular nerve, commonly C5 and C6, arises from the upper trunk of the brachial plexus and joins the corresponding artery in its course. The dorsal scapular nerve, arising from the fifth cervical just outside the intervertebral fora-

men, has a deeper course, passing through the scalenus medius, and then through or deep to the levator scapulae (to which it may give a branch) to come into apposition with the deep branch of the transverse cervical artery close to the superior angle of the scapula.

If the transverse cervical artery arises as a single stem, it divides under cover of the trapezius into a superficial ("ascending") and a deep (descending or dorsal) branch. In the more common instances in which what should be the superficial branch arises from the thyrocervical trunk and what should be the deep branch arises directly from the subclavian, there is, of course, no division and officially no transverse cervical artery; instead, the two arteries are then named the "superficial cervical" and the "dorsal (or descending) scapular" artery. The origin and therefore the technically correct terminology would usually not be known when one is working on the back of the shoulder, but the courses of the

Fig. 4-62. Origin of nerves to the shoulder (*shaded*).

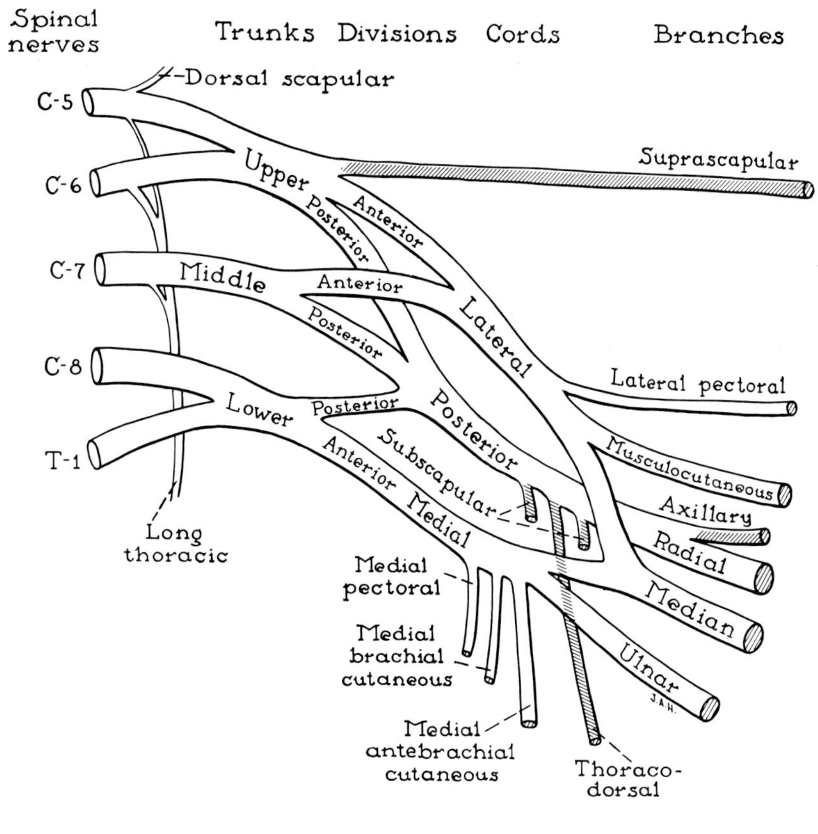

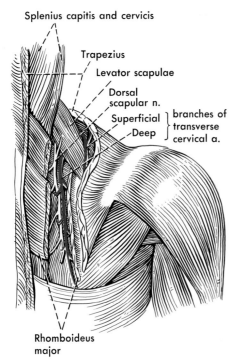

Splenius capitis and cervicis

Trapezius

Levator scapulae

Dorsal
scapular n.

Superficial ⎤ branches of
            ⎥ transverse
Deep ⎦ cervical a.

Rhomboideus
major

Fig. 4-63. Transverse cervical artery and dorsal scapular nerve.

vessels here remain the same. The superficial branch, or the superficial cervical artery, passes lateral to the levator scapulae and runs downward on the deep surface of the trapezius with the accessory nerve, sometimes also giving off an ascending branch that may anastomose above with twigs of the occipital. The deep branch of the transverse cervical, or the dorsal scapular artery, passes to the deep side of the levator scapulae and, joined by the dorsal scapular nerve, descends deep to the insertions of the levator scapulae and of the rhomboideus muscles, closely paralleling the medial border of the scapula (Fig. 4-63). The nerve ends in the two rhomboideus muscles and the artery anastomoses across the medial border of the scapula with the suprascapular and scapular circumflex arteries, and also with posterior branches of the intercostals that help to supply the rhomboids.

The origin of the suprascapular artery makes no difference in either its name or its course in the shoulder. The suprascapular vessels and nerve, upon reaching the superior border of the scapula, usually separate briefly, for the vessels typically pass above the superior transverse scapular ligament, the nerve beneath the ligament and therefore through the scapular notch, where it may be entrapped (Clein). Thereafter they run together again (Fig. 4-64). They lie in the supraspinatous fossa between the supraspinatus muscle and the bone, and give off branches into the muscle. They then pass deep to the root of the acromion into the infraspinatous fossa, where the nerve ends in the infraspinatus muscle; the artery helps to supply this muscle and anastomoses with branches of the scapular circumflex and transverse cervical arteries.

SUBSCAPULAR ARTERY AND NERVES, THORACODORSAL NERVE

These all arise in the axilla (Figs. 4-23 and 4-32), where the axillary artery and the posterior cord of the brachial plexus are against the costal surface of the subscapularis muscle. The nerves arise close together from the posterior cord, the elements forming it, or its branches, and usually consist of fifth and sixth cervical nerve fibers. The upper subscapular nerve, sometimes double and usually accompanied by a subscapular ramus from the axillary artery, goes directly into the upper part of the subscapularis muscle. The thoracodorsal, next to arise, emerges from behind the axillary vessels to run downward and laterally across the posterior wall of the axilla and enter the deep (costal) surface of the latissimus dorsi. It at first approximately parallels the subscapular artery, and below the origin of the scapular circumflex artery is closely accompanied by the branch of the thoracodorsal artery to the latissimus. The lower subscapular nerve divides soon after its origin to send one branch into the lower part of the subscapularis muscle, the other downward to the teres major.

The subscapular artery gives branches into the subscapularis muscle, and after a short course downward on the costal surface of the muscle divides into scapular circumflex and thoracodorsal branches.

The scapular circumflex artery turns around the lower border of the subscapularis

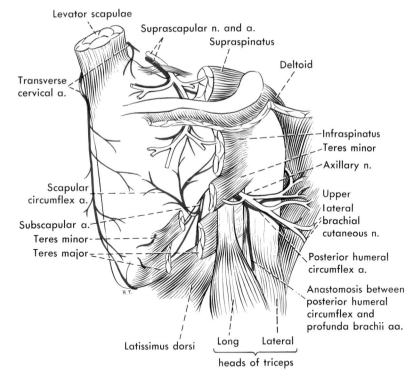

Fig. 4-64. Nerves and vessels of the posterior aspect of the shoulder.

muscle, through the triangular space, and gives branches to the teres major and minor, including a branch that descends between these muscles to the inferior angle of the scapula. It then turns around the lateral (axillary) border of the scapula, through the origin of the teres minor muscle, to lie between the infraspinatus muscle and the bone of the infraspinatous fossa (Fig. 4-64) and anastomose here with the suprascapular and transverse cervical arteries.

The thoracodorsal artery runs downward on the subscapularis muscle in company with the thoracodorsal nerve. It gives branches to this muscle and ends in branches that are distributed to the lateral thoracic wall and to the teres major and latissimus dorsi muscles.

HUMERAL CIRCUMFLEX ARTERIES AND AXILLARY NERVE

The anterior and posterior humeral circumflex arteries, usually the lowest branches of the axillary, arise at about the same level, sometimes by a common stem, sometimes

from opposite sides of the axillary artery. The anterior humeral circumflex is small; it passes anteriorly around the surgical neck of the humerus, deep to the two heads of the biceps and to the coracobrachialis, supplying adjacent muscles and ending by anastomosing with the posterior humeral circumflex. It sends a rather constant branch to the capsule of the shoulder joint, that passes upward just deep to the long head of the biceps.

The posterior humeral circumflex artery is much larger than the anterior, and travels with the axillary nerve. The latter, one of the terminal branches of the posterior cord, or sometimes with an independent origin from the posterior divisions of the brachial plexus, runs downward and laterally on the subscapularis muscle, behind the axillary artery, to the lower border of this muscle. Here it comes into contact with the posterior humeral circumflex artery. It typically contains fibers from the fifth and sixth cervical nerves. At the lower border of the subscapularis muscle the posterior humeral circumflex vessels and the axillary nerve turn posteriorly around the edge of the muscle and against the surgical neck of the humerus, through the quadrangular space. Artery and nerve then wind around the humerus under cover of the deltoid muscle, giving branches into it (Fig. 4-64). They usually lie no more than about 2 inches below the level of the acromion, hence longitudinal incisions in the deltoid for approaches to the shoulder joint have to be limited to this distance or less if the more anterior part of the deltoid is not to be denervated.

The artery supplies mainly the deltoid muscle. It sends a twig to the acromion and usually communicates with the deltoid branch of the profunda brachii, which ascends along the long head of the triceps, and it ends by anastomosing with the anterior humeral circumflex.

As the axillary nerve emerges from the quadrangular space it gives off a branch to the teres minor, which forms the upper border of this space, and then divides. One branch, the superior lateral brachial cutaneous nerve, passes around the posterior border of the deltoid muscle or through its posterior fibers, to be distributed to skin over the deltoid muscle; the remainder of the nerve circles the humerus with the posterior humeral circumflex artery, ending in the deltoid. The most posterior muscular branches may arise before the nerve divides or may arise from the cutaneous branch of the nerve.

OTHER BLOOD VESSELS AND NERVES

In the interval between the anterior border of the deltoid muscle and the lateral border of the clavicular head of the pectoralis major the cephalic vein, a convenient landmark for this interval, ascends and passes deeply to join the axillary vein. In its upper part it is paralleled by the deltoid branch of the thoracoacromial artery, which gives branches into both muscles, and usually gives rise to the acromial branch. The latter crosses the distal end of the clavicle to ramify subcutaneously on the upper surface of the acromion and here form, with twigs from the suprascapular and posterior humeral circumflex arteries, the acromial rete.

The *accessory* (spinal accessory) *nerve* arises both from the medulla and the spinal cord, but that portion arising from the medulla is distributed with the vagus, of which it actually forms the most caudal rootlets, and the part distributed to the sternocleidomastoid and trapezius muscles arises entirely from the spinal cord (Fig. 4-65). After making its exit from the jugular foramen it normally passes laterally between the internal jugular vein and the internal carotid artery, then crosses the lateral side of the vein to pass under cover of the sternocleidomastoid muscle. Here it receives a communication from the second cervical nerve, some of the cervical fibers entering the sternocleidomastoid, others accompanying the accessory nerve to the trapezius. The accessory nerve usually penetrates the sternocleidomastoid but may give off a separate branch into it and thereafter run deep to it. In either case it emerges at the posterior border of the muscle to cross the posterior triangle of the neck; it runs inferiorly and posteriorly on the outer surface of the levator scapulae (and superficially, since it is here enclosed only between two layers of the deep fascia of the neck, and covered by no muscle) to attain the deep or anterior surface of the trapezius muscle. Here it is commonly joined by branches from the third and fourth cervical nerves, some of which may go directly to the trapezius instead of joining the accessory. The nerve then travels down on the deep surface of the muscle in company with the superficial branch of the transverse cervical artery.

It has already been noted that, although the cervical nerve fibers have been thought to include motor ones, experimental evidence indicates that they are entirely afferent, and most clinical evidence indicates that if they do contain any motor fibers these are not an important part of the innervation of the trapezius.

The *long thoracic nerve*, or nerve to the serratus anterior, commonly arises from the posterior surfaces of the fifth, sixth, and seventh cervical nerves shortly after they have emerged through the intervertebral foramina. The roots emerge through, or less commonly

Fig. 4-65. The grouping of the cranial (bulbar) rootlets of the accessory nerve as the internal ramus that joins the vagus, and the continuation of the spinal rootlets as the external ramus, or accessory nerve of gross anatomy.

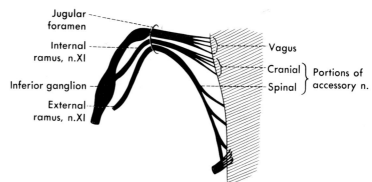

behind or in front of, the scalenus medius muscle. Almost as soon as it is formed, the nerve lies on the external surface of the serratus anterior muscle, the upper part of which appears in the base of the neck.

The long thoracic nerve passes downward on the serratus anterior muscle, largely hidden by the brachial plexus until it emerges from behind the plexus and the axillary vessels in the axilla (Figs. 4-22 and 4-23). It continues down along the medial wall of the axilla, on the serratus anterior, to supply the entire muscle; in the lower part of its course, especially, it is likely to be accompanied by branches of the thoracodorsal (subscapular) artery. It is said that the fibers of this nerve are distributed to the serratus anterior in a segmental fashion, the upper part of the muscle being innervated through fibers derived from the fifth cervical nerve when this rootlet is present and, if not, through fibers from the sixth, while the lower part is innervated through fibers derived from the seventh when this rootlet is present.

LYMPHATICS

The superficial lymphatics of the shoulder region and upper part of the arm converge toward the axilla. Those from the back of the shoulder and upper part of the arm pass around the posterior axillary fold to reach lower members of the group of axillary nodes, those from the anterolateral surfaces of the shoulder and arm pass forward; the lower ones pass below the edge of the pectoralis major to join the lateral axillary nodes, the upper ones pass between the pectoralis major and the deltoid to join infraclavicular nodes.

The deep lymphatics of the shoulder follow in general the blood vessels and therefore drain largely into axillary nodes; some of the deep lymphatics from the shoulder, however, follow the transverse cervical and suprascapular vessels, and therefore drain into outlying nodes of the lower deep cervical group (subclavian nodes).

References

ABBOTT LC, SAUNDERS JB DE CM: Acute traumatic dislocation of the tendon of the long head of the biceps brachii: A report of six cases with operative findings. Surgery 6:817, 1939

ADAMS JC: Recurrent dislocation of the shoulder. J Bone Joint Surg 30-B:26, 1948

ANDREASEN AT: Congenital absence of the humeral head: Report of two cases. J Bone Joint Surg 30-B:333, 1948

ARMSTRONG JR: Excision of the acromion in treatment of the supraspinatus syndrome: Report of ninety-five excisions. J Bone Joint Surg 31-B:436, 1949

BAILEY RW: Acute and recurrent dislocation of the shoulder. J Bone Joint Surg 49-A:767, 1967

BANKART ASB: The pathology and treatment of recurrent dislocation of the shoulder-joint. Brit J Surg 26:23, 1938

BARLOW RN: The sternalis muscle in American whites and Negroes. Anat Rec 61:413, 1935

BASMAJIAN JV, BAZANT FJ: Factors preventing downward dislocation of the adducted shoulder joint: An electromyographic and morphological study. J Bone Joint Surg 41-A:1182, 1959

BEARN JG: Direct observations on the function of the capsule of the sternoclavicular joint in clavicular support. J Anat 101:159, 1967

BEATON LE, ANSON BJ: Variations in the origin of the m. trapezius. Anat Rec 83:41, 1942

BEEVOR CE: The Croonian Lectures on Muscular Movements and Their Representation in the Central Nervous System. London, Adlard and Son, 1904

BISHOP WA, JR: Calcification of the supraspinatus tendon: Cause, pathologic picture and relation to the scalenus anticus syndrome. Arch Surg 39:231, 1939

BOYD HB, HUNT HL: Recurrent dislocation of the shoulder: The staple capsulography. J Bone Joint Surg 47-A:1514, 1965

BOYD HB, SISK JD: Recurrent posterior dislocation of the shoulder. J Bone Joint Surg 54-A:779, 1972

BROOME HL, BASMAJIAN JV: The function of the teres major muscle: An electromyographic study. Anat Rec 170:309, 1971

BUSCHMAKIN N: Die Lymphdrüsen der Achselhöhle, ihre Einteilung und Blutversorgung. Anat Anz 41:3, 1912

CALDWELL GA, UNKAUF BM: Results of treatment of subacromial bursitis in three hundred forty cases. Ann Surg 132:432, 1950

CAVE AJE, BROWN RW: On the tendon of the subclavius muscle. J Bone Joint Surg 34-B:466, 1952

CLARK E: Congenital variation of the pectoral muscles, with report of a case. J Anat Physiol 49:155, 1915

CLEIN LJ: Scapular entrapment neuropathy. J Neurosurg 43:337, 1975

CODMAN EA: The Shoulder. Boston, T Todd Co, 1934

CORBIN KB, HARRISON F: The sensory innervation of the spinal accessory and tongue musculature in the rhesus monkey. Brain 62:191, 1939

CRERAR JW: Note on the absence of the subclavius muscle. J Anat Physiol 26:554, 1892

CYRIAX EF: A second brief note on 'floating clavicle.' Anat Rec 52:97, 1932

DANESE C, HOWARD JM: Postmastectomy lymphedema. Surg Gynecol Obstet 120:797, 1965

DEBEYRE J, PATTE D, ELMELIK E: Repair of ruptures of the rotator cuff of the shoulder. With a note on advancement of the supraspinatus muscle. J Bone Joint Surg 47-B:36, 1965

DE GARIS CF, SWARTLEY WB: The axillary artery in white and Negro stocks. Am J Anat 41:353, 1928

DEHNE E, HALL RM: Active shoulder motion in complete deltoid paralysis. J Bone Joint Surg 41-A:745, 1959

DENHAM RH, JR, DINGLEY AF, JR: Epiphyseal separation of the medial end of the clavicle. J Bone Joint Surg 49-A:1179, 1967

DEPALMA AF: Recurrent dislocation of the shoulder joint. Ann Surg 132:1052, 1950

DEPALMA AF, CALLERY G, BENNETT GA: Variational anatomy and degenerative lesions of the shoulder joint. Am Acad Orth Surgeons Instructional Course Lectures 6:255, 1949

DEWAR FP, HARRIS RI: Restoration of function of the shoulder following paralysis of the trapezius by fascial sling fixation and transplantation of the levator scapulae. Ann Surg 132:1111, 1950

DICKSON JA, O'DELL HW: A phylogenetic study of recurrent anterior dislocation of the shoulder joint. Surg Gynecol Obstet 95:357, 1952

DUCHENNE GB: Physiology of Motion Demonstrated by Means of Electrical Stimulation and Clinical Observation and Applied to the Study of Paralysis and Deformities. Translated and edited by E B Kaplan. Philadelphia, JB Lippincott, 1949

DUGGAN N: Recurrent dislocation of the sternoclavicular cartilage. J Bone Joint Surg 13:365, 1931

DURMAN DC: An operation for paralysis of the serratus anterior. J Bone Joint Surg 27:380, 1945

ELLIS VH: The diagnosis of shoulder lesions due to injuries of the rotator cuff. J Bone Joint Surg 35-B:72, 1953

FICK R: Notiz über einen M. sternalis. Arch f Anat u Physiol (Anat Abt) p 193, 1899

FREEDMAN L, MUNRO RR: Abduction of the arm in the scapular plane: Scapular and glenohumeral movements. A roentgenographic study. J Bone Joint Surg 48-A:1503, 1966

GALLIE WE, LE MESURIER AB: Recurring dislocation of the shoulder. J Bone Joint Surg 30-B:9, 1948

GARDNER E: The innervation of the shoulder joint. Anat Rec 102:1, 1948

GIBSON JMC: Rupture of the axillary artery. J Bone Joint Surg 44-B:114, 1962

GRANT JCB, SMITH CG: Age incidence of rupture of the supraspinatus tendon. Anat Rec 100:666, 1948 (abstr)

HARMON PH: Surgical reconstruction of the paralytic shoulder by multiple muscle transplantations. J Bone Joint Surg 32-A:583, 1950

HAUSER EDW: Avulsion of the tendon of the subscapularis muscle. J Bone Joint Surg 36-A:139, 1954

HAYES WM: Rupture of the pectoralis major muscle: Review of the literature and report of two cases. J Internat Coll Surgeons 14:82, 1950

HITCHCOCK HH, BECHTOL CO: Painful shoulder: Observations on the role of the tendon of the long head of the biceps brachii in its causation. J Bone Joint Surg 30-A:263, 1948

HORAN FT, BONAFEDE RP: Bilateral absence of the trapezius and sternal head of the pectoralis major muscles: A case report. J Bone Joint Surg 59-A:133, 1977

HORN JS: The traumatic anatomy and treatment of acute acromio-clavicular dislocation. J Bone Joint Surg 36-B:194, 1954

HORWITZ MT, TOCANTINS LM: An anatomical study of the role of the long thoracic nerve and the related scapular bursae in the pathogenesis of local paralysis of the serratus anterior muscle. Anat Rec 71:375, 1938a

HORWITZ MT, TOCANTINS LM: Isolated paralysis of the

serratus anterior (magnus) muscle. J Bone Joint Surg 20:720, 1938b

HOWARD FM, SHAFER SJ: Injuries to the clavicle with neurovascular complications: A study of fourteen cases. J Bone Joint Surg 47-A:1335, 1965

HUELKE DF: Variation in the origins of the branches of the axillary artery. Anat Rec 135:33, 1959

HUNTINGTON GS: The derivation and significance of certain supernumerary muscles of the pectoral region. J Anat Physiol 39:1, 1904

IMATANI RJ, HANLON JJ, CADY GW: Acute, complete, acromioclavicular separation. J Bone Joint Surg 57-A:328, 1975

INMAN VT, SAUNDERS JB DE CM, ABBOT LC: Observations on the function of the shoulder joint. J Bone Joint Surg 26:1, 1944

JEANNOPOULOS CL: Congenital elevation of the scapula. J Bone Joint Surg 34-A:883, 1952

JOHNSON EW, JR, COLLINS HR: Nonunion of the clavicle. Arch Surg 87:963, 1963

JOHNSTON GW, LOWRY JH: Rupture of the axillary artery complicating anterior dislocation of the shoulder. J Bone Joint Surg 44-B:116, 1962

JOHNSTON TB: The movements of the shoulder-joint: A plea for the use of the "plane of the scapula" as the plane of reference for movements occurring at the humero-scapular joint. Brit J Surg 25:252, 1937

JONES V: Recurrent posterior dislocation of the shoulder. Report of a case treated by bone block. J Bone Joint Surg 40-B:203, 1958

KAPLAN T, KATZ A: Thrombosis of the axillary vein: Case report with comments on etiology, pathology and diagnosis. Am J Surg 37:326, 1937

KENNEDY JC: Retrosternal dislocation of the clavicle. J Bone Joint Surg 31-B:74, 1949

KERR AT: The brachial plexus of nerves in man, the variations in its formation and branches. Am J Anat 23:285, 1918

LAMBERT AE: A rare variation in the pectoralis minor muscle. Anat Rec 31:193, 1925

LANDRY SO, JR: The phylogenetic significance of the chondro-epitrochlearis muscle and its accompanying pectoral abnormalities. J Anat 92:57, 1958

LESLIE JT, JR, RYAN TJ: The anterior axillary incision to approach the shoulder joint. J Bone Joint Surg 44-A:1193, 1962

LEWIS OJ: The coraco-clavicular joint. J Anat 93:296, 1959

LIBERSON F: Os acromiale—A contested anomaly. J Bone Joint Surg 19:683, 1937

LOCKHART RD: Movements of the normal shoulder joint and of a case with trapezius paralysis studied by radiogram and experiment in the living. J Anat 64:288, 1930

LORHAN PH: Isolated paralysis of the serratus magnus following surgical procedures: Report of a case. Arch Surg 54:656, 1947

MARMOR L, BECHTOL CO, HALL CB: Pectoralis major

muscle. Function of sternal and mechanism of rupture of normal muscle: Case reports. J Bone Joint Surg 43-A:81, 1961

MARTIN CP: The movements of the shoulder-joint, with special reference to rupture of the supraspinatus tendon. Am J Anat 66:213, 1940

MCCALLY WC, KELLY DA: Treatment of fractures of the clavicle, ribs and scapula. Am J Surg 50:558, 1940

MCCORMACK LJ, CAULDWELL EW, ANSON BJ: Brachial and antebrachial arterial patterns: A study of 750 extremities. Surg Gynecol Obstet 96:43, 1953

MCLAUGHLIN HL: Lesions of the musculotendinous cuff of the shoulder: III. Observations on the pathology, course and treatment of calcific deposits. Ann Surg 124:354, 1946

MEYER AW: Spontaneous dislocation and destruction of tendon of long head of biceps brachii: Fifty-nine instances. Arch Surg 17:493, 1928

MEYER AW: Chronic functional lesions of the shoulder. Arch Surg 35:646, 1937

MILCH H: The treatment of recent dislocations and fracture-dislocations of the shoulder. J Bone Joint Surg 31-A:173, 1949

MILLER RA: Observations upon the arrangement of the axillary artery and brachial plexus. Am J Anat 64:143, 1939

MOORE DC, BRIDENBAUGH LD, EATHER KF: Block of the upper extremity: Supraclavicular approach versus axillary approach. Arch Surg 90:68, 1965

MORTENSEN OA, WIEDENBAUER M: An electromyographic study of the trapezius muscle. Anat Rec 112:366, 1952 (abstr)

MOSELEY HF, OVERGAARD B: The anterior capsular mechanism in recurrent anterior dislocation of the shoulder: Morphological and clinical studies with special reference to the glenoid labrum and gleno-humeral ligaments. J Bone Joint Surg 44-B:913, 1962

NEER CS II: Anterior acromioplasty for the chronic impingement syndrome of the shoulder: A preliminary report. J Bone Joint Surg 54-A:41, 1972

NEUHOF H: Excision of the axillary vein in the radical operation for carcinoma of the breast. Ann Surg 108:15, 1938

NICOLA T: Recurrent dislocation of the shoulder. Am J Surg 86:85, 1953

OSMOND-CLARKE H: Habitual dislocation of the shoulder: The Putti-Platt operation. J Bone Joint Surg 30-B:19, 1948

OWEN R: Bilateral glenoid hypoplasia: Report of five cases. J Bone Joint Surg 35-B:262, 1953

PRESCOTT MU, ZOLLINGER RW: Alar scapula: An unusual surgical complication. Am J Surg 65:98, 1944

READ WT, TROTTER M: The origins of transverse cervical and of transverse scapular arteries in American whites and Negroes. Am J Phys Anthropol 28:239, 1941

REISS FP, DE CAMARGO AM, VITTI M, DE CARVALHO

CAF: Electromyographic study of the subclavius muscle. Acta Anat 105:284, 1979

ROWE CR, PATEL D, SOUTHMAYD WW: The Bankart procedure. A long-term end-result study. J Bone Joint Surg 60-A:1, 1978

SAUNDERS WH, JOHNSON EW: Rehabilitation of the shoulder after radical neck dissection. Ann Otol Rhinol Laryngol 84:812, 1975

SCHEVING LE, PAULY JE: An electromyographic study of some muscles acting on the upper extremity of man. Anat Rec 135:239, 1959

SEIB GA: The m. pectoralis minor in American whites and American Negroes. Am J Phys Anthropol 23:389, 1938

SELDEN BR: Congenital absence of trapezius and rhomboideus major muscles. J Bone Joint Surg 17:1058, 1935

SHEEHAN D: Anatomical notes. Bilateral absence of trapezius. J Anat 67:180, 1932

SHEVLIN MG, LEHMANN JF, LUCCI JA: Electromyographic study of the function of some muscles crossing the glenohumeral joint. Arch Phys Med Rehab 50:264, 1969

STAPLES OS, WATKINS AL: Full active abduction in traumatic paralysis of the deltoid. J Bone Joint Surg 25:85, 1943

STEINDLER A: The reconstruction of upper extremity in spinal and cerebral paralysis. Am Acad. Oth. Surgeons, Instructional Course Lectures 6:120, 1949

SYMEONIDES PP: The significance of the subscapularis muscle in pathogenesis of recurrent anterior dislocation of the shoulder. J Bone Joint Surg 54-B:476, 1972

TAYLOR S: Clavicular dysostosis: A case report. J Bone Joint Surg 27:710, 1945

TELFORD ED, MOTTERSHEAD S: Pressure at the cervico-brachial junction: An operative and anatomical study. J Bone Joint Surg 30-B:249, 1948

TROTTER M, HENDERSON JL, GASS H, BRUA RS, WEISMAN S, AGRESS H, CURTIS GH, WESTBROOK ER: The origins of branches of the axillary artery in whites and in American Negroes. Anat Rec 46:133, 1930

URIST MR: Complete dislocation of the acromioclavicular joint. J Bone Joint Surg 45-A:1750, 1963

VARE AM, INDURKAR GM: Some anomalous findings in the axillary musculature. J Anat Soc India 14:34, 1965

WATERSTON D: Variations in the teres minor muscle. Anat Anz 32:331, 1908

WEATHERSBY HT: The valves of the axillary, subclavian and internal jugular veins. Anat Rec 124:379, 1956 (abstr.)

WILLIAMS GA: Pectoral muscle defects: Cases illustrating three varieties. J Bone Joint Surg 28:417, 1930

WILSON CL, DUFF GL: Pathologic study of degeneration and rupture of the supraspinatus tendon. Arch Surg 47:121, 1943

WILSON JC, MCKEEVER FM: Traumatic posterior (retroglenoid) dislocation of the humerus. J Bone Joint Surg 31-A:160, 1949

WILSON JT: Further observations on the innervation of axillary muscles in man. J Anat Physiol 24:52, 1889

WOODWARD JW: Congenital elevation of the scapula: Correction by release and transplantation of muscle origins. A preliminary report. J Bone Joint Surg 43-A:219, 1961

WRIGHT WG: Muscle Function New York, Hoeber, 1928: Hafner, 1962

Chapter 5

ARM, ELBOW, AND FOREARM

The Arm

The musculature (Fig. 5-1), nerves, and vessels of the arm are relatively simple. The muscles arise from the scapula or from the humerus and insert into the humerus or cross the elbow joint to insert upon the radius or the ulna. The muscles on the anterior surface of the arm are innervated by the musculocutaneous nerve, a derivative of the anterior portion of the brachial plexus, while the muscles on the posterior surface of the arm are innervated by the radial nerve, from the posterior cord of the plexus. The other nerves, the median and ulnar, simply traverse the arm on their way to the forearm and hand, and commonly give off no branches in the arm except perhaps articular ones to the elbow joint. The brachial artery, the direct continuation of the axillary, normally undergoes no appreciable diminution in size during its course through the arm, and ends by dividing into radial and ulnar arteries in the forearm.

HUMERUS

Form

The upper end of the humerus (Figs. 5-2 to 5-4) is formed by the head and the two tubercles. The rounded head, covered by articular cartilage, is directed upward, medially, and somewhat backward. A narrow rim of bone at the edge of the articular surface is the ana-

tomical neck, distinct only where it separates the head from the tubercles. The greater tubercle (tuberosity) is on the lateral side of the upper end of the humerus, the lesser tubercle (tuberosity) is on the anterior side; between the two and their downward extensions, or crests, is the intertubercular (bicipital) groove. The tubercles and their crests serve for the attachment of a number of muscles.

The lesser tubercle may be prolonged upward as far as the epiphyseal line by a supratubercular ridge (Fig. 5-3), which predisposes to fraying of the biceps tendon as this curves medially above the tubercle (Meyer, '28). Hitchcock and Bechtol found such a ridge in 67% of humeri, and described it as "markedly developed" in 8%. Spurs projecting from the lesser tubercle, increasing the possibility of damage to the biceps tendon, are often associated with supratubercular ridges.

The narrowest part of the body or shaft of the humerus lies a little below the tubercles and is called the surgical neck because of the relative frequency of fractures here. The body below the tubercles is roughly cylindrical until it begins to be flattened in an anteroposterior direction in its lower part, but is described as having three surfaces and three borders. The anterior border, between anterolateral and anteromedial surfaces, begins with the crest of the greater tubercle, includes the medial edge of the deltoid tuberosity, becomes indistinct on the more flattered lower

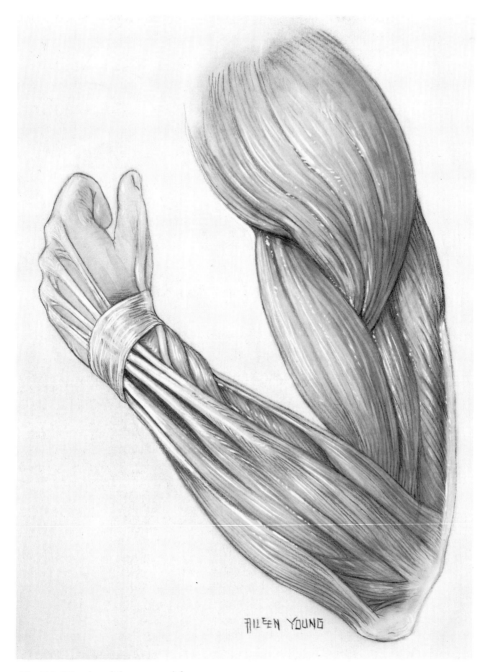

Fig. 5-1. Muscles of the arm and forearm.

part of the body, and ends at the sharp lateral edge of the coronoid fossa. The medial border begins with the crest of the lesser tubercle, often fades out in the middle of the bone, and ends as the medial supracondylar ridge. The lateral border, poorly defined above but cor-

responding to the posterior border of the greater tubercle, ends below as the lateral supracondylar ridge.

The anterolateral surface of the humerus presents the deltoid tuberosity above; on the lower part of the body, where the anterior border becomes poorly defined, anterolateral and anteromedial surfaces tend to blend as they both give rise to the brachialis muscle. The anteromedial surface has no markings particularly worthy of note. The posterior surface is crossed obliquely by a shallow groove, the groove of the radial nerve (radial or spiral groove) that runs downward and laterally from behind the deltoid tuberosity.

According to Krahl's findings, the medial torsion that the humerus undergoes between its upper and lower ends occurs primarily at the upper epiphyseal plate and is not complete until the nineteenth or twentieth year.

The lower end or condyle of the humerus bears two articular surfaces, a medial trochlea for articulation with the ulna and a lateral capitulum for articulation with the radius. The trochlea is pulleylike, with its articular surface extending around the lower border of the bone from the anterior to the posterior aspect. The convex articular surface of the ca-

Fig. 5-2. Anterior views of the humerus. In *b* origins of muscles are shown in *red*, insertions in *black*.

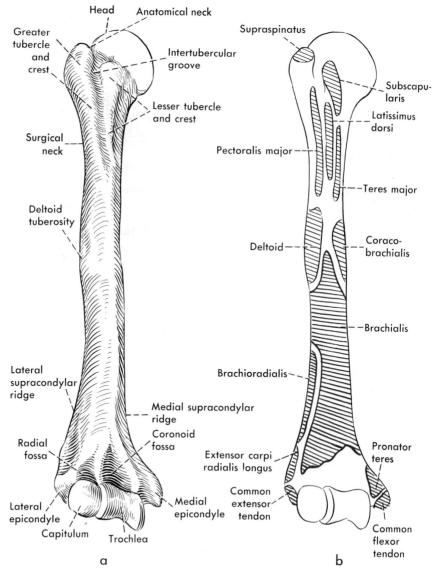

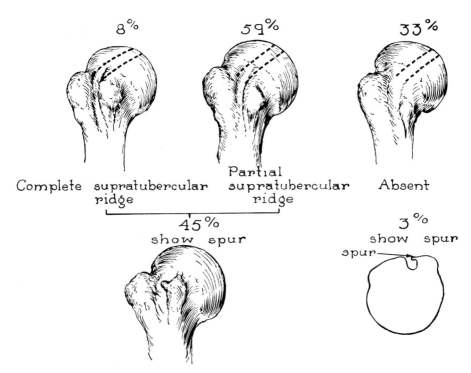

8%   59%   33%

Complete supratubercular ridge   Partial supratubercular ridge   Absent

45% show spur   3% show spur

spur

Fig. 5-3. Variations in the supratubercular ridge. (Hitchcock HH, Bechtol CO: J Bone Joint Surg 30-A:263, 1948)

pitulum is limited to the anterior and inferior aspects of the bone. Above the trochlea there is a pit, the coronoid fossa, in the anterior surface of the humerus, for the reception of the coronoid process of the ulna when the forearm is flexed, and on the posterior surface there is a larger pit, the olecranon fossa, for the reception of the olecranon when the forearm is extended. The bone between these two pits is very thin and may be replaced by membrane, thus leaving a gap, the septal aperture, in the dried bone. Trotter found septal apertures in 8% of humeri. The radial fossa is a depression on the anterior surface of the humerus above the capitulum, for reception of the head of the radius during flexion of the forearm.

The sharp edges of the flattened lower end of the humerus above the condyle are usually called the supracondylar ridges. The medial and lateral intermuscular septa of the arm are attached to these ridges, and the lateral one gives origin on its anterior surface to some of the extensor muscles of the forearm. The lat-

eral supracondylar ridge ends below in the lateral epicondyle, less prominent than the medial. The medial supracondylar ridge ends below in the prominent medial epicondyle, which has posteroinferiorly a smooth sulcus for the ulnar nerve, and a roughened anteromedial surface for the attachment of some of the flexor muscles of the forearm.

VARIATIONS, ANOMALIES, AND FRACTURE

Common variations in the upper end of the humerus are the supratubercular ridge, already noted, and differences in the depth of the intertubercular groove and in the angle that the medial wall (lesser tubercle) forms with the floor. The results of Hitchcock and Bechtol's investigation of the latter are discussed in connection with the shoulder joint and illustrated in Figure 4-49. As pointed out there, the lesser tubercle functions as a pulley for the long tendon of the biceps during flexion-extension movements, and shoulders with

Fig. 5-4. Posterior views of the humerus. In *b* origins of muscles are shown in *red,* insertions in *black.*

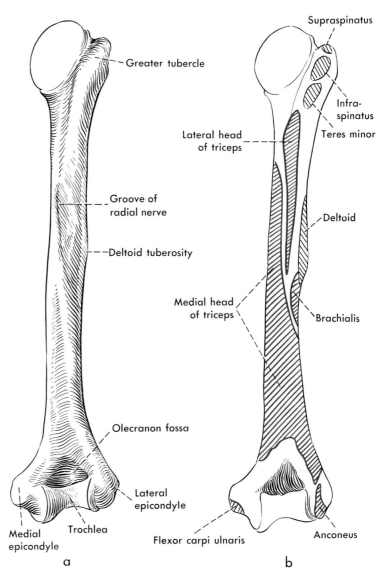

Greater tubercle

Supraspinatus

Infra-spinatus

Lateral head of triceps

Teres minor

Groove of radial nerve

Deltoid

Deltoid tuberosity

Medial head of triceps

Brachialis

Olecranon fossa

Lateral epicondyle

Medial epicondyle   Trochlea

Flexor carpi ulnaris

Anconeus

a

b

shallower grooves and lesser angles are more disposed to dislocation of the long head of the biceps than are those with deeper grooves and greater angles.

A relatively common variation in the lower part of the humerus is the presence of a su-pracondylar (supracondyloid) process, a projection from the anteromedial surface of the humerus about 2 inches (5 cm) above the medial epicondyle (Fig. 5-5). The supracondylar process varies in size and form, for it may be a small nubbin, it may be in part cartilaginous, or it may be a decided bony hook.

Terry concluded that 1% probably represents the incidence in which there is a bony spine of 3 mm or more in height. Apparently the process is very rare in colored races.

A well-developed supracondylar process is often connected to the medial epicondyle by a fibrous band (ligament of Struthers), and upper fibers of the pronator teres typically arise from the process and the band; usually then the median nerve, and often the brachial artery, pass behind the process and then forward between the fibrous band and the bone, and the ulnar nerve has been reported

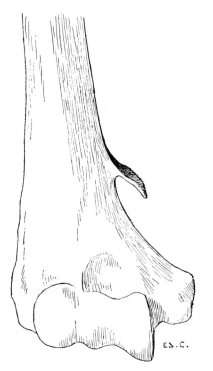

Fig. 5-5. A supracondylar process. (Poirier P, Charpy A: Traité d'Anatomie Humaine [ed 3], Vol 1. Paris, Masson et Cie, 1911)

as lying against it posteriorly (Kessel and Rang). In spite of this relationship, most supracondylar processes are symptomless, but compression of the median nerve or of the artery, or both, especially in extension and supination, may require operative intervention (Kessel and Rang). Kolb and Moore noted that it may be important to recognize fracture of the process, since the symptoms subside without treatment if there is no entrapment of nerve and vessel. Smith and Fisher reported a case in which the nerve was entrapped by a "Struther's ligament" that arose from the humerus in the absence of a supracondylar process.

Isolated absence of the humerus is, according to O'Rahilly, extremely rare if it exists at all, for he could find no well-authenticated case of it; he himself reported a case in which one humerus was very rudimentary, but the radius was missing bilaterally.

Humerus varus is a rare congenital abnormality of the upper end of the humerus. According to Lloyd-Roberts this is defined as a reduction in the head-body angle to less than 140°, protrusion of the greater tubercle above the level of the upper margin of the neck, and reduction of the distance between the articular surface of the head and the cortex of the lateral side of the humerus. In the case he reported, abduction was limited to about 80° and was mostly carried out by the scapula because the greater tubercle became impacted against the acromion after about 20° of abduction at the scapulohumeral joint; lateral rotation of the humerus was also limited. Removal of the acromion relieved the pain and increased the mobility of the joint.

Fracture

Because the humerus is closely surrounded by muscle, nonunion of fractures may result as a consequence of muscle being interposed between the bony ends.

In fracture of the humerus displacement of the ends may result from the pull of the muscles inserting into the parts: thus in fracture of the surgical neck the upper fragment may be abducted and externally rotated through the actions of the muscles inserting on the greater tubercle, while the lower fragment is adducted and internally rotated through the pull of such muscles as the pectoralis major, latissimus, and teres major. Fracture at other levels also may lead to angulation (Fig. 5-6) and, as in fracture of any long bone, overriding of the ends of the fracture may result from muscular pull (Fig. 5-7), hence the necessity for traction. Holstein and Lewis found the radial nerve to be commonly injured in humeral fractures only by gross displacement of the fragments, therefore typically in spiral fractures of the distal third with proximal displacement of the distal fragment and lateral displacement of the proximal one; injury is then close to the passage of the nerve through the lateral intermuscular septum, where the nerve is least mobile, and of course spares the triceps. Supracondylar fractures may injure the median or ulnar nerves, or blood vessels, and Lipscomb and Burleson found that in

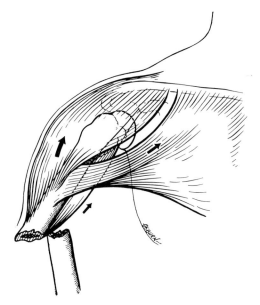

Fig. 5-6. Angulation of a fractured humerus as a result of muscle pull. (Bickel WH. In Morris GM (ed): The Cyclopedia of Medicine, Surgery, Specialties, Vol 5, p 843. Philadelphia, FA Davis, 1956)

their referral series of more severe injuries vessels, nerves, or both were involved in 22% of 108 such fractures. Spinner and Schreiber reported cases of paralysis of the anterior interosseous branch of the median nerve following supracondylar fractures in children, and suggested this might result from traction on a nerve held closely against the ulna.

BLOOD SUPPLY

The blood supply of the humerus is from branches of the brachial artery, including one or more nutrient arteries and the vessels intimately associated with the ends of the bone. The nutrient branch of the brachial artery, sometimes called either the nutrient or the chief nutrient artery of the humerus, is a slender branch that arises from the brachial (or sometimes from the superior ulnar collateral) artery about the middle of the arm and enters the anteromedial surface of the humerus. The foramen through which this vessel travels in the bone is directed very obliquely downward; its external end is usually located in the middle third of the bone, but varies consider-

ably, from a position above the lower end of the deltoid tuberosity to one at about the junction of middle and lower thirds of the bone.

The profunda brachii artery also usually sends a nutrient branch into the body of the humerus as it starts its course around the bone. This nutrient vessel usually enters the bone above and medial to the groove of the radial nerve, and may be the largest nutrient artery of the humerus.

The upper end of the humerus recieves its blood supply primarily from the posterior humeral circumflex vessels, and nutrient foramina for these vessels are particularly numerous. The lower end of the humerus receives its blood supply from the vessels that form the anastomosis around the elbow, but particularly posteriorly, and therefore especially from the anastomoses formed by the superior ulnar collateral and the posterior ulnar recurrent vessels on the medial side, and the profunda brachii (middle collateral) and interosseous recurrent vessels laterally. Nutrient foramina through which these vessels enter or leave are especially prominent on the posterior surface of the lateral epicondyle.

Fig. 5-7. Shortening of a fractured humerus as a result of muscle pull. (Bickel WH. In Morris GM (ed): The Cyclopedia of Medicine, Surgery, Specialties, Vol 5, p 843. Philadelphia, FA Davis, 1956)

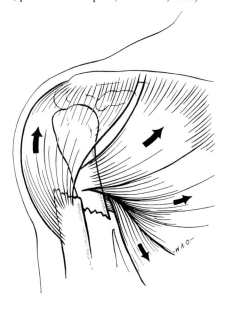

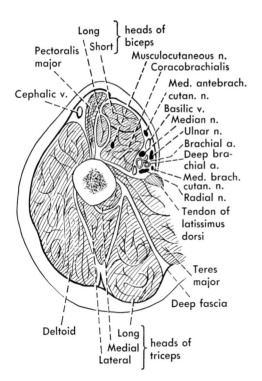

Fig. 5-8. Cross section through the left arm at the level of the lower border of the axilla. (Redrawn from Eycleshymer AC, Schoemaker DM: A Cross-Section Anatomy. New York, Appleton, 1923)

FASCIA; SUPERFICIAL VESSELS AND NERVES

Fascia

The superficial fascia, or tela subcutanea, of the arm contains a variable amount of fat, parts of the cephalic and basilic veins and their tributaries, the terminal ramifications of the cutaneous nerves of the arm, and, in its lower part, the upper stems of the antebrachial cutaneous nerves.

The brachial fascia or deep fascia of the arm (Figs. 5-8 to 5-10) is continuous above with the fascia over the deltoid and pectoral muscles and with the axillary fascia, and below with the deep fascia of the forearm. It is attached to the medial and lateral epicondyles of the humerus, and above these sends medial and lateral intermuscular septa between the flexor and extensor muscles to attach to the humerus.

The *medial intermuscular septum* (Figs. 5-9 and 5-10) begins above at the lower border of the pectoralis and latissimus dorsi muscles, with the fascia of which it is continuous. It lies between the coracobrachialis and brachialis muscles anteriorly and the triceps posteriorly, and gives origin to some of the fibers of the brachialis and triceps muscles. In the lower part of the arm it is attached to the medial supracondylar ridge, and it ends at the medial epicondyle. It is pierced by the ulnar nerve and the superior ulnar collateral artery as these pass backward toward the medial epicondyle.

The *lateral intermuscular septum* (Figs. 5-9 and 5-10) begins above at the insertion of the deltoid muscle and is attached to the humerus, including the lateral supracondylar ridge and the lateral epicondyle; it lies between the brachialis, the brachioradialis, and the extensor carpi radialis longus muscles anteriorly and the triceps posteriorly, and gives origin to

Fig. 5-9. Cross section through the middle third of the left arm. (Redrawn from Eycleshymer AC, Schoemaker DM: A Cross-Section Anatomy. New York, Appleton, 1923)

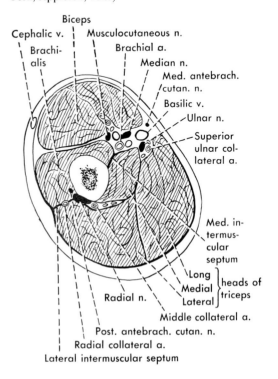

some of the fibers of all these muscles. It is pierced by the radial nerve and the radial collateral (anterior descending) branch of the profunda brachii artery as these pass anteriorly to reach the anterior aspect of the elbow joint.

The intermuscular septa divide the major part of the arm into two compartments, of which the anterior contains the flexor muscles, the posterior the extensor (triceps). In the posterior compartment the triceps is firmly attached to the brachial fascia, but in the anterior one the fascia lies loosely over the biceps, and the brachialis is attached only to the anterior surfaces of the intermuscular septa.

Superficial Vessels

The *cephalic vein* lies in the subcutaneous tissue of the arm, passing in front of the elbow and upward in the lateral bicipital sulcus or groove at the lateral border of the biceps, and then in the groove between the deltoid and pectoral muscles (Fig. 5-11). In the lower part of the arm it passes superficial to the lateral antebrachial cutaneous nerve. It receives no major branches in the arm.

The *basilic vein* also ascends in front of the elbow, communicating here with the cephalic through the vena mediana cubiti. In the lower part of the arm it lies in the medial bicipital groove, where it is accompanied by the medial antebrachial cutaneous nerve. At about the junction of the middle and lower thirds of the arm the basilic vein penetrates the brachial fascia (at about the same point at which the medial antebrachial cutaneous nerve emerges from deep to this fascia). The exact pattern of the veins at the elbow varies considerably (Charles).

The *superficial lymphatics* of the arm are especially numerous on the medial side, where they run upward to penetrate the brachial fascia and end in axillary nodes (Fig. 5-12); there may be also a lymphatic along the cephalic vein that bypasses the axillary nodes, ending first in a deltopectoral node and then continuing to cervical nodes (Danese and Howard). They form a number of almost parallel trunks that are largely formed by, and represent the upward continuation of,

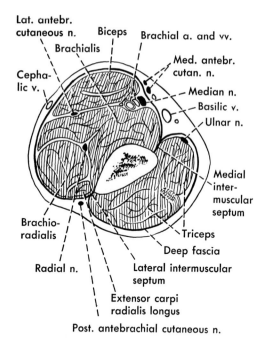

Fig. 5-10. Cross section through the lower third of the left arm. (Redrawn from Eycleshymer AC, Schoemaker DM: A Cross-Section Anatomy. New York, Appleton, 1923)

the superficial lymphatics of the forearm. In their course they receive lymphatics from the back of the arm; the latter vessels diverge from each other along a line on the posterior surface of the arm (as do the vessels in the upper part of the forearm), those on the outer side passing around the lateral border of the limb, those on the inner side passing medially, to join the lymphatic trunks along the medial side of the arm.

Some lymphatics from the ulnar side of the forearm end in one or two cubital (supratrochlear) nodes that lie alongside the basilic vein above the medial epicondyle, before the vein disappears through the deep fascia. At and above the cubital fossa there are said to be numerous communications between the superficial lymphatics and the deep group that accompanies the brachial vessels.

Cutaneous Nerves

Cutaneous nerves are rather numerous (Fig. 5-13). The medial brachial cutaneous nerve from the medial cord of the brachial plexus,

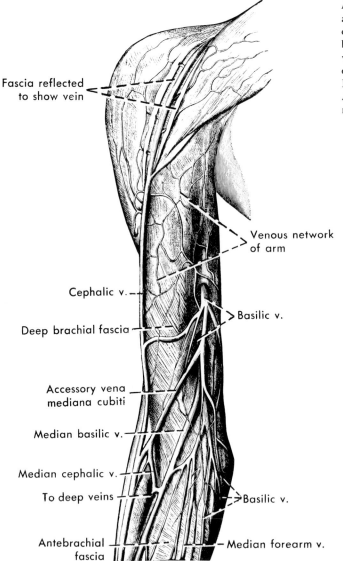

Fascia reflected
to show vein

Venous network
of arm

Cephalic v.

Basilic v.

Deep brachial fascia

Accessory vena
mediana cubiti

Median basilic v.

Median cephalic v.

To deep veins

Basilic v.

Antebrachial
fascia

Median forearm v.

Fig. 5-11. Superficial veins of the arm, anterior view. The cephalic vein is here doubled in the forearm, and the median basilic and median cephalic veins are variations from the usual pattern at the elbow. For a more common pattern, see Fig. 5-44. (Toldt C: An Atlas of Human Anatomy [ed 2], Vol 2. New York, Macmillan, 1928)

and the intercostobrachial nerve from the second or second and third intercostal nerves, supply skin of much of the medial side of the arm; the lower part is supplied by twigs from the medial antebrachial cutaneous, which supplies also most of the skin over the anterior surface of the biceps muscle. The superior lateral brachial cutaneous branch of the axillary nerve supplies an upper part of the lateral and anterolateral surface, largely the skin over the deltoid muscle. The inferior lateral brachial cutaneous nerve, arising from the radial (with the posterior antebrachial cutaneous in about 30% of cases [Kasal] and sometimes called the upper branch of the posterior antebrachial), supplies skin of the lower part of the lateral surface. The posterior brachial cutaneous nerve, a branch from the radial before it leaves the axilla, supplies an upper part of the posterior surface and extends a variable distance toward the elbow; and there may be unnamed branches of the radial that pierce the lateral or long heads of the triceps, or run between them, to reach posterior skin.

MUSCLES OF THE ARM

The muscles of the arm are the biceps brachii, coracobrachialis, and brachialis muscles on the anterior or flexor surface, the triceps and the anconeus posteriorly. Of these muscles, both heads of the biceps, the coracobrachialis, and the long head of the triceps arise from the scapula, while the remainder arise from the humerus; only the coracobrachialis inserts on the humerus.

Innervation

The three muscles of the front of the arm are innervated by the musculocutaneous nerve (Fig. 5-14); through this nerve the biceps and brachialis regularly receive fibers derived from the fifth and sixth cervical nerves; the coracobrachialis is usually said to receive fibers from the sixth and seventh, but it is said that it may receive fibers from C5 also, or receive only seventh cervical fibers. The musculocutaneous nerve apparently receives fibers from the seventh cervical nerve in only about two thirds of cases, but in some 50% of cases the nerve or nerves to the coracobrachialis arise in part from the lateral cord or from C7, and thus can receive seventh cervical fibers even when the musculocutaneous has none.

The triceps and the anconeus are both innervated by the radial nerve. The triceps is variably reported to receive its segmental innervation from the sixth through the eighth cervical nerves, from the seventh and eighth alone, and from the sixth and seventh alone; Harris reported contraction of it upon stimulation of the fifth cervical nerve, but Frohse and Fränkel doubted that any of the fibers of C5 to the radial nerve are motor ones. Further, fibers of the sixth cervical nerve are frequently said to go to the lateral head only. The most important innervation therefore seems to be from the seventh and eighth cervical. The anconeus is said to receive fibers from the seventh and eighth cervical nerves only.

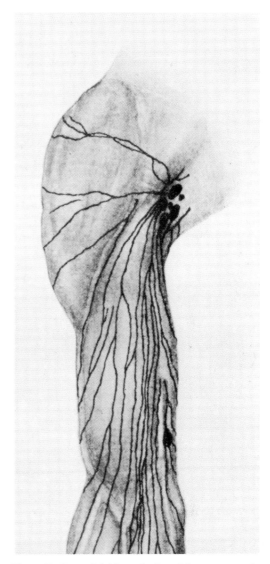

Fig. 5-12. Superficial lymphatics of the arm, anterior view. A supratrochlear node and several axillary ones are shown. (Rouvière H: Anatomie des Lymphatiques de l'Homme. Paris, Masson et Cie, 1932)

BICEPS

The biceps brachii is the most superficial muscle of the front of the arm, and almost entirely covers the brachialis; the musculocutaneous nerve lies between the two muscles. The median nerve and the brachial artery run in the medial bicipital sulcus, approximately at the level of the biceps' posterior border.

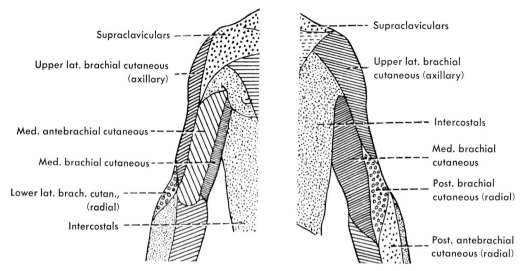

Fig. 5-13. Distribution of nerves to the anterior and posterior surfaces of the arm. (Flatau E: Neurologische Schemata für die ärztliche Praxis. Berlin, Springer, 1915)

The biceps typically arises by two heads (Fig. 5-15; see also Fig. 5-17), as implied by its name; its long head arises from the supraglenoid tubercle (and from the adjacent glenoidal labrum) by means of a long tendon that traverses the shoulder joint, covered by a reflection of synovial membrane, and passes downward between the two humeral tubercles surrounded by a diverticulum of the cavity of the shoulder joint (the intertubercular synovial sheath). Here it is held in place by the weak transverse humeral ligament and is crossed anteriorly by the tendon of the pectoralis major muscle. It emerges from the capsule of the shoulder joint still entirely tendinous, and begins to form its muscular belly behind the lower part of the insertion of the pectoralis major. This head of the muscle unites with the short head at a somewhat variable level, but commonly at about the junction of middle and lower thirds of the arm.

The long head of the biceps has usually been said to develop outside the shoulder joint and gradually sink into it, but Neale found it deep to the capsule in the youngest fetuses in which it could be identified. Its relationship to the synovial membrane varies somewhat: most of the intra-articular part of the tendon usually lies free within the shoulder joint, surrounded by a tubular sheath (visceral layer) of synovial membrane, but a mesotendon may persist for various lengths, connecting the sheath about the tendon to the lining of the joint cavity. DePalma and coworkers observed also, among 100 shoulders, one case in which the long tendon lay between the synovial membrane and fibrous capsule of the shoulder, and another in which it apparently lay entirely outside the capsule.

The short head of the biceps is also tendinous at its origin, but flattened rather than rounded; it arises with the coracobrachialis from the tip of the coracoid process, forming the anterior tendinous part of this common origin. It is here concealed by the anterior fibers of the deltoid. Separating from the coracobrachialis, it forms the medial of the two muscular bellies of the biceps.

The insertion of the biceps is through two tendinous attachments, the rounded and strong tendon proper and the rather thin, broad and flattened bicipital aponeurosis (lacertus fibrosus). The tendon of insertion of the biceps passes deeply between the flexor and extensor muscles into the cubital fossa and attaches to the posterior part of the tuberosity of the radius, a bursa (bicipitoradial) separating the tendon from the anterior part of the tuberosity. The bicipital aponeurosis arises

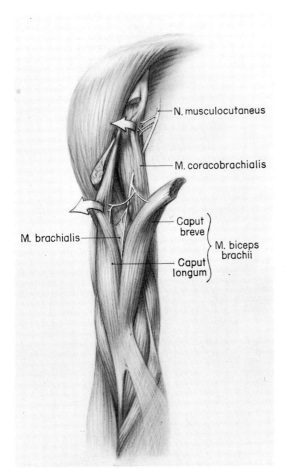

Fig. 5-14. Innervation of the muscles of the front of the arm by the musculocutaneous nerve. (Hollinshead WH, Markee JE: J Bone Joint Surg 28:721, 1946)

from the medial border of the tendon and from the lower medial muscular portion of the muscle, and passes downward and medially across the front of the cubital fossa to join the deep fascia over the upper ends of the flexor muscles of the forearm. It is usually described as ending by attaching into this fascia, but Congdon and Fish said that it can be traced, in part directly and in part indirectly, to an attachment into the ulna; they regarded this attachment as providing additional leverage for the biceps in flexion of the forearm. While the bicipital aponeurosis is flat and generally thin, its upper edge is somewhat thickened and can be easily felt when the forearm is flexed.

Innervation and Action

The biceps receives its innervation from the musculocutaneous nerve (Fig. 5-14), which sends a branch into each head of the muscle; these may arise separately or together and are usually composed of fibers derived from the fifth and sixth cervical nerves.

As discussed in connection with movements of the shoulder, contraction of both heads of the biceps during flexion at the shoulder and of the long head during abduction has been said both to aid these movements and to be primarily a protection for the elbow; Basmajian and Latif regarded their electromyographic studies as confirming the action of the biceps in flexion but casting doubt upon its action in abduction. If extension at the elbow and flexion at the shoulder are required simultaneously, as in pushing up from a chair, the biceps does not contract; contraction would interfere with extension at the elbow far more than it would aid flexion at the shoulder.

Fig. 5-15. Muscles of the front of the arm.

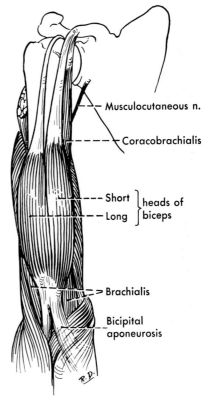

At the elbow, the biceps is an important flexor and also an important supinator, since in the pronated forearm the tendon of the biceps is wound around the radius. Because flexion and supination ordinarily accompany each other, supination is less powerful when the forearm is extended, and the biceps will participate only against resistance. Similarly, the biceps plays little or no part in flexion of a pronated forearm (Basmajian and Latif).

Variations

A not uncommon anomaly of the biceps is for it to present three or more heads. Greig, Anson, and Budinger found such variations in 28 (about 21.5%) of 130 biceps muscles studied, one or more accessory heads from the middle third of the humerus representing the commonest type (Fig. 5-16). In 2 instances an accessory head arose from the intertubercular groove of the humerus, and in 5 cases one arose from the pectoralis major. DePalma and his co-workers reported a case of doubling of the intra-articular part of the long head.

In 3 of the 130 specimens of Greig and his colleagues the long head of the biceps arose from the intertubercular groove rather than the supraglenoid tubercle, and had also a small origin from the capsule of the shoulder joint. Meyer ('28) reported a number of cases of this type and attributed them to the tendon's having attained a secondary attachment as its surface became frayed by the wear

Fig. 5-16. Some variations in the biceps brachii muscle. Accessory heads of origin are marked with crosses. The asterisks on the coracobrachialis muscle identify bundles arising from the short head of the biceps rather than with it from the coracoid process. (Greig HW, Anson BJ, Budinger JM: Quart Bull Northwestern Univ M. School 26:241, 1952)

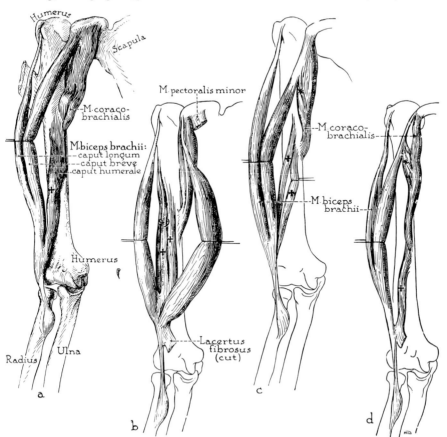

and tear on it; the suprahumeral portion of the tendon then tends to disappear.

As pointed out by Meyer ('28, '37), and others, the long tendon of the biceps may, especially in elderly people, be very much frayed where it lies in the bicipital groove; fraying of the under surface usually results from contact with a cartilaginous margin of the head of the humerus, with slight irregularities in the floor of the sulcus, or with a supratubercular ridge; lateral fraying results similarly from irregularities in the walls of the groove.

Rupture

Stimson found about 100 cases of acute rupture of the tendon of the long head of the biceps reported by 1935, and noted that rupture may occur in young and healthy individuals as a result of sudden muscular effort. Michele and Krueger found this not unusual in the United States Marine Hospital where they worked; and Waugh and co-workers noted that the biceps brachii has been reported to stand third in order of frequency of rupture, preceded only by ruptures of the muscles of the calf and of the extensor muscles of the leg. In their own experience they found rupture of the biceps to be most frequent.

Rupture of the biceps tendon at its insertion is much less common, but apparently not really rare, for although Dobbie could find only 24 reported cases he was able to obtain reports of 51 more by correspondence. Rupture of the tendon of insertion usually occurs at or very near the insertion on the radial tuberosity. Boyd and Anderson suggested that after finding the end of the ruptured tendon in an anterior approach it can be most easily reattached through a posterior one.

Dislocation of the long head of the biceps is discussed in the preceding chapter.

CORACOBRACHIALIS

The coracobrachialis muscle (Figs. 5-17 and 5-18) arises with the short head of the biceps from the coracoid process, but unlike the biceps is fleshy rather than tendinous at its origin. It is penetrated by the musculocutaneous

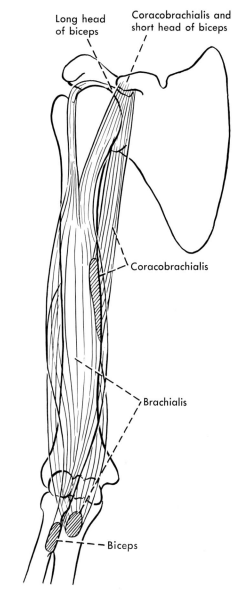

Fig. 5-17. Origins (*red*) and insertions (*black*) of the muscles of the front of the arm.

nerve and inserts into the medial border of the shaft of the humerus about its middle. Its nerve supply (usually C6 and C7) arises from the musculocutaneous nerve, commonly before this nerve pierces the muscle, or from the lateral cord, or from both sources; there may be a single nerve, but there are often several twigs to the muscle. The coracobrachialis is a flexor and adductor of the arm. According to Wright, it contracts more strongly for adduc-

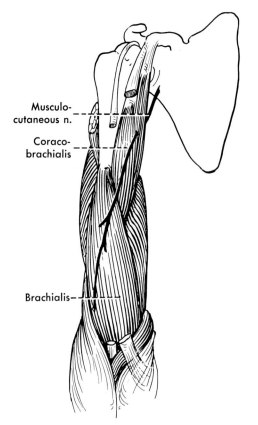

Musculo-
cutaneous n.

Coraco-
brachialis

Brachialis

Fig. 5-18. Brachialis and coracobrachialis muscles after removal of the biceps.

tion than for flexion; Duchenne regarded this as in part for protection to the shoulder joint, helping to prevent subluxation from the downward pull of the more powerful adductors, the pectoralis major and latissimus dorsi.

Variations

Some variations in the coracobrachialis are said to result from the fact that the part normally found in man represents only two parts (separated by the musculocutaneous nerve) of a muscle that in many animals has three parts. Perhaps the commonest variation, however, is for the most superficial part to extend downward farther than usual, sometimes even as far as the medial epicondyle.

The original third head of the muscle, termed the coracobrachialis brevis, occurs rather rarely. Beattie recorded a case in which this muscle was present bilaterally, arising

from the coracoid process and inserting into the crest of the lesser tubercle about a centimeter below this tubercle. He said the muscle has been reported as inserting into the capsule of the shoulder joint, into the tendon of the latissimus dorsi, into the skin and fascia of the axilla, or into the subscapularis tendon. A coracobrachialis brevis may rub against the lesser tubercle during rotation of the arm, and cause pain over this tubercle (Bechtol). Chouké reported a case in which what he called a coracobrachialis muscle arose from the root of the coracoid process and slightly from the conoid ligament, and was entirely separate from the normal muscle; it inserted by a thin tendon into the medial intermuscular septum about the middle of the arm, and its nerve supply was by a special twig from the *posterior* cord.

BRACHIALIS

The brachialis muscle (Figs. 5-17 and 5-18) arises from about the lower half to two thirds of the front (anteromedial and anterolateral surfaces) of the body of the humerus and from the adjacent medial and lateral intermuscular septa; the upper part of its origin is divided into two parts by the insertion of the deltoid muscle. The broad and strong tendon of the muscle is closely applied to the front of the elbow joint; it inserts into the lower part of the coronoid process and into the tuberosity of the ulna.

Because of its origin from the humerus and insertion into the ulna, the brachialis is limited in its action to flexion of the elbow. Its nerve supply is from the fifth and sixth cervical nerves through the musculocutaneous nerve. The lower lateral part of the muscle, where the radial nerve is adjacent to it, may be penetrated by a branch from the radial nerve, but there is disagreement as to whether this is commonly motor to the muscle or not. Linell and some texts state that stimulation of the nerve has indicated that it is probably not motor and, since a branch from the radial nerve to the elbow joint commonly traverses the muscle, the nerve may be, usually, purely one to the joint. It has also been stated (Kirklin) that contraction of the brachialis can sometimes be provoked by stimulation of this branch of the radial nerve, and some authors have regarded the brachialis muscle as being of double origin, a lateral part of it being derived from the posterior premuscle mass of the embryonic limb. Ip and Chang supported this view, for they found a branch from the radial nerve rather regularly, and said that while its fibers did go to muscle spindles, they also formed many motor endings.

Jones ('19) found only one individual with a severed musculocutaneous nerve in whom any contraction of the brachialis could be palpated, and this contraction was not strong enough to produce flexion of the elbow. It seems unlikely that the brachialis muscle receives enough innervation from the radial nerve to allow it to act effectively as a flexor at the elbow after interruption of the musculocutaneous nerve; rather, flexion at the elbow under this circumstance is brought about by the radial nerve because it innervates the brachioradialis, an excellent flexor of the elbow, and the other more anteriorly arising forearm extensors that can also flex the elbow.

The brachialis muscle has the biceps in front of it and shares the intermuscular septa on each side with the triceps; the musculocutaneous nerve lies between it and the biceps, and the brachial artery and median nerve come to lie on the front of the muscle at the elbow. In its lower part the brachialis is related laterally to the extensor muscles of the forearm, especially the brachioradialis, and to the radial nerve; this nerve runs between the brachialis and the origins of the brachioradialis and extensor carpi radialis longus muscles, but from the front is covered by the lateral bulge of the brachialis.

TRICEPS BRACHII

The triceps (Figs. 5-19 and 5-20) occupies almost the whole of the back of the arm, being superficially placed after it emerges from under cover of the posterior part of the deltoid. The long head arises by a strong tendon from the infraglenoid tubercle on the lateral border of the scapula and passes downward in front of the teres minor but behind the teres major. The axillary nerve and the posterior humeral circumflex vessels run in the space (quadrangular space) between the long head of the triceps and the surgical neck of the humerus.

The lateral head of the triceps arises from the posterior surface of the humerus above the radial groove, and in part also from the lateral intermuscular septum; it and the long head form both the lateral and the medial portions of the upper and superficial parts of the muscle and unite first to form almost the entire portion of the muscle visible superficially. The fibers of the long head are directed downward, those of the lateral head downward and medially; they cover the groove of the radial nerve (spiral groove), the radial nerve itself and the accompanying profunda brachii vessels, and most of the posterior surface of the medial head.

The medial head has an extensive origin, from the entire posterior surface of the humerus below the radial groove, from the medial intermuscular septum, and from the lower part of the lateral intermuscular septum below the point at which the radial nerve passes through this septum. The medial head of the triceps therefore appear not only on the medial side of the arm but also on the lateral side. It attaches to the deep surface of the combined long and lateral heads, and the entire muscle forms a very strong, broad tendon that inserts into the olecranon, a bursa (subtendinous bursa of the triceps) separating the tendon from the posterior capsule of the elbow joint. Expansions from the edges of the tendon attach into the fascia of the forearm, and a few deep muscle fibers attach into the synovial membrane of the elbow joint to form the articularis cubiti muscle.

The long head of the triceps, since it passes across the shoulder joint, can assist other muscles in extension at this joint. Wright said it does so only against resistance. It is also an adductor at the shoulder joint, but Duchenne surmised that much of its contraction during forcible adduction is, like the contraction of the coracobrachialis, to resist subluxation by the more powerful adductors. The entire

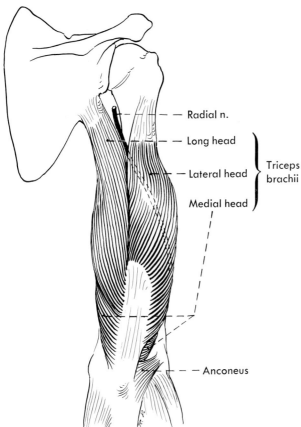

Fig. 5-19. Triceps and anconeus muscles.

Radial n.

Long head

Lateral head   } Triceps brachii

Medial head

Anconeus

muscle is the only strong extensor at the elbow joint, but it is assisted in this action by the anconeus. Duchenne found the long head to be a much weaker extensor at the elbow than the other two heads, and Travill showed that the medial head usually exhibits the greatest activity during extension, the lateral less, and the long head none unless additional strength is needed.

The nerve supply to the triceps is from the radial nerve, each head receiving one or more branches; these are said to derive their fibers from the sixth through the eighth, or the seventh and eighth cervical nerves. A branch to the long head normally leaves the radial nerve while it lies in the axilla, and so does the first branch to the medial head; the latter nerve runs downward on the medial surface of the medial head, just posterior to the ulnar nerve, and is sometimes therefore known as the ulnar collateral nerve. As the radial nerve

winds around the humerus it usually supplies an additional branch to the long head, one or two branches to the lateral head, and at least one additional branch to the medial head (Figs. 5-21 and 5-22). A part of the latter nerve extends through the medial head in its lateral part to end in the anconeus muscle.

ANCONEUS

The anconeus is a small triangular muscle that arises from the distal end of the back of the lateral epicondyle of the humerus and fans out to insert on the lateral side of the olecranon and an upper part of the posterior surface of the body of the ulna (Figs. 5-19 and 5-20); its upper border is close to the lower border of the lateral portion of the medial head of the triceps, and it partly covers the posterior capsule of the elbow joint, the annu-

the elbow joint in other movements of the extremity; they regarded its activity during pronation as stabilizing the elbow against the flexor action of the pronator teres. Basmajian and Griffith, reporting only moderate activity in the anconeus during both pronation and supination, agreed with this.

VESSELS

BRACHIAL ARTERY

The brachial artery is the continuation of the axillary, the change in name occurring at the lower border of the teres major muscle. The artery lies at first on the medial side of the arm in front of the long head of the triceps, and then on the medial head of the triceps; as it passes downward it inclines forward along the medial border of the biceps, to lie

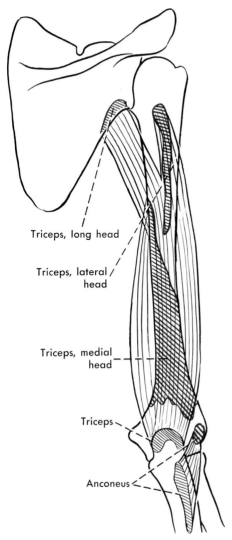

Fig. 5-20. Origins (*red*) and insertions (*black*) of the triceps and anconeus muscles.

lar ligament of the radius, and the upper part of the ulna.

Duchenne reported that isolated stimulation of the anconeus produced better extension than did stimulation of the long head of the triceps. Ray and co-workers adduced evidence that its chief action is that of producing abduction of the ulna, amounting to about 8°, as the forearm is moved from extreme supination to extreme pronation. However, Travill found it most active during extension, and Pauly and co-workers reported that it initiates elbow extension and acts to stabilize

Fig. 5-21. Upper branches of the radial nerve to the triceps muscle. (Hollinshead WH, Markee JE: J Bone Joint Surg 28:721, 1946)

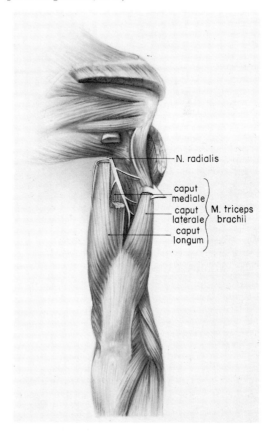

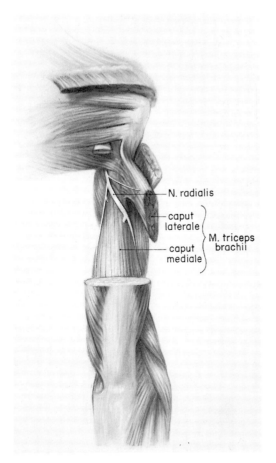

Fig. 5-22. Deeper branches of the radial nerve to the triceps muscle. (Hollinshead WH, Markee JE: J Bone Joint Surg 28:721, 1946)

on the anterior surface of the brachialis muscle, in which position it disappears into the cubital fossa (Fig. 5-23). As it enters the cubital fossa it lies deep to the aponeurosis of the biceps (lacertus fibrosus).

In the upper part of the arm the radial nerve lies behind the artery, the ulnar nerve lies medial to it, and the median nerve lies anterolateral or lateral to it. The artery loses its relationship to the radial nerve as the latter starts its course around the humerus, and the ulnar nerve gradually diverges from the artery by penetrating the medial intermuscular septum and passing toward the back of the medial epicondyle. As the median nerve runs down the arm it crosses the artery, almost always superficially, to lie on the medial side of the vessel at the elbow. Among 34 limbs Linell found 2, both from the same body, in

which the median nerve crossed deep to the brachial artery, but this relationship is much rarer than this finding might indicate.

The brachial artery ends by dividing into radial and ulnar arteries in the cubital fossa. It is usually accompanied by two brachial veins, one lying medial and one lateral to it, but with numerous anastomoses between them.

The named branches of the brachial artery above its terminal ones are the profunda brachii, the nutrient artery of the humerus, and the superior and inferior ulnar collateral arteries. The artery also gives branches to the muscles of the front and medial side of the arm.

Fig. 5-23. Chief vessels and nerves of the arm.

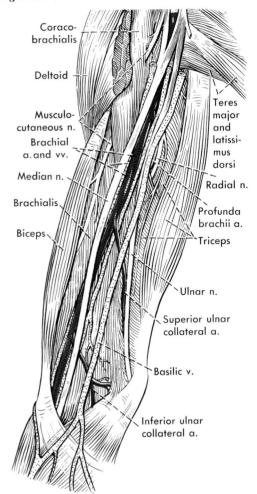

Profunda Brachii

The profunda brachii artery is the largest branch of the brachial in the arm. It usually arises from the posteromedial side of the brachial artery in the upper part of the arm, and runs downward and laterally behind the brachial artery to parallel the radial nerve in a spiral around the posterior surface of the humerus, passing at first deep to the long head of the triceps and then deep to the lateral head, and thus between it and the medial head of the muscle. Before or as it reaches the lateral intermuscular septum the profunda divides into two branches, one of which, the middle collateral artery, runs on or in the medial head of the triceps, supplying this, and either anastomosing broadly with the interosseous recurrent artery or taking part in a rete to which the latter artery also contributes. The other terminal branch of the profunda is the radial collateral artery, which passes through the lateral intermuscular septum with the radial nerve and descends with this between the origins of the brachialis muscle and the upper extensor muscles of the forearm. It gives off branches to neighboring muscles, and ends by anastomosing with the radial recurrent artery.

During its course the profunda brachii gives branches to the triceps and sends a deltoid branch upward under cover of the deltoid to supply this muscle and anastomose with the posterior humeral circumflex artery. Enlargement of this anastomotic branch with disappearance of the stem arising from the brachial artery leaves the profunda arising from the posterior humeral circumflex. The profunda usually gives off also a small nutrient branch to the humerus; this sometimes largely replaces the nutrient branch of the branchial artery.

The profunda brachii varies somewhat in its origin. Charles and his co-workers found a single deep brachial arising alone from the brachial artery in only 54.7% of 300 arms, double deep brachials with this origin in 0.7%, and triple deep brachials in one case (about 0.3%). The deep brachial arose from the brachial artery with the superior ulnar collateral in 22.3%, at the junction of brachial and axillary in 8%, and directly from the axil-

lary in 8.7%. In 4% of their series it arose from the posterior humeral circumflex, in 0.7% from the subscapular, and in 0.7% it was absent.

Nutrient Artery

The nutrient branch of the brachial artery arises in about the middle of the arm and enters the anteromedial surface of the humerus, its exact point of entrance varying considerably. It may be small or missing.

Ulnar Collateral Arteries

The *superior ulnar collateral artery* (Figs. 5-23 and 5-24) typically arises below the profunda, but may share a common trunk with it; it arises from the posteromedial side of the bra-

Fig. 5-24. Diagram of the lower part of the brachial artery and its branches.

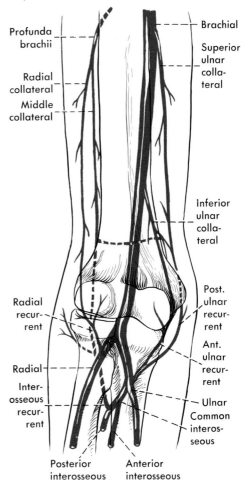

chial artery, usually at about the middle of the arm. It passes downward and backward with the ulnar nerve through the medial intermuscular septum to the back of the medial epicondyle, where it anastomoses with the posterior ulnar recurrent artery and often with the inferior ulnar collateral.

The *inferior ulnar collateral artery* (supratrochlear artery) arises from the posteromedial surface of the brachial about 1.5 inches (3.8 cm) above the medial epicondyle. It runs downward on the brachialis muscle, and under cover of the pronator teres anastomoses in front of the medial epicondyle with the anterior ulnar recurrent artery. Typically, shortly beyond its origin, it gives off a posterior branch that pierces the medial intermuscular septum and passes between the humerus and the medial head of the triceps to anastomose with the posterior ulnar recurrent artery.

Polonskaja pointed out that the smaller branches of the brachial artery, especially those vessels that anastomose around the elbow to form the collateral circulation, have no constant patterns; he said he was never able to find the same pattern even on the two sides of one body.

Involvement of Brachial Artery in Injuries at Elbow

Rupture of the brachial artery as a consequence of supracondylar fracture of the humerus or dislocation of the elbow is apparently rare (Spear and Janes), although perhaps not as rare as reports in the literature indicate. It may result in so much hemorrhage into the cubital fossa that a large portion of the collateral blood supply to the forearm is blocked and either gangrene or Volkmann's ischemic contracture follows. According to Spear and Janes, a similar result may ensue from rupture of the ulnar artery, the radial artery, or both vessels, or even from involvement of a vein.

"SUPERFICIAL" ARTERIES OF ARM

The term "superficial brachial" is sometimes applied to any artery that arises either in the axilla or the arm and courses anterior (super-

ficial) to the median nerve; whether it arises in the axilla or the arm, there is also a deeper artery, but the specific vessels contributed to by the superficial and deep arteries, respectively, vary from case to case. For this reason a more descriptive terminology is obtained if the designation "superficial brachial" is restricted to superficial arteries that end by bifurcating into radial and ulnar arteries (the deep artery then typically ends in the forearm as the common interosseous), while terms such as "superficial radial" and "superficial ulnar" are used when they are more appropriate. Superficial arteries of the arm have essentially similar courses and relations in the arm, whether they are of axillary or brachial origin and regardless of their distribution; if they arise from the axillary, however, they nearly always emerge between the two heads of the median nerve.

Superficial arteries of the arm are relatively common. McCormack and co-workers found some type in 30.77% of 364 bodies in which both limbs were investigated. They were commonly unilateral, so the incidence among limbs was much less (18.53% of 750 limbs).

Superficial Brachial Artery

A superficial brachial artery, in the sense of a superficial vessel that bifurcates into radial and ulnar arteries, has been reported as arising from the axillary in 17 of 512 limbs (De Garis and Swartley), but Miller found only 1 case in 960 limbs, and McCormack and co-workers only 1 among 750 limbs. Origin of a superficial brachial artery in the arm rather than in the axilla is more common, for McCormack and his co-workers found 7 of brachial arterial origin among the 750 arms that they dissected, an incidence of superficial brachial artery in 1.07% of their series, and representing 5.75% of the total number of vascular variations they observed in these limbs.

In the one case in McCormack's series in which the superficial brachial artery arose in the axilla, it passed between the two heads of the median nerve to lie anterior and medial to the median nerve, following the medial border of the biceps muscle. In the cases in which it arose in the arm, the median nerve in turn

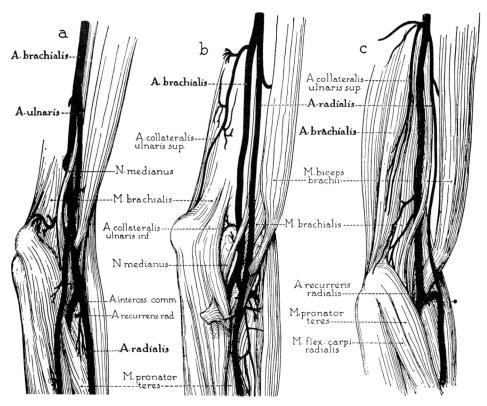

Fig. 5-25. Superficial arteries arising from the brachial artery. In *a* there is a superficial ulnar artery, in *b* and *c* there are superficial radials. The asterisk in *c* marks a large anastomosis between the brachial artery and the small superficial radial. (McCormack LJ, Cauldwell EW, Anson BJ: Surg Gynecol Obstet 96:43, 1953 [by permission of Surgery, Gynecology & Obstetrics])

lay anterior to the normally positioned or "deep" brachial artery; this "deep" brachial artery in the forearm formed the two interosseous arteries. The lowest level of origin of a superficial brachial artery in their series was 10.5 cm above the intercondylar line.

Superficial Radial Artery

McCormack and his co-workers found a high origin of the radial artery to be far the most common variation of the arteries of the arm, for it represented 77% of all the variations they observed, and was found in 14.27% of all specimens; in 2.13% of limbs a radial artery arose from the axillary, and in 12.14% from some part of the brachial artery (Fig. 5-25*b* and *c*). De Garis and Swartley reported a superficial radial of axillary origin in about 2.7% of white, 5% of Negro arms; Miller

found it bilaterally in 15 of 480 bodies (about 3%).

All the radial arteries that McCormack and his co-workers found arising from the axillary artery, and four that came from the proximal portion of the brachial, arose from the anterior aspect of the main artery and passed between the lateral and medial heads of the median nerve; in the remainder, the artery arose from the anteromedial or medial surface of the brachial artery, most frequently from the upper part of this vessel. When the vessel passed between the two heads of the median nerve, it lay anterior and very close to that nerve throughout the arm; when it was not related to the heads of the median nerve, it passed medially anterior to the nerve, which then lay between the superficial radial and brachial arteries for a distance of 6 cm to 8 cm, finally crossing in front of the brachial artery to occupy its usual more medial position in the cubital fossa.

Of the 107 aberrant radial arteries, 20 had an anastomotic connection with the brachial artery or (1 case) with the median. These connections were usually in the region of the cubital fossa, and in 14 instances passed anterior to the biceps tendon. Those passing anterior to the tendon were usually of relatively large size, and the caliber of the radial artery proximal to the anastomosis was usually particularly small. Anastomoses passing behind the biceps tendon were typically slender.

Superficial Ulnar

McCormack and his co-workers also reported instances of superficial ulnar arteries, which occurred in 2.26% of their series; 7 arose from the axillary artery, 10 from the brachial (Fig. 5-25a). De Garis and Swartley reported a superficial ulnar artery of axillary origin in about 0.8% of cases. Miller found none on either side of 480 bodies.

The arteries of axillary origin in McCormack's series passed, as is usual, between the two heads of the median nerve, and one vessel of brachial origin had a similar course; in the remaining cases of brachial origin the superficial ulnar remained anterior to both the brachial artery and the median nerve, and just posterior to the medial border of the biceps brachii muscle. In the lower part of the arm the artery tended to be medial to both the median nerve and the brachial artery, the nerve lying between the two arteries after crossing, in its usual manner, the anterior aspect of the brachial. In contrast to superficial radial arteries, which tend to have the normal course of the radial in the forearm, superficial ulnar ones frequently run superficially across the flexor muscle mass in the forearm.

Other Variations

A superficial median artery is rare. De Garis and Swartley found only 1, arising from the axillary, in 512 limbs, Miller only 1 in 960 limbs, and there were apparently none in McCormack's series. When it is of axillary origin, the vessel tends to pass between the heads of the median nerve, and in any case has a superficial course in the arm.

McCormack and his co-workers encountered a single example of the rare accessory brachial artery, or doubling of the brachial. In their case doubling occurred 21 cm above the intercondylar line; the accessory artery lay medial to the main brachial artery throughout its course, but about the middle of the arm passed deep to the median nerve and, 4 cm proximal to its ending, crossed back over the median nerve, rejoining the brachial artery in the cubital fossa.

VEINS

The deep veins of the arm need little description. There are typically two brachial veins, which anastomose freely around the brachial artery, by which they are otherwise separated from each other. The medial one frequently ends by joining the basilic vein, the lateral usually joins the axillary. The larger branches of the brachial artery also have two venae comitantes each.

The basilic vein lies deep to the deep brachial fascia in the upper half or two thirds of the arm, just posterior or medial to the brachial artery and veins. It may receive the medial of the two brachial veins, but otherwise usually has no connection with the deep veins in the arm until after it has become the axillary vein at the lower border of the teres major.

LYMPHATICS

The deep lymphatic vessels of the arm accompany the brachial artery and are continuations upward of the lymphatics about the radial and ulnar vessels and their branches. They are said to receive numerous communications from the superficial lymphatics and end largely in the lateral group of axillary nodes.

Deep lymph nodes in the arm are usually very small and apparently inconstant. There may be some in the cubital fossa, and along the brachial artery in the arm; most of the deep lymphatic vessels reach the axillary nodes without interruption in lower nodes (as do also most of the superficial ones).

NERVES

MUSCULOCUTANEOUS NERVE

The musculocutaneous nerve is one of the terminal branches of the lateral cord and derives its fibers especially from the fifth and sixth cervical nerves, but more often than not contains some fibers from the seventh. It diverges laterally from the lateral root of the median nerve to lie between the coracobrachialis muscle and the axillary artery; in this position it usually gives off one or more branches to the coracobrachialis muscle, and then penetrates the muscle to lie between the biceps and the brachialis muscles (Fig. 5-26). It supplies a branch or branches to both heads of the biceps and a branch to the brachialis, and thereafter is purely cutaneous, passing downward between biceps and brachialis muscles to emerge lateral to the tendon of the biceps and continue into the forearm as the lateral antebrachial cutaneous nerve. Linell found that in an average arm of 30.5 cm length, measured from the tip of the acromion to the tip of the lateral epicondyle of the humerus, the nerve to the coracobrachialis arose, on an average, 4.76 cm below the tip of the acromion, and entered the muscle 7.35 cm below this tip; the nerve to the biceps arose 12.99 cm below the acromion and entered the muscle 15.28 cm below, while the nerve to the brachialis arose 17.32 cm below the acromion and entered the muscle 20.27 cm below. Thus a lesion of the nerve much below the middle of the arm will probably interrupt no fibers to muscles, but only the cutaneous portion of the nerve.

The nerve or nerves to the coracobrachialis may arise from the lateral cord rather than from the musculocutaneous nerve, and Linell found this in 4 of 26 dissections; Kerr reported an origin from the musculocutaneous alone in only 49.4% of 109 limbs, and said that in the

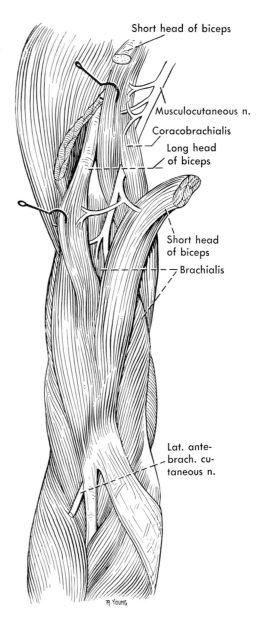

Fig. 5-26. The musculocutaneous nerve in the arm.

remainder there was some origin from the lateral cord, the lateral root of the ulnar nerve (C7, probably) or the seventh cervical nerve directly.

The nerve to the biceps typically arises as a single trunk, which then branches to supply both bellies of the muscle, but there may be a separate one for each belly. The nerve to the brachialis is typically single.

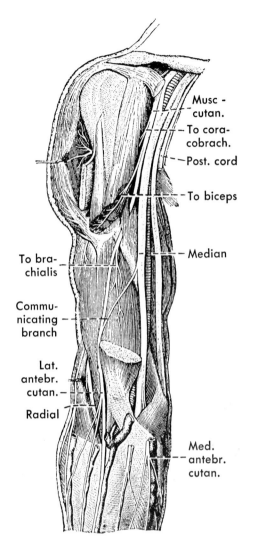

Musc-
cutan.

To cora-
cobrach.

Post. cord

To biceps

Median

To bra-
chialis

Commu-
nicating
branch

Lat.
antebr.
cutan.

Radial

Med.
antebr.
cutan.

Fig. 5-27. Connection from the median nerve to the musculocutaneous. A branch in the opposite direction is much more common. (After Sappey, from Ranney AL: The Applied Anatomy of the Nervous System. New York, Appleton, 1881)

Variations

It is not particularly uncommon to find a nerve trunk of considerable size leaving the musculocutaneous, while this lies behind the biceps, and passing distally and medially to join the median nerve. This is to be regarded as a result of median nerve fibers from the lateral cord passing into the musculocutaneous rather than into the lateral root of the median, and then rejoining the median nerve at

a lower level; when this occurs, the lateral root of the median nerve is typically abnormally small. The communication may be a tiny filament rather than a real nerve trunk. Kerr found that a branch from the musculocutaneous nerve to the median had been reported in from 8.1% to 36.19% of different series, and estimated its incidence as somewhat less than the 24% found in his series. A branch from the median to the musculocutaneous (Fig. 5-27) is much rarer; Kerr quoted an incidence of less than 2% in a series in which the reverse communication was found in 36.19%. Both Testut and Villar described the various types in considerable detail.

Very occasionally, the musculocutaneous nerve fails to separate from the median, and the latter therefore gives off the branches that should arise from the musculocutaneous (for example, Villar). In a similar fashion, the musculocutaneous has been reported to go only to the muscles of the arm, while the lateral antebrachial cutaneous arose from the median. Hari Rao and Ramachandra Rao reported 28 cases, found among 300 specimens, in which the musculocutaneous nerve did not pierce the coracobrachialis but instead passed between it and the biceps (in 8 of these the nerve was at first bound to the median), and 6 cases in which the nerve split, one part going over, the other through, the muscle.

MEDIAN AND ULNAR NERVES

Neither the median nor the ulnar nerve normally supplies any muscles of the arm, and typically neither gives off a branch to a muscle of the forearm above the level of the elbow (Fig. 5-23). In rare instances, the median nerve instead of the musculocutaneous may contain the fibers for the coracobrachialis (Testut), or even have the entire musculocutaneous nerve united with it (Villar).

The median nerve is formed anterior or anterolateral to the brachial artery, and it descends lateral or anterolateral to this artery to about the middle of the arm, thereafter crossing the artery, almost always in front, to lie on its medial side and accompany it into the cubital fossa. In the lower part of its course it lies not only medial to the artery but also behind the bicipital aponeurosis (lacertus fibrosus); in the forearm it disappears between the two heads of the pronator teres muscle. The median nerve most commonly gives off no branches in the arm, its uppermost branches (to the pronator teres and to the elbow joint) typically arising in the forearm; however, a nerve to the pronator teres may arise as high as 7 cm above the level of the epicondyles.

At its origin, the ulnar nerve lies between the axillary artery and the axillary vein, with the medial antebrachial cutaneous nerve in front of it. In the upper half of the arm it lies medial or posterior to the brachial artery, but at about the middle of the arm it begins to diverge from this vessel and pierces the medial intermuscular septum to lie on the front of the medial head of the triceps muscle; it is joined in this course by the superior ulnar collateral artery, with which it passes downward to the back of the medial epicondyle. The nerve enters the forearm by passing behind the epicondyle, between the humeral and ulnar heads of origin of the flexor carpi ulnaris muscle.

The ulnar nerve typically gives off no muscular branches in the arm, although occasionally the highest branch to the flexor carpi ulnaris may arise above the level of the epicondyle; Linell found 2 cases among 26 in which this occurred, the level of origin of this first muscular branch being 1 cm and 0.5 cm respectively above the level of the lateral epicondyle (Linell's point of measurement). Rather regularly, however, the ulnar nerve does give off an articular branch to the elbow above the epicondylar level, although Linell said this origin was seldom more than 1 cm above that. The articular nerve usually ran almost horizontally to reach the elbow joint, but in one of his cases it arose particularly high and paralleled the main trunk of the ulnar nerve for a distance of 12 cm.

RADIAL NERVE

The radial nerve is the larger of the two terminal branches of the posterior cord and, as a branch of that cord, may contain fibers from all the nerves entering the brachial plexus. However, the contribution from the lower trunk of the brachial plexus, C8 and T1, to the posterior cord is typically small, and is said to contain often only eighth cervical fibers. Walsh traced fibers from T1 into the radial nerve in only 6 of 74 special dissections of the brachial plexus. The radial nerve therefore consists usually of fifth, sixth, and seventh cervical fibers, with fewer fibers from the eighth, and with the contribution from the seventh being usually noticeably large. Frohse and Fränkel regarded the contribution of the fifth cervical nerve to the radial as being probably afferent; Harris reported contraction of the triceps upon stimulation of the fifth nerve. Most authors do not list any muscles as being supplied by first thoracic fibers through the radial, but Harris expressed the opinion that such fibers go to the special extensors of the fingers and to the extensors of the thumb.

After separating from the axillary nerve, which it may do before the contribution of the lower trunk has joined the posterior cord (in which case this contribution joins the radial nerve directly), the radial nerve lies behind the axillary artery, successively on the front of the subscapularis, teres major, and latissimus dorsi muscles. In the upper part of the arm it lies behind the brachial artery, and on the front of the long head of the triceps muscle. It then passes laterally and posteriorly with the profunda brachii artery, deep to the long head of the triceps muscle and, as it attains the posterior aspect of the humerus, deep to the lateral head and therefore between the lateral and medial heads of the triceps (Fig. 5-28). In this position it lies close to but apparently not actually in the groove of the radial nerve (spiral groove) of the humerus according to Whitson; further, the radial nerve does not usually lie against the humerus, but on the upper part of the medial head of the triceps.

As the nerve reaches the lateral border of the medial head of the triceps it pierces the lateral intermuscular septum, in company with the radial collateral branch of the pro-

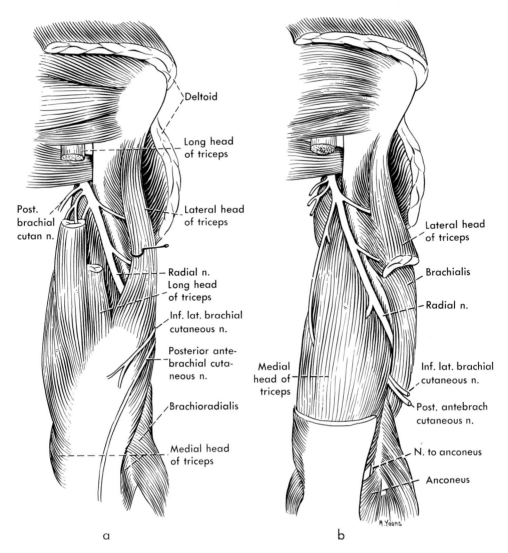

Fig. 5-28. The radial nerve in the back of the arm.

funda brachii artery, and thereafter loses its association with the triceps. Continuing its spiral course, it passes forward in close association with the lateral side of the brachialis muscle, lying first in the groove between this and the brachioradialis, and subsequently between the brachialis and the extensor carpi radialis longus, to pass in front of the lateral epicondyle into the forearm. After it becomes associated with the brachialis muscle, it lies close to the origin of this muscle from the humerus; consequently, since this muscle bulges laterally from its origin, the radial nerve ac-

tually lies behind the lower part of the brachialis muscle, a position that must be taken into consideration if the lateral part of this muscle is split in an approach to the elbow joint.

Branches

During its course in the arm the radial nerve gives off two cutaneous branches, a series of muscular branches to the triceps and the anconeus, and nerves to the two uppermost of the extensor forearm muscles—those that arise above the lateral epicondyle.

While the radial nerve is still in the axilla it commonly gives off the posterior brachial cutaneous nerve, which may arise independently or with one of the muscular branches to the triceps. This pierces the fascia near the axilla to supply a limited area of skin on the posterior surface of the proximal third of the arm, close to that area supplied by the medial brachial cutaneous and intercostobrachial nerves, with one or both of which it anastomoses. It varies inversely in size with the sizes of these other two nerves.

Also while it is still in the axilla (or just as it enters the arm) the radial nerve gives off two or more muscular branches to the triceps, at least one to the long head and another to the medial part of the medial head. The latter branch, once called the ulnar collateral nerve because it parallels the ulnar nerve for some distance in the arm, usually gives off several branches into the upper part of the medial head, but a part of it passes with the ulnar nerve as far as the lower third of the arm before it disappears into this head.

As the radial nerve passes between the lateral and long heads of the triceps it may give an additional branch to the long head, and typically supplies one branch to the lateral head. Then, as it lies on the medial head, it gives its main supply to this head; a part of this nerve travels down through the substance of the muscle to reach the anconeus and supply that muscle also. Linell regarded the branches to the triceps as being typically four in number, the nerve to the long head arising, according to his observations, about 7.1 cm below the tip of the acromion, the "ulnar collateral" nerve arising about 9.5 cm below the acromial tip, the nerve to the lateral head arising about 10.1 cm, and the main nerve to the medial head about 11.2 cm below the acromial tip. Sunderland ('46) found the branches to the triceps to be variable; there were from 5 to 10, arising variably in the axilla, as the nerve reached the arm, and as it spiralled around the humerus. Linell said that all the branches of the radial nerve to the triceps except the main one to the medial head are given off before the nerve becomes associated with the radial groove; as Sunderland

noted, this is by no means true of all specimens. However, it is true that injury to the nerve from a fracture of the humerus is unlikely to produce marked paralysis of the triceps muscle, because so many of the nerves to this muscle arise high.

As the radial nerve nears the lateral intermuscular septum it gives off two cutaneous branches in succession, or by a common stem in about 30%; the first branch, the *inferior lateral brachial cutaneous* (also called the upper branch of the posterior antebrachial cutaneous) pierces the deep fascia close to the lateral intermuscular septum, and lies close to the lower brachial part of the cephalic vein, supplying skin over about the lower half of the lateral and anterior aspects of the arm. The lower and larger branch, the *posterior antebrachial cutaneous*, penetrates the fascia of the arm a little below the level at which the lateral brachial cutaneous emerges, and descends to the back of the forearm to supply skin there as far as the wrist or onto the dorsum of the hand.

Lotem and co-workers reported cases of neuropraxis of the radial nerve produced by exercise, and involving the nerve below its branches to the triceps but above or at the level of the origin of the posterior antebrachial cutaneous. On the basis of dissection of cadavers, they attributed this to pressure from a tendinous band associated with the origin of the lateral head of the triceps but attached to the humerus both above and below the nerve. Manske reported a case of permanent paralysis of the radial nerve resulting from compression as it passed deep to the lateral head.

While it lies lateral and anterior to the humerus, the radial nerve supplies a branch, sometimes two, to the brachioradialis muscle, and a branch to the extensor carpi radialis longus (Fig. 5-29). It may also supply a branch to the extensor carpi radialis brevis, but this branch commonly arises below the elbow, usually from the deep branch of the radial rather than from the radial trunk itself.

As the radial nerve becomes closely related to the brachialis muscle, it often gives off a branch into that muscle; Sunderland ('46) found one or more such branches in 18 of 20 limbs. As already discussed in connection

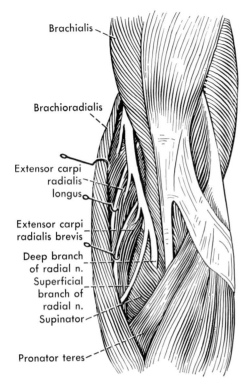

Brachialis

Brachioradialis

Extensor carpi
radialis
longus

Extensor carpi
radialis brevis

Deep branch
of radial n.

Superficial
branch of
radial n.

Supinator

Pronator teres

Fig. 5-29. The radial nerve at the front of the elbow.

with the brachialis muscle, there is a difference of opinion as to whether this nerve commonly helps to supply the brachialis, or whether it is usually an articular branch only. Gardner apparently regularly traced this branch to the elbow joint, but in only one case did he find it branching into the brachialis muscle. Whatever may be the condition here, it seems certain that the clinician should not rely on the presence of this radial twig to maintain a functioning brachialis muscle.

The Elbow Joint

The elbow joint is formed by the articulation of the distal end of the humerus with both the radius and the ulna, flexion and extension of the forearm taking place here, and by the articulation of the radius with the ulna, pronation and supination occurring here. The humeroulnar, humeroradial, and proximal radioulnar joints are all enclosed within a common synovial and fibrous capsule.

Surgical incisions for approaching various parts of the joint range from anterolateral and anteromedial ones through lateral, medial, posterolateral, posteromedial, and posterior ones. In the anterolateral approach the position of the radial nerve between the brachialis and brachioradialis muscles must be borne in mind; in posteromedial approaches the ulnar nerve is encountered as it passes behind the medial epicondyle. Strachen and Ellis reported that the radial nerve close to the elbow moved as much as 1 cm medially during pronation, and thus advised a posterolateral approach with the forearm in this position for removal of the head of the radius.

Articular Surfaces

The lower end of the humerus presents the rather rounded capitulum for articulation with the slightly concave upper end of the head of the radius and, medial to the capitulum and separated from it by a groove, the concave and pulleylike trochlea that articulates with the trochlear notch of the ulna. The two surfaces are covered by a continuous layer of articular cartilage, but the capitulum is limited to the anterior and lower surfaces of the end of the humerus, while the trochlea is present anteriorly, inferiorly, and posteriorly. Posteriorly, the lips of the trochlea extend obliquely upward and laterally, so that when the forearm is extended it is not in exact line with the arm but deviates laterally, thus forming the so-called carrying angle of the arm.

The trochlear notch of the ulna is a deep concavity on the front of the upper end of this bone, between the olecranon and the coronoid process; the middle of this notch is elevated, in order to fit the concavity of the trochlea. When the forearm is flexed, the coronoid process fits into the coronoid fossa of the lower end of the humerus, and when the forearm is extended the olecranon fits into the olecranon fossa on the posterior surface of the distal end of the humerus. The trochlear notch is commonly covered with a continuous layer of articular cartilage, but at about its middle the cartilage is ordinarily deficient medially and laterally, and the bone is roughened; sometimes these rough places extend completely across the bone, thus leaving superior and inferior articular surfaces separated by a shallow roughened trough. The articular cartilage of the coronoid process extends onto the lateral surface of the process to line the radial notch, the smooth concavity for articulation with the head of the radius.

The disklike head of the radius, larger in diameter than the rest of the bone and therefore united to it by a neck, is covered by articular cartilage both on its slightly concave upper surface and on its periphery; this circumferential articular surface articulates with the radial notch of the ulna and rotates within the annular ligament, thus forming the proximal radioulnar joint.

Fibrous Capsule and Ligaments

The articular capsule of the elbow joint (Fig. 5-30) is thin anteriorly and posteriorly, where the joint is protected by the brachialis and triceps muscles, respectively, but, as in all typical hinged joints, is strengthened medially and laterally by special ligaments.

The anterior part of the capsule attaches to the front of the humerus immediately above the radial and coronoid fossae, below to the anterior border of the coronoid process of the ulna and to an anterior part of the annular ligament of the radius. The posterior part of the capsule is especially weak in its middle part, where it is closely related to the tendon of the triceps. It is attached above to the upper border and sides of the olecranon fossa, close to the sides of the posterior part of the trochlea, and to the back of the lower edge of the lateral epicondyle. Inferiorly, it attaches to the upper and lateral margins of the trochlear notch of the ulna, to the roughened area on the lateral side of the ulna just distal to the radial notch, and to the annular ligament of the radius, with which it largely blends but sends a few fibers to attach to the neck of the radius.

Both anterior and posterior parts of the capsule are blended laterally and medially with the radial and ulnar collateral ligaments.

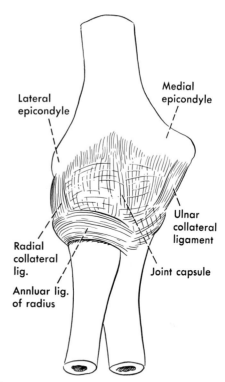

Fig. 5-30. Anterior view of the ligaments of the elbow joint.

The thickened medial portion of the fibrous capsule of the elbow joint is the *ulnar collateral ligament,* attached above to the anterior and inferior surfaces of the medial epicondyle and, behind the epicondyle, to the groove between the trochlea and the epicondyle. From this origin as an apex the ligament extends downward and its parts diverge, so that it becomes triangular (Fig. 5-31a). It is commonly described as consisting of three bands, continuous with each other: the anterior part, from the epicondyle, runs distally and anteriorly to attach to the medial edge of the coronoid process; the middle part, also from the epicondyle, descends and spreads out to attach along the ridge between the coronoid and olecranon processes and blend with the transverse band (oblique ligament of Cooper) that stretches between the olecranon and coronoid processes; and the posterior part, from behind the epicondyle, runs slightly backward to attach to the medial side of the olecranon. The anterior and posterior bands are strong, the middle part is very thin.

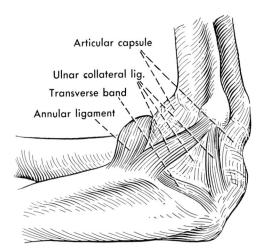

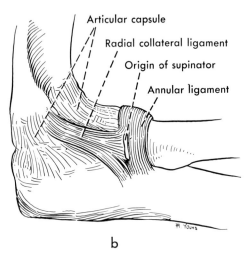

Fig. 5-31. Medial, *a*, and lateral, *b*, views of the ligaments of the elbow joint.

The *radial collateral ligament*, or lateral ligament of the elbow joint (Fig. 5-31*b*), is attached to the anterior and inferior aspects of the lateral epicondyle, deep to the common extensor tendon that also arises from this epicondyle. From this origin it extends distally to attach into the annular ligament of the radius. Some of its fibers run with the annular ligament to the ulna, others blend with the origin of the supinator muscle.

The part of the fibrous capsule between the radial notch on the ulna and the medial side of the neck of the radius is called the quadrate ligament (Fig. 5-32); Martin ('58a) did not regard it as being a true ligament.

The *annular ligament* (lig. anulare radii, orbicular ligament) is a well-marked band of fibers attached both anteriorly and posteriorly to the margins of the radial notch of the ulna; it receives most of the attachment of the radial collateral ligament, and some of that ligament's fibers follow the annular ligament to the ulna (Martin, '58a); thus, although the annular ligament forms about four fifths of a circle, it is not readily displaced downward. The circumference of the circle formed by the upper edge of the annular ligament and the upper border of the radial notch is greater than that of the circle formed by the lower edge of ligament and notch, so that the whole resembles somewhat a cup with the bottom broken out. Since the head and neck fit neatly into this cup and its broken bottom, distal displacement of the head of the radius is discouraged. However, the head of the radius can escape from the annular ligament in children without tearing it (Steindler, '35). This has been attributed to the head being regularly no larger or only a little larger than the neck in infants and children, but Ryan found the proportions between the two to be similar in both fetal and adult bones. The deep surface of the annular ligament is connected by a few fibers of the

Fig. 5-32. Frontal section through the elbow joint. The synovial membrane is in *red*.

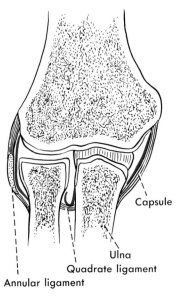

capsule of the joint to the neck of the radius, but they are too few and too loose to interfere with rotation of the bone.

Synovial Membrane

The synovial membrane of the elbow joint lines the fibrous membrane of the capsule, including that part supported by the annular ligament, and is attached to the margins of the articular surfaces of the humerus, ulna, and radius (Figs. 5-32 and 5-33). On the humerus it is reflected into the bottom of the coronoid, radial, and olecranon fossae, where it is separated from the bone by connective tissue and a little fat. Between the synovial and fibrous capsules, in front of the coronoid and radial fossae and behind the olecranon fossa, are pads of fat that move in and out of the joint cavity with extension and flexion.

The cavity of the elbow joint does not communicate with any of the bursae about this joint, but the synovial membrane may pouch slightly beneath the lower border of the transverse ligament (oblique ligament of Cooper) connecting the olecranon and the coronoid process medially. Ehrlich reported on two patients with rheumatoid arthritis who had developed antecubital cysts, similar to Baker's or popliteal cysts.

Vessels and Nerves of the Elbow Joint

The blood supply to the elbow joint is from all those vessels (Fig. 5-24) that form the collateral circulation about the elbow (thus, on the medial side, from the superior and inferior ulnar collateral arteries from above and the two ulnar recurrents below; and on the lateral side from the radial and middle collateral arteries of the profunda above, and the radial and interosseus recurrent arteries from below).

The nerves to the elbow joint typically arise from all four nerves that pass across the joint (Fig. 5-34), although none are absolutely constant, and in general the distribution of any one nerve to the joint is overlapped by the distribution of branches from other nerves (Gardner). Gardner found a small branch from the musculocutaneous nerve, derived from the nerve to the brachialis muscle and descending along the medial part of this mus-

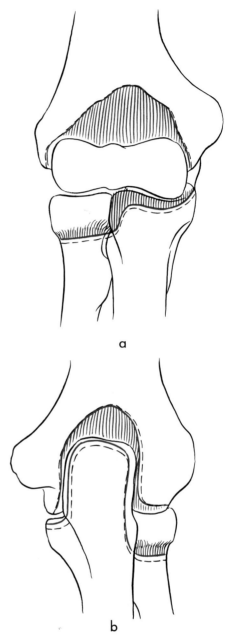

a

b

Fig. 5-33. Attachments of the capsule of the elbow joint from the front, *a,* and from behind, *b.* The attachment of the fibrous capsule is indicated by *broken black lines,* the lines along which the synovial membrane reaches the bone, and its areas of reflection over the bone, by *red.*

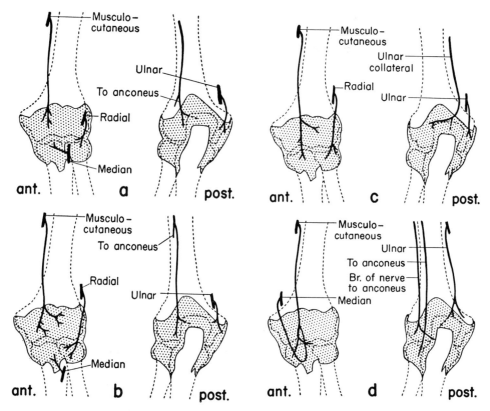

Fig. 5-34. Four variations in the nerves to the elbow joint. (Gardner E: Anat Rec 102:161, 1948)

cle to the anterior part of the capsule, to be most constant. The median nerve usually gives off a branch, just before it passes through the pronator teres, to run recurrently to the anteromedial part of the capsule; and in sections of fetuses Gardner often found a branch of the median descending along the brachialis muscle to be distributed in the same region as is the musculocutaneous branch, and another from the anterior interosseous branch that ran recurrently to supply the medial and posteromedial part of the capsule.

A branch or several fine filaments usually arise from the ulnar nerve as it passes behind the medial epicondyle, to supply the posteromedial part of the joint; a branch may arise higher and descend with the nerve, but Linell said it seldom arose more than a centimeter above the epicondyle. Gardner usually found several branches from the radial nerve, which were distributed to the posterolateral, the lateral, and the anterior parts of the capsule, and

gave this nerve the widest distribution to the joint of any of the nerves.

Bateman found the articular branch or branches of the ulnar nerve to be the easiest to find at operation. Because of the difficulty in identifying all articular branches, he advocated denervating the elbow joint by exposing the main nerve trunks and stripping them of their articular branches; he said that while the articular branches usually arise above the motor ones, it is advantageous to pinch a nerve branch gently, to be sure there is no motor response, before dividing it.

The lymphatics from the elbow joint, according to Tanasesco, in part follow the blood vessels, in part run separately from them. They drain into nodes in the cubital fossa or along the brachial artery or into a node that he found in the groove of the radial nerve and called the radial or retrohumeral node, and into the lateral (brachial) or central groups of axillary nodes.

Anomalies, Dislocation, and Fracture at the Elbow

Bones or bony fragments have been found within the elbow joint: Atsatt reported the removal of one that lay adjacent to the lateral epicondyle and finally locked the joint; in this same joint another fragment lay just above the tip of the olecranon, was rounded, and seemed to fit into a curved defect in the olecranon. Simril and Trotter reported articulated rather than loose bones at the elbow joint bilaterally in a male Negro, but were unable to decide whether the anomalous bones resulted from the moderate arthritis that was present, were of congenital origin, or represented centers of ossification of the lateral epicondyles that were avulsed at an early age.

Fractures of the humerus in the medial epicondylar region are particularly likely to affect the ulnar nerve, because of its close association with the medial epicondyle. Fracture of the capitulum, according to McLaughlin, usually occurs in young individuals, the average age being about 15 years, and is usually produced by falling with the forearm extended so that the head of the radius is jammed against the capitulum, which is then dislocated forward. Fractures of the radial head usually result from the same type of fall; these and other fractures of the elbow have been discussed by Keon-Cohen. Radin and Riseborough recommended that displaced fractures involving more than two thirds of the radial head, and all comminuted ones, should be treated by early total excision.

Dislocation of the elbow is said to be only second in frequency to that of the shoulder (Eliason and Brown). By far the greater number of dislocations are posterior ones, anterior dislocation being rare. Dislocations of the radius alone, however, are commonly anterior (Steindler, '35).

Either supracondylar fracture or dislocation of the elbow may result in rupture of vessels; Spear and Janes, in reviewing the literature, found that the brachial artery, the ulnar, the radial, or both radial and ulnar arteries may be involved, and in one case a vein was. Eliason and Brown reported a case of rupture of the radial and ulnar arteries consequent to posterior dislocation of the elbow, and Kilburn and co-workers reported 3 cases of rupture of the brachial, but such cases are rare. The median nerve is usually spared, although muscle damage is present; Kilburn and his colleagues regarded simple ligation of the artery to be as satisfactory as grafting or suture. Jones ('30) reported paralysis of the radial nerve in 2 of 32 posterior dislocations of the elbow, paralysis of the median in 1 of 29 posterolateral dislocations, and paralysis of the ulnar in 12 of 16 lateral dislocations, but found no paralysis resulting from dislocations in other directions. With combined dislocation of the head of the radius and fracture of the ulna, according to Stein and co-workers, only one case of nerve injury had been reported; in contrast, among their 11 patients so afflicted, 6 had nerve lesions, all involving the radial or deep radial nerves, and 3 involving also the ulnar.

King found that relatively few cases of recurrent dislocation of the elbow had been reported; he listed among the causes fracture of the epicondyles, torn collateral ligaments, fracture of the coronoid process at its base with retraction upward of the brachialis muscle, a congenitally shallow trochlear notch of the ulna, osteochondritis dissecans, and fracture of the medial part of the humeral condyle.

According to Buxton, ossification in the ligaments of the elbow, and sometimes in adjacent muscles, following dislocation, is probably more common than ossification in any other ligaments of the body. He attributed this to the frequency with which bone is avulsed with the ligaments, and to lack of proper care. He said that proper reduction and immobilization will prevent or minimize this abnormal ossification, and that immobilization will even lead to resorption of bone already deposited in the muscles, although little resorption from the ligaments occurs.

MOVEMENTS AT THE ELBOW

Movements at the elbow are of two types, flexion-extension and pronation-supination. The former takes place between the humerus and both the radius and the ulna; the latter, a

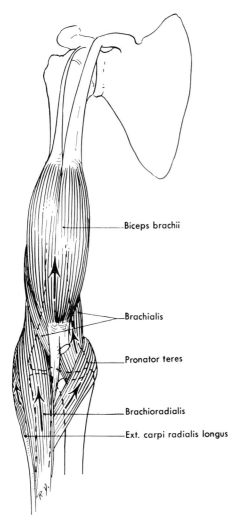

Fig. 5-35. Flexors at the elbow.

rotatory movement of the forearm and hand, between the radius and the ulna. Both sets of movements are carried out in part by muscles of the arm, in part by muscles of the forearm.

Flexion and Extension

During flexion and extension of the forearm the trochlear notch of the ulna rotates around the pulleylike trochlea of the humerus, and the upper articular surface of the head of the radius slides upon the capitulum of the humerus. The trochlea is so formed that in flexion the bones of the forearm are held in line with the humerus, but upon extension the oblique margins of the posterior part of the

trochlea so engage the ulna that a lateral deviation of the forearm results. Measured on the medial side, the angle between the arm and the extended forearm (the "carrying angle") varies from about 3° to 29° (Atkinson and Elftman).

Flexion of the forearm is normally limited by contact between the soft tissues of the front of the forearm and those of the arm; extension, by the contact of the olecranon with the fat in the olecranon fossa of the humerus. The average range of movement permitted by the articular surfaces is about 140° (Steindler, '35); the most useful part of this range is said to be between 60° and 120° (Steindler, '49).

The chief flexors of the forearm (Fig. 5-35) are the biceps and the brachialis, the one acting upon the radius and the other upon the ulna, and the brachioradialis. The brachialis is equally effective whether the forearm is in pronation or supination and, as shown by Basmajian and Latif, commonly participates in all flexion. The biceps acts more promptly when movement is started from the supine or semiprone position; if flexion is started from pronation, the muscle participates only when there is resistance to the movement (Beevor; Basmajian and Latif). Flexion of the forearm is also the chief function of the brachioradialis muscle; in slow movements it apparently participates only against resistance, and mostly then from the prone or semiprone position, but it participates in all fast movements of flexion (Basmajian and Latif). The brachioradialis can be aided in flexion by adjacent muscles of the extensor group, notably the extensor carpi radialis longus; even the extensor digitorum can sometimes aid in flexion if the forearm is pronated (Wright). However, these accessory flexors are at a disadvantage until flexion is started, because they lie so close to the joint. The pronator teres is also a flexor at the elbow, although usually a weak one. Wright said that it usually assists in flexion only when the forearm is pronated; Basmajian and Travill found it contributing from no position unless flexion was resisted. Jones ('19) showed that the brachioradialis (and adjacent muscles) can produce immediate and strong flexion of the forearm when

Table 5-1
Innervation of the Muscles Flexing the Forearm

| | | | SEGMENTAL INNERVATION | |
	MUSCLE	NERVE	*Reported Range*	*Probable Most Important Segment(s)*
Chief flexors	Biceps	Musculocutaneous	C5-C6	C5, 6
	Brachialis	Musculocutaneous	C5-C6	C5, 6
	Brachioradialis	Radial	C5-C6	C5, 6
Accessory flexors	Pronator teres	Median	C5-C7	C6, 7
	Radial muscles of extensor group	Radial	C5-C8	C6, 7

both the biceps and brachialis are paralyzed; he also showed that the pronator teres alone can flex the forearm, but said that this movement was neither so strong nor so complete as that produced by the brachioradialis, and varied in strength from one not quite sufficient to raise the hand to the mouth to a strong and useful movement of the pronated forearm.

Thus flexion at the elbow can be brought about by muscles innervated by the musculocutaneous, radial, and median nerves (Table 5-1). However, Hendry said that active flexion by forearm muscles after injuries to the brachial plexus is possible only when the flexion is started passively (most injuries of the brachial plexus would simultaneously affect the flexors in the arm and the extensor forearm muscles that act as elbow flexors). He, Steindler ('49), Bunnell, and several others have discussed the surgical problems connected with attempts to provide function to a limb in which the muscles about the elbow are paralyzed.

Active extension at the elbow is brought about by the triceps muscle, particularly the medial head, with the lateral and long heads usually being recruited in that order as needed to overcome more resistance (Travill). Extension is also brought about by the anconeus. Although the latter muscle cannot contribute strength comparable to the triceps, Travill found it active in all movements of extension, and Pauly and co-workers, confirming this, said it initiates extension and acts to stabilize the joint, its activity decreasing with increase in activity of the triceps.

Among the flexors, it is the brachialis that

offers the necessary resistance to slow extension of the elbow, but the brachioradialis regularly contracts when there is quick extension (Basmajian and Latif). Extension, save through gravity, is therefore entirely dependent upon the radial nerve, but this nerve must be injured near the axilla if the injury is to affect the entire triceps.

Pronation and Supination

Pronation and supination involve a rotation of the head of the radius within the circle formed by the annular ligament and the radial notch of the ulna, the rotation of the radius carrying the hand into pronation or supination. There is also, therefore, a movement between the entire lengths of the radius and ulna, so that from the parallel position of supination the radius comes to cross the ulna in pronation; this movement is provided for by the interosseous membrane and the distal radioulnar joint. Braune and Flügel found the total excursion of the radius in pronation-supination to be from 150° to 160°, and this can be slightly supplemented by rotation at the radiocarpal joint. Steindler regarded the most useful part of the pronation-supination movement as being the range between 80° pronation and 45° supination (zero position being the thumb-up one).

The movements of pronation and supination have usually been said to occur around an axis denoted by a line passing from the center of the capitulum of the humerus through the distal end of the ulna and hence through the little finger, that is, to involve no movement of the ulna, but only a rotation of the radius. However, Ray and co-workers

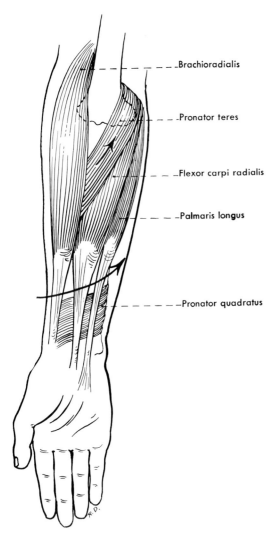

Fig. 5-36. Pronators of the forearm.

Pronation is brought about especially by the pronator teres and quadratus muscles (Fig. 5-36). Slow unresisted pronation is carried out entirely by the pronator quadratus; the pronator teres is also employed if the movement is fast or more strength is needed (Basmajian and Travill). In spite of the fact that the major part of the pronator teres is also a potential flexor of the elbow, the position of the joint did not affect activity of the muscle in pronation. Other muscles, which are often listed as potential pronators because of their slant from a medial position at the elbow to a more lateral position at the wrist, are the flexor carpi radialis and, against resistance, the palmaris longus. Duchenne did not obtain pronation from stimulating the palmaris longus, but this muscle does contract during forceful pronation. To these, the brachioradialis is often added as pronating from extreme supination, but its contribution under normal circumstances must be regarded as questionable; Duchenne said the brachioradialis pronates relatively little when it alone is stimulated; Wright said it never contracts for pronation when that, rather than flexion, is the willed movement; and Basmajian and Latif were not sure whether its activity during pronation (against resistance only) was a synergistic one or that of a prime mover. Severing the median nerve above the elbow typically paralyzes all potential pronators except the brachioradialis, however, so the finding of Boswick and Stromberg that pronation is not abolished by such a lesion indicates that the brachioradialis must be included among the pronators.

Supination is brought about primarily by the supinator and biceps muscles (Fig. 5-37); the supinator alone produces good supination and, if the movement is unresisted and slow, acts alone whether the elbow is flexed or extended. The biceps is recruited for added strength or speed (Travill and Basmajian); considerable resistance is necessary before the biceps will attempt to supinate the extended forearm. The extensor and abductor pollicis longus muscles are sometimes said to supinate, but apparently cannot (Duchenne), and

have pointed out that pronation and supination often are more nearly in an axis passing through the index than the little finger, in which case there is necessarily during pronation a lateral abduction of the distal end of the ulna. They regarded this abduction as being the chief function of the anconeus, but Travill noted that the muscle is active in both pronation and supination if the movement is resisted, and Pauly and co-workers regarded its greater activity during pronation as a part of its usual activity in stabilizing the joint against flexion.

the extensor carpi radialis longus and the brachioradialis can conceivably contribute to its initiation from extreme pronation. In spite of its old name, supinator longus, the brachioradialis is not a good supinator, if it supinates at all. Duchenne was unable to obtain supination by stimulating it, Wright noted that, although it tends to put the forearm into the midposition when it functions as a flexor, it does not contract when either supination or pronation alone is carried out; and Basmajian and Latif expressed the same doubts about its activity during resisted supination that they did of its role in pronation.

Pronation is, for practical purposes, almost entirely a function of the median nerve (Table 5-2), supination a function of the musculocutaneous and radial nerves combined (Table 5-3).

The Forearm

The skeleton of the forearm consists of the radius and ulna.

The majority of the muscles of the forearm arise from the humerus, and thus pass across the elbow joint; the majority pass also across the wrist joint, to gain an insertion upon the metacarpals or the phalanges. The muscles on the flexor side of the forearm, arising largely from the medial aspect of the humerus or from the ulna, are innervated by both the median and the ulnar nerves, the former nerve supplying by far the larger number. The muscles of the extensor surface of the

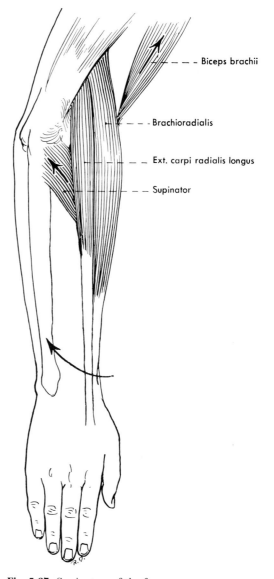

Fig. 5-37. Supinators of the forearm.

Table 5-2
Innervation of the Muscles Pronating the Forearm

	MUSCLE	NERVE	Reported Range	Probable Most Important Segment(s)
			SEGMENTAL INNERVATION	
Chief pronators	Pronator teres	Median	C5-C7	C6, 7
	Pronator quadratus	Median	C6-T1	C8, T1
Accessory pronators	Flexor carpi radialis	Median	C6-C8	C6, 7
	Palmaris longus	Median	C6-T1	C7, 8(?)
	Brachioradialis	Radial	C5-C6	C5, 6

Table 5-3
Innervation of the Muscles Supinating the Forearm

	MUSCLE	NERVE	SEGMENTAL INNERVATION	
			Reported Range	*Probable Most Important Segment(s)*
Chief supinators	Supinator	Radial	C5-C7	C6
	Biceps	Musculocutaneous	C5-C6	C5, 6
Accessory supinators	Extensor carpi radialis longus	Radial	C5-C8	C6, 7
	Brachioradialis (?)	Radial	C5-C6	C5, 6

forearm are innervated exclusively by the radial nerve.

The chief nerves of the forearm are therefore the median and ulnar nerves, the former being much the more important, on the anterior side, the radial nerve on the posterior side. The chief arteries, the radial and ulnar, lie on the anterior side, and much of the blood supply to the extensor muscles enters their upper ends; the posterior interosseous artery, the longitudinally running artery of the posterior aspect, is small after it has given off its upper branches in the forearm.

RADIUS AND ULNA

The radius and ulna parallel each other in the forearm, but are so articulated that a rotation of the head of the radius permits a spiraling of this bone from the position of supination, in which the bones are parallel, to one of pronation in which the distal end of the radius is carried medially and hence the radius crosses the ulna. This movement is provided for by the free rotation of the head of the radius within the annular ligament at the proximal radioulnar joint, by the flexibility of the interosseous membrane, and by the distal radioulnar joint.

Articulations

The proximal radioulnar joint is a part of the elbow joint and has already been described with that; the distal radioulnar joint, although separated from the radiocarpal joint by an articular disk, is one of the several joints at the wrist and is therefore described with these (Chap. 6).

In addition to their two synovial articulations, the radius and the ulna are connected by fibrous tissue, most of it in the form of an interosseous membrane (Figs. 5-38 and 5-39); above the level of the interosseous membrane there is a small band of fibrous tissue, the oblique cord. The oblique cord arises from the tuberosity of the ulna and runs obliquely downward and laterally to attach to the radius just distal to the radial tuberosity. There is a gap between the lower border of the oblique cord and the upper border of the interosseous membrane and through this gap, usually, the posterior interosseous artery passes from the flexor to the extensor aspect of the forearm. Martin ('58b) regarded the oblique cord as probably representing a fibrous part of the supinator muscle and said it becomes slack in pronation and tense in supination.

The interosseous membrane is a broad sheet of fibrous tissue that stretches between the sharp lateral edge of the ulna and the similarly sharp medial edge of the radius. Its uppermost fibers attach to the radius a little below the radial tuberosity; the fibers are directed laterally and upward from the ulna to the radius, so that they tend to resist independent upward movement of the radius. In addition, there are weaker fibers at a right angle to those just described, and at the lower end of the membrane there may be some very strong tranversely placed fibers. Christensen and co-workers found the interosseous membrane taut in the neutral position, largely lax in both pronation and supination.

The lower end of the interosseous membrane reaches about as far as the distal radioulnar joint. The anterior interosseous ar-

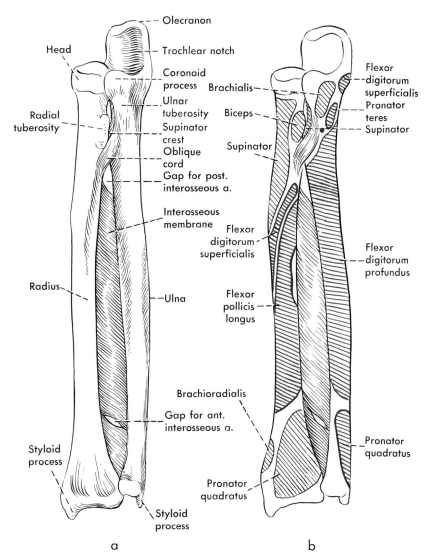

Labels in figure:

- Olecranon
- Head
- Trochlear notch
- Coronoid process
- Brachialis
- Radial tuberosity
- Ulnar tuberosity
- Biceps
- Supinator crest
- Oblique cord
- Gap for post. interosseous a.
- Supinator
- Flexor digitorum superficialis
- Pronator teres
- Supinator
- Interosseous membrane
- Flexor digitorum superficialis
- Radius
- Ulna
- Flexor pollicis longus
- Flexor digitorum profundus
- Brachioradialis
- Gap for ant. interosseous a.
- Styloid process
- Pronator quadratus
- Styloid process
- Pronator quadratus

a   b

Fig. 5-38. Anterior view of the bones of the forearm. In *b* origins of muscles are shown in *red*, insertions in *black*. The leader to the origin of the supinator crosses the occasional origin of the flexor pollicis longus (unlabeled) from the ulna.

tery sends its largest terminal branch through the interosseous membrane to the dorsum of the wrist and hand, perforating the membrane about 0.75 inch (1.9 cm) above the distal end.

Radius

Both the shallowly concave upper surface and the circumference of the disklike head of the radius are covered with articular cartilage, for articulation at the elbow joint; the head and the narrowed neck are embraced by the annular ligament and the radial notch of the ulna. Just below the neck, on the medial side of the body of the bone, is the radial (bicipital) tuberosity for the insertion of the biceps muscles. Below this the body gradually increases in diameter and is somewhat triangular in shape, so that it presents three borders and three surfaces. Of the borders, the interosseous or medial is the sharpest and most obvious, and gives attachment to the interosseous

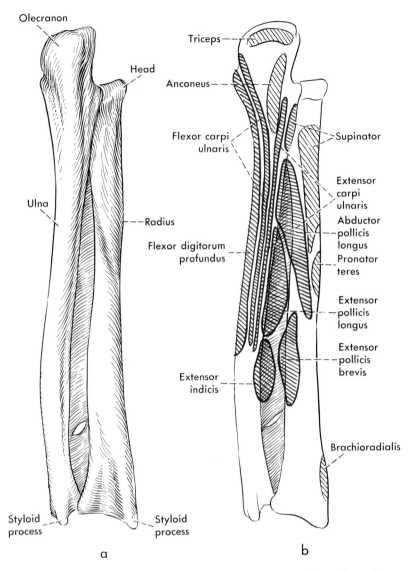

Olecranon

Triceps

Head

Anconeus

Flexor carpi
ulnaris

Supinator

Extensor
carpi
ulnaris

Abductor
pollicis
longus

Pronator
teres

Ulna

Radius

Flexor digitorum
profundus

Extensor
pollicis
longus

Extensor
pollicis
brevis

Extensor
indicis

Brachioradialis

Styloid
process

Styloid
process

a

b

Fig. 5-39. Posterior view of the bones of the forearm. In *b* origins of muscles are shown in *red,* insertions in *black.*

membrane. The anterior border begins above at the tuberosity, runs obliquely across the front of the bone to its lateral side, and descends to the base of the styloid process. At about the junction of the upper and middle thirds of the anterior surface, which lies between these borders, is the chief nutrient foramen. The posterior border is distinct only in the middle third of the bone, but is described as beginning above at the back of the radial tuberosity, ending below at the dorsal tubercle on the distal end of the radius. The lateral surface, between the anterior and posterior borders, therefore extends onto the anterior aspect of the bone above, its posterior aspect below. The posterior surface, between the posterior and interosseous borders, needs no particular comment.

The expanded distal end of the radius is slightly concave anteriorly but presents no special markings on this surface. On the lateral surface there is a shallow groove for the tendons of the long abductor and short extensor of the thumb. The short and thick styloid process projects from the anterolateral aspect of the distal end. On the posterior surface there is a relatively prominent ridge, the dorsal or radial tubercle (of Lister); the tendons of the extensor carpi radialis muscles pass lateral to the tubercle, that of the extensor pollicis longus passes close against its medial side. The medial surface of the distal end of the radius consists of the shallow ulnar notch for articulation with the head of the ulna. The cartilage of the concave distal surface, which articulates with the carpals, extends also onto the anteromedial surface of the styloid process.

Ulna

The deep trochlear (semilunar) notch, bordered above by the olecranon and below by the coronoid process, is the prominent feature of the upper end of the ulna. The articular surface of the notch is narrow where the olecranon and coronoid process meet, and the articular cartilage is sometimes completely interrupted at this level. On the lateral side of the coronoid process is the radial notch for the reception of the head of the radius; the crest of the supinator muscle extends from the posterior border of the notch toward the interosseous border of the bone. Anteromedially, just distal to the base of the coronoid process, is the roughened ulnar tuberosity on which the brachialis inserts.

From its large upper end the ulna tapers almost to its distal end, where the head is slightly larger than the adjacent part of the body. The body is usually described as having three borders, interosseous or lateral, anterior, and posterior, and three surfaces, anterior, medial, and posterior; the interosseous border is the only one really well marked, and on the rounded distal part of the bone, especially, the three surfaces blend.

The head of the ulna, separated from the wrist joint by an articular disk, is covered on its inferior surface and the anterolateral part of its periphery by articular cartilage; this forms a part of the distal radioulnar joint. The styloid process, projecting from the posteromedial aspect of the head, is partly separated from it on the posterior surface by the deep groove for the tendon of the extensor carpi ulnaris.

ABNORMALITIES AND FRACTURE

Absence

Absence or partial absence of either the radius or ulna is rare, but either or both may be missing or deficient.

Congenital absence of the radius (Fig. 5-40), usually associated with absence of the radial side of the wrist and hand, is certainly rare, but more common than congenital absence of the ulna. O'Rahilly listed the radius as being second among the long bones (following the fibula) in incidence of congenital absence, and found that more than 253 cases had been reported as long ago as 1924. Riordan found complete absence to be only a little more frequent than partial absence and, in contrast to previous reports, bilateral absence to be approximately as frequent as unilateral. Delorme and Bora and co-workers described their methods of treatment for the condition. Laurin and co-workers reported a case in which bilateral absence of both the radius and tibia was accompanied by bilateral duplication of the ulna and fibula.

Fig. 5-40. Almost complete absence of the radius; note also the absence of the first metacarpal and the clubbing of the hand. (Rocher H-L: Anomalies congénitales du poignet. In Ombrédanne L, Mathieu P (eds): Traité de Chirurgie Orthopédique, Vol 3. Paris, Masson et Cie, 1937)

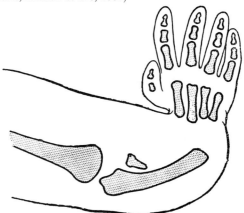

Absence or deficiency of the ulna, which tends to have associated with it absence of the ulnar part of the wrist and hand, is according to O'Rahilly less common than absence of any long bone, other than the humerus; he said that partial absence is more common than complete absence, and found that more than 50 cases of ulnar deficiency had been reported by 1932. Pringle reported a case in which both ulnas were absent.

Humeroradial Synostosis

Humeroradial synostosis is a congenital deformity that is extremely rare. Murphy and Hanson reported a case in which the condition was present bilaterally, and said that apparently only 24 cases had been reported previously.

Madelung's Deformity

Madelung's deformity is an anterior bowing of the distal end of the radius; Anton and co-workers collected reports of 171 cases in 1938, but expressed the opinion that the condition may not be as rare as indicated by the paucity of reports of it, since it is frequently not recognized except by roentgenographic examination. They regarded it as being brought about by abnormal growth of the distal radial epiphysis: the deformity appears in persons in the age group of 10 to 14 years, and dyschondroplasia is often evidenced by premature disappearance of the medial portion of the distal epiphyseal line. In the fully developed condition, the radius may be shortened in addition to being bowed, and the ulna may also be somewhat bowed.

Fracture

Fracture of either the radius or ulna, or both, may occur as a result of direct trauma, but since the radius is the weight-bearing bone between the hand and the arm, fractures of the radius are much more common than are those of the ulna. In turn, the most common fracture of the radius is Colles fracture, defined by Bacorn and Kurtzke as being a complete transverse fracture within the distal inch (2.5 cm) of the radius *with dorsal displacement of the distal fragment;* it is this dorsal displacement that gives rise to the typical "silver fork deformity" characteristic of this fracture (Fig. 5-41).

In the extensive study of fractures of the forearm and wrist reported in Workmen's Compensation files, Bacorn and Kurtzke found that true Colles fracture, as they defined this condition, constituted 60% of all fractures of the radius; the next most common fracture, that of its head, occurred in 14% of radial fractures. Fractures of the body between the head and the distal inch accounted for 6.2%. Fractures in the distal inch that were not Colles accounted for 19.8%. In this last group, 6% were fractures of the radial styloid; 1% were fractures similar to Colles but with anterior displacement of the distal fragment (Smith's fracture, or reversed Colles fracture); and the remaining 12.8% were a heterogeneous group, including, among others, chip fractures and transverse fractures.

MacAusland said that in Colles fracture either the tip of the ulna is broken off or the ulnar collateral ligament is torn, and that in children the fracture may be represented by a separation of the lower radial epiphysis.

Colles fracture can be caused by a direct blow (8.4% of the cases reviewed by Bacorn and Kurtzke), but is more commonly caused by a fall in which the weight is thrown on the outstretched forearm and hand. According to Lewis such a fall cannot produce a Colles fracture unless the hand is markedly dorsiflexed; he produced it in the cadaver only by so forcefully hyperextending the hand that its dorsum came in contact with the forearm, and then pushing the hand and lower fragment backward and upward. As he expressed it, the fracture is produced not by falling on the extended hand but rather by "falling over" the hand, with the forced hyperextension that results from this. The concept that the mechanism is something other than direct transmission of force through the carpal bones to the radius is strongly supported by Bacorn and Kurtzke's findings that fracture of any carpals in association with a Colles fracture is rare; they found associated fractures of no carpals other than the scaphoid (navicular), and only 0.5% of those, in their series of more than 2,000 Colles fractures.

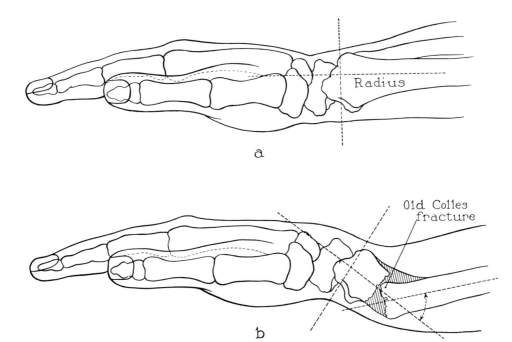

Fig. 5-41. Normal alignment of the forearm and hand, *a*, and the angulation ("silver fork deformity"), *b*, in a Colles fracture. The one represented here is an old one, the *shaded area* indicating the bone formed in healing. (Meyerding HW, Overton LM: Minnesota Med 18:84, 1935)

MacAusland emphasized that if the normal palmar tilt of the lower end of the radius is not restored the wrist joint will remain weak; Bacorn and Kurtzke reported that, regardless of the treatment employed, and in spite of the physical therapy that was almost universally used, only 2.9% of the Colles fractures that they studied apparently resulted in no permanent disability. They found a defect of pronation in 28.2%, of supination in 36.9%, of palmar flexion in 94.5%, of dorsiflexion in 80%, and of lateral mobility at the wrist in 49.4%; sometimes also there was impairment of flexion and extension of the fingers.

Abbott and Saunders expressed the opinion that many injuries to the median nerve resulting from fractures of the lower end of the radius or from the treatment of such fractures go unrecognized. They regarded fixation of the wrist in acute flexion, thus forcing the edge of the flexor retinaculum (transverse carpal ligament) against the nerve, as being the most frequent cause of damage to the nerve, and said that if this position is used for fixation the fingers and thumb should be ex-

amined frequently for possible sensory or motor deficits. Lewis implicated hematomas either in the carpal canal or above this, deep to the antebrachial fascia, as one cause of median nerve compression following a Colles fracture. Zoëga reported three cases of injury to the ulnar nerve and expressed surprise that this has not been noted more often.

Fractures of the body of the radius that fall between the insertion of the supinator and that of the pronator teres may show the upper fragment supinated, the lower one pronated (Fig. 5-42). Those below the pronator show less displacement.

FASCIA, SUPERFICIAL VEINS AND LYMPHATICS

The superficial fascia, or tela subcutanea, is thicker in the upper part of the forearm, where it contains more fat than in the lower, and in it are embedded the antebrachial nerves, the superficial veins, and the superficial lymphatic vessels.

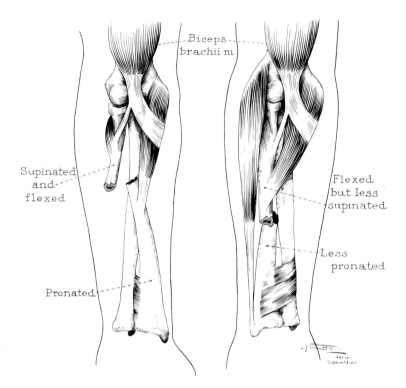

Fig. 5-42. Displacement of the radius by muscle pull in fracture above the pronator teres, between it and the supinator, and below it, between it and the pronator quadratus and brachioradialis.

The veins show a variable pattern, but the main ones below the elbow are the basilic, the cephalic, and the median vein of the forearm; sometimes an accessory cephalic, paralleling the cephalic, is also named. The cephalic vein begins on the dorsum of the hand on the radial side (Fig. 5-43), and curves around the radial side of the forearm to reach the anterior surface and travel in this position to the elbow; here it often receives an accessory cephalic vein, which arises more on the ulnar side of the hand and travels up the radial border of the forearm. At the elbow it has an oblique connection to the basilic vein; this is the median cubital vein, which passes upward and medially from the cephalic to the basilic; here likewise there is usually a prominent connection from the cephalic vein to deep veins of the forearm. After leaving the elbow the cephalic vein passes upward in the lateral bicipital groove, and thence in the groove between the pectoralis major and deltoid muscles.

The basilic vein likewise begins on the back of the hand, but from the ulnar side of the venous network; it passes upward on the ulnar side of the forearm (Fig. 5-44), and as it continues up the arm lies with the medial antebrachial cutaneous nerve in the medial bicipital groove. A little distance above the elbow it receives the median cubital vein from the cephalic, and the median forearm vein; at about the junction of middle and lower thirds of the arm it penetrates the brachial (deep) fascia to course deep to this to the axilla.

The median forearm vein is a usually small vein that begins on the palmar surface of the hand, and runs up the anterior surface of the forearm, in about the midline, to join the basilic vein or the vena mediana cubiti at about the level of the elbow. Variations in the venous pattern at the elbow are numerous.

It is important to realize that there may be a superficially located artery lying in the cubital fossa, and that an aberrant ulnar artery may course superficial to rather than deep to

aspect pass around the radial border of the limb, those on the ulnar aspect around the ulnar border of the limb, to converge upon the lymphatic vessels of the anterior aspect of the forearm (Fig. 5-45), and with these give rise to the less numerous superficial lymphatics of the arm. Danese and Howard divided the lymphatics of the forearm into three groups: radial, median, and ulnar; although all three groups end in axillary nodes, they said, members of the ulnar group may end first in the cubital (trochlear, epitrochlear) node or nodes, while it is some member of the radial group that may travel with the cephalic vein, end in a deltopectoral node, and continue to a deep cervical node, thus bypassing the axillary nodes.

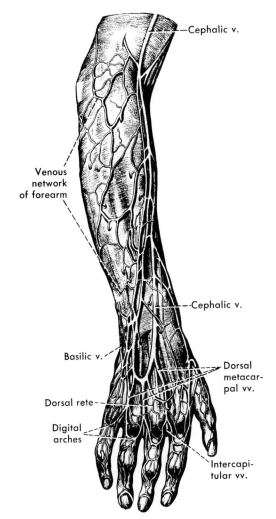

Fig. 5-43. Superficial veins of the posterior aspect of the forearm. (Toldt C: An Atlas of Human Anatomy [ed 2], Vol 2. New York, Macmillan, 1928)

Fig. 5-44. Anterior view of the superficial veins of the forearm. (Toldt C: An Atlas of Human Anatomy [ed 2], Vol 2. New York, Macmillan, 1928)

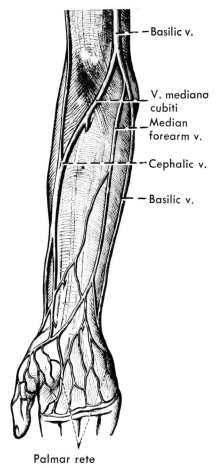

most of the flexor muscles of the forearm. Hager and Wilson, among others, have warned of the danger of gangrene of the hand if material intended for an intravenous injection is delivered by mistake into an artery.

The superficial lymphatic vessels of the forearm originate from the superficial lymphatic network on the hand and are joined by the efferent vessels originating from the cutaneous plexus of the forearm. The channels, especially numerous on the posterior aspect, tend at first to parallel each other. Higher in the forearm the vessels on the posterior side diverge from each other; those on the radial

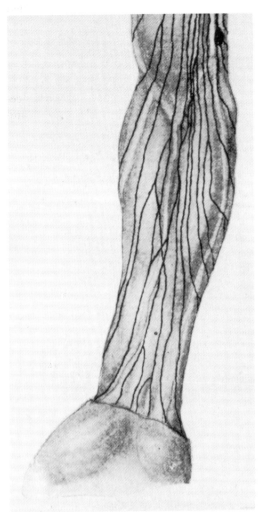

Fig. 5-45. Superficial lymphatics of the front of the forearm. Note the vessels from the dorsum rounding the borders of the limb to run up the medial side of the arm. (Rouvière H: Anatomie des Lymphatiques de l'Homme. Paris, Masson et Cie, 1932)

Cutaneous Innervation

The cutaneous nerve supply to the forearm is through three nerves, the lateral antebrachial cutaneous, the medial antebrachial cutaneous, and the posterior (dorsal) antebrachial cutaneous (Fig. 5-46).

The *lateral antebrachial cutaneous nerve* is the continuation of the musculocutaneous. It emerges from behind the biceps brachii by passing lateral to the tendon of this muscle, and penetrates the deep fascia in front of the elbow. It typically divides into two branches,

an anterior and a posterior one, which pass downward on the lateral aspects of the anterior and posterior surfaces of the forearm. The anterior branch usually communicates with the superficial branch of the radial nerve at the wrist and extends onto the thenar eminence to supply skin there. The posterior branch usually stops at the wrist, but may continue onto the dorsum of the hand to supply skin over the first or first and second metacarpals. In anomalous arrangements of the nerves on the dorsum of the hand, rather rare but more often encountered than those on the palmar surface, the posterior branch of the lateral antebrachial cutaneous may take over much of the distribution of the superficial branch of the radial nerve.

The *medial antebrachial cutaneous nerve* is a branch from the medial cord of the brachial plexus and contains fibers derived from the first thoracic nerve. As it passes downward from the axilla it lies superficial to the brachial artery, and at about the junction of the middle and distal thirds of the. arm it pierces the deep fascia in company with the basilic vein, which it then accompanies to the elbow. It gives off twigs to the front of the arm, but is primarily distributed to the forearm through two terminal branches, an anterior and a posterior one. The anterior branch runs on the flexor surface of the medial side of the arm, supplying skin about as far as the wrist; it and the anterior branch of the lateral antebrachial cutaneous nerve supply the entire anterior surface of the forearm. The posterior branch passes over the upper part of the flexor muscle mass to supply skin on the ulnar side of the posterior surface of the forearm almost to the wrist, but usually does not reach as far down as does the anterior branch. It tends to anastomose with the dorsal branch of the ulnar nerve.

The posterior branches of the lateral and medial antebrachial cutaneous nerves are separated in their distribution to the skin on the back of the forearm by the *posterior antebrachial cutaneous nerve,* a branch of the radial. This nerve arises in the arm at about the point at which the radial nerve pierces the lateral intermuscular septum, penetrates the brachial

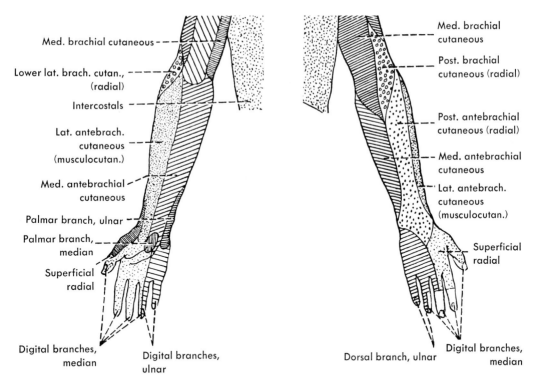

Med. brachial cutaneous

Lower lat. brach. cutan.,
(radial)

Intercostals

Lat. antebrach.
cutaneous
(musculocutan.)

Med. antebrachial
cutaneous

Palmar branch, ulnar

Palmar branch,
median

Superficial
radial

Digital branches,
median

Digital branches,
ulnar

Med. brachial
cutaneous

Post. brachial
cutaneous (radial)

Post. antebrachial
cutaneous (radial)

Med. antebrachial
cutaneous

Lat. antebrach.
cutaneous
(musculocutan.)

Superficial
radial

Dorsal branch, ulnar

Digital branches,
median

Fig. 5-46. Distribution of nerves to the forearm and hand. (Flatau C: Neurologische Schemata für die ärtzliche Praxis. Berlin, Springer, 1915)

fascia, and passes behind the lateral epicondyle to run a course down the middle of the posterior aspect of the forearm, supplying skin here and overlapping in its distribution the distributions of the posterior branches of the medial and lateral antebrachial cutaneous nerves.

Deep Fascia

The deep fascia of the forearm (antebrachial facia) encloses both the flexor and extensor muscles in a common cylindrical sheath (Figs. 5-47 and 5-48). This sheath is continuous above with the fascia of the arm, below with that of the hand; in the forearm it is attached to the posterior surface of the olecranon and to the subcutaneous border of the ulna. In its upper part, where it covers both flexor and extensor muscles close to their origins, it is firmly fused to these muscles, and some of the muscle fibers take origin from it; also, it sends septa between the muscles, these septa likewise contributing to the origins of the mus-

cles, so that tendons and septa together form the common flexor and common extensor tendons. The extensor and flexor groups of muscles in the upper part of the forearm are separated from each other by the attachment of the antebrachial fascia to the ulna and by a septum that it sends to the radius.

In the lower part of the forearm the antebrachial fascia of the anterior surface is split into two layers, a thin one covering the palmaris longus, flexor carpi ulnaris, and flexor carpi radialis muscles, and a stronger one between these and the other flexor muscles. At the wrist both of these fascias are strengthened by transverse fibers. The transverse fibers in the more superficial layer largely blend with those of the deep layer, but on the ulnar side attach to the front of the pisiform bone and form, with the posteriorly lying deep fibers (the flexor retinaculum) a tunnel (Guyon's canal) enclosing the ulnar nerve and vessels. These anterior fibers may or may not be well developed; they have been called the

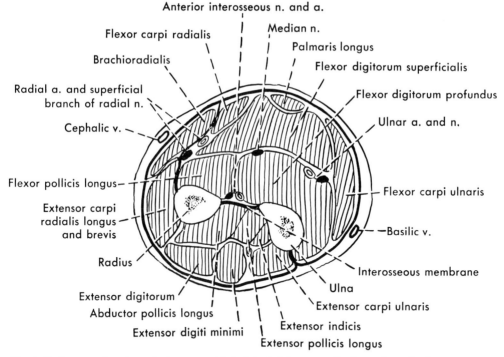

Anterior interosseous n. and a.

Flexor carpi radialis

Median n.

Brachioradialis

Palmaris longus

Flexor digitorum superficialis

Radial a. and superficial
branch of radial n.

Flexor digitorum profundus

Cephalic v.

Ulnar a. and n.

Flexor pollicis longus

Extensor carpi
radialis longus
and brevis

Flexor carpi ulnaris

Basilic v.

Radius

Interosseous membrane

Ulna

Extensor digitorum

Extensor carpi ulnaris

Abductor pollicis longus

Extensor indicis

Extensor digiti minimi

Extensor pollicis longus

Fig. 5-47. Cross section through the upper third of the left forearm. (Redrawn from Eycle-shymer AC, Shoemaker DM: A Cross-Section Anatomy. New York, Appleton, 1923)

Fig. 5-48. Cross section through the lower third of the left forearm. (Redrawn from Eycle-shymer AC, Shoemaker DM: A Cross-Section Anatomy. New York, Appleton, 1923)

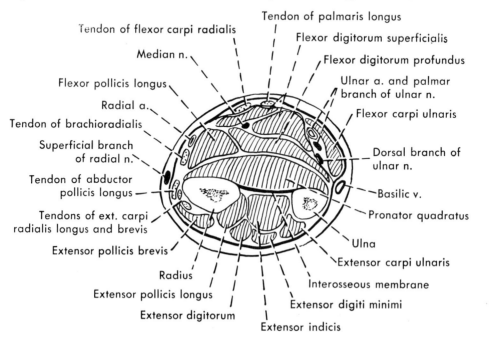

Tendon of palmaris longus

Tendon of flexor carpi radialis

Flexor digitorum superficialis

Median n.

Flexor digitorum profundus

Flexor pollicis longus

Ulnar a. and palmar
branch of ulnar n.

Radial a.

Flexor carpi ulnaris

Tendon of brachioradialis

Superficial branch
of radial n.

Dorsal branch of
ulnar n.

Tendon of abductor
pollicis longus

Basilic v.

Tendons of ext. carpi
radialis longus and brevis

Pronator quadratus

Extensor pollicis brevis

Ulna

Radius

Extensor carpi ulnaris

Extensor pollicis longus

Interosseous membrane

Extensor digitorum

Extensor digiti minimi

Extensor indicis

superficial part of the retinaculum and the
"volar carpal ligament," but should not be
confused with the retinaculum proper. The
transverse fibers of the deeper fascia give rise
to the strong flexor retinaculum (transverse
carpal ligament), described in the following
chapter; the ulnar vessels and nerve pass in
front of the retinaculum, but the median
nerve and the long flexor tendons of the fingers and thumb pass behind it.

On the dorsum of the wrist there are
oblique fibers, a thickening of the posterior
fascia, that extend from the radius to the styloid process of the ulna and the ulnar side of
the carpals, and these form the extensor retinaculum (dorsal carpal ligament). This retinaculum sends septa to the radius and ulna so
that there are a number of separate compartments for tendons on the back of the wrist.

FLEXOR MUSCLES OF FOREARM

The anterior or flexor muscles of the forearm
form a prominent mass on the medial side of
the upper part of the forearm, since many of
them arise from the medial epicondyle or in
its immediate vicinity. Some of the more superficial muscles arise from the medial epicondyle through a common flexor tendon;
this is formed by the fusion of tendinous fibers
of the pronator teres, flexor carpi radialis,
palmaris longus, flexor digitorum superficialis
(sublimis) and flexor carpi ulnaris with each
other, with the fibrous septa that the deep fascia sends between the muscles, and with the
overlying deep fascia. Other members of the
flexor group arise lower, from the ulna, the radius, and the interosseous membrane.

The muscles of the flexor surface of the
forearm are innervated by two nerves, the
median and the ulnar; of these, the ulnar
nerve supplies only the flexor carpi ulnaris
and an ulnar portion of the flexor digitorum
profundus, and the median nerve therefore
supplies all the remaining muscles.

The segmental innervation of these muscles
is not accurately known in detail, and presumably varies somewhat. However, the

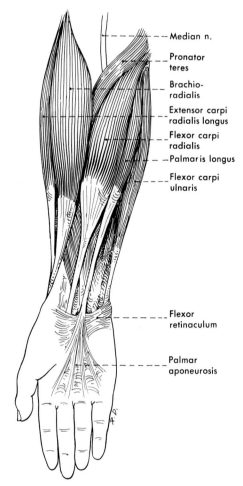

Fig. 5-49. Superficial muscles of the front of the forearm. These include the more anterior muscles of the
extensor group.

pronator teres and the closely associated
flexor carpi radialis are apparently usually
supplied through C6 and C7, the palmaris
longus through C7 and C8, and the other
muscles, except for the flexor carpi ulnaris,
through C7 to T1. The flexor carpi ulnaris is
usually said to receive fibers from C8 and T1
only, but fibers from C7 have been thought to
reach the muscle in the many cases in which
the ulnar nerve receives fibers from at least as
high as this spinal nerve (Harris).

For convenience of description, the flexor
muscles are often divided into three layers: a
superficial, consisting of the muscles that help
form the common flexor tendon (Fig. 5-49);

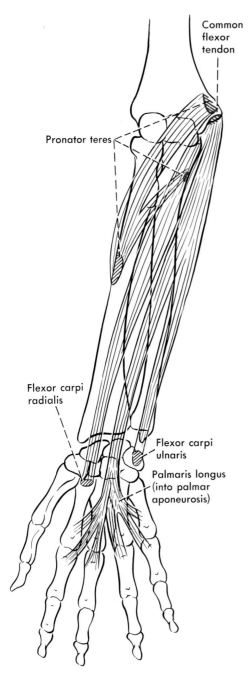

Fig. 5-50. Origins (*red*) and insertions (*black*) of the four muscles of the superficial layer of the flexor aspect of the forearm. The ulnar origin of the flexor carpi ulnaris, on the posterior surface of the ulna, is not visible here.

an intermediate, the flexor digitorum superficialis (sublimis); and a deep, the flexor digitorum profundus, the flexor pollicis longus, and the pronator quadratus.

SUPERFICIAL GROUP

Pronator Teres

The pronator teres commonly arises by two heads: the humeral head takes origin from the lower part of the medial supracondylar ridge and the medial intermuscular septum, and from the common flexor tendon, while the much smaller ulnar head arises from the coronoid process of the ulna and joins the deep surface of the humeral head (Fig. 5-50). The median nerve typically leaves the cubital fossa by passing between the two heads of the pronator teres. After the two heads of the muscle are joined, it continues its downward and lateral direction to pass under cover of the brachioradialis and to insert on the lateral side of the radius at about its middle point.

The pronator teres forms the medial boundary of the cubital fossa. At its origin it is somewhat overlapped by the flexor carpi radialis, and at its insertion, under cover of the brachioradialis, it is crossed by the radial vessels; in between, and along much of its upper border, it is subcutaneous.

The pronator teres usually receives two or more nerves (Fig. 5-51), but may receive a single one; they typically arise from the median nerve just before this nerve passes between the two parts of the muscle, and usually contain fibers derived from the sixth and seventh cervical nerves. The muscle is a good pronator and a weak flexor of the forearm.

The most common variation of the pronator teres is absence of its ulnar head; Jamieson and Anson found this in 8.7% of 300 limbs. Other, rarer, variations include doubling of the humeral head, occurrence of a third head arising from the medial intermuscular septum, and variations in the insertion of the muscle.

Flexor Carpi Radialis

The flexor carpi radialis arises from the common flexor tendon and from the fascia overlying its upper portion, and runs downward and slightly laterally. About the middle of the forearm its relatively broad belly gives rise to a flattened tendon that continues in the direction of the muscle and, becoming rounded, runs first across the front of the flexor retinaculum (transverse carpal ligament), but then enters a compartment formed by the splitting of the radial attachment of the retinaculum. Here it lies in a groove on the trapezium (greater multangular), within a tendon sheath; it inserts into the base of the second metacarpal and, rather commonly, into the third, either by a splitting of the tendon or by a short medial process given off from the base of the tendon.

The flexor carpi radialis is bound in its upper part to the pronator teres and to the palmaris longus by the intermuscular septa from which they take common origin, and partially overlaps the upper parts of both these muscles. It covers a part of the flexor digitorum superficialis in its course down the arm, and a part of it is subcutaneous from the epicondyle to the wrist, where its tendon can be palpated.

Abnormalities of the muscle are apparently rare. It has been reported missing (Dwight). It may apparently, rarely, be involved in a stenosing tendovaginitis at the wrist (Burman).

The radial flexor is a good flexor of the wrist. Although its chief insertion is into the second metacarpal, its origin from the medial epicondyle gives it an oblique course across the forearm and lessens its power of radial abduction; Duchenne could not obtain radial deviation from stimulating it, and some writers have denied that it is a radial abductor; however, he could also not obtain ulnar abduction from stimulation of the flexor carpi ulnaris, which is certainly an ulnar abductor. He admitted that his failure to obtain lateral deviation from stimulating either of these muscles may have been because the major action of both is flexion, and that lateral deviation of the wrist from a flexed position is difficult. Beevor and Wright both reported that the radial flexor normally contracts during radial deviation. Although it apparently does not usually aid in pronation, strong stimulation of it produces pronation following flexion (Duchenne). This muscle receives one or sometimes two branches (C6 and 7) arising from the median nerve either separately or in common with nerves to the palmaris longus and flexor digitorum superficialis.

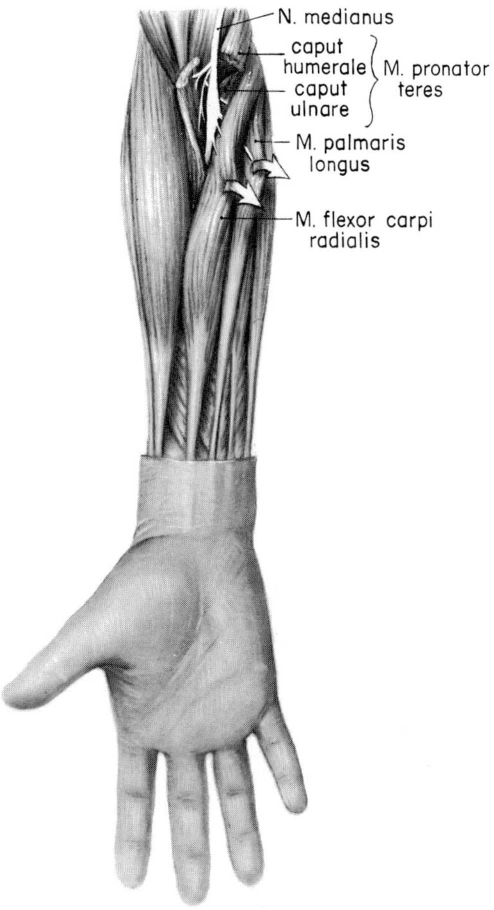

Fig. 5-51. Branches of the median nerve to superficial muscles of the flexor surface of the forearm. (Hollinshead WH, Markee JE: J Bone Joint Surg. 28:721, 1946)

Palmaris Longus

The palmaris longus likewise arises from the medial epicondyle of the humerus by means of the common flexor tendon, from the deep

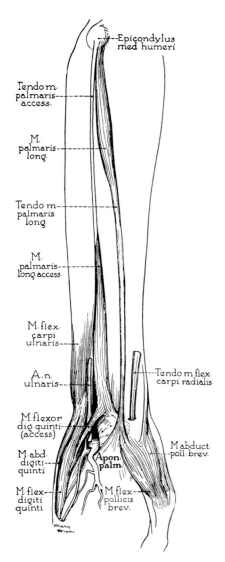

Fig. 5-52. Doubling of the palmaris longus muscle, with a distally placed muscle belly on the duplicate. (Reimann AF, Daseler EH, Anson BJ, Beaton LE: Anat Rec 89:495, 1944)

fascia overlying it, and from the two septa that it shares respectively with the flexor carpi radialis and the humeral head of the flexor carpi ulnaris. The fleshy part of the muscle tends to be short, but is variable; commonly it extends only about a third of the length of the forearm, and ends in a rather long, sharp tendon. At the wrist, this tendon passes superficial to the flexor retinaculum and is therefore a particularly good landmark; it is attached

to the surface of this ligament as it reaches the hand, and in the palm becomes continuous with the palmar aponeurosis.

The palmaris longus muscle is innervated usually by a single branch (probably composed of fibers of about C7 and 8) from the median nerve, this branch often penetrating the flexor carpi radialis (Fig. 5-51), and frequently arising in common with one of the nerves to the flexor radialis or with nerves to other muscles. It is a good flexor at the wrist, and apparently can assist also in pronation; while Duchenne obtained only flexion upon stimulating it, Beevor noted that it contracts, with the flexor carpi radialis, if pronation is against resistance.

Variations in the palmaris longus muscle, both in form and size and in its presence, are frequent. The most common variation of the muscle is its absence, which may be either unilateral or bilateral, and can be readily appreciated in the living individual because of the usual prominence of its tendon. Thompson, McBatts, and Danforth made a study of the genetics of the palmaris longus in several racial groups, and apparently showed that the absence of the muscle is a mendelian characteristic.

Reimann and his co-workers sought the palmaris longus muscle in 1,600 limbs, and found it absent in 12.9% of these. Among 362 bodies in which they examined both limbs, it was absent equally often bilaterally and unilaterally (8.3% in each). In addition, about 9% (46) of 530 palmaris longus muscles showed noticeable variations, of which about half were variations in form such as a centrally or distally placed belly instead of a proximal one, and duplication or splitting of the tendon. More marked variations were abnormal origins or insertions, and accessory slips (Fig. 5-52).

Abnormal origins were from the bicipital aponeurosis (lacertus fibrosus), the flexor carpi radialis, the flexor carpi ulnaris, and the flexor digitorum superficialis. Abnormal insertions included attachment into the antebrachial fascia at about the middle of the forearm, attachment wholly or partly to the fascia of the thenar eminence, and attachment to some of the bones of the carpal canal; apparently, origin from the flexor superficialis and insertion into the hamate bone are the commonest variations in the attachments of this muscle.

While duplication of the entire muscle is not particularly frequent (0.8% in the series of Reimann and his co-workers, but reported to be as high as 3.1% in another series), accessory slips arising from an otherwise normal muscle accounted for 15 of the 46 abnormalities in Reimann's series. These insert into muscles of the little finger or into ligaments, bones, or tendons at the wrist.

Flexor Carpi Ulnaris
The flexor carpi ulnaris arises by two heads: the humeral one originates from the medial epicondyle of the humerus (by way of the common flexor tendon), the fascia covering the muscle, and the intermuscular septum on its radial side; the ulnar one originates, through an aponeurosis that it shares with the profundus, from the medial border of the olecranon and the posterior border of about the upper three fifths of the ulna. The humeral and ulnar heads are joined together just below the epicondyle by a tendinous arch, deep to which the ulnar nerve enters the forearm to lie behind the muscle and on the flexor digitorum profundus. The ulnar flexor forms a broad tendon that lies on the medial border of the anterior surface of the wrist and inserts into the pisiform bone. Its action across the wrist is through the pisohamate and pisometacarpal ligaments to the hook of the hamate bone and the proximal end of the fifth metacarpal. The flexor carpi ulnaris largely covers the ulnar nerve and vessels, but these emerge at the wrist to lie on the lateral side of its tendon.

The muscle is innervated by the ulnar nerve, commonly by two or three branches (Fig. 5-53), the uppermost of which usually leaves the ulnar nerve very close to its passage between the two heads of the muscle, but may arise above the medial epicondyle. Linell found two nerves to the muscle to be the rule, but sometimes found four; he described the two constant branches as arising close to-

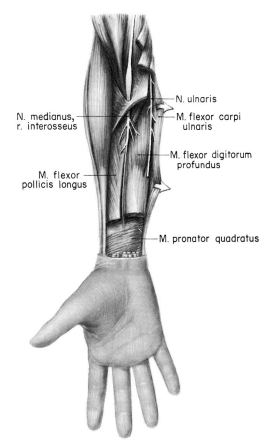

Fig. 5-53. Innervation of the flexor carpi ulnaris and of the deep muscles of the flexor aspect of the forearm. (Hollinshead WH, Markee JE: J Bone Joint Surg. 28:721, 1946)

gether, just before the ulnar trunk disappears between the two heads of the flexor carpi ulnaris. The segmental innervation is from C8 and T1, and perhaps often also from C7. The muscle is both a flexor and ulnar abductor at the wrist.

INTERMEDIATE LAYER

Flexor Digitorum Superficialis
The flexor digitorum superficialis (formerly "sublimis") stretches across the forearm between the medial epicondyle and the radius (Fig. 5-54), and separates the superficial group of muscles from the deep group throughout much of the forearm. The large and somewhat thick humeroulnar head arises

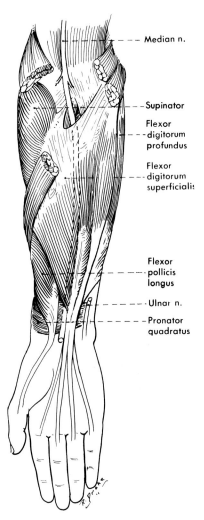

Fig. 5-54. Flexor digitorum superficialis (sublimis) and some related muscles.

finger (middle digit), the other for the ring finger; the two posterior or deeper tendons are destined for the index and little fingers, respectively (Fig. 5-55). If the tendons are traced upward, it may be seen that the hu-

Fig. 5-55. Origins (red) and insertions (black) of the flexor digitorum superficialis.

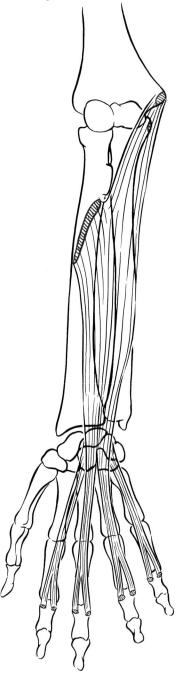

from the medial epicondyle by means of the common flexor tendon, and usually also from the ulnar collateral ligament and the medial border of the base of the coronoid process. The radial head arises from about the upper two thirds of the anterior border of the radius, and is broad but also thin. As the two heads converge to unite high in the forearm their angle of junction is bridged by a dense membrane behind which pass the median nerve and the ulnar artery.

In the lower part of the forearm the muscle gives rise to four tendons, which are arranged in two layers: the radially situated one of the anterior two tendons is destined for the long

meroulnar head divides into two laminae, a superficial and a deep, of which the deep goes to the index and little fingers, while the superficial goes to the ring finger and is joined by the radial head in forming the tendon of the long finger.

The flexor digitorum superficialis is largely covered by the superficial group of the flexor muscles and their tendons, and in turn almost completely covers the flexor digitorum profundus and an upper part of the flexor pollicis longus. Behind it the median nerve gives off its anterior interosseous branch and then continues downward between it and the deep flexor to emerge about 1.5 inches (3.8 cm) above the flexor retinaculum on the radial side of the superficial flexor. Behind it, also, the ulnar artery gives off the large common interosseous branch and, running obliquely medially and therefore leaving the median nerve, emerges from behind the medial border of the muscle to join the ulnar nerve in its course between the flexor carpi ulnaris and the flexor digitorum profundus.

At the wrist the tendons enter a tendon sheath, common to them and the deep flexor tendons; still arranged in two laminae, and with the deep flexor tendons behind them, they pass behind the flexor retinaculum, the two sets of long flexor tendons largely filling the carpal canal (carpal tunnel). The median nerve, on the radial side of the tendons of the superficial flexor above the wrist, also passes behind the retinaculum, but in front of the tendon sheath that houses the long flexor tendons.

In the palm, at the distal end of the carpal canal, the tendons diverge toward the four fingers, and lie anterior or superficial to the similarly diverging tendons of the profundus muscle. The three lateral tendons usually leave the common tendon sheath of the flexor tendons and attain new tendon sheaths, with the accompanying profundus tendons, on the digits; a tubular derivative of the common flexor tendon sheath usually surrounds the superficial and deep tendons to the little finger and follows them out on the digit. Within the tendon sheath on the finger each superficial tendon divides to allow the underlying deep one to pass through, and then goes to an

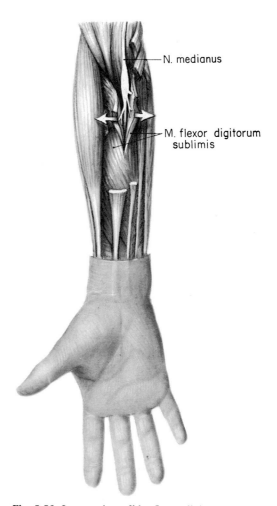

Fig. 5-56. Innervation of the flexor digitorum superficialis (sublimis). (Hollinshead WH, Markee JE: J Bone Joint Surg 28:721, 1946)

insertion on the base of the middle phalanx of the associated finger. The details of the tendons and their sheaths on the fingers are described in connection with the hand.

The flexor digitorum superficialis is *innervated* by several branches, containing fibers from C8 and T1, and perhaps C7, from the median nerve (Fig. 5-56). The uppermost usually arises before the median passes through the pronator teres, and goes to the humeroulnar head of the muscle as the median nerve passes behind it. While the median nerve lies behind the muscle, it gives one or more branches to the radial head and usually an additional branch or so into the humeroulnar head.

The primary *action* of the muscle is flexion of the proximal interphalangeal joints; by continued action it also helps to flex the metacarpophalangeal joints, but its action is weak because the annular ligament (pulley) on the basal phalanx holds it closely to that bone and thus keeps its pull largely along the longitudinal axis of that phalanx. In weakness of the intrinsic muscles of the hand (the chief flexors at the metacarpophalangeal joints) the action of the superficial flexor at these joints can be increased by cutting proximal parts of the annular ligaments (on both sides, not in the midline) so as to allow the tendons to move farther forward in relation to the joints. If the fingers are extended, the superficial flexor can aid in flexion of the wrist, but neither it nor the profundus can shorten sufficiently to produce simultaneous flexion of the fingers and wrist. Consequently, they produce the best flexion of the fingers when the wrist is somewhat dorsiflexed. Both the superficial and the deep flexors of the fingers tend to adduct the fingers as they flex them.

Major variations in the flexor digitorum superficialis are apparently rare. Occasionally, muscular slips derived from it may replace the palmaris longus or form a reduplicate of that muscle, terminating in the fascia of the forearm or the palmar aponeurosis; a lumbrical may arise from it; or there may be muscle slips passing from it to other adjacent muscles, one from the ulnar head of the superficial flexor to the long flexor of the thumb being relatively common. The length of the attachment of the radial head to the bone varies markedly in different specimens, and it is said that this attachment may be missing. Du Bois-Reymond reported an origin of the part of the flexor digitorum superficialis for the index finger so low from the ulna that it crossed behind the tendon from the long finger within the carpal canal; Mainland reported a case in which the part of the superficial flexor to the index finger was split in the forearm into two separate bellies with separate tendons, and had still a third small belly in the palm of the hand; and Chowdhary reported a bilateral condition in which the tendon to the ring finger passed superficial to the flexor retinaculum and that to the little finger

arose in the hand, from the fourth metacarpal.

DEEP GROUP

Flexor Digitorum Profundus

The flexor digitorum profundus (deep flexor of the fingers—Figs. 5-57 and 5-58) is a large muscle that arises from about one half to two thirds of the medial and anterior surfaces of the ulna, from a medial part of the interosseous membrane, and from an aponeurosis common to it and the flexor carpi ulnaris, by which the latter muscle obtains its origin from the posterior border of the ulna. There may also be a small slip of origin from the radius below the radial tuberosity. The muscle tends to divide early into two parts; the lateral part, arising mostly from the interosseous membrane, forms an independent tendon for the index finger, while the larger medial part tends to form a broad common tendon which may not divide into its three definitive tendons until the muscle lies in the carpal canal. Presumably, the degree of independence of the tendons of the three medial fingers accounts for the fact that some persons can flex the terminal phalanges of these fingers independently, others cannot.

At the wrist, the tendons of the flexor digitorum profundus enter the same synovial tendon sheath as do those of the superficial flexor, lying behind these tendons; the relations of the tendon sheaths of the wrist and fingers are the same for both sets of tendons. Within the fibro-osseous tunnels on the fingers each profundus tendon passes through the split in the superficial tendon, and goes to an insertion on the distal phalanx of the digit.

The flexor digitorum profundus is primarily a flexor of the distal interphalangeal joints, but in so doing, as discussed in the following chapter, it also flexes the proximal interphalangeal joints and therefore is an entirely adequate substitute for the superficial flexor. In attempting to flex the metacarpophalangeal joints or the wrist joint the deep flexor is subject to the same limitations as is the superficial flexor.

The flexor digitorum profundus has a double *innervation*, from both the median and the

dency for the two nerves to share in the innervation of the muscle about equally, the median nerve supplying that portion of the muscle going to the tendons of the index and long fingers, the ulnar nerve those portions

Fig. 5-58. Origins (*red*) and insertions (*black*) of the muscles of the deep layer on the flexor aspect of the forearm.

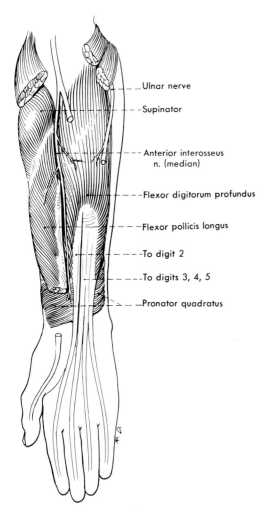

Fig. 5-57. Deep layer of flexor muscles of the forearm; part of the supinator is also visible.

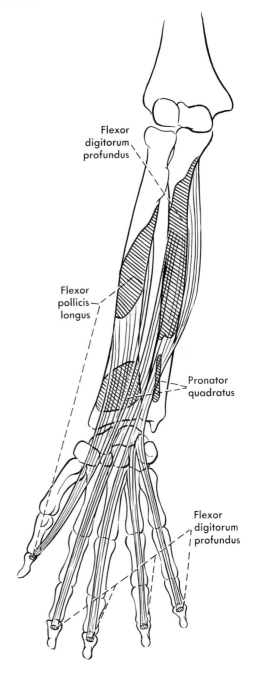

ulnar nerve (Figs. 5-53 and 5-57); the segmental innervation apparently includes fibers of C8 and T1 that reach it from both nerves, and perhaps also fibers of C7, probably by way of the median nerve. The branch or branches from the median nerve typically arise from the anterior interosseous branch rather than from the median nerve itself, and enter the radial portion of the muscle, while the ulnar nerve, as it lies on the front of the profundus a little distance below the elbow, gives rise to a branch that enters the ulnar portion of the muscle. The two branches may anastomose. As discussed later, the exact distribution of the branches to the profundus apparently varies; however, there is a ten-going to the tendons of the ring and little fingers. That part of the muscle to the index finger apparently is always innervated exclusively by the median nerve. The median nerve is more apt to invade ulnar nerve territory, helping to innervate the muscle to the ring or even the little finger, than is the ulnar nerve to invade median nerve territory.

The most common variation in the flexor digitorum profundus is in the level at which its tendons become independent, that for the index finger sometimes failing to become independent until the wrist is reached, while those for the remaining fingers may be independent well above the wrist. Occasional muscle bundles from the medial epicondyle, the ulna, or the neighboring flexor digitorum superficialis and flexor pollicis longus may join the muscle; an accessory bundle arising from the coronoid process and forming a tendon that joins the tendon of the long or index finger is said to be found in about 20% of bodies, and is called the accessory tendon of the flexor digitorum profundus. Rarely, a lumbrical may arise from the deep flexor in the forearm.

Flexor Pollicis Longus
The flexor pollicis longus, or long flexor of the thumb, lies in the same plane as the flexor digitorum profundus, and arises from the lateral part of about the middle half of the interosseous membrane and from a large part of the radius below the radial tuberosity; the upper part of this attachment is below and medial to the insertion of the supinator muscle (the line along which the muscles adjoin each other sometimes being known as the oblique line of the radius). As the muscle is traced down the origin expands to cover the entire width of the anterior surface of the radius. The tendon of this muscle originates high in the forearm from the more vertically descending fibers, and the muscle fibers arising lower enter its posterior surface and its edges; a short distance above the wrist the tendon becomes rounded and free of muscle attachment, and then passes behind the flexor retinaculum on the radial side of the closely grouped tendons of the superficial and deep

flexor digitorum muscles. Within the carpal canal the flexor pollicis longus has a tendon sheath, which may be completely independent or may communicate with the common tendon sheath of the flexor tendons; this tendon sheath usually passes along the flexor tendon almost to its insertion, and on the thumb is enclosed within a fibro-osseous tunnel similar to those of the fingers. The flexor pollicis longus inserts on the distal phalanx of the thumb.

The flexor pollicis longus is a flexor of the distal phalanx of the thumb, and by continued action may act on the metacarpophalangeal joint. Duchenne was unable to produce any movement of the metacarpal by stimulating it, but Jones ('19) described it as being very important in producing opposition after the thenar muscles are paralyzed. Presumably it can help flex the wrist if the thumb is kept extended. It is innervated through the anterior interosseous branch of the median nerve, commonly by two branches; its segmental innervation has been reported to be C6 and 7 and also C8 and T1, but is probably usually regarded as being C7, C8, and T1.

A common variation of the flexor pollicis longus is the presence of an accessory head of origin derived from the common flexor mass in the upper part of the forearm, and ultimately taking origin from either the medial humeral epicondyle or the coronoid process of the ulna, or both. Dykes and Anson found such an accessory head or tendon in 80 of 150 extremities (about 53%); of 21 tendons that they traced to a bony origin, only one arose solely from the ulna, 11 arose both from the humerus and the ulna. In all cases, a part of the origin was from the articular capsule of the elbow joint. Very rarely, the long flexor of the thumb may give rise to a lumbrical muscle that originates in the forearm.

Pronator Quadratus
The pronator quadratus is a flat, quadrilateral muscle in the lower part of the forearm, lying behind the flexor digitorum profundus and the flexor pollicis longus, and is therefore sometimes regarded as a fourth layer of the flexor forearm muscles. It arises from about

the distal fourth of the anterior surface of the ulna, and its fibers run almost transversely, but somewhat downward and laterally, to insert into about the distal fourth of the anterior surface of the radius. This muscle participates in all pronation, and carries out the movement alone unless the need for more speed or power requires the participation of the pronator teres (Basmajian and Travill). It is supplied by the anterior interosseous branch of the median nerve, which usually passes behind the muscle and in front of the interosseous membrane, and sends a branch or branches into the deep surface of the muscle. Its segmental innervation is said to be from C8 and T1, with perhaps a contribution from C7 or even C6.

Deep to (behind) the muscle, the larger terminal branch of the anterior interosseous artery passes through a foramen in the lower part of the interosseous membrane to attain the dorsum of the hand; terminal twigs of the anterior interosseous artery and nerve descend behind the muscle to the flexor surface of the wrist joint, to supply this.

The muscle varies somewhat in the length of its attachment to the radius and ulna; it is said that it is sometimes split into two laminae by the interosseous branch of the median nerve.

ISCHEMIC CONTRACTURE OF FOREARM MUSCLES

Ischemic contracture of the muscles of the forearm, often known as Volkmann's contracture, is a disabling condition that affects especially the flexor muscles of the forearm; it is particularly likely to occur in children as a complication of fracture of the lower end of the humerus (Meyerding and Krusen, and others). As a result of interruption of their blood supply, the muscles become contracted and fibrotic and produce a permanent flexion deformity of the wrist and fingers.

According to Meyerding and Krusen, the mechanisms leading to interruption of the circulation after fracture of the lower end of the humerus are these: the force that breaks the humerus carries the distal fragment back-ward and strips the periosteum away from the posterior surface of the upper fragment, this space immediately filling with blood and therefore exerting pressure; at the same time the lower end of the proximal fragment is carried forward and pierces the periosteum, so that it comes to lie against soft tissues in front of the elbow. Thus blood vessels and nerves become compressed and hemorrhage fills the cubital fossa, further interfering with the blood supply, especially the venous drainage, of the forearm. The nerves may, of course, be severed, and thus there may be a paralysis before ischemia sets in.

Meyerding and Krusen emphasized that all these effects are increased by placing the arm in acute flexion, and that, if there is a large hematoma, incision and drainage of it are indicated. They said that reduction and internal fixation of the fracture may prevent the contracture, and that proper splinting and physical therapy may also help prevent contracture. Parkes described the extensive plastic surgical treatment necessary to restore usefulness to a hand in which ischemic contracture has become established.

VESSELS AND NERVES OF FLEXOR FOREARM

ARTERIES

The brachial artery continues into the forearm on the front of the brachialis muscle, with the biceps on its lateral side and the median nerve on its medial side; it passes deep to the aponeurosis of the biceps muscle (lacertus fibrosus) and in the cubital fossa divides into the radial and the ulnar arteries (Fig. 5-59).

Radial Artery

The radial artery is in its course the more direct continuation of the brachial; it runs downward and slightly laterally, crossing in front of the tendon of the biceps, a part of the supinator muscle, and the pronator teres close to that muscle's insertion; here it lies behind the medial border of the brachioradialis muscle. As it continues downward, crossing the front of the flexor digitorum superficialis, the flexor pollicis longus, and pronator quadratus, it comes to lie medial to the tendon of the brachioradialis and lateral to that of the flexor carpi radialis; it is here, on the pronator quadratus as it inserts on the radius, that the vessel is commonly palpated by compression against the bone. Just above the wrist the artery turns laterally and posteriorly to wind around the wrist; it passes first deep to the tendons of the abductor pollicis longus and extensor pollicis brevis, here in close contact with each other, then through the depression, the anatomic snuffbox, on the radial aspect of the wrist between tendons of the thumb, and then deep to the tendon of the extensor pollicis longus to attain the dorsum of the hand.

Soon after its origin the radial artery gives off the radial recurrent artery, which runs laterally to disappear deep to the brachioradialis muscle. This branch gives off muscular branches to the more anterior muscles of the extensor group as well as to the flexor muscles here related to the radius; the remainder of it ascends, paralleling the course of the radial nerve and therefore passing upward in the groove between the brachialis and the brachioradialis, to end by anastomosing with the radial collateral branch of the profunda brachii artery.

Below the level of origin of the radial recurrent artery, the radial gives off a series of unnamed muscular branches to the muscles on the radial side of the flexor surface of the forearm, and just before it leaves the palmar side of the wrist gives off two named branches in fairly close succession. The first of these is the superficial palmar branch of the radial, which runs downward into the muscles of the thenar eminence, and either ends by supplying these or continues to join the superficial palmar arch. The second branch is the palmar carpal branch, a small twig that passes medially deep to the long flexor tendons to reach the front of the wrist joint.

Ulnar Artery
The ulnar artery, larger at its origin than the radial, has a somewhat more complicated course and bears more named branches. As it leaves the brachial and radial arteries it is directed downward and medially, and passes behind the entire pronator teres muscle while the median nerve, closely associated with it at its origin, passes between the two heads of the muscle. It then rejoins the median nerve, and nerve and artery pass together between the two heads of the flexor digitorum superficialis, under cover of which the artery gives off most of its named branches.

Deep to the superficial flexor of the fingers, the ulnar artery continues its medial and downward direction to emerge from behind the medial border of this muscle and pass under cover of the flexor carpi ulnaris, where it lies between this and the flexor digitorum profundus; here it gives off muscular branches and descends more vertically, in company with the ulnar nerve. As it nears the wrist it begins to appear at the lateral side of the flexor carpi ulnaris muscle, and as it reaches the wrist it and the ulnar nerve, which lies medial to it, emerge from behind the flexor carpi ulnaris to pass across the front of the flexor retinaculum just lateral to the pisiform bone. At this location the artery and nerve are covered by the transverse thickening of the anterior lamina of the deep fascia at the wrist sometimes known as the volar carpal ligament. Toward the distal end of the flexor retinaculum, at the base of the hypothenar eminence, the ulnar artery ends by dividing into superficial and deep branches; the distribution of these is described in connection with the hand.

The first branch of the ulnar artery is typically an *ulnar recurrent artery* (Figs. 5-59 and 5-60), usually a short stem arising while the ulnar artery lies behind the flexor digitorum superficialis, that divides into anterior and posterior branches. The anterior branch of the ulnar recurrent artery (also called "anterior ulnar recurrent artery") is ordinarily a small vessel, which runs upward to pass deep to the humeral head of the pronator teres and other muscles attached to the common flexor tendon and ascend anterior to the medial epicondyle and deep to the brachialis muscle, anastomosing in front of the elbow with the inferior ulnar collateral artery especially. It gives off twigs to the muscles adjacent to it in

Fig. 5-59. Chief vessels and nerves of the anterior aspect of the forearm.

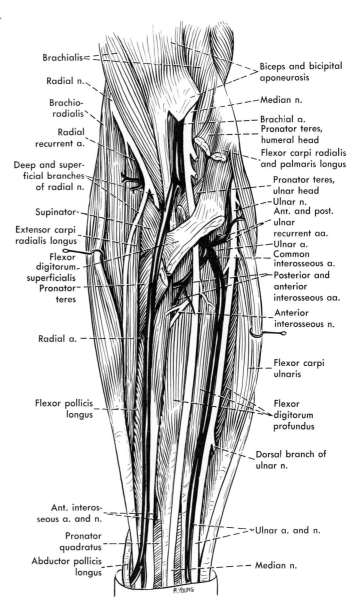

Brachialis

Radial n.

Brachio-radialis

Radial recurrent a.

Deep and super-ficial branches of radial n.

Supinator

Extensor carpi radialis longus

Flexor digitorum superficialis

Pronator teres

Radial a.

Flexor pollicis longus

Ant. interos-seous a. and n.

Pronator quadratus

Abductor pollicis longus

Biceps and bicipital aponeurosis

Median n.

Brachial a.
Pronator teres, humeral head
Flexor carpi radialis and palmaris longus

Pronator teres, ulnar head
Ulnar n.
Ant. and post. ulnar recurrent aa.
Ulnar a.
Common interosseous a.
Posterior and anterior interosseous aa.

Anterior interosseous n.

Flexor carpi ulnaris

Flexor digitorum profundus

Dorsal branch of ulnar n.

Ulnar a. and n.

Median n.

its course, and its anastomosis with the inferior ulnar collateral (and sometimes with the superior) may either be obvious or be by tiny twigs not visible in the usual dissection.

The posterior branch of the ulnar recurrent artery (posterior ulnar recurrent artery) passes laterally and upward between the flexor digitorum superficialis and the flexor digitorum profundus to lie deep to the flexor carpi ulnaris, and pass upward along the ulnar nerve between the two heads of origin of the latter muscle. It gives off twigs to the adjacent muscles and to the elbow joint, and anastomoses especially with the superior ulnar collateral artery, a direct anastomosis here being usually observable. In addition, twigs of communication with the inferior ulnar collateral, the interosseous recurrent, and the middle collateral branch of the profunda brachii, unite close to the olecranon to form the rete articulare cubiti (formerly rete olecrani).

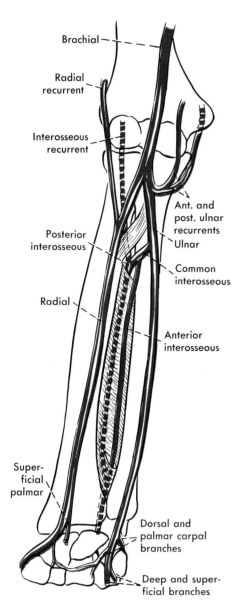

Brachial

Radial recurrent

Interosseous recurrent

Posterior interosseous

Radial

Ant. and post. ulnar recurrents

Ulnar

Common interosseous

Anterior interosseous

Superficial palmar

Dorsal and palmar carpal branches

Deep and superficial branches

Fig. 5-60. Diagram of the arteries of the forearm. The *broken lines* indicate vessels on the dorsal aspect.

The *common interosseous artery* is a short, stout trunk, as large as or larger than the continuation of the ulnar, which arises from the dorsolateral surface of the ulnar artery about an inch (2.5 cm) or so below the origin of this vessel. It passes downward and somewhat backward in the groove between the flexor pollicis longus and the flexor digitorum profundus, and divides into posterior and ante-

rior (dorsal and volar) interosseous arteries. The posterior interosseous artery turns backward between the radius and the ulna above the upper border of the interosseous membrane, usually in the gap between this and the oblique cord, but sometimes above the oblique cord; its course on the dorsum of the forearm is decribed in connection with that part.

The anterior interosseous artery passes downward in front of the interosseous membrane, between the flexor digitorum profundus and the flexor pollicis longus, by which it is usually completely hidden. In this position it gives branches into both muscles, and is accompanied by the anterior interosseous branch of the median nerve, this nerve normally lying on its lateral, or radial, side. Besides its muscular branches, the anterior interosseous artery gives off a median artery that arises from its proximal part, and is usually a small twig that passes outward between the flexor digitorum profundus and flexor pollicis longus to reach the median nerve and descend with this to the wrist, supplying this nerve. (The median artery represents a part of the original axial stem into the limb, and when it is of large size, as it may be, it participates in the formation of the superficial palmar arch, or supplies digital vessels directly to the fingers.)

In addition, the anterior interosseous artery gives rise to the nutrient arteries of both the radius and the ulna, and ends by dividing into anterior and posterior terminal branches. This division is usually close to the upper border of the pronator quadratus or behind the muscle; the anterior terminal branch continues down on the front of the interosseous membrane, behind the pronator quadratus muscle, to the front of the wrist joint. The posterior terminal branch, considerably larger than the anterior, passes through the gap in the lower part of the interosseous membrane and appears in the lower part of the posterior aspect of the forearm, under cover of the extensor muscles here. Its further course is described in connection with the posterior aspect of the forearm.

At the wrist the ulnar artery gives off palmar and dorsal carpal branches, either by a common stem or separately, which pass deep to the tendons to help supply the palmar and dorsal surfaces of the wrist.

Arterial Variations in the Forearm

The exact point of branching of the brachial artery into radial and ulnar vessels in the forearm varies somewhat, but commonly lies at about the level of the upper end of the radius. As already noted in connection with the description of the brachial artery in the arm, the common stem for the radial and ulnar arteries may leave the brachial or even the axillary artery at various levels to form a superficial brachial artery, the deeper brachial artery then continuing into the forearm as the common interosseous artery; also, either the radial or the ulnar artery may arise from the brachial in the arm, or from the axillary artery, and thus form a superficial radial or superficial ulnar artery.

Hofer and Hofer described a case in which the brachial artery passed between the heads of the pronator teres instead of dividing above it, but were unable to find a similar case in the literature. With the exception of high origin of one of its branches, however, the lower end of the brachial artery in the forearm very rarely shows any real variations.

Charles (1894) recorded a case of absence of the radial artery; its place at the wrist was taken by the anterior interosseous artery, which wound around the radial border of the wrist deep to the long tendons of the thumb and entered the palm of the hand by passing between the first and second metacarpal bones in the fashion of a normal radial artery. Real anomalies of the radial artery in the forearm are apparently rare, however.

McCormack and his co-workers found a *superficial radial artery* (one arising high, above the intercondylar line) in 14.27% of 750 upper extremities. Yet, regardless of whether this origin was low in the arm or as high as the axillary artery, the course of the aberrant artery in the forearm was like that of a normal vessel (Fig. 5-61). The most important variation these workers found in a radial artery of high origin was that it had, in a minority of

cases, an anastomosis with the "deep" or usual brachial artery in the cubital region (Fig. 5-61c), this anastomosis passing anterior to the biceps tendon more commonly than posteriorly, and in one case being from the median artery rather than from the "deep" brachial. In many of the cases in which there was an anastomotic branch to the aberrant radial from the "deep" brachial, the radial recurrent artery arose from this anastomotic branch rather than from the radial itself. In addition, among the 107 radial arteries that arose abnormally high, McCormack and his co-workers found one in which the branch to the dorsum was apparently lacking; one in which the artery was duplicated throughout its entire length; and five in which the radial artery divided in the distal fourth of the forearm into two large branches. One ran superficial to the tendons of the long dorsal muscles of the thumb to enter the palm between the first interosseous muscle and the head of the second metacarpal bone, while the other formed the radial portion of the deep palmar arch. Superficial arteries on the dorsum of the wrist have been mistaken for veins and injected with disastrous results (Hager and Wilson).

Superficial ulnar arteries, that is, ulnar arteries arising unusually high (Fig. 5-61a), are far less common, apparently, than are superficial radial ones: McCormack and his co-workers found only 17 superficial ulnar arteries, a percentage of 2.26 of the vessels examined, in contrast to the 107 superficial radial arteries. While the courses of these superficial ulnar arteries in the arm were rather constant, 12 of the 17 showed an abnormal course in the forearm: In 11 of these, the ulnar artery in the forearm had a superficial course immediately deep to the antebrachial fascia (such vessels have also been said to run subcutaneously sometimes) running obliquely downward over the origins of the pronator teres and the other superficial muscles to reach the lateral border of the flexor carpi ulnaris at about the middle of the forearm, where it either passed under cover of this muscle or paralleled it. By the time the wrist was reached, the ulnar nerve and artery had the usual relationship.

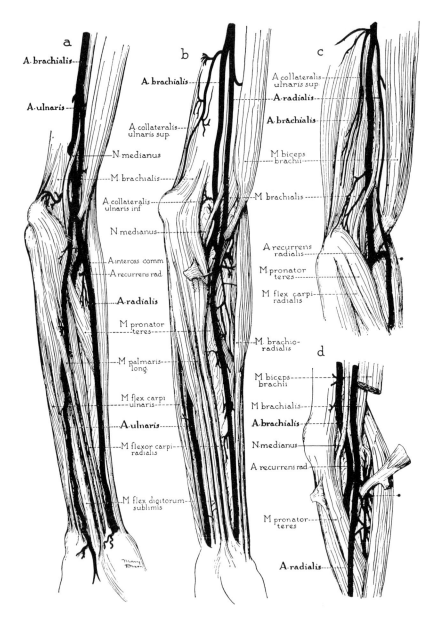

Fig. 5-61. Superficial ulnar and superficial radial arteries in the forearm. The former usually passes superficial to the palmaris longus, instead of deep to it as it does in *a*. (McCormack LJ, Cauldwell EW, Anson BJ: Surg Gynecol Obstet 96:43, 1953 [by permission of Surgery, Gynecology & Obstetrics].)

The twelfth abnormal course was superficial to most of the flexor group of muscles but deep to the palmaris longus, the artery thereafter following the course described for the other 11 cases. Weathersby reported 3 cases of superficial ulnar artery in which the vessel ran on the lateral side of the palmaris longus almost to the wrist, therefore not at all under cover of the flexor carpi ulnaris. Although all three turned medially deep to the palmaris tendon, only one crossed the flexor retinaculum in the usual position; the other two crossed it while lying immediately medial to the palmaris tendon.

As already noted, the *median artery* varies in size, and may contribute significantly to the superficial palmar arch or to palmar digital vessels directly. McCormack and co-workers found a prominent median artery in 4.43% of 750 extremities. Among 80 specimens in which the relations of the ulnar, radial, and median arteries to the superficial arch in the hand were studied, McCormack and his co-workers found that a median artery made some contribution to the superficial palmar arch or to the digits in 16 (more than 20%). Misra found a median artery contributing to the circulation of the hand less often, in 11 hands from 66 bodies; in 4 cases the condition was bilateral.

VEINS

The larger branches of the brachial artery on the flexor side of the forearm (radial and ulnar, the common interosseous, and the anterior interosseous) are accompanied by paired veins; these typically lie one on either side of the corresponding artery and, like the brachial veins, have numerous anastomoses with each other. They are so valved as to aid in directing the flow of blood upward.

LYMPHATICS

The superficial lymphatics of the forearm have already been described. Deep lymphatic vessels accompany the radial, ulnar, and anterior interosseous arteries, and end in the deep lymphatics of the arm. Occasional tiny lymph nodes, by which some of the vessels are interrupted, are said to be present on these arteries, and similar nodes are associated with the bifurcation of the brachial artery in the cubital fossa.

NERVES

Other than the cutaneous nerves, already described, the nerves of the flexor surface of the forearm are two, the median and the ulnar. The ulnar nerve supplies only the flexor carpi ulnaris muscle and a medial or ulnar part of the flexor digitorum profundus; the median nerve supplies all the remaining flexor muscles.

Median Nerve

The median nerve contains fibers from the last three cervical and first thoracic nerves, and usually (in all but 8 of Walsh's 74 special dissections) from the fifth cervical also. Frohse and Fränkel regarded the fibers of C5 in the median nerve as being probably entirely afferent, but Harris and others have listed them as being part of the motor innervation of the pronator teres.

The median nerve enters the forearm medial to the brachial artery (Fig. 5-59), lying on the anterior surface of the brachialis muscle. As it reaches the upper border of the pronator teres it passes between the two heads of this muscle (Fig. 5-62) and parts company with the artery, which passes deep to the muscle. After emerging from between the two heads of the pronator teres the nerve passes, with the ulnar artery on its medial side, between the two heads of the flexor digitorum superficialis, and runs deep to this muscle almost to the wrist. A little above the wrist it appears on the radial side of the tendons of the superficial flexor, behind and slightly to the radial side of the tendon of the palmaris longus; as it passes with the flexor tendons behind the flexor retinaculum (transverse carpal ligament) it assumes a more anterior position, between the tendons and the retinaculum.

The common variation in the course of the median nerve in the forearm is its relation to the pronator teres. When the ulnar head of this muscle is missing (8.7% of Jamieson and Anson's 300 specimens), the nerve cannot, of course, pass between heads of this muscle, but simply passes deep to the humeral head; in the presence of a normal muscle, the nerve may pass through the humeral head or deep to the entire muscle (2% and 6%, respectively, in Jamieson and Anson's series).

The branches of the median nerve in the forearm are almost entirely muscular ones, but one or more articular branches arise at or a little below the level of the elbow joint to supply this. Also, the palmar (palmar cutane-

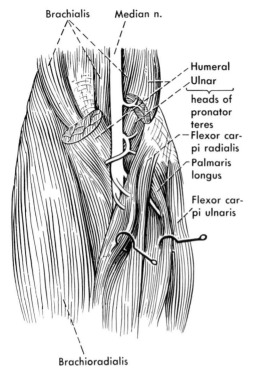

Fig. 5-62. The median nerve as it enters the forearm.

upper branches to other flexor muscles, frequently arise either by common stems, in varying combinations, or very close together (Linell; Sunderland and Ray). The branches to the flexor carpi radialis, palmaris longus, and flexor digitorum superficialis arise for the

Fig. 5-63. Further course and branches of the median nerve in the forearm. Most of the branches shown in Fig. 5-62 have been cut away in the present figure.

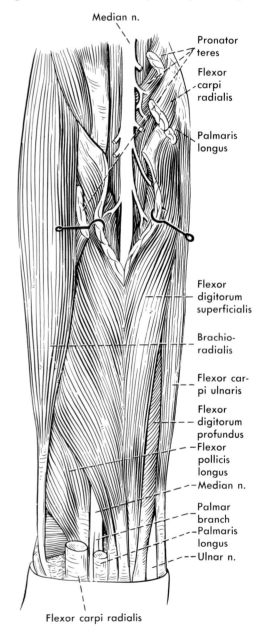

ous) branch arises a little above the wrist and runs on the anterior surface of the nerve to about the upper border of the flexor retinaculum; here the palmar branch becomes subcutaneous and the median nerve passes behind the retinaculum.

The first muscular branch of the median nerve is usually one to the pronator teres (Figs. 5-62 and 5-63), but there is otherwise a great deal of variation in the muscular branches, both in regard to their levels of origin and their number. Linell noted that the uppermost nerve to the pronator usually arises 1 cm to 2 cm below the level of the tip of the lateral epicondyle (his point of measurement), but varies considerably, and Sunderland and Ray found a range in its origin from 7 cm above to 2.3 cm. below the medial epicondyle (their point of measurement). In 6 of Sunderland and Ray's 20 specimens there was only 1 nerve to the pronator teres, but there were 2 in 9 specimens, 3 in 4, and 4 in 1. The uppermost of these nerves usually arises alone, but other nerves to this muscle, and

most part from the median nerve before it gives off its anterior interosseus branch; however, the lowest of these branches, which typically supplies the part of the superficial flexor going to the index finger, may arise as much as 20 cm below the level of the medial epicondyle, from the trunk of the median nerve below the origin of the anterior interosseous or from the latter nerve (Linell; Sunderland and Ray). Sunderland and Ray found a single nerve to the flexor carpi radialis in 18 of 20 specimens, and 2 to 7 nerves to the flexor digitorum superficialis in 18 of 20 specimens. The nerve to the palmaris longus is almost always single.

The anterior interosseous nerve is the largest muscular branch of the median, and typically supplies the flexor digitorum profundus (radial part), the flexor pollicis longus, and the pronator quadratus (Fig. 5-64). It arises from the posterior surface of the median nerve some 2 cm to 8 cm below the level of the medial epicondyle, averaging about 5 cm (Linell; Sunderland and Ray), and runs downward at first on the flexor digitorum profundus. It then passes between this muscle and the flexor pollicis longus and runs downward on the interosseous membrane, with the anterior interosseous artery, to pass behind the pronator quadratus. Isolated paralysis of the anterior interosseous nerve is apparently rare; Lake, reporting three such cases, found only 26 previous ones in the literature. Although no cause was found in the one instance in which Lake explored the area surgically, fibrous bands across the nerve have been found in some cases.

In its course the anterior interosseous nerve usually gives off multiple nerves to both the flexor pollicis longus and the radial part of the flexor digitorum profundus. Sunderland and Ray found from 2 to 6 branches to each muscle in 17 of 20 specimens, and 1 branch only to each muscle in the remainder; the branches arise in various combinations with each other, or sometimes with a branch to the superficial flexor. (Either the flexor pollicis or the radial side of the profundus may also receive a branch from the median nerve that arises above the origin of the interosseous.)

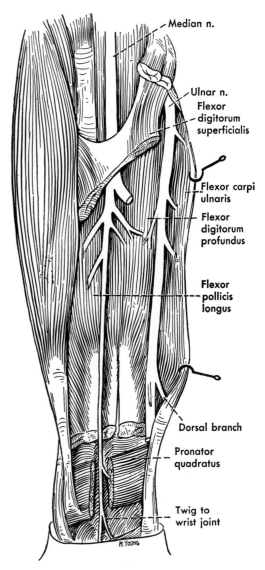

Fig. 5-64. Branches of the ulnar and anterior interosseous (median) nerves in the forearm. The median nerve itself has been removed from a little below the level of origin of its anterior interosseous branch.

The larger terminal branch of the interosseous nerve ends in the deep surface of the pronator quadratus, and the smaller one extends down behind the muscle to reach the front of the wrist joint and help supply this.

Blunt described the median nerve as possessing good longitudinal arterial anastomoses derived from the median artery in the forearm, the ulnar artery just above the flexor retinaculum, and the superficial palmar arch in the hand.

Communications Between Median and Ulnar Nerves

It has long been known that communications between the median and ulnar nerves in the forearm are sometimes present, and several of the earlier authors who studied these and other connections between the nerves of the upper limb pointed out that their presence accounted for the then-known fact that complete section of a nerve was not always followed by the expected anesthesia and paralysis. These facts were largely forgotten until relatively recently, when interest in the occurrence of anomalous innervation of the muscles of the hand (Chap. 6) emphasized the importance of the communications.

Connections between the median and ulnar nerves in the forearm were found by Gruber (1870) bilaterally in 10 and unilaterally in 18 of 125 cadavers, an incidence of more than 20% of bodies, about 15% of limbs. Thomson collected reports on 406 limbs, in which connections between the nerves were present in 63, or about 15%.

These connections usually arise in the upper part of the forearm, and generally originate from the median or its anterior interosseous branch and pass distally between the two long flexor muscles of the digits to join the ulnar. In 33 of Thomson's 63 cases the communicating branch arose from the upper part of the anterior interosseous nerve, in 14 from the median, and in 16 there was a loop between the two nerves supplying the flexor profundus; in the latter cases the anastomosis may be of no importance, since the muscle is supplied by both nerves anyway. Villar quoted evidence that looping anastomoses between the two nerves regularly supply the profundus. Neither Gruber nor Thomson found connections from the ulnar nerve to the median in the forearm, although it is said that they do occur rarely.

The connections from the median to the ulnar nerve are usually rather slender filaments; the opinion has been expressed that they usually contain median nerve fibers that

rejoin that nerve in the anastomosis between the ulnar and the digital branch of the median to the adjacent sides of the ring and middle fingers. This may be true in part, but stimulation of these branches at operation, and a great deal of clinical evidence, indicate that they may also contain fibers for muscles of the hand, either median nerve fibers that reach the thenar muscles by way of the ulnar nerve, or ulnar nerve fibers that travel to the forearm with the median nerve. Cliffton's finding that the sensory distribution may be normal in hands with anomalous innervation of muscles indicates that these communications are not necessarily of any real importance so far as sensory fibers are concerned. Further, the contribution of the ulnar nerve to the adjacent sides of the ring and middle fingers is not dependent upon its receiving fibers from the median in the forearm, for while Gruber found a connection from the ulnar to the median in the hand in 14 out of 15 limbs in which a connection from the median to the ulnar was present in the forearm, he also found it in 44 out of 50 limbs in which there was no connection in the forearm.

Ulnar Nerve

The ulnar nerve has a simple course and relatively few branches in the forearm. It enters the forearm by passing between the humeral and ulnar heads of the flexor carpi ulnaris muscle, and thus comes to lie deep to this muscle on the surface of the flexor digitorum profundus. It passes practically straight distally, being joined toward the middle of the forearm by the ulnar artery (Fig. 5-59), which courses obliquely medially to reach it; thereafter, the nerve and vessel travel distally together, emerging from under cover of the flexor carpi ulnaris just above the wrist. The nerve, being medial to the artery, is more protected than is the artery; however, both nerve and artery pass lateral to the pisiform bone, across the front of the flexor retinaculum, where they emerge from under cover of the "volar carpal ligament." At the base of the hypothenar eminence the ulnar nerve ends by dividing into deep and superficial palmar branches, the distribution of which is described in connection with the hand.

The ulnar nerve consists primarily of fibers derived from the eighth cervical and first thoracic nerves. Through its frequent but variable lateral root it can also obtain fibers from C7 or higher. Harris expressed the opinion that such fibers go to the flexor carpi ulnaris.

The articular branch or the several branches of the ulnar nerve to the elbow joint are usually given off in the lowest part of the arm, at about the level at which the nerve lies behind the medial epicondyle. The first muscular branch arises only a little lower; it usually goes to the flexor carpi ulnaris, but in 1 of 20 limbs investigated by Sunderland and Hughes it went to both this and the flexor digitorum profundus, and in another to the profundus alone. There is usually a second nerve to the flexor carpi ulnaris, and there may be three or four in all (Linell; Sunderland and Hughes). They usually arise in fairly close succession, but in their 20 specimens Sunderland and Hughes found a range in levels of origin of the highest and lowest branches from 4 cm above to 10 cm below the medial epicondyle. The upper branches usually go directly into the muscle; the lowest one may run down some distance before entering it.

In one of their specimens, Sunderland and Hughes found a twig from the ulnar nerve into the flexor pollicis longus, but normally the only other muscular branch in the forearm is to the ulnar side of the flexor digitorum profundus. This nerve is usually single (Sunderland and Hughes found none in 1 specimen, 1 in 16, and more than 1 in 3), and often arises below or with the lowest branch to the flexor carpi ulnaris. Linell found it arising typically about 3 cm below the epicondyle, and running downward about 2.5 cm before entering the muscle.

Since both the median and the ulnar nerves usually supply the deep flexor, the distribution of these nerves within the muscle is of interest. Sunderland ('45), from an examination of 38 patients with lesions of the median or ulnar nerves, concluded that in about 50% of cases the ulnar nerve supplies the ulnar half of

the muscle, the median the radial half. However, he found that the median nerve is much more likely to invade ulnar nerve territory (even as far as the part to the little finger) than the ulnar nerve is to invade median nerve territory, and the part of the muscle to the index finger is apparently regularly supplied by the median nerve alone. Other clinical studies, as indicated in the following section, have indicated much variation in the effects of paralysis of the ulnar and median nerves on the profundus.

The cutaneous branches of the ulnar nerve in the forearm are two, the dorsal branch to the hand and a small palmar branch. The former arises deep to the flexor carpi ulnaris, an average 7 cm above the level of the tip of the radial styloid process (his point of measurement) according to Linell, and becomes subcutaneous by appearing between the medial border of the flexor ulnaris and the ulna, an average 3 cm above the radial styloid. If it arises high, it almost parallels the ulnar nerve before it becomes subcutaneous, and if it arises very low it winds around the styloid process of the ulna. This branch supplies skin on the ulnar side of the dorsum of the hand, and on the dorsum of one and one-half or more fingers.

The small palmar (cutaneous) branch of the ulnar nerve supplies some of the skin over the hypothenar eminence. It arises in the lower third of the forearm, or as high as the middle, sometimes above and sometimes below the origin of the dorsal branch, and reaches the hand by running downward immediately in front of the ulnar nerve.

Because of the high level at which the muscular branches of the ulnar nerve arise, *lesions* of this nerve must be close to the level of the medial epicondyle if they are to produce paralysis of forearm muscles. It is, of course, where the nerve is in contact with the medial epicondyle that it is more commonly injured. It may be involved in fractures of the humerus, or it may be injured purely through contact with the medial epicondyle, either by repeated mild trauma against this or as a result of friction in a roughened groove. Chang and co-workers found that almost half of 400

nerves showed some enlargment here, produced by an increase in the connective tissue;
marked enlargement was found only after age
35. According to Bennett, bony deposits in
the tissue of the elbow joint deep to the ulnar
nerve are an occupational hazard of baseball
players.

When the nerve was simply trapped at the
elbow by the tendon uniting the two heads of
the flexor carpi ulnaris, Wilson and Krout
found that dividing the tendon over the nerve
gave good results. Macnicol agreed that this
was true if the symptoms were of not more
than 3 months' duration, the nerve was definitely constricted, and no adhesions were
present. Otherwise, he said, better results
were obtained by anterior transplantation.
Occasionally the nerve is compressed at the
elbow by some unusual strand: Ho and Marmor found two instances in which a fibrous
band between the medial epicondyle and the
olecranon was responsible, and Mittal and
Gupta found a band that stretched from just
behind a supracondylar process (that had affected the median nerve) to the median epicondyle.

McGowan reported 42 cases in which the
ulnar nerve had been transferred anterior to
the medial epicondyle (Fig. 5-65) because of
injury at the epicondylar level. She emphasized that if transplantation is to be suc-

cessful, it should be done before there is too
much motor damage; when there were paralysis of the interosseous muscles of the hand
and general gross wasting of the muscles supplied by the ulnar nerve, full recovery of
motor function was seldom regained in her
experience, although in all cases the interossei
apparently showed some recovery.

Because the motor branches of the ulnar
nerve arise close to the level of the medial
epicondyle and parallel the main nerve, there
should be no damage to these branches from
transplanting the nerve from a posterior to an
anterior location; however, the articular
branch to the elbow joint would probably
usually have to be severed. Neblett and Ehni
avoided any possible damage to the nerve by
removing the epicondyle and supracondylar
ridge subperiosteally.

Paralysis of the Median and Ulnar Nerves

Combined paralysis of both nerves above the
elbow abolishes most flexion of the wrist, and
paralyzes all the intrinsic muscles of the hand.
A lesion of the median nerve alone, depending upon its level, may affect primarily the
thumb, or affect also movements of the other
digits, of the wrist, and of the elbow. Boswick
and Stromberg found no loss of pronation, although this was much weakened, in any of

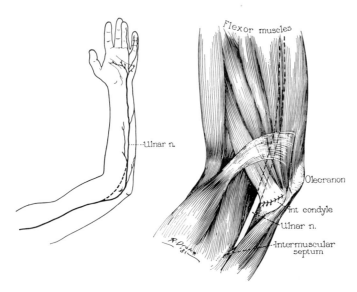

Ulnar n.

Flexor muscles

Olecranon

Int condyle

Ulnar n.

Intermuscular septum

Fig. 5-65. Transplantation of the ulnar nerve to the front of the elbow joint. (Craig WMcK. In Walters W (ed): Lewis' Practice of Surgery, Vol 3, Chap 6. Hagerstown, Prior, 1955)

their 13 cases in which the median nerve was severed at or as high as 6 cm above the elbow; thus, whether the brachioradialis normally participates in pronation or not, it can produce this movement. In only 5 of these was the ability to rotate the thumb to a position of grasp lost; in 4 thumb rotation was normal; in 4 others rotation occurred but the metacarpal was abducted less than normal. The only constant motor deficit that these workers found was loss of flexion in the interphalangeal joint of the thumb and those of the index finger. (According to Linell, injury to the median nerve in the middle of the forearm usually paralyzes the flexor superficialis tendon to the index finger; and injury to the anterior interosseous nerve typically produces loss of flexion of the distal phalanx of the thumb and that of the index finger—Warren.) Thus the profundus was less affected than would be expected, although in 3 of their patients there was less than complete flexion of the interphalangeal joints of the long finger. Boswick and Stromberg were inclined to believe that sparing of the long finger indicated a very frequent ulnar innervation to the part of the profundus to this digit, although they could not exclude the possibility that it was flexion of more ulnar digits that, because of low separation of the tendons, produced flexion of the long finger. Similarly, Kirklin concluded from clinical observations that the ulnar nerve constantly activates the profundus to the long finger (and in 100% of cases, he said, the median nerve activates the profundus to the three radial fingers and in some 30% or more, that to all four). It would thus seem that, regardless of the details of innervation, lesions of neither the median nor the ulnar nerve can be expected to have a constant effect on the three ulnar digits.

Lesions of the ulnar nerve alone produce their chief effect through paralysis of the short muscles in the hand; paralysis of the flexor carpi ulnaris is apparently hard to detect (Jones, '19) and, as indicated in the preceding paragraph, there is no constancy in the effect on the distal phalanges of the fingers.

Since the gravest effects of lesions of the median and ulnar nerves, either alone or combined, are in the hand, they and the tendon transfers that have been used to compensate for them will be discussed in connection with the hand. It can be pointed out here, however, that combined lesions of the two nerves do not necessarily abolish flexion at the wrist, for the abductor pollicis longus is frequently a strong flexor of the wrist (Jones, '19).

EXTENSOR ASPECT OF FOREARM

The superficial vessels and the cutaneous nerves of the extensor or posterior aspect of the forearm have already been described. The extensor muscles of the forearm arise from the lateral and posterior sides of the limb; the uppermost ones arise well above the level of origin of the highest muscle of the flexor group, and all those muscles arising from the humerus arise from the anterior rather than the posterior surface of the bone. Thus the upper members of the extensor group actually function as flexors at the elbow. The muscles of the extensor part of the forearm are divisible into two layers, a superficial one that arises mostly from the humerus, and a deep one that arises mostly from the radius and ulna.

The chief artery on the extensor side of the forearm is the posterior interosseous, which supplies the more posteriorly placed muscles, running downward between the superficial and deep groups. Just above the wrist it is reinforced by a terminal branch of the anterior interosseous artery, which passes through a gap in the lower part of the interosseous membrane. The descending (middle and radial collateral) branches of the profunda brachii are intimately related to the origins of the higher arising (anterolateral) muscles of the extensor group, and the radial artery gives off the chief blood supply to this group in the upper part of the forearm.

All muscles of the extensor group are supplied by the radial nerve, the branches to the muscles of higher origin arising while this nerve is anteriorly situated in the arm and

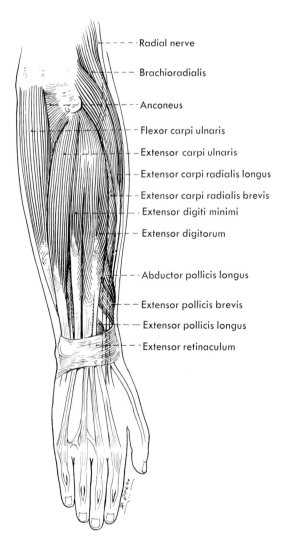

Radial nerve

Brachioradialis

Anconeus

Flexor carpi ulnaris

Extensor carpi ulnaris

Extensor carpi radialis longus

Extensor carpi radialis brevis

Extensor digiti minimi

Extensor digitorum

Abductor pollicis longus

Extensor pollicis brevis

Extensor pollicis longus

Extensor retinaculum

Fig. 5-66. Muscles of the extensor surface of the forearm.

forearm, the remainder arising after it has reached the back of the forearm.

As in the case of the flexor group, their segmental innervations are probably not accurately known. A general statement is that the brachioradialis, a flexor at the elbow, is innervated by C5 and C6 like the other flexors of the elbow; the supinator is probably innervated primarily by C6; the radial extensors of the wrist through C6 and 7, and the long muscles of the thumb through these or C6, 7, and 8; the common and proper extensors of the fingers through C6, 7, and 8; and the ulnar extensor of the wrist by C7 and 8.

SUPERFICIAL MUSCLES

Brachioradialis

The brachioradialis (Figs. 5-66 and 5-67) has a lengthy origin from the anterior surface of about the upper two thirds of the lateral supracondylar ridge, where it lies between the triceps and the brachialis muscles, with the radial nerve intervening between it and the brachialis. It partly overlies the extensor carpi radialis longus, and passes down on the radial side of the limb well in front of the axis of flexion at the elbow joint. The muscular belly tapers to form a flat tendon that is crossed a little above the wrist by the abductor pollicis longus and extensor pollicis brevis, and inserts into the lateral side of the lower end of the radius. As the most anterior of the extensor muscles of the forearm, the brachioradialis forms the lateral border of the cubital fossa.

As already noted in the discussion of movements at the elbow, the chief and perhaps only action of the brachioradialis is to flex the forearm; it is employed both for added speed and power and also contracts synergistically ("antagonistically") in quick extension of the forearm. There is disagreement as to whether it normally participates at all in either pronation or supination, both of which it is in a position to do, but all investigators have agreed that any such contribution is at the most a very minor one. The muscle receives one, sometimes two, branches from the radial nerve, which arise while the nerve is in the groove between the brachioradialis and the brachialis (see Fig. 5-69), but commonly run down some distance, an inch (2.5 cm) or so, before entering the deep surface of the muscle. Its segmental innervation is commonly C5 and C6.

Extensor Carpi Radialis Longus

The extensor carpi radialis longus originates in line with the brachioradialis, from the lower third of the front of the lateral supracondylar ridge of the humerus, and as it passes downward is overlapped by the brachioradialis and in turn overlaps the extensor carpi radialis brevis. Like the brachioradialis, it passes in front of the axis of motion through the elbow, but not so far in front as the first-named muscle. Also, like the brachioradialis, its belly tapers and gives rise to a flattened tendon about the middle of the forearm. In company with the extensor carpi radialis brevis, at first largely overlapping the latter and then on its radial side, it descends on the posterior surface of the radius, passes deep to the abductor pollicis longus and the extensor pollicis brevis, and then enters a tendon sheath common to it and the short radial extensor, with which it also shares a compartment in the extensor retinaculum (dorsal carpal ligament). It inserts on the radial side of the dorsal surface of the base of the second metacarpal bone, a small bursa sometimes occurring between the tendon and the bone close to its insertion.

The extensor carpi radialis longus is presumably a flexor of the forearm, although not as efficient a one as the brachioradialis; Duchenne apparently did not obtain flexion of the forearm upon stimulating it, and perhaps it can aid in this only when the forearm is pronated. It is both an extensor and a radial abductor of the wrist but, according to Tournay and Paillard, is largely inactive in extension (in contrast to the extensor carpi radialis brevis) unless speed is needed. It is typically innervated by a single branch (containing especially fibers from C6 and 7) from the radial nerve. This branch arises from the anterior surface of the radial nerve about 0.5 inch (1.3 cm) below the origin of the nerve to the brachioradialis, and runs downward to enter the deep surface of the muscle at about the level of the lateral epicondyle.

Extensor Carpi Radialis Brevis
The extensor carpi radialis brevis is the most radially placed of the several muscles that arise by the common extensor tendon from the front of the lateral epicondyle (Figs. 5-67 and 5-68); it arises also from the underlying radial collateral ligament, from intermuscular septa, and from its covering fascia. In its upper part it is largely covered by the extensor carpi radialis longus, and it is bordered on its ulnar side by the extensor digitorum; its tendon begins to form about the middle of the

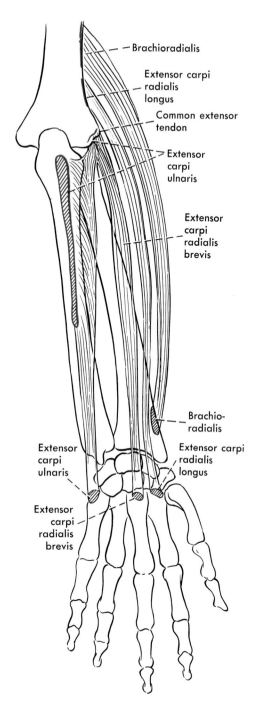

Fig. 5-67. Origins (*red*) and insertions (*black*) of some of the superficial muscles of the extensor group in the forearm.

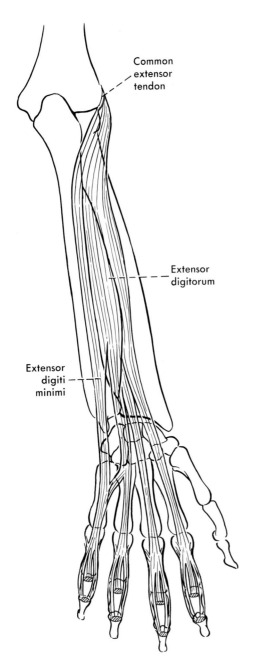

Fig. 5-68. Origin (*red*) and insertions of the extensor digitorum and extensor digiti minimi muscles.

forearm but muscle fibers continue to insert into the tendon to about the lower third. The muscle closely accompanies the extensor carpi radialis longus, but in the lower third of the forearm it is separated from the extensor digitorum by the abductor pollicis longus and the extensor pollicis brevis; these, arising deeply, pass between the short radial extensor and the extensor digitorum, and then cross superficial to the tendons of both radial carpal extensors and of the brachioradialis to reach the radial side of the wrist.

At the wrist, the tendon of the extensor carpi radialis brevis lies immediately on the ulnar side of that of the longus, occupies the same tendon sheath and compartment deep to the extensor retinaculum, but goes to an insertion on the base of the third rather than the second metacarpal. There is typically a small bursa between its tendon and the most proximal part of the base of the metacarpal.

The extensor carpi radialis brevis presumably assists the two previously described muscles in flexing the elbow, but probably contributes relatively little. Its chief action is to extend the wrist, which it does with neither radial nor ulnar deviation (Duchenne); it acts alone in extension unless more speed or force is necessary (Tournay and Paillard). It has been both stated and denied that it is also a radial abductor. Wright said it contracts in radial deviation against strong resistance, and Tournay and Paillard found it regularly active electromyographically during radial abduction. The muscle typically receives a single branch from the radial nerve, which leaves this nerve in the neighborhood of the division into superficial and deep branches; the nerve to the extensor brevis may arise from either the deep or the superficial branch of the radial nerve, occasionally from above the division of the nerve, and sometimes exactly at the angle of its branching. It usually consists of fibers derived from the sixth and seventh cervical nerves.

Extensor Digitorum

The extensor digitorum (extensor digitorum communis), the extensor or common extensor of the fingers, occupies the central part of the dorsum of the forearm. It arises by the common extensor tendon from the front of the lateral epicondyle and soon separates from the extensor carpi radialis brevis, on its radial side; however, it usually shares not only a common origin but a common muscular belly with the extensor digiti minimi, which separates from it some distance down in the forearm, and lies on its ulnar side (Fig. 5-68). The common extensor ends in either three or four tendons, which pass together (in a tendon sheath common to them and the proper extensor of the index finger) through a compartment formed by the extensor retinaculum and the radius, and then fan out toward the bases of the digits. The tendon to the little finger is frequently missing, and when present is usually smaller than the others; on the hand, as discussed later, its place may be taken by one of the intertendinous connections that usually unite the common extensor tendons here, or by an insertion of the tendon to the ring finger into the little finger also.

The details of the extensor tendon on the digits are discussed in the following chapter, along with the functional importance of its attachments. It need only be recalled here that each tendon typically receives attachments of the lumbricals and interossei and thereafter divides into three bands, a central and two lateral ones, of which the central inserts on the middle phalanx while the two lateral come together to insert on the distal phalanx.

The chief function of the extensor digitorum, which is extension of the digits, involves also intrinsic muscles of the hand, and details of function are therefore discussed in the following chapter. The muscle spreads the fingers apart (abducts them) as it extends. Duchenne said this movement is centered about the long finger, as it is when the dorsal interossei produce it, and that the long finger does not move; Jones ('19) said it is primarily a radial spreading from the little finger. While it does not usually assist in flexion at the elbow, Wright described a case in which a hypertrophied extensor digitorum produced flexion at the elbow when the forearm was pronated. It is innervated by the deep branch of the radial nerve, which emerges through the supinator to lie immediately deep to the extensor digitorum. Typically, some three or four nerves, constituting a large part of the bulk of the deep radial nerve, are given into the deep surface of the extensor and supply it

with fibers derived from the sixth through the eighth cervical nerves.

Extensor Digiti Minimi

The extensor digiti minimi (extensor digiti quinti proprius, extensor or proper extensor of the little finger) is sometimes described as arising from the lateral epicondyle with the extensor digitorum, but arises largely from the septum between it and the latter muscle, and in small part from the septum between it and the exensor carpi ulnaris; it occupies a superficial position on the dorsum of the forearm between these two muscles. Separating from the extensor digitorum, this slender muscles gives rise to a tendon, which at the wrist passes through a separate compartment deep to the extensor retinaculum, and usually divides here into two or sometimes more slips, of which the radial one is joined on the hand by the tendon of the common extensor to the little finger, if there is one, both tendons then joining to form the extensor tendon on the dorsum of the little finger.

The extensor of the little finger acts with the common extensor in extending the fingers, but because of its presence the little finger can be extended independently of the others; the muscle is also a powerful abductor of that finger (Jones, '19). It is innervated by one or more branches of the radial nerve that may arise with those to the extensor digitorum, presumably bringing in fibers from C6, 7, and 8.

Extensor Carpi Ulnaris

The extensor carpi ulnaris is the most medial of the superficial group of extensor muscles of the forearm. It arises by two heads: one is from the lateral epicondyle of the humerus by the common extensor tendon, and the other is from the posterior border of the ulna by means of an aponeurosis and the covering deep fascia that closely adheres to the muscle. The latter origin extends for about the middle half of the length of the bone, beginning close to the lower end of the insertion of the anconeus into the ulna. The tendon of the muscle passes through a special compartment deep to the extensor retinaculum, being situated here

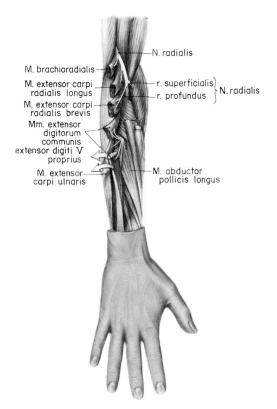

M. brachioradialis

M. extensor carpi
radialis longus

M. extensor carpi
radialis brevis

Mm. extensor
digitorum
communis

extensor digiti V
proprius

M. extensor
carpi ulnaris

N. radialis

r. superficialis
r. profundus } N. radialis

M. abductor
pollicis longus

Fig. 5-69. Innervation of the superficial group of extensor muscles in the forearm. (Hollinshead WH, Markee JE: J Bone Joint Surg 28:721, 1946)

in a groove on the lower end of the ulna, and inserts into the medial side of the base of the fifth metacarpal bone.

The extensor carpi ulnaris is both an extensor of the wrist and an ulnar abductor. It is supplied by a branch, sometimes more than one, of the deep radial nerve, arising in common with or close to the branches to the extensor digitorum and the extensor of the little finger (Fig. 5-69), and usually containing fibers from C6 through C8, or C7 and 8.

Variations and Abnormalities of the Superficial Extensor Muscles

Marked variations from the normal are rarely seen in the superficial group of extensors. Du Bois-Reymond reported unusual origins of some of them: a case in which both of the radial extensors of the wrist arose high from the humerus, close to the deltoid tuberosity; and a case in which the brachioradialis arose from

the extensor digitorum low in the forearm. Variations in the length of the origin of the brachioradialis from the humerus are relatively common, and the muscle may sometimes extend almost halfway up the body of the humerus. The two radial extensors may blend to form a common muscle, and more blending than usual of the extensor of the little finger with the common extensor is relatively frequent. Any of the muscles may be lacking, but this is rare.

Variations in the insertions of the muscles are somewhat more common: The tendon of the brachioradialis may divide, but all the slips nevertheless insert into the radius; it may be double throughout most or all its length. The tendon of the extensor carpi radialis longus may send a slip to the third metacarpal, to the trapezium (greater multangular), or to some adjacent tendon; that of the short radial extensor may attach to the second metacarpal as well as to the third. The tendons of the extensor digitorum vary in number above the wrist, since two or more tendons may remain fused at this level; it is relatively common to see three tendons instead of four, that to the little finger usually being the missing one. These tendons also vary considerably in the way in which they are united on the dorsum of the hand. Du Bois-Reymond reported a case in which the common extensor had six tendons, the first going to the thumb, the second to the index finger, the third and fourth to the middle digit, and the fifth and sixth partly uniting and then spreading out to the ring and little fingers.

Occasionally, aberrant muscle slips are present among the superficial group of the extensors; the brachioradialis may be connected by a muscle slip to some one of the muscles of the arm or to one of the more radially situated muscles of the forearm. A slip known as the extensor carpi radialis intermedius may arise from one or both of the radial extensors and insert into the second or the third metacarpal bone or both, and one known as the extensor carpi radialis accessorius, with a similar origin, may be present (Gruber, 1877), attaching to the metacarpal or first phalanx of the thumb, or to one of the tendons of the thumb. Rarely, a slip arising from the dorsal surface of the ulna and termed the "ulnaris digiti quinti" may be present, inserting into the base of the proximal phalanx of the little finger.

Rupture of the superficial group of extensor muscles is rare. Hamilton reported a case of rupture of the brachioradialis muscle, at the junction of muscle and tendon, and said it was apparently the first such case to be reported. Gladstone reported two cases of rupture of tendons of the extensor digitorum that occurred years after ununited and malunited fractures of the lunate bone and the distal end of the radius, respectively. Vaughan-Jackson has also reported rupture of extensor tendons in two cases in which the cause was apparently attrition at a markedly arthritic distal radioulnar joint.

"Tennis elbow," or epicondylitis, is the occurrence of pain, exacerbated by radial extension of the wrist, over the lateral aspect of the elbow. Carp regarded it as caused by bursitis of the radiohumeral bursa, an inconstant bursa that lies between the common extensor tendon and the radiohumeral joint. Capener suggested it may be caused by compression of the deep branch of the radial nerve within the supinator muscle, and Werner estimated that this is the cause in 5% of cases, but van Rossum and co-workers found the deep radial nerve to be unaffected, and pointed out that a lesion here would be below the level of origin of the nerve to the epicondyle. However, Roles and Moudsly reported cases in which the pain was apparently due to entrapment of the radial nerve or its deep branch at the point at which the radial nerve penetrates the lateral intermuscular septum or between this and the entrance of the deep radial into the supinator muscle. Bosworth regarded it as resulting from any of several different conditions, and said that section of the annular ligament of the radius, or of it and the common tendon, will relieve the pain; Boyd and McLeod both sectioned the tendon and removed a strip from the proximal part of the annular ligament. Garden later agreed that it is probably degenerative changes in this liga-

ment that give rise to the pain, but said it is the pull of the extensor carpi radialis brevis (on the radial collateral ligament and therefore indirectly on the annular ligament) that produces the pain. Z-lenghtening of the tendon of this muscle at the wrist, he found, will relieve symptoms that do not yield to conservative measures.

Coonrad and Hooper, pointing out that the condition sometimes involves the medial epicondyle (and that only a small percentage of their patients played tennis) reported finding tears or scar tissue in a tendon, usually that of the extensor carpi radialis brevis or flexor carpi radialis, in 26 of the 28 patients upon whom they operated. They obtained successful results by a V section of the torn or scarred tissue. In a larger series (82 patients) Nirschl and Pettione "routinely" found a lesion in the origin of the short radial extensor. Thus tears of this muscle seem to be the common cause of "tennis elbow."

Repetitive movements, involving repeated contraction of the muscles with movement of the tendons over the radius where they lie in their synovial sheaths, or even direct trauma, may produce *stenosing tendovaginitis* of some of the extensor tendons. This condition is more common in the tendons and sheath of the abductor pollicis longus and extensor pollicis brevis, the active radial abductors, and is then usually referred to as de Quervain's disease. However, Burman called attention to the fact that other tendons may also be involved: he said that stenosis of the sheath of the long extensor of the thumb is more common than that of stenosis of the radial extensors of the wrist, but that these may be involved; and that stenosis of the common extensor tendon more commonly involves the index and little fingers, which diverge at the wrist and therefore take a bend, while the tendons to the middle and ring fingers, which run straighter, usually escape. Stenosis about the tendon of the extensor of the little finger and also that of the extensor carpi ulnaris has apparently been reported a few times; thus, apparently, most of the tendons of the dorsum of the wrist may on occasion be involved by this process.

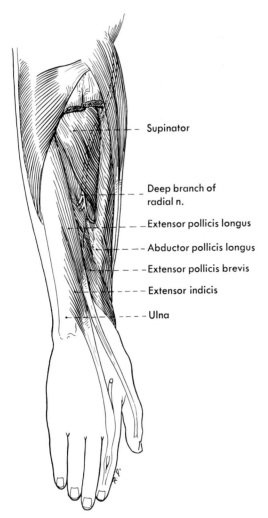

Fig. 5-70. Deep muscles of the extensor group in the forearm.

DEEP GROUP OF EXTENSOR MUSCLES

The deeper-lying muscles of the extensor group (Fig. 5-70) are the supinator, the abductor pollicis longus, the extensor pollicis brevis, the extensor pollicis longus, and the extensor indicis. In the upper part of the forearm these muscles are completely covered by the superficial group, but most of them appear between members of the superficial group at varying levels between the middle of the forearm and the wrist.

Supinator

The supinator is a flat, rhomboid muscle in the upper part of the forearm, completely hidden by the members of the superficial group. It arises in part from the lateral epicondyle of the humerus, in part from the radial collateral and annular ligaments, and in part from the ulna, thus having an origin that runs obliquely downward and backward. Its muscle fibers sweep downward and radially, to be closely applied to the posterior and lateral aspects of the radius, and insert into the lateral surface of this bone in the area in which the lateral surface is also anterior (see Figs. 5-72 and 5-73). The insertion extends from about the level of the radial tuberosity to the attachment of the pronator teres.

The muscle has two laminae, between which the deep radial nerve passes from the flexor to the extensor surface of the forearm. The fibers of the superficial lamina are longer and have a more vertical direction than do those of the deep, and according to Davies and Laird the deep portion may insert on the radius in such an oblique fashion that the radial nerve lies directly upon the bone rather than on the deep portion of the muscle.

The supinator muscle is, as indicated by its name, a supinator of the forearm. Against no resistance it carries out supination by itself, but if more power is needed it is assisted in this action by the biceps. It has a mechanical advantage over the biceps, in that it is almost equally effective whether the forearm is flexed or not; however, it is by no means as powerful as the biceps, and is usually said to have only about half the latter's power. The supinator muscle typically receives several branches (probably largely composed of sixth cervical fibers) from the deep branch of the radial nerve, which arise from this before and as it passes between the two layers of the muscle (Fig. 5-71).

Abductor Pollicis Longus

The origin of the abductor pollicis longus (long abductor of the thumb) is in part from the posterior surface of the ulna distal to the supinator muscle, but spreads across the interosseous membrane to include an origin from this and from about the middle third of the posterior surface of the radius, distal to the insertion of the supinator (Fig. 5-73). From this wide origin the fibers converge and

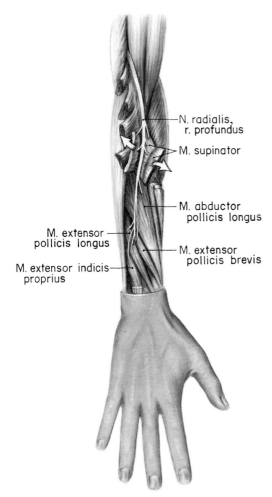

Fig. 5-71. Innervation of the deep muscles of the extensor group in the forearm. (Hollinshead WH, Markee JE: J Bone Joint Surg 28:721, 1946)

N. radialis, r. profundus
M. supinator
M. abductor pollicis longus
M. extensor pollicis longus
M. extensor indicis proprius
M. extensor pollicis brevis

terolateral side (Fig. 5-72), but very frequently is double or triple and has other attachments also—see a following section.

The chief action of the abductor pollicis longus is not pure abduction of the thumb;

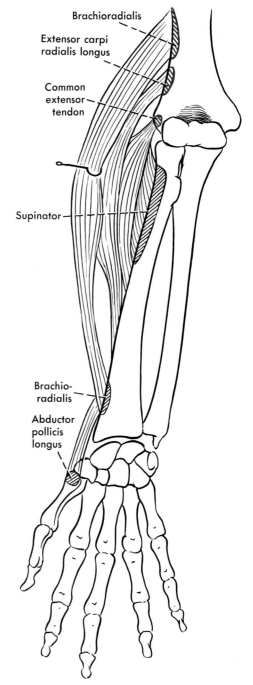

Fig. 5-72. Those origins (*red*) and insertions (*black*) of the extensor group of forearm muscles that are visible in an anterior view of the limb.

Brachioradialis
Extensor carpi radialis longus
Common extensor tendon
Supinator
Brachioradialis
Abductor pollicis longus

the muscle passes obliquely downward and radially, to emerge between the extensor carpi radialis brevis and the common extensor of the fingers, with the extensor pollicis brevis immediately inferomedial to it; these two muscles thus become superficial in the distal part of the forearm, and cross superficial to both radial extensors. As the two closely associated muscles become tendinous they also cross the tendon of the brachioradialis, and usually share a common tendon sheath and compartment at the wrist. The two tendons lie superficial to the radial artery as this passes toward the dorsum of the hand. The tendon of the long abductor attaches to the base of the first metacarpal bone on its an-

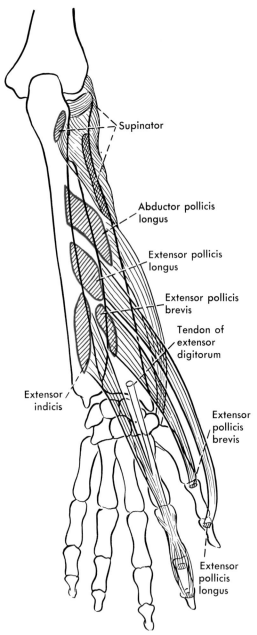

Supinator

Abductor pollicis longus

Extensor pollicis longus

Extensor pollicis brevis

Tendon of extensor digitorum

Extensor indicis

Extensor pollicis brevis

Extensor pollicis longus

Fig. 5-73. Origins (*red*) and insertions (*black*) of the deep group of extensor muscles of the forearm. The insertion of the abductor pollicis longus, seen in Fig. 5-72, is too far on the anterior surface to be seen here.

the hand. It is also an important radial abductor at the wrist and, by virtue of its anterior insertion here, also a flexor of the wrist. The muscle is innervated by one or more branches from the posterior interosseous (radial) nerve, containing fibers from C6 and 7 or C6, 7, and 8.

Extensor Pollicis Brevis
The extensor pollicis brevis (short extensor of the thumb) arises from the posterior surface of the radius just distal to the radial origin of the long abductor, and from the adjacent interosseous membrane, and appears between the two radial carpal extensors and the extensor digitorum in close contact with the abductor longus. It therefore crosses superficial to the radial extensors and the brachioradialis tendon, and on the radial side of the wrist helps to form the anterior boundary of the anatomic snuffbox, where the radial artery passes deep to it and the abductor longus. Its tendon continues along the back of the first metacarpal bone to be inserted into the base of the proximal phalanx of the thumb; commonly, also, it sends a slip to join the long extensor tendon, or shows other variations in its insertion. It is an extensor of at least the basal phalanx of the thumb (and may extend the terminal one) and of the first metacarpal, moving this directly radially. It can also radially abduct the wrist, but according to Wright is used in this action only when there is strong resistance to the movement. It is commonly innervated by a single branch arising from the posterior interosseous branch of the radial nerve, probably containing fibers from C6 and C7.

Extensor Pollicis Longus
The extensor pollicis longus arises from about the middle third of the posterior surface of the ulna and from the interosseous membrane (Fig. 5-73), where it is in close contact with the abductor pollicis longus and partly overlaps the extensor pollicis brevis. It extends downward deep to the more superficial muscles of the forearm, but the lowest part of its muscular belly as it changes into tendon appears on the radial side of the tendons of the extensor digitorum just above the wrist. The

rather, it is that of removing this member from the position of opposition: that is, it draws the first metacarpal obliquely radially and backward (extending and abducting) and at the same time rotates it externally, so that the flexor surface of the thumb more nearly corresponds to the plane of the palm of

tendon runs obliquely across the wrist in a separate compartment formed by the extensor retinaculum and a groove on the radius just medial to the radial tubercle; emerging on the dorsal side of the radial aspect of the wrist, it forms the dorsal boundary of the anatomic snuffbox. It continues distally on the dorsal aspect of the first metacarpal, almost paralleling but gradually converging toward the tendon of the extensor brevis, and passes to an insertion on the base of the distal phalanx of the thumb; not uncommonly it receives some of the insertion of the extensor pollicis brevis, and usually slips from both the abductor pollicis brevis and the adductor pollicis join it.

The long extensor of the thumb is an extensor of both phalanges of the thumb, and carries these and the metacarpal dorsally and medially. It is sometimes also regarded as a supinator, but Duchenne was unable to obtain supination by stimulating it; Beevor listed it as a radial abductor at the wrist, but Wright could find no evidence of this, and neither, apparently, did Duchenne. Apparently it can help somewhat in extension at the wrist (Beevor). It is innervated by one or more branches from the posterior interosseous nerve, given off at approximately the point at which this nerve passes deep to the muscle to lie against the interosseous membrane, and probably containing fibers from the seventh and eighth cervical nerves.

Extensor Indicis

The extensor indicis (extensor indicis proprius, extensor of the index finger) arises, distal to the origin of the long extensor of the thumb, from the posterior surface of the ulna, and may also arise in part from the interosseous membrane. Like the long extensor of the thumb, it is largely covered by the superficial muscles, but a small portion of it is visible above the wrist between the diverging tendons of the extensor carpi ulnaris and the extensor of the little finger. The tendon of the extensor indicis passes deep to the tendon of the extensor digiti minimi, and comes to lie deep to the tendons of the extensor digitorum; it shares with these tendons the same tendon sheath and compartment on the dorsum of

the hand. On the back of the hand it lies on the ulnar side of the extensor digitorum tendon to the index finger; it inserts into the expansion of that tendon on the back of the proximal phalanx.

The extensor indicis assists in extension of the index finger, its independent action, however, allowing extension of this finger without that of the remaining fingers. According to Jones ('19) it is also an excellent adductor of the digit. It is supplied by the posterior interosseous branch of the deep radial nerve; the nerve to this muscle (usually from C7 and C8) is commonly the last muscular branch of the radial, which then continues as a twig to the dorsum of the wrist joint.

Variations of the Deep Muscles

The supinator muscle varies in the distinctness of its separation into two laminae, there sometimes being an obvious fibrous layer between the two parts, while sometimes they seem to blend largely except where they are separated by the radial nerve.

Any of the muscles may have unusual slips of origin, connecting them with other muscles, or there may be fusion between two adjacent muscles. Parsons and Robinson reported that among 131 limbs all the long muscles of the thumb were separate in only 81 (61.8%); in 36.6% the bellies of the long abductor and short extensor were partially or completely fused; in one case all these muscle bellies were fused; and in one case the long extensor of the thumb and the extensor indicis were fused. They also reported absence of the extensor brevis in 8 of 126 limbs, and of the extensor longus (the brevis also being absent) in 2 of 131 limbs.

More common, however, and also of more clinical interest, are doubling and abnormal insertion of the tendons on the hand, the long abductor and the short extensor of the thumb being particularly variable in this regard.

Lacey and co-workers reported that among 38 limbs the *tendon of the long abductor* was single and inserted on the base of the first metacarpal exclusively in only 7; of the remaining number, 1 aberrant tendon was present in 19 limbs, 2 were found in 10, and 3 were found in 2; thus one or more aberrant tendons were

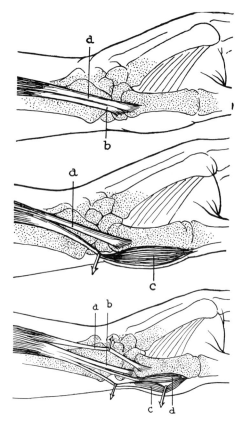

Fig. 5-74. Varieties of anomalous insertion of the abductor pollicis longus. *a* is the normal tendon, *b*, a duplicate inserting into the trapezium, *c* is the abductor pollicis. (Lacey T II, Goldstein LA, Tobin CE: J Bone Joint Surg 33-A:347, 1951)

present in 82% of this series. The majority of the accessory tendons inserted into the trapezium (greater multangular), most of the remainder into the short abductor of the thumb (Fig. 5-74); in only 5 did an accessory tendon have an insertion other than one of these, and such an accessory tendon was very small, inserting into the fascia of the wrist or into the opponens pollicis.

Stein in a similar study found 57 wrists among 84 dissections (68%) which showed one or more accessory tendons of insertion for the long abductor of the thumb, with no significant differences as to either sex or race. In only 2 of his cases, however, was more than 1 accessory tendon present. Twenty-four of these tendons inserted into the trapezium, and 24 inserted into muscles or fascia of the thenar eminence; the remaining 11 had both

insertions, into the bone and into the muscles or the fascia of the thenar eminence. It is well known to most surgeons who work on tendons at the wrist that doubling of the tendon of the abductor pollicis longus is rather common, and the two series of observations just quoted indicate that doubling may be expected in about 70% or more of limbs. Baba, in a later report, found accessory tendons even more frequently, in 98.5% of 134 hands.

The closely adjacent *tendon of the extensor pollicis brevis* also often shows some doubling, either at the wrist or on the dorsum of the thumb. Stein found only 7 among the 84 wrists he studied that showed doubling of this tendon, and in 3 of these the tendon was doubled as it lay in the osseofibrous canal at the wrist. In 5 of these cases the accessory tendon of the short extensor was attached to the tendon of the long extensor, its more common attachment. In 6 wrists he was unable to find an individual extensor brevis tendon, but a single tendon, apparently representing those of both the short extensor and the long abductor, inserted into the metacarpal and gave off a slip that extended to an attachment on the base of the proximal phalanx of the thumb. In 9 of the 84 wrists the tendons of the long abductor and short extensor, instead of sharing a common compartment, occupied separate compartments deep to the dorsal carpal ligament.

Parson and Robinson reported that among 118 cases the extensor brevis inserted entirely into the proximal phalanx (or partly into the adjacent joint capsule) in 72%; inserted entirely on the distal phalanx with the extensor longus in 6.8%; and had both insertions in 21.2% (Fig. 5-75).

Cauldwell and co-workers studied the *extensor indicis* muscle in 263 forearms, and in all except 3 of these the muscle arose primarily from the ulna; there may be no origin from the interosseous membrane. They also found that in the majority of 80 tendons studied for this point, some muscle fibers extended down into the compartment deep to the extensor retinaculum, rather than tendon alone being here.

Cauldwell and his co-workers found no case of absence of the extensor indicis, although in

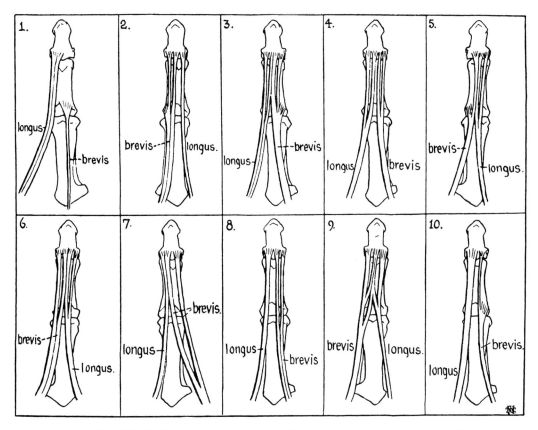

Fig. 5-75. Variations in the insertion of the long and short extensors of the thumb. (McGregor AL: J Anat 60:259, 1926)

3% of their series the tendon was markedly reduced in size. In 10.3% of 263 limbs supernumerary tendons were associated with muscles of normal origin: 6.5% had an accessory tendon that inserted into the middle digit, so that the muscle acted as an extensor of both the index and long fingers (Fig. 5-76); in only 2 cases, however, was there complete duplication of the muscle, so that the accessory tendon and its muscular belly could be considered as forming a separate muscle, the extensor of the middle digit. In 2.7% there was complete duplication of the tendon to the index finger, in 1 case there were tendons going to the index, long and ring fingers, and in 2 cases there were tendons to the index finger and the thumb.

Tendovaginitis

The long abductor and short extensor tendons are particularly apt to be involved at the wrist as they lie in their tendon sheath or sheaths on the radius, constant stretching and contraction of the muscles in radial and ulnar deviation of the hand leading to a condition variously known as tenosynovitis, stenosing tendovaginitis, or de Quervain's disease. Loomis said this disability is most common in persons in their third or fourth decade, but occasionally occurs in the very young or in the aged, and that it is usually said also to be more common in men than in women. Patterson and Jones noted that it has been confused with sprain, arthritis, neuritis, and the like, but that conservative treatment (i.e., rest and splinting) is frequently followed by recurrence of the condition.

As already pointed out, the condition may affect tendons other than those of the long abductor and short extensor of the thumb; Rhodes said that it may even involve flexor tendons, and that the condition is more com-

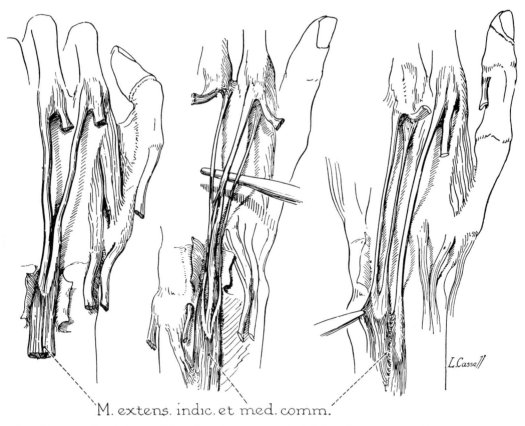

M. extens. indic. et med. comm.

Fig. 5-76. Examples of abnormal insertion of the extensor indicis into the extensor tendon of the middle finger. (Cauldwell EW, Anson BJ, Wright RR: Quart Bull Northwestern Univ M School 17:267, 1943)

mon among automobile mechanics and others whose work requires powerful grip and whose hands are frequently subject to vibration or jolt.

Rupture

While the long abductor and short extensor of the thumb are more likely to be involved in tendovaginitis, it is the long extensor tendon of the thumb that is more commonly ruptured as a result of Colles fracture. Trevor reported 9 patients who presented rupture of this tendon after fracture of the lower end of the radius; however, he found that there is often no evidence that the fracture surfaces were displaced enough to tear the tendon, or that the tendon had become frayed by acting over a sharp fragment. He supported the concept that in such cases tearing of the mesoten-

don at the time of the initial injury interfered with the blood supply to the tendon and caused avascular necrosis, subsequent strain on the tendon causing it to rupture.

BLOOD VESSELS

The posterior (dorsal) interosseous artery is the chief vessel to the extensor muscles of the forearm, but toward the wrist it is reinforced by the anterior interosseous artery (Fig. 5-77). The posterior interosseous artery takes origin from the common interosseous on the front of the forearm, passes between the oblique cord and the upper edge of the interosseous membrane, or above the oblique cord, and emerges between the lower border of the supinator and the upper border of the abductor pollicis longus to branch deep to the superficial muscles of the extensor group. As it emerges between the deep muscles it gives off the recurrent interosseous artery, which runs upward on or through the supinator, and deep to the anconeus muscle, to give off branches to adjacent muscles and to the elbow joint; it then passes behind the lateral epicondyle to anastomose with the middle collateral branch of the profunda brachii and with twigs of the inferior ulnar collateral artery.

Most of the muscular branches of the posterior interosseous artery are given off soon after it reaches the posterior aspect of the forearm, hence its continuation downward is relatively small. As it crosses the abductor pollicis longus it runs with the posterior interosseous nerve, but thereafter it continues downward between the superficial and deep muscles, while the nerve passes deep to the extensor pollicis longus to lie directly on the interosseous membrane. As it runs down the forearm the posterior interosseous artery gives off additional muscular branches, and twigs to the skin, and at the lower end of the forearm is reduced to a mere twig. At this level it is reinforced by the posterior branch of the anterior interosseous artery, which comes through a gap in the lower part of the interosseous membrane and runs downward against this membrane, appearing between the extensor pollicis longus and extensor indicis muscles and usually forming an anastomotic loop with the posterior interosseous deep to the extensor indicis tendon. From this loop twigs are given off to the wrist joint and to the dorsal carpal rete, or the anterior interosseous itself may continue to the rete, the chief contributor to which is usually the radial artery.

It may be noted that the more anteriorly placed muscles of the extensor group receive their blood supply primarily from the radial and radial recurrent arteries, on the anterior surface of the forearm.

RADIAL NERVE

The course of the radial nerve in the lower part of the arm has already been described; it lies here between the brachialis and the bra-

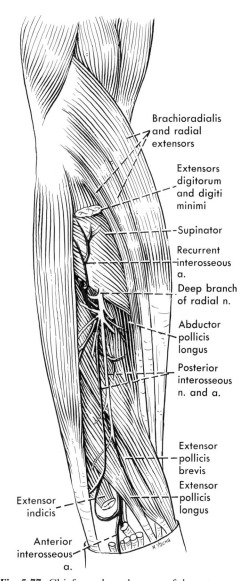

Fig. 5-77. Chief vessels and nerves of the extensor aspect of the forearm.

chioradialis and extensor carpi radialis longus muscles, and in this position gives off the nerves to these muscles (Fig. 5-78). It enters the front of the forearm under cover of these muscles, and at about the level of the tip of the lateral epicondyle divides into superficial and deep branches. At about this point, also, the radial nerve gives off the nerve to the extensor carpi radialis brevis.

Salsbury found that among 50 limbs the nerve to the short radial extensor arose from

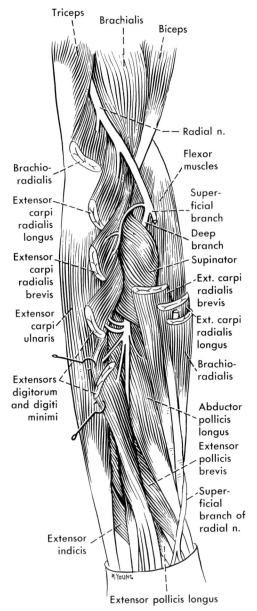

Fig. 5-78. Course and muscular branches of the radial nerve in the forearm.

the superficial branch of the radial nerve in 56% of cases, from the deep branch in 36%, and from the angle formed by the two in 8%. Linell regarded it as commonly arising from the deep branch of the radial nerve, and placed its average origin at 1.82 cm below the level of the lateral epicondyle; Sunderland ('46), among 20 limbs, found the nerve or nerves arising below the lateral epicondyle in 14 specimens, above it in 4, and both above and below in 2 (from the radial itself, from the superficial or deep radial, or from a combination of these.) Origin from the radial nerve itself seemed to be related to a low division of the radial into superficial and deep branches, and origin from the deep branch seemed to be related to a higher division of the nerve, but he noted no apparent relationship between origin from the superficial radial and the site of division of the nerve. He found the range for the origin of the nerve or nerves to this muscle to be from 2.2 cm above the epicondyle to 4.8 cm below it.

Superficial Branch

The superficial branch of the radial nerve, sometimes referred to in the older literature as *the* radial nerve when the main nerve was called the musculospiral, is the smaller of the two terminal branches of the radial nerve. It arises at a variable level: Linell found the bifurcation of the radial nerve to vary from 4.5 cm above the lateral epicondyle to 4 cm below, 9 of the specimens showing the nerve divided above it, 3 at the level of the epicondyle, and 11 below; Sunderland ('46) found among 20 specimens no divisions above the epicondylar level, 4 at this level, and the remainder from 1 cm to 3 cm below. Thus an origin below the level of the lateral epicondyle is apparently slightly more common.

Unless the superficial radial nerve supplies the extensor carpi radialis brevis, which it does fairly often, it gives off no muscular branches at all. It descends under cover of the brachioradialis on the anterior and then the lateral aspect of the forearm, gradually working backward deep to the tendon of the brachioradialis to emerge on the ulnar side of this tendon at about the junction of the middle and distal thirds of the forearm; from here it descends to supply skin on the lateral part of the dorsum of the wrist and hand. Variations in the distribution of this nerve are discussed in the following chapter.

Anomalies of the superficial branch of the radial nerve are occasionally found. Linell reported a case in which the nerve was absent on both sides of one body; and he also reported, apparently among only 26 extremities, 2 cases of abnormal course of the superficial radial: in one, the abnormality was actually a fusion between the brachioradialis and extensor carpi radialis longus muscles, and the superficial radial nerve pierced the tendon formed by the two muscles; in the other, the nerve wound around the superficial aspect of the brachioradialis at about the level of the elbow joint and passed down the forearm superficial rather than deep to this muscle.

Deep Branch

The deep branch of the radial nerve (often called the posterior or dorsal interosseous nerve) has no cutaneous distribution, but is distributed to the posterior muscles of the extensor group and to the wrist joint. It may give off one or more branches to the extensor carpi radialis brevis, but if the nerve or nerves to this muscle arise from the radial above its bifurcation or from the superficial branch of the radial, the first muscular branches of the deep radial are to the supinator muscle.

The deep radial nerve arises under cover of the brachioradialis muscle and enters the upper border of the supinator muscle, coursing obliquely around the radius within this muscle and helping to divide the muscle into superficial and deep parts (Fig. 5-79). Capener described a case of compression of the nerve by a lipoma as it lay in the muscle and suggested that it may be compressed between thickened aponeurotic sheets that may separate it from the muscle fibers. Spinner reported that it may be compressed by a tendinous band ("arcade of Frohse") that sometimes develops in association with the upper edge of the superficial lamina of the muscle. Werner found such a band in 80 of 90 cases he investigated, and said the nerve was compressed upon passive pronation in 83 of 90. Roles and Moudsly said it can also be compressed at a higher level by a firm band that holds it against the elbow joint or, during pronation, by either the extensor carpi radialis brevis or a fascial band that blends with the deep fascia over the forearm flexors. Nor-

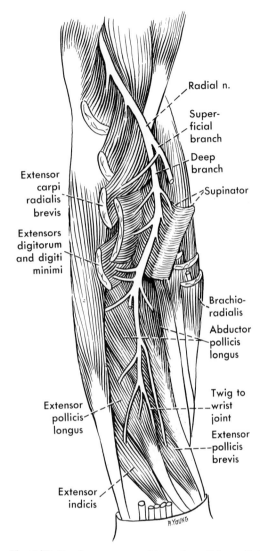

Fig. 5-79. Further course and branches of the radial nerve after some of the overlying muscles have been removed.

mally it does not rest directly against the bone, but rather on the deep part of the muscle; however, as already noted, the insertion of the deep portion may be such as to parallel the direction of the radial nerve, so that the nerve lies on the bone below this insertion.

The deep branch of the radial regularly supplies the supinator, usually by several branches that arise before or as the nerve courses through the supinator muscle. Linell described the branches to the supinator as

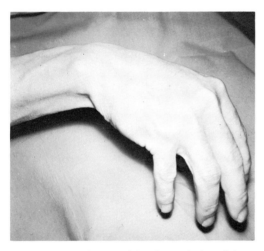

Fig. 5-80. Wristdrop resulting from a lesion of the radial nerve. (Courtesy of Dr. E. D. Henderson)

being numerous short ones arising while the nerve coursed through the muscle; Sunderland ('46) found the supinator receiving its supply partly from the radial itself in 4 specimens, but never from the superficial radial nerve. In his specimens one or more nerves to the supinator were apparently given off by the deep radial before it entered the muscle, often accompanying the nerve for some distance between the strata of the muscle before they penetrated one of these; additional branches usually arose while the deep radial nerve lay in the muscle.

It is in its course around the radius that the radial nerve is most apt to be injured in fractures of the upper third or half of the radius, thus producing a paralysis of all the deep and most of the superficial muscles of the extensor forearm, the brachioradialis and the two radial extensors being spared. The nerve is flat and thin in this location, and suture of it is undoubtedly difficult; however, Mayer and Mayfield expressed the belief that nerve repair should be attempted before tendons are transplanted. They advocated following the radial nerve from the elbow to the site of injury and attempting end-to-end suture.

As the deep radial nerve emerges at the lower border of the supinator it gives off a group of branches that turn superficially into the superficial extensor muscles, the number and arrangement of these branches varying from one individual to another. As pointed out by Sunderland ('46), repair of the nerve at this level would be particularly difficult because of this leash of branches. Sunderland found the order and manner of branching to the extensor digitorum, the extensor digiti minimi, and the extensor carpi ulnaris to be variable; sometimes branches radiated from one level of the nerve, sometimes they arose by a common stem at or just below the site of the emergence of the nerve from the supinator, and sometimes they left the nerve not together but in rapid succession; an additional supply, from farther distally, to the extensor digitorum and to the extensor of the little finger was fairly frequent.

After these branches to the superficial muscles are given off, the posterior interosseous nerve, the continuation of but considerably smaller than the deep radial, descends across the superficial surface of the abductor pollicis longus; it typically gives off one or more branches into the superficial surface of each of the deep extensor muscles before continuing, deep to the extensor pollicis longus and extensor indicis, to the wrist joint. Sunderland described multiple branches to these muscles as being frequent.

Paralysis of Radial Nerve
Paralysis of the radial or deep radial nerve that is not attended by regeneration may be particularly disabling, because of the wristdrop (Fig. 5-80) and interference with the ability to extend the fingers. However, Jones ('19) said that the wristdrop may not be very obvious, because flexion of the metacarpophalangeal joints may produce forceful extension at the wrist by pulling upon the tendons of the extensor digitorum; Jones also said that the interossei extend the two distal phalanges after paralysis of the radial nerve, but Bunnell said that they can do so only if the metacarpophalangeal joints are fixed in extension. The latter statement is made suspect by the finding (Chap. 6) that the interossei normally extend the interphalangeal joints only when the metacarpophalangeal joint is flexed or being flexed.

In his measurements of the lengths of the radial nerve fibers from the level of the epicondyle to the point at which they entered the muscle, Sunderland ('46) found the extensor digitorum was supplied before the extensor indicis in all but one specimen, and here the lengths were equal; and that in only 2 of 20 specimens were the shortest fibers to the extensor digiti minimi significantly shorter than those to the extensor digitorum; thus, in nerve regeneration, the extensor digitorum would be expected to recover before or at least no later than the other two muscles mentioned. He found the extensor of the little finger to be supplied before the extensor of the index in most cases, and the long abductor of the thumb to be supplied before the short extensor in most cases. In some cases, the distance to the extensor carpi ulnaris was less than that to the extensor digitorum, in others it was greater; thus, in regeneration to the extensor muscles, either the extensor of the fingers or the ulnar extensor of the wrist may first show recovery.

If repair and regeneration of the radial nerve do not succeed, it is possible to transfer flexor tendons to such attachments that they can produce extension of the wrist and fingers. The chief purpose of such transfers is to produce extension of the fingers and extension and abduction of the thumb. The relative scarcity of wrist flexors is increased when the palmaris longus is absent, yet hyperextension (dorsiflexion) of the wrist must be prevented if there is to be good extension of the fingers. Young and Lowe quoted evidence that extension of the fingers may amount to no more than about 21% when no flexors of the wrist are left, to as much as 97% when only one is left. Many workers have emphasized the desirability of preserving one wrist flexor, usually the palmaris longus or the flexor carpi radialis, in tendon transfers for paralysis of the radial nerve.

References

ABBOTT LC, SAUNDERS JB DE CM: Injuries of the median nerve in fractures of the lower end of the radius. Surg Gynecol Obstet 57:507, 1933

ANTON JI, REITZ GB, SPIEGEL MB: Madelung's deformity. Ann Surg 108:411, 1938

ATKINSON WB, ELFTMAN H: The carrying angle of the human arm as a secondary sex character. Anat Rec 91:49, 1945

ATSATT RF: Loose bodies in the elbow joint: An unusual location and form. J Bone Joint Surg 15:1008, 1933

BABA MA: The accessory tendon of the abductor pollicis longus muscle. Anat Rec 119:541, 1954

BACORN RW, KURTZKE JF: Colles' fracture: A study of two thousand cases from the New York State Workmen's Compensation Board. J Bone Joint Surg 35-A:643, 1953

BASMAJIAN JV, GRIFFIN RW, JR: Function of anconeus muscle: An electromyographic study. J Bone Joint Surg 54-A:1712, 1972

BASMAJIAN JV, LATIF A: Integrated actions and functions of the chief flexors of the elbow: A detailed electromyographic analysis. J Bone Joint Surg 39-A:1106, 1957

BASMAJIAN JV, TRAVILL A: Electromyography of the pronator muscles in the forearm. Anat Rec 139:45, 1961

BATEMAN JE: Denervation of the elbow joint for the relief of pain: A preliminary report. J Bone Joint Surg 30-B:635, 1948

BAUMAN GI: Rupture of the biceps tendon. J Bone Joint Surg 16:966, 1934

BEATTIE PH: Description of bilateral coracobrachialis brevis muscle, with a note on its significance. Anat Rec 97:123, 1947

BECHTOL CO: Coracobrachialis brevis. Clin Orthopaed 4:152, 1954

BEEVOR CE: The Croonian Lectures on Muscular Movements and Their Representation in the Central Nervous System. London, Adlard and Son, 1904

BENNETT GE: Shoulder and elbow lesions distinctive of baseball players. Ann Surg 126:107, 1947

BLUNT MJ: The vascular anatomy of the median nerve in the forearm and hand. J Anat 93:15, 1959

BORA FW, JR, NICHOLSON JT, CHEEMA HM: Radial meromelia: The deformity and its treatment. J Bone Joint Surg 52-A:966, 1970

BOSWICK JA, JR, STOMBERG WB, JR: Isolated injury to the median nerve above the elbow: A review of thirteen cases. J Bone Joint Surg 49-A:653, 1967

BOSWORTH DM: The role of the orbicular ligament in tennis elbow. J Bone Joint Surg 37-A:527, 1955

BOYD HB, ANDERSON LD: A method for reinsertion of the distal biceps brachii tendon. J Bone Joint Surg 43-A:1041, 1961

BOYD HB, MCCLEOD AC, JR: Tennis elbow. J Bone Joint Surg 55-A:1183, 1973

BRAUNE W, FLÜGEL A: Ueber Pronation und Supination des menschlichen Vorderarmes und der Hand. Arch f Anat u Physiol (Anat Abt) p 169, 1882

BUNNELL S: Tendon transfers in the hand and forearm. Am Acad Orth Surgeons Instructional Course Lectures 6:106, 1949

BURMAN M: Stenosing tendovaginitis of the dorsal and volar compartments of the wrist. AMA Arch Surg 65:752, 1952

BUXTON ST JD: Ossification in the ligaments of the elbow joint. J Bone Joint Surg 20:709, 1938

CAPENER N: The vulnerability of the posterior interosseous nerve of the forearm. A case report and an anatomical study. J Bone Joint Surg 48-B:770, 1966

CARP L: Tennis elbow (epicondylitis) caused by radiohumeral bursitis: Anatomic, clinical, roentgenologic and pathologic aspects, with a suggestion as to treatment. Arch Surg 24:905, 1932

CAULDWELL EW, ANSON BJ, WRIGHT RR: The extensor indicis proprius muscle: A study of 263 consecutive specimens. Quart Bull Northwestern Univ M School 17:267, 1943

CHANG KSF, LOW WD, CHAN ST, CHUANG A, POON KT: Enlargement of the ulnar nerve behind the medial epicondyle. Anat Rec 145:149, 1963

CHARLES CM: On the arrangement of the superficial veins of the cubital fossa in American white and American Negro males. Anat Rec 54:9, 1932

CHARLES CM, PENN L, HOLDEN HF, MILLER RA, ALVIS EB: The origin of the deep brachial artery in American white and in American Negro males. Anat Rec 50:299, 1931

CHARLES JJ: A case of absence of the radial artery. J Anat Physiol 28:449, 1894

CHOUKÉ KS: Variation of the coracobrachialis muscle. Anat Rec 27:157, 1924

CHOWDHARY DS: A rare anomaly of m. flexor digitorum sublimis. J Anat 85:100, 1951

CHRISTENSEN JB, ADAMS JP, CHO KO, MILLER L: A study of the interosseous distance between the radius and ulna during rotation of the forearm. Anat Rec 160:261, 1968

CLIFFTON EE: Unusual innervation of the intrinsic muscles of the hand by median and ulnar nerve. Surgery 23:12, 1948

CONGDON ED, FISH HS: The chief insertion of the bicipital aponeurosis is on the ulna. A study of collagenous bundle patterns of antebrachial fascia and bicipital aponeurosis. Anat Rec 116:395, 1953

COONRAD RW, HOOPER WR: Tennis elbow: Its course, natural history, conservative and surgical management. J Bone Joint Surg 55-A:1177, 1973

DANESE C, HOWARD JM: Postmastectomy lymphedema. Surg Gynec Obst 120:797, 1965

DAVIES F, LAIRD M: The supinator muscle and the deep radial (posterior interosseous) nerve. Anat Rec 101:243, 1948

DE GARIS CF, SWARTLEY WB: The axillary artery in white and Negro stocks. Am J Anat 41:353, 1928

DELORME TL: Treatment of congenital absence of the radius by transepiphyseal fixation. J Bone Joint Surg 51-A:117, 1969

DEPALMA AF, CALLERY G, BENNETT GA: Variational anatomy and degenerative lesions of the shoulder

joint. Am Acad Orth Surgeons Instructional Course Lectures 6:255, 1949

DOBBIE RP: Avulsion of the lower biceps brachii tendon: Analysis of fifty-one previously unreported cases. Am J Surg 51:662, 1941

DU BOIS-REYMOND R: Beschreibung einer Anzahl Muskelvarietäten an einem Individuum. Anat Anz 9:451, 1894

DUCHENNE GB: Physiology of Motion: Demonstrated by Means of Electrical Stimulation and Clinical Observation and Applied to the Study of Paralysis and Deformities. Translated and edited by E B Kaplan. Philadelphia, Lippincott, 1949

DWIGHT T: Notes on muscular abnormalities. J Anat Physiol 22:96, 1887

DYKES J, ANSON BJ: The accessory tendon of the flexor pollicis longus muscle. Anat Rec 90:83, 1944

EHRLICH GE: Antecubital cysts in rheumatoid arthritis—a corollary to popliteal (Baker's) cysts. J Bone Joint Surg 54-A:165, 1972

ELIASON EL, BROWN RB: Posterior dislocation at the elbow with rupture of the radial and ulnar arteries. Ann Surg 106:1111, 1937

FROHSE F, FRÄNKEL M: Die Muskeln des menschlichen Armes. Vol. 2, Sect 2, Pt 2. In von Bardeleben K (ed): Handbuch der Anatomie des Menschen. Jena, Fischer, 1908

GARDEN RS: Tennis elbow. J Bone Joint Surg 43-B:100, 1961

GARDNER E: The innervation of the elbow joint. Anat Rec 102:161, 1948

GLADSTONE H: Rupture of the extensor digitorum communis tendons following severely deforming fractures about the wrist. J Bone Joint Surg 34-A:698, 1952

GREIG HW, ANSON BJ, BUDINGER JM: Variations in the form and attachments of the biceps brachii muscle. Quart Bull Northwestern Univ M School 26:241, 1952

GRUBER W: Ueber die Verbindung des Nervus medianus mit dem Nervus ulnaris am Unterarme des Menschen und der Säugethiere. Arch f Anat Physiol u wissensch Med p 501, 1870

GRUBER W: Ueber den Musculus radialis externus accessorius. Arch f Anat u Physiol (Anat Abt) p 388, 1877

HAGER DL, WILSON JN: Gangrene of the hand following intra-arterial injection. Arch Surg 94:86, 1967

HAMILTON AT: Subcutaneous rupture of the brachioradialis muscle. Surgery 23:806, 1948

HARI RAO GRK, RAMACHANDRA RAO V: Musculocutaneous nerve of the arm. J Anat Soc India 4:48, 1955

HARRIS W: The true form of the brachial plexus, and its motor distribution. J Anat Physiol 38:399, 1904

HENDRY AM: The treatment of residual paralysis after brachial plexus injuries. J Bone Joint Surg 31-B:42, 1949

HITCHCOCK HH, BECHTOL CO: Painful shoulder: Observations on the role of the tendon of the long head of the biceps brachii in its causation. J Bone Joint Surg 30-A:263, 1948

HO HC, MARMOR L: Entrapment of the ulnar nerve at the elbow. Am J Surg 121:355, 1971

HOFER K, HOFER G: Ueber den Verlauf der Arteria brachialis mit dem Nervus medianus zwischen den beiden Köpfen des Musculus pronator teres. Anat Anz 36:510, 1910

HOLSTEIN A, LEWIS GB: Fractures of the humerus with radial-nerve paralysis. J Bone Joint Surg 45-A:1382, 1963

IP MC, CHANG KSF: A study on the radial supply of the human brachialis muscle. Anat Rec 162:363, 1968

JAMIESON RW, ANSON BJ: The relation of the median nerve to the heads of origin of the pronator teres muscle: A study of 300 specimens. Quart Bull Northwestern Univ M School 26:34, 1952

JONES FW: Voluntary muscular movements in cases of nerve lesions. J Anat 54:41, 1919

JONES RW: Primary nerve lesions in injuries of the elbow and wrist. J Bone Joint Surg 28:121, 1930

KASAL T: About the N. cutaneous brachii lateralis inferior. Am J Anat 112:305, 1963

KEON-COHEN BT: Fractures at the elbow. J Bone Joint Surg 48-A:1623, 1966

KERR AT: The brachial plexus of nerves in man, the variations in its formation and branches. Am J Anat 23:285, 1918

KESSEL L, RANG M: Supracondylar spur of the humerus. J Bone Joint Surg 48-B:765, 1966

KILBURN P, SWEENEY JG, SILK FF: Three cases of compound posterior dislocation of the elbow with rupture of the brachial artery. J Bone Joint Surg 44-B:119, 1962

KING T: Recurrent dislocation of the elbow. J Bone Joint Surg 35-B:50, 1953

KIRKLIN JW: Personal communication to the author.

KOLB LW, MOORE RD: Fractures of the supracondylar process of the humerus: Report of two cases. J Bone Joint Surg 49-A:532, 1967

KRAHL VE: The bicipital groove: A visible record of humeral torsion. Anat Rec 101:319, 1948

LACEY T, II, GOLDSTEIN LA, TOBIN CE: Anatomical and clinical study of the variations in the insertions of the abductor pollicis longus tendon, associated with stenosing tendovaginitis. J Bone Joint Surg 33-A:347, 1951

LAKE PA: Anterior interosseous nerve syndrome. J Neurosurg 41:306, 1974

LAURIN CA, FAVREAU JC, LABELLE P: Bilateral absence of the radius and tibia with bilateral reduplication of the ulna and fibula: A case report. J Bone Joint Surg 46-A:137, 1964

LEWIS MH: Median nerve compression after Colles's fracture. J Bone Joint Surg 60-B:195, 1978

LEWIS RM: Colles fracture—causative mechanism. Surgery 27:427, 1950

LINELL EA: The distribution of nerves in the upper limb, with reference to variabilities and their clinical significance. J Anat 55:79, 1921

LIPSCOMB PR, BURLESON RJ: Vascular and neural complications in supracondylar fractures of the humerus in children. J Bone Joint Surg 37-A:487, 1955

LLOYD-ROBERTS GC: Humerus varus: Report of a case treated by excision of the acromion. J Bone Joint Surg 35-B:268, 1953

LOOMIS LK: Variations of stenosing tenosynovitis at the radial styloid process. J Bone Joint Surg 33-A:340, 1951

LOTEM M, FRIED A, LEVY M, SOLZI P, NAJENSEN T, NATHAN H: Radial palsy following muscular effort: A nerve compression syndrome possibly related to a fibrous arch of the lateral head of the triceps. J Bone Joint Surg 53-B:500, 1971

MACAUSLAND AR: Colles's fracture. Am J Surg 36:320, 1937

MACNICOL MF: The results of operation for ulnar neuritis. J Bone Joint Surg 61-B:159, 1979

MAINLAND D: An uncommon abnormality of the flexor digitorum sublimis muscle. J Anat 62:86, 1927

MANSKE PR: Compression of the radial nerve by the triceps muscle: A case report. J Bone Joint Surg 59-A:835, 1977

MARTIN BF: The annular ligament of the superior radio-ulnar joint. J Anat 92:473, 1958a.

MARTIN BF: The oblique cord of the forearm. J Anat 92:609, 1958b

MAYER JH, JR, MAYFIELD FH: Surgery of the posterior interosseous branch of the radial nerve: Analysis of 58 cases. Surg Gynecol Obstet 84:979, 1947

MCCORMACK LJ, CAULDWELL EW, ANSON BJ: Brachial and antebrachial arterial patterns: A study of 750 extremities. Surg Gynecol Obstet 96:43, 1953

MCGOWAN AJ: The results of transposition of the ulnar nerve for traumatic ulnar neuritis. J Bone Joint Surg 32-B:293, 1950

MCLAUGHLIN EF: Fracture of the capitellum: Report of a case successfully treated by closed reduction. Ann Surg 112:122, 1940

MEYER AW: Spontaneous dislocation and destruction of tendon of long head of biceps brachii: Fifty-nine instances. Arch Surg 17:493, 1928

MEYER AW: Chronic functional lesions of the shoulder. Arch Surg 35:646, 1937

MEYERDING HW, KRUSEN FH: The treatment of Volkmann's ischemic contracture. Ann Surg 110:417, 1939

MICHELE AA, KRUEGER FJ: Tenodesis of biceps tendons: A preliminary report. Surgery 29:555, 1951

MILLER RA: Observations upon the arrangement of the axillary artery and brachial plexus. Am J Anat 64:143, 1939

MISRA BD: The arteria mediana. J Anat Soc India 4:48, 1955

MITTAL RL, GUPTA BR: Median and ulnar-nerve palsy:

An unusual presentation of the supracondylar process. J Bone Joint Surg 60-A:557, 1978

MURPHY HS, HANSON CG: Congenital humeroradial synostosis. J Bone Joint Surg 27:712, 1945

NEALE RM: The alleged transmigration of the long tendon of the biceps brachii. Anat Rec 67:205, 1937

NEBLETT C, EHNI G: Medial epicondylectomy for ulnar nerve palsy. J Neurosurg 32:55, 1970

NIRSCHL RN, PETTRONE FA: Tennis elbow: The surgical treatment of lateral epicondylitis. J Bone Joint Surg 61-A:832, 1979

O'RAHILLY R: Morphological patterns in limb deficiencies and duplication. Am J Anat 89:135, 1951

PARKES A: The treatment of established Volkmann's contracture by tendon transplantation. J Bone Joint Surg 33-B:359, 1951

PARSONS FG, ROBINSON A: Eighth report of the Committee of Collective Investigation of the Anatomical Society of Great Britain and Ireland, for the year 1897–98. J Anat Physiol 33:189, 1898

PATTERSON DC, JONES EK: DeQuervain's disease: Stenosing tendovaginitis at the radial styloid. Am J Surg 67:296, 1945

PAULY JE, RUSHING JL, SCHEVING LE: An electromyographic study of some muscles crossing the elbow joint. Anat Rec 159:47, 1967

POLONSKAJA R: Zur Frage der Arterienanastomosen im Gebiete der Ellenbogenbeuge des Menschen. Anat Anz 74:303, 1932

PRINGLE JH: Notes of a case of congenital absence of both ulnae. J Anat Physiol 27:239, 1893.

RADIN EL, RISEBOROUGH EJ: Fractures of the radial head. A review of eighty-eight cases and analysis of the indications for excision of the radial head and non-operative treatment. J Bone Joint Surg 48-A:1055, 1966

RAY RD, JOHNSON RJ, JAMESON RM: Rotation of the forearm: An experimental study of pronation and supination. J Bone Joint Surg 33-A:993, 1951

REIMANN AF, DASELER EH, ANSON BJ, BEATON LE: The palmaris longus muscle and tendon: A study of 1600 extremities. Anat Rec 89:495, 1944

RHODES RL: Tenosynovitis of the forearm. Am J Surg 73:248, 1947

RIORDAN DC: Congenital absence of the radius. J Bone Joint Surg 37-A:1129, 1955

ROLES NC, MOUDSLY RH: Radial tunnel syndrome: Resistant tennis elbow as a nerve entrapment. J Bone Joint Surg 54-B:499, 1972

VAN ROSSUM J, BURUMA OS, KAMPHUISEN HA, ONVLER GJ: Tennis elbow—a radial tunnel syndrome? J Bone Joint Surg 60-B:197, 1978

RYAN JR: The relationship of the radial head to radial neck diameters in fetuses and adults with reference to radial-head subluxation in children. J Bone Joint Surg 51-A:781, 1969

SALSBURY CR: The nerve to the extensor carpi radialis brevis. Brit J Surg 26:95, 1938

SIMRIL WA, TROTTER M: Anomalous elbow joints in a cadaver. Anat Rec 100:775, 1948 (abstr.)

SMITH RV, FISHER RG: Struthers' ligament: A source of median nerve compression above the elbow. Case Report. J Neurosurg 38:778, 1973

SPEAR HC, JANES JM: Rupture of the brachial artery accompanying dislocation of the elbow or supracondylar fracture. J Bone Joint Surg 33-A:889, 1951

SPINNER M: The arcade of Frohse and its relationship to posterior interosseous nerve paralysis. J Bone Joint Surg 50-B:809, 1968

SPINNER M, SCHREIBER SN: Anterior interosseousnerve paralysis as a complication of supracondylar fractures of the humerus in children. J Bone Joint Surg 51-A:1584, 1969

STEIN AH, JR: Variations of the tendons of insertion of the abductor pollicis longus and the extensor pollicis brevis. Anat Rec 110:49, 1951

STEIN F, GRABIAS SL, DEFFER PA: Nerve injuries complicating Monteggia lesions. J Bone Joint Surg 53-A:1432, 1971

STEINDLER A: Mechanics of Normal and Pathological Locomotion in Man. Springfield, Ill, Thomas, 1935

STEINDLER A: The reconstruction of upper extremity in spinal and cerebral paralysis. Am Acad Orth Surgeons Instructional Course Lectures 6:120, 1949

STIMSON H: Traumatic rupture of the biceps brachii. Am J Surg 29:472, 1935

STRACHEN JCH, ELLIS BW: Vulnerability of the posterior interosseous nerve during radial head resection. J Bone Joint Surg 53-B:320, 1971

SUNDERLAND S: The innervation of the flexor digitorum profundus and lumbrical muscles. Anat Rec 93:317, 1945

SUNDERLAND S: Metrical and non-metrical features of the muscular branches of the radial nerve. J Comp Neurol 85:93, 1946

SUNDERLAND S, HUGHES ESR: Metrical and non-metrical features of the muscular branches of the ulnar nerve. J Comp Neurol 85:113, 1946

SUNDERLAND S, RAY LJ: Metrical and non-metrical features of the muscular branches of the median nerve. J Comp Neurol 85:191, 1946

TANASESCO J GH: Lymphatiques de l'articulation du coude. Anat Anz 40:602, 1912

TERRY RJ: A study of the supracondyloid process in the living. Am J Phys Anthropol 4:129, 1921

TESTUT JL: Recherches anatomiques sur l'anastomose du nerf musculo-cutané avec le nerf médian. J de l'anat et de la physiol 19:103, 1883

THOMPSON IM: Anomaly of median nerve and flexor digitorum sublimis muscle. Anat Rec 23:375, 1922

THOMPSON JW, MCBATTS J, DANFORTH CH: Hereditary and racial variation in the musculus palmaris longus. Am J Phys Anthropol 4:205, 1921

THOMSON A: Third Annual Report of the Committee of Collective Investigation of the Anatomical Society of Great Britain and Ireland for the year 1891–92. J Anat Physiol 27:183, 1893

TOURNAY A, PAILLARD J: Électromyographie des muscles radiaux à l'état normal. Révue neurol 89:277, 1953

TRAVILL AA: Electromyographic study of the extensor apparatus of the forearm. Anat Rec 144:373, 1962

TRAVILL A, BASMAJIAN JV: Electromyography of the supinators of the forearm. Anat Rec 139:557, 1961

TREVOR D: Rupture of the extensor pollicis longus tendon after Colles fracture. J Bone Joint Surg 32-B:370, 1950

TROTTER M: Septal apertures in the humerus of American whites and Negroes. Am J Phys Anthropol 19:213, 1934

VAUGHAN-JACKSON OJ: Rupture of extensor tendons by attrition at the inferior radio-ulnar joint: Report of two cases. J Bone Joint Surg 30-B:528, 1948

VILLAR F: Quelques recherches sur les anastomoses des nerfs du membre superieur. Bull de la Soc anat de Paris, p 607, 1888

WALSH JF: The anatomy of the brachial plexus. Am J M Sci 74:387, 1877

WARREN JD: Anterior interosseous nerve palsy as a complication of forearm fractures. J Bone Joint Surg 45-B:511, 1963

WAUGH RL, HATHCOCK TA, ELLIOTT JL: Ruptures of muscles and tendons: With particular reference to rupture (or elongation of long tendon) of biceps brachii with report of fifty cases. Surgery 25:370, 1949

WEATHERSBY HT: Unusual variation of the ulnar artery. Anat Rec 124:245, 1956

WERNER C-O: Lateral elbow pain and posterior interosseous nerve entrapment. Acta orthop scandinav suppl 174, 1979

WHITSON RO: Relation of the radial nerve to the shaft of the humerus. J Bone Joint Surg 36-A:85, 1954

WILSON DH, KROUT R: Surgery of ulnar neuropathy at the elbow: 16 cases treated by decompression without transposition: Technical note. J Neurosurg 38:780, 1973

WOLF-HEIDEGGER G: Contribution à l'étude des anomalies des muscles biceps brachial et brachial antérieur. Arch d'anat d'histol et d'embryol 23:207, 1937

WRIGHT WG: Muscle Function. New York, Hoeber, 1928; Hafner, 1962

YOUNG HH, LOWE GH, JR: Tendon transfer operation for irreparable paralysis of the radial nerve: Long term follow-up of patients. Surg Gynecol Obstet 84:1100, 1947

ZOËGA H: Fracture of the lower end of the radius with ulnar nerve palsy. J Bone Joint Surg 48-B:514, 1966

Chapter 6

THE WRIST AND HAND

The tendons of most of the long muscles of the forearm and its major nerves and vessels continue across the wrist into the hand. On the dorsum of the hand, which is typically not provided with any intrinsic musculature, the tendons and vessels are essentially subcutaneous. The palm, however, not only transmits the long tendons and the chief nerves and vessels of the digits but also presents special fascial layers and a number of intrinsic muscles that are particularly important in controlling movements of the fingers and thumb (Fig. 6-1); it is, therefore, especially complicated.

Bones and Joints

Eight carpal bones form the wrist, five metacarpal bones form the hand proper, and fourteen phalanges form the skeleton of the digits (Figs. 6-2 to 6-5).

CARPUS

The eight bones of the wrist are held rather tightly together by ligaments, and the carpus as a whole is concave anteriorly, convex posteriorly.

In the *proximal row,* the scaphoid (navicular), the lunate (semilunar), and the triquetral bones, in that order from the radial to the ulnar side, articulate proximally at the radiocarpal joint, transversely with each other, and distally with the distal row of carpals. The fourth member of the proximal row, the pisiform, articulates directly only with the anterior surface of the triquetral bone, although it is bound to other bones also by ligaments.

Important features of the scaphoid are its large convex proximal articular surface for articulation with the radius, and the tubercle on its lateral surface; this extends also onto its palmar surface, where it serves for the attachment of part of the flexor retinaculum (transverse carpal ligament) and is the prominent part of the bone visible and palpable at the wrist. The lunate gets its name from its convex proximal articular surface, through which it is in contact with the radius and the articular disk of the wrist joint, and its highly concave distal surface for articulation with the capitate. The triquetrum has only a small articular surface entering into the wrist joint; this is in contact with the ulnar collateral carpal ligament when the wrist is straight, and with the articular disk of the wrist joint during ulnar abduction. The pisiform is the only bone of the wrist that regularly receives a tendon of insertion from a muscle of the forearm; like the scaphoid, it gives attachment to a part of the flexor retinaculum. Kadasne and Bansal have denied that the pisiform is a sesamoid bone, as usually described; they found

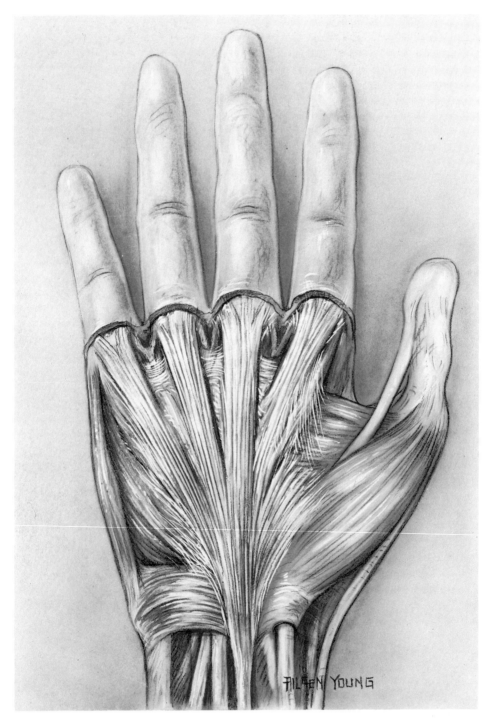

Fig. 6-1. The palm of the hand after removal of the superficial fascia. Most of the long tendons at the wrist disappear behind the flexor retinaculum, but the palmaris longus runs in front of it to become continuous with the superficial longitudinal fibers of the palmar aponeurosis.

that it does not develop in the tendon of the flexor carpi ulnaris but independently, and that the tendon attains a secondary attachment to the bone.

The *distal row* of carpals, composed of the trapezium (greater multangular), trapezoid (lesser multangular), capitate (os magnum), and hamate (unciform) bones, in that order from the radial side, articulates proximally with the proximal row and distally with the metacarpals, while the members of the row are so tightly articulated transversely with each other that this row moves as a unit.

Important features of the trapezium are its large saddle-shaped distal articular surface for articulation with the metacarpal of the thumb; the tubercle on the lateral part of its anterior surface, for attachment of a part of the flexor retinaculum; and the groove, just

medial to the tubercle, that houses the tendon of the flexor carpi radialis muscle. The trapezoid is the smallest of the carpals, except for the pisiform, and the capitate is the largest. The hamate gets its name from the prominent hook (hamulus or uncinate process) that projects forward from its anterior surface, and serves primarily for the attachment of a part of the flexor retinaculum.

The trapezium may receive the insertion of one part of a doubled abductor pollicis longus tendon (Chap. 5). Other than this and the attachment of the flexor carpi ulnaris to the pisiform, muscular attachments to the carpals (Figs. 6-3 and 6-5) are limited, for the muscles of the thumb and little finger arise primarily from the flexor retinaculum, with usually only a few fibers arising from the bones of the wrist.

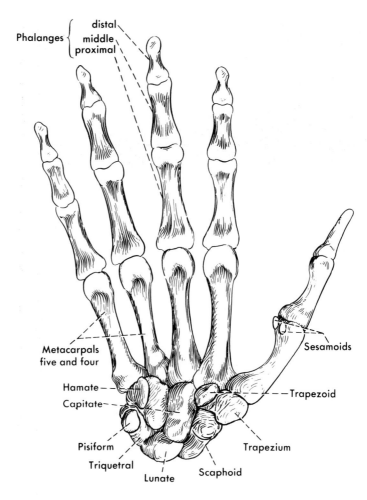

Fig. 6-2. Skeleton of the wrist and hand, palmar view.

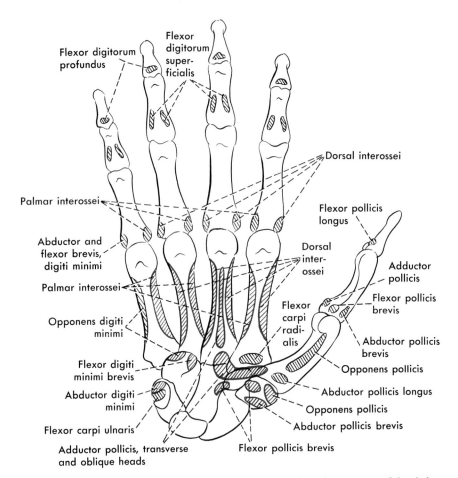

Fig. 6-3. Origins (*red*) and insertions (*black*) of muscles on the palmar aspect of the skeleton of the hand. The palmar interossei often have no insertion into bone, but if they do they attach as indicated here.

ANOMALIES, DISLOCATION, AND FRACTURE

Anomalies of the bones of the wrist are not common, but usually consist of departures from the normal number: fusions between carpals diminish the number of individual bones, while bipartite or even tripartite bones and accessory carpals increase the number. The carpal most frequently fractured is the scaphoid; that most frequently dislocated is the lunate.

Fusion

While fusion may occur between any of the carpal bones, the most common fusion is usually reported as being between the lunate and the triquetral (Waugh and Sullivan; O'Rahilly). Minnaar, O'Rahilly, and others have noted that congenital fusion of these two bones is more common among Negroes than other races, and it has been reported to occur, according to Minnaar, in 5.9% of West African Negroes. Among white persons, fusion of the lunate and triquetral is apparently rare; O'Rahilly found only one such case among 743 wrists that he investigated. In these same wrists he found two instances of fusion between the capitate and the hamate, and two of fusion between the trapezium and the trapezoid, the total incidence of fusion in this series therefore amounting to 0.7%. Cockshott reported that he had seen five cases of the rare pisiform-hamate fusion.

According to Minnaar, fusion of the lunate and triquetral bones never causes symptoms, cannot be recognized by clinical examination, and is usually recognized incidentally in roentgenograms. While limitation of motion does occur in some intercarpal arthrodeses, Peterson and Lipscomb found these often preferable to total arthrodesis in relieving pain and weakness; they suggested, however, that fusion between the proximal and distal rows should involve more than one bone in each row.

Supernumerary Bones

Each of the carpal bones usually develops from a single center of ossification, but it is said that the scaphoid and the capitate, especially, may each develop from two or even three centers, and bipartite or tripartite bones are thought to arise as a result of failure of fusion of such centers. In the bipartite scaphoid, the most commonly observed example of bipartite carpal, the line of division between the two parts is usually at the narrowed waist of the bone, which is commonly also the site of fracture of a normal bone. Waugh and Sullivan expressed the opinion that bipartite scaphoid is often mistaken for an ununited fracture; O'Rahilly suggested that a mistake is frequently made in the opposite direction, and that many cases of so-called bipartite scaphoid are actually traumatic. Since a fractured scaphoid may occur bilaterally and a congenital bipartite scaphoid may be unilateral, according to O'Rahilly, there seems to be no possibility of distinguishing between the two conditions on this basis. Bizarro found among roentgenograms of the wrists of 100 persons only one case of possible bipartite scaphoid—unilateral, but with no history of trauma.

Partridge described a case in which both the trapezoid (lesser multangular) and the hamate were each partially divided into two

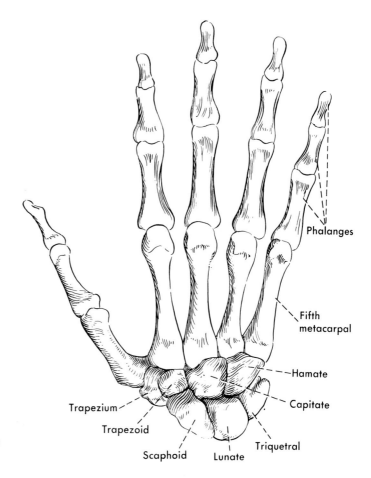

Phalanges

Fifth metacarpal

Hamate

Capitate

Trapezium

Trapezoid

Triquetral

Scaphoid

Lunate

Fig. 6-4. Dorsal view of the skeleton of the hand.

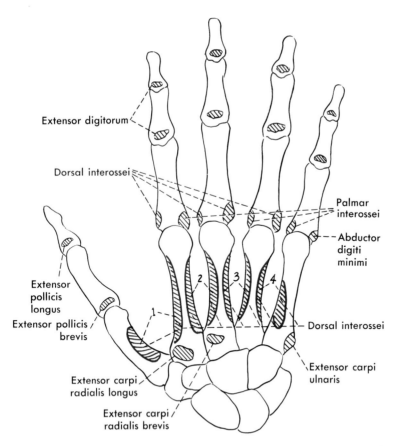

Fig. 6-5. Origins (*red*) and insertions (*black*) of muscles on the dorsal aspect of the skeleton of the hand. The numbers identify the origins of the dorsal interossei.

parts; he said that while the trapezoid had been described as the carpal most frequently found divided, this appearance may sometimes be due to an os centrale at the back of the trapezoid. Obviously, it may often be impossible to draw a firm line between wrists that have bipartite bones and those that have true accessory ones.

The hamulus (uncinate process) of the hamate may arise from a separate center of ossification, and occasionally this fails to unite with the rest of the bone. It can be mistaken for a fracture, but is usually bilateral (Wilson).

Accessory bones of the wrist are apparently far less common than are accessory bones of the ankle; excluding a case of possible bipartite scaphoid, Bizarro found only three small accessory bones in his study of the roentgenograms of the wrists of 100 individuals. O'Ra-

hilly was able to find accounts of more than 20 apparently different accessory bones of the wrist (Fig. 6-6) some of which were interpreted as being accessory carpal bones and others as sesamoid bones. The most common and the best known of the accessory carpals is the *os centrale*, which sometimes appears as a separate bone on the back of the hand, in approximately a central position between the scaphoid, the trapezoid, and the capitate. The os centrale is in many animals a separate bone, but in man it is usually said to fuse typically with the scaphoid during the cartilaginous stage; Waugh and Sullivan, however, found that it has also been reported simply to disappear, to degenerate and be represented only by a ligament between adjacent bones, or to merge with the capitate instead of the scaphoid.

Rushforth noted that four accessory centers

of ossification have been seen in association with the trapezium, and described a case in which an extra bone associated with the joint between the trapezium and the first metacarpal, bilaterally, may have represented persistence of one of these accessory centers. In his case, the condition was marked by a prominence at the base of the first metacarpal and by limited abduction and extension at the carpometacarpal joint. Wilson stressed the advantages of profile views of the wrist in identifying injuries or developmental disturbances of the carpus; he noted that the supernumerary carpals not seen in routine roentgenograms are usually well visualized by this technique.

Dislocation and Fracture

Simple dislocation at the radiocarpal joint is apparently rare, for the violence necessary to produce extensive tearing of the ligaments of this joint usually produces a Colles fracture or a fracture-dislocation of the carpus. Dislocation of the bones of the wrist may or may not be accompanied by fracture, and may involve a single bone or a group of bones (Fig. 6-7). Isolated anterior dislocation of the lunate is frequently said to be the most common carpal dislocation, and this was true in the series of Campbell

and co-workers; in contrast, Watson-Jones and Russell both regarded transscaphoperilunar dislocation (a posterior dislocation of all the other carpals, with the lunate and a part of the scaphoid only remaining in place) as more common. Both dislocations are typically caused in the same way, by a fall upon the hyperextended hand; in the one case, all the bones except the lunate and a part of the scaphoid are carried backward by the force of the blow, while in the other the compression of the carpals upon forced hyperextension—they move closer together with dorsiflexion—so squeezes the narrow posterior part of the lunate as to force the bone forward.

Associated injuries may include other carpals (MacConaill found either the scaphoid or triquetral bones, or both, fractured in 6 of 9 cases of lunate dislocation), the styloid process of the radius; the median nerve is frequently involved in the carpal canal by anterior dislocation of the lunate, but in the series of Campbell and his co-workers there was never more than transient damage to this nerve. MacConaill reported a technique that he regarded as particularly useful in restoring a displaced lunate to normal position; Campbell and co-workers suggested that in intractable cases, excision of the entire proximal row

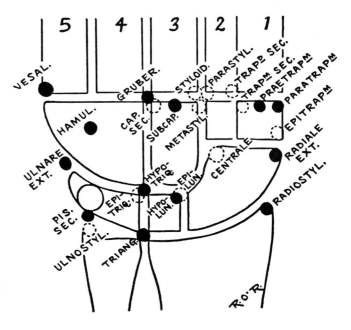

Fig. 6-6. Diagram of the accessory carpal bones in a palmar view of the hand. Those shown in *broken outline* are more dorsally situated ones. (O'Rahilly R: J Bone Joint Surg 35–A: 626, 1953)

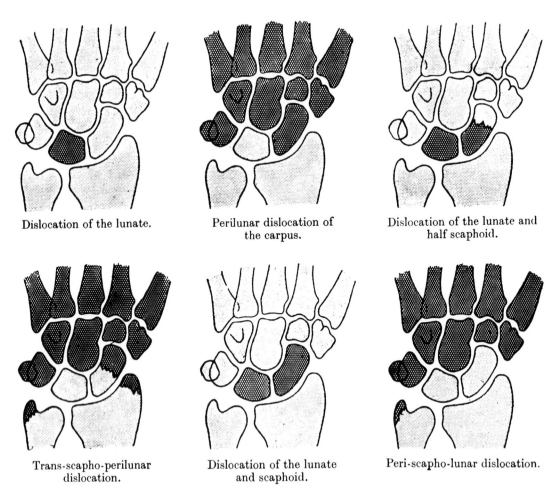

Dislocation of the lunate.

Perilunar dislocation of the carpus.

Dislocation of the lunate and half scaphoid.

Trans-scapho-perilunar dislocation.

Dislocation of the lunate and scaphoid.

Peri-scapho-lunar dislocation.

Fig. 6-7. Types of carpal dislocations. The heavily shaded bones and parts are those that are dislocated (Watson-Jones R: Fractures and Joint Injuries [ed 4] Vol 2. Baltimore, Williams & Wilkins, 1955)

of bones may be preferable to arthrodesis or to removal of only part of this row.

While any of the bones of the wrist may be fractured and, as already noted, dislocations are often accompanied by fractures, the scaphoid is the most commonly fractured bone. According to Cleveland, indirect violence to the forearm and hand fractures the scaphoid some four to five times as frequently as it does the radius in young individuals with strong muscles; he said Colles fracture predominates in older individuals with flabby muscles. In a review of a large series of workmen's compensation cases, however, Bacorn and Kurtzke found Colles fracture to be about three times as frequent, and all fractures of the radius

about five times as frequent, as carpal fractures. Of all the carpal fractures of which they studied records, scaphoid ones formed 70.8%; the triquetral was fractured in 14.3%, the lunate in 5.6%, the capitate in 2.4%, the pisiform, trapezium, and hamate in 2.1% each, and the trapezoid in 0.5%.

Simultaneous fracture of one or more carpals and the radius is not rare, but the most common fracture of the radius, Colles, is seldom associated with carpal fracture: Bacorn and Kurtzke found the scaphoid fractured in only 0.5% of 2,100 Colles fractures, and no other carpal fractures at all in this group.

Fracture of the scaphoid is commonly through the narrowed waist of the bone, and according to Cleveland healing usually occurs within 8 to 10 weeks if the proximal fragment has an adequate blood supply and the fracture is properly immobilized; otherwise it may required up to 20 weeks. The usual quotation that only 13% of scaphoids have no blood supply entering proximal to the waist is probably incorrect in view of the careful studies of Taleisnik and Kelly. These workers found only one specimen among 11 in which an artery entered the proximal end of the bone, and it was very small and had a similarly limited distribution; otherwise the supply was through three groups of arteries, two of which penetrated the region of the waist, while the third entered the tuberosity. Whether the first two groups mentioned might still supply the proximal fragment would seem to depend on the exact position of the fracture and perhaps the amount of displacement.

Friedenberg pointed out that both dorsiflexion and radial deviation move the carpals closer together, and showed experimentally that these same maneuvers bring the two parts of a fractured scaphoid together. He therefore advised maintaining the wrist in at least 30° of dorsiflexion, and using radial deviation also if better closure of the fracture line is necessary. There is little agreement as to the best method of treating ununited fractures of the carpal scaphoid (Lipscomb, '68). Stewart stressed immobilization for as long as a year or more if necessary, saying that even if the proximal fragment undergoes avascular necrosis proper immobilization will usually lead to revascularization; London, in contrast, found no good evidence that prolonged immobilization produces union.

JOINTS AT THE WRIST

Radiocarpal Joint

The radiocarpal joint, or wrist joint proper, is formed by the three major bones of the proximal row of carpals (the scaphoid, lunate, and triquetral) distally, and proximally by the distal surface of the radius and by the articular disk intervening between the ulna and the wrist joint (Fig. 6-8). As a whole, the articular surface of the radius faces somewhat palmarward, usually said to vary from about 1° to 23°, and somewhat ulnarward, about 15° to 30°.

The articular disk binds the radius to the ulna and separates the distal radioulnar joint

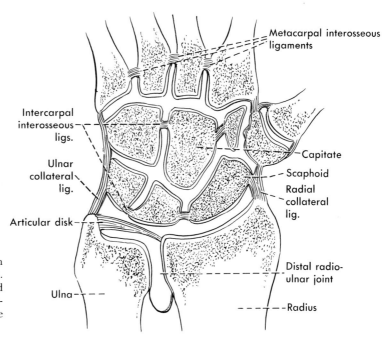

Fig. 6-8. A frontal section through the joints at the wrist. Synovial membrane is indicated by *red;* that bulging upward between radius and ulna forms the sacciform recess.

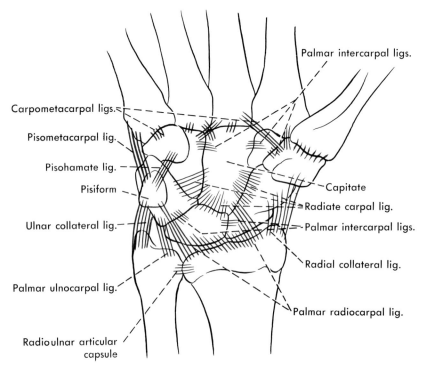

Fig. 6-9. Diagram of the palmar ligaments attaching to the carpal bones. Intercarpal ligaments are *black*, others *red.*

from the radiocarpal joint. The distal wall of the joint is the articular cartilages of the three bones concerned, and the two interosseous ligaments between these bones; the proximal surfaces of these ligaments are set flush with the articular cartilage, so that the radiocarpal joint extends not at all between the carpals of the proximal row. Rarely, the articular disk may be perforated, and the wrist joint may therefore communicate with the distal radioulnar joint; also rarely, one of the interosseous ligaments may be defective and allow communication between the radiocarpal joint and the midcarpal one.

The articular capsule of the wrist joint is especially thickened medially and laterally, these thickenings being described as the *radial* (lateral) *collateral carpal ligament* and the *ulnar* (medial) *collateral carpal ligament* (Figs. 6-9 and 6-10). The radial collateral ligament is attached to the styloid process of the radius and to the tubercle of the scaphoid with a few fibers extending to the trapezium; similarly, the ulnar collateral ligament is attached to

the styloid process of the ulna and to the nonarticular part of the medial surface of the triquetrum, extending also to the pisiform.

The *palmar radiocarpal ligament* forms most of the anterior part of the articular capsule. It arises from the palmar surface of the distal edge of the radius, and extends both distally and medially to attach to all the bones of the proximal row, some of the fibers also extending to the capitate in the distal row. The most medial fibers blend with the *palmar ulnocarpal ligament,* which arises from the distal end of the ulna and from the articular disk and attaches into the lunate and triquetral bones; this in turn blends with the ulnar collateral ligament.

Mayfield and co-workers described ligaments deep to those just mentioned as being more important for stabilizing the wrist. The intrascapular ligaments associated with the palmar radiocarpal ligament are the radiocapitate, radiotriquetral, and radioscaphoid; those associated with the palmar ulnocarpal ligament are the ulnolunate and the ulnotriquetral. Since the distal row of carpals is connected to the radius on the palmar side by the radiocapitate ligament only, this ligament may be ruptured in forced hyperextension, thus separating the two rows. Hyperextension also puts the radioscaphoid ligament under maximal tension, and both it and the radiocapitate ruptured when perilunar dislocation was produced.

On the dorsum of the joint the capsule is represented by the *dorsal radiocarpal ligament;* the fibers of this ligament originate largely from the dorsal edge of the distal end of the radius and, like those of the palmar ligament, pass distally and medially to attach to the scaphoid, lunate, and triquetral bones. The prevailing direction of the fibers of both radiocarpal ligaments is such that movement of the radius in either pronation or supination immediately tightens one of them; during pronation, the dorsal radiocarpal ligament helps to assure that the hand also is carried into pronation, while in supination this duty falls upon the palmar radiocarpal ligament.

The loose synovial membrane of the wrist joint lines the deep surface of the ligamentous capsule as a whole, and covers the proximal surfaces of the scapholunate and lunotriquetral interosseous ligaments. It is attached above to the edges of the articular surface of the radius and of the articular disk, below to the edges of the proximal articular surfaces of the scaphoid, lunate, and triquetral bones.

Distal Radioulnar Joint

Properly speaking, this joint is not a part of the wrist joint. However, it is more closely related topographically to the radiocarpal joint than it is to the proximal radioulnar joint with which it works in pronation and supination.

This joint is a very simple one. Its distal border is a fibrocartilaginous disk that is attached to the lateral side of the styloid process of the ulna and to the sharp edge of the distal end of the radius between that bone's articular surfaces for the ulna, on the one hand, and the carpal bones on the other. This disk

Fig. 6-10. Diagram of the dorsal ligaments attaching to the carpal bones. Intercarpal ligaments are *black*, others *red*.

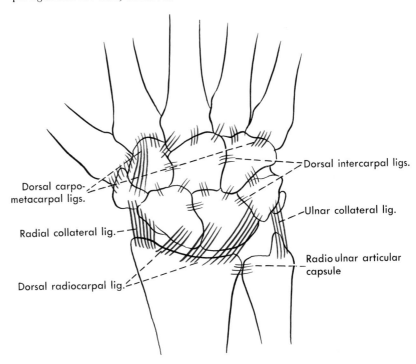

blends with the capsules of both the radiocarpal and distal radioulnar joints, and normally completely separates the two joint cavities. The capsule of the distal radioulnar joint is rather thin; its anterior and posterior parts are formed by sparse fibers that stretch transversely between the radius and the ulna. The cavity of the joint is L-shaped, since it consists both of a vertical portion between the radius and ulna and a horizontal portion between the distal end of the ulna and the articular disk; the upper part of the synovial membrane between the radius and the ulna forms a small pouch that extends upward beyond the articular surfaces to form the sacciform recess.

Intercarpal Joints

The intercarpal joints are of two types, those between the bones of a single row and those between the bones of the proximal and distal rows. Relatively little movement is permitted, in general, between the bones composing a single row, but much more movement occurs between the proximal and distal rows; the articulation between the rows is therefore particularly important, and is usually designated the *midcarpal (mediocarpal) joint*, or the transverse carpal joint. It usually extends from one side of the wrist to the other, between the proximal and distal rows of bones, but is sometimes interrupted by the occurrence of an interosseous ligament between the scaphoid and the capitate.

The intercarpal joints between the scaphoid, lunate, and triquetral bones, sealed off proximally from the radiocarpal joint by two *intercarpal interosseous ligaments,* open distally into the midcarpal joint. Those of the distal row are interrupted nearer their middles by three interosseous ligaments so that they open proximally into the midcarpal joint and distally into the common carpometacarpal joint of the fingers (Fig. 6-8). Occasionally, one of these interosseous ligaments, more commonly that between the trapezium and trapezoid, is incomplete, and thus allows communication between the midcarpal and the carpometacarpal joint.

Besides the interosseous ligaments, the bones in each row are bound to each other and to members of the other row by superficially placed *palmar and dorsal intercarpal ligaments* (Figs. 6-9 and 6-10). Most of these pass from one bone to the next contiguous one and have no special names, but palmar ones between the proximal and distal rows are grouped together as the *radiate carpal ligament*, because they tend to radiate from the capitate bone.

The *pisotriquetral joint* has, of course, no communication with the other intercarpal joints. It is a small synovial cavity, with its own articular capsule, between the pisiform and triquetral bones. The pisiform is strongly anchored proximally to the wrist by the ulnar collateral and palmar radiocarpal ligaments and distally by two strong bands, the pisohamate and pisometacarpal ligaments. The pisohamate ligament is attached to the hamulus on the hamate bone, the pisometacarpal to the base of the fifth metacarpal; these ligaments transmit the pull of the flexor carpi ulnaris muscle to the distal row of carpals and to the metacarpals, and thus actually serve as the lower end of the tendon of this muscle; indeed, the attachment of the pisometacarpal ligament to the fifth metacarpal has often been regarded as the true insertion of the flexor carpi ulnaris.

Carpometacarpal Joints

The carpometacarpal joints of the four medial digits form a continuous cavity stretching between the distal row of carpals and the second through the fifth metacarpals. The second metacarpal articulates with the trapezium, the trapezoid, and in small part with the capitate; the third with the capitate; the fourth with the hamate and a small surface on the capitate; and the fifth with the hamate alone. The sinuous cavity thus formed between the carpals and the metacarpals is continuous proximally with the distal parts of the intercarpal joints of the distal row of bones, and distally with the intermetacarpal joints between the four medial metacarpals. The carpometacarpal joint of the thumb, between the trapezium and the first metacarpal, is usually completely separated from the joint between the trapezium and the second meta-carpal by an interosseous ligament.

The common carpometacarpal joint of the four medial digits is bounded anteriorly and posteriorly by *palmar* and *dorsal carpometacarpal ligaments*. For the most part these stretch merely between the adjacent parts of the distal row of carpals and the bases of the metacarpals, but the pisometacarpal ligament is an exception to this. The common arrangement of these ligaments is shown in Figures 6-9 and 6-10.

The carpometacarpal joint of the thumb, so important for the movement of this digit, is usually described as saddle-shaped. It has its own articular capsule, which is, however, somewhat loose in order to allow free movement of the metacarpal. Special ligaments of this capsule are usually not named; however, Haines described five—radial, anterior oblique, and posterior oblique carpometacarpal ones, and anterior and posterior intermetacarpal ones (Fig. 6-11). He regarded them as being responsible for the rotation occurring at this joint when the metacarpal is flexed or extended. Napier agreed in general with Haines' description and described movement at the joint in opposition as being of two types: for precision, the thumb is first fully abducted, then circumduction and medial rotation follow (this rotation being contributed to by the ligaments); for power, the metacarpal is moved directly into medial rotation, flexion, and adduction by the opponens, and the ligaments of the joint remain relaxed until the movement is almost completed.

According to Waugh and Yancey, dislocations at the carpometacarpal joints, except that of the thumb, are rare; those of the thumb are accompanied by marked swelling and deformity, the latter varying according to the direction of displacement. Waugh and Yancey also discussed dislocation of the bases of the fourth and fifth metacarpals simultaneously.

Blood Supply and Innervation

The ligaments and the bones of the wrist are supplied by arterial twigs from the dorsal and palmar carpal retes, which lie immediately

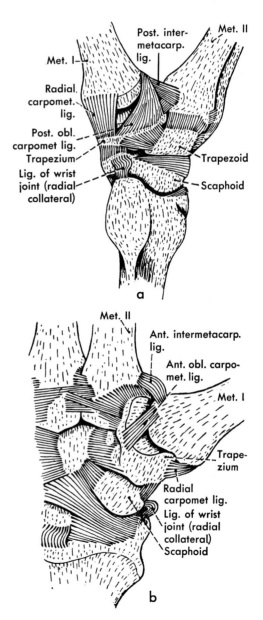

Fig. 6-11. Deep ligaments at the carpometacarpal joint of the thumb. *a* is a lateral view, *b* a palmar one. (Haines RW: J Anat 78:44, 1944)

adjacent to the ligaments. The dorsal rete is formed by the dorsal carpal branches of the radial and ulnar arteries and the posterior terminal branch of the anterior interosseous artery. The palmar rete is similarly formed by the palmar carpal branches of the radial and ulnar arteries and the anterior terminal branch of the anterior interosseous artery.

The distal radioulnar joint is said to be supplied by twigs of both the anterior and posterior interosseous nerves (median and radial). Although they found some variation, Gray and Gardner described the radiocarpal joint as supplied primarily from the anterior and posterior interosseous nerves; the intercarpal and midcarpal joints by these and the median and ulnar, including the latter's deep branch; the intermetacarpal joints by the deep branch of the ulnar; and the carpometacarpal joints by the ulnar nerve and its deep branch, plus nerves of the dorsum. The superficial branch of the radial helped supply some of the lateral intercarpal and carpometacarpal joints.

MOVEMENTS AT THE WRIST

The usually described movements of the hand upon the forearm are flexion (palmar flexion), dorsiflexion and hyperextension, and radial and ulnar deviation. These movements involve, to a varying degree, motion at both the radiocarpal joint and the midcarpal joint, but the carpometacarpal joints contribute almost nothing to them. The hand can also be rotated at the wrist; according to Cyriax the total amount of passive rotation at the wrist usually approximates 45°.

Range of Movements
The range of motion in flexion-extension and in radial and ulnar abduction varies considerably among individuals and between the two hands of the same individual (and even, if active motion alone is considered, in the same hand on successive days); it is also influenced by whether the hand is in pronation or supination. Horwitz regarded the average amount of palmar flexion as being 68°, of dorsiflexion, 69°; Steindler ('35) quoted an average of 84° of palmar flexion, 64° of dorsiflexion; and Hewitt found an average of 69.4° palmar flexion for the right hand, slightly less for the left, with the hand supinated, but one of 80.4° for the right hand, slightly more for the left, with the hand pronated; dorsiflexion averaged about 64° with the hand supinated, about 72° when it was pronated. These averages hardly tell the story, however, for Hewitt found that the maximal flexion of the right hands in her series varied from 40° to 101°, that of the left hands from 31° to 116°, when the hands were supinated; there was a lessened range, about 50° on both sides, when movement was carried out from pronation.

Horwitz said ulnar abduction averages 56°, radial 18°. Steindler quoted ulnar abduction as ranging from 30° to 50°, radial abduction up to 30°. Hewitt found an average of slightly more than 61° for ulnar deviation when the hand was supinated, but a range of from 35° to 88°; radial deviation averaged about 27° with the hand supinated, but ranged from 4° to 70°. Pronation, which increases the maximal palmar flexion by about 10° or more and the maximal dorsiflexion by between 4° and 9°, also increases radial abduction some 10° or more, according to Hewitt; however, it markedly decreases ulnar abduction, from an average of about 61° in supination to one of about 32° in pronation.

Radiocarpal and Intercarpal Movements
The exact mechanisms of movements of the wrist are necessarily complex, involving as they do the variously shaped articular surfaces at the radiocarpal and midcarpal joints; there is sufficient disagreement among various workers to indicate that the question is not completely understood. For instance it has, according to Horwitz, been both stated and denied that flexion involves not only flexion at the radiocarpal and midcarpal joints, but also, because of the obliquity of these joints, a radial abduction in the radiocarpal joint and an ulnar abduction in the midcarpal, these two canceling each other out to give pure flexion. Aside from such questions as this, however, there are a number of facts that are generally agreed upon.

Most movements of the hand at the wrist involve both the radiocarpal and the midcarpal joints, but in varying amount depending upon the movement: Wright ('35) and Horwitz, for instance, agreed that in palmar flexion there is more motion in the midcarpal joint than there is in the radiocarpal one, but that in dorsiflexion the reverse is true, the larger amount of movement occurring at the radiocarpal joint and the lesser amount at the midcarpal. They also said that in radial abduction all or almost all the movement occurs at the midcarpal joint, while in ulnar abduction four fifths or more of the movement is at the radiocarpal joint (according to Horwitz, this distribution is typical of abduction with the hand pronated, but if the hand is supinated about a third of the movement in radial abduction is at the radiocarpal joint, and the two joints share about equally in ulnar abduction). Bradley and Sunderland, comparing movements in a wrist in which bony fusions prevented movement at the radiocarpal joint with those of the unaffected wrist, concluded that the midcarpal joint contributes most to both palmar flexion and dorsiflexion, almost nothing to ulnar deviation, and is responsible for all radial deviation.

Horwitz' study emphasizes that the scaphoid moves much more freely upon the radius than does the lunate-triquetral complex; he reported an average total range of movement, from flexion to hyperextension, of 88° at the radioscaphoid part of the radiocarpal joint, and of only 43° at the radiolunate part. MacConaill reported that the scaphoid moves at the midcarpal joint with other bones of the proximal row during flexion mostly, when the two rows of bones are relatively loosely in contact with each other; in dorsiflexion, he said, the capitate is brought more tightly against the scaphoid so that the two move increasingly together in regard both to the radius and to the lunate. Regardless of the exact mechanism, the tighter packing of the bones of the wrist both during hyperextension and radial deviation, and their loosening both during flexion and ulnar deviation, have apparently been demonstrated by MacConaill and by Friedenberg.

Muscles Acting at the Wrist

The chief flexors of the wrist (Fig. 6-12) are the flexor carpi radialis and the flexor carpi ulnaris; the palmaris longus, not always present, is also a primary flexor of the wrist. The abductor pollicis longus, although largely situated on the extensor and radial sides of the

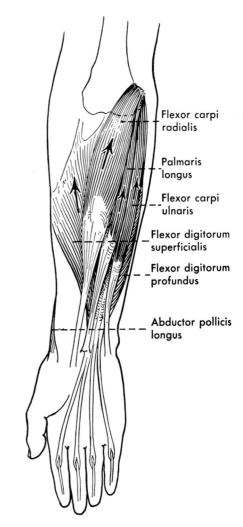

Flexor carpi radialis

Palmaris longus

Flexor carpi ulnaris

Flexor digitorum superficialis

Flexor digitorum profundus

Abductor pollicis longus

Fig. 6-12. Flexors of the wrist.

forearm, attains the flexor side of the wrist at its insertion, and according to Jones ('19) can flex the wrist with considerable force. The long flexors of the fingers are effective in flexing the wrist only when the digits are kept extended; similarly, their length is such that they can effectively flex the fingers only when the wrist is extended. The flexors of the wrist are innervated by the median, ulnar, and radial nerves (Table 6-1).

The chief extensors of the wrist are the extensor carpi radialis longus and brevis and the extensor carpi ulnaris (Fig. 6-13). As noted in Chapter 5, the short radial extensor participates in all extension of the wrist, and the

Table 6-1
Innervation of the Muscles Flexing the Hand at the Wrist

	MUSCLE	NERVE	SEGMENTAL INNERVATION	
			Reported Range	*Probable Most Important Segments*
Chief flexors	Flexor carpi radialis	Median	C6-C8	C6, 7
	Flexor carpi ulnaris	Ulnar	C7-T1	C8, T1
	Palmaris longus	Median	C6-T1	C7, 8
Accessory flexors	Abductor pollicis longus	Radial	C6-T1	C6, 7
	Flexor digitorum superficialis	Median	C7-T1	C7, 8, T1
	Flexor digitorum profundus	Median and ulnar	C7-T1	C8, T1
	Flexor pollicis longus	Median	C6-T1	C7, 8 (?)

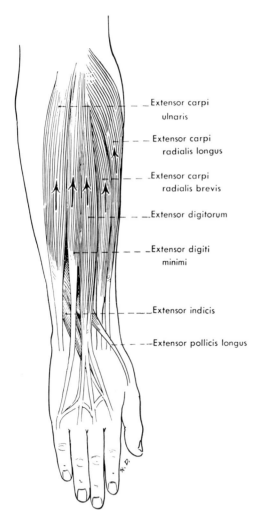

Fig. 6-13. Extensors of the wrist.

- Extensor carpi ulnaris
- Extensor carpi radialis longus
- Extensor carpi radialis brevis
- Extensor digitorum
- Extensor digiti minimi
- Extensor indicis
- Extensor pollicis longus

other two muscles, which tend to extend and abduct at the same time, are recruited for added speed or power. The long extensors of the digits can assist in this movement only when the digits are flexed; since they do not have sufficient length to allow simultaneous flexion of the wrist and fingers, a person with paralysis of the extensors—wristdrop—can extend his wrist by first making a fist. Extension at the wrist is entirely dependent upon the radial nerve (Table 6-2), hence the well-known wristdrop following lesions of this nerve.

Radial abduction at the wrist (Fig. 6-14) is carried out primarily by thumb muscles, the abductor pollicis longus and the extensor pollicis brevis, and hence is dependent especially upon the radial nerve (Table 6-3). The flexor carpi radialis, primarily a flexor, is regarded by some workers as being able to assist in radial abduction, while others deny that it can do so. Similarly, both radial extensors are sometimes described as assisting in radial abduction (*e.g.*, Wright, '28), or the short extensor only is regarded as so assisting. Ulnar abduction is normally carried out by the combined action of the flexor and the extensor carpi ulnaris, hence through both the ulnar and radial nerves (Table 6-4).

Since the major flexors of the wrist and all of the flexors of the digits are innervated by the median and ulnar nerves, and the extensors of the wrist and of the digits are all innervated by the radial, all flexors or all extensors may be involved simultaneously. Tendons of various extensors of the wrist can be trans-

Table 6-2
Innervation of the Muscles Extending the Hand at the Wrist

	MUSCLE	NERVE	SEGMENTAL INNERVATION	
			Reported Range	*Probable Most Important Segments*
Chief extensors	Extensor carpi radialis longus	Radial	C5-C8	C6, 7
	Extensor carpi radialis brevis	Radial	C5-C8	C6, 7
	Extensor carpi ulnaris	Radial	C6-C8	C7, 8
Accessory extensors	Extensor digitorum	Radial	C6-C8	C7, 8
	Extensor indicis	Radial	C6-T1	C7, 8
	Extensor digiti minimi	Radial	C6-T1	C7, 8
	Extensor pollicis longus	Radial	C6-T1	C7, 8

Fig. 6-14. Radial and ulnar abductors at the wrist. *a* is a posterior, *b* an anterior view. Whether any muscles other than the abductor pollicis longus and extensor pollicis brevis actually assist in radial abduction has been questioned.

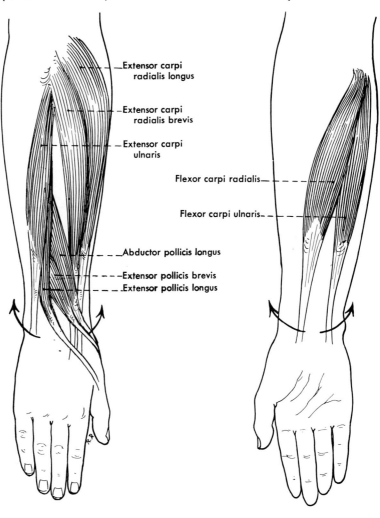

Table 6-3
Innervation of the Radial Abductors of the Hand

| | MUSCLE | NERVE | SEGMENTAL INNERVATION | |
			Reported Range	Probable Most Important Segments
Chief radial abductors	Abductor pollicus longus	Radial	C6-T1	C6, 7
	Extensor pollicis brevis	Radial	C6-T1	C6, 7
Possible accessory radial abductors	Extensor carpi radialis longus	Radial	C5-C8	C6, 7
	Flexor carpi radialis	Median	C6-C8	C6, 7
	Extensor carpi radialis brevis	Radial	C5-C8	C6, 7
	Extensor pollicis longus	Radial	C6-T1	C7, 8

ferred to the flexor surface to produce flexor movement of the digits and, similarly, flexor tendons can be transferred to the dorsum in order to produce extensor movement. It is, however, important that not all wrist flexors or wrist extensors be so moved, since the stabilizing action of at least one wrist flexor is needed to allow good extension of the digits and, similarly, the stabilizing action of a wrist extensor is needed to allow good flexion. If stabilization of the wrist cannot be obtained by the residual muscles, arthrodesis of the wrist joint in a position of slight hyperextension and ulnar deviation, thus permitting full action of the important flexor movement of the digits, has been carried out. While arthrodesis of one wrist may be advisable, that of both wrists is contraindicated, since an individual so handicapped cannot take care of daily hygienic needs.

METACARPALS AND PHALANGES

METACARPALS

The five metacarpals, which are numbered from the radial side, are essentially similar to each other (Figs. 6-2 and 6-4). Each has a somewhat expanded and cuboidal base (Fig. 6-15) that presents one or more facets for articulation with the carpus and, except for the first, facets for the adjacent metacarpal bones; the palmar and dorsal surfaces are roughened for the attachment of ligaments. The base of

the third metacarpal bears a proximally projecting styloid process.

The bodies of the metacarpals, that of the first being shorter and stouter than those of the others, tend to be curved concavely in the palmar direction and to be somewhat triangular in cross section. The flattened dorsal surface is subcutaneous except where it is covered by extensor tendons; the medial and lateral surfaces give rise to the interossei.

The head, or expanded distal extremity, of a metacarpal presents a convex articular surface for articulation with the proximal phalanx. This surface extends more onto the palmar than the dorsal aspect of the bone, and very little onto the sides, each of which is roughened and presents a tubercle for the attachment of the collateral ligaments of the metacarpophalangeal joint.

Fractures of the Metacarpals

Owen, and McNealy and Lichtenstein have discussed the anatomy and treatment of fractures of the long bones of the hand and emphasized the role that the muscular attachments have in influencing the displacements of the fracture fragments. Thus the common displacement in fractures of any of the four medial metacarpals is a flexion of the distal fragment, because of the flexor action of the interossei upon the proximal phalanges, and hence a dorsal bowing. Fractures of the first metacarpal may result in adduction of the distal fragment by the short muscles, and abduction or extension of the proximal fragment by the long muscles.

Table 6-4
Innervation of the Ulnar Abductors of the Hand

| | MUSCLE | NERVE | SEGMENTAL INNERVATION | |
			Reported Range	Probable Most Important Segments
Chief ulnar abductors	Extensor carpi ulnaris	Radial	C6-C8	C7, 8
	Flexor carpi ulnaris	Ulnar	C7-T1	C8, T1
Accessory ulnar abductors	None			

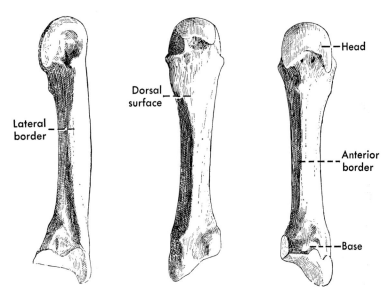

Fig. 6-15. Lateral, dorsal, and palmar views of the third metacarpal. (Poirier P, Charpy A: Traité d' Anatomie Humaine [ed 3] Vol 1. Paris, Masson et Cie, 1911)

PHALANGES

The body of a phalanx tends to be slightly curved palmarward in its long axis (Fig. 6-16), like a metacarpal, but its dorsal surface is convex, its palmar surface flat or slightly concave, from side to side, so that it tends to be semilunar in cross section. The articular surface of the base of each proximal phalanx is shallowly concave for articulation with the head of the metacarpal, but its distal end or head, and the head of each middle phalanx, has a pulley-shaped articular surface. The articular surfaces of the bases of the middle and distal phalanges, which must fit these pulleys, therefore consist of two shallow concavities

separated by a ridge. The free end of each distal phalanx bears a roughened tuberosity on its palmar surface.

The proximal phalanges receive at least parts of the insertions of many of the intrinsic muscles of the hand. The middle and distal phalanges receive the insertions of the flexor tendons and of the extensor aponeurosis (long extensor tendons and associated intrinsic muscles).

Fractures of the Phalanges
Fractures of the phalanges are, like fractures of the metacarpals, relatively common, and as is true of the latter the displacement of the fragments is often influenced not so much by

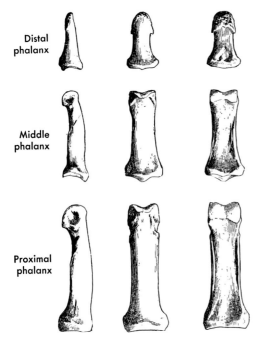

Distal phalanx

Middle phalanx

Proximal phalanx

Fig. 6-16. Phalanges of a finger in lateral, dorsal, and palmar views. (Poirier P, Charpy A: Traité d' Anatomie Humaine [ed 3] Vol 1. Paris, Masson et Cie, 1911)

the direction of the fracturing force as by the pull of muscles.

Fractures of the proximal phalanges, regardless of their level, usually show a flexion of the proximal fragment brought about by the short muscles of the hand, and a hyperextension of the distal fragment through the pull of the long extensor tendons, hence a V-shaped deformity with the open end of the V directed dorsally (Owen; McNealy and Lightenstein). Fractures of the proximal phalanx of the thumb usually show a similar deformity, the proximal segment being adducted and flexed by the adductor and flexor brevis, the distal extended by the long extensor.

Muscular displacement in fractures of the middle phalanges may not occur, but when it does it may vary according to the level of fracture. If the fracture is distal to the insertion of the superficial flexor, the proximal fragment may be flexed by this tendon and the distal extended by the long extensor tendon, so that a V-shaped deformity results. If,

however, it is proximal to the insertion of the superficial flexor the distal fragment may be flexed by this muscle, and the deformity is then the reverse, an inverted V—but this deformity is rarely seen, apparently because the middle slip of the extensor tendon neutralizes the action of the superficial flexor (Lipscomb, '68). It has been said that a straight dorsal splint must be used to correct an inverted-V deformity, but many workers treat almost all phalangeal fractures by fixing the digits in slight flexion, the position of function.

In fractures of the terminal portion of a distal phalanx, whether of the four medial digits or the thumb, the distal segment is free of muscle action and displacement may not occur at all. Fractures of the middle or distal portion of this phalanx are particularly likely to be associated with hematoma beneath the nail, however (McNealy and Lichtenstein), and it may be necessary to relieve pressure here by boring a small hole through the nail. Fractures of the base of the terminal phalanx involve the region of insertion of both the long flexor and long extensor tendons, and the displacement therefore varies, for the proximal fragment may be displaced either dorsally or palmarward, the distal portions sharply flexed or hyperextended.

METACARPAL AND PHALANGEAL JOINTS

The carpometacarpal joints have already been described, and the intermetacarpal joints need little comment. These three small synovial cavities lie between the bases of the four medial digits and are continuous proximally with the carpometacarpal joint cavity (Fig. 6-8). Each synovial pouch is strengthened by a dorsal and a palmar metacarpal ligament that pass from the base of one metacarpal to the base of the next one (Figs. 6-17 and 6-18). The fibrous capsule is completed distally by an interosseous metacarpal ligament between the distal parts of the bases of two contiguous metacarpals. Where there are two joint surfaces between a pair of metacarpals, the interosseous ligament penetrates deeply between them.

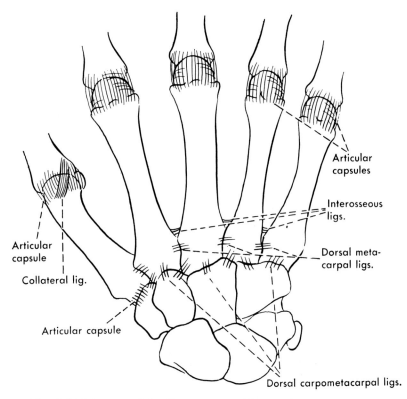

Fig. 6-17. Diagram of the ligaments attaching to the metacarpals, dorsal view.

Since the metacarpals diverge from each other and their distal ends are appreciably separated from each other, with intervening soft tissues, the heads of the metacarpals do not articulate as do the bases. However, the four medial metacarpals are held to each other and kept from spreading apart by three deep transverse metacarpal (deep intermetacarpal, transverse capitular) ligaments; these are attached to the sides of the heads of adjacent metacarpals, close to the palmar surface, and blend with the palmar ligaments and with the lateral capsules of the metacarpophalangeal joints. The lack of a similar ligament between forefinger and thumb facilitates free movement of the first metacarpal.

The palmar digital nerves and vessels and the tendons of the lumbrical muscles pass on the palmar side of the deep transverse metacarpal ligaments. The tendons of all the interossei, palmar and dorsal, pass across the dorsal aspect of these ligaments.

Metacarpophalangeal Joints

Although the metacarpophalangeal joints function somewhat differently from the interphalangeal ones, the ligamentous arrangement at all three joints is similar: the metacarpophalangeal, proximal interphalangeal, and distal interphalangeal joints all have heavy ligaments on their palmar surfaces and strengthening collateral ligaments on their sides; dorsally they are protected by the extensor tendon and its expansions. Details of the arrangement of the extensor tendon at the various joints of the digit are described later in this chapter.

The metacarpophalangeal joints allow not only movements of flexion and extension of the proximal phalanges upon the metacarpals, but also side-to-side movement and some passive rotation; they are in general of the condyloid type, with the rounded head of the metacarpal fitting into a small concavity of the base of the phalanx. Lateral movement

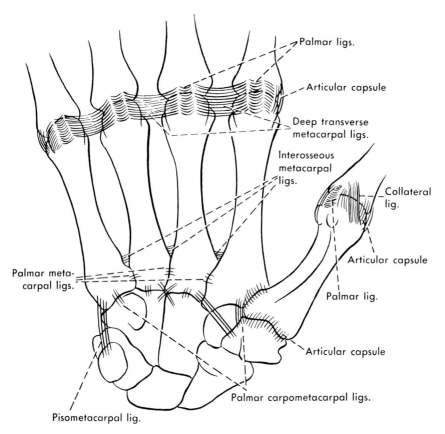

Fig. 6-18. Diagram of the ligaments attaching to the metacarpals, palmar view.

and rotation at the metacarpophalangeal joint of the thumb are more restricted, and this joint tends to be more nearly of the hinge type. It also almost always has a sesamoid imbedded in the palmar aspect of the fibrous capsule of the joint on both the radial and ulnar sides.

On their dorsal aspects, the metacarpophalangeal joints are protected by thin articular capsules reinforced by the expansion of the long extensor tendons over these joints. On each side the capsule is reinforced by a collateral ligament, which runs obliquely across the side of the joint, being attached to the dorsum of the head of the metacarpal, but to the palmar aspect of the base of the proximal phalanx. While it is generally stated that these ligaments are taut in flexion but relaxed in extension, thus allowing abduction–adduction in the latter position, Landsmeer ('55) said they are tense in both positions and that

rotation of the proximal phalanx (not possible in flexion) is necessary for abduction to take place.

A number of explanations have been offered for the ulnar deviation of the digits typical of the rheumatoid hand. Hakstian and Tubiana and Smith and Kaplan agreed that because of the shape of the metacarpal heads the radial collateral ligaments allow ulnar deviation more readily than the ulnar ones allow radial deviation, and attributed the rheumatoid condition to an exaggeration of the normal ulnar deviation. Landsmeer ('55) wrote that the greater length and obliquity of the radial collateral ligaments contribute to the ease of ulnar deviation; Backhouse ('72) said that in the majority of cases the insertions of the ulnar interossei into the extensor tendons are stronger than those of the radial ones, so that the radial fibers are usually stretched. The thesis that all the ulnar interosseous muscles are stronger than the radial ones is apparently not valid; Matheson and co-workers found that this varied with the finger involved.

The palmar surface of the joint is protected by the dense fibrocartilaginous palmar ligament (volar plate, volar ligament, accessory volar ligament), grooved on its free surface for the long flexor tendons (Fig. 6-18). The grooved surface is covered by a portion of the synovial tendon sheath of the long flexor tendons, and the edges of the groove represent the attachment of the fibrous sheath of these tendons. Each palmar ligament is firmly attached laterally to the collateral ligaments and distally to the proximal phalanx; their attachment to the metacarpals is less strong and is, according to Gad, confined to its sides with the middle part having no attachment at all.

Alldred described a case in which there was a tear in the capsule of a metacarpophalangeal joint, between the collateral ligament and the palmar ligament, and the proximal portion of the torn ligament had rolled into the joint so as to prevent its extension beyond about 40° of flexion.

Murphy and Stark, among others, emphasized that in dorsal dislocation at the metacarpophalangeal joint a palmar incision is necessary for reduction. They listed the obstacles to reduction as three:

1. The palmar ligament is torn from the metacarpal and folded into the joint.
2. The flexor tendons lie dorsal to the metacarpal head.
3. The metacarpal head, forced forward, is wedged between the superficial transverse metacarpal ligament and the transverse fasciculi of the palmar aponeurosis.

The metacarpophalangeal joints have a variable innervation. Their chief nerve supply is from the palmar digital nerves, but they typically also receive branches from the dorsal digital nerves, and a variable number are supplied also by the deep branch of the ulnar (Gray and Gardner). Stopford ('21) investigated loss of appreciation of passive movement following division of various nerves, and found the clinically important innervation to

be as follows: for the metacarpophalangeal joint of the thumb, both the median and radial nerves, but sometimes either alone; for the joints of the second and third digits, the median principally or entirely as a rule, but occasionally the radial; for the ring finger, the ulnar primarily, usually with a supplementary supply from the median; and for the little finger, the ulnar alone.

Interphalangeal Joints

Although the interphalangeal joints are pure ginglymoid or hinge joints and thus differ from the metacarpophalangeal joints which are, except for that of the thumb, condylar, the ligaments of the interphalangeal joints are exactly similar to those of the metacarpophalangeal ones. Thus the capsules are thin dorsally, where they are largely replaced by the expansions of the long extensor tendons over the joints; they are strengthened on each side by collateral ligaments, which arise from the dorsal aspect of the side of the head of one phalanx and extend distally and in a palmar direction to insert on the palmar aspect of the side of the base of the next phalanx; and on the palmar surface the capsules blend with and are much reinforced by the heavy palmar ligaments, which provide part of the deep surfaces over which the flexor tendons glide and which separate the synovial membrane of the tendon sheath from those of the joints (Figs. 6-19 and 6-20). The fibrocartilaginous palmar ligaments are like those of the metacarpophalangeal joints and are more firmly attached distally than proximally.

The collateral ligaments of the interphalangeal joints are usually described as being taut in all positions; regardless of that, the difference in shape between the heads of the metacarpals and those of the phalanges would prevent abduction–adduction as long as the collateral ligaments were intact. Redler and Williams found prolonged disability resulting from complete rupture of a collateral ligament, but reported that suturing to soft tissue rather than bone provided satisfactory repair.

Although a torn palmar ligament may become wedged into an interphalangeal joint

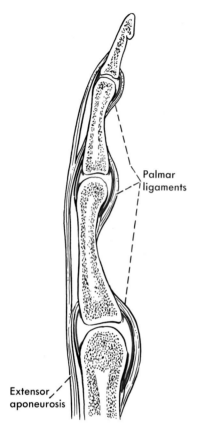

Palmar
ligaments

Extensor
aponeurosis

Fig. 6-19. Longitudinal section of a finger to show the metacarpophalangeal and interphalangeal joints. Synovial membrane is depicted by *red*.

and thus lock it in hyperextension, hyperextension deformities of the proximal interphalangeal joints probably arise more frequently from anterolateral tears that abolish the slight flexion in which the joints are usually maintained. Kaplan ('36) found that in such circumstances the profundus tendon, as it flexes the distal interphalangeal joint, hyperextends the proximal joint, and that repair of the anterolateral part of the capsule is necessary to restore the slight flexion that allows the profundus to flex this joint. An important part of the deformity is that the lateral bands of the extensor aponeurosis gradually slip dorsally after such tears.

Lipscomb ('67) pointed out that the results of synovectomy of the metacarpophalangeal joints of the fingers in rheumatoid arthritis may often be enhanced by a similar synovec-

tomy at the metacarpophalangeal joint of the thumb (Fig. 6-21) or of the proximal interphalangeal joint of one or more fingers (Fig. 6-22). He approached the joint from one side only, and dislocated it in order to perform a complete synovectomy.

Most of the dorsal digit nerves are so short that the interphalangeal joints are necessarily supplied primarily by the palmar digital nerves, but Gray and Gardner reported that the interphalangeal joint of the thumb, the proximal one of the index finger, and both those of the little finger are usually supplied also by the dorsal digital nerves. Stopford ('21) concluded that the interphalangeal joint of the thumb is typically supplied chiefly by the median nerve, to a minor degree by the radial, and in 42% by the median alone; that the joints of the index and the long or middle fingers are typically supplied by the median only; that those of the ring finger are typically supplied by the ulnar nerve supplemented by the median, but by the ulnar alone in about 25%; and that the joints of the little finger are supplied by the ulnar nerve alone.

ANOMALIES OF HAND

There are numerous reports of isolated cases or short series of cases of congenital anomalies of the hand; more general discussions of this subject include those of Barsky ('51) and of Kelikian and Doumanian.

Anomalies of the Digits

According to Barsky, *syndactylism* is the most common congenital deformity of the hand; he quoted an estimated occurrence of 1 in 3,000 births, and said that it is most frequent between the middle and ring fingers, and is sometimes said to be twice as frequent in males as in females. Apparently, syndactylism is often associated with polydactylism, and symphalangism (fusion of interphalangeal joints, Fig. 6-23) occasionally also accompanies syndactylism. Syndactylism is said to vary markedly in the degree of fusion; the most common type, according to Barsky, is a simple webbing of the skin, which may or

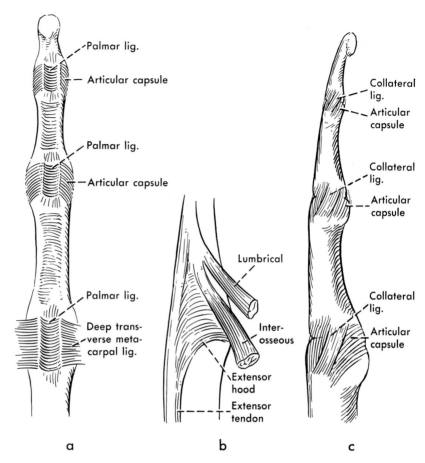

Fig. 6-20. Palmar and lateral views of the ligaments of the metacarpophalangeal and interphalangeal joints.

may not extend the entire length of the digits; in the most severe type the bones may be fused and sometimes even the tendons and nerves are, and there is a double nail at the end of the finger.

Polydactylism is usually a reduplication of the little finger or of the thumb, rather than of any of the intervening fingers, and the condition is said to be more common on the radial side of the limb; as a rule, if the reduplicate contains a skeleton this articulates with the enlarged head of an otherwise normal metacarpal, or with the bifurcated head of a metacarpal; an additional metacarpal for the extra digit is said to be rare. The reduplicate may vary from a fleshy mass to a well-formed digit.

Handforth found an incidence of 2.4 cases

of polydactylism of the hand per 1,000 among a group of Chinese whom he examined; most commonly the duplication was of the first digit. He expressed the opinion that this incidence is a rather high one. Among the 14 cases of polydactylism he found, only one individual showed an associated congenital abnormality (polydactylism of the feet), and in only one case was there a family history of the occurrence of supernumerary digits. Cooperman reported a case of a supernumerary metacarpal on the ulnar side of the hand, fused with the fifth metacarpal, but unassociated with a supernumerary digit; in this case the patient had six toes on each foot.

Most congenital abnormalities of the digits are acknowledged to have a hereditary basis, and *brachydactylism* is said to be the first exam-

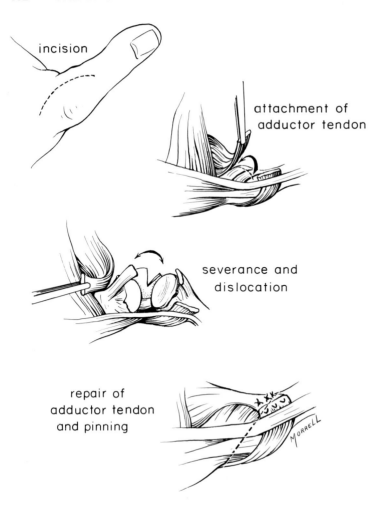

incision

attachment of
adductor tendon

severance and
dislocation

repair of
adductor tendon
and pinning

MORRELL

Fig. 6-21. Synovectomy of the metacarpophalangeal joint of the thumb. (Lipscomb PR: J Bone Joint Surg 49-A: 1135, 1967)

ple of mendelian inheritance that was demonstrated in man. According to Barsky, the shortening of the digits may involve one or all of them, and the shortening may be the result of decreased length of the phalanges, absence of a phalanx, or even a shortened metacarpal. In a familial series studied by Drinkwater, in which the index and middle fingers were short, some of the digits were said to have an increased number of segments. Barsky stated that in cases of brachydactylism the tendon of the flexor profundus may attach to the distal phalanx and be able to flex it, it may fail to extend to this phalanx, or there may be a rigid joint here.

Annular grooves around one or more digits, the wrist, or the hand, and *congenital amputation*, two different expressions of the same un-

derlying defect, have already been discussed (Chap. 3).

Macrodactylism was said by Barsky to be rare; the second and third digits are more commonly involved, and the condition occurs more often in males than in females.

Triphalangeal Thumb

Triphalangeal thumb is very rare, although apparently not as rare as clefthand. Milch, reporting 4 cases, said that the extra phalanx usually takes the form of a small ossicle lying between the nearly normal proximal and distal phalanges, and that unequal growth of the ossicle and interference with the growth of the adjacent epiphysis of the distal phalanx may result in an abnormal curvature of the thumb. He said that if the diagnosis is made in infancy or in childhood the ossicle should be removed surgically, but that operation is said to be contraindicated for an adult. Lapidus and co-workers reported six persons with triphalangeal thumbs who were encountered in roentgenographic examination of 75,000.

The occasional occurrence of an extra phalanx on the thumb has been variably explained: Lapidus and his co-workers favored the theory that it represents the remains of one of the phalanges of an incompletely developed bifid thumb (Fig. 6-24), but noted that a triphalangeal thumb has been said to represent a doubling of the index finger on a hand that failed to develop a true thumb. McGregor described a thumb with three phalanges that was also reduplicated laterally, and he expressed the opinion that triphalangeal thumb is an atavism; the normal distal phalanx of the thumb, according to this view,

actually represents fusion of the middle and distal phalanges of a primitive triphalangeal thumb.

Minor Anomalies and Sesamoids

Minor skeletal anomalies of the hand include two epiphyses for the metacarpal of the thumb, a proximal and a distal (Joseph); a proximal epiphysis in one or more of the second to fifth metacarpals (Wakeley); the development of a separate center of ossification in the styloid process of the third metacarpal, and failure of this process to unite with the rest of the bone, thus forming a styloid bone; and supernumerary sesamoid bones at the metacarpophalangeal or interphalangeal joints.

Accessory sesamoid bones in the hand (Fig. 6-25), lying in the capsule of the joint and always having an articular surface with the underlying bone (Joseph), are not uncommon,

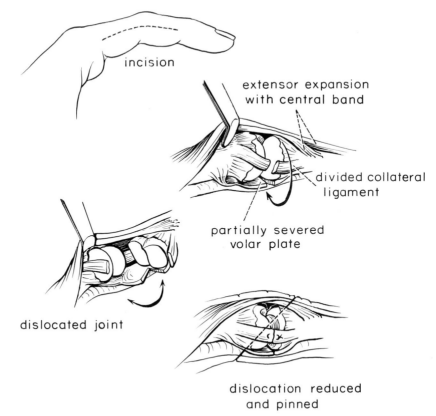

incision

extensor expansion
with central band

divided collateral
ligament

partially severed
volar plate

dislocated joint

dislocation reduced
and pinned

Fig. 6-22. Synovectomy of the proximal interphalangeal joint of a finger. (Lipscomb PR: J Bone Joint Surg 49-A: 1135, 1967)

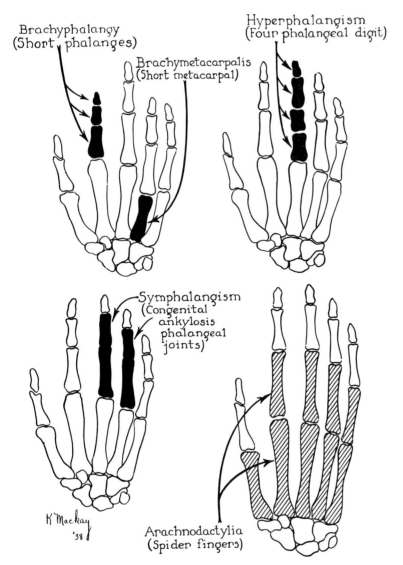

Fig. 6-23. Some congenital deformities of the fingers. (Meyerding HW, Dickson DD: Am J Surg 44:218, 1939)

although they may be difficult to demonstrate by roentgenogram because of their sizes and positions. Figure 6-26a shows the locations and incidence of sesamoid bones in one series of hands, as determined by dissection. Bizarro found fewer, and a lower incidence, by roentgenogram (Fig. 6-26b). The most common interphalangeal sesamoid occurs at the interphalangeal joint of the thumb, although Bizarro found it much less frequently than had former workers; Joseph found it bilaterally in 69% of males and 79% of females. He also reported that if a sesamoid is present at the metacarpophalangeal joint of the index finger there is usually one also at the joint of the little finger. Flatt, reporting a case in which the index finger was recurrently locked at the metacarpophalangeal joint by a particularly large sesamoid, regarded one here, one at the corresponding joint of the little finger, and one at the interphalangeal joint of the thumb as being almost constant. Typically, all the interphalangeal sesamoids, and the metacarpophalangeal ones except for the nor-

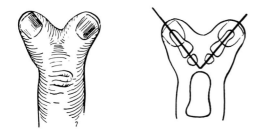

Fig. 6-24. Bifid thumb. (Meyerding HW, Dickson DD: Am J Surg 44:218, 1939)

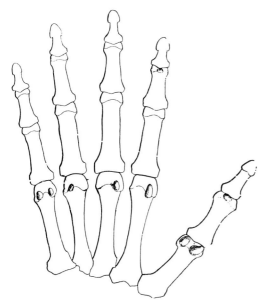

Fig. 6-25. A hand with numerous accessory sesamoids. (Poirier P, Charpy A: Traité d' Anatomie Humaine [ed 3], Vol 1. Paris, Masson et Cie, 1911)

mally present double ones at the metacarpophalangeal joint of the thumb, are single bones, but both Joseph and Bizarro reported cases in which there were two at a joint.

Clubhand

Absence of one or more digits is commonly associated with a deficiency of the radius or ulna (hemimelia) and of the carpals and metacarpals on the border of the limb (according to O'Rahilly, 80% or more of cases in which the radius is absent present also deficiencies of the radial side of the hand); congenital absence of the digits is therefore fre-

quently associated with clubhand. In the case described by Forbes, the affected hand was deflected so far to the radial side that its original radial border lay adjacent to the radial border of the forearm, and hand and forearm were united by a cutaneous bridge.

It has already been noted that congenital absence of the radius, in part or in whole, is more common than congenital absence of the ulna; in accordance with this, clubhand involving a deformity on the radial side (see Fig. 5-40) is more common than is one on the ulnar side. Jones and Roberts reported one of the rare cases of clubhand involving a deformity on the ulnar side. Barsky said that 253 examples of congenital absence of the radius or at least a distal part of it, and generally associated with clubhand, were collected as long ago as 1923.

Fig. 6-26. Incidence of sesamoids in the hand, a, in a series (Pfitzner's) in which the bones were carefully cleaned, b, in a series of roentgenograms. a is schematized as a palmar view of the left hand, b, as one of the right hand. (Bizarro AH: J Anat 55:256, 1921)

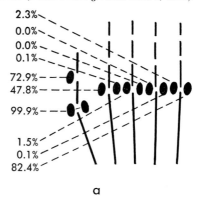

a

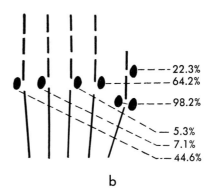

b

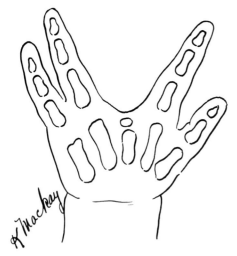

Fig. 6-27. Cleft or lobster-claw hand. (Meyerding HW, Dickson DD: Am J Surg 44:218, 1939)

Clefthand

The severe defect known as clefthand or lobster-claw hand (Fig. 6–27) is, fortunately, a very rare congenital anomaly. Barsky ('64), reporting on the reconstructive surgery involved in 19 cases and reviewing the literature on the subject, said it is often familial. A common form is absence of the phalanges of the middle digit, or of those and the metacarpal.

Palmar Anatomy

Subcutaneous Tissue

The subcutaneous tissue of the palm of the hand, often referred to as superficial fascia, is a particularly tough layer that intervenes between the skin and the deep fascia (the palmar aponeurosis and the thenar and hypothenar sheaths). As witnessed by the relative immobility of the skin on the palmar surface, the subcutaneous tissue is permeated by tough bands of connective tissue that unite the deeper fascial layers to the skin, and hence anchor the skin rather firmly. While the subcutaneous tissue contains fat, especially over the fingers, the fat is markedly subdivided by the connective tissue septa. Septa attaching to the palmar and digital creases are especially

strong and serve to help prevent slipping of the skin when something is held firmly in the grip; on the sides of the fingers special septa (the cutaneous ligaments of the digits) form imperfect canals in which run the digital nerves and vessels.

The subcutaneous tissue of the palm is in marked contrast to the thin and loose subcutaneous tissue or superficial fascia of the dorsum, where the skin is mobile and can be torn off somewhat like a glove, but is continuous with this around the borders of the hand and digits. The palmar subcutaneous tissue of the digits, especially on the distal phalanges, is frequently referred to as the pulp of the fingers; similarly, this tissue in the palm is sometimes referred to as the palmar pulp.

Clinical aspects of infections of the hand have been discussed by Kanavel, by Scott and Jones, and by Robins. Of interest here is the fact that a large proportion of infections of the hand are in the pulp of either the fingers or the palm; Robins reported that among 1,000 cases of infection of the hand 233 were infections around the nail (paronychia, the most frequent complication of which is an extension into the pulp of the finger), but 244 were infections of the pulp of the distal segment of the digits, and 412 were subcutaneous infections elsewhere, only 43 of these being on the dorsum. Similarly, Scott and Jones analyzed infections of the hand in 1,211 patients, and said that in 263 there was paronychia, but in 413 there was infection of the palmar pulp of the finger, most commonly of the distal pulp, in 76 infection of the distal palmar pulp, and in 16 "subcutaneous" infection of the palm.

Because of the arrangement of septa over the distal phalanx, infections of the pulp space here are particularly likely to spread transversely between the septa, thus either toward the skin or toward the bony phalanx, and such infections are known as felons. Subcutaneous infections of the hand other than in the distal pulp space are, according to Robins, essentially similar, and their main complication is necrosis of the skin. Although they may, it seems, sometimes spread through the palmar fascia, they apparently do not usually pass the flexor tendons so as to involve the deep spaces of the palm.

The subcutaneous tissue of the palm contains the limited palmar venous network, which drains mostly toward the sides of the digits and of the palm; small cutaneous twigs of nerves and arteries also course in it, after having penetrated the deep fascia of the wrist or hand. In the fingers, the large palmar digital nerves and vessels lie in the subcutaneous tissue on the sides of the flexor tendons; they lie, however, not loosely in this tissue, but rather in a sort of tunnel formed by septa stretching from the phalanges to the skin (Fig. 6-28); these septa have been described as the "cutaneous ligaments of the digits" (Grayson); the septa dorsal to the digital nerves and vessels were originally called "Cleland's ligaments."

Superficial Veins and Lymphatics

The superficial veins of the palmar surface are rather small and inconspicuous. The palmar digital veins of each finger tend to form paired channels that communicate with each other across the surface of the digit; these veins, however, drain only in part toward the palm of the hand, for they send communications around the borders of the finger with which they are associated to reach the dorsal digital veins; as they reach the bases of the fingers, they unite to form a small superficial arch, which in turn drains in part to the dorsal venous network by intercapitular veins (Fig. 6-29) and veins on the borders of the hand. Proximally running vessels from the superficial arch join other vessels in the palm of the hand. The palmar veins, like the digital ones, drain largely around the medial and lateral borders of the hand, but in part give rise to the small median forearm vein that runs up the flexor surface of the forearm to a junction with a vena mediana cubiti, the basilic, or the cephalic vein (Chap. 5). The drainage of much of the blood around the borders of the fingers and of the hand to the dorsal veins serves an obviously useful purpose; it allows free circulation from the superficial aspect of the palm even though the longitudinally running veins may be occluded by grasping something in the hand.

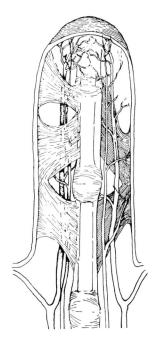

Fig. 6-28. Cutaneous ligaments of the digits, palmar aspect. Palmar ligaments have been removed on the viewer's right to show the dorsal ones. Note the position of the digital nerve and artery, between the two sets of ligaments. (Grayson J: J Anat 75:164, 1941)

The lymphatic cutaneous plexuses on the palmar surface of the digits and in the palm of the hand are particularly dense; as is true of the veins, the digital lymphatics drain for the most part around the sides of the digits to join the dorsal digital lymphatics. Similarly, while a few efferent vessels from the palm of the hand run up the flexor surface of the forearm, the great majority either turn around the ulnar border of the hand to join lymphatics associated with the little finger, turn around the radial border of the hand to join those arising from the thumb, or pass distally in the palm to reach the interdigital clefts, turn dorsally in these, and join vessels on the dorsum of the hand (Fig. 6-30). The lymphatic drainage from the palmar surface of the hand is thus largely to the dorsum; both because of this and of the difference in the subcutaneous tissue of the palmar and dorsal surfaces of the hand, infections of the palm may be first evidenced by edema on the dorsum of the hand.

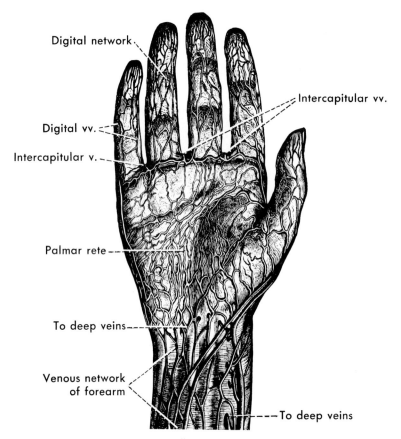

Digital network

Intercapitular vv.

Digital vv.

Intercapitular v.

Palmar rete

To deep veins

Venous network
of forearm

To deep veins

Fig. 6-29. Superficial veins of the palm of the hand. (Toldt C: An Atlas of Human Anatomy [ed 2], Vol 2. New York, Macmillan, 1928)

CUTANEOUS NERVES AND ARTERIES

Palmar Nerves and Vessels

The cutaneous nerves of the palm of the hand consist of several small twigs derived primarily from the median and ulnar nerves. However, the anterior branch of the medial antebrachial cutaneous nerve usually sends a terminal twig to join the palmar branch of the ulnar nerve, and hence helps to supply a limited portion of skin over the proximal portion of the hypothenar eminence, and the anterior branch of the lateral antebrachial cutaneous nerve typically supplies skin over the lateral part of the thenar eminence.

The palmar branch of the median nerve supplies a larger area of skin in the palm; this nerve arises from the median above the wrist, pierces the antebrachial fascia or the upper edge of the flexor retinaculum (transverse carpal ligament), and descends superficially into the palm where it divides into medial and lateral branches. The lateral branch helps to supply the skin over the thenar eminence, communicating with the lateral antebrachial cutaneous nerve here; the medial branch supplies much of the remaining skin of the palm, extending medially about to the line of the fourth metacarpal, where it overlaps the distribution of or communicates with the palmar cutaneous branch of the ulnar.

The palmar branch of the ulnar nerve arises at a variable level in the lower half of the forearm; it accompanies the ulnar nerve and the corresponding artery downward toward the wrist to become subcutaneous, run superficially across the flexor retinaculum (transverse carpal ligament), and supply an ulnar portion of the hand corresponding to the width of about one and one-half fingers. The digital branch of the ulnar nerve to the ulnar border of the little finger frequently gives off additional twigs to skin of the ulnar border of the hand, and digital branches of both the median and ulnar nerves may contribute twigs to skin of the distal part of the palm of the hand.

The cutaneous arteries of the palm are tiny twigs derived largely from the digital arteries and therefore pierce the deep fascia of the hand, deep to which the digital arteries lie.

Digital Nerves and Arteries

The palmar digital nerves are not strictly cutaneous, for they supply the deep structures such as tendons and joints (this is also true of the nerves of the hand, regularly regarded as "cutaneous"); however, these nerves are the sole nerve supply to the digital skin on the palmar surface, and a major supply also to the skin of the dorsal surface of the digits. Similarly, the palmar digital arteries are the chief supply to the entire digit, for the dorsal arteries are very small (Fig. 6-31).

The palmar digital nerves are derived from both the median and ulnar nerves; their origins and course are described in more detail in the descriptions of these nerves. In brief, they take origin in the proximal part of the hand at about the distal border of the flexor retinaculum (transverse carpal ligament) and run forward in the hand under cover, except on the hypothenar eminence, of the palmar aponeurosis (see Figs. 6-34 and 6-39). There are three common digital nerves in the palm for the adjacent sides of the four fingers, and two proper digitals for the radial side of the index finger and the ulnar side of the little finger. The digital nerves to the thumb may arise separately or by a short common stem. There are, similarly, three common digital arteries and a proper digital for the little finger, arising from the superficial palmar arch and joined by branches from the deep arch; the proper digital artery to the radial side of the little finger, and the two for the thumb, arise separately or together from the radial ar-

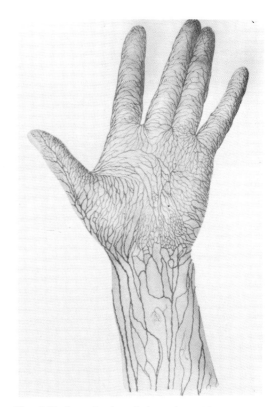

Fig. 6-30. Lymphatics of the palm. (After Sappey, from Rouvière H: Anatomie des Lymphatiques de l'Homme. Paris, Masson et Cie, 1932)

tery as it enters the palm. These vessels are described in more detail in following sections.

The branches of the median nerve to the thumb and to the radial side of the index finger emerge from behind the lateral part of the aponeurosis and are joined as they start out on the digits by the corresponding arteries, which typically have a deeper origin. The digital nerves and vessels to the adjacent sides of the four fingers appear in the distal part of the palm in the fatty connective tissue that lies between the longitudinal fasciculi that the palmar aponeurosis sends to the digits. At the level at which they appear, the common digital nerves may or may not have divided into their proper digital branches; the arteries usually have not, for they typically divide distal to the nerves. Before they reach the bases of the digits, the nerves and vessels lie between the long flexor tendons with their associated lumbrical muscles. Each proper digital

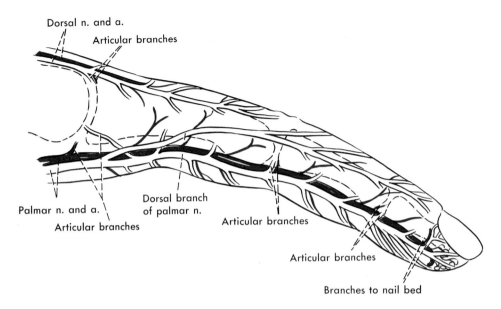

Dorsal n. and a.

Articular branches

Palmar n. and a.

Articular branches

Dorsal branch
of palmar n.

Articular branches

Articular branches

Branches to nail bed

Fig. 6-31. Digital nerves and vessels in a lateral view of a finger.

nerve proceeds onto the digit, immediately in front of the proper digital artery, and both nerve and artery lie on the sides of the flexor tendons at about the level of the palmar surfaces of the phalanges. They are thus largely protected by the flexor tendons from undue pressure when an object is grasped in the hand. On the digits, each nerve and its corresponding artery pass between the palmar and dorsal cutaneous ligaments of the fingers.

As they pass along the digit, the nerve and artery give branches to the palmar surface of the digit, many of the nerve branches ending in large lamellated (Pacinian) corpuscles; they also give branches to the side of the fingers, a branch to each of the interphalangeal joints, and, at about the level of the middle of the distal phalanx, each sends a large branch straight dorsally, close around the lateral border of the phalanx, to supply the nail bed of the finger. The remainder of the nerve then breaks up into numerous branches for the distal end of the finger, while the two arteries on a digit anastomose to form an arch across the distal phalanx; from this arch a fine network of vessels extends into the pulp of the distal phalanx.

In its course the digital artery gives numerous branches to the dorsum; the nerve tends to behave differently in this respect. Each palmar digital nerve of the second, third, and fourth digits typically gives off, as it runs along the proximal portion of the basal phalanx, a dorsal cutaneous branch. This branch winds dorsally around the side of the proximal phalanx, sometimes passing deep to and sometimes superficial to the palmar digital artery, and attains the lateral side of the dorsum of the finger at about the proximal interphalangeal joint; it may or may not grossly anastomose with the small dorsal digital nerve, but is distributed more distally, to the skin over the dorsum of the middle and distal phalanges. Thus it comes about that the median nerve, rather than the radial, is the nerve supply to a distal part of the dorsum of the index and middle fingers and radial side of the ring finger. In accordance with this, the branches of the radial nerve to the dorsum of the finger are usually not traceable much beyond the proximal interphalangeal joint, if that far.

The branches of the median nerve to the thumb typically give several twigs, rather than one major branch, to the skin of the dorsum of the distal phalanx. The proper digital branches of the ulnar nerve to the adjacent sides of the ring and little fingers often, but not always, have dorsal digital branches like the median nerve; the proper digital branch of the ulnar nerve to the medial border of the little finger usually does not have a dorsal branch, for the dorsal digital nerve typically runs the length of the finger. The presence or absence of dorsal branches from the palmar digital nerves of the ulnar makes no real difference, since both palmar and dorsal surfaces of the medial one and one half digits are in either event supplied by the ulnar nerve. As is true of the median nerve, however, it is the palmar digital branches, not the dorsal ones, that supply the nail bed.

FLEXOR RETINACULUM AND PALMAR APONEUROSIS

Flexor Retinaculum and Carpal Canal

The flexor retinaculum or transverse carpal ligament, also known as the annular or anterior annular ligament, is a very heavy band of fibers that stretches across the concavity of the carpal arch, thus forming with the carpals a compartment, the carpal canal or tunnel, through which the long flexor tendons and the median nerve reach the palm of the hand (Fig. 6-32). The retinaculum is attached medially to the pisiform bone and to the hamulus of the hamate bone, laterally to the tubercle of the scaphoid (navicular) and to the trapezium (greater multangular). At its attachment to the trapezium it divides into two parts, one of which attaches to the tubercle of the trapezium, the other to the medial part of the palmar surface of the bone dorsal to the groove that lodges the tendon of the flexor carpi radialis; this attachment converts the groove for the flexor radialis tendon into a special tunnel through which the tendon proceeds to the floor of the carpal canal to reach its insertion on the base of the second metacarpal.

At its upper border the flexor retinaculum is continuous with the deeper layer of deep fascia at the wrist, that which lies behind the flexor carpi ulnaris, flexor carpi radialis, and

Fig. 6-32. Cross section at the wrist showing the flexor retinaculum (*red*) and the carpal canal behind it.

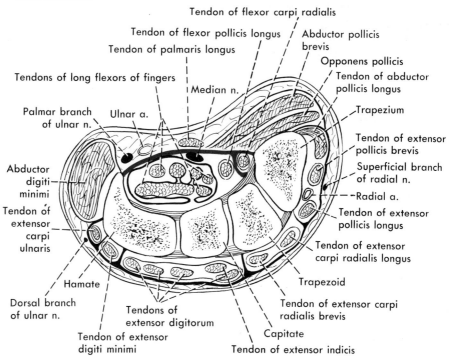

Tendon of flexor carpi radialis
Tendon of flexor pollicis longus
Tendon of palmaris longus
Abductor pollicis brevis
Opponens pollicis
Tendons of long flexors of fingers
Tendon of abductor pollicis longus
Median n.
Palmar branch of ulnar n.
Ulnar a.
Trapezium
Tendon of extensor pollicis brevis
Abductor digiti minimi
Superficial branch of radial n.
Tendon of extensor carpi ulnaris
Radial a.
Tendon of extensor pollicis longus
Hamate
Tendon of extensor carpi radialis longus
Dorsal branch of ulnar n.
Tendons of extensor digitorum
Trapezoid
Tendon of extensor digiti minimi
Tendon of extensor carpi radialis brevis
Capitate
Tendon of extensor indicis

palmaris longus muscles, but in front of the other muscles. Superficially, it has fused to it the distal edge of the "volar carpal ligament," except where the ulnar nerve and vessels separate these two layers, and on the superficial surface of its distal part it has attached to it fibers of the palmar aponeurosis, and of the palmaris longus tendon as these contribute to the palmar aponeurosis. Medially and laterally, it gives origin to many of the muscles of the hypothenar and thenar eminences, respectively (see Fig. 6-39). In appearance, it is more tendinous than fascial.

The "volar carpal ligament," mentioned in the preceding paragraph, is a variable thickening of transverse fibers in the superficial layer of the deep fascia of the forearm at the wrist (the layer of fascia lying in front of the palmaris longus and the flexor carpi radialis and ulnaris muscles) and is not listed in present anatomical terminology. Sometimes referred to as a superficial part of the retinaculum, it lies largely proximal to the flexor retinaculum and more superficially; it attaches laterally to the styloid process of the radius, medially to the styloid process of the ulna and to the pisiform bone; its lower border is not distinct, since it fuses with the anterior surface of the upper part of the flexor retinaculum except over the ulnar nerve and vessels. The part over the ulnar nerve and vessels, the only distinct part, forms with the upper part of the flexor retinaculum a tunnel ("Guyon's canal") from which these nerves and vessels emerge to run into the palm superficial to the distal part of the retinaculum.

It is the flexor retinaculum or transverse carpal ligament, not the volar carpal ligament, that with the carpals forms the carpal canal (tunnel). The carpal canal, in the broad sense of the space between retinaculum and carpals, actually contains two compartments: the major compartment or canal proper lies behind the flexor retinaculum and in front of the carpals, and is a large space that transmits the long flexor tendons of the fingers and of the thumb, and the median nerve; this compartment can be opened by a longitudinal incision through the flexor retinaculum. The

second compartment is very small, just large enough to accommodate the tendon of the flexor carpi radialis and its surrounding synovial tendon sheath; it lies at first on the radial side of the major portion of the canal, between the two attachments of the flexor retinaculum to the trapezium, but distally is more on the dorsal aspect (floor) of the canal. This compartment is not opened by the usual section of the flexor retinaculum, which leaves the flexor carpi radialis held closely against the carpus by the radial attachment of the retinaculum.

RELATIONS AT THE WRIST

Since the relations of structures just above the wrist and at the level of the carpal canal are of particular surgical importance, they will be briefly reviewed here.

Flexor Carpi Ulnaris, Ulnar Nerve and Vessels

The flexor carpi ulnaris lies on the ulnar border of the wrist, but ends with its attachment to the pisiform bone. The ulnar nerve and artery, the latter with two accompanying veins, descend in the lower part of the forearm under cover of the flexor carpi ulnaris, the nerve lying medial to the artery, and emerge from behind this muscle close to the level of the pisiform bone (see Fig. 6-39); they pass around the lateral aspect of the prominence of the pisiform bone, emerge from under cover of the volar carpal ligament, and pass distally into the palm of the hand in front of the flexor retinaculum but behind (deep to) the palmaris brevis. As the ulnar nerve crosses the flexor retinaculum it divides, at about the level of the distal end of the pisiform bone, into its terminal superficial and deep branches, which will be described later; the ulnar artery, at about the same level, gives off a usually small deep palmar branch, while the main branch of the artery, the superficial palmar arch, continues with the superficial branch of the ulnar nerve across the retinaculum and then turns laterally across the palm to run deep to the palmar aponeurosis.

It is while the nerve is in or close to the canal of Guyon, that is, between the volar carpal ligament and the flexor retinaculum, that it is most commonly injured below the epicondylar level. Shea and McLain found that lesions proximal to or in the canal may produce both sensory and motor deficits, those in or distal to the canal either sensory or motor deficit only, depending on whether the superficial or deep branch is involved.

Flexors of the Fingers, and Palmaris Longus

Lateral to the ulnar nerve and vessels at the wrist are the tendons of the two flexor digitorum muscles and, superficially, the tendon of the palmaris longus. As they lie at the wrist and as they enter the carpal canal, the flexor tendons are arranged in three layers: the tendons of the flexor digitorum superficialis (sublimis) to the middle and ring fingers form the most anterior layer; behind these, forming the middle layer, are the tendons of the superficial flexor to the index and little fingers; and the most posterior layer is formed by the tendons of the flexor profundus, which may be four but are often only two in number at this level, one for the index finger, the remaining one for the other three fingers. Just before they pass behind the flexor retinaculum all these flexor tendons, superficial and deep, enter a common synovial tendon sheath, which surrounds them and is continued beyond the distal border of the flexor retinaculum into the palm of the hand.

The palmaris longus tendon lies almost exactly in the midline of the wrist, and in front of the radial border of the common flexor tendons. It is a prominent landmark at the wrist, except when the palmaris is missing; just above the wrist the median nerve lies almost behind the tendon, very slightly to the radial side. The tendon of the palmaris continues into the palm of the hand, where it blends with the palmar aponeurosis across the front of the flexor retinaculum, some of its deeper fibers attaching to the latter. Like the ulnar nerve and vessels, therefore, it does not traverse the carpal canal.

Median Nerve

For much of its course in the forearm the median nerve lies behind the flexor digitorum superficialis (sublimis), but just above the wrist it emerges on the radial side of the superficial flexor and passes slightly forward and medially around this side of the tendons to lie in front of them in the carpal canal. The small palmar branch of the median nerve, given off as this nerve appears on the radial side of the flexor tendons, pierces the deep fascia or sometimes the upper edge of the flexor retinaculum to run superficial to the retinaculum into the palm of the hand, but the large median nerve itself passes behind the retinaculum. At the wrist, the median nerve lies at first on the radial side of the superficial flexor tendons and almost directly behind, but slightly to the radial side of, the palmaris longus tendon; in the carpal canal it lies on the front of the flexor tendon sheath, and immediately behind the flexor retinaculum. In the limited space of the carpal canal the median nerve is subject to damage if it is forced against the relatively unyielding flexor retinaculum. Robbins reported that the canal is narrowest at its middle and that both flexion and extension decrease the space in the canal, apparently through rotation of the lunate; and Tanzer studied pressures in the canal and found that both flexion and extension increase the pressure in the proximal part of the canal, and that extension also increases that in the distal part.

At about the distal border of the flexor retinaculum the median nerve divides under cover of the retinaculum or of the palmar aponeurosis into, usually, five or six branches, the exact method of division varying somewhat; these branches are the muscular branch into the thenar muscles (Fig. 6-33), a common or two proper digital nerves for the thumb, the proper digital for the radial side of the index finger, and two common digitals for the adjacent surfaces of index and middle, and middle and ring, fingers, respectively.

Compression of the median nerve within the carpal canal, commonly called "carpal tunnel syndrome," is a recognized cause of in-

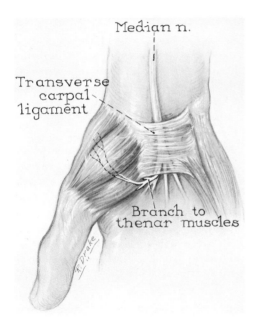

Median n.

Transverse
carpal
ligament

Branch to
thenar muscles

Fig. 6-33. The median nerve and its muscular branch in the hand. (Love JG: North Carolina MJ 16:463, 1955)

jury to the nerve at the wrist (Zachary; Cannon and Love; Brain, and co-workers). Zachary reported that the condition was first described in 1909, as isolated atrophy of muscles of the thenar eminence, and has been variously regarded as resulting from pressure on the muscular branch only of the median nerve (as a result of its curving too sharply around the distal edge of the flexor retinaculum), or from compression of the median nerve as a whole behind the retinaculum; the sensory fibers of the nerve may or may not be obviously involved. Many of the cases reported, especially the earlier ones, were associated with advanced arthritis or with fracture or dislocation of the carpus, which might be expected to crowd the carpal canal, but others showed no such association. Cannon and Love reported 38 patients with median nerve palsy from compression in the carpal canal, in 9 of whom surgical division of the retinaculum was carried out with satisfactory results; included among these were 3, apparently the first reported, in which the compression was not associated with previous injury to the wrist.

Brain and his co-workers stressed the importance, in differential diagnosis, of the distribution of loss of sensation even though the sensory findings are likely to be slight, pointing out that in this respect in particular the symptoms differ from those of compression at the level of the brachial plexus. Subsequent to these early studies, it has been recognized that the syndrome occurs very often when there is no preceding injury to the wrist. (Smith, '71, Butler and Bigley, and Schultz and co-workers all reported cases associated with an anomalous muscle belly or tendon in the canal.) Phalen described thickening or fibrosis of the synovial membrane of the tendon sheath as the common cause and said that in only 16% of 644 hands could a preceding injury have contributed. He reported that hormonal treatment for rheumatoid conditions often gave relief and found section of the retinaculum necessary in only 40%. Fullerton adduced evidence that an important effect of pressure on the nerve may be the production of ischemia. Thomas and co-workers found conduction studies of the median nerve, especially its afferent fibers, particularly useful in confirming the diagnosis of carpal tunnel syndrome, but not predictive of the amount of recovery to be expected after operation.

Section of the flexor retinaculum produces no ill effects, because all forceful movements of flexion of the digits are carried out with the wrist in moderate hyperextension, and therefore the tendons remain against the floor of the carpal canal.

Flexor Pollicis Longus, Flexor Carpi Radialis

On the radial side of the median nerve above the wrist in a more posterior plane (that of the flexor digitorum profundus) is the tendon of the flexor pollicis longus; this tendon acquires a synovial tendon sheath just above the proximal edge of the flexor retinaculum, and runs, in this sheath, behind the retinaculum into the hand. A communication between the tendon sheath of the flexor pollicis longus and the adjacent common tendon sheath of the flexor superficialis and profundus behind the flexor retinaculum is common; because of the relationships of these two tendon sheaths within the carpal canal the sheath of the flexor pollicis longus is sometimes referred to as the "radial bursa," that of the common flexor tendons as the "ulnar bursa."

Almost directly in front of the tendon of the flexor pollicis longus, and so covering the latter tendon that it cannot be palpated, is the tendon of the flexor carpi radialis. This tendon passes superficial to the uppermost fibers of the flexor retinaculum (attached to the scaphoid), and then dives into a special compartment in the attachment of the retinaculum to the trapezium, where it lies in the groove on the trapezium and separates the radial attachment of the retinaculum into two parts; from an anterolateral position in the carpal canal it quickly attains the floor of the canal, to reach its attachment on the base of the second metacarpal. In contrast to the long flexor tendons of the digits, the tendon of the flexor carpi radialis thus does not lie in the compartment behind the flexor retinaculum, but lies at first external to the retinaculum, and then between the two parts of its radial attachment and the trapezium (Fig. 6-32). Just before or as it pierces the retinaculum it enters a tendon sheath that accompanies it almost to the insertion on the metacarpal.

Radial Artery

Lateral to the tendon of the flexor carpi radialis, above the wrist, is the radial artery with its two venae comitantes, lying superficially and here palpable against the underlying pronator quadratus muscle and the radius. However, this artery is not associated with the flexor retinaculum in any way, for above the level of the retinaculum it winds laterally and backward deep to the tendons of the abductor pollicis longus and extensor pollicis brevis to lie subcutaneously in the triangular area (anatomic snuff box) between these tendons and the tendon of the extensor pollicis longus, on the radial side of the wrist (see Fig. 6-86), and pass onto the dorsum of the hand deep to the latter tendon. Rarely, it has a superficial course to the dorsum. Its further course will be described later.

Before the radial artery leaves the anterior surface of the wrist, it gives off a small palmar carpal branch which, with similar branches of the ulnar and anterior interosseous arteries and recurrent branches from the deep palmar arch, supplies the tissue of the floor of the carpal canal and the carpal bones. At about the same level, the radial artery also gives off its superficial palmar branch, which passes into or sometimes over the muscles of the thumb on the lateral side of the prominence of the scaphoid; it helps to supply these muscles, and may anastomose with the ulnar artery in the palm of the hand to complete the superficial palmar arch.

PALMAR APONEUROSIS

The palmar aponeurosis (Figs. 6-1 and 6-34) represents a special thickening of the central portion of the superficial layer of the deep fascia of the hand. Superficially, it is continuous laterally and medially with the thinner deep fascia over the muscles of the thenar and hypothenar eminences, and deeply, medial to muscles of the thenar eminence and lateral to those of the hypothenar, with septa that pass dorsally to attach to the first and fifth metacarpals, respectively. There are also deep connections between the palmar aponeurosis and all four metacarpals in the distal part of the hand.

The palmar aponeurosis is a very strong, somewhat triangular membrane that occupies the central part of the palm and expands from a blunted apex at the distal edge of the flexor retinaculum to a serrated border at about the level of the heads of the metacarpals. A majority of its fibers run longitudinally, and are in part continuous with the tendon of the palmaris longus when that is present, but are for the most part attached to the distal border of the flexor retinaculum; deep to the longitudinal fibers, where these begin to sort themselves into longitudinal fasciculi for the four fingers, there appear heavy transverse fibers, the *transverse fasciculi*. According to Kaplan ('38), the deeper longitudinal fibers that arise from the flexor retinaculum have actually not so much a longitudinal

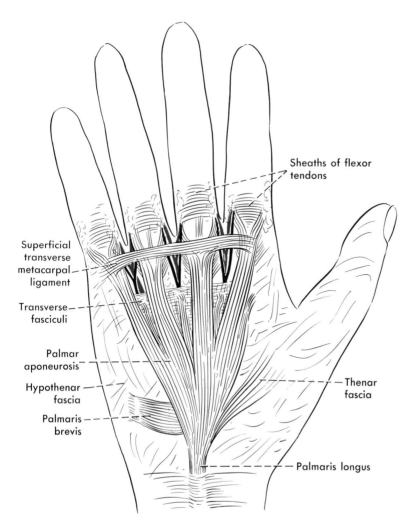

Fig. 6-34. Diagram of the palmar aponeurosis. Digital arteries (*red*) and digital nerves (*black*) appear between the aponeurotic digitations in the distal part of the palm.

as an oblique direction: the fibers arising from the lateral side of the retinaculum, he said, run toward the metacarpophalangeal joint of the little finger, while those from the medial side of the retinaculum run toward the meta-carpophalangeal joint of the index finger, so that an X is formed.

The palmar aponeurosis gives off from its superficial surface fibers that attach into the skin, these attachments being particularly marked at the palmar and digital creases; as the region of the heads of the metacarpal bones is reached, the transverse fibers cease, but the longitudinal fibers run forward to-

ward each of the four fingers. The three sets of digital nerves and vessels to the adjacent sides of the four fingers appear in the fatty connective tissue between the longitudinal slips to the fingers. As these diverging slips reach the region of the interdigital webs, they are again united to each other by less aponeurotic, superficially lying, transverse fibers, which attach to the front of the fasciculi and pass in front of the digital nerves and vessels between the fasciculi. These fibers constitute the *superficial transverse metacarpal ligament* (natatory ligament, superficial transverse ligament of the fingers, interdigital ligament), and are quite distinct from the transverse fasciculi of the palmar aponeurosis, already mentioned.

The distal attachments of the palmar aponeurosis are usually regarded as of particular importance because of their relationship to the flexor tendons of the fingers and the involvement of the palmar fascia in Dupuytren's contracture.

As the four longitudinal fasciculi to the fingers pass distally they are closely applied to the long flexor tendons; some of the more superficial fibers, often termed the pretendinous bands, continue distally in front of the tendons onto the fingers, but the major portion of each digitation splits into two bands which are joined by fibers from the transverse fasciculi (Bojsen-Møller and Schmidt) and then pass deeply on each side of the pair of flexor tendons associated with that digitation, to attach to the fascia over the interossei and through that to the metacarpal bone; the heaviest fibers attach to the deep transverse metacarpal ligaments. These bands and septa form tunnels about the tendons; between the four tunnels are the three areas of fibrous and fatty tissue in which lie the digital nerves and vessels and the distal parts of the lumbrical muscles (Fig. 6-35). The fibers about the sides of the tendons blend with the fibrous flexor tendon sheaths on the digits, and are often described as attaching to the proximal, and sometimes even the middle, phalanges. Fibers from the deep surface of the pretendinous bands can often be seen attaching to the front of the flexor tendon sheath, but the more superficial fibers turn forward to the skin, especially at the proximal digital creases. Bojsen-Møller and Schmidt have emphasized the importance of the aponeurosis in anchoring the skin to the skeleton of the hand.

Kaplan ('53) expressed the opinion that the palmar aponeurosis normally has no attachment at all to bone and does not continue into the fingers. The septa of the palmar aponeurosis which, according to most descriptions, form the proximal parts of the tunnels for the flexor tendons, and blend distally with the fibrous flexor digital sheaths are, according to him, not parts of the aponeurosis but rather parts of the sheaths; admitting that in Dupuytren's contracture, bands from the palmar aponeurosis can be traced to the metacarpals and onto the fingers, he said such bands are entirely abnormal, and do not represent an exaggeration of the normal attachments of the palmar aponeurosis.

Fig. 6-35. Diagrammatic cross section of the hand close to the metacarpophalangeal joints, showing the fibrous tunnels formed about the flexor tendons by the palmar aponeurosis (*red*).

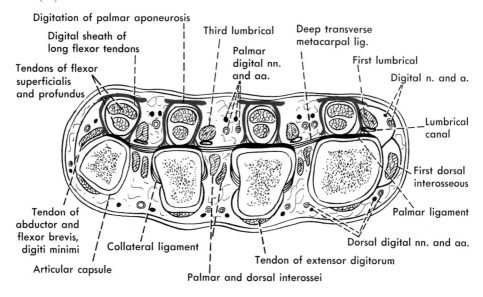

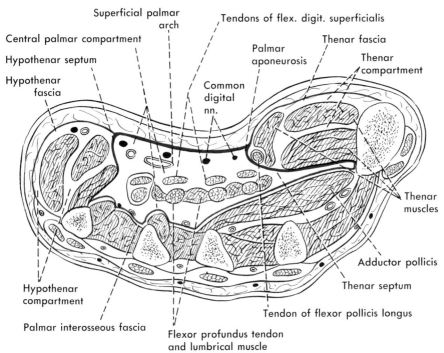

Fig. 6-36. Palmar aponeurosis and intermuscular septa in a cross section of the palm. These, the thenar and hypothenar fascia, and the fascia that helps to bound the central compartment of the palm are *red*.

From its lateral border the palmar aponeurosis gives off a variably developed slip toward the metacarpophalangeal joint of the thumb; this slip, even when well developed, usually blends with the thenar fascia proximal to the metacarpophalangeal joint; it may, rarely, proceed far enough out to contribute to the formation of the sheath of the long flexor tendon of the thumb (Harper).

Compartments of the Palm

Superficially, the palmar aponeurosis is continuous with the less dense fascias that cover the musculature of the thenar and hypothenar eminences, and through these the deep fascia of the palm becomes continuous with the deep fascia of the dorsum. Along the medial border of the thenar eminence, at the junction of the palmar aponeurosis and thenar fascia, there arises a septum from the deep surface; this septum, usually called the thenar or the lateral intermuscular septum, separates the muscles of the thenar eminence

from the long flexor tendons of the fingers, and passes between the muscles of the thenar eminence and the adductor pollicis to an attachment upon the first metacarpal (Fig. 6-36). Similarly, at the lateral border of the hypothenar eminence, where the palmar aponeurosis becomes continuous with the hypothenar fascia, another septum, the hypothenar or medial intermuscular septum, leaves the deep surface to run lateral to the hypothenar muscles and to attach to the fifth metacarpal. The thenar and hypothenar septa are sometimes regarded as derivatives of the palmar aponeurosis, sometimes as continuations of the thenar and hypothenar fascia respectively. They are relatively thin fascial layers, comparable in structure to the thenar and hypothenar fascia, rather than aponeurotic like the palmar aponeurosis.

Through these attachments of the palmar fascia, the palm of the hand is divided into three major compartments: the central (middle) compartment of the hand (not to be confused with the midpalmar fascial space, which lies within the central compartment) lies behind the palmar aponeurosis, and contains the tendons of the superficial and deep flexors of the digits, most of the digital nerves in the palm, the palmar arterial arches, and the two major fascial spaces (midpalmar and thenar); its dorsal boundary is the anterior surfaces of the interossei and of the metacarpal bones. The lateral or thenar compartment, not to be confused with the thenar fascial space in the central compartment of the palm, contains the short muscles of the thumb except for the adductor pollicis (the long flexor tendon of the thumb, and the nerves and vessels to the thumb, lie at first in the central palmar compartment, but then leave this to enter the thenar one); it is subdivided by thin fascial layers that separate the various muscles. The medial or hypothenar compartment contains the short muscles of the little finger.

Dupuytren's Contracture

The distal attachments of the palmar aponeurosis are particularly important in the surgical treatment of Dupuytren's contracture (Figs. 6-37 and 6-38). According to Meyer-ding and co-workers, the pathologic basis of this contracture is an inflammatory reaction with proliferation of capillaries and fibroblasts in the skin and subcutaneous tissue, and with proliferation of fibroblasts, but no other signs of inflammation, among the fibers of the palmar aponeurosis. The proliferation of fibroblasts in the aponeurosis leads to the deposition of additional collagen fibers, and the subsequent contraction of these newly formed fibers presents, in the terminal stages, a picture of heavy avascular scar tissue.

Mason ('52) also said that the disease usually begins as a nodule in the palm of the hand, and that this is commonly in line with the ring finger; following this, a fibrous cord develops in the palm, and as this develops it tends to pull the finger distal to it into flexion, first at the metacarpophalangeal joint and later at the proximal interphalangeal joint. There are differences of opinion as to whether or not trauma is a precipitating cause of the condition; Mason regarded the evidence in respect to trauma as being somewhat inconclusive, but said the condition is almost always familial and presumably often hereditary, although the exact hereditary pattern is unknown. The condition has been regarded

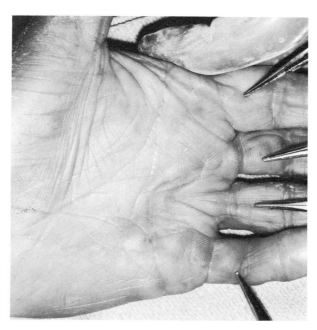

Fig. 6-37. Dupuytren's contracture with attachment of fascia to skin in the palm. The points of the instruments are on contracting bands or nodules in the fingers; the thumb is also involved. (Courtesy of Dr. P.R. Lipscomb)

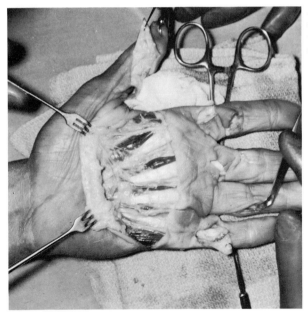

Fig. 6-38. Photograph at time of operation shows the hand seen in Fig. 6–37 after a subtotal fasciectomy. (Courtesy of Dr. P.R. Lipscomb)

as some type of fibroma, or as fibrositis associated with arthritis; Burch suggested that it is an autoimmune disease and analyzed its probable genetic basis.

According to Mason the ring, little, and middle fingers are usually involved in that order, either singly or together, and the index finger and the thumb are more rarely affected. Davis reported that both hands were affected in 58% of 31 patients. Among these, the ring and little fingers were affected in 58%; these plus the middle finger in 13%; the middle and ring fingers only, in 3%; the ring finger only, in 13%; and the little finger only, in 13%.

Davis, in 1932, reported that radical excision of the diseased fascia gave cures in his series in only 38% of cases, and that while improvement was noted in 32% more, the remainder were no better or were even worse. Kaplan, in 1936, suggested that sectioning the longitudinal fibers alone might suffice to cure the condition, but radical and complete removal has usually been regarded as being necessary; for example, both Mason and Richards have emphasized that the removal must include not only all the aponeurosis, whether it is apparently diseased or not, but also any attachments to the fingers and all the

deep vertical sheaths that pass down toward the metacarpals. McFarlane and Jamieson varied their technique, sometimes doing radical and sometimes limited excision of the palmar fascia, or even subcutaneous fasciotomy, but they described their overall results as comparable to those obtained by others who used one technique only.

SUPERFICIAL ARCH, DIGITAL VESSELS AND NERVES IN PALM

Behind the palmar aponeurosis in the central part of the hand there is a layer of connective tissue surrounding the long flexor tendons and their sheaths; this is often not named but, when it is, it is called the deep palmar membrane (Grodinsky and Holyoke), or the subaponeurotic fascia (Flynn, '42). The tissue in front of the long flexor tendons may contain an appreciable amount of fat, and regularly contains the superficial palmar arterial arch and its branches, the median nerve and its branches, and the superficial branch of the ulnar nerve (Fig. 6-39). These vessels and nerves lie at first entirely in front of the long flexor tendons, but as these tendons diverge toward the digits the fascial layers superficial and deep to the tendons come together between them, and the digital branches of the nerves and vessels sink down into the common layer between the tendons, to assume positions alongside the tendons as the fingers are approached.

SUPERFICIAL PALMAR ARCH

The superficial palmar arch is the chief branch of the ulnar artery in the hand. Distal to the pisiform bone, over the origin of the hypothenar muscles, the ulnar artery gives off its small deep palmar branch, which passes

Fig. 6-39. Superficial palmar arch and digital arteries and nerves. In this case there is a communication between the median and ulnar nerves just after they enter the palm.

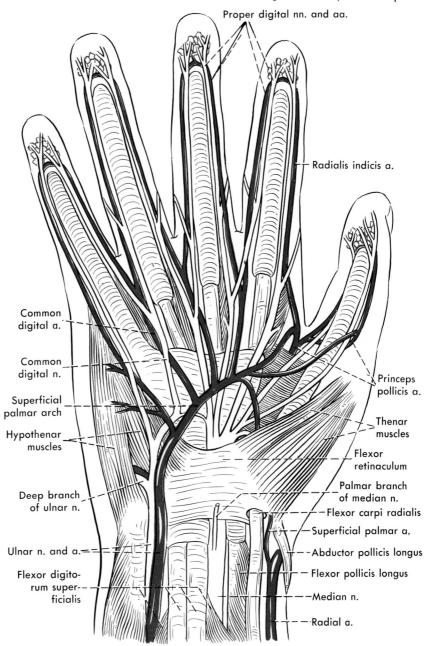

Proper digital nn. and aa.

Radialis indicis a.

Common digital a.

Common digital n.

Superficial palmar arch

Hypothenar muscles

Deep branch of ulnar n.

Ulnar n. and a.

Flexor digitorum superficialis

Princeps pollicis a.

Thenar muscles

Flexor retinaculum

Palmar branch of median n.

Flexor carpi radialis

Superficial palmar a.

Abductor pollicis longus

Flexor pollicis longus

Median n.

Radial a.

dorsally, usually between the abductor digiti minimi and the flexor digiti minimi, into the muscles of the hypothenar eminence. The continuation of the ulnar artery beyond the origin of the deep palmar branch is the superficial palmar arch.

The superficial arch continues distally across the flexor retinaculum and the origins of the hypothenar muscles from the retinaculum, but deep to the palmaris brevis; distal to the retinaculum it turns laterally and penetrates the septum separating the hypothenar muscles from the central palmar compartment. It then arches across the palm immediately behind the palmar aponeurosis, in front of the branches of the median nerve and the long flexor tendons. The superficial palmar arch has its concavity proximally and lies at about the level of the proximal transverse palmar crease. The arch varies somewhat in shape and position, especially on its radial side, since it may curve smoothly back toward the origins of the thumb muscles, or may run more distally as it is traced to the radial side; this is correlated, in part at least, with the manner in which the arch is completed on the radial side.

In its course in the hand, the arch, which is accompanied by small paired veins that form a superficial venous arch, gives off the proper digital branch to the ulnar side of the little finger while it is under cover of the palmaris brevis muscle; as it lies behind the palmar aponeurosis it gives rise to three common digital arteries; beyond the origin of the common digital to the index and middle fingers the arch often sends a branch to the interspace between the thumb and index finger and commonly thereafter curves into the thumb muscles, where it may either end or anastomose with the superficial palmar branch of the radial.

The superficial palmar arch or its branches give off a number of small cutaneous and muscular twigs, but its main branches are the four palmar digital arteries. The first digital artery given off is the proper palmar digital one to the ulnar side of the little finger, given off from the arch while it is under cover of the palmaris brevis, and before it has entered the

central compartment of the hand. This artery runs distally on the hypothenar muscles, accompanied by the branch of the ulnar nerve to this side of the little finger. The artery lies at first medial to the nerve, but before it reaches the digit crosses behind the nerve and maintains a position behind the nerve as it runs along the digit.

The three remaining palmar digital arteries are common digital ones, destined for the adjacent sides of the four fingers. (Coleman and Anson found one so frequently to the adjacent sides of the thumb and index, as in Figure 6-40B and D, that they regarded four common digitals as the normal number.) Each of these runs forward toward the interdigital cleft, at first in a plane anterior to the digital nerve destined for these clefts. As the common digital nerves divide into their proper digital branches, the common digital arteries come to lie between these branches and on a plane slightly dorsal to them; here the digital nerves and vessels emerge from behind the transverse fasciculi of the palmar aponeurosis, and lie in the loose connective tissue between the tunnels formed by the palmar aponeurosis about the long flexor tendons. At about the level of the heads of the metacarpals each common digital artery is usually joined by a palmar metacarpal artery from the deep palmar arch; shortly thereafter, at about the level of the base of the proximal phalanx, it divides into its two proper digital arteries for the adjacent sides of two fingers. These pass onto the fingers, behind the corresponding nerves, and on the sides of the sheaths of the long flexor tendons; their course and branches on the finger have already been described.

Variations in the Arch

The superficial palmar arch varies much in its connections and branches. Both Weathersby ('54) and Coleman and Anson agreed that it is completed by the superficial branch of the radial in about 35% of hands, and Coleman and Anson found it ending in the thumb muscles in another 37%. In only 12% in Weathersby's series, however, was there no connection of the arch to branches of the radial artery on the radial side (Fig. 6-40A);

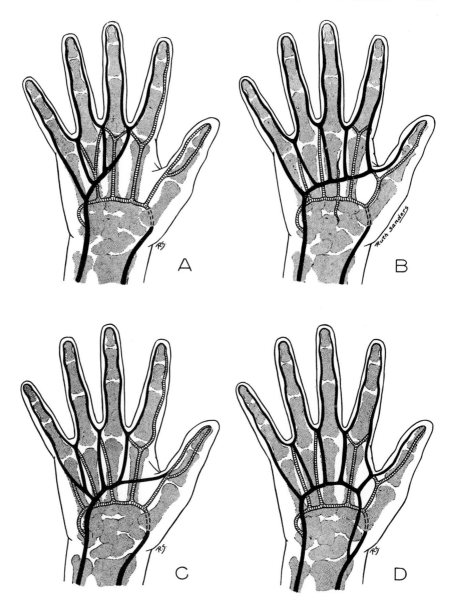

Fig. 6-40. Some variations in the superficial palmar arch: an incomplete arch in *A*, one connected to the princeps pollicis and radialis indicis in *B*, one connected to the princeps pollicis in *C*, and one to the superficial palmar branch of the radial and the stem of the princeps pollicis and radialis indicis in *D*. (*B* is from the original of Fig. 3, Weathersby HT: Anat Rec 122:57, 1955; *A*, *C*, and *D*, courtesy of Dr. Weathersby)

sometimes it was with a branch of the princeps pollicis alone (Fig. 6-40C), but most often it was with the radialis indicis or that and the princeps pollicis (Fig. 6-40B and D), this being the connection that Coleman and Anson regarded as the first common digital artery. In 10% of Weathersby's ('55) series

and in 18% of Coleman and Anson's, the first dorsal metacarpal artery passed between the thumb and forefinger to join the superficial arch or its branches. Weathersby found a persistent median artery joining the arch in 10%; Coleman and Anson found an arch formed by the median and ulnar, or an arch receiving a

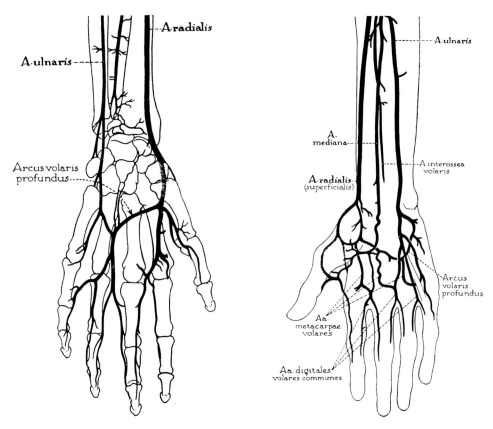

Fig. 6-41. Absence of the superficial arch, and an incomplete arch in which digital arteries arise from both ulnar and median arteries. (McCormack LJ, Caudwell EW, Anson BJ: Surg Gynecol Obstet 96:43, 1953 [by permission of Surgery, Gynecology & Obstetrics])

significant contribution from a median artery, in 5%. The arch is rarely completely absent (Fig. 6-41); Coleman and Anson, moreover, listed the arch as incomplete when, as in Figure 6-40A, it did not send a branch to the thumb and index finger; more than half of their 21.5% of incomplete arches were of this type.

From the foregoing, it is obvious that the most frequent variation of the branches of the arch are on its radial side, where it may or may not send a branch to the radial side of the index finger or this and the thumb. Weathersby ('55) found this to be the only origin of the radialis indicis in 13% of hands. Common digital branches to any of the other interspaces may, however, be small or absent, with the proper digital arteries then deriving their blood supply from the deep arch through the corresponding metacarpal artery.

Palmar Branches of the Radial Artery

The palmar branches of the radial artery are typically derived from this artery both while it lies on the anterior surface of the wrist and after it reaches the dorsum of the hand. The first of these branches is the superficial palmar branch of the radial, variable in size, which arises from the radial before it starts a backward course around the wrist, and descends into the thumb muscles. As already noted, it may end in these muscles, fairly frequently continues through them to join the superficial palmar arch, and occasionally (3.2% in Coleman and Anson's series) continues into the palm to supply one or more of the radial digits without having a connection to the arch.

The other branches of the radial artery in the palm are derived from the deep palmar arch, described later. Of these branches, the metacarpal arteries (Fig. 6-42) typically end by turning forward, proximal to the heads of the metacarpal bones, to join the common digital arteries, but if the latter are small or missing the metacarpal arteries themselves divide and continue as proper digital arteries. The other important branches are the princeps pollicis and the radialis indicis; both appear on the digits, the branch of the princeps

to the radial side of the thumb passing behind the tendon of the flexor pollicis longus to attain its definitive position dorsal to the digital nerve here; both may be connected to the superficial arch, and the radialis indicis may arise entirely from that arch, but otherwise they have a deep course until they reach the digits. They are described further with the deep arch.

Fig. 6-42. Diagram of the arteries of the palm.

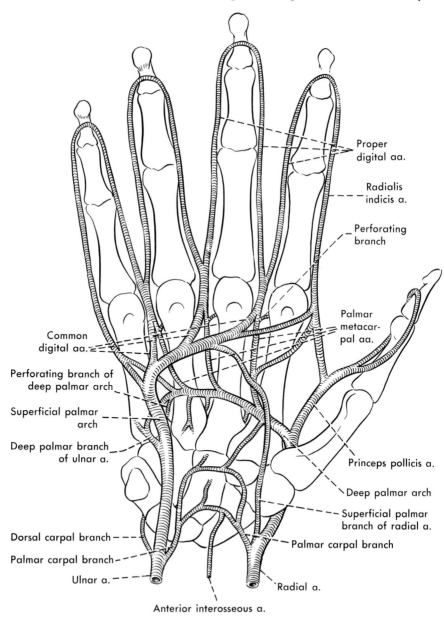

MEDIAN NERVE

As the median nerve emerges from behind the lower edge of the flexor retinaculum, it and its branches (except for the palmar cutaneous branch, given off above the wrist and running subcutaneously) lie immediately behind the palmar aponeurosis. At about the level of the distal border of the flexor retinaculum the median nerve breaks up into its terminal branches, which are its muscular branch into the thenar muscles and, variably, stems that give rise to a common digital nerve for the thumb, a proper digital nerve for the radial side of the index finger, and two common digital nerves for the adjacent sides of index and ring, and ring and middle, fingers, respectively (Fig. 6-39). The exact level of branching of the median nerve varies, for it may be in the distal part of the carpal canal or it may be in the proximal portion of the palm of the hand; similarly, the manner in which the branches arise varies. Sometimes the nerve consists of two large bundles at the wrist (*e.g.,* Ogden), and Wood and Frykman found three in one instance.

Muscular Branch

The branch of the median nerve to the thenar muscles ("recurrent branch") may arise from the anterior or the radial aspect of the median above the level of origin of the other branches, it may arise with the first common digital branch of the median, or it may be one of several simultaneously arising branches of the median. (Sunderland and Ray, in a series of only 20 hands, found the first-mentioned origin in 3, the last-mentioned in 10.) It is a short, relatively stout nerve that turns anteriorly and laterally into the muscles of the thenar eminence, leaving the central compartment of the palm by piercing the septum that the palmar aponeurosis sends down to the first metacarpal bone. (Papathanassiou reported two cases in which the motor branch to the thenar muscles passed through the flexor retinaculum instead of behind and below it; Johnson and Shrewsbury found this in 8 of 10 cases, the branch having its own canal in the distal part of the retinaculum.) Brooks reported a case in which there was no branch from the median nerve into the thenar muscles, and Linburg and Albright reported one in which there were two branches, both of which pierced the retinaculum.

Depending on its origin, the motor branch may curve forward and laterally immediately below the lower border of the flexor retinaculum, or may have no close relationship to this border; wasting of the muscles of the thenar eminence has been attributed to pressure upon the muscular branch by the lower edge of the flexor retinaculum, but is now conceded to be often, if not always, a result of pressure upon the median nerve as a whole while this nerve lies in the carpal canal.

After it penetrates the septum between the central palmar compartment and the thenar muscles, the muscular branch of the median nerve typically gives off a branch into the superficial head of the flexor pollicis brevis (by far the larger portion of this muscle), then passes across or through this muscle to send a branch into the abductor pollicis brevis, and end in the opponens (Fig. 6-43). With the addition of the two radial lumbricals, these three thumb muscles normally receive all the muscular branches of the median nerve in the hand; the median nerve thus typically supplies the superficial or major portion of the flexor pollicis brevis, the abductor pollicis brevis, the opponens, and the first two lumbricals, usually described as four and one-half muscles. It is now apparent, however, that there is much more variation in the distribution of the median and ulnar nerves to the muscles of the hand than was formerly realized.

Digital Branches

A common pattern of the terminal division of the median nerve, regardless of the origin of the muscular branch, is for it to divide into a number of branches simultaneously, and Sunderland and Ray found this in 10 of 20 specimens; another is for it to divide first into .2 branches, a medial and a lateral, and they found this in 7 specimens. In the latter case, the lateral branch typically gives off the digital nerves to the thumb and the radial side of the index finger, the other gives off the com-

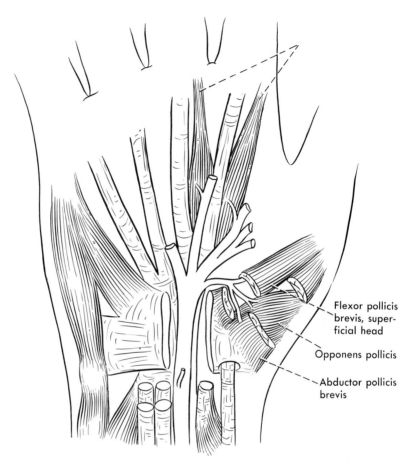

Fig. 6-43. Muscular branches of the median nerve in the palm. The tendons of the superficial flexor and the flexor pollicis, a part of the flexor retinaculum, and the digital branches of the nerve, have been cut away. See also Fig. 6–39.

mon digital branches to the adjacent sides of the index, middle, and ring fingers. Instead of either, the digital nerves may arise in various combinations in which they share common stems; there is, obviously, no fixed pattern of branching of the median nerve. The level of the branching also varies much; Sunderland and Ray gave it as being from 19 mm to 60 mm below the level of the radial styloid in their cases.

As the digital branches of the median nerve pass distally they typically course behind the superficial palmar arch, but then pass slightly palmarward so that they or their proper digital branches lie anterior to the arteries before reaching the digits. Occasionally a digital nerve will split as it passes an artery, allowing

that vessel to pass through it. The proper digital nerve to the radial side of the index finger gives off a twig to the first lumbrical as it passes superficial to this muscle, and the common digital nerve to the adjacent sides of the index and middle fingers similarly supplies the second lumbrical.

The two proper digital nerves to the thumb penetrate the intermuscular septum to enter the thenar compartment and run distally along the thumb, the one to the radial side crossing in front of the tendon of the flexor pollicis longus. The nerve to the radial side of the index finger runs outward along the lateral side of the flexor tendons to that finger, in front of the radialis indicis artery. The common digital nerves to the adjacent sides of the index and middle, and middle and ring, fingers, respectively, pass into the loose connective tissue between the flexor tendons as these diverge toward the fingers, and here divide into proper digital nerves. The common digital arteries, which usually divide later than the nerves, typically appear between the proper digital nerves before dividing to lie dorsal to them; in this position, proper digital nerves and arteries run distally on the sides of the flexor tendon sheaths. The most medial common digital nerve or its proper digital branch to the radial side of the ring finger not infrequently receives a branch of communication from the ulnar nerve.

ULNAR NERVE

As the ulnar nerve passes distally across the superficial surface of the medial border of the flexor retinaculum, in company with (medial to) the ulnar artery, it lies close against the origin of the hypothenar muscles; distal to the pisiform bone the ulnar nerve divides into deep and superficial branches, the deep branch turning dorsally between the heads of origin of the abductor digiti minimi and the flexor digiti minimi, the superficial branch continuing its course downward with the superficial arch toward the palm of the hand (Fig. 6-39). At a variable level, but usually proximal to the distal edge of the flexor retinaculum, the superficial branch of the ulnar nerve divides into two terminal branches, a proper digital branch for the ulnar side of the little finger and a common digital for the adjacent sides of little and ring fingers. This division is usually under cover of or just distal to the palmaris brevis, and the nerve to the palmaris is commonly derived from the proper digital branch to the ulnar side of the little finger—although it may be derived from the superficial branch before this divides or from the undivided ulnar nerve, and Shrewsbury and co-workers found two branches in all of their 10 dissections.

The proper digital branch to the ulnar border of the little finger runs distally on the anterior surface of the flexor digiti minimi, being crossed by the corresponding proper digital artery, which originates lateral to the

nerve but crosses medial to it to assume a position posterior to the nerve on the digit. The branch to the palmaris brevis is normally the only muscular branch of the superficial portion of the ulnar nerve; the common digital branch of the ulnar nerve therefore contains no fibers to voluntary muscles. It passes almost straight distally toward the cleft between the two fingers that it supplies, and is stimilar in its branches and relations to the common digital branches of the median nerve. It may, as already mentioned, have one or more communicating branches with the most medial common digital branch of the median nerve; otherwise its branches are distributed entirely to the little finger and to the radial side of the ring finger.

CUTANEOUS DISTRIBUTION OF THE MEDIAN AND ULNAR NERVES

Except for the usually small communication that the ulnar nerve may send to the median, the palmar digital branches of the median nerve usually are distributed lateral to the midline of the ring finger, those of the ulnar nerve medial to this line: that is, the median has seven digital branches that together supply the palmar skin of three and one-half digits, while the three palmar digital branches of the ulnar nerve typically supply the palmar skin of one and one-half digits. Departures from this, which can be revealed by dissection, are not often reported, in contrast to the well-recognized variability in the dorsal digital nerves. In only 1 hand among 166 did Ming-tzu recognize a variability in the distribution of the palmar digital nerves: in this, each nerve supplied two and one-half digits. Among only 20 hands, however, Linell found 1 in which the ulnar nerve was distributed to both sides of the ring finger, as well as to the little one, and 3 in which the ulnar nerve was distributed to the ulnar side of the middle finger—whether alone or through a communicating branch with the median is not made clear, but the high incidence suggests the latter, for more than three palmar proper digital branches of the ulnar nerve are unusual.

As determined clinically the distribution of

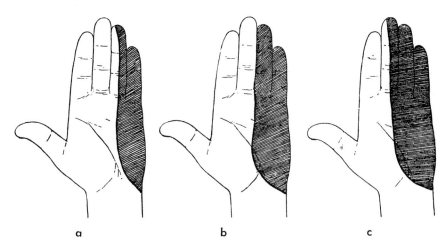

Fig. 6-44. Variations in the distribution of the median and ulnar nerves on the palmar surface of the hand determined on the basis of loss of sensation following complete interruption of the ulnar nerve. The *shaded* areas represent the distribution of the ulnar, the *unshaded* ones, therefore, that of the median. (Stopford JSB: J Anat 53:14, 1918)

the two nerves to the digits varies more than would be supposed from most anatomical observations (Fig. 6-44). This is presumably to be explained, as is anomalous innervation of muscles of the hand, upon aberrant courses of nerve fibers and upon connections between the two nerves in the forearm and hand. Thus Stopford ('18) concluded that the superficial branch of the ulnar nerve supplies exactly one and one-half digits in only about 70% to 80% of cases: in 20% of cases in which the median nerve had been divided there was no anesthesia on the radial side of the ring finger, although there was diminution of sensibility; in 4% there was loss of sensibility on the ulnar side of the ring finger; and in 4% there was loss on the radial side of the little finger. Similarly, after section of the ulnar nerve, the radial side of the ring finger was anesthetic in 14% of cases, the ulnar side of the middle digit was in 3% (Fig. 6-44b and c).

LONG FLEXOR TENDONS AND THEIR SHEATHS

As the long flexor tendons of the fingers enter the carpal canal (carpal tunnel), the four tendons of the flexor digitorum superficialis (sublimis) lie in front of the tendons of the profundus; moreover, the tendons of the su-

perficial flexor are in two layers, for the tendons to the middle and ring fingers lie directly in front of those to the index and little fingers. As the superficial tendons reach the distal end of the carpal canal and diverge in the palm of the hand, the tendons to the index and little fingers appear from behind those of the middle and ring fingers, and all four tendons assume the same plane in the hand. There may be four tendons of the profundus entering the carpal canal, immediately behind the second layer of the superficial tendons, but if so they are all arranged in the same plane, rather than forming two layers as do the superficial tendons; rather commonly, the tendons of the profundus as they enter the carpal canal are only two, a rounded one for the index finger and a broader, flattened one destined for the other three fingers. In that case, the common flexor tendon mass for the three medial fingers divides into three tendons in the canal; distally these tendons fan out in the palm of the hand, toward the digits, and lie exactly dorsal to the four tendons of the superficial flexor.

COMMON FLEXOR TENDON SHEATH

Just before they enter the carpal canal, the tendons of both the superficial and deep flexors enter the upper end of a tendon

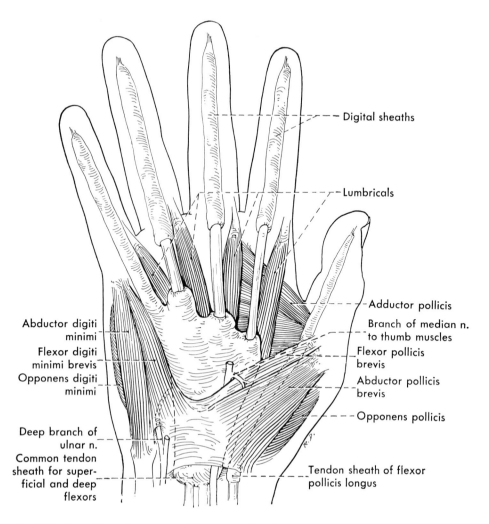

Fig. 6-45. Synovial tendon sheaths of the palm of the hand.

sheath, the common flexor tendon sheath or "ulnar bursa" (Fig. 6-45), the upper end of which extends as a rule about 0.5 cm to 2.5 cm above the upper border of the flexor retinaculum. The walls of the tendon sheath are thin, for they consist only of the parietal synovial membrane, with no special supporting ligaments; however, part of this sheath lies within the carpal canal, and is therefore supported posteriorly by the floor of the canal, and anteriorly by the flexor retinaculum. The median nerve lies on the anterior surface of the sheath, on the radial side, and thus intervenes here between the sheath and the retinaculum; in this position the nerve is embedded in some loose connective tissue, and sometimes invaginates somewhat the anterior wall of the tendon sheath.

On its ulnar side, as well as posteriorly, the common flexor tendon sheath is in contact with the ligaments of the wrist; on its radial side, however, it is immediately adjacent to the tendon sheath of the flexor pollicis longus, and the two sheaths may or may not communicate with each other here. Scheldrup found a communication between them in 85% of 367 hands in which he investigated the pattern of the sheaths, thus a pattern allowing spread of infection from one tendon sheath to the other in the majority of hands; however, he said that spread from the common flexor tendon sheath to the sheath of the long flexor of the thumb has been said to be twice as common as spread in the opposite direction.

Beyond the carpal canal, the major portion of the common flexor tendon sheath continues distally around all eight flexor tendons to about the middle of the palm. It lies deep to the palmar aponeurosis, from which it is separated by the loose connective tissue and fat of the anterior part of the deep palmar membrane or subaponeurotic fascia, in which lie the digital nerves and the superficial palmar arch. Posteriorly, the common flexor tendon sheath and its continuation or continuations to the digits, and the bare areas of the long flexor tendons, rest upon a layer of connective tissue (part of the deep palmar or subaponeurotic fascia) situated behind them and forming the anterior wall of the two chief fascial spaces of the palm. This relationship continues until the tendons to the fingers become bound down to the metacarpals by the septa sent by the palmar aponeurosis around each pair of tendons. The tunnels formed about the tendons by these septa are continuous with the fibro-osseous tunnels on the phalanges, which are lined with the synovial digital tendon sheaths. Between the tunnels for the flexor tendons in the distal part of the palm are the digital nerves and vessels.

Continuity with Digital Sheaths

The common flexor tendon sheath typically ends blindly, at about the middle of the palm, around the flexor tendons to the index, middle, and ring fingers, so that these tendons are bare for a variable distance in the distal part of the palm before they enter the digital tendon sheaths; the tendons to the little finger typically have no bare area in the palm, however, for the common flexor tendon sheath is usually continuous around these tendons with the digital sheath of this finger. Scheldrup found this pattern in 71.4% of 367 hands that he investigated, but seven different patterns in the remainder (Fig. 6-46). Among these hands, the digital tendon sheath of the little finger was continuous with that of the common flexor tendon sheath for a total of 80.8%;

that of the index finger was continuous with the common flexor tendon sheath in 5.1%, that of the middle finger in 4%, and that of the ring finger in 3.5%; and in 17.4% none of the digital sheaths was continuous with the common sheath.

Tenosynovitis of any of the flexor digital sheaths can therefore sometimes spread proximally to involve the common sheath, but Scheldrup said that such spread apparently does not occur in the exact proportion indicated by the anatomical findings. Swelling of the common flexor tendon sheath as a result of infection is usually evidenced above the upper border of the flexor retinaculum. Infections within digital sheaths with blind proximal endings, typically therefore those of the index, middle, and ring fingers, may lead to rupture of their thin proximal ends, with involvement of deep fascial spaces of the hand.

Visceral Layer and Mesotendons

The exact arrangement of the synovial membrane about the tendons of the superficial and deep flexors within their common tendon sheath varies. Usually, the tendons of the flexor digitorum profundus are surrounded by a common layer of visceral synovial membrane, rather than by an individual layer about each tendon, and this common visceral layer is attached by a rather heavy mesotendon to the radial and dorsal aspect of the parietal layer of the tendon sheath (Fig. 6-32). In contrast to the deep ones, the superficial tendons tend to have an individual visceral layer of synovial membrane around each tendon (although they may be enclosed by a common layer), and may or may not have mesotendons. When there is a mesotendon, it attaches to the anterior surface of the visceral layer around the deep flexor tendons; one tendon may have a mesotendon, the next one have none; a mesotendon may run the length of the tendon within the tendon sheath, and may be either a continuous membrane or fenestrated, or mesotendon may be confined to one part of the superficial tendon.

These various arrangements indicate that while the tendons of the deep flexor can receive a blood supply by way of their mesoten-

(*Text continues p. 494.*)

Fig. 6-46. Variations in the connections of the common flexor and digital sheaths. The percentages are those found by Scheldrup. (Scheldrup EW: Surg Gynecol Obstet 93:16, 1951 [by permission of Surgery, Gynecology & Obstetrics])

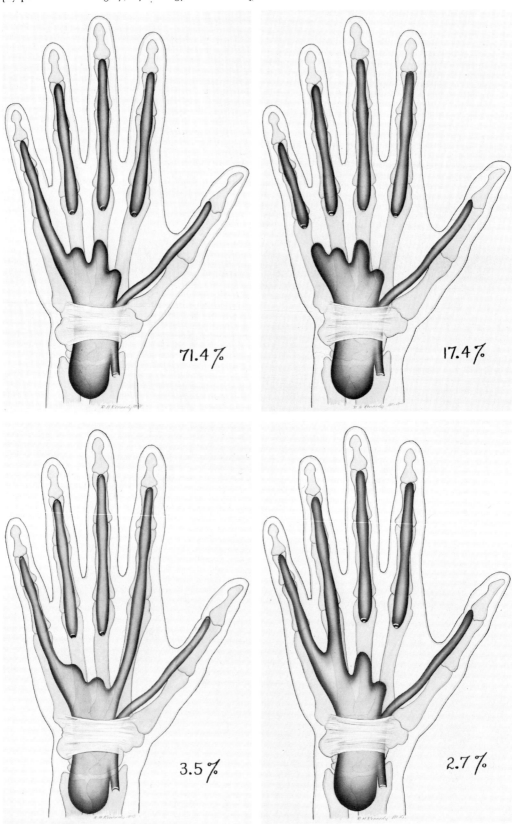

71.4%

17.4%

3.5%

2.7%

Fig. 6-46. (Continued).

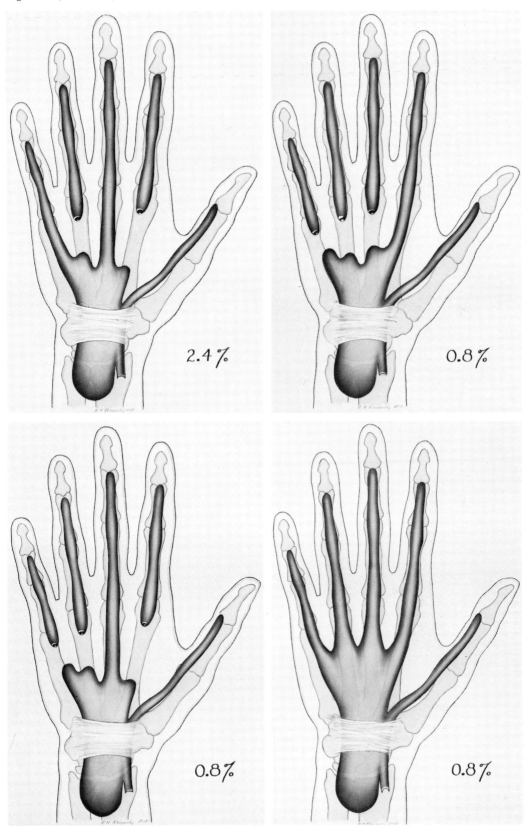

493

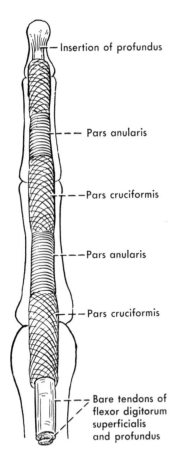

Insertion of profundus

Pars anularis

Pars cruciformis

Pars anularis

Pars cruciformis

Bare tendons of
flexor digitorum
superficialis
and profundus

Fig. 6-47. A digital tendon sheath. The annular parts shown here are the number 2 and 4 parts of Doyle and Blythe (1975).

don, the blood supply of the tendons of the superficial flexor within the common flexor tendon sheath must often descend along the tendon from above the upper border of this sheath, or pass recurrently into the tendon from vessels entering it either through its bare area in the palm or, if a bare area does not exist, through the short vinculum lying at the insertion of the tendon. Brockis found frequent vessels, one artery to two veins, entering the long flexor tendons where they are bare, that is, above the wrist and in the distal part of the palm; he described the flexor profundus tendons in the common flexor tendon sheath as being supplied by vessels that enter both the proximal and distal ends of the sheath, and run through the mesotendon to

course between the visceral layer of the tendon sheath and the tendon tissue itself. He described the vessels to the superficial flexor as continuing into its mesotendons and making arcades here but, as already pointed out, the vessels cannot enter except at the proximal and distal ends of the sheath when mesotendons are lacking.

DIGITAL TENDON SHEATHS

Where they lie in the fibro-osseous tunnels on the fingers, the superficial and deep tendons to each finger are enclosed by the synovial lining of these tunnels. As already noted, the synovial sheath for the tendons of the little finger is typically continuous with the common flexor tendon sheath, while the tendon sheaths for the other three fingers typically are not; they begin slightly proximal to the level of the metacarpophalangeal joints, and continue out along each finger to about the level of the distal interphalangeal joint. The synovial tendon sheaths of the fingers are supported by and fused to the walls of the fibro-osseous tunnels; dorsally, these are the heads of the metacarpals proximally and, distalward, the palmar ligaments and the palmar surfaces of the proximal and middle phalanges. Anteriorly and on the sides the palmar aponeurosis at first protects the synovial sheaths, but this is succeeded on the fingers by special fibrous sheaths that represent thickenings of the outer layer of the digital sheaths. The fibrous sheaths of the digits on the phalanges blend with and are reinforced by the slips of the palmar aponeurosis that pass to the heads of the metacarpals.

The fibrous flexor tendon sheaths of the digits have long been described as divisible into two components, according to whether the sheath is over the body of the phalanx or over a joint (Fig. 6-47). According to this description, each fibrous sheath is thinner in front of the joint and many of the fibers run obliquely either from the radial to the ulnar side of the digit or from the ulnar to the radial side; this part is known as the pars cruciformis. In front of the bodies of the proximal and middle phalanges the fibrous flexor sheath is different, for it takes the form of very heavy bands in which the fibers run transversely, arching across the flexor tendons and attaching to the margins of the phalanges; these parts, each known as a pars anularis, are the pulleys of the flexor tendons on the digits, and serve to keep these tendons applied closely to the bone during flexion of the fingers. Surgeons have for a good many years described a pars annularis or "pulley" proximal to that over the proximal phalanx. According to Doyle and Blythe ('75) this begins about 5 mm proximal to the metacarpophalangeal joint and is attached mostly to the palmar ligament, in small part to the proximal phalanx. These authors also described another or third annular part attached largely to the palmar ligament over the proximal interphalangeal joint, and therefore between the two annular parts shown in Figure 6-47. (They thus recognized only very short cruciform parts.) According to their experiments, the second and fourth annular parts (the traditional ones of anatomists) are most important. Small synovial ganglia sometimes protrude through a proximal part of the fibrous sheath, and if they do not disappear spontaneously can be torn and ruptured by a needle (Bruner).

Infections of the tendons and tendon sheaths on the digits are not particularly common among infections of the hand as a whole; Scott and Jones found only 16 among 1,211 patients treated for infections of the hand, and Robins found only 3 primary ones among 1,000 consecutive cases. Flynn ('43a) said that infection usually enters the tendon sheaths in front of the joints, where the sheath is thin. He, Kanavel ('39), and others have discussed clinical aspects of tenosynovitis of the fingers. This condition may lead to involvement of the common flexor tendon sheath or, through rupture, of the deep fascial spaces of the palm. Drainage of a tendon sheath is usually carried out by opening it laterally, the incision passing behind the digital nerves and vessels so as not to interfere with the important palmar branches of the nerves.

Narrowing of the first annular portion of a

fibrous digital tendon sheath, or local swelling of the tendons, is the common cause of "trigger finger." The disproportion between the tendon or tendons and the sheath leads to such a tight fit that while the tendons will slide proximally under the pull of the powerful long flexors, they hang when the less powerful extensors attempt to extend the digit and pull them distally again; thus the digit, after flexion, tends to remain in a flexed position, as if pulling a trigger. The condition can be treated by sectioning the annular part of the sheath through a small incision on the palm.

LONG FLEXOR TENDONS ON THE FINGERS

Within the synovial sheath on the digit the tendon of the superficial flexor lies at first anterior to that of the profundus, just as it does in the palm. While it lies on the proximal phalanx, however, the superficial tendon splits into two equal portions, and as it proceeds to its insertion on the middle phalanx these two parts diverge to allow the profundus tendon to pass through and go to its insertion on the distal phalanx (Fig. 6-48).

As the superficialis tendon divides into its two component bands, each of these bands, as it diverges from the other and passes around the side of the flexor profundus tendon, undergoes a gradual rotation away from the midline: the fibers that are at first most medial in the band gradually become the most palmar portion of it, and with further rotation the most lateral component of the tendon in regard to the midline; similarly, the fibers that form the lateral portion of the tendon at the level of the split become the most medial portion at the insertion of the tendon. This rotation of each band of the superficial tendon through 180° around the profundus tendon forms a tunnel in which the profundus tendon fits exactly. However, since the original medial fibers of the superficial tendon insert not only most laterally but also more proximally on the middle phalanx, while the original lateral fibers insert most medially and continue most distally, the tunnel cannot

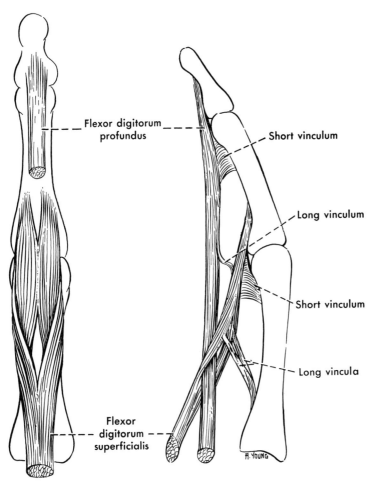

Fig. 6-48. Arrangement of the superficial and deep flexor tendons within a digital tendon sheath. *a* is a palmar, *b* a lateral view.

be closed by pulling upon the tendon, even when the profundus tendon has been removed from this tunnel. In consequence, the deep tendon has a smooth gliding surface on the superficial one, yet contraction of the latter tendon does not grip the deep tendon and hence interfere with its action.

As the two bands of the superficial tendon approach each other behind the profundus tendon, over the distal part of the proximal phalanx and the proximal interphalangeal joint, they are united by synovial membrane and by a few tendon fibers that cross to the opposite side, but nevertheless retain their identity as tendinous bands so that each superficial tendon has a double insertion on the base of the middle phalanx. The profundus tendon continues distally in contact with the middle phalanx, and goes to an insertion on the base of the distal phalanx; the synovial tendon sheath of the digit ends over the head of the middle phalanx.

Vincula and Blood Supply

Within the synovial tendon sheath of each finger the tendons of the superficial and deep flexors are each separately covered by a closely adherent tube of synovial membrane; this visceral layer of membrane is continuous at the ends of the sheath and tendons with the parietal layer by vincula, which represent remains of mesotendons.

The short vincula are relatively thick triangular membranes that unite the terminal parts of the superficial and deep flexor tendons to the dorsal wall of the tendon sheath.

The short vinculum of the superficial flexor tendon arises from the dorsal aspect of the tendon and attaches to the distal end of the proximal phalanx and to the palmar ligament of the proximal interphalangeal joint; that of the profundus tendon similarly arises from the dorsal aspect of this tendon and attaches to the distal half of the middle phalanx, tapering to its apex where the tendon becomes closely applied to the palmar ligament of the distal interphalangeal joint.

The long vincula are placed proximal to the short ones, and are narrow, rather translucent, bands. They vary in number, but there are usually two long vincula from the superficial tendon, one from each of the two parts after it has split; these attach to the proximal phalanx. The profundus tendon usually has a single long vinculum, which may pass between the two bands of the superficial tendon to reach the phalanx, but usually attaches instead to the palmar surface of the synovial membrane uniting the terminal parts of these two bands.

The short vincula are especially important in the blood supply of the superficial and deep flexor tendons within the digital sheaths. The longitudinally running channels, an artery and two veins, on the dorsal aspect of the superficial tendon are derived from the branches that the digital vessels give off to the proximal interphalangeal joint, and reach the tendon by way of the short vinculum; the arteries of the two sides anastomose with each other, and each sends a twig distally along the insertion of the tendon, a larger branch proximally along the tendon slip of its side (Brockis). The deep flexor tendon typically has a blood supply from both ends of its sheath, for a longitudinal channel enters it in the palm and follows its dorsal surface, while branches of the digital arteries to the distal interphalangeal joint reach the tendon through the short vinculum, supply its termination, and run proximally along the tendon to anastomose with the vessel arising in the palm (Fig. 6-49). The long vincula contribute very tiny vessels that connect with the main longitudinally running channels.

Transection of the Long Flexor Tendons

Successful repair with full restoration of function of severed flexor tendons of the hand and digits is notoriously more difficult than similar repair of extensor tendons, being especially complicated by the extensive gliding mechanisms, the synovial sheaths, associated with the flexor tendons, and the limitation of the tendons' blood supply to vessels entering at the ends of the sheaths. As would be expected, repair of flexor tendons severed in the palm has usually been more successful than repair of tendons sectioned at the wrist, within the flexor tendon sheath, or on the fingers within the fibrous digital sheaths (Bunnell, '56, and others); Siler reported "excellent" results (recovery of 80% or more of normal function) in 37% of cases in which tendons severed in the palm of the hand were treated by primary suture, but similar results in only 9% of tendons severed on a finger.

Strictly surgical aspects, such as the type of suture and the placement of sutures, and the very important postoperative care, have been discussed by most of the authors referred to here, but are inappropriate to the present text. Anatomico-clinical problems relate to approximation of the ends of the tendons and

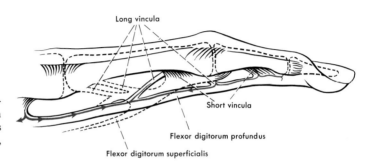

Fig. 6-49. Diagram of the blood supply (*red*) of a profundus tendon on the finger. (Redrawn from Brockis JG: J Bone Joint Surg 35-B:131, 1953)

Long vincula

Short vincula

Flexor digitorum profundus

Flexor digitorum superficialis

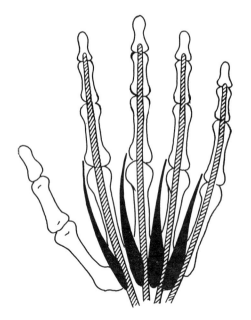

Fig. 6-50. Diagram of the lumbrical muscles (*black*). They are shown here as arising from the flexor profundus tendons (*shaded*) and inserting into the extensor aponeurosis (not shown). (Eyler DL, Markee JE: J Bone Joint Surg 36-A:1, 1954)

to providing the repaired tendons with both a gliding surface and a blood supply if they are severed on the digits. Traditionally, primary suture of tendons has been preferred only when the lesion is not within a digital tendon sheath (the "no man's land" of Bunell). Repair within the digital tendon sheath typically gave poor results because of constriction of the healing tendon by the sheath, adhesions between tendon and sheath, or between the two tendons if the injury was on the proximal phalanx. Thus the sheath was usually cut away over the suture line and fatty tissue placed over the tendon. Since the profundus alone can flex the digit completely, the superficial tendon was usually removed—always if it also was injured.

The development of microsurgery has allowed successful repair in digital sheaths, and Verdan, reviewing the subject in 1972, stated that better results are obtained by repair of both tendons, when both are injured, rather than removal of the superficial one, and that an intact superficial tendon should not be removed. A little later Green and Niebauer·

agreed that primary repair within the sheath can give good results in carefully selected cases, and said that even secondary repairs can be successful; however, while they never removed an intact superficial tendon, they sometimes removed one when one of its two slips was severed, and always removed one that was completely severed.

Providing a gliding surface for a tendon repaired within a digital tendon sheath is extremely important, since the profundus tendons are said to move about 0.75 inch (1.9 cm) over the proximal phalanx in flexing all the phalanges. When tendon grafts are necessary to replace a profundus tendon they usually extend from the base of the hand to the insertion of the tendon, thus placing the suture lines outside the reconstructed tendon sheath. The sheath can be more easily reconstructed over a small tendon, so that of the palmaris longus or plantaris is often preferred (Peacock). Peacock has replaced badly damaged tendons and sheaths with a tendon and complete sheath graft from a cadaver; Chacha has used the tendon and sheath from the big toe; and Urbanisk and co-workers have produced a sheath around a flexible silicone rod, and then replaced the rod with a free tendon graft. Parkes found it useful to section the lumbrical when putting in a profundus graft, just in case the graft should be too long; in this case, the origin of the lumbrical may be shifted proximally enough so that attempts to flex the digit produce extension instead.

LUMBRICAL MUSCLES

The lumbrical muscles arise from the tendons of the flexor digitorum profundus, and are therefore closely associated with the long flexor tendons in the palm of the hand (Figs. 6-45 and 6-50). Like these tendons, they have the palmar aponeurosis and the digital nerves and vessels in front of them. As the lumbrical muscles and the long flexor tendons pass toward the fingers, each lumbrical is closely associated with the radial side of the long flexor tendons of its finger; where the flexor tendons enter the tunnel formed by the septa from the palmar aponeurosis, each lumbrical tendon becomes separated from the flexor tendons by this tissue, and lies on the ulnar side of the interspace in which the digital nerves and vessels also lie (see Fig. 6-52). Proximally, the lumbricals overlie the deep fascial spaces of the palm, as do the long flexor tendons; while these spaces stop distally, in general, at the level at which the long flexor tendons enter their tunnels, a special prolongation extends distally deep to each lumbrical muscle, between it and the deep transverse metacarpal ligament.

There are typically four lumbrical muscles, the first one arising from the radial side of the flexor digitorum profundus tendon to the index finger, the second arising from the profundus tendon to the middle finger, or from that and the tendon to the index finger in some 20% to 30%, the other two arising from both of the profundus tendons between which they lie. As the long flexor tendons pass across the palmar surface of the metacarpal head and proceed out onto the digit, the tendon of the associated lumbrical passes on the radial side of the metacarpal head (and in the case of the medial three, across the palmar surface of the deep transverse metacarpal ligament) and around the radial side of the digit to attach into the extensor aponeurosis distal to the attachment of the interossei into the aponeurosis.

The first two lumbricals (from the radial side) are typically innervated by the median nerve; the nerve twigs enter the superficial surfaces of the muscles from the digital branches of the median nerve that cross them. The third and fourth lumbrical muscles are typically innervated by the deep branch of the ulnar nerve, and their muscular branches therefore enter their deep surfaces.

The lumbricals are primarily extensors of the interphalangeal joints, and participate in this during both flexion and extension of the metacarpophalangeal joints. A secondary action is flexion of the latter joints.

Variations

Braithwaite and his colleagues quoted evidence that more than four lumbricals occur more frequently than less than four and that,

although absence of one is rare, it is the third that is most frequently missing. Among a total of 147 hands studied in the two series of Basu and Hazary and of Mehta and Gardner, however, there were no instances of an additional lumbrical muscle, and in all five cases (in three bodies) in which a lumbrical was absent it was the fourth, not the third.

As brought out especially by Mehta and Gardner, variations in origin or insertion of the lumbricals are very common; they found occasional origin in the forearm or from a metacarpal, origin from the superficial instead of the deep flexor tendons, and origin of the third and fourth lumbricals from a single tendon instead of two. They also reported frequent insertion into a phalanx, in addition to or in place of the usual insertion into the extensor tendon, or even into the flexor tendon sheath. Both they and Basu and Hazary reported fairly numerous instances in which a lumbrical tendon split to go to the adjacent sides of two fingers or, less frequently, inserted entirely on the ulnar side of the adjacent digit.

The third lumbrical is particularly variable in its innervation. Brooks (1887) found this muscle innervated in whole or in part by the median nerve in 12 of 21 hands that he dissected, Sunderland and Ray found it so innervated in 5 of 20 dissected hands, and Mehta and Gardner found it innervated exclusively by the median in two of 75 hands. The ulnar nerve may, but rarely does, invade median nerve territory in regard to the lumbricals, and help supply even the first lumbrical (Mehta and Gardner). Occasionally, the lumbrical branches of the ulnar nerve arise from its superficial rather than its deep branch (Brooks; Mehta and Gardner).

FLEXOR POLLICIS LONGUS

The tendon of the flexor pollicis longus passes through the carpal canal on the radial side of the tendons of the superficial and deep flexors of the digits. About 2.5 cm to 0.5 cm above the proximal border of the flexor retinaculum (transverse carpal ligament) this tendon enters a synovial tendon sheath, which typically surrounds the tendon to about the inter-phalangeal joint of the thumb (Fig. 6-45). The tendon sheath of the flexor pollicis at the wrist is therefore usually continuous with the digital tendon sheath; Grodinsky and Holyoke quoted a lack of continuity in only about 5%. The tendon sheath of the flexor pollicis longus is sometimes known as the radial bursa; as already noted, Scheldrup found a communication between this tendon sheath and that of the flexors of the fingers in 85% of hands.

Until just before it reaches the metacarpophalangeal joint, the tendon sheath about the flexor pollicis longus is entirely synovial; over the proximal phalanx there are, as on the other digits, two layers to the sheath, synovial and fibrous. The fibrous digital sheath is similar in composition to the fibrous digital sheaths of the fingers, but since there is no middle phalanx of the thumb the arrangement of the pulleys is necessarily different. Doyle and Blythe ('77) found three pulleys: an annular one attached to the palmar ligament at the metacarpophalangeal joint; an oblique one beginning on the ulnar side of the base of the proximal phalanx and extending obliquely to the radial side close to the interphalangeal joint; and a second annular one centered over the palmar ligament of the interphalangeal joint. They reported that section of the oblique pulley lessened movement at the metacarpophalangeal joint, and increased it at the interphalangeal joint.

As the tendon of the long flexor of the thumb, surrounded by its tendon sheath, emerges from the carpal canal, it lies behind the origin of the superficial or main head of the flexor pollicis brevis from the flexor retinaculum; if there is a deep head to the flexor brevis (arising from the floor of the carpal canal) the long flexor tendon passes between the two heads. In any case it is at first somewhat under cover of the flexor pollicis brevis as it runs distally, and becomes superficial in front of the metacarpophalangeal joint. Proximal to the joint it is usually crossed superficially by the proper digital nerve to the radial side of the thumb, but the proper digital artery to the radial side of the thumb (from the princeps pollicis) crosses behind the flexor tendon, between this and the bone.

Within its tendon sheath the long flexor tendon may have proximally a more or less well-developed mesotendon; in the digital part of its course it may have a vinculum longum, and regularly has a well-developed short vinculum, these being similar to the vincula connected with the tendons of the fingers. "Trigger thumb," a locking of the thumb in flexion, is essentially similar to "trigger finger," according to Sprecher; he said that at operation the annular or pulley portion of the digital sheath (presumably the oblique pulley of Doyle and Blythe) is found to be thickened and constricted and the tendon shows a fusiform enlargement, these being, apparently, the result of traumatic inflammation.

FASCIAL SPACES

THENAR AND MIDPALMAR SPACES

The thenar and midpalmer fascial spaces are regions of loose connective tissue lying in the central compartment of the hand, therefore behind the common flexor tendon sheath and its tendons, and in front of the adductor pollicis and of the interosseous muscles on the ulnar side of the middle metacarpal (Fig. 6-51). Of clinical importance because they are sometimes the location of abscesses, their exact boundaries are not agreed upon—a usual finding in descriptions of "fascial spaces," which are, after all, difficult to define by their very nature, since their walls are often also rather delicate laminae of connective tissue.

A valid and understandable concept of the fascial spaces of the palm is that they arise as a result of splitting of the connective tissue deep to the common flexor tendon sheath and the flexor tendons and lumbricals (Anson and Ashley). This connective tissue is actually a part of the rather loose, fat-containing tissue, often called the subaponeurotic fascia or the deep palmar membrane, which intervenes between the palmar aponeurosis and the heavier fascia on the anterior surface of the adductor pollicis and the interossei and

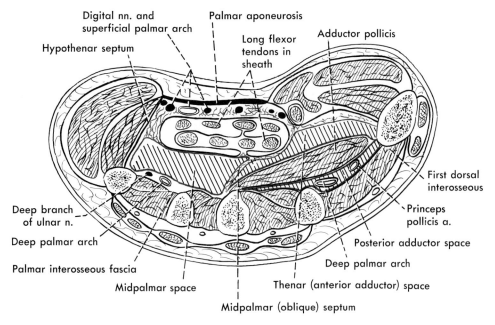

Digital nn. and
superficial palmar arch

Palmar aponeurosis

Long flexor
tendons in
sheath

Adductor pollicis

Hypothenar septum

First dorsal
interosseous

Deep branch
of ulnar n.

Princeps
pollicis a.

Deep palmar arch

Posterior adductor space

Palmar interosseous fascia

Deep palmar arch

Midpalmar space

Thenar (anterior adductor) space

Midpalmar (oblique) septum

a

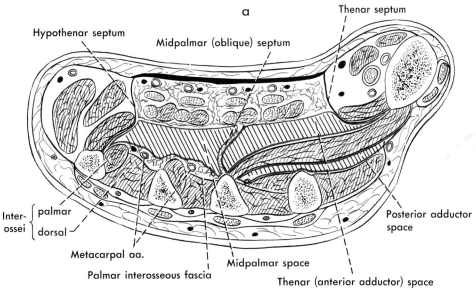

Thenar septum

Hypothenar septum

Midpalmar (oblique) septum

Inter-
ossei

palmar

dorsal

Posterior adductor
space

Metacarpal aa.

Palmar interosseous fascia

Midpalmar space

Thenar (anterior adductor) space

b

Fig. 6-51. Diagrammatic cross sections through proximal, *a*, and distal, *b*, parts of the hand to show the palmar fascial spacies (shaded). They are here exaggerated. The connective tissue in which they develop is shown in *red*.

metacarpals; through the flexor tendons and the splitting behind the tendons it is divided, in much of the palm, into three layers: a superficial one, immediately behind the palmar aponeurosis and in front of the common flexor tendon sheath and the flexor tendons, which contains the superficial palmar arch and the proximal portions of the digital nerves and vessels; an intermediate lamina, immediately behind the flexor tendon sheath and tendons; and a deep lamina, in front of the fascia of the adductor and interosseous

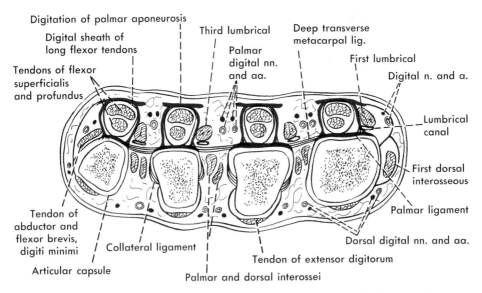

Fig. 6-52. Cross section of the hand at about the level of the metacarpophalangeal joints. The lumbrical canals, forward extensions of the fascial spaces, appear between the lumbrical muscles and the deep transverse metacarpal ligaments.

muscles. The latter two laminae form the immediate anterior and posterior walls of the fascial spaces of the palm. The anterior of these two layers is thin, and supported by the flexor tendons in front of it; the posterior is also flimsy, and rests upon the adductor, and upon the interossei of the ulnar side of the hand, but usually contains fat where it covers the interossei. The looser connective tissue between, not in the form of laminae since it is not applied to a surface, forms the fascial spaces.

All three laminae of this connective tissue come together distally between the flexor tendons, thus becoming continuous with the connective tissue in the webs of the fingers. Deep to the long flexor tendons the intermediate and deep layers come together as these tendons are surrounded by the septa that the palmar aponeurosis sends to the metacarpals, but deep to the lumbricals they remain separate over the deep transverse metacarpal ligaments, thus forming the lumbrical canals (Fig. 6-52); these canals are distal extensions of the palmar spaces that are otherwise obliterated at about the level of the distal palmar crease. According to Anson and Ashley, and Grodinsky and Holyoke, the septa about the

long flexor tendons extend proximally for a variable distance, subdividing the distal part of the midpalmar space. The intermediate and deep laminae also come together proximally, at or in the carpal canal, and are continuous with each other at the margins of the central palmar compartment, where the more delicate connective tissue of the subaponeurotic fascia lines the inner surfaces of the thenar and hypothenar septa.

Midpalmar Septum

On clinical grounds, Kanavel first described two fascial spaces in the palm of the hand, separated by a septum attaching to the third metacarpal: the one on the ulnar side of this, in front of the second and third palmar and the third and fourth dorsal interossei, is called the midpalmar space; the one on the radial side, in front of the adductor pollicis, is usually called the thenar space (anterior adductor is a better term, since the space is not among the muscles of the thenar eminence). Kanavel described the septum between the two spaces as arising behind the flexor tendons of the index finger and coursing obliquely posteriorly to its attachment on the third metacarpal; thus the midpalmar space extends radially in front of the thenar space, and the septum between the two is usually termed the midpalmar septum or the oblique septum. Accepting Kanavel's description, and disregarding the delicate connective tissue laminae behind the long flexor tendons and in front of the fascia on the anterior surfaces of the adductor pollicis and the interossei, the boundaries of the thenar space (Fig. 6-51) are, laterally, the lateral or thenar intermuscular septum; posteriorly, the adductor pollicis and its covering fascia; medially, the midpalmar or oblique septum; and, anteriorly, this same septum on the medial side, the tendons of the index finger on the radial side. Similarly, the boundaries of the midpalmar space are, anteriorly, the tendons of the three ulnar digits; medially, the medial or hypothenar intermuscular septum; posteriorly, from the ulnar side, the fourth dorsal and third palmar interosseous muscles, the fourth metacarpal bone, the third dorsal and second palmar interossei, and the ulnar side of the third metacarpal; and posteriorly and laterally, the midpalmer or oblique septum.

Flynn ('42, '43b), on the basis of dissections of normal hands and examinations of infected ones, modified this description slightly: he described the midpalmar septum as passing to the third metacarpal from the radial side of the tendons to the middle finger, and said these tendons are therefore related to both spaces, and that infections of their tendon sheath may break into either; among four such infections that had involved the palmar spaces, he found the thenar space involved in two and the midpalmar space in two.

The midpalmar and thenar spaces, and especially the midpalmar or oblique septum, can often be demonstrated by dissection, and the septum can be seen by simply lifting the flexor tendons, as Flynn illustrated. However, there are other interpretations of the septum and of the thenar space: Grodinsky and Holyoke described the septum as attaching on the radial side of the tendons to the index finger, and said that all four sets of tendons are related to the midpalmar space; the thenar space, as they described it, sometimes lies directly against the adductor pollicis muscle.

It is not clear, from their description and figures, that they recognized the usually described midpalmar or oblique septum. Anson and Ashley's interpretation of the thenar space and of the septum is very similar to that of Grodinsky and Holyoke; however, they also described and illustrated an "intermediate septum" which seems to correspond to the midpalmar or oblique septum of other authors.

The midpalmar and thenar spaces, as defined here, have been said to communicate freely with each other, sometimes, because the midpalmar septum is deficient proximally. Flynn, however, demonstrated a complete septum, extending at least as far proximally as the distal edge of the flexor retinaculum (transverse carpal ligament) and sometimes higher, in all of 100 hands that he dissected, and concluded that the two spaces never communicate; further, among 100 infections of the deep spaces of the hand, there were only two in which, he said, an infection in one space had spread to the other, and this had occurred because of necrosis of the septum. According to Flynn, the thenar space stops proximally at the distal edge of the flexor retinaculum, but the midpalmar space extends upward into the carpal canal. Through the connective tissue between the posterior walls of the flexor tendon sheaths and the floor of the carpal canal, continuity can be established between the palmar fascial spaces and the deep fascial space of the forearm (Parona's space, between the deep flexor tendons and the pronator quadratus); the space in the carpal canal not occupied by the tendons is very limited, however, and Flynn found that infections of the fascial spaces of the palm do not tend to spread into the canal.

A very different interpretation of the fascial spaces has been offered by Bojsen-Møller and Schmidt. They regarded all the connective tissue in the central compartment between the palmar aponeurosis and the fascia over the palmar surfaces of the interossei and the adductor pollicis, thus encompassing the entire central compartment, as being a fascial space. They therefore described it as single proximally, where it contains, for instance, the superficial palmar arch, the diverging branches of the median nerve, and the flexor tendons, but subdivided distally by the septa that the palmar aponeurosis sends around the flexor tendons into four extensions along these tendons, four extensions for the lumbricals, and three, separated by thin septa from the lumbrical canals, for the digital nerves and vessels. They may of course be correct in their interpretation that it is primarily the para-tendinous septa that prevent spread from one side of the hand to the other, but their defini-tion of the fascial space as consisting of all the loose connective tissue around the flexor ten-dons allows no comparison with the descrip-tions of the midpalmar and thenar spaces.

POSTERIOR ADDUCTOR AND POSTERIOR INTEROSSEOUS SPACES

The attachment of the connective tissue of the radial wall of the thenar space to the dis-tal edge of the adductor pollicis in the web between thumb and forefinger is said (Lan-non) to separate the thenar space from an-other potential space lying behind the adduc-tor pollicis muscle, between this and the first dorsal interosseous. Lannon called this the posterior adductor space (it has also been called the accessory thenar space) and de-scribed the fascia at the free edge of the ad-ductor as being firmly attached to the muscle before continuing back around the first dorsal interosseous. He regarded this attachment as preventing spread of infection between the thenar and posterior adductor spaces, and vice versa, around the muscle.

Lannon also emphasized an additional space, best mentioned here even though it is a dorsal rather than a palmar space, as being of particular importance in infections of the hand; this is the potential space dorsal to the first dorsal interosseous muscle, between this and the part of the deep dorsal fascia that stretches between the first and second meta-carpal bones (recognizable, but not labeled, in Fig. 6-51a). Lannon called this the poste-rior interosseous space.

He described the posterior adductor space as being completely closed except around the

radial artery, along which infection can ex-tend dorsally into the posterior interosseous space, or anteriorly into the thenar space; among 10 infections, however, he found only 1 in which there had been spread between any of these 3 spaces, and in 8 injections into the posterior adductor space he failed to get spread into other spaces. He said the posterior interosseous space is continuous with the sub-fascial tissues of the thumb and index finger, and therefore infections in this tissue can spread proximally and involve the space.

INFECTIONS OF PALMAR SPACES

Infections of the deep spaces of the hand can be particularly serious because of the relation of these spaces to the long flexor tendons and to the bones and intrinsic muscles of the palm; fortunately, they have in the past formed only a small percentage of all infec-tions of the hand, and are at present of di-minishing clinical importance, like infections of other fascial spaces, because of the antibiot-ics. Robins found only 6 cases of deep palmar infection among 1,000 cases of infections of the hand, and Scott and Jones found only 17 in 1,211. Pemberton regarded them as being usually secondary to infection elsewhere, but Flynn ('42, '43b) said they most commonly originate from deep puncture wounds. Ap-parently first emphasized by Kanavel, they are discussed in greater or less detail in all surgical works on the hand (e.g., Kanavel; Bunnell, '56).

Secondary infections of the deep spaces may result from spread of subcutaneous in-fections of a digit or an interdigital web, from rupture of an infected flexor tendon sheath of the wrist, or from rupture of an infected digi-tal tendon sheath, most commonly of the index, middle, or ring finger. Infections of the tendon sheath of the little finger or of the long flexor of the thumb are most likely to spread within the tendon sheaths to the wrist, since both of these digital sheaths normally are continuous with those at the wrist. The me-chanical reason favoring rupture of a digital tendon sheath into a deep palmar space is very apparent: the blind proximal end of the sheath is thin, and when it ruptures spread occurs more easily in the loose connective tissue deep to the flexor tendons than in the more limited connective tissue between the tendons and the dense palmar aponeurosis.

In line with the fact that perforating wounds on the thumb side of the hand are more common than wounds on the ulnar side is the finding that infection of the thenar (anterior adductor) space has been more common than infection of the midpalmar space. Among the 100 cases of infection of one of these spaces that Flynn collected there were 70 in which the thenar space was said to have been involved, 30 in which the midpalmar space was said to have been involved; of Scott and Jones' 17 cases, the infection was said to be in the thenar space in 13 and in the midpalmar space in 4; and of Robins' 6 cases, the infection was said to be in the thenar space in 5 and in the midpalmar in 1. Lannon agreed that infections of the deep spaces are more common on the radial side, but said that in his experience they were more often posterior to the adductor pollicis muscle than anterior to it (in the posterior adductor rather than in the thenar or anterior adductor space).

With involvement of either of the chief palmar spaces one of the chief symptoms is tenderness over the palm, according to Flynn; if the thenar space is involved there is ballooning of the tissue between the thumb and the radial longitudinal crease of the palm, while if the midpalmar space is involved there is obliteration of the concavity of the palm. There is usually extensive swelling over the dorsum of the hand in both instances. With involvement of the thenar space, also, the index finger may be flexed, while the middle and ring fingers may be flexed when the midpalmar space is involved.

Lannon said that the clinical signs of infection of the dorsal interosseous and posterior adductor spaces are essentially similar, and that a differential diagnosis can be made only at operation. He described a marked ballooning on the dorsum between the thumb and the index finger, far exceeding the dorsal swelling usually associated with infections of the palmar spaces proper. In contrast to the condition in infections of the thenar space, the index finger can be extended either actively or passively without pain, he said, and there is no tenderness over the thenar space when it is palpated from the palmar aspect; rather, the tenderness is over the posterior adductor space, the maximum being at the distal border of the interosseous muscle.

Surgical drainage of the midpalmar and thenar spaces, discussed by many of the authors cited, was originally through palmar incisions, special care having to be taken, in opening the thenar space, to avoid damaging the branch of the median nerve to the thenar muscles. The thenar space has also been opened from the dorsum, through longitudinal incisions between the first and second metacarpals or transverse ones on the web. Lannon preferred the last-mentioned approach for opening the posterior adductor or dorsal interosseous spaces, but advocated an incision on the palmar side of the web for approach to the thenar space.

INTEROSSEI

The interossei form the deepest layer of muscles of the palm of the hand, and, as their name implies, lie largely between the metacarpal bones (Fig. 6-53). They are divided into two sets, palmar and dorsal; the palmar interossei, however, fit into concavities on the palmar surface of the dorsal interossei, so that the anterior surfaces of both dorsal and palmar interossei are in part in the same plane. The interossei are covered on both their palmar and dorsal surfaces by relatively strong fascia that bridges the spaces between the metacarpals and represents, in both locations, the deep layer of the deep fascia. The palmar interosseous fascia is continuous, at the first and fifth metacarpals, with the thenar and hypothenar septa, and hence with the superficial layer of the deep fascia. On the third metacarpal it is continuous with the fascia of the adductor pollicis. On the ulnar side of the third metacarpal, therefore, the palmar interosseous fascia forms the posterior wall of the midpalmar fascial space; on the radial

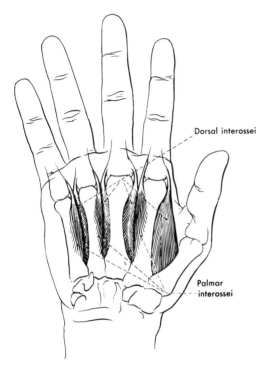

Fig. 6-53. The interossei of the right hand.

side, however, the adductor pollicis muscle intervenes between the posterior wall of the thenar (anterior adductor) space and the interossei, and the interosseous fascia lateral to the third metacarpal forms the posterior wall of the posterior adductor (accessory thenar) space. Proximally, this fascia fuses with the ligaments of the palmar surface of the wrist, while distally it is continuous with the deep transverse metacarpal ligaments; these are usually regarded as special thickenings of the palmar interosseous fascia. In contradistinction to the lumbricals, which send their tendons onto the fingers across the palmar surface of the deep transverse metacarpal ligaments, the interossei therefore sent their tendons posterior or dorsal to these ligaments (but anterior to the axis of motion at the metacarpophalangeal joints).

PALMAR INTEROSSEI

There are three distinct palmar (volar) interossei, although some accounts describe four of them; under the latter system, the first palmar interosseous is a small group of muscle fibers that takes origin from the ulnar side of the first metacarpal and blends with the oblique head of the adductor pollicis to insert with it on the ulnar side of the thumb. The continuity of this slip with the origin of the adductor pollicis from the bases of the second and third metacarpal bones, and its insertion with the adductor, seem to be sufficient reasons for calling it a part of the adductor pollicis rather than a first palmar interosseous; this same slip has been called, by some authors, the deep head of the flexor pollicis brevis. Functionally, of course, the entire adductor pollicis is similar to a palmar interosseous in that it, like the palmar interossei proper, adducts the digit to which it is attached and flexes the proximal phalanx.

The three palmar interossei, as they are numbered here, are associated with the index, ring, and little fingers (Fig. 6-54). Each arises from the more anterior part of the metacarpal body of the digit with which it is associated, in front of the origin of the dorsal interosseous from that same metacarpal, passes dorsal to the deep transverse metacarpal ligament, and inserts primarily into the extensor aponeurosis. Since the palmar interossei adduct, the

Fig. 6-54. Origins (*red*) and insertions of the palmar interossei.

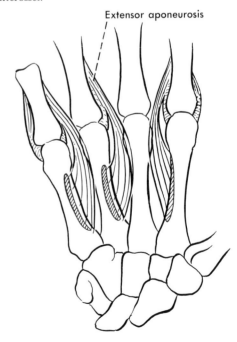

first one arises from the ulnar side of the second metacarpal and passes on the ulnar side of the metacarpophalangeal joint, while the second and third arise from the radial sides of the fourth and fifth metacarpals, respectively, and pass dorsally on the radial sides of the metacarpophalangeal joints. As the tendon of each muscle crosses the joint it is bound to the long extensor tendon by fibers that help to form the extensor hood, and over the proximal phalanx it joins the edge of the extensor tendon and contributes fibers both to the middle band of the extensor aponeurosis and to the lateral band on its side.

In addition to its insertion into the extensor aponeurosis, a palmar interosseous may have an insertion into the proximal phalanx of the digit with which it is associated. The third interosseous occasionally inserts entirely into bone (Salsbury; Eyler and Markee); usually, however, an insertion into the phalanx involves only a very small part of the total insertion of the muscle, and even this occurs so seldom in the case of the first and second palmar interossei that the percentage of the total mass of these muscles that inserted into bone was negligible (Fig. 6-55) in the more than 30 hands investigated by Eyler and Markee.

DORSAL INTEROSSEI

Each of the four dorsal interossei arises from the two metacarpals between which it lies, and the converging heads form a bipennate muscle (Fig. 6-56). Between the proximal ends of the two heads there is a small gap through which an artery passes—the radial artery reaching the palm by passing between the two heads of the first dorsal interosseous, the perforating branches from the deep palmar arch to the dorsal metacarpal arteries between the heads of the remaining three interossei. Since the dorsal interossei abduct, the first one, filling the first intermetatarsal space, inserts on the radial side of the index finger; the second and third insert on the radial and ulnar sides, respectively, of the middle finger; and the fourth inserts on the ulnar side of the ring finger. Their tendons of insertion pass dorsal to the deep transverse metacarpal ligament, and as they pass across the

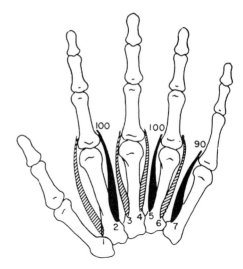

Fig. 6-55. The approximate percentage of insertion of the palmar interossei (*black*) into the extensor aponeurosis; the remaining insertion is into the proximal phalanx. The dorsal interossei are here *shaded*. (Eyler DL, Markee JE: J Bone Joint Surg 36-A:1, 1954)

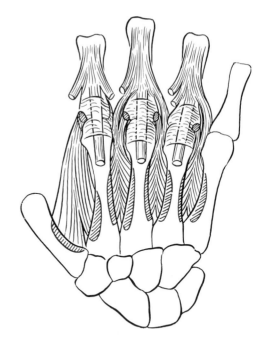

Fig. 6-56. Origins (*red*) and insertions (*bony ones black*) of the dorsal interossei.

metacarpophalangeal joints each tendon tends to divide into a palmar and a dorsal slip. When this occurs, the palmar slip inserts into the base of the proximal phalanx, while the dorsal slip becomes associated with the

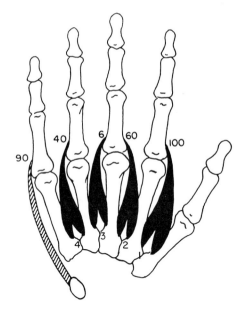

Fig. 6-57. The approximate percentage of insertion of the dorsal interossei (*black*) into bone of the proximal phalanges; the remainder is into the extensor aponeurosis. The large insertion of the abductor digiti minimi (*shaded*) into bone is also indicated. (Eyler DL, Markee JE: J Bone Joint Surg 36-A:1, 1954)

extensor hood and the remainder of the extensor aponeurosis in exactly the same manner as do the tendons of the palmar interossei.

There is considerable variation in the amount of insertion into the proximal phalanx and into the extensor aponeurosis among similarly numbered dorsal interossei in different hands, but a still greater variation in the proportion of insertion into bone and into aponeurosis among the interossei in a single hand. Salsbury and Eyler and Markee estimated, as they did for the palmar interossei, the amount of muscle inserting into the bone of the proximal phalanx or the capsule of the metacarpophalangeal joint, and the amount continuing into the extensor aponeurosis: it appears that the first dorsal interosseous inserts almost entirely into bone, the second only slightly more into bone than into aponeurosis, the third almost entirely into aponeurosis, and the fourth slightly more into aponeurosis than into bone (Fig. 6-57). In contrast, Smith and Kaplan reported "frequent" insertion of the first dorsal muscle into the aponeurosis.

Truly anomalous arrangements of the interossei are apparently rare. Griffith described what was apparently an accessory first dorsal interosseous muscle, and du Bois-Reymond described a hand in which the interossei were arranged as they are in the foot, around the second rather than around the middle digit.

ACTIONS AND INNERVATION

When the fingers are extended and the metacarpophalangeal joints are fixed in extension, the interossei act only as abductors and adductors of the digits: the dorsal interossei, acting with the abductor pollicis brevis and the abductor digiti minimi, spread the digits apart (the second and third interossei abducting the middle digit in either direction from the midline); and the palmar interossei, acting with the adductor pollicis, bring the digits together again. Both sets of interossei have usually been regarded as flexors of the metacarpophalangeal joints; Long and Brown concluded that the dorsal ones, at least, contribute little to this, but Backhouse found all interossei active in flexion here. By virtue of their insertions into the extensor aponeurosis all except the first dorsal (which may lack such insertion) are extensors of the interphalangeal joints. Long and Brown found that the dorsal ones (they did not test the palmar ones) do this only when the metacarpophalangeal joint is being flexed or is held in a flexed position.

All the interossei are normally innervated by the deep branch of the ulnar nerve (C8, T1) through twigs which this nerve gives off to them as it passes across their palmar surfaces near the bases of the metacarpals. Sunderland ('46) reported, however, that among 100 hands that were dissected the first dorsal interosseous was supplied exclusively in 3 cases by the median nerve, and partially by this nerve in a fourth; that among 41 clinical cases with complete interruption of the ulnar nerve the first dorsal interosseous was apparently unimpaired in 3, only partially impaired in 2 more; and that among 17 cases of complete interruption of the median nerve the muscle appeared to be normal in 15 but was significantly weak in 2. Clifton estimated from his clinical findings that about 10% of first dorsal interossei are innervated by the median nerve. A very few first dorsal interossei have been reported to receive a twig from the radial nerve, but this is assumed to be an afferent one; and, as discussed in the section on anomalous innervation of the muscles of the hand, some isolated clinical cases have been reported in which all the interossei seemed to be innervated by the median nerve.

THE INTEROSSEI AND DEFORMATION OF THE HAND

Paralysis or overaction of the interossei contribute much to functional deformities of the hand. The common deformity produced by an ulnar nerve paralysis is a clawing of the hand (Fig. 6-58), in which the long extensors, without the resistance offered by the interossei, tend to hyperextend the digits at the metacarpophalangeal joints; they produce this through their attachments to the palmar ligaments, so their distal parts are relaxed (Mulder and Landsmeer), and they lose part of their power of extension of the interphalangeal joints. The long flexors, with lessened opposition from the long extensors and none from the interossei, then tend to flex the middle and distal phalanges. Since clawing is, however, dependent upon hyperextension at the metacarpophalangeal joints, it varies markedly with the individual; it is minimal if the joint capsules are such as to prevent appreciable hyperextension under normal circumstances. In the usual ulnar paralysis, the less extensive involvement of the index and long fingers results from the action of the first two lumbricals, which are usually innervated by the median, not the ulnar, nerve.

The clawhand, or *main en griffe,* is still more evident, and the thumb also is involved, if there is a lesion of both the median and ulnar nerves at the wrist and the metacarpophalangeal joints are hypermobile; it may then involve all four fingers equally. Further, scar tissue formed at the level of the lesion may produce flexion at the wrist, and hence increase the pull upon the long extensor tendons and exaggerate the hyperextension at the metacarpophalangeal joints (Goldner). In

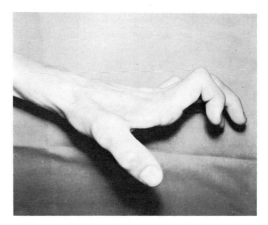

Fig. 6-58. Clawhand resulting from ulnar nerve paralysis. (Courtesy of Dr. E.D. Henderson)

the same way, the clawing that may result from ischemic (Volkmann's) contracture, because of the flexion of the wrist and interphalangeal joints by the long flexors and the passive insufficiency of the long extensors at the metacarpophalangeal joints, is accentuated by paralysis of the ulnar nerve (Steindler, '32). On the other hand, high paralysis of both the median and ulnar nerves regularly results in minimal clawing (Goldner), because the pull of the long flexors is absent.

In paralysis of the radial nerve the unopposed pull of the long flexors tends to put the wrist and fingers into moderate flexion, but the tendency to hyperextension at the metacarpophalangeal joints is resisted by the interossei and lumbricals; without the brake of the extensor digitorum on the metacarpophalangeal joints, however, these short muscles are not good extensors of the interphalangeal joints.

Good return of muscle function in the hand following injury to its nerves may not occur, leaving a hand that is almost useless. In these cases, the deformity caused by paralysis of the ulnar nerve can be overcome by attaching a part of a superficial flexor tendon to the extensor aponeurosis, to prevent the hyperextension of the proximal phalanx (Bunnell, '49; Littler). Restoring abduction to the index finger is of particular importance in order to allow a pinch between thumb and forefinger, and can be obtained by passing the superficial flexor tendon around the radial side of the digit to an attachment on the dorsal apo-neurosis (Bunnell), or by attaching the ten-don of the extensor indicis or of the extensor pollicis brevis to the tendon of the paralyzed first dorsal interosseous muscle (Bunnell; Littler). Riordan also has discussed tendon transfers in paralysis of the nerves of the hand.

Contractures of the intrinsic muscles of the hand produce a deformity just the reverse of that which follows their paralysis (Harris and Riordan); there is flexion at the metacarpo-phalangeal joints and extension at the inter-phalangeal ones, or at least at the proximal joints; the distal ones may be in slight flexion because of the pull of the profundus tendons. According to Harris and Riordan, the condi-tion has been treated both by extensive strip-ping of the interosseous muscles from their origins, and by sectioning parts of their ten-dons. Differential section of the oblique fibers continuing onto the extensor tendon, with sparing of the transverse fibers at the extensor hood, would seem to be particularly suitable, since it should abolish the power of these muscles to extend the interphalangeal joints and weaken but not abolish their power of flexion at the metacarpophalangeal joints—thus relieving the flexion at the latter joints, but at the same time preventing hyperexten-sion here and therefore leaving the long ex-tensor free to extend the interphalangeal joints.

DEEP PALMAR ARCH AND DEEP ULNAR NERVE

Both of these pass across the palm in front of the proximal ends of the bodies of the meta-carpals and of the palmar and dorsal interos-sei which fill the spaces between the metacar-pals (Fig. 6-59). The deep branch of the ulnar nerve emerges from the hypothenar muscles to run across the palm in a radial direction, giving off branches as it goes, and disappears between the two heads of the adductor pol-licis, while the deep palmar arch appears be-tween the two heads of the adductor pollicis and runs in an ulnar direction, being typi-

cally completed on this side by the deep pal-mar branch of the ulnar artery emerging through the muscles of the hypothenar emi-nence. Both structures therefore lie primarily behind the posterior wall of the mid-palmar fascial space; Kaplan ('53) suggested that the fat pad normally found in the posterior wall of this space may have a function in protect-ing the ulnar nerve.

The deep branch of the ulnar nerve may lie either proximal or distal to the deep arch, but when it or its branches cross the arteries they typically do so superficial to them. In contrast to the superficial arch which, although some-what variable in its position, often extends distally as far as a transverse line drawn across the palm from the metacarpophalangeal joint of the thumb, the deep arch and the deep branch of the ulnar nerve lie somewhat more proximally in the palm, only slightly distal to the lower edge of the flexor retinaculum.

DEEP PALMAR ARCH

The deep palmar arch is the chief continua-tion of the radial artery into the hand and is described as being, with the smaller princeps pollicis, one of the terminal branches of the radial, or as giving rise to the princeps. As the radial artery leaves the dorsum of the hand by passing between the two heads of the first dorsal interosseous muscle, the deep palmar arch and the princeps pollicis artery separate from each other, the princeps running distally and the deep arch turning medially, between the first dorsal interosseous and the adductor pollicis, across the metacarpals and interossei. On the ulnar side, the deep arch is usually completed by a branch of the ulnar artery. The *deep branch* of the ulnar, typically much smaller than the superficial branch, usually arises just distal to the pisiform bone and passes with the deep branch of the ulnar nerve between the heads of origin of the short abductor and flexor of the little finger; it sup-plies the muscles of the little finger, and fairly often (in somewhat more than half of Weath-ersby's series, '55, somewhat less than half in Coleman and Anson's) emerges to join the deep arch. In 34% of Weathersby's series the

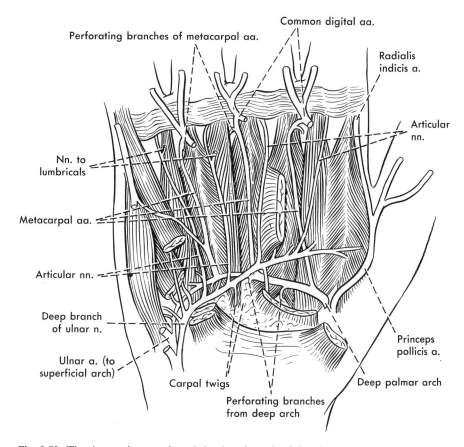

Fig. 6-59. The deep palmar arch and the deep branch of the ulnar nerve.

arch was completed by a branch, called the inferior deep branch by Coleman and Anson, that ran across the palmar surface of the short flexor of the little finger and passed medial to the hypothenar muscles to join the deep arch; Coleman and Anson found the deep arch completed by this vessel alone in 49% of hands. When there was no branch from the ulnar artery to the deep arch, in Weathersby's series, a perforating artery from a dorsal metacarpal completed it.

Marked variations in the deep arch are uncommon. Coleman and Anson reported a few instances in which the arch was incomplete, with the radial artery supplying the thumb and radial side of the index finger, or doing this and ending by anastomosing with a perforating artery from a dorsal metacarpal, while the ulnar end of the arch was formed by the deep branch of the ulnar and another

perforating artery. They found 1 case also in which the deep arch was formed entirely by the deep ulnar and a perforating artery, there being no deep branch of the radial. Charles reported a case in which there was no radial artery in the forearm and the lower end of the anterior interosseous artery replaced it. According to Griffith, the arch sometimes runs through the oblique head of the adductor pollicis instead of between the two heads of that muscle.

Branches

The *princeps pollicis,* when not considered a terminal branch of the radial artery, is the first branch of the deep arch and has also been called the first palmar metacarpal artery. It is rarely absent (2% in Coleman and Anson's series) and typically runs distally between the first dorsal interosseous and the adductor pollicis muscles, close to the first meta-carpal. It may give rise to or communicate with the radialis indicis, but otherwise it divides at about the distal edge of the adductor pollicis into two branches, one of which continues along the ulnar side of the thumb, while the other crosses behind the tendon of the flexor pollicis longus to attain the radial side. As noted in connection with the superficial arch, this arch frequently communicates with the princeps.

The *radialis indicis artery* is particularly variable (Weathersby, '55). It is usually described as originating from the deep arch, but Weathersby found this to be its sole origin in only 45%, and in about a fifth of these it arose with the princeps rather than independently; in 13% it arose entirely from the superficial arch, and in the remainder from both arches. When it arises from the deep arch it runs distally along the second metacarpal, and while it may supply only the radial side of the index finger, it sent a branch to the common digital of the second and third fingers in 75% of cases, and in a fifth of all hands it was the chief or sole supply here. Because of this, Weathersby suggested it should be called a metacarpal artery, and Coleman and Anson, regarding the princeps pollicis as the first metacarpal, called this the second.

The deep arch usually sends perforating (proximal perforating) branches between the heads of the second, third, and fourth dorsal interossei to join the dorsal metacarpal arteries; they are usually an important or the sole source of blood to these vessels, but occasionally, as already noted, a large perforating artery joins a part of the deep arch and carries blood to it. The arch also gives off two or three small recurrent branches to the front of the carpals and a variable number of *metacarpal arteries*. Counting the radialis indicis as one, there may be only three for the interspaces between the four fingers, there may be fewer, or there may be more than one for an interspace. Typically, each joins the common digital artery of its interspace and may even replace it, but sometimes it divides to join two common digitals or ends in the metacarpophalangeal joint (Coleman and Anson).

Usually the palmar metacarpal artery sends a perforating (distal perforating) branch to the dorsal metacarpal proximal to the heads of the metacarpal bones. Beyond the metacarpal of the interspace between the fourth and fifth digits there is frequently another branch of the deep arch that runs distally along the lateral side of the hypothenar muscles, supplying them, and ending by joining the proper digital artery to the ulnar border of the little finger.

DEEP BRANCH OF THE ULNAR NERVE

At the level of or just distal to the pisiform bone, the deep and superficial branches of the ulnar nerve separate (Fig. 6-39); the ulnar nerve here lies on the front of the flexor retinaculum (transverse carpal ligament) medial to the ulnar artery. The superficial branch of the ulnar nerve continues its course distally, across the front of the flexor retinaculum and the muscles of the little finger, and gives rise to digital branches in the hand; normally, its only muscular branch is a twig to the palmaris brevis muscle. This nerve is, therefore, almost exclusively cutaneous and articular. The deep branch of the ulnar, in contrast, is distributed to the great majority of the muscles of the hand; it has no cutaneous branches.

The deep branch of the ulnar nerve (Fig. 6-60) leaves the superficial branch just distal to the pisiform bone, where it runs dorsally and slightly medially to enter the muscles of the hypothenar eminence, passing between the origins of the abductor of the little finger (from the pisiform) and of the flexor of the little finger (from the flexor retinaculum). In company with the deep branch of the ulnar artery, it then either pierces the dorsal border of the opponens close to its origin or passes around this border, and turns more medially to run across the palm. In its course through the hypothenar muscles the ulnar nerve supplies one or more branches to each of these three muscles; the exact origins of these muscular branches vary, and upper ones may be derived from the ulnar nerve before its terminal branching, but there is a tendency for the

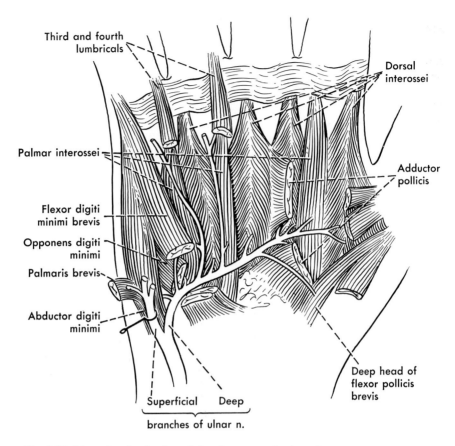

Third and fourth lumbricals

Dorsal interossei

Palmar interossei

Adductor pollicis

Flexor digiti minimi brevis

Opponens digiti minimi

Palmaris brevis

Abductor digiti minimi

Deep head of flexor pollicis brevis

Superficial   Deep

branches of ulnar n.

Fig. 6-60. Muscular distribution of the ulnar nerve in the palm.

muscles to be supplied in the order of abductor, flexor, and opponens.

While pressure from carpal ganglia usually involves both branches of the ulnar nerve, the deep branch alone may be affected (Richmond), and a few cases have been reported in which the nerve was injured distal to the hypothenar muscles (Sedon; Bakke and Wolfe). Similarly, fractures of the ulnar carpals or metacarpals may injure the deep branch alone (Howard, '61).

In the palm of the hand the deep branch of the ulnar nerve runs approximately parallel to the deep palmar arch, until it disappears with this arch between the two heads of the adductor pollicis. Thereafter the terminal part of the nerve diverges from the arch, for the latter curves back to its entrance through the first dorsal interosseus, while the ulnar nerve travels more transversely to end in the

first dorsal interosseous muscle, supplying a branch to each head of the adductor before it does so. Fenning reported a case in which the branch to the first dorsal interosseous apparently arose while the deep ulnar nerve was among the hypothenar muscles: a branch given off here presented a neuroma where it curved sharply around the hook of the hamate, and the first dorsal interosseous was much atrophied.

Branches

In addition to the hypothenar muscles, the deep ulnar nerve typically supplies the two medial lumbricals, all the interossei, the adductor of the thumb, and the deep head of the short flexor of the thumb (however this deep head may be defined); it also usually has articular branches to the metacarpophalangeal joints of at least the little and ring fingers, and sometimes to those of the middle and index ones also. The exact order of origin of these branches varies, and so does their number, for branches to two or more muscles frequently arise by a common stem, a branch that has already supplied hypothenar muscles may innervate other of the muscles on the ulnar side, and muscular branches may leave the articular branches as the latter run distally toward the joints.

Sunderland and Hughes investigated the branches of the ulnar nerve in 20 hands, and found that the palmar and dorsal interossei occupying a single intermetacarpal space usually were supplied by a single stem that supplied first the palmar, then the dorsal, interosseous, but that occasionally each muscle had an independent branch. The nerves to the third and fourth interspaces frequently arose together, and the nerves to the third and fourth lumbricals (which curve forward into the deep surfaces of these muscles) most frequently arose with the nerves to the third and fourth dorsal interossei; similarly, the nerves to the lumbricals may arise from the articular branches to the metacarpophalangeal joints. The branch to the interossei of the second space usually arises independently, according to Sunderland and Hughes; it may arise under cover of the adductor, or medial to the adductor, and in the latter case it may accompany the major portion of the ulnar nerve between the two heads or pierce the transverse head.

The terminal branches of the nerve, typically given off under cover of the adductor, are usually three, one to each head of the adductor and one to the first dorsal interosseous. When the deep head of the short flexor is present and supplied by the ulnar nerve, its nerve supply usually arises from the nerve to the oblique head of the adductor. The branch to the first dorsal interosseous may be missing (this was true of one of the 20 hands investigated by Sunderland and Hughes), for this muscle is sometimes supplied by the median nerve.

The articular branches of the ulnar nerve also vary. As the deep branch passes across the palm it gives off tiny recurrent twigs to

the intercarpal joints, but the large articular branches run distally across the front of the interosseous muscles to reach the metacarpophalangeal joints. Two or more of the articular branches may arise together, they may arise from nerves to the interossei, or they may arise independently and may give rise to the nerves to the lumbricals. Articular branches to the metacarpophalangeal joints of the little and ring fingers are found regularly, and usually there is also one to this joint of the middle finger; less commonly, but fairly frequently, there is also a branch to this joint of the index finger. As already discussed, Stopford's ('21) findings indicate that an ulnar innervation to the metacarpophalangeal joints of the two more radial fingers is commonly of little clinical importance because of their usual innervation from the median and radial nerves.

THENAR MUSCULATURE

The superficial short muscles of the thumb, the abductor pollicis brevis and the superficial head of the flexor pollicis brevis, overlie the opponens pollicis and with this form the prominence of the thenar eminence (Fig. 6-61). Since all three of these muscles are normally supplied by the median nerve, atrophy in the thenar eminence is a typical indication of involvement of this nerve. Su and co-workers reported a case of absence of these three muscles, with intact adductor and deep head of the flexor brevis (innervated by the ulnar nerve).

These muscles are covered by the thenar fascia, continuous medially with the palmar aponeurosis but much less densely fibrous than this; a slip from the palmar aponeurosis may run out onto this fascia toward the metacarpophalangeal joint of the thumb, but is usually not well developed and usually does not join the long flexor tendon sheath, so that Dupuytren's contracture rather rarely involves the thumb. The thenar fascia is attached laterally and dorsally to the dorsum of the first metacarpal; at its junction with the palmar aponeurosis, on the medial side of the

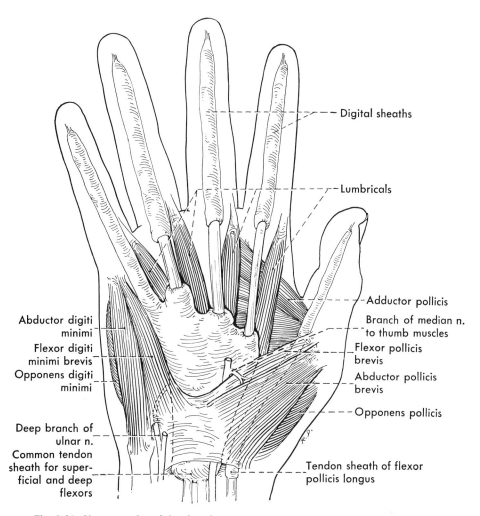

Fig. 6-61. Short muscles of the thumb.

muscles of the thenar eminence, there arises from the deep surface of the fascia the thenar or lateral intermuscular septum, which passes dorsally, and then laterally between the opponens and the adductor, to attach to the first metacarpal; thus the abductor, the flexor, and the opponens are enclosed within a special compartment, the thenar compartment, while the adductor lies in the central palmar or midpalmar compartment. (The term "compartment" is preferable to that of "space" because of the easy confusion between the *fascial* spaces of the hand, which are areas of loose connective tissue between layers of muscles or of muscles and tendons, and the larger areas, whether called compartments or spaces, that contain the muscles and tendons and the fascial spaces, and are limited by relatively heavy layers of fascia.)

Within the thenar compartment, between and around the muscles it contains, there are potential fascial spaces which, properly speaking, should collectively be termed the thenar space, as Grodinsky and Holyoke did; however, following Kanavel, the term "thenar space" has been so long applied to the more radially situated of the two major fascial spaces in the midpalmar compartment that only confusion can result from an attempt to use more appropriate terms. The fascial spaces within the thenar compartment apparently have no particular clinical significance.

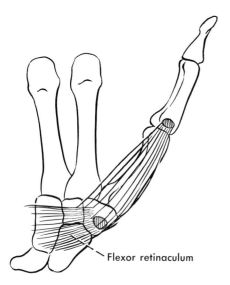

Fig. 6-62. Origin (*bony origin red*) and insertion (*black*) of the abductor pollicis brevis. The main origin is from the flexor retinaculum.

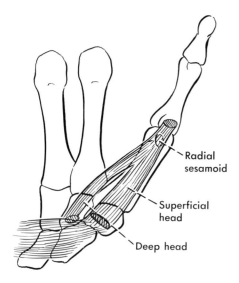

Fig. 6-63. Origin (*bony origin red*) and insertion (*black*) of the flexor pollicis brevis. The main origin of the superficial head is from the flexor retinaculum.

Abductor Pollicis Brevis

The abductor pollicis brevis arises (Fig. 6-62) largely from the proximal portion of the radial border and superficial surface of the flexor retinaculum (transverse carpal ligament), with some deep origin from the tubercle of the trapezium and often, it is said, with some origin also from the scaphoid. It often receives a portion of the insertion of a divided tendon of the abductor pollicis longus, and some of its fibers may arise from such a tendon. The muscle lies on the radial side of the thumb, therefore parallel to the palm of the hand when the thumb is in the normal resting condition, and the contracted muscle is easily palpated, for it is the most superficial muscle here. The abductor is somewhat triangular, its fibers converging upon a flattened tendon that inserts largely into the radial side of the base of the proximal phalanx of the thumb, but usually gives off a superficial lamina that extends around the radial border of the thumb to attach to the tendon of the extensor pollicis longus (an insertion which recalls that of the dorsal interossei, also abductors).

Flexor Pollicis Brevis

The flexor pollicis brevis is usually described as having two heads, a superficial and a deep one.

The superficial head of the short flexor, usually by far the larger component, arises from the distal portion of the flexor retinaculum and may have also a bony origin from the tubercle (crest) of the trapezium (Fig. 6-63). This muscle forms the superficial part of the more medial portion of the thenar eminence, and can be easily palpated when it contracts. It passes distally along the flexor side of the thumb, to insert into the base of the proximal phalanx on the radial side of the flexor surface; before it crosses the metacarpophalangeal joint, the tendon of insertion attaches in part into the radial sesamoid of this joint.

The deep head of the short flexor has been variably described, but is here regarded as consisting of fibers arising in conjunction with the oblique head of the adductor pollicis and passing obliquely distally deep to the tendon of the flexor pollicis longus to join the superficial head in inserting into the radial side of the thumb. It varies much in size and may be missing, and in the past there has been much disagreement as to the identity of the deep head. For instance, Brooks (1886) regarded the fibers just described as part of the superficial head, while Flemming regarded them as part of the adductor; in contrast, Cunningham described fibers arising from the first metacarpal and inserting into the ulnar side of the thumb (apparently the same ones that have been called the first palmar interosseous when four of these are described) as the deep head of the short flexor. Day and Napier have more recently found a deep head as described here in 41 of 53 hands; in 3, they said, there was no deep head; in 8 it divided to insert into both sides of the thumb; and in 1 it inserted only into the ulnar side. Like the adductor with which it is associated, this deep head is typically innervated by the deep branch of the ulnar nerve but, according to Day and Napier, may also be innervated by the median, or even by the latter alone.

Opponens Pollicis

The opponens pollicis is in part covered by the short abductor and in part by the short flexor. Like the preceding muscles, it arises largely from the flexor retinaculum and has also an origin from the tubercle of the trapezium (Fig. 6-64); it runs obliquely distally and laterally deep to the foregoing muscles to insert upon the radial surface and radial half of the anterior surface of the whole length of the body of the first metacarpal bone.

Adductor Pollicis

The adductor pollicis arises by both transverse and oblique heads. The transverse head arises from the ridge on the palmar surface of the third metacarpal, but some fibers, it is said, may take origin from the fascia over the adjacent interossei and even from the ligaments of the metacarpophalangeal articulations. This head is triangular, but its distal border passes almost transversely across the palm toward the ulnar side of the metacarpophalangeal joint of the thumb, where the two heads join (Fig. 6-65).

The oblique head of the adductor pollicis has, as already indicated, been the subject of much controversy; fibers associated with it but arising from the first metacarpal have been called the first palmar interosseous, making four of these, and the fibers that are

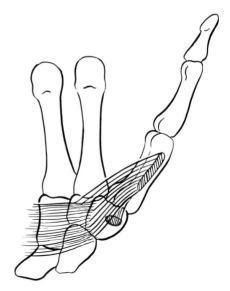

Fig. 6-64. Origin (*bony origin red*) and insertion (*black*) of the opponens pollicis. The main origin is from the flexor retinaculum.

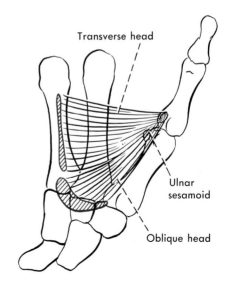

Fig. 6-65. Origin (*red*) and insertion (*black*) of the adductor pollicis.

described in this text as the deep head of the short flexor have been called a part of the oblique head of the adductor, thus giving it an insertion on both sides of the proximal phalanx. In the present text, the view adopted is that the fibers from the first metacarpal are part of the adductor, but that those inserting with the short flexor are part of the short flexor. The oblique head of the adductor pollicis, as thus defined, arises from the bases of the first, second, and third metacarpal bones and from the ligamentous floor of the carpal canal over the distal border of the capitate and perhaps the trapezoid, these fibers being closely associated with the origin of the deep head of the short flexor. This head also is triangular, and its fibers converge as it runs distally toward the metacarpophalangeal joint; the oblique and transverse heads join before they reach the joint, and the common tendon of insertion, before it attaches to the ulnar side of the base of the proximal phalanx of the thumb, has an attachment to the ulnar sesamoid of the metacarpophalangeal joint of the thumb. Some fibers may be traced to an insertion into the extensor aponeurosis of the thumb.

The adductor, especially the transverse head, separates the thenar (anterior adductor) space of the palm of the hand from the dorsal (posterior) adductor space, and passes in front of all the interossei lateral to the middle digit. The first part of the deep palmar arch lies between this muscle and the interossei, and the terminal branches of the ulnar nerve lie also in this position; both the deep arch and the deep ulnar nerve usually pass between the two heads of the adductor.

Actions of the Short Thumb Muscles

Movements of the thumb as a whole, like movements of other digits, are best discussed after all the muscles and tendons attaching to them have been described.

In the present discussion, it should be remembered that abduction and adduction of the thumb are at an almost right angle to the palm; that flexion and extension therefore involve moving the thumb parallel to the palm of the hand; and that although all these movements involve the metacarpal as well as the phalanges, opposition, or placing the tip of the thumb in contact with the tip of a finger, especially involves the metacarpal, whose movements are governed not only by the direction of muscular pull but also by the special ligaments of the carpometacarpal joint. A further complication in discussing the action of a specific muscle is that most of the short muscles of the thumb are active, in varying degree, in almost every movement (Weathersby and colleagues; Forrest and Basmajian), and therefore even electromyographic evidence is difficult to interpret.

The abductor pollicis brevis, lying as it does in a plane parallel to the plane of the palm, is the only pure abductor of the thumb, raising the thumb away from the palm of the hand at an approximate right angle to the plane of the palm. The muscle should be able to assist in flexion of the proximal phalanx; Weathersby and colleagues found it active not only in flexion but in all movements, including those at the metacarpophalangeal and interphalangeal joints, except adduction, and in many of these movements it must have been acting only to stabilize the thumb; Forrest and Basmajian found it largely inactive when all three joints of the thumb are flexed simultaneously. Through its attachment to the long extensor tendon it should be able to assist in extension of the distal phalanx, and Weathersby and colleagues reported that it contracts first in unopposed extension. It apparently participates with variable strength in opposition, which it might be expected to assist by abducting and rotating the metacarpal and flexing the proximal phalanx.

The flexor pollicis brevis flexes and rotates medially the proximal phalanx of the thumb, and acting indirectly upon the metacarpal does the same to that; it becomes particularly active when firm opposition of the thumb is demanded.

The opponens acts as its name implies and is active whether the opposition is one of light touch or pressure. It also acts, like the short abductor, to stabilize the thumb, since it becomes active in all movements except flexion of the interphalangeal joint (Weathersby and colleagues).

The adductor pollicis adducts the metacarpal and helps to flex the metacarpophalangeal joint; thus if the metacarpal is rotated for opposition, it should be expected to assist in opposition by bringing the metacarpal palmarward and helping to flex the metacarpophalangeal joint. Through its partial inser-

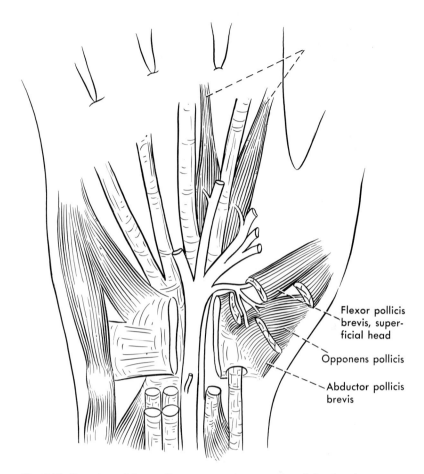

Flexor pollicis
brevis, super-
ficial head

Opponens pollicis

Abductor pollicis
brevis

Fig. 6-66. Branches of the median nerve to short muscles of the thumb.

tion into the tendon of the long extensor of the thumb, the adductor should also be able to help extend the distal phalanx of this digit, but it apparently is not active during this movement (Weathersby and colleagues).

Innervation of Short Thumb Muscles

The muscular branch of the median nerve into the muscles of the thumb is often said to consist of fibers derived from the sixth and seventh cervical nerves, but there is a good deal of clinical evidence that all these muscles, whether innervated by the median or by the ulnar nerve, receive their innervation primarily from C8 and T1. The branch into the thumb muscles leaves the median nerve either slightly proximal or slightly distal to the lower edge of the flexor retinaculum, and curves laterally or laterally and distally into

the thenar muscles (Fig. 6-66). In 3 of 20 hands investigated by Sunderland and Ray the muscular branch of the median was the first branch given off by the nerve in the palm, while in the remaining cases it either arose with the terminal digital branches or arose from one of these branches. Variations in the course of the nerve have already been mentioned.

This branch of the median nerve is commonly described as being distributed only to the abductor pollicis brevis, the superficial head of the flexor pollicis brevis, and the opponens pollicis. It may give off branches into the superficial surfaces of the flexor and the abductor, or may pass between them and give off branches to them there; in any event, the remainder of the nerve passes between the short abductor and the short flexor to reach the opponens and innervate this muscle. As indicated in a following section on anomalous innervation, anatomical and clinical evidence indicate that any of these muscles may be innervated also, or exclusively, by the ulnar nerve, although there is as yet little agreement as to the incidence of this.

The manner of branching of the median nerve in the thenar muscles is variable. Sunderland and Ray found that the short flexor usually received two or three branches, which sometimes arose with branches to other muscles, and that the branches to the short abductor and to the opponens frequently ran together, with each muscle then receiving a single branch; sometimes, however, two or three nerves were found going into the abductor, and, rarely, two into the opponens. They measured the length of nerve from the level of the radial styloid process to each of the thenar muscles and differences in length were so small that, for practical purposes, regenerating fibers might be expected to reach all the muscles at about the same time.

The abductor pollicis is typically supplied by the deep branch of the ulnar nerve, as is the deep head of the flexor pollicis brevis, but partial or complete supply by the median nerve has been reported.

HYPOTHENAR MUSCULATURE

The hypothenar muscles consist of one superficial and three deeper muscles. The superficial muscle is the *palmaris brevis* (Fig. 6-60), which arises from the ulnar side of the palmar aponeurosis and extends medially to attach into the skin along the medial border of the palm. The ulnar artery and the ulnar nerve pass deep to it, the nerves usually dividing into superficial and deep branches while in this position; it receives its nerve supply from the ulnar nerve on its deep surface. The branch to the palmaris brevis usually arises from the superficial branch of the ulnar nerve (Sunderland and Hughes found this in 10 of 14 cases), but may have a different origin; in 2 of their 14 cases Sunderland and Hughes described it as coming from the point of division

of the ulnar into the superficial and deep branches, in 1 as arising by 2 stems, 1 from the superficial and 1 from the deep branch, and in 1 as arising from the deep branch. As already mentioned, Shrewsbury and co-workers regularly found two nerves to the muscle. The palmaris brevis draws the skin of the medial border of the hand laterally, thus deepening the hollow of the palm and assisting the grip on the ulnar border of the hand. Shrewsbury and co-workers suggested that its main function may be to protect the ulnar nerve and vessels by drawing the hypothenar pad over them.

The three deeper muscles (Figs. 6-60 and 6-61) are covered by the hypothenar fascia, a thin layer of fascia that is continuous laterally with the palmar aponeurosis, and attaches around the medial border of the hypothenar muscles to the dorsal aspect of the fifth metacarpal. At the junction of the hypothenar fascia and the palmar aponeurosis a septum is given off from the deep surface to extend to the fifth metacarpal; this and the hypothenar fascia enclose the muscles of the hypothenar eminence in a hypothenar compartment.

These three muscles are all muscles of the little finger and, in accordance with their actions, are named the abductor digiti minimi, the flexor digiti minimi brevis, and the opponens digiti minimi.

The *abductor digiti minimi* is the most superficial and medial of the three. It arises from the distal surface of the pisiform bone, and often from the lower end of the tendon of the flexor carpi ulnaris and runs along the medial border of the hand to insert (with the flexor) into the medial side of the base of the proximal phalanx of the little finger (Fig. 6-67). Usually this tendon of insertion sends a slip superficially around the ulnar border of the little finger to attach into the extensor tendon; Eyler and Markee estimated that an average of about 10 per cent of the insertion of the muscle is into that tendon.

The *flexor digiti minimi brevis* arises just distal to the origin of the abductor, from the flexor retinaculum and the hook of the hamate bone. Like the abductor, it is a flat muscle, but its surface is approximately in the plane of the palm rather than in that of the medial border of the hand; as it runs distally its fibers converge and it diverges medially to join the abductor and be inserted with it, but more on the palmar than the medial surface of the phalanx.

The *opponens digiti minimi* arises under cover of the abductor and the flexor, from the flexor retinaculum and the hook of the hamate (Fig. 6-68). Its fibers run distally and medially at the same time to insert into about the distal three fourths of the medial border of the fifth metacarpal, and a corresponding length along the medial half of the palmar surface of the bone.

Of these three muscles, the opponens is apparently constantly present, although it may be fused with the short flexor or the abductor; the abductor is rarely absent, but may be completely fused with the flexor, or have an accessory head arising at the wrist or in the forearm. The flexor is usually fused partially with the abductor and may be fused with the opponens, and is said to be lacking in about 15 to 20% of hands.

Actions of the Hypothenar Muscles

In contrast to the thenar musculature, the muscles of the hypothenar eminence have simple actions that are also of much less clinical importance. The abductor abducts the little finger at the metacarpophalangeal joint, and thus acts with the dorsal interossei in abduction or spreading of the four fingers; it also assists the flexor in flexing the metacarpophalangeal joint of the little finger and, because of the slip it sends to the extensor tendon, can help slightly in extending the interphalangeal joints. The direction of pull of the flexor is such that it cannot aid the abductor in abduction of, but only flexes, the metacarpophalangeal joint. The opponens draws the fifth metacarpal forward, thus deepening the hollow of the palm and thereby increasing the power of the grip along the ulnar border of the palm, and allowing better opposition between thumb and little finger. Forrest and Basmajian found little activity in any of the muscles during extension, but all were active in flexion and in abduction, with the abduc-

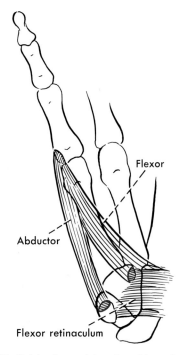

Fig. 6-67. Origin (*bony origin red*) and insertion (*black*) of the abductor and flexor digiti minimi. The flexor brevis arises chiefly from the flexor retinaculum.

Fig. 6-68. Origin (*bony origin red*) and insertion (*black*) of the opponens digiti minimi. The main origin is from the flexor retinaculum.

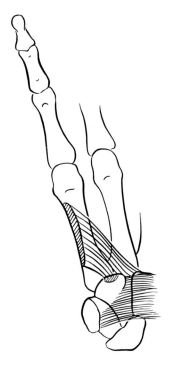

tor predominating in the latter movement. All were likewise active during firm opposition of the thumb, with the opponens particularly active during thumb-little finger opposition.

Innervation

The hypothenar muscles are typically innervated by the deep branch of the ulnar nerve (usually said to consist of fibers derived from the eighth cervical and first thoracic nerves), which starts its deep course by passing between the origins of the abductor and flexor of the little finger, and continues downward to pass behind the major portion of the opponens as it curves laterally in the palm of the hand (Fig. 6-60); it may pass around the deep border of the origin of the opponens, but commonly penetrates the muscle, separating a small upper and deep portion of the origin from the major portion.

The branches to the muscles may arise as the deep branch of the ulnar passes between the muscles, or may arise earlier; Sunderland and Hughes found that the abductor digiti minimi usually receives no more than two branches, of which the uppermost branch arose 3 times (among 20 hands) from the ulnar nerve before its division, 5 times at the point of division, and 12 times from the deep branch to the ulnar 17 mm to 39 mm below the origin of this branch. They described the innervation of the flexor as being from either one or two branches, occurring in about equal proportions; a branch arose from the ulnar nerve before its division once, but all were from the deep branch of the ulnar in the remaining 19. The opponens usually received a single nerve, which was usually the last branch to the hypothenar muscles to be given off, but sometimes arose between or even above branches to the other muscles.

The nerves to the hypothenar muscles often share a common stem, which supplies two or even all three muscles, and sometimes continues with the deep branch of the ulnar to supply the fourth lumbrical or the third palmar and fourth dorsal interossei.

ANOMALOUS INNERVATION OF MUSCLES OF THE HAND

The short muscles of the hand consist of the short muscles of the thumb, the lumbricals and the interossei, and the hypothenar muscles. As a whole, these muscles are innervated by the median nerve and the deep branch of the ulnar nerve (palmaris brevis usually by the superficial branch of the ulnar), but these nerves participate unequally in the innervation of the muscles of the hand. The innervation of the various muscles has already been described in some detail.

In summary, the muscular distribution usually described for the median nerve in the hand is to four and one-half muscles—the two lateral lumbricals, the abductor pollicis brevis and the opponens pollicis, and the superficial or main head of the flexor pollicis brevis. If this distribution holds, the distribution of the ulnar nerve is then to all the hypothenar muscles; to the two medial lumbricals; to all the interossei, both palmar and dorsal; to the deep head of the flexor pollicis brevis, if there is one; and to the adductor pollicis.

ANATOMIC OBSERVATIONS

Anatomically, as seen in dissections of the nerves in the hand, the distribution just quoted can usually apparently be confirmed, and variations from it seem to be relatively infrequent. In regard to the *thumb muscles,* it is known that there is some variation in the gross distribution of the nerves to them, especially the short flexor. Apparently the deep branch of the ulnar nerve did not extend into the thenar eminence in any of the 20 hands Sunderland and Hughes dissected, but Brooks (1886) reported that among 31 hands he dissected the flexor pollicis brevis was supplied exclusively by the ulnar nerve in 5 and exclusively by the median nerve in 7; among the latter, the deep head, normally supplied by the ulnar, was said to be missing in 1 and very poorly developed in another, but no note was made concerning the remainder. Similarly, Day and Napier found the superficial head to be supplied exclusively from the median nerve, as expected, in 17 of 30 hands, but by the ulnar alone in 6, and by both in 7; of 24 deep heads investigated, 16 were supplied, as expected, by the ulnar nerve alone, but 5 were supplied by both nerves, and 3 by the median nerve alone. Brooks also reported 2 cases in which the oblique head of the adductor was partially supplied by the median, and 1 in which the ulnar nerve supplied not only the flexor brevis but also the opponens and the abductor, there being no branch from the median nerve into the thumb muscles.

To what extent the observations quoted reveal the true incidence of anomalous innervation of the thumb muscles is of course debatable. Occasional fairly obvious connections between the deep branch of the ulnar and the median branch into the thenar eminence have been mentioned, and the author has seen one (in which it was impossible to tell which nerve was extending beyond its own territory); Frohse and Fränkel said that there are regularly one or more anastomoses between the two nerves among or in the muscles of the thumb. If so, they are presumably small since they have not been noticed by most observers, but they do suggest that some limited overlap between the two nerves may not be as rare as usually supposed.

The *hypothenar muscles* apparently always derive their nerves from the ulnar nerve in the hand. The *interossei* apparently regularly do this also, except for the first dorsal interosseous, which was found by Sunderland ('46) to be innervated exclusively by the median nerve in 3 of 100 hands, and partially by that nerve in a fourth. Of the *lumbricals,* the third is apparently most variable in its innervation, for Mehta and Gardner found it innervated exclusively by the median nerve instead of the ulnar in 2 of 75 hands, Brooks (1887) found it partially or completely innervated by the median in 12 of 21 hands, and Sunderland and Ray found this in 5 of 20 hands. The ulnar nerve rarely extends to the first two lumbricals, but Mehta and Gardner found cases in which both were innervated in part by the ulnar.

A limitation of the observations just cited is that they are based upon the distribution within the hand and cannot thereby disclose the course of the nerve fibers as they run to the hand. It is well known, however, that fibers not uncommonly leave the brachial plexus in one nerve and subsequently join another (Chap. 5). Thus a communication from the musculocutaneous to the median has been reported in from about 8% to 36% of arms and, more important to the present discussion, one from the median to the ulnar in the forearm in about 15% of limbs (the reverse communication is rare). This would indicate that the ulnar nerve at the wrist may fairly frequently contain fibers that run at first with the median nerve, and Murphey and coworkers have shown that in one case these fibers were distributed to the first dorsal interosseous and flexor pollicis brevis muscles, while Cliffton showed that in another case they were distributed only to those muscles usually innervated by the ulnar nerve. Thus in the first instance, based upon their distribution in the hand, some of the fibers in the communicating branch to the ulnar nerve would be thought of as true median nerve fibers which had an anomalous course into the hand, while in the second all would be regarded as true ulnar nerve fibers that ran at first with the median nerve.

These findings emphasize that the level of the lesion may be crucial to the clinical determination of anomalous innervation: in the arrangement described by Murphey and his colleagues the median nerve would have to be severed below the communication, or the ulnar nerve below it, to reveal the aberrant course of the fibers, while in Cliffton's case one of the two nerves would have to be severed above the communication. Since the communication is almost always from the median to the ulnar nerve, they also explain why dissections of the hand or lesions of the ulnar nerve at the wrist cannot be expected to reveal anomalous innervation of hypothenar muscles, although dissection or lesions of either the median or ulnar nerves at the wrist may reveal anomalous innervation of thenar muscles.

CLINICAL STUDIES

The possibility of anomalous innervation of the muscles of the hand was largely ignored by modern anatomists and clinicians until the latter rediscovered it and convincingly proved during and after the Second World War that it does occur (Murphey and co-workers; Cliffton; Rowntree). In these studies the lesion of the nerve was shown to be a complete one, and "trick" or substitution movements were ruled out as completely as possible; Murphey and his colleagues eliminated such movements entirely in their work by considering only those muscles that they could palpate. In all the studies, also, electrical stimulation of nerves and muscles was frequently employed.

It should be pointed out that these studies, even though extensive, cannot be expected to establish accurately the incidence of anomalous innervation of muscles of the hand, for there are, as already indicated, too many variables in the anatomy involved. With their severe criteria, Murphey and his colleagues proved the existence of anomalous innervation, but since they accepted as examples of this only those cases in which they were definitely able to prove it by inspection, stimulation, and palpation of the muscles concerned, they reported relatively few examples. Thus, among 698 cases of ulnar nerve injuries, they found only 4 in which the first dorsal interosseous was definitely shown to be innervated by the median nerve, although, as already noted, the incidence of such an innervation as determined by dissection is considerably higher than that. Interestingly, in one of these cases of innervation of the first dorsal interosseous by the median nerve they were able to prove that the innervation was not by the thenar branch of the median but was, rather, through fibers from the median that joined the ulnar nerve below the level of the lesion. This would indicate that the anatomic finding of 3% to 4% of first dorsal interosseous muscles innervated entirely or partly by the median may be too low, and Cliffton found 10% with this innervation in his clinical series. In addition to anomalous innervation of the first dorsal interosseous, Murphey and his co-workers reported cases in which the flexor pollicis brevis was innervated by the ulnar nerve exclusively; a case in which the abductor digiti minimi was innervated by the median nerve; and a case in which the opponens pollicis was innervated by the ulnar nerve.

Cliffton examined the records on 250 injuries of the ulnar nerve, 150 of the median, and 151 of both the median and the ulnar nerve, and among these found 4 cases in which, although the ulnar nerve was completely interrupted in the arm or the upper part of the forearm, muscles in the hand normally supplied by the ulnar nerve were found to show partial or total function, both on clinical examination and on stimulation of the median nerve at operation. Among these records he also found 6 cases in which the median nerve was divided yet the thumb muscles as a whole seemed to be functioning, and cited other instances in which "recovery" of function was apparently a result of anomalous innervation rather than of regeneration of severed fibers.

Muscles of Thumb

The studies of Cliffton and of Murphey and his colleagues thus agree with the anatomical evidence that muscles of the thumb may be anomalously innervated and indicate why neither dissection nor clinical findings after a nerve is severed can be expected to reveal all of them. Rowntree attempted to analyze the innervation of the individual muscles of the thumb in more detail, and the results of his study are shown in the accompanying tables.

The cases that Rowntree studied were ones in which only the median or the ulnar nerve had been severed, and the lesion was shown to be complete either by examination at operation or by electrical stimulation of the nerve. His series may include cases of "trick" movements, but it is said that all examinations were done by skilled workers who took pains to exclude these, so the results probably represent typical clinical findings. In summary, the data indicate that in only 35 of 226 hands (about 15%) in which either the median or the ulnar nerve was sectioned did these nerves have the usually described distribution to the muscles of the thumb (Table 6-5), if one as-

Table 6-5
Apparent Innervation of Thumb Muscles Deduced from Paralysis Following Lesions of Median and Ulnar Nerves*
(102 median nerve injuries)

MUSCLE	NUMBER OF PATIENTS										
	4	*3*	*13*	*14*	*38*	*16*	*10*	*1*	*1*	*1*	*1*
Abductor brevis	U	MU	M	M	M	M	M	M	U	MU	MU
Opponens	U	U	U	MU	M	M	M	MU	MU	M	M
Flexor brevis	U	U	U	U	U	MU	M	MU	MU	MU	U
Adductor	U	U	U	U	U	U	U	U	U	U	U

(124 ulnar nerve injuries)

MUSCLE	NUMBER OF PATIENTS					
	34	*19*	*64*	*1*	*5*	*1*
Abductor brevis	M	M	M	M	M	U
Opponens	M	M	M	M	M	MU
Flexor brevis	U	MU	M	M	M	M
Adductor	U	U	U	MU	M	U

* Adapted from Rowntree, T.: J Bone Joint Surg 31-B:505, 1949.
U denotes innervation by the ulnar nerve, M by the median, MU by both.

Table 6-6
Apparent Incidence of Abnormal Innervation of Thumb Muscles.*

MUSCLE	INNERVATION				
	EXPECTED	DEDUCED			
		M	*U*	*MU*	*Per Cent Abnormal*
Abductor brevis	M	215	6	5	4.9
Opponens	M	189	20	17	16.4
Flexor brevis	MU	81	107	38	83.2†
Adductor	U	5	220	1	2.7

* Rearrangement of the data in Table 6-5.
† This figure may be too high; see text.

sumes that the data on the flexor pollicis brevis are accurate and correctly interpreted. They are in accord with other observations, already cited, that this muscle does have an anomalous innervation more frequently than any other muscle, but the incidence of this given in Table 6-6 may be too high for several reasons. First, there is the obvious difficulty in assessing by clinical examination the activity of a specific muscle of the thumb, perhaps particularly of the short flexor since it is only one of several muscles important in flexion and opposition. Second, more than two thirds of the cases of apparent sole innervation by the ulnar nerve were in the series of median nerve injuries, where a normally innervated deep head should still have been active, and more than three fourths of the cases of ap-

parent sole innervation by the median nerve were in the series of ulnar nerve injuries, where a normally innervated superficial head should have been active. This suggests that it may often have been impossible to diagnose paralysis of one head only. Finally, since these workers were apparently reporting on the superficial head only, the "expected innervation" becomes the median nerve, and the number of hands showing normal innervation of all thumb muscles rises to 74, or 33%, while the incidence of abnormal innervation of the flexor pollicis brevis falls to 64.1%, somewhat closer to the 43.3% demonstrated by dissection in Day and Napier's short series.

In his series as a whole, Rowntree found 4 patients with lesions of the median nerve in which all the muscles of the hand seemed to have an intact innervation (compare the report of Brooks, 1886, on a median nerve with no muscular branch in the hand) and another 3 in which the only motor loss was reported to be a weakness of the abductor pollicis brevis; in 5 of these the lesion of the median nerve was at the level of the wrist, and in a sixth procaine block of the ulnar nerve at the elbow produced complete paralysis of the hand. Thus the fibers were in the ulnar nerve above the level of the wrist.

Rosen stimulated both nerves at the elbow and wrist in 96 cases and recorded electromyographically the responses of the opponens pollicis muscles. He found none that was not innervated by the median nerve, but 16 that were innervated also by the ulnar nerve; in 11 of these the fibers were in the ulnar nerve at the levels of both elbow and wrist, while in the remainder they joined the ulnar nerve between these points.

Interossei and Hypothenar Muscles

The innervation of the interossei and of hypothenar muscles as a whole seems to be less variable than that of the thenar muscles, although this may be an illusion: as pointed out by Cliffton, the ulnar nerve in the lower part of the forearm regularly contains the fibers for most of these muscles; therefore many lesions of the ulnar nerve cannot be expected to reveal an anomalous innervation. Nevertheless, Cliffton found among 250 complete lesions of the ulnar nerve 4 cases in which these muscles were functioning. Among these muscles the first dorsal interosseous, by far the easiest in-

terosseous to test, apparently has also the most variable innervation, being supplied by the median nerve in about 10% of cases, according to Cliffton; it is also the member of this group in which it has been shown that fibers to it may traverse either the ulnar or the median nerve in the hand. Rosen investigated only the abductor among the hypothenar muscles, and found it innervated by the ulnar nerve in all 96 cases; in 4 of these, fibers to the muscle were also in the median nerve at the levels of both elbow and wrist.

Rowntree attempted an analysis of the interossei and of the hypothenar muscles in only a few instances, and in none of these was there an attempt to separate the individual hypothenar muscles. The results, shown in Table 6-7, again emphasize the fact that any or all of these muscles may have an anomalous innervation.

PARALYSIS OF ULNAR AND MEDIAN NERVES

Ulnar Nerve

Paralysis of the ulnar nerve alone seriously affects movements of the digits only, regardless of the level at which the lesion lies. Section of the nerve at or above the level of the elbow typically denervates not only small muscles of the hand, but also the flexor carpi ulnaris and an ulnar part of the flexor digitorum profundus. However, flexion of the wrist is not seriously interfered with, because of the other flexors; Jones found paralysis of the flexor

Table 6-7
Apparent Innervation of Interossei and Hypothenar Muscles*

	MUSCLES							
	DORSAL INTEROSSEI				PALMAR INTEROSSEI			HYPOTHENAR
LEVEL OF LESION	*1*	*2*	*3*	*4*	*1*	*2*	*3*	
Elbow	M	MU	MU	U	?	?	?	MU
Elbow	M	M	M	M	M	M	M	M
Elbow	MU	MU	MU	MU	MU	MU	U	U
Elbow	U	U	U	U	U	U	U	U
Wrist	MU	MU	MU	U	?	?	?	U
Wrist	U	U	U	U	U	U	U	U

* In six patients with lesions of the ulnar nerve. Adapted from Rowntree T: J Bone Joint Surg 31-B:505, 1949.

carpi ulnaris difficult to detect, but more evident on attempted ulnar abduction than upon flexion. Unless clawing of the fingers occurs, there may be also relatively little loss of movement in the hand: the fingers can be abducted by the common extensor, adducted by the long flexors; further, the little finger can be independently abducted by the extensor digiti minimi, and the index finger independently adducted by the extensor indicis (Jones). Because of the involvement both of the intrinsic muscles of the hand and the ulnar part of the profundus, there is weakness of flexion in the little and ring fingers especially, and the little finger is usually extended at the metacarpophalangeal joint; however, Jones found that this joint could usually be flexed fairly well by the superficial flexor, which is not involved by ulnar paralysis, and that this joint of the ring finger is readily flexed. Until hyperextension develops at the metacarpophalangeal joints, as it may or may not, the long extensors can replace the interossei and lumbricals in extending the middle and distal phalanges.

The problem of tendon transfers in high ulnar paralysis is essentially similar to that where the interossei are the chief muscles involved. In the former case, however, as Bunnell ('49) pointed out, the superficial flexor tendons of the little and ring fingers cannot be attached to the extensor tendons to replace the intrinsic muscles, since they are necessary as flexors of the proximal interphalangeal joints. Bunnell has restored muscle balance at the metacarpophalangeal joints by slitting the annular ligament (pulley) of the proximal phalanx on its sides, beginning proximally and carrying the cuts just far enough distally to allow the long flexor tendons to pass far enough in front of the joint to flex it. Zweig and colleagues attached the two tendons of the extensor digiti minimi to the insertions of the adductor and the first dorsal interosseous to restore pinch.

Low lesions of the ulnar nerve may or may not involve the dorsal cutaneous branch of that nerve, but usually do involve both the superficial and deep branches of the palm. A few cases have been reported in which only the deep branch of the nerve distal to the hypothenar muscles was involved. Bakke and Wolff reported one such instance, with wasting of the interossei, which they said was apparently an example of an occupational neuritis resulting from constantly recurring intermittent pressure over the hypothenar eminence; Seddon found that in four similar cases the immediate cause of the paralysis was a "ganglion" that protruded against the nerve from the palmar aspect of a carpal joint. Jeffery reported a case, apparently the sixth, in which the deep branch was compressed by an anomalous muscle, and Gore reported one in which it was injured by dislocation of the bases of the fourth and fifth metacarpals. Richmond reported compression of the nerve above the hypothenar eminence by "ganglia," usually involving the superficial and deep branches, but sometimes the deep branch alone.

Median Nerve

High paralysis of the median nerve is much more crippling than is an interruption of this nerve at the wrist, since the latter affects primarily only opposition of the thumb; the former may abolish not only this but also flexion of the distal phalanx of the thumb and flexion of the two or three radial fingers. Bunnell ('49) listed, as available for transfer, the tendons of one of the radial carpal extensors, the part of the flexor profundus supplied by the ulnar nerve, the brachioradialis, and the extensor carpi ulnaris, fully sufficient for the three movements needed. He suggested that the tendons of the flexor profundus of the paralyzed fingers can be attached to those of the ulnar fingers, and the strength of the grip further increased by attaching also the brachioradialis; the extensor carpi radialis brevis can be attached to the long flexor of the thumb, the extensor carpi ulnaris used for opposition of the thumb (see the following paragraph).

Paralysis of the median nerve at the wrist may necessitate no tendon transfers at all, for while Kirklin and Thomas found that regeneration into the muscles of the thumb rarely occurs, about 65% of patients with a paralyzed opponens and abductor pollicis brevis can substitute satisfactorily by using the long abductor (radial nerve) and the short flexor (usually supplied in part, sometimes entirely, by the ulnar nerve); Jones said that the long abductor and the adductor (regularly innervated by the ulnar nerve) can together produce opposition; and, as discussed in Chapter 5, Boswick and Stromberg found rotation of the thumb to a position of grasp to be more frequently retained than lost after section of the median nerve at the elbow. In other cases, however, opposition of the thumb must be provided for, and even when a substitute movement is possible, better strength and pinch can usually be provided by tendon transfer (Lipscomb, '68). Fitch, in 1930, reported that fixation of the first metacarpal in a dorsally dislocated position allowed the flexors of the thumb to be more effective in opposing it, but the usual method of obtaining opposition is by a tendon transfer.

According to Littler, any active wrist flexor or any extensor tendon can be used as the prime mover; a tendon of the superficial flexor, expecially that to the ring or little finger, is also available, but many workers feel it is best not to depend upon the tendon to the little finger because of its small size. Henderson has used various wrist extensors and the brachioradialis when flexors were not available, Burkhalter and co-workers used the extensor indicis, and Littler and Cooley have used the abductor digiti minimi. It is necessary that the tendon pass obliquely from the ulnar side and have a firm pulley to maintain the obliquity; therefore extensor tendons are carried around the ulnar side of the wrist, flexor tendons are looped around the flexor carpi ulnaris, held by a fascial pulley attached to the pisiform bone, or otherwise held firmly to the ulnar side. The tendon of the mover, prolonged by a tendon graft if necessary, has been attached around the radial side of the thumb to the dorsal aspect of the proximal phalanx, to the tendon of the extensor pollicis brevis, to the tendon of the abductor pollicis brevis, or to the short abductor and the long extensor. Jacobs and Thompson attached it sometimes to the metacarpal neck and the

base of the proximal phalanx, and expressed the opinion that the precise location of attachment is not as important as the selection of a good "motor" and attachment of the tendon under proper tension. Makin has reported good results from translocating the tendon of the flexor pollicis longus so that it winds medially and dorsally around the thumb.

Although there is no agreement as to whether the action of the abductor, of the opponens, or of both and perhaps the short flexor is being replaced by such operations, Hendry attributed eventual failure of a successful operation to stretching of the ligaments of the carpometacarpal and metacarpophalangeal joints under the strain imposed upon them. If Napier's analysis of opposition is correct, that opposition by the abductor depends on the ligaments of the metacarpophalangeal joint, while these become taut only at the end of opposition by the opponens, reproducing the action of the latter as closely as possible should put less strain on the ligaments.

Combined Lesions of Median and Ulnar Nerves

When both nerves are injured above the level of innervation of the forearm flexors there is a dearth of available muscles to substitute for the wrist and finger flexors as well as the long muscles of the thumb and the intrinsic muscles of the hand. Bunnell ('49) said that in such a case the wrist should be arthrodesed; the available extensors are then all three extensors of the wrist and the brachioradialis. Instead of arthrodesis, others prefer to preserve a wrist extensor to hyperextend the wrist and thus assist in flexion of the fingers even though this limits still more the tendons available for transfer. The problems of tendon transfer are to restore muscle balance at the metacarpophalangeal joints to prevent their hyperextension, to restore finger flexion, and to restore opposition of the thumb. Hendry regarded the attainment of the first of these goals as being most difficult; he reported failure in transferring the tendons of the superficial flexor in place of the lumbricals, and in activating this flexor by the brachioradialis or a radial extensor.

If the lesion of both nerves is at the wrist, the problem is simplified, because three extensors of the wrist are still available and so are superficial flexor tendons; the correction then involves only transfers for the intrinsic muscles, as discussed with the interossei and in the preceding section.

Anatomy of the Dorsum

Superficial Fascia

The superficial fascia of the dorsum of the hand is thin and, except over the phalanges, allows rather free movement of the overlying skin. As usual, it consists of fatty and fibrous layers (see Fig. 6-74). It loosely unites the skin to the underlying superficial layer of deep fascia over the long extensor tendons, and in it are imbedded the cutaneous nerves of the dorsum of the hand, the dorsal venous network, and most of the efferent lymphatic trunks from the fingers and hand.

CUTANEOUS NERVES

The cutaneous nerves on the dorsum of the hand are typically the dorsal branch of the ulnar and the superficial branch of the radial (Fig. 6-69), these being supplemented, over the middle and distal phalanges, by dorsal branches from the palmar digital nerves—therefore by the median for the first three and one-half digits.

Ulnar

The dorsal branch of the ulnar nerve arises from this nerve under cover of the flexor carpi ulnaris, usually somewhat below the middle of the forearm, and gradually diverges from the ulnar nerve as the two run distally. A little above the wrist, where the ulnar nerve lies under cover of the lateral border of the flexor carpi ulnaris, the dorsal branch of the ulnar turns medially and dorsally to emerge from under cover of the muscle and pass around the medial side of the wrist onto the dorsum

of the hand, piercing the deep fascia as it does so. On the dorsum of the hand it divides into at least two and usually three digital branches: a proper digital for the ulnar border of the little finger; a common digital for the adjacent sides of the little and ring fingers; and, usually, a common digital branch for the adjacent sides of the ring and long fingers that may be the sole supply to these, or join a branch of the superficial radial nerve. The proper digital nerves of the little finger can usually be traced for most of the length of the digit, and therefore usually supply skin over the middle and distal phalanges; the palmar digital branches of the ulnar nerve on the little finger typically do not have dorsal branches to the skin over the two distal phalanges, but do supply the nail bed. The branch to the ulnar side of the ring finger varies, but is frequently reinforced or replaced over the middle and distal phalanges by a dorsal branch from the palmar digital nerve. This is of no clinical significance, since in most nerve lesions the differences in distribution cannot be apparent; in contrast to the nerves to the radial fingers, both the palmar and dorsal branches here are from the same nerve. As the branches of the ulnar proceed toward the bases of the digits, they give off smaller branches to supply the skin on the ulnar side of the dorsum of the hand.

Radial

The superficial branch of the radial nerve originates on the flexor surface of the forearm a little below the elbow, where the deep branch of the nerve starts its course toward the dorsum of the forearm by entering the supinator muscle. The radial nerve is at this level under cover of the brachioradialis, and its superficial branch runs distally deep to this muscle, between it and the extensor carpi radialis longus. It lies just lateral to the radial artery in the middle third of the forearm, but leaves the artery to pass onto the dorsum of the forearm, emerging from under cover of the tendon of the brachioradialis. It then penetrates the deep fascia and divides into its terminal branches, the dorsal digital nerves. Linscheid noted that painful neuromas may

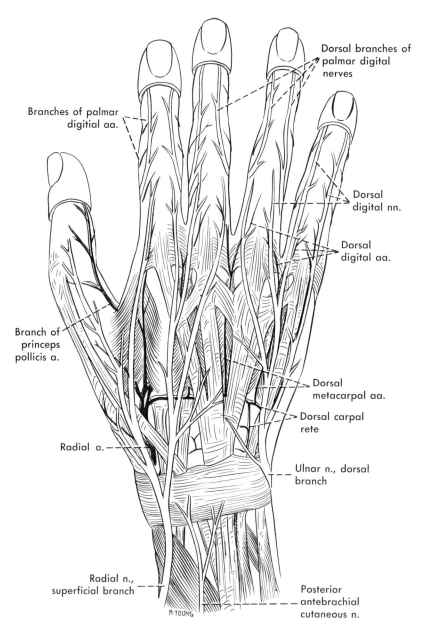

Branches of palmar digitial aa.

Dorsal branches of palmar digital nerves

Dorsal digital nn.

Dorsal digital aa.

Branch of princeps pollicis a.

Dorsal metacarpal aa.

Dorsal carpal rete

Radial a.

Ulnar n., dorsal branch

Radial n., superficial branch

R. YOUNG

Posterior antebrachial cutaneous n.

Fig. 6-69. Dorsum of the hand, with the deep fascia except for the extensor retinaculum shown as removed, but with the cutaneous nerves left in place.

arise if the nerve is sectioned as it passes onto the hand, and suggested that care should be taken to avoid it in such operations as opening stenotic tendon sheaths.

There are usually five dorsal digital branches of the radial nerve. The origins of the branches of the superficial radial on the hand show no regular arrangement, but usually the first branch to arise is to the radial side of the thumb; this branch typically anastomoses with the dorsal branch of the lateral antebrachial cutaneous nerve, to supply skin on the lateral portion of the thenar eminence, and on the radial side of the dorsum of the proximal phalanx. The second dorsal digital nerve is distributed to the ulnar side of the thumb, and typically does not extend much beyond the interphalangeal joint on this side; the third dorsal digital nerve supplies the dorsum of the radial side of the index finger, and the fourth the adjacent sides of the index and middle fingers, to about as far as the proximal interphalangeal joints, the median nerve supplying the distal portion of the dorsum. The fifth branch, not always present, is called the ramus communicans with the ulnar nerve, and supplies a proximal part of the adjacent sides of the middle and ring digits or, if it does not anastomose with the ulnar nerve, part of the long finger alone.

The dorsal digital nerves may give branches to the metacarpophalangeal joints, and even to the proximal interphalangeal ones; as already noted, the branches of the radial nerve usually help to supply both the metacarpophalangeal and interphalangeal joints of the thumb, rather frequently help to supply the metacarpophalangeal joint of the second digit, and sometimes supply also this joint of the third one.

Variations

The nerves on the dorsum of the hand show considerable variability, the most common being in the way in which the digital nerves to the adjacent surfaces of the ring and long fingers are formed. The nerve to these fingers may be from the superficial radial, or be formed by the union of branches from the ulnar and superficial radial (Fig. 6-70b); the branch to the radial side of the ring finger is more often from the ulnar alone (Fig. 6-69); and sometimes it is from the radial alone (Fig. 6-70a). Linell apparently found, among 16 hands, no examples in which the two nerves anastomosed upon the dorsum, but in only 2 did the ulnar nerve fail to extend beyond the ulnar side of the ring finger; in 2 it supplied both sides of the ring and little fingers, in 11 it supplied the little, ring, and ulnar side of the long finger, and in 1 it supplied also the ulnar side of the index. Ming-tzu found the radial nerve innervating three and one-half digits, the ulnar only one

and one-half, in about 29% of 166 Chinese hands; each innervating two and one-half digits in about 65%; anastomotic branches between the nerves to the middle digits in about 5%; and the radial restricted to one and one-half digits, while the ulnar supplied three and one-half, in less than 1%. He pointed out that there is some difference in percentages according to race.

In addition to these nerves, the posterior antebrachial cutaneous nerve, arising from the radial in the arm, frequently extends a variable distance beyond the wrist to supply skin on a proximal portion of the dorsum of the hand between the distribution of the ulnar and superficial radial nerves.

Because of overlap between the nerves, the loss of sensation following section of a nerve to the dorsum of the hand is usually more limited than anatomic description would suggest; Stopford ('18), for instance, found that loss of sensation on the dorsum of the hand following section of the ulnar nerve was usually confined largely to the one and one-half medial fingers, although not infrequently there was some loss over the base of the ring finger, or even the ulnar side of the middle one. However, the loss extended the entire length of the radial side of the ring finger in only about 14% of 102 cases, and the entire length of the ulnar side of the middle finger in only 3%. Similarly, clinical experience has shown that the area of total anesthesia following section of the superficial branch of the radial nerve is likely to be a small one, mostly in the area over and between the first and second metacarpals.

More marked variations in the distribution of nerves to the dorsum of the hand have been reported. These include replacement of the dorsal branch by the radial (Fig. 6-70c); replacement of the superficial branch of the radial by the lateral antebrachial cutaneous (Fig. 6-70e); replacement of the superficial radial by both the lateral antebrachial cutaneous and the dorsal branch of the ulnar (Fig. 6-70d); and replacement of the superficial radial and part of the dorsal branch of the ulnar by the lateral and posterior antebrachial cutaneous nerves (Fig. 6-70f).

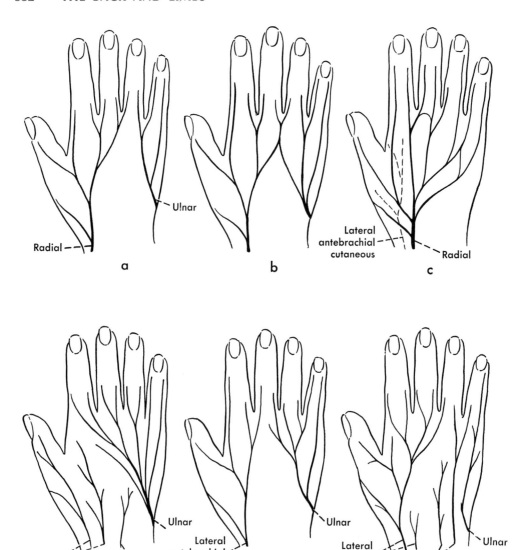

Fig. 6-70. Common variations, *a* and *b*, and unusual distributions, *c* to *f*, of nerves to the dorsum of the hand. (*c* is redrawn from Learmonth JR: J Anat 53:371, 1919; *d* and *f* from Hutton WK: J Anat Physiol 40:326, 1906; *e* from Appleton AB: J Anat Physiol 46:89, 1911)

SUPERFICIAL VEINS OF THE DORSUM

The veins on the dorsum of the digits tend to take the form of two longitudinally running channels, one on either side, but anastomose freely with each other across the dorsum, and receive, from around the borders of the fingers, many of the veins that originate on the palmar surface. At about the level of the heads of the metacarpals the veins on the adjacent sides of two digits tend to join to form a dorsal metacarpal vein, and these metacarpal veins, plus the digital ones from the free borders of the little and index fingers, unite with each other in an extremely variable pattern to form the dorsal venous rete (Fig. 6-71). Veins from the palmar surface pass around the medial border of the hand to join the veins on the dorsum, and similar veins pass from the palmar surface to the dorsum around the base of the thumb and, as intercapitular veins, between the bases of the digits. The veins from the thumb join the radial side of the dorsal network or proceed independently toward the cephalic vein.

The major superficial venous drainage from the hand is through the veins on the dorsum; the basilic vein takes origin from the ulnar side of the dorsal venous plexus, and the cephalic vein arises usually by several chan-nels both from the radial and the major remaining part of the plexus.

LYMPHATICS

The lymphatics on the dorsum of the fingers and of the hand receive most of the lymphatic drainage from the palm; a majority of the efferents from the palmar surface pass around the sides of the fingers and the hand to join the lymphatics originating on the dorsum. Although some of the efferents from the hand run upward on the anterior surface of the forearm, most of them originate on the dorsum and therefore start upward along the dorsum of the forearm (Fig. 6-72). In lym-

Fig. 6-71. Veins of the dorsum of the hand. (Toldt C: An Atlas of Human Anatomy [ed 2], Vol 2. New York, Macmillan, 1928)

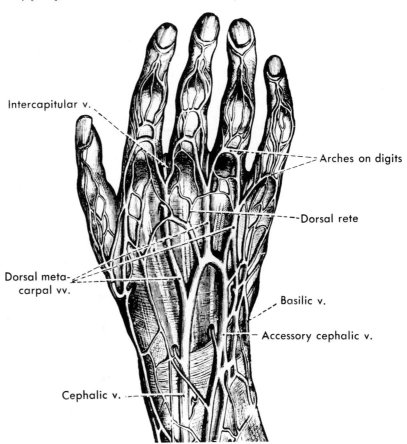

Intercapitular v.

Arches on digits

Dorsal rete

Dorsal meta-carpal vv.

Basilic v.

Accessory cephalic v.

Cephalic v.

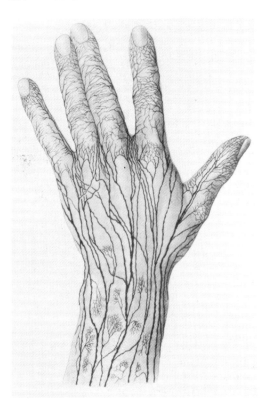

Fig. 6-72. Lymphatics of the dorsum. (Rouvière H: Anatomie des Lymphatiques de l'Homme, Paris, Masson et Cie, 1932)

phangiography of the upper limb, radiopaque material is injected into a lymphatic of the dorsum after it has been visualized by the injection of supravital dyes, usually into the subcutaneous tissue of the web. Before these vessels reach the elbow, however, most of them pass superficially around either the radial or the ulnar border of the forearm to run upward along the medial side of the arm.

EXTENSOR RETINACULUM AND DEEP FASCIA

At the level of the wrist the deep fascia of the forearm is thickened by oblique fibers that form a band about an inch (2.5 cm) wide, extending from the distal part of the anterior border of the radius around the dorsum and over the wrist and lower part of the forearm to the styloid process of the ulna, the ulnar

collateral ligament of the wrist, and the ulnar carpal bones (Figs. 6-69 and 6-73). This band, the extensor retinaculum (dorsal carpal ligament), holds the tendons of the extensor muscles closely against the bones as they cross the wrist. It differs from the flexor retinaculum in one important respect, however: it does not extend from its radial to its ulnar attachments as a free band, but rather sends septa from its deep surface to the bones that it crosses, between various tendons or groups of tendons.

Thus there is on the dorsum no equivalent of the carpal canal or tunnel; rather, there are a number of separate compartments deep to the extensor retinaculum, each compartment occupied by a single tendon or by two or more closely associated tendons. Typically there are six compartments, but fairly frequently there are seven: from the radial side, these compartments contain (1) the abductor pollicis longus and the extensor pollicis brevis; (2) the extensor carpi radialis longus and brevis; (3) the extensor pollicis longus; (4) the extensor digitorum and the extensor indicis; (5) the extensor digiti minimi, and (6) the extensor carpi ulnaris. The common variation from this arrangement is in the first compartment, for the abductor pollicis longus and the extensor pollicis brevis may each have its own compartment rather than share a common one.

Tendon Sheaths of the Dorsum
As each tendon or group of tendons lies in one of the compartments deep to the extensor retinaculum it is surrounded by a tendon sheath; thus there are separate tendon sheaths for the extensor pollicis longus, the extensor digiti minimi, and the extensor carpi ulnaris; on the other hand, the extensor carpi radialis longus and brevis share a single tendon sheath, and so do the extensor digitorum and extensor indicis, while the abductor pollicis longus and the extensor pollicis brevis usually do.

The tendon sheaths of the extensor tendons typically begin a very short distance above the proximal edge of the extensor retinaculum, but continue a greater distance onto the dorsum of the hand below the distal border of this retinaculum. Since there are no dorsal

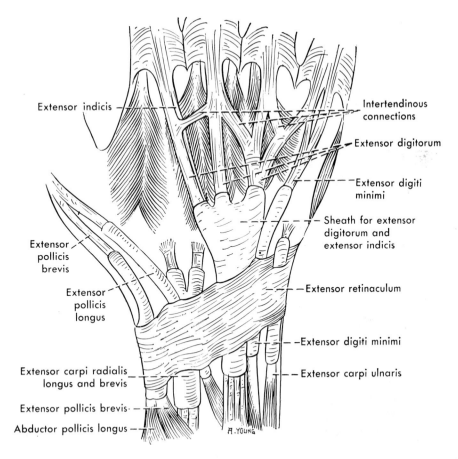

Extensor indicis

Intertendinous connections

Extensor digitorum

Extensor digiti minimi

Sheath for extensor digitorum and extensor indicis

Extensor pollicis brevis

Extensor pollicis longus

Extensor retinaculum

Extensor digiti minimi

Extensor carpi radialis longus and brevis

Extensor carpi ulnaris

Extensor pollicis brevis

Abductor pollicis longus

A. YOUNG

Fig. 6-73. The extensor retinaculum and the sheaths of the extensor tendons.

digital tendon sheaths, none of the sheaths at the wrist is continued onto the fingers; most of them extend around the tendons only a little beyond the bases of the metacarpals, but the sheath of the long extensor of the thumb frequently extends almost to the distal end of the metacarpal, while the sheaths of the tendons inserting on a metacarpal—the abductor pollicis longus, the two radial extensors, and the ulnar extensor—stop short of the attachment of these tendons to the base of the metacarpal. Between the proximal end of the base of the metacarpal and the tendon as it goes to its insertion there is often, in the case of the two radial extensors (particularly the short one) and of the ulnar extensor, a small bursa intervening; this has no connection with the tendon sheath.

When the long abductor and the short ex-

tensor of the thumb have a common tendon sheath, this is often bifurcated at its distal end, where the sheath follows each individual tendon for a short distance; this is also true of the common tendon sheath for the radial extensors, and there is usually some subdivision of the distal end of the tendon sheath around the common extensor tendons. Normally, four of the usual six tendon sheaths at the wrist have no communication with each other, but the sheaths around the radial extensors and the long extensor of the thumb usually do communicate: the tendon of the extensor pollicis longus, after rounding the radial tubercle, passes obliquely across the wrist toward the dorsum of the first metacarpal, and in so doing crosses superficial to the radial extensors and their sheath; in this location the adjacent walls of the two sheaths are usually de-

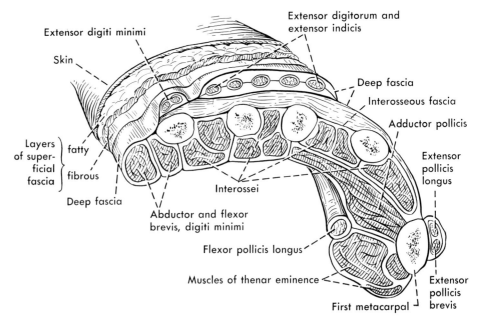

Fig. 6-74. Fascia of the dorsum of the hand; the several layers have been cut at different levels in order to show them more clearly. (Redrawn from Anson, BJ, Wright RR, Ashley FL, Dykes J: Surg Gynecol Obstet 81:327, 1945 [by permission of Surgery, Gynecology & Obstetrics])

ficient, so that they communicate freely with each other here.

Tenosynovitis (tendovaginitis) of the extensor tendons has already been discussed (Chap. 5).

Deep Fascia of the Dorsum

The deep fascia of the dorsum of the hand is thin. It is usually described as consisting of a superficial and a deep layer, between which lie the extensor tendons and their sheaths (Fig. 6-74). The superficial or supratendinous layer of the deep fascia is continuous proximally with the distal border of the extensor retinaculum; the deep or infratendinous layer lies over the dorsal interosseous fascia. On the sides of the extensor tendons the two layers come together so as to form a compartment about these tendons. Anson and his co-workers described the deep fascia as being fused to the second and fifth metacarpal bones, and thereafter extending around the borders of the hand to become continuous with the thenar and hypothenar fascias, respectively; sometimes, they said, the infratendinous layer is fused also to the third and fourth metacar-

pals. Distally the two layers blend into the webs of the fingers and with the extensor aponeurosis at the metacarpophalangeal joints.

Anson and his colleagues described also a very thin layer of connective tissue that connects the extensor tendons within the compartment on the dorsum of the hand, and Lucien described this layer in the embryo as giving rise to both the tendons and the connective tissue connecting them, and thus forming an intermediate layer on the dorsum. Lucien expressed the opinion that variations in the development of this layer explain the variations in the tendons of the dorsum of the hand.

The dorsal interosseous fascia is a thin and discontinuous layer that directly covers the dorsal interossei, extending from one metacarpal to the next.

EXTENSOR TENDONS

The cutaneous nerves of the dorsum of the hand, as well as the superficial veins, cross the tendons of the dorsum superficially; in contrast, the radial artery and its branches lie deep to these tendons.

Abductor Pollicis Longus and Extensor Brevis

Above the wrist, the abductor pollicis longus and the extensor pollicis brevis cross superficial to the tendons of the extensor carpi radialis longus and brevis and enter the most radial compartment formed by the extensor retinaculum, on the radial side of the wrist (Fig. 6-73). The closely associated tendons pass distally well toward the palmar side of the radial surface of the wrist, the extensor brevis tendon coming to somewhat overlap the long abductor tendon, but lying slightly to its dorsal side. Here the two tendons form the palmar boundary of the anatomic snuff box, the depression on the radial side of the wrist seen when the thumb is abducted and extended.

The radial artery passes onto the dorsum of the hand deep to the tendons of the long abductor and short extensor of the thumb, and hence lies in the anatomic snuff box. Distally, the two tendons separate, the short extensor passing onto the dorsum of the thumb, the long abductor passing still more palmarward to attach to the anterior surface of the base of the first metacarpal. Attention has already been called to the fact that the insertion of the long abductor is often by more than one tendon, and that this muscle therefore frequently has an additional insertion other than into the metacarpal; variations in the insertion of the extensor brevis have likewise been described.

Radial Extensors

The tendons of the extensor carpi radialis longus and brevis are largely covered in the lower part of the forearm by the lower parts of the muscular bellies of the abductor pollicis longus and extensor pollicis brevis. Just as they appear below the lower border of the extensor brevis they pass into their compartment deep to the extensor retinaculum; distal to the retinaculum the two tendons diverge, the longus attaching to the radial side of the base of the second metacarpal, the brevis to the radial side of the base of the third meta-

carpal. It has already been noted that the tendon sheath around these tendons typically communicates with the overlying tendon sheath of the extensor pollicis longus.

Extensor Pollicis Longus

The extensor pollicis longus extends obliquely downward in the forearm, under cover of the common extensor, and barely appears on the radial side of the common extensor before it enters its compartment deep to the extensor retinaculum. The radial side of this compartment rounds the radial tubercle, and the tendon of the long extensor works around this as a pulley, turning somewhat more sharply in a radial direction as it passes the tubercle. The tendon thus runs obliquely across the lower part of the wrist and the upper part of the hand onto the dorsum of the metacarpal, crossing the tendons of the extensor carpi radialis longus and brevis toward the distal end of the extensor retinaculum; distal to the retinaculum, it forms the dorsal border of the anatomic snuff box.

Ulnar Extensor

The tendon of the extensor carpi ulnaris is in the most medial (ulnar) of the compartments formed by the extensor retinaculum. It lies in a groove on the lower end of the ulna, and passes along the dorsal aspect of the ulnar border of the wrist to insert into the ulnar border of the base of the fifth metacarpal. Its tendon sheath ends just before it inserts, and deep to the tendon as it crosses the base of the fifth metacarpal there is typically a small bursa.

Extensor Digitorum Brevis

This anomalous muscle (Fig. 6-75), the equivalent of the similarly named but normally occurring muscle on the dorsum of the foot, varies from a tiny muscular slip on the dorsum to a well-formed muscle. In its best-developed form it arises from the ulnar side of the dorsum of the wrist and divides into four slips whose tendons attach to the tendons of the extensor digitorum just proximal to the metacarpophalangeal joints. Well-developed specimens are usually reported sporadically as isolated cases. Smith (1896) reported find-

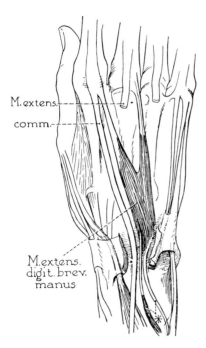

M.extens.----

comm.----

M.extens.
digit.brev.
manus

Fig. 6-75. An example of an extensor digitorum brevis of the hand. (Cauldwell EW, Anson BJ, Wright RR: Quart Bull, Northwestern Univ M School 17:267, 1943)

ing some trace of it in 35 of 50 hands, more commonly as a slip in the second metacarpal space. It may be a source of pain in the living person, in which case it should be excised (Ross and Troy; Reef and Brestin).

Long Extensors of the Fingers

The common extensor tendons and the proper extensor tendons of the index and little fingers occupy a central position at the wrist. The *extensor indicis* lies deep to the common extensor, and at the wrist shares the compartment within the extensor retinaculum, and a common tendon sheath, with the tendons of this muscle. The musculotendinous junction frequently extends into the compartment, and clenching the fingers, which draws the muscle deeper into the compartment, may produce a tenosynovitis here (Ritter and Inglis). Distal to the extensor retinaculum the proper extensor tendon of the index finger appears between the common extensor tendons to the middle and index fingers, and runs distally immediately on the

ulnar border of the latter tendon to join it at the level of the extensor hood.

The *extensor digiti minimi* arises in common with the extensor digitorum, and lies in the same plane as this muscle, superficially, in the forearm; the tendon at the wrist separates, however, from the common extensor tendons and passes through its own compartment deep to the extensor retinaculum; distal to the retinaculum it diverges toward the metacarpophalangeal joint of the little finger. On the dorsum of the hand it lies on the ulnar side of the extensor digitorum tendons, including that to the little finger when there is one; Schenck found the extensor digitorum tendon to the little finger absent in 56% of 57 hands and very small in 21.5% more. The extensor digiti minimi may thus have to form the extensor hood at the metacarpophalangeal joint largely or entirely by itself. It is typically split into two parts on the hand, a medial and a lateral, and the lateral tendon receives the attachment of the common extensor tendon; both parts unite in helping to form the extensor aponeurosis. Schenck found a single tendon in only 4 of 57 hands, 2 tendons in 48, and 5 instances in which there were more than 2.

The three or four tendons of the *extensor digitorum,* or common extensor, occupy the central compartment on the dorsum of the wrist, which they share with the deeper-lying extensor indicis tendon. They begin to diverge from each other toward the distal border of the extensor retinaculum, and pass distally over the metacarpals and the dorsal interossei toward the metacarpophalangeal joints. The common extensor tendon of the index finger lies on the radial side of the extensor indicis tendon; the common extensor tendon to the little finger, when present, joins the lateral of the two parts of the extensor digiti minimi in forming the extensor hood.

Distally on the hand, the tendons of the extensor digitorum are usually joined together by obliquely placed bands, termed "intertendinous connections," or "junctura"; these are somewhat variable. Bands that connect the tendon of the ring finger to those of the long and little fingers are apparently most common; these arise from the tendon of the ring finger somewhat proximal to the head of the metacarpal and pass obliquely distally, one going laterally to join the tendon to the long finger, the other medially to join that to the little finger or, sometimes, the proper extensor tendon to the little finger. Sometimes the tendon of the extensor digitorum to the little finger is closely bound to the tendon to the ring finger until the level at which the intertendinous connection should appear, and replaces this connection by turning medially to join the tendon of the extensor digiti minimi. Schenck found a tendon going to both fingers and dividing usually not far from the metacarpal heads in 12.5% of his cases. He also said that the smaller the tendon to the little finger, the more likely there will be an intertendinous connection to it.

A connection between the tendons of the index and long fingers is less constant, but often present. Such a connection may pass obliquely from the tendon of the long finger, distally and laterally, to join the common extensor tendon of. the index, or it may arise from the tendon of the index finger and pass distally and medially to join the tendon of the middle finger; in either case it usually passes superficial to the tendon of the extensor indicis.

Extensor Hood and Extensor Aponeurosis
On the fingers, the extensor tendons are joined by the lumbricals and in varying degree by the interossei, and the flattened complex tendon thus formed is commonly referred to as the extensor aponeurosis. There is little subcutaneous tissue between it and the skin, but bursae may occur in this tissue. Howard ('39) reported 26 cases of digital bursitis, usually occurring as a result of a cut or scratch over the dorsum of the joint, and emphasized that since these bursae are subcutaneous they have no connection with the joint cavity; thus opening of an infected one leads to prompt healing with no other involvement. Howard described these small bursae as being fairly constant over the proximal interphalangeal joints and occurring occasionally over the distal ones; bursae over the metacarpo-phalangeal joints were said to be rare and to occur more commonly on the fifth digit. Intermetacarpophalangeal bursae, between the heads of the metacarpals dorsal to the deep transverse metacarpal ligaments, were said to occur fairly frequently.

The formation and the attachments of the extensor aponeurosis and the functional implications of these details have been the subject of much investigation. The chief contributors to this aponeurosis are the long extensor tendons just discussed, and the following description applies to all four fingers; although the extensor aponeuroses of these fingers vary in regard to the insertion of interossei into them, and those of the little and index fingers differ from the others in that the proper extensors form or help form the aponeuroses, these differences do not affect the basic arrangement.

As the flattened but relatively narrow tendon of the extensor digitorum (communis) reaches the distal portion of the head of the metacarpal, it expands markedly so as to cover not only the entire dorsum of the metacarpophalangeal joint but also much of both sides of this joint; this expansion, the central portion of which is thicker and continues the extensor tendon, is usually known as the extensor hood (Figs. 6-20*b* and 6-76). The more proximal fibers of the hood, as they leave the sides of the extensor tendon, arch around the metacarpophalangeal joint in a palmar direction, covering the sides of the joint capsule and attaching, according to Braithwaite and co-workers, to the deep transverse metacarpal ligaments. Distal to the deep transverse metacarpal ligament the hood consists of additional transverse fibers that unite the extensor tendon to the lumbrical and interosseous tendons, which lie on the flexor aspect of each side of the joint and form the palmar edges of the hood. According to Landsmeer ('49) these intertendinous fibers consist of two layers, more superficial ones that pass across the dorsum of the extensor tendon to unite the laterally lying interossei and lumbricals, and deeper ones that pass from the edge of the extensor tendon to these lateral tendons (Figs. 6-77 and 6-78). Some of the deeper fibers of

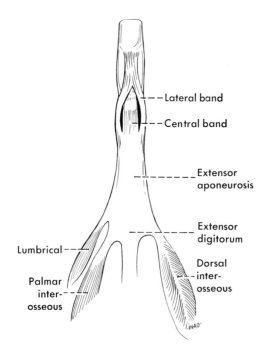

Fig. 6-76. Diagram of the basic structure of the extensor aponeurosis of a finger.

this part of the hood also attach to the capsule of the joint, and have in general a palmar and distal direction. The attachments of the hood to the joint serve as a hinge, about which the hood can slip distally over the dorsum of the joint as it is flexed, proximally as it is extended.

Because the chief direction of the attaching fibers of the hood is distally and palmarward, movement of the hood in a distal direction is more free than is movement in the proximal direction: the latter tends to tighten the fibers and move the phalanx into extension or hyperextension. It is through these attachments that the extensor tendons can act upon the proximal phalanges alone, as when these phalanges are extended or hyperextended while the other two phalanges are at the same time flexed. This action has been said to be due to an insertion of the extensor tendon upon the dorsum of the proximal phalanx just distal to the metacarpophalangeal joint; however, Kaplan ('50) reported that a slip from the deep surface of the tendon to the proximal phalanx occurs in only about a third of digits, and moreover, is never tendinous in nature;

Landsmeer agreed that it is neither constant nor tendinous. Thus it can hardly be regarded as a true insertion of the muscle.

Proximal movement of the hood is brought about by contraction of the common or proper extensor muscles; distal movement occurs passively, through the pull exerted on the extensor aponeurosis when the long flexors act on the interphalangeal joints. This distal movement gives the lumbrical and interosseous components of the hood a more favorable leverage on the proximal phalanx, and presumably contributes to their efficiency in flexing this phalanx.

As the extensor tendon passes across the dorsal aspect of the metacarpophalangeal joint it is often said to have loose connective tissue between it and the capsule of the joint. Braithwaite and his colleagues said, however, that a bursa intervenes between the two, thus facilitating movement of the hood over the joint.

Over the body of the proximal phalanx the extensor hood narrows as the interosseous and lumbrical tendons converge toward the centrally situated extensor tendon. As these join this extensor tendon, but before it reaches the level of the proximal interphalangeal joint, the tendon divides into three bands, a middle and two lateral ones (Figs. 6-76 to 6-78). In contrast to some earlier descriptions, it is now agreed that the tendons of the lumbricals and of those interossei that insert into the extensor apparatus contribute fibers (labeled medial and lateral interosseous bands in Figure 6-77) to both the middle and the lateral band and therefore obtain, like the long extensor, an insertion on both the middle and distal phalanges.

As the extensor tendon divides into middle and lateral bands and these are joined by the interossei and lumbrical, the three bands are united by fibers that have been referred to as the triangular membrane or ligament; as they pass across the proximal interphalangeal joint, the median and lateral slips together with the triangular membrane form a hood over this joint that protects it both dorsally and on the sides. The central band of the aponeurosis is fused to the dorsal aspect of the

Fig. 6-77. Details of the attachment of the extensor aponeurosis, lateral view. (Redrawn from original of Fig. 2, Landsmeer JMF: Anat Rec 104:31, 1949)

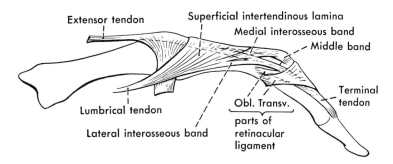

Extensor tendon

Superficial intertendinous lamina

Medial interosseous band

Middle band

Lumbrical tendon

Lateral interosseous band

Obl. Transv.

parts of retinacular ligament

Terminal tendon

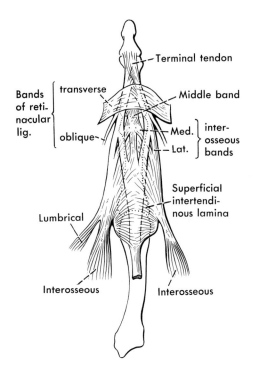

Terminal tendon

Bands of retinacular lig.

transverse

oblique

Middle band

Med.

Lat.

interosseous bands

Superficial intertendinous lamina

Lumbrical

Interosseous

Interosseous

Fig. 6-78. Details of the attachment of the extensor aponeurosis, dorsal view. (Redrawn from original of Fig. 1, Landsmeer, JMF: Anat Rec 104:31, 1949)

capsule of the joint, and the lateral bands, which lie dorsolateral to the joint, are prevented from slipping too far forward by the triangular membrane. Finally, each lateral band is firmly anchored to the lateral part of the joint capsule by a retinacular ligament (Landsmeer, '49) that allows proximal and distal movement at the joint but limits dorsal displacement of the lateral band. It consists of two sets of fibers (Figs. 6-77 and 6-78), a thin layer of superficial ones that pass almost transversely to attach to the flexor tendon sheath on the proximal phalanx, and deeper fibers forming a slender but strong oblique band or cord that passes proximally to insert into the lateral border of the proximal phalanx.

Landsmeer ('49, '63), Haines ('51), and Stack ('62) have all emphasized the importance of the oblique cords of the retinacular ligaments, now often called Landsmeer's ligaments, in producing flexion of the middle phalanx concomitant with that of the distal phalanx; as the profundus flexes the distal phalanx, the lateral bands are drawn distally, and therefore both relax the middle band and tense the oblique cords which, because of their palmar attachment at the proximal interphalangeal joint, produce flexion at that joint. Thus, the more the distal joint is flexed, the more the proximal one is. Conversely, as the proximal joint is extended the oblique cords are said to help to extend the distal one. Harris and Rutledge, however, denied this entire concept, stating that in their experiments the oblique cords were relaxed in full extension of the proximal joint and became taut only when the distal joint was flexed to 70°; they also found that if the lateral bands were cut proximal to the oblique cords while both joints were in full extension, the distal phalanx dropped to 70° of flexion. Thus, according to their findings, the oblique cords can neither contribute to the first 70° of flexion of the proximal joint, nor to the extension at the distal joint. Sarrafian and co-workers, in contrast, interpreted their measurements of strain in various elements of the extensor complex (they did not investigate the oblique cords) as confirming the concept of flexion at the proximal interphalangeal joint.

After passing across the proximal interphalangeal joint, the middle band of the extensor aponeurosis, consisting, as already described, of fibers from both the extensor tendon and the interossei (and a lumbrical) inserts upon the base of the middle phalanx, while the two lateral bands, also consisting of extensor and interosseous fibers (and, on the radial side, of lumbrical fibers) converge over the dorsal aspect of the middle phalanx to unite into a single tendon that crosses the distal interphalangeal joint, fusing here with the capsule of that joint, and goes to its insertion on the base of the distal phalanx.

MOVEMENTS OF THE DIGITS

Movements of the thumb are complex and different enough to demand a special description for this digit, but the movements of the other four digits can be discussed together, since they are essentially similar. As already noted, the only real difference in movements of the four fingers is in the special although slight mobility of the fifth metacarpal, which can be appreciably flexed in the movement of opposition so useful in gripping objects in the palm of the hand. This movement, brought about largely by the opponens digiti minimi, usually involves both movement at the carpometacarpal joint of this digiti and increase in the curvature of the carpal arch, the latter presumably aided by the origin of the flexor and opponens from the flexor retinaculum (transverse carpal ligament).

MOVEMENTS OF THE FINGERS

The primary movements of the four medial digits at the metacarpophalangeal joints are those of abduction-adduction and of flexion-extension. Circumduction, also permitted, is simply a combination of these fundamental movements; active rotation at the joint is not appreciable, but a limited amount of passive rotation can be produced by twisting the digit.

Abduction and Adduction

The abductors of the fingers are the four dorsal interossei and the abductor digiti minimi (Fig. 6-79), which represents functionally a dorsal interosseous on the ulnar border of the hand; the dorsal interossei vary in their relative amount of insertion into the proximal phalanx and into the extensor aponeurosis, respectively, but this affects primarily the extensor rather than the abductor action of these muscles. In brief, the insertions of the dorsal interossei are such that the first one abducts (moves radially) the index finger; the second and third abduct the long finger or middle digit, through which the midline of the hand runs, in either direction; the fourth abducts or moves ulnarward the ring finger; and the abductor digiti minimi acts similarly on the fifth digit. The extensor digitorum also abducts the fingers as it extends them, but cannot abduct one finger independently of the others. The extensor digiti minimi (proprius) is a strong abductor of the little finger (Jones).

In an exactly similar fashion, the three palmar interossei (Fig. 6-80) adduct the index, ring, and little fingers toward the middle digit, which needs no adductor of its own since it is supplied with abductors that will move it in either direction. The long flexors of the fingers adduct these as they flex them, but like the extensor cannot exert an effect on each finger separately. However, the extensor indicis (proprius) is said to be able to adduct the index finger independently (Jones).

The four lumbricals, which typically pass to the extensor aponeurosis on the radial side of each finger, might be expected to abduct the index and middle fingers, adduct the ring and little fingers—that is, produce radial deviation of all four digits. Braithwaite and his colleagues regarded this as an important action of the lumbricals: they said that the first dorsal interosseous cannot maintain a good grip between the thumb and forefinger, that is, cannot produce good radial deviation or prevent ulnar deviation against pressure, but that the lumbrical can. However, neither

Fig. 6-79. Abductors of the digits, dorsal view. The extensor digiti minimi, not shown, is also an abductor of that finger.

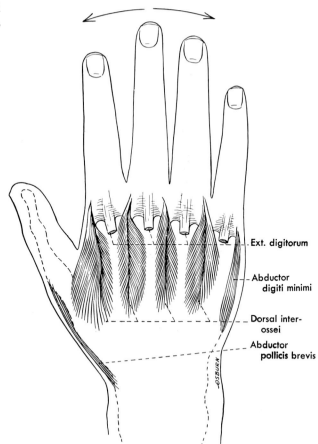

Ext. digitorum

Abductor digiti minimi

Dorsal inter-ossei

Abductor pollicis brevis

Eyler and Markee nor Backhouse ('68) could demonstrate any tendency toward radial deviation of the digits upon stimulation of the lumbricals. Sunderland did report that when the first two lumbricals are paralyzed the digits to which they are attached rotate so that the dorsum is turned to the ulnar side, and rotate back again after regeneration of the nerve supply to the lumbricals.

The expected innervation of all the interossei and of the abductor digiti minimi is through the ulnar nerve; usually, therefore, this nerve is entirely responsible for abduction and adduction of the digits. However, Sunderland ('46) reported that he found, both upon dissection and upon clinical examination, that about 3% of first dorsal interossei are supplied exclusively by the median nerve rather than by the ulnar, and that occasionally this muscle is supplied by both the me-

dian and the ulnar. Clifton placed the incidence of innervation by the median nerve at 10%.

Flexion at the Interphalangeal Joints

Flexion here is apparently well understood. The only muscle capable of acting upon the distal phalanges is the flexor digitorum profundus, therefore it, through its tendons, is the sole flexor at the distal interphalangeal joints. As already noted, flexion at the distal interphalangeal joint, by drawing the lateral bands of the extensor tendon distalward and thus tightening the oblique cords that attach to the lateral bands, produces concomitant flexion at the proximal interphalangeal joint. Thus action by the obvious flexor of the proximal joint, the flexor digitorum superficialis, while adding strength to the flexion, is not necessary to produce it. (However, as noted

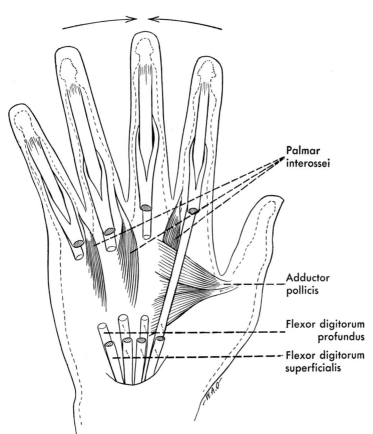

Fig. 6-80. Adductors of the digits, palmar view.

Palmar interossei

Adductor pollicis

Flexor digitorum profundus

Flexor digitorum superficialis

by Kaplan, '36, and later by Landsmeer, '63, this is true only if the proximal joint is not hyperextended; in such a case the lateral and oblique bands are drawn dorsally so that tightening them by flexing the distal phalanx further hyperextends at the proximal interphalangeal joint.)

The fact that the flexor digitorum profundus alone can produce satisfactory flexion of both the distal and the middle phalanges was long made use of in repairing flexor tendons severed in the digital sheath; in such cases, best results were usually obtained by removing the tendon of the superficial flexor from the finger, thus avoiding the difficulty of trying to restore both the normal gliding mechanism between the tendons of the superficial and deep flexors and an adequate gliding mechanism between these and their surroundings.

Flexion at the Metacarpophalangeal Joints

The usual concept of flexion at the metacarpophalangeal joints has been that it is brought about primarily by the intrinsic muscles of the hand, particularly the interossei (Fig. 6-81); this is logical, since it is these and the lumbricals (and for the little finger the flexor digiti minimi) that prevent the joints from being hyperextended, as they are when the intrinsic muscles are paralyzed, by flexion at the proximal interphalangeal joints and the pull of the extensor digitorum. The exact part normally played by the various muscles in movements of flexion is not yet completely clear, however. In regard to the lumbricals, Eyler and Markee regarded them as weak flexors of the proximal phalanx, but as important in stabilizing it against the pull of the extensor digitorum. Long and Brown and subsequently Backhouse ('68), studied the muscles electromyographically and found that they contract during metacarpophalangeal flexion only when the interphalangeal joints are either extended or being extended; Backhouse added that when they are stimulated they first extend the interphalangeal joints, and only thereafter, if the stimulation is intense enough, do they flex the metacarpophalangeal joints.

Although the interossei are usually regarded as being flexors of the metacarpophalangeal joints as well as extensors of the interphalangeal joints, Stack expressed the opinion that only those parts of the interossei attaching to the extensor aponeurosis can flex the joint, while that part attaching to the proximal phalanx extends. Long and Brown, investigating the activity of the two dorsal interossei of the long finger, which typically

have both attachments, concluded that these muscles do not participate at all in movement at the metacarpophalangeal joint alone, but only when flexion here is combined with interphalangeal extension. Thus they said that in the case of this finger, at least, flexion of the metacarpophalangeal joint simultaneously with flexion of the interphalangeal joints is brought about entirely by the long flexor tendons. Backhouse, however, while noting that it is the long flexors that are chiefly responsible for the power grip, found activity in both dorsal and palmar interossei whenever the metacarpophalangeal joints were flexed, regardless of the position of or direction of movement at the interphalangeal joints. (In a later paper, Long and co-workers agreed that the interossei are active in the power grip.) It thus seems that the interossei are indeed the

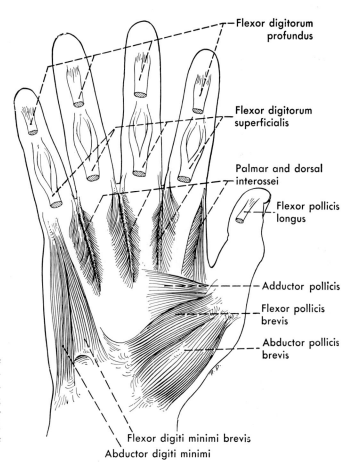

Fig. 6-81. Flexors of the digits. The lumbricals, which help flex the metacarpophalangeal joints if the interphalangeal ones are extended, are not shown; and there has been doubt as to how much the interossei contribute when all the joints are flexed simultaneously (see text).

Flexor digitorum profundus

Flexor digitorum superficialis

Palmar and dorsal interossei

Flexor pollicis longus

Adductor pollicis

Flexor pollicis brevis

Abductor pollicis brevis

Flexor digiti minimi brevis
Abductor digiti minimi

prime flexors at the metacarpophalangeal joints, participating in every movement that requires their flexion or stabilization in flexion.

Extension at the Metacarpophalangeal Joints

Although, as already noted, Stack regarded the parts of the interossei attaching into the proximal phalanges as being extensors at the metacarpophalangeal joints, Long and Brown found no participation by the dorsal interossei during this movement, even when the interphalangeal joints were extended at the same time. Active extension of the proximal phalanges at the metacarpophalangeal joints is apparently carried out solely by the long extensors—the extensor digitorum for all fingers and the extensor indicis and extensor digiti minimi for the index and little fingers. (Long and Brown regarded these muscles as being primarily responsible for extension of all the joints of a digit when they are extended simultaneously.) Two factors contribute to extension of this joint with simultaneous flexion of the interphalangeal joints; there is a passive distal pull on the extensor tendon when the proximal interphalangeal joint is flexed, thus tending to extend the proximal phalanx (e.g., Landsmeer, '63), and at the same time the extensor digitorum also contracts (Long and Brown).

Extension at the Interphalangeal Joints

The exact mechanism of extension of the interphalangeal joints, that is, of the middle and distal phalanges, has been the subject of considerable diversity of opinion. The argument centers around the role played by the long extensor tendon on the one hand and the lumbrical and interosseous muscles on the other, in extension of these joints; the deductions concerning the relative importance of these roles have been based upon anatomic dissection alone, this combined with pulling upon the tendons at anatomic or surgical dissection, analysis of the movements after various nerve lesions, experimental observations on the effect of nerve blocks and of stimula-

tion of the muscles, and electromyographic investigation.

Willan, for instance, expressed the view that the extensor digitorum has no appreciable effect on the middle and distal phalanges, extension of these joints being caused primarily by the interossei and lumbricals. Braithwaite and his colleagues supported the opposite view, that the extensor digitorum is responsible for all natural movements of extension at the interphalangeal joints, the interossei assisting only by maintaining the extensor tendon in position over the knuckles, and the lumbricals by drawing forward and relaxing the profundus tendons. Even extension of the interphalangeal joints with flexion of the metacarpophalangeal ones, usually attributed to the interossei and lumbricals, was explained by them as being a result of the passive pull backward on the extensor tendon caused by angulation of this tendon over the large metacarpal head (Figs. 6-82 and 6-83).

The observations and experiments of numerous workers indicate the the truth lies between these two concepts, and that the long extensors and the intrinsic muscles both normally contribute to interphalangeal extension, although there may be some small disagreement as to how much each contributes under various conditions. Kaplan ('50), and Eyler and Markee, for instance, agreed that either the extensor digitorum or the intrinsic muscles can extend the interphalangeal joints, and that they normally act together in so doing. Kaplan added that the extensor digitorum alone cannot fully and strongly extend at the interphalangeal joints against the pull of the long flexors, and Stack said that the extensor alone cannot fully extend the distal joint because the bowed lateral bands move dorsally and closer together unless they are retained by the pull of the intrinsic muscles, and thereby become so functionally lengthened that they are still not completely tight by the time the middle band stops the proximal movement of the tendon. Long and Brown found a range from no activity to major contraction of the extensor digitorum when the interphalangeal joints were ex-

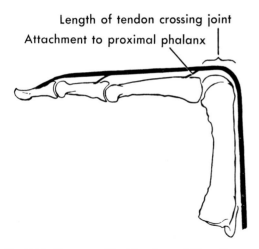

Length of tendon crossing joint
Attachment to proximal phalanx

Fig. 6-82. Concept of Braithwaite and his colleagues as to the mechanism of extension of the interphalangeal joints with flexion of the metacarpophalangeal one: because of the greater length that the tendon must traverse in crossing the metacarpophalangeal joint (compare the brackets in this and Fig. 6–83) the extensor hood does not shift appreciably forward, as shown by the attachment to the proximal phalanx; therefore, the tendon distal to this is passively tensed and extends the interphalangeal joints. (Braithwaite F, Channell GD, Moore FT, Whillis J: Brit J Plastic Surg 2:175, 1949)

tended simultaneously with flexion at the metacarpophalangeal joints (this indicates that the intrinsic muscles, and the passive factors analyzed by Braithwaite and his colleagues, may often be most important during this particular movement), but they said it averaged definite participation without major contraction.

Stack apparently regarded the role of the lumbricals and interossei in extension of the interphalangeal joints as being primarily concerned with the distal phalanx, where they pull upon the lateral bands and therefore aid the extensor digitorum in producing complete extension of this phalanx. Kaplan regarded the lumbricals as being particularly good extensors of the distal phalanges. Eyler and Markee found, after appropriate nerve blocks, that the interossei alone can extend the interphalangeal joints completely, and reported that stimulation of the first and second lumbricals also produced extension of the interphalangeal joints, although not complete ex-

tension. In contrast to this, Backhouse reported strong and complete extension when a stimulating current was applied to the lumbricals, and regarded them as the chief extensors of the middle and distal phalanges; the action of the interossei in this extension was thought to be a minor effect, secondary to their action in flexing the metacarpophalangeal joints, while the lumbricals were active regardless of the position of the latter joint. Long and Brown also found the lumbricals active in all extension of the interphalangeal joints, but apparently contributing less when the extensor digitorum was extending all the joints; the interossei appeared to contract to reinforce the lumbricals in extending the interphalangeal joints only when the metacarpophalangeal joint was flexed or being flexed.

In summary, therefore, it appears that extension at the interphalangeal joints is partly dependent upon mechanical factors determined by the attachments of the extensor tendon, but that both the extensor digitorum and the lumbricals actively participate in most of this extension, and the interossei also aid if the metacarpophalangeal joint is either being flexed or held in flexion.

Fig. 6-83. Distal movement of the extensor hood with flexion of the interphalangeal joints, as shown by the reversed direction of the exaggerated attachment to the proximal phalanx. (Braithwaite F, Channell GD, Moore FT, Whillis J: Brit J Plastic Surg 2:175, 1949)

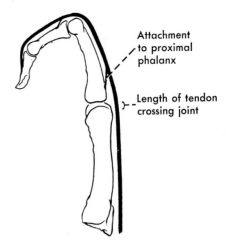

Attachment to proximal phalanx

Length of tendon crossing joint

MOVEMENTS OF THE THUMB

These typically involve simultaneous movement, in varying degree, of the two phalanges and the metacarpal, and the following analysis of movement at the various joints is therefore necessarily somewhat artificial. Further, movements are steadied by the contraction of other muscles, including often opposing ones, as brought out particularly by Weathersby and his colleagues.

At the Interphalangeal Joint

The interphalangeal joint of the thumb, a hinge (ginglymoid) one like the other interphalangeal joints, allows flexion to about 90°, and a variable amount of hyperextension that probably depends upon the laxity of the anterior part of the capsule (Harris and Joseph). The only flexor at this joint is the flexor pollicis longus, innervated by the median nerve, but a number of other muscles, including the opposing extensor pollicis longus, contract to steady the movement. The chief extensor is the extensor pollicis longus. However, a part of the tendon of the extensor pollicis brevis frequently joins or parallels the long extensor tendon to gain an insertion on the distal phalanx, and both the short abductor and the adductor typically send tendinous slips to the long extensor tendon; thus four muscles, two innervated by the radial nerve, one by the median, and one by the ulnar may conceivably help to extend the distal phalanx. Weathersby and colleagues reported that in unopposed extension the short abductor typically contracts before the long extensor does, but said the adductor and the extensor brevis are inactive unless the movement is opposed, in which case all the muscles of the thumb except the long flexor become active.

At the Metacarpophalangeal Joint

Because the head of the first metacarpal is more flattened anteroposteriorly than are the heads of the other metacarpals, and the articular surface usually does not extend as far onto the palmar aspect of the head, movements of abduction-adduction, of rotation, and of flexion are often somewhat limited at this joint. Coonrad and Goldner reported that

among 1,000 hands flexion varied from 10° to 100° and averaged 75°; abduction-adduction, with the joint slightly flexed, ranged from 0° to 20° and averaged 10°; and extension, or hyperextension, ranged from 0° to 90° and averaged only 20°. Hyperextension is apparently influenced not only by the degree of laxity of the anterior part of the capsule but also by the variable shape of the articular surface (Harris and Joseph).

It should be noted that movements of flexion and extension and of abduction and adduction at both the metacarpophalangeal and the carpometacarpal joints of the thumb are defined in terms of the normal position of the thumb. Since the dorsum of the thumb lies almost at a right angle to the dorsum of the fingers, a movement of flexion of the thumb, that is, a bending of the thumb toward its flexor or palmar surface, is necessarily at an approximate right angle to a movement of flexion of the fingers—while the fingers move toward the palm, the thumb moves in flexion across and parallel to the surface of the palm. Similarly, while abduction and adduction of the fingers are movements in the plane of the palm, abduction at the carpometacarpal and metacarpophalangeal joints moves the thumb away from contact with the radial border of the index finger (at an angle of 80° to the plane of the palm, according to Napier), and adduction brings it back into contact with the index finger.

The chief flexors at the metacarpophalangeal joint of the thumb are the flexor pollicis brevis and the adductor pollicis, but both the short abductor and the long flexor are usually regarded as assisting in this action. During flexion of all three joints simultaneously, however, the short abductor is largely inactive (Forrest and Basmajian); it therefore apparently contributes little to metacarpophalangeal flexion. The long flexor does not contract for this movement (Weathersby and colleagues), although the long abductor does, obviously to steady the metacarpal. While most of the flexors are innervated by the median nerve, the adductor and the deep head of the short flexor (not always present) are innervated by the ulnar nerve, and Jones ('19) found that good flexion of this joint occurred after interruption of the median nerve. Abduction and adduction are carried out, primarily, by the short abductor and the adductor, respectively. The chief extensor is the extensor pollicis brevis, assisted secondarily by the extensor longus. Both the opponens and the short abductor are also active during this movement (Weathersby and colleagues, Forrest and Basmajian), probably simply for stabilization.

At the Carpometacarpal Joint

Movements at the carpometacarpal joint, so important to the usefulness of the human thumb, are usually complex, for movements of the metacarpal are rarely exclusively in a plane parallel to the palm (the plane of flexion-extension) or in one at an approximate right angle to the palm (the plane of abduction-adduction). Rather, abduction usually accompanies both flexion and extension, and any extensive movement of flexion, extension, or abduction is accompanied by rotation of the metacarpal (a medial rotation with flexion and with abduction, a lateral one with extension). This rotation is facilitated by the attachments of the muscles, but is apparently not exclusively a product of the angle of muscular pull; it occurs also upon passive movement (Haines) if the ligaments of the carpometacarpal joint (see Fig. 6-11) are intact, but disappears if they are cut.

Abduction of the first metacarpal is brought about primarily by the abductor pollicis brevis, hence through the median nerve; however, the long abductor, primarily an antagonist of the opponens in that while abducting it extends the metacarpal and rotates it laterally, and innervated by the radial nerve, can compensate fairly adequately for the short abductor (Jones). The opponens, innervated by the median nerve, has been regarded as both an abductor and an adductor, but primarily flexes and rotates the metacarpal. Weathersby and co-workers recorded it as more active in abduction than adduction, but active also in all other movements of the thumb except flexion at the interphalangeal joint; Forrest and Basmajian did not test it in

adduction, but suggested that its activity in abduction may be primarily for stabilization rather than movement of the metacarpal. The former workers reported that both the short extensor and the short flexor are active during abduction; the latter (who did not test the extensor) reported that the activity of the flexor was slight.

Adduction is brought about primarily by the adductor, which also tends to flex the proximal phalanx, and is innervated by the ulnar nerve. The short and long flexors can assist in the adduction that accompanies opposition, the long extensor adducts as it extends, and the opponens is sometimes listed as an adductor (see above). The flexor brevis and the long but not the short extensor are also active during adduction (Weathersby and colleagues). The first dorsal interosseous, although acting largely on the index finger, is important in the pinch between the adducted thumb and the forefinger.

Flexion of the metacarpal, typically preceded by abduction, is brought about primarily by the adductor and the short flexor as they flex the proximal phalanx; pure flexion is largely a useless movement, since in this the thumb is dragged across the palm of the hand. When this is done, however, the short abductor and the opponens are both active and the long extensor is also, obviously steadying the movement. Pure *extension* from the resting position is limited, and apparently brought about primarily by the extensor pollicis brevis. The long abductor both extends and radially rotates.

As already noted, *rotation* of the metacarpal typically accompanies other movements. Total rotation amounts to about 45°, according to Haines, of which 30° of medial rotation is contributed during flexion, 15° of lateral rotation is contributed during extension. According to Haines, rotation during passive movement occurs primarily toward the end of the movement, as the guiding ligaments tauten. The direction of muscle pull is apparently also important, however, and Hendry expressed the opinion that most operations for producing opposition of the thumb fail after a time because the exact pull of the muscles cannot be duplicated by tendon grafts, and the strain upon the ligaments stretches them.

Napier said *opposition* takes place in two ways: for power, the opponens is the prime mover, medially rotating, flexing, and adducting the metacarpal at the same time; for precision, the short abductor first stabilizes the thumb in abduction, after which the thumb is circumducted and medially rotated, this being the movement that is guided primarily by the ligaments of the carpometacarpal joint. Thus, in the more common movement of opposition in which the thumb is brought in contact with a fingertip, most of the short muscles of the thumb might be expected to take part in the abduction, rotation, flexion, and adduction of the metacarpal and proximal phalanx, while the flexor pollicis longus contracts to flex the distal phalanx. Weathersby and his colleagues reported that the long flexor and all four short muscles are active in opposition; the long abductor and sometimes the short extensor, which oppose the movement, were also active.

Forrest and Basmajian investigated in detail the activity of the opponens, the short flexor, and the short abductor in making light and firm contact between the thumb and both the side and the palmar surface of each fingertip. Although all three muscles were active in each movement and became more active as movement progressed to the little finger, the opponens was most active during light touch, the short flexor least active; with firm contact to the side of each finger the short flexor became more active than the opponens, while in firm contact with the palmar surface the activity of the opponens increased to equal or exceed that of the short flexor, the abductor being the least active of the three muscles in these movements. Although they did not test the adductor, it also obviously may be active in opposition, as Weathersby found it, especially when contact is between the thumb and index finger.

Since many of the long and short muscles of the thumb typically work together in producing opposition, this movement may be carried out fairly normally when some of them are not working. For instance, Murphey and his co-workers said that the short flexor can flex and rotate the metacarpal almost as well as the opponens, and Jones said that the long abductor and long flexor working together can produce good rotation and opposition. He, Murphey and his colleagues, and others, have thus emphasized that the activity of the various short muscles cannot be assessed by simply observing the movement of the digit.

Just as flexion and medial rotation usually go together in movement of the first metacarpal, so do extension and lateral rotation, a combined movement best referred to as *reposition*. The muscles used in this movement are the long abductor and the long extensor, both of which externally rotate the metacarpal as they extend it, and the short extensor, which helps the long extensor to extend it. These muscles are all innervated by the radial nerve, and their unopposed action following paralysis of the median and ulnar nerves produces the extension and external rotation of the thumb that is designated as "ape hand."

INJURIES TO EXTENSOR MUSCLES AND TENDONS

Radial Paralysis

Paralysis of the long extensor muscles as a result of injury to the radial nerve not only produces wristdrop but, because these muscles are the only ones that can produce extension at the metacarpophalangeal joints, badly cripples the hand. Since the interossei and lumbricals are efficient extensors of the interphalangeal joints only when the metacarpophalangeal joints are stabilized in partial extension, none of the digital joints can be actively extended: extension of the fingers is limited to that obtainable by gravity or by sharply flexing the wrist to produce a pull on the extensor tendons. When regeneration of the radial nerve does not occur, transfer of flexor tendons to the dorsum can do much in restoring a functionally useful hand.

The muscles commonly used to replace the extensors are the flexors of the wrist: the flexor carpi radialis, flexor carpi ulnaris, or palmaris longus. Scuderi reported best results from suturing the tendon of the palmaris longus to the tendon of the extensor pollicis longus, and that of the flexor carpi ulnaris to the extensor digitorum; Young and Lowe reported satisfactory results from a number of different combinations (*e.g.,* attachment of the flexor carpi radialis to tendons of the thumb, and of the flexor carpi ulnaris to the common extensor). These transfers leave a flexor of the wrist when all three are present; if the palmaris longus is absent, however, necessitating transfer of both the flexors present, the efficiency of the operation is much reduced because the wrist can no longer be stabilized. Young and Lowe quoted 97% of extension of the fingers as being obtainable when one flexor of the wrist is left, only 21% when there is no such flexor.

Bunnell ('49) emphasized that the normal excursion of the movers of the digits is so much greater than that of movers of the wrist—about 2 to 2.87 inches (7.3 cm), as opposed to about 1.25 inches (3.2 cm)—that no mover of the wrist can be expected to function adequately in moving both the wrist and the fingers. He also said that if full excursion in extensor of the thumb is expected, the same tendon should not be attached to both the long extensor and long abductor tendons, since the latter, having a shorter excursion, will markedly limit the movement of the former. Bunnell regarded four movers—for dorsiflexion of the wrist, extension of the fingers, extension of the thumb, and stabilization of the thumb in dorsiflexion—as being essential if best results are to be obtained from tendon transfers following radial palsy. As examples of desirable transfers, he listed: transfer of the flexor carpi ulnaris to the extensor tendons of the three medial fingers, of the flexor carpi radialis to the long extensors of the thumb and the index finger, of the pronator teres to the extensor carpi radialis, and either a slip from the latter tendon to the abductor pollicis longus or transfer of the palmaris longus to the abductor, where it can still flex the wrist; or use of the flexor carpi ulnaris to extend all four fingers, of the palmaris longus to extend

the thumb, of the pronator teres to extend the wrist, and of the flexor carpi radialis to abduct the thumb and flex the wrist.

If flexors are also involved, a proper number for transfer may not be available. Arthrodesis of the wrist in about 30° of dorsiflexion abolishes the necessity for both an extensor and a flexor of the wrist, and the metacarpophalangeal joint of the thumb can be arthrodesed to stabilize the thumb. Tenodesis of the extensor tendons to the radius and ulna, to produce extension of the fingers with flexion of the wrist, has also been used when no flexors were available for transplant.

In contrast, if only the deep branch of the radial nerve is injured, the problem is relatively simple: a procedure found satisfactory by Bunnell is transfer of the extensor carpi radialis brevis to the extensor tendons of all the digits and transfer of the flexor carpi radialis or palmaris longus to the long abductor of the thumb.

Injuries to Extensor Tendons on Hand

Suture of severed extensor tendons on the dorsum of the hand or on the fingers can usually be expected to give good results, in contrast to the results frequently obtainable when flexor tendons must be sutured. The extensor tendons need no smooth gliding surface as do the flexor ones, and the connective tissue of the dorsum, which normally allows free movement of the tendons, is too scanty to form the dense adhesions that may involve flexor tendons. Further, the direct contact between the tendons and the surrounding tissue provides an adequate local blood supply.

The tendon of the extensor pollicis longus is sometimes ruptured in connection with Colles' fracture; if necessary, the extensor indicis can be used to replace this tendon (Pulvertaft). Apparently, only a very few cases of rupture of the tendons of the extensor digitorum as a result of fracture at the wrist have been reported (Gladstone); however, rupture of the more ulnar tendons of the extensor in rheumatoid arthritis with dorsal subluxation of the distal end of the ulna, thus a consequence of fraying over a roughened area, is not especially uncommon (Lipscomb, '68).

Burman reported three cases of tendinitis of the extensor digitorum, apparently as a result of repeatedly stretching the tendon by flexing the proximal interphalangeal joint with the metacarpophalangeal joint extended, or flexing the latter with the former extended.

Tenosynovitis, or tendovaginitis, of extensor tendons at the wrist has been discussed in the preceding chapter.

Injuries of Extensor Apparatus on Fingers

Injuries affecting the tendons on the dorsum of the fingers may be to some part of these tendons, or may involve some other part of the extensor apparatus and alter the function of the extensor tendons. Thus a number of different deformities can result. Injuries over the metacarpophalangeal and proximal interphalangeal joints are particularly likely to involve the complicated and rather delicate retinacula that hold the various components of the extensor aponeurosis in place, and Kaplan, who has studied the extensor apparatus of the fingers intensively, suggested in his 1950 paper that if surgical repair at these joints is to be successful it must restore, essentially, these finer connections. Since displacements of the tendons, or parts of them, at these joints can be in either a dorsal or a palmar direction, some injuries cause an extensor deformity, others a flexor deformity.

"Trigger finger," in which a finger can be flexed voluntarily but cannot be extended past a certain point without assistance, is most commonly due to swelling of the flexor tendon or narrowing of its sheath, but may be caused by anterior displacement of the extensor tendon at a metacarpophalangeal joint (Straus). This displacement is said to occur usually following a direct blow over the joint or a marked contraction of the extensor tendon against resistance, with resulting detachment of the tendon from its continuity with the extensor hood. A tendon so detached lies in its normal position when the fingers are extended, but flexion of the proximal phalanx to about 45°, with consequent increasing ten-

sion on the tendon as it crosses the pulley of the metacarpophalangeal joint, tends to make the tendon slide off the prominence of the knuckle and lie on the side of the joint rather than on the dorsum; in consequence, voluntary contraction of the extensor digitorum cannot thereafter produce complete extension of that metacarpophalangeal joint.

Elson reported a case in which the radial side of the hood on the little finger showed no tear, but was thin and lax, and allowed the tendon to slip so far forward on the ulnar side of the metacarpal during flexion that attempts to extend the finger then produced flexion at the metacarpophalangeal joint. Kettlekamp and co-workers reported five cases in which the tendon to the long finger was dislocated as a result of trauma. Displacement to the ulnar side, with flexion of the metacarpophalangeal joints, hyperextension of the proximal interphalangeal joints, and perhaps slight flexion of the distal interphalangeal joints together with ulnar deviation of the digits is more commonly seen as a deformity associated with rheumatoid arthritis. Backhouse ('68) suggested that a number of factors may contribute to the ulnar displacement of the tendons, including the stretching of the aponeurosis and collateral ligaments on the weaker radial side, and the stronger attachment of the ulnar interossei into the aponeurosis. The interossei, he reported, are usually less active than normal, and never spastic.

Braithwaite and his co-workers also reported that the tendons are prone to slip laterally when the interossei, which help to anchor them, are paralyzed. Further evidence of the importance of the attachments of the extensor tendon to the extensor hood is found in the statement of Braithwaite and his colleagues that if the hood becomes displaced dorsally the metacarpophalangeal joint is held in hyperextension; apparently, the lateral attachments of the extensor tendon through the hood are necessary to maintenance of a proper balance at the metacarpophalangeal joint. The importance of mobility of the tendon at both the metacarpophalangeal and proximal interphalangeal joints (where the hood and the retinaculum, respec-

Anterior displacement of lateral band

Proximal displacement of hood

Destruction of middle band

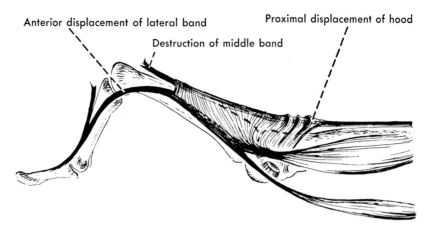

Fig. 6-84. Diagram of the mechanism of hyperextension of the metacarpophalangeal and distal interphalangeal joints with flexion of the proximal interphalangeal joint, as a result of rupture of the median band and forward displacement of the lateral bands of the extensor aponeurosis. (Braithwaite F, Channell GD, Moore FT, Whillis J: Brit J Plastic Surg 2:175, 1949)

tively, allow it to slip distally during flexion, proximally during extension) is well illustrated by the loss of movement at these joints following burns or other trauma that produce dense adhesions here.

Displacements of the lateral bands of the extensor aponeurosis at the proximal interphalangeal joints also can result in deformities of the digits. Tears of the ligaments at these joints, presumably involving especially the oblique parts (Landsmeer's ligaments) of the retinacular ligament, may disrupt the normal linkage by which flexion at the distal joint produces flexion at the proximal one and also allow a gradual dorsal displacement of the lateral bands of the extensor tendon; as a result, these bands pass over the dorsum rather than the sides of the phalangeal trochlea here, and flexion of the distal phalanx so tightens the displaced tendon as to produce hyperextension instead of flexion of the middle phalanx.

On the other hand, tears of the triangular membrane may allow the lateral bands to slip so far forward that they lie in front of the axis of motion; attempts to extend the interphalangeal joints then produce flexion at the proximal joint, instead of extension (Kaplan, '39). In the same way, rupture of the central band of the extensor tendon over the proximal interphalangeal joint (the buttonhole tear, or boutonnière defect) puts all the passive pull on the tendon during flexion upon the lateral bands; since they lie on the sides of the joint, they tend to move farther forward and hence to tear the triangular ligament, and if they become displaced enough anteriorly they hold the proximal interphalangeal joint in flexion. With disruption of the central band, the extensor hood is displaced more proximally than usual by contraction of the long extensor, therefore the lateral bands are pulled upon more than usual; in assuming the straightest course possible, under these abnormal conditions, they therefore produce flexion at the proximal interphalangeal joint and hyperextension at the distal one (Fig. 6-84). The hyperextension is contributed to also by the tension on the oblique parts of the retinacular ligaments and often (Littler and Eaton) by hypertrophy and shortening of these ligaments.

From these observations it follows that the integrity of both the triangular membrane and the retinacular ligaments at the proximal interphalangeal joint 'is particularly important. In the normal condition the triangular membrane allows the lateral bands to slip palmarward enough to permit better flexion at both interphalangeal joints, yet it prevents such forward displacement of these bands that they themselves become flexors at the

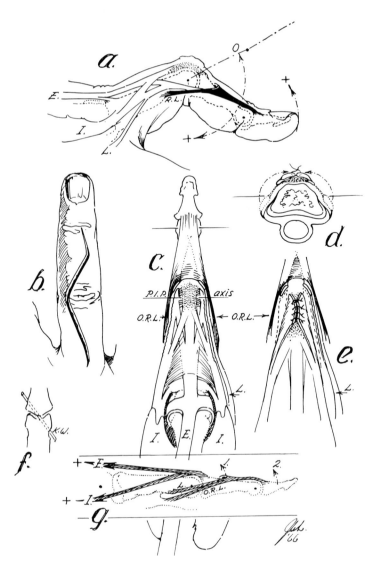

Fig. 6-85. Correction of the boutonnière deformity. *a.* The characteristic flexion of the middle phalanx and hyperextension of the distal phalanx. *b.* The dorsal incision to expose the extensor mechanism. *c.* Separation of the long extensor and interosseous system from the lumbrical tendon and the oblique retinacular ligament, *O.R.L. d.* and *e.* Detaching palmarly displaced lateral bands, rolling them together dorsally, and suturing them to each other and to the central slip with the proximal joint in full extension, thus concentrating the force of the long extensor and interossei on the middle phalanx. *f.* Transfixion of the proximal interphalangeal joint with a Kirschner wire. *g.* The long extensor and interossei now extend the middle phalanx; this in turn tenses the retinacular ligaments which, aided by the lumbrical, extend the distal phalanx. (Littler JW, Eaton RG: J Bone Joint Surg 49-A:1267, 1967)

proximal interphalangeal joint; the retinacular ligaments aid in flexion at the proximal interphalangeal joint through pull upon them when the distal phalanx is flexed, and prevent the dorsal displacement of the lateral bands that both weakens their action on the distal phalanx and reverses the usual effect of flexion of that phalanx.

Souter found conservative treatment of the boutonnière deformity superior to operative treatment, usually end-to-end suture, if the patients are seen within 6 weeks of injury. Littler and Eaton reported obtaining better results in selected cases by splitting the lumbrical portion of the radial lateral band and its attached oblique retinacular ligament away from the remainder of the band, sectioning the latter, sectioning the ulnar lateral band proximal to the attachment of its reinacular ligament, and bringing the two parts together dorsally so as to concentrate the power of the long extensor and the interossei on the middle phalanx, while allowing the lumbrical and the retinacular ligaments to extend the distal one (Fig. 6-85).

"Mallet finger," in which the distal phalanx is held in unopposed flexion, results from rupture of the terminal tendon (the apposed lateral bands) at the distal interphalangeal joint or over the middle phalanx. According to Kaplan ('39), the extensor tendon in mallet finger is not usually torn at its insertion, but rather over the dorsum of the middle phalanx; Smillie said that it results from sudden passive flexion nipping the extensor tendon against the dorsal aspect of the base of the distal phalanx and thus rupturing it. Roemer regarded this injury (also referred to as baseball finger) as being commonly due to a forced hyperextension that jams the proximal end of one phalanx against the dorsum of the more proximal one, and pinches off the insertion of the tendon from the bone, and a similar hyperextension as a cause of disruption of the middle band as it inserts on the middle phalanx.

ARTERIES OF THE DORSUM

Radial Artery

While it lies on the radial side of the flexor surface of the wrist, lateral to the tendon of the flexor carpi radialis muscle, the radial artery gives off a small palmar carpal branch, and a larger superficial palmar branch that passes into the thenar muscles and frequently communicates with the superficial palmar arch to complete this arch. At the level of the distal end of the radius the radial artery turns dorsally and runs obliquely across the radial side of the wrist, around the scaphoid and the trapezium, to reach the dorsum of the hand (Fig. 6-86). In this course it passes superficial

Fig. 6-86. Radial aspect of the wrist.

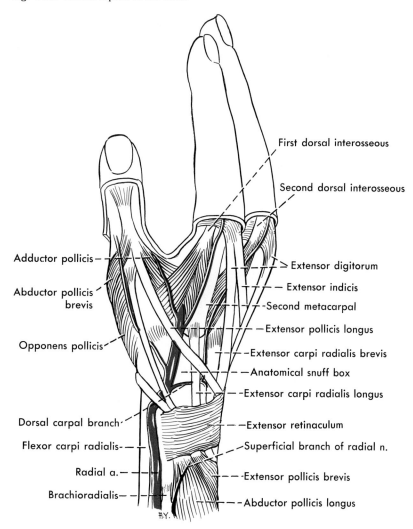

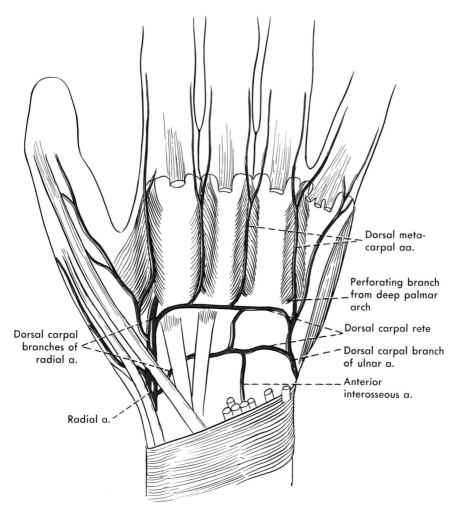

Fig. 6-87. Arteries of the dorsum of the hand.

Labels on figure:
- Dorsal meta-carpal aa.
- Perforating branch from deep palmar arch
- Dorsal carpal rete
- Dorsal carpal branch of ulnar a.
- Anterior interosseous a.
- Dorsal carpal branches of radial a.
- Radial a.

to the radial collateral carpal ligament, but deep to the adjacent tendons of the abductor pollicis longus and extensor pollicis brevis; it thus enters the anatomic snuff box, which it leaves by passing deep to the tendon of the extensor pollicis longus, to attain the interspace between the first and second metacarpal bones. Here the radial artery turns deeply, passing between the two heads of the first dorsal interosseous muscle to attain the palm of the hand; its branches there have already been described.

Before the radial artery reaches the extensor pollicis longus it usually gives off one or more dorsal carpal branches and a small dorsal digital artery to the radial border of the thumb;

just before it disappears between the heads of the first dorsal interosseous, it gives off a metacarpal artery to the ulnar side of the thumb and the radial side of the index finger.

Dorsal Carpal Rete, Metacarpal and Digital Arteries

The dorsal carpal branch or branches of the radial artery pass transversely across the wrist, deep to the extensor tendons, to unite with the dorsal carpal branch of the ulnar and with the posterior interosseous artery, the dorsal terminal branch of the anterior interosseous, or both, to form a dorsal carpal rete (Fig. 6-87); or the rete may arise from any combination of these branches.

The first dorsal metacarpal artery, usually the largest, arises either from the radial artery or the radial side of the dorsal rete (and in Fig. 6-86 its two branches arise separately), and divides to go to the adjacent sides of the thumb and forefinger. The dorsal carpal branch of the ulnar artery usually gives off a dorsal digital artery to the ulnar border of the little finger, and the distal arching part of the rete typically gives rise to three more dorsal metacarpal arteries; these run distally on the second, third, and fourth dorsal interossei. They may be very tiny at their origins and become suddenly larger over the interossei, for each receives a perforating (proximal perforating) branch from the deep palmar arch; this, rather than the rete, may be the origin of the metacarpal artery. Each artery gives off small twigs to the tissues of the hand as it runs distally, usually receives a perforating (distal perforating) branch from the appropriate palmar metacarpal, and ends by dividing at the levels of the heads of the metacarpals into dorsal digital arteries. The dorsal digital arteries, whether arising from the radial and ulnar or from the metacarpal arteries, are in all cases very small, and supply only tissue of the proximal phalanx; the much larger palmar digital arteries supply the larger part of the dorsum of the finger through numerous dorsal branches. Sometimes a dorsal metacarpal artery or one of its digital branches receives still another communicating or perforating vessel from the common digital artery through the tissue of the web distal to the deep transverse metacarpal ligament; and it has already been noted that in some 10% to 18% of cases an enlarged first dorsal metacarpal artery passes palmarward around the distal edge of the adductor pollicis to join the superficial arch or one of its branches to the index finger or thumb.

Variations

Variations of the radial artery on the dorsum of the hand are rare; a few cases have been described in which a part of the radial artery passed superficial rather than deep to the tendons of the long muscles of the thumb on the radial side of the wrist (Fig. 6-88), and such a

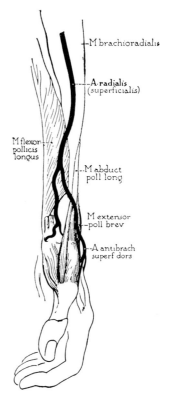

Fig. 6-88. A superficial radial artery with a large branch passing onto the dorsum superficial to the long extensor tendons of the thumb. (McCormack LJ, Cauldwell EW, Anson BJ: Surg Gynecol Obstet 96:43, 1953 [by permission of Surgery, Gynecology & Obstetrics].)

vessel may enter the palm more distally than usual, instead of passing between the two heads of the first dorsal interosseous muscle.

References

ALLDRED A: A locked index finger. J Bone Joint Surg 36-B:102, 1954

ANSON BJ, ASHLEY FL: The midpalmar compartment, associated spaces and limiting layers. Anat Rec 78:389, 1940

ANSON BJ, WRIGHT RR, ASHLEY FL, DYKES J: The fascia of the dorsum of the hand. Surg Gynecol Obstet 81:327, 1945

BACKHOUSE KM: The mechanics of normal digital control in the hand and an analysis of the ulnar drift of rheumatoid arthritis. Roy Coll Surg Eng 43:154, 1968

BACKHOUSE KM: Extensor expansion of the rheumatoid hand. Ann Rheumatic Dis 31:112, 1972

BACORN RW, KURTZKE JF: Colles' fracture: A study of two thousand cases from the New York State Workmen's Compensation Board. J Bone Joint Surg 35-A:643, 1953

BAKKE JL, WOLFF HG: Occupational pressure neuritis of the deep palmar branch of the ulnar nerve. Arch Neurol Psychiat 60:549, 1948

BARSKY AJ: Congenital anomalies of the hand. J Bone Joint Surg 33-A:35, 1951

BARSKY AJ: Cleft hand: Classification, incidence, and treatment. Review of the literature and report of nineteen cases. J Bone Joint Surg 46-A:1707, 1964

BASU SS, HAZARY S: Variations of the lumbrical muscles of the hand. Anat Rec 136:501, 1960

BIZARRO AH: On sesamoid and supernumerary bones of the limbs. J Anat 55:256, 1921

DU BOIS-REYMOND R: Beschreibung einer Anzahl Muskelvarietäten an einem Individuùm. Anat Anz 9:451, 1894

BOJSEN-MØLLER F, SCHMIDT L: The palmar aponeurosis and the central spaces of the hand. J Anat 117:55, 1974

BRADLEY KC, SUNDERLAND S: The range of movement at the wrist joint. Anat Rec 116:139, 1953

BRAÍN WR, WRIGHT AD, WILKINSON M: Spontaneous compression of both median nerves in the carpal tunnel: Six cases treated surgically. Lancet 1:277, 1947

BRAITHWAITE F, CHANNELL GD, MOORE FT, WHILLIS J: The applied anatomy of the lumbrical and interosseous muscles of the hand. Guy's Hosp Rep 97:185, 1948

BROCKIS JG: The blood supply of the flexor and extensor tendons of the fingers in man. J Bone Joint Surg 35-B:131, 1953

BROOKS H ST J: Variations in the nerve supply of the flexor brevis pollicis muscle. J Anat Physiol 20:641, 1886

BROOKS H ST J: Variations in the nerve-supply of the lumbrical muscles in the hand and foot, with some observations on the innervation of the perforating flexors. J Anat Physiol 21:575, 1887

BRUNER JM: Treatment of "sesamoid" synovial ganglia of the hand by needle rupture. J Bone Joint Surg 45-A:1689, 1963

BUNNEL S: Tendon transfers in the hand and forearm. Am Acad Orth Surgeons Instructional Course Lectures 6:106, 1949

BUNNEL S: Surgery of the Hand (ed. 3). Philadelphia, JB Lippincott, 1956

BURCH PRJ: Dupuytren's contracture: An auto-immune disease? J Bone Joint Surg 48-B:312, 1966

BURKHALTER W, CHRISTENSEN RC, BROWN P: Extensor indicis proprius opponensplasty. J Bone Joint Surg 55-A:725, 1973

BURMAN M: Tendinitis of the insertion of the common extensor tendon of the fingers. J Bone Joint Surg 35-A:177, 1953

BUTLER B JR, BIGLEY EC: Aberrant index (first) lumbrical tendinous origin associated with carpal-tunnel syndrome. J Bone Joint Surg 53-A:160, 1971

CAMPBELL RD JR, LANCE EM, YEOH CB: Lunate and perilunar dislocations. J Bone Joint Surg 46-B:55, 1964

CANNON BW, LOVE JG: Tardy median palsy; median neuritis; median thenar neuritis amenable to surgery. Surgery 20:210, 1946

CHACHA P: Free autologous composite tendon grafts for division of both flexor tendons within the digital theca of the hand. J Bone Joint Surg 56-A:960, 1974

CHARLES JJ: A case of absence of the radial artery. J Anat Physiol 28:449, 1894

CLEVELAND M: Fracture of the carpal scaphoid. Surg Gynecol Obstet 84:769, 1947

CLIFFTON EE: Unusual innervation of the intrinsic muscles of the hand by median and ulnar nerve. Surgery 23:12, 1948

COCKSHOTT WP: Pisiform hamate fusion. J Bone Joint Surg 51-A:778, 1969

COLEMAN SS, ANSON BJ: Arterial patterns in the hand based upon a study of 650 specimens. Surg Gynecol Obstet 113:409, 1961

COONRAD RW, GOLDNER JL: A study of the pathological findings and treatment in soft-tissue injury of the thumb metacarpophalangeal joint. With a clinical study of the normal range of motion in one thousand thumbs and a study of post mortem findings of ligamentous structures in relation to function. J Bone Joint Surg 50-A:439, 1968

COOPERMAN MB: An unusual congenital deformity of the hand combined with supernumerary toes: A case report. J Bone Joint Surg 12:956, 1930

CUNNINGHAM DJ: The flexor brevis pollicis and the flexor brevis hallucis in man. Anat Anz 2:186, 1887

CYRIAX EF: On the rotary movements of the wrist. J Anat 60:199, 1926

DAVIS AA: The treatment of Dupuytren's contracture: A review of 31 cases, with an assessment of the comparative value of different methods of treatment. Brit J Surg 19:539, 1932

DAY MH, NAPIER JR: The two heads of flexor pollicis brevis. J Anat 95:123, 1962

DOYLE JR, BLYTHE W: The finger flexor tendon sheath and pulleys: Anatomy and reconstruction. AAOS Symposium on tendon surgery in the hand, pp. 81–87. St. Louis, C. V. Mosby, 1975

DOYLE JR, BLYTHE WF: Anatomy of the flexor tendon sheath and pulleys of the thumb. J Hand Surg 2:149, 1977

DRINKWATER H: Hereditary abnormal segmentation of the index and middle fingers. J Anat 50:177, 1916

ELSON RA: Dislocation of the extensor tendons of the hand: Report of a case. J Bone Joint Surg 49-B:324, 1967

EYLER DL, MARKEE JE: The anatomy and function of the intrinsic musculature of the fingers. J Bone Joint Surg 36-A:1, 1954

FENNING JB: Deep ulnar-nerve paralysis resulting from an anatomical abnormality: A case report. J Bone Joint Surg 47-A:1381, 1965

FITCH RR: An operation to improve the condition of hands disabled by paralysis of the opponens pollicis. J Bone Joint Surg 28:190, 1930

FITZGERALD RR: Habitual dislocation of the digital extensor tendons. Ann Surg 110:81, 1939

FLATT AE: Recurrent locking of an index finger. J Bone Joint Surg 40-A:1128, 1958

FLEMMING W: Über den Flexor brevis pollicis und hallucis des Menschen. Anat Anz 2:68, 1887

FLYNN JE: Clinical and anatomical investigations of deep fascial space infections of the hand. Am J Surg 55:467, 1942

FLYNN JE: Acute suppurative tenosynovitis of the hand. Surg Gynecol Obstet 76:227, 1943a

FLYNN JE: Surgical significance of the middle palmar septum of the hand. Surgery 14:134, 1943b

FORBES G: A case of congenital clubhand with a review of the aetiology of the condition. Anat Rec 71:181, 1938

FORREST WJ, BASMAJIAN JV: Functions of human thenar and hypothenar muscles. An electromyographic study of twenty-five hands. J Bone Joint Surg 47-A:1585, 1965

FRIEDENBERG ZB: Anatomic considerations in the treatment of carpal navicular fractures. Am J Surg 78:379, 1949

FROHSE F, FRÄNKEL M: Die Muskeln des menschlichen Armes, Vol 2, Section 2, Part 2. In von Bardeleben K (ed): Handbuch der Anatomie des Menschen. Jena, Fischer, 1908

FULLERTON PM: The effect of ischaemia on nerve conduction in the carpal tunnel syndrome. J Neurol Neurosurg Psychiat 26:385, 1963

GAD P: The anatomy of the volar part of the capsules of the finger joints. J Bone Joint Surg 49-B:362, 1967

GLADSTONE H: Rupture of the extensor digitorum communis tendons following severely deforming fractures about the wrist. J Bone Joint Surg 34-A:698, 1952

GOLDNER JL: Deformities of the hand incidental to pathological changes of the extensor and intrinsic muscle mechanisms. J Bone Joint Surg 35-A:115, 1953

GOLDNER JL, IRWIN CE: An analysis of paralytic thumb deformities. J Bone Joint Surg 32-A:627, 1950

GORE DR: Carpometacarpal dislocation producing compression of the deep branch of the ulnar nerve. J Bone Joint Surg 53-A:1387, 1971

GRAY DJ, GARDNER E: The innervation of the joints of the wrist and hand. Anat Rec 151:261, 1965

GRAYSON J: The cutaneous ligaments of the digits. J Anat 75:164, 1941

GREEN WL, NIEBAUER JJ: Results of primary and secondary flexor-tendon repairs in No man's land. J Bone Joint Surg 56-A:1216, 1974

GRIFFITH TW: An abnormal muscle of the hand, with remarks on the course of the radial artery. J Anat Physiol 31:283, 1897

GRODINSKY M, HOLYOKE EA: The fasciae and fascial spaces of the palm. Anat Rec 79:435, 1941

HAINES RW: The mechanism of rotation at the first carpo-metacarpal joint. J Anat 78:44, 1944

HAINES RW: The extensor apparatus of the finger. J Anat 85:251, 1951

HAKSTIAN RW, TUBIANA R: Ulnar deviation of the fingers. The role of joint structure and function. J Bone Joint Surg 49-A:299, 1967

HANDFORTH JR: Polydactylism of the hand in southern Chinese. Anat Rec 106:119, 1950

HARPER WF: The distribution of the palmar aponeurosis in relation to Dupuytren's contraction of the thumb. J Anat 69:193, 1935

HARRIS C JR, RIORDAN DC: Intrinsic contracture in the hand and its surgical treatment. J Bone Joint Surg 36-A:10, 1954

HARRIS C, RUTLEDGE CL JR: The functional anatomy of the extensor mechanism of the finger. J Bone Joint Surg 54-A:713, 1972

HARRIS H, JOSEPH J: Variation in extension of the metacarpo-phalangeal and interphalangeal joints of the thumb. J Bone Joint Surg 31-B:547, 1949

HENDERSON ED: Transfer of wrist extensors and brachioradialis to restore opposition of the thumb. J Bone Joint Surg 44-A:513, 1962

HENDRY AM: The treatment of residual paralysis after brachial plexus injuries. J Bone Joint Surg 31-B:42, 1949

HEWITT D: The range of active motion at the wrist of women. J Bone Joint Surg 10:775, 1928

HIPPS HE: A safe method of inducing local digital anesthesia. Am J Surg 80:216, 1950

HORWITZ T: An anatomic and roentgenologic study of the wrist joint: Observations on a case of recurrent radiocarpal dislocation complicating Madelung's deformity and its surgical correction. Surgery 7:773, 1940

HOWARD FM: Ulnar-nerve palsy in wrist fractures. J Bone Joint Surg 43-A:1197, 1961

HOWARD NJ: Subcutaneous dorsal digital bursitis. Surgery 5:939, 1939

JACOBS B, THOMPSON TC: Opposition of the thumb and its restoration. J Bone Joint Surg 42-A:1015, 1960

JEFFERY AK: Compression of the deep palmar branch of the ulnar nerve by an anomalous muscle. Case report and review. J Bone Joint Surg 53-B:718, 1971

JOHNSON RK, SHREWSBURY MM: Anatomical course of the thenar branch of the median nerve—usually in a separate tunnel through the transverse carpal ligament. J Bone Joint Surg 52-A:269, 1970

JONES FW: Voluntary muscular movements in cases of nerve lesions. J Anat 54:41, 1919

JONES HW, ROBERTS RE: A rare type of congenital club hand. J Anat 60:146, 1926

JOSEPH J: The sesamoid bones of the hand and the time of fusion of the epiphyses of the thumb. J Anat 85:230, 1951

KADASNE DK, BANSAL PG: Origin of human pisiform bone. J Anat Soc India 14:23, 1965

KANAVEL AB: Infections of the Hand: A Guide to the Surgical Treatment of Acute and Chronic Suppurative Processes in the Fingers, Hand, and Forearm (ed. 7). Philadelphia, Lea and Febiger, 1939

KAPLAN EB: Extension deformities of the proximal interphalangeal joints of the fingers: An anatomical study. J Bone Joint Surg 18:781, 1936

KAPLAN EB: The palmar fascia in connection with Dupuytren's contracture. Surgery 4:415, 1938

KAPLAN EB: Pathology and operative correction of finger deformities due to injuries and contractures of the extensor digitorum tendon. Surgery 6:35, 451, 1939

KAPLAN EB: Embryological development of the tendinous apparatus of the fingers: Relation to function. J Bone Joint Surg 32-A:820, 1950

KAPLAN EB: Functional and Surgical Anatomy of the Hand. Philadelphia, JB Lippincott, 1953

KELIKIAN H, DOUMANIAN A: Congenital anomalies of the hand. J Bone Joint Surg 39-A:1002; 1249, 1957

KETTLEKAMP DB, FLATT AE, MOULDS R: Traumatic dislocation of the long finger extensor tendon. A clinical anatomical and biomechanical study. J Bone Joint Surg 53-A:229, 1971

KIRKLIN JW, THOMAS CG JR: Opponens transplant: An analysis of the methods employed and results obtained in seventy-five cases. Surg Gynecol Obstet 86:213, 1948

LANDSMEER JMF: The anatomy of the dorsal aponeurosis of the human finger and its functional significance. Anat Rec 104:31, 1949

LANDSMEER JMF: Anatomical and functional investigations on the articulation of the human fingers. Acta Anatomica 25 (Suppl 24-2):5, 1955

LANDSMEER JMF: The coordination of finger-joint motions. J Bone Joint Surg 45-A:1654, 1963

LANNON J: The posterior adductor and posterior interosseous spaces of the hand. South African MJ 22:283, 1948

LAPIDUS PW, GUIDOTTI FP, COLETTI CJ: Triphalangeal thumb: Report of six cases. Surg Gynecol Obstet 77:178, 1943

LINBURG RM, ALBRIGHT JA: An anomalous branch of the median nerve. A case report. J Bone Joint Surg 52-A:182, 1970

LINELL EA: The distribution of nerves in the upper limb, with reference to variabilities and their clinical significance. J Anat 55:79, 1921

LINSCHEID RL: Injuries to radial nerve at wrist. Arch Surg 91:942, 1965

LIPSCOMB PR: Synovectomy of the distal two joints of the thumb and fingers in rheumatoid arthritis. J Bone Joint Surg 49-A:1135, 1967

LIPSCOMB PR: Personal communication to the author, 1968

LITTLER JW: Tendon transfers and arthrodeses in combined median and ulnar nerve paralysis. J Bone Joint Surg 31-A:225, 1949

LITTLER JW, COOLEY SGE: Opposition of the thumb and its restoration by abductor digiti quinti transfer. J Bone Joint Surg 45-A:1389, 1963

LITTLER JW, EATON RG: Redistribution of forces in the correction of the boutonnière deformity. J Bone Joint Surg 49-A:1267, 1967

LONDON PS: The broken scaphoid bone. The case against pessimism. J Bone Joint Surg 43-B:237, 1961

LONG C, BROWN ME: Electromyographic kinesiology of the hand: Muscles moving the long finger. J Bone Joint Surg 46-A:1683, 1964

LONG C II, CONRAD PW, HALL EA, FURLER MS: Intrinsic–extrinsic muscle control of the hand in power grip and precision handling. An electromyographic study. J Bone Joint Surg 52-A:853, 1970

LUCIEN M: L'aponévrose dorsale moyenne de la main. Acta anat 4:188, 1947

MACCONAILL MA: The mechanical anatomy of the carpus and its bearing on some surgical problems. J Anat 75:166, 1941

MAKIN M: Translocation of the flexor pollicis longus tendon to restore opposition. J Bone Joint Surg 49-B:458, 1967

MASON ML: Dupuytren's contracture. AMA Arch Surg 65:457, 1952

MATHESON AB, SINCLAIR DC, SKENE WG: The range and power of ulnar and radial deviation of the fingers. J Anat 107:439, 1970

MAYFIELD JK, JOHNSON RP, KILCOQUE RF: The ligaments of the human wrist and their functional significance. Anat Rec 186:417, 1976

MCFARLANE RM, JAMIESON WG: Dupuytren's contracture. The management of one hundred patients. J Bone Joint Surg 48-A:1095, 1966

MCNEALY RW, LICHTENSTEIN ME: Fractures of the bones of the hand. Am J Surg 50:563, 1940

MEHTA HJ, GARDNER WU: A study of lumbrical muscles in the human hand. Am J Anat 109:227, 1961

MEYERDING HW, BLACK JR, BRODERS AC: The etiology and pathology of Dupuytren's contracture. Surg Gynecol Obstet 72:582, 1941

MILCH H: Triphalangeal thumb. J Bone Joint Surg 33-A:692, 1951

MING-TZU P'AN: The cutaneous nerves of the Chinese hand. Am J Phys Anthropol 25:301, 1939

MINNAAR AB DE V: Congenital fusion of the lunate and triquetral bones in the South African Bantu. J Bone Joint Surg 34-B:45, 1952

MULDER JD, LANDSMEER JMF: The mechanism of claw finger. J Bone Joint Surg 50-B:664, 1968

MURPHEY F, KIRKLIN JW, FINLAYSON AI: Anomalous innervation of the intrinsic muscles of the hand. Surg Gynecol Obstet 83:15, 1946

MURPHY AF, STARK HH: Closed dislocation of the metacarpophalangeal joint of the index finger. J Bone Joint Surg 49-A:1579, 1967

NAPIER JR: The form and function of the carpo-meta-carpal joint of the thumb. J Anat 89:362, 1955

OGDEN JA: An unusual branch of the median nerve. J Bone Joint Surg 54-A:1779, 1972

O'RAHILLY R: A survey of carpal and tarsal anomalies. J Bone Joint Surg 35-A:626, 1953

OWEN HR: Fractures of the bones of the hand. Surg Gynecol Obstet 66:500, 1938

PAPATHANASSIOU BT: A variant of the motor branch of the median nerve in the hand. J Bone Joint Surg 50-B:156, 1968

PARKES A: The "lumbrical plus" finger. J Bone Joint Surg 53-B:236, 1971

PARTRIDGE EJ: Anomalous carpal bones. J Anat 57:378, 1923

PEACOCK EE, JR: Some technical aspects and results of flexor tendon repair. Surgery 58:330, 1965

PEMBERTON PA: Infection of fascial spaces of the palm. Am J Surg 50:512, 1940

PETERSON HA, LIPSCOMB PR: Intercarpal arthrodesis. Arch Surg 95:127, 1967

PHALEN GS: The carpal-tunnel syndrome. Seventeen years' experience in diagnosis and treatment of six hundred fifty-four hands. J Bone Joint Surg 48-A:211, 1966

PULVERTAFT RG: Repair of tendon injuries in the hand. Hunterian Lecture. Ann Roy Coll Surg England 3:3, 1948

REDLER I, WILLIAM JT: Rupture of a collateral ligament of the proximal interphalangeal joint of the fingers: Analysis of eighteen cases. J Bone Joint Surg 49-A:322, 1967

REEF TC, BRESTIN SG: The extensor digitorum manus and its clinical significance. J Bone Joint Surg 57-A:704, 1975

RICHARDS HJ: The surgical treatment of Dupuytren's contracture. J Bone Joint Surg 36-B:90, 1954

RICHMOND DA: Carpal ganglion with ulnar nerve compression. J Bone Joint Surg 45-B:513, 1963

RIORDAN DC: Tendon transplantations in median-nerve and ulnar-nerve paralysis. J Bone Joint Surg 35-A:312, 1953

RITTER MA, INGLIS AE: The extensor indicis proprius syndrome. J Bone Joint Surg 51-A:1645, 1969

ROBBINS H: Anatomical study of the median nerve in the carpal tunnel and etiologies of the carpal-tunnel syndrome. J Bone Joint Surg 45-A:953, 1963

ROBINS RHC: Infections of the hand: A review based on 1,000 consecutive cases. J Bone Joint Surg 34-B:567, 1952

ROEMER FJ: Hyperextension injuries to the finger joints. Am J Surg 80:295, 1950

ROSEN AD: Innervation of the hand: An electromyographic study. Electromyography Clinical Neurophysiol 13:175, 1973

ROSS JA, TROY CA: The clinical significance of the extensor digitorum brevis manus. J Bone Joint Surg 51-B:473, 1969

ROWNTREE T: Anomalous innervation of the hand muscles. J Bone Joint Surg 31-B:505, 1949

RUSHFORTH AF: A congenital abnormality of the trapezium and first metacarpal bone. J Bone Joint Surg 31-B:543, 1949

RUSSELL TB: Inter-carpal dislocations and fracture-dislocations: A review of fifty-nine cases. J Bone Joint Surg 31-B:524, 1949

SALSBURY CR: The interosseous muscles of the hand. J Anat 71:395, 1937

SARRAFIAN SK, KAZARIAN LE, TOPOUZIAN LP, SARRA-FIAN VK, SIEGELMAN A: Strain variation in the components of the extensor apparatus of the finger during flexion and extension: A biomechanical study. J Bone Joint Surg 52-A:980, 1970

SCHELDRUP EW: Tendon sheath patterns in the hand: An anatomical study based on 367 hand dissections. Surg Gynecol Obstet 93:16, 1951

SCHENCK RR: Variations of the extensor tendons of the fingers. Surgical significance. J Bone Joint Surg 46-A:103, 1964

SCHULTZ RJ, ENDLER PM, HUDDLESTON HD: Anomalous median nerve and an anomalous muscle belly of the first lumbrical associated with carpal-tunnel syndrome. J Bone Joint Surg 55-A:1744, 1973

SCOTT JC, JONES BV: Results of treatment of infections of the hand. J Bone Joint Surg 34-B:581, 1952

SCUDERI C: Tendon transplants for irreparable radial nerve paralysis. Surg Gynecol Obstet 88:643, 1949

SEDDON HJ: Carpal ganglion as a cause of paralysis of the deep branch of the ulnar nerve. J Bone Joint Surg 34-B:386, 1952

SHEA JD, MCCLAIN EJ: Ulnar-nerve compression syndromes at and below the wrist. J Bone Joint Surg 51-A:1095, 1969

SHREWSBURY MM, JOHNSON RK, OSTERHOUT DK: The palmaris brevis—A reconsideration of its anatomy and possible function. J Bone Joint Surg 54-A:344, 1972

SILER VE: Primary tenorrhaphy of the flexor tendons in the hand. J Bone Joint Surg 32-A:218, 1950

SMILLIE IS: Mallet finger. Brit J Surg 24:439, 1937

SMITH EB: Some points in the anatomy of the dorsum of the hand, with special reference to the morphology of the extensor brevis digitorum manus. J Anat Physiol 31:45, 1896

SMITH RJ: Anomalous muscle belly of the flexor digitorum superficialis causing carpal tunnel syndrome. Report of a case. J Bone Joint Surg 53-A:1215, 1971

SMITH RJ, KAPLAN EB: Rheumatoid deformities at the metacarpophalangeal joints of the fingers. A correlative study of anatomy and pathology. J Bone Joint Surg 49-A:31, 1967

SOUTER, WA: The boutonnière deformity. A review of 101 patients with division of the central slip of the extensor expansion of the fingers. J Bone Joint Surg 49-B:710, 1967

SPRECHER EE: Trigger thumb in infants. J Bone Joint Surg 31-A:672, 1949

STACK HG: Muscle function in the fingers. J Bone Joint Surg 44-B:899, 1962

STEINDLER A: The mechanics of muscular contractures in wrist and fingers. J Bone Joint Surg 14:1, 1932

STEINDLER A: Mechanics of Normal and Pathological Locomotion in Man. Springfield, Ill., Thomas, 1935

STEWART MJ: Fractures of the carpal navicular (scaphoid). A report of 436 cases. J Bone Joint Surg 36-A:998, 1954

STOPFORD JSB: The variations in distribution of the cutaneous nerves of the hand and digits. J Anat 53:14, 1918

STOPFORD JSB: The nerve supply of the interphalangeal and metacarpo-phalangeal joints. J Anat 56:1, 1921

STRAUS FH: Luxation of extensor tendons in the hand. Ann Surg 111:135, 1940

SU CT, HOOPES JE, DANIEL R: Congenital absence of the thenar muscles innervated by the median nerve. Report of a case. J Bone Joint Surg 54-A:1087, 1972

SUNDERLAND S: The innervation of the first dorsal interosseous muscle of the hand. Anat Rec 95:7, 1946

SUNDERLAND S: Rotation of the fingers by the lumbrical muscles. Anat Rec 116:167, 1953

SUNDERLAND S, HUGHES ESR: Metrical and non-metrical features of the muscular branches of the ulnar nerve. J Comp Neurol 85:113, 1946

SUNDERLAND S, RAY LJ: Metrical and non-metrical features of the muscular branches of the median nerve. J Comp Neurol 85:191, 1946

TALEISNIK J, KELLY PJ: The extraosseous and intraosseous blood supply of the scaphoid bone. J Bone Joint Surg 48-A:1125, 1966

TANZER RC: The carpal-tunnel syndrome. A clinical and anatomical study. J Bone Joint Surg 41-A:626, 1959

THOMAS JE, LAMBERT EH, CSEUZ KA: Electrodiagnostic aspects of the carpal tunnel syndrome. Arch Neurol 16:635, 1967

URBANIAK JR, BRIGHT DS, GILL LH, GOLDNER JL: Vascularization and the gliding mechanism of free flexor-tendon grafts inserted by the silicone-rod method. J Bone Joint Surg 56-A:473, 1974

VERDAN CE: Half a century of flexor-tendon surgery. Current status and changing philosophies. J Bone Joint Surg 54-A:472, 1972

WAKELEY CPG: Bilateral epiphysis at the basal end of the second metacarpal. J Anat 58:340, 1924

WATSON-JONES R: Fractures and Joint Injuries. Vol 2 (ed. 4). Baltimore, Williams and Wilkins, 1955

WAUGH RL, SULLIVAN RF: Anomalies of the carpus: With particular reference to the bipartite scaphoid (navicular). J Bone Joint Surg 32-A:682, 1950

WAUGH RL, YANCEY AG: Carpometacarpal dislocations: With particular reference to simultaneous dislocation of the bases of the fourth and fifth metacarpals. J Bone Joint Surg 30-A:397, 1948

WEATHERSBY HT: The volar arterial arches. Anat Rec 118:365, 1954 (abstr.)

WEATHERSBY HT: The artery of the index finger. Anat Rec 122:57, 1955

WEATHERSBY HT, SUTTON LR, KRUSEN UL: The kinesiology of muscles of the thumb: An electromyographic study. Arch Phys Med 44:321, 1963

WERSCHKUL JD: Anomalous course of the recurrent motor branch of the median nerve in a patient with carpal tunnel syndrome. J Neurosurg 47:113, 1977

WILLAN R: The action of the extensor, lumbrical, and interosseous muscles in the hand and foot. Anat Anz 42:145, 1912

WILSON JN: Profiles of the carpal canal. J Bone Joint Surg 36-A:127, 1954

WOOD VE, FRYKMAN GY: Unusual branching of the median nerve at the wrist. A case report. J Bone Joint Surg 60-A:267, 1978

WRIGHT RD: A detailed study of movement of the wrist joint. J Anat 70:137, 1935

WRIGHT WG: Muscle Function. New York, Hoeber, 1928; Hafner, 1962

YOUNG, HH, LOWE GH, JR: Tendon transfer operation for irreparable paralysis of the radial nerve: Long term follow-up of patients. Surg Gynecol Obstet 84:1100, 1947

ZACHARY RB: Thenar palsy due to compression of the median nerve in the carpal tunnel. Surg Gynecol Obstet 81:213, 1945

ZWEIG J, ROSENTHAL S, BURNS H: Transfer of the extensor digiti quinti to restore pinch in ulnar palsy of the hand. J Bone Joint Surg 54-A:51, 1972

Chapter 7
GENERAL SURVEY OF THE LOWER LIMB

Anatomical details of the lower limb are discussed in the following chapters in a regional manner. The present chapter is therefore simply a summary of those features of the lower limb that can best be discussed together rather than regionally.

Development

The early development of the lower limb closely parallels that of the upper (Chap. 3), but the lower limb bud appears slightly later than the upper one, and in some respects lags slightly behind it in development. As in the upper limb, the blastemal swelling becomes cylindrical; the distal end becomes flattened as the forerunner of the foot; constrictions appear, to indicate the regions of the ankle and knee; and ridges on the distal flattened portion give rise to the toes (Fig. 7-1).

In an early stage of development the lower limb bud projects slightly caudally, but soon comes to project at about a right angle to the body. As development proceeds it is obvious that the big-toe or tibial side is directed cranially, while the fibular side is directed caudally. At this stage, the borders and surfaces of the lower limb correspond very nicely to those of the upper limb: the cephalic, preaxial, or tibial border, marked by the big toe, corresponds to the radial or thumb border of

the upper limb; the caudal, postaxial, or fibular border is that marked by the little toe; and the ventral and dorsal surfaces, which are flexor and extensor ones, correspond to the planes of the trunk. In later development, however, the lower limb undergoes a medial rotation (beginning prenatally but not completed until postnatally) of about 90° around its long axis; this alone would result in the knee pointing cranially rather than dorsally, but combined adduction and extension have the effect of another 90° of medial rotation, and therefore direct the knee forward (ventrally). Thus when the limb is straightened and in line with the trunk, the tibial or preaxial side is medial rather than cephalic, and the original extensor surface of the limb is now directed largely anteriorly, the original flexor surface largely posteriorly (Fig. 7-2). The rotation that the lower limb has undergone is plainly illustrated by the direction of the fibers of the capsule of the hip joint and by the somewhat spiral direction of the dermatomes, particularly marked in Keegan's dermatomal charts.

The development of the lower limb of man has been investigated especially by Bardeen and Lewis, and by Bardeen ('05).

Skeleton and Musculature
At a very early stage a mesenchymal condensation at the base of each limb bud represents the pelvic girdle, and a similar condensation

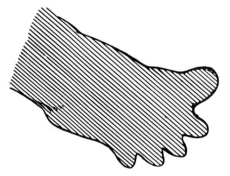

Fig. 7-1. Right leg and foot of a 21-mm human fetus. (After Strauss, from Coventry, MB: Am Acad Orth Surgeons, Instructional Course Lectures 6:218, 1949)

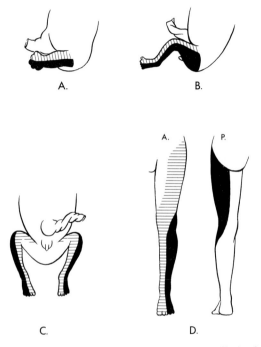

Fig. 7-2. Rotation of the lower limb. *A* is the limb of a 7-week embryo, and *B* that of an 8-week one, seen from the side; *C* shows the limbs at about the end of the third month; and *D* shows anterior (A) and posterior (P) views of the adult limb. The original extensor or dorsal surface, directed laterally in *A*, is *shaded* to differentiate the preaxial part (*lined*) from the postaxial part (*solid black*); the original ventral part of the limb, visible in *C* and *D*, is *unshaded.*

in the center of the free limb is the forerunner of the skeleton there. The blastemal condensation representing the girdle becomes chondrified except in the region of the future obturator foramen, and expands so as to come in

contact with the upper three sacral vertebrae; similarly, the forerunners of the various bones of the free limb are formed in cartilage from the central blastemal condensation of the bud (the patella, a sesamoid bone, is the exception, since it develops within the tendon of the quadriceps). All the bones of the lower limb are cartilage bones, that is, are first represented by cartilage. The first center of ossification in the lower limb is in the body of the femur; this appears during the seventh week, making the femur the second bone (following the clavicle) to show ossification. Centers for many of the other long bones appear toward the end of the second fetal month.

The development of the musculature of the lower limb closely parallels that of the upper limb, and lags only a little behind it. Condensations of mesenchyme form premuscle masses, which as they differentiate become subdivided into dorsal and ventral, or extensor and flexor, groups, from which the individual muscles are differentiated. As in the upper limb, proximal muscles appear sooner than distal ones do, and at any given level the extensor muscles tend to appear sooner than do the flexors.

Nerves

The nerves contributing to the lower limb correspond, as is the case also with the upper limb, to those at the level of the originally wide limb bud. They form two connected plexuses, a lumbar and a sacral, at the base of this bud, and as the elements from each plexus grow out into the limb they become subdivided (in the case of the lumbar plexus, by the developing girdle) into dorsal and ventral components, for the dorsal (extensor) and ventral (flexor) musculature. The large dorsal element of the lumbar plexus is the femoral nerve, the ventral element the obturator. The large dorsal component of the sacral plexus is the common peroneal nerve, the ventral the tibial; these two components go downward close together to form the sciatic (ischiadic) nerve. With rotation of the limb around its long axis, the original dorsal nerves become laterally placed, while the original ventral nerves become medially placed, thus occu-

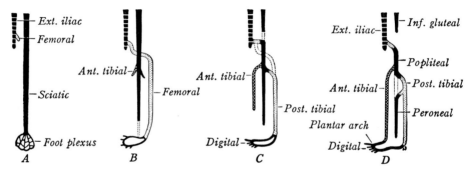

Fig. 7-3. Four stages in the development of the arteries of the lower limb. (Arey LB: Developmental Anatomy [ed 6]. Philadelphia, WB Saunders, 1954)

pying the relative positions that they have in the adult. Thus in the thigh, where in the adult the original ventral surface is now largely medial, the obturator nerve supplies the anteromedial muscles, the tibial part of the sciatic the posteromedial ones; similarly, the femoral nerve supplies most of the anterolateral, originally dorsal, musculature, but the common peroneal supplies the one posterior muscle arising lateral to the posterior midline, where original dorsal and ventral surfaces meet.

Arteries

The original artery to the lower limb is the sciatic (ischiadic), a branch given off from that part of the umbilical artery which will later become the internal iliac or hypogastric (Senior). It grows down the flexor aspect of the limb to feed the developing foot (Fig. 7-3). A little later the femoral artery arises from the external iliac and grows down the dorsal surface of the thigh, but passes through the developing musculature to the ventral aspect, where it eventually establishes connections with and takes over the lower part of the sciatic artery. Parts of the original sciatic stem then disappear, an uppermost part persisting as the inferior gluteal artery, an intermediate part as the popliteal and a lower part, having lost its connection with the digital vessels, as the peroneal. This method of development obviously explains the very rare persistence of a sciatic artery (ischiadic artery) as the main stem to the leg and the occasional occurrence of a peroneal artery that gives rise to the plantar arteries.

OSSIFICATION OF THE SKELETON

Centers of ossification appear in the cartilaginous bodies of the femur, the tibia, and the metatarsals during the seventh and eighth fetal weeks; centers in the body of the ilium and the body of the fibula during the eighth and ninth weeks; centers for the bodies of the phalanges about the tenth week; one in the body (superior ramus) of the ischium during the third month; one in the superior ramus of the pubis during the fourth or fifth month; one for the calcaneus during the sixth month; one for the talus during the seventh month; and, usually, centers for the distal epiphysis of the femur and the proximal one of the tibia during the ninth month.

At birth, the three bony elements of the coxal (innominate) bone, although represented by bone, are widely surrounded by areas of cartilage, which are later ossified both by extension of the primary centers and by the appearance of secondary or epiphyseal centers. The acetabulum is entirely cartilaginous, as are the inferior ramus of the pubis and the ramus (inferior ramus) of the ischium, and the margins of all three elements. The bodies of the femur, tibia, and fibula are well represented by bone, but except for the distal end of the femur and the proximal end of the tibia, in which small centers of ossification are present, their ends are entirely cartilaginous. Only two or three tarsals, the calcaneus, the talus, and often the cuboid, show centers of ossification. The bodies of the metatarsals and phalanges are well formed in

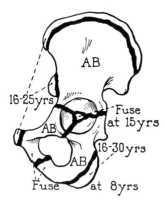

Fig. 7-4. Ossification of the hip (coxal, innominate) bone. *AB* indicates that centers for the bones so marked are present at birth. Two numbers (for instance, *16–25 yrs.*) indicate respectively the age at which the center so identified appears and the age at which it fuses with the rest of the bone. (Camp JD, Cilley EIL: Am J Roentgenol 26:905, 1931)

bone, but the proximal ends of the first metatarsal and of all the phalanges and the distal ends of the four lateral metatarsals are still cartilaginous. As in the upper limb, there is a good deal of variability in regard to the times of appearance and of fusion of the epiphyses, and the following account, with the accompanying figures, should be taken as giving only approximate dates.

Hip

Expansion of the primary centers of ossification of the *ilium, ischium,* and *pubis* (Fig. 7-4) gradually gives shape to these bones, and at about the age of 7 or 8 years the inferior ramus of the pubis and the ramus of the ischium, ossified by extension from the primary centers (and therefore separated until then by a gradually narrowing plate of cartilage) unite. Flecker found this union as early as age 3 years, 11 months in a girl, 4 years in a boy, and said that while it may not be completed until 11 or after, the majority are fused by the age of 8 in girls, 7 in boys. This seems to be an exception to the usual rule that skeletal maturation in females typically precedes that·in males. Expansion of the primary centers also involves the acetabulum and narrows the cartilage here.

Ossification of the coxal bone is completed by secondary centers, somewhat variable in number but usually five or more. The earliest are in the cartilage of the acetabulum; they are difficult to study by roentgenogram, but are said to be usually two or more, of which one, sometimes called the os acetabuli or the cotyloid bone, first (at about the age of 12 years) fuses with the pubis to form much of the acetabular part of this bone. Through growth of the three bones into the acetabulum this is largely converted into bone, and the three elements narrow the uniting cartilage to a Y-shaped piece. One or more secondary centers appear in this Y-cartilage by at least the twelfth to fourteenth year, and complete the union of the three bones—the ilium and pubis first fusing, then the ilium and ischium, and finally the pubis and ischium. Union at the acetabulum is sometimes said to occur usually about the eighteenth year, but Flecker found it as early as the eleventh year in a girl, the fourteenth in a boy, and said a majority of fusions were completed by the age of 13 in girls, 15 in boys.

The secondary centers for the hip bone in addition to the acetabular ones are epiphyseal centers, and apparently vary. In general, they appear in about the thirteenth to fifteenth years, and regularly include one for the iliac crest, one for the ischial tuberosity and, probably, one for the anterior inferior iliac spine; Flecker said he had never demonstrated this by roentgenogram. Apparently less frequent are epiphyses for the pubis, either for the pecten or for the body at the symphysis, and an epiphysis for the anterior superior iliac spine seems to be rare (Flecker). Irregular separate centers for the rim of the acetabulum have apparently also been described as sometimes present. The epiphyses are usually all fused at about the ages of 20 to 21 in both sexes.

The *upper end of the femur* typically presents three centers of ossification, one for the head and one for each trochanter (Fig. 7-5). According to Paterson ('29), the center for the head of the femur is never present at birth (but it has been reported, Flecker said, in a fraction of 1% within 48 hours after birth), usually appears during the second half of the first year, and is always present at the end of the first year. By the end of the third year the head as seen in roentgenograms has the form of the adult head. It is united to the neck of the femur, developed from the body, between the ages of 14 and 18 years; Flecker found the great majority fused at age 14 in the female, 17 in the male.

Both trochanters develop in part from the body, but they are completed by epiphyses. Paterson found the center of ossification for the epiphysis of the greater trochanter sometimes present during the fourth year, but not constant until the fifth year in the male, usually appearing about a year earlier in the female. A center for the lesser trochanter he found to be apparently inconstant, but usually appearing between the ages of 9 and 12 years if it appears at all. He described the epiphyses of the trochanters as joining the body at about the same time as does the head.

Knee

As already noted, the center for the *lower end of the femur* is typically present at birth, and the lower end of the femur develops from this single center (Fig. 7-6). It begins to take the shape of the condyles at the age of 2 years and unites with the body at about age 16 in the female, 18 in the male; Flecker found a range in time of union of from 14 years in girls to later than 19 in both sexes.

The center for the *upper end of the tibia* also is usually present at birth, but according to Paterson it sometimes does not appear until about the age of 1 month. At the age of 10 or 11 years the center expands downward to form the tibial tuberosity, the tip of which may occasionally have a separate center of ossification appearing between the ages of 7 and 15 years. The epiphysis of the upper end of the tibia is said to be usually fused with the body by age 16 (Paterson) or 15 (Flecker) in the female, age 18 in the male, although nonfusion may still be evident in either sex at 19 years (Flecker). The lower part of the tibial tuberosity is the last portion to unite.

A single center of ossification appears for the *upper end of the fibula* usually during the last of the third or the early part of the fourth year; it unites with the body by about the age

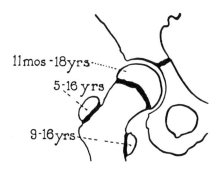

Fig. 7-5. Ossification of the upper end of the femur. The numbers indicate respectively the age at which the epiphyseal center can be expected to appear and the age at which it can be expected to fuse. (Camp JD, Cilley EIL: Am J Roentgenol 26:905 1931)

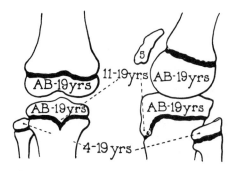

Fig. 7-6. Ossification at the knee. *AB* indicates centers present at birth, and the first of two numbers indicates the age at which the center so marked appears; the second number indicates the age at which it usually fuses to the rest of the bone. (Camp JD, Cilley EIL: Am J Roentgenol 26:905, 1931)

of 16 or 17 years in the female, 18 or 19 years in the male.

A center for the *patella* (sometimes two) is variable in its time of appearance. Paterson found it sometimes as early as 3 years of age, Flecker as early as 2 years, 6 months, but both agreed it is not present in a majority of boys until about the age of 5. This bone apparently reaches full growth by the age of 16. When there are two centers, the second is at the lower extremity of the patella according to Paterson, either here or at the upper and lateral aspect according to Flecker. Nonfusion of the two centers gives rise to bipartite patella.

The lateral *fabella*, a sesamoid bone associated with the tendon of origin of the lateral head of the gastrocnemius, may also first ap-

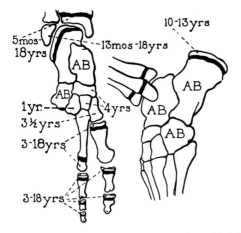

Fig. 7-7. Ossification of the ankle and foot. *AB* indicates a center of ossification present at birth; a single number, or the first of two, indicates the age at which the center for the bone or bones so marked appears; the second of two numbers indicates the age at which the center fuses with the rest of the bone. (Camp JD, Cilley EIL: Am J Roentgenol 26:905, 1931)

pear in roentgenograms of knees between the sixteenth and twentieth years. Flecker found it in 8.5% of persons 16 to 29 years old.

Ankle and Foot

Paterson found the center for the *lower end of the tibia* (Fig. 7-7) appearing earlier than it is usually said to do, being often visible at 6 months, and always before the end of the first year. At about the age of 8 years it starts to project downward to form the medial malleolus; it unites with the body at about the age of 14 years (Flecker) or 16 years (Paterson) in the female, 17 or 18 years in the male. Flecker reported an instance in which the tips of the medial malleoli showed separate centers of ossification.

Paterson found a center for the *lower end of the fibula* present at the end of the first year, although it is usually said to appear during the second. He found it usually fused to the body by the age of 15 years in the female, 17 in the male, and always fused by 16 in the female, 18 in the male. Flecker said a majority were fused by age 14 in the female, 17 in the male, but reported nonfusion in a girl of 16 years, 4 months, and in a boy of 18.

The *tarsals* typically develop from single

centers of ossification, although double centers have been observed for at least the talus, the navicular, and the medial cuneiform, and the calcaneus regularly has an epiphysis. McDougall said that the talus typically has two epiphyses, one for each tubercle, that appear and fuse within a year. Separate centers of ossification for some of the tarsals account for several of the more common supernumerary tarsal bones.

As already noted, the primary center of ossification for the calcaneus appears before birth, and so do the centers for the talus and, usually, the cuboid; however, Paterson said the one for the cuboid may appear shortly after birth. He found the center for the lateral cuneiform appearing usually during the first year, that for the navicular during the third, those for the intermediate and medial cuneiforms during the third year in the female, the fourth in the male. The epiphysis for the posterior end of the calcaneus appears between the ages of about 7 and 9 years in the female, 9 and 11 years in the male, and fuses with the remainder of the bone at about the age of 15 in the female, 17 to 18 years in the male.

Centers of ossification for the epiphyses of the second through the fifth *metatarsals* appear at the distal ends of these bones at about the age of 2 years in the female, 3 years in the male, while one for the first metatarsal commonly appears at the proximal end of this bone a few months earlier; their order of appearance is not constant. A majority were said by Flecker to have fused by the age of 14 in girls, 17 in boys, and Paterson said that in his series all were fused by the age of 15 in the female, 18 in the male. The tuberosity of the fifth metatarsal may ossify separately; Flecker found a separate center in 15% of 100 subjects between 12 and 17 years old. If it fails to unite to the rest of the bone, it forms the os vesalianum.

Centers of ossification for the epiphyses of all the *phalanges* appear at the proximal ends of these elements almost simultaneously, between the ages of 2 and 2.5 years in females, 3 years in males. Paterson said they were usually fused at the end of the fifteenth year in females, while in males 75% were fused at age 17, all at age 18. Flecker reported a good deal of variation in the times of appearance and fusion of the phalangeal epiphyses.

Sesamoids of the foot appear rather late; Flecker found those at the head of the first metatarsal constant after the age of 12 years, but found none earlier than the tenth year. He noted that the medial sesamoid often, and the lateral one sometimes, may have double centers and thus be split as if it had been fractured.

ANOMALIES

The lower limb is in general subject to the same developmental anomalies that may affect the upper limb, and the same classification used there, such as paraxial and transverse hemimelia, complete and partial adactylia, and the like (Chap. 3), is applicable also to the lower limb. In addition, because the lower limbs develop close together, malformation of the caudal end of the body may lead to their fusion with each other (*sympodia*), producing a mermaidlike or sireniform monster. Multiple anomalies, especially of the urogenital system, regularly accompany sympodia.

A few rare instances of supernumerary lower limbs have been reported (*e.g.,* Smillie and Murdoch), and Strivastava and Garg reported an apparently unique case in which there were two femurs, only one tibia but apparently two fibulas, and the foot had eight toes.

Congenital dislocation of the hip (Chap. 8) may result in maldevelopment of the acetabulum and the femoral head. Clubfoot (Chap. 9) has been variably said to be a primary defect in the development of the skeleton or in the development of soft tissues; its rather frequent association with spina bifida, and the finding that the musculature may show pathologic changes, have led to increasing belief that the causative factors may lie in muscular or other soft tissue defect rather than in the skeleton per se. There seems to be a hereditary factor involved in the production of clubfoot.

General Anatomy

SKELETON

The chief differences in the skeleton of the upper and lower limbs result from the functional adaptations of a common plan, the upper limb being adapted especially for freedom of movement, the lower limb (Fig. 7-8) for support. Thus the pelvic girdle articulates firmly with the sacrum, instead of being largely attached to the axial skeleton through

Fig. 7-8. Skeleton of the lower limb.

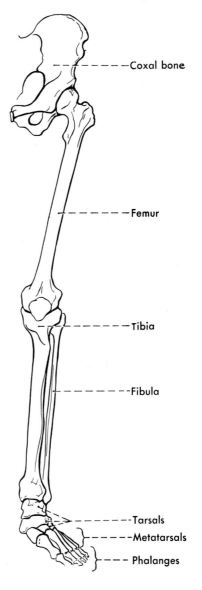

muscles as is the pectoral girdle; the pelvic girdle also attains better stability and strength through fusion of its three paired elements, ilium, ischium, and pubis, into paired single bones, and the firm union of these with each other at the pubic symphysis. Similarly, the deep acetabulum, receiving the head of the femur to form the hip joint, gives strength to this joint, and the heavy ligaments of the joint add more; in contrast to the shoulder joint, therefore, the strength of which depends primarily upon the surrounding musculature, the hip joint is intrinsically strong.

The femur, corresponding to the humerus, is succeeded by the tibia and fibula, corresponding to the radius and ulna; the knee joint owes its strength both to the muscles that pass across it and to the special ligaments connected with it. In the leg, the tibia is the chief weight-bearing bone at both ankle and knee, and the fibula serves largely simply for the attachment of muscles. The tarsals and metatarsals, the equivalent of the carpals and metacarpals, are obviously arranged for weight bearing, with the backward-projecting calcaneus forming the posterior end of the longitudinal arch of the foot and the heads of the metatarsals forming the anterior end.

Most of the phalanges of the foot are rudimentary as compared to the hand, and because of the weight-bearing function of the first metatarsal the big toe remains in the plane of the other digits, and is not opposable.

Miller has discussed important anatomical aspects of paracentesis of the joints of the lower limb. As in the upper limb, there is no fixed relationship between the attachment of the capsule of a joint to a long bone and the epiphyseal line or lines at that end of the bone (Fig. 7-9).

MUSCLES

The muscles of the lower limb (Figs. 7-10 and 7-11) are approximately equal in number to those of the upper limb (about 57 or so, depending upon how they are counted), but are in general more simple because of the limited mobility demanded of the lower limb as compared with the upper. In consequence of the firm attachment of the pelvic girdle to the axial skeleton, muscles passing between the axial skeleton and the lower limb are very few—primarily the quadratus lumborum, the piriformis, the psoas minor, and the iliopsoas,

Fig. 7-9. Anterior and posterior views of the ends of the large long bones of the lower limb, showing relations of epiphyseal lines and the attachments of joint capsules. The epiphyseal lines are the *heavy broken ones. Solid lines* and *stipple* indicate lines and areas of attachment of heavy ligaments, *dotted lines* the lines of attachment of thin parts of the capsules. (Mainland D: Anatomy. New York, Hoeber, 1945)

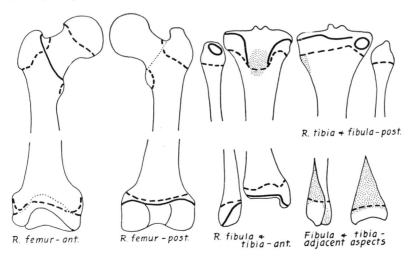

many muscles, such as those of the abdominal wall, of the pelvic diaphragm, and of the perineum, that are not muscles of the limb.

As a result of the rotation and adduction that the lower limb has undergone at the hip, the originally dorsal musculature of the free limb lies anteriorly and laterally, while that at the buttocks remains largely dorsal. This musculature is still innervated by the dorsal

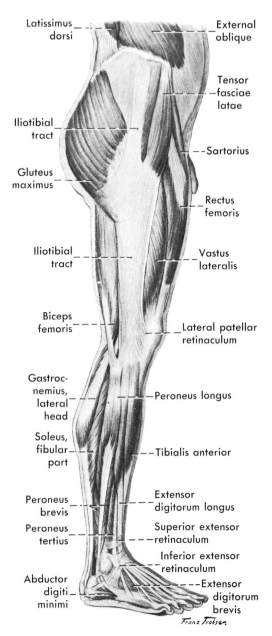

Fig. 7-10. Lateral view of muscles of the lower limb. (Frohse F, Fränkel M: Vol 2, Section 2, Part 2B in K von Bardeleben's Handbuch der Anatomie des Menschen. Jena, Fischer, 1913)

Fig. 7-11. Medial view of muscles of the lower limb. (Frohse F, Fränkel M: Vol 2, Section 2, Part 2B in K von Bardeleben's Handbuch des Anatomie des Menschen. Jena, Fischer, 1913)

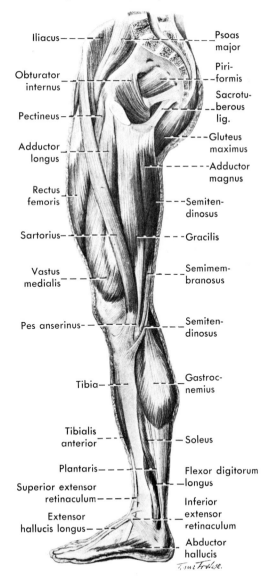

and the first of these is more properly regarded as a lateral muscle of the back. This is in contrast to the numerous extrinsic muscles associated with the upper limb. Because of its relationship to the caudal end of the body, the pelvic girdle does receive attachments of

Table 7-1
Innervation of the Muscles of the Lower Limb

MUSCLE GROUP	INDIVIDUAL MUSCLE	PERIPHERAL NERVE SUPPLY	APPROXIMATE SEGMENTAL NERVE SUPPLY*
Extrinsic at hip	Quadratus lumborum	Several twigs from anterior rami	T12-L3
	Iliopsoas	Twigs from lumbar plexus	L2-L4
	Psoas minor	,, ,, ,, ,,	L1, L2
In buttock	Gluteus maximus	Inferior gluteal	L5-S2
	Gluteus medius	Superior gluteal	L4-S1
	Gluteus minimus	,, ,,	,,
	Tensor fasciae latae	,, ,,	,,
	Piriformis	N. to piriformis	S1, S2
	Obturator internus	N. to obturator internus and superior gemellus	L5-S2
	Superior gemellus	N. to obturator internus and superior gemellus	,,
	Inferior gemellus	N. to quadratus femoris and inferior gemellus	L4-S1
	Quadratus femoris	N. to quadratus femoris and inferior gemellus	,,
Anterior of thigh	Sartorius	Femoral	L2, L3
	Quadriceps	,,	L2-L4
	Articularis genus	,,	L3, L4
	Pectineus	Femoral sometimes obturator	L2, L3
Anteromedial (adductor) in thigh	Adductor longus	Obturator	,,
	Adductor brevis	,,	L3, L4
	Gracilis	,,	,,
	Adductor magnus	Obturator and sciatic (tibial)	L3-L5
	Obturator externus	Obturator	L3, L4
Posterior in thigh (hamstrings)	Semitendinosus	Sciatic (tibial)	L5-S1
	Semimembranosus	,, ,,	,,
	Biceps, long head	,, ,,	L5-S2
	Biceps, short head	Sciatic (common peroneal)	L5, S1
Anterolateral in leg	Tibialis anterior	Deep peroneal	L4-S1
	Extensor digitorum longus	,, ,,	,,
	Peroneus tertius	,, ,,	,,
	Extensor hallucis longus	,, ,,	,,
	Peroneus longus	Superficial peroneal (and deep?)	,,
	Peroneus brevis	Superficial peroneal	,,
Of calf	Gastrocnemius	Tibial	S1, S2
	Soleus	,,	,,
	Plantaris	,,	L4-S1
	Popliteus	,,	L5, S1
	Tibialis posterior	,,	,,
	Flexor digitorum longus	,,	,,
	Flexor hallucis longus	,,	L5-S2
On dorsum of foot	Extensor hallucis brevis	Deep peroneal	L5, S1
	Extensor digitorum brevis	,, ,,	,,
In sole of foot	Abductor hallucis	Medial plantar	L5, S1
	Flexor hallucis brevis	,, ,,	,,
	Adductor hallucis	Lateral plantar	S1, S2
	Flexor digitorum brevis	Medial plantar	L5, S1

Table 7-1 (continued)

MUSCLE GROUP	INDIVIDUAL MUSCLE	PERIPHERAL NERVE SUPPLY	APPROXIMATE SEGMENTAL NERVE SUPPLY*
In sole of foot *(cont.)*	Quadratus plantae	Lateral plantar	S1, S2
	Flexor digiti minimi brevis	" "	"
	Abductor digiti minimi	" "	"
	First lumbrical	Medial plantar	L5, S1
	Lateral three lumbricals	Lateral plantar	S1, S2
	All interossei	" "	"

* While many authors would add a segment either above or below, sometimes both, to the segmental innervations shown here, there are few real points of disagreement.

or posterior branches of the lumbosacral plexus (for the buttock, several small nerves, for the thigh the femoral nerve and the common peroneal part of the sciatic, and for the anterolateral side of the leg and the dorsum of the foot the common peroneal). Of the originally ventral musculature, that of the thigh is then innervated by the obturator nerve and the tibial portion of the sciatic, and that of the calf of the leg and the plantar side of the foot by the tibial nerve. The innervation of the muscles of the lower limb is shown in Table 7-1, and the motor points of the muscles in Fig. 7-12 and Fig. 7-13.

As is true for the upper limb, the segmental innervations given for the various muscles must be assumed to be only approximate. If the usually given segmental innervations are approximately correct, however, there is, as pointed out by Last and discussed briefly in Chapter 1, a general pattern of innervation of the muscles of the lower limb if they are considered in regard to their placement: the joint across which they chiefly act, and whether they lie in front of or behind that joint. Thus Last pointed out that, if one assumes that the important innervation to any muscle is usually from two segments rather than more, it is unnecessary to memorize segmental innervations for the lower limb. Instead, the general pattern, possibly accurate enough for clinical use, is that the anteriorly lying muscles at a given joint in the lower limb are innervated by nerves that lie segmentally above those for the posteriorly lying muscles: for instance, the anteriorly lying muscles at the hip,

which produce flexion and adduction, are supplied largely by the second and third lumbar nerves, while the posteriorly lying muscles, which produce extension and abduction, are supplied largely by the fourth and fifth lumbar nerves—thus, assuming only a bisegmental innervation for each group of muscles, by the two succeeding nerves. Further, if one moves down either the anterior or posterior surface of the lower limb to the next joint, the muscles acting across this joint are found to be innervated by nerves one segment distal to those innervating the joint above: for instance, hip flexors are probably innervated primarily by L2 and L3, extensors of the knee by L3 and L4, dorsiflexors at the ankle by L4 and L5; similarly, extensors at the hip are probably innervated primarily by L4 and L5, flexors at the knee by L5 and S1, plantar flexors at the ankle by S1 and S2.

Bursae and Tendon Sheaths

Most of the bursae in the lower limb, as in the upper, are associated with tendons of muscles, but there are several rather constant subcutaneous bursae. These include the subcutaneous trochanteric bursa over the greater trochanter of the femur; the prepatellar subcutaneous bursa in front of the lower part of the patella; the subcutaneous infrapatellar related to the upper part of the patellar ligament; a subcutaneous bursa of the tibial tuberosity over the front of this eminence; subcutaneous bursae of the lateral and the medial malleoli over these prominences, and a subcutaneous calcaneal bursa between the skin of the sole and

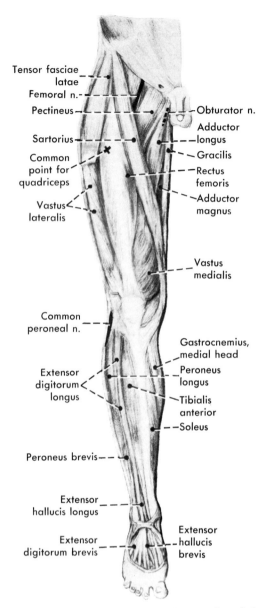

Fig. 7-12. Motor points of anterior muscles of the lower limb. (After Cohn, from Frohse F, Fränkel M: Vol 2, Section 2, Part 2B in K von Bardeleben's Handbuch der Anatomie des Menschen. Jena, Fischer, 1913)

chanteric bursa of the gluteus maximus, a large bursa between the tendon of the gluteus maximus and the posterolateral surface of the greater trochanter; two trochanteric bursae of the gluteus medius, both small, one lying be-

Fig. 7-13. Motor points of posterior muscles of the lower limb. (After Cohn, from Frohse F, Fränkel M: Vol 2, Section 2, Part 2B in K von Bardeleben's Handbuch der Anatomie des Menschen. Jena, Fischer, 1913)

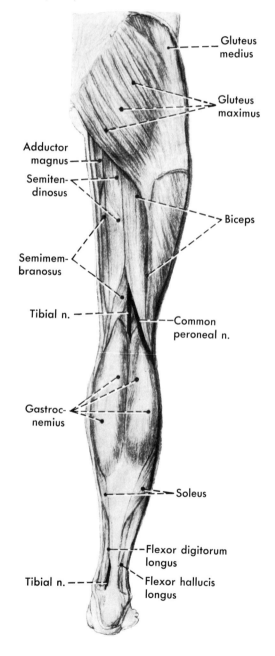

the plantar surface of the calcaneus. In the region of the knee, in addition to the subcutaneous bursae already mentioned and the subtendinous bursae yet to be recounted, there may also be a subfascial prepatellar bursa, in the connective tissue covering the lower part of the patella.

Deeper bursae around the hip are the tro-

tween the tendon of the gluteus medius and the lateral surface of the greater trochanter, the other between the tendon of the gluteus medius and the piriformis muscle; the trochanteric bursa of the gluteus minimus, a fairly large one usually present between the tendon of the gluteus minimus and the edge of the greater trochanter; the bursa of the piriformis muscle, between the tendon of the piriformis and the femur; the subtendinous bursa of the obturator internus, deep to the tendon of this muscle as it crosses the lesser sciatic notch; variable bursae (intermuscular bursae of the gluteus muscles) between the gluteus muscles; the ischiadic bursa of the gluteus maximus, sometimes present between the ischial tuberosity and the gluteus maximus; a similar ischiadic bursa of the obturator internus, which may lie between this muscle and the upper border of the ischial tuberosity; an iliopectineal (iliopsoas) bursa, deep to the iliopsoas tendon over the front of the hip joint, and communicating with the hip joint in about 15% of adults; a subtendinous iliac bursa, a small one between the tendon of insertion of the iliopsoas and the lesser trochanter; and the superior bursa of the biceps femoris muscle, a fairly constant one that lies between the common tendon of origin of the long head of the biceps and semitendinosus, expanding also beneath the tendon of the semimembranosus, and the ischial tuberosity. A rather large number of other bursae, which are both small and inconstant, have also been described in the region of the hip.

The deeper bursae around the knee are particularly numerous. They include the following:

The subtendinous prepatellar bursa, not constant, between the tendon of the quadriceps and the upper end of the patella (thus three prepatellar bursae in all, a subcutaneous, a subfascial, and a subtendinous, but rarely all three simultaneously)

The suprapatellar bursa, between the tendon of the quadriceps and the anterior surface of the lower part of the femur, but usually communicating with the cavity of the knee joint so as to form in the adult not a bursa, but simply a suprapatellar recess of the knee joint

The deep infrapatellar bursa, between the patellar ligament and the tibia

One or more subtendinous bursae of the sartorius, deep to the tendon of this muscle and lying between it and the tendons of the semitendinosus and gracilis, and often communicating with the following bursa

The anserine bursa, a rather large and complicated one, lying over the tibial collateral ligament of the knee joint deep to the tendons of the sartorius, gracilis, and semitendinosus muscles

The inferior bursa of the biceps femoris, between the tendon of insertion of this muscle and the fibular collateral ligament

The lateral subtendinous bursa of the gastrocnemius, between the tendon of origin of the lateral head of the muscle and the capsule of the knee joint

The medial bursa of the gastrocnemius, between the tendon of origin of the medial head of the gastrocnemius and the capsule of the knee joint

The bursa of the semimembranosus, between the semimembranosus and the medial head of the gastrocnemius muscle, and often communicating with the bursa of the medial head, which in turn may communicate with the knee joint.

Many of these bursae are somewhat variable, and additional bursae have been described. In addition to the bursae, some of which, as noted, may communicate with the knee joint, the popliteus has between it and the posterior aspect of the lateral condyle of the tibia a prolongation of the synovial membrane of the knee joint, the subpopliteal recess; like most apparent extensions of the synovial membrane of joints to serve as bursae beneath tendons, the subpopliteal recess is developmentally a bursa that is secondarily fused with the synovial membrane of the knee joint.

Bursae in the ankle and foot are relatively few, because most of the tendons reaching the ankle are provided with synovial tendon sheaths. Those named in the NA, fairly constant, are only two: the subtendinous bursa of the tibialis anterior, between this tendon and the medial surface of the medial cuneiform bone, and the bursa of the tendo calcaneus (tendon of Achilles), between this tendon and the upper part of the back end of the calcaneus. Other bursae sometimes described as being related to the region of the ankle are a subtendinous bursa of the extensor hallucis longus, deep to the tendon of this muscle at about the level of the tarsometatarsal joint; the bursa of the sinus tarsi, in the sinus tarsi and extending over the lateral surface of the neck of the talus and its covering ligaments to lie deep to the tendons of the extensor digitorum longus; and a subtendinous bursa of the tibialis posterior, between this tendon and the navicular bone just above its attachment.

As is indicated by this account, the bursae of the lower extremity tend to vary even more than do those of the upper extremity; presumably the more inconstant ones, at least, are purely adventitial bursae, developing because of a particularly close association between a tendon and bone, or unusual use of a muscle. Adventitial subcutaneous bursae are also particularly likely to develop in the lower limb with abnormal weightbearing and friction of the parts: for instance, subcutaneous bursae may appear over the medial condyles of the two femurs in cases of genu valgum (knock knee), as a result of friction between the two knees, and in talipes (clubfoot), bursae may appear deep to the skin most subject to friction during walking.

The synovial tendon sheaths associated with muscles of the lower limb lie anteriorly, posterolaterally, and posteromedially at the ankle, where they allow the tendons to glide freely between the underlying bones and ligaments and the overlying retinacula. Details of the tendon sheaths are discussed in a following chapter.

NERVES

The chief nerves of the lower limb are shown in Figure 7-14. They are discussed and illustrated in detail in the following chapters; in the present discussion their origins, varia-

tions, and general distribution, rather than detailed relations, are dealt with.

The nerve supply to the lower limb is derived from both the lumbar and the sacral plexuses, these frequently being grouped together as the lumbrosacral plexus. Since the lumbar plexus is the cranial one of the two, its distribution to the limb is to anteromedial structures, that is, to the parts developed along the original cranial border of the limb; further, since the plexus is divisible into originally dorsal and ventral branches, its chief posterior (dorsal) branch, the femoral nerve, innervates the adult anterior (originally dorsal) aspect of the thigh, with anterolateral skin being supplied by another posterior branch, the lateral femoral cutaneous; and its chief anterior (ventral) branch, the obturator nerve, innervates the anteromedial aspect of the thigh. It has also been said that the medial cutaneous branch of the femoral nerve, distributed to skin of the medial side of the thigh and often partly replacing the cutaneous branch of the obturator here, is actually a ventral derivative of the plexus in contrast to the major portion of the femoral nerve.

The distribution of the lumbar plexus to the lower limb is limited to the thigh except for the cutaneous distribution of the saphenous branch of the femoral nerve, along the anteromedial surface of the leg and the medial border of the foot.

The sacral plexus, the caudal of the two plexuses, has larger branches and is distributed to more of the limb than is the lumbar plexus, but nevertheless supplies expecially the originally more caudal parts of the limb. As in the case of the lumbar plexus, its branches are subdivided into posterior and anterior (dorsal and ventral) ones. The posterior ones include most of the nerves to the buttock, and the common peroneal portion of the sciatic nerve, the latter being distributed to the dorsal part of the postaxial portion of the limb—to the one muscle (short head of biceps) arising lateral to the posterior midline of the thigh, the postaxial border along which the original dorsal and ventral surfaces met, and anterolaterally to skin and muscles of the leg and the dorsum of the foot. The tibial nerve, the largest anterior or ventral branch

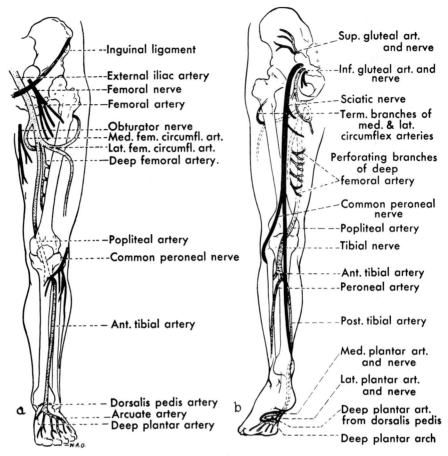

Labels (left, a):
- Inguinal ligament
- External iliac artery
- Femoral nerve
- Femoral artery
- Obturator nerve
- Med. fem. circumfl. art.
- Lat. fem. circumfl. art.
- Deep femoral artery.
- Popliteal artery
- Common peroneal nerve
- Ant. tibial artery
- Dorsalis pedis artery
- Arcuate artery
- Deep plantar artery

Labels (right, b):
- Sup. gluteal art. and nerve
- Inf. gluteal art. and nerve
- Sciatic nerve
- Term. branches of med. & lat. circumflex arteries
- Perforating branches of deep femoral artery
- Common peroneal nerve
- Popliteal artery
- Tibial nerve
- Ant. tibial artery
- Peroneal artery
- Post. tibial artery
- Med. plantar art. and nerve
- Lat. plantar art. and nerve
- Deep plantar art. from dorsalis pedis
- Deep plantar arch

Fig. 7-14. Diagrams of the nerves and arteries of the leg in anterior, *a*, and posterior, *b*, views.

of the sacral plexus, is then distributed to tissue associated with the original flexor surface of the postaxial portion of the limb, namely, to muscles of the thigh arising medial to the posterior midline and to skin and muscles of the calf of the leg and of the plantar surface of the foot.

The cutaneous innervation of the posterior aspect of the thigh is through the posterior femoral cutaneous nerve, which, like the sciatic, lies in the posterior midline of the thigh, is derived from both dorsal and ventral parts of the plexus, and is distributed to skin as the sciatic nerve is to muscles: the fibers it receives from the dorsal part of the plexus are distributed lateral to the posterior midline, those from the ventral part medial to the midline.

LUMBOSACRAL PLEXUS

All except the very lowest nerves caudal to the last thoracic, and frequently a part of that, enter into the formation of the lumbosacral plexus. The lumbar plexus is that portion of the lumbosacral plexus lying in the abdomen, formed within the substance of the psoas major muscle; the sacral plexus, in contrast (although it receives fibers from lumbar nerves), is formed in the pelvis, on the front of and lateral to the sacrum. The major portions of both the lumbar and sacral plexuses pass to the lower limb, but an upper part of the lumbar plexus goes largely to the anterior abdominal wall, and nerves from the lower end of the sacral plexus pass to the perineum and to skin and soft tissue around the coccyx.

The sacral plexus was once defined as that

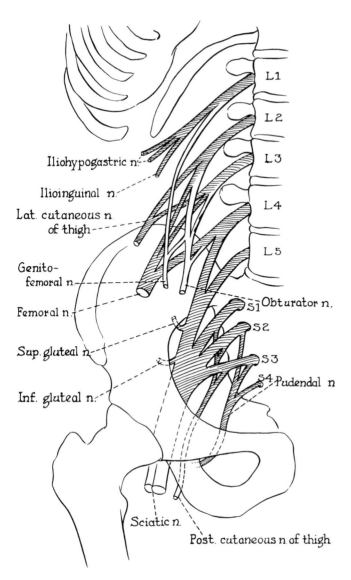

Fig. 7-15. Diagram of the lumbosacral plexus. (Sections of Neurology and the Section of Physiology, Mayo Clinic and Mayo Foundation: Clinical Examinations in Neurology [ed 2]. Philadelphia, WB Saunders, 1963)

part of the spinal nerve plexus in the pelvis that is distributed to the lower limb, and succeeding parts have therefore been described as the pudendal and coccygeal plexuses; present terminology includes the pudendal and coccygeal nerves in the sacral plexus. The lumbar and the sacral plexuses will be discussed separately in detail, following the general description here.

General Composition

The lumbosacral plexus (Fig. 7-15) is formed by ventral (anterior) rami of spinal nerves, typically all those from the first lumbar downward, and often with a contribution also from the last thoracic nerve. The ventral rami of the lumbar nerves increase in size on down through the fifth, which is therefore the largest lumbar nerve contributing to the plexus; the first sacral nerve is typically but not always larger than the fifth lumbar nerve, and the ventral rami of the remaining sacral nerves diminish in size to the tiny coccygeal.

As already noted, the branches of both the lumbar and sacral portions of the plexus are divisible into anterior ones that supply musculature originally on the ventral or flexor surface of the limb, and posterior ones that supply muscle originally on the dorsal or extensor surface of the limb. However, this division is not particularly obvious except upon careful dissection of each plexus; in relation to the adult limb, which has undergone a medial torsion that has distorted its original relations, both the anterior and the posterior branches of the lumbar plexus leave the abdomen to pass anteriorly into the thigh, while all the branches, anterior and posterior, of the sacral plexus leave the pelvis to pass posteriorly into the buttock. The obvious subdivision therefore is between the lumbar or cranial part of the plexus and the sacral or caudal part. Typically, these two plexuses share only one spinal nerve in common, and this is usually the fourth lumbar; therefore in a reasonably normal plexus a part of the fourth lumbar nerve (usually less than half) leaves the part going into the lumbar plexus and runs caudally in the psoas muscle, immediately lateral to the vertebral bodies. It is joined or paralleled by all the ventral ramus of the fifth lumbar nerve, the two forming a lumbosacral trunk; this emerges between the diverging psoas muscle and the lateral surface of the vertebral column, crosses the pelvic brim and, as the highest component of the sacral plexus, joins the upper sacral nerves.

While the lumbosacral plexus as a whole may contain fibers from the eleventh or twelfth thoracic through the last sacral and the coccygeal nerve, that is, from a total of 11 spinal nerves, the two extremes of the plexus are, as already indicated, distributed to structures other than the limb. Thus, in what is the most common and therefore presumed to be the "normal" plexus, much of the most cranial contribution to the lumbar plexus, whether L1 alone or T12 and L1, is distributed to the lower part of the abdominal wall, and secondarily to skin of the thigh; similarly, the most caudal nerves of the sacral plexus, typically the fourth and fifth sacral and the coccygeal, pass to the pudendal nerve and to the coccygeal plexus instead of to the limb. In consequence, the usual number of spinal nerves contributing to the chief nerves of the limb (the most cranial ones being the lateral femoral cutaneous and the femoral, the most caudal the tibial) is most commonly eight, the twenty-first through the twenty-eighth—with a normal vertebral column, the first lumbar

through the third sacral (Bardeen and Elting). There may be, however, as few as six or as many as nine. If one excludes from the "chief nerves" of the limb the lateral femoral cutaneous nerve, which often receives all the first lumbar fibers going to the lower limb (although some may be distributed, probably only to skin, through the femoral nerve), the femoral nerve is then regarded as the most cranially arising principal nerve of the limb; in this circumstance, the nerves contributing to the plexus are most commonly seven in number, the second lumbar through the third sacral.

Variations in Formation
Variations in the most cranial part of the lumbar plexus (see Table 7-4) in which the roots contribute to the origin of the iliohypogastric, ilioinguinal, and genitofemoral nerves, are usually not included in speaking of variations in the lumbosacral plexus; these nerves contribute little to the innervation of the limb, and all their contribution is cutaneous. However, Bardeen ('02, '07) reported that the uppermost nerve, the iliohypogastric, may arise from as high as T11 or contain fibers of L1 only. Thus, in the strict sense, the upper border of the lumbosacral plexus may vary between T11 and L1, although the uppermost fibers are probably always distributed to the abdominal wall rather than to the thigh. The three uppermost branches of the plexus, since they supply skin at the border between abdomen and thigh, are sometimes referred to as the "border nerves."

The lumbosacral plexus as a whole shows many variations, although a number of them, including those just described, are minor ones. Fibers from either higher (more cranial) or lower (more caudal) nerves than usually participate in the formation of the plexus may enter it, thus expanding the plexus, or an upper or lower nerve that usually contributes may fail to do so, thus leading to contraction of the plexus; or the plexus may be shifted in either direction through loss of a usual contribution at one end and an addition of one at the other end. Degrees of shifting that are slighter, but often more important, occur when a given spinal nerve contributes fewer

Table 7-2
*Variations in the Formation of the Lumbrosacral Plexus**

TYPE OF PLEXUS	RANGE IN COMPOSITION	FURCAL NERVE	VARIATIONS IN COMPOSITION†
Normal	L1-S3	L4	L1-S3; L2-S3
Prefixed	T12-S3	L4 or L3 and L4	L1-S1; L1-S2; L2-S2; T12-S3; L1-S3; L2-S3
Postfixed	L1-S4	L4 or L5 or both	L1-S3; L2-S3; L1-S4; L2-S4

* Based upon the observations and classification of Bardeen and Elting. The iliohypogastric, ilioinguinal, and genitofemoral nerves at the cranial end, and the pudendal and coccygeal nerves at the caudal end, are not included here.

† Normal, prefixed, and postfixed plexuses may have the same composition; the classification is then based upon the furcal nerve: if L3 is a furcal nerve or most of the fibers of L4 go to the sacral plexus, the plexus is prefixed; if L5 is a furcal nerve the plexus is postfixed.

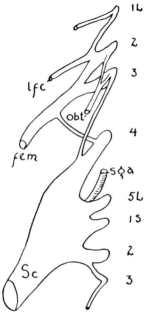

Fig. 7-16. A prefixed plexus in which the third and fourth sacral nerves both contribute to the lumbar and sacral plexuses. *lfc, obt, fem,* and *Sc* identify the lateral femoral cutaneous, obturator, femoral, and sciatic nerves, respectively; *sga* is the superior gluteal artery. (From part of original of Fig. 1, Horwitz MT: Anat Rec 74:91, 1939)

fibers than usual to one branch of the plexus, more fibers than usual to another; thus the fourth lumbar nerve, which normally sends the majority of its fibers into the femoral nerve, may instead send the majority to the sacral plexus and the sciatic nerve. It is rela-

tively rare for the plexus to both gain an upper nerve and lose a lower one, or vice versa, at the same time. Thus the plexus seldom shifts a complete segment in either direction; more commonly, it is expanded or contracted by additions or subtractions at either or both ends.

Obviously, variations in the plexus as a whole can be of many different types, so as almost to defy classification. The major types of plexus are usually described as being three: normal, prefixed, and postfixed (Table 7-2). The latter two terms cannot be accurately defined, but prefixation implies a tendency on the part of the plexus to be shifted cranially through a more cranial contribution than usual reaching it, through loss of its usual caudal contribution, or through an entirely internal shift in which the cranial and caudal limits of the plexus are normal, but some important nerve trunk derives fibers from a higher source than usual (Fig. 7-16). Similarly, postfixation implies a tendency for the plexus to shift caudally, by gaining fibers caudally, losing fibers cranially, or an internal shift.

Bardeen and Elting used three criteria in defining a normal lumbosacral plexus: they regarded a plexus as normal if (1) the most cranial nerve to supply the lower limb (other than through the "border nerves" already mentioned) is either L1 of L2; (2) the most caudal nerve (to the sciatic—they disregarded the pudendal nerve) is S3; and (3) L4, dividing to go to both the lumbar and the sacral plexuses, sends a majority of its fibers into the lumbar plexus. On this basis they found normal plexuses in 42.3% of 256. The remainder they divided into either prefixed or postfixed (the terms they preferred were "proximal" and "distal"), and subdivided each of these into three groups according to the degree of prefixation and postfixation. They did not regard the highest entering nerve as being important in this classification, but based the classification upon the most caudal nerve entering the plexus and on the bifurcation of the nerve that goes to both plexuses (since this nerve forks, it is often called the nervus furcalis or furcal nerve; it is usually L4, but L3 and L4 may both bifurcate, or L5 may be the furcal nerve as in Figs. 7-16 and 7-17).

Among their prefixed types they found only one example, the most prefixed, in which the twenty-first spinal nerve, normally the first sacral, was the most caudal to enter the sciatic nerve; their second type, constituting about 10.2% of all plexuses, was one in which the second sacral nerve was the lowest to enter the plexus; and their third type was one in which the third sacral was the lowest to enter, just as it is in a normal plexus, but most of the fibers of L4 went to the sacral rather than to the lumbar plexus. They divided postfixed plexuses also into three types, the least postfixed, constituting 7.3% of all plexuses, being one in which, while most of L4 goes to the lumbar plexus, L5 also contributes to this plexus; the next postfixed type occurred in 6.1%, and was characterized by the fact that the twenty-ninth spinal nerve, S4, rather than the twenty-eighth, S3, was the last nerve to contribute to the limb; and in the most postfixed type in their series, 8.1% of the total, not only did S4 contribute to the nerves of the limb, but L5 went in part to the lumbar plexus.

Obviously, the exact percentages of the different types of plexus described by different workers will vary not only with the material (by chance variation in the samples, rather than according to sex and race), but also with the criteria adopted as distinguishing prefixed and postfixed plexuses from normal. In Bardeen and Elting's series, however, 42.3% were

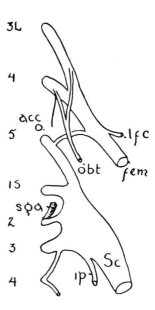

Fig. 7-17. A postfixed plexus in which the fifth lumbar nerve forks to contribute to both the lumbar and the sacral plexus. Labels as in Fig. 7-16; in addition, *acc o* is the accessory obturator nerve, *ip* the pudendal. (From part of original of Fig. 2, Horwitz MT: Anat Rec 74:91, 1939)

classified as normal plexuses, slightly more than 36% as prefixed, and 21.5% as postfixed. Horwitz used different criteria, and found rather different percentages. He defined a normal plexus as one in which the most cranial root entering the femoral nerve was from L2, the furcal nerve was L4, and the sacral plexus (defined as that giving rise to both the sciatic and pudendal nerves) received fibers from L4 through S4. On this basis, he found 71.93% of 228 plexuses (both sides of 114 bodies) to be normal, 28.07% to be prefixed or postfixed. Of the latter, slightly more than half were classified as prefixed, and both the prefixed and postfixed groups were subdivided into a number of different subgroups.

Actually, although there is a general tendency for a major branch of a prefixed plexus to receive more fibers from cranially situated nerves than it usually does, or less fibers from caudally situated nerves than it usually does, this does not necessarily hold true for each individual nerve, and in two plexuses of the same general type a given peripheral nerve may have different origins. For instance,

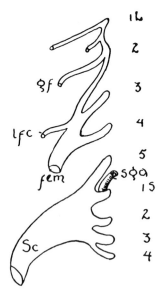

Fig. 7-18. A lumbosacral plexus in which there is no nerve contributing to both plexuses. Labels as in Fig. 7-17; in addition, *gf* is the genitofemoral nerve. (From part of original of Fig. 2, Horwitz MT: Anat Rec 74:91, 1939)

among Horwitz' prefixed plexuses there were two groups in which nerves L3 and L4 both contributed to the lumbar and sacral plexuses; in one group the femoral nerve had its normal origin, yet the sacral-pudendal complex received fibers from S5 and thus, while prefixed on the basis of its highest entering fibers, was postfixed according to its lowest entering fibers. In the second group the first lumbar nerve contributed to the femoral nerve (which in contrast to this nerve in the first group was therefore prefixed), and the sacral-pudendal complex received its lowest entering fibers from S3 and was therefore prefixed regardless of which end is considered. There are thus variations in the general form of the plexus, and other variations, not always strictly paralleling them, in the formation of the branches of the plexus.

Variations in the branches of the lumbar and sacral plexuses will be considered in further sections. In summary of the variations of the plexus as a whole, the highest nerve contributing to the limb (other than the border nerves) may be T12, but is more commonly L1, or L2; the lowest nerve contributing to the sacral-pudendal plexus may be S3, S4, or S5 (of if the pudendal nerve is omitted, the lowest spinal nerve contributing to the limb, that is, to the sciatic nerve, may vary from the first sacral through the fourth sacral, according to Bardeen and Elting); the furcal nerve, the one that goes to both plexuses, may be L4 as it is commonly, in which case, however, it may send most of its fibers to the lumbar plexus as usual, or most to the sacral; instead of L4, L3 and L4 together may be furcal nerves (Fig. 7-16); there may be no furcal nerve, all of L4 going into the femoral and all of L5 into the sciatic (Fig. 7-18); or L5 may be the furcal nerve, sending some of its fibers into the femoral and some into the sciatic nerve (Fig. 7-17).

In the discussions and tables of the lumbosacral plexus here presented the nerves are all named as they would be with a normal vertebral column: that is, the twenty-first spinal nerve is called the first lumbar regardless of the number of thoracic vertebrae, the twenty-sixth is called the first sacral even when it emerges between fifth and sixth lumbar vertebrae, and, when, because of a sacralized fourth lumbar vertebra, it is really the second nerve to emerge through the sacrum. Actually, a normal plexus, in terms of the serial number of its nerves, may be associated with an abnormal vertebral column, and vice versa, but Bardeen and Elting found that prefixed plexuses tend to be associated with shortening of the lumbosacral column, postfixed plexuses with lengthening of it. Horwitz, with his more exclusive definitions of prefixation and postfixation, found an even better correlation: prefixed plexuses, he found, tend to occur with any shortening of the vertebral column—a reduction in the total number of segments, or sacralization of the fifth lumbar vertebra; postfixed plexuses tend to be associated with any lengthening of the column, whether in total length or in the lumbar or sacral regions alone without increase in total length; further, the more abnormal the plexus, the greater the incidence of an associated abnormality of the vertebral column.

Pudendal Nerve and Coccygeal Plexus

These are of no importance insofar as the limb is concerned, and can be disposed of briefly.

The *pudendal nerve* is the largest of the branches from the caudal part of the sacral plexus that are not distributed to the limb. It goes to the perineum, and the nerves contributing to it were once described as forming the pudendal plexus. It is usually derived from the second and third, frequently from the third and fourth, sacral nerves; or occasionally it receives fibers either from as high as the first sacral nerve or as low as the fifth sacral, or both simultaneously. The third sacral contribution is usually the most important, but sometimes the nerve is derived exclusively from the second sacral.

The visceral or splanchnic branches of the pelvic nerves (pelvic splanchnic nerves, nervi erigentes) arise from the same ventral rami (usually S2, 3, and 4, sometimes S2 and 3, or S3 and 4) that contribute to the pudendal nerve. Although these nerves are composed of afferent and autonomic fibers, they are conveniently grouped with the pudendal nerve since they control the smooth musculature of the bladder and rectum, while the pudendal nerve is concerned with the voluntary muscle of the outlet of the bladder and rectum through the perineum. It need only be noted here that, because both the pudendal and the pelvic splanchnic nerves typically arise from the second to the fourth sacrals, bilateral damage to the cord or the spinal nerves at this level can be expected to have serious consequences in regard to the emptying of both the bladder and the bowel.

Other caudal branches of the sacral plexus include muscular ones to the levator ani and the coccygeus muscles, and to a part of the sphincter ani externus. They tend to arise from the fourth sacral nerve.

The *coccygeal plexus* is formed by two or more of the most caudal nerves, thus always including the coccygeal nerve. Commonly it is formed by the union of a part of the anterior branch of S4 with both S5 and the coccygeal nerve; its branches of distribution, the anococcygeal nerves, pass dorsally through the sacrotuberous ligament to supply skin in the region of the coccyx. It may receive fibers from as high as the third sacral, or only part of the fifth sacral.

LUMBAR PLEXUS

The lumbar plexus is formed in the substance of the psoas major muscle, and its branches emerge through the muscle laterally, anteriorly, and medially. Its usual branches are the iliohypogastric, the ilioinguinal, the genitofemoral, the lateral femoral cutaneous, the femoral, the obturator, and a branch to the lumbosacral trunk and the sacral plexus. In addition, there are muscular branches to the adjacent psoas, quadratus lumborum, and iliacus muscles; these branches may arise from the lumbar plexus itself or from the femoral nerve and, depending upon their destination, may or may not appear outside of the psoas muscle. Of the main branches, the iliohypogastric, the ilioinguinal, the lateral femoral cutaneous, and the femoral appear in that order, from above downward, at the lateral border of the psoas muscle; the genitofemoral appears on the anterior surface of the psoas, usually being formed here by two parts that pierce the muscle before uniting; and the obturator nerve and the contribution to the lumbosacral trunk appear just above the pelvic brim medial to the psoas. If there is an accessory obturator nerve, it also appears in this position.

Formation

The lumbar plexus is usually formed by the union of the ventral (anterior) rami of the first, second, and third lumbar nerves, and a major part of the ventral ramus of the fourth (Fig. 7-19); in the majority of cases it also receives some fibers from the twelfth thoracic that usually go, however, into the lower nerves of the abdominal wall, not to the limb, and in any case apparently never supply musculature of the limb.

A common form of the plexus is for the iliohypogastric and the ilioinguinal nerves to

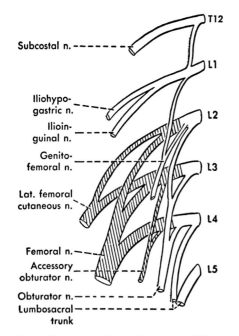

Fig. 7-19. Diagram of the lumbar plexus. The nerves derived from the posterior divisons are *shaded.*

arise close together, sometimes by a long common stem, from the first lumbar nerve after this has received a twig from the last thoracic; for part of or all the rest of the first lumbar to join a small part of the second lumbar to form the genitofemoral; and for the remainder of the plexus to divide into anterior and posterior parts, comparable to the division of the brachial plexus into anterior and posterior divisions. The posterior divisions of the lumbar plexus give rise to the lateral femoral cutaneous and the femoral nerves, while the smaller anterior divisions give rise to the obturator nerve (and according to many workers, to the accessory obturator, when it is present); the lumbosacral trunk, a part of the sacral plexus, does not divide into anterior and posterior divisions until it reaches the pelvis.

The lateral femoral cutaneous nerve most commonly arises from the posterior divisions of the first three lumbar nerves, or from the second and third, while the femoral commonly arises from the posterior divisions of the second, third, and fourth lumbar nerves; the obturator nerve most commonly arises

from the anterior divisions of the same nerves that contribute to the femoral, the second through the fourth lumbar; the accessory obturator, when it is present, arises most frequently from the third and fourth lumbar nerves. The accessory obturator nerve is sometimes described and figured as an anterior, sometimes as a posterior branch of the plexus; Paterson (1893–94) was among the early workers who argued that it is really a posterior branch, in which case it might more appropriately have been called the accessory femoral nerve.

Variations

As indicated by the account of the variations in the lumbosacral plexus as a whole, the lumbar plexus shows a number of variations (Tables 7-3, 7-4, and 7-5). The most cranial nerve entering it is usually a part of the twelfth thoracic, but is occasionally the eleventh thoracic, and not uncommonly the first lumbar. The most caudal contribution to it is usually from L4, and includes considerably more than half the fibers in this nerve. The commonest variation in its caudal boundary is in the relative number of fibers that the fourth lumbar nerve sends to it and to the sa-

Table 7-3
Variations in the Formation of the Lumbar Plexus*

TYPE OF PLEXUS	MOST COMMON COMPOSITION	VARIATIONS IN COMPOSITION
Normal	L1-L4	L2-L4
Prefixed	L1-L4	T12-L4; L2-L4
Postfixed	L2-L5	L1-L4; L1-L5; L2-L4

* Excluding the "border nerves."

Table 7-4
Variations in the Uppermost Branches of the Lumbar Plexus

NERVE	MOST COMMON COMPOSITION	VARIATIONS IN COMPOSITION
Iliohypogastric	T12, L1	T11, T12; T12; L1
Ilioinguinal	L1	T12; T12, L1
Genitofemoral	L1, L2	T12-L2; L1; L1-L3; L2, L3

Table 7-5
Variations in the Branches of the Lumbar Plexus*

NERVE	COMMON COMPOSITION	REPORTED VARIATIONS
To psoas major	L2-L4	L2, 3; L1-L3; L1-L4; L2-L5
To psoas minor	L1 or L1, 2	T12; T12, L1; L2; L2, 3
To quadratus lumborum	T12, L1 (?)	T12-L3; T12-L4; L1-L3; L1-L4
To iliacus	L2, 3	L2-L4; L3, 4
Lateral femoral cutaneous	L1-L3	T12-L2; L1, 2; L2; L2, 3; L3; L3, 4
Femoral	L2-L4	L1-L4; L1-L5; L2-L5
Obturator	L2-L4 or L3, 4	L1-L4; L2-L5; L3; L3-L5; L4; L4, 5
Accessory obturator	L3 or L3, 4	L1-L3; L2; L2, 3; L4

* Excluding the "border nerves," already listed in Table 7-4.

cral plexus, but occasionally both the fourth and fifth lumbars contribute to both plexuses (Fig. 7-20) or the fifth lumbar alone does so (Fig. 7-17), so that the lumbar plexus receives fibers from L5. Horwitz reported this as occurring in about 12% of plexuses. Occasionally, the plexus receives fewer fibers than usual from the fourth lumbar, and only a part of the third lumbar (Fig. 7-16), both nerves forking to contribute to both plexuses. Horwitz found this in slightly more than 5% of plexuses. Apparently, the lumbar plexus always receives at least some fourth lumbar fibers.

Iliohypogastric, Ilioinguinal, and Genitofemoral Nerves

These nerves, the highest to reach the limb, are distributed primarily to the lower abdominal wall, hence Bardeen ('02, '07) called them "border nerves." Reported variations in the origins of these nerves are summarized in Table 7-4. Bardeen found the iliohypogastric arising rarely (about 2%) from T11 and T12, arising from T12 in 32%, from T12 and L1 in 34%, and from L1 alone in 32%. The chief origin of the ilioinguinal is from L1 and Bardeen found it receiving fibers from this nerve only in slightly more than half (51.5%) of cases, receiving fibers from T12 only in 3.5%, and from T12 and L1 in 38.3%. In 6.6% of cases it was absent. He found the genitofemoral arising from L1 alone in 19%, from L1 and L2 in 79%, and from L1 to L3, or L2 and L3, in 2%; Paterson said it often contains fibers from T12.

The iliohypogastric and the ilioinguinal

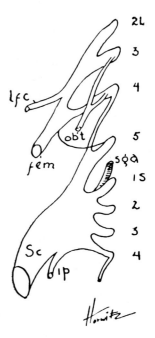

Fig. 7-20. A lumbosacral plexus in which the fifth lumbar nerve contributes a small branch to the lumbar plexus. Labels as in Fig. 7-16 and 7-17. (From part of original of Fig. 1, Horwitz MT: Anat Rec 74:91, 1939)

nerves emerge through the lateral border of the psoas muscle and run obliquely downward and laterally across the quadratus lumborum, thus on the posterior abdominal wall, to enter the transversus abdominis a little above the iliac crest. The two nerves may have separate origins, or may be combined into a single trunk until they reach the abdominal musculature. In like fashion, the iliohypogastric may run with the twelfth thoracic, which lies above it, or the ilioinguinal may contain the fibers that should be in the nerve below it, the genitofemoral (Bardeen, '07).

After the iliohypogastric nerve has entered the abdominal musculature it gives off a lateral cutaneous (iliac) branch, similar to the lateral cutaneous branches of the intercostal nerves, which is distributed to a posterolateral portion of the skin of the buttock. Its remaining portion helps to supply the transversus abdominis and the internal oblique muscles as it runs forward between them, and ends as the anterior cutaneous (hypogastric) branch by supplying skin above the symphysis pubis.

The ilioinguinal nerve usually communicates with the iliohypogastric in the anterior abdominal wall. It penetrates the internal oblique to reach the superficial inguinal ring, emerges through this, and is distributed to skin of the root of the penis and scrotum in the male, to the mons pubis and labium majus in the female.

The genitofemoral nerve emerges on the anterior surface of the psoas muscle, sometimes being formed in the muscle, sometimes on' its anterior surface by branches from L1 and L2 that do not fuse until they have separately emerged from the muscle. The nerve runs downward on the anterior surface of the muscle to divide into genital (external spermatic) and femoral (lumboinguinal) branches. The genital branch then passes through the inguinal canal and supplies twigs to the cremaster muscle and to the skin of the scrotum in the male, or to the skin of the labium majus in the female. The femoral branch runs along the external iliac artery, with which it passes behind the inguinal ligament into the thigh. In the thigh it lies lateral to the femoral artery, and is usually distributed to an upper area of skin in the middle of the anterior surface of the thigh. Bardeen found it somewhat variable. It was absent in 1.2% of cases, had its common distribution in slightly more than 60%; but in about 19% it was distributed more medially than usual, in about 20% more laterally. He also found a good deal of variation in the amount of skin it supplies.

Although these upper branches of the plexus bring first lumbar and sometimes lower thoracic fibers into the limb, they are purely cutaneous and have such a limited distribution that their segmental contribution to the limb is of little moment and usually ignored.

Muscular Branches

The small muscular branches of the lumbar plexus are a number of twigs that arise mostly from the several nerves as they form the plexus, but sometimes from the femoral nerve; they are distributed directly to the psoas major, the psoas minor if this is present, the quadratus lumborum, and the iliacus. The psoas major usually receives fibers from about L2 to L4, may apparently receive them from L1 or L5 also; the psoas minor usually receives fibers, apparently, from not more than two of the upper three or four lumbar nerves; the quadratus lumborum is said to receive fibers from T12 and L1, or from as low as L3 or L4; the iliacus receives its nerves from the femoral nerve rather than directly from the plexus, and these are apparently commonly composed of fibers of the second and third lumbar, perhaps also some from the first or the fourth.

Lateral Femoral Cutaneous

The lateral femoral cutaneous nerve arises most commonly (in 43% of 287 cases, according to Bardeen, '02) from the first three lumbar nerves (Table 7-5). Bardeen described it as arising from the twelfth thoracic and the first two lumbars in 39%; Paterson, in a detailed dissection of 23 plexuses, apparently found no origin from the twelfth thoracic, but did find it arising, in addition to its usual origin, from L1 and L2 only, from L2 and L3 only, and from L2 only. The nerve may be at first bound with the femoral and therefore arise from that; Bardeen found this in 18% of cases, Horwitz in 65.3%. This apparent discrepancy is probably largely a result of differences in dissection of the surrounding connective tissue.

This nerve is derived from the posterior divisions of the ventral rami to the lumbar plexus. It emerges from the lateral border of the psoas muscle, and runs downward and laterally across the anterior surface of the iliacus; it enters the thigh a little below and medial to the anterior superior iliac spine, passing behind the inguinal ligament and appearing just medial to the upper end of the sartorius muscle. It is the highest of the nerves with a reasonably large distribution to the thigh, and reaches a subcutaneous position by passing in front of, behind, or through the sartorius muscle. It supplies skin of the lateral aspect of the thigh, usually from about the level of the greater trochanter to the level of the knee (see Fig. 7-28); Bardeen found it varying considerably in its distribution, often (in about 30%) replacing some of the more lateral cutaneous branches of the femoral nerve.

Femoral Nerve

The femoral nerve, like the lateral femoral cutaneous, arises from the posterior divisions of the nerves to the lumbar plexus. Its most common origin is from the second, third, and fourth lumbar nerves, with most of the fibers derived from the latter two; in plexuses that tend to be prefixed, however, it may receive fibers from L1, and in those that tend to be postfixed it may receive fibers from L5. Bardeen and Elting found a contribution from the first lumbar nerve to the femoral in 15 (60%) of the plexuses that they classified as prefixed, but said these fibers were distributed through the cutaneous branches only, not to muscular branches.

The femoral nerve emerges through the lateral border of the psoas muscle, usually below the lateral femoral cutaneous but sometimes having a common stem with it, and passes downward in the groove between the psoas and iliacus muscles. It gives off branches to the iliacus muscle and passes into the thigh on the surface of the iliopsoas muscle, running with the muscle behind the inguinal ligament and emerging into the femoral triangle. Shortly below the inguinal ligament it divides into a number of terminal branches, the chief

distribution of which is diagrammed in Figure 7-21.

Obturator

The obturator nerve is derived from the anterior divisions of the lumbar plexus. Paterson described it as arising most commonly from the second, third, and fourth lumbar nerves, that is, having the same roots of origin as does the femoral; Bardeen and Elting apparently always found origin from these three nerves in their series, but said that when the plexus was prefixed the obturator nerve usually also received fibers from L1, and that when it was postfixed it usually received fibers from L5. As in the case of the femoral nerve, fibers from the first lumbar nerve that reach the obturator are apparently destined for its cutaneous branch. Horwitz described a much more variable origin: in his series of 228 plexuses it arose from the third and fourth lumbar nerves in 175, and from the second through the fourth lumbars in only 23; in the remainder it arose from the first through the fourth lumbars twice, from the third through the fifth 10 times, from the third alone 3 times, from the fourth alone 12 times, and from the fourth and fifth, 3 times.

The obturator nerve emerges from the medial side of the psoas muscle just above the brim of the pelvis and runs downward between the muscle and the lateral aspect of the vertebral column parallel to the lumbosacral trunk, typically being separated from this by the iliolumbar artery. Upon entering the pelvis it diverges laterally from the lumbosacral trunk, passing along the pelvic wall lateral to the internal iliac vessels and the ureter, therefore along the upper part of the internal surface of the obturator internus muscle, to the obturator canal. Through this it passes into the thigh. The nerve may divide or give off one of its smaller branches while it is in the pelvis, but most commonly its branching occurs while it lies in the canal. Larochelle and Jobin found the division of the nerve occurring in the canal in 80 of 106 dissections, above the canal in 10, and below it in 16. It is distributed to anteromedial muscles of the

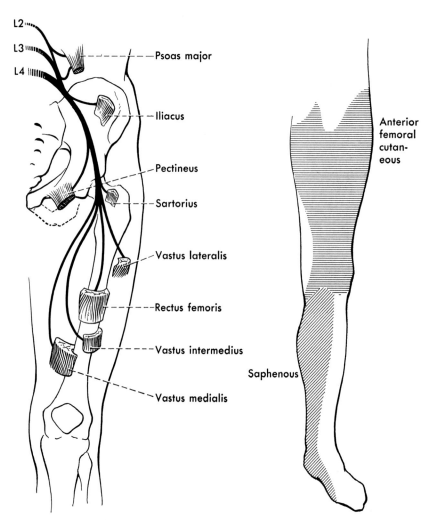

L2

L3

L4

Psoas major

Iliacus

Pectineus

Sartorius

Vastus lateralis

Rectus femoris

Vastus intermedius

Vastus medialis

Anterior femoral cutaneous

Saphenous

Fig. 7-21. Muscular and cutaneous distribution of the femoral nerve.

thigh, and usually to some of the skin of the medial side (Fig. 7-22).

Accessory Obturator

An accessory obturator nerve was found in 8.4% of 250 sides by Bardeen and Elting, in 8.7% of 550 sides by Woodburne, and almost twice as frequently, 17.1% of 278 sides, by Horwitz. Bardeen and Elting expressed the opinion that it is perhaps more commonly associated with prefixed than with other types of plexuses. It typically arises between the divisions that contribute to the femoral and obturator nerves and may be bound to either of these nerves by connective tissue close to its origin. Its origin is regularly less extensive

than that of the obturator or the femoral nerve. Bardeen and Elting found it arising most commonly from the third and fourth lumbars; Horwitz found it arising most commonly from either the obturator or the femoral, rather than from the divisions of the plexus, but described the origin in the remaining cases as being most commonly from L3, about half as frequently from L3 and L4, and sometimes from L2 alone, L2 and L3, or L4 alone.

The accessory obturator, arising between posterior (femoral) and anterior (obturator) parts of the plexus, is sometimes classed with one, sometimes with the other. It emerges from the medial surface of the psoas muscle

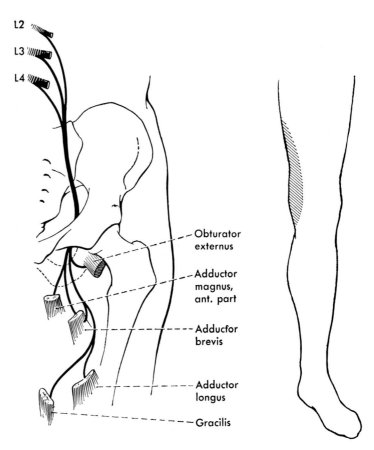

Fig. 7-22. Muscular and cutaneous distribution of the obturator nerve.

L2
L3
L4

Obturator externus

Adductor magnus, ant. part

Adductor brevis

Adductor longus

Gracilis

close to the obturator nerve, but instead of following this toward the obturator foramen it descends along the medial border of the muscle and passes anterior to the pubis, as does the femoral nerve. On entering the thigh it passes around the lateral edge of the pectineus muscle to lie deep to this muscle, usually supplying only this and a part of the hip joint and perhaps communicating with the obturator nerve. Paterson recorded an unusual case in which it sent branches of communication to several branches of the obturator nerve: those to the adductor brevis, gracilis, obturator externus, adductor magnus, and knee joint.

SACRAL PLEXUS

The pudendal nerve and coccygeal nerve and plexus have already been discussed; here we are concerned only with the contribution of the sacral plexus to the limb.

The upper border of the sacral plexus varies with the lower border of the lumbar plexus. Typically, therefore, the highest nerve contributing to it is the fourth lumbar (Table 7-6), which sends the major proportion of its fibers into the lumbar plexus, but a smaller proportion to the sacral plexus. In the most common form of prefixed sacral plexus, the fourth lumbar still contributes to both plexuses but sends fewer fibers than usual into the lumbar plexus. Bardeen and Elting's prefixed

Table 7-6
Variations in the Formation of the Sacral Plexus*

TYPE OF PLEXUS	MOST COMMON COMPOSITION	VARIATIONS IN COMPOSITION
Normal	L4-S3	
Prefixed	L4-S3	L3-S2; L3-S3; L4-S1; L4-S2
Postfixed	L4-S4	L4-S3; L5-S3; L5-S4

* Excluding the pudendal nerve.

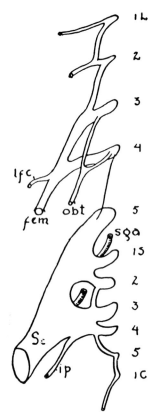

Fig. 7-23. A lumbosacral plexus in which the fourth lumbar nerve contributes only a twig to the sacral plexus, and the fifth sacral joins the caudal end of the plexus. Labels as in Fig. 7-16 and 7-17. (From part of original of Fig. 1, Horwitz MT: Anat Rec 74:91, 1939)

plexuses were all of this type, but in 6 (of 256) there was also a small twig from L3 going to the sacral plexus; Horwitz found the third lumbar as well as the fourth contributing to the sacral plexus (Fig. 7-16) in about 5%. In postfixed plexuses the fourth lumbar nerve may contribute very little to the sacral plexus (Fig. 7-23), or both the fourth and fifth lumbar nerves may contribute to both plexuses. The latter was true in 6.1% of Bardeen and Elting's cases, and of slightly more than 3% of Horwitz's cases. In another variety of postfixation the fourth nerve contributes exclusively to the lumbar plexus, the fifth exclusively to the sacral plexus (1.75% of Horwitz' cases) or, more commonly, the fifth lumbar nerve divides to go to both plexuses (Fig. 7-24; about

7% of Horwitz' cases, 9.7% of Bardeen and Elting's).

The most caudal fibers that enter the limb typically go into the tibial nerve, and if the caudal end of the sacral plexus is defined thus, it varies from the first sacral through the fourth (Bardeen and Elting); in prefixed plexuses the most caudal contribution to the sacral plexus is most commonly S3 as it is in the normal plexus, but may be higher, while in postfixed plexuses the most caudal contribution is most commonly S4. Since L4 may contribute to the sacral plexus in either type, the usual range of the sacral plexus is L4 to S2 and L4 to S4. Horwitz, defining the caudal border of the sacral plexus as the pudendal nerve, found one additional segment usually contributing to it.

The sacral plexus is formed in the concavity of the pelvis by the union of the lumbosacral trunk (typically receiving a smaller por-

Fig. 7-24. A postfixed sacral plexus, in which the fifth lumbar is the highest nerve contributing to it, and the fifth sacral joins its caudal end. Labels as in Fig. 7-16 and 7-17. (From part of original of Fig. 2, Horwitz, MT: Anat Rec 74:91, 1939)

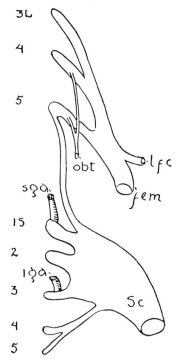

tion of the ventral ramus of the fourth lumbar nerve, all that of the fifth lumbar nerve) and the ventral rami of the first three sacral nerves. The lumbosacral trunk is directed primarily downward into the pelvis, and the sacral nerves are directed laterally and downward in the grooves extending from the upper pelvic sacral foramina; the plexus takes form on the anterior surface of the piriformis muscle, behind (deep to) branches of the internal iliac vessels and the special layers of pelvic fascia associated with these. The superior gluteal artery, in leaving the pelvis, usually runs between the lumbosacral trunk and the first sacral nerve (Fig. 7-20), or between the first and second sacral nerves (Fig. 7-17), before these join the plexus; sometimes, instead, it divides the lumbosacral trunk, passing between the parts contributed by the fourth and by the fifth lumbar nerve. The inferior gluteal artery likewise varies in its exact relation to the plexus, sometimes passing between two of the sacral nerves before they join (Fig. 7-24), sometimes piercing the plexus after most of the components have come together (Fig. 7-23).

The branches of the plexus pass through the greater sciatic foramen, the superior gluteal nerve passing above the piriformis muscle, which largely fills this foramen, and the remainder of the plexus typically passing below the muscle. For the most part the branches of the plexus take origin just as the plexus passes through the foramen, therefore a part of the plexus appears in the pelvis, while a part can be seen only in a dissection of the buttock. Because of its relationship to the viscera and the blood vessels, surgical approach to the sacral plexus in the pelvis is particularly difficult; in an approach through the buttock, the plexus can be displayed in practically its entirety by removal of bone and ligaments at the margins of the sciatic foramen—a part of the sacrotuberous ligament and a lateral part of the sacrum medially, ilium and sacrum above.

As is true of the ventral rami going to the brachial and the lumbar plexuses, those contributing to the sacral plexus divide into anterior and posterior divisions, from which the various nerves arise (Fig. 7-25); because of the connective tissue surrounding the plexus, the divisions are not, however, as obvious as they are in either of the other plexuses. The nerves to the larger muscles of the buttock (the superior and inferior gluteal nerves), the nerve or nerves to the piriformis, and the common peroneal portion of the sciatic nerve are derived from the posterior divisions of the sacral plexus; the tibial nerve, and the nerves to the short rotators in the buttock (the nerve of the obturator internus and superior gemellus, and the nerve of the quadratus femoris and inferior gemellus) are derived from the anterior divisions of the plexus. The posterior femoral cutaneous nerve, running along the original postaxial line and therefore between flexor and extensor surfaces, receives fibers from both parts of the plexus; the fibers from the posterior divisions of the plexus are distributed laterally from this nerve, and the fibers from the anterior divisions are distributed medially.

Gluteal Nerves

The *superior gluteal nerve* arises from the posterior surface of the upper nerves contributing to the sacral plexus, therefore typically the lumbosacral trunk and the first sacral nerve. In consequence of this origin, it ordinarily receives fibers from L4, L5, and S1 (Table 7-7); in those cases in which the furcal nerve is L5, this is of course the highest nerve contributing to the superior gluteal. When L5 is the furcal nerve and L4 fails to contribute, the superior gluteal nerve may also receive some fibers from S2. Its strongest origin is typically from L5.

The superior gluteal nerve passes almost directly dorsally between the upper border of the greater sciatic notch and the upper border of the piriformis muscle, to run deep to the gluteus medius in the buttock.

The *inferior gluteal nerve* also arises from the posterior divisions of the plexus, typically receiving fibers from L5, S1, and S2; its chief origin is often from S1. In prefixed plexuses it may apparently sometimes receive fibers from L4, and in postfixed ones it may receive a few

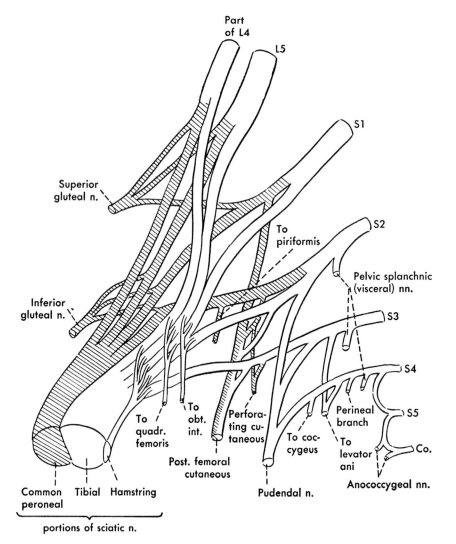

Fig. 7-25. Diagram of the sacral plexus. The elements derived from the posterior divisions of the plexus are *shaded.*

from S3. This nerve takes origin just as the plexus is passing through the sciatic foramen, and from the pelvic side is largely hidden by the roots of origin of the sciatic nerve; it passes with the sciatic nerve beneath the lower border of the piriformis muscle, and turns sharply backward lateral to the sacrotuberous ligament to enter the gluteus maximus.

Smaller Muscular Branches

As the sacral plexus is taking form on the anterior or pelvic surface of the piriformis muscle, one or two branches derived from the pos-

terior surfaces of the first and second sacral nerves (or either of these) pass directly into the piriformis to supply it.

As the sciatic nerve takes form from the major mass of nerves contributing to the sacral plexus it passes into the buttock beneath the lower border of the piriformis muscle, accompanied by two small nerves lying on its anterior surface: the nerve to the quadratus femoris and inferior gemellus, and the nerve to the obturator internus and superior gemellus. These nerves may arise independently from the plexus, but usually appear to arise

Table 7-7
Variations in the Branches of the Sacral Plexus

NERVE	COMMON COMPOSITION	REPORTED VARIATIONS
Superior gluteal	L4-S1	L4, 5; L4-S2; L5; L5-S1; L5-S2
Inferior gluteal	L5-S2	L4-S1; L5, S1; L5-S3; S1-S3
To piriformis	S1, 2	L4, 5; L5, S1; S2
To quadratus femoris	L4-S1	L4, 5; L5, S1
To obturator internus	L5-S2	L4-S1; L4-S2; L4-S3; L5-S3; S1, 2; S1-S3; S2
Common peroneal	L4-S2	L3?-S1; L4-S1; L4-S3; L5-S1; L5-S2; L5-S3
Tibial	L4-S3	L3?-S2; L3?-S3; L4-S1; L4-S2; L4-S4; L5-S2; L5-S3; L5-S4
Posterior femoral cutaneous	S1-S3	L4-S1; L4-S2; L5-S3; S1, 2; S2, 3; S2-4
Pudendal	S2-S4	S2, 3; S3, 4

directly from the upper part of the sciatic nerve. When traced to their origins by careful dissection (*e.g.,* Paterson) the nerve to the quadratus femoris and inferior gemellus is usually found to arise from the anterior divisions of the fourth and fifth lumbar and first sacral nerves, and the nerve to the obturator internus and superior gemellus from the anterior divisions of the fifth lumbar and first two sacrals. In the case of the former nerve, Paterson found the first sacral contribution sometimes missing, so that this nerve consisted of fourth and fifth lumbar fibers; when the fourth lumbar was missing from the sacral plexus, it consisted of fifth lumbar and first sacral fibers. He found that the nerve to the obturator internus and superior gemellus rarely receives fibers from the fourth lumbar nerve, and in postfixed plexuses sometimes receives fibers from only the first three sacrals.

Sciatic Nerve

The sciatic (ischiadic) nerve, the largest nerve in the body, is actually two nerves, the common peroneal and the tibial, which are usually bound together by a common sheath of connective tissue from their origin to the upper end of the popliteal space. In spite of their common connective tissue sheath, these two nerves do not interchange fibers; in about 10% to 15% of cases they separate at their origins, the anteromedial or tibial component then typically entering the buttock by passing below the piriformis muscle, as the entire nerve usually does, the peroneal or posterolateral component passing through or some-

times above the piriformis muscle. As a whole, the sciatic nerve is derived from all the nerves contributing to the sacral plexus, except that the last nerve contributing to the pudendal nerve typically does not contribute to the sciatic.

The *common peroneal* division of the sciatic nerve is derived from the posterior divisions of the nerves entering the sacral plexus, and typically contains fibers from the last two lumbar and the first two sacral nerves. In extreme prefixation of the plexus it may receive some fibers from L3 (which Bardeen and Elting described as going into its cutaneous branches); Bardeen and Elting found one case in which it apparently received fibers only from L4 and L5, and in postfixed plexuses it may receive fibers from S3. As already noted, it usually passes with the tibial nerve, as a component of the sciatic, below the lower border of the piriformis muscle into the buttock and down the posterior aspect of the thigh; it lies largely lateral but also a little dorsal to the tibial part of the nerve. It gives off its first branch, the only one arising from the lateral aspect of the sciatic nerve, in about the middle of the thigh; this goes to the short head of the biceps femoris. It is thereafter distributed to the muscles and skin on the anterolateral aspect of the leg and the dorsum of the foot (Fig. 7-26).

The *tibial* portion of the sciatic nerve, derived from the anterior divisions of the sacral plexus, typically, has a wider segmental origin than does the common peroneal, arising in the usual type of plexus from the fourth lumbar through the third sacral. It com-

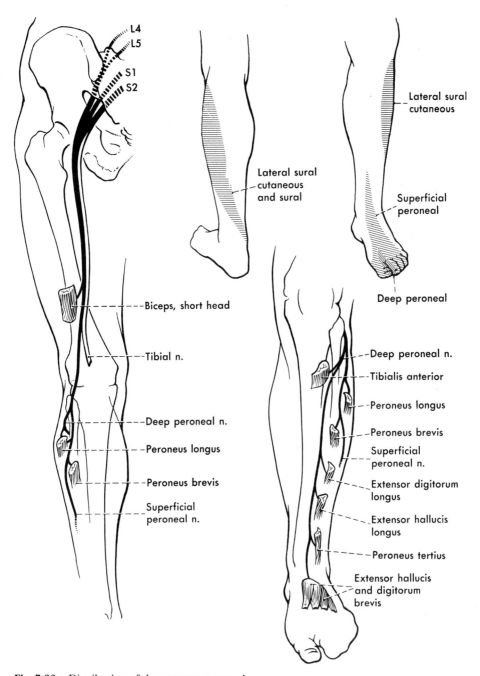

Fig. 7-26. Distribution of the common peroneal nerve.

monly, therefore, receives fibers from one segment lower than does the common peroneal nerve. In prefixed plexuses it is more likely to receive third lumbar fibers (apparently destined for its cutaneous branches) than is the common peroneal, and in postfixed plexuses

its caudal limit is usually the fourth sacral nerve rather than the third.

The tibial component of the sciatic nerve typically passes into the buttock with the common peroneal portion, but gives off branches to the upper parts of those hamstring muscles arising from the ischial tuberosity while it is still in the buttock and upper part of the thigh. It continues into the leg and foot to supply muscles and skin of the posterior aspect of the leg and sole of the foot (Fig. 7-27).

As pointed out by Paterson and others, the part of the tibial nerve to the hamstring mus-cles is fairly easily separable by dissection from the rest of the nerve. It forms a medial part of the tibial, and may either separate from it in the lower part of the buttock or re-main a component of the tibial until its final branches (to the lower parts of the ham-strings) have been given off. Its origin is from all the roots that normally contribute to the

Fig. 7-27. Distribution of the tibial nerve.

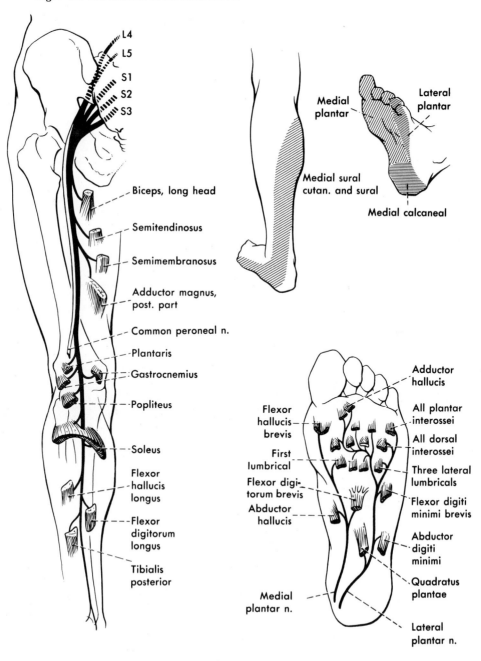

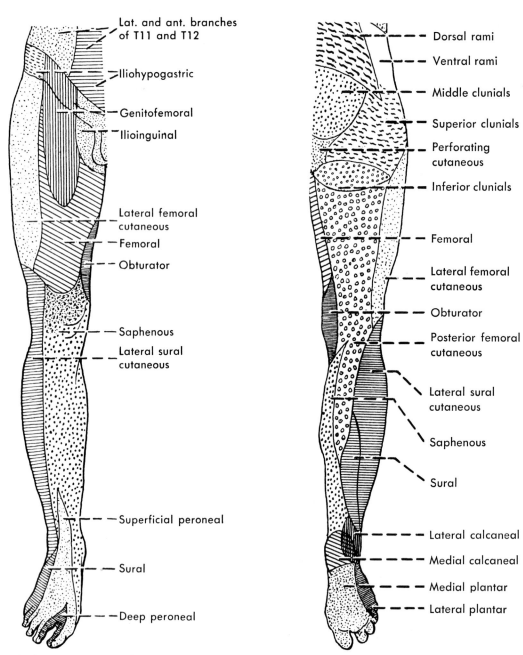

Fig. 7-28. Distribution of cutaneous nerves to the lower limb. There is a good deal of variation in the sizes of the areas to which the nerves are distributed. (Flatau E: Neurologische Schemata für die ärtzliche Praxis. Berlin, Springer, 1915)

origin of the tibial nerve, that is, the fourth lumbar through the third sacral.

Posterior Femoral Cutaneous

The posterior femoral cutaneous nerve arises from both the anterior and the posterior divi-

sions of the sacral plexus, typically from the first three sacral nerves. Its range of variation is from as high as the fourth lumbar to as low as the fourth sacral. It passes into the buttock beneath the piriformis muscle, posteromedial to the sciatic nerve, and gives off its first branches (to skin of the buttock and perineum) soon after it enters the buttock. In the thigh, its fibers are distributed according to whether they are derived from the posterior or the anterior divisions of the sacral plexus, the fibers from the posterior divisions being distributed laterally from the nerve, those from the anterior divisions being distributed medially.

In somewhat more than half of bodies there is present a small twig, the perforating cutaneous or inferior medial clunial nerve, that arises from the posterior surface of the second and third sacral nerves. It pierces the lower part of the sacrotuberous ligament to pass around the lower border of the gluteus maximus muscle and be distributed to skin over this muscle. Although representing a posterior part of the plexus, it may be bound with the pudendal nerve and therefore arise from it; when it is lacking, its place is typically taken by a branch of the posterior femoral cutaneous.

PERIPHERAL NERVES

These have already been discussed largely in connection with the plexuses. Details of

their anatomy are given in the following chapters.

Cutaneous Distribution

The lateral femoral cutaneous and genitofemoral nerves, from the lumbar plexus, and the posterior femoral cutaneous nerve, from the sacral plexus, supply part of the skin of the lower limb, and cutaneous branches from all the major nerves—the femoral, the obturator, the tibial, and the common peroneal—supply the remainder. The general distribution of the cutaneous nerves of the lower limb is shown in Figure 7-28; average areas of complete and partial loss of sensibility found clinically following lesions of the various cutaneous nerves are shown in Figures 7-29 and 7-30.

As in the case of the upper limb, cutaneous innervation must be considered both from the standpoint of the distribution of peripheral nerves and from the standpoint of dermatomes, since lesions affecting sensibility may involve either the peripheral nerves or the nerve roots or spinal cord segments contributing to the innervation of the limb.

As in the case of the upper limb, it is impossible to follow the fibers of the various spi-

Fig. 7-29. Average areas of sensory loss following interruption of various cutaneous nerves to the anterior and medial surfaces of the leg. *Black* areas represent ones with total sensory loss, and *immediately surrounding areas* those in which sensation of light touch is lost but some sensation of pain and temperature is preserved. (After Lewandowsky M, Foerster O, from Woltman HW, Kernohan JW, Goldstein NP, in Baker AB ed: Clinical Neurology [ed 2], Vol 4. New York, Hoeber, 1962)

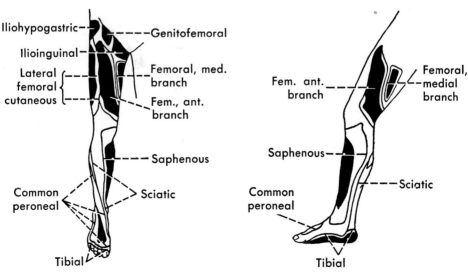

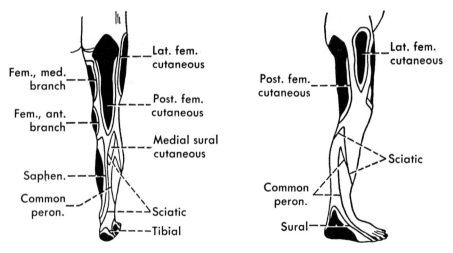

Fig. 7-30. Average areas of sensory loss following interruption of nerves distributed to the posterior and lateral surfaces of the lower limb. See the legend for Fig. 7-29. (After Lewandowsky M, Foerster O, from Woltman HW, Kernohan JW, and Goldstein NP, in Baker AB, (ed): Clinical Neurology [ed 2], Vol 4. New York, Hoeber, 1962)

Fig. 7-31. Dermatomes of the lower limb from the side, according to Foerster. (After Foerster, from Haymaker W, Woodhall B: Peripheral Nerve Injuries [ed 2]. Philadelphia, WB Saunders, 1953)

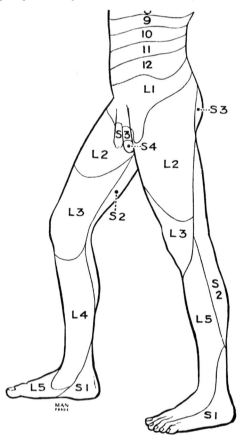

nal nerves as they are distributed to the skin, and the embryology of the limb, likewise, gives no definitive information on the arrangement of the dermatomes. If one follows Foerster's schema of the dermatomes (Figs. 7-31 to 7-33), one assumes that the more centrally placed nerves are drawn out exclusively into the distal part of the limb, while if one follows Keegan's (Fig. 7-34) one assumes that all the nerves retain a dermatomal distribution at the base of the limb.

In the lower limb, because of the rotation it has undergone, the original preaxial border is now anteromedial, therefore it is on the anteromedial surface of the thigh that one would expect to encounter the distribution of the uppermost nerves reaching the limb. If one then proceeds around the original dorsal surface of the lower limb, now its front, one should encounter the dermatomes in numerical order unless they have been drawn entirely out into the limb, and Keegan's charts show the dermatomes in order, with no dorsal axial line. Other charts show a dorsal axial line, where the middle nerves to the limb have been drawn farther distally, as extending from the posterior superior iliac spine over the buttock and along the posterolateral side of the thigh, and thence part way down the leg over the fibula. The upper part of this line is visible in Figure 7-33. There is necessarily a

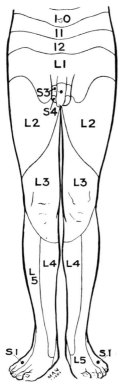

Fig. 7-32. Dermatomes from the front. (After Foerster, from Haymaker W, Woodhall B: Peripheral Nerve Injuries [ed 2]. Philadelphia, WB Saunders, 1953)

line on the ventral surface of the limb where the most cranial and the most caudal nerves supplying the limb meet, and there is therefore an abrupt change in the segmental innervation. Because of the rotation of the lower limb, this line, the ventral axial line, lies medially in the thigh, then runs along the back of the leg to the heel.

The differences between these two dermatomal schema probably lie in part, at least, in the method by which the dermatomes were defined. Woodhall expressed the opinion that Foerster's charts are probably more useful to the neurosurgeon in determining the nerve root involved by a protruding lumbar disk than are Keegan's, and said that in his experience it has been involvement of nerve L5, not L4, that typically produces sensory changes over the dorsum of the foot and the big toe; he did note, however, that the proximal extension of the dermatomes toward the base of the limb, depicted by Keegan but not by Foers-

ter, was evident in at least some cases of compression of a nerve by a disk.

A general rule (Herringham's rule) long regarded as valid is that of two spots at the same level on the limb, the one nearer the preaxial border tends to be supplied by the higher nerve; of two spots along the preaxial border, the higher one tends to be supplied by the higher nerve; and of two spots along the postaxial border, the lower one tends to be supplied by the higher nerve. The first part of this statement holds true according to all dermatomal schema; the latter two are true only if centrally placed nerves are drawn distally in the limb, or the dermatomes are obliquely placed along these borders.

Muscular Distribution
The distributions of the larger nerves to the muscles of the lower limb are diagrammed in

Fig. 7-33. Posterior view of the dermatomes. (After Foerster, from Haymaker W, Woodhall B: Peripheral Nerve Injuries [ed 2]. Philadelphia, WB Saunders, 1953)

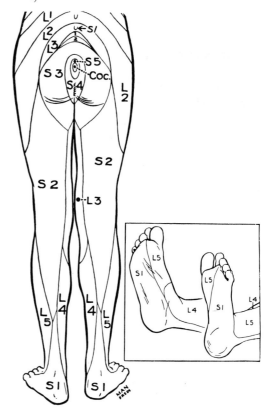

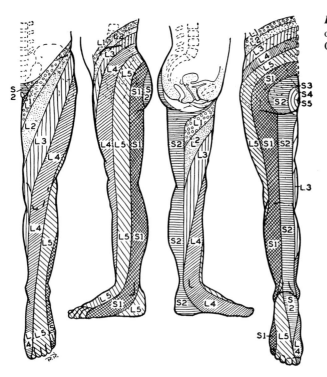

Fig. 7-34. Dermatomes of the lower limb according to Keegan. (From Fig. 7, Keegan JJ, Garrett FD: Anat Rec 102:409, 1948)

Figures 7-21, 7-22, 7-26, and 7-27, and the muscular and cutaneous distributions of the nerves of the lower limb are listed in Table 7-8, which also shows the usual segmental composition of these nerves. In summary, a number of small nerves arising from the sacral plexus supply the musculature of the buttock; the femoral nerve supplies the anterior muscles of the thigh; and the obturator nerve supplies the anteromedial or adductor group of muscles. The tibial component of the sciatic nerve supplies the muscles on the posterior side of the thigh arising medial to the posterior midline, that is, from the ischial tuberosity, while the only muscle arising lateral to this line, the short head of the biceps, is innervated by the common peroneal component; the tibial nerve also supplies the muscles of the calf of the leg and the sole of the foot, and the common peroneal nerve supplies the anterolateral muscles of the leg and the dorsal muscles of the foot. The approximate segmental innervation of the muscles of the limb is indicated in Table 7-1, and is further discussed and tabulated in the following chapters.

VESSELS

Arteries

Of the three major arterial stems into the lower limb, two of them, the superior and inferior gluteal arteries (Fig. 7-35), pass through the greater sciatic foramen and their distribution is confined largely to the buttock, even though the inferior gluteal was at one stage of development the stem vessel supplying the thigh, leg, and foot. Thus the femoral artery, the continuation of the external iliac, is the chief artery to the free limb. The largest branch of the femoral artery is the profunda femoris, given off in the upper part of the thigh; it runs deeply in the anteromedial portion of the thigh to send branches to the posteriorly lying structures, which after the disappearance of the primitive sciatic artery lack a longitudinally running vessel on their side. After giving off the profunda, the remainder of the femoral artery, sometimes called the superficial femoral, runs more superficially (but under cover of the sartorius muscle) along the anteromedial side of the thigh, and perforates the tendon of the adductor magnus muscle to reach the popliteal fossa and pass behind the knee; the femoral artery changes its name to popliteal as the vessel passes through the adductor hiatus.

The popliteal artery ends a little below the knee by dividing into the anterior and posterior tibial arteries. The posterior tibial gives off a deep branch, the peroneal artery (corresponding to the anterior interosseous in the forearm) and continues down the calf between the superificial and deep groups of muscles, to pass around the posteromedial aspect of the ankle and help supply expecially the more superficial tissues of the sole of the foot. The anterior tibial, corresponding to the posterior interosseous artery in the forearm, passes between the tibia and fibula to run down the front of the leg, and continue onto the dorsum of the foot as the dorsalis pedis artery; the dorsalis pedis artery behaves like the radial artery in the hand, sending its main branch between the first and second metatar-

Table 7-8
Composition and Distribution of the Nerves to the Lower Limb

NERVE	USUAL SEGMENTAL COMPOSITION	MUSCULAR DISTRIBUTION	CUTANEOUS DISTRIBUTION
From Lumbar Plexus			
Ilioinguinal	(T12), L1	(To abdominal wall?)	(Scrotum or labia); upper thigh
Genitofemoral	L1, L2	(Cremaster)	(Scrotum or labia); upper thigh
Lateral femoral cutaneous	L2, L3	None	Thigh
Femoral	L2-L4	Sartorius Quadriceps Articularis genus Pectineus	Thigh, leg, and foot
Obturator	L2-L4	Pectineus (sometimes) Adductor longus Adductor brevis Adductor magnus Gracilis Obturator externus	Thigh
From Sacral Plexus			
N. to quadr. fem. and inf. gemel.	L4-S1	Quadratus femoris Inferior gemellus	None
N. to obt. int. and sup. gemel.	L5-S2	Obturator internus Superior gemellus	None
N. to piriformis	(S1), S2	Piriformis	None
Superior gluteal	L4-S1	Gluteus medius Gluteus minimus Tensor fasciae latae	None
Inferior gluteal	L5-S2	Gluteus maximus	None
Posterior femoral cutaneous	S1-S3	None	Thigh
Sciatic (tibial and peroneal)	L4-S3	See below	See below
Tibial	L4-S3	Semitendinosus Biceps, long head Semimembranosus All muscles of calf of leg and sole of foot	Leg and foot
Common peroneal	L4-S2	Biceps, short head All lateral and anterior muscles of leg, dorsal muscles of foot	Leg and foot

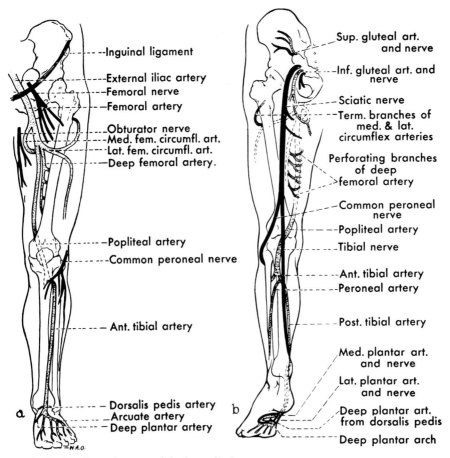

Labels on figure (left, a):
- Inguinal ligament
- External iliac artery
- Femoral nerve
- Femoral artery
- Obturator nerve
- Med. fem. circumfl. art.
- Lat. fem. circumfl. art.
- Deep femoral artery.
- Popliteal artery
- Common peroneal nerve
- Ant. tibial artery
- Dorsalis pedis artery
- Arcuate artery
- Deep plantar artery

Labels on figure (right, b):
- Sup. gluteal art. and nerve
- Inf. gluteal art. and nerve
- Sciatic nerve
- Term. branches of med. & lat. circumflex arteries
- Perforating branches of deep femoral artery
- Common peroneal nerve
- Popliteal artery
- Tibial nerve
- Ant. tibial artery
- Peroneal artery
- Post. tibial artery
- Med. plantar art. and nerve
- Lat. plantar art. and nerve
- Deep plantar art. from dorsalis pedis
- Deep plantar arch

Fig. 7-35. Arteries and nerves of the lower limb.

sal bones to reach the plantar surface of the foot and become the main contributor to the plantar (deep) arch.

VEINS

The veins of the lower extremity, like those of the upper, are divided into deep and superficial ones. The deep veins generally follow the arteries, and are typically two in number for each artery, taking the name of the artery they accompany. They need no special description here; their regional anatomy is described in the following chapters.

The superficial veins of the lower extremity begin in the foot and form two chief channels running up·the leg, the great or greater (long) saphenous and the small or lesser (short) saphenous. The great saphenous vein originates along the medial border of the dorsum of the foot and runs up the medial aspect of the leg and the medial aspect of the thigh. It ends by penetrating the deep fascia through a gap, the saphenous hiatus (fossa ovalis), in the deep fascia of the thigh a little below the level of the inguinal ligament, there joining the femoral vein. The small saphenous vein originates on the lateral aspect of the dorsum of the foot and passes up the posterior aspect of the calf. It usually enters the popliteal fossa to join the popliteal vein. There are typically connections between the great and small saphenous veins as they run upward, and both communicate also with the deep veins of the limb, especially those of the leg, through communicating veins that perforate the deep fascia. The valves in the communicating veins (commonly called "perforators") of the leg and thigh are so arranged that blood normally passes only from the superficial to the deep veins, never in the reverse direction (*e.g.,* Kosinski). Incompetence of some of these valves is, however, regularly associated with varicose veins, and under these conditions the blood flow is reversed, passing from the deep to the superficial veins. In the foot the communicating veins either have no valves, or have valves that direct blood from the deep to the superficial veins (Lofgren and co-workers, Kuster and co-workers).

Superficial Veins of Foot

The superficial veins of the plantar surface of the foot are small, and join to form a plantar venous network. This receives distally the superficial veins from the plantar surfaces of the toes and drains into the dorsal venous system by numerous small channels that pass around the lateral and medial borders of the foot or between the heads of the metatarsals (these are often called intercapitular veins). The plantar venous network also communicates with the deep veins of the plantar surface through slender vessels that run vertically upward along the intermuscular septa of the foot.

On the dorsum, the dorsal digital veins are joined by veins from the plantar surface to form four dorsal metatarsal veins that quickly unite to form a dorsal venous arch at about the level of the heads of the metatarsals (Fig. 7-36). The medial end of the dorsal venous arch receives vessels from the medial side of the plantar surface of the foot and cutaneous veins from the dorsum of the foot and continues to form the great saphenous vein; the lateral end of the dorsal arch similarly receives veins from the lateral part of the plantar surface of the foot and veins of the dorsum, and typically continues to form the small or lesser saphenous vein. Proximal to the dorsal venous arch there is an irregular dorsal venous network or rete, which receives some of the cutaneous veins from the dorsum of the toes and foot, some vessels from the medial and lateral sides of the sole that run around the borders of the foot but pass superficial to the dorsal arch and its saphenous continuations, and veins from the inner borders of the dorsal arch and saphenous veins; the veins draining this plexus converge toward the front of the ankle and continue upward as a few small channels that connect to the saphenous system at a higher level. Lateral parts of this plexus drain the dorsal venous arch when the small saphenous vein begins more posteriorly.

Parts of both the great and small saphenous systems lie more deeply than do others; the more superficial parts plainly lie in the loose connective tissue of the superficial fascia of the limb, but the deeper parts usually run beneath a decidedly fibrous layer of tissue. Excluding the cases in which the upper part of the small saphenous definitely lies deep to the deep fascia of the leg, opinion varies as to whether the membranous layer covering the deeper-lying veins is a part of the superficial or of the deep fascia; some clinicians have supported the latter concept, but in most anatomical studies the former is implied or stated. It is of no particular importance if one realizes that parts of the system do lie more deeply than others, deep to a tougher fibrous layer.

Small Saphenous Vein

The small, lesser, or short saphenous vein begins along the lateral side of the dorsum of the foot, usually as a continuation of the dorsal venous arch but sometimes (Kuster and co-workers) originating in a perforating vein at the base of the fifth metatarsal; it passes upward behind the lateral malleolus, along the lateral border of the tendo calcaneus. On the foot and the lower part of the calf it is accompanied by the sural nerve; higher in the calf it usually lies directly superficial to the medial sural nerve, but in about 30% of cases is medial to it, in about 12% lateral (Kosinski). The lower end of the posterior femoral cutaneous nerve may also parallel its upper part. As it ascends it crosses superficial to the tendo calcaneus and the gastrocnemius muscle to run straight up the posterior midline of the calf (Fig. 7-37). According to Kosinski, the vein lies in the superficial fascia in the foot and lower part of the leg, usually deep to this

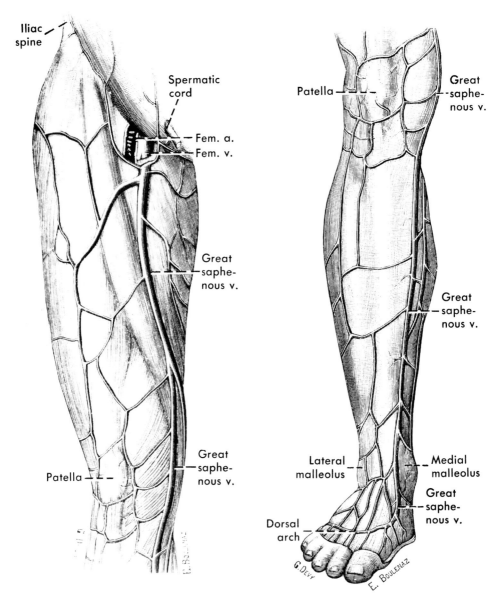

Fig. 7-36. Anterior views of the veins of the lower limb. The great saphenous close to its upper end receives a large lateral accessory saphenous vein. (Testut L: Traité d' Anatomie Humaine, Vol 2. Paris, Doin, 1891)

fascia in the upper half of the leg, but in about 25% of cases deep to the deep fascia here; Mullarky described it as usually lying in the crural (deep) fascia of the upper third of the leg, and Shah and Srivastava described this relation as existing also in the lower part of the leg where the tunnel, with its superficial layer thickest, encloses also the sural nerve. At the popliteal fossa it typically passes

between the two heads of the gastrocnemius and empties into the popliteal vein. In so doing, it arches very close to the tibial nerve, more often lateral to it, but very frequently medial to it.

About 60% of small saphenous veins have from 7 to 11 *valves* (Kosinski), the remainder have as few as 4 or as many as 13; it is said that the number does not vary either with sex or age. The highest valve is typically found close to the termination of the small saphenous, the other valves just below junctions of its tributaries.

In its course up the leg the small saphenous vein receives a number of tributaries from the calf, some of which connect also with veins draining into the great saphenous. Kosinski found from two to four direct connections between the small and great saphenous, usually toward the middle of the leg, and emphasized that the valves in these are so arranged that they always drain from the small into the great saphenous (Fig. 7-38). Just before it ends, the small saphenous may give off a further connection, of variable size, to the great saphenous; this may join the great saphenous almost directly, or run up the medial side of the thigh behind the latter to join it close to its upper end, thus contributing to or forming what is sometimes called the medial accessory saphenous (medial superficial femoral) vein, discussed in a following section. The small saphenous vein also has connections with the deep veins of the foot and leg; these are best discussed in connection with the similar perforating or communicating veins of the great saphenous, and are therefore described in a following section.

Variations in the manner of ending of the small saphenous vein are fairly common. It sometimes fails to pierce the deep fascia of the leg or gives off only a minor branch that does so, and instead proceeds superficially up the posteromedial aspect of the thigh to join the great saphenous vein; or it may, after piercing the deep fascia, either fail to communicate at all with the popliteal vein or bifurcate and send only one branch to the popliteal. In the former case the entire small saphenous, in the latter one part of it, typically runs higher up the thigh, usually then entering the deep femoral vein. Kosinski found the small saphenous vein ending in the popliteal in only 57% of cases, and in about a fourth of these it had a higher ending also; other types of ending included ending below the popliteal level, ending in the veins of the gastrocnemius or in the profunda femoris, and high junction with the great saphenous (Fig. 7-39). Mullarky found

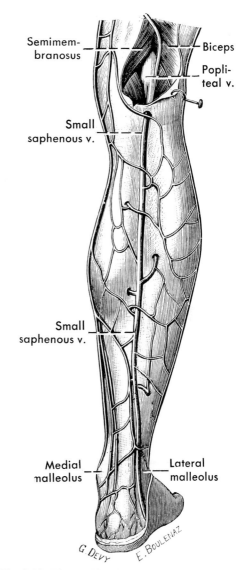

Fig. 7-37. The small or lesser saphenous vein; in this case it sends a connection from its upper end to the great saphenous above the knee. (Testut L: Traité d' Anatomie Humaine, Vol 2. Paris, Doin, 1891)

somewhat different percentages, with termination in the popliteal only in 49.3%, in the great saphenous only in 21.4%, and communicating with both in 9.3%.

Great Saphenous Vein

The great, greater, or long saphenous vein originates on the medial side of the dorsum of the foot. It passes upward in front of the medial malleolus and along the medial side of

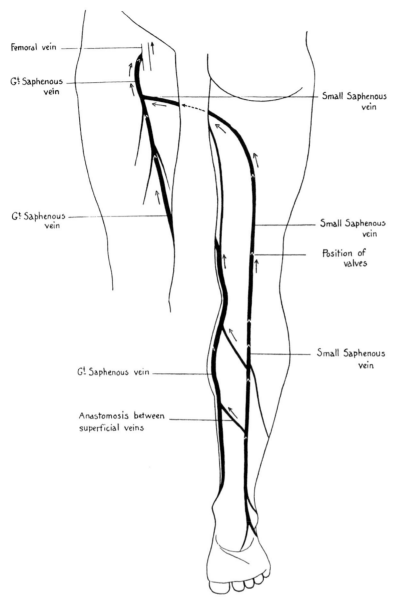

Fig. 7-38. Diagram of a small saphenous vein that joins the great close to the latter's upper end. *Arrows* indicate the only normal directions of flow. (Kosinski C: J Anat 60:131, 1926)

the leg, gradually inclining posteriorly and typically passing behind the medial condyles of the tibia and the femur. During its course in the leg it is accompanied by the saphenous branch of the femoral nerve which may be avulsed in stripping the vein (Garnjobst). It receives veins from the leg, and anastomotic channels, as noted, from the small saphenous;

also, as discussed in a following section, it has many communications with deep veins of the leg.

As the great saphenous vein ascends in the thigh it runs gradually more anteriorly, typically receiving additional branches and becoming larger as it ascends. According to Sherman ('44) it usually consists of two major

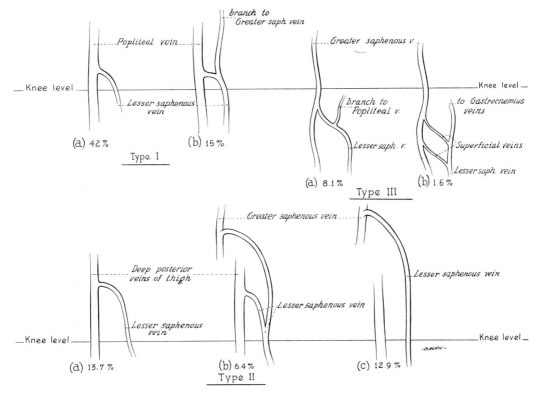

Fig. 7-39. Various types of ending of the lesser saphenous in the greater saphenous. Percentages are those of Kosinski. (Courtesy of Dr. T.T. Myers)

tributaries rather than of a single trunk in the lower third or more of the thigh (and may be doubled for most of its length), and one of these lies more deeply than the other; he described this deeper tributary and much of the main saphenous channel as lying deep to a superficial layer of the deep fascia, in contrast to other tributaries and to the upper end of the great saphenous. Mullarky did not remark upon two channels at different depths, but he did describe the great saphenous as lying in a tunnel in the crural fascia and fascia lata in the upper part of the leg and lower third of the thigh. The upper part of the great saphenous vein is usually described as lying between two layers of superficial fascia. A little below the inguinal ligament the great saphenous vein enters the saphenous hiatus (fossa ovalis), penetrates the lower end of the femoral sheath, and empties into the femoral vein. Just before it ends it is usually joined by the superficial epigastric, superficial iliac cir-

cumflex, and superficial external pudendal veins (small veins having the general course of the arteries of the same name), and often one or two larger and more important, but inconstant, superficial tributaries from the thigh.

Daseler and co-workers, calling these larger upper tributaries the lateral and medial accessory saphenous veins, pointed out that many names have been applied to them. The vein they designated as lateral accessory saphenous has apparently been variously called the external superficial femoral vein, the lateral femoral circumflex vein, the accessory saphenous vein, and the anterior saphenous vein (Daseler and his co-workers). The vein they called the medial accessory saphenous vein, in addition to being called simply accessory saphenous vein, has also been called the internal femoral cutaneous vein, the medial superficial femoral vein, the medial femoral circumflex vein, and the posterior saphenous vein (a name also applied to the small saphenous).

As described by Daseler and his co-workers, the lateral accessory saphenous vein originates in the suprapatellar venous network, and as it ascends along the front of the thigh runs also medially, receiving branches from both the lateral and anterior surfaces of the thigh; Sherman ('44) stressed that it usually communicates with the great saphenous stem in the region of the knee. It empties into the upper end of the saphenous vein, sometimes being joined by other veins, especially the superficial iliac circumflex, before so doing.

The medial accessory saphenous vein typically originates along the posteromedial border of the thigh according to Daseler and his co-workers, and receives most of its branches from the posterior surface; it runs around the medial aspect of the thigh as it ascends, to end in the upper part of the great saphenous vein. It frequently has connections below with lower tributaries of the great saphenous; and it is the upper end of the small saphenous when this vessel joins the great saphenous high in the thigh. Sometimes the medial accessory saphenous has been defined as the femoral part of the small saphenous vein.

The *valves* in the great saphenous vein vary considerably in number. Klotz reported from 6 to 25; Kampmeier and Birch found fewer, from 6 to 14, in 34 limbs. They are variably placed, but typically one is located in the upper part of the great saphenous. According to Kampmeier and Birch there was, in a series of 100 veins investigated, a valve at the mouth of the great saphenous in 82, one varying from 1 cm to 13 cm lower down in 16, and no valve at all present in the upper part in two.

Van Cleave and Holman presented evidence, which they regarded as tentative because of the rather large possible margin of error, that varicose veins have fewer valves than do normal veins. They quoted Kampmeier and Birch as having found an average of one valve per 6.6 cm of length in the great saphenous vein above the knee (without reference to the very irregular spacing of these valves), and themselves found an average of one valve per 8.81 cm of length of normal veins (but with a variation from no valves at all in this part of the vein to as many as 11). In contrast, they were able to find an average of one valve to only 16.8 cm of vein in varicose veins removed at operation. They suggested that the number and distribution of valves at birth may therefore be a factor in the subsequent development of varicosities.

The important variations in the great saphenous vein are in relation to the veins entering it close to its upper end. Daseler and co-workers and Mansberger and co-workers have reported that any or all of the upper tributaries that usually enter the great saphenous separately may fuse before they terminate in the saphenous, or may enter the femoral vein or an accessory saphenous vein. Various patterns, and the percentage in which they were found by Daseler and his co-workers in 550 specimens, are shown in Figure 7-40.

Daseler and his co-workers found the lateral accessory saphenous to be almost constant in its presence, usually of larger size than the other upper tributaries of the saphenous, and sometimes larger than the great saphenous. The medial accessory saphenous was an unusual tributary, apparently occurring in only 8% of their specimens. (This is in contrast to Kosinski's smaller series, in which the small saphenous ended or sent one branch to end high in the medial side of the greater saphenous in almost 20%.) Only rarely were both accessory saphenous veins present at the same time. The superficial epigastric and superficial iliac circumflex veins tended to empty into the lateral accessory saphenous instead of into the great saphenous or femoral veins, while the superficial external pudendal often entered the great saphenous, but sometimes the medial accessory saphenous or other tributaries, or the femoral vein. Any of these upper tributaries may be doubled.

Glasser, reporting upon a dissection of both sides of 50 bodies, said that the "usually pictured" pattern was found in only 37% and that prominent accessory saphenous (apparently medial accessory) and lateral superficial femoral (apparently lateral accessory)

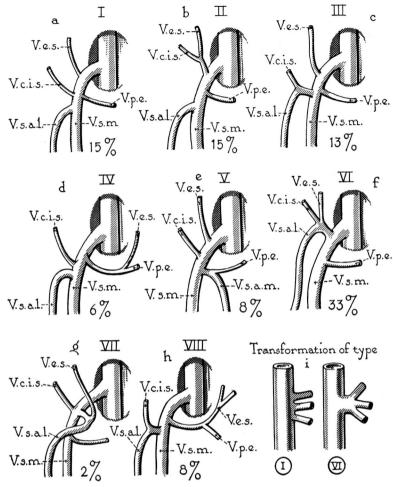

Fig. 7-40. Various arrangements of the veins entering the upper end of the great saphenous. *V.c.i.s.* is the superfical iliac circumflex vein, *V.e.s.* the superficial epigastric, *V.p.e.* the external pudendal, *V.s.a.l.* the lateral accessory saphenous, *V.s.a.m.* the medial accessory saphenous, and *V.s.m.* the great saphenous (Daseler EH, Anson BJ, Reismann AF, Beaton, LE: Surg Gynecol Obstet 82:53, 1946 [by permission of Surgery, Gynecology and Obstetrics])

veins were found often, at least one of the two being found in more than 50% of the dissections. He found a double saphenous vein in 3% and expressed the opinion, also held by other workers, that variations in the upper tributaries of the vein are a frequent cause of failure to achieve the expected results following primary ligation for the treatment of varicose veins; he was inclined to blame especially the variations in the lateral vein.

Mansberger and his co-workers said that the more unusual variations are for the great saphenous and both a lateral and a medial ac-cessory saphenous to be present simultaneously and to be of approximately equal size; for the great saphenous to enter the femoral vein below the saphenous hiatus (fossa ovalis); and for the lateral accessory saphenous to drain directly into the femoral vein above the hiatus.

"Perforators"

Of particular importance in relation to the surgery of incompetent saphenous veins are the connections, commonly termed "perforators" or "perforator veins," which these veins

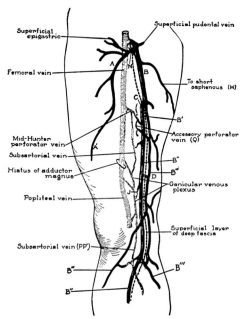

Fig. 7-41. Perforating veins of the medial side of the thigh; the superficial veins are *black,* the deep ones *shaded. A, A'* is a tributary, apparently the lateral accessory saphenous, of *B,* the great saphenous. *B'* and *B''* are deeper lying parts of the great saphenous, emerging from beneath the "superficial layer of the deep fascia" at *C* after communicating with *B''',* another part of the great saphenous, at *D. J* is an inconstant perforator. Line *M* calls attention to the *junction* of *A* and *B,* and the necessity of distinguishing *A* as a tributary, not the main vessel. (Sherman RS: Ann Surg 130:218, 1949)

usually have to deep veins. Linton pointed out that a distinction should really be made between perforating veins and communicating veins, for two or more veins may perforate the deep fascia some distance apart, but join a single vein that communicates with the deep veins, and sometimes a perforating vein is simply a drainage from a muscle, with no connection to the larger deep veins.

The communicating veins *of the thigh,* connecting the great saphenous to the deep veins, are far fewer than are those of the leg, and apparently less important clinically; Sherman ('49) reported finding incompetent perforators in the thigh in only 9% of 901 operations for varicose veins.

Kosinski found anastomoses between the great saphenous and the deep veins of the

thigh to be usually few in number, but one or more occurred rather regularly at about the middle of the thigh. Sherman ('44) found from 1 to 6, with an average of 1.94 in 101 dissections. The constant communication which he found, usually leaving the saphenous in about the middle of the thigh, joined the femoral vein at about the middle of the adductor canal (Fig. 7-41); the superficial end of this connection was often doubled, thus consisting of two perforating veins of which both sometimes originated from the great saphenous or one originated from the lateral accessory saphenous. Deep to the fascia lata this communication usually joined a vein from the sartorius muscle and through this vein it often had connections to communicating veins between the femoral and saphenous at the knee. Sherman also found rather constant connections between the great saphenous and the popliteal or lower end of the femoral veins at the knee, through a plexus of veins connecting to the genicular veins (his "genicular plexus"). Sherman noted that any of the communicating veins may receive tributaries from perforators that do not originate from the main trunk of the great saphenous, and Mullarky found several perforators on the medial side and a whole series on the lateral side of this kind.

The perforating or communicating veins of the leg are considerably more numerous and complicated and much more frequently incompetent—Sherman ('49) found from 1 to 14 incompetent ones—when the saphenous systems is varicose. Linton pointed out that the communicating veins regularly run between muscles, hence often along intermuscular septa in the leg (Fig. 7-42), and that they are fairly constant in position even though the levels at which the perforators pass through the deep fascia may vary somewhat. Sherman emphasized that many of them connect to tributaries rather than the main stems of the saphenous veins. Most of them are between the great saphenous or its tributaries and the deep veins, but some are from the small saphenous and its tributaries, or from veins connecting the two saphenous vessels.

Both Sherman ('49) and Linton have described the communicating and perforating veins of the leg. Their reports differ somewhat, and comparison is made more difficult by the difference in groupings and in terminology they employed, but both agreed that the veins are relatively constant. They appear in several positions on the leg, and there tend to be several members arranged linearly in each position. Mullarky has not classified the veins in such detail, but figured perforating veins corresponding to most of those described by Sherman and by Linton.

The medial group of veins, along the course of the great saphenous, is most important, for some member of this group is most commonly incompetent (Linton; Sherman). The larger members of this group connect deeply to the posterior tibial vein, superficially to the great saphenous or, especially in the lower part of the leg, to its tributaries; in the upper part of the leg they course between the soleus muscle and the tibia, sometimes penetrating a small part of the muscle, and in the lower part of the leg they course between the soleus and the flexor digitorum longus. Linton noted that several perforators often joined one communicating vein, and Sherman said the upper veins, usually joining the great saphenous, typically have collaterals that join its tributaries. Linton found usually two sets of communicating veins in the upper third of the leg, one or two in the middle third, and three or four in the lower third. Sherman found an almost constant six, arranged at about 5 cm intervals at levels from about 13.5 cm to 40 cm above the sole of the foot, and described the second vein in this series (Fig. 7-43) as being incompetent in 35.2% of more than 900 operations, the third as being incompetent in 19.6%.

Linton also described other communicating veins, apparently not recognized by Sherman, on the anteromedial side of the leg. He described them as connecting the greater saphenous to the anterior tibial, perforating the interosseous membrane to run posterior to the tibia close against the periosteum. He found three or four in the middle two thirds of the leg, but said they are very rarely incompetent.

A lateral or anterolateral group of veins

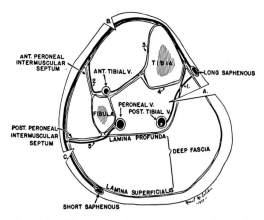

Fig. 7-42. Diagrammatic cross section of the leg, showing the relation of the communicating or perforating veins to the fascial septa. (Linton RR: Ann Surg 107:582, 1938)

also connects the great saphenous to the anterior tibial. Linton described two sets of these: three or four pairs of small veins, very rarely incompetent, in the middle two thirds of the leg, that pass between the tibialis anterior muscle and the tibia, perforating the deep fascia at the lateral border of this bone; and a larger, more important, lateral set of five or six members, of which the upper ones pass between the tibialis anterior and the extensor digitorum longus muscles, the lower between the latter muscle and the peroneus longus. This lateral set corresponds to one of Sherman's two groups of lateral perforators; he found five typically present, at levels of from 16 cm to 39 cm above the level of the sole, and reported that some member of this group was incompetent in 2.9% of his operations.

A posterolateral group of veins, Linton's peroneal communicating veins and the other of Sherman's two lateral groups of perforators, connects the superficial veins, tributaries of the great saphenous and especially the small saphenous, to the peroneal veins. According to Linton, they are usually six or seven in number, and run between the soleus and the peroneus longus muscles in the upper part of the leg, between the latter muscle and the flexor hallucis longus in the lower part of the leg. Sherman said that they often also connect deeply to the veins draining the gastrocnemius and and soleus muscles, and that

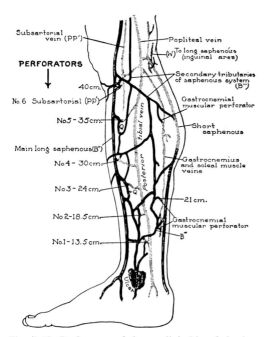

Fig. 7-43. Perforators of the medial side of the leg. The most important ones are numbered, and the approximate distances they appear above the sole of the foot are indicated. *B'''* indicates tributaries of the great saphenous; *Q* indicates accessory perforators that connect to smaller tributaries of the saphenous system. Deeper-lying veins and parts of veins are *lightly shaded.* (Sherman RS: Ann Surg 130:218, 1949)

line, the other two are 3 cm or more lateral to the midline. Linton apparently did not find most of this group, or included them with other groups; he did describe, as inconstant but frequent, a popliteal communicating vein that communicated deeply with the popliteal vein and the veins of the gastrocnemius, and joined the lesser saphenous vein in the upper part of the middle third of the leg. He said that often perforating veins here were simply a drainage from the gastrocnemius muscle and therefore not truly communicating ones. Sherman listed incompetent "perforators in calf" as having been found in 11.2% of his operations.

From this summary it should be obvious that the medial and the lateral or posterolateral groups of communicating veins and perforators in the leg are of the most importance clinically, one or more members of the medial

Fig. 7-44. Perforators on the lateral side of the leg. Most of them emerge between the peroneal muscles and the gastrocnemius-soleus complex; measurements indicate expected levels above the sole of the foot. *B'''* is a tributary of the great saphenous; *L* is a communicating vein that passes through the lateral border of the gastrocnemius or soleus; *Q* is an accessory perforator that connects with the same communicating vein as the unlabeled one below it. (Sherman, RS: Ann Surg 130:218, 1949)

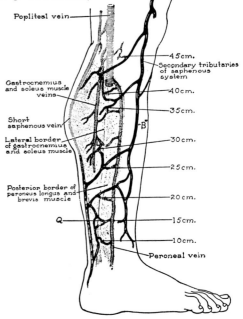

while they vary considerably, they commonly occur at about 5 cm intervals between levels 10 cm to 45 cm above the sole of the foot (Fig. 7-44). Both Linton and Sherman emphasized the importance of this group of veins; Sherman listed involvement of some of the group as having occurred in 24% of the operations in his series.

Among posterior perforators, Sherman listed the various endings of the small saphenous vein, already mentioned in connection with the description of that vein. He also described a group of perforators of the calf which, he said, usually connect tributaries of both the small and great saphenous veins to the muscular veins of the gastrocnemius and soleus muscles. He found usually four rows of these (Fig. 7-45), with from one to four elements in each row, spaced from 10 cm to 45 cm above the sole of the foot; two of the rows, he said, are within 2 cm of the posterior mid-

group being especially frequently incompetent; also, that stripping of the main stems of the great and small saphenous veins will not eradicate many of the perforators, since so many of these connect superficially with tributaries of the saphenous system.

Perforators of *the ankle and foot* have not been studied extensively, but have been figured by Linton and by Mullarky, and discussed briefly by Sherman, and designated as either competent or incompetent. More recently, Kuster and co-workers found communicating veins between the deep venous system and the superficial veins of the dorsum in as many as 12 locations (Fig. 7-46), although some were more constant than others, and they averaged 9 per foot. The dorsal venous arch, the great saphenous, and the small saphenous all tended to have one connection with the dorsalis pedis veins, and the remainder were between the great saphenous vein and the medial plantar veins, or the small saphenous and the lateral plantar veins. In sharp contrast to the communicating veins in the leg, those of the foot were found either to have no valves (49 of 91 veins) or to have valves so placed that blood could flow only from the deep to the superficial system. Since flow in this direction is apparently normal in the foot, it would appear impossible to describe any of these veins as incompetent. Myers emphasized the importance of removing dilated veins of the feet, especially those associated with dilated perforating veins.

Varicose Veins

Neither the circulatory nor the anatomic factors leading to the development of varicose veins are clearly understood. Often stressed is the concept that incompetent valves in the saphenous system lead to increased venous pressure in this system: however, it is not known whether the incompetence of the valves precedes or is a result of the varicosities (*e.g.*, Van Cleave and Holman). Further, the pressure in the superficial veins of the ankle or on the dorsum of the foot in a normal individual standing quietly is, on the average, sufficient to support a column of blood to about the level of the third intercostal space or the

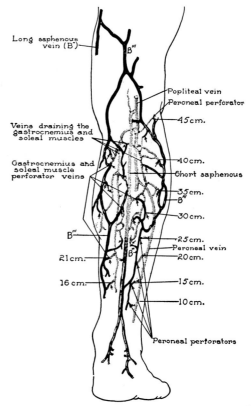

Fig. 7-45. Composite drawing of the communicating and perforating veins on the back of the leg. B''' indicates tributaries of the great saphenous vein; arrows indicate the perforators; and the measurements are from the sole of the foot. (Sherman RS: Ann Surg 130:218, 1949)

manubrium sterni (Pollack and Wood; Boyd, and co-workers). The valves therefore seem to contribute little or nothing to the support of the blood as high as the heart, and incompetence or lack of them could hardly increase the venous pressure at the ankle during quiet standing, when it is highest. Heller said, however, that there is a difference in circulation, for he described a downward rush of blood into incompetent varicose veins when a subject stands, and said that it may take as much as 10 minutes for the saphenous trunk to empty itself.

The poorer venous circulation (the arterial flow in limbs with varicose veins is said to be at least normal, according to Abramson and Fierst) is particularly obvious during walking. In the normal individual walking reduces the

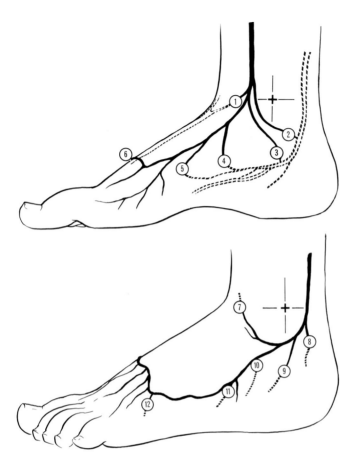

Fig. 7-46. Perforating veins on the medial and lateral sides of the foot. Those numbered *1* to *5* open into the great saphenous vein, *2* and *5* being less constant than the others; *6* opens into the dorsal venous arch; and those numbered *7* to *12* open into the small saphenous vein. *8* was present in only half the cases, *12* in only about a third. (Kuster G, Lofgren EP, Hollinshead WH: Surg Gynecol Obstet 127:817, 1968 [by permission of Surgery, Gynecology and Obstetrics])

blood pressure in the saphenous system very appreciably, since the contracting muscles massage the blood upward in the deep veins, and thus increase the rate of flow from the superficial to the deep veins. In consequence, the blood pressure in the superficial veins drops, usually to a level sufficient to support a column of blood only as high as the knee (Pollack and Wood). In persons with varicose veins, exercise augments the blood flow less (Heller) and decreases the venous pressure less (Boyd and co-workers) than it does in the normal individual; this would seem to implicate incompetent perforating or communicating veins, allowing blood to be forced into the superficial venous system instead of drained from it, rather than incompetence of valves in the superficial system. Blood pressure in the femoral and common iliac veins apparently changes little during exercise (Boyd and his co-workers), hence lack or incompetence of

valves at the upper end of the great saphenous vein should not lead to increased pressure in this vein; however, they would allow increased pressure on straining, when the abdominal venous pressure is increased.

In summary, it would appear that persons with varicose veins have normal venous pressures when they are inactive, but that during activity their pressures drop less than do those of normal persons. Hence, once varicosities and incompetence are established, their superficial venous system, obviously already weakened, is subjected to more stress during activity, in terms of venous pressure, than is that of a normal individual.

In regard to anatomic etiologic factors, probably even less is known. There seems to be a decided hereditary factor in many cases of varicose veins, suggesting therefore that a congenital deficiency may contribute to the development of varicosities and incompetence. Van Cleave and Holman, on the basis of their tentative findings that upper segments of incompetent varicose veins apparently have fewer valves than do similar lengths of normal veins, suggested a paucity of valves as a factor.

Wagner and Herbut suggested that abnormal thinning of the normally thinned wall of the vein at the level of a valve, the valve sinus (see Fig. 1-21) may be a predisposing cause. They investigated histologically the upper end of the great saphenous in the region of its uppermost valve and found what they considered to be a normal wall of the valve sinus in only 16 of 100 specimens; in the remainder it was from one half to one fifth of its normal thickness, contained no internal elastic membrane, and either no tunica media or only a few small bundles of smooth muscle. This was true not only of all the veins that were varicose, but of a majority of those that were not, and was found also in 2 children and 3 infants; they regarded it as allowing dilation of the wall and hence the occurrence of varicosities, and suggested that such defects may also occur in the perforator veins, and predispose to dilatation under the constant stress of the blood flow.

A very different view was expressed by Edwards and O'Connor; they suggested that many varices, especially in young people, are caused by their connection to a congenital hemangioma that lies deep to the deep fascia.

It is because of the obvious contribution that incompetent communicating veins can make to pressure in the superficial veins that surgeons have increasingly emphasized the necessity of locating and ligating incompetent perforators. Linton and Sherman both stressed that these vessels connect not only with the main stems of the saphenous system but also with their tributaries, and that incompetent communicating veins should be ligated at and removed from well below the level of the deep fascia. Grady and Colvin described ascending phlebography as being useful in locating incompetent perforating veins.

Common causes of the appearance of varicosities after high ligation and stripping are said to be failure to ligate all the superficial veins because of the variations in the region of the saphenous opening, and failure to ligate incompetent perforators; stripping, without ligation of the perforators, is usually said to be inadequate (e.g., Linton; Sherman; Summers; and others); Myers has expressed the opinion that incompetent superficial veins must be removed, not merely ligated and injected, if control of the condition is to be expected. Politowski and co-workers reported a technique of electrocoagulation which they apparently preferred to stripping.

LYMPHATICS

The lymphatics and the lymph nodes of the lower limb are generally divided into superficial and deep ones, and as in the case of the upper limb the superficial ones are more important. The deep lymphatic vessels follow the blood vessels, and the deep lymphatic nodes are both small and scarce. A tiny node, the anterior tibial, is apparently usually present on the upper part of the anterior surface of the interosseous membrane, but is usually too small to be detected by dissection; small popliteal lymph nodes lie for the most part in the fat around the popliteal vessels, with usually one node lying just deep to the deep fascia close to the upper end of the small saphenous vein; they receive superficial lymphatics that have run with this vein, and some of the deep lymphatics of the leg. Deep inguinal lymph nodes lie on the medial side of the upper end of the femoral vein, but are again usually too small to be demonstrated easily upon dissection, and typically number not more than two or three. One or more of the nodes of this group usually lie in the femoral canal.

The superficial lymphatic vessels are numerous (Figs. 7-47 and 7-48). Those from the plantar surface of the foot run largely around the borders of the toes or the medial and lateral borders of the foot or ankle to join the lymphatics of the dorsum, which for the most part drain upward along the course of the great saphenous vein. Vessels from the posteromedial aspect of the calf pass mostly around the medial aspect of the leg to join the

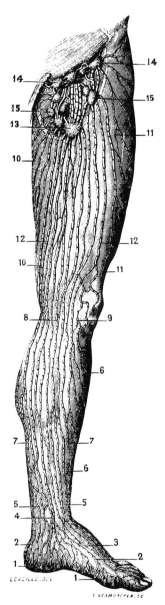

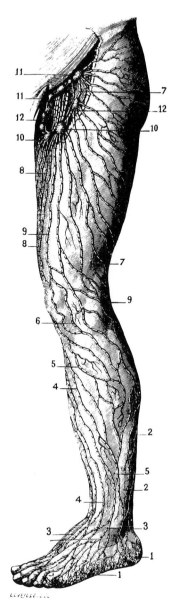

Fig. 7-47. Medial view of the superficial lymphatics of the lower limb. The identifying numbers are for the most part of no importance, for they indicate the obvious: for instance, *1, 1* is the plexus on the medial side of the foot, *7, 7* are vessels of the medial side of the leg, *12, 12* vessels of the medial side of the thigh. *13, 14,* and *15* are superficial inguinal nodes. (After Sappey, from Poirier P, Charpy A: Traité d' Anatomie Humaine [ed 2] Vol 2, Fasc 4. Paris, Masson et Cie, 1909)

Fig. 7-48. Anterolateral view of the superficial lymphatics of the lower limb. As in Fig. 7-47, the numbers largely identify the obvious; however, it might be noted that vessels *2, 2* are diverging posteriorly, to join the lymphatics along the small saphenous vein and end in popliteal nodes, while most of the vessels are running toward the medial side of the leg. All the drainage from the thigh is toward the superficial inguinal nodes. (Poirier P, Charpy A: Traité d' Anatomie Humaine [ed 2] Vol 2, Fasc 4. Paris, Masson et Cie, 1909)

vessels along the great saphenous, and those from the posterolateral, lateral, and anterior surfaces of the calf pass anteriorly and medially to join these vessels. Some drainage from the foot and the posterior aspect of the calf is, however, carried out by several lymphatics that follow the small saphenous vein along the posterior aspect of the calf and penetrate the fascia lata over the popliteal fossa with it, to join the deep lymphatics here. Above the level of disappearance of these vessels there are no longitudinally running channels on the back of the lower limb, so the vessels of the back of the thigh separate as if from a watershed, those originating posterolaterally and anteriorly passing around the anterior surface of the thigh to join vessels along the greater saphenous, those originating posteromedially passing around the medial surface of the thigh.

The superficial inguinal lymph nodes lie in the fat between the skin and the fascia lata and are typically divided into two sets, an upper one that approximately parallels the inguinal ligament and a lower one whose elements lie around the upper end of the great saphenous vein. These lymph nodes receive not only the lymphatic drainage from the free limb, but also that from the buttock, from the skin of the anterior abdominal wall below the level of the umbilicus, from the anal canal, from the scrotum and from the major part of the penis (or, in the female, from the vulva and the lower end of the vagina), and even, in the female, one or two vessels that pass outward from the uterus along the round ligament, accompanying this through the inguinal canal. The superficial inguinal lymph nodes and their connections are described in more detail in the following chapter.

References

ABRAMSON DI, FIERST SM: Arterial blood flow in extremities with varicose veins. Arch Surg 45:964, 1942

BARDEEN CR: A statistical study of the abdominal and border-nerves in man. Am J Anat 1:203, 1902

BARDEEN CR: Studies of the development of the human skeleton: (A) The development of the lumbar, sacral and coccygeal vertebrae. (B) The curves and the proportionate regional lengths of the spinal column during the first three months of embryonic development. (C) The development of the skeleton of the posterior limb. Am J Anat 4:265, 1905

BARDEEN CR: Development and variation of the nerves and the musculature of the inferior extremity and of the neighboring regions of the trunk in man. Am J Anat 6:259, 1907

BARDEEN CR, ELTING AW: A statistical study of the variations in the formation and position of the lumbo-sacral plexus in man. Anat Anz 19:124; 209, 1901

BARDEEN CR, LEWIS WH: Development of the limbs, body-wall and back in man. Am J Anat 1:1, 1901

BOYD AM, CATCHPOLE BN, JEPSON RP, ROSE SS: Some observations on venous pressure estimations in the lower limb. J Bone Joint Surg 34-B:599, 1952

DASELER EH, ANSON BJ, REIMANN AF, BEATON LE: The saphenous venous tributaries and related structures in relation to the technique of high ligation: Based chiefly upon a study of 550 anatomical dissections. Surg Gynecol Obstet 82:53, 1946

EDWARDS EA, O'CONNOR JF: Ordinary varicose veins as an expression of congenital hemangioma. Surg Gynecol Obstet 122:1245, 1966

FLECKER H: Time of appearance and fusion of ossification centers as observed by roentgenographic methods. Am J Roentgenol 47:97, 1942

GARNJOBST W: Injuries to the saphenous nerve following operations for varicose veins. Surg Gynecol Obstet 119:359, 1964

GLASSER ST: An anatomic study of venous variations at the fossa ovalis. The significance of recurrences following ligations. Arch Surg 46:289, 1943

GRADY ED, COLVIN EM: Treatment of venous insufficiency of the lower extremities with a note on the use of ascending phlebography. Am Surgeon 19:936, 1953

HELLER RE: The circulation in normal and varicose veins. Surg Gynecol Obstet 74:1118, 1942

HORWITZ MT: The anatomy of (A) the lumbosacral nerve plexus—its relation to variations of vertebral segmentation, and (B), the posterior sacral nerve plexus. Anat Rec 74:91, 1939

KAMPMEIER OF, BIRCH C LA F: The origin and development of the venous valves, with particular reference to the saphenous district. Am J Anat 38:451, 1927

KLOTZ K: Untersuchungen über die Vena saphena magna beim Menschen, besonders rücksichtlich ihrer Klappenverhältnisse. Arch f Anat u Physiol (Anat Abt) 1887, p. 159

KOSINSKI C: Observations on the superficial venous system of the lower extremity. J Anat 60:131, 1926

KUSTER G, LOFGREN EP, HOLLINSHEAD WH: Anatomy

of the veins of the foot. Surg Gynecol Obstet 127:817, 1968

LAROCHELLE J-L, JOBIN P: Anatomical research on the innervation of the hipjoint. Anat Rec 103:480, 1949 (abstr)

LAST RJ: Innervation of the limbs. J Bone Joint Surg 31-B:452, 1949

LINTON RR: The communicating veins of the lower leg and the operative technic for their ligation. Ann Surg 107:582, 1938

LOFGREN EP, MYERS TT, LOFGREN KA, KUSTER G: The venous valves of the foot and ankle. Surg Gynecol Obstet 127:289, 1968

MANSBERGER AR, YEAGER GH, SMELSER RM, BRUMBACK FM: Saphenofemoral junction anomalies. Surg Gynecol Obstet 91:533, 1950

MCDOUGALL A: The os trigonum. J Bone Joint Surg 37-B:257, 1955

MILLER JA JR: Joint paracentesis from an anatomic point of view: II. Hip, knee, ankle and foot. Surgery 41:999, 1957

MULLARKY RE: The Anatomy of Varicose Veins. Springfield, Ill., Thomas, 1965

MYERS TT: Results and technique of stripping operation for varicose veins. JAMA 163:87, 1957

PATERSON AM: The origin and distribution of the nerves to the lower limb. J Anat Physiol 28:84;169, 1893–94

PATERSON RS: A radiological investigation of the epiphyses of the long bones. J Anat 64:28, 1929

POLITOWSKI M, SZPAK E, MARSZALEK Z: Varices of the lower extremities treated by electrocoagulation. Surgery 56:355, 1964

POLLACK AA, WOOD EH: Venous pressure in the saphenous vein at the ankle in man during exercise and changes in posture. J Appl Physiol 1:649, 1949

SENIOR HD: An interpretation of the recorded arterial anomalies of the human leg and foot. J Anat 53:130, 1919

SHAH AC, SRIVASTAVA HC: Fascial canal for the small saphenous vein. J Anat 100:411, 1966

SHERMAN RS: Varicose veins: Anatomic findings and an operative procedure based upon them. Ann Surg 120:772, 1944

SHERMAN RS: Varicose veins: Further findings based on anatomic and surgical dissections. Ann Surg 130:218, 1949

SMILLIE IS, MURDOCH JH: Man with three legs. J Bone Joint Surg 34-B:630, 1952

SRIVASTAVA KK, GARG LD: Reduplication of bones of lower extremity. J Bone Joint Surg 53-A: 1445, 1971

SUMMERS JE: Highlights in the treatment of varicose veins and ulcers. Am J Surg 86:443, 1953

VAN CLEAVE CD, HOLMAN RL: A preliminary study of the number and distribution of valves in normal and varicose veins. Am Surgeon 20:533, 1954

WAGNER FB JR, HERBUT PA: Etiology of primary varicose veins: Histologic study of one hundred saphenofemoral junctions. Am J Surg 78:876, 1949

WOODBURNE RT: The accessory obturator nerve and the innervation of the pectineus muscle. Anat Rec 136:367, 1960

WOODHALL B: Sensory patterns in the localization of disc lesions. J Bone Joint Surg 29:470, 1947

Chapter 8

BUTTOCK, HIP JOINT, AND THIGH

The pelvic girdle articulates very strongly with the sacrum, and no muscles are needed at this joint. The muscles of the buttock, almost entirely covered by the largest member of the group, the gluteus maximus (Fig. 8-1), insert for the most part on the upper end of the femur, but the gluteus maximus and the tenson fasciae latae gain additional leverage by their insertions into the strong iliotibial tract that runs down the lateral side of the thigh to attach across the knee joint on the tibia. The musculature of the thigh assists that of the buttock in movements at the hip joint and is also primarily responsible for movements at the knee. The longer muscles of the thigh are biarticular ones, crossing both the hip and knee joints.

Nerves and vessels enter the thigh both anteriorly and posteriorly. The anterior artery, the femoral, continues on down to supply the leg and foot but the anterior nerves, chiefly the femoral and obturator, are largely confined in their distribution to the thigh. The reverse is true of the posterior vessels and nerves: the arteries entering the buttock are largely confined in their distribution to the buttock, while the largest nerve, the sciatic, continues into the leg to supply all the muscles of the leg and foot.

Pelvic Girdle

The two coxal or hip bones (innominate bones) articulate firmly with the sacrum to form the pelvic girdle (Fig. 8-2). Each coxal bone is in turn composed of three fused elements, the ilium, ischium, and pubis, and each of these has its own named parts. In addition, the coxal bone as a whole presents two prominent features, the acetabulum, for accommodation of the head of the femur, and the obturator foramen.

The ilium, ischium, and pubis all enter into the formation of the acetabulum, which faces not only outward but also somewhat downward and anteriorly. Its weight-bearing upper and posterior walls are particularly heavy, while its inferior wall is deficient, presenting the acetabular notch; this leads into the acetabular fossa, the rough area in the center of the articular or lunate surface of the acetabulum. In life, the acetabular cavity is deepened by the fibrocartilaginous acetabular labrum (glenoidal labrum) attached to its lip; the labrum crosses the acetabular notch as the transverse ligament of the acetabulum.

The obturator foramen, closed by the obturator membrane except anterosuperiorly,

619

where the obturator canal transmits the obturator nerve and vessels, is surrounded by the pubis and ischium (Fig. 8-3). Wakeley ('39) described the several directions that a hernia may take as it emerges through the foramen, including a rare type in which the

hernia passes between two layers of membrane. Wakeley said that only about 10% to 15% of obturator hernias occur in males; Pernworth found 420 cases reported by 1938, the hernial sac most commonly containing a segment of small bowel.

Fig. 8-1. Some superficial muscles of the buttock and thigh. The gluteus maximus and the tensor fasciae latae can be seen converging on the iliotibial tract, and some of the gluteus medius appears between them; anterior to the tract is a part of the quadriceps, posterior to it the biceps femoris. On the right thigh, the prominent muscles seen are the sartorius, gracilis, and semitendinosus, converging toward their insertions on the tibia.

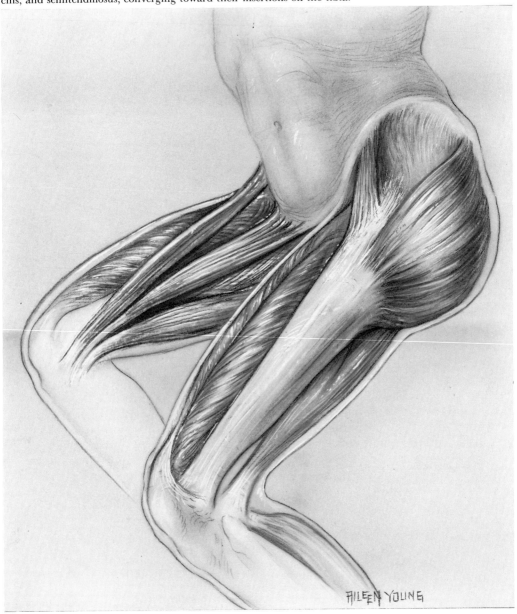

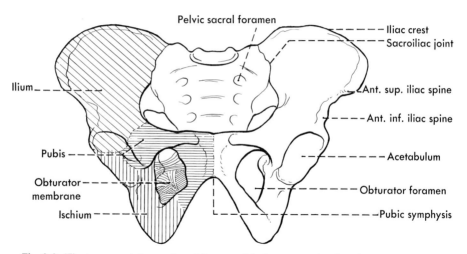

Fig. 8-2. The two coxal (innominate) bones, with the sacrum in place between them. The *shading* on the reader's left identifies the three elements that form the adult bone.

Fig. 8-3. Lateral view of the right coxal bone.

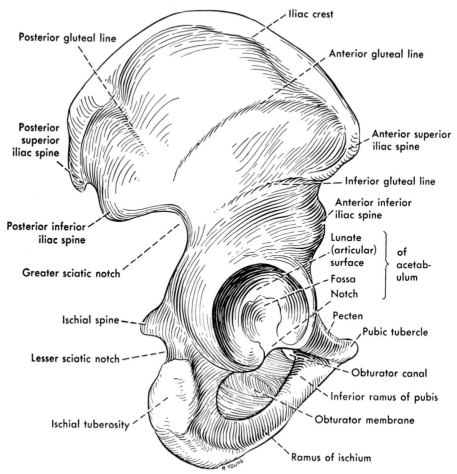

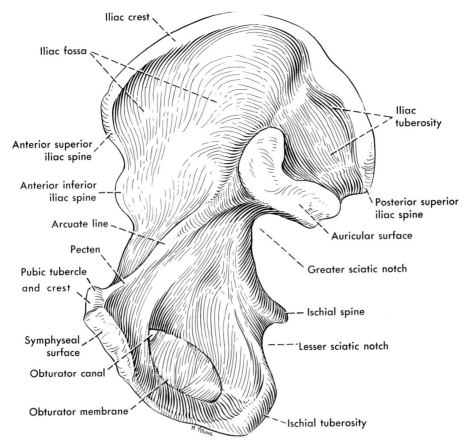

Iliac crest

Iliac fossa

Anterior superior
iliac spine

Anterior inferior
iliac spine

Arcuate line

Pecten

Pubic tubercle
and crest

Symphyseal
surface

Obturator canal

Obturator membrane

Iliac
tuberosity

Posterior superior
iliac spine

Auricular surface

Greater sciatic notch

Ischial spine

Lesser sciatic notch

Ischial tuberosity

Fig. 8-4. Medial view of the right coxal bone.

Ilium, Pubis, and Ischium

The *ilium* (os ilium) has a thick body that enters into the formation of the acetabulum, and an expanded, thinner, ala or wing that ends above in the iliac crest. The iliac crest, curving between the anterior superior and posterior superior iliac spines, is subcutaneous, but other parts of the bone are covered by muscles. The crest is described as presenting an external lip or labium, an intermediate line, and an internal lip, marking the attachment of the anterolateral abdominal and other muscles and of the fascia lata and iliac fascia to it. On the anterior border of the ilium, a little above the acetabulum and therefore at about the region of junction of the wing and body of the bone, is the anterior inferior iliac spine. On the posterior border the greater sciatic (ischiadic) notch extends from the posterior inferior iliac spine to the ischial spine.

The only markings of importance on the outer or gluteal surface of the wing of the ilium are the anterior, posterior, and inferior gluteal lines. The first-mentioned one marks the posterosuperior border of the origin of the gluteus minimus and therefore the anteroinferior border of the origin of the gluteus medius; the second the posterior border of the origin of the medius and the anterior border of the origin of the gluteus maximus; and the third marks the inferior border of the origin of the minimus. The larger part of the inner surface of the wing of the ilium forms the iliac fossa, but posteriorly the sacropelvic surface is divisible into an auricular surface and the iliac tuberosity. The auricular surface, the facet for the synovial joint with the sacrum, extends approximately horizontally along the upper border of the greater sciatic notch and then turns upward at its anterior end (Fig. 8-4). The roughened iliac tuberosity, above the auricular surface, serves for the attachment of ligaments. An arcuate line, the upper major part of the linea terminalis, runs downward from the anterior end of the auricular surface toward the pecten of the pubis and helps mark the border between the major and the minor pelvis.

The *pubis* (os pubis) needs little description. Its body has a symphyseal surface for articulation with the bone of the other side, its superior ramus runs above the obturator foramen to take part in the formation of the acetabulum, and its inferior ramus curves inferiorly and posteriorly to fuse with the ischium. The paired inferior rami form the pubic arch, with its variable subpubic angle. The upper border of the body presents anteriorly a thickened pubic crest; at the lateral end of the crest is the prominent pubic tubercle, from which a raised line, the pecten, extends upward toward the arcuate line of the ilium. Anterior to the pecten, the junction of the superior ramus of the pubis with the ilium is marked by the iliopubic (iliopectineal) eminence; and on the inner surface of this ramus the obturator sulcus, bordered above by the obturator crest, leads toward the obturator canal.

On the posterior border of the body of the *ischium* (os ischii) is the ischial spine, and at the region of junction of the body and ramus (once called superior and inferior rami, respectively) the large ischial tuberosity projects posterolaterally. The lesser sciatic notch lies between the spine and tuberosity.

Ligamentous and muscular attachments on the hip bone are particularly numerous. The chief ones are shown in Figures 8-5 and 8-6.

BLOOD SUPPLY

As evidenced by its numerous vascular foramina, typically occurring in the thicker parts of the bone, the hip bone has an abundant vascular supply; this is derived from branches of most of the neighboring smaller vessels. Branches from the deep iliac circumflex and the iliolumbar arteries enter the ilium near the crest and in the iliac fossa, and branches of the obturator penetrate especially the area of the arcuate line. On the outer surface, numerous small branches from the superior and inferior gluteal arteries penetrate the body of the ilium just above the acetabulum. The ischium receives branches, especially numerous on the outer surface between the acetabulum and the tuberosity, from the obturator and the medial and lateral femoral circumflex arteries. The chief blood supply of the pubis is also derived from these three arteries. An acetabular branch, usually from the obturator artery, sends branches both into the soft tissues and the bone of the acetabular fossa.

CONGENITAL ANOMALIES

Although the pelvis as a whole, and therefore the hip bones, may become much distorted during the growth period in response to abnormal postures, congenital abnormalities other than those that may occur in connection with congenital dislocation of the hip (discussed in a following section) are apparently rare.

Abnormal depth of the acetabulum, presenting a picture of apparent protrusion of the head of the femur into the pelvis, is probably, according to Rechtman, a result of a congenital, possibly hereditary, defect. Apparently it results from overgrowth of the elements entering into the formation of the acetabulum.

The hip bone normally descends somewhat during its early growth, to lie primarily alongside the sacral rather than the lumbar portion of the vertebral column with which it is first associated. Levinthal and Wolin described a case of apparent incomplete descent, in which the right ilium, smaller than the left, articulated with the right transverse process of the fifth lumbar vertebra instead of with the pars lateralis of the sacrum (which appeared to be partially absent).

FRACTURE

Direct force applied to the pelvis may fracture it anywhere. The fractures that involve the pelvic ring are typically produced by compression and may be either single or multiple. Since a fracture tends to be across a weaker part of the ring, it is especially likely to be at

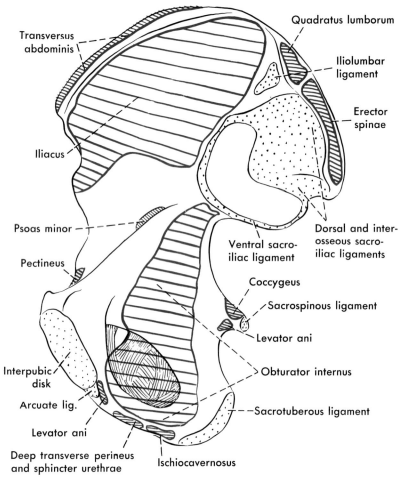

Fig. 8-5. Origins (*red*) and insertions (*black*) of muscles on the medial surface of the coxal bone. Some important ligamentous attachments are also shown.

the obturator foramen, involving one or both pubic rami, or across the wing of the ilium from the crest to the margin of the greater sciatic foramen (Fig. 8-7). A single fracture in either location allows little separation of parts, and treatment is simple (Watson-Jones). Fractures across the pubic rami of both sides, or unilateral fracture with separation at the symphysis, is usually caused by lateral compression, according to Watson-Jones; since the anterior part of the ring does not support the weight of the body, and the parts tend to be held in position by muscles, displacement is limited to that originally produced by the injury. In contrast, fracture of the pubis or pubic separation with fracture of the ilium or dislocation at the sacroiliac joint

(typically produced by anteroposterior compression, according to Watson-Jones) may be accompanied by great displacement of the fragment.

Fractures of the anterior part of the pelvis are particularly dangerous if there is displacement, because they may be associated with damage to the viscera, especially the bladder and urethra; Wakeley ('29) reviewed 100 cases of fracture of the pelvis and found that in only 11 were there visceral complications, but this report was made when automobile accidents were fewer. Posterior fractures of the pelvis involving the ilium or the sacrum may be associated with injury to the rectum, great vessels, or the sacral plexus, but Wakeley found no such involvement in any of his cases as a direct result of injury—although later there was injury to the plexus in one case, as the result of callus formation at the sacral foramina. Burman reported that tears of the sacrotuberous ligament may be associated with fracture of the inferior ramus of the ischium or of the ischial tuberosity; and that the sacrospinous ligament may be avulsed at its origin or at its insertion on the ischial spine.

Lesser fractions of the hip bone include those of the crest or spinous processes of the ilium, or of the acetabular rim. More serious is fracture of the acetabular floor, which may

result from a force directed primarily along the femur so as to drive its head into the acetabulum. In this case, the thin floor of the acetabulum may give way entirely, and the head of the femur, protruding into the pelvic cavity, may carry bony fragments into close contact with the rectum.

HEMIPELVECTOMY

Carcinoma that has involved the coxal bone has been treated by interinnominoabdominal amputation (hindquarter amputation, hemipelvectomy). Saint said the first operation of

Fig. 8-6. Lateral view of the right coxal bone showing muscular attachments (origins *red*, insertions *black*) and chief ligamentous attachments.

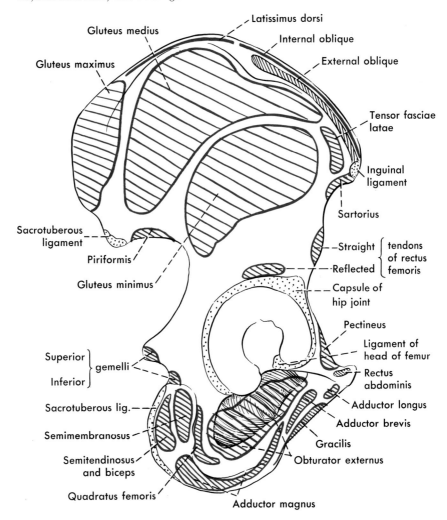

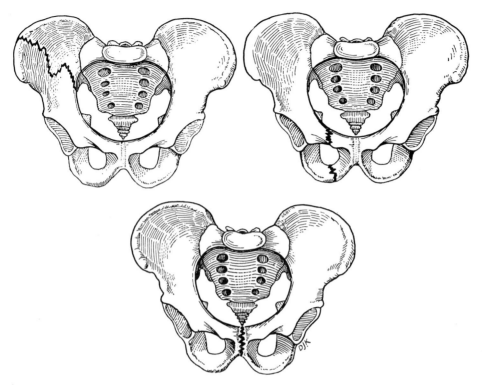

Fig. 8-7. Typical fractures of the pelvis that show little displacement; a combination of any two of these readily allows displacement. (Watson-Jones R: Brit J Surg 25:773, 1938)

this type was in 1891; most of the authors writing on the subject have mentioned the greatly lowered mortality of this drastic operation with modern technics, and Saint, in 1950, estimated the mortality as being about 15%. Among others who have discussed the operation are Gordon-Taylor and colleagues; Beck and Bickel; and Brittain.

PUBIC SYMPHYSIS

The pubic symphysis is simple. The bodies of the two pubic bones are united at their symphyseal surfaces by a fibrocartilaginous interpubic disk, reinforced above by a superior pubic ligament and below by an arcuate pubic ligament. The superior pubic ligament extends across the upper borders of the bodies of the two pubes, blending with the disk between, and is thick enough over the bone to hold sutures. The arcuate pubic ligament follows the pubic arch, running between the two inferior pubic rami and blending with the disk between the pubes. The dorsal vein of the

penis (or clitoris) enters the pelvis immediately beneath it.

The interpubic disk may be solid throughout, but often presents a narrow fissure that apparently results from degeneration; Sutro ('36) noted that pregnancy, which causes relaxation and stretching of the ligamentous tissues of the pelvis, accentuates the degenerative process in an interpubic disk.

The arterial supply of the interpubic joint is by a number of tiny twigs from the internal pudendal, the pubic branches of the obturator and inferior epigastric arteries, and ascending branches of the medial femoral circumflex and external pudendal arteries.

The lymphatic drainage of the pubic symphysis is partly into deep inguinal lymph nodes, partly into external iliac ones (Tanasesco).

SACROILIAC JOINT

The union between a hip bone and the sacrum is particularly strong, for the rather limited synovial joint is strongly reinforced above and behind by heavy ligaments, and the union is further strengthened by accessory ligaments—the sacrotuberous, sacrospinous, and iliolumbar ligaments.

Accessory Ligaments

The *sacrotuberous* (NA, sacrotuberal) *ligament* (Fig. 8-8) is attached at its upper end to the posterior surfaces of the lower three sacral vertebrae, and in part also to a posterior portion of the iliac crest and the region of the two posterior iliac spines; it blends in part with the dorsal sacroiliac ligaments. From this broad upper attachment the fibers converge and run laterally and inferiorly, forming a strong band that attaches to the medial border of the ischial tuberosity. An extension from its attachment on the tuberosity is sent along the ramus of the ischium to form the falciform process.

The *sacrospinous* (NA, sacrospinal) *ligament* is both thinner and narrower than the sacrotuberous; it is somewhat triangular, its expanded base being attached to the lateral border of the lower part of the sacrum and to the coccyx, while its apex is attached to the pelvic surface of the ischial spine. On its anterior surface it blends with the coccygeus muscle, and it has been said to represent a degenerated posterior part of this muscle. The sacrospinous ligament forms the lower border of the greater sciatic foramen, and the upper border of the lesser sciatic foramen; the sacrotuberous ligament forms the posteromedial border of the greater sciatic foramen and the postero-inferior border of the lesser foramen. These two ligaments together especially brace the pelvis against rotation of the sacrum between the two hip bones—the weight of the body on the upper end of the sacrum tends to force this downward and forward, hence the coccygeal end would move upward and backward were it not for these two ligaments.

The *iliolumbar ligament* arises from the antero-inferior part of the transverse process of the fifth lumbar vertebra and passes downward and laterally, expanding as it does so, to blend in part with the anterior sacroiliac ligament at the base of the sacrum and to attach also to the inner surface of the ilium just anterior to the sacroiliac joint.

Synovial Cavity and Ligaments

The sacroiliac joint has a relatively small synovial cavity between the auricular surfaces of the ilium and sacrum, at the levels of the first and second sacral vertebrae, and involving the third vertebra more commonly in the male than in the female (Derry). The articular surfaces of the two elements are usually described as being hyaline cartilage; Schunke

Fig. 8-8. The chief bracing or accessory ligaments of the sacroiliac articulation. Posterior view, with the three elements entering into the adult coxal bone *shaded* on the right side.

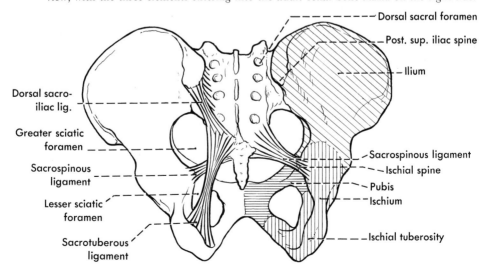

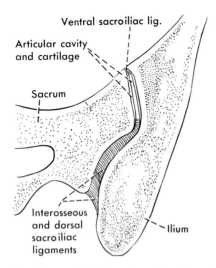

Fig. 8-9. Schematic horizontal section through the sacroiliac joint.

described the sacral cartilage as being either hyaline or fibrous before birth, but invariably hyaline after birth, while he found the iliac auricular cartilage to be hyaline until the fourth fetal month, but nearly always fibro-cartilage thereafter. The joint is so frequently involved in degenerative changes, especially osteo-arthritic ones and bony ankylosis, that Sashin maintained that the normal anatomy of the joint can be understood only if joints of individuals less than 30 years old are examined.

The cavity of the sacroiliac joint (Fig. 8-9) is a narrow cleft, with only enough fluid to moisten the surfaces of the articular cartilages. These surfaces are typically not actually plane but sinuous, so that they resist movement. Further, they slant medially as they are traced from above downward and, except for a lower portion where the sacrum is slightly wider on its pelvic side, also slant a little medially as they are traced from behind forward; thus they tend to be forced even closer together in weight bearing, for the weight of the body tends to force the upper end of the sacrum downward and forward in the pelvis. The very strong interosseous sacroiliac ligaments resist the widening of the interiliac space necessary for descent and forward rotation of the sacrum; the dorsal sacroiliac ligaments also resist this, and with

the sacrotuberous and sacrospinous ligaments resist rotation of the sacrum.

The synovial membrane of the joint cavity simply passes from the edge of one articular cartilage to that of the other. The fibrous capsule, thin anteriorly but thick posteriorly where strength is needed, is closely applied to the outer surface of the synovial membrane, and consists of several special ligaments.

The anterior part of the fibrous capsule, *the ventral,* formerly anterior, *sacroiliac ligaments,* is both thin and weak, and is a broad membrane separating the joint cavity from the pelvic cavity. The posterior part of the capsule is enormously thickened as the interosseous and dorsal sacroiliac ligaments. The *interosseous sacroiliac ligaments* are short but very thick and strong; they attach to the rough iliac and sacral tuberosities and thus fill the narrow gap above and behind the auricular surfaces of the two bones. The *dorsal* (posterior) *sacroiliac ligaments,* sometimes divided into long and short ones, are continuous on their deep surface with the interosseous sacroiliac ligaments, but are longer and more obliquely placed. The uppermost fibers, nearly horizontal in direction, pass between the tuberosity of the ilium and the lateral sacral crest (tubercles) of the first and second segments of the sacrum, while the lower and longer fibers pass downward, some of them almost vertically, from the region of the posterior superior spine of the ilium to the lateral crest of the third segment of the sacrum, blending here with some of the origin of the sacrotuberous ligament.

Lichtblau reported a case of marked dislocation of the sacroiliac joint following removal of a posterior part of the iliac crest, thus emphasizing the importance of the interosseous and posterior sacroiliac ligaments in preventing subluxation at this joint.

Accessory sacroiliac synovial joints, usually between the posterior superior iliac spine and the sacrum at the level of the second sacral foramen, but sometimes between the iliac and sacral tuberosities at the level of the first sacral foramen, are fairly common. Trotter ('40) found them in 36% of 958 skeletons, more frequently in older than in younger age groups;

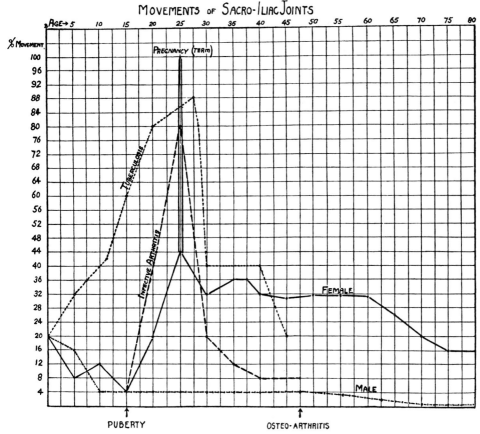

Fig. 8-10. Movement at the sacroiliac joint at different ages and under different conditions of the joint. (Brooke R: J Anat 58:299, 1924)

Hadley found about the same frequency (33.5%) in 200 patients in whom roentgenograms were taken because of low back pain.

Mobility

Sashin described mobility at the sacroiliac joint as being very slight at the most, and consisting of a gliding up and down with a slight anteroposterior movement; these together allow a limited amount of rotation, which is usually what is measured. In individuals of less than 30 years he was able to produce by manual pressure on the lower end of the sacrum a movement of an average 4°, the minimum being 2° and the maximum, occurring in women who died shortly after parturition, 8°. Brooke found the movement of the sacroiliac joint difficult to measure, and reported it (Fig. 8-10) in comparative terms. It

decreases in both sexes between birth and puberty, and then increases in the female until age 25; in both sexes, barring parturition, the mobility remains low (but that of the female is much higher than that of the male) and tends to diminish with age. Sashin found that it begins to disappear in men about the fourth decade and in women usually toward the end of the fifth; Brooke's graph indicates a slightly later occurrence.

The softening of the ligaments brought about by the hormones of pregnancy increases movement at the joints, as both Brooke and Sashin have shown. Sashin further reported that while forcible separation of the two bones usually resulted in tearing the interosseous ligament from one of the two, the ligaments were so lax in four females dying after parturition that the joint surfaces could be separated about 0.25 inch (6 mm) by manual traction.

Sashin found gross pathological changes in the articular cartilages in 91% of male and 77% of female joints among 111 bodies of persons between the ages of 30 and 59, and microscopic pathologic changes in all. Osteoarthritic changes in the joints were found in 91% of males and 53% of females, ankylosis in 51% of males and 5.8% of females. After the age of 60, osteoarthritic changes were even more marked, especially in males, and bony ankylosis of the joint was said to be present in 82% of males and 30% of females. Brooke found ankylosis present in 37% of all (105) the males of all ages whose joints he examined, and described the joint as being typically ankylosed in the male after the age of 50.

Sacroiliac Sprain

In spite of the fact that the sacroiliac joint appears to be the strongest and most stable joint in the body, diagnosis of sacroiliac sprain (or strain), either alone or combined with sacrolumbar sprain, was once common, and these disabilities were widely accepted as being common causes of low back pain. For instance, Miltner and Lowendorf, in 1931, recorded 525 cases of what they believed to be sacroiliac, sacrolumbar, or combined sprain, and said that in so doing they had rejected 850 more cases in which the original diagnosis was that but later was changed because of the obvious development of arthritis in the lumbosacral region. Yeoman did not agree that strain of the sacroiliac joint was a common cause of low back pain or sciatica, but did claim that arthritis of the sacroiliac joint accounted for a very large percentage of cases of sciatica, 36% in his series. The symptoms, he said, were due to periarthritis that involved the ventral sacroiliac ligaments and the piriformis muscle, and therefore secondarily the roots of the sciatic nerve. Miltner and Lowendorf reported that sciatic pain occurred in 64.1% of their cases in which a diagnosis of sacroiliac sprain was made; in such cases, they said, straight-leg raising causes pain.

Albee was apparently one of the earlier workers to abandon the diagnosis, pointing out that severance of the pubic symphysis, removal of the pubis, or fracture of the pelvis should all place a much greater strain on the sacroiliac joint than could stooping or lifting, and yet in his experience symptoms referable to the joint were missing after such severe injuries; he claimed that in at least 90% of cases of low back pain, sacroiliac strain, and the like, the etiologic factor was what he called "myofascitis." Most of the literature on sacroiliac strain or sprain appeared in the days before protruded lumbar intervertebral disks were clearly recognized as a cause of back pain and sciatica; undoubtedly, many cases of what was diagnosed as sacroiliac strain were actually those of protruded disk, the symptoms of which are largely identical with those described as being typical of sacroiliac derangement. Naffziger and co-workers reported in 1938 that they had found a very high incidence of protruded disk in patients whose condition previously had been diagnosed as a disorder of the lumbosacral and sacroiliac joints and who had been treated for such a disorder. Consonant with the anatomic strength of the sacroiliac joint, strain of it is not widely recognized as a prominent source of low back or sciatic pain at the present time; differences of opinion on this subject still exist, however.

Bickel and Romness reported a rare case of nontraumatic separation of both sacroiliac joints in which subluxation of the ilia was demonstrated by roentgenogram; the joints were surgically fused because of back pain, and at operation it was found that the ilia could be moved fairly easily upward, downward, or laterally for a distance of 1 cm.

Femur

The femur (Figs. 8-11 to 8-16), the largest of the long bones, is mostly covered by the muscles of the thigh, but is palpable toward its upper and lower ends. The greater trochanter, largely subcutaneous, is the most prominent projection at the hip, and the lateral and medial surfaces of the lateral and medial condyles, respectively, are subcutaneous at the knee. The patella lies in front of and between the condyles. A subcutaneous bursa typically lies over the greater trochanter; there may be one over each of the raised projections (epicondyles) on the sides of the condyles; and there is one in front of the patella.

Evans has reported in considerable detail upon the behavior of the femur subjected to various loads as determined by precise engineering technics.

Upper End

The upper end of the femur includes the head, the neck, and the two trochanters. The globular head is covered by articular cartilage except at the fovea for the attachment of the ligament of the head (ligamentum teres). The neck, considerably smaller in diameter than the head, is somewhat flattened anteroposteriorly, generally thinner in its middle than at either end, and flares out very considerably as it attaches to the body (shaft) in the region of the trochanters. Since the capsule of the hip joint is attached at the base of the neck everywhere except posteriorly, where it is somewhat higher, fractures of the neck are likely to interrupt the blood supply to the upper fracture segment, derived largely from vessels that turn upward along the bone from the capsular attachment.

The neck projects markedly medially, as well as upward, from the body, and also somewhat forward. The angle of the medial and upward projection, measured as the angle between the long axes of the neck and of the body, is the *angle of inclination;* the angle of the anterior projection, measured by the angle that the longitudinal axis of the neck makes with a line drawn through the centers of the two femoral condyles, is the *angle of torsion* or *angle of declination.* Humphry found the angle of inclination to be about 141° in fetuses of 2.5 months, about 128° in those of 9 months; it is usually said to average about 127° in adults, and tends to be slightly smaller in women than in men. Pick and co-workers ('41) recorded a number of measurements of the femoral head and neck of adults,

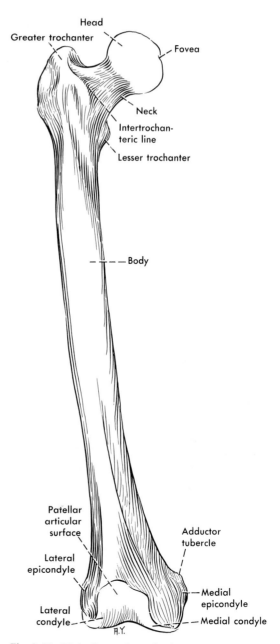

Fig. 8-11. Right femur from the front.

and some of their findings are shown in Figures 8-12 and 8-13. Among their 152 specimens, of which about 5% were from females, the angle of inclination varied from 104° to 147°, with a mean of 126.4°. There was a very wide variation in the angle of torsion, from 1° to 41°, but the mean was 14.01°. Dunlap, Shands, and their co-workers reported what

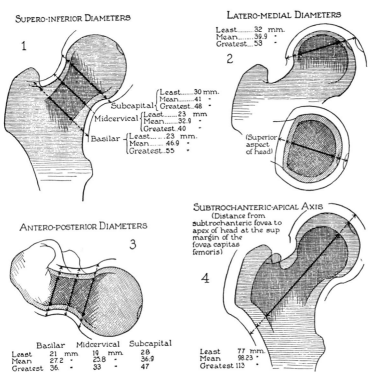

SUPERO-INFERIOR DIAMETERS

1

Subcapital { Least........30 mm.
Mean........41   "
Greatest..48   "

Midcervical { Least......23 mm
Mean......32.9   "
Greatest.40   "

Basilar { Least......23 mm.
Mean........46.9   "
Greatest..55   "

LATERO-MEDIAL DIAMETERS

Least........32 mm.
Mean.......39.9   "
Greatest...53   "

2

(Superior
aspect
of head)

ANTERO-POSTERIOR DIAMETERS

3

	Basilar	Midcervical	Subcapital
Least	21 mm	19 mm.	28
Mean	27.2 "	25.8 "	36.9
Greatest	36. "	33 "	47

SUBTROCHANTERIC-APICAL AXIS
(Distance from
subtrochanteric fovea to
apex of head at the sup
margin of the
fovea capitas
femoris)

4

Least	77 mm.
Mean	98.23 "
Greatest	113 "

Fig. 8-12. Some measurements of the head and neck of the femur. (Pick JW, Stack JK, Anson BJ: Quart Bull Northwestern Univ M School 15:281, 1941)

they regarded as an improved method for measuring the latter angle in the living individual, and Shands and Steele found that coxa valga is often accompanied by an increase in torsion, coxa vara by a decrease.

The *greater trochanter,* projecting above the junction of the neck and the upper end of the body, has on its medial side a small deep depression, the trochanteric fossa, that marks the insertion of the obturator externus, and smooth surfaces representing the attachments of gluteal muscles on its upper end. The intertrochanteric crest extends obliquely downward and medially, on the posterior surface, from the greater to the posteromedially situated *lesser trochanter.* The intertrochanteric line, on the front of the femur, originates also at the greater trochanter but passes just below the lesser one, spiraling onto the back of the femur (this portion has also been called the spiral line) to become continuous with the medial lip of the linea aspera.

The internal structure of the femur, especially of its complicated upper end, has been studied in considerable detail and has already been discussed briefly in Chapter 1; there Koch's analysis of lines of compression and tension was quoted as showing that the femur's structure closely follows engineering principles and is adapted very precisely to its function. Tobin described the changes in the trabeculae with coxa vara and valga.

The heaviest cortex of the neck is inferior, where the neck joins the body (Tobin; Harty, '57); hence, a nail against this has much more support than one driven through the trabeculae of the center of the neck (Tobin). Two other structural features of the upper end of the femur are often mentioned. The weakest place in the neck appears to be a triangular area bounded medially by heavier trabeculae that run from the lower surface of the neck upward and medially into the head, above by trabeculae from the greater trochanter to the head, and below by trabeculae between the two trochanters (*e.g.,* Tobin); this has been called the *internal femoral trigone* or *Ward's triangle.* The *calcar femorale,* according to Harty ('57, '65), is a dense cortical plate of bone that extends laterally toward the greater tro-

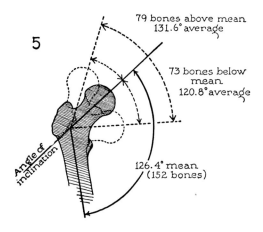

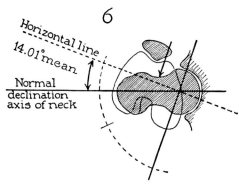

Fig. 8-13. Angles of inclination and declination of the neck of the femur. (Pick JW, Stack JK, Anson BJ: Quart Bull Northwestern Univ M School 15:281, 1941)

coming the effect of the angle of inclination of the neck by shifting the weight-bearing articular surfaces of the knee closer to the center of gravity. The angle of obliquity is measured as that between the longitudinal axis of the femur and a line drawn perpendicular to the horizontal one across the lower surfaces of the femoral condyles. Among the 152 specimens of Pick, Stack, and Anson ('41) this angle var-

Fig. 8-14. Posterior view of the right femur.

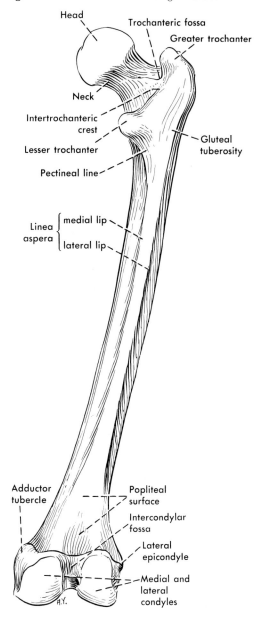

chanter and the gluteal tuberosity from the cortex of the posteromedial part of the femoral body and the posterior surface of the femoral neck; it is most marked deep to the lesser trochanter and, because of its disposition, can be seen only in lateral roentgenograms (Harty). It apparently develops as a special support in response to stress produced by angulation of the neck and is not present in early infancy.

Body

The body or shaft of the femur is slightly bowed anteriorly, and below the trochanters departs only a little from a tubular shape. In the normal position of standing the body of the femur inclines medially, this inclination, the angle of obliquity, partially over-

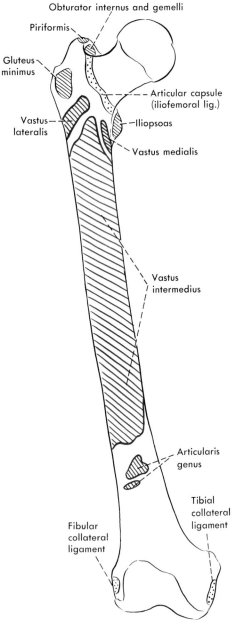

Fig. 8-15. Muscular attachments (origins *red*, insertions *black*) and chief ligamentous attachments on the front of the femur.

ied from 3 to 15°, and had a mean of 9.56°.

The body of the femur is generally smooth, but posteriorly, in about the middle half of the bone, there is a prominent longitudinal ridge, the linea aspera, with roughened medial and lateral labia or lips (Fig. 8-14). Above, the lips of the linea aspera diverge, the

medial one, if well marked, spiraling forward below the lesser trochanter to become continuous with the intertrochanteric line, the lateral one running upward to end in the broader gluteal tuberosity (for the insertion of the gluteus maximus), which may project markedly enough to form a third trochanter.

Fig. 8-16. Muscular attachments (origins *red*, insertions *black*) and chief ligamentous attachments on the back of the femur.

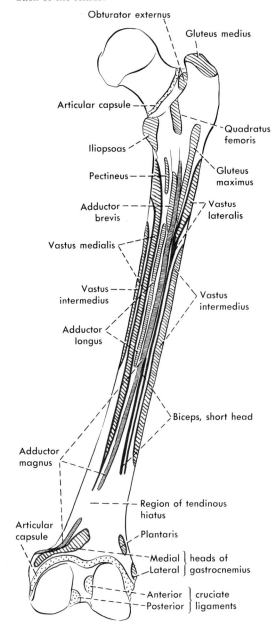

The lips of the linea aspera also diverge, toward the condyles, as the line is traced downward, and are here often called the medial and lateral supracondylar lines. They enclose between them the smooth popliteal surface of the femur, bordered below by the intercondylar line. The only other particular marking on the body of the femur is the pectineal line, which runs downward from the lesser trochanter between the gluteal tuberosity and the upper part of the medial lip of the linea aspera; it may blend with this lip, or reach the middle part of the linea.

The body of the femur is surrounded by and gives attachment to a large number of muscles (Figs. 8-15 and 8-16).

Lower End

The rounded condyles project only a little in front of the body of the femur, but markedly behind. Their long axes are not quite parallel, being separated more widely posteriorly than anteriorly, and the medial condyle projects about 0.5 cm below the lateral one when the femur is held vertically; this is a compensation for the obliquity of the femur. Anteriorly, both condyles participate in the formation of the patellar articular surface, although this is larger on the lateral than on the medial condyle. Posteriorly, they are separated from each other below the intercondylar line by the deep intercondylar fossa.

The articular surfaces of the two condyles differ somewhat from each other: that of the lateral condyle is wide, but that of the medial condyle is longer; also, the medial condyle is more highly curved posteriorly than is the lateral one. As discussed in the following chapter (see Movements at the Knee Joint), this difference in curvature produces a difference in the movement of the two condyles and accounts for the medial rotation of the femur on the tibia that occurs on full extension at the knee.

Except for the differences mentioned, the two condyles are essentially similar. Their articular surfaces, viewed from the side, are complex curves, highly curved posteriorly, much more flattened inferiorly. No parts of the curve, however, are actually segments of a

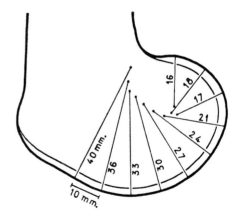

Fig. 8-17. The irregular curve of the femoral condyles. Note that the ends of the radii, instead of meeting, form part of a spiral. (Pouzet F: Genou: I. Généralités. In Ombrédanne L, Mathieu P (eds): Traité de Chirurgie Orthopédique, Vol 4. Paris, Masson et Cie, 1937)

circle; in consequence, there are no common radii for any two parts of the curve, and there is no real axis of rotation for the condyles (Fig. 8-17).

Of the epicondyles, the medial is more prominent than the lateral and bears on its upper border the adductor tubercle, which receives the lowest part of the insertion of the adductor magnus. Above and behind the lateral epicondyle is a pit for the attachment of the lateral head of the gastrocnemius, and below the epicondyle is a pit marking the origin of the popliteus muscle; from this a well-marked groove for the accommodation of the popliteus tendon when the knee is flexed curves upward toward the posterior surface of the condyle.

BLOOD SUPPLY

The blood supply of the femur, especially that of the upper end, particularly important in fractures of the neck, has been studied in considerable detail.

BODY

As is true of other long bones, the body of the femur can derive a blood supply from small

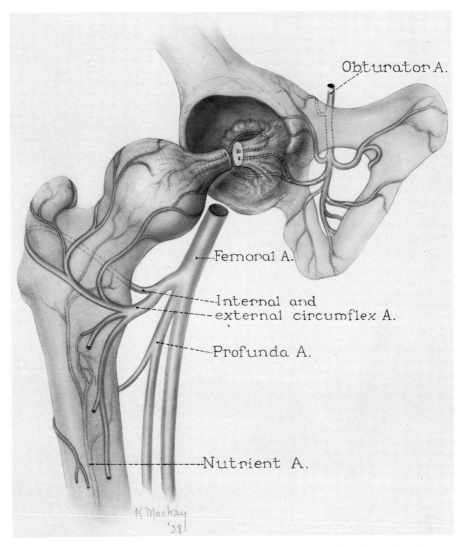

Fig. 8-18. Schematic drawing of the blood supply to the upper end of the femur. (Henderson MS: S Clin North America [Aug] 1939, p 927)

periosteal vessels, but its usual supply is from the nutrient artery or arteries. Laing found 2 nutrient arteries in 11 of 17 children less than 1 year of age, only 1 in the remaining six: however, there was only 1 in 6 of 10 adult limbs, and in 3 of the femurs with 2 vessels both entered the upper half of the femur. The lower part of the body of the femur seems, therefore, to be usually supplied in adults by a long descending branch of the nutrient artery (as is the lower part of the tibia), and Laing suggested that femoral fracture can sometimes lead to impaired osteogenesis or avascular necrosis of the lower fragment, just

as tibial fracture sometimes does. He also warned that operative stripping of muscular attachments from the linea aspera may interfere with the femur's blood supply, since the nutrient artery or arteries usually enter at or close to this line. They are usually derived from the first or second perforating branches of the deep femoral artery.

TROCHANTERS, NECK, AND HEAD

The blood supply of the upper end of the femur is through a number of branches derived from the vessels in the immediate neighborhood (Fig. 8-18). While upper branches of the nutrient artery are said (*e.g.,* Wolcott; Tucker) to anastomose with the vessels that enter the upper end directly, this source of blood supply to the upper end is apparently of no great importance: Howe and his colleagues described the nutrient artery as terminating at the upper end of the hollow part of the femoral shaft in numerous small branches, and expressed the opinion, based upon their findings at dissection, that it contributes relatively little to the blood supply of the trochanters, the neck, or the head. Trueta and Harrison were unable to demonstrate any anastomosis between the nutrient artery and the vessels at the base of the head.

Origin

The extracapsular arteries to the upper end of the femur (entering the trochanters and the base of the neck) arise from both femoral circumflex arteries, from the superior gluteal but probably not the inferior gluteal, from the obturator, and from the first perforating branch of the profunda.

Howe and his co-workers described the lateral femoral circumflex artery as sending branches lateral to the iliopsoas muscle to reach the femur at the intertrochanteric line, supplying twigs to the front of the capsule, and piercing the base of the femoral neck extracapsularly along this line. The lateral femoral circumflex artery also supplies two or three trochanteric branches to the anterior and lateral surfaces of the greater trochanter, one of them continuing around to the posterior surface of the trochanter to pierce this along with branches from the first perforating artery.

The medial femoral circumflex artery, as it passes around the femur proximal to the lesser trochanter, gives off two or three small vessels to this trochanter; as it runs between the trochanters it also supplies branches to the posterior surface of the base of the neck, and as it passes more laterally it gives off two or three branches into the upper surface of the neck near its junction with the greater trochanter (Fig. 8-19).

Howe and his co-workers found twigs from

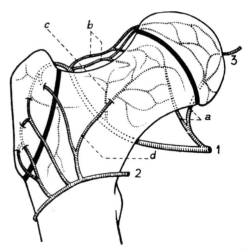

Fig. 8-19. Arteries of the femoral head and neck. *1* is the medial femoral circumflex artery, with *a* identifying the posteroinferior arteries to the femoral head and neck, *b* the posterosuperior ones, and *c* a posterior one to the neck; *2* is the lateral femoral circumflex artery, *d* an anterior branch to the neck; *3* is the artery of the ligament of the head. (After Nussbaum and Funck-Brentano, from Mathieu P: Lesions traumatiques du hanche. Fractures du col du femur. In Ombrédanne L, Mathieu P (eds): Traité de Chirurgie Orthopédique, Vol 4. Paris, Masson et Cie, 1937)

the superior gluteal artery to the upper and lateral surfaces of the greater trochanter, but apparently none to the neck. They found no branches of the inferior gluteal artery actually going into the femur, but did describe a good-sized branch from the first perforating artery as ascending along the femur deep to the insertion of the gluteus maximus and helping to supply the posterior surfaces of both the greater and lesser trochanters.

The arteries to the head and to the major portion of the neck are derived from both femoral circumflexes and, to a variable degree, from an acetabular branch that is usually described as arising from the obturator artery. Weathersby found the acetabular ramus arising solely from the obturator artery in 54.5% of 134 hip joints; arising solely from the medial femoral circumflex in 14.9%; double, one from each source, with both sending a branch to the femoral head in 6.7%; and arising from an anastomotic connection between the medial femoral circumflex and the obturator in 23.9%. Regardless of its origin, the acetabular ramus passes through the acetabular notch to supply the soft tissue in the acetabular fossa, send branches into the hip bone, and send one or more branches (artery of the ligament of the head, or of the ligamentum teres) to the head of the femur through the ligament of the head (formerly ligamentum teres).

Of the four arteries, superior gluteal, first perforating, and both femoral circumflexes, that supply the extracapsular part of the upper end of the femur, only the femoral circumflexes supply the intracapsular part. Their branches have similar courses, for they all pierce the fibrous capsule of the joint at its attachment to the femur (the intertrochanteric line anteriorly, the neck of the femur posteriorly) and run upward toward the head on the surface of the neck, deep to the synovial membrane that is reflected upward around the neck from the attachment of the fibrous capsule to the rim of the cartilage covering the head (see Fig. 8-25). Because of this course, they are liable to interruption in any intracapsular fracture.

Artery of the Ligament of the Head

The importance of this artery, the only one entering the head without first running along or through the neck, is variable, for, regardless of the origin of the acetabular branch from which it is derived, the artery of the ligament of the head is inconstant in size and distribution.

Wolcott said that the artery of the ligament of the head (artery of the ligamentum teres, also called the foveolar artery) contributes to the growth of the head in children only in the immediate area of the fovea, and does not anastomose with other arteries of the head until ossification is practically complete, at approximately the age of 15; and according to Trueta the artery does not contribute to the blood supply of the head until the age of about 8 or 9 years. In about 20% of cases Wolcott was unable to demonstrate anastomoses between this vessel and the other vessels of the head even in specimens from

adults. In these cases, therefore, the contribution from this artery was still limited to a small area adjacent to the fovea. Even when anastomotic connections are present there may be a great deal of variation in the size of the vessel, and opinion differs as to its potential importance.

Chandler and Kruescher studied 114 ligaments of the head and said that while in four of these the vessels were of precapillary size, in all the others the vessels were of significant size and presumably contributed to the circulation in the head of the femur. Tucker also described the artery as being present in every ligament, but varying so considerably in size that in about 30% of adults it could not be expected to supply any appreciable amount of the head if the circulation from below were interrupted, while in the remaining 70% it could presumably support portions varying from a very limited one near the fovea to almost all the head. Howe and his colleagues likewise found one or two small arteries constantly present in the ligament of the head, but commented that they were very small. In investigating 11 cases in which they had been unable to inject vessels in the ligament of the head in otherwise successful injections, Wertheimer and Lopez also regularly found very small vessels, usually multiple, in the ligament. Trueta and Harrison said that the artery of the ligament reached no more than one fifth of the head in 7 of 15 specimens, slightly less than half in 1, and perhaps as much as two thirds in 7. Smith ('59) found the proximal fragment in cases of fracture through the head to be apparently avascular in 4 or 24 cases, while 12 bled sluggishly and 8 bled briskly. In the latter cases, rotation in either direction, or extreme valgus, stopped the bleeding.

Ascending Vessels

The arteries that ascend along the neck to enter the neck and head of the femur have often been called "capsular vessels," although they are to the bone rather than the capsule; "capital arteries" (Howe and colleagues); "retinacular arteries" (Tucker), because they are often associated with folds or retinacula of the synovial membrane as they turn upward deep to this membrane; "ascending cervical branches" (Crock); and "metaphyseal" and "epiphyseal" arteries (Trueta and Harrison). Tucker, and Howe and his colleagues divided them into three general groups, posterosuperior and posteroinferior, both from the medial femoral circumflex, and an anterior artery or group, from the lateral femoral circumflex. (According to Ogden, the two femoral circumflexes at first participate equally in this supply, each furnishing half, but the supply from the lateral circumflex gradually regresses.) Trueta and Harrison divided the posterosuperior group into lateral epiphyseal arteries, two to six in number, entering the head superiorly and posterosuperiorly above the epiphyseal line (their "medial epiphyseal" artery is the artery of the ligament of the head), and superior metaphyseal, derived from the same vessels that continue as the lateral epiphyseal arteries, but piercing the superior aspect of the femoral neck; they called the posteroinferior group inferior metaphyseal arteries, saying they entered the bone close to the rim of the articular cartilage but below the epiphyseal line; and they apparently did not observe the anterior artery or group, which according to Howe enters the neck but not the head. After the disappearance of the epiphyseal cartilage, the metaphyseal arteries anastomose with the epiphyseal arteries so that the distinction between the two sets of vessels in the adult is not primarily one of distribution, but one of where they pierce the bone; and according to Trueta branches of the metaphyseal vessels help supply the femoral head from birth to the age of 3 or 4 years, after which they disappear. (The two to six relatively large arteries designated "upper capital arteries" by Wertheimer and Lopez and described as entering the upper surface of the femoral neck apparently correspond to Trueta and Harrison's lateral epiphyseal arteries, and the fewer and smaller "lower capital arteries," described as entering along the lower surface of the femoral neck, to the inferior metaphyseal arteries.)

Crock, pointing to the difficulties of injecting and examining these small vessels, followed none of these classifications, and particularly objected to the inferior arteries being called "metaphyseal." Although he apparently found the superior or lateral group of arteries to be particularly numerous, he found arteries ascending on all sides of the femoral neck, and all behaving similarly. They arise at fairly regular intervals from a ring of vessels at the inferior attachment of the capsule, formed posteriorly by a branch of the medial femoral circumflex, anteriorly by branches of the lateral femoral circumflex; they ascend along the femoral neck, anastomosing with each other through side branches as they do so, and each, regardless of position, ends in exactly the same fashion by dividing into a metaphyseal artery that enters the neck and an epiphyseal one that continues to enter the head. This was true both in the infant (in which Trueta also described the "metaphyseal" arteries as helping to supply the head) and in the adult. Crock suggested that the arterial supply to the neck and head of the femur be described as consisting of the arterial ring, ascending cervical arteries derived from the ring and giving rise to both metaphyseal and epiphyseal branches, and the artery of the ligament of the head. Regardless of how they are otherwise classified, there seems to be general agreement that the superior retinacular arteries form the major supply to the femoral head; Sevitt and Thompson found that they could inject nearly all the vessels of the head through these vessels.

LOWER END

Rogers and Gladstone investigated the blood supply of the distal end of the femur: Arteries to the medial condyle arise from the deep branch of the descending (highest) genicular and from the medial superior genicular, arteries to the lateral condyle chiefly from the lateral superior genicular, and all three of these vessels supply the anterior and posterior supracondylar regions. The vessels entering the intercondylar fossa are derived chiefly from the middle genicular. Since the vessels enter laterally, anteriorly, and posteriorly,

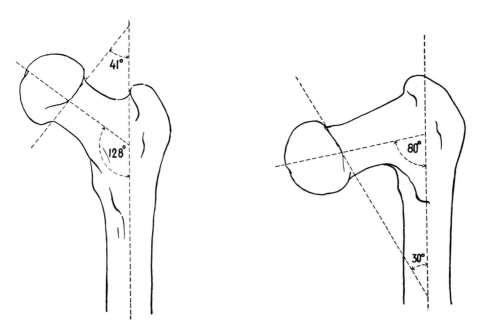

Fig. 8-20. Femur showing coxa vara (*right*) compared with a normal femur. (Lance M: Coxa vara. In Ombrédanne L, Mathieu P (eds): Traité de Chirurgie Orthopédique, Vol 4. Paris, Masson et Cie, 1937)

and are very numerous, Rogers and Gladstone concluded that it is unlikely that any fracture here could cause necrosis.

ANOMALIES, FRACTURE, AND EPIPHYSEAL SEPARATION

Congenital anomalies of the femur are relatively infrequent.

Absence
Absence or partial absence of the femur occurs occasionally. Among the major long bones of the upper and lower limbs, O'Rahilly placed the femur third in order of frequency in regard to deficiencies, following the fibula and the radius, but preceding the tibia, the ulna, and the humerus.

Coxa Vara
Coxa vara is a decrease in the angle of obliquity that the femoral neck makes with the shaft, so that the normal angle of 125° to 130° (measured inferiorly) may in the adult approach 90°, or be even less (Fig. 8-20). Since

this produces a shortening of the distance between the acetabulum and the intercondylar plane on the lower surface of the femur, the condition leads to a limp not unlike that of congenital dislocation; as the head sinks, the greater trochanter rises until abduction becomes impossible because of impingement of the trochanter against the ilium.

The most common type of coxa vara apparently develops as a result of disease conditions that weaken the femoral neck; the rarer type, usually reported as congenital coxa vara, becomes evident during the early years of life (Le Mesurier; Babb and co-workers). Babb and his colleagues agreed with others that the origin of so-called congenital coxa vara is unknown, and suggested that there might be two types included under this clinical entity, one being a true congenital lesion accompanied by multiple deformities elsewhere, the other a disturbance in ossification confined to the upper portion of one or both femurs. They noted that although the condition is frequently associated with an abnormally short femur, it is not always so associated, and expressed the opinion that it has not been proved to be congenital. Le Mesurier, and Babb and his colleagues discussed methods of treatment for this crippling condition.

FRACTURE

Fracture of the femur may occur at any level, but fracture of the neck, anywhere between the head and the base, is particularly frequent both because this is the narrowest part of the femur and because of the strain upon this part in all weight-bearing positions. Fracture of the neck is more likely to occur in elderly people, either as a result of a fall or, perhaps, as fatigue fractures. Griffiths and co-workers produced fatigue fractures in bones from elderly persons by applying loads within the range of normal activity, and thus supported the concept that at least sometimes a fracture leads to a fall, rather than the reverse.

Fractures of the upper end of the femur have been variously classified. The simplest classification is to divide them into extracapsular and intracapsular ones, although a few fractures, either because they are sufficiently oblique or are so placed on the neck as to be intracapsular anteriorly, extracapsular posteriorly, fall strictly into neither category. Intracapsular fractions, in turn, may be close to the head (subcapital), or through the neck proper (transcervical or midcervical). The highest extracapsular fracture is at the very base of the neck along the intertrochanteric line and hence is called "intertrochanteric" or "cervicotrochanteric." These three types are shown in Figure 8-21. Fractures through the greater trochanter extending through, proximal, or distal to the lesser trochanter are called "pertrochanteric" or "transtrochanteric."

Evans has reviewed the work in which he and his colleagues produced the various subtypes of fracture of the upper end of the femur—subcapital, oblique and transverse fractures of the neck (see Fig. 8-23) and trochanteric region—by applying dynamic or static loading to the greater trochanters of femurs supported by their heads and medial condyles. They also investigated deformation

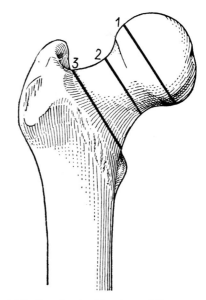

Fig. 8-21. Simple classification of fractures of the femoral neck. *1* is a subcapital, *2* a transcervical, and *3* an intertrochanteric or cervicotrochanteric fracture. (Mathieu P: Lesions traumatiques du hanche. Fractures du col du femur. In Ombrédanne L, Mathieu P (eds): Traité de Chirurgie Orthopédique, Vol 4. Paris, Masson et Cie, 1937)

of the femur under vertical and transverse loading. They emphasized that variations in the position of the bone as it was subjected to force could produce all the variations in fractures found clinically and that no tortional forces, which have been said to be responsible for certain fractures, were present in their experiments. In accordance with the fact that the tensile strength of bone is regularly less than its compressive strength, they found that failure of the bone under load occurred on the side of tensile strain—the convexity of the bone as it began to bend.

In unimpacted fractures through the upper end of the femur, whether through the neck or the trochanteric region, there is, of course, a tendency for the lower fragment to override the upper fragment. Thus a varus deformity persists after healing if the bone has not been properly set and, as in dislocation of the hip, the upper end of the greater trochanter tends to ride above its normal level. The amount of shortening consequent to upward displacement of the greater trochanter, whether from

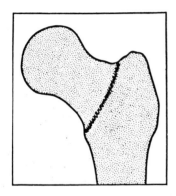

Fig. 8-22. Transtrochanteric fracture of the femur. (Perkins G: Fractures. London, Oxford University Press, 1940)

fracture or dislocation, is often gauged by the relationship of the tip of the trochanter to Nélaton's line, Bryant's line, or Bryant's triangle.

Nélaton's line is a line drawn between the anterior superior iliac spine and the ischial tuberosity; since this line usually crosses the tip of the greater trochanter, the distance of the trochanteric tip above it is a measure of the shortening of the limb. *Bryant's line* is a vertical one dropped to the highest point on the greater trochanter from a horizontal plane through both anterior superior iliac spines; it is usually about 5 cm long, and shortening of it on one side of the body indicates upward displacement of the trochanteric tip.

The use of *Bryant's triangle* does not differ essentially from the use of Bryant's line, except that it is constructed with the patient supine. The base of Bryant's triangle is actually Bryant's line—in this case, because of the position of the patient, a horizontal line drawn from the tip of the greater trochanter to a perpendicular line erected to the anterior superior iliac spine; the hypotenuse of the triangle is the part of Nélaton's line between the anterior superior iliac spine and the upper tip of the trochanter. With upward displacement of the trochanter, it is the base of the triangle that is shortened.

Extracapsular Fractures
Into this class fall fractures of the trochanters, and fractures at the base of the femoral neck

or the uppermost part of the body, whether they are called trochanteric, trans-, inter-, or subtrochanteric. They are characterized by the fact that they occur in spongy bone that has an abundant blood supply from several directions, and therefore present no particular problem in healing if apposition of the parts is obtained.

Fracture of the greater trochanter in adults is usually a result of a direct blow or of a fall on the trochanter, although it may apparently be caused by muscular action alone. It may involve various-sized pieces of the bone; in adolescents, separation is likely to occur at the epiphyseal line. In either case, since the muscles inserting on the trochanter tend to pull the fragment upward and inward, apposition can usually best be obtained by abducting the limb. Because of the abundant blood supply of the greater trochanter, there is typically no difficulty in healing.

Fractures of the lesser trochanter are usually not isolated ones, but occur rather in fracture of the trochanteric region. Isolated fracture of the lesser trochanter may apparently occur as a result of hyperextension and abduction at the hip with unusual strain thrown on the iliopsoas muscle. The injury in adolescents may be at the epiphyseal line, as in the case of the greater trochanter; Lapidus said that while epiphyseal separation of the lesser trochanter is comparatively rare, it is most commonly seen in boys of 13 to 17 years of age. Apposition is best accomplished by flexing the hip and rotating it slightly outward.

Fractures at the base of the neck (Fig. 8-22) or through the trochanters also lie in spongy bone, and therefore are likely to have an adequate blood supply. They may present difficulties in securing and maintaining reduction, however, and Boyd and Griffin have classified trochanteric fractures into four types, according to the ease or difficulty of these procedures.

Intracapsular Fractures
It is evident from the course of the vessels to the neck and head of the femur that intracapsular fractures always endanger the blood supply of the upper fragment. Claffey

could find no relationship between tear of the "inferior metaphyseal vessels" associated with displaced fracture of the femoral neck and avascular necrosis, but said fracture across the line of entrance of the lateral epiphyseal vessels into the bone always produced necrosis. Clinical experience apparently agrees with anatomic observations that the blood supply through the vessels of the ligament of the head varies. Even in subcapitate fractures, where the tissue to be supported is limited to the head, the vessels entering the fovea may be entirely inadequate to support the head, with consequent necrosis of it.

Both because of the variability in the size and distribution of the artery of the ligament of the head, and because it may be impossible to know from the usual roentgenograms the extent to which capsular vessels have been interrupted, it is often impossible to estimate the chances of reunion versus nonunion of the upper fragment. Rook found arteriograms of the region of the hip joint useful for establishing the adequacy or inadequacy of the blood supply to the proximal fragment (from capsular vessels: among six ligaments of the head examined by dissection he found no macroscopic vessels, and apparently saw none in arteriograms). Boyd and Calandruccio have used the uptake of radioactive phosphorus by the head as an indication of its vascularity and the possibility of subsequent avascular necrosis.

Catto's study of avascular necrosis indicated that although a partially necrotic femoral head can be revascularized by growth of arteries from the artery of the ligament of the head and across the fracture line, completely nectrotic heads do not revascularize completely and are subject to late collapse. Gordon, however, expressed the belief that improper weight distribution resulting from malalignment of the capital fragment, rather than avascular necrosis, is the usual cause of late collapse. Frangakis emphasized the importance of avoiding rotation or other malposition in fixing the fracture in order to avoid damage to the artery of the head, and the experiments of Hayes and Groth on dogs indicate that marked rotation of the head pro-

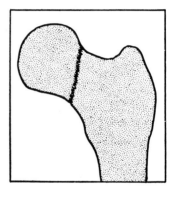

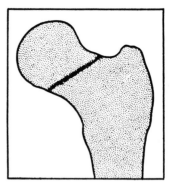

Fig. 8-23. Transcervical fractures; the upper is usually said to be the adduction type, the lower the abduction type. (Perkins G: Fractures. London, Oxford University Press, 1940)

duces avascular necrosis and fibrous nonunion.

A common subdivision of fractures of the neck of the femur is into abduction fractures and adduction fractures. The original classification was presumably upon the basis of the position of the limb at the time of fracture; the term has apparently been extended to include the position of the upper fragment at the time of examination. Perkins said that in adduction fractures the fracture is complete, runs more or less vertically across the neck, and there is no impaction, so that displacement can be expected (Fig. 8-23 *top*). He said that in abduction fractures the line of fracture runs more or less horizontally (Fig. 8-23 *bottom*), there is impaction, and repair usually presents no particular problem. (Pauwels' classification of transcervical fractures into three groups is similarly based upon the relation of the line of fracture to the thrust on the femoral head.) Trueta and Harrison explained the reported greater incidence of necrosis of the femoral head following unimpacted adduction fractures of the femoral neck as compared with abduction fractures as being due to the fact that the "lateral" (posterosuperior) group of arteries, which are the chief supply to the femoral head in most persons, are most likely to be torn in adduction fractures, while in abduction ones they have a better chance of remaining uninjured.

Linton said that so-called abduction, adduction, and intermediate fractures are actually stages in one process and that the type of injury and the direction of force producing the injury have nothing to do with the final positions of the fragments. He said, further, that impaction is simply a first stage of displacement and that one type of fracture may pass into another: a head that is at first in slight abduction, for instance, may gradually rotate within the socket until it lies posterior to the femoral neck, in adduction.

Fractures of the Body

In contrast to fractures of the neck of the femur, fractures of the body are usually brought about by external violence, and the majority occur in young adults, especially males, and in children. Since they are usually produced by violence, there is likely to be a great deal of damage also to soft tissues.

Fracture of the body of the femur is most commonly in the middle third, but may be either higher or lower. Overriding of the fragments and shortening of the limb usually occur, and the direction of displacement of the parts may be the result primarily of muscular pull. In general, there is a tendency for the proximal fragment to be rotated outward both as a result of gravity and of the pull of the external rotator muscles, which are stronger than the internal rotators; other than this, the displacement, if it results from muscular pull, varies with the level of fracture.

In fracture through the upper third the upper fragment tends to be flexed, primarily through the pull of the iliopsoas muscle; abducted, especially by the gluteus medius and minimus; and externally rotated by the pull of the short rotators and by gravity. The greatest degree of flexion and abduction is likely to occur in a subtrochanteric fracture, where the upper segment is very short. The lower segment may be adducted by the pull of the adductor muscles (Fig. 8-24).

Displacement in fracture of the middle third of the shaft varies much, since fractures here do not primarily fall betweeen attachments of various groups of muscles, but rather occur at a site where the fracture line lies across the attachments of a number of muscles. There is, therefore, no characteristic muscle pull, except for overriding. It is said that either fragment may lie in front of the other, and that either fragment may lie lateral to the other. Also, there is some tendency for the proximal fragment to be abducted (by the glutei) when the fracture is in the upper half of the middle third, and for it to be adducted (by the adductor group of muscles) when the fracture is in the lower half of the middle third.

In fracture of the lower third, the chief displacement other than shortening is brought about by the muscles arising from the posterior surface of the lower fragment, therefore especially by the gastrocnemius. The pull of this muscle, the plantaris, and the popliteus tends to flex the lower fragment at the knee, with the obvious danger that the jagged edge may be forced back against the popliteal vessels (the artery, lying in front of the vein, would be most endangered). Ischemic contracture of the lower extremity (Volkmann's contracture) is apparently usually due to extensive injury in the region of the knee (Horwitz, '40), although it is not common; Seddon reported and discussed the treatment of 15 cases of ischemia of the limb as a result of femoral fracture and other causes. Neer and co-workers found only one instance of arterial injury among 110 cases of supracondylar fracture. Fracture of the lower third of the femur may or may not involve the condyles; the T fracture is one in which a vertical intercondylar fracture joins a transverse one above the condyles.

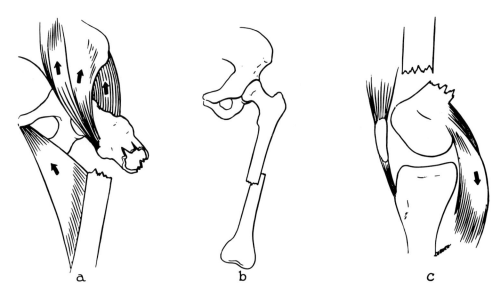

Fig. 8-24. Typical displacement in fractures of the upper, middle, and lower parts of the shaft of the femur. In *a* the upper fragment is flexed, abducted, and externally rotated, the lower fragment adducted, and the limb shortened; in *b* overriding of the fragments is shown; and in *c*, a lateral view, the lower fragment is flexed. (Ivins JC: Modern Medicine Annual, 1951, p 222)

EPIPHYSEAL SEPARATION AND SLIPPING

Separation of the greater or lesser trochanter from the body of the femur at the epiphyseal line is actually a type of fracture, and has already been mentioned with fracture of these parts.

Separation of the head at the upper femoral epiphyseal plate may likewise be purely traumatic, resulting from severe injury, but is probably very often the result of a minor strain superimposed upon an abnormal epiphyseal line: that is, it is the end result of slipping of the upper femoral epiphysis, discussed in the following section.

Separation at the lower femoral epiphyseal plate again belongs in the category of fracture. Tees said that it usually results from hyperextension at the knee, and the lower fragment is commonly displaced forward. Since the separation is below the attachments of the gastrocnemius and associated muscles, there is no muscle pull to displace the lower fragment backward.

Bellin reported that traumatic separation of the lower epiphysis of the femur is rather rare, but found approximately 200 cases recorded by 1937. As in the case of supracondylar fracture of the femur, there may be severe injury to the vessels or nerves in the popliteal fossa, but in this case, it is through the displacement of the lower end of the body. Flexion of the knee, by reducing the pull of the gastrocnemius muscle on the lower part of the body, is said to allow ready apposition of the parts.

Slipping of Upper Femoral Epiphysis

Slipping of the upper femoral epiphysis may occur in children and adolescents before the upper epiphyseal cartilage has disappeared; the weight on the head produces downward and posterior sliding of the head upon the neck through a weakened cartilage. As in fracture of the neck, this allows a relative upward displacement of the tip of the greater trochanter, and produces a varus deformity. Howorth ('49), among others, has discussed surgical aspects of this question.

As also in fractures of the neck, slipping of any great degree may interrupt the vessels to the head that lie on the posterosuperior or lateral surface of the neck, leaving for circulation to the head only the less important postero-inferior vessels and the vessels of the ligament of the head. Acute separation may occur as the result of trauma, but probably in a great majority of cases such separation occurs only after the basic pathologic changes essential to slipping have been present.

The exact pathologic basis of slipping has apparently not been determined although it is generally agreed that slipping results from weakening of the cartilage. Howorth ('41) suggested that decalcification and softening of the cartilage result from circulatory changes produced by synovitis occurring during a period of rapid growth; Lacroix and Verbrugge said that the essential lesion is a transformation of epiphyseal cartilage into fibrous tissue instead of into bone; Harris pointed out that slipping of the upper femoral epiphysis rather regularly appears only in children who show abnormalities of growth that are apparently caused by some endocrine disorder, and regarded an endocrine disturbance as being responsible for the condition.

Alexander regarded clinical slipping as being an exaggeration of a usual process of slow posterior migration of the femoral head brought about by the weight that is borne by the femur in the sitting posture.

Hip Joint

Although both are ball-and-socket joints, the hip joint stands in strong contrast to the shoulder joint in a number of respects, all of them designed to increase the strength of the hip joint even though they necessarily limit mobility. The head of the femur forms considerably more than half a sphere, in contrast to the small fraction of a sphere formed by the head of the humerus, and is moreover deeply embedded in the acetabular socket; the fibrous capsule of the hip joint is particularly strong, in contrast to that of the shoulder joint; and the muscles that cross the hip joint, instead of inserting close to the head, insert some distance from the head, and thus obtain greater leverage in strengthening the joint.

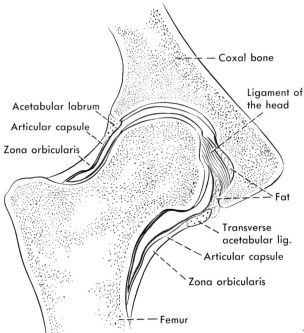

Coxal bone

Ligament of the head

Acetabular labrum

Articular capsule

Zona orbicularis

Fat

Transverse acetabular lig.

Articular capsule

Zona orbicularis

Femur

Fig. 8-25. Frontal section through the hip joint. The synovial membrane is represented by *red.*

The bony acetabulum is deepened by the acetabular (once glenoidal) labrum, attached to its rim; this is a fibrocartilaginous lip to the acetabulum (Fig. 8-25). Its free edge is slightly smaller than its base, so that the acetabulum accommodates slightly more than half of a sphere, and the head of the femur cannot be withdrawn without stretching or rupture of the labrum.

At the acetabular notch in the lower part of the acetabulum the labrum itself ceases, but the ring is completed by the transverse ligament of the acetabulum, a heavy fibrous band continuous with the labrum. The transverse ligament is attached only at its ends, for its proximal or deep edge does not reach the bottom of the notch; thus a foramen is left between the ligament and the bone. Through this foramen the acetabular artery enters to supply the fat in the floor of the acetabulum and give rise to the vessels of the ligament of the head of the femur.

The flattened ligament of the head of the femur (once known as the ligamentum teres or round ligament, in spite of its shape) attaches to both sides of the acetabular notch and to the floor of the acetabular fossa deep to the transverse ligament. From this origin the fibers of the ligament converge to form a flattened band that passes medially and upward around the articular surface of the head to attach into the upper part of the fovea of the head. As this ligament traverses the hip joint it is surrounded by a funnel-like sheath of synovial membrane, continued upward from the synovial membrane covering the fat in the acetabular fossa.

The development of the hip joint has been described by Gardner and Gray. Surgical approaches to the joint have varied markedly, from anterolateral ones through lateral, posterolateral, and posterior ones (*e.g.,* Osborne; Campbell; Naffziger and Norcross; Horwitz, '52; and Burwell and Scott).

ARTICULAR CAPSULE

Synovial Membrane
The synovial membrane, where it is in contact with the fibrous capsule, is closely attached to this capsule, but at both the acetabular and the femoral ends of the joint it separates in part from the fibrous capsule. Proximally it is, for the most part, attached to the acetabular rim, but it passes across the inner surface of the transverse acetabular ligament to cover the fat extending inward from the floor of the acetabular notch, and to attach to the edges of the acetabular fossa. It is this portion that is reflected up loosely around the ligament of the head of the femur to attach to the edges of the fovea capitis, the nonarticular portion of the head. Distally, on the neck of the femur, the synovial membrane and the fibrous capsule reach anteriorly, superolaterally, and inferomedially to the very base of the neck, so that on these aspects the entire neck is intracapsular; posteriorly, however, they reach less far distally, and approximately the lower third of the neck posteriorly is extracapsular (Fig. 8-26).

As the synovial membrane reaches the neck of the femur it is reflected proximally along it as far as the articular cartilage covering the head. Some of the inner fibers of the fibrous capsule are likewise reflected upward for a variable distance upon the neck, but since these inner fibers are not uniformly so reflected they form retinacula, projections of the fibrous capsule covered by the synovial layer, which are visible when the joint is opened. The vessels for the neck and head of the femur tend to run in these retinacula, hence the name sometimes applied to them, retinacular vessels.

Fibrous Capsule and Associated Ligaments
The fibrous capsule is composed of fibers that for the most part run approximately longitudinally between the pelvis and the femur, but there are also some deeper fibers that run circularly. On the posterior aspect of the capsule, at the back of the femoral neck, these circular fibers become better marked and also superficially located, since the longitudinal fibers are deficient here; the circular fibers constitute the *zona orbicularis,* usually regarded as one of the four special ligaments at the hip joint—but, as noted, readily demonstrable only posteriorly.

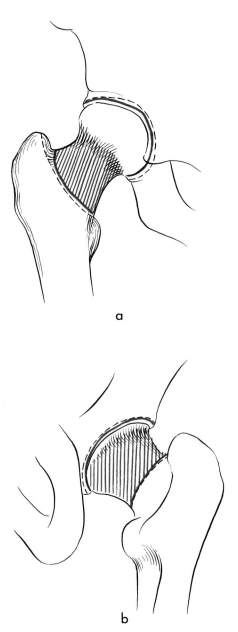

a

b

Fig. 8-26. Anterior and posterior views of the attachment of the capsule of the hip joint. The broken *black* lines represent the attachment of the fibrous capsule, and the *red* lines and the *red shaded* areas represent the lines along which the synovial membrane first reaches the bones and its reflection over the bone to the articular cartilage.

The other three special ligaments at the hip joint are marked thickenings of the longitudinal fibers of the capsule. According to their attachments to the pelvis, they are known as the iliofemoral, ischiofemoral, and pubofem-

oral ligaments (Fig. 8-27), sometimes also called iliocapsular, ischiocapsular, and pubo-capsular.

The *iliofemoral ligament* is rather thick, and probably one of the strongest ligaments in the body. It is somewhat triangular, with its apex attached to the lower part of the anterior inferior iliac spine and to the body of the ilium between this spine and the acetabular rim; its base is attached along the intertrochanteric-line, therefore across the whole anterior aspect of the hip joint. The fibers of this ligament spiral somewhat medially as they pass downward and tend to separate into two bands with a relatively weak place between them; for this reason the iliofemoral ligament may resemble an inverted Y, and it has been called the Y ligament of Bigelow. In consequence of its attachments the iliofemoral ligament occupies the entire anterior aspect of the hip joint, except for an upper medial portion. It strongly resists hyperextension at the hip joint, a function particularly important because in the usual standing position the weight of the body tends to roll the bony pelvis backward upon the heads of the two femurs.

The *ischiofemoral ligament*, the thinnest of the three, arises from the body of the ischium behind and below the acetabulum. Its upper fibers are directed horizontally, its lower fibers spiral upward and laterally, and all attach on the upper and posterior part of the neck at the junction with the greater trochanter. The uppermost fibers blend with the uppermost fibers of the iliofemoral ligament. In flexion at the hip the fibers of the ischiofemoral ligament tend to be relaxed, since the fibers then run almost transversely, but in extension, as the trochanter is carried upward, they assume their spiral course and are therefore tightened. Increased stability of the hip joint in extension has been said to be caused by an inward movement of the head of the femur produced by this tightening of the ischiofemoral ligament; Roberts, however, denied that there is an inward movement. He found that the femur is abducted and medially rotated during extension, and this tightens both the ischiofemoral and iliofemoral ligaments.

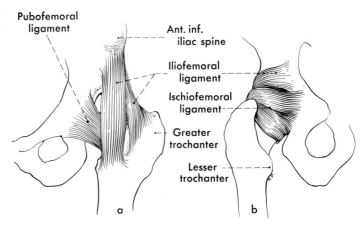

Fig. 8-27. Chief ligaments of the capsule of the hip joint from in front, *a*, and behind, *b*.

The *pubofemoral ligament* arises from the body of the pubis close to the acetabulum and from an adjacent part of the superior ramus. It passes distally in front of the lower part of the head of the femur, and below it, to blend anteriorly with the lower limb of the iliofemoral ligament and attach to the lower surface of the femoral neck. The pubofemoral ligament is especially put under stress in abduction of the thigh, and therefore assists the adductor muscles in preventing excess abduction.

The capsule of the joint is rather thin between the medial border of the lower limb of the iliofemoral ligament and the upper border of the pubofemoral ligament; the tendinous deep surface of the iliopsoas is adjacent to it here. Between the iliopsoas and the capsule there is typically a bursa (iliopectineal, iliopsoas) and in approximately 15% of adults the thin capsule is perforated, so that the cavity of the hip joint and the bursa are continuous with each other.

The other particularly thin place in the capsule of the hip joint is situated posteriorly. Because the posterior fibers of the pubofemoral ligament insert inferiorly on the neck, while the inferior fibers of the ischiofemoral ligament, arising adjacent to those of the pubofemoral, insert superiorly on the neck, the posterior surface of the neck receives few longitudinally coursing fibers; this is the location in which the fibers of the zona orbicularis appear. Between the zona orbicularis and the back of the femoral neck there may be a slight outpouching of the capsule; this serves as a

bursa for the obturator externus muscle as it crosses the posterior surface of the neck (the lower part of which is extracapsular) to reach its insertion in the trochanteric fossa.

NERVES AND VESSELS

Nerves

The innervation of the hip joint is from a number of different nerves, including the femoral nerve or one or more of its muscular branches, the obturator, the accessory obturator when present, sometimes one from the superior gluteal, and one or two from the nerve to the quadratus femoris (Figs. 8-28 and 8-29). Gardner commented that there tends to be less overlap than is found in other major joints, and that many of the branches seem to be primarily vascular ones. The femoral branches are distributed largely to the region of the lower part of the iliofemoral ligament, with some distribution also to a posterosuperior part of the capsule and the region of the pubofemoral ligament; the obturator and accessory obturator are distributed also to the region of the pubofemoral ligament; the branch of the superior gluteal goes to a superior and lateral part of the capsule; and the branch from the nerve to the quadratus femoris goes to the sparsely innervated posterior part of the capsule. Peterson and co-workers, reporting on the nerve endings in the capsule of the joint in the cat, found the highest density of endings in the upper part of the capsule, but no parts that could be described as sparsely innervated.

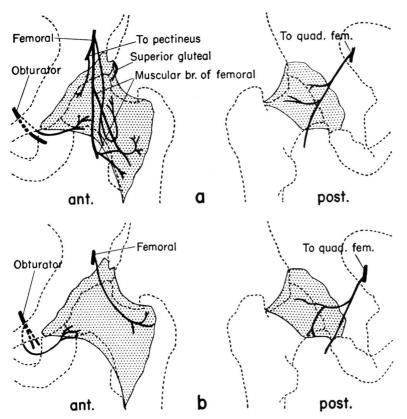

Fig. 8-28. Anterior and posterior views of two types of distribution of nerves to the hip joint. In *a*, *ant.*, are seen twigs to the joint from the nerve to the pectineus and from the superior gluteal nerve. (Original of Fig. 1, *a-b*, Gardner E: Anat Rec 101:353, 1948)

Wertheimer apparently found no femoral branches going directly to the capsule of the joint, but did find twigs accompanying vessels; among 50 hips, he found a branch from an accessory obturator nerve in 1, 1 or more branches from the obturator in the remainder; 1 from the nerve to the quadratus femoris in all of 53 cases in which he sought it; and in 2 of the latter, an additional branch directly from the sciatic nerve. Larochelle and Jobin found a branch from the nerve to the quadratus in only 90 of 106 specimens.

The origin of the branch or branches from the obturator is reported to be particularly variable. Gardner found the usual origin to be from the main nerve as it lay in the obturator canal. Kaiser found it arising more frequently in the pelvis than in the canal. Wertheimer found it proximal to the canal in 40%, and

arising from the obturator trunk in almost 72%, from the posterior branch in 25%, and from the anterior in 3%. In contrast, Larochelle and Jobin found origin from the main trunk in only 10 of 106 dissections; there was an origin from the anterior branch in 70%, one from the posterior branch in 17%, and no obturator branch in 16%.

Neurectomy for Painful Hip

Attempts to relieve pain from the hip joint by neurectomy have been largely directed at the obturator nerve alone, or the obturator nerve and the nerve to the quadratus femoris, with apparently somewhat variable results (perhaps not to be wondered at, since the operation ignores the rather inaccessible branches of the femoral nerve to the hip joint, and it is apparently often assumed that the nerve supply from the obturator is by far the most important one). The nerve is more easily found and the articular branch more likely to be severed if the operation is an intrapelvic (extraperitoneal) one.

Liebolt and co-workers reported prompt relief from pain in 10 of 12 patients in whom the obturator nerve had been divided, and said that only 4 of these were aware of the weakness in the adductor muscles postoperatively; these authors were unwilling to commit themselves on the permanent value of the operation. Obletz and co-workers reported that satisfactory immediate relief of pain, lasting for at least as long as 20 months after operation, was obtained in 28 of 42 patients in whom both the obturator nerve and the nerve to the quadratus femoris were sectioned. Mulder reported that section of the posterior branch of the obturator nerve or re-

section of the obturator trunk, combined with resection of the nerve to the quadratus femoris, had given good results in the relief of painful osteo-arthritic hip joints in 11 of 44 patients, fair results in another 11, and no improvement in 22; however, apparently only 2 of the men so operated upon had been able to resume their work, and he said he had discarded denervation of the hip joint as an independent operation.

Blood Supply

The blood supply to the hip joint is from branches of most of the vessels in its neighborhood, especially from the medial femoral circumflex, the lateral femoral circumflex, the obturator artery, and the superior and inferior gluteals; the first perforating branch of the profunda femoris usually also sends a branch upward to the hip joint. The blood

Fig. 8-29. Other types of innervation of the hip joint. In *a, ant.,* a branch to the joint from the accessory obturator nerve is shown. (Original of Fig. 1, *c-d,* Gardner E: Anat Rec 101:353, 1948)

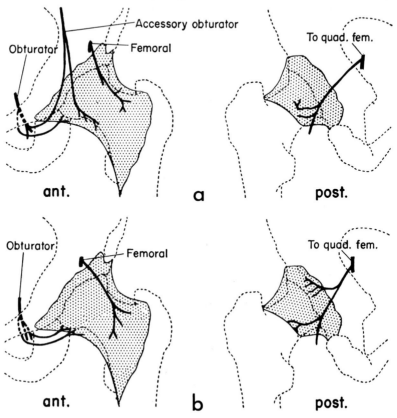

supply to the hip bone and that to the upper end of the femur have already been described, and most of the vessels to the capsule of the joint are twigs from the branches that supply the upper end of the femur.

Briefly described, the vessels to the upper end of the femur, primarily from the two femoral circumflex arteries, give small twigs to the lower part of the capsule of the joint; and the acetabular branch (of the obturator or medial femoral circumflex, or both), besides its branch to the head of the femur, supplies the tissue of the acetabular fossa. Branches from the superior gluteal supply an upper portion of the acetabulum, an upper portion of the fibrous capsule of the hip joint, and a small portion of the greater trochanter; branches of the inferior gluteal artery supply inferior and posterior portions of the acetabular rim and the adjacent fibrous capsule, but apparently do not enter the femur. Twigs from the branch of the first perforating artery to the hip joint supply the posterior surfaces of the two trochanters and the adjacent capsule.

STABILITY AND MOVEMENTS AT THE HIP JOINT

The entire weight of the trunk is supported by the two rounded femoral heads; in slight flexion of the hip joints, when the line through the center of gravity lies in front of them, the pelvis is prevented from rolling downward on the femoral heads through the action of the posterior hamstrings; the more powerful gluteus maximus apparently does not function at all as a postural muscle (Joseph and Williams). The stable position of the hip joint is reached only when the thigh is extended; then the line of gravity of the body passes slightly behind the hip joint, so that the posterior muscles can relax, and the strain is thrown primarily on structures at the front of the joint, especially the iliofemoral ligament and the covering iliopsoas muscle. The iliacus part of the iliopsoas apparently does function as a postural muscle, for it shows continuous slight to moderate activity during standing (Basmajian, '58); however, the psoas major does not

(Keagy and colleagues). Stability of the hip joint in extension is increased by the tightening of the ligaments that occurs as the limb is extended.

As a ball-and-socket joint, the hip joint allows all the movements typical of such a joint: flexion and extension, abduction and adduction (therefore also circumduction), and external and internal rotation. Because of the structure of the hip joint and its covering muscles, however, some of these movements are much more limited than others, for in the standing position the thigh is extended, adducted, and inwardly rotated from its midposition of movement. Thus, from the extended position, flexion is rather free, amounting perhaps to about 120°, and is primarily checked, if the hamstrings are relaxed through flexion at the knee, by apposition of the soft tissues. Hyperextension is usually said to be limited to 20°. However, Mundale and co-workers found that the maximal extension of the femur on the pelvis, even in marked hyperextension of the trunk, was 178°, therefore less than complete extension. Extension is checked by the pressure of the head of the femur against the structures lying in front of the joint—the iliofemoral ligament primarily, and secondarily the tendon of the iliopsoas muscle.

Adduction in the extended position with the heels together is of course impossible, and with the hip flexed is limited to perhaps 20° to 25°, as in crossing the knees, by the apposition of soft tissues, and possibly by the upper fibers of the iliofemoral and ischiofemoral ligaments. From its anatomy, the only ligamentous function of the ligament of the head (ligamentum teres) should be to help check this action. However, this ligament appears to have little function in the adult, for it is said that it never becomes very taut even in a skeletal preparation in which adduction is halted by contact of the femoral neck and the acetabular brim. It appears, however, that excessive length of the ligament in the fetus and newborn may be a factor in the instability that sometimes results in congenital dislocation (Crelin).

Medial rotation is also very limited, for the

iliofemoral and ischiofemoral ligaments run on such a spiral that the former particularly resists medial rotation from the extended position, the latter medial rotation from the flexed position.

In contrast, lateral rotation and abduction are relatively free; varying data have been given for these movements, but Boscoe found a range in abduction from 59.6° to 23.6°, with an average of 35° to 36°, in a group of university students, and a range in rotation from 69.3° to only 25.6°, with an average of from 50.8° to 51.3°. Presumably, abduction is limited by the pubofemoral ligament and the adductor muscles, and possibly by contact between the neck of the femur and the acetabular rim; external rotation is probably stopped by contact of the posterior aspect of the neck of the femur with the rim of the acetabulum.

Although, unlike the humerus, the femur cannot rotate around its long axis, there is necessarily rotation of the head of the femur within the acetabulum; because of the angle between the neck and the body, this rotation is greater in flexion and extension than it is in other movements.

Muscles Acting at Hip Joint

The muscles acting at the hip joint are numerous. Their actions are discussed in more detail in the following descriptions of the individual muscles and in the final section of this chapter on movements at the hip and knee.

Of the *extensors* (Fig. 8-30), the gluteus maximus and the sciatic or hamstring part of the adductor magnus (that arising from the ischial tuberosity and innervated by the sciatic nerve) are equally effective regardless of the position of the knee; the hamstrings proper (semimembranosus, semitendinosus, biceps) contribute most when the knee is extended, but this contribution apparently varies with the degree of flexion at the hip. Waters and co-workers found that the strength of extension at the hip was greatest when the hip was flexed to 90°, and that while flexion at the knee produced no change in strength in this position, it did produce a

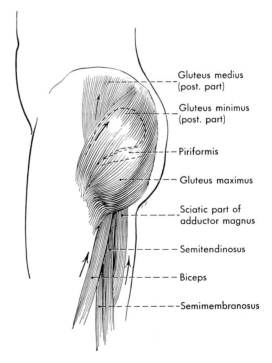

Fig. 8-30. Extensors of the thigh. According to recent evidence, the gluteus minimus may not belong here.

loss of from 18% to 12% when hip flexion was 45° or less. Block of the sciatic nerve indicated that the hamstrings contribute from 31% to 48% of the total strength of extension, depending on the angle at which the hip is flexed.

Joseph and Williams emphasized that the hamstrings become active as soon as the center of gravity is shifted forward, as in swaying forward from a standing position or bending over, and remain active throughout the flexion and extension of the hip during toe touching; in contrast, the powerful gluteus maximus becomes active only during the final flexion of toe touching and relaxes shortly before the erect posture is reached. Posterior parts of the gluteus medius and minimus have been said to assist in extension, but the latter apparently assists only from a flexed position, as is possibly true also of the small piriformis. The innervation of these muscles is shown in Table 8-1.

The chief *flexor* at the hip joint is the iliopsoas, and the pectineus is also primarily a

Table 8-1
Innervation of the Extensors of the Thigh

	MUSCLE	NERVE	SEGMENTAL INNERVATION Reported Range	Probable Most Important Segments
Chief extensors	Gluteus maximus	Inferior gluteal	L4-S3	L5, S1, 2
	Adductor magnus (post. part)	Tibial	L3-S3	L4, 5
Accessory extensors	Semimembranosus	Tibial	L4-S3	L5, S1
	Semitendinosus	Tibial	L4-S3	L5, S1
	Biceps femoris (long head)	Tibial	L4-S3	L5, S1, 2
	Gluteus medius	Superior gluteal	L4-S2	L4, 5, S1
	Gluteus minimus (from flexion)	Superior gluteal	L4-S1	L4, 5, S1
	Piriformis	N. to piriformis	S1-S3	S1, 2

flexor here. Most of the other muscles attaching to the front of the pelvis can also help in flexion—the tensor fasciae latae, the sartorius, the adductor brevis, the adductor longus, the anterior part of the adductor magnus (that part innervated by the obturator nerve), the gracilis (if the leg is not also flexed) and the anterior part of the gluteus minimus (Fig. 8-31). The rectus femoris is a much better flexor of the thigh when the knee also is flexed. The innervation of the flexors is shown in Table 8-2.

The *adductors* of the thigh are in part anteriorly placed, in part posteriorly (Fig. 8-32); the best ones are the three named "adductor" and the gracilis. The pectineus has a slight adductor action, but is a better flexor. The potential posterior adductors (except for the sciatic part of the adductor magnus) apparently contribute relatively little. The gluteus maximus apparently acts in adduction only when there is strong resistance in the abducted position, and this is true also of the posterior hamstrings, the semimembranosus, semitendinosus, and biceps. The obturator externus, and sometimes the quadratus femoris, which are primarily rotators, have been said to adduct slightly also. These muscles are innervated through several nerves (Table 8-3).

The *abductors* are severely limited in number (Fig. 8-33), the really strong ones being the gluteus medius and minimus; the tensor fasciae latae, innervated through the nerve to the two glutei, has been said to assist in this action, and the piriformis and sartorius pre-

sumably assist. Duchenne obtained little outward movement of the free limb upon stimulating the tensor, apparently more on stimulating the sartorius; he and Kaplan said the tensor is not really an abductor, but a flexor and internal rotator. McKibbin reported that in the infant, at least, the iliopsoas becomes a good abductor toward the extreme of abduction, although Pauly and Scheving could find no evidence that it either abducts or adducts. With the thigh flexed the obturator internus also may assist in abduction, and the inferior fibers of the gluteus maximus may do so when the thigh is flexed to 90°. The important abductors, the gluteus medius and minimus, are innervated by the superior gluteal nerve, typically receiving their fibers from the last two lumbar and the first sacral nerve.

Rotation of the femur is not a true rotation (around its long axis) for the angle between the head and the long axis of the body makes this impossible. Instead, the femur primarily swings forward (internal rotation) or backward (external rotation) like a gate on its hinges. Hooper and Ormond have apparently confirmed that the axis of "rotation" extends between the head of the femur and any fixed point, be it knee, heel, or great toe.

The chief *internal* or *medial rotators* at the hip are usually said to be the anterior parts of the gluteus medius and minimus, and the tensor fasciae latae (Fig. 8-34); the semitendinosus and the semimembranosus are presumably weak internal rotators, and the gracilis is said to become active in internal rotation (Wheatley and Jahnke). Although Duchenne reported that stimulation of the anterior part of the gluteus medius first produced strong internal rotation, followed by a movement of flexion and abduction, electromyography has indicated that although the tensor fasciae latae and the gluteus minimus do normally participate in internal rotation, the gluteus medius does not. The role of the iliopsoas and the adductors in rotation has only recently been settled. Traditionally, they have more commonly been regarded as internal rotators, although Duchenne reported they rotate externally. Evidence now indicates that the three named adductors all internally rotate (Basmajian, '74) and so also does the pectineus (Takebe and co-workers), but the iliopsoas apparently does rotate externally.

The chief *external* or *lateral rotators* are posteriorly placed (Fig. 8-35) and include the gluteus maximus, the so-called short rotators in the buttock (the piriformis, the obturator internus and associated gemelli, the obturator externus, and the quadratus femoris) and posterior fibers of the gluteus medius. Pauly and Scheving found the medius active only against resistance, and the minimus, usually listed as assisting, inactive. The sartorius on the front of the thigh becomes active in external rotation (Wheatley and Jahnke) and is usually said to have an external rotatory action; Duchenne found it rotated only a little, and weakly. The iliopsoas and the adductor group of muscles, including the pectineus, have been described, as just mentioned, as internal rotators; the iliopsoas apparently is not, but the others do internally rotate (see the preceding paragraph). The long head of the biceps apparently has a slight external rotatory action. The numerous rotators of the thigh are supplied by a number of different nerves (Table 8-4).

Contractures of muscles about the hip are usually treated by tenotomy or by nerve section. Problems and technics have been discussed by, among others, Barr; Freeman; and Peterson. One of the more common deformities of the hip is the flexion deformity that is a part of the syndrome known as "contracture of the iliotibial band," discussed later.

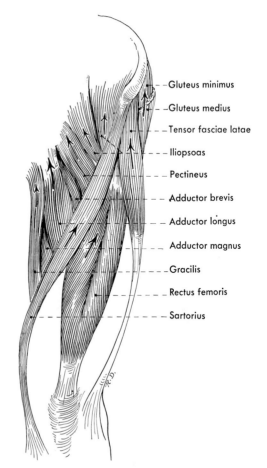

Fig. 8-31. Flexors of the thigh. The gluteus medius apparently does not belong here.

Labels: Gluteus minimus — Gluteus medius — Tensor fasciae latae — Iliopsoas — Pectineus — Adductor brevis — Adductor longus — Adductor magnus — Gracilis — Rectus femoris — Sartorius

Because of the numerous muscles that usually are able to assist in any one movement of the hip, paralysis usually must be extensive to interfere seriously with movement here. Abduction is, however, an exception, for the only two good abductors are innervated by the superior gluteal nerve. In bilateral abductor paralysis the attempt to balance the weight of the body leads to an especially awkward gait, and muscle transfers are frequently used for this (Thomas and co-workers).

FRACTURE AND DISLOCATION

Fractures of the hip have already been discussed under the separate headings of fracture of the hip bone and fracture of the upper end of the femur.

Table 8-2
Innervation of the Flexors of the Thigh

	MUSCLE	NERVE	SEGMENTAL INNERVATION	
			Reported Range	*Probable Most Important Segments*
Chief flexors	Iliopsoas	Nn. to iliopsoas	L2-L5	L2, 3, 4
	Pectineus	Femoral or obturator	L2-L4	L2, 3
	Tensor fasciae latae	Superior gluteal	L4-S1	L4, 5
	Adductor brevis	Obturator	L2-L5	L3, 4
	Sartorius	Femoral	L2-L4	L2, 3
Accessory flexors	Adductor longus	Obturator	L2-L4	L2, 3
	Adductor magnus (ant. part)	Obturator	L2-L5	L3, 4
	Gracilis	Obturator	L2-L5	L3, 4
	Gluteus minimus	Superior gluteal	L4-S1	L4, 5, S1

Fig. 8-32. Adductors of the thigh from the front, *a*, and behind, *b*.

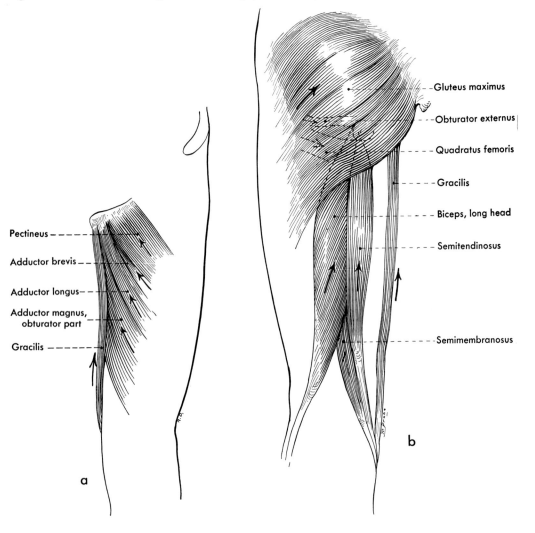

Table 8-3
Innervation of the Adductors of the Thigh

| | MUSCLE | NERVE | SEGMENTAL INNERVATION | |
			Reported Range	*Probable Most Important Segments*
Chief adductors	Adductor brevis	Obturator	L2-L5	L3, 4
	Adductor longus	Obturator	L2-L4	L2, 3
	Adductor magnus	Obturator and tibial	L2-S3	L3, 4, 5
	Gracilis	Obturator	L2-L5	L3, 4
Accessory adductors	Gluteus maximus	Inferior gluteal	L4-S3	L5, S1, 2
	Pectineus	Femoral or obturator	L2-L4	L2, 3
	Obturator externus	Obturator	L2-L5	L3, 4
	"Hamstrings"	Tibial	L4-S3	L5, S1

Fig. 8-33. Abductors of the thigh.

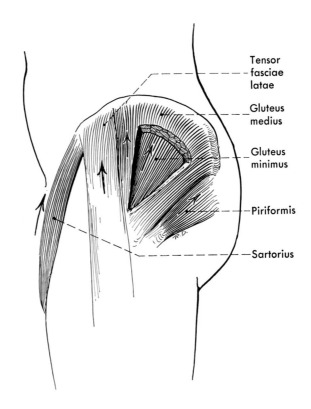

Tensor fasciae latae

Gluteus medius

Gluteus minimus

Piriformis

Sartorius

Dislocation

Dislocation of the hip may be the result of direct trauma or may be of congenital origin. According to Gordon and Freiberg, traumatic dislocation of the hip constitutes between 2% and 5% of all traumatic dislocations. Because of the deep position of the head of the femur in the acetabulum, a very severe force is usually necessary to produce dislocation, and this commonly is a blow upon the knee while the hip is in flexion. This probably accounts primarily for Walker's finding that posterior traumatic dislocation at the hip is seven times as frequent as is anterior dislocation, although the weakness of the posterior part of the capsule as compared with the anterior

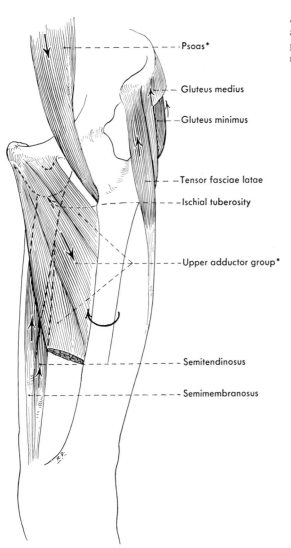

Psoas*

Gluteus medius

Gluteus minimus

Tensor fasciae latae

Ischial tuberosity

Upper adductor group*

Semitendinosus

Semimembranosus

Fig. 8-34. Some of the rotators of the thigh. Most are internal rotators, but both the iliopsoas and the gluteus medius have now been shown to be external rather than internal rotators.

may also be a factor. Dislocation of the hip of a child with cerebral palsy is also usually a posterior one (Samilson and co-workers).

Posterior dislocation at the hip always involves tearing of the posterior part of the capsule, and frequently a fracture of the posterior rim of the acetabulum. Associated fracture of the head of the femur is apparently rare (Gordon and Freiberg), however, although when it does occur the fracture is said to be usually in the inferior or antero-inferior part of the head; the fragment may or may not remain within the acetabulum, and may perhaps be attached to the ligament of the head

of the femur (ligamentum teres). This ligament is usually torn away at its attachment to the head, but may be torn from its attachment to the acetabulum. The short rotators on the back of the hip joint are of course usually injured by the head of the femur, and the sciatic nerve may be, but commonly is not. Usually the head rests on the posterior surface of the ilium (Fig. 8-36a), but sometimes it is downwardly dislocated and lies upon the posterior surface of the ischium close to the acetabulum.

In anterior dislocations at the hip the head of the femur usually passes around the medial aspect of the strong iliofemoral ligament, therefore through the lower part of the proximal portion of the anterior capsule; it usually lies either on the body of the pubis (Fig. 8-36b) or in the obturator foramen, depending upon whether it moves upward or downward. It is said that anterior dislocation has, rarely, produced rupture of femoral vessels; and that occasionally either the obturator nerve or the femoral may be somewhat bruised, but usually the damage is not severe.

Aseptic necrosis of the head of the femur may occur following dislocation, as a result of injury of the blood vessels to the neck and head; Ghormley and Sullivan quoted evidence that early reduction, before there has been excessive damage to the capsular vessels, is important if best results are to be obtained. Kleinberg reported two cases of aseptic necrosis in young adults limited to a small portion of the femoral head, apparently as a result of rupture of the ligament of the head of the femur with the cutting off of the blood supply usually afforded the region of the fovea by the vessels in this ligament.

Congenital dislocation of the hip has been thought to result from a dysplasia of the acetabulum, but Harrison, from experiments on rats, concluded that the presence of a nor-

mally placed and normally growing femoral head is necessary for proper acetabular growth, and that acetabular dysplasia is a result rather than a cause of congenital dislocation. Stanisavljevic and Mitchell concluded from their dissections that neither the femoral head and neck nor the acetabulum grow properly when dislocation occurs early in development, and Ponseti ('66) concurred. In a later ('78) paper he reported a study of the acetabulum of six newborn dead infants with congenital displacement, and (by x-ray) of the development of the acetabular roof of 180 children after reduction of dislocation. The six hips examined at autopsy all showed a ridge across the acetabulum dividing it into two parts; this ridge was formed by the acetabular cartilage alone in three cases, while in the others an inverted acetabular labrum lay over the cartilaginous ridge. (Ponseti emphasized that a click alone upon moving the femoral head is not sufficient to diagnose congenital dislocation. He said the click must be accompanied by a jolt as the head passes over the acetabular ridge.) Perhaps the acetabulum is more easily affected than the femur, for Laurenson described a case of bilateral dislocation in a 26-week fetus in which the femoral heads were normal but the acetabula were

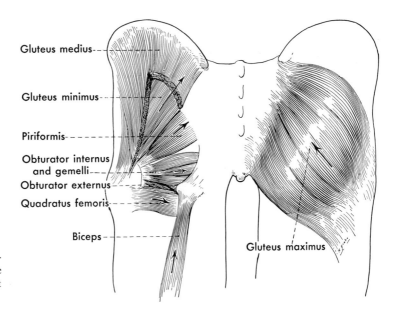

Fig. 8-35. Posteriorly placed external rotators of the thigh. The gluteus minimus is now said not to belong here.

Table 8-4
Innervation of the Rotators of the Thigh

	MUSCLE	NERVE	SEGMENTAL INNERVATION	
			Reported Range	*Probable Most Important Segments*
Chief internal rotators	Gluteus minimus	Superior gluteal	L4-S1	L4, 5, S1
	Tensor fasciae latae	Superior gluteal	L4-S1	L4, 5
Accessory internal rotators	Gracilis	Obturator	L2-L5	L3, 4
	Adductor brevis	Obturator	L2-L5	L3, 4
	Adductor longus	Obturator	L2-L4	L2, 3
	Adductor magnus (ant. part)	Obturator	L2-L5	L3, 4
	Adductor magnus (post. part)	Tibial	L3-S3	L4, 5
	Pectineus	Femoral or obturator	L2-L4	L2, 3
	Semitendinosus	Tibial	L4-S3	L5, S1
	Semimembranosus	Tibial	L4-S3	L5, S1
Chief external rotators	Gluteus maximus	Inferior gluteal	L4-S3	L5, S1
	Piriformis	To piriformis	L4-S2	S1, 2
	Obturator externus	Obturator	L2-L5	L3, 4
	Obturator internus Superior gemellus	N. to obturator internus and superior gemellus	L4-S3	L5, S1, 2
	Inferior gemellus Quadratus femoris	N. to inferior gemellus and quadratus femoris	L4-S1	L5, S1
Accessory external rotators	Gluteus medius	Superior gluteal	L4-S2	L4, 5, S1
	Sartorius	Femoral	L2-L4	L2, 3
	Iliopsoas	Nn. to iliopsoas	L1-L5	L2, 3, 4
	Biceps, long head	Tibial	L4-S3	L5, S1

shallow pits. However, the heads do not grow normally when dislocation persists (Fig. 8-37).

Von Rosen denied that most congenital dislocation results from dysplasia, and Stanisavljevic and Mitchell suggested that there may be several causes of congenital dislocation; Ponseti ('66) said that a large proportion of dislocations appear during the last few weeks of intrauterine life. An important factor has been said to be a temporary relaxation of the ligaments in utero through the action of hormones (Wilkinson; Carter and Wilkinson; von Rosen) or a familial general laxity of joints (Carter and Wilkinson); the former might account for the preponderance of congenital dislocation in females. Crelin implicated an excessively long ligament of the head of the femur as the specific cause of the disability. Breech presentation may also be a factor (Wilkinson; Barlow).

Although the femoral head in congenital dislocation may, after the limb has borne weight, be either anterosuperior or posterosuperior to the joint, evidence indicates that the original dislocation is an anterosuperior one. Howorth found such dislocations to be typical, and posterior dislocations to be few and usually in older children. Somerville, in explorations of 32 congenital dislocations for the purpose of open reduction, found that in all except 2, the head of the femur, when the hip was extended, lay against the ilium anterosuperior to the acetabulum; the exceptions, in which it lay superior to the joint, were in children older than 4 years.

Howorth ('47), Somerville, MacKenzie and co-workers, Scaglietti and Calandriello, and Ferguson, among others, have described the variable anatomy of the soft tissues as seen in open reductions of congenital dislocation of the hip, and therefore the obstacles that may make closed reduction difficult or impossible. Thus there may be a short or hypertrophied psoas muscle, a constriction of the capsule between the femoral head and the acetabulum, an acetabular labrum that is doubled into the acetabulum, an enlarged or long ligament of the head, or an acetabulum that is largely filled with fibrous tissue.

Early reduction contributes to more normal development of the hip joint, and Schwartz reported much better results in hips reduced before the age of 20 months. Barlow preferred to wait until the child was about a year old, but von Rosen and Ponseti ('66) both emphasized diagnosis and treatment of the newborn. Ponseti said the preferable time for treatment is during the first week, and that if treatment is delayed for several months closed reduction may be difficult or impossible. In his '78 study of the development of the acetabulum in children whose hips had been reduced at various ages, he found only 3 (about 6.7%) instances of abnormal development among the 45 children in whom the reduction had been done before the age of 1 year; there were, however, 26 such instances (30.3%) among 59 in whom the reduction had been done between the ages of 1 and 2 years, and 23 (44.1%) among 76 when the reduction had been done after the age of 2.

The Buttock

The superficial fascia (tela subcutanea) of the gluteal region or buttock is typically thick and loaded with fat, thereby contributing to the contour of the buttock. It is continuous with the superficial fascia of the abdomen and back, with that of the perineum, and with that of the thigh.

The *deep fascia* of the gluteal region is continuous inferiorly and anteriorly with the fascia lata, the deep fascia of the thigh. Posteriorly and laterally it is divided into two laminae, which pass superficial and deep to the gluteus maximus and to the tensor fasciae latae. The external layer on the gluteus maximus is attached to the crest of the ilium and the back of the sacrum; it is a thin layer that is

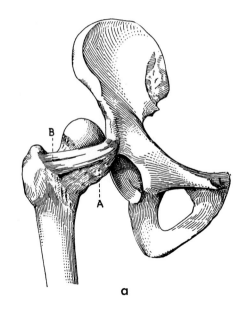

a

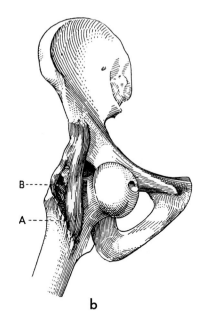

b

Fig. 8-36. Iliac, *a*, and pubic, *b*, dislocation of the head of the femur. *A* and *B* are the more vertical and the more transverse parts, respectively, of the iliofemoral ligament. (Mathieu P: Luxations traumatiques de la hanche. In Ombrédanne L, and Mathieu P (eds): Traité de Chirurgie Orthopédique, Vol 4. Paris, Masson et Cie, 1937)

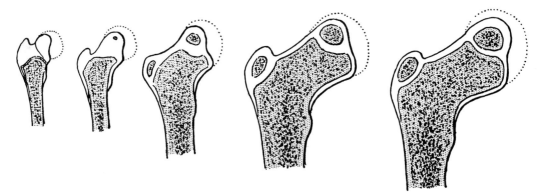

Fig. 8-37. Shapes of the femoral head in congenital dislocation of the femur at the ages of 1 month, 1 year, 3 years, 4.5 years, and 7 years; the *broken outlines* indicate the expected development at those ages. (After Ducroquet, from Lance M: Malformation luxante: Luxation congénital de la hanche. In Ombrédanne L, and Mathieu P (eds): Traité de Chirurgie Orthopédique, Vol 4. Paris, Masson et Cie, 1937)

firmly bound both to the superficial fascia and to the muscle, sending septa into the muscle. The deep layer here lies between the gluteus maximus and the deeper muscles, and most of the nerves and vessels, of the buttock; it is attached especially to the sacrotuberous ligament and to the ischium.

At the inferior border of the gluteus maximus these two layers come together and extend downward as a part of the fascia lata. On the superolateral border of the gluteus the layers also come together to form a single layer between this border and the iliac crest; on its deep surface this part gives origin to some of the fibers of the gluteus medius, so it is referred to as the gluteal aponeurosis. More anteriorly the deep fascia splits again to surround both surfaces of the tensor fasciae latae, being united to it by septa, and comes together anterior to it to continue as the fascia lata on the front of the thigh. Below the gluteal aponeurosis the deep fascia is much thickened by a large part of the insertion of the gluteus maximus and all the tendon of insertion of the tensor fasciae latae; these tendons fuse with the fascia lata, and fascia and tendons together give rise to the iliotibial tract or band, the particularly strong lateral part of the fascia lata.

The superficial vessels of the buttocks are unnamed twigs, but the *superficial nerves* of the buttock are larger. The most laterally lying

nerve, descending across the crest of the ilium, is typically the lateral branch of the twelfth thoracic nerve, and just medial to that, paralleling it, is the lateral or iliac branch of the iliohypogastric. These nerves supply an upper and lateral portion of the skin over the gluteal region. A major middle portion of the skin of the buttock is supplied by the superior clunial nerves, typically three in number, which are the cutaneous continuations of the lateral branches of the dorsal rami of the upper three lumbar nerves. In a similar fashion, the lateral branches of the dorsal rami of the first three sacral nerves supply the medial aspect of the buttock, and form the three middle clunial nerves; and the posterior femoral cutaneous nerve, as it leaves the cover of the gluteus maximus, sends inferior clunial nerves around the lower border of this muscle to supply skin over a lower part of the buttock.

MUSCLES

The large gluteus maximus muscle covers most of the structures in the buttock; hence approaches to the structures here, or posterior approaches to the hip joint, involve dividing this muscle either at its insertion or by splitting it in the direction of its fibers. The gluteus medius muscle appears above and lateral to the superolateral border of the gluteus maximus, and still more laterally and anteriorly is the tensor fasciae latae, actually a muscle of the buttock although it arises very close to the anterior superior iliac spine. The gluteus minimus is entirely concealed by the gluteus medius, and the small muscles of the buttock (the piriformis, the obturator internus and gemelli, and the quadratus femoris) are concealed by the gluteus maximus (Fig. 8-38).

Gluteus Maximus

The gluteus maximus is a particularly heavy and coarse-fibered muscle, running diagonally downward and laterally, and largely contributing to the shape of the buttock. It arises from the relatively small part of the outer surface of the ilium behind the posterior gluteal line, and more extensively from the dorsal surface of the sacrum and coccyx, blending here with tendons of origin of the erector spinae, and from the adjacent part of the sacrotuberous ligament (Fig. 8-39).

The insertion of the gluteus maximus is in two parts, a superficial and a deep one. All of approximately the upper half of the muscle and a superficial half of the lower half—about three fourths of the muscle—give rise to tendinous fibers that continue the downward and lateral direction of the muscle and cross the greater trochanter to insert into the fascia lata and help form the iliotibial tract. A subcutaneous trochanteric bursa usually lies superficial to the tendon over the greater trochanter, and the large trochanteric bursa of the gluteus maximus lies between the tendon and this bony protuberance. The deep part of the lower portion of the gluteus maximus is inserted almost entirely into the gluteal tuberosity of the femur, some of the fibers, however, attaching into the fascia lata and the lateral intermuscular septum which passes inward from the deep surface of the fascia.

The chief *action* of the gluteus maximus is to extend the hip joint. It is inactive in standing (Joseph and Williams), and persons with atrophy of it can walk on a flat surface with no apparent limp (Duchenne); it is of greater importance in walking up a grade or ascending stairs, and in jumping, running, or standing up from a sitting position. Stimulation of it produces primarily extension of the thigh and only a little external rotation (Kaplan, '58),

Fig. 8-38. Musculature of the buttock. Most of the left gluteus maximus and a posterior part of the corresponding gluteus medius are shown removed.

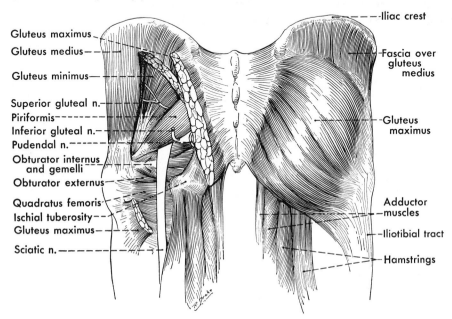

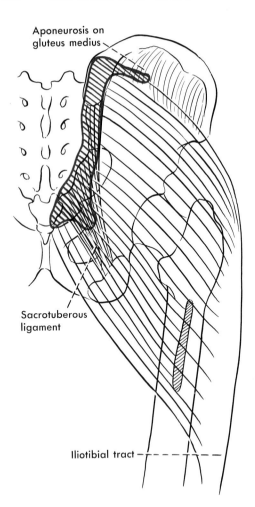

Fig. 8-39. Origin and insertion (origin *red*, bony insertion *black*) of the gluteus maximus.

but electromyography has indicated that it participates in external rotation, adduction against resistance in the abducted position, and, when the thigh is flexed to 90 °, abduction against strong resistance (Wheatley and Jahnke).

The gluteus maximus is *innervated* by the inferior gluteal nerve, which turns around the lower border of the piriformis muscle and the lateral edge of the sacrotuberous ligament to enter the buttock. The branches of the nerve are accompanied by branches of the inferior gluteal artery; the nerve is typically composed of fibers from the last lumbar and first two sacral nerves.

Gluteus Medius and Minimus

These two muscles are closely associated anatomically, have largely similar actions, and are innervated by the same nerve.

The gluteus medius arises from that large area of the wing of the ilium between the iliac crest and the anterior gluteal line, as far back as the posterior gluteal line (Fig. 8-40). Its origin is thus largely above and posterior to the gluteus minimus, and it completely covers that muscle from behind. In front of the upper border of the gluteus maximus, the gluteus medius also obtains origin from the deep fascia covering its surface.

From this wide origin the muscle fibers converge to insert by a short tendon into the posterior and lateral surfaces of the upper part of the greater trochanter. A bursa typically lies between the anterior fibers and the anterior part of the trochanter as they cross this to reach their insertion.

Between this muscle and the gluteus min-

Fig. 8-40. Origin (*red*) and insertion (*black*) of some of the muscles of the buttock.

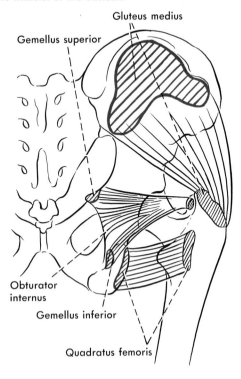

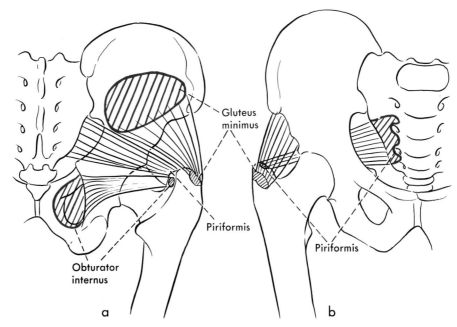

Fig. 8-41. Origin (*red*) and insertion (*black*) of some of the muscles of the buttock. *a* is a posterior, *b* an anterior, view.

imus are the major branches of the superior gluteal nerve and vessels.

The gluteus minimus, also a fan-shaped muscle, arises from the ilium between the anterior and inferior gluteal lines (Fig. 8-41), therefore under cover of the gluteus medius. Its fibers converge to insert on the uppermost part of the anterior surface of the greater trochanter. A bursa lies between the anterior part of the muscle and the trochanter.

The gluteus medius and minimus have essentially the same *functions,* the primary one being abduction. While other muscles can aid somewhat in abduction, they are the only two strong ones, and are therefore particularly important in preventing the pelvis of the opposite side from sagging markedly from the horizontal when all the weight is put upon one limb. Inman regarded the tensor fasciae latae as being the only other abductor of any particular importance at all, and said that if these three muscles act with forces proportional to their masses, the ratio is 1:2:4 for the tensor, the gluteus minimus, and the gluteus medius. Inman stated that if the unsupported

side of the pelvis drops much more than 15°, the fascia lata assumes all the support; that in the level position the abductor muscles and the fascia share the support about equally; and that raising the unsupported side much above 15° throws all the support on the abductor muscles. (Stimulation of the tensor fasciae latae indicates that it is not effective in abduction of the free limb, but this does not negate a contribution to the support of the pelvis.) Merchant used a model in which three gauges represented these muscles, and found that not only did the total stress on them vary both with the tilt of the pelvis and rotation of the thigh, but so did the distribution of the stress among the three. Thus with external rotation, he said, more weight is borne by the tensor and the iliotibial tract, while in internal rotation little or none is.

Because of their origins anterior, above, and posterior to the hip joint, the gluteus medius and the gluteus minimus together have other actions in addition to abduction; indeed they very much resemble the deltoid muscle, at the shoulder, in their relations to the joint and in their actions. Duchenne described stimulation of the anterior third of the gluteus medius as producing strong internal rotation followed by an oblique movement of flexion and abduction, and stimulation of the posterior third as producing an oblique movement of extension and abduction combined with external rotation; he said the latter movements were much less powerful than those produced by stimulation of the anterior part of the muscle. In contrast, Pauly and Scheving reported that the gluteus medius is active in hyperextension and in forced external rotation, but not in internal rotation, while the gluteus minimus, assumed to have similar actions although not tested by Duchenne, is active in internal rotation and flexion. The minimus was also reported to act with the medius in extending the partially flexed thigh.

These two muscles are *innervated* by the superior gluteal nerve, which enters the buttock at the upper border of the piriformis muscle, thus at the lower border of the two gluteal muscles, and turns upward and forward between them, supplying both muscles and continuing also into the deep surface of the tensor fasciae latae. The nerve is accompanied in its course by the deep or major branches of the superior gluteal vessels; the superficial branches of these vessels help supply the upper part of the gluteus maximus.

Tensor Fasciae Latae

The tensor fasciae latae arises from the anterior part of the iliac crest as far forward as the level of the anterior superior iliac spine (see Fig. 8-56). Its posterior border is closely related to the anterior border of the gluteus medius and it is the most anterior member of the abductor group. It is a short, rather straplike muscle that passes downward and slightly backward to insert into the anterior part of the iliotibial tract a little below the level of the greater trochanter. As already noted, it is enclosed within the fascia lata, which splits to go upon both sides of it.

The tensor fasciae latae has been described as a flexor, internal rotator, and abductor of the thigh, and Pauly and Scheving found it

participating in all these actions. Both Duchenne and Kaplan ('58) reported that stimulation of it produces flexion and internal rotation but denied that it can be considered an abductor. Its chief action is flexion. Wright said it does not contract for internal rotation alone, but only when this is combined with flexion, and Duchenne regarded its internal rotatory action as being primarily the prevention of external rotation; however, Pauly and Scheving found it the most active of all muscles during internal rotation. It apparently acts to help overcome the tendency of other flexors of the hip to rotate the thigh outward, and Duchenne said that when it is paralyzed the limb rotates outward during its forward swing in walking. A conjectured action at the knee through the iliotibial tract, sometimes said to be extension, sometimes flexion, has no evidence to support it.

The tensor fasciae latae is *innervated* by a continuation of the superior gluteal nerve. This passes forward from between the gluteus medius and minimus to enter the deep surface of the tensor. It is usually described as receiving fibers from all three nerves, L4 to S1, that typically contribute to the superior gluteal nerve.

Short Rotators

The piriformis, the obturator internus and the associated gemelli, the quadratus femoris, and the deeper and more anteriorly lying obturator externus are frequently referred to as the short rotators of the femur. The last-named muscle belongs to the anteromedial rather than the gluteal group, and is described with the anteromedial muscles; it passes across the postero-inferior aspect of the hip joint to an insertion in the trochanteric fossa.

The *piriformis* is the uppermost of the small muscles of the gluteal region, and the key to the arrangement of the nerves and vessels in the buttock. It largely fills the greater sciatic (ischiadic) foramen, through which the branches of the sacral plexus and the branches of the internal iliac (hypogastric) vessels to the gluteal and pudendal regions leave the pelvis; therefore, the vessels and nerves that enter the buttock necessarily are closely related to this muscle. The superior gluteal nerve and vessels are typically the only ones appearing at its upper border, while the pudendal nerve and vessels, the inferior gluteal nerve and vessels, the sciatic (ischiadic) nerve, the posterior femoral cutaneous, and the nerves to the small rotators except the piriformis typically appear at its lower border. In more than 10% of cases the piriformis muscle is perforated by one or both parts of the sciatic nerve, and, less often, a part of the nerve enters the buttock above the piriformis muscle and courses downward on its posterior surface (see a following section for details).

The piriformis arises from the pelvic surface of the pars lateralis (lateral mass) of the sacrum (Fig. 8-41), at the levels of the second through the fourth sacral segments. The muscle forms, at its origin, a portion of the posterior wall of the pelvis, and is related anteriorly to a part of the sacral plexus and to the rectum. As it leaves the greater sciatic foramen it usually receives some additional fibers that arise from the upper margin of the greater sciatic notch and from the pelvic surface of the sacrotuberous ligament. In the buttock the fibers converge to form a rounded tendon that inserts into the medial side of the upper border of the greater trochanter under cover of the insertion of the gluteus medius.

The piriformis is usually described as both an external rotator and abductor. Duchenne said that isolated stimulation of it produced simultaneously external rotation and a slight amount of extension and abduction—thus a movement similar to that produced by the posterior fibers of the gluteus medius, which the piriformis so closely parallels. The muscle receives its innervation on its pelvic surface, usually from both the first and second sacral nerves, but sometimes from either alone, as the sacral plexus is taking form there.

The *obturator internus* (Figs. 8-40 and 8-41) has an extensive origin, not visible from the buttock. This origin is from the internal surface of the bony pelvis, and includes all the margin of the obturator foramen except the region of the obturator groove; the inner surface of the obturator membrane; and a broad strip on the pelvic surface of the coxal bone above and behind the obturator foramen. Some fibers also arise from the covering obturator fascia. The levator ani obtains origin from the pelvic surface of the obturator internus; thus an upper part of the latter helps to form the lateral wall of the pelvic cavity, while a lower part, below the levator, forms the lateral wall of the ischiorectal fossa.

From its wide origin the fibers of the obturator internus converge to pass through the lesser sciatic (ischiadic) foramen, making a sharp turn around the smooth bone of the lesser sciatic notch of the ischium; as it makes this turn its deep surface is tendinous, and separated from the bone by a large bursa. The fibers of the muscle continue to converge after the muscle has entered the buttock, and the muscle as a whole runs almost transversely across the posterior aspect of the hip joint to insert into the medial surface of the greater trochanter just above the trochanteric fossa. In the buttock the muscle is closely associated with the gemelli.

The *superior and inferior gemelli* lie immediately above and below the obturator internus muscle in the buttock. The superior gemellus takes origin from the posterior surface of the ischial spine, the inferior gemellus from the upper part of the ischial tuberosity; they converge upon the intervening obturator internus to attach to its tendon and be inserted with it. The gemelli vary a good deal in their sizes; not infrequently one consists of only a few muscle bundles, and one or both may be lacking.

The obturator internus and the associated gemelli are primarily external rotators at the hip. It is usually said that they are abductors when the hip is flexed, but Duchenne found them producing external rotation without abduction when the hip was flexed unless the foot was allowed to touch the ground, in which case abduction was carried out.

The obturator internus and the superior gemellus are supplied by the same nerve, the "nerve to the obturator internus and superior gemellus." It is usually derived from L5 to S2 or S1 to S3, and appears on the medial side of the sciatic nerve below the piriformis muscle.

It crosses the superior gemellus superficially, close to the muscle's origin, and gives off a twig to it; then it turns more medially below the ischial spine to follow the free internal surface of the obturator internus muscle toward the origin of the muscle. The nerve to the inferior gemellus is a branch of the nerve to the quadratus femoris and enters its deep surface.

The *quadratus femoris* (Figs. 8-38 and 8-40) arises from the lateral border of the ischial tuberosity and inserts on the posterior aspect of the femur on the intertrochanteric crest and a line leading down from it, about midway between the two trochanters. A bursa lies on its deep surface as it crosses the lesser trochanter and the tendon of insertion of the obturator externus just above this trochanter.

The quadratus femoris is primarily an external rotator at the hip joint; because of its position below the head of the femur it is usually also said to be an adductor, although Duchenne apparently did not observe this action. The muscle is innervated by the nerve to the quadratus femoris and inferior gemellus (usually from L4 to S1). After emerging below the lower border of the piriformis muscle this nerve passes downward deep to (on the anterior surfaces of) the two gemelli and the associated tendon of the obturator internus, giving off a twig to the inferior gemellus and to the posterior aspect of the hip joint before entering the deep surface of the quadratus.

Variations and Anomalous Muscles

Variations in the musculature of the buttock are apparently not common. Occasionally there are fusions among the muscles or between one and a neighboring muscle of another group (*e.g.*, between the quadratus femoris and the adductor magnus). Occasionally, also, one of the smaller muscles is missing; this is most commonly true of the superior gemellus, but may be true of the quadratus femoris or the inferior gemellus. The piriformis muscle, sometimes pierced by the sciatic nerve and divided into upper and lower parts, is, rarely, divided close to its origin into anterior and posterior planes by the roots of origin of the sacral plexus (Celli). As noted in the discussion of anomalous muscles of the thigh, aberrant slips of muscle may arise in the buttock, commonly close to or with the piriformis muscle, and pass into the thigh, where they usually join the biceps femoris.

NERVES AND VESSELS

The buttock contains not only the nerves and vessels destined for its own structures, but also ones of passage—particularly the sciatic nerve, but also the posterior femoral cutaneous, largely destined for the thigh, and the pudendal nerve and vessels to the perineum. Since all the nerves and vessels appearing in the buttock leave the pelvis by way of the greater sciatic foramen, they are closely related to the piriformis muscle which largely fills this foramen; the superior gluteal nerve and vessels typically appear at the upper border of the piriformis muscle, all the other structures at the lower border.

Sciatic Nerve

The large sciatic or ischiadic nerve (commonly from L4 to S3) typically appears in the buttock at the lower border of the piriformis muscle and courses downward posterior (superficial) to the obturator internus and gemelli and the quadratus femoris (Fig. 8-42), passing about midway between the ischial tuberosity and the greater trochanter, and continuing downward in the midline of the posterior aspect of the thigh. In the buttock it is under cover of the gluteus maximus and of the upward continuation from the fascia lata that lies deep to the maximus; as it passes into the upper part of the thigh it lies at first deep only to the fascia lata, but soon passes under cover of the hamstring muscles. It gives off no branches to the buttock, but its uppermost branch or branches to the hamstrings frequently arise above the ischial tuberosity. As it lies in the buttock, it is related anteriorly to the nerve to the quadratus femoris and superior gemellus, but separated from this nerve by the short rotator muscles in the buttock. The posterior femoral cutaneous and the infe-

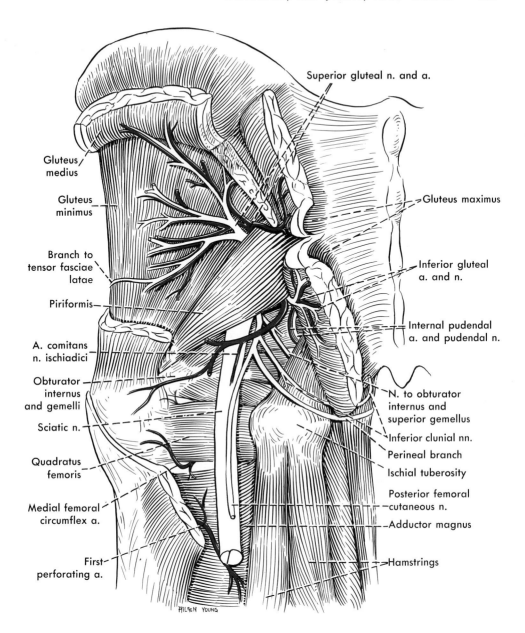

Fig. 8-42. Nerves and arteries of the buttock.

rior gluteal nerves appear below the piriformis immediately medial to it, and the former nerve in leaving the buttock and passing into the thigh typically comes to lie directly behind the sciatic nerve. The nerve to the obturator internus and superior gemellus, and the pudendal nerve and vessels, lie more medially, close to the ischial spine.

Since the lower end of the sacral plexus and the upper end of the sciatic nerve are essentially inapproachable surgically from the pelvis, because of the overlying viscera, blood vessels, and autonomic nerves, they can best be approached from the posterior aspect. Section of the piriformis muscle gives access to a higher part of the nerve and also allows liga-

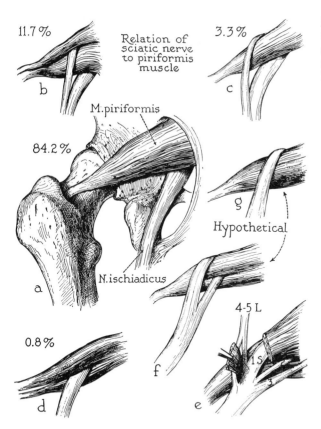

11.7%   Relation of sciatic nerve to piriformis muscle   3.3%

b   c

M.piriformis

84.2%

g

Hypothetical

N.ischiadicus

a

4-5 L

0.8%

f

1 S

3

d   e

Fig. 8-43. Varying relations between the sciatic nerve and piriformis muscle. Percentages are those found by Beaton and Anson. (Original of Plate 1, Beaton LE, Anson BJ: Anat Rec 70:1, 1937)

tion of the stems of the gluteal vessels if this is necessary, and still greater exposure can be obtained by operative enlargement of the greater sciatic foramen—sectioning the sacrotuberous ligament, and if necessary removing bone bordering the foramen.

The relation just described between the sciatic nerve and the piriformis muscle holds in approximately 85% of cases, while in the remainder part or all of the sciatic nerve courses through, or a part courses above, the piriformis muscle. The more common variations depend upon the facts that the common peroneal and tibial divisions of the sciatic nerve may separate from each other at their origins and take different courses into the buttock, and that the piriformis muscle may be divided into two parts by one or both components of the sciatic nerve. Apparently, passage of the entire nerve above the muscle has not been reported, nor has passage of one part of the nerve above and one through the muscle. Thus the variations from the normal usually

reported are only three: passage of the entire nerve through the muscle; passage of a part of it through and the other part below the muscle; and passage of one part below and the other above the muscle.

Statistics on the occurrence of the varying relations of the sciatic nerve to the piriformis muscle differ, apparently in part because of racial factors (*e.g.,* Ming-tzu, '41; Misra). Beaton and Anson found the nerve passing below the muscle, therefore with normal relations, in 84.2% of cases (Fig. 8-43). The most common variation is for the tibial nerve to pass below the piriformis, the common peroneal through it, this having been found in 11.7%; Ming-tzu found among Chinese a very high percentage, 32.9%. Beaton and Anson reported passage of the common peroneal nerve above, the tibial below, the piriformis muscle in 3.3%, passage of the entire nerve through the piriformis in only 0.8% of cases.

Coccygodynia and sciatic pain have been attributed to abnormal relations between the piriformis muscle and the sciatic nerve. Robinson described the "piriformis syndrome" as consisting of pain and tenderness over the lower part of the sacroiliac joint, the greater sciatic notch, and the piriformis muscle, and sometimes pain in the hip and gluteal atrophy. As with "sacroiliac strain," the diagnosis of "piriformis syndrome" is not often made nowadays.

Posterior Femoral Cutaneous Nerve

This nerve appears at the lower border of the piriformis muscle just medial to the sciatic nerve, and runs downward, assuming a position posterior to the sciatic, with a branch of the inferior gluteal artery. At the lower border of the gluteus maximus the nerve gives inferior clunial branches around the border of the muscle to the skin of the buttock, and a perineal branch that curves forward around the medial surface of the thigh below the ischial tuberosity. The remainder of the nerve continues down the posterior aspect of the thigh in almost exactly the posterior midline, lying deep to the fascia lata and sending small branches through that to the skin.

Superior Gluteal Nerve and Vessels

The superior gluteal nerve is commonly derived from the fourth and fifth lumbar and first sacral nerves, and may receive fibers from S2; it innervates three muscles, the gluteus medius, minimus, and tensor fasciae latae. The superior gluteal artery is a branch of the internal iliac (hypogastric) artery; it passes through the sacral plexus in order to reach the buttock, usually between the lumbosacral trunk and the first sacral nerve, but sometimes between the first and second sacral nerves or between the fourth and fifth lumbars. The accompanying vein is a tributary of the internal iliac vein.

The superior gluteal nerve and vessels enter the buttock together, between the upper border of the piriformis muscle and the lower border of the gluteus medius and minimus muscles. As the artery and vein appear in the buttock each divides into a superficial and a deep branch; the superficial branch divides quickly into a number of subsidiary branches, while the deep one runs upward and laterally between the gluteus medius and minimus, with the superior gluteal nerve. Many of the twigs from the superficial branch supply an upper part of the gluteus maximus, others pass through it to the skin, and still others anastomose with branches of neighboring vessels, particularly the inferior gluteal and internal pudendal.

The superior gluteal nerve and the deep branches of the superior gluteal vessels turn upward between the gluteus medius and minimus just as they emerge into the buttock between the glutei and the piriformis; they are, therefore, hardly visible unless the gluteus medius is reflected. The deep branch of the artery gives a branch into the ilium and then divides into upper and lower branches, accompanied by corresponding veins.

The upper branch runs laterally and forward along the upper border of the gluteus minimus, supplying this and the medius, and continuing beyond their anterior borders to reach the deep surface of the tensor fasciae latae; here it anastomoses with the ascending branch of the lateral femoral circumflex artery, and with twigs of the deep iliac circumflex.

The inferior ramus of the deep branch of the artery also passes laterally and forward between the gluteus medius and the minimus, running almost transversely and accompanied by the major part of the superior gluteal nerve; it gives branches to both muscles and emerges to continue over the greater trochanter, where it anastomoses with the ascending branch of the lateral femoral circumflex artery. It typically supplies a branch to the hip joint, and before it passes the trochanter gives rise to a branch that anastomoses with the inferior gluteal artery and with ascending branches of the medial femoral circumflex and first perforating arteries.

The superior gluteal nerve runs almost transversely with the inferior ramus of the deep branch of the artery. It ends largely in the gluteus medius and minimus as it runs between them, but part of it continues forward beyond the anterior borders of these muscles to enter the deep surface of the tensor fasciae latae.

Inferior Gluteal Nerve and Vessels

The inferior gluteal nerve, artery, and vein enter the buttock at the lower border of the piriformis muscle. The nerve, usually derived from the posterior divisions of the fifth lumbar and the first two sacral nerves, is closely associated in its emergence with the posterior femoral cutaneous; it immediately breaks up into a number of branches to the gluteus maximus, which perforate the fascia on the deep surface of the muscle, some of them passing immediately into the muscle, others running some distance laterally between fascia and muscle fibers before they enter.

The inferior gluteal artery arises from the internal iliac artery, often by a stem common to it and the internal pudendal; in leaving the pelvis it passes through the sacral plexus, typically between the first and second or the second and third sacral nerves, and branches as it enters the buttock. It gives muscular branches especially to the deep surface of the gluteus maximus and to the superficial surface of the short rotators and the upper ends of the hamstring muscles. It also gives off one or two twigs that pierce the sacrotuberous ligament to be distributed to tissue around the posterior surface of the coccyx, and twigs that pass around the lower border of the gluteus maximus to reach the skin over the lower part of this muscle. A small branch may descend for a variable distance with the posterior femoral cutaneous nerve; another small branch, the arteria comitans nervi ischiadici (artery accompanying the sciatic nerve) runs downward on the dorsal surface of the sciatic nerve or disappears into its substance, as the chief arterial supply of the upper end of the nerve.

The inferior gluteal artery usually also gives off articular twigs to the posterior aspect of the hip joint, these passing downward deep to the gemelli and the obturator internus, and two anastomotic branches that pass laterally, either in front of or behind the sciatic nerve. One of the latter typically anastomoses medial to the greater trochanter with the superior gluteal artery, with the ascending branch of the medial femoral circumflex (usually appearing between the inferior gemellus and the

quadratus), and with an ascending branch of the first perforating artery. The other anastomoses with branches of both femoral circumflex arteries and the ascending branch of the first perforating artery.

The anastomoses between the two gluteal arteries, the two femoral circumflexes, and the first perforating branch of the profunda are a major source of blood supply to the femoral artery after ligation of the external iliac or of the femoral proximal to the profunda. They are sometimes grouped together as the *crucial* or *cruciate anastomosis,* or sometimes this anastomosis is defined as being roughly cruciate in shape and formed as follows: a branch of the inferior gluteal artery descends toward the posterolateral surface of the greater trochanter; the transverse branch of the medial femoral circumflex typically appears between the lower border of the quadratus femoris and the upper border of the abductor magnus; the ascending branch of the first perforating artery ascends on the posterior aspect of the adductor magnus; and the transverse branch of the lateral femoral circumflex reaches the posterolateral surface of the greater trochanter by traveling through the upper part of the vastus lateralis muscle. The anastomoses between these vessels are typically through numerous small branches, but sometimes a gross anastomosis between the inferior gluteal and the medial femoral circumflex occurs.

Nerves of the Short Rotators

The *nerves to the piriformis,* derived directly from the posterior surfaces of the first and second sacral nerves, or either, as they lie on the anterior surface of this muscle, do not appear in the buttock.

The small *nerve of the obturator internus and superior gemellus* (Fig. 8-44), usually containing fibers from the fifth lumbar and first two sacrals, but with a variation from the fourth lumbar to the third sacral, appears in the buttock below the piriformis muscle toward the medial side of the sciatic, posterior femoral cutaneous, and inferior gluteal nerves. It runs downward across the base of the ischial spine, gives a twig into the posterior surface of

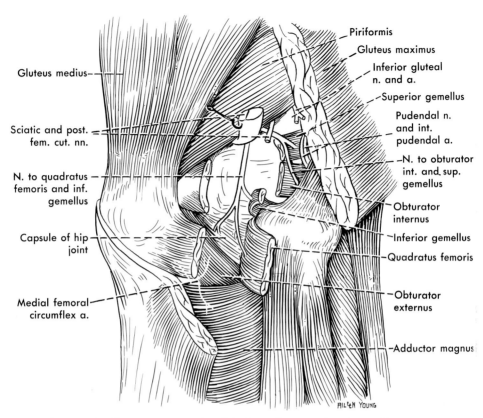

Fig. 8-44. Smaller nerves to muscles of the buttock, as seen in a deeper dissection than that shown in Fig. 8-43.

the superior gemellus, and then turns medially below the ischial spine to follow the superficial surface of the obturator internus muscle through the lesser sciatic notch, thus disappearing deep to the sacrotuberous ligament.

The small *nerve of the quadratus femoris and inferior gemellus,* said to consist usually of fibers from the fourth and fifth lumbars and first sacral, has a deep course in the buttock. It leaves the lower border of the piriformis muscle in front of (deep to) the sciatic nerve, and then runs deep to the superior gemellus, obturator internus, and inferior gemellus. It gives off a branch to the deep surface of the inferior gemellus, another to the posterior aspect of the hip joint, and ends in the deep surface of the quadratus femoris. (Wilson recorded a case in which this nerve continued to supply an uppermost anterior part of the adductor magnus, above the dis-

tribution of the obturator to this muscle; Smith reported a case in which it helped supply the superior gemellus.) This is the nerve usually cut in attempts to denervate the posterior aspect of a painful hip joint; it is best found by reflecting the gemelli and obturator internus.

Pudendal Nerve and Vessels

The pudendal nerve and internal pudendal vessels appear for a very short distance in the buttock, entering it between the lower border of the piriformis and the spine of the ischium, and passing downward and medially, around the spine of the ischium or the attachment of the sacrospinous ligament to the spine, to disappear through the lesser sciatic foramen under cover of the sacrotuberous ligament. In their brief course through the buttock the nerve is most medial, and the artery tends to lie between the nerve and the vein.

The Thigh

Most of the muscles of the thigh arise from the hip bone and insert into the femur, but the psoas part of the iliopsoas muscle arises from the vertebral column and is situated primarily in the posterior abdominal wall; other muscles of the thigh cross both the hip and knee joints; and still others, notably the largest part of the quadriceps femoris, arise from and largely cover the femur, crossing the knee joint only.

All the nerves and vessels in the thigh are continued into it from the abdominal or pelvic cavities. The posterior nerves, the posterior femoral cutaneous and the sciatic, are continued down through the buttock, as are small branches from the inferior gluteal artery; the anteriorly located femoral artery is a continuation of the external iliac and enters the thigh behind the inguinal ligament, as does the femoral nerve; and the other major anterior nerve of the thigh, the obturator, has a deeper entrance into the thigh, passing along the lateral wall of the true pelvis and entering the thigh through the obturator foramen.

SUPERFICIAL STRUCTURES

The superficial fascia or tela subcutanea of the thigh contains a variable amount of fat but presents few features of special interest. It is continuous posteriorly and above with the superficial fascia of the gluteal region, medially with the superficial fascia of the perineum, and anteriorly and laterally with the superficial fascia of the lower abdominal wall. Immediately below the inguinal ligament it tends to be separated into two layers by the superficial inguinal lymph nodes, the upper end of the great saphenous vein, and the stems of the several small subcutaneous vessels entering here. The deeper part is usually described as being membranous, and a continuation of the deep or membranous layer (Scarpa's fascia) of the superficial layer of fascia of the lower abdominal wall; it is fused ,to

the deep fascia of the thigh (fascia lata) a little below the inguinal ligament, and attached around the margin of the saphenous hiatus (fossa ovalis) to form the fascia cribrosa. The other part of the superficial fascia of the lower abdominal wall is regularly described as passing across the inguinal ligament and becoming continuous with the remainder of the superficial fascia of the thigh.

Around the knee the superficial fascia becomes thin, and blends with the deep fascia and its special expansions at the knee.

FASCIA LATA

The fascia lata is the deep fascia of the thigh; it enwraps the entire thigh as a more or less strong, membranous layer, particularly well developed laterally. In the distal two thirds of the thigh it gives rise on its deep surface to two septa, the lateral and medial intermuscular septa (Fig. 8-45).

The *lateral intermuscular septum* arises from the strong lateral portion of the fascia lata, the iliotibial tract, and extends inward toward the femur to attach along the lower part of the linea aspera and its diverging lower lateral lip (supracondylar line); it passes between the vastus lateralis and intermedius, anteriorly, and the short head of the biceps, posteriorly, and gives origin to some of the fibers of these muscles.

The *medial intermuscular septum* is best developed in the middle third of the thigh, where it attaches to the medial lip of the linea aspera and lies between the adductor muscles and the vastus medialis. As it passes outward toward the fascia lata it divides into two layers, one of which follows the vastus, the other the adductors, to enclose between them the femoral vessels and the sartorius muscle. Deep to the sartorius the two layers are united by an especially thick layer across the femoral vessels, and this forms the roof of the adductor (subsartorial, Hunter's) canal. In the lower third of the thigh the medial intermuscular septum is not well marked, but blends with the anterior surface of the tendon of insertion of the adductor magnus muscle. Except where

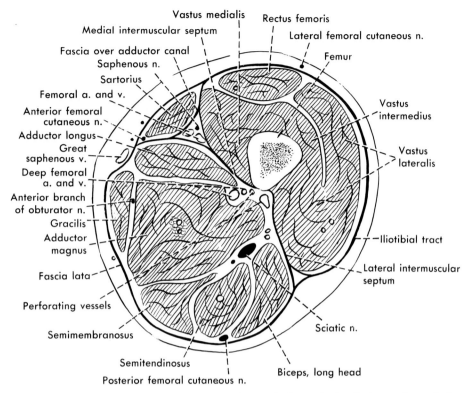

Fig. 8-45. Cross section through the middle third of the thigh. (Redrawn from Eycleshymer AC, Schoemaker DM: A Cross-section Anatomy. New York, Appleton, 1923)

it is bound down by the intermuscular septa, the fascia lata rather loosely ensheaths the muscles of the thigh.

Sometimes the underlying muscles may herniate through the fascia lata, giving rise to a circumscribed swelling that usually disappears as the muscle is contracted vigorously; Simon and Sacchet said that this occurs almost exclusively in males, and is usually not associated with a history of direct trauma. They found that the muscles most commonly involved are the adductors, the rectus femoris, and the vastus lateralis.

The fascia lata varies much in strength as it is traced circumferentially around the thigh: it is particularly poorly developed medially, over the adductor muscles; is moderately well developed anteriorly and posteriorly, where it consists of interwoven fibers directed both transversely and longitudinally, but especially the former; and it becomes particularly

thick laterally where a strong longitudinally running middle layer, derived from the tendons of the gluteus maximus and tensor fasciae latae, forms the iliotibial tract (Figs. 8-45 and 8-47).

Because the lateral part of the fascia lata, in the region of the iliotibial tract, is the strongest part, fascial strips for hernial repair or other uses are usually obtained here. Gratz, reporting upon the strength of fascia lata (presumably in the region of the iliotibial tract), said that it had little strength in resisting a transverse pull, but that its ultimate tensile strength longitudinally was about 7,000 lb/in^2; he calculated that on this basis the safe tension that could be put upon a strip of average thickness and about ³/₁₆ in. (0.48 cm) wide is approximately 8 lb (3.6 kg), and the safe tension for one about 0.5 in. (1.27 cm) wide is 21.5 lb (9.8 kg).

Foshee reported that regenerated portions

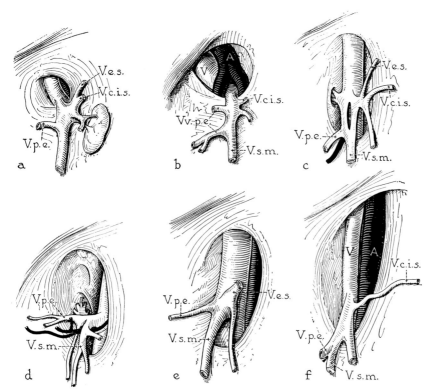

Fig. 8-46. Various types of saphenous hiatus (fossa ovalis). *a* is the smallest hiatus found among 400 examined, *f* the largest. Although the hiatus is small and almost circular in *b*, the femoral artery is as fully exposed as it is in the large hiatus in *f*; in *a, c,* and *d* the femoral vein is fully or partly exposed, but the artery is not; and in *e* the artery is partly exposed. *A* and *V* are the femoral artery and vein, respectively; *V.e.s.* is the superficial epigastric vein, *V.c.i.s.* the superficial iliac circumflex, *V.p.e.* and *Vv.p.e.* the external pudendal vein or veins, and *V.s.m.* the great saphenous vein. (Daseler EH, Anson BJ, Reimann AF, Beaton LE: Surg Gynecol Obstet 82:53, 1946 [by permission of Surgery, Gynecology and Obstetrics])

of the fascia lata (again presumably in the region of the iliotibial tract) typically consist of only two layers, inner and outer transverse ones, with the stronger longitudinal middle layer, typical of the fascia in this region, being missing. He said that after 6 months to a year, regenerated fascia is of sufficient density and strength to be used successfully when normal fascia lata might be used, but it seems obvious from his description that transverse rather than longitudinal strips of regenerated fascia should have the greater strength.

Saphenous Hiatus

Just below the inguinal ligament, on the anteromedial side of the thigh, the fascia lata has special relations at the saphenous opening (saphenous hiatus, fossa ovalis) through which the great saphenous vein passes more deeply to enter the femoral vein (Fig. 8-46). Here a part of the fascia lata, arising from the lower end of the inguinal ligament, curves laterally and downward across the front of the femoral vein (enclosed with the artery in the femoral sheath) and then curves medially across the vein once again just below the entrance of the great saphenous vein into it. Thus a falciform margin, with superior and inferior cornua, demarcates the saphenous hiatus above, laterally, and below. On the medial side of the femoral vein and the saphenous hiatus, however, there is no sharp margin, for the origin of the fascia lata as it is traced laterally from the pubic tubercle is along the pecten of the pubis, where it is continuous behind the femoral vein and the femoral sheath with the fascia on the anterior surface of the pectineus muscle.

The deeper layer of the superficial fascia covering the saphenous hiatus is penetrated by a number of small vessels, and is called the cribriform fascia (fascia cribrosa).

Continuities

As already mentioned in connection with the gluteal region, the fascia lata on the posterior aspect of the thigh attains an attachment above by splitting and passing, as the gluteal fascia, on both surfaces of the gluteus maximus muscle (being closely attached to this muscle by the septa which it sends into it) and thereby reaching the crest of the ilium, the posterior aspect of the sacrum, and the sacrotuberous ligament. In the interval above and lateral to the upper border of the gluteus maximus the fascia lata extends to the ilium over the gluteus medius, where it is aponeurotic and gives rise to superficial fibers of the muscle; anterior to this, it splits to go on both sides of the tensor fasciae latae. In the region of and below the greater trochanter the tendon of the tensor fasciae latae, directed downward and slightly posteriorly, converges with the tendon of the major portion of the gluteus maximus, directed laterally and inferiorly, and the two contribute numerous longitudinal bundles to the fascia lata, so that it becomes thickened here into a heavy band, the iliotibial tract or band, that runs downward to an attachment on the tibia.

Medially, the fascia lata is attached to the ischial tuberosity and hence forward along the conjoined rami of the ischium and pubis to the body of the pubis; from here its line of attachment runs upward and laterally along the anterior surface of the inguinal ligament to the anterior superior iliac spine and the iliac crest.

At the knee the fascia lata is attached laterally, medially, and anteriorly to the upper ends of the fibula and tibia and to the patella, blending with expansions from the quadriceps tendon to form the medial and lateral

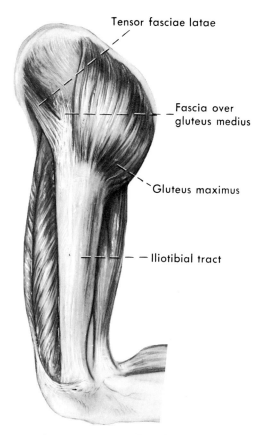

Fig. 8-47. Iliotibial tract of the left thigh. Its three-pronged attachment above (to the gluteus maximus posteriorly, the tensor fasciae latae anteriorly, and the aponeurosis of the gluteus medius in between) is clearly visible.

retinacula of the patella; below the knee it is continuous with the deep fascia of the leg.

Iliotibial Tract

This strong lateral part of the fascia lata begins in the region of the greater trochanter and extends down the thigh to cross the lateral aspect of the knee joint. Above the trochanter both the tensor fasciae latae, anteriorly, and the gluteus maximus, posteriorly, are enclosed in the continuation of the fascia lata to the iliac crest, and the converging tendons of these two muscles, meeting at about the level of the trochanter, give rise to the heavy longitudinally running fibers characteristic of the iliotibial tract (Fig. 8-47). Between these two muscles the fascia lata is continued upward to the iliac crest over the gluteus medius muscle and is tightly bound to the muscle through the origin of the superficial muscle fibers from it. Thus the upper end of the iliotibial tract can be thought of as having a three-pronged attachment to the iliac crest: an anterior attachment through the tensor fasciae latae, a middle one through the aponeurosis over the gluteus medius, and a posterior one through the gluteus maximus.

As it extends down the lateral aspect of the thigh the iliotibial tract is firmly attached on its deep surface to the linea aspera and its diverging lateral lip on the posterior aspect of the femur. Through this attachment, both the tensor fasciae latae and the gluteus maximus gain an indirect insertion into the femur. At its lower end the thickened posterior border of the iliotibial tract passes across the anterior part of the lateral aspect of the knee joint to attach to a smooth facet on the upper part of the lateral tibial condyle. (When the knee is flexed, the tract is carried over the lateral femoral condyle to a posterior position; Renne described a friction syndrome involving the tract that may develop after marching or running.) Expansions from the anterior border of the tract at the knee join expansions from the quadriceps muscle to form the lateral patellar retinaculum, while the posterior part of the tract blends with the more lateral part of the capsule of the knee.

The attachments of the iliotibial tract have been thought to be of particular importance in some deformities of the limb associated with poliomyelitis and with spastic paralysis. In a postpoliomyelitic syndrome referred to as "contracture of the iliotibial band" the more grave deformities are flexion of the hip joint, perhaps with abduction and external rotation, genu valgum (knock knee), and perhaps external rotation and flexion at the knee (Johnson, '53, and others); these may be accompanied by a number of secondary deformities such as discrepancy in leg length, equinovarus deformity of the foot, increased obliquity (downward rotation) of the pelvis, and, in consequence of the latter, increased lumbar lordosis. Section, a few inches above the knee, of the iliotibial tract and of the underlying lateral intermuscular septum that attaches it to the femur, combined with section of the tract and the anterolateral part of the fascia lata at the level of the greater trochanter, is said to relieve the primary deformities; the mechanism of their production, especially whether it is an involvement of the tract or of the muscles associated with it, is not entirely clear, however.

If contracture of the tensor fasciae latae is responsible for the flexion at the hip, section of its attachment to the iliotibial tract, the tendon of insertion of the muscle, should relieve this; the tensor is not a good abductor, however, and certainly cannot produce external rotation, although the gluteus maximus can do both. It is difficult to believe that either muscle can have any effect at the knee, since the tract is fastened firmly to the femur above the knee. Thus the genu valgum is probably due either to direct involvement of the iliotibial tract or to contracture of the muscle attaching to that tract and the lateral intermuscular septum—the short head of the biceps femoris. Contracture of the latter muscle can, of course, produce both genu valgum and flexion at the knee. Flexion at the knee relieved by section of the iliotibial tract at the insertion of the tensor fasciae latae is presumably produced by the passive action of the hamstrings, made more taut by the flexion at the hip.

Kaplan ('58) reported that pulling on the tract does not produce movement at the knee, and that stimulation of neither the tensor fasciae latae nor of the gluteus maximus produced such movement; neither does section of it interfere with the actions of these muscles, for their action is transmitted to the femur through the lateral intermuscular septum. Kaplan expressed the opinion that the hip flexion may be caused by the gluteus minimus and anterior part of the medius, the knee flexion by the short head of the biceps, and that the tensor and tract may not be associated at all with this deformity.

In cerebral palsy, Eggers ('50) reported, the prolonged flexion contracture of the knee stretches the patellar ligament more than it does the patellar retinacula, so that when the flexor spasm is released the quadriceps is restricted in its action by the retinacula. Eggers said that division of the iliotibial tract and section of both retinacula improved the function at the knee, and ('52) that this combined with transplantation of hamstring muscles to the femoral condyles, so that they could not produce flexion at the knee as a result of flexion at the hip, produced a decided improvement in walking and stance in 40 patients.

CUTANEOUS NERVES

There are usually four or five main superficial nerves to the skin of the thigh; two are derived from the femoral, two, the lateral femoral cutaneous and the posterior femoral cutaneous, are independent branches from the lumbar and sacral plexuses respectively, and the fifth, rather variable, is derived from the obturator nerve. In addition to these the femoral branch of the genitofemoral nerve supplies a variably sized area of skin on the front of the thigh beginning immediately below the inguinal ligament, while the genital branch of the genitofemoral and the scrotal or labial branches of the ilioinguinal supply the adjacent skin of the base of the penis and scrotum, or of the mons. Areas of distribution of these nerves, as indicated by clinical findings, are shown in Figures 7-29 and 7-30.

Femoral Branches

The two anterior femoral cutaneous branches of the femoral nerve have also been called the intermediate and medial femoral cutaneous nerves. The former, typically subdivided into two branches, pierces the fascia lata approximately in the anterior midline of the thigh over the sartorious muscle, one branch often passing through the sartorius to reach the fascia; these two branches are then distributed downward over the front of the thigh as far as the knee (Fig. 8-48). The medial of the two anterior cutaneous branches typically gives off a few twigs that pierce the fascia lata close to the saphenous hiatus (fossa ovalis), and supply skin on the upper medial side of the thigh; others pierce the fascia lata in the middle and lower thirds of the thigh to become

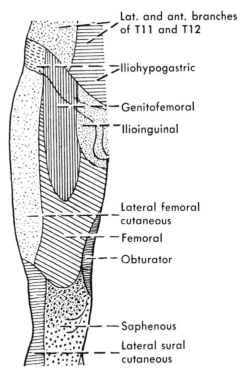

Fig. 8-48. Distribution of cutaneous nerves on the front of the thigh. The genitofemoral has here a rather large distribution. (Flatau E: Neurologische Schemata für die ärtzliche Praxis. Berlin, Springer, 1915)

subcutaneous close to the great saphenous vein. The longer branches tend to descend along the sartorius muscle deep to the fascia lata, one of them piercing the fascia lata in the lower third of the thigh and supplying the more medial part of the anterior surface as low as the knee, while another pierces the fascia more medially, usually gives a twig of communication to the saphenous nerve, and supplies the skin of the medial side in about the distal third or more of the thigh.

Obturator Branch

The skin of the middle third of the medial side of the thigh may be supplied by an anterior femoral cutaneous nerve, but is typically supplied by the cutaneous branch of the obturator nerve. This is not constant, and is of variable size. When it is present, it usually arises from the anterior division of the obturator nerve and pierces the fascia lata after

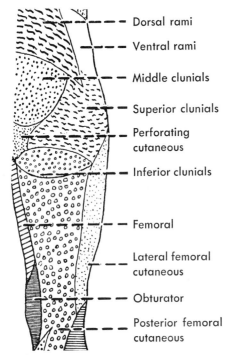

Dorsal rami

Ventral rami

Middle clunials

Superior clunials

Perforating cutaneous

Inferior clunials

Femoral

Lateral femoral cutaneous

Obturator

Posterior femoral cutaneous

Fig. 8-49. Distribution of cutaneous nerves on the posterior aspect of the thigh. (Flatau E: Neurologische Schemata für die ärtzliche Praxis. Berlin, Springer, 1915)

passing between the gracilis and adductor longus muscles in about the upper part of the middle third of the thigh. It supplies skin in its area of emergence, sometimes reaching as far distally as the knee.

Lateral Femoral Cutaneous Nerve

The anterolateral and lateral aspects of the thigh are supplied by the lateral femoral cutaneous nerve (Figs. 8-48 and 8-49). This nerve, usually derived from the first three lumbar nerves, courses across the anterior surface of the iliacus muscle where this forms a part of the wall of the false pelvis, and enters the thigh between the lateral border of the iliacus and the upper end of the inguinal ligament; thereafter it pierces the fascia lata, commonly just below and medial to the anterior superior iliac spine. It may, however, course through the fascia lata obliquely, and attain a subcutaneous position several inches below the spine, giving off smaller subcutaneous branches before it itself attains this position.

The larger part of the nerve runs downward on the lateral side of the anterior aspect of the thigh to supply skin here as far as the knee; a smaller posterior branch runs more laterally, supplying skin below the greater trochanter but usually not reaching the knee. The larger branch of the lateral femoral cutaneous and the branches of the femoral nerve that reach the knee may enter into a plexus over the front of the patella, usually with an infrapatellar branch of the saphenous nerve (the cutaneous branch of the femoral nerve to the leg).

Posterior Femoral Cutaneous

The posterior femoral cutaneous nerve, usually derived from the first three sacrals, sometimes from the first and second or the second and third, runs down the posterior aspect of the thigh in almost exactly the posterior midline, and supplies a posterior strip of skin extending from the buttock to, usually, below the knee (Fig. 8-49). Below the gluteus maximus it lies at first on the posterior aspect of the sciatic nerve deep to the fascia lata, but soon separates from this nerve, since the sciatic nerve passes deep to the biceps femoris while the posterior femoral cutaneous passes superficial to this muscle. In its course deep to the fascia lata it gives off the inferior clunial branches, already mentioned, to skin of the lower part of the buttock, a perineal branch, and other branches that pierce the fascia lata to supply an upper part of the thigh. The terminal branches of the nerve usually become subcutaneous by piercing the fascia lata in the upper part of the popliteal fossa, and continue downward to supply skin over this fossa, including some of the upper part of the leg.

SUPERFICIAL ARTERIES

The main subcutaneous arteries in the thigh arise from the upper part of the femoral artery, and penetrate the fascia lata at or close to the saphenous hiatus. These arteries are the superficial epigastric, the superficial iliac circumflex, and the superficial external pudendal—the first and last, as indicated by their names, being distributed primarily to structures other than the thigh. These vessels all arise close together, sometimes by a stem common to two or all three of them, from the upper few centimeters of the femoral artery.

The *superficial epigastric artery* tends to arise from the anterior surface of the femoral artery. It perforates the femoral sheath and, usually, the superficial fascia (fascia cribrosa) overlying the saphenous opening, and turns upward superficial to the inguinal ligament to run in the superficial fascia of the lower part of the abdomen toward the umbilicus. Before it leaves the thigh it usually gives twigs to adjacent superficial inguinal lymph nodes, but it is distributed primarily to the abdominal wall.

The *superficial iliac circumflex artery* is usually the smallest of the three branches; if it arises independently from the femoral artery it tends to come from the lateral aspect of this, at the level of or only slightly distal to the origin of the superficial epigastric. It also pierces the femoral sheath and either emerges through the saphenous hiatus or pierces the fascia lata to run laterally and upward in a subcutaneous position in the upper part of the thigh, paralleling the inguinal ligament. It supplies branches to the skin of the upper part of the thigh and to the superficial inguinal lymph nodes, connects with the deep iliac circumflex by twigs that pass deep to the inguinal ligament, and anastomoses with the superior gluteal and lateral femoral circumflex arteries below the anterior part of the iliac crest.

The *external pudendal arteries*, usually two, arise from the femoral artery and are frequently described as superficial and deep external pudendals. Both give off inguinal branches and end as anterior scrotal or labial branches. The distinction between the two is that the superficial external pudendal emerges through or close to the saphenous hiatus and runs medially, while the deep external pudendal runs medially deep to the fascia lata and pierces this on the medial side of the thigh close to the scrotum or labium majus.

SUPERFICIAL VEINS

The larger superficial veins of the thigh are all parts of the saphenous system of veins. The details of this system, beginning in the foot and ending in the thigh, have been discussed in the preceding chapter on the limb as a whole, and only a brief mention of the basic anatomy is necessary here.

The *small (lesser) saphenous vein* usually barely extends into the thigh, for after passing up the lateral side of the leg to a posterior position behind the knee it typically penetrates the fascia lata over the popliteal fossa to end in the popliteal vein.

The *great (greater) sapehnous vein* (Fig. 8-50) crosses the medial side of the knee in front of the medial femoral condyle and runs obliquely up the anteromedial aspect of the thigh, gradually assuming a more anterior position, to enter the saphenous hiatus (fossa ovalis), penetrate the lower end of the femoral sheath, and end in the femoral vein a little below the inguinal ligament. In its course in the thigh it receives additional tributaries, but its more constant ones enter it close to its upper end. These may include the veins accompanying the superficial arteries (the superficial epigastric, superficial iliac circumflex, and external pudendal, Fig. 8-46) and typically include at least one, sometimes two, larger veins. These larger veins entering the upper end of the great saphenous are the medial and lateral accessory saphenous veins. The lateral accessory saphenous vein is apparently more constant (Chap. 7), and frequently receives the superficial epigastric and superficial iliac circumflex veins, which otherwise enter the great saphenous or the femoral vein directly. The medial accessory saphenous is much less constant; it sometimes receives the external pudendal, or the latter may go directly into the great saphenous or the femoral.

LYMPHATIC VESSELS AND NODES

As noted in Chapter 7, the superficial lymphatic vessels of the lower limb are much more numerous than are the deep ones. In the thigh, the greatest number course upward in

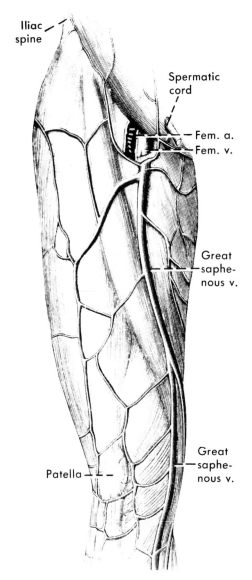

Iliac spine

Spermatic cord

Fem. a.
Fem. v.

Great saphenous v.

Great saphenous v.

Patella

Fig. 8-50. The great saphenous vein in the thigh; a prominent lateral accessory saphenous vein opens into its upper end. (Testut L: Traité d'Anatomie Humaine, Vol 2. Paris, Doin, 1891)

the region of and approximately parallel to the great saphenous vein; they are a continuation upward of vessels of the leg and are joined by the superficial lymphatics draining the thigh. The latter originate close to the posterior midline of the thigh, where there is a "lymph-shed"; vessels on the lateral side of the posterior midline pass laterally and upward around the lateral aspect of the thigh to-

ward the great saphenous vein and the superficial iliac lymph nodes; those on the medial side pass around the medial side of the thigh.

The deep lymphatics of the thigh accompany the femoral vessels.

Superficial Inguinal Nodes

The superficial inguinal lymph nodes lie in the superficial fascia of the upper part of the thigh, arranged in general slightly below and parallel to the inguinal ligament, and extending downward on both sides of the upper end of the great saphenous vein (Fig. 8-51). As a whole, therefore, they form a somewhat distorted T. They vary much in both number and size. It is said that when their afferents are injected so that even the smallest ones can be recognized, some 12 to 20 are usually found; Daseler, Anson, and Reimann found from 4 to 25 per limb among 450 extremities (apparently of regular dissecting room material), with an average of 8.25. Variation in size is not necessarily inversely correlated with their number, for these nodes are so frequently infected that hyperplasia of them is common. Daseler and his colleagues said that in their material the surface area of the nodes varied from 0.6 cm^2 to 10 cm^2.

These nodes have numerous connections with each other, and since they vary so much in number they necessarily vary in regard to the afferent lymphatics that they receive. Any classification of them into subgroups is therefore quite arbitrary, and certainly not of functional importance. For purposes of description, however, they are usually subdivided; the simplest subdivision is into the two obvious although interlocking groups, the proximal group approximately paralleling the inguinal ligament (sometimes the term "superficial inguinal nodes" is restricted to these), and the distal group around the upper end of the great saphenous vein (sometimes these are called "superficial subinguinal nodes"). Another way of subdividing them is into five groups, according to their relation to two lines, one drawn horizontally across the saphenofemoral junction, the other vertically along and above the upper end of the great saphenous; by this method the groups in the

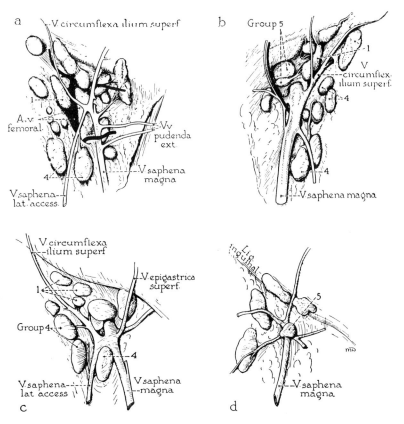

Fig. 8-51. Varying arrangements of the superficial inguinal lymph nodes. *a, c,* and *d* are drawn from a right limb, *b* from a left one. The grouping shown here by numerals is on the basis of quadrants about the saphenofemoral junction, Groups *1, 2, 3,* and *4* being nodes in the upper lateral, upper medial, lower medial, and lower lateral quadrants, respectively, and *5* the central group over the venous junction. (Daseler EH, Anson BJ, Reimann AF: Surg Gynecol Obstet 87:679, 1948 [by permission of Surgery, Gynecology & Obstetrics])

four quadrants become superior medial and superior lateral, inferior medial and inferior lateral, those at about the junction of the quadrants become "central" nodes.

The superficial inguinal lymph nodes receive superficial lymphatics from the entire lower limb, the majority but not all of these going into the distal or "superficial subinguinal" nodes; they also receive superficial lymphatics from the infraumbilical part of the abdominal wall, the buttock, the perineum, the anal canal, and the penis and scrotum or the labia, the majority of these, but not all, going into the proximal or "superficial inguinal" nodes. According to some workers lymphatics from the glans penis or glans clito-

ris may also join the superficial nodes, but they are usually said to go to the deep nodes.

The efferent lymphatics of the superficial inguinal nodes go in part to the deep inguinal nodes (especially from the lower nodes), but mostly to the external iliac nodes.

Deep Nodes

The *popliteal nodes* are tiny and difficult to demonstrate. They lie in the connective tissue of the popliteal space, deep to the fascia lata. They are said to include one on the terminal portion of the lesser saphenous vein, two to four along the sides of the popliteal vessels, and usually one in front of the vessels, on the back of the capsule of the knee joint. However, Riveros and Cabanas injected lymphatics about the lateral malleolus and found that the popliteal nodes not only vary in size and number, but are not always present, so that lymphatics from the region of the heel may drain directly into inguinal nodes.

The *deep inguinal nodes* (sometimes also known as deep subinguinal nodes) likewise lie deep to the fascia lata. They tend to be smaller than the superficial nodes, and are usually described as varying in number from one to three. Daseler and his colleagues described them as forming an almost continuous chain along the upper ends of the femoral and deep femoral vessels. In Poirier and Charpy it is said that when there are as many as three, the lowest is below the saphenofemoral junction, the next is in the femoral canal, and the highest (node of Cloquet or of Rosenmuller) lies at the femoral ring. It is said that the middle node is the least constant, and that while none of them is constant, complete absence of all deep nodes is very rare.

The afferent vessels of the deep inguinal nodes are the lymphatics accompanying the femoral and deep femoral vessels, some vessels from the superficial inguinal nodes, and vessels from the glans penis or glans clitoris. Their efferent lymphatics pass into the pelvic cavity to terminate mostly in the lowest medial member of the external iliac chain of nodes.

Dissection of Inguinal Nodes
Because of the wide field of drainage of their afferent lymphatics, the inguinal nodes may be involved by metastasis from carcinoma not only of the lower limb but also of the buttock, the perineum including the anal canal and genitalia, and the infra-umbilical portion of the abdominal wall. Daseler and his colleagues, Baronsofsky, and Mendelsohn and Mansfield have discussed the technic of radical dissection of inguinal nodes. In general, it includes removal of all the superficial fascia in a quadrilateral block extending slightly above, farther below, the inguinal ligament, with sacrifice of the great saphenous vein; removal of the fascia lata over the femoral triangle, with clean dissection about the femoral vessels and nerve; and, usually, an extra-peritoneal abdominal incision with removal of all connective tissue and nodes about the external iliac vessels. Daseler and his co-workers stressed also the importance of opening the inguinal canal and stripping the connective tissue from the ductus deferens and the accompanying vessels, in order to eliminate the lymphatic vessels that take this course to the pelvis.

The relatively new technic of lymphangiography has made possible preoperative studies of the lymphatic drainage of the limb and the extent of involvement of lymph nodes by metastases. The superficial inguinal and the external and common iliac nodes, as well as the lumbar nodes and the thoracic duct, can be opacified through a lymphatic of the dorsum of the foot after the lymphatics have been visualized by injection of dye into the web space between the first and second toes (Fig. 8-52).

ANTERIOR MUSCLES

The muscles of the anterior aspect of the thigh (Fig. 8-53; see also Fig. 8-62) are conveniently divided into three groups: the iliopsoas, largely in the abdomen; the anterior muscles proper, innervated by the femoral nerve; and the anteromedial muscles, innervated by the obturator nerve. Closely associated with the iliopsoas in the abdomen are the psoas minor and the quadratus lumborum, attached both to the axial skeleton and to the pelvic girdle. The latter muscle has already been described (Chap. 2).

The anterior group includes the sartorius and the four parts of the quadriceps muscle; the pectineus also belongs to this group, but sometimes receives an innervation from the obturator nerve. The tensor fasciae latae, also appearing on the anterior aspect of the upper part of the thigh, actually is a member of the gluteal group of muscles, and is described with those muscles. The anteromedial muscles include the adductor brevis, the adductor longus, the upper and anterior part of the adductor magnus (a part of this muscle also belongs with the hamstrings), and the gracilis.

The most superficial anterior and the most

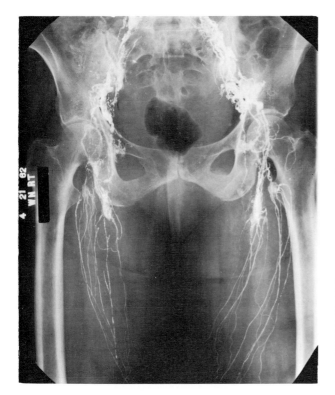
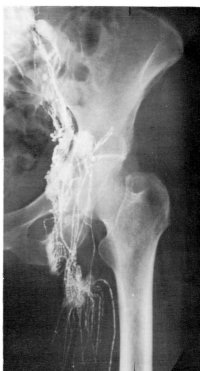

Fig. 8-52. Two lymphangiograms showing superficial inguinal nodes and their drainage into iliac nodes. (Courtesy of Dr. W.E. Miller)

superficial anteromedial muscles, the sartorius and the adductor longus, converge toward each other from their origins close to the two ends of the inguinal ligament, and therefore with this ligament form a triangle, the femoral (Scarpa's) triangle, in the upper part of the thigh. Further, the anterior and anteromedial muscles, as they lie close together in the thigh, shelter the femoral vessels after these leave their almost superficial position in the femoral triangle, and thus form a canal about these vessels. These two structures, the femoral triangle and the adductor canal, merit special description.

Femoral Triangle

The upper border or base of the femoral or Scarpa's triangle (Fig. 8-54) is the inguinal ligament, its lateral border is the medial edge of the sartorius muscle, and its medial border is usually defined as the medial edge of the adductor longus muscle, but sometimes as its lateral edge. The sartorius runs obliquely downward and medially across the front of

the thigh to pass across the front of and then onto the medial side of the adductor muscles, and the apex of the triangle therefore lies at the point at which the sartorius and the adductor longus meet.

The fascia lata covers, or forms the roof of, the femoral triangle, and is in this position perforated by the saphenous hiatus (fossa ovalis) to permit the junction between the great saphenous and femoral veins. The medial part of the floor of the triangle is the anterior surface of the adductor longus muscle. Lateral to the adductor longus, the pectineus forms a major part of the floor; a small part of the adductor brevis may appear between the adjacent borders of the adductor longus and pectineus and thus in the floor of the triangle, or these borders may be contiguous, and the brevis entirely hidden behind the longus. The most lateral part of the floor is the iliopsoas muscle.

The femoral artery, vein, and nerve all lie in the femoral triangle, the great saphenous vein enters the triangle to end in the femoral

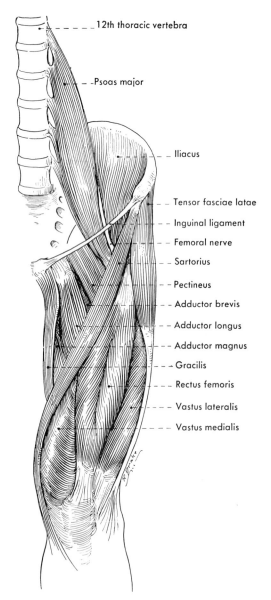

12th thoracic vertebra

Psoas major

Iliacus

Tensor fasciae latae

Inguinal ligament

Femoral nerve

Sartorius

Pectineus

Adductor brevis

Adductor longus

Adductor magnus

Gracilis

Rectus femoris

Vastus lateralis

Vastus medialis

Fig. 8-53. Anterior muscles of the groin and thigh.

vein, and the deep inguinal lymph nodes also lie in this triangle. As the femoral artery and vein lie in the triangle just below the inguinal ligament, they are enclosed within the femoral sheath, which contains a compartment for each vessel, but also a third, most medial, compartment, the femoral canal. The arrangement of the more important structures immediately below the inguinal ligament is therefore, beginning medially, the femoral

canal, the femoral vein, and the femoral artery, all enclosed within the femoral sheath, and then, most laterally, the femoral nerve, lying outside the sheath. This exact relationship is not maintained for much more than 2 cm or 3 cm, however, for the femoral canal ends a short distance below the inguinal ligament, through fusion of its walls to each other, and the femoral nerve begins to branch extensively. The great saphenous vein pierces the femoral sheath to enter the femoral vein, and lymphatics and the smaller upper branches of the femoral vessels also pierce the sheath.

All the major branches of the femoral artery leave it while it lies in the femoral triangle, all the major tributaries of the femoral vein join it in this position, and the femoral nerve divides into its terminal branches in the triangle. The femoral artery and vein, accompanied by two branches of the femoral nerve (the saphenous nerve and the nerve to the vastus medialis) run straight downward, the artery gradually passing in front of the vein, to the apex of the triangle, where they disappear under cover of the sartorius into the adductor canal.

Adductor Canal

The adductor canal, also known as the subsartorial or Hunter's canal, begins at the apex of the femoral triangle and extends downward and medially deep to the sartorius muscle to end at the hiatus in the tendon of the adductor magnus, close to the medial aspect of the femur, a little above the medial epicondyle. It is essentially triangular in cross section, for it lies between the converging medial surface of the vastus medialis and the anterior surfaces of the adductor muscles, adductor longus above, magnus below. The sartorius covers it, but not directly; after that muscle is removed the actual roof of the canal is seen to be a dense membrane, a derivative of the medial intermuscular septum that passes between the muscles forming the two sides of the canal (Fig. 8-45).

The adductor canal contains the femoral (superficial femoral) artery and vein, with the artery anteromedial (superficial) to the vein,

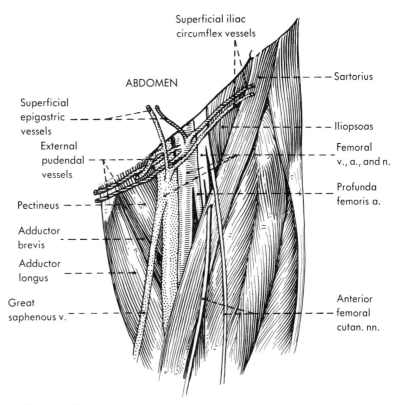

Fig. 8-54. The femoral triangle.

the saphenous nerve, and, in its upper part, the nerve to the vastus medialis.

PSOAS GROUP

Iliopsoas

The iliopsoas as it appears in the thigh is a blending of two distinct muscles of the abdominal wall, the iliacus and the psoas major (Fig. 8-55).

The *iliacus* is a fan-shaped muscle, fleshy at its origin, which is primarily from the inner surface of the wing of the ilium; it thus forms the lateral wall and floor of the false pelvis or iliac fossa. Some few fibers arise also from the pelvic surface of the lateral part of the sacrum, and from adjacent ligaments—the ventral sacroiliac, the iliolumbar, and the lumbosacral ligaments. It passes behind the inguinal ligament lying close on the lateral side of the psoas major, and blending with that muscle as the two descend into the thigh; some of its fibers insert into the la-

teral side of the tendon of the psoas, others descend parallel to the psoas to insert into the lesser trochanter with it and into the femur for approximately 2 cm below the trochanter.

The *psoas major* arises from the side of the vertebral column, from about the lower border of the twelfth thoracic through the upper border of the fifth lumbar vertebra, by fleshy origins from the adjacent borders of the vertebrae and the intervening disks; it also arises from the membranous arches that span the middle of the bodies of the upper four lumbar vertebrae, allowing passage for the lumbar vessels and, usually, the rami communicantes; it likewise has a deep origin from the transverse processes of all the lumbar vertebrae. Its upper end is crossed by the medial arcuate ligament (lumbocostal arch) of the diaphragm, so that it lies largely but not entirely in the abdominal cavity. The muscle narrows as it lies along the lateral brim of the pelvis; it passes behind the inguinal ligament, with the

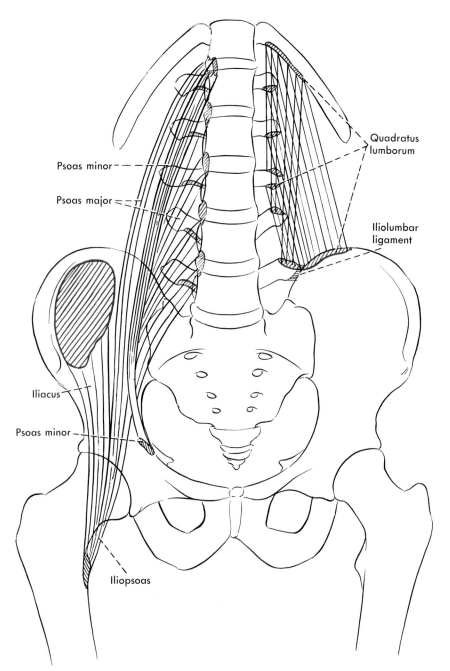

Fig. 8-55. Origin (*red*) and insertion (*black*) of the iliopsoas (psoas major and iliacus). The psoas minor and the quadratus lumborum, which attach to the coxal bone, are also shown.

iliacus muscle joining it on its lateral side, and crosses the front of the hip joint to insert into the lesser trochanter. Below the inguinal ligament the iliopsoas forms a part of the floor of the femoral triangle. The deep surface of the muscle is tendinous as it crosses the hip joint,

and a bursa (see a following section) lies deep to the muscle here.

The *innervation* of the psoas major is usually from about the second to the fourth lumbar nerves, probably fairly frequently also from the first, and sometimes from the fifth; the twigs to it are typically derived from the nerves as they are forming the lumbar plexus in the substance of the muscle. The iliacus usually receives its nerve or nerves as a direct branch or branches from the femoral nerve; it is usually supplied through the second and third or third and fourth lumbar nerves.

There is universal agreement that the chief *function* of the iliopsoas is that of flexion of the thigh; it is the most powerful flexor at the hip. It has been said that it is used during flexion only when power is demanded, but Keagy and co-workers found the psoas major participating regularly in the forward swing of the limb during walking. Basmajian ('58) found the iliacus active during quiet standing, but Keagy and co-workers found no activity in the psoas major under these conditions, although it became markedly active upon leaning back; thus the iliacus may be a postural muscle at the hip, but the psoas major apparently is not. In its action on the vertebral column, the psoas major does not participate in lateral flexion to its own side, but the contralateral muscle becomes markedly active.

The iliopsoas has been listed also as an adductor, especially when the thigh is flexed, and as a rotator. Its adductor action, if any, is unimportant; Duchenne obtained none from stimulation of the iliacus, and Keagy and co-workers, describing some activity in the psoas major during both abduction and adduction, concluded that this was an isometric stabilizing action only. McKibbin, however, said it becomes a strong abductor as the limb reaches the extreme of this movement. It is the supposed rotatory action of the iliopsoas that has aroused the most controversy, for the muscle has variously been regarded as internally rotating, externally rotating, or rotating in either direction depending on the amount of flexion or extension at the hip. The controversy is perhaps now settled, for while Keagy and his colleagues found slightly more activity in the psoas major during external than internal rotation, implying that the iliopsoas was acting largely as a stabilizer, Basmajian and Greenlaw reported that neither the psoas nor the iliacus is active during internal rotation, but that both often become active in ex-

ternal rotation. Basmajian ('71) added that the iliopsoas is regularly active in external rotation. Although this apparent external rotatory action may be of no great importance in the adult, McKibbin has described the muscle as having a strong external rotatory effect in the infant, particularly when the limb is abducted, and noted that the muscle may have to be severed in order to correct an external rotatory deformity.

Psoas Minor

The psoas minor does not appear in the thigh, but it is closely associated with the anterior surface of the psoas major muscle, and is best considered here. It is a thin and inconstant muscle, typically arising from the adjacent borders of the last thoracic and first lumbar vertebrae and from the intervertebral disk between; it descends in contact with the anterior surface of the psoas major to give rise to a long flat tendon that inserts into the iliopubic (iliopectineal) eminence and the arcuate line.

Seib reported upon the incidence of the muscle in 500 American bodies, and summarized his findings and those already reported in the literature as indicating that among 2,627 cadavers the muscle as absent bilaterally in 51.5%, present bilaterally in 35.8%, and present unilaterally in 12.7%; among 5,903 sides it was absent in 57.3%. Some racial differences were found to exist, the muscle apparently being most often absent in Negroes, least often absent in yellow races, and intermediate in its absence (57.1%) in the white race.

The innervation of the psoas minor is apparently variable; Thomson reported that among 108 cases in which its nerve supply was determined this was from the twelfth thoracic in 2, from a loop between the twelfth thoracic and the first lumbar in 2, from the first lumbar in 36, from a loop between the first and second lumbars in 28, from the genitofemoral (typically itself from L1 and L2) in 7, from the second lumbar in 27, and from the loop between the second and third lumbars in 6. The psoas minor presumably helps to tilt the pelvis upward.

Anomalies

Anomalous arrangements of the psoas group, other than the very common absence of the psoas minor already noted, are apparently very rare. Occasionally a slip of the iliacus, called the iliacus minor or the iliocapsularis, arises from the anterior inferior iliac spine and inserts either into the intertrochanteric line of the femur or into the iliofemoral ligament. Clarkson and Rainy reported a case in which on one side there were four psoas muscles: the usual psoas major and minor, and two others that fused with the tendon of the psoas major at about the level of the inguinal ligament.

Iliopsoas Fascia and Abscess

The iliopsoas muscle is covered by a distinct and fairly strong, but thin, layer of fascia. This fascia is attached above to the medial arcuate ligament of the diaphragm, and below this to the vertebral column at the origins of the muscle; laterally, it is continuous above the iliac crest with the fascia of the quadratus lumborum, below the crest with the fascia on the inner surface of the iliacus muscle. The iliac fascia, in turn, although attached to the crest of the ilium, is continuous above the crest with the transversalis fascia lining a major part of the abdominal cavity. As the iliac and psoas fascia come together they form the iliopsoas fascia, and as the muscle starts across the pubis the iliopsoas fascia becomes continuous also with the fascia over the upper part of the pectineus muscle, thus forming an iliopectineal fascia.

The lateral part of the inguinal ligament is attached more or less firmly to the anterior surface of the fascia over the iliacus. Medial to this, behind the inguinal ligament, a special thickened band of the fascia, the iliopectineal arch, passes from the junction of the inguinal ligament and the iliac fascia to run across the front of the femoral nerve, as this lies in the groove between the iliacus and the psoas major, and blend with the anterior layer of the psoas fascia. It continues behind the femoral artery and vein to attach to the pubis between the psoas major and the pectineus, and thence becomes continuous with the anterior layer of fascia of the pectineus.

Because of the iliopectineal arch the space behind the inguinal ligament is subdivided into two compartments, a lacuna musculorum, situated primarily laterally and containing the iliacus and psoas major muscles and the femoral nerve, and a lacuna vasorum, situated primarily medially but also overlapping anteriorly the lacuna musculorum, and containing the femoral vessels and the femoral sheath which encloses them and the femoral canal. In the thigh, the iliopsoas and the pectineal fascias are continued to the insertions of these muscles, thereby helping to form the fascial floor of the femoral (Scarpa's) triangle.

As the femoral nerve enters the thigh it emerges through and then lies anterior to the iliopsoas fascia, and contains no special fascial investment, but the femoral sheath follows the vessels for a short distance into the thigh.

Because the fascia of the psoas is continuous from the vertebral column to the insertion of the muscle in the thigh, infections that gain entrance into the muscle can travel down or up within the fascia, and thus progress from the abdominal cavity to the thigh or vice versa. Many iliopsoas abscesses are tuberculous, originating from involvement of the adjacent vertebral column, and Weinberg has discussed diagnosis and treatment of these. Norrish said that acute psoas abscess may arise from intestinal infection, from a perinephric abscess, from involvement of lumbar retroperitoneal nodes, and the like. Freiberg and Perlman pointed out that acute infections of the hip joint can give rise to abscesses in the abdominal cavity through passage upward of the infection in the iliacus or psoas muscles. Norrish described the typical symptom of psoas abscess as being the maintenance of the thigh in a position of flexion and external rotation.

Iliopectineal Bursa

The iliopectineal (iliopsoas) bursa lies deep to the iliopsoas muscle and tendon as it crosses the anterior surface of the hip joint. Hucherson and Denman found the bursa to be usually about 3 cm to 7 cm long, 2 cm to 4 cm wide. Since it is over the thin anterior portion of the capsule between the iliofemoral and ischiofemoral ligaments, communications between the bursa and the hip joint occur fairly frequently. Chandler found it absent in only 3 of 206 adult limbs that he investigated, and it communicated with the hip joint in almost 15% of cases, a percentage quoted by Hucherson and Denman as having been found also among 700 adult specimens. However, there was no communication in the 27 fetuses and still-born specimens that Chandler investigated, and the communication appears therefore to be a result of attritional changes produced by the action of the iliopsoas tendon.

O'Connor, Finder, and Hucherson and Denman all expressed the opinion that iliopectineal bursitis is probably not as rare as it has seemed to be, because it has frequently not been recognized. The latter workers described the typical signs of iliopectineal bursitis as being a characteristic area of tenderness over the region of the bursa, with the thigh held in flexion, adduction, and moderate external rotation. Stephens, reporting two cases of cystic tumor of the iliopectineal bursa in which the bursa had to be removed, said that in the one of the two in which there was a connection to the joint there were arthritic and atrophic changes there.

ANTERIOR (FEMORAL) GROUP

These are the sartorius, the quadriceps (rectus femoris and three vasti) and the closely associated articularis genus, and the pectineus. The pectineus extends from the pelvic girdle to the femur, and therefore crosses only one joint, that of the hip; the sartorius and the rectus femoris cross both the hip and the knee joints; and the three vastus muscles cross the knee joint only. The muscles are innervated by the femoral nerve.

Sartorius

The sartorius (Figs. 8-53 and 8-56) is a long muscle, straplike both in general form and in the direction of its fiber bundles, that runs obliquely across the anteromedial aspect of the thigh. It arises from the anterior superior iliac spine and runs downward and medially, in a superficial position immediately deep to the fascia lata, to the medial surface of the

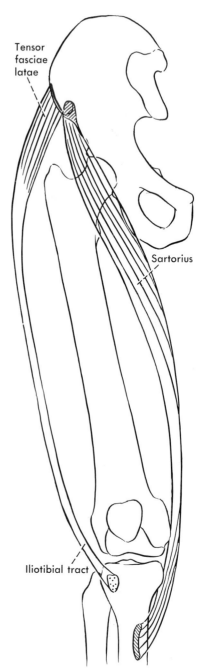

Fig. 8-56. Origin (*red*) and insertion (*black*) of the sartorius. The tensor fasciae latae, closely associated at its origin, is also shown.

thigh, passing across the knee somewhat behind the axis of motion of the joint. Below the knee it turns forward and becomes tendinous; the tendon of insertion expands to attach along the upper part of the medial surface of

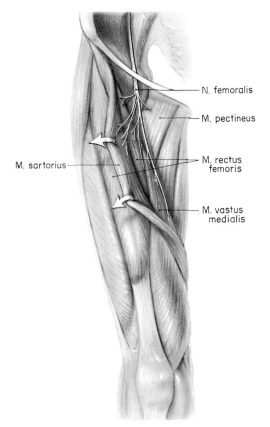

N. femoralis

M. pectineus

M. sartorius

M. rectus femoris

M. vastus medialis

Fig. 8-57. Innervation of the sartorius, pectineus, and parts of the quadriceps by the femoral nerve. (Hollinshead WH, Markee JE: J Bone Joint Surg 28:721, 1946)

the body of the tibia, in front of and largely above the insertions of the gracilis and the semitendinosus. With these two tendons, the tendon of the sartorius is sometimes described as forming the "pes anserinus" tendon. Like most of the tendons at the knee, the tendon of the sartorius gives off expansions that join the fascia of the leg and the superficially lying ligaments on the medial side of the knee.

In its upper part the sartorius forms the lateral boundary of the femoral triangle. As it runs downward and medially on the front of the thigh it comes to fill the angle between the vastus medialis and the adductor group of muscles, and in the middle of the thigh lies immediately upon the fascial roof of the adductor (subsartorial, Hunter's) canal. The chief stems of the femoral artery and femoral

vein (below the profunda femoris, and therefore sometimes described as the superficial femoral artery and vein) lie in the adductor canal, with the muscular branch of the femoral nerve to the vastus medialis, and the saphenous branch of the femoral nerve to skin of the leg.

Deep to the tendon of insertion of the sartorius there is usually a subtendinous bursa (subtendinous bursa of the sartorius, tibial intertendinous bursa) that tends to separate it from the insertions of the gracilis and semitendinosus; it may communicate with, and thus form a part of, the anserine bursa.

The sartorius muscle is frequently pierced by one of the cutaneous branches of the femoral nerve; it receives its innervation by, as a rule, two muscular branches that enter the upper third of the muscle not very far apart (Fig. 8-57). Both muscular branches may leave the femoral with the cutaneous branch that pierces the muscle. As indicated by its name, "the tailor muscle," the complicated series of actions involved in assuming a sitting posture and crossing the legs has been attributed to the sartorius—flexion, abduction, and external rotation at the hip joint, and flexion at the knee joint. The particularly long fiber bundles that are characteristic of the sartorius indicate that it is adapted for simultaneous flexion at the hip and knee, and Duchenne obtained this upon stimulation of it; however, Johnson and co-workers found it active in both flexion and extension of the knee (usually more so in flexion than extension, but in some persons the reverse). Its greatest activity, they reported, is in flexion at the hip, but it also participates more weakly in abduction and external rotation; they regarded it as a regulator of flexion and lateral rotation of the thigh during the swing phase of gait.

Quadriceps Femoris

The quadriceps is composed of four muscles or heads of origin—the rectus femoris, and the vastus lateralis, intermedius, and medialis—which blend together at their insertions (Figs. 8-53, 8-58 and 8-59).

The *rectus femoris* is the anterior member of this group; it expands in a spindle-shaped

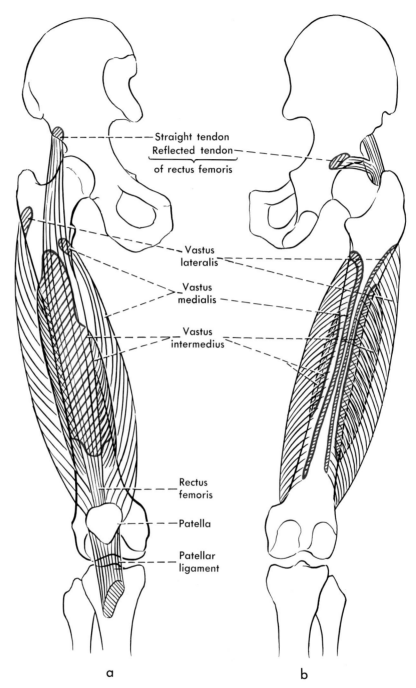

Straight tendon
Reflected tendon
of rectus femoris

Vastus
lateralis

Vastus
medialis

Vastus
intermedius

Rectus
femoris

Patella

Patellar
ligament

a   b

Fig. 8-58. Origin (*red*) and insertion (*black*) of the quadrices in anterior, *a*, and posterior, *b*, views.

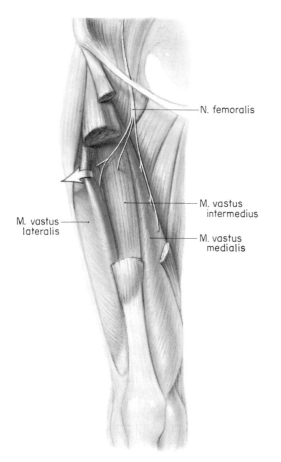

M. vastus lateralis

N. femoralis

M. vastus intermedius

M. vastus medialis

Fig. 8-59. Distribution of the femoral nerve to the vastus muscles. See also Fig. 8-57. (Hollinshead WH, Markee JE: J Bone Joint Surg 28:721, 1946)

distance in the muscle, giving rise to muscle fibers on its sides. The muscle at first lies under cover of the sartorius and of the lateral part of the iliopsoas, but appears superficially between the upper part of the sartorius and the tensor fasciae latae.

Toward its lower end the rectus femoris again becomes tendinous, and the tendon narrows and thickens; it expands as it reaches and attaches to the upper border of the patella, receiving at the same time some attachment of the muscle fibers of the vastus medialis and vastus lateralis. The patellar ligament, actually the lower end of the tendon of the quadriceps, attaches the rectus and the other members of the quadriceps group, through the mediation of the patella (a sesamoid bone), to the tibial tuberosity.

The *vastus lateralis* is a large muscle, partly covered anteriorly by the rectus femoris and the tensor fasciae latae, and itself covering much of the anterior and lateral surface of the vastus intermedius. It arises anteriorly, laterally, and posterolaterally along a line that extends from the lower part of the front of the greater trochanter, below the trochanter, and lateral to the gluteal tuberosity, to the lateral lip of the linea aspera, and posteriorly, by tendinous fibers that help form the lateral intermuscular septum, from most of the length of the lateral lip. The upper part of its origin is largely muscular. The fibers run downward and forward to insert into the lower part of the tendon of the rectus femoris and the upper and lateral borders of the patella; some of the most lateral tendinous fibers at the knee continue to an attachment on the front of the lateral condyle of the tibia, thereby forming a part of the lateral patellar retinaculum.

The *vastus medialis* likewise has some of its origin on the anterior aspect of the femur, but most of it posteriorly; its upper fibers arise from the lower part of the anterior portion of the intertrochanteric line, and the origin of the muscle follows the medial and posterior continuation of this line, formerly called the spiral line, to an attachment on the medial lip of the linea aspera; the origin of this muscle extends lower than that of the vastus lateralis, for it occupies the whole length of the linea

fashion between its origin and insertion, forming the prominence of the anterior part of the thigh; it is a particularly strong muscle, of a general bipenniform type but with the fibers also describing a curve as they originate posteriorly and insert anteriorly. It is the only member of the quadriceps group that crosses the hip joint; it arises by two tendons, the larger, rounded, straight tendon arising from the anterior inferior iliac spine, and a thinner and flattened tendon with a lower lacunar border, the reflected tendon, arising from the ilium just above the acetabulum and joining the deep (posterior) aspect of the straight tendon. A bursa typically lies deep to the reflected tendon. The single tendon formed by the junction of these two descends for some

aspera and extends onto the lower diverging portion (supracondylar line) to occupy about the upper two thirds of this. In its origin from the linea aspera it is tightly bound to the adductor muscles by the medial intermuscular septum, from which some of its fibers arise; superficial to this union between it and the adductor muscles, the vastus medialis forms the medial wall of the adductor canal. The fibers of this muscle are directed downward and laterally, the lowest muscle fibers being inserted more nearly horizontally; most of the fibers attach, with no obvious tendon, into the medial aspect of the tendon of the rectus femoris and the upper and medial borders of the patella; some are continued by a thin tendon to the front of the medial condyle of the tibia, here forming a part of the medial patellar retinaculum.

The *vastus intermedius* covers a major portion of the front and sides of the body of the femur, arising from approximately the upper two thirds of the front and lateral side of the body, and so bulging as to cover much of the medial side, even though it has little attachment on this side. It is completely covered by the vastus lateralis and medialis on the sides and by the rectus femoris in front. In the lower half of the thigh its origin extends posteriorly around the femur to the lateral lip of the linea aspera, including an upper portion of the diverging part of this lip (lateral supracondylar line), where it is closely fused with the vastus lateralis and is usually described as obtaining, like the latter muscle, some origin from the lateral intermuscular septum. The fiber bundles of the muscle descend almost vertically, the more anterior ones becoming tendinous some distance above the knee, to fuse more or less with the adjacent vastus medialis and vastus lateralis muscles and insert especially into the deep part of the upper border of the patella, and into the deep surface of the tendon of insertion of the vastus lateralis. Between its lower part and the anterior surface of the femur is the suprapatellar bursa, typically appearing as an upward prolongation of the cavity of the knee joint.

The *articularis genus* (formerly genu), a derivative of the vastus intermedius, consists of a few bundles of muscle fibers that arise from the front of the femur at about the junction of the upper three fourths with the lower fourth, under cover of the vastus intermedius, and insert into the upper and posterior aspect of the synovial membrane of the knee joint.

The *innervation* of the quadriceps muscle is by several different nerves, all branches of the femoral (Figs. 8-57 and 8-59). They arise high in the femoral triangle, only a little below the inguinal ligament, as the femoral nerve breaks up into a number of muscular and cutaneous branches. The rectus femoris typically receives two nerves, one entering the deep surface of its upper third, the other tending to continue along the medial surface of the muscle and enter it at about the junction of its upper and middle thirds. The vastus lateralis receives from one to three nerves, all of them typically entering its upper third; sometimes a branch of a nerve to the vastus lateralis is continued into the substance of the lateral part of the vastus intermedius, to help supply this muscle.

The vastus medialis receives one or sometimes two nerves that run downward along the medial surface of this muscle. The lower part of the nerve lies in the adductor canal close to the saphenous nerve, to which it may be bound by a common connective tissue wrapping; it gives off a series of branches into the muscle, the terminal one entering above the midpoint of the muscle. It is said that fibers from the nerve to the vastus medialis usually also supply a medial part of the vastus intermedius. The nerve to the vastus intermedius is usually single, passing into the anterior surface of the muscle, but continuing down through the muscle to supply also the articularis genus. As already noted, the vastus intermedius may be also supplied from the nerves of the vastus lateralis and vastus medialis.

The various nerves to the quadriceps consist largely of fibers derived from the third and fourth lumbar nerves, usually with a minor proportion derived from L2; in some cases, they apparently contain fibers derived also from L5.

The *action* of the quadriceps as a whole is

extension of the knee. The rectus femoris, because it passes in front of the hip joint, is at the same time a flexor of the thigh; it is, however, so closely related to the front of the hip joint that it is not as good a flexor of the thigh as its strength would indicate. Duchenne said it flexed the thigh only weakly if the leg was extended, but flexed it more strongly when the leg was flexed; Wright said that upon palpation the rectus appeared to contract not at all in unresisted hip flexion, but did contract, although weakly, when the movement was strongly resisted. Further evidence of its rather weak flexor action at the hip is afforded by Wright's observation that if extension of the knee is carried out with the hip already sharply flexed, the rectus does not even maintain it in flexion, for the pull of the hamstrings produces some extension at the hip. Wheatley and Jahnke found the upper, but not the lower, part of the rectus femoris apparently participating in abduction of the thigh.

While the rectus femoris pulls the patella proximally only, the vastus lateralis and vastus medialis pull both proximally and laterally, and proximally and medially, respectively; Duchenne obtained subluxation of the patella by isolated stimulation of the vastus lateralis, but could not obtain subluxation by isolated stimulation of the vastus medialis. As might be expected, the rectus is less efficient as an extensor of the leg when the thigh is flexed. It is usually said (e.g., Steindler) that the rectus femoris alone cannot produce complete extension at the knee, and that about the last 10° to 15° of extension is carried out by the vasti. Neither Lieb and Perry nor Basmajian and co-workers, however, could confirm the often-stated belief that it is the vastus medialis that is responsible for the last 15° of extension. Wheatley and Jahnke's electromyographic records seem to indicate that the rectus may initiate the movement of extension (against gravity, with the subject seated and the leg hanging) and continue to contract throughout the whole movement, whereas the vastus lateralis and medialis may not become active until the movement is about half completed. Under the same conditions, Basmajian and co-workers found much variation, but reported that the rectus and the vastus lateralis usually contracted first, with the other two heads starting their contraction only some 5° later; with a load, however, all started the movement.

The articularis genus has the function of pulling upward on the synovial capsule of the knee joint as the knee is extended by the quadriceps, thus preventing a redundant fold from getting caught within the knee joint.

Rupture of the quadriceps tendon is rare; Carlucci reported that in a 10-year series there were 318 cases of fracture of the patella to only 4 cases of rupture of the quadriceps tendon. Sloane and Sloane said that muscular pull rarely ruptures the quadriceps tendon, for in young people this is stronger than the patella; however, after the fifth decade the tendon may give way. Scuderi and Shrey found the mechanism of rupture to be usually somewhat similar to that of fracture of the patella, resulting from indirect violence applied during a time when the knee is in a semiflexed position; it apparently differs in that when the tendon ruptures it is because the patella is held firmly against the anterior aspect of the femur by the pull of the quadriceps and its upper pole cannot then be fractured across the patellar surface of the femur. Conway pointed out that, as is typical of tendons, the normal quadriceps tendon does not rupture in its middle but at its attachment into bone or its attachment into muscle. He said that rupture usually occurs at either the superior or inferior borders of the patella, but if it is high in the tendon it may be either limited to the rectus femoris or involve also the vasti; it may also be at the tibial insertion of the tendon.

Pectineus

The pectineus (Figs. 8-53 and 8-60) lies somewhat intermediate between the anterior and the anteromedial muscles; because of its similarity to and close relationship with the adductor brevis, a muscle of the anteromedial group, it is frequently classed with these muscles; however, it is more commonly innervated by the femoral nerve, and in this re-

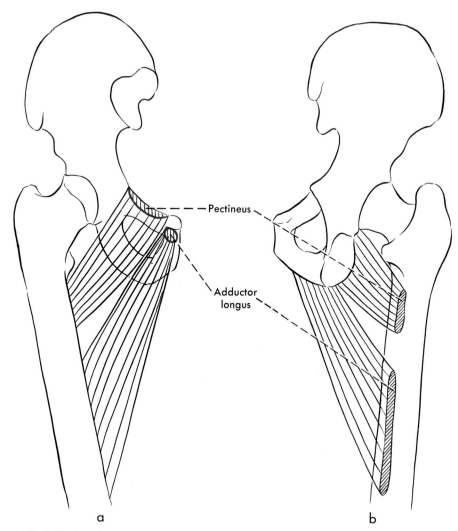

Fig. 8-60. Anterior and posterior views, *a* and *b*, of the origins (*red*) and insertions (*black*) of the pectineus and adductor longus muscles.

spect therefore should be classed with the anterior muscles. It arises from the pecten of the pubis and the bone anterior to this, and with its covering fascia forms a part of the floor of the femoral triangle, medial to the psoas major. It inserts by a broad but thin tendon into the upper half of a line (pectineal line) leading downward from the lesser trochanter to the linea aspera.

The pectineus muscle is usually innervated by the femoral nerve (Fig. 8-57), but may be supplied by the obturator or the accessory obturator nerve (Paterson, 1891); according to Woodburne it always receives a branch from the femoral when it receives one from the accessory obturator. Paterson regarded the accessory obturator as being really an accessory femoral, but also regarded the possibility of innervation through the obturator nerve as indicating that the muscle is really a composite: an anterior or outer part, constant, being innervated by the femoral, while a deeper or inner part, very inconstant, is innervated by the obturator nerve. A belief in the double origin of the pectineus muscle, that is, that one part is derived from the extensor mass and another from the flexor mass of the thigh, is of course predicated upon the assumption that fibers in the lumbar plexus separate accurately into posterior and anterior divisions to supply these two developing muscle masses. The nerve fibers reaching the pectineus are usually said to be from the second and third lumbars or, if it is supplied by the obturator nerve, the third and fourth lumbars.

The pectineus has been said to adduct, but is primarily a flexor of the thigh. According to Wright, it helps to start the movement from the extended position and continues to contract until the limb is flexed at a right angle. Duchenne described it as producing flexion of the thigh and lateral rotation, but only slight adduction; Takebe, Vitti and Basmajian found it active in flexion, adduction, and *medial* rotation.

Variations and Anomalous Muscles

Other than being sometimes more, sometimes less, fused together, the quadriceps group shows little variation. A common variation of the pectineus is for it to have two distinct strata, a superficial and a deep one, corresponding to the double innervation (the superficial from the femoral nerve, the deep from the obturator) which this muscle may receive (*e.g.,* Ochiltree). It may be partly fused with the adductor brevis or the adductor longus, or have aberrant muscle fascicles connecting it with the lesser trochanter, the hip joint, or neighboring muscles. The sartorius may have two heads of origin, the second head arising from the iliopubic (iliopectineal) eminence, the anterior inferior iliac spine, or other neighboring points, and joining the normal head at a variable distance downward, so that the muscle is split over a part of its length (Schaeffer); such slips may have anomalous distal attachments instead of joining the major part of the muscle.

Truly anomalous muscles of the front of the thigh are apparently very rare. Tyrie described an interesting one, a saphenous muscle, which he found bilaterally in one specimen. It arose from the lateral end of the inguinal ligament, passed downward and medially just deep to the fascia lata to cross the front of the femoral vein just below the sa-

phenofemoral junction and ascended again to attach to the medial end of the inguinal ligament.

ANTEROMEDIAL (OBTURATOR) GROUP

The pectineus, often considered one of this group, is described with the anterior muscles. The muscles constituting this group are then the adductor brevis, adductor longus, a large part of the adductor magnus, the gracilis, and the obturator externus, all innervated by the obturator nerve (Fig. 8-61; see also Fig. 8-64). The gracilis is the only member of this group to cross the knee joint as well as the hip.

Adductor Longus and Brevis

The *adductor longus,* the most anteriorly placed of the adductor muscles in the upper part of the thigh, is a triangular muscle (Fig. 8-60), arising by a strong tendon from the front of the pubis just below the pubic tubercle and expanding as it passes down, to insert into the medial lip of the linea aspera between the attachments of the vastus medialis and adductor magnus to this lip. It lies medial to the pectineus, and with this and the iliopsoas forms the floor of the femoral triangle when the medial border of the muscle is regarded as the medial boundary of the triangle. As it and the sartorius converge at the apex of the femoral triangle it is covered by the latter, and forms an upper part of the floor of the adductor canal. It lies in front of and usually completely covers the adductor brevis, which may, however, appear in the narrow interval between the pectineus and the adductor longus, and thus form a part of the floor of the femoral triangle; its medial border is adjacent to the gracilis.

The *adductor brevis* (Fig. 8-62; see also Fig. 8-65) arises by a relatively broad tendon from the body and inferior ramus of the pubis; it expands in a triangular fashion as it goes to its insertion on about the upper half of the linea aspera and a line leading up from this toward the lesser trochanter. Although it is a shorter muscle than the adductor longus, it is also a thicker and apparently more powerful one. It

usually derived from the second, third, and fourth lumbar nerves.

The results obtained by Duchenne on stimulating the adductor longus (and magnus) have not been confirmed by electromyography. As summarized by Basmajian ('74), the adductors are not used for the most part as prime movers, but are primarily activated through reflexes of gait. Thus they usually become active in flexion of the knee, and also in extension if that is against resistance; similarly, their upper parts often become active during flexion of the hip against resistance, and always become active during extension against resistance. The adductor longus (brevis not mentioned) is said to become active in adduction, as expected, but the magnus was frequently inactive unless the

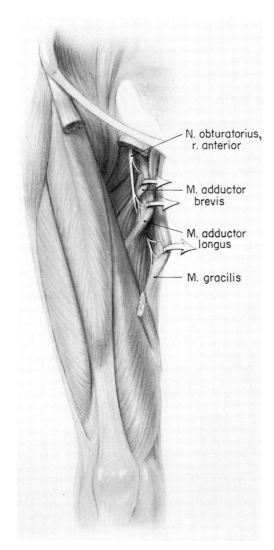

Fig. 8-61. Distribution of the anterior branch of the obturator nerve to the adductor longus and other muscles of the adductor group. (Hollinshead WH, Markee JE: J Bone Joint Surg 28:721, 1946)

Fig. 8-62. Deeper anterior muscles of the thigh. See also Fig. 8-53.

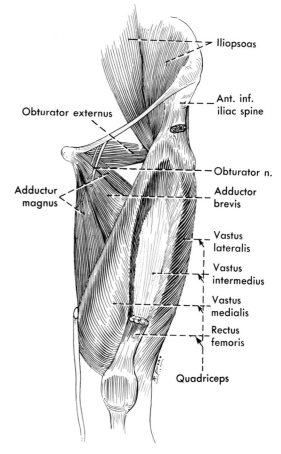

lies for the most part between the adductor longus and the adductor magnus, with the gracilis on its medial side.

The adductor longus is *innervated* by the anterior branch of the obturator nerve, which passes between the longus and the brevis to reach the gracilis (Fig. 8-61). The adductor brevis is usually supplied by the anterior branch, but sometimes by the posterior branch as this passes behind it to be distributed to the adductor magnus. The fibers are

movement was resisted. Finally, it now seems certain that the adductors as a group are active in internal or medial rotation of the thigh, not the external rotation described by Duchenne.

Adductor Magnus

The adductor magnus is a very large muscle, triangular like the other adductor muscles. It arises from the lower part of the inferior pubic ramus, and from the entire length of the ramus of the ischium as far as the lower part of the ischial tuberosity (Fig. 8-63). From this extensive origin it spreads out, its more anteriorly arising fibers inserting on the upper part of the linea aspera and a portion of the femoral body above this up to the insertion of the quadratus femoris; its intermediately arising fibers inserting along the remainder of the length of the linea aspera; and its most posteriorly arising fibers inserting on the diverging medial lower lip of the linea (medial supracondylar line) for most of its length, and onto the adductor tubercle at the upper border of the medial epicondyle. A little above the adductor tubercle the tendon of insertion is interrupted by a gap, the tendinous or adductor hiatus, for the passage of the femoral vessels into the popliteal fossa. This muscle lies behind the members of the adductor group already described, and has the gracilis on its medial side. Below the adductor longus it forms the posterior wall of the adductor canal, which ends at the tendinous hiatus.

This is a complex muscle, apparently actually representing at least two different muscles that have fused. The more anteriorly arising fibers, which also insert higher (and are sometimes divisible into two groups), belong with the other adductor muscles both from the standpoint of their innervation and their action. However, the more posteriorly arising fibers, especially those that run almost straight downward from the ischial tuberosity and insert into the adductor tubercle, not only act with the hamstrings but are innervated by the nerves that also supply the hamstrings. Thus the adductor magnus is divisible, functionally but not anatomically, into two distinct portions, the upper and more anterior portion, called the adductor or obturator portion, and the more posterior and lower-inserting portion, referred to as the hamstring or sciatic portion.

The upper portion of the muscle, viewed anteriorly, is innervated by the posterior branch of the obturator nerve (Fig. 8-64); this apparently brings into it fibers from the third and fourth lumbar nerves, as a rule. This part of the muscle is an adductor at the hip joint and is more powerful than the adductor longus. This portion is also a flexor at the hip (the more posterior portion is an extensor at the hip, with the hamstrings). Both parts are apparently internal rotators, although Duchenne claimed that only the lower (posterior) part internally rotates, and that when that was atrophied the remaining adductors externally rotate as they adduct.

Gracilis

The gracilis is a long, straplike muscle on the medial side of the thigh, where it lies immediately adjacent to the adductor longus and adductor magnus. It arises (Fig. 8-63) from a lower part of the body of the pubis close to the symphysis, and from an upper part of the inferior ramus; it tapers as it approaches the knee, first lying between the sartorius and the semimembranosus, then between the sartorius and the semitendinosus. Below the knee its thin tendon curves forward and expands to insert into the medial surface of the upper end of the tibia below the medial condyle, between the insertions of the sartorius and the semitendinosus. With these tendons it forms the so-called pes anserinus or "goose-foot" tendon. A rather complicated bursa, the anserine bursa, intervenes between these three associated tendons and the tibia; it may become painful and much distended as a result of acute or chronic trauma to the medial side of the knee (Meyerding and Chapman; Sutro, '48).

The gracilis is innervated by the anterior branch of the obturator nerve. It adducts and medially rotates the thigh, and flexes it if the knee is kept from flexing (Wheatley and Jahnke), and it flexes and medially rotates the leg. Wheatley and Jahnke found some activ-

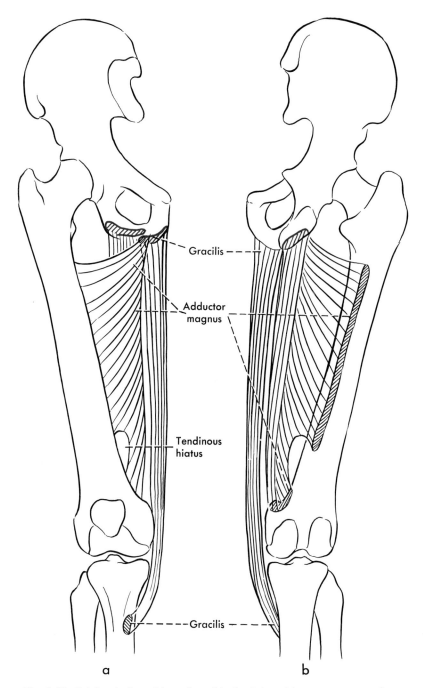

Fig. 8-63. Origins (*red*) and insertions (*black*) of the adductor magnus and gracilis, in anterior and posterior views (*a* and *b*).

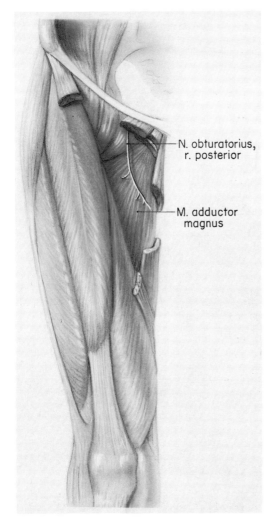

Fig. 8-64. Innervation of the anterior part of the adductor magnus. (Hollinshead WH, Markee JE: J Bone Joint Surg 28:721, 1946)

ity during extension of the thigh also, but were not sure the muscle was aiding the extension.

Obturator Externus

This muscle is the most deeply placed of the adductor group. It arises from the external surface of the obturator membrane and from the external surface of the pubis and ischium around the obturator foramen (Fig. 8-65). Its fibers converge, so that the muscle takes the form of a somewhat twisted cone; it passes below the capsule of the hip joint and then upward across the lower part of the back of this joint, and inserts into the trochanteric fossa. As it crosses the hip joint it may be separated from it by a bursa, or a protrusion of the thin capsule here may serve as a bursa for it.

The nerve supply to the obturator externus is by a branch of the obturator nerve, derived either from the undivided nerve or from its posterior branch. It may arise within the pelvis or from the posterior branch as this passes through the muscle, and is usually said to contain fibers from the third and fourth lumbar nerves.

The obturator externus is primarily a lateral rotator; it is often also said to be an adductor, but such an action is assuredly weak, if it exists at all.

Variations and Anomalous Muscles

The muscles of the adductor group may show some fusion among themselves, especially between the adductor brevis and longus, or with the pectineus muscle. Doubling of the origin of the adductor longus or adductor brevis sometimes occurs, and Ochiltree reported a case in which the brevis divided into two quite separate insertions into the linea aspera. The superior fibers of the adductor magnus, arising from the inferior ramus of the pubis and adjacent part of the ischial ramus and inserting largely above the linea aspera, may form a distinct muscle, the adductor minimus (*e.g.,* Ochiltree); and the two parts of the muscle, the lower adductor part and the posterior hamstring part, may be more distinct than usual.

ANTEROMEDIAL NERVES AND VESSELS

There are only two nerves associated with the anterior and anteromedial muscles of the thigh, the femoral nerve and the obturator. The femoral nerve, although the more anterior in the adult, is a posterior derivative of the lumbar plexus, while the obturator nerve is an anterior one (Chap. 7); the altered rela-

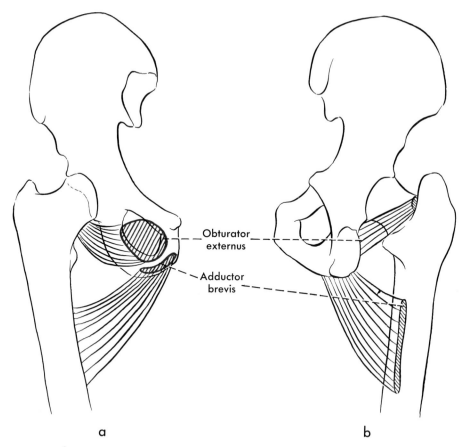

Fig. 8-65. Origins (*red*) of the insertions (*black*) of the obturator externus and adductor brevis. *a* and *b* are anterior and posterior views, respectively.

tionship has been brought about through the rotation of the lower limb.

Femoral Nerve

The femoral nerve, typically derived from the second, third, and fourth lumbar nerves, with fibers from the latter two predominating, passes into the thigh behind the inguinal ligament, on the anterior surface of the iliopsoas muscle. It is immediately lateral to the femoral artery (see Fig. 8-70), from which it is separated behind the inguinal ligament by the iliopectineal arch, and below the ligament by the femoral sheath. Soon after entering the upper part of the femoral triangle it breaks up in a somewhat variable manner into a number of cutaneous and muscular branches.

The *cutaneous branches to the thigh*, the anterior femoral cutaneous nerves, have already

been described, as have the *articular branches* to the hip joint, which arise either directly from the nerve or from its muscular branches.

The *muscular branches* (Fig. 8-66) include ones to the sartorius, to all four members of the quadriceps group, and to the pectineus, arising in no regular order. There are typically two nerves or two groups of nerves to the sartorius; usually one branch to the deep surface of the rectus femoris and one into its lateral surface; usually two branches to the vastus lateralis; one or two into the anterior surface of the lastus intermedius; and one, sometimes two, that continue downward with the saphenous nerve and femoral vessels into the adductor canal to end on the medial side of the vastus medialis. The nerve to the pectineus (usually but not always present) is one of the higher branches of the femoral nerve,

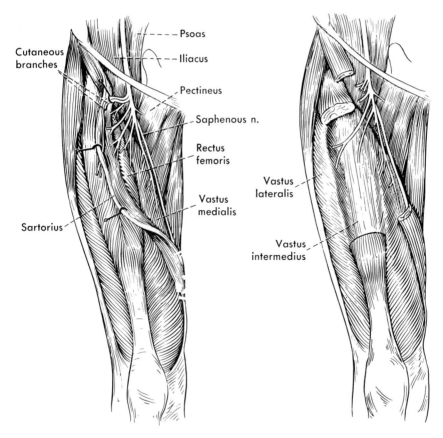

Cutaneous branches

Psoas

Iliacus

Pectineus

Saphenous n.

Rectus femoris

Vastus medialis

Sartorius

Vastus lateralis

Vastus intermedius

Fig. 8-66. Muscular branches of the femoral nerve. Cutaneous branches have been cut away in the figure on the left, these and some superficial muscular branches on the right.

and derived from its medial side; it passes medially behind the femoral vessels to enter the superficial surface of the muscle.

The nerves for the vastus lateralis and vastus medialis are said to continue often to help supply also the vastus intermedius; the nerve to the vastus intermedius continues to supply the subjacent articularis genus; and the nerves of all three vasti are said usually to continue below the muscles to assist in supplying the knee joint.

Through the branches just recounted, the femoral nerve supplies skin on the anterior and anteromedial part of the thigh, twigs to both the hip and knee joints, and all the anterior group of muscles of the thigh. One further branch, the saphenous nerve, remains to be described.

As the femoral nerve gives off its various

branches, the *saphenous nerve* continues the course of the femoral nerve to the apex of the femoral triangle, usually with the nerve to the vastus medialis just lateral to it, and at the apex of the triangle disappears into the adductor canal, where it descends in company with the femoral vessels; it usually lies at first lateral to the artery but then crosses its front to lie along its medial side. At the lower end of the canal, where the femoral vessels pass through the tendinous hiatus, the saphenous nerve emerges between the sartorius and adductor magnus muscles (therefore posterior to the sartorius) and descends along the medial side of the knee in company with the descending genicular (supreme genicular) artery. At the level of the knee it becomes subcutaneous by perforating the fascia between the sartorius and gracilis muscles, and runs downward along the medial side of the leg to supply skin here. The level of emergence of the saphenous nerve is a convenient landmark for the lower end of the adductor canal when the popliteal vessels are to be explored (Johnson, '68). The distribution of this nerve will be further described in the following chapter; one or more infrapatellar branches usually arise at the distal end of the adductor canal, pierce or run around the posterior border of the sartorius muscle, and curve anterolaterally to supply skin in the infrapatellar region. These are to be avoided in the usual incisions for anteromedial approaches to the knee joint.

Obturator Nerve

The obturator nerve is, like the femoral, derived usually from the second through the fourth lumbar nerves, but with most of the fibers coming from the third and fourth. Its course in the abdomen and pelvis has been described in connection with the lumbar plexus. The obturator nerve enters the thigh through the obturator canal, therefore deep to the superior ramus of the pubis and under cover of the pectineus muscle. As already noted in connection with the description of the innervation of the hip joint, it may give off, while it lies in the pelvis, both an articular branch to this joint and the nerve to the obturator externus; it may also divide into its two branches, anterior and posterior, while it is in this location, but more commonly these branches arise as the nerve enters the obturator canal. On emerging from the obturator canal the anterior branch usually appears in front of the obturator externus muscle, between this and the covering pectineus, while the posterior branch usually passes through the upper part of the obturator externus.

The first muscular branch, whether derived from the undivided nerve or from the posterior branch, is to the obturator externus. In their further courses the two branches separate: The anterior branch passes downward and slightly medially in front of the adductor brevis, between this and the adductor longus, to supply regularly one or more branches

into the deep surface of the adductor longus and continue to end in the medial surface of the gracilis; it often supplies also the adductor brevis (Fig. 8-67a). Usually the anterior branch also gives rise to a cutaneous branch of variable size, which supplies skin of the middle third of the medial side of the thigh or may even continue as far as the upper part of the calf. It also, usually, gives rise to a twig to the femoral artery, some of this apparently ending on the artery, some of it following the artery and its branches to reach the knee joint. If an accessory obturator nerve is present it typically communicates with the anterior branch. The posterior branch passes downward and slightly medially behind the adductor brevis, typically giving off a branch into the deep surface of this muscle and ending by giving a series of branches into the anterior surface of the upper part of the adductor magnus (Fig. 8-67b). The posterior branch often also gives rise to a branch to the knee joint; this may follow the femoral artery to the popliteal space or pierce the adductor magnus independently.

Banks and Green sectioned both the anterior branch of the obturator nerve and the adductor longus and brevis, sometimes the gracilis, or even anterior fibers of the adductor magnus muscle to relieve adduction contraction resulting from spastic cerebral palsy. Pollock reported that adductor tenotomy may leave the limb so abducted, laterally rotated, and flexed that the child cannot walk; in such cases he transferred the tendons of insertion of the posterior hamstring muscles to a medial attachment on the lower end of the femur and moved their origins as close as possible to the pubic symphysis.

Accessory Obturator

An accessory obturator nerve, arising from the second and third or third and fourth lumbar nerves or from the third alone, has been reported in 8 per cent or more of sides; it differs from the obturator nerve in its entrance into the thigh, for this is anterior to or above the superior ramus of the pubis, between the iliopsoas and pectineus muscles. It typically

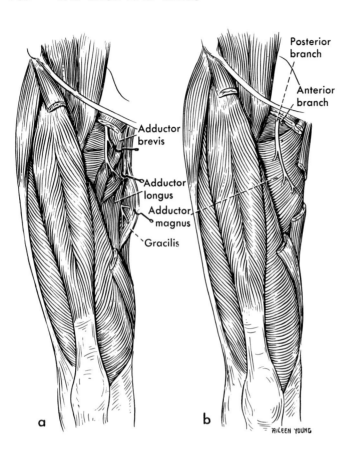

Fig. 8-67. Distribution of the obturator nerve. That of the anterior branch is shown in *a*, that of the posterior branch in *b*.

disappears deep to the pectineus muscle. Woodburne found a branch to the hip joint, one to the pectineus (which also received a femoral branch), and a communication to the anterior branch of the obturator to be typical, although its branches may be either fewer or more numerous. Paterson (1893–94) reported a case in which it not only supplied the pectineus and gave twigs to the hip joint, but sent contributions also to the nerve to the gracilis; to the nerve to the adductor brevis; to the cutaneous branch of the obturator; and to the deep branch of the obturator, which was then distributed to the obturator externus, the adductor magnus, and the knee joint—a distribution that suggests that in this case, at least, it was more closely related to the obturator than to the femoral nerve. Paterson argued in another paper that the accessory obturator is really from the posterior divisions of the plexus and should be called the "accessory femoral" nerve.

FEMORAL VESSELS

Femoral Sheath

The femoral artery and vein pass between abdomen and thigh side by side, behind the inguinal ligament in the medial compartment (lacuna vasorum), which is separated from the lateral compartment (lacuna musculorum) by the thickened fascia forming the iliopectineal arch (Fig. 8-68). The vessels therefore are anteromedial to the iliopsoas muscle, from which they are separated by the iliopectineal arch, and lie on the anterior surface of the pectineus muscle and its fascia. The two vessels bring down around them a special reflection of the fascia of the abdomen, the femoral sheath, a funnel-shaped diverticulum of the abdominal fascia that ends by blending with the adventitia of the vessels about 3 cm to 4 cm below the inguinal ligament. Within the sheath there are three compartments, a lateral one for the femoral artery (and the femoral or lumboinguinal branch of the genitofemoral nerve, which descends on the artery), an intermediate one for the femoral vein, and a medial, essentially empty one, the femoral canal (Fig. 8-69). The femoral canal extends a shorter distance below the inguinal ligament than the other two compartments, being closed by the fusion of its medial wall to the septum between it and the femoral vein. Its upper end, the femoral ring, opens into the extraperitoneal space of the abdominal cavity. The canal contains only a little loose connective tissue, some lymphatic vessels, and a lymph node or two, and thus can offer little resistance to femoral hernias.

The sequence of structures at the level of the inguinal ligament is, beginning laterally, the iliopsoas muscle and the femoral nerve, iliopectineal arch, femoral artery, femoral vein, and femoral ring. Below the inguinal ligament the sequence is similar, but there is no heavy layer of fascia between the nerve and the vessels, here enclosed in the femoral sheath, so the sequence from lateral to medial is femoral nerve and femoral sheath, with the artery, vein, and femoral canal enclosed within the sheath.

Cutaneous Branches

These small arteries, usually the superficial epigastric, superficial iliac circumflex, and one or more external pudendal arteries (Fig. 8-70), have already been discussed.

Muscular Branches in Femoral Triangle

The femoral artery may also give off, in the femoral triangle, one or more direct muscular branches that run to adjacent muscles, especially to the sartorius and to the adductors, and the femoral vein may receive corresponding tributaries. However, most of the muscular branches in the anterior part of the thigh are indirect derivatives of the femoral vessels, coming from or into the circumflexes and the profunda femoris.

Profunda Femoris

The profunda femoris or deep femoral artery (Figs. 8-70 to 8-72) is the largest branch of the

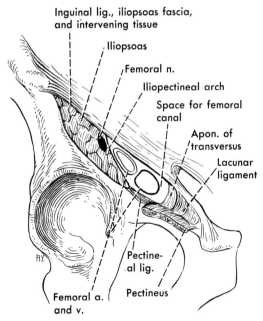

Fig. 8-68. Relations deep to the inguinal ligament, seen from below.

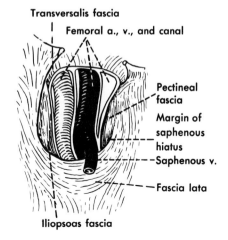

Fig. 8-69. A right femoral sheath opened from the front to show the three compartments.

femoral artery, and is usually given off from the posterolateral aspect of this artery fairly high in the femoral triangle, approximately 4 cm or 5 cm below the inguinal ligament. It curves posteriorly, medially, and downward, on the anterior surfaces of the iliopsoas and pectineus muscles, passing behind the femoral artery and vein and then assuming a deeper position than these: it passes deeply on the

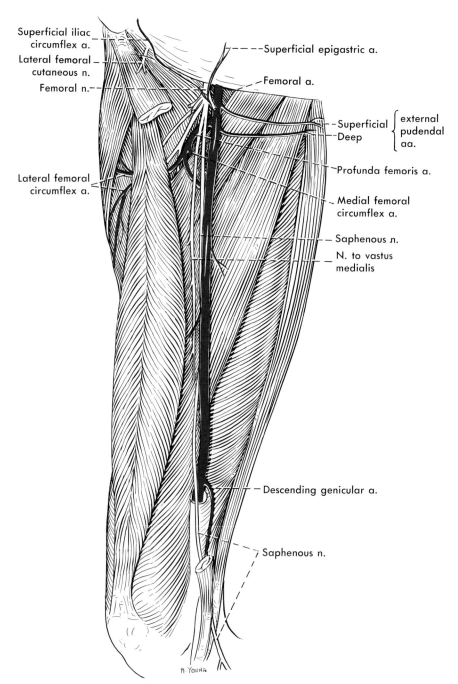

Superficial iliac circumflex a.
Lateral femoral cutaneous n.
Femoral n.
Superficial epigastric a.
Femoral a.
Superficial
Deep
external pudendal aa.
Profunda femoris a.
Lateral femoral circumflex a.
Medial femoral circumflex a.
Saphenous n.
N. to vastus medialis
Descending genicular a.
Saphenous n.

Fig. 8-70. The femoral artery and related nerve branches in the thigh. The sartorius muscle and the fascial roof of the adductor canal have been removed, and the artery can be seen disappearing through the tendinous (adductor) hiatus.

ductor magnus. As it runs downward the profunda femoris vein is in front of it. The artery has a number of muscular branches, most of which end in the adductors, but some of which continue through the adductor magnus to anastomose with other vessels of the posterior side of the thigh.

Among the branches of the profunda femoris vessels are the *perforating branches,* typically four in number (the terminal branch being the fourth perforating). As implied by their names, the perforating branches perforate the insertions of the adductor muscles close to the femur to appear in the posterior side of the thigh behind the adductor magnus and supply the hamstring muscles (see Fig. 8-81). There is some variation in the number and position of the perforating arteries and of

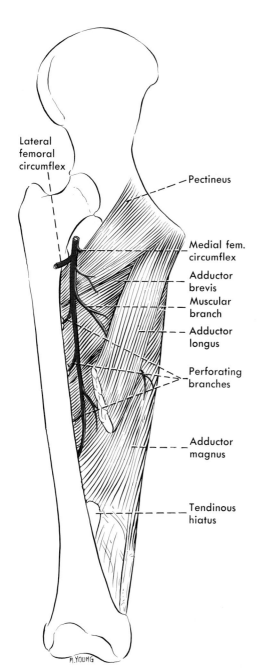

Fig. 8-71. The deep femoral artery.

front of the adductor brevis to lie close to the femur, and runs downward on the muscle in this position. On reaching the upper border of the adductor longus it passes behind this muscle, by which it is then separated from the femoral vessels, to lie on the front of the ad-

Fig. 8-72. Diagrammatic representation of various types of origin of the femoral circumflex arteries. *D* is the profunda, *De* the descending branch of the lateral femoral circumflex, *F* the femoral, and *L* and *M* the lateral and medial circumflexes, respectively. Percentages shown are those of Williams and his coworkers. (Adapted from original of Fig. 1, Williams GD, Martin CH, McIntire LR: Anat Rec 60:189, 1934)

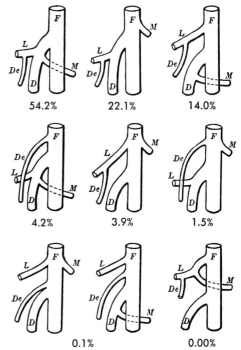

their accompanying veins, but the following description of the arteries generally applies to both.

The first perforating artery usually penetrates the adductor brevis and then the adductor magnus, branching on the posterior side of the thigh primarily into ascending and descending branches; these run on the posterior surface of the adductor magnus, give rise to muscular branches, and anastomose with adjacent vessels. The ascending branch usually runs upward and crosses the posterior surface of the quadratus femoris to anastomose medial to and over the greater trochanter with both gluteal arteries and both femoral circumflex arteries (see the description in connection with the buttock). The descending branch tends to anastomose with the second perforating artery.

The second perforating artery also usually pierces both the adductor brevis and the adductor magnus, and on appearing on the posterior aspect of the thigh gives rise to muscular branches and to ascending and descending branches that tend to anastomose with the first and third perforating arteries. The third perforating artery usually arises below the adductor brevis and perforates the adductor magnus only; it tends to anastomose with the perforating vessels above and below it, and one of its muscular branches usually penetrates the short head of the biceps to supply also a posterior part of the vastus lateralis. The fourth perforating or terminal branch of the profunda behaves like the others; it tends to anastomose with the third perforating and with the upper muscular branches of the popliteal artery; among its muscular branches one or more usually penetrate the short head of the biceps to help supply the vastus lateralis.

The *nutrient artery* of the femur is commonly given off from the second perforating artery, but when there are two nutrient arteries they may arise from any two of the perforating arteries.

Important variations of the profunda femoris artery are not common. Very rarely, it has an abnormal course in its upper part.

Schrutz described a case in which the profunda passed medially across the fronts of the femoral artery and vein to reach its normal position, thus forming an arterial circle about the saphenofemoral junction. Johnston described a similar case in which the profunda first ran laterally across the fronts of the femoral vessels. In such instances as these, exposure of the saphenous vein for ligation of it at its junction with the femoral carries the risk of damage to a large and unexpected arterial channel.

Blankfein recorded an unusual case in which the first two perforating arteries arose from the medial femoral circumflex instead of from the profunda. The profunda was apparently represented by a large vessel arising below the origin of the lateral femoral circumflex (the medial circumflex arose still higher) that gave rise to the last two perforating arteries but was not accompanied by a vein.

A relatively common variation in the *deep femoral vein*, of particular clinical importance, is the presence of a large connection between its lower end and either the femoral vein or the popliteal. Small anastomotic channels regularly exist, as they do in the case of the arteries, but in some cases the popliteal vein divides, with one part ending in the femoral, the other in the deep femoral, and in others there may be a fairly large connection between the deep femoral and the lower part of the femoral vein. Mavor and Galloway found the profunda femoris connecting directly to the popliteal vein in 38% of 22 limbs, and with a tributary of the popliteal in 48% more, and Edwards and Robuck reported an "obvious" communication between the lower part of the deep femoral vein and the femoral vein in 10% of 61 extremities. These findings suggest that ligation of the femoral vein below the entrance of the profunda may often be ineffective in preventing extension of a thrombus from the popliteal vein. Edwards and Robuck found the entrance of the deep femoral into the femoral to average 8 cm below the inguinal ligament; Basmajian ('52) found it to average 9 cm.

Femoral Circumflex Vessels

The lateral and medial femoral circumflex arteries are typically the largest branches of the profunda femoris, although they do not always arise from this vessel; either or both may arise from the femoral artery above the origin of the profunda. In any case, they originate in the femoral triangle. Whether they arise from the profunda or from the femoral artery, they may share a common stem or have separate origins.

Statistics on the origins of the circumflex arteries differ somewhat, as reported and reviewed by several workers (*e.g.,* Ming-tzu, '37; Lipshutz; Williams and his co-workers); the incidence of different types apparently varies somewhat with race. Lipshutz found that the vessels on the two sides of the same body differed in origin in 60% of cases.

Williams and his colleagues described six major patterns of branching and three subtypes, based upon the varying origins of the two circumflexes from the profunda and the femoral, and the separate origin of the descending branch of the lateral femoral circumflex. Including their own cases, they summarized the findings on 979 white persons as presenting the several different types of origin in the percentages shown in Figure 8-72; a common stem from the femoral for both circumflexes has been reported in Negroes and Japanese, but not white persons. Most commonly, in all races reported upon, both vessels arise from the profunda, this being apparently slightly more common in Negroes and Japanese than in white persons.

Baird and Cope found that the terminations of the circumflex veins usually did not correspond to the origins of the arteries. In contrast to the most common origin of the arteries, from the deep femoral, the most common ending of the veins was for both of them to enter the femoral; this occurred in 86.14%, while in only 1.85% did both terminate in the deep femoral. In the remainder, either the lateral or the medial entered the femoral, the other the deep femoral. In more than 20% of all their cases one or both of the circumflex veins were doubled as they joined the femoral

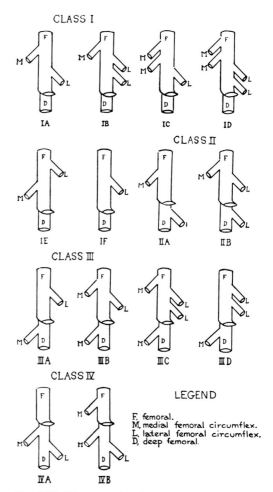

Fig. 8-73. Diagram of various patterns of ending of the femoral circumflex veins. In *Class I,* both circumflexes terminate, although with varying patterns, in the femoral; in *Class II* a lateral, and in *Class III* a medial, circumflex terminates in the deep femoral; and in *Class IV* both do. (Fig. 1, Baird RD, Cope JS: Anat Rec 57:325, 1933)

system; there was a very strong tendency in all their types for the medial circumflex to terminate higher than the lateral (Fig. 8-73).

The *lateral femoral circumflex artery* (Fig. 8-74) therefore typically arises from the lateral side of the upper end of the profunda, or less frequently from the femoral artery above the origin of the profunda. It runs laterally across the front of the iliopsoas muscle and between branches of the femoral nerve to disappear, at the lateral border of the femoral triangle, by

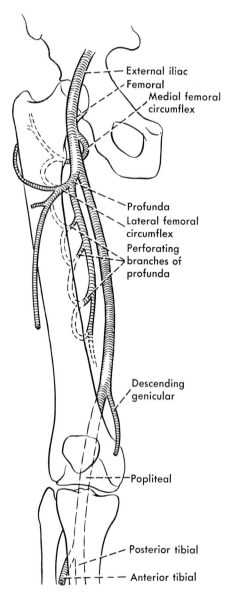

Fig. 8-74. Diagram of the femoral artery and its chief branches.

passing posterior to the sartorius and the rectus femoris muscles. It gives off muscular branches to adjacent muscles and typically divides into three terminal branches, an ascending, a transverse, and a descending (it is the last one that may arise separately from the femoral artery or from the profunda).

The ascending branch runs laterally and upward, deep to the rectus femoris and the tensor fasciae latae, to reach the interval be-

tween the latter and the gluteus medius and minimus muscles; it supplies branches to adjacent muscles, including these, and anastomoses with twigs of the superior gluteal and deep iliac circumflex arteries. It also typically supplies a branch to the front of the hip joint.

The transverse branch of the lateral femoral circumflex is the smallest of the three terminal branches. It passes laterally between the rectus femoris and the vastus intermedius and enters the upper part of the vastus lateralis; a part of it often emerges from the muscle posterolateral to the greater trochanter to participate, usually rather weakly, in the cruciate anastomosis.

The descending branch of the lateral femoral circumflex runs almost straight downward behind the rectus femoris and in the groove where the vastus lateralis and vastus intermedius meet anteriorly, giving off branches to adjacent muscles. Some of these are said to anastomose with the transverse branch. At its lower end the descending branch anastomoses with twigs of the lower perforating arteries, with upper branches of the lateral superior genicular artery, and with the descending (supreme) genicular.

The *medial femoral circumflex artery* (Fig. 8-74) typically arises from the medial or posteromedial aspect of the profunda or the femoral and runs posteriorly, or downward and then posteriorly, to leave the femoral triangle between the iliopsoas and pectineus muscles. Before leaving the femoral triangle it gives off usually a single muscular or superficial branch that supplies the upper ends of the pectineus and adductor muscles; the remaining and larger part of the artery is the deep branch. It winds around the medial side of the femur between the obturator externus and the adductor brevis. As this branch passes below the neck of the femur it may give rise to an acetabular branch that enters the acetabulum to supply the fat in the acetabular fossa and give rise to the artery of the ligament of the head of the femur. Weathersby found an acetabular branch of the medial femoral circumflex in 21.3%, and in almost a third of these there was another acetabular branch from the obturator; and in 23.9% the acetabular branch arose from an anastomotic connection between the medial femoral circumflex and obturator arteries.

As the deep branch reaches the anterior surface of the quadratus femoris muscle it divides into two terminal branches, an ascending and a transverse. The ascending branch typically appears posteriorly above the upper border of the quadratus femoris, the transverse branch between the quadratus femoris and adductor magnus muscles, but it may pierce the upper part of the magnus; both branches typically anastomose with the gluteal and first perforating arteries.

Important variations in the femoral circumflex vessels, other than those of origin, are apparently rare. Blankfein's case, in which the medial femoral circumflex gave rise to the first two perforating arteries, has already been mentioned.

Femoral Artery Below Profunda

After it gives off the profunda in the femoral triangle, the continuing stem of the femoral artery (Figs. 8-70 and 8-74) is sometimes known as the superficial femoral, to distinguish between it and the "common" femoral above the origin of the profunda. It continues to the apex of the triangle where it passes under cover of the sartorius muscle to lie between the vastus medialis and the adductor muscles in the adductor canal, which is roofed deep to the sartorius by a strong fascia passing between the adductor and vastus muscles. Within the canal the artery lies in front of the vein; the nerve to the vastus medialis lies lateral to the artery in the upper part of the canal, but leaves the canal to be distributed to the vastus medialis; the saphenous branch of the femoral nerve also lies at first lateral to the artery but typically crosses in front of it to lie medial to it at the lower end of the canal. The femoral artery leaves the adductor canal a little above the medial epicondyle of the femur, passing here through the gap in the tendon of insertion of the adductor magnus muscle (the tendinous or adductor hiatus), and changing its name to popliteal as it emerges into the popliteal fossa. It is here subject to damage against the tendon at the

hiatus (see the following section). Just before it penetrates the adductor magnus it gives off its last branch, usually the only one arising in the adductor canal. This is the descending genicular (supreme genicular) artery.

The *descending genicular artery* gives off one or more articular branches, which join the anastomosis around the knee, and continues as the saphenous branch. This gives twigs to the distal ends of the vastus lateralis and the adductor magnus muscles, and then becomes subcutaneous. (These muscular branches sometimes arise directly from the lower end of the femoral artery, rather than from the stem of the descending genicular.) The saphenous branch leaves the adductor canal to pass between the sartorius and gracilis muscles with the saphenous nerve, perforate the deep fascia with this nerve, and supply skin on the upper part of the medial side of the leg. It anastomoses with vessels entering the plexus about the knee, especially the medial inferior genicular artery; it commonly also gives muscular twigs to the lower part of the sartorius and gracilis muscles.

Rarely, the saphenous artery is a large one and contributes to the blood supply of the deeper part of the leg, thus presenting the surgeon with an important and unexpected superficially lying artery. Popowsky described a case in which the saphenous artery arose directly from the femoral and was almost as large as the popliteal; it ran superficially down the medial aspect of the leg, gave off a branch to the knee as it passed this, and about a third of the way down the leg divided into two branches. One of these continued superficially downward with the great saphenous vein and the saphenous nerve as far as the medial malleolus, here giving rise to a branch that passed laterally onto the dorsum of the foot, deep to the extensor tendons, to anastomose with the dorsalis pedis. The other branch ran deeply to anastomose with the posterior tibial at about the middle of the leg. Aasar described a somewhat similar case in which the artery was apparently smaller, and after running superficially to about the middle of the leg penetrated the deep fascia to join the posterior tibial. An example of an

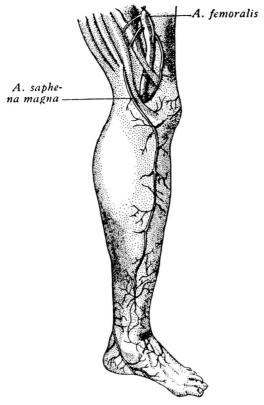

A. femoralis

*A. saphe-
na magna*

Fig. 8-75. A great saphenous artery. (After Hyrtl, from Senior HD: J Anat 53:130, 1919)

anomalous saphenous artery is shown in Figure 8-75.

Anomalies of the femoral artery are apparently rare; Musgrove reported a case in which it was doubled throughout the length of the adductor canal, dividing into parallel vessels at the apex of the femoral triangle but uniting again to a single vessel just before it passed through the adductor magnus.

Occlusion of Femoral Vessels

Holden reported that arteriosclerosis of the vessels of the lower limb has been said usually to involve first the lower half of the femoral artery. Palma described damage of the artery, as a result of its pulsations bringing it into contact with the tendon of the adductor magnus at the hiatus, as tending to produce pathologic changes and narrowing at the femoral-popliteal junction; he regarded this as a common cause of intermittent claudication,

and of thrombosis beginning at this level, in relatively young and muscular males.

Thrombosis of the deep veins of the lower limb apparently more commonly originates in the deep veins of the leg or at the level of the knee: Fine and Sears quoted evidence that of 133 cases of venous thrombosis investigated by dissection, the thrombosis was distal to the opening of the profunda into the femoral vein in 85%, with 64% being in the popliteal vein or its tributaries, 18% in the profunda, and 3% in the femoral distal to the opening of the profunda (superficial femoral).

Ligation of the femoral vein in order to prevent ascending thrombosis and pulmonary embolism is a not uncommon measure; the preferable level for ligation is, however, disputed. Fine and Sears preferred as a rule to divide and ligate the femoral vein just below the profunda, inspecting the terminal portion of the profunda; if there was evidence of thrombosis along it they ligated higher, above the profunda. Edwards and Robuck called attention to the frequency of large communications between the popliteal vein or the lower end of the femoral and the profunda, and suggested that the femoral ligation should always be above the level of entrance of the deep femoral. Mavor and Galloway regarded the collateral channels available after high ligation of the femoral vein as being usually inadequate in varying degree.

Homans suggested that advancing thrombosis in the veins of the lower limb can best be treated by ligation of the common iliac vein, pointing out that the anastomotic channels available after this ligation are much more abundant than those available following ligation of the femoral itself.

A variable amount of usually transitory edema may follow ligation of the femoral vein, or there may be persistent edema with pain or functional disability. Fine and Sears were not convinced that these more serious disturbances were actually a consequence of the interruption of the vein, maintaining that the phlebitis already present might be at least equally responsible.

O'Keeffe and co-workers, from measurements and venous pressure, concluded that interruption of the femoral vein below the deep femoral produces no demonstrable change in the venous circulation from the limb provided other vessels are normal.

Valves

The number, arrangement, and apparent competence of the valves in the femoral venous system have been investigated by several groups of workers.

It has been generally supposed that deficiencies of the valves in the deep veins, consequent to phlebitis, may be responsible for the poor circulation, resulting in ulcers and other complications, sometimes seen in a postphlebitic limb. Luke found a variable degree of backflow of radiopaque material injected into the femoral vein in 15 persons with no evidence of phlebitis, and found no real difference in this respect between apparently normal individuals and those persons who had had phlebitis. He concluded that the competency of the valves of the deep veins varies tremendously from one person to another, and that they are usually not so competent as to prevent all backflow. Since congenital weakness or absence of valves does not, in the average and presumably normal individual, lead to the changes seen in the postphlebitic leg and often attributed to valvular incompetence, Luke regarded ligation of the deep veins of the postphlebitic leg as being a questionable procedure.

Kampmeier and Birch found a valve in about 35% of external iliac veins; Basmajian ('52), combining his observations with those of others, estimated that a valve is present in about 25% of veins; and Powell and Lynn found one in 33%. Kampmeier and Birch described this valve as being usually incompetent, while Basmajian said that it is competent in more than two thirds of cases.

One or more valves are usually present in the femoral vein above the opening of the profunda (common femoral vein), most commonly above the opening of the great saphenous. Kampmeier and Birch found a valve in the latter location in 72% of 100 veins and 2 valves in 5%; Powell and Lynn found one at about the level of the inguinal liga-

ment in 72%; and Basmajian found one or more, usually within 1 cm of the inguinal ligament, in about 67%. In only one of Basmajian's cases was there a valve immediately above the opening of the profunda femoris.

The femoral vein below the opening of the profunda (superficial femoral vein) apparently almost always has one or more valves in it. Although Powell and Lynn reported finding a great variation in number, even between the two sides of a body, they found no evidence of a progressive reduction in the number of valves with increasing age; among their specimens there was only 1 valve in that part of the femoral in 10%, there were 2 in 33%, and there were 3 or more in 57% (they also found a valve in the upper part of the popliteal vein in 96%). Kampmeier and Birch found valves always present in their specimens, usually in two sets. Basmajian found, among 76 veins, two which had no valve in the part below the deep femoral (one of these had no valve anywhere) but there was 1 valve in 17.2%, there were 2 in 30.4%, 3 in 3.17%, and 4 or 5 in the remainder. Basmajian said the most constant valve in his series (found in 89.5%, usually bilaterally) lay in the femoral just below the mouth of the deep femoral. The valves are usually bicuspid, sometimes tricuspid, and in bicuspid valves one cusp is typically placed anteriorly, one posteriorly (Basmajian).

OBTURATOR VESSELS

The obturator artery typically arises in the pelvis from the anterior division of the internal iliac (hypogastric) artery, and the vein enters the internal iliac vein. The obturator vessels run forward and downward along the lateral pelvic wall with the obturator nerve, a little below the pelvic brim (with the nerve above, the vein below the artery) and enter the obturator canal in the upper part of the obturator foramen. In its pelvic course the artery gives off branches to adjacent muscles and to the ilium, and its terminal branches emerge into the thigh.

In some 25% to 30% of cases the obturator artery has an anomalous origin from the infe-

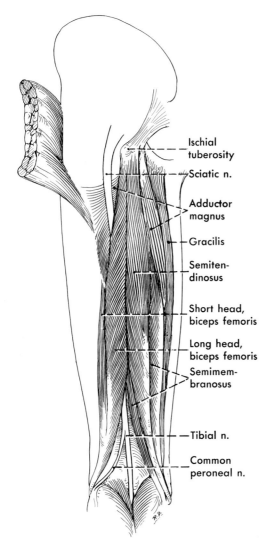

Fig. 8-76. Posterior muscles of the thigh.

Labels on figure: Ischial tuberosity; Sciatic n.; Adductor magnus; Gracilis; Semitendinosus; Short head, biceps femoris; Long head, biceps femoris; Semimembranosus; Tibial n.; Common peroneal n.

thigh, but rather simply circle the obturator foramen: the anterior branch runs anteriorly, the posterior one posteriorly, around the margin of the foramen, and the two anastomose at the lower border of the foramen. Both branches give off twigs to the obturator externus and other adjacent muscles, and anastomose with similar twigs from the medial femoral circumflex. The posterior branch of the artery commonly gives rise to an acetabular branch that enters the hip joint, as it did in 62% of Weathersby's series; next most frequently the acetabular branch arises from a communication between the obturator and the medial femoral circumflex.

POSTERIOR MUSCLES

These muscles (Fig. 8-76), because of the relations of their tendons at the knee, are frequently referred to as the "hamstring muscles"; they include not only the semitendinosus, the biceps, and the semimembranosus, but also that part of the adductor magnus that arises from the ischial tuberosity and inserts into the adductor tubercle.

Except for the short head of the biceps, all these muscles arise from the ischial tuberosity, therefore on the medial or flexor side of the postaxial line (marked by the posterior femoral cutaneous and sciatic nerves). Again with the exception of the short head of the biceps, they are innervated by the tibial component of the sciatic nerve, derived from the anterior divisions of the sacral plexus. The short head of the biceps is innervated by the common peroneal component, derived from the posterior divisions of the plexus.

Semitendinosus

The semitendinosus arises from the posteromedial surface of the distal part of the ischial tuberosity, its tendon of origin being fused with that of the long head of the biceps. The muscle is somewhat flattened, but fusiform when viewed posteriorly, and reaching its greatest width about the middle of the thigh; the muscular belly is divided into upper and

rior epigastric or, less often, from the external iliac. While this does not affect its distribution in the thigh, it is of importance in operation for femoral hernia: the aberrantly arising obturator artery, on its way to the obturator foramen, courses very close to (across, medial, or lateral to) the femoral ring, and hence is likely to be intimately related to the neck of the sac of a femoral hernia.

As the obturator vessels pass through the obturator canal, they divide into anterior and posterior branches; these branches do not, like the similarly named branches of the obturator nerve, continue on downward into the

lower parts by a tendinous intersection that runs obliquely across it close to its middle. Its tendon of insertion becomes free of muscle fibers only a little above the medial condyle of the femur but extends up into the muscle substance to about the middle of the thigh. As the muscle runs downward behind the knee it lies at first on the surface of the semimembranosus; below the knee its tendon curves forward around the medial side to insert into the medial aspect of the body of the tibia just posterior to the insertions of the gracilis and sartorius, giving off an expansion to the fascia of the leg before it reaches its insertion. It forms a part of the tendon complex (sartorius, gracilis, and semitendinosus) sometimes referred to as the pes anserinus, or anserine tendon, with which the anserine bursa, already mentioned, is associated.

The semitendinosus is supplied by two or more nerve branches (Figs. 8-77 and 8-79) arising from the medial (tibial) side of the sciatic nerve, or a single branch that divides. At least one nerve regularly enters the muscle above the tendinous intersection, another enters below this intersection. These nerves usually contain fibers derived from the fifth lumbar and first and second sacral nerves.

Since it crosses two joints, the semitendinosus can produce simultaneous flexion of the leg at the knee and extension of the thigh at the hip, but acts more strongly when it does either alone. Its adductor and internal rotatory actions at the hip are probably of no great importance, but when the knee is flexed the muscle becomes a good internal rotator of the leg.

Biceps Femoris

The long head of the biceps femoris arises from the posteromedial aspect of the lower part of the ischial tuberosity in common with the semitendinosus (Fig. 8-78). Some of its tendon also arises from the lower part of the sacrotuberous ligament. The long head runs downward and laterally, separating from the semitendinosus; it is fusiform in shape and its muscle bundles extend almost to the knee, but the tendon of insertion begins higher and is joined in the lower third of the thigh by the

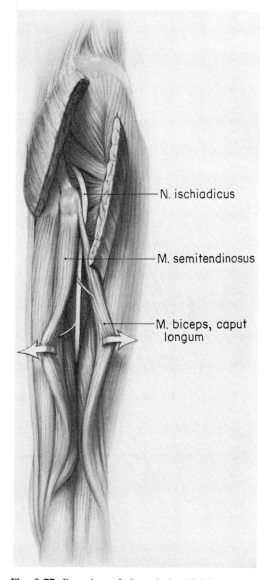

Fig. 8-77. Branches of the sciatic (tibial) nerve to hamstring muscles. See also Fig. 8-79. (Hollinshead, WH, Markee JE: J Bone Joint Surg 28:721, 1946)

fiber bundles of the short head of the biceps. This head arises from the lateral lip of the linea aspera, sometimes as high as the lower end of the insertion of the gluteus maximus, and from the diverging lower part of this lip (supracondylar line), as well as from the lateral intermuscular septum between it and the vastus lateralis. The biceps forms the upper and lateral border of the popliteal fossa. After the tendon has received the fiber bundles

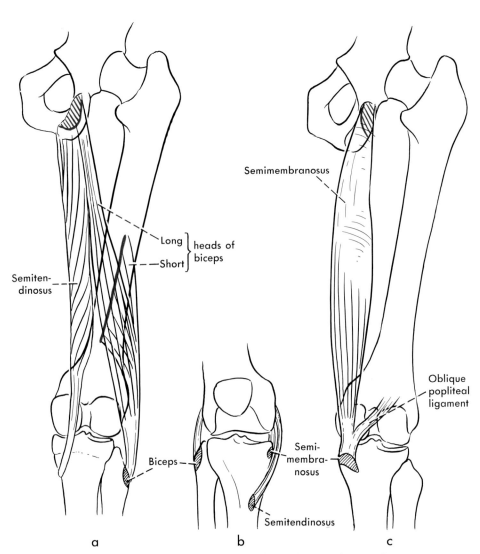

Fig. 8-78. Origins (*red*) and insertions (*black*) of the hamstring muscles. *a* and *c* are posterior views, *b* an anterior one.

from both muscular bellies it becomes rounded, crosses the most lateral part of the posterior aspect of the knee joint, and inserts into the head of the fibula, giving expansions to the lateral condyle of the tibia and the deep fascia of the leg. Sneath described the part derived from the long head as being partly wrapped around the lateral, posterior, and medial side of the attachment of the fibular collateral ligament to the fibula, and that derived from the short head as being attached primarily to the tibia and to the fibular collateral ligament.

The long head of the biceps receives a nerve from the medial or tibial side of the sciatic nerve, which enters its upper end, and may receive a second nerve lower down; its segmental nerve supply is usually from the last lumbar and first two sacral nerves. The short head receives its nerve from the lateral or peroneal part of the sciatic nerve; the fibers to it are derived from the fifth lumbar through the second sacral.

The long head of the biceps can both extend the thigh and flex the knee, while the short head can only flex the knee; the long head also laterally rotates the thigh, in small degree, and both heads strongly laterally rotate the leg when the knee is flexed. Like the other hamstrings, the long head does not have sufficient length to permit free simultaneous flexion at the hip and extension at the knee, nor sufficient contractability to fully flex the knee when the hip is extended. The short head is a more efficient flexor at the knee than is the long head; Paterson ('17) reported that when the short head is paralyzed the long head can be seen to assist flexion only until the knee is semiflexed; if flexion is continued, he said, the long head relaxes.

Semimembranosus

This muscle arises by a long and flat tendon from the posterolateral part of the lower portion of the ischial tuberosity, lateral to the origin of the common tendon of the biceps and semitendinosus; the tendon runs down in front of the latter tendon, and on the posterior surface of the adductor magnus. In the middle third of the thigh the tendon of origin, now wide, gives place to muscle fibers, which arise on an oblique line beginning on the medial margin of the muscle and extending downward toward the lateral margin. The muscle becomes larger until it begins to taper as it forms the upper part of the medial border of the popliteal fossa. It becomes entirely tendinous behind the medial condyle of the femur, the tendon being rather short but heavy and inserting into the back of the medial tibial condyle. The straight part of the tendon attaches to the medial meniscus as it crosses that, and hence can help move the meniscus posteriorly (Chap. 9). As it goes to its insertion it gives off an expansion that runs anteriorly to attach to the tibia deep to the tibial collateral ligament, another that runs obliquely upward and laterally across the posterior aspect of the knee joint, reinforcing and largely forming the oblique popliteal ligament, and still another that joins the fascia over the posterior surface of the popliteal muscle.

The semimembranosus usually receives two or more branches from the tibial part of the sciatic nerve (Fig. 8-79), typically containing

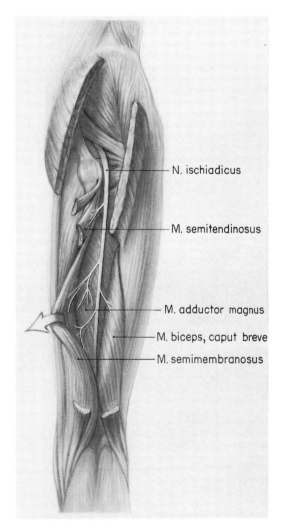

Fig. 8-79. Nerve supply to the hamstring muscles. See also Fig. 8-77. (Hollinshead WH, Markee JE: J Bone Joint Surg 28:721, 1946)

fibers from L5, S1, and S2; one or more of these nerves usually branch to supply also the posterior and lower portion of the adductor magnus. It is like the semitendinosus in its action, extending and to a slight extent medially rotating and adducting the thigh, flexing the leg and medially rotating it. It, the semitendinosus, and the biceps apparently help to adduct the thigh only when movement from an abducted position is resisted (Wheatley and Jahnke), and with them it is also a postural muscle, resisting flexion at the hip joint.

Popliteal Fossa

Lateral to the lower part of the semimembranosus in the thigh is the diamond-shaped popliteal space or fossa, the upper medial border of which is the semimembranosus, the upper lateral border the biceps femoris, the two lower borders the two heads of the gastrocnemius. The popliteal fossa is covered by the fascia lata as this passes downward to become continuous with the deep fascia of the leg. The floor of the fossa is largely the posterior aspect of the femur and of the capsule of the knee joint, but on the lateral side the plantaris appears above the lateral head of the gastrocnemius, and in the lower part of the fossa the popliteus muscle and its covering fascia form a part of the floor.

The tibial nerve and the popliteal vessels run almost directly vertically through the fossa, the nerve lying at first on the lateral aspect of the vessels but diverging somewhat medially, while the popliteal vessels, the vein lying posterior to the artery, diverge slightly laterally; as these structures disappear deep to the gastrocnemius, at the lower end of the fossa, the nerve lies posterior or medial to the vein, and the vein, posterior to the artery, largely hides that. The common peroneal nerve appears along the upper lateral border of the fossa, close against the medial margin of the biceps femoris, and leaves the fossa by following the tendon of this muscle across the upper part of the lateral head of the gastrocnemius.

It might be noted that Baker's or popliteal cysts, which commonly involve the semimembranosus bursa (Chap. 9), may not only come to fill the popliteal fossa, thereby distorting the anatomy here, but may extend downward for a variable distance deep to the fascia of the calf.

Adductor Magnus

The adductor magnus has already been described with the medial muscles of the thigh, with which the anterior and upper part of the muscle belongs. The most posterior part of the muscle arises from the ischial tuberosity, in front of the origins of the other hamstring muscles; these fibers descend lowest to insert by a flattened tendon into the medial lip of the linea aspera, to be interrupted by the tendinous hiatus for the passage of the femoral-popliteal vessels, and end by a strong tendon of insertion into the adductor tubercle.

One of more branches from the tibial nerve, usually arising with the nerves to the semimembranosus, enter the lower posterior part of the adductor magnus; thus this part of the muscle, from the ischial tuberosity, has a completely separate innervation from that of the more anterior and upper part, which is innervated by the obturator nerve. The segmental nerve supply of the posterior part of the adductor magnus is said to be through the fourth and fifth lumbar nerves and often the first sacral, or sometimes also through other sacrals; it typically is about one segment lower than the nerve supply to the anterior and upper part (L3 and L4).

The hamstring or sciatic portion of the adductor magnus was said by Duchenne to have an action quite different from that of the other part of the muscle. He said that both parts together adduct with no rotation, flexion, or extension, but that acting separately the upper part rotates laterally and flexes as it adducts, while the posterior and lower part rotates medially and extends the thigh as it adducts. Basmajian ('74) adduced electromyographic evidence that both parts of the adductor magnus rotate medially, as do the other adductors.

Variations and Anomalous Muscles

Minor variations in the hamstring muscles include some degree of fusion among them, insertion of a muscular slip of the biceps into the fascia of the leg to form a tensor fasciae suralis muscle, insertion of a slip of the semimembranosus into the fascia of the thigh, and accessory heads of origin of one of the muscles. The most striking and apparently most common accessory origins and insertions are in connection with the long head of the biceps femoris. This muscle may have a head arising from the anterior aspect of the sacrum with or close to the piriformis, and passing out through the greater sciatic foramen beneath the piriformis to join the long head, or the combined long and short heads, in the thigh (Moore; Seelaus). According to Moore, most anomalous slips of origin of the biceps arising from the sacrum, coccyx, or pelvis join the long head, most of those arising from the femur join the short head. He noted that insertion of slips of the biceps into the gastrocnemius muscle and into the tendo calcaneus has been reported.

POSTERIOR NERVES AND VESSELS

The posterior nerves of the thigh are the posterior femoral cutaneous nerve and the sciatic nerve. The posterior femoral cutaneous has already been described in connection with the cutaneous nerves of the thigh; and the upper end of the sciatic nerve has been described in connection with the buttock.

Sciatic Nerve

In entering the thigh, the sciatic nerve curves just lateral to the ischial tuberosity and then descends vertically to the middle of the popliteal fossa, thus running very close to the middle of the posterior aspect of the entire thigh (Figs. 8-80 and 8-81). As it emerges below the lower border of the gluteus maximus it passes under cover of the long head of the biceps, and lies deep to this and the lateral part of the semimembranosus until it divides, usually just above the popliteal fossa, into its tibial and common peroneal branches. The tibial nerve continues down through the approximate middle of the popliteal fossa, at first lateral and then posterior to the popliteal vessels, and disappears between the two heads of the gastrocnemius; the common peroneal branch follows the lateral border of the fossa, the medial border of the biceps, to course laterally in the uppermost part of the leg.

As already noted, the sciatic nerve is sometimes divided into its two components, tibial and common peroneal nerves, even as it enters the buttock, and in this case the two nerves simply course downward side by side. Otherwise they are bound together by a common sheath of connective tissue although they interchange no nerve fibers; within this common sheath the tibial nerve lies largely medially but a little anteriorly, and the common peroneal nerve lies largely laterally but a little posteriorly. All the branches of the sciatic nerve with the exception of that to the short head of the biceps are given off from the medial or tibial aspect of the nerve and are derived from the tibial nerve; thus the only laterally arising branch of the sciatic nerve appears about the middle of the thigh. It carries fibers usually derived from the fifth lumbar and first, or first two, sacral nerves to the short head of the biceps.

The other branches of the sciatic nerve in the thigh are all medial ones, derived from the tibial, and usually said to carry fibers derived from the fourth and fifth lumbar and the first sacral nerves to the hamstring portion of the adductor magnus, and from the fifth lumbar and first two sacrals to the semitendinosus, semimembranosus, and long head of the biceps. These branches typically take the form of a number of separate ones that arise at intervals from the medial aspect of the nerve between the level of the ischial tuberosity and the middle of the thigh; sometimes, however, all of them have a common stem of origin that arises independently from the sacral plexus, and simply parallels the sciatic nerve, lying on its medial side. In most cases, it is said, the fibers to the hamstrings are easily separable by careful dissection from the medial side of the tibial nerve.

The upper of the two nerves to the semitendinosus may be the first branch given off by the sciatic to the hamstring muscles, arising close to (above, at the level of, or below) the ischial tuberosity; the nerve, or the upper of two, to the long head of the biceps often arises with this or precedes it, after which are found two or more nerves to the semimembranosus and the adductor magnus, and further nerves to the semitendinosus and semimembranosus. Sunderland and Hughes found among 20 carefully studied specimens a great deal of variation in the order in which the branches to the various muscles arose, and in the lengths of the nerve fibers (hence the expected order of regeneration) between fixed

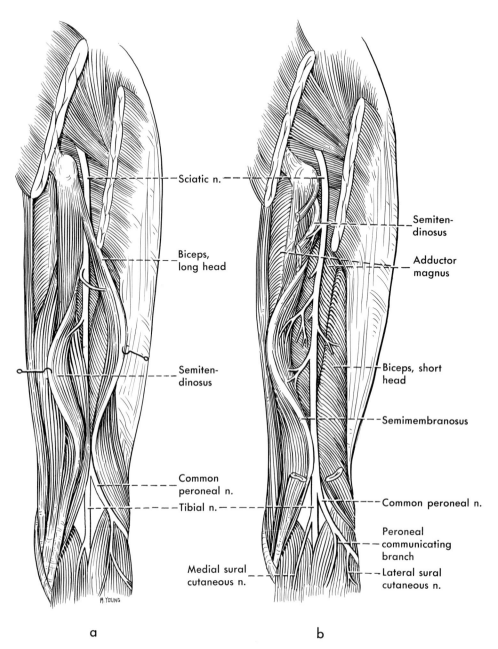

Fig. 8-80. Course and branches of the sciatic nerve.

skeletal landmarks and the entrance into the muscle. In general, however, the shortest lengths of nerves were to the semitendinosus, with the nerves to the long head of the biceps, the adductor magnus, the semimembranosus, and the short head of the biceps following in that order. They found a single branch to the long head of the biceps in 10 cases, 2 nerves in 8, and 3 in 2; 1 to the short head in 14, 2 in 6; a single nerve to the semitendinosus in only 2 cases, 2 or sometimes 3 in the remainder; and 1 to 4 branches, sharing a common trunk in all but 1 specimen, to the adductor magnus and semimembranosus. In 3 cases all the

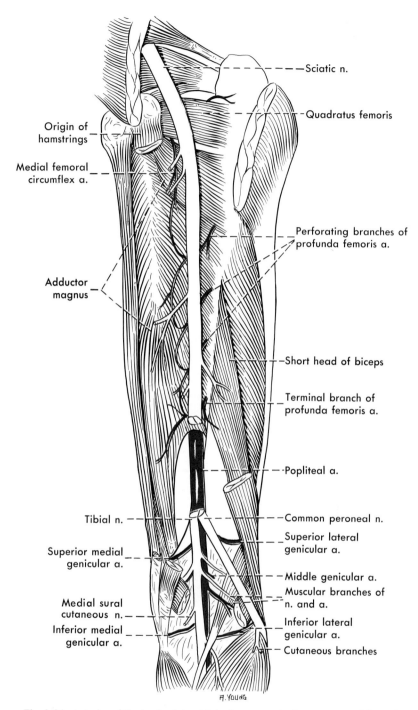

Sciatic n.

Quadratus femoris

Origin of
hamstrings

Medial femoral
circumflex a.

Perforating branches of
profunda femoris a.

Adductor
magnus

Short head of biceps

Terminal branch of
profunda femoris a.

Popliteal a.

Tibial n.

Common peroneal n.

Superior lateral
genicular a.

Superior medial
genicular a.

Middle genicular a.

Muscular branches of
n. and a.

Medial sural
cutaneous n.

Inferior lateral
genicular a.

Inferior medial
genicular a.

Cutaneous branches

A. YOUNG

Fig. 8-81. Arteries of the back of the thigh. A segment is shown removed from the
sciatic nerve, and the cut ends shifted somewhat medially, in order to show the
arteries better.

nerves to the hamstrings arose at the same level; in other specimens nerves to muscles other than the adductor magnus and semimembranosus shared a common stem or arose at the same level.

As it lies in the popliteal fossa the tibial part of the sciatic nerve gives off its first branches to the leg; usually these are the medial sural cutaneous nerve, articular branches to the knee joint, and muscular branches to the two heads of the gastrocnemius and to the plantaris. The common peroneal likewise gives rise to a cutaneous branch, the lateral sural cutaneous, usually to the peroneal communicating branch, which helps form the sural nerve, and to an articular branch; its other branches arise more distally, after it lies laterally in the leg.

Popliteal Vessel

The proximal ends of the popliteal artery and vein are at the posterior aspect of the tendinous hiatus in the adductor magnus, through which these vessels become continuous with the femorals. They pass downward and slightly laterally, at first under cover of the semimembranosus, but soon entering the popliteal fossa; the artery (Fig. 8-81) is in front of the vein; the lower end of the sciatic nerve, or the tibial nerve, is posterolateral or posterior to the vessels, and the tibial nerve gradually crosses them posteriorly to attain their medial side. An aneurysm of the popliteal artery cannot expand anteriorly, since the vessel lies here against the bone, and in expanding posteriorly it exerts pressure on its posterior relations, the vein and nerve; hence the pain that may be associated with such an aneurysm. (Hara urged that even asymptomatic and small popliteal aneurysms be removed and replaced by a graft as soon as they are diagnosed because of the danger that they will thrombose and occlude the artery.) At the lower end of the popliteal fossa the popliteal vessels disappear between the two heads of the gastrocnemius (at which level they may be injured by contact with a fascial band between the heads—Chap. 9), subsequently dividing into their terminal branches farther down in the leg. Their major branches are all in the leg; the only ones arising in the popli-

teal fossa are the paired superior geniculars, the unpaired middle genicular, and muscular branches (primarily to the two heads of the gastrocnemius). They and their connections are described in more detail in the following chapter.

The two *superior geniculars,* lateral and medial, arise from these two aspects of the popliteal artery and run laterally and medially respectively, across the bony floor of the popliteal space; they circle the femur just above the epicondyles, uniting with other genicular vessels to form the plexus around the patella, and therefore being a part of the collateral circulation to the leg.

The *middle genicular artery* is small, and arises opposite the back of the knee joint; it quickly pierces the capsule to supply the deeper ligaments and the synovial membrane of the joint.

The *muscular branches* of the popliteal artery include several small ones to the lower ends of the hamstring muscles. The large muscular branches, arising from the popliteal artery just before it disappears between the two heads of the gastrocnemius, are the sural arteries. These are two large branches arising from the sides or from the posterior surface of the artery, and branching almost immediately to disappear into and deep to the two heads of the gastrocnemius, supplying this muscle, the soleus, and the small plantaris.

The *popliteal vein* lies behind the artery, between it and the tibial nerve. The largest veins it receives above the level of the knee joint are those from the two heads of the gastrocnemius, which usually enter it separately some 2 cm to 4 cm above the joint. The varying manner of formation of the popliteal vein is described in the following chapter. However, Williams found it doubled to within about 1 cm of the level of the joint in almost three fourths of cases, often doubled for several centimeters above the joint. He regarded ligation of the vein as being of doubtful efficacy in improving the circulation of a limb in which the circulation was already damaged, and suggested that perhaps ligation produced deleterious results no more often than has been reported because only one of two channels was ligated. The popliteal vein may also, more rarely, divide at its upper end, one part continuing as the femoral, the other as the lowest tributary to the profunda femoris.

Other Vessels

Other than in exceptional circumstances, where the lower part of the inferior gluteal artery persists as a stem of considerable size (the sciatic or ischiadic artery) accompanying the sciatic nerve and connecting with the popliteal artery at the knee joint, there is no longitudinally running artery in the posterior aspect of the thigh. Rather, the vessels here include a lower part of the inferior gluteal, terminal branches of the medial and lateral femoral circumflex arteries, usually some muscular twigs that pierce the adductor magnus to appear posteriorly, the perforating branches of the profunda (Fig. 8-81) and, at the knee, the descending (supreme) and superior genicular arteries. The inferior gluteal has been described with the buttock, the superior geniculars in a preceding paragraph, and the remaining arteries in connection with the anterior aspect of the thigh, where they take origin. It might be noted that through the anastomosis of the first perforating with the femoral circumflexes and inferior gluteal above, the anastomoses of the perforating vessels with each other, and the anastomosis of the lowest perforating artery (the termination of the profunda) with upper muscular branches from the popliteal artery and perhaps with the lateral superior genicular, a longitudinally running arterial channel of sorts can be made out on the posterior surface of the adductor magnus close to the femur. The anastomotic connections are small, however, consisting of a series of small loops, or even of intramuscular communications. They therefore may or may not be grossly visible.

Stability and Movement at the Hip and Knee

The structure of the hip joint has already been considered, and that of the knee joint follows in the next chapter; all the musculature acting upon the hip has likewise been de-scribed, and so has almost all that acting at the knee. In these descriptions the functions of the individual muscles have been stressed. They should also be considered from the standpoint of their contribution to standing and walking. In the movement of walking, hip and knee act together, in the sequence of flexion at the hip and the knee, extension at the knee as the heel touches the ground, then flexion again at the knee followed by extension at the knee (Saunders and co-workers) and extension at the hip.

The only stable position of the hip and of the knee is the extended one. In extension at both these joints the major ligaments—the iliofemoral and ischiofemoral at the hip, the two collateral and two cruciate ligaments at the knee—are all tightened, and the knee is further stabilized through the fact that in extension the flattened anterior portions of the femoral condyles are brought into contact with the only slightly concave surfaces of the menisci and the tibial condyles. Both the hip and the knee are further stabilized in the position of extension by gravity, since the center of gravity lies slightly behind the hip joint and in front of the knee joint.

Thus the weight of the body tends to keep the hip in extension, while the strong iliofemoral ligament tends to prevent hyperextension; most of the anterior muscles of the thigh are then relaxed, although the iliacus, but not the psoas major, is said to show some activity. Similarly, the weight of the body tends to keep the lower limb extended without muscular effort by the quadriceps, and further extension is resisted by the ligaments. With a slight sway forward, both flexion at the hip and hyperextension at the knee are resisted by the hamstrings (Wheatley and Jahnke), and knee hyperextension and ankle flexion by the triceps surae (Houtz and Walsh, and others).

Extension at the knee merits further comment because there are differences of opinion as to whether any muscle other than the quadriceps can contribute to this. The gastrocnemius and soleus have been said to assist in extension at the knee, or at least in preventing further flexion from a slightly flexed condition, by pulling backward upon the lower end of the femur and the upper end of the tibia and fibula. Their real contribution to maintaining extension at the knee (or helping to produce it) is obviously, however, through their role of plantar flexors at the ankle. The weight-supporting limb cannot be flexed at the knee without simultaneous dorsiflexion at the ankle, which these muscles resist; similarly, when the ankle is dorsiflexed and the knee flexed, contraction of the muscles of the calf, to overcome the dorsiflexion and restore the tibia to a vertical position, is a necessary concomitant of extension at the knee if the balance is to be maintained. Certainly the contraction of the muscles of the calf during normal standing and their greater contraction when one bends at the hips with the knees straight have nothing directly to do with producing or maintaining extension at the knee. Gravity does this, and the calf muscles are therefore resisting dorsiflexion at the ankle, also produced by gravity, and, in the case of the gastrocnemius, resisting hyperextension at the knee; when the foot is free, contraction of the gastrocnemius tends to flex, not extend, the knee.

The tensor fasciae latae has often, the gluteus maximus less frequently, been said to extend the leg, or at least help keep it extended; the argument has been that they attach into the iliotibial tract, and that this passes in front of the axis of motion of the knee joint. Evidence has already been cited that neither stimulation of the tensor fasciae latae nor pulling upon the iliotibial tract produces movement at the knee, so this theory would seem to be disproved. The gluteus maximus could contribute to knee extension through its effect on the hip joint, since when the limb bears weight the hip must flex as the knee does, but even this is problematical; electromyography has indicated that it is the hamstrings, not the gluteus maximus, that act as postural muscles in resisting flexion at the hip.

In walking, the strongest hip flexor, the iliopsoas, has been said not to assist in flexion, but Keagy and co-workers found the psoas major active during almost all of the forward swing. Of course this may be primarily to stabilize the vertebral column. Whether the anterior part of the gluteus minimus, and the adductor muscles, also participate is apparently not known, although, according to Basmajian ('74), the upper part of the adductor magnus is active during most of the cycle, while the brevis and longus show triphasic activity. Wright said that flexion is apparently started by the tensor fasciae latae, the pectineus, and the sartorius, and that the iliopsoas and the adductor muscles typically contract only against resistance. The rectus femoris presumably helps only slightly, since it must relax to allow flexion of the knee as the hip is flexed. According to Wright, the tensor fasciae latae alone can produce flexion, although it cannot carry the thigh through the complete arc of motion, and if the other flexors are paralyzed it becomes enormously hypertrophied; similarly, if it is paralyzed, the action of the other flexors is said to produce outward rotation simultaneously with flexion. This apparently indicates that flexion is then carried out primarily by the iliopsoas or sartorius, since all the other flexors are internal rotators.

As the hip is flexed, gravity and the passive pull upon the hamstring muscles as they are tightened by flexion at the hip is sufficient to produce also flexion at the knee. This seldom needs to be a powerful movement, and the hamstring muscles are said to be only about a third as strong as the quadriceps. The action of the sartorius and gracilis on the knee, in any flexion, is weaker than that of the hamstrings, but the sartorius does help in flexing the knee simultaneously with the hip.

As the heel touches the ground, the knee is partly extended by the action of the quadriceps, but this muscle relaxes before the weight is shifted forward, and further extension of the knee is apparently brought about by the plantar flexors controlling the rate of dorsiflexion of the foot while momentum and the anterior leg muscles carry the weight of the body forward over the foot (Sutherland).

As the weight is shifted onto one limb, the abductors of that side, primarily the gluteus medius and minimus, necessarily contract to stabilize the pelvis and prevent the unsupported side from dropping very far below the horizontal; in the usual movement of walking, the unsupported side drops only about 5° (Saunders and co-workers). Extension of the thigh during the last phase of the step is probably brought about primarily by the gluteus maximus and the hamstring portion of the adductor magnus. The other hamstrings contract also, but probably not so much to assist in extension at the hip as to prevent hyperextension at the knee. Although the gluteus maximus has been described as not being used in extension of the hip during walking on the level, it becomes active as soon as the heel touches the ground (Sutherland). However, patients with atrophy of the gluteus maximus can walk without apparent limp on a horizontal surface, although they have difficulty in walking up a hill (Duchenne), indicating the lesser efficiency of the hamstrings as compared with that of the gluteus maximus.

If the hamstrings are paralyzed a hyperextension deformity at the knee develops; if the medially lying flexors at the knee, probably especially the gracilis, are weak, or the biceps is hypertrophied or overactive, a knock-knee deformity can be expected to develop.

As implied in the foregoing, the triceps surae becomes active as the heel touches the ground, controlling the forward shift over the foot; other leg muscles become active to aid in this forward shift and steady the foot, and as the weight is shifted forward the plantar flexors of the foot also become active. Liberson has analyzed the somewhat complicated sequence in which muscles of the thigh and leg become active during walking; Joseph and Watson have reported similarly on the major muscles involved in walking up and down stairs.

References

AASAR YH: The saphenous artery. J Anat 73:194, 1938

ALBEE FH: Myofascitis: A pathological explanation of many apparently dissimilar condidtions. Am J Surg 3:523, 1927

ALEXANDER C: The etiology of femoral epiphysial slipping. J Bone Joint Surg 48-B:299, 1966

BABB FS, GHORMLEY RK, CHATTERTON CC: Congenital coxa vara. J Bone Joint Surg 31-A:115, 1949

BAIRD RD, COPE JS: On the terminations of the circumflex veins of the thigh and their relations to the origins of the circumflex arteries. Anat Rec 57:325, 1933

BANKS HH, GREEN WT: Adductor myotomy and obturator neurectomy for the correction of adduction contracture of the hip in cerebral palsy. J Bone Joint Surg 42-A:111, 1960

BARLOW TG: Early diagnosis and treatment of congenital dislocation of the hip. J Bone Joint Surg 44-B:292, 1962

BARONOFSKY ID: Technique of inguinal node dissection. Surgery 24:555, 1948

BARR JS: Muscle transplantation for combined flexion-internal rotation deformity of the thigh in spastic paralysis. Arch Surg 46:605, 1943

BASMAJIAN JV: The distribution of valves in the femoral, external iliac, and common iliac veins and their relationship to varicose veins. Surg Gynecol Obstet 95:537, 1952

BASMAJIAN JV: Personal communication to the author, 1971

BASMAJIAN JV: Muscles Alive. Their Functions Revealed by Electromyography, ed. 3. Baltimore, Williams and Wilkins, 1974

BASMAJIAN JV, GREENLAW RK: Electromyography of iliacus and psoas with inserted fine-wire electrodes. Anat Rec 160:310, 1968 (abstr.)

BASMAJIAN JV, HARDEN TP, REGENOS EM: Integrated actions of the four heads of quadriceps femoris. Anat Rec 172:15, 1972

BEATON LE, ANSON BJ: The relation of the sciatic nerve and of its subdivisions to the piriformis muscle. Anat Rec 70:1, 1937

BECK NR, BICKEL WH: Interinnomino-abdominal amputations: Report of twelve cases. J Bone Joint Surg 30-A:201, 1948

BELLIN H: Traumatic separation of epiphysis of lower end of femur. Am J Surg 37:306, 1937

BICKEL WH, ROMNESS JO: True diastasis of the sacroiliac joints with hypermobility. J Bone Joint Surg 39-A:1381, 1957

BLANKFEIN E: An example of dissociation of the branches of the a. profunda femoris. Anat Rec 21:329, 1921

BOSCOE AR: The range of active abduction and lateral rotation at the hip joint of men. J Bone Joint Surg 14:325, 1932

BOYD HB, CALANDRUCCIO RA: Further observations on the use of radioactive phosphorus (P^{32}) to determine the viability of the head of the femur. Correlation of clinical and experimental data in 130 patients with fractures of the femoral neck. J Bone Joint Surg 45-A:445, 1963

BOYD HB, GRIFFIN LL: Classification and treatment of trochanteric fractures. Arch Surg 58:853, 1949

BRITTAIN HA: Hindquarter amputation. J Bone Joint Surg 31-B:404, 1949

BROOKE R: The sacro-iliac joint. J Anat 58:299, 1924

BURMAN M: Tear of the sacrospinous and sacrotuberous ligaments. J Bone Joint Surg 34-A:331, 1952

BURWELL HN, SCOTT D: A lateral intermuscular approach to the hip joint for replacement of the femoral head by a prosthesis. J Bone Joint Surg 36-B:104, 1954

CAMPBELL WC: Surgery of the hip joint from the physiologic aspect. Surgery 7:167, 1940

CARLUCCI GA: Rupture of the quadriceps extensor tendon: A case report. J Bone Joint Surg 16:456, 1934

CARTER C, WILKINSON J: Persistent joint laxity and congenital dislocation of the hip. J Bone Joint Surg 46-B:40, 1964

CATTO M: A histological study of avascular necrosis of the femoral head after transcervical fracture. J Bone Joint Surg 47-B:749, 1965

CELLI E: Sulla morfologia del M. piriformis. Anat Anz 44:551, 1913

CHANDLER SB: The iliopsoas bursa in man. Anat Rec 58:235, 1934

CHANDLER SB, KREUSCHER PH: A study of the blood supply of the ligamentum teres and its relation to the circulation of the head of the femur. J Bone Joint Surg 14:834, 1932

CLAFFEY TJ: Avascular necrosis of the femoral head. An anatomical study. J Bone Joint Surg 42-B:802, 1960

CLARKSON RD, RAINY H: An unusual arrangement of the psoas muscle. J Anat Physiol 23:504, 1889

CONWAY FM: Rupture of the quadriceps tendon: With a report of three cases. Am J Surg 50:3, 1940

CRELIN ES: An experimental study of hip stability in human newborn cadavers. Yale J Biol Med 49:109, 1976

CROCK HV: A revision of the anatomy of the arteries supplying the upper end of the human femur. J Anat 99:77, 1965

DASELER EH, ANSON BJ, REIMANN AF: Radical excision of the inguinal and iliac lymph glands: A study based upon 450 anatomical dissections and upon supportive clinical observations. Surg Gynecol Obstet 87:679, 1948

DENNIS C: Disaster following femoral vein ligation for thrombophlebitis: Relief by fasciotomy; clinical case of renal impairment following crush injury. Surgery 17:264, 1945

DERRY DE: The influence of sex on the position and composition of the human sacrum. J Anat Physiol 46:184, 1912

DUCHENNE GB: Physiology of Motion: Demonstrated by Means of Electrical Stimulation and Clinical Observation and Applied to the Study of Paralysis and Deformities. Translated and edited by E. B. Kaplan, Philadelphia, JB Lippincott, 1949

DUNLAP K, SHANDS AR JR, HOLLISTER LC JR, GAUL JS JR, STREIT HA: A new method for determination

of torsion of the femur. J Bone Joint Surg 35-A:289, 1953

EDWARDS EA, ROBUCK JD JR: Applied anatomy of the femoral vein and its tributaries. Surg Gynecol Obstet 85:547, 1947

EGGERS GWN: Surgical division of the patellar retinacula to improve extension of the knee joint in cerebral spastic paralysis. J Bone Joint Surg 32-A:80, 1950

EGGERS GWN: Transplantation of hamstring tendons to femoral condyles in order to improve hip extension and to decrease knee flexion in cerebral spastic paralysis. J Bone Joint Surg 34-A:827, 1952

EVANS FG: Stress and Strain in Bones. Their Relation to Fractures and Osteogenesis. Springfield, Ill., Thomas, 1957

EVANS FG, HAYES JF, POWERS JE: "Stresscoat" deformation studies of the human femur under transverse loading. Anat Rec 116:171, 1953

FERGUSON AB JR: Primary open reduction of congenital dislocation of hip using a median adductor approach. J Bone Joint Surg 55-A:671, 1973

FINDER JG: Iliopectineal bursitis. Arch Surg 36:519, 1938

FINE J, SEARS JB: The prophylaxis of pulmonary embolism by division of the femoral vein. Ann Surg 114:801, 1941

FOSHEE JC: Fascia lata regeneration: Final report. Surgery 21:819, 1947

FRANGAKIS EK: Intracapsular fractures of the neck of the femur. Factors influencing non-union and ischaemic necrosis. J Bone Joint Surg 48-B:17, 1966

FREEMAN LW: A simple combined approach to the obturator and femoral nerves. Surgery 24:968, 1948

FREIBERG JA, PERLMAN R: Pelvic abscesses associated with acute purulent infection of the hip joint. J Bone Joint Surg 18:417, 1936

GARDNER E: The innervation of the hip joint. Anat Rec 101:353, 1948

GARDNER E, GRAY DJ: Prenatal development of the human hip joint. Am J Anat 87:163, 1950

GHORMLEY RK, SULLIVAN R: Traumatic dislocation of the hip. Papers of Am Assoc Surg Trauma 1952, p 298

GORDON EJ, FREIBERG JA: Posterior dislocation of the hip with fracture of the head of the femur. J Bone Joint Surg 31-A:869, 1949

GORDON RS: Malreduction and avascular necrosis in subcapital fractures of the femur. J Bone Joint Surg 53-B:183, 1971

GORDON-TAYLOR G, WILES P, PATEY DH, WARWICK WT, MONRO RS: The interinnomino-abdominal operation: Observations on a series of fifty cases. J Bone Joint Surg 34-B:14, 1952

GRATZ CM: Tensile strength and elasticity tests on human fascia lata. J Bone Joint Surg 13:334, 1931

GRIFFITHS WE, SWANSON SAV, FREEMAN MAR: Experimental fatigue fracture of the human femoral neck. J Bone Joint Surg 53-B:136, 1971

HADLEY LA: Accessory sacro-iliac articulations. J Bone Joint Surg 34-A:149, 1952

HARA M: The hazards of popliteal aneurysms. Surg Gynecol Obstet 124:358, 1967

HARRIS WR: The endocrine basis for slipping of the upper femoral epiphysis: An experimental study. J Bone Joint Surg 32-B:5, 1950

HARRISON TJ: The influence of the femoral head on pelvic growth and acetabular form in the rat. J Anat 95:12, 1961

HARTY M: The calcar femorale and the femoral neck. J Bone Joint Surg 39-A:625, 1957

HARTY M: The significance of the calcar femorale in femoral neck fractures. Surg Gynecol Obstet 120:340, 1965

HAYES AG, GROTH HE: The influence of rotational malpositions on intracapsular fracture of the femoral neck. Surg Gynecol Obstet 124:40, 1967

HOLDEN WD: Reconstruction of the femoral artery for arteriosclerotic thrombosis. Surgery 27:417, 1950

HOMANS J: Deep quiet venous thrombosis in the lower limb: Preferred levels for interruption of veins; iliac sector or ligation. Surg Gynecol Obstet 79:70, 1944

HOOPER AC, ORMOND DJ: A radiographic study of hip rotation. Irish J Med Science 144:25, 1975

HORWITZ T: Ischemic contracture of the lower extremity. Arch Surg 41:945, 1940

HORWITZ T: The posterolateral approach in the surgical management of basilar neck, intertrochanteric and subtrochanteric fractures of the femur: A report of its use in 36 acute fractures. Surg Gynecol Obstet 95:45, 1952

HOUTZ SJ, WALSH FP: Electromyographic analysis of the function of the muscles acting on the ankle during weight-bearing with special reference to the triceps surae. J Bone Joint Surg 41-A:1469, 1959

HOWE WW JR, LACEY T, SCHWARTZ RP: A study of the gross anatomy of the arteries supplying the proximal portion of the femur and the acetabulum. J Bone Joint Surg 32-A:856, 1950

HOWORTH MB: Slipping of the upper femoral epiphysis. Surg Gynecol Obstet 73:723, 1941

HOWORTH MB: Congenital dislocation of the hip. Ann Surg 125:216, 1947

HOWORTH MB: Slipping of the upper femoral epiphysis. J Bone Joint Surg 31-A:734, 1949

HUCHERSON DC, DENMAN FR: Non-infectious iliopectineal bursitis. Am J Surg 72:576, 1946

HUMPHRY PROF: The angle of the neck with the shaft of the femur at different periods of life and under different circumstances. J Anat Physiol 23:273, 1889

INMAN VT: Functional aspects of the abductor muscles of the hip. J Bone Joint Surg 29:607, 1947

JAHNKE WD: An electromyographic study of some of the superficial thigh and hip muscles in normal individuals. Anat Rec 106:206, 1950 (abstr.)

JOHNSON CE, BASMAJIAN JV, DASHER W: Electromyography of sartorius muscle. Anat Rec 173:127, 1972

JOHNSON EW JR: Contractures of the iliotibial band. Surg Gynecol Obstet 96:599, 1953

JOHNSON EW JR: Personal communication to the author, 1968

JOHNSTON TB: A rare anomaly of the arteria profunda femoris. Anat Anz 42:269, 1912

JOSEPH J, WATSON R: Telemetering electromyography of muscles used in walking up and down stairs. J Bone Joint Surg 49-B:774, 1967

JOSEPH J, WILLIAMS PL: Electromyography of certain hip muscles. J Anat 91:286, 1957

KAISER RA: Obturator neurectomy for coxalgia: An anatomical study of the obturator and the accessory obturator nerves. J Bone Joint Surg 31-A:815, 1949

KAMPMEIER OF, BIRCH C LA F: The origin and development of the venous valves, with particular reference to the saphenous district. Am J Anat 38:451, 1927

KAPLAN EB: The iliotibial tract. Clinical and morphological significance. J Bone Joint Surg 40-A:817, 1958

KEAGY RD, BRUMLIK J, BERGAN JJ: Direct electromyography of the psoas major muscle in man. J Bone Joint Surg 48-A:1377, 1966

KLEINBERG S: Aseptic necrosis of the femoral head following traumatic dislocation: Report of two cases. Arch Surg 39:637, 1939

LACROIX P, VERBRUGGE J: Slipping of the upper femoral epiphysis: A pathological study. J Bone Joint Surg 33-A:371, 1951

LAING PG: The blood supply of the femoral shaft: Anatomical study. J Bone Joint Surg 35-B:462, 1953

LAPIDUS PW: Epiphyseal separation of the lesser femoral trochanter. J Bone Joint Surg 12:548, 1930

LAROCHELLE J-L, JOBIN P: Anatomical research on the innervation of the hipjoint. Anat Rec 103:480, 1949 (abstr.)

LAURENSON RD: Bilateral anomalous development of the hip joint. Post mortem study of a human fetus, twenty-six weeks old. J Bone Joint Surg 46-A:283, 1964

LE MESURIER AB: Developmental coxa vara. J Bone Joint Surg 30-B:595, 1948

LEVINTHAL DH, WOLIN I: Unilateral congenital elevation of the ilium or congenital dislocation of the sacro-iliac joint. J Bone Joint Surg 21:193, 1939

LIBERSON WT: Biomechanics of gait: a method of study. Arch Phys Med 46:37, 1965

LICHTBLAU S: Dislocation of the sacro-iliac joint. A complication of bone-grafting. J Bone Joint Surg 44-A:193, 1962

LIEB FJ, PERRY J: Quadriceps function. An anatomical and mechanical study using amputated limbs. J Bone Joint Surg 50-A:1535, 1968

LIEBOLT FL, BEAL JM, SPEER DS: Obturator neurectomy for painful hip. Am J Surg 79:427, 1950

LINTON P: Types of displacements in fractures of the femoral neck and observations on impaction of fractures. J Bone Joint Surg 31-B:184, 1949

LIPSHUTZ BB: Studies on the blood vascular tree: I. A composite study of the femoral artery. Anat Rec 10:361, 1916

LUKE JC: The deep vein valves: A venographic study in normal and postphlebetic states. Surgery 29:381, 1951

MACKENZIE IG, SEDDON HJ, TREVOR D: Congenital dislocation of the hip. J Bone Joint Surg 42-B:689, 1960

MAVOR GE, GALLOWAY JMD: Collaterals of the deep venous circulation of the lower limb. Surg Gynecol Obstet 125:561, 1967

MCKIBBIN B: The action of the iliopsoas muscle in the newborn. J Bone Joint Surg 50-B:161, 1968

MENDELSOHN SN, MANSFIELD RD: Radical groin dissection for carcinoma: A simplified operative procedure. Surg Gynecol Obstet 92:432, 1951

MERCHANT AC: Hip abductor muscle force. An experimental study of the influence of hip position with particular reference to rotation. J Bone Joint Surg 47-A:462, 1965

MEYERDING HW, CHAPMAN JP: Anserina bursitis, a painful condition of the knee (cystic hygroma of horsemen). S Clin North America (Aug.), p 987, 1947

MILTNER, LJ, LOWENDORF CS: Low back pain: A study of 525 cases of sacro-iliac and sacrolumbar sprain. J Bone Joint Surg 13:16, 1931

MING-TZU P'AN: Origin of deep and circumflex femoral group of arteries in the Chinese. Am J Phys Anthropol 22:417, 1937

MING-TZU P'AN: The relation of the sciatic nerve to the piriformis muscle in the Chinese. Am J Phys Anthropol 28:375, 1941

MISRA BD: The relation of the sciatic nerve to the piriformis in Indian cadavers. J Anat Soc India 3:44, 1954

MOORE AT: An anomalous connection of the piriformis and biceps femoris muscles. Anat Rec 23:307, 1922

MULDER JD: Denervation of the hip joint in osteoarthritis. J Bone Joint Surg 30-B:446, 1948

MUNDALE MO, HISLOP HJ, RABIDEAU RJ, KOTTKE FJ: Evaluation of extension at hip. Arch Phys Med 37:75, 1956

MUSGROVE J: Bifurcation of the femoral artery with subsequent re-union. J Anat Physiol 26:239, 1892

NAFFZIGER HC, INMAN V, SAUNDERS JB DE CM: Lesions of the intervertebral disc and ligamenta flava: Clinical and anatomical studies. Surg Gynecol Obstet 66:288, 1938

NAFFZIGER HC, NORCROSS NC: The surgical approach to lesions of the upper sciatic nerve and the posterior aspect of the hip joint. Surgery 12:929, 1942

NEER CS II, GRANTHAM A, SHELTON ML: Supracondylar fracture of the adult femur: A study of one hundred and ten cases. J Bone Joint Surg 49-A:591, 1967

NORRISH RE: Acute ilio-psoas abscess. Brit J Surg 24:55, 1936

OBLETZ BE, LOCKIE LM, MILCH E, HYMAN I: Early effects of partial sensory denervation of the hip for relief of pain in chronic arthritis. J Bone Joint Surg 31-A:805, 1949

OCHILTREE AB: Some muscular anomalies in the lower limb. J Anat Physiol 47:31, 1912

O'CONNOR DS: Early recognition of iliopectineal bursitis. Surg Gynecol Obstet 57:674, 1933

OGDEN JA: Changing patterns of proximal femoral vascularity. J Bone Joint Surg 56-A:941, 1974

O'KEEFE AF, WARREN R, DONALDSON GA: Venous circulation in lower extremities following femoral vein interruption. Surgery 29:267, 1951

O'RAHILLY R: Morphological patterns in limb deficiencies and duplications. Am J Anat 89:135, 1951

OSBORNE RP: The approach to the hip-joint: A critical review and a suggested new route. Brit J Surg 18:49, 1930

PALMA EC: Stenosed arteriopathy of the Hunter canal and loop of the adductor magnus. Am J Surg 83:723, 1952

PATERSON AM: The pectineus muscle and its nerve-supply. J Anat Physiol 26:43, 1891

PATERSON AM: The origin and distribution of the nerves to the lower limb. J Anat Physiol 28:84; 169, 1893–94

PATERSON AM: The action of the biceps flexor cruris. J Anat 51:362, 1917

PAULY JE, SCHEVING LE: An electromyographic study of some hip and thigh muscles in man. Electromyography (Suppl 1) 8:131, 1968

PAUWELS F: Der Schenkelhalsbruch, ein mechanisches Problem. Grundlagen des Heilungsvorganges Prognose und kausale Therapie. Stuttgart, Ferdinand Enke, 1935

PERKINS G: Fractures. London, Oxford University Press, 1940

PERNWORTH P: Obturator hernia: Report of an operation for irreducible incarceration. Am J Surg 71:539, 1946

PETERSON HA, WINKELMANN RK, COVENTRY MB: Nerve endings in the hip joint of the cat: Their morphology, distribution, and density. J Bone Joint Surg 54-A:333, 1972

PETERSON LT: Tenotomy in the treatment of spastic paraplegia: With special reference to tenotomy of the iliopsoas. J Bone Joint Surg 32-A:875, 1950

PICK JW, STACK JK, ANSON BJ: Measurements on the human femur: I. Lengths, diameters and angles.

Quart Bull Northwestern Univ M School 15:281, 1941

PICK JW, STACK JK, ANSON BJ: Measurements on the human femur: II. Lengths, diameters and angles (concl'd). Quart Bull Northwestern Univ M School 17:121, 1943

POIRIER P, CHARPY A: Traité d'Anatomie Humaine Vol. 2, Fasc. 4 (ed 2). Paris, Masson et Cie, 1909

POLLOCK GA: Treatment of adductor paralysis by hamstring transposition. J Bone Joint Surg 40-B:534, 1958

PONSETI IV: Non-surgical treatment of congenital dislocation of the hip. J Bone Joint Surg 48-A:1392, 1966

PONSETI IV: Morphology of the acetabulum in congenital dislocation of the hip. Gross, histological and roentgenographic studies. J Bone Joint Surg 60-A:586, 1978

POPOWSKY J: Ueberbleibsel der Arteria saphena beim Menschen. Anat Anz 8:580, 1893

POWELL T, LYNN RB: The valves of the external iliac, femoral, and upper third of the popliteal veins. Surg Gynecol Obstet 92:453, 1951

RECHTMAN AM: Etiology of deep acetabulum and intrapelvic protrusion. Arch Surg 33:122, 1936

RENNE JW: The iliotibial band friction syndrome. J Bone Joint Surg 57-A:1110, 1975

RIVEROS M, CABANAS R: A lymphangiographic study of the popliteal lymph nodes. Surg Gynec Obst 134:227, 1972

ROBERTS WH: The locking mechanism of the hip joint. Anat Rec 147:321, 1963

ROBINSON DR: Pyriformis syndrome in relation to sciatic pain. Am J Surg 73:355, 1947

ROGERS WM, GLADSTONE H: Vascular foramina and arterial supply of the distal end of the femur. J Bone Joint Surg 32-A:867, 1950

ROOK FW: Arteriography of the hip joint for predicting end results in intracapsular and intertrochanteric fractures of the femur. Am J Surg 86:404, 1953

SAINT JH: The hindquarter (interinnomino-abdominal) amputation. Am J Surg 80:142, 1950

SAMILSON RL, TSOU P, AAMOTH G, GREEN WM: Dislocation and subluxation of the hip in cerebral palsy. Pathogenesis, natural history and management. J Bone Joint Surg 54-A:863, 1972

SASHIN D: A critical analysis of the anatomy and the pathologic changes of the sacro-iliac joints. J Bone Joint Surg 12:891, 1930

SAUNDERS JB DE CM, INMAN VT, EBERHART HD: The major determinants in normal and pathological gait. J Bone Joint Surg 35-A:543, 1953

SCAGLIETTI O, CALANDRIELLO B: Open reduction of congenital dislocation of the hip. J Bone Joint Surg 44-B:257, 1962

SCHAEFFER JP: On two muscle anomalies of the lower extremity. Anat Rec 7:1, 1913

SCHRUTZ A: Zu Zaaijer's Artikel: Seltene Abweichung

(Schligenbildung um die V. cruralis) der A. profunda femoris. Anat Anz 9:727, 1894

SCHUNKE GB: The anatomy and development of the sacro-iliac joint in man. Anat Rec 72:313, 1938

SCHWARTZ DR: Acetabular development after reduction of congenital dislocation of the hip. A follow-up study of fifty hips. J Bone Joint Surg 47-A:705, 1965

SCUDERI C, SCHREY EL: Ruptures of the quadriceps tendon: Study of fourteen tendon ruptures. Arch Surg 61:42, 1950

SEDDON HJ: Volkmann's ischaemia in the lower limb. J Bone Joint Surg 48-B:627, 1966

SEELAUS HK: On certain muscle anomalies of the lower extremity. Anat Rec 35:185, 1927

SEIB GA: Incidence of the m. psoas minor in man. Am J Phys Anthropol 19:229, 1934

SEVITT S, THOMPSON RG: The distribution and anastomoses of arteries supplying the head and neck of the femur. J Bone Joint Surg 47-B:560, 1965

SHANDS AR JR, STEELE MK: Torsion of the femur. A follow-up report on the use of the Dunlap method for its determination. J Bone Joint Surg 40-A:803, 1958

SIMON HE, SACCHET HA: Muscle hernias of the leg: Review of literature and a report of twelve cases. Am J Surg 67:87, 1945

SLOANE D, SLOANE MF: Rupture of the quadriceps tendon. Am J Surg 29:470, 1935

SMITH FB: Effects of rotatory and valgus malpositions on blood supply to the femoral head. Observations at arthroplasty. J Bone Joint Surg 41-A:800, 1959

SMITH GE: An account of some rare nerve and muscle anomalies, with remarks on their significance. J Anat Physiol 29:84, 1894

SNEATH RS: The insertion of the biceps femoris. J Anat 89:550, 1955

SOMERVILLE EW: Development of congenital dislocation of the hip. J Bone Joint Surg 35-B:568, 1953

STANISAVLJEVIC S, MITCHELL CL: Congenital dysplasia, subluxation, and dislocation of the hip in stillborn and newborn infants. An anatomical-pathological study. J Bone Joint Surg 45-A:1147, 1963

STEINDLER A: Mechanics of Normal and Pathological Locomotion in Man. Springfield, Ill., Thomas, 1935

STEPHENS VR: Cystic tumor of the iliopectineal bursa: Report of two cases. Arch Surg 49:9, 1944

SUNDERLAND S, HUGHES ESR: Metrical and non-metrical features of the muscular branches of the sciatic nerve and its medial and lateral popliteal divisions. J Comp Neurol 85:205, 1946

SUTHERLAND DH: An electromyographic study of the plantar flexors of the ankle in normal walking on the level. J Bone Joint Surg 48-A:66, 1966

SUTRO CJ: The pubic bones and their symphysis. Arch Surg. 32:823, 1936

SUTRO CJ: Trauma to the region of the bursa anserina. Am J Surg 75:489, 1948

TAKEBE K, VITTI M, BASMAJIAN JV: Electromyography of pectineus muscle. Anat Rec 180:281, 1974

TANASESCO JGH: Lymphatiques de la symphyse pubienne. Anat Anz 41:415, 1912

TEES FJ: Fractures of the lower end of the femur. Am J Surg 38:656, 1937

THOMAS LI, THOMPSON TC, STRAUB LR: Transplantation of external oblique muscle for abductor paralysis. J Bone Joint Surg 32-A:207, 1950

THOMPSON A: Third annual report of the committee of collective investigation of the Anatomical Society of Great Britain and Ireland for the year 1891–92. J Anat Physiol 27:183, 1893

TOBIN WJ: The internal architecture of the femur and its clinical significance. The upper end. J Bone Joint Surg 37-A:57, 1955

TROTTER M: A common anatomical variation in the sacro-iliac region. J Bone Joint Surg 22:293, 1940

TRUETA J: The normal vascular anatomy of the human femoral head during growth. J Bone Joint Surg 39-B:358, 1957

TRUETA J, HARRISON MHM: The normal vascular anatomy of the femoral head in adult man. J Bone Joint Surg 35-B:442, 1953

TUCKER FR: Arterial supply to the femoral head and its clinical importance. J Bone Joint Surg 31-B:82, 1949

TYRIE CCB: Musculus saphenous. J Anat Physiol 28:288, 1894

VON ROSEN S: Diagnosis and treatment of congenital dislocation of the hip joint in the newborn. J Bone Joint Surg 44-B:284, 1962

WAKELEY CPG: Fractures of the pelvis: An analysis of 100 cases. Brit J Surg 17:22, 1929

WAKELEY CPG: Obturator hernia: Its aetiology, incidence, and treatment with two personal operative cases. Brit J Surg 26:515, 1939

WALKER WA: Traumatic dislocations of the hip joint. Am J Surg 50:545, 1940

WATERS RL, PERRY J, MCDANIELS JM, HOUSE K: The relative strength of the hamstrings during hip extension. J Bone Joint Surg 56-A:1592, 1974

WATSON-JONES R: Dislocations and fracture-dislocations of the pelvis. Brit J Surg 25:773, 1938

WEATHERSBY HT: The origin of the artery of the ligamentum teres femoris. J Bone Joint Surg 41-A:261, 1959

WEINBERG JA: The surgical excision of psoas abscesses resulting from spinal tuberculosis. J Bone Joint Surg 39-A:17, 1957

WERTHEIMER LG: The sensory nerves of the hip joint. J Bone Joint Surg 34-A:477, 1952

WERTHEIMER LG, LOPEZ S DE LF: Arterial supply of the femoral head. A combined angiographic and histological study. J Bone Joint Surg 53-A:545, 1971

WHEATLEY MD, JAHNKE WD: Electromyographic study of the superficial thigh and hip muscles in normal individuals. Arch Phys Med 32:508, 1951

WILKINSON JA: Prime factors in the etiology of congenital dislocation of the hip. J Bone Joint Surg 45-B:268, 1963

WILLIAMS AF: The formation of the popliteal vein. Surg Gynecol Obstet 97:769, 1953

WILLIAMS GD, MARTIN CH, MCINTYRE LR: Origin of the deep and circumflex femoral group of arteries. Anat Rec 60:189, 1934

WILSON JT: Abnormal distribution of the nerve to the quadratus femoris in man, with remarks on its significance. J Anat Physiol 23:354, 1889

WOLCOTT WE: The evolution of the circulation in the developing femoral head and neck: An anatomic study. Surg Gynecol Obstet 77:61, 1943

WOODBURNE RT: The accessory obturator nerve and the innervation of the pectineus muscle. Anat Rec 136:367, 1960

WRIGHT WG: Muscle Function. New York, Hoeber, 1928; Hafner, 1962

YEOMAN W: The relation of arthritis of the sacro-iliac joint to sciatica: With an analysis of 100 cases. Lancet 2:1119, 1928

ZADEK I: Congenital coxa vara. Arch Surg 30:62, 1935

Chapter 9

KNEE, LEG, ANKLE, AND FOOT

The bones of the leg are the tibia and fibula, the former being developmentally the preaxial bone and therefore corresponding to the radius of the forearm, the latter corresponding to the ulna. Unlike the forearm, however, where the ulna forms the chief articulation at the elbow, the radius the chief one at the wrist, the tibia forms the chief articulation at both the knee and ankle. The fibula does not at all enter into the articulation with the femur at the knee, and its articulation at the ankle is on the side, so that it is poorly equipped to bear weight. Its chief importance lies in the facts that it gives origin to numerous muscles, and that at its lower end it helps the tibia so to grasp the talus that the talocrural or ankle joint is an almost strictly hinge one. The tibia, in spite of its larger size, gives rise to fewer muscles than does the fibula; its chief function is the support of weight. The tarsal bones of the foot are homologous with the carpals of the hand, and the metatarsals and the phalanges of the toes are very similar to the metacarpals and the phalanges of the fingers.

The muscles of the leg are divisible into two general groups, the posterior (originally flexor) muscles, or muscles of the calf, and the anterolateral muscles (Fig. 9-1). In turn, each of these groups is divisible into two chief subgroups: the muscles of the calf are conveniently divided into a superficial and a deep group, the anterolateral muscles into a lateral and an anterior group. Certain muscles of the

leg are the obvious equivalents of muscles of the forearm, and there is an even closer resemblance between the muscles of the foot and those of the hand.

The arteries of the leg are branches of the popliteal, and the nerves of the leg are all branches of the sciatic, with the exception of the saphenous nerve (a cutaneous branch of the femoral). However, as elsewhere in the limbs, both anterior and posterior branches of the nerve plexus supplying the limb are continued into the leg, and these two components have typically different distributions. Thus the tibial nerve, the anterior or ventral component of the sciatic nerve, is distributed to the calf and the plantar surface of the foot, the original ventral or anterior part of the limb; and the common peroneal nerve, the posterior or dorsal component of the sicatic, is distributed anterolaterally on the leg and onto the dorsum of the foot, thus to the original dorsal or posterior part of the leg and foot. Many comparisons are also possible between the vessels and nerves of the leg and foot and those of the forearm and hand; perhaps the greatest similarity is between the medial and lateral plantar nerves in the foot and the median and ulnar nerves in the hand: these have almost identical distributions.

Fascia

The superficial fascia or tela subcutanea of the leg is continuous above with that of the

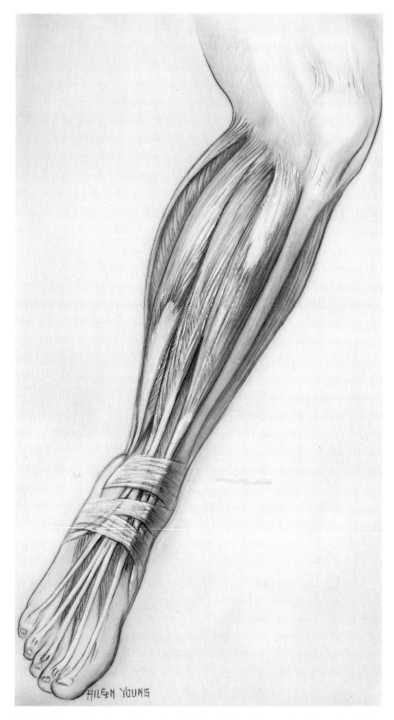

Fig. 9-1. Anterior view of muscles of the leg. The wider parts of muscles of the
calf are visible; anteriorly, the tendons of the tibialis anterior, extensor hallucis
longus, extensor digitorum longus, and peroneus tertius, beginning medially
(the viewer's right) can be recognized as they course deep to the extensor retinacula. The peronei are also visible, just lateral to the extensor digitorum
longus.

thigh, below with that of the foot. As usual, it consists of mingled fat and fibrous connective tissue in which are embedded most of the superficial nerves and vessels. Its deep surface, in contact with the deep fascia of the leg, tends to be more fibrous than its outer surface, and the lesser saphenous vein, in the fascia in the lower part of the leg, lies deep to it (and sometimes deep to the deep fascia) in the upper part of the leg. The superficial fascia of the dorsum of the foot is thin, that of the plantar surface much thicker and tougher, especially under the heel where it forms a particularly heavy protective pad.

The deep fascia of the leg, the crural fascia, is more complicated. It is continuous above with the fascia lata of the thigh, but since this fascia is largely fastened down around the knee, the crural fascia is likewise. Thus it is attached above to the patella, the patellar ligament, and the various bony projections about the knee—the tuberosity of the tibia, the femoral and tibial condyles and the head of the fibula; in passing from the patella and the patellar ligament to the bones of the leg, the fascia therefore helps to form the patellar retinacula, formed also in part by the fascia lata and by tendinous fibers derived from the vasti muscles. Over the popliteal fossa the fascia is more directly continuous with the fascia lata. The upper part of the crural fascia is strengthened by receiving expansions from the tendons of the biceps femoris, gracilis, semitendinosus, and sartorius muscles. Muscle hernias through it, more commonly of the tibialis anterior or the gastrocnemius, sometimes appear to be the result of trauma, but are probably more often located at the sites of congenital defects in the fascia (Simon and Sacchet).

Over the subcutaneous surface of the tibia the crural fascia blends with the periosteum of this bone; lateral to its attachment to the tibia it gives rise on its deep surface to some of the fibers of the anterior muscles of the leg, especially the tibialis anterior. As it passes around the lateral side of the leg, it gives off from its deep surface an anterior and a posterior intermuscular septum that pass to the fibula in front of and behind the lateral muscles,

the peroneus longus and brevis (Fig. 9-2). Thus, in consequence of the attachment of the crural fascia to the tibia, there are three major compartments of the leg: an anterior one for the anterior or extensor muscles, between the anterior intermuscular septum and the tibia; the lateral one, containing the peroneus longus and brevis; and a posterior one, between the posterior intramuscular septum and the tibia, and containing the muscles of the calf. The posterior compartment is usually described as being subdivided by a septum, derived from the deep fascia, that passes almost transversely across the calf and separates the superficial muscles from the deep muscles here. This is usually called the transverse crural septum or deep transverse fascia of the leg.

The deep fascia of the plantar surface of the foot is rather similar to that of the hand, being thinner over the special muscles of the big and little toes, but forming in the central part of the foot a heavy plantar aponeurosis. The fascia of the dorsum is for the most part thin; it and the plantar fascia are described in more detail in the section on the foot.

The deep fascia of the leg and foot is strengthened about the ankle by additional fibers stretching between bony attachments to form retinacula for the tendons crossing the ankle. Since there are three sets of tendons, there are three sets of retinacula: extensor, peroneal, and flexor.

The *extensor retinaculum* (Figs. 9-3 and 9-5) is in two parts: one, the superior extensor retinaculum (transverse crural ligament) lies in the lower part of the leg; the other, the inferior extensor retinaculum (cruciate ligament) lies on the proximal portion of the dorsum of the foot. The superior extensor retinaculum stretches from the tibia to the fibula, as does the deep fascia here, and is represented by so few additional fibers that its upper and lower borders are indistinct. The tendons of the tibialis anterior, extensor hallucis longus, and extensor digitorum longus (and muscle fibers of the peroneus tertius) pass deep to this retinaculum, but usually only the tendon of the tibialis anterior is provided with a tendon

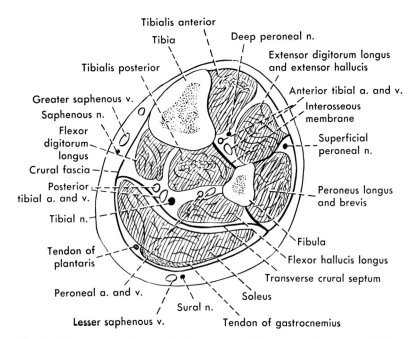

Fig. 9-2. Cross section through the lower part of the leg. (Redrawn from Eycleshymer AC, Schoemaker DM: A Cross-section Anatomy. New York, Appleton, 1923)

sheath at this level. The inferior extensor retinaculum, formerly called the cruciate crural ligament, is much better defined than is the superior one. It resembles a Y, with a lateral stem attached mostly to the calcaneus, and two diverging bands, arising from this stem on the front of the ankle, to represent the two limbs of the Y.

The lateral stem can be divided into three parts or roots (Smith, '58). The lateral or superficial blends with the deep fascia on the lateral side of the foot; the strong intermediate root descends to attach to the upper surface of the calcaneus and is here separated from the lateral root by fibers of the extensor digitorum brevis that also arise from the upper surface of the calaneus; and the slender medial root runs medially between talus and calcaneus (in the sinus tarsi) to insert into the sulcus on the upper surface of the calcaneus (see Fig. 9-67). According to Cahill, both the medial and intermediate roots also send fibers to the talus. On the medial side, the upper limb proceeds medially and upward to attach to the medial malleolus, while the lower limb extends medially and somewhat downward to

blend with the fascia of the medial side of the sole of the foot.

The inferior extensor retinaculum does not simply lie in front of the extensor tendons, as does the superior retinaculum, but rather is divided into deep and superficial laminae by the tendons, and each set of tendons lies in its own compartment in the inferior retinaculum. Thus the stem of the retinaculum splits into superficial and deep layers on the lateral side of the closely associated tendons of the peroneus tertius and extensor digitorum longus muscles, and these two laminae unite again on the medial border of these tendons; within this compartment the tendons are provided with a common synovial tendon sheath. Similarly, the upper limb of the Y splits on the lateral border of the extensor hallucis longus, and comes together again on its medial border, the anterior lamina here being especially thick. In contrast, most of the fibers of the upper limb pass behind the tendon of the tibialis anterior on their way to the tibia, and there may even be no fibers passing in front of it. It is because of the thinness or absence of the upper limb of the extensor retinaculum in front of the tibialis anterior that this tendon becomes prominent upon dorsiflexion of the foot. The lower limb of the inferior extensor retinaculum passes largely in front of both the extensor hallucis longus and the tibialis anterior tendons, or sometimes mostly behind the latter.

The *tendon sheaths* on the front of the ankle are three in number: a sheath common to the closely associated extensor digitorum longus and peroneus tertius tendons, and separate ones for the extensor hallucis longus and the tibialis anterior. The common tendon sheath of the extensor digitorum longus and peroneus tertius begins usually between the levels of the superior and inferior extensor retinacula, and extends downward about 3 inches (7.5 cm) onto the dorsum of the foot. As the extensor digitorum tendon widens and splits into its terminal tendons and the peroneus tertius starts to diverge from these, the sheath widens correspondingly. The tendon of the peroneus tertius leaves the lateral side of the sheath, which usually continues down a little farther on the long extensor tendons. The upper end of the sheath of the extensor hallucis longus lies usually only a little above the upper border of the inferior retinaculum, but its lower end extends to or beyond the cuneometatarsal joint of the big toe. The tendon sheath of the tibialis anterior begins much higher, behind or above the superior extensor retinaculum, and also ends higher, usually at about the upper border of the inferior limb of the retinaculum.

The *flexor retinaculum* or laciniate ligament (Fig. 9-4) is a thickening of the transverse crural septum (deep transverse crural fascia) at the medial side of the ankle; the tendons, nerve, and vessels that run from the calf of the leg into the plantar aspect of the foot pass deep to it. The flexor retinaculum is attached above to the tibia, below to the calcaneus, and between these points sends three strong septa down to the tibia; thus there are four compartments under cover of this retinaculum. The upper and most anterior compartment, immediately against the posterior and inferior surface of the medial malleolus, transmits the tendon of the tibialis posterior

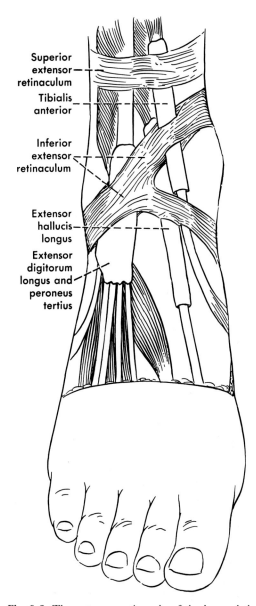

Fig. 9-3. The extensor retinacula of the leg and the tendon sheaths of the extensor muscles.

muscle; the next compartment contains the tendon of the flexor digitorum longus; the third the tibial nerve and posterior tibial vessels; and the most posterior and lowest compartment the tendon of the flexor hallucis longus.

Each of the tendons has, within its compartment deep to the retinaculum, its own *tendon sheath*. That of the tibialis posterior commonly begins highest, some 2 inches (5

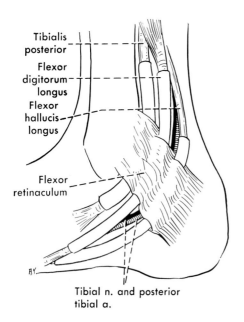

Fig. 9-4. The flexor retinaculum and the tendon sheaths of the deeper layer of muscles of the calf.

cm) or more above the tip of the medial malleolus, but extends only a little below the lower edge of the retinaculum. The sheath of the flexor digitorum longus usually begins below that of the tibialis posterior, but extends onto the plantar surface of the foot to about the point at which the tendons of the flexor digitorum longus and flexor hallucis longus cross each other. The sheath of the flexor hallucis longus usually begins just barely above the upper edge of the retinaculum, and extends to about the same level or slightly beyond that of the flexor digitorum longus.

While these tendon sheaths are necessarily separate as they lie in their fibrous compartments, they may communicate with each other below the retinaculum; Lovell and Tanner quoted evidence that the tendon sheaths of the flexor hallucis longus and of the flexor digitorum longus communicate with each other at the point at which the tendons cross in about 20% of cases, and said that sometimes the sheath of the tibialis posterior communicates with that of the flexor digitorum longus. Grodinsky found communications between the first-mentioned sheaths in "about half" of his specimens, and said he

had observed once a communication between the upper ends of the sheaths of the tibialis posterior and the flexor digitorum.

There are two *peroneal retinacula,* a superior and an inferior (Fig. 9-5). The superior is attached above to the lateral malleolus, below to the lateral surface of the calcaneus, and has deep to it a single compartment in which run the tendons of the peroneus longus and brevis muscles. The inferior is attached to the lateral surface of the calcaneus above and below the two peroneus tendons, and sends a septum between them to attach to the peroneal trochlea, a more or less obvious ridge on the lateral surface of the calcaneus.

Behind the lateral malleolus and deep to the superior peroneal retinaculum the peroneus longus and brevis muscles share a common tendon sheath. The upper end of the sheath is bifurcated, one component extending upward around the peroneus longus tendon for perhaps an inch (2.5 cm), the other extending around the peroneus brevis for about half this distance. Similarly, the lower end of the tendon sheath is bifurcated, from

Fig. 9-5. The peroneal retinacula and the sheath of the peroneal tendons.

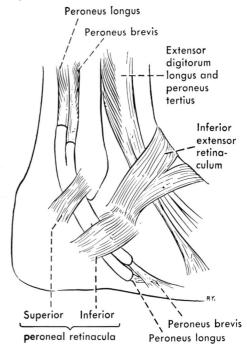

at least the level of the peroneal trochlea downward; the part of the tendon sheath for the peroneus brevis usually follows this tendon to within about an inch of the attachment of the muscle, while that of the longus extends farther downward: on the superficial surface of the tendon it usually stops about a half inch (1 cm to 1.5 cm) beyond the peroneal trochlea, but on the deep surface of the sheath is continued to the level at which the tendon of the peroneus longus rounds the lateral and plantar surface of the cuboid bone. Here it may communicate with a second, plantar, sheath associated with the tendon in its rather long course across the plantar surface of the foot; Lovell and Tanner cited one report that such a communication is rare, another that it exists in about 33% of feet. Grodinsky found such a communication in 1 of 7 feet, and said that while the adjacent ends were contiguous in 2 more, in the other 4 they were from 0.25 cm to 1.0 cm apart.

SUPERFICIAL NERVES AND VESSELS

The cutaneous nerves of the leg are derived from the femoral and both divisions of the sciatic, and typically continue to supply also skin of the foot; in addition, the posterior femoral cutaneous nerve, which supplies a broad strip down the posterior aspect of the thigh, continues into the upper part of the leg to supply skin here.

Except for the usual minute arterial twigs to the skin, the superficial vessels are the veins belonging to the saphenous system, which begins on the foot and ends in the thigh. The digital arteries, necessarily superficially placed on the toes, are described with the arterial system of the foot.

Cutaneous Innervation

Skin of the anteromedial side of the leg is typically supplied by the *saphenous nerve*. This nerve becomes superficial just above the knee, where it pierces the fascia lata behind the tendon of the sartorius to pass downward along the medial side of the leg in company with the great saphenous vein. Just before it becomes subcutaneous, the saphenous nerve

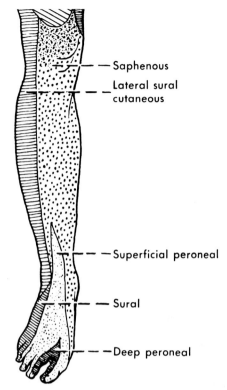

Fig. 9-6. Distribution of cutaneous nerves on the anterior surface of the leg. (Flatau E: Neurologische Schemata für die ärztliche Praxis. Berlin, Springer, 1915)

gives off a large infrapatellar branch, which pierces the sartorius and is distributed to skin in front of the knee; this branch in particular is endangered in medial approaches to the knee joint, and most of the saphenous stem below the knee may be avulsed in "stripping" a varicose great saphenous vein. Its other branches are known collectively as the medial crural cutaneous branches, and are distributed to the anterior and medial sides of the leg (Fig. 9-6). The longest terminal branch passes in front of the medial malleolus and along the medial side of the foot for a variable distance; it may reach to about the level of the metatarsophalangeal joint of the great toe. The nerve should be identified and avoided during surgical approaches to this region. Kosinski reported that the saphenous nerve never normally supplies skin of this toe; in the one instance he found in which it did, the cutaneous part of the superficial peroneal nerve

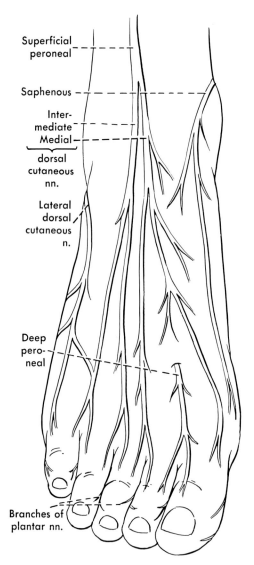

Superficial
peroneal

Saphenous

Inter-
mediate
Medial

dorsal
cutaneous
nn.

Lateral
dorsal
cutaneous
n.

Deep
pero-
neal

Branches of
plantar nn.

Fig. 9-7. Nerves of the dorsum of the foot.

(normally supplying skin of the big toe) was lacking.

The skin of the anterolateral side of the leg is supplied by three nerves, the lateral sural cutaneous, the superficial peroneal, and the sural; the first-mentioned and the last are also distributed posteriorly. The *lateral sural cutaneous nerve* arises from the common peroneal, usually some 1 to 2 inches (2.5 cm to 5 cm) above the level of the knee joint, and passes down for a short distance with the common peroneal before piercing the fascia lata. Through its branches it supplies skin over the

upper part of the lateral and posterolateral aspects of the leg, but it varies considerably. Kosinski found it missing in 1.7% of cases, and said that in another 9.4% it had no cutaneous branches of its own, but joined the medial sural cutaneous; most workers would probably regard the nerve in the latter case as being the peroneal communicating rather than the lateral sural cutaneous. Huelke found the lateral sural cutaneous absent in 22% of his cases, much of its posterior distribution being taken over by the posterior femoral cutaneous and the saphenous nerves, and in 13% it was partly replaced by branches of the peroneal communicating. Both workers reported that sometimes the lateral sural cutaneous nerve replaces the sural in forming the lateral dorsal cutaneous nerve of the foot.

The *superficial peroneal nerve* is one of the terminal branches of the common peroneal. It passes down the leg between the peroneus longus and brevis muscles, and then between these and the extensor digitorum longus to pierce the crural fascia at about the junction of the upper two thirds with the lower third. Soon after piercing the fascia, or sometimes before, it divides into its two main branches for the dorsum of the foot; as these continue the downward course of the nerve they diverge slightly, and give off twigs that supply skin of the anterolateral leg and the dorsum of the foot.

The medial terminal branch of the superficial peroneal nerve is the medial dorsal cutaneous nerve (Fig. 9-7). On the dorsum of the foot it divides into two dorsal digital nerves, of which the medial is distributed to the medial side of the big toe, where it may communicate with the saphenous nerve and with the dorsal digital branch of the deep peroneal. The lateral branch of the medial dorsal cutaneous nerve runs toward the interspace between the second and third toes and divides into dorsal digital branches for the adjacent sides of these two toes.

The lateral terminal branch of the superficial peroneal nerve, the intermediate dorsal cutaneous, also divides into two branches: the medial one divides into dorsal digital branches for the adjacent sides of the third and fourth toes, the lateral one into branches for the adjacent sides of the fourth and fifth toes. In about 60% of cases (Kosinski) the lateral branch receives a communication from the lateral dorsal cutaneous nerve, the continuation of the sural nerve onto the dorsum of the foot. These nerves should be avoided in the relatively frequent operations done through dorsolateral incisions on the foot.

The *deep peroneal nerve* supplies skin in the foot only. Its terminal branch emerges from under cover of the flexor hallucis brevis close to the interspace between the first and second toes, and divides into dorsal digital branches for the adjacent sides of these toes, usually receiving a twig of communication from the medial digital branch of the medial dorsal cutaneous nerve.

Although the superficial peroneal nerve is the chief supply to skin of the dorsum of the foot, there is a great deal of variation in the distribution of the cutaneous nerves here. Kosinski described 3 main types of distribution, and some 40 subtypes. The lateral dorsal cutaneous nerve may replace part or all of the intermediate dorsal cutaneous, supplying the adjacent sides of the fourth and fifth toes, or these and the adjacent sides of the third and fourth. Similarly, either the medial or the intermediate dorsal cutaneous nerves may show expansion or contraction of their usual areas of distribution, replacing or being replaced by branches of each other, the lateral dorsal cutaneous, or the deep peroneal.

As in the hand, the dorsal digital nerves are smaller than the plantar digital ones, and do not run the entire lengths of the digits. The innervation of the distal skin of the digits, and the nail bed, is supplied by the plantar digital nerves.

Skin of the lateral side of the calf, of the lower part of the lateral surface of the leg, and of the lateral and dorsolateral aspect of the foot is usually supplied by the *sural nerve* (Figs. 9-7 and 9-8). This nerve is formed by the union of the medial sural cutaneous nerve and the peroneal communicating branch from the common peroneal. The medial sural cutaneous nerve arises from the tibial in the popliteal fossa, typically a little below the

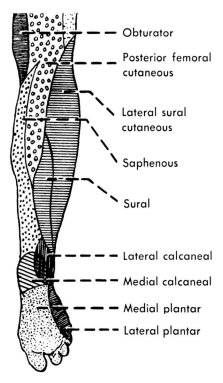

Fig. 9-8. Distribution of nerves on the posterior surface of the leg. (Flatau E: Neurologische Schemata für die ärztliche Praxis. Berlin, Springer, 1915)

level of the knee joint. The peroneal communicating branch arises from the common peroneal nerve in the popliteal fossa, either with the lateral sural cutaneous (58.5% of Huelke's series) or alone. It may be very short and join the medial sural cutaneous to form the sural nerve in the popliteal fossa, or it may run down over the lateral head of the gastrocnemius and join the medial sural cutaneous at almost any level above the ankle. In 75% of Huelke's series the union was in the lower half of the leg. Kosinski found the nerve present in less than half of limbs, Huelke in 80%; when it is absent the medial sural cutaneous usually continues onto the foot as the lateral dorsal cutaneous nerve, normally a continuation of the sural. The medial sural cutaneous nerve, or the sural if it is formed in the popliteal fossa, runs downward deep to the deep fascia and between the two heads of the gastrocnemius muscle; in consequence of this position, it is easily injured in posterior approaches to the popliteal space. It typically does not pierce the crural fascia until it reaches the middle of the leg, or even the junction of the upper two thirds and lower third, but after piercing the fascia it lies alongside the lesser saphenous vein, usually medial to it but sometimes lateral. The sural nerve usually supplies skin of at least the lower third of the lateral surface of the leg and the lateral side of the calf, and after giving off lateral calcaneal branches to the heel it continues onto the foot as the lateral dorsal cutaneous nerve; as already noted, this supplies at least the dorsal digital branch to the lateral side of the little toe.

The remaining nerve to supply cutaneous twigs to the leg, the *posterior femoral cutaneous,* may fail to do so; Kosinski found it apparently stopping at about the level of the knee joint in 5.5% of cases, but usually extending some distance into the leg, most commonly from 40% to 60% of the length of the leg. In its course down the leg it accompanies the lesser saphenous vein if that is superficial, usually lying on the medial side of this vein. It supplies a variable amount of skin of the posterior surface of the upper part of the calf, its distribution typically overlapping with that of the saphenous medially, the lateral sural cutaneous laterally, and the sural nerve inferiorly.

Except for the calcaneal branches of the sural nerve, cutaneous branches to the sole of the foot are entirely from the tibial nerve. As it lies in the flexor retinaculum or as it passes onto the sole of the foot the tibial nerve or its chief terminal branches give off *medial calcaneal branches* that pierce the retinaculum or the origin of the abductor hallucis and supply skin on the posterior and medial sides of the heel and most of its plantar surface. The remainder of the skin of the sole is supplied by the *medial and lateral plantar nerves,* the terminal branches of the tibial. Their digital branches, described in more detail in connection with the deep structures of the foot, are distributed exactly as are the palmar cutaneous nerves in the hand: the medial plantar nerve supplies the medial three and one-half toes, thus corresponding to the median, and the lateral

plantar nerve supplies the lateral one and one-half toes, or more, thus corresponding to the ulnar; as is true also in the hand, the nerves to the adjacent sides of the third and fourth digits often, but not always, are formed by fibers from both nerves. As these nerves pass forward, they give off branches that penetrate the deep fascia and supply skin of the anterior two thirds of the sole corresponding to their distribution on the digits.

Superficial Veins

The superficial veins of the lower limb, and their communications with each other and with the deep veins, have been described in some detail in Chapter 7; thus they need little description here.

Both of the two large superficial veins of the leg originate on the dorsum of the foot. The *great saphenous vein* (Fig. 9-9) originates from the medial side of the dorsal venous arch and passes upward in front of the medial malleolus; it is relatively easily located here, and this part of the vein is therefore one of the favored sites for administering intravenous therapy, especially in infants. It continues upward along the medial side of the leg, running slightly backward to cross the knee joint close to the posterior border of the medial surface of the limb. Throughout its course in the leg it is accompanied by the saphenous nerve. In addition to the communicating veins with deep veins of the leg, described in Chapter 7, it receives veins from the subcutaneous network of the leg, and communications from the small saphenous vein which are said to be so valved that blood can pass only from it to the great saphenous.

The *small saphenous vein* (lesser saphenous) arises from the lateral side of the dorsal venous arch and runs upward behind the lateral malleolus (Fig. 9-10). As it continues straight up the leg and the calf widens, it comes to lie at about the middle of the posterior surface of the calf. In the lower part of its course it is accompanied by the sural nerve and usually, in the upper part of its course, by the posterior femoral cutaneous nerve. Both these nerves tend to lie medial to the vein, but either may lie lateral. The small saphenous vein typically

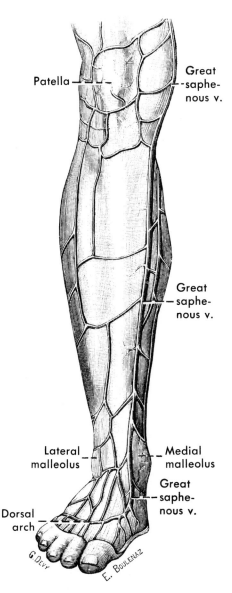

ends largely or entirely in the popliteal vein; variations in its ending have been mentioned in Chapter 7.

A very few instances of a superficially placed (great saphenous) artery accompanying the great saphenous vein, even as far as the big toe, have been reported (Fig. 8-75, and Senior, '19). Similarly, there have been a few reports of an artery (great sural artery) of

variable size accompanying the lesser saphenous vein (Senior).

Superficial Lymphatics

The superficial lymphatics have also been described in Chapter 7. Briefly summarized, most of the superficial lymphatic vessels from the plantar surface of the foot pass around the margins of the toes or of the foot to join the lymphatics of the dorsum. These in turn tend

Fig. 9-10. The small saphenous and other superficial veins of the calf. (Testut L: Traité d'Anatomie Humaine, Vol 2. Paris, Doin, 1891)

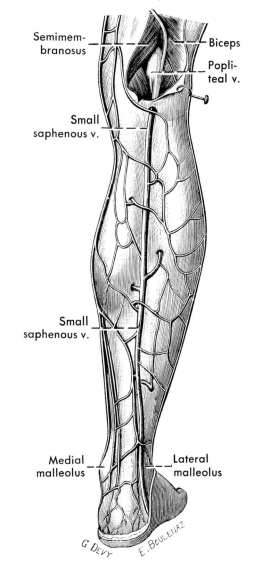

to converge toward and follow the course of the great saphenous vein, but some from the lateral side of the foot run upward with the small saphenous vein. In the leg, also, the lymphatics originating here tend to converge toward and join the lymphatics accompanying the great saphenous vein: vessels from the posteromedial side of the leg pass around the medial border toward the great saphenous, those from the lateral run anteriorly, are joined by vessels originating anteriorly, and then pass around the medial side of the leg toward the vessels accompanying the great saphenous. Not all the lymphatic vessels of the leg do this, however; the vessels running upward with the small saphenous vein receive the lymphatic drainage from a small area on each side of the vein, so that there is a strip of skin along the posterior aspect of the calf that does not send its efferents along the great saphenous vein to the superficial inguinal lymph nodes. The lymphatics accompanying the lesser saphenous vein penetrate the deep fascia of the leg with this vein and end usually in the popliteal lymph nodes, a variable number of very small nodes in the popliteal fossa.

The Knee Joint

The knee joint (Fig. 9-11; see also Figs. 9-14 to 9-21) is the articulation between the femoral and tibial condyles; the fibula does not enter into this articulation. The joint resembles a hinge one more closely than it does a ball-and-socket one, although rotation at the knee does occur, and the joint is therefore not strictly of the hinge type. Like hinge joints, the knee joint is largely protected anteriorly and posteriorly by muscles, and the special ligamentous thickenings associated with its capsule lie mostly on the sides of the joint. Since the anterior muscle associated with the joint is the quadriceps, the tendon of which contains the patella, this bone also enters into the articulation at the knee.

The lower end of the femur has already been described (Chap. 8). Features particularly important in the present discussion include the facts that the two rounded femoral condyles are eccentrically curved, being more curved posteriorly than anteriorly; that the curve of the lateral condyle is slightly greater, especially anteriorly, than is that of the medial, but at the same time is also slightly shorter; and that the two articular surfaces are confluent anteriorly, through the articular surface for the patella, but widely separated postero-inferiorly and posteriorly by the intercondylar fossa.

Pertinent features of the upper end of the tibia can be quickly summarized. The medial and lateral tibial condyles form the upper, expanded end of the tibia, the lateral condyle presenting posterolaterally an articular surface for the head of the fibula. The upper surface of each condyle is an articular surface, approximately ovoid in shape, and almost flat but very slightly concave. The two condyles and their articular surfaces are separated anteriorly by a slight depression, the anterior intercondylar area (fossa) that runs from the anterior to the superior surface of the bone, and posteriorly by a more marked depression, the posterior intercondylar area (fossa) that rounds the posterior margin of the upper surface. Between the anterior and posterior intercondylar areas the condyles and their articular surfaces are separated by a raised area, the intercondylar eminence, that bears medial and lateral tubercles.

Between the femoral and tibial condyles, attached primarily to the latter, are fibrocartilaginous menisci, thick peripherally but tapering to thin free edges as they approach the intercondylar areas and eminence. While these menisci, because of their curvatures, provide a slightly more congruous surface for the femoral condyles, they do not deepen the joint very much. In consequence, the strength of the knee joint is derived entirely from the muscles and ligaments associated with it, and it is particularly vulnerable to a force delivered at an approximately right angle to the long axis of the limb.

The femur and the tibia are not in a direct line with each other, for the femur slants inward toward the knee, departing from a right angle in relation to the infracondylar plane by about 9°, while the tibia is more nearly at a right angle to this plane, passing almost

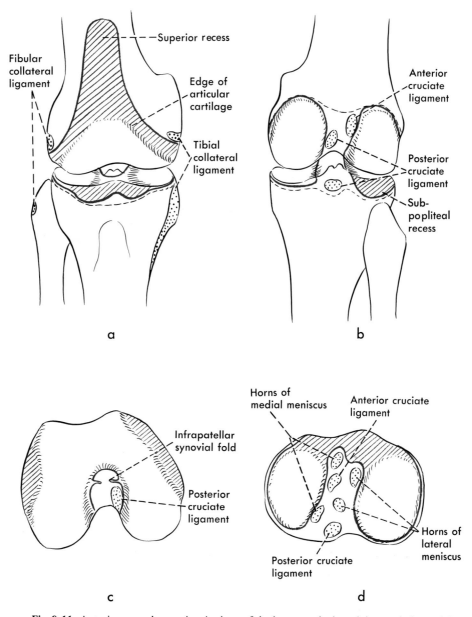

Fig. 9-11. Anterior, *a*, and posterior, *b*, views of the bones at the knee joint, and views of the articular surfaces of the femur, *c*, and tibia, *d*. The *broken* lines indicate the attachment of the fibrous capsule, the *red* lines and areas indicate the synovial membrane and its reflections on the surfaces of the bones.

vertically downward in the male, with a slight lateral inclination in the female. Thus there is a normal valgus deviation at the knee of about 10° to 12° (Steindler).

There is a difference of opinion as to which femoral condyle at the knee bears the greater weight. Steindler said the lateral one does, because of the natural genu valgum. Pick and co-workers described the weight-bearing surface of the medial condyle as usually slightly lower than that of the lateral, and therefore bearing the greater weight; they said this opinion seems to be supported by the much higher incidence of attritional and hypertro-

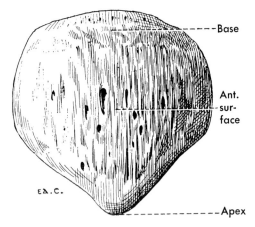

Fig. 9-12. Anterior view of the right patella. (Poirier P, Charpy A: Traité d'Anatomie Humaine [ed 3], Vol 1. Paris, Masson et Cie, 1911)

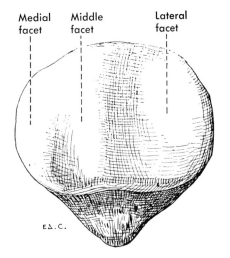

Fig. 9-13. Posterior view of the right patella. (Poirier P, Charpy A: Traité d'Anatomie Humaine [ed 3], Vol 1. Paris, Masson et Cie, 1911)

phic changes found in the medial femoral condyle as compared with the lateral one. De Palma ('54) apparently agreed with Steindler; he also agreed that changes in the normal angle necessarily shift the stress, increase of genu valgum throwing more weight upon the lateral condyle, genu varum shifting weight abnormally onto the medial condyle.

In regard to the weight-bearing surfaces, Kettelkamp and Jacobs reported that in the extended position the area of contact between the medial condyles of the femur and tibia is 1.6 times that between the lateral condyles, and although both areas decrease steadily with flexion, this ratio is maintained. Maquet and co-workers agreed that the area of contact decreases during flexion, for in their experiments they found it decreasing from an average of 20.13 cm^2 at extension to one of 11.6 cm^2 between 90° and 110° of flexion; they said, however, that the area of contact was about equal on the medial and lateral condyles. Removal of the menisci markedly reduced the weight-bearing area, to 12 cm^2 at extension, and 6 cm^2 at 90° of flexion.

PATELLA

The somewhat triangular patella is a sesamoid bone, the largest of the body, and is developed in the tendon of the quadriceps. Its rough anterior surface (Fig. 9-12) receives fibers of the quadriceps tendon and ligamentum patellae that spread over it, but the chief attachment of the tendon is on the upper border (base) of the bone and on its sides, while the chief attachment of the ligament is on both the anterior and posterior surfaces of its apex. The patellar retinacula, formed by expansions of the quadriceps tendon and the fascia lata, extend between the sides of the bone and the femoral and tibial condyles. Between the bone and the covering skin is the prepatellar bursa.

The articular surface of the patella (Fig. 9-13) is unequally divided into larger lateral and smaller medial parts by a longitudinal ridge and occupies most of the posterior surface of the bone. The blood vessels to the patella are derived from a vascular ring formed by anastomoses between the genicular arteries; most of them enter the anterior surface, but those to the apex enter posteriorly, below the articular surface (Scapinelli).

Dislocation, Fracture, and Anomalies

Although the patella can be forcibly dislocated either laterally or medially, the more gentle slope of its lateral articulation makes a lateral dislocation easier than a medial one. Duchenne found that stimulation of the vastus lateralis could produce lateral subluxation of the patella, but failed to obtain medial subluxation by stimulating the vastus medialis, in spite of the lower attachment of this muscle into the quadriceps tendon and patella.

Recurrent dislocation of the patella is apparently always a lateral dislocation (Thompson and Bosworth) and may be in part hereditary. Heywood regarded ligamentous laxity as being the common cause, and said that transplanting the tibial tuberosity downward and medially in young adults gives good results, but should not be used in children, and that removal of the patella is better in older patients. Brattström, through careful measurements on roentgenograms, adduced evidence that femoral dysplasia is a usual finding, with a lessened anterior projection of the lateral femoral condyle beyond the floor of the patellar groove and a diminished depth of the groove.

Fracture of the patella is relatively common (Michele and Krueger) and traumatic arthritis may result (Paschall and co-workers). Fracture may be a result of a blow delivered to the knee cap, but often it is a clean break of the bone across the fulcrum of the lower end of the femur. With the knee semibent, only a small middle part of the patella's articular surface is engaged with the articular surface of the femur: thus a sudden powerful contraction of the quadriceps, or a force delivered so as to produce further sudden flexion of the knee, can result in the patella (held below by the patellar ligament and pulled upon above by the quadriceps muscle) being snapped across the femur. Fracture of the lower pole may be distal to the attachment of the synovial membrane and hence extraarticular. There may be wide separation of fragments if the retinacula are torn.

Scapinelli found that marginal fractures of the patella rarely unite, presumably because the blood vessels to most of the bone radiate from their entrance anteriorly. Since the lower part of the patella has vessels entering it posteriorly also, he suggested that if part of the patella is to be removed after transverse fractures, it should be the upper fragment.

Bipartite patella should not be confused with fracture of the patella. According to Oetteking, about 3% of patellae have a defect in the upper and outer margin of the bone, as a result of the development of a secondary center of ossification and its incomplete fusion with the rest of the bone. The secondary center therefore forms an accessory bone that fills the defect but is either unfused or incompletely fused to the patella.

Patellectomy

Because the patella holds the quadriceps tendon farther away from the axis of motion of the knee, it increases the efficiency of the quadriceps in the last part of its extension of the knee. Haxton found that in the cadaver removal of the patella increased the quickness of movement of the leg from a fully flexed position, when acted upon by a constant force, but that the movement stopped about 30° short of full extension; he therefore agreed that the patella particularly improves the efficiency of extension as the extended position is approached.

In consequence of this function, there is some difference of opinion as to when fracture of the patella should be treated by total or partial excision, and when it should be treated conservatively. Cave and Rowe, noting that it has been said that the knee functions as well without as with the patella, quoted evidence that only about 5% of 101 patients in whom the patella had been excised believed that the knee was normal. Scott also reviewed this question, and found in a follow-up on patients that many of them did not consider the knee to be normal following total removal of the patella. However, there is no way of knowing how much of the disability is actually due to the removal of the patella and how much to concomitant injury of the knee at the time the patella was fractured. Obtaining complete extension after a good repair is not so much of a problem as is the loss of power to maintain extension against resistance when the knee is flexed to about 30° or 40° (Bickel).

Horwitz and Lambert regarded successful results from patellectomy as depending upon eliminating the slack in the quadriceps ten-

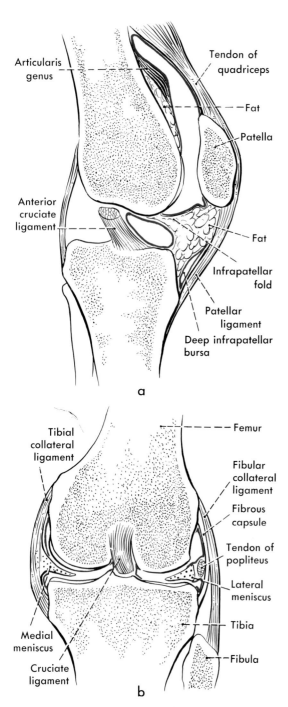

Articularis genus

Tendon of quadriceps

Fat

Patella

Anterior cruciate ligament

Fat

Infrapatellar fold

Patellar ligament

Deep infrapatellar bursa

a

Tibial collateral ligament

Femur

Fibular collateral ligament

Fibrous capsule

Tendon of popliteus

Lateral meniscus

Tibia

Medial meniscus

Fibula

Cruciate ligament

b

Fig. 9-14. Sagittal, *a*, and frontal, *b*, sections through the knee joint. The synovial membrane is shown in *red*.

don produced by removal of the patella; they stated that this amounts to about 1.6 cm when the knee is in full extension. They reported that after the slack was removed the preoperative range of movement was obtained, in their experience, although patients with a limited motion because of damage to the knee joint before operation did not regain normal movement. Kaufer, also, emphasized the necessity of eliminating the slack. He reported that when this was not done following patellectomy about 30% more force was required to extend the knee completely, thus demanding an increase in quadriceps force beyond the strength of some individuals. (Sutton and co-workers reported a loss of 49% in the strength of extension following a complete patellectomy.) Kaufer stated that tuberoplasty alone, moving the tendon forward by a bone block, reduced the 30% decrease to 15%, as did transverse repair of the defect (sewing the tendon to the patellar ligament); the two together eliminated the need for increased force.

SYNOVIAL MEMBRANE

The cavity of the knee joint, lined by the synovial membrane (Figs. 9-11 and 9-14), is complicated, and partly subdivided into a number of communicating compartments. In front, the synovial membrane is attached around the border of the articular surface of the patella. From the sides of the patella it extends laterally and medially around the joint in close contact with the retinacula. From the lower end of the patella it extends downward and backward, separated from the patellar ligament by the infrapatellar fat pad. Traced from the apex of the patella, the synovial membrane forms a central fold, the infrapatellar synovial fold (sometimes called ligamentum mucosum), which is joined by two lesser folds, the alar folds (plicae alares), that extend down from the sides of the patella. The edge of the infrapatellar fold swings upward again to attach to the anterior margin of the intercondylar fossa of the femur. Its sides attach, above, to the lateral and medial margins of this fossa, so that it covers the front and sides of the upper end of the posterior cruciate ligament; they attach, below, to the margins of the anterior intercondylar area of the tibia, and thus cover also the lower attachment of the anterior cruciate ligament. Because the fold attaches to both the femur and the tibia, it divides the part of the knee joint between these bones into medial and lateral synovial cavities, separated by the extrasynovial space that houses the cruciate ligaments. The infrapatellar fold is said to represent the remains of a septum that completely divided the knee joint into medial and lateral cavities.

From the upper border of the patella the synovial membrane passes upward a variable distance, closely applied to the deep surface of the quadriceps muscle, and is then reflected onto the front of the femur. The suprapatellar bursa, between the quadriceps and the front of the femur, is usually fused with the upper end of this part of the cavity of the knee joint but may be separate. When it is fused, the area of fusion may be marked by a constriction or may be indistinguishable. Whether or not this fusion has occurred, the upper reflection of the synovial membrane receives on its posterior surface the insertion of the articularis genus muscle, and as it passes downward on the front of the femur it is at first separated from the bone by a layer of fat. It then becomes closely applied to the front of the femur, and ends at the margins of the articular surface for the patella. The lateral walls of the synovial pouch in front of the femur swing laterally as they are followed downward, attaching to the sides of the femoral condyles above the edges of the articular surfaces. From these attachments the synovial membrane covers the sides of the condyles to the edges of the articular cartilages.

Traced posteriorly along the lateral and medial sides, respectively, of the lateral and medial condyles, the attachment of the synovial membrane to the femur reaches the edges of the articular surfaces at the back of the condyles, and then follows these edges around the posterior aspects of the two condyles; it then runs forward along the lateral and medial lips of the intercondylar fossa, to become

continuous here with the synovial membrane of the infrapatellar fold. From the sides of the femoral condyles the synovial membrane passes downward, that from the edges of the intercondylar fossa attaching to the internal borders of the articular surfaces of the two tibial condyles, that from the external periphery of the femoral condyles attaching to the periphery of the menisci; from the lower edges of the periphery of the menisci the membrane is continued to attach on the periphery of the articular surfaces of the two tibial condyles.

As a result of these attachments, the cavity of the knee joint, viewed posteriorly (Fig. 9-15), is completely divided into medial and lateral parts by this deep forward pouching of the synovial membrane. The intersynovial space, extending from the femoral intercondylar fossa above to the intercondylar areas of the tiba below, houses the cruciate ligaments. These ligaments are therefore covered by synovial membrane anteriorly and on their sides, but not posteriorly. Also, because of the menisci, the medial and lateral synovial cavities between the femoral and tibial condyles are each partially subdivided into an upper portion, above the meniscus, and a lower one below it. The two parts of the cavity communicate centrally around the free edge of the meniscus.

A further complication of the synovial membrane of the knee joint is introduced by its relation to the popliteus muscle. The tendon of this muscle, arising from the lateral aspect of the lateral condyle deep to the attachment of the fibrous capsule to the femur, runs obliquely across the lateral and posterior aspect of the joint between the fibrous and synovial capsules, and as it emerges through the fibrous capsule has deep to it a diverticulum of the synovial membrane; this forms the subpopliteal recess, from the lateral of the two synovial cavities. The medial of the two cavities normally has no diverticula; however, only a thin fibrous capsule separates it from the bursa lying deep to the medial head of the gastrocnemius, and sometimes the bursa is continuous with the cavity of the joint. (This bursa typically lies behind the capsule of the joint and the lower end of the femur, deep to

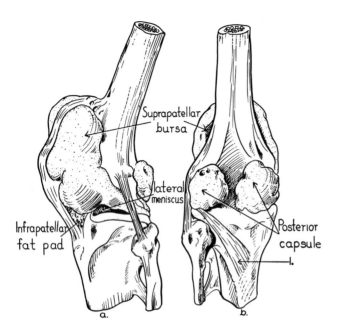

Fig. 9-15. Lateral and posterior views of a knee joint filled with plaster and then stripped of most soft tissues. In *b, l* indicates a part of the popliteus muscle. (Brantigan OC, Voshell AF: J. Bone Joint Surg 23:44, 1941)

the medial head of the gastrocnemius, and extends to lie between this head and the tendon of insertion of the semimembranosus.) It is said (Fullerton) that when the two do communicate the contents of the bursa can be easily emptied into the knee joint while the knee is flexed, but that the overlying muscle usually closes the opening between the two when the knee is extended. There is also a bursa between the lateral head of the gastrocnemius and the lateral epicondyle, but this very rarely communicates with the knee joint.

The lower end of the subpopliteal recess is very close to the synovial sac of the tibiofibular joint, and Weeks found a communication between the two cavities in 4 of 12 joints that he dissected. He quoted a series of 80 dissections in which there was said to be communication in 11, and concluded that it may be expected in about 15% of joints; he pointed out that removal of the head of the fibula in such cases necessarily involves opening the knee joint at the subpopliteal recess.

FIBROUS CAPSULE AND ASSOCIATED LIGAMENTS

The fibrous capsule of the knee joint is simple, in contrast to the synovial capsule. Anteriorly, it is entirely replaced by the quadriceps and its tendon, by the patella itself, and by the patellar ligament and retinacula (Fig. 9-14).

Laterally and medially the capsule is attached to the femur just outside the synovial layer, and shows only a little thickening. On the lateral side, the longitudinally running fibers that thicken and strengthen the capsule arise from the arcuate ligament, but the chief lateral ligament of the joint is the fibular collateral ligament, which stands well away from the capsule. On the medial side, the longitudinally running fibers that strength the capsule are attached, between femur and tibia, to the medial meniscus; the fibers between the femur and the middle third of the meniscus are relatively strong (this part was once called the short internal lateral ligament), but the continuation from the meniscus to the tibia is weaker. The chief ligament here, applied to the capsule but blending with it only posteriorly, is the tibial collateral ligament.

The posterior part of the capsule (Fig. 9-16) is attached above to the lower border of the popliteal surface of the femur, so that the intercondylar fossa of the femur lies within the fibrous capsule although it lies outside the synovial one; and it is attached below to the posterior surface of the upper end of the tibia. It is thin over the back of the condyles, where it is further supported by the two heads of origin of the gastrocnemius; laterally, along its lower border, there is a gap through which the popliteus muscle escapes from its position between the fibrous and synovial capsules of the joint. The posterior part of the capsule is, however, strengthened by two special bands of fibers, the oblique popliteal ligament and the arcuate popliteal ligament.

The *oblique popliteal ligament* is largely formed by an expansion from the semimembranosus tendon that turns upward and laterally to run across the joint as far as the attachment of the capsule to the lateral condyle of the femur. It resists hyperextension at the knee. The *arcuate popliteal ligament* is best marked laterally and inferiorly, where it is attached to the head of the fibula; as it runs upward on the lateral side of the popliteus tendon a lateral part of it extends almost vertically to attach to the femur, while other fibers arch over the posterior surface of the popliteus as it emerges from the capsule. They spread out to blend with the other fibers of the posterior part of the capsule. According to Last ('50) the tendon of the popliteus muscle is firmly attached to the deep surface of the lateral portion of the arcuate ligament, and some of the superficial fibers of the popliteus muscle attach to the arching edge of the thinner medial portion of the ligament.

Kaplan ('61) found the straight part of the arcuate ligament, which has apparently also been called the short (external) lateral ligament, poorly represented as a rule; however, in those knees in which a fabella is present in association with the lateral head or the gastrocnemius there was a well-developed ligament extending from the fabella to the fibula. Kaplan called this the fabellofibular ligament, and expressed the opinion that it, rather than part of the arcuate ligament, represents the short lateral ligament. Although it apparently blends with the arcuate ligament above, it is separated below by the lateral inferior genicular vessels which run superficial to the arcuate ligament but deep to the fabellofibular ligament.

The *coronary ligament,* sometimes named, is composed of those portions of the knee joint capsule that attach the menisci to the tibia; it

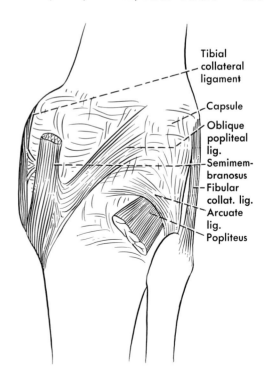

Fig. 9-16. Posterior view of the knee joint.

is relatively lax, thus allowing movement of the menisci on the tibia.

Popliteal cysts, also called Baker's cysts and popliteal hernia, are cystlike swellings in the popliteal fossa that are usually adherent to the medial head of the gastrocnemius. They have been said to develop from the bursa deep to the medial head of the gastrocnemius, and from the semimembranosus bursa. Rauschning described a bursa, the gastrocnemico-semimembranosus bursa, that lies between the two, and said that while the opening of the cyst always lies deep to the medial head of the gastrocnemius, it is the gastrocnemico-semimembranosus bursa that is expanded to form the major part of the cyst.

Gristina and Wilson expressed the belief that in children the condition arises from direct inflammation of a normal bursa, but that in adults it often represents an extension from the joint. They found communications between the cyst and a diseased joint in 66% of their adult patients and suggested that absence of communication may result from obliteration of a preexisting one through repeated compression during motion. Childress

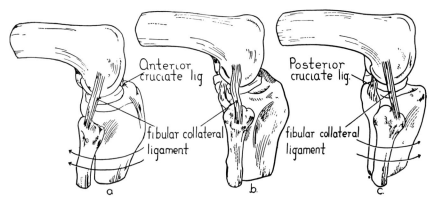

Fig. 9-17. The fibular collateral ligament. Relaxed by flexion, *b*, it allows rotation of the leg in either direction, but is also tightened by rotation, *a* and *c*, and thus helps to check it. (Brantigan OC, Voshell AF: J Bone Joint Surg 23:44, 1941)

agreed that most popliteal cysts in children are primary, but stressed that almost all cysts in young and middle-aged adults are secondary to an intra-articular lesion; of 36 patients who had cysts and a torn medial meniscus, meniscotomy alone led to the disappearance of the cyst in 21. Wolfe and Coloff also found tears of the medial meniscus to be the most frequent cause, and attributed that to the fact that such tears are the most frequent cause of effusions into the knee joint. Pinder quoted evidence that in rheumatoid arthritis the cyst is usually a protrusion of the posterior wall of the capsule of the knee. Rather than removing the cyst, he performed an anterior synovectomy to reduce the effusion.

In addition to the swelling and the consequent discomfort on flexion of the knee resulting from a popliteal cyst, there may be intermittent swelling of the leg and ankle, apparently from pressure on the popliteal vein; or there may be compression also of the artery and the tibial nerve, and the cyst may become so large that it extends far into the calf.

Collateral Ligaments

The *fibular collateral ligament* (Figs. 9-16 and 9-17) is a strong rounded cord attached above to the lateral epicondyle of the femur and below to the head of the fibula. It stands well away from the lateral part of the capsule, and the lateral inferior genicular vessels run deep

to it. Last ('48) and others have regarded it as the superficial part of a bipartite lateral ligament, the deep part of which is the slightly thickened lateral part of the articular capsule. The fibular collateral ligament is tight in extension, further tightened by hyperextension; it is relaxed by flexion, tightened by rotation in either direction (Fig. 9-17). Sneath reported that a fibular collateral ligament relaxed by flexion can be bowed back, and therefore tightened, by pull on it by the biceps tendon, which attaches in part into its posterior border. He also described the lower end of the ligament as provided with a bursa that surrounds all except its posterior part, separating it from the attachments of the biceps that reach the tibia.

Kaplan ('57b) emphasized also the roles of the iliotibial tract and the popliteal tendon, noting that these can adequately stabilize a knee in which the head of the fibula, and therefore the fibular collateral ligament, has been removed.

The *tibial collateral ligament* is a broad band closely applied to the capsule. It is attached above to the whole of the medial epicondyle of the femur from the adductor tubercle down; below, it attaches to the tibia as far as the level of the tuberosity. The anterior part of the ligament extends straight down, and is separated from the capsule by loose connective tissue in which there is usually at least 1 bursa (De Palma, '54, found one in 94%) and sometimes as many as 3 (Brantigan and Voshell, '43). The semimembranosus tendon sends an expansion forward between the ligament and the tibia, and the medial inferior genicular vessels pass between this expansion and the ligament.

While the anterior part of the ligament is straight, the posterior part runs obliquely downward and backward to the level of the medial meniscus, then obliquely forward to its attachment on the tibia, so that the ligament is widest at the level of the medial meniscus (Fig. 9-18). The posterior part blends with the capsule and is itself firmly attached to the medial meniscus. Last ('48) regarded the tibial collateral ligament as being, like the fibular, the superficial part of a bipartite ligament of which the deep part is represented by the medial part of the capsule, also attached to the medial meniscus.

The posterior part of the tibial collateral ligament relaxes during flexion, but the anterior part remains taut in all positions (Brantigan and Voshell, '41; De Palma, '54).

Lamb reported 2 cases in which calcium salts had been deposited in the tibial collateral ligament, and said that this was apparently rare; he suggested that it might result from changes in the blood supply. Calcification of the tibial collateral ligament is sometimes known as Pellegrini-Stieda disease. The roentgenographic appearance has apparently been interpreted as that of a chip fracture of the femur, or new bone formation from a torn periosteum, but Wetzler and Elconin found no evidence of this; they regarded the condition as resulting from metaplasia of the fibrous tissue, apparently resulting from trauma, with calcification appearing in the ligament from 7 weeks to 7 months following this. Brantigan and Voshell ('43) suggested that the calcification may start in a bursa deep to the ligament. Oxford noted that recurrence of the deposition after its operative removal has been reported, and he himself reported apparent success in arresting the progress of the disease by treating the condition medically and with immobilization.

Fig. 9-18. The tibial collateral ligament. Tense in extension, *a*, it slides backward in flexion, *b*, and its posterior but not its anterior fibers relax. (Brantigan OC, Voshell AF: J Bone Joint Surg 23:44, 1941)

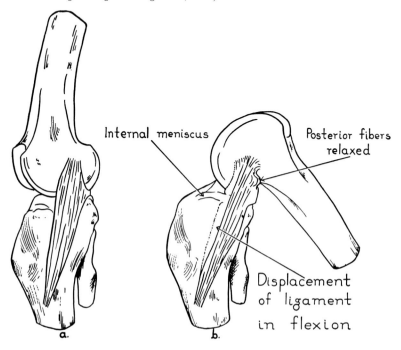

Internal meniscus

Posterior fibers relaxed

Displacement of ligament in flexion

a.   b.

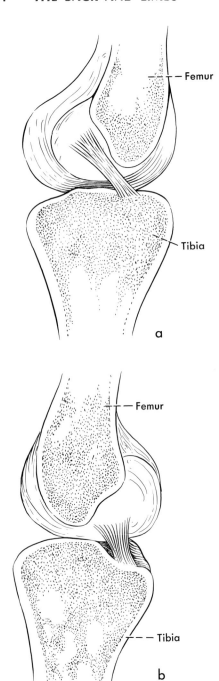

Fig. 9-19. The anterior, *a*, and posterior, *b*, cruciate ligaments of the left knee.

Cruciate Ligaments

The cruciate ligaments (Fig. 9-19) lie within the fibrous capsule of the knee joint but outside the synovial membrane, occupying the deep extrasynovial cul-de-sac projecting for-

ward into the posterior aspect of the knee joint. They therefore lie behind the posterior wall of the synovial cavity at the front of the knee joint, and between the two compartments of the infrafemoral portion of the cavity. They are strong, rather rounded bands that cross each other between their tibial and femoral attachments, and are named from their tibial attachments.

The anterior cruciate ligament arises from the anterior (nonarticular) intercondylar area of the tibia adjacent to the medial condyle. It extends obliquely upward, backward, and laterally to attach into the medial side of the lateral condyle of the femur. The anterior cruciate ligament is typically more slender and longer than the posterior; Steindler said the ratio in length of the two is about 5:3. In contrast, De Palma reported that the anterior cruciate ligament measured an average of 2.7 cm long with the knee flexed to 90 °, while the posterior cruciate measured an average of 3.3 cm with the knee extended. This indicates that the anterior cruciate ligament becomes shorter by about 50%, as a result of the twisting of its fibers, when the knee is flexed to a right angle.

The posterior cruciate ligament arises from the posterior intercondylar area and passes obliquely upward, forward, and medially, crossing behind the anterior cruciate ligament, to attach at the front of the intercondylar fossa to the lateral surface of the medial condyle. The femoral attachment of the posterior cruciate ligament is thus more anteriorly placed, that of the anterior cruciate ligament more posteriorly placed (see Fig. 9-21).

Widely varying statements have been made concerning the tenseness of the cruciate ligaments in various positions and their consequent role in stability at the knee; it is usually granted that they are reasonably tense in all positions, and sometimes said that the anterior becomes particularly tense in full extension, the posterior in full flexion. De Palma ('54) said both ligaments are "taut" in both extremes, some part "tense" but no part "taut" in all positions between; Brantigan and Voshell ('41) said some part of each is "taut" in all positions of flexion and extension, and they are both "tense" in hyperextension. The difference lies in how one defines "taut" and "tense."

MENISCI

The menisci or semilunar cartilages of the knee (Fig. 9-20) lie within the knee joint between the femur and the tibia, and somewhat adapt the tibial and femoral condyles to each other; as discussed in Chapter 1, one function may be that of providing the mechanical factors for lubrication, but they are particularly important also in allowing rotation at the knee joint, in which movement they move with the femur on the tibia. Each meniscus is a somewhat semilunar piece of fibrocartilage with a thickened outside edge and a thin inner edge, so that it is wedge-shaped in cross section. As is usual with fibrocartilaginous disks, the periphery is largely fibrous tissue.

The fibrous and synovial capsules of the knee joint attach to the thickened convex edges of the menisci. Since the fibrous capsule on the medial side is fused to the posterior part of the tibial collateral ligament, the medial meniscus is usually regarded as being relatively firmly anchored on its periphery. In contrast, the lateral meniscus gains no attachment to the fibular collateral ligament, since this stands away from the joint; it is attached posteriorly both to the capsule and to fibers of the popliteus muscle (for instance, Kaplan, '57b; Heller and Langman).

The two menisci normally differ in shape. The medial meniscus tends to be C-shaped or semicircular in form, with a posterior extremity or horn wider than its anterior one; the two horns of the lateral meniscus tend to be the same size, and this meniscus more nearly approaches the form of an incomplete O. Chandler found, however, a good deal of variation in the menisci of 55 knee joints that he examined, and in 12 of the joints there was little difference between the two.

Discoid menisci are nearly always lateral ones (Ross and co-workers, reporting a medial one, found only four cases in the literature); they have been regarded as representing the retention of an embryological disklike shape, but Kaplan ('57a) and Ross and co-workers could

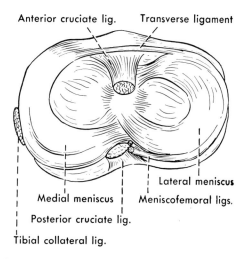

Fig. 9-20. The menisci of the right knee joint.

find no developmental stage in which the menisci are disklike. Kaplan thought they are not congenital; he described discoid lateral menisci as regularly lacking attachment of their posterior horns to the tibia and having particularly short and strong meniscofemoral ligaments. This combination of circumstances, he said, leads to abnormal movement of the meniscus that produces secondarily the thickening reported as discoid: as the meniscus moves forward in extension, this movement is checked by the meniscofemoral ligaments, and the meniscus is displaced posteromedially into the intercondylar space, from which it moves out again on flexion.

Berson and Herman found no report of torn discoid menisci in adults, saying that this usually occurs in childhood. However, they reported three cases of a torn lateral discoid meniscus, and one of a torn medial discoid meniscus, in adults.

At the anterior and posterior extremities, or horns, of each meniscus, the fibrocartilaginous tissue gives way entirely to bands of fibrous tissue through which the menisci attain a firm attachment to the tibia: The anterior and posterior horns of the more highly curved lateral meniscus attach close together, one just in front of and one just behind the tibial intercondylar eminence. The anterior horn of the medial meniscus attaches to the anterior intercondylar area anterior to the attachment of the anterior cruciate ligament, while its

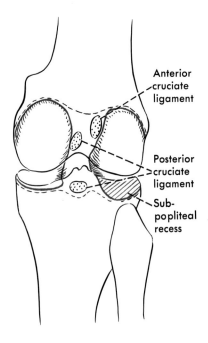

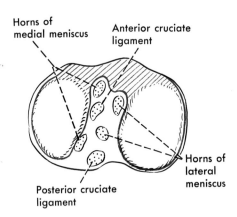

Fig. 9-21. Posterior view of the bones at the knee, and the upper end of the tibia, to show the attachments of the menisci and the cruciate ligaments. Synovial membrane is in *red*.

posterior horn attaches posteriorly, between the attachments of the posterior cruciate ligament and the posterior horn of the lateral meniscus (Fig. 9-21).

The posterior horn of the lateral meniscus, closely related to the lateral aspect of the posterior cruciate ligament, often gives off a band of fibers, a *meniscofemoral ligament,* that follows the posterior cruciate ligament to its attach-

ment on the femur; it may follow its anterior surface, its posterior surface, or split and follow both. Heller and Langman found an anterior meniscofemoral ligament (of Humphry) present in only 36% of 140 knees, a posterior (of Wrisberg) in only 35%, and both present in only 6 knees. The meniscofemoral ligament or ligaments are particularly important in lateral rotation of the femur, when they pull the posterior horn of the meniscus medially and slightly anteriorly, accommodating it to the femoral condyle. They normally limit anterior movement of the meniscus (*e.g.,* Kaplan, '57a), but according to Heller and Langman they move the posterior horn slightly medially and anteriorly when they and the posterior cruciate ligament become tense in the weight-bearing flexed knee. Under this condition, the activity of the popliteal muscle would help to limit both anterior movement of the femur and of the meniscus.

Between the anterior horns of the two menisci there may be a band of transverse fibers joining the menisci together; this is the *transverse ligament* of the knee. Chandler found a well-formed one in only 2 of 55 knee joints, a very small ligament in 6, and a few thin shreds apparently representing the ligament in 8.

Rotation at the knee joint, relatively free when the knee is flexed, takes place largely between the menisci and the tibia, for in this movement they, and especially the more freely movable lateral one, move with the femur; in flexion and extension without rotation the movement is largely between the femoral condyles and the menisci, but as the limb is completely extended rotation occurs with consequent increased movement of the menisci.

Both menisci move slightly forward with extension at the knee, slightly backward with flexion (Fig. 9-22). During extension they are also somewhat spread by the flatter anterior surfaces of the femoral condyles. Forward movement of the menisci during extension is probably produced largely by the roll of the femoral condyles, but aided also (Kaplan, '62) by the quadriceps through thickenings of the patellar retinacula ("patellomeniscal liga-

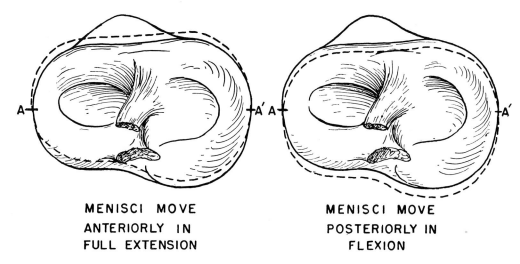

MENISCI MOVE
ANTERIORLY IN
FULL EXTENSION

MENISCI MOVE
POSTERIORLY IN
FLEXION

Fig. 9-22. Normal anteroposterior movement of the menisci with extension and flexion. *A* and *A'* call attention to the fact that there is practically no lateral movement. (Smith FB, Blair HC: J Bone Joint Surg 36-A:88, 1954)

ments") between the patella and the menisci. Posterior movement during flexion, also brought about in part by the femoral condyles, is aided in the case of the lateral meniscus by the popliteus, in that of the medial meniscus by the contraction of the semimembranosus (Kaplan, '62, demonstrated both the attachment of the semimembranosus to the meniscus and its role in moving this).

VESSELS AND NERVES

The *arteries* of the knee joint are branches of the vessels that enter into the anastomosis around the knee, described in a following section. Thus twigs to the capsule are derived directly or indirectly from the descending genicular and the descending branch of the lateral femoral circumflex, from the superior medial and lateral genicular arteries, from the inferior medial and lateral genicular arteries, and from the anterior tibial recurrent artery. Occasionally there is also, or instead, a posterior tibial recurrent which helps to supply the joint. In addition to these, the fibrous capsule is perforated posteriorly by the middle genicular artery, which supplies especially the tissue of the intercondylar region, therefore the cruciate ligaments and the attachments of the menisci. The menisci are largely avascular,

but are more vascular at their extremities than elsewhere.

The *lymphatic network* about the knee joint is apparently better developed anteriorly, laterally, and medially than it is posteriorly; according to Tanasesco the lymphatics leaving the joint are divisible into two general groups, superficial ones that run toward and follow the course of the great saphenous vein and eventually end in superficial inguinal nodes, and deep ones that in general follow the articular arteries; he described the deep lymphatics as being the more important. Many of the deep lymphatics end in the small popliteal nodes; others, and the drainage from these nodes, run upward with the popliteal and femoral vessels to enter the deep inguinal nodes or to pass on into the pelvis.

Nerves

The innervation of the knee joint, as described by Gardner, is by means of a number of twigs from the femoral, the obturator, and both parts of the sciatic nerve; Gardner found evidence in the literature that the accessory obturator nerve may, although apparently rarely, also help to supply the knee joint.

The articular branches of the femoral nerve to the knee joint arise from the saphenous nerve and from the nerves to the three vasti

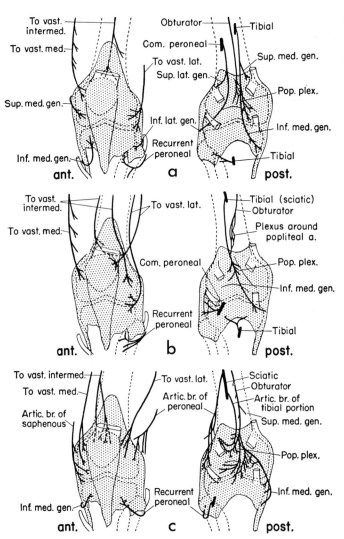

Fig. 9-23. Anterior and posterior views of some different types of distribution of nerves to the knee joint. (Fig. 2, Gardner E: Anat Rec 101:109, 1948)

(Fig. 9-23). The branch from the saphenous is distributed to the anteromedial part of the joint; that from the nerve to the vastus medialis to a medial part of the joint; that from the nerve to the vastus intermedius to the suprapatellar part; and that from the nerve to the vastus lateralis to an anterolateral part. There is a good deal of overlap and frequently there are anastomoses between these branches.

The branch from the obturator, usually derived from its posterior division, follows the femoral and popliteal arteries to the knee joint and is distributed especially to the posteromedial part of the capsule. Gardner often

found a single large branch from the tibial nerve instead of the several usually described; its subsidiary branches accompany various genicular vessels or run directly to the capsule to supply a posterior part of that. The branch from the common peroneal nerve was also usually single, and distributed to an anterolateral part of the capsule.

The recurrent branch of the peroneal nerve, arising about at the point of division of the common peroneal into its superficial and deep branches, is largely distributed to the periosteum of the anterolateral surface of the tibia and to the tibiofibular joint, but some of its subsidiary branches follow blood vessels to the knee joint and apparently supply the infrapatellar fat pad and the adjacent capsule.

O'Connor and McConnaughey found that in the cat the meniscal horns, but not the body of the meniscus, have a rich innervation, as does the posterior meniscofemoral ligament.

Surgical Exposure of the Knee Joint

The knee joint can be exposed anteromedially or anterolaterally by an incision paralleling the lower part of the rectus femoris, skirting the edge of the patella, and paralleling the patellar ligament; the joint can be opened still more widely by incising the capsule on both sides of the patella and its ligament. For wide exposure of either the medial or the lateral side of the joint, Cave recommended a skin incision carried downward from just behind the epicondyle and curving forward to the patellar ligament about one-quarter inch below the joint line; the anterior part of the capsule is then opened by an incision beginning just anterior to the medial or lateral ligament, as the case may be, and curved forward to the patellar ligament, and when necessary the capsule is opened posteriorly by a vertical incision just behind the collateral ligament. A posteromedial approach is usually made by retracting the sartorius muscle posteriorly; a posterolateral one by retracing the iliotibial tract anteriorly, the biceps and the lateral head of the gastrocnemius posteriorly. A still more posterior approach can be made, following an appropriate popliteal incision, by reflecting the gastrocnemius backward, the appropriate hamstring muscles forward. In a posterolateral approach, the superficial position of the common peroneal nerve must be borne in mind. Abbott and Carpenter have discussed and illustrated various approaches to the knee joint, and Abbott and Gill discussed approaches particularly to the epiphyseal cartilages.

MOVEMENTS AT THE KNEE JOINT

The chief movements at the knee joint are those of flexion and extension but, as already noted, the joint is not strictly a hinge one and

some rotation can take place here. In starting from a flexed position, the posterior and more highly curved parts of the articular surfaces of the condyles of the femur are in contact with the posterior parts of the menisci, and the superior part of the articular surface of the patella is engaged against the lowest portion of the patellar articular surface of the femur. The fibular and the posterior part of the tibial collateral ligaments are relaxed; in full flexion both cruciate ligaments are taut, but the tension on them is decreased by a movement of extension from this position. Because of the very small area of contact between the incongruous surfaces of the femur and the menisci and tibial condyles, and because of the relative laxity of most of the ligaments, the leg is relatively easily rotated in most positions of flexion.

Wang and Walker found a great deal of variation in the amount of rotation allowed in the semiflexed knee. Meniscectomy increased rotation in about half of the limbs tested, but caused no change in the others. Cutting both cruciate ligaments increased rotation by an average 25%, but cutting both collateral ligaments increased it by an average 49%. The collateral ligaments are therefore the most important restraint to rotation at the knee.

As *extension* is carried out through the activity of the quadriceps the two femoral condyles roll forward and skid backward (Fig. 9-24) so that their more anterior parts engage the more anterior parts of the menisci and tibial condyles (in this respect also the knee joint departs from the hinge joint type, since the movement involves a constant shifting of the axis of motion). However, the shorter curve of the lateral condyle is exhausted faster by this motion than is the longer and less curved articular surface of the medial condyle; consequently, as full extension approaches and the lateral condyle nears completion of its forward movement (and its posterior skidding begins to be checked by the anterior cruciate ligament), the medial condyle both rolls forward and at the same time slips or skids posteriorly faster than does the lateral one, in order to engage its more anterior part. In this movement the medial condyle tends to form

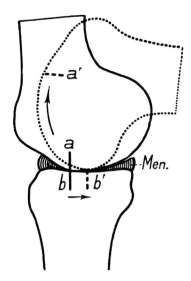

Fig. 9-24. Schema indicating the necessity for gliding or skidding of the femoral condyles on the tibia as the knee is flexed and extended. In extension, the femur and tibia are in contact along line *a-b*, but in flexion they are in contact along line *b'*. The femur therefore rolls backward only a little, from *b* to *b'*, during flexion, yet at the same time the contact point in extension is carried to *a'*; the difference of the distances between *a* and *a'* and *b* and *b'* thus indicates the amount of gliding that accompanies the rolling movement of the condyles. *Men.* is the meniscus. (Pouzet F: Genou I. Généralités. In Ombrédanne L, Mathieu P (eds): Traité de Chirurgie Orthopédique, Vol 4. Paris, Masson et Cie, 1937)

part of the arc of a circle around the lateral condyle. (If the anterior cruciate ligament is cut, this "screw-home" pattern is irregular, according to Shaw and Murray, for the non-weight-bearing tibia shifts forward as it rotates—thus, in standing, the femur should skid backward. Cutting the posterior cruciate ligament had little additional effect.) Finally according to Barnett, the quadriceps contributes to the medial rotation, promoting the backward glide of the medial condyle and rotating the lateral condyle and the lateral meniscus so that the anterior horn of the latter glides downward on the front of the tibial condyle; this then allows further extension of the lateral condyle.

At the end of extension, the anterior and flatter surfaces of both femoral condyles are in contact with the menisci and tibial condyles.

Because of this lesser curvature, the surfaces of the extended joint are more congruent and therefore the joint is more stable; in addition, the faster backward skidding of the medial condyle results in a further tightening of the tibial and fibular collateral ligaments, already tightened by extension, so that the joint is generally tightened. Both cruciate ligaments are also tightened by extension, and help to resist hyperextension (Brantigan and Voshell, '41; De Palma, '54).

The terminal medial rotation of the femur on the tibia tends to unwind the cruciate ligaments slightly from each other, and slightly reduce the tension on the anterior cruciate ligament. If the leg is free rather than supporting the weight of the body, some of the rotation may be a lateral rotation of the tibia on the femur.

At the same time that these changes are occurring, the patella has moved up until its lower articular surface is in contact with the upper part of the patellar articular surface of the femur.

In the last phase of extension, as the lateral condyle of the femur completes its anterior roll and the medial condyle moves backward, the two condyles move the menisci with them, although until this time they have been for the most part moving upon the menisci. Thus as the lateral condyle completes its roll it moves the lateral meniscus slightly farther forward, and at the same time further tightens the anterior cruciate ligament; as the femur is rotated medially by the posterior sliding of the medial condyle, the medial meniscus is carried somewhat backward and laterally, while the lateral meniscus is carried slightly medially.

In contrast to most workers, who have regarded stability of the knee during standing (when the center of gravity is in front of the knee joint) as depending primarily on articular structures, Smith ('56) concluded that they contribute relatively little; in standing on both legs, he said, the activity of the knee flexors contributes about 30% of the resistance to further extension, and the passive resistance of other extraarticular structures con-

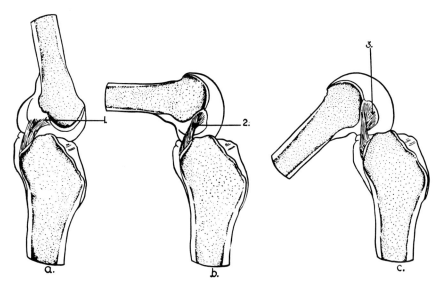

Fig. 9-25. View of the medial half of a knee joint and the posterior cruciate ligament. Note that as the knee is flexed the ligament is twisted on itself. The same thing occurs to the anterior cruciate ligament. (Brantigan OC, Voshell AF: J Bone Joint Surg 23:44, 1941)

tributes about 50%, while in standing on one leg the muscles contributes 65% and other extraarticular structures 20%.

In moving from complete extension into *flexion*, the terminal medial rotation of the femur has to be undone. This is accomplished by the popliteus muscle, an excellent external rotator of the femur or internal rotator of the tibia, and thus important in starting flexion even though it is apparently a weak flexor. As the femur rotates laterally as a result of pull of the popliteus, the menisci again move with the femur. Since the medial meniscus is much more firmly attached at its periphery to the femur than it is to the tibia, it can rotate with the femur while its horns remain fixed to the tibia. The lateral meniscus has a more controlled movement with external rotation of the femur, for the popliteus is attached in part to it and therefore tends to draw it backward and downward; and, according to Last ('50), Heller and Langman, and others, the meniscofemoral ligaments, made more tense by flexion, help to move the posterior part of the lateral meniscus medially and anteriorly, thus helping to rotate it. After the initial rotation, the femoral condyles then roll backward on

the tibial condyles and the menisci, while the patella, held at a fixed distance from the tibial tuberosity by the inelastic patellar ligament, is brought into contact with the lower part of the patellar surface of the femur as the knee bends. Flexion relaxes the posterior, oblique, part of the tibial collateral ligament and also relaxes the fibular collateral ligament, and in complete flexion both actually buckle out in a ligamentous preparation. However, in complete flexion the cruciate ligaments become rather taut again. Because of their lines of attachment to the femur, both cruciate ligaments are twisted by flexion (Fig. 9-25). It is the twisting and untwisting between the extremes of extension and flexion that shift the tension among the fibers of the ligaments.

Except for the slight terminal rotation involved in complete extension and the slight initial rotation involved as flexion is started, *rotatory movements* of the leg at the knee joint occur only in a partly flexed joint. In partial flexion all the ligaments, with perhaps the exception of the anterior part of the tibial collateral, are less taut than in other positions, thus favoring rotation. It is usually said that

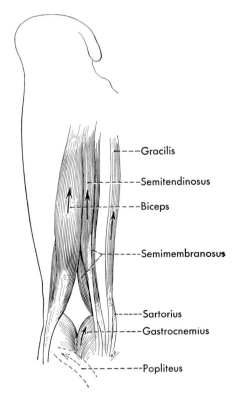

Fig. 9-26. Flexors of the knee joint.

Gracilis

Semitendinosus

Biceps

Semimembranosus

Sartorius

Gastrocnemius

Popliteus

rotation of the leg at the knee is greatest when the limb is flexed to about 90 °; however, Ross tested the amount of total rotation at the knee in 100 individuals and found an average of about 40 ° when the knee was flexed to 45 °, one of 35 ° when it was flexed to 90 °, and slightly more, 39 °, when it was flexed to 135 °.

Rotation of the femur in either direction tends to tighten the collateral ligaments, especially the lax fibular one. Medial rotation of the tibia on the femur tends to wind the cruciate ligaments around each other, and thus make both more taut; they therefore help to check this movement. Lateral rotation is apparently checked by the medial part of the capsule and by the collateral ligaments; in forced rotation it is the capsule that is torn first (Slocum and Larson, '68a). During rotation, the femur and the menisci move together on the tibia, or the tibia moves on the menisci and the femur; in either case, the latter pivot on their horns attached to the tibia.

Muscles Moving the Knee Joint

The *flexors* of the knee are primarily the hamstrings (semimembranosus, semitendinosus, and biceps, especially its short head) and the closely associated gracilis and sartorius muscles (Fig. 9-26). The popliteus is usually regarded as important in early phases of flexion because of its rotatory action upon the femur or tibia. Electromyographic studies have indicated (Barnett and Richardson, Basmajian and Lovejoy) that it is a weak flexor, its chief action during flexion being to draw the lateral meniscus posteriorly.

When the limb is free of weight, the gastrocnemius can also help to flex the knee. These muscles are innervated by a number of different peripheral nerves (Table 9-1).

The chief *extensor* of the knee is the quadriceps. As noted in the previous chapter, the tensor fasciae latae and the gluteus maximus have been listed as extensors at the knee, on the basis that they can exert sufficient pull on the iliotibial tract to produce movement here. However, stimulation of neither muscle produces extension of the knee, nor does pulling upon the iliotibial tract, and the tensor cannot be regarded as contributing to this movement. Nevertheless, because the knee of the weight-supporting limb necessarily moves with the hip and the ankle (in the combination of flexion at the knee, flexion at the hip, and dorsiflexion at the ankle, or extension at the knee, extension at the hip, and plantar flexion at the ankle) both the gluteus maximus and the triceps surae (gastrocnemius and soleus) help extend the weight-bearing knee as they act at hip and ankle (Fig. 9-27). These muscles, like the flexors, are innervated through several different nerves (Table 9-2).

Rotators of the leg act primarily when the leg is flexed, at which time relaxation of the ligaments of the knee allows more rotation and most of the muscles have a favorable leverage. The external rotator of the free, nonweight-bearing, leg is the biceps femoris; the internal rotators are the semimembranosus, semitendinosus, gracilis, sartorius, and popliteus.

The most important *tendon transplant* at the

Table 9-1
Innervation of the Flexors of the Knee

| | MUSCLE | NERVE | SEGMENTAL INNERVATION | |
			Reported Range	*Probable Most Important Segments*
Chief flexors	Semimembranosus	Tibial	L4-S3	L5, S1
	Semitendinosus	Tibial	L4-S3	L5, S1
	Biceps, short head	Peroneal	L4-S2	L5, S1
	Gracilis	Obturator	L2-L5	L3, 4
	Sartorius	Femoral	L2-L4	L3, 4
Accessory flexors	Popliteus	Tibial	L4-S3	L4, 5, S1
	Gastrocnemius	Tibial	L4-S3	S1, 2
	Biceps, long head	Tibial	L4-S3	L5, S1, 2

knee is transference of one or more hamstrings to replace a paralyzed quadriceps (Mayer). According to Mayer, the biceps and sartorius make a good combination, but the operation probably should not be done unless there are four good hamstring muscles, and the semimembranosus is best not used because of its importance in preventing hyperextension at the knee. The biceps is often used alone.

INJURIES TO THE KNEE JOINT

Acute injuries to the knee joint typically result from forces applied laterally or medially, often also with a rotatory component, to the partly flexed knee. The knee is ill equipped to withstand forced abduction or adduction in any position, but during flexion the ligaments are as relaxed as they ever get, rotation is possible, and hence the joint is particularly vulnerable. The extent of injury varies considerably, depending probably both on the direction and on the strength of the injuring force.

McMurray said that the two collateral ligaments are usually injured together, and that the anterior cruciate ligament is frequently stretched or torn by forced hyperextension, but that the posterior ligament or both of them together are less frequently torn. Meyers reported 14 cases in which the only tear of ligaments of the knee was avulsion of the posterior cruciate ligament from its tibial attachment.

Fig. 9-27. Extensors of the knee joint. The gluteus maximus, gastrocnemius, and soleus extend only when the limb is supporting weight, and then only incidentally as a result of their actions at the hip and ankle.

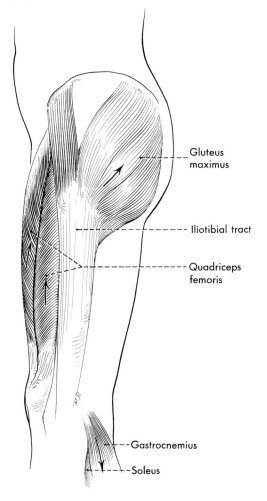

Gluteus maximus

Iliotibial tract

Quadriceps femoris

Gastrocnemius

Soleus

Table 9-2
Innervation of the Extensors of the Knee

| | MUSCLE | NERVE | SEGMENTAL INNERVATION | |
			Reported Range	*Probable Most Important Segments*
Chief extensor	Quadriceps	Femoral	L2-L5	L3, 4
Accessory extensors	Gluteus maximus	Inferior gluteal	L4-S3	L5, S1, 2
	Gastrocnemius	Tibial	L4-S3	S1, 2
	Soleus	Tibial	L4-S3	S1, 2

Injuries to Cruciate Ligaments

Apparently the most common statement in the past concerning the functions of the cruciate ligaments has been that the anterior cruciate becomes particularly tight in extension, less tight (or according to some authors actually slack) in flexion, while the posterior cruciate becomes especially tight in flexion, less tight (or slack, according to some) during extension. However, as already mentioned, Brantigan and Voshell, and De Palma, described them as both being especially tense or taut in full flexion and extension, and some fibers of each as being tense or taut in the range of movement between these extremes.

Regardless of some variation in the tenseness of the ligaments with various positions at the knee, it is clear that in flexion the anterior cruciate ligament is so slanted as to resist particularly anterior displacement of the tibia on the femur, while in the weight-bearing flexed limb the posterior cruciate ligament resists anterior displacement of the femur on the tibia, and thus also will resist passive posterior movement of the leg at the knee. Thus a test for the integrity of the cruciate ligaments has been an attempt to move the unsupported and flexed tibia forward and backward on the femur: if the tibia could be moved forward abnormally, this was considered evidence of rupture of the anterior cruciate ligament, while if it could be moved abnormally backward it was considered evidence of rupture of the posterior cruciate ligament. (Katz and co-workers reported that in the fetus less than 28 weeks old only the cruciate ligaments are capable of preventing anterior dislocation of the leg, and said that in the congenitally dislocated knees they explored the cruciate liga-

ments were either absent or hypoplastic. They reported good results from reconstructing an anterior cruciate ligament.)

Horwitz ('38a), De Palma ('54), and others noted that extensive anteroposterior displacement of the leg on the femur is typically indicative of tears of both the anterior cruciate ligament (or both cruciates) and of the collateral ligaments, especially the tibial one; De Palma reported that although cutting the anterior cruciate ligament allowed some 2 mm to 5 mm of anterior displacement when the knee is extended, cutting the tibial collateral ligament much increased the possible displacement. In contrast, Butler and co-workers stated that the anterior cruciate ligament contributes an average of 86% of the resistance to anterior movement, and the other ligaments—the iliotibial tract, the middle thirds of the medial and lateral parts of the capsule, and the tibial and fibular collateral ligaments—contribute only about 3% each. (The posterior cruciate ligament, they found, contributes about 95% of the resistance to posterior movement.)

Slocum and Larson ('68a) described the "anterior drawer" test as fairly reliable for rupture of the anterior cruciate ligament only when forward dislocation amounts to three quarters of an inch or more; they said the anterior cruciate ligament is probably never injured alone, and that in the usual injury caused by abduction and external rotation of the flexed tibia the sequence of tearing is the medial part of the capsule, the tibial collateral ligament, and the anterior cruciate ligament. (Kennedy and Fowler reported the same sequence of injury in tests on a machine. Wang and co-workers, however, reported a case in which the only injury was said to be rupture of the anterior cruciate ligament.) The most sensitive sign of such injury, Slocum and Larson found, was rotatory instability, tested with the knee flexed to 90° and the foot supported. With the tibia externally rotated about 15°, the extent of damage to the medial structures and the anterior cruciate ligament can be gauged by the amount of forward movement and external rotation that occurs as the leg is drawn forward (but in mild injury to the medial structures, they found, the posterior part of the medial meniscus may limit the rotation and tend to mask the tear). In contrast to the movement when the tibia is externally rotated, internal rotation to about 30° so tightens the lateral structures that if they are intact they prevent forward displacement even if the medial ligaments and the anterior cruciate are ruptured.

In regard to lateral movement, the cruciate and collateral ligaments apparently all contribute to stability here. Brantigan and Voshell ('41) found no lateral instability at the extended knee when both cruciate ligaments were cut, but there was abnormal movement when the knee was flexed and, since the fibular collateral ligament was relaxed, only the tibial one could resist the movement.

While some workers have tended to emphasize the contribution of the tibial collateral ligament to stability at the knee, and others the contribution of the cruciate ligaments, Brantigan and Voshell pointed out that all the ligaments (both cruciates, the collaterals, and the posterior capsule of the joint) work together, and found it difficult to assign a single specific function to any one ligament. Their results on the effect of cutting one ligament or various combinations of ligaments indicate that some increased lateral motion and rotation can be expected, in some position of the joint, when any one ligament is ruptured, but that the effect is absent or minimal in hyperextension, when all other ligaments are tightened; cutting almost any combination of ligaments increased the abnormal motion. (As already noted, Wang and Walker found that cutting both collateral ligaments increased the amount of rotation almost twice

as much as did cutting both cruciate ligaments.) Brantigan and Voshell expressed the opinion that when the anterior cruciate and medial collateral ligament are both torn, as they often are, both probably should be repaired, but that repair of either should give a satisfactorily functioning knee. Some workers have advocated repair of only the tibial collateral ligament.

O'Donoghue, who has dealt much with athletes in whom stability of the knee is extremely important, has repeatedly emphasized (for instance, '55, '63) that when both ligaments are ruptured both should be repaired, and that an unrepaired anterior cruciate ligament is a prime cause of instability of the knee. A number of procedures have been used in this repair; among the more recent ones are replacement by a strip of the patellar ligament (O'Donoghue, '55; Jones, '63), by the lower part of the iliotibial tract (O'Donoghue, '63), or by the semitendinosus tendon (Cho).

Other workers have suggested that a well-developed quadriceps muscle can compensate to a variable degree for lax or torn cruciate ligaments, especially the anterior cruciate. For instance, Michele described laxity of the anterior ligament as being a cause of anteroposterior instability at the knee, but said that strengthening the quadriceps rather than repair of the ligament may suffice if the patient is not particularly active; and McMurray recommended conservative treatment especially for rupture of the anterior cruciate ligament, saying that much can often be accomplished by strengthening the quadriceps. Other workers who have regarded loss of the anterior cruciate ligament alone as being sometimes insignificant if there is a good quadriceps include Horwitz ('38a), Hauser, and Quigley.

Tears of Collateral Ligaments

As already noted, there are differences of opinion as to when repair of the anterior cruciate ligament is necessary, but there seems to be general agreement that when it and the tibial collateral ligament are both torn, at least the latter should be repaired. O'Donoghue, as already mentioned, insisted that

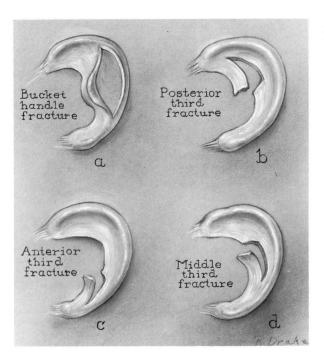

Fig. 9-28. Tears of the menisci. (Courtesy of Dr. P.R. Lipscomb)

both should be. O'Donoghue ('55) has used various methods of repair, including simple suture, suturing to bone through drill holes, and moving its tibial attachment down or replacing the ligament by fascia lata. Some workers (*e.g.,* Bosworth, Fenton) have preferred to use the semitendinosus tendon, imbedding it in the medial femoral condyle.

Starr, and later Slocum and Larson ('68b), stated that reconstruction of both the tibial collateral and anterior cruciate ligaments often does not provide sufficient lateral rotatory stability to allow an abrupt change in the line of progression. The latter authors described substantial improvement in a majority of cases through transposition of the semitendinosus part of the "pes anserinus" tendon to a higher position, where at 45° of knee flexion it functions primarily as a medial rotator rather than as a flexor. The lower part of the tendon, contributed especially by the semitendinosus, is cut away from its insertion on the tibia and folded up over the upper attached part; its end is then reattached to the side of the quadriceps tendon, and its upper border sewed into position just below the medial condyle.

There has been relatively little discussion of the fibular collateral ligament, since it alone is not often torn. Kaplan ('57b) suggested that stability at the knee is not impaired so long as two of the three lateral supporting structures—the fibular collateral ligament, the popliteus muscle, and the iliotibial tract—are intact.

Tears of Menisci

Tearing (fracture) of the menisci (Fig. 9-28) sometimes occurs, especially as a result of rotational injuries during flexion while the tibia is supporting the weight of the body. The torn cartilage is more commonly the medial one, although the ratios reported vary widely. In 62 of 64 cases investigated by Ferguson and Thompson it was the medial cartilage that was torn rather than the lateral. In one series of tears reported over a period of 10 years, Lipscomb and Henderson found, there were 10 tears of the medial meniscus to one of the lateral, but in the following 10 years there were only 5 of the medial to 1 of the lateral; they regarded this apparent increase in damage to the lateral cartilage as resulting from better diagnosis.

The more common incidence of tears of the medial meniscus over those of the lateral is usually attributed to lesser mobility of the medial meniscus as a result of its attachment to the tibial collateral ligament. Last ('50) maintained, however, that the medial meniscus is only loosely attached at its margin to the tibia, and is actually as mobile as is the lateral meniscus. He blamed the greater incidence of tears of the medial meniscus on the fact that it is moved entirely passively by the femoral condyle, while the lateral meniscus, although moved by the femoral condyle in medial rotation of the femur, is provided with two mechanisms, the popliteus muscle, and its attachment by the meniscofemoral ligaments to the femur; they move it out of harm's way during lateral rotation of the femur. There may also be a factor of tensile strength, for Mathur and co-workers found that among 32 joints the medial meniscus was weaker than the lateral in 27 (the medial meniscus had an average tensile strength of 56 pounds [25.4 kg]; the lateral, one of 74 pounds [33.6 kg]). In one joint there was equal strength, and in only four were the medial menisci stronger than the lateral ones.

Volk and Smith reported a case of fracture of the medial meniscus in a 5-year-old boy, and said that this was apparently the earliest age at which it had been reported to occur. According to Mathur and his co-workers, there is evidence that the menisci may become progressively weaker with age.

Tears of the menisci are usually accompanied by locking of the joint, as a result of rolling of the fractured portion, and pain and swelling at the knee. Complete removal of a torn meniscus has been advocated by some workers, but Lipscomb and Henderson advocated removing the torn part only, expressing the opinion that complete removal does not improve the end result. McGinty and co-workers agreed, and reported that partial meniscectomy gave much better results than total meniscectomy. Further, Lutfi reported that in the monkey total medial meniscectomy produces thinning and erosion of the articular cartilage, and Jones and co-workers found osteoarthritis so prevalent on the

operated side in patients older than 40 years that they advised not operating when there were what they called "degenerative tears."

Regeneration of the menisci is said to occur in dogs (King, '36a and b); the ingrowth apparently takes place from the synovial membrane, and grossly resembles fibrocartilage but was found on histologic study to be only connective tissue. Further, the regeneration apparently did not prevent a good deal of roughening and degeneration of the articular surfaces of the femur and tibia.

Miscellaneous Injuries

Smith and Blair described a number of instances of "chronic tibial collateral ligament strain" that failed to respond to conservative therapeutic measures; in the patients so diagnosed some deformity of the medial meniscus, resulting in its being displaced against or even herniated through the tibial collateral ligament, was found at operation.

Other derangements of or in the knee joint include a whole variety of things; degenerative changes of the menisci (Burman and Sutro); cysts in the menisci, far more commonly the lateral one, and sometimes described as being caused by direct trauma (Taylor), sometimes as being a developmental defect (Willard and Nicholson); ossification of the meniscus (Weaver); ossification in a cruciate ligament (Garden); damage to the articular cartilage of the joint from various causes (Ghormley); bursitis in some of the numerous bursae around the knee (De Palma, '48); and the apparently rare calcification in bursae about the knee (Norley and Bickel).

The Leg

TIBIA AND FIBULA

The tibia articulates with the femur at the knee joint, and both tibia and fibula enter into the formation of the ankle joint. The two bones are united to each other by the tibiofibular synovial joint, the interosseous membrane, and the tibiofibular syndesmosis.

The *tibiofibular synovial joint* lies between the

head of the fibula and a posterolateral part of the lower surface of the lateral tibial condyle. Its small cavity sometimes communicates with the cavity of the knee joint through the subpopliteal recess. Anterior and posterior ligaments of the head of the fibula run upward and medially to the tibia and strengthen the capsule. The nerve supply of the joint is from twigs of the common peroneal and tibial nerves, those from the latter being derived from the nerve to the popliteus muscle.

The heaviest fibers of the *interosseous membrane* pass downward and laterally from the interosseous margin (crest) of the tibia to that of the fibula. Above the upper end of the membrane is an oval aperture for passage of the anterior tibial vessels to the front of the leg, and close to its lower end a smaller aperture for the perforating branch of the peroneal artery. Between the closely adjacent lower ends of the tibia and fibula the membrane becomes very much thicker, and this part is often called the interosseous ligament.

The *tibiofibular syndesmosis,* between the lower ends of the tibia and fibula, is sometimes converted into a synovial joint by a diverticulum from the talocrural (ankle) joint. In addition to the lower thickened part of the interosseous membrane, the ligaments of this syndesmosis are the anterior and posterior tibiofibular ligaments, and transverse fibers that are often described as the inferior transverse ligament. The anterior tibiofibular ligament is somewhat triangular, and extends laterally and downward from the lower end of the tibia to a narrower attachment on the fibula. The posterior tibiofibular ligament is smaller and quadrilateral in shape, but anteriorly it is continuous with very heavy transverse fibers in front of and below it, these being the ones that are sometimes described as forming a separate ligament, the inferior transverse one. These fibers extend from the posterior border of the articular surface of the tibia to the lateral malleolus; they are related, through synovial membrane, to the articular surface of the talus, and form a part of the articular surface of the ankle joint.

Tibia

The expanded upper end or superior articular surface of the tibia is formed largely by the two condyles. The articular surfaces for the femoral condyles are separated posteriorly by the shallow depression of the posterior intercondylar area (fossa), anteriorly by the anterior intercondylar area, and between these two by the intercondylar eminence with its medial and lateral intercondylar tubercles (Figs. 9-29 and 9-30). The tibial ends of the cruciate ligaments and the extremities of the menisci are attached to the anterior and posterior intercondylar areas, but the intercondylar eminence receives no ligamentous attachments.

On its anterolateral surface the lateral condyle usually presents a raised area that marks the attachment of the strongest and most posterior part of the iliotibial tract; on its curved inferior or distal surface is the facet for articulation with the head of the fibula. The chief marking of the medial condyle, other than its articular surface, is the posteriorly placed deep transverse groove for the major part of the insertion of the semimembranosus muscle.

The body of the tibia is described as having three borders, anterior, medial, and interosseous, and three surfaces, medial, posterior, and lateral. The anterior border and the medial surface are largely subcutaneous. The tibial tuberosity, on the anterior border, receives the attachment of the ligamentum patellae; according to Hughes and Sunderland, the strongest attachment of this ligament is into the groove demarcating the tuberosity above and laterally, but there is also strong attachment into the superior margin and the inferior angle of the lower roughened area of the tuberosity. Neither the medial nor the lateral surface of the tibia presents any special markings, but the posterior surface is marked in its upper third by the line of the soleus muscle (soleal line, popliteal line); this, the tibial origin of the soleus muscle, extends obliquely medially and downward.

The lower end of the tibia bears the medial malleolus, the articular surface of which is

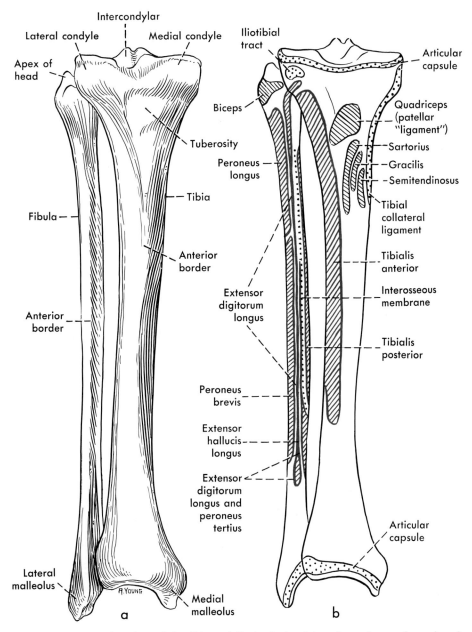

Fig. 9-29. Anterior views of the tibia and fibula. In *b*, origins of muscles are shown in *red*, insertions in *black*.

continuous with that of the inferior surface and enters with it into articulation with the trochlea tali. Behind the medial malleolus is the malleolar sulcus, which accommodates the tendons of the tibialis posterior and flexor digitorum longus muscles; farther laterally there may be a less distinct groove for the tendon of the flexor hallucis longus. The lateral surface of the lower end of the tibia presents the fibular notch for articulation with that bone.

The blood supply of the tibia has been described in considerable detail by Nelson and co-workers.

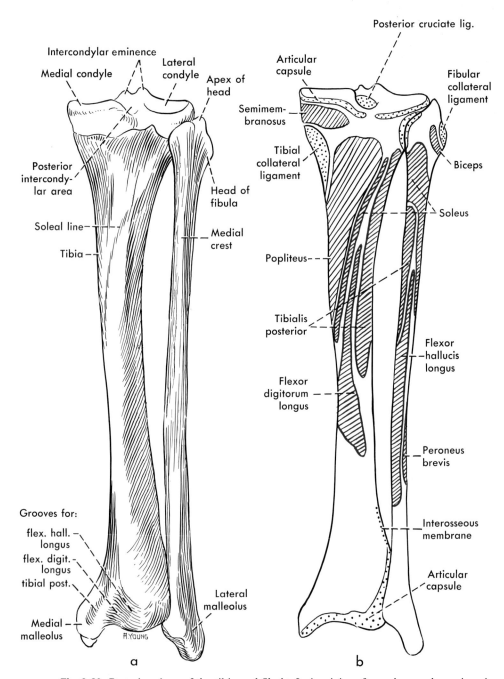

Fig. 9-30. Posterior views of the tibia and fibula. In *b*, origins of muscles are shown in *red*, insertions in *black*.

Fibula

The head of the fibula is subcutaneous and rises laterally to its apex (formerly called the styloid process). Medially, it bears the facet for articulation with the tibia. The region of junction of head and body is sometimes re-

ferred to as the neck; the common peroneal nerve here passes around the lateral side of the fibula and is particularly vulnerable in this location.

The long slender body of the fibula is usually described as having three borders, interosseous, anterior, and posterior, and three surfaces, medial, lateral, and posterior, but the borders are difficult to identify, and borders and surfaces spiral laterally as they are traced downward (Fig. 9-31). Further, the large posterior surface of the bone is marked, especially in its middle part, by the prominent medial crest; this so divides the posterior surface that four surfaces of the bone are sometimes described.

The expanded lower end of the fibula, the lateral malleolus, is subcutaneous like the head. Its medial surface is largely occupied by the malleolar articular surface for the talus; behind and below this is the roughened malleolar fossa into which the posterior talofibular ligament attaches. There may be a groove on the posterior surface of the malleolus where the peroneus brevis tendon passes behind it.

Abnormalities and Fracture

Of all the long bones the fibula is apparently most frequently deficient or absent; O'Rahilly found some 297 cases tabulated by 1935. Deficiency or absence of the tibia is said to be less common; many writers list the tibia as following the fibula in incidence of deficiency, but O'Rahilly said that both the radius and the femur are defective more often than is the tibia. He found 122 cases of tibial defect reported by 1924. Although only a part of the fibula may be absent, Harmon and Fahey said usual reports have been of complete replacement of the bone by a fibrous band, or at most the presence of a small piece of bone in the expected site, and O'Rahilly agreed that total rather than partial absence of the tibia is more frequent.

Associated with absence of the fibula is usually absence or malformation of some of the tarsals and absence of some of the toes. Harmon and Fahey suggested that in many cases the deficient metatarsals and toes are the second and third; O'Rahilly expressed the opinion that, as one would expect, the fifth metatarsal and toe are most frequently lacking, the second least frequently. Characteristic features of a limb with congenital absence of the fibula are shortening of the leg, and pronounced forward bowing (convex side

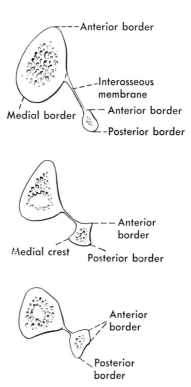

Fig. 9-31. Cross sections through the upper, middle, and lower parts of the bones of the leg, to illustrate especially the varying shape of the fibula. The interosseous border of both bones can be identified by the attachment of the interosseous membrane.

forward) of the tibia (Fig. 9-32). Harmon and Fahey described delay in the appearance of the center of ossification and slow development of the lower tibial epiphysis and of some of the bones of the ankle and foot as characteristic of the development of the deformity. Fusion of the developing calcaneus and talus, with obliteration of the important subtalar joint, is very frequent. While any of the anterior tarsal bones are occasionally absent, the cuboid is said to be almost constantly present; it or other tarsals may be involved in the talocalcaneal fusion. Because of the abnormality of the ankle joint in absence of the fibula, the foot tends to be subluxated posteriorly and laterally; tight calcaneal and peroneal tendons may hold the foot, to a varying degree, in the equinovalgus position (plantar flexed, and everted and abducted—pronated). Wood and co-workers recommended total removal of the foot, and described the technic that leaves the most satisfactory stump.

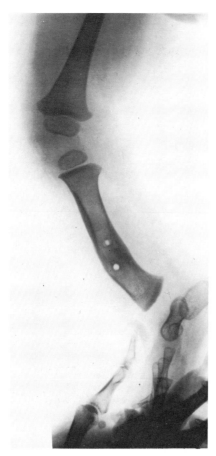

Fig. 9-32. Roentgenogram of a leg with congenital absence of the fibula. Note the anterior bowing of the tibia. (Courtesy of Drs. M.B. Coventry and E.W. Johnson Jr)

In congenital absence of the tibia, also, the limb is markedly distorted. Gray dissected a leg from a 3-year-old child presenting this anomaly, and in his case, although the leg was straight, the foot was so completely inverted that the sole faced upward, and so markedly adducted that the toes pointed toward the opposite heel. In this instance there were only two toes, associated with the lateral three metatarsals; the only tarsals present were the calcaneus and the cuboid, both deformed. (O'Rahilly found in his survey that while any of the medial bones [talus, navicular, the first two cuneiforms, or metatarsals and phalanges] may be absent, any one is more frequently present than it is absent. However, he found a high incidence of talo-calcaneal and talonavicular fusion.) The fibula in Gray's specimen was straight and projected below the calcaneus so as to form the lowest part of the limb. Among the soft tissues, a number of muscles and nerves of the leg and foot were absent, and several anomalous muscles were found.

Dankmeijer described a somewhat similar case from a near-term fetus in which the fibula was markedly curved, there were a number of maldevelopments of the muscles, and although the foot had five metatarsals and toes it was in a marked varus position (adducted and inverted—supinated). Roberts reported a case in which the tibia tapered to a point above the ankle joint, and did not enter into formation of the joint; in this case the lower end of the fibula expanded and entered into an articulation with a misshapen talus, but there was talipes equinovarus in spite of the attempt to form an ankle joint.

According to Heyman and Herndon, anterior tibial bowing when the fibula is absent is associated with shortening of the muscles of the calf, from which a severe talipes equinus ensues. They said that syndactylism and congenital dislocation of the hip are also frequently associated with this type of bowing. Heyman and co-workers described also a much rarer posterior bowing of the tibia, apparently associated with tightness of the anterior muscles of the leg and weakness of the triceps surae (gastrocnemius and soleus).

Fracture of the upper ends of the tibia and fibula may result from a blow on the knee, and fracture of the tibia may be produced by a force that drives the femur violently down against it.

Transverse fracture of the body of either the tibia or the fibula alone is typically not accompanied by much displacement since the sound bone acts as a splint for the fractured one; however, they are often fractured simultaneously (Fig. 9-33). Fracture of the tibia resulting from indirect violence is most likely to occur at about the junction of the middle and lower thirds of the bone and is often oblique; in this case, regardless of whether or not the fibula is fractured, there is likely to be considerable overriding of the two parts of the bone.

The pull of the strong calf muscles tends also to produce an anterior angulation at the line of fracture. It is particularly important in reducing a tibial fracture that the parts be restored to their normal alignment in order that the weight shall be transmitted in as normal a fashion as possible to the ankle joint.

Fracture of a malleolus may be produced by a direct blow, but is more commonly produced by a fall that produces a severe twisting at the ankle, the type of fracture varying somewhat (see Fig. 9-76)) according to whether the strain was primarily one of eversion or abduction (Pott's fracture), or inversion or adduction (reversed Pott's fracture).

According to Devas, "shin splints," a painful condition of the tibia well known to track men, are actually stress fractures of the tibia.

MUSCLES OF THE CALF

The muscles of the calf of the leg lie in two compartments within the deep fascia, separated from each other by a layer usually called the transverse crural septum or deep transverse fascia of the leg. The gastrocnemius, soleus, and plantaris muscles occupy the superficial compartment; the popliteus, the flexor hallucis longus, the flexor digitorum longus, and the tibialis posterior occupy the deep compartment, and arise in part from the covering fascia. The nerves and vessels lie at first in the superficial compartment, but penetrate the transverse septum to run thereafter in the deep compartment.

SUPERFICIAL GROUP

The two large muscles of this group, the gastrocnemius and soleus, share a common tendon of insertion on the calcaneus (Fig. 9-34), and are together called the triceps surae; the tendon of the plantaris, the third member of this group, typically inserts also on the calcaneus, but sometimes ends otherwise.

Triceps Surae
The *gastrocnemius,* the superficial member of this complex, forms most of the prominence

Fig. 9-33. Fracture of the tibia and fibula.

of the calf. It arises by two heads, both of which take some origin from the posterior aspect of the capsule of the knee, but largely arise by strong tendons from the femur. The medial and larger head arises from the popliteal surface of the femur just above the medial condyle (Fig. 9-35), and the lateral head arises farther forward, from the posterior part of the lateral surface of the lateral femoral condyle and from a little of the lateral lip of the linea aspera just above the condyle. Treanor has noted that this anterior origin of the gastrocnemius allows the muscle, acting from its insertion, to produce the hyperextension of genu recurvatum. A bursa typically lies deep to each of these heads at its origin; that of the medial head, which is larger, may communicate with the cavity of the knee joint

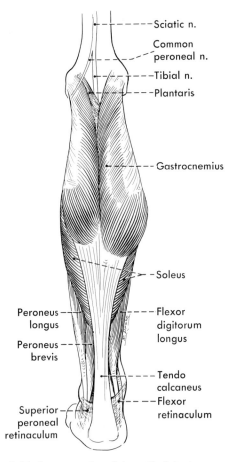

Fig. 9-34. Some muscles of the calf of the leg.

ported as the fabella, but is sometimes called the lateral fabella. Sutro and co-workers found shadows of at least one fabella in 11.5% of roentgenograms of 700 patients. Since sesamoids may be fibrocartilaginous or cartilaginous instead of bony, a higher incidence has been reported from dissection: Parsons and Keith summarized reports indicating that the

Fig. 9-35. Origin, *red*, and insertion, *black*, of the gastrocnemius muscle.

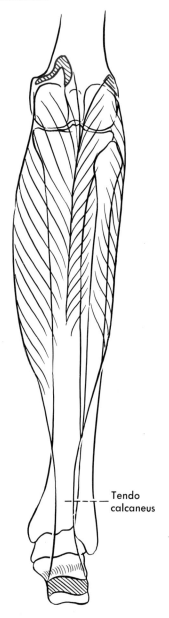

(and may be involved in popliteal or Baker's cyst); that of the lateral head very rarely does.

As the two heads of the gastrocnemius run downward, the popliteal vessels and the tibial nerve disappear between them. Additional fibers arise from a heavy aponeurosis that extends down on the superficial surface of each belly, and as the bellies come together they form a tendinous raphe that in turn becomes continuous with a heavy aponeurosis on the anterior surface of the muscle. This aponeurosis receives the attachment of the lower-inserting fibers and subsequently the tendon of the soleus to form the tendo calcaneus (tendon of Achilles).

The lateral head especially, the medial head less frequently, may bear on its deep surface a sesamoid bone or cartilage. A bony sesamoid of the lateral head is usually re-part of the medial border of that bone. The popliteal vessels and the tibial nerve pass between the tibial and fibular heads of the soleus, deep to the tendinous arch, to attain their deeper position in the leg.

Muscular fibers of the soleus end in a broad aponeurosis applied to the posterior surface of the muscle and therefore immediately adja-

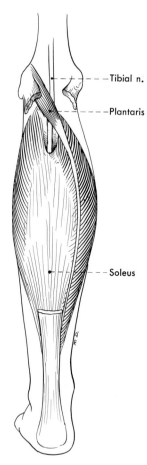

Fig. 9-36. Soleus and plantaris muscles. The cut ends are parts of the gastrocnemius.

lateral head of the gastrocnemius has a sesamoid in about 27% to 29%, the medial head one in about 12% to 15%, but that only about a third of these sesamoids are bony. Kaplan ('61) regularly found a well-developed ligament (fabellofibular ligament) extending from the fabella to the fibula.

The *soleus* (Figs. 9-36 and 9-37) lies immediately deep to the gastrocnemius but is visible on the sides of and below that muscle, since it is wider and its muscular fibers extend lower to attach into the tendo calcaneus. It arises by two heads united by a tendinous arch (arcus tendineus of the soleus muscle) from which additional fibers arise. The lateral origin is from the head and about the upper third of the body of the fibula, the other from the soleal (popliteal) line on the tibia and from a

Fig. 9-37. Origins, *red*, and insertions, *black*, of the plantaris and soleus muscles.

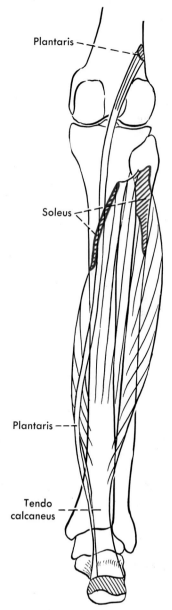

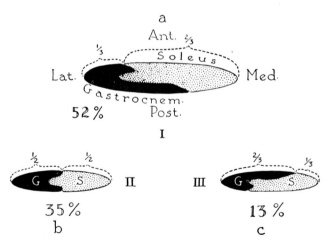

Fig. 9-38. Varying amounts of twisting of the tendo calcaneus at its insertion, diagrammed as of the left limb. (Cummins EJ, Anson BJ, Carr BW, Wright RR, Hauser EDW: Surg Gynecol Obstet 83:107, 1946 [by permission of Surgery, Gynecology and Obstetrics])

cent to the aponeurosis of the gastrocnemius on the anterior surface of this muscle; the two parallel each other for a distance, but as they both narrow and thicken they unite to form the tendo calcaneus about 6 inches (15 cm) above the heel.

The tendo calcaneus begins superficially in the middle third of the leg where the muscular belly of the gastrocnemius ends, but it continues to receive muscle fibers on its anterior surface, from the soleus, almost to the malleolar level. The two components of the calcaneal tendon can be separated fairly well almost to the insertion on the bone, but Cummins, Anson, and colleagues noted a good deal of variation in the number and strength of fibers that these two parts interchange. White pointed out that the fibers of the tendon take an approximate 90° lateral twist so that the tendinous fibers connected with the gastrocnemius insert largely laterally into the posterior aspect of the calcaneus, those connected with the soleus largely medially. Because of this twist, he advocated cutting the anterior half just above the calcaneus and the medial half just below the muscles when tendon lengthening is necessary. Cummins and his colleagues found that the amount of rotation varies, sometimes being more, sometimes less, than 90° (Fig. 9-38). They expressed the opinion that a better lengthening procedure is to section the posterior or gastrocnemius portion of the tendon above, the medial or soleal one below.

Hall and co-workers reported, as congenital short tendo calcaneus, a condition in which children habitually walked on their toes, and the only abnormality was a short tendo calcaneus, with the muscle fibers extending farther down than usual.

Attrition and calcification of the tendon with advancing age may lead to its rupture, and it may be ruptured in vigorous individuals as a result of the sudden strain thrown upon it by landing on the ball of the foot in a jump from a height. Lea and Smith advised against surgical treatment, saying the tendon will heal itself if the foot is placed in a gravity-equinus boot.

A bursa typically lies between the upper part of the posterior surface of the calcaneus and the tendon; Dickinson and co-workers reported that bursitis here is typically associated with wearing high-heeled pumps.

The triceps surae is the important plantar flexor of the foot. According to Duchenne, the soleus alone acts if the movement is unopposed. The gastrocnemius, because of its origin from the femur, is more powerful when the knee is extended; also, because of this origin, it can act as a flexor at the knee joint when the limb is free of weight-bearing, and a short gastrocnemius interferes with simultaneous normal dorsiflexion at the ankle and extension at the knee such as occurs in walking (Nutt). Simon and co-workers found that the triceps functions during walking primarily to restrain forward movement as the weight is shifted during the stance phase from the heel to the ball of the foot; in the absence of triceps function, they said, the forward swing of the normal limb is both faster and shorter. Murphy and co-workers also observed this in a woman whose triceps had been excised. Other than this, they said, walking was almost normal; however, there was excessive dorsiflexion of the foot of the affected limb, and the speed could not be increased beyond that of normal walking. Sutherland and co-workers added that paralysis of the plantar flexors also interferes with extension of the knee during midstance; further, they agreed with other workers that the muscles do not contract for "push-off" in normal walking.

The gastrocnemius is innervated by two branches, which may arise together, from the tibial nerve; these go into each head (Fig. 9-39). The soleus receives one or two branches also from the tibial nerve, usually a chief one entering the superficial surface of the muscle and a smaller one to the deep surface of its medial head. Both muscles are typically supplied with fibers derived from the first two sacral nerves.

Plantaris

The plantaris arises from the lateral prolongation of the linea aspera toward the lateral condyle, a little above the origin of the lateral head of the gastrocnemius (Fig. 9-37). The tapering muscle belly is usually no more than 3 or 4 inches (about 7.5 cm to 10 cm) long, and its slender tendon therefore begins high in the leg. This tendon runs downward between the gastrocnemius and soleus muscles to assume a position along the medial border and insert just medial to the tendo calcaneus on the os calcaneus. Daseler and Anson reported that this muscle was absent in 6.67% of 750 lower extremities that they examined, and found absence recorded in 6.2% and 7.5% in other series.

While minor variations in the origin and insertion of the plantaris muscle are apparently common, marked variations such as failure of the tendon to reach the level of the heel are apparently uncommon; Daseler and

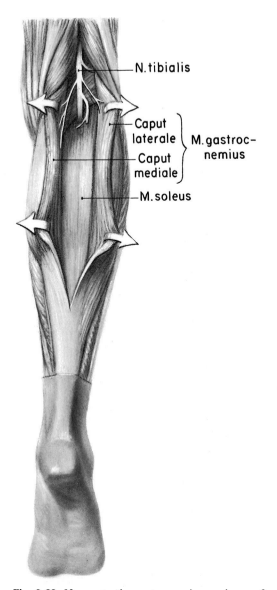

Fig. 9-39. Nerves to the gastrocnemius, and one of two to the soleus. See also Fig. 9-43. (Hollinshead WH, Markee JE: J Bone Joint Surg 28:721, 1946)

Anson encountered variations in the insertion at the heel, but no variations in origin and no truly anomalous insertions. They found the tendon inserting in what is described as the normal fashion, into the bone immediately medial to the calcaneal tendon, in 48%; in other cases it inserted anterior to the tendo calcaneus or fanned out so as partly to embrace this tendon; in 5% it inserted chiefly into the tendo calcaneus. In some of their

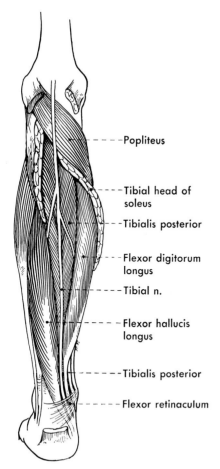

Fig. 9-40. Deeper muscles of the calf.

cases there was also an insertion into the flexor retinaculum and adjacent fascia, an insertion that is occasionally encountered with no attachment of the plantaris tendon to the calcaneus.

The plantaris is a rudimentary muscle. It is equivalent to the palmaris longus of the forearm and in some animals extends beyond the calcaneus to become continuous with the plantar aponeurosis, in the same manner that the palmaris longus become continuous with the palmar aponeurosis. The muscle is a plantar flexor of the foot and a flexor of the leg, although obviously with no appreciable power; rupture of it may occur as a result of extreme plantar flexion, as in toe dancing, and is said to be very painful. The muscle is innervated by a twig from the tibial nerve,

apparently usually containing fibers from the fourth and fifth lumbar and first sacral nerves.

DEEP GROUP

These muscles, the popliteus above and the flexor hallucis longus, flexor digitorum longus, and tibialis posterior below (Fig. 9-40), lie anterior to the deep transverse fascia or transverse septum of the leg. This fascia, continuous with the deep fascia that encircles the leg, is stretched from the tibia to the fibula; it is thick above, over the popliteus muscle, where it receives an expansion from the tendon of the semimembranosus muscle; it is thinner and more transparent in the middle of the leg, but behind the medial malleolus is thickened by additional transverse fibers to form the flexor retinaculum, already described. The popliteal vessels and the tibial nerve, after passing deep to the tendinous arch of the soleus, penetrate this fascia below the popliteus muscle to run a deeper course in the leg; the continuations of the vessels and nerve into the foot therefore pass deep to the flexor retinaculum, along with the tendons that continue into the foot. Matsen and Clawson described a syndrome of the deep compartment, in which there is hypesthesia over the distribution of the tibial nerve, usually pain, and weakness of the flexors of the toes. As for the anterior tibial compartment, the treatment is prompt decompression.

Popliteus

The popliteus arises (or inserts) by a strong tendon attached well forward on the lateral surface of the lateral condyle of the femur (Fig. 9-41); this tendon runs posteriorly, medially, and inferiorly across the femur, between the fibrous and synovial capsules of the knee joint, and leaves the posterior aspect of the joint by passing below a more or less well-defined fibrous arch, the arcuate popliteal ligament. The muscle is attached to the lateral meniscus of the knee joint and to the arcuate ligament (Last, '50, Heller and Langman, and others); Lovejoy and Harden described it as arising also from the head of the fibula, by a tendon that is usually described as a part of the arcuate ligament. The flattened triangular muscle belly outside the joint has a diverticulum of the joint cavity, the subpopliteal recess, deep to it. The muscle forms the lower part of the floor of the popliteal fossa, and is attached into much of the posterior surface of the tibia above the soleal (popliteal) line. The popliteal vessels and the tibial nerve cross the posterior surface of the muscle and pierce the transverse crural septum below the muscle; occasionally, however, the anterior tibial artery arises high and in this case it may cross the upper border of the muscle to run anterior (deep) to it.

The popliteus muscle is innervated by a twig of the tibial nerve, said to contain usually fibers from the fifth lumbar and first sacral nerves. Its early action in flexion is apparently not to flex the knee but to rotate the femur and the lateral meniscus laterally; and in standing on the flexed knee it also resists anterior displacement of the femur and the meniscus. (Barnett and Richardson; Basmajian and Lovejoy). Mann and Hagg found that it is active during any exercise that involves medial rotation of the tibia; further, in walking it becomes active shortly before heel-strike, and maintains its activity through about three fourths of the stance phase. They therefore concluded that its chief function is to bring about and maintain medial rotation of the tibia.

Flexor Hallucis Longus

The flexor hallucis longus arises chiefly from the posterior surface of about the lower two thirds of the fibula (Fig. 9-42), lateral to the medial crest, with an additional origin from the fascia that covers it and from the intermuscular septa on its sides. It is a bipennate muscle; its tendon begins high in the muscle, and the muscle fibers insert obliquely on this tendon to about the level at which it enters the flexor retinaculum. Here it is provided with a tendon sheath and occupies its own special compartment deep to the retinaculum. This compartment is the most posterolateral of the four compartments and is bor-

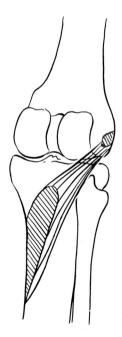

Fig. 9-41. Origin, *red*, and insertion, *black*, of the popliteus muscle.

dered anteromedially by the compartment containing the posterior tibial vessels and the tibial nerve. The tendon passes across the posterior process of the talus in the groove between the medial and lateral tubercles, then across the medial suface of the calcaneus in a groove just below the sustentaculum tali. Its course in the foot, to be described in more detail later, can be summarized by saying that it crosses the upper or deep surface of the flexor digitorum longus, typically giving off a slip that joins tendons of that muscle, and farther forward lies on the superficial surface of the flexor hallucis brevis; thereafter, it enters its special tendon sheath on the big toe and goes to an insertion on the base of the distal phalanx. Doubling of the tendon, with one part inserting into the calcaneus, has been reported (Burkard).

The muscle is supplied by one or more branches from the tibial nerve, usually containing fibers from the fifth lumbar and first two sacral nerves. It is primarily a flexor of the distal phalanx of the great toe and secondarily a supinator (adductor and inverter) of the foot. Duchenne said it adducts with

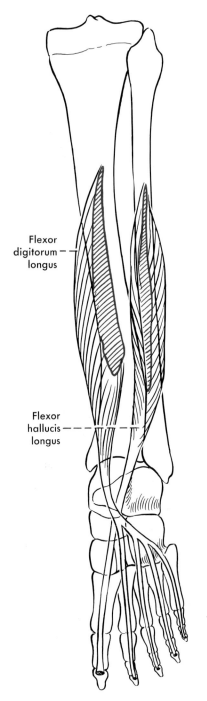

Flexor
digitorum —
longus

Flexor
hallucis — — —
longus

Fig. 9-42. Origins, *red*, and insertions, *black*, of the flexor hallucis longus and flexor digitorum longus muscles.

force, but inverts weakly. It is a very weak plantar flexor of the foot.

Flexor Digitorum Longus

This muscle arises from much of the middle three fifths of the posterior surface of the tibia (Fig. 9-42), the upper part of the origin tapering to a point between the soleal line and the origin of the posterior tibial muscle from the medial side of the posterior surface. As in the case of the flexor hallucis longus, there is also origin from the fascia covering the muscle and from the intermuscular septa adjacent to it. As the tendon takes form in the lower part of the leg this muscle becomes separated from the tibia by the tendon of the tibialis posterior, which crosses between the flexor longus and the tibia to assume the most anteromedial position at the ankle. The flexor digitorum longus, although the most medial of the three muscles in the leg, therefore assumes an intermediate position at the ankle: its compartment deep to the flexor retinaculum is the second, counting backward from the medial side; the compartment for the tibialis posterior lies anteromedially, that for the tibial nerve and posterior tibial vessels lies posterolaterally. The tendon sheath within this compartment has already been described.

After passing onto the sole of the foot around the medial margin of the sustentaculum tali the tendon of the flexor digitorum longus extends forward below (superficial to) the tendon of the flexor hallucis longus, which it crosses, but above (deep to) the flexor digitorum brevis; at about the point at which the tendon of the flexor digitorum longus crosses that of the flexor hallucis longus, it divides into separate tendons for the four toes. These give rise to the lumbrical muscles of the foot, and after they have entered the digital tendon sheaths behave, in respect to the tendons of the flexor digitorum brevis, in exactly the way that the tendons of the flexor digitorum profundus do in respect to those of the flexor superficialis in the hand: the short flexor of the toes, representing the superficial flexor of the forearm and hand, has tendons that divide to let the tendons of the long flexor, the equivalent of the deep flexor of the forearm and hand, pass through to their insertions on the bases of the distal phalanges. The tendon of the flexor digitorum longus in the foot also receives the insertion of a muscle of the foot, the quadratus plantae (accessory flexor), that is not represented by a corresponding muscle of the hand.

The flexor digitorum longus is innervated by one or more branches from the tibial nerve (Fig. 9-43), usually containing fibers from the fifth lumbar and the first sacral nerves. It is primarily a flexor of the distal phalanges of the four lateral toes, secondarily a supinator (inverter and adductor) and weak plantar flexor of the foot.

Tibialis Posterior

The tibialis posterior, the deepest lying of the three deep muscles of the calf, is partly overlapped on its superficial or posterior surface by the flexor hallucis longus and the flexor digitorum longus. It arises (Fig. 9-44) from the posterior surface of the fibula between the medial crest and the interosseous border of the bone, from the medial part of the posterior surface of the tibia below the soleal line, from the posterior surface of the interosseous membrane, and from the covering fascia and intermuscular septa. As the muscle passes downward and converges onto its tendon it also passes medially to run in front of the flexor digitorum longus, between this and the tibia, and enter the most medial and anterior of the compartments deep to the flexor retinaculum. Here the tendon, enclosed in its own tendon sheath, lies against the posterior surface of the medial malleolus. Below the medial malleolus it turns forward under (superficial to) the plantar calcaneonavicular ("spring") ligament to expand and divide into bands that insert into the tuberosity of the navicular bone, into the plantar surfaces of the three cuneiforms and of the bases of the second, third, and fourth metatarsal bones, and into the cuboid bone. As the tendon crosses the calcaneonavicular ligament it usually contains a sesamoid cartilage or bone. If its tendon sheath is short, a bursa may occur between the tendon and the ligament (Lovell and Tanner).

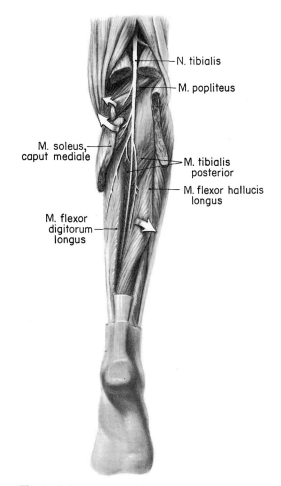

Fig. 9-43. Innervation of the deep muscles of the calf. A deep nerve to the soleus is also shown. (Hollinshead WH, Markee JE: J Bone Joint Surg 28:721, 1946)

Ghormley and Spear reported anomalous tendons within the tendon sheath of the tibialis posterior in 4 of 11 patients operated upon for tenosynovitis of this sheath: in 3 the tendon of the muscle was doubled, in the fourth there was a tendon that represented an accessory flexor of the toes. Key reported partial rupture of the tendon within its sheath.

The tibialis posterior is supplied by one or more branches of the tibial nerve, typically containing fibers from the fifth lumbar and first sacral nerves. It is an inverter and adductor (supinator) of the foot, and secondarily a weak plantar flexor; Duchenne said it adducts with force, but inverts weakly.

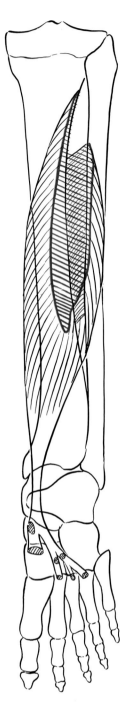

Fig. 9-44. Origin, *red*, and insertion, *black*, of the tibialis posterior.

Functions of the Deep Flexors

The chief functions of these three muscles, as they act on the foot which is not supporting weight or engaged in walking, have already been given. All three of them are, it will be noted, plantar flexors and supinators of the foot, but it is obvious that they actually contribute relatively little under normal circumstances to plantar flexion of the weight-bearing foot; most of this movement is carried out by the massive triceps surae, which also, because of its insertion on the calcaneus, has an additional advantage of leverage. Jones ('41) calculated from their size and leverage that they contribute no more than 5% of the power normally used in plantar flexion; similar calculations by Sutherland ('66) gave a somewhat different, but still low, ratio of about 6 to 1 in favor of the triceps surae. However, these muscles have an importance in standing and walking that is out of proportion to their effect as plantar flexors of the foot, for they and the anterior ones help to keep the weight properly distributed on the foot. It is also true that, in spite of their mechanical inefficiency, they (with the peroneus longus) may be able to substitute fairly well for the triceps surae; for instance, Boyd and his co-workers advocated section of the tendo calcaneus as being sometimes the most effective way to relieve the pain of intermittent claudication, and said that most of their patients so treated could, after 4 days, walk a mile or more, and could still rise on their toes. However, the ability to stand on the toes is frequently lost with paralysis of the triceps or section of its tendon.

ANOMALOUS CALF MUSCLES

Variations in the arrangement of the tendo calcaneus and in the insertion of the plantaris, as well as absence of that muscle, have already been noted.

Real anomalies of muscles in the calf are rare. As mentioned in Chapter 8, there is occasionally a muscular slip, the tensor fasciae suralis, that arises from a lower part of one of the hamstrings, frequently the biceps femoris, or from the deep surface of the fascia lata, and passes down to insert into the deep surface of the fascia of the leg (*e.g.,* Barry and Bothroyd); similarly, there are, rarely, abnormal slips of origin or insertion associated directly with the calf muscles. Extra slips of origin associated with the gastrocnemius and soleus muscles usually join those muscles or the tendo cancaneus; Parsons, however, recorded an instance of a muscle slip that passed transversely between the two heads of origin of the gastrocnemius. Rarely, a muscle slip associated with the popliteus, but arising from the posterior aspect of the femur and inserting into the posterior part of the capsule of the knee joint, may be present as a popliteus minor; more frequently, a muscle slip lying deep to the popliteus may take origin from the head of the fibula to insert on the tibia under cover of the popliteus. The rare peroneocalcaneus is sometimes regarded as being associated with the tibialis posterior, sometimes with the flexor hallucis longus. It arises from the lower end of the fibula below the origin of the flexor hallucis, passes downward and medially with the tendon of this muscle, and usually inserts into the calcaneus but may join the tendon of the flexor hallucis longus.

One of the more common anomalous muscles of the leg is the accessory flexor digitorum longus; this arises variably in the leg, from the deep fascia or either of the two bones, and on reaching the foot inserts either into the quadratus plantae or into the tendons of the flexor digitorum longus (Driver and Denison).

VESSELS AND NERVES OF THE CALF

Popliteal and Posterior Tibial Vessels

The popliteal vessels, with the artery in front of the vein, emerge from under cover of the medial hamstring muscles to enter the popliteal space; they and the tibial nerve cross each other obliquely, and they first lie medial to, then in front of, and finally lateral to, the nerve (Fig. 9-45). At the lower end of the popliteal space these three structures disap-

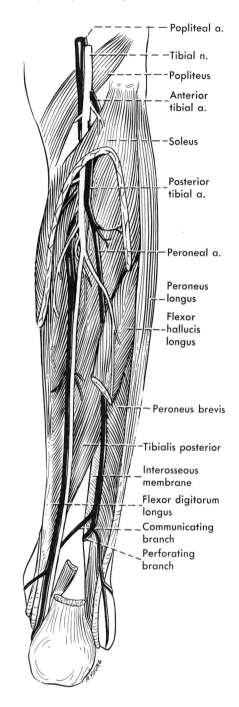

Fig. 9-45. Arteries and nerves of the calf.

pear together between the two heads of the gastrocnemius muscle. Boyd and co-workers said that intermittent claudication is sometimes due to primary thrombosis of the artery that results from injury to it, as it passes between the two heads of the gastrocnemius, by a fibrous band uniting the heads. The variable connections of the popliteal vein with the deep femoral vein or its tibutaries have been mentioned in the preceding chapter.

In the popliteal space the popliteal artery gives off muscular, cutaneous, and genicular branches. The upper muscular branches are twigs to the lower ends of the posterior muscles of the thigh, and anastomose with similar twigs from the profunda artery; they arise from the upper part of the popliteal. The lower muscular branches, arising close to the level of the knee joint, are typically two large ones, the sural arteries, that are distributed primarily to the gastrocnemius and soleus muscles. They anastomose with the lower genicular arteries and with branches of the posterior tibial.

The cutaneous branches of the popliteal artery are twigs, derived either from that artery or its branches, that supply skin over the popliteal space.

There are five genicular arteries from the popliteal, paired superior and inferior and an unpaired middle. The medial superior genicular artery passes medially, above the femoral condyle and deep to the semimembranosus, semitendinosus, and tendon of the adductor magnus, to supply the vastus medialis and the knee joint. It anatomoses across the front of the knee with the lateral superior genicular, above with the descending (supreme) genicular, and below with the inferior medial genicular. The lateral superior genicular passes above the lateral femoral condyle, deep to the biceps, to supply the vastus lateralis and a lateral part of the knee joint; it anastomoses with its fellow of the medial side, with the descending branch of the lateral femoral circumflex, and with the lateral inferior genicular artery.

The middle genicular artery, unpaired, arises opposite the knee joint and pierces the oblique popliteal ligament to supply the cru-

ciate ligaments and the synovial membrane around them.

The medial inferior genicular artery arises under cover of the medial head of the gastrocnemius and winds medially around the tibia deep to this head, passing below the medial tibial condyle and deep to the tibial collateral ligament. It gives twigs to the popliteus muscle, the upper end of the tibia, and the capsule of the knee joint, and anastomoses with the lateral inferior genicular and the superior medial genicular. The lateral inferior genicular artery arises deep to the lateral head of the gastrocnemius and passes laterally around the lateral tibial condyle, above the head of the fibula and deep to the fibular collateral ligament and tendon of the biceps. It anastomoses with the lateral superior genicular, the medial inferior genicular, and the anterior tibial recurrent.

Through the anastomotic connections just recounted, the superior and inferior genicular arteries form an important part of the collateral circulation around the knee. (Fig. 9-46). As noted in Chapter 1, the incidence of gangrene following sudden occlusion of the popliteal artery is very high, thus giving evidence that this collateral circulation is not very often sufficiently developed to compensate for the loss of the great flow through the popliteal artery.

The popliteal vessels continue, deep to the soleus, across the posterior surface of the popliteus muscle, and at about the lower border of this muscle the popliteal artery ends by dividing into anterior and posterior tibial arteries; the popliteal vein may be formed at this same level by the confluence of the paired posterior and anterior tibial veins accompanying the arteries, but varies considerably (see a following paragraph) in the manner and level of its formation. The anterior tibial vessels pass laterally and forward between the lower border of the popliteus and the upper border of the tibialis posterior to pass above the upper edge of the interosseous membrane and reach the anterior aspect of the leg. The posterior tibial vessels continue downward with the tibial nerve, disappear between the two heads of the soleus, and penetrate the deep transverse fascia of the leg; here they continue downward and somewhat medially in a fascial canal at about the overlapping borders of the flexor digitorum longus and the tibialis posterior. Approximately an inch (2.5 cm) below its origin the posterior tibial artery gives off under cover of the soleus muscle a lateral branch, the peroneal artery, that has a somewhat deeper course and is described in a following section.

Most of the branches of the posterior tibial artery in the leg are muscular ones, but the large nutrient artery of the tibia arises close to the upper end of the posterior tibial, and after supplying a few twigs to adjacent muscles enteres the bone. One of the upper muscular branches is the fibular circumflex artery, which pierces the fibular origin of the soleus to wind around the neck of the fibula, help supply the peroneal muscles arising from the lateral aspect of this bone, and anastomose with the sural and the lateral inferior genicular arteries. This branch may arise from the lower end of the popliteal artery, or occasionally from the anterior rather than the posterior tibial.

A little above the ankle joint the posterior tibial artery receives the communicating branch of the peroneal (Fig. 9-45; see also Fig. 9-48) and a little lower it gives off one or more medial malleolar branches that join with twigs from the anterior tibial and peroneal arteries in forming a rete over the medial malleolus. Under cover of the flexor retinaculum, just before the posterior tibial artery ends by dividing into medial and lateral plantar arteries, it gives off one or more medial calcaneal branches that pierce the retinaculum to supply the medial side of the heel and form here a calcaneal rete with twigs from the malleolar branches of the anterior tibial and peroneal arteries.

In its course down the leg the posterior tibial artery is accompanied by two veins. The tibial nerve at first lies medial to it but then crosses the artery to lie on its lateral side throughout most of the length of the leg. Nerve and vessels pass together into a special compartment within the flexor retinaculum, situated between the compartments for the

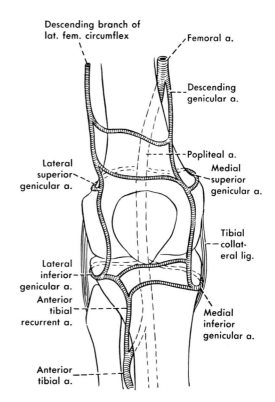

Fig. 9-46. Diagram of the anastomoses around the knee joint.

flexor digitorum longus, medially, and the flexor hallucis longus, laterally.

The most common variation of the popliteal artery is in its level of division. Trotter found a high ending of the popliteal artery, with the division occurring on the posterior surface of or above the upper border of the popliteus muscle, in 4.9% of 246 extremities from white persons and 6.2% of 338 extremities from Negro persons. Parsons and Robinson collected reports on the level of ending of the popliteal artery in 106 specimens, and said that in 82% of these the division of the artery was either at, or within 0.25 inch (about 0.6 cm) of, the lower border of the popliteus, but that in 8.2% it divided more than 0.5 inch (1.3 cm) above this border. Trotter found the anterior tibial artery passing distally on the anterior surface of the popliteus muscle in about one third of the instances in which there was high division of the artery, and Parsons and Robinson found this in about one fourth of their cases of high division. Parsons and Robinson, and Trotter, also found that high division of the popliteal artery is often associated with origin of the peroneal artery from the anterior rather than the posterior tibial. Senior ('29) has discussed the various types of abnormal branching of the popliteal artery and shown that most of the variations here result from arrest of normal development of the vessels.

Occasionally, the posterior tibial artery is small in the leg after giving off the peroneal, but attains its usual size at the ankle after being joined by the communicating branch of the peroneal. Rarely, the posterior tibial artery is absent, it and the peroneal being replaced in the leg by an artery, the great peroneal, that contributes to the peroneal and otherwise normally disappears during later development; or both it and the peroneal are replaced by persistent anomalous vessels representing incomplete development from the embryologic condition (Pierson). According to Senior, true absence of the peroneal artery has never been reported. When it replaces the posterior tibial on the foot, it typically passes superficial to the flexor hallucis longus tendon to assume the position of the posterior tibial at the ankle, and usually continues into the foot as the lateral plantar artery; the medial plantar is then usually absent.

The *popliteal vein* is usually described as being formed at the lower border of the popliteus muscle by the union of the anterior and posterior tibial veins. Williams, however, found a single large trunk representing the popliteal vein at the level of the knee joint in only 13 of 50 limbs that he investigated, while in 33 limbs there were two venous trunks at this level, and in 4 there were three. Further, he found a very considerable variation in the manner in which the veins came together. In his specimens, the paired veins accompanying the posterior tibial joined to form a single trunk in about 50%, and in the remainder joined or were joined by the peroneal veins in a variable manner; the peroneal veins formed a common trunk in only 10 cases before ending, and in the other cases joined in a variable manner either the anterior or the posterior

tibial veins. The anterior tibial veins formed a single trunk only 5 times, and when they did not (sometimes there were three anterior tibial veins) they usually joined different venous trunks. Some of the various patterns of formation he found are shown in Figure 9-47. In the minority of instances in which a single trunk was present at the level of the knee joint, the posterior tibials usually united first, then were joined by the peroneals, and then by the anterior tibials. In those cases in which there was more than one trunk present at the level of the knee joint, there was great inconstancy in their formation; however, one was usually smaller than the other.

Williams found the muscular branches emptying into the veins close to the level of the knee to be apparently more constant than are the tibial and peroneal veins in their entrance. He described veins from the soleus as usually joining the peroneal veins, and tending to make these larger than the posterior tibials, some 3 cm or 4 cm below the level at which the anterior tibials appear; and he found the veins from the gastrocnemius typically joining the single or the larger of two popliteal trunks separately, sometimes at the level of the knee joint but more commonly 2 cm or 3 cm above it.

Peroneal Vessels

The peroneal artery (Figs. 9-45 and 9-48), normally the largest branch of the posterior tibial in the leg, typically arises about an inch (2.5 cm) below the origin of that vessel. Accompanied by its two veins, it passes obliquely downward and laterally across the posterior surface of the upper part of the tibialis posterior and enters a canal formed by the fibula anterolaterally, the tibialis posterior in front and on the medial side, and the flexor hallucis longus behind; sometimes it actually runs in the substance of the latter muscle. As it passes downward in this deep position it gives branches to the adjacent muscles, some of these passing through the interosseous membrane to supply the anterior muscles of the leg; it typically gives rise to a nutrient artery to the fibula, and about an inch (2.5 cm) above the inferior tibiofibular

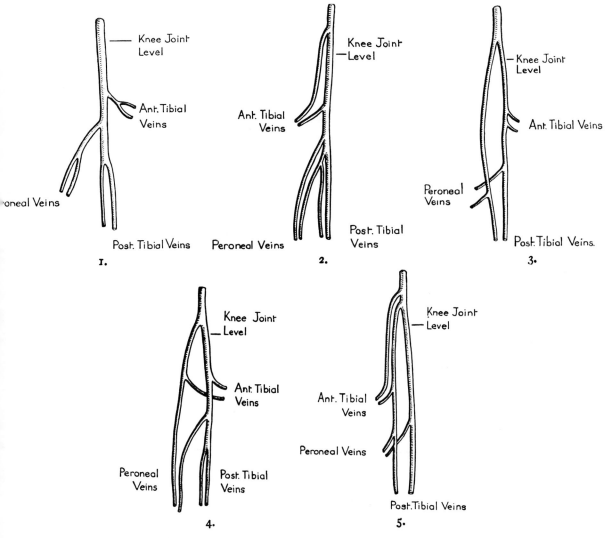

Fig. 9-47. Five different patterns of formation of the popliteal vein. (Williams AF: Surg Gynecol Obstet 97:769, 1953 [by permission of Surgery, Gynecology and Obstetrics])

joint, while it is still under cover of the flexor hallucis longus, it gives off a more or less transverse branch that runs between this muscle and the tibia to join the posterior tibial artery. Close to the same level, either before or after the communicating branch arises, the peroneal gives rise to a perforating branch that passes forward between the tibia and the fibula, through the gap near the distal border of the interosseous membrane, to reach the front of the ankle joint; it anastomoses on the dorsum of the foot with the lateral malleolar branch of the anterior tibial artery

and with the tarsal branch of the dorsalis pedis. The terminal branches of the peroneal artery are a malleolar one and usually several calcaneal ones that take part in the anastomosis on the lateral side of the malleolus and the heel.

It has already been noted that the peroneal artery may arise from the anterior tibial rather than from the posterior tibial, and Parsons and Robinson reported that in 5% of 106 cases it arose at the level of division of the popliteal artery into the anterior and posterior tibials, so that the popliteal artery ac-

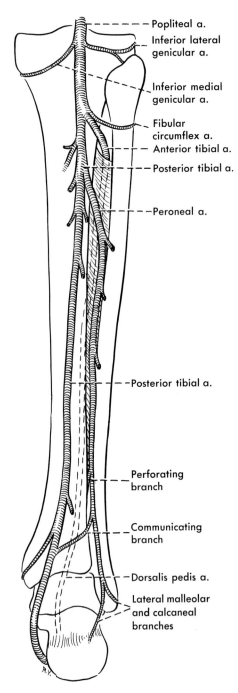

- - - - Popliteal a.
Inferior lateral
genicular a.

Inferior medial
genicular a.

Fibular
circumflex a.
- - - Anterior tibial a.
- - - Posterior tibial a.

- - - Peroneal a.

- - Posterior tibial a.

- Perforating
branch

- Communicating
branch

- Dorsalis pedis a.

Lateral malleolar
and calcaneal
branches

Fig. 9-48. Diagram of the posterior tibial and peroneal arteries.

tually ended in a trifurcation. Branches of the peroneal artery may partially replace those of the posterior tibial, and if the latter is small or missing the lower end of the peroneal is typically continued into the foot to form the lateral plantar artery (see the discussion of the tibial artery). Similarly, if the anterior tibial artery is small or missing, the perforating branch of the peroneal usually supplements or replaces the continuation of the anterior tibial onto the dorsum of the foot, forming the dorsalis pedis artery, or giving rise to its tarsal and arcuate branches but not necessarily the deep plantar branch.

LYMPHATICS

As elsewhere in the limbs, the deep lymphatics of the calf are few as compared with the superficial lymphatics. They consist of channels that accompany the arteries and veins already described. Most of these lymphatics, as well as those accompanying the anterior tibial vessels, and lymphatics from the knee joint accompanying the genicular vessels, empty into the popliteal lymph nodes, approximately a half dozen very small nodes located along the popliteal vessels and rarely seen because of their minute size and the fact that they are buried in fat. One member of this group usually lies just deep to the popliteal fascia on the small saphenous vein, and receives the lymphatics that penetrate the fascia with this vein. Thus the efferents leaving the popliteal nodes represent drainage both from the deep and a part of the superficial aspect of the leg, as well as from the knee joint. These efferents pass upward along the femoral artery to end in the deep inguinal nodes, described in the preceding chapter. Occasionally, as noted previously, the popliteal nodes are lacking so that the lymphatics which usually enter them drain instead into inguinal nodes.

TIBIAL NERVE

The tibial nerve passes straight down through the popliteal fossa and lies at first lateral to the popliteal vessels; as these run obliquely laterally, the nerve crosses behind them to their medial side; it passes with them between the two heads of the gastrocnemius and then deep to the tendinous arch of the soleus to lie on the posterior surface of the tibialis posterior muscle, in front of the deep transverse fascia of the leg. The nerve at first lies medial to the posterior tibial artery, but as the artery runs slightly medially the nerve crosses behind it to lie therafter on its lateral side. The nerve passes downward with the posterior tibial vessels on the tibialis posterior muscle and then on the posterior aspect of the tibia, and disappears with the vessels into the neurovascular compartment of the flexor retinaculum, between the compartments for the flexor digitorum longus and the flexor hallucis longus. It ends by dividing, under cover of the flexor retinaculum, into the medial and lateral plantar nerves.

A number of the branches of the tibial nerve arise while this nerve is in the popliteal fossa, others commonly arise after it has passed under cover of the soleus. As is usual, there is some variation in the levels of origin of the branches, and in regard to whether they are separate branches or arise in common with others. The articular branches to the knee joint are often described as two or three in number, but Gardner found usually only a single large one that sometimes arose high in the thigh and descended with the tibial nerve, at other times arose from it in the popliteal fossa. Whether the articular branches are single or multiple, some of the smaller branches pass directly to the capsule of the joint, others accompany genicular vessels.

The tibial also gives off one cutaneous branch, the medial sural cutaneous, in the popliteal fossa. This arises from the posterior aspect of the tibial nerve and courses downward and superficially between the two heads of the gastrocnemius and then on the superficial surface of this muscle, but deep to the crural fascia, to penetrate the fascia at about the junction of the upper two thirds and lower third of the leg. It has already been described with the cutaneous nerves of the leg.

The *muscular branches* of the nerve that typi-cally arise before it passes deep to the soleus are a branch to each head of the gastrocnemius, one to the plantaris, one into the superficial aspect of the soleus, and one to the popliteus (Fig. 9-49). The nerves to the two heads of the gastrocnemius may arise separately from the tibial, or by a common stem. Sunderland and Hughes found the nerves to the two heads sharing a common stem in only 2 of 20 specimens, but the nerve or nerves to the lateral head frequently sharing a common stem with other muscular branches; while the nerve to the medial head may arise at the same level as other nerves, it usually does not share a common stem with any of them unless it be the nerve to the lateral head. Either head of the muscle may receive more than one branch. The nerves to the gastrocnemius may arise high in the popliteal fossa or close to its middle.

The nerve to the plantaris is small, and frequently arises with that to the lateral head of the gastrocnemius; it tends, at any rate, to be one of the higher branches. The upper of the two nerves to the soleus (the second arises usually under cover of this muscle) typically arises below the origins of the nerves to the gastrocnemius, but may share a common stem with the nerve to the lateral head; it courses downward between the gastrocnemius and soleus to enter the superficial surface of the upper end of the soleus. The nerve to the popliteus usually arises from the anterior surface of the tibial nerve, runs downward over the posterior surface of the popliteus muscle, and turns around the lower border of this muscle to enter its anterior surface. It usually gives off the interosseous nerve of the leg (n. interosseus cruris) that runs downward on the posterior aspect of the interosseous membrane as far as the tibiofibular syndesmosis (inferior tibiofibular joint); it also gives a twig to the (superior) tibiofibular joint, and a branch to the tibialis posterior may arise with it.

The remaining muscular branches of the tibial nerve (to the tibialis posterior, flexor digitorum longus, and flexor hallucis longus muscles) usually arise after the nerve has passed under cover of the soleus muscle, but a branch to any of these may arise with or at

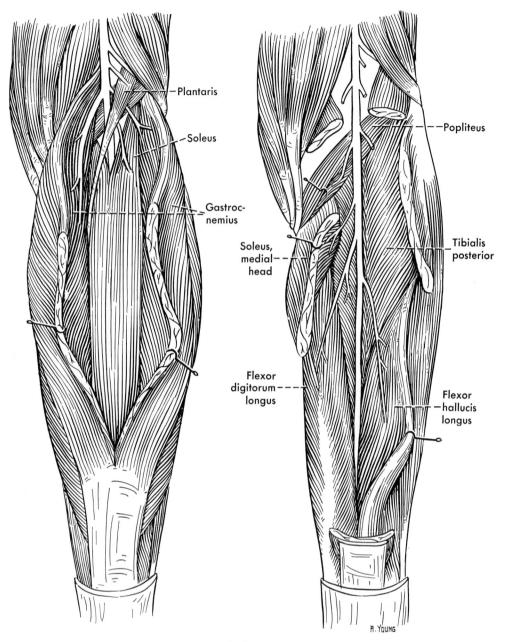

Fig. 9-49. Distribution of the tibial nerve in the leg.

the level of the nerve to the popliteus. The second of the usual two nerves to the soleus (Sunderland and Hughes found 2 in 18 of 20 specimens) typically arises from 6 cm to 8 cm below the first; in Sunderland and Hughes' specimens it had an independent origin from the tibial in 6 cases, and arose with other muscular branches in the remainder.

Sunderland and Hughes found 1 nerve to the posterior tibial muscle in 9 specimens, from 2 to 4 nerves in 11; these usually arose at the level of or in common with branches to the other deep muscles of the leg. The nerve to the flexor digitorum longus tends to be fairly long when it is single, passing down on the muscle's posterior surface to divide into two or more branches before entering it; Sunderland and Hughes found a single branch in 11, two or three branches in the remainder of their specimens. The nerve to the flexor hallucis longus may be the last of the muscular branches to arise, or may arise in common with other nerves; it has a particularly long course downward on the superficial surface of the muscle, and usually sends several branches into this muscle before it finally ends. There may be two nerves to this muscle also.

The nerve to the flexor hallucis longus usually accompanies the upper part of the peroneal artery, and gives twigs to this artery; the nerve to the tibialis posterior is said to supply twigs to the posterior tibial artery.

Just before its division into medial and lateral plantar nerves, the tibial nerve gives off two or three medial calcaneal branches that perforate the flexor retinaculum to supply skin on the medial side and much of the plantar surface of the heel.

ANTEROLATERAL MUSCLES OF LEG

The two lateral muscles of the leg, the peroneus longus and the peroneus brevis (Fig. 9-50), lie in a special compartment, separated from the anterior muscles by the anterior intermuscular septum of the leg, and from the muscles of the calf by the posterior intermus-

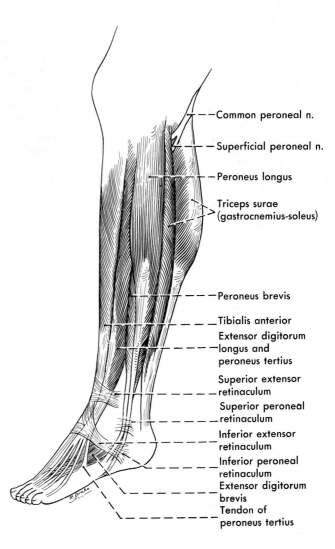

—Common peroneal n.

—Superficial peroneal n.

—Peroneus longus

Triceps surae
(gastrocnemius-soleus)

—Peroneus brevis

Tibialis anterior

Extensor digitorum
longus and
peroneus tertius

Superior extensor
retinaculum

Superior peroneal
retinaculum

Inferior extensor
retinaculum

Inferior peroneal
retinaculum

Extensor digitorum
brevis

Tendon of
peroneus tertius

Fig. 9-50. Lateral view of the muscles of the leg.

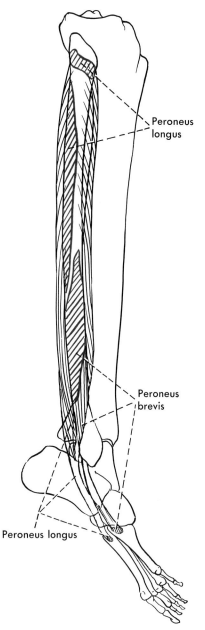

Fig. 9-51. Origins, *red*, and insertions, *black*, of the peroneus longus and brevis.

cular septum. The common peroneal nerve enters the upper end of this compartment and divides into the superficial and deep peroneal nerves here; the superficial peroneal nerve then supplies the two lateral muscles and subsequently becomes cutaneous by piercing the crural fascia, while the deep peroneal

nerve continues forward through the anterior intermuscular septum to supply the anterior muscles of the leg.

LATERAL MUSCLES

These both arise from the lateral surface of the fibula, and are supplied by the superficial peroneal nerve; the peroneus longus frequently also gets a branch from the common peroneal or the deep peroneal nerve.

Peroneus Longus

The peroneus longus (fibularis longus) arises from the lateral surface of the head and about the upper two thirds of the body of the fibula (Fig. 9-51), from the fascia that covers the muscle, and from the intermuscular septa. Its origin from the bone is not continuous from the head to the body of the fibula, but presents a gap through which the common peroneal nerve runs between the muscle and bone in its course around the lateral aspect of the leg. The muscle fibers give way to a flattened tendon that lies superficial to the peroneus brevis, and becomes rounded as it approaches the ankle.

The tendon of the peroneus longus passes deep to the superior peroneal retinaculum along with the peroneus brevis, which separates it from contact with the lateral malleolus; the two tendons here share a common tendon sheath. It then passes across the lateral surface of the calcaneus, where the bone may present a slight groove to accommodate it, and is held in place here by the inferior peroneal retinaculum; it has its own compartment within this retinaculum, posteroinferior to the compartment for the peroneus brevis. The tendon then crosses the lateral surface of the cuboid bone and turns medially across its plantar surface, lying here against the posterior wall of the sulcus for the tendon of the peroneus longus (the anterior smooth surface of the tuberosity of the cuboid). The long plantar ligament converts the groove here into a canal, and the tendon is provided, at least on its deep surface, with a synovial sheath. It runs obliquely, medially and distally, across the sole of the foot to insert into the lateral side of the medial cuneiform and the base of the first metatarsal bone.

Where it lies in contact with the smooth surface of the tuberosity of the cuboid the tendon usually has within it a sesamoid fibrocartilage; Parsons and Keith found this sesamoid reported as ossified in about 20% of specimens.

Peroneus Brevis

The peroneus brevis arises from about the lower two thirds of the lateral surface of the body of the fibula (Fig. 9-51), and from the two intermuscular septa adjacent to it. Its upper fibular fibers arise anterior to the lower fibers of the peroneus longus. It passes straight down, deep to the peroneus longus, and shares a common tendon sheath with this muscle behind the lateral malleolus and deep to the superior peroneal retinaculum. Here the peroneus brevis tendon is more anterior and lies in the groove on the posterior surface of the lateral malleolus. As the tendon curves forward below the lateral malleolus it crosses the lateral surface of the calcaneus above the tendon of the peroneus longus, and here lies in its own compartment and a subdivison of the common tendon sheath. The tendon continues downward and forward across the lateral surface of the foot to insert into the dorsolateral surface of the base of the fifth metatarsal.

Innervation and Action

The peroneus longus receives one or more nerves that arise close to the division of the common peroneal (Fig. 9-52); the uppermost branch, when there are two, sometimes comes from the common peroneal, sometimes from the superficial, and sometimes from the deep; usually at least one branch is from the superficial peroneal nerve. The peroneus brevis usually receives a single branch from the superficial peroneal. The nerves to both muscles are said to contain fibers derived from the fourth and fifth lumbar and first sacral nerves.

The peroneus longus and brevis are the chief evertors of the foot; since they insert into the forepart of the foot they simultaneously

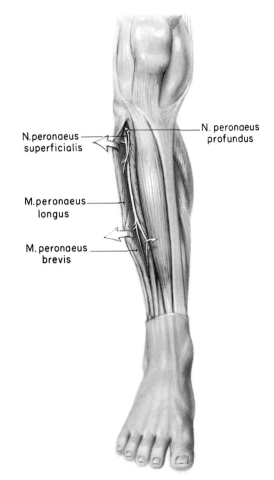

N.peronaeus superficialis

N. peronaeus profundus

M.peronaeus longus

M.peronaeus brevis

Fig. 9-52. Innervation of the peroneus longus and brevis muscles. (Hollinshead WH, Markee JE: J Bone Joint Surg 28:721, 1946)

abduct this, and are therefore pronators of the foot. Duchenne found that the brevis is a more powerful abductor than the longus. He described the peroneus longus as a weak flexor, and said that while the brevis will plantar flex from sharp dorsiflexion, it will dorsiflex from plantar flexion. However, both peronei are said to be inactive during pure dorsiflexion (Basmajian); although both are simultaneously active during flexion, their chief action appears to be on the transverse tarsal rather than the talocrural joint. O'Connell, recording from the peroneus longus, found it sometimes initiating unopposed plantar flexion. In contrast, Jones ('19) pointed out that the important action of

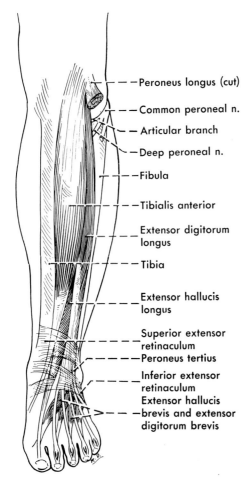

Peroneus longus (cut)

Common peroneal n.

Articular branch

Deep peroneal n.

Fibula

Tibialis anterior

Extensor digitorum longus

Tibia

Extensor hallucis longus

Superior extensor retinaculum

Peroneus tertius

Inferior extensor retinaculum

Extensor hallucis brevis and extensor digitorum brevis

Fig. 9-53. Anterior muscles of the leg.

pronation is best carried out in combination with dorsiflexion; he found that some patients in whom the tibial nerve had been divided could produce plantar flexion through the peronei, but that most of the patients whom he examined could not because they were unable to contract the peronei without also contracting the anterior muscles of the leg, thus producing dorsiflexion. Nevertheless, the movement can be learned, just as movements can be learned after tendon transfers. Bickel and Moe, for instance, reported obtaining satisfactory plantar flexion from the peroneus longus by moving its tendon backward and fitting it into a groove in the calcaneus to give it better leverage for plantar flexion; it then substituted for a paralyzed gastrocnemius-soleus complex.

With the deep muscles of the calf and the

anterior muscles of the leg the peronei also play an important part in helping to assure proper weight distribution in the foot. When they are paralyzed, the foot is so inverted that the weight is thrown primarily on the lateral border of the foot.

Variations

Rarely, the two muscles may be fused. Also rarely, a slip of the peroneus brevis may join the long extensor tendon to the little toe and thus form a peroneus digiti minimi muscle. More common is an accessory peroneal muscle, variably known as the peroneus accessorius, the peroneus quartus, or the peroneus of the tarsi; Hecker reported an incidence of 13% of this muscle. He found that it could be associated with either the peroneus longus or the brevis, and could arise in either the lower or the upper third of the leg; he said it inserted, variably, into the lateral surface of the calcaneus, into the cuboid, or with the peroneus longus.

ANTERIOR MUSCLES

These muscles (Fig. 9-53) occupy the anterior compartment of the leg, deep to the deep fascia between the tibia and the anterior intermuscular septum. The four muscles of this group are innervated by the deep peroneal nerve.

Tibialis Anterior

The tibialis anterior is the largest of the anterior muscles of the leg. It arises from the inferior surface of the lateral condyle and approximately the upper two thirds of the lateral surface of the body of the tibia (Fig. 9-54), from the interosseous membrane, from a septum between it and the extensor digitorum longus, and from the deep surface of the crural fascia. The muscle fibers give way to a strong tendon in the lower third of the leg; this passes downward on the anterior surface of the tibia behind the superior and in the most medial compartment of the inferior extensor retinaculum, over the front of the ankle joint, and the medial side of the dorsum of the foot, to insert into the medial and lower sides of the medial cuneiform bone and of the base

Fig. 9-54. Origins, *red*, and insertions, *black*, of the tibialis anterior, extensor digitorum longus, and peroneus tertius muscles. *a* is an anterior view of the leg and foot, *b* a plantar view of the foot, and *c* a dorsal view of a digit to show the details of the extensor tendon here.

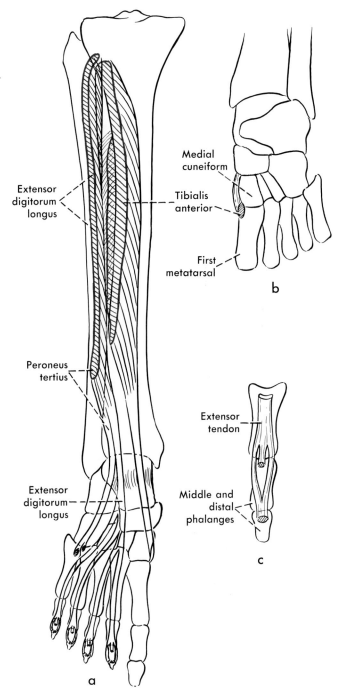

Medial cuneiform

Tibialis anterior

First metatarsal

b

Extensor digitorum longus

Peroneus tertius

Extensor tendon

Extensor digitorum longus

Middle and distal phalanges

c

a

of the first metatarsal. The lower end of the tendon may be split, as it goes to its insertion, into a part for each bone and, rarely, is doubled for a greater length; occasionally, it inserts only on the metatarsal or sends a slip to attach to the dorsum of the proximal phalanx of the big toe or to the distal part of the first metatarsal (Hallisy).

In the upper part of the leg, this muscle covers the anterior tibial vessels and the deep peroneal nerve. As it passes through the extensor retinacula it is provided with its own tendon sheath. As already noted, it may or may not have an appreciable part of the inferior retinaculum in front of it; the strongest band of this retinaculum passes behind it.

Ischemia restricted to or centered in the tibialis anterior is sometimes reported as anterior tibial compartment syndrome. The tibialis anterior has been said to be especially susceptible to ischemia because it derives all its blood supply from the anterior tibial artery, while the other anterior muscles of the leg receive some supply also from branches of the posterior vessels that penetrate the interosseous membrane. However, Getzen and Carr suggested that the condition results from block of either lymphatic or venous drainage, or both, which in turn interferes with intramuscular arterial flow. They and Leach and co-workers stressed that fasciotomy to decompress the compartment should be done as early as possible to prevent severe muscle damage. Bradley expressed the opinion that venous obstruction is not a cause, and that while exercise, trauma, and arterial obstruction may all be causes, they have one thing in common: they affect the microcirculation, causing sludging of the red corpuscles and thus an increase in extravascular fluid.

Rupture of the tibialis anterior tendon is apparently rare; Mensor and Ordway found reports of only 10 instances of this, and reported two of their own; they said it usually resulted from sharp plantar flexion of the foot such as may occur from a fall on the foot in plantar flexion.

This muscle receives its nerve supply, typically containing fibers from the fourth and fifth lumbar and first sacral nerves, from the deep peroneal nerve; there are usually two or more branches to the muscle. It is the most important dorsiflexor of the foot, and by virtue of its insertion on the medial border of the foot it adducts and inverts (supinates) the foot; because it inserts relatively far posteriorly on the forepart of the foot, it is not a very powerful adductor.

Extensor Digitorum Longus
This muscle, the lateral one of the group, arises from the lateral side of the lateral condyle of the tibia, and from approximately the upper two thirds or more of the anterior surface of the body of the fibula (Fig. 9-54), from the anterior intermuscular septum of the leg and from a septum between it and the tibialis anterior, from the deep surface of the crural fascia, and from a part of the interosseous membrane adjacent to its fibular origin. The muscle is penniform, with its tendon beginning on the medial border at about the junction of the upper two thirds and lower third of the leg. The lower part of the muscle is typically continuous with the peroneus tertius; this, not always present, is a split-off part of the extensor digitorum and its tendon accompanies that of the extensor through the intermediate compartment in the inferior extensor retinaculum, and shares with it a common tendon sheath in this position. As the long extensor tendon enters the inferior extensor retinaculum it usually broadens and divides into four slips, one for each of the four lateral toes.

At the level of the metatarsophalangeal joints the tendons of the extensor digitorum longus to the second, third, and fourth toes are joined laterally by tendons of the extensor digitorum brevis. On the toes the combined tendons of the extensor digitorum longus and brevis behave essentially as do the tendons of the extensor digitorum in the hand: each expands over the metatarsophalangeal joint to form a hood over the dorsum and sides of the capsule, where it may receive a small contribution from the interossei associated with that toe (although these insert primarily into bone rather than joining the extensor tendon, as do those of the hand), regularly receives the insertion of the lumbrical tendon, and then divides into three slips, a middle one and two lateral ones; the middle slip attaches to the base of the middle phalanx, the two lateral slips converge to an insertion on the distal phalanx.

The extensor digitorum longus receives its innervation from the deep peroneal nerve (Fig. 9-55); there are usually two or more nerves entering it, containing fibers from the fourth and fifth lumbar and first sacral nerve. It is an extensor of the four lateral toes, but is unable to overcome the pull of the more powerful flexors and therefore acts primarily in hyperextending the toes at the metatarsophalangeal joints. In addition, it works with the peroneus tertius in dorsiflexing and pronating the foot; Duchenne said that these muscles produce less powerful dorsiflexion than does the tibialis anterior, but that they abduct the anterior part of the foot much more strongly than the tibialis anterior adducts it.

Peroneus Tertius

The peroneus tertius, indistinguishable at its origin from the extensor digitorum longus, arises from approximately the lower third of the anterior surface of the fibula, from the adjacent interosseous membrane, and from the anterior intermuscular septum, and gives rise to a tendon that lies on the lateral side of the tendon of the extensor digitorum longus and accompanies that through the extensor retinacula. On the dorsum of the foot it passes slightly more laterally than do the tendons of the extensor digitorum and inserts into the dorsal surface of the base of the fifth metatarsal bone.

The peroneus tertius, really a part of the extensor digitorum longus that has been split off to assist specifically in dorsiflexing and raising the outer border of the foot, is sometimes absent, and shows considerable variation in size, although there appear to be no quantitative data on its variations. The strength of the peroneus tertius is a factor in the results achieved by tendon transplant to replace a paralyzed anterior tibial muscle (Bickel): attachment of the tendon of the extensor hallucis longus so that it can function better as a dorsiflexor does not markedly increase its power of inversion, which it shared with the tibialis anterior; thus if the peroneus tertius is strong it will evert the foot against the pull of the transplanted extensor hallucis, while helping to dorsiflex the foot, and in such a case it itself must be transplanted or cut. If, however, it is weak, it can be disregarded.

The nerve supply to this muscle is typically continued from one of the branches of the deep peroneal nerve that supplies the extensor digitorum longus. As already noted, the muscle works with the extensor digitorum longus in dorsiflexing and pronating the foot.

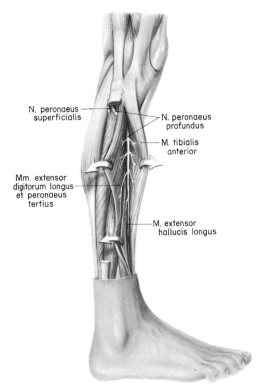

Fig. 9-55. Innervation of the anterior muscles of the leg. (Hollinshead WH, Markee JE: J Bone Joint Surg 28:721, 1946)

Extensor Hallucis Longus

The extensor hallucis longus is largely covered by the tibialis anterior and the extensor digitorum longus, but begins to appear between the two muscles at about the middle of the leg. It arises from approximately the middle three fifths of the anterior surface of the fibula, medial to the origin of the extensor digitorum longus from this bone, and from an adjacent part of the interosseous membrane (Fig. 9-56). This muscle is also penniform, with its tendon beginning along its medial border. As it passes through the inferior extensor retinaculum it has a separate compartment and is provided with a tendon sheath. The tendon then passes distally over the foot to insert into the distal phalanx of the big toe; it sometimes has a slip of insertion into the proximal phalanx (this occurred in 72 of 290 feet that Hallisy studied), but usually the insertion into this phalanx is that of the extensor hallucis brevis.

Like the other anterior muscles of the leg,

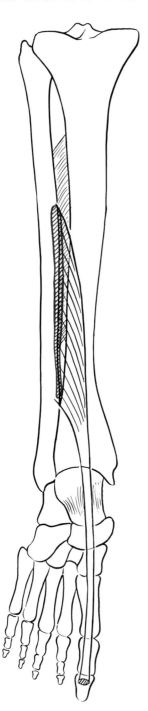

Fig. 9-56. Origin, *red*, and insertion, *black*, of the extensor hallucis longus.

the extensor hallucis longus typically receives several branches from the deep peroneal nerve, containing fibers from the fourth and fifth lumbar and first sacral nerves. It is primarily an extensor of the big toe, and according to Duchenne it is only a weak supinator and dorsiflexor of the foot. He said that it contracts strongly in dorsiflexion if the tibialis anterior is paralyzed, but it is not as strong a dorsiflexor as is the extensor digitorum longus, nor is its power of supination as great as that of the extensor digitorum longus and peroneus brevis in pronating; thus, a foot deprived of the dorsiflexing and supinating power of the tibialis anterior gradually goes into pronation.

Variations

Variations of the anterior muscles of the leg other than those already mentioned are rare. There may be a small muscle, the tibiofascialis anterior, apparently a separated portion of the tibialis anterior, arising from the lower part of the tibia and inserting into fascia of the lower part of the leg or of the foot. A tendon of the extensor digitorum longus to one of the toes is sometimes doubled; the muscle varies a good deal in the extent of its origin. Stevenson reported an unusual case in which it arose only from an upper part of the fibula while the peroneus tertius arose from most of the length of this bone and sent a slip of insertion to the proximal phalanx of the little toe, and the extensor digitorum longus sent tendons to the second, third, and fourth toes only. The extensor hallucis longus occasionally inserts partly into the metatarsal or into the proximal phalanx.

ANTEROLATERAL VESSELS AND NERVES OF LEG

ANTERIOR TIBIAL VESSELS

The anterior tibial artery (Fig. 9-57) is one of the terminal branches of the popliteal, and the variations in its origin and course on the back of the leg have already been discussed with that vessel. The artery and its two accompanying veins pass through the gap at the upper end of the interosseous membrane to lie on the anterior surface of that membrane, at first under cover of the tibialis anterior muscle, then of that and the flexor hallucis longus. The deep peroneal nerve, completing its course around the fibula, comes to lie just lateral to the anterior tibial artery and descends with it on the interosseous membrane. The nerve lies lateral to the artery as they start their course, and may so lie throughout the entire length of the leg; however, Horwitz ('38b) said that in 95 of 100 specimens that he examined the nerve passed anterior to the artery about 4 inches (10 cm) above the ankle but in the lower 2 inches (5 cm) of the leg it lay again lateral or anterolateral to the vessels. Perlow and Halpern also described the nerve as lying anterior to the artery and vein 10 cm (4 inches) above the medial malleolus. In four of Horwitz' 100 cases the nerve passed posterior to the artery and then lay medial to it for the rest of the length of the leg, while in one it lay first anterior and then medial to it in the lower part of the leg. In about the lower third of the leg the artery passes from under cover of the tibialis anterior and extensor hallucis longus to lie between the two muscles and then, with the nerve, passes deep to the extensor hallucis longus. Both nerve and artery pass deep to the inferior extensor retinaculum and as the artery crosses the ankle joint and becomes subcutaneous its name changes to dorsalis pedis.

The upper branches of the anterior tibial artery are variable. The anterior tibial recurrent artery is a vessel that commonly arises from the anterior tibial shortly after this has reached the anterior aspect of the leg, and runs upward to enter into the anastomosis around the knee joint. Parsons and Robinson reported that this was present in 84.5% of specimens that they surveyed; a vessel of similar distribution (then called posterior tibial recurrent) arose from the popliteal artery in 10.4%, and from the posterior tibial in 3.4%. (The circumflex fibular or fibular circumflex artery, to the upper ends of the anterolateral muscles, is also variable in its origin; it is usually a branch from the posterior tibial, but may arise from the popliteal or from the ante-

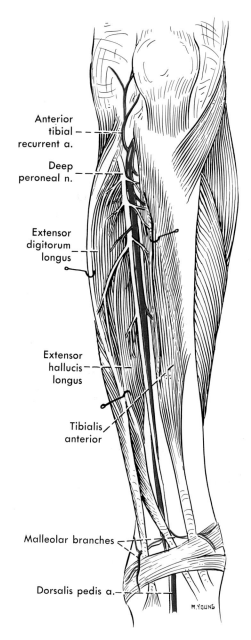

Fig. 9-57. Anterior tibial artery and deep peroneal nerve.

rior tibial; in any case, it is the curve around the neck of the fibula that is characteristic of it.) Most of the remaining branches of the anterior tibial artery in the leg are muscular or cutaneous; the muscular ones are distributed to the adjacent muscles on the front of the leg or penetrate the interosseous membrane to supply the deep muscles of the calf; the small cutaneous branches supply the skin of the front of the leg.

Close to the level at which it passes in front of the ankle joint to assume the name of dorsalis pedis, the artery may give off medial and lateral malleolar arteries. These are usually regarded as typically arising from the anterior tibial, but Huber found them more frequently arising a little below the joint, therefore really from the dorsalis pedis. They arise at about the same level, a little above the upper border of the inferior extensor retinaculum as a rule. The lateral anterior malleolar artery, usually larger than the medial, passes laterally behind the tendons of the extensor digitorum longus and peroneus tertius to anastomose with the perforating branch of the peroneal artery and with the tarsal branch of the dorsalis pedis, and with these supply the lateral side of the ankle. The medial anterior malleolar artery runs medially behind the tibialis anterior tendon and anastomoses with branches of the posterior tibial artery over the medial malleolus, helping to supply the skin and the ankle joint on this side.

In approximately 3.5% of limbs the anterior tibial artery either fails to reach the dorsum of the foot or is reduced to a very slender channel by this level, and the dorsalis pedis artery is a continuation of the perforating branch of the peroneal. In about 50% of feet, Huber found a branch, arising from the anterior tibial about 5 cm above the ankle, that ran laterally to reinforce or replace the perforating peroneal, and in 1.5% of feet the anterior tibial artery took this lateral course to the foot instead of running straight down in its usual fashion.

The *anterior tibial veins* need no further comment; the variations in their manner of ending in the popliteal vein have already been described with the latter vein.

The deep *lymphatics* of the anterior aspect of the leg accompany the anterior tibial vessels. There is said to be usually a very tiny anterior tibial lymph node on the upper part of the interosseous membrane; apparently one of the deep lymphatics ends here, while the others bypass it to pass above the interosseous membrane with the anterior tibial vessels and end in popliteal nodes.

PERONEAL NERVES

The *common peroneal nerve* usually contains fibers from the last two lumbar and first two sacral nerves, and distributes fibers from the last two lumbar and the first sacral nerve to the muscles that it and its branches supply. After giving off its lone branch in the thigh, to the short head of the biceps femoris, it diverges from the tibial nerve to run downward along the lateral border of the popliteal fossa (close to the medial border of the biceps femoris) and then leaves the fossa to pass downward and laterally between the biceps and the lateral head of the gastrocnemius. (Mangieri reported a rare case in which the nerve was stretched by a large fabella here.) It is subcutaneous just behind the head of the fibula, where it is most vulnerable to injury, and then winds laterally and forward around the neck of the fibula deep to the peroneus longus, in the gap between the origin of this muscle from the head and body of the bone. As it starts its course around the neck of the fibula it divides into its two major terminal branches, the superficial and deep peroneal nerves. Stack and co-workers reported a number of cases of compression of the common peroneal nerve by ganglion cysts in the region of the knee; in three instances the cyst communicated with the tibiofibular joint (Muckart reported eight similar cases, in five of which the ganglion lay in the peroneus longus muscle); in one of Stack's cases there was a communication with the knee joint, and in another, apparently, with the prepatellar bursa.

While it is in the popliteal fossa, or before it has separated from the tibial, the common peroneal nerve gives off an articular branch that descends to supply an anterolateral portion of the joint capsule of the knee; it also gives off the lateral sural cutaneous nerve, distributed to a part of the lateral surface of the leg; and either separately or as a branch of the latter it usually gives off the peroneal communicating nerve that joins the medial sural cutaneous branch of the tibial to form the sural nerve. These cutaneous nerves have been described in a previous section of this chapter.

At about the level at which it divides into superficial and deep branches, the common peroneal nerve gives rise to a recurrent articular branch. This curves around the fibula and is distributed in part to the periosteum and interosseous membrane of the upper part of the leg, in part to the (superior) tibiofibular joint and the knee joint; it sometimes arises with a branch to the upper part of the tibialis anterior muscle. Sometimes the recurrent branch is almost as large as the superficial and deep peroneal nerves, and the common peroneal nerve therefore seems to trifurcate. Not uncommonly, also, the common peroneal nerve gives rise to a branch to the peroneus longus muscle. Sunderland and Hughes found a branch to this muscle arising from the common peroneal above the level of its ending in four of 20 specimens, and one arising at the level of its division in nine.

The *superficial peroneal nerve* (Fig. 9-58), arising as one of the terminal branches of the common peroneal as this nerve comes in contact with the fibula, runs downward at first between the peroneus longus and the fibula, later between the peroneus longus and the peroneus brevis. It typically gives off a branch into the peroneus longus, which may be the first or only nerve supplying this muscle or may be the second of two or more, in which case the first branch may have been derived from either the common or the deep peroneal; Sunderland and Hughes found from two to seven branches, of variable origin, to the peroneus longus. While it lies on the peroneus brevis it gives off a branch into this muscle, and thereafter it is cutaneous; emerging from between the peroneus longus and brevis, it runs downward on the superficial surface of the latter muscle to penetrate the deep fascia of the leg at about the junction of the middle and lower thirds, soon thereafter dividing into its two terminal branches for the skin of the lower part of the leg and the foot.

The *deep peroneal nerve* (Fig. 9-57) passes farther around the fibula than does the common peroneal, and penetrates the anterior intermuscular septum to reach the anterior compartment of the leg. Here it runs largely lat-

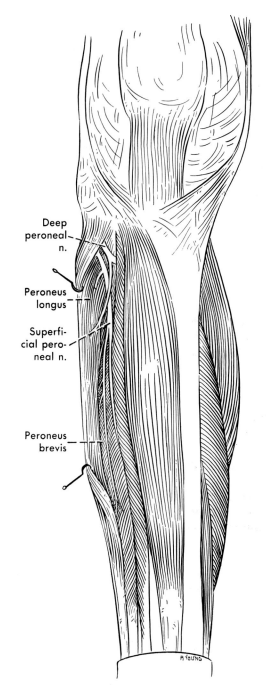

Fig. 9-58. Muscular branches of the superficial peroneal nerve, and a part of the deep peroneal nerve.

eral to, but for a short distance in the lower half of the leg anterior to, the tibial vessels, and in its course gives off numerous muscular branches to the anterior muscles of the leg. A small branch descends to help supply the ankle joint, and the remainder of the nerve passes onto the dorsum of the foot, usually lying lateral to the anterior tibial artery.

The nerves to the several anterior muscles of the leg are commonly multiple, except that the peroneus tertius is usually supplied by a single branch that either is a continuation of a branch into the extensor digitorum longus or is a separate branch to this muscle. Sunderland and Hughes found two or more branches to the tibialis anterior in 19 of 20 cases. The first branch to this muscle was sometimes derived from the common peroneal or from its bifurcation. The other branches, however, were from the deep peroneal. Similarly, these workers found that the first branch to the extensor digitorum longus arose from the common peroneal or its level of division, rather than the deep peroneal, in 9 of 20 cases and from the deep peroneal in the remainder, and that in all but one instance there were two or more branches to the muscle. They also reported that the extensor hallucis longus frequently received multiple branches, and that its uppermost branch usually arose below the level of origin of the branches to the tibialis anterior and extensor digitorum longus. They described the peroneus tertius as being supplied by a single branch in the 14 specimens in which they were able to trace it; the nerve arose alone from the deep peroneal, with branches to the extensor digitorum longus, or with those to the extensor hallucis longus.

Ankle and Foot

BONES AND JOINTS

The foot is, to a large extent, comparable to a pronated hand. The soft structures of the hand and foot are similar in many details, and the bony structure is fundamentally similar, but modified in the foot because of its weight-bearing duties. Thus the bones of the ankle and foot consist of tarsals, metatarsals, and phalanges, the latter two groups of bones being essentially similar to the metacarpals and phalanges of the hand; the tarsals, al-

though homologous with the carpals, depart more from these in their shape, size, and arrangement.

TARSUS

Seven bones—the talus, the calcaneus, the navicular or tarsal scaphoid, three cuneiforms (medial, intermediate, and lateral, or first, second, and third), and the cuboid—form the tarsus. They are arranged in two rows, with one bone between. In the posterior row the talus sits upon the calcaneus; in the distal row the cuneiforms and the cuboid lie side by side; and the navicular lies between the two rows on the medial side. Only one of these bones, the talus, enters into the articulation with the bones of the leg, and therefore all the weight on the foot is transmitted through this bone to the others and to the points of contact with the ground. Normally, the calcaneus and the heads of the five metatarsals are the weight-bearing points of the foot; between these points the skeleton of the foot is arched, the arch being much higher on the medial side than it is laterally (Fig. 9-59).

Talus
The body of the talus (astragulus), bearing the pulleylike trochlea tali, is separated from the smooth convex head by the slightly constricted, roughened neck. The trochlea is slightly wider in front than it is behind; on its lateral surface, below the articular cartilage, is the projecting lateral process, which gives rise to the lateral talocalcaneal ligament, and on its medial surface, below the smaller area of articular cartilage here, is a roughened depression for the attachment of the deltoid ligament.

The narrower part of the body projecting behind the trochlea is the posterior process. It ends in medial and lateral tubercles, which are separated by the groove for the tendon of the flexor hallucis longus.

On the inferior surface of the body is the oblique sulcus tali, which with the corresponding sulcus calcanei forms the sinus tarsi or canalis tarsi. (The sinus tarsi is often described as consisting of two parts, the narrow

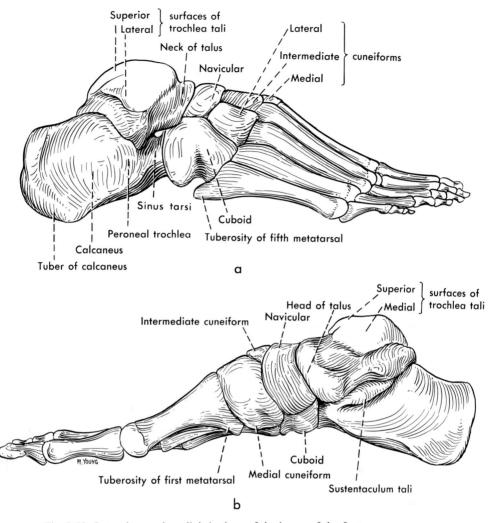

Superior ⎱ surfaces of
Lateral ⎰ trochlea tali
Neck of talus
Navicular
Lateral ⎱
Intermediate ⎰ cuneiforms
Medial
Sinus tarsi
Cuboid
Peroneal trochlea
Tuberosity of fifth metatarsal
Calcaneus
Tuber of calcaneus

a

Superior ⎱ surfaces of
Medial ⎰ trochlea tali
Head of talus
Navicular
Intermediate cuneiform
A.YOUNG
Tuberosity of first metatarsal
Medial cuneiform
Cuboid
Sustentaculum tali

b

Fig. 9-59. Lateral, *a*, and medial, *b*, views of the bones of the foot.

medial part between the sulci just mentioned, which is then called the tarsal canal, and the wider lateral space between talus and calcaneus that leads into the tarsal canal, to which the term "sinus tarsi" is then restricted.) Posterior to the sulcus tali is the large oval posterior calcaneal articular surface, and anterior to it is the middle calcaneal articular surface (which rests upon the sustentaculum tali of the calcaneus).

The anterior calcaneal articular surface is on the inferior surface of the head of the talus, and is continuous with the convex articular surface of the head; it may be, but usually is not, continuous across the neck with the mid-

dle articular surface. The head articulates mostly with the navicular bone, but a lower medial part lies directly on the plantar calcaneonavicular (spring) ligament.

Calcaneus

The calcaneus, the largest of the tarsal bones, lies beneath the talus and supports it (Figs. 9-59, 9-60, and 9-61). The posterior talar articular surface is on the upper surface about the middle of the bone; anterior to this is the sulcus calcanei, which helps to form the canal portion of the sinus tarsi; anterior to this, on the upper surface of the medially projecting sustentaculum tali, is the middle talar articu-

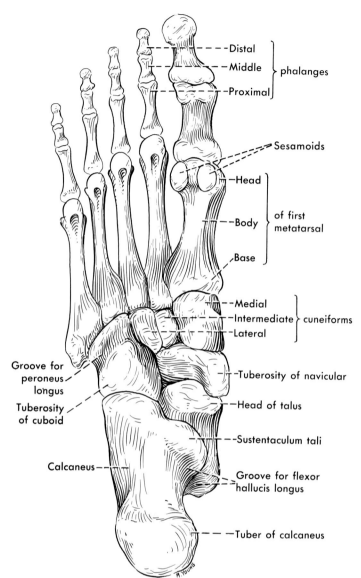

Fig. 9-60. Bones of the foot, plantar view.

-Distal
-Middle } phalanges
-Proximal

Sesamoids

-Head]
-Body } of first metatarsal
-Base]

-Medial
-Intermediate } cuneiforms
-Lateral

Groove for peroneus longus
Tuberosity of cuboid

-Tuberosity of navicular

-Head of talus

-Sustentaculum tali

Calcaneus-

Groove for flexor hallucis longus

-Tuber of calcaneus

lar surface; and most anteriorly is the anterior talar articular surface, sometimes continuous with the middle one. The anterior end of the bone forms the cuboid articular surface.

On the medial surface, behind and below the sustentaculum tali, is the groove for the tendon of the flexor hallucis longus. On the lateral surface, in well-developed specimens, there is a small projection, the peroneal trochlea, for the attachment of the inferior peroneal retinaculum between the tendons of the peroneus longus and peroneus brevis. Be-

hind and below this there is often a slight groove for the tendon of the peroneus longus.

The downward projection at the back end of the inferior surface is the tuber calcanei. It carries rounded medial and lateral tuberal processes, which support the weight put upon the heel.

The calcaneus and talus are frequently referred to as the hindfoot, the remainder of the foot then being called the forefoot. All the muscular attachments into the hindfoot are into the calcaneus (Figs. 9-62 and 9-63).

Navicular Bone

The navicular (scaphoid) bone is distinguished by its concave posterior articular surface, for accommodation of the head of the talus, and its convex anterior articular surface for the three cuneiforms. Medially and inferiorly, the rough tuberosity of the navicular bone marks the attachment of a major portion of the tibialis posterior tendon.

Cuneiform Bones

These three bones form, with the cuboid, the distal row of tarsals. The medial or first cunei-

form is the largest, the intermediate or second the smallest. While all three bones are somewhat wedge-shaped in cross section, the medial one differs from the others in that its blunt apex is directed dorsally, while the apices of the other two bones are directed plantarward. The slightly concave posterior surfaces of all three bones articulate with the navicular, and the anterior surface of each articulate, respectively, with the bases of the three medial metatarsals. Their sides are partly roughened for attachments of the interosseous ligaments, but dorsally (except on

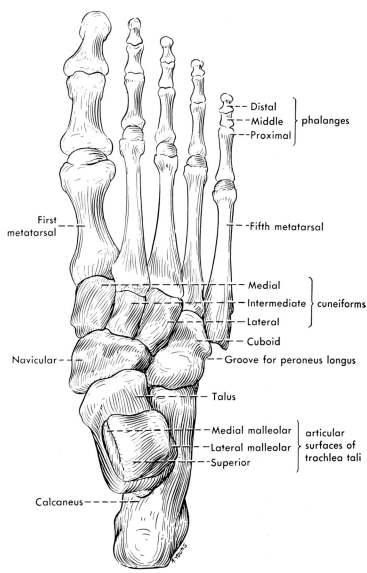

Distal
Middle } phalanges
Proximal

First metatarsal

Fifth metatarsal

Medial
Intermediate } cuneiforms
Lateral

Cuboid

Navicular

Groove for peroneus longus

Talus

Medial malleolar
Lateral malleolar } articular surfaces of trochlea tali
Superior

Calcaneus

Fig. 9-61. Dorsal view of the bones of the foot.

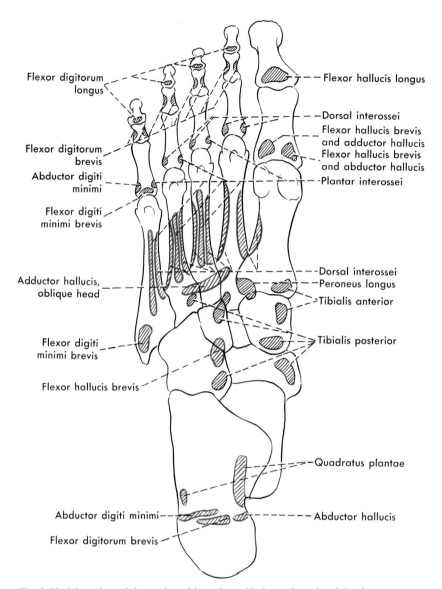

Fig. 9-62. Muscular origins, *red*, and insertions, *black*, on the sole of the foot.

the medial side of the medial cuneiform) there are articular facets for articulation with each other and with the sides of the bases of the metatarsals, and, on the lateral cuneiform, one for the cuboid bone.

Cuboid Bone

The posterior surface of this bone articulates with the calcaneus and its anterior surface articulates with the fourth and fifth metatarsals. On its medial surface there is an articular facet for the lateral cuneiform. The tuberosity of the cuboid is largely on the inferior surface

but extends also onto the lateral surface; in front of it is the groove for the tendon of the peroneus longus muscle.

METATARSALS AND PHALANGES

Except for size, these closely resemble the metacarpals and the phalanges of the hand, respectively.

Metatarsals

The metatarsals are essentially similar, and each consists of a base, a body, and a head.

The proximal surface of the base articulates with a tarsal bone—the first three metatarsals articulate with the three cuneiforms, the fourth and fifth with the cuboid. The bases of the first and second metatarsals are usually united to each other by an interosseous ligament only, but the contiguous surfaces of the bases of the four lateral metatarsals (II to V) bear one or more articular facets for articulation with each other; there are usually two joints, separated by an interosseous ligament, between the second and third metatarsals, and one each between the remainder. The base of the second metatarsal has also on its sides articular facets for the medial and lateral cuneiforms, respectively. Singh found a good deal of variation in the facets of the metatarsals. The bases of the first and fifth metatarsals bear tuberosities, not present on the others. That of the first, on its plantar surface, is simply a small oval projection for the insertion of the peroneus longus tendon, but that of the fifth is an obvious posterolateral projection. It sometimes develops from a separate center of ossification, in which case the epiphyseal line can be mistaken for a fracture; if this separate center fails to unite with the rest of the bone it forms an accessory bone, the os vesalianum.

The bodies of the first four metatarsals tend to be somewhat triangular in cross section, especially proximally, while that of the fifth tends to be flattened in a dorsolateral to a medioplantar direction. The inferior convexity of the body of each metatarsal is exaggerated by the projection of the base and head below the level of the body. As a metatarsal tapers toward its head, its dorsal surface tends to pass onto its medial side.

The convex articular surface of the head of the metatarsal extends more onto the plantar than the dorsal surface of the bone, and is usually grooved on its plantar aspect for accommodation of the flexor tendons; the head of the first metatarsal likewise bears a pair of articular facets for the two sesamoid bones of the metatarsophalangeal joint. On the dorsal aspect of the flattened sides of the head of each metatarsal are tubercles for the attachment of the collateral ligaments, and usually

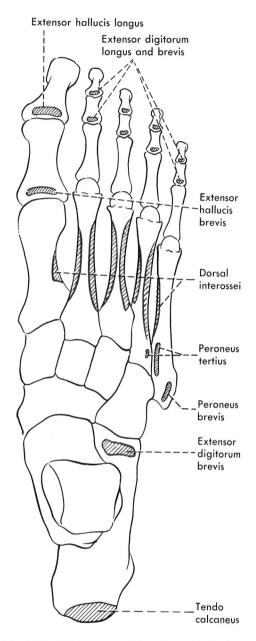

Fig. 9-63. Origins, *red*, and insertions, *black*, of muscles on the dorsum of the foot.

there is a groove for the ligament extending from the tubercle.

Shortness of the first metatarsal, hypermobility of it, and a more proximal position than usual of its sesamoids, were said by Morton ('35) to throw more weight than usual onto the second metatarsal, or this and the third, with consequent distortion of the arch and, in

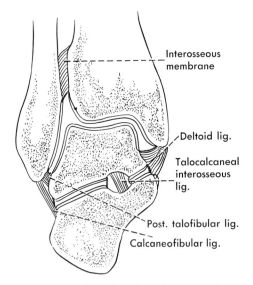

Interosseous
membrane

Deltoid lig.

Talocalcaneal
interosseous
lig.

Post. talofibular lig.

Calcaneofibular lig.

Fig. 9-64. Frontal section through the ankle (talo-crural) and subtalar joints. Synovial membrane is shown by *red*. The interosseous ligament shown here is also called the ligament of the tarsal canal.

some cases, pain as a result of functional disorder of the foot. Harris and Beath, in an investigation of the feet of 3,619 Canadian soldiers, found a short first metatarsal and variations in the positions of the sesamoids to be rather common, but found no correlation between these and evidence of excess pressure on the more lateral metatarsals, nor any correlation between them and the development of objective and clinical symptoms of derangement of the longitudinal arch. They were unable to obtain evidence of a hypermobile first metatarsal.

Bunion formation at the metatarsophalangeal joint of the big toe is accompanied by a valgus position of the toe. According to Haines and McDougall, the essential lesion in hallux valgus is a stretching of the ligaments of the medial side of the metatarsophalangeal joint and an erosion of the ridge on the metatarsal head that separates the grooves for the two sesamoids.

Phalanges

Although the proximal phalanges are much longer and larger than the middle and distal ones, they are much smaller than those of the fingers. The middle phalanges of the toes are

very short, and may have little form; the distal ones, except that of the big toe, are tiny nodules of bone.

The base of each proximal phalanx is concave proximally for articulation with the head of the metatarsal; the body tends to be concave in a plantar direction; and the head is pulley-shaped, with an articular surface that extends well onto the plantar surface but hardly at all onto the dorsum. The base of each middle phalanx is somewhat saddle-shaped, and its head is pulley-shaped like that of the proximal phalanx. Each distal phalanx has a somewhat saddle-shaped base, almost no body, and a terminal expanded tuberosity.

ANKLE (TALOCRURAL) JOINT

The ankle joint is formed by the tibia above and medially, the fibula laterally, and the trochlea of the talus below the tibia and between the malleoli (Fig. 9-64). The inferior surface of the tibia articulates with the superior surface of the trochlea tali, and the weight of the body is transmitted mostly through these surfaces; however, Lambert ('71) has apparently shown that about a sixth of the load is transmitted to the talus by the fibula. The medial malleolus of the tibia projects downward on the medial side of the ankle to articulate with the medial surface of the trochlea, the lateral malleolus of the fibula projects downward to articulate with the lateral surface of the trochlea, and these give lateral stability to the ankle joint.

Since the trochlea resembles a segment of a cylinder, the ends of which are embraced by the malleoli, the ankle joint is almost strictly a hinge one. However, the medial border of the superior trochlear surface has a smaller arc anteriorly than it does posteriorly, while the lateral border presents a single arc (Barnett and Napier); thus the free foot tends to pronate or evert slightly at the talocrural joint during dorsiflexion, and to supinate slightly during plantar flexion (Barnett and Napier; Hicks, '53).

Although the trochlea is gripped between the two malleoli, the tightness of the grip varies with the position of the joint, for both the trochlea and the inferior articular surface of the tibia are wider anteriorly than posteriorly, and the distance between the medial and lateral malleoli is greater anteriorly than posteriorly. In consequence, when the foot is plantar flexed the narrower posterior part of the talus is brought into contact with the wider anterior part between the malleoli, and a small amount of lateral movement is then permitted when the foot is not bearing weight. As the foot is returned to the normal position at a right angle to the leg, and then dorsiflexed, the broader part of the trochlea is brought into contact with the narrower portion between the malleoli, and the trochlea is therefore gripped increasingly tightly, and according to Weinert and co-workers the fibula moves upward to accommodate this broader surface. A further consequence of the narrower dimension of the posterior part of the ankle joint is that the joint tends to resist anterior subluxation: since the center of gravity of the body falls somewhat in front of the ankle joint, there should be a tendency for the tibia and fibula to dislocate forward on the talus, and this is accentuated by standing upon the toes; however, any such movement drives the narrower space between the malleoli toward the wider part of the talus, so that the talus is gripped ever more firmly between these; the malleoli cannot be separated appreciably since the tibia and fibula are closely bound together by the inferior tibiofibular ligaments. The joint is further tightened, during running, by a downward movement of the fibula at the strike phase; this also tightens the interosseous membrane, and thus transfers more force from the fibula to the tibia (Weinert and co-workers).

The mediolateral axis of the hinge joint of the ankle is usually not quite parallel to that of the knee (Hutter and Scott), because of torsion between the upper and lower ends of the tibia. The torsion is typically an external one, of about 20° according to Hunter and Scott, but they expressed the opinion that a variation of 20° in either direction, from no torsion to an external one of 40°, is compatible with normal appearance and function. Apparently, most of the torsion develops between birth and the age of 7 years; abnormal internal torsion, which is more common, may result from an infant sleeping prone with legs curled under him; abnormal external, especially in paraplegic children, from gravity and the weight of bedclothes.

Since most persons toe out about 10°, the axis of the talocrural joint is normally at about a 120° angle, measured externally, to the line of progression, instead of what would seem to be a more logical 90° angle. The fact that the ankle joint is not normally at a right angle to the line of progression can easily be demonstrated by standing on both feet and bending the knees: in most individuals, unless they are standing pigeon-toed, the knees diverge as they are bent.

Ligaments

The capsule of the ankle joint attaches to the edges of the articular cartilages. In the usual fashion of hinge joints, the capsule is thin and weak anteriorly and posteriorly, but is strengthened by special ligaments on the sides (Figs. 9-64 and 9-65).

The strong *medial or deltoid ligament* presents somewhat the form of a triangle with its base below on the tarsals, and a very blunted apex above attached to the medial malleolus; its fibers radiate from the tibia to the navicular, the calcaneus, and the talus. Although all the fibers of the deltoid ligament arise together from the tibia, four parts of this are distinguished according to the lower attachment of the fibers. The most anterior fibers pass downward and forward to attach into the upper and medial surface of the navicular bone, and constitute the pars tibionavicularis of the deltoid ligament; at their lower border these fibers blend anteriorly with the plantar calcaneonavicular ("spring") ligament. Deep to these fibers there are others that pass from the tibia to the neck of the talus, and these constitute the pars tibiotalaris anterior. Succeeding the tibionavicular ligament posteriorly on the surface of the deltoid ligament, and arising from the apex of the medial malleolus, are fibers, the pars tibiocalcanea, that attach to the sustentaculum tali; the anterior tibiotalar fibers attaching to the medial side

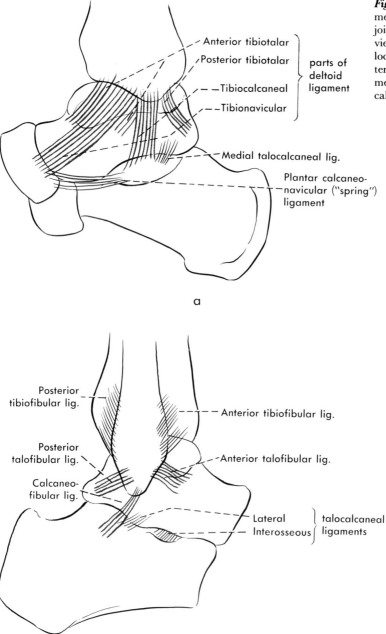

Anterior tibiotalar ⎤
Posterior tibiotalar ⎬ parts of deltoid ligament
Tibiocalcaneal ⎥
Tibionavicular ⎦

Medial talocalcaneal lig.

Plantar calcaneo-navicular ("spring") ligament

a

Posterior tibiofibular lig.

Anterior tibiofibular lig.

Posterior talofibular lig.

Anterior talofibular lig.

Calcaneo-fibular lig.

Lateral ⎤
Interosseous ⎦ talocalcaneal ligaments

b

Fig. 9-65. Diagrams of the ligaments of the ankle and subtalar joints in medial and lateral views. The ligaments of the talocrural joint are *red*. The interosseous talocalcaneal ligament shown here is the one also called the cervical ligament.

of the talus are continued also deep to this part. The posterior part of the deltoid ligament consists of superficially placed fibers that attach to the posterior part of the medial surface of the talus, including the medial tubercle (the projection medial to the groove for the tendon of the flexor hallucis longus muscle); these form the pars tibiotalis posterior.

The *lateral ligaments* of the ankle consist of three distinct bands instead of a broad one with several parts, and each of these bands is therefore named as a separate ligament. These three ligaments are the posterior talofibular, the calcaneofibular, and the anterior talofibular ligaments.

The posterior talofibular ligament arises from the back part of the lateral malleolus and runs backward and slightly downward to attach into the lateral tubercle of the posterior process of the talus (lateral to the sulcus for the flexor hallucis longus tendon). This is the strongest of the three lateral ligaments and helps to resist forward dislocation of the leg on the foot. The calcaneofibular ligament is a narrow bundle of fibers that runs downward and slightly posteriorly to insert into a small tubercle present on the upper part of the lateral surface of the calcaneus. This ligament is the longest of the three. The anterior talofibular ligament arises from the anterior border of the lateral malleolus and passes forward and somewhat medially to attach to the neck of the talus.

The medial and lateral ligaments all contribute to the lateral stability of the ankle joint, although probably in varying degree. The deltoid ligament, being the only one on the medial side of the ankle, tends to resist eversion, and some part of it is torn in eversion sprains of the ankle. Sectioning it alone in the cadaver produces lateral instability of the ankle (Anderson and co-workers). Anderson and his colleagues, and Leonard agreed that many workers have regarded the calcaneofibular ligament as the most important of the lateral ones, but they themselves regarded the anterior talofibular as the most important, for they found that cutting it, although leaving the foot fairly stable when it is at 90° to the leg (but even in that position, according to Anderson and his co-workers, allowing some anterior displacement of the foot on the leg) deprives it of a good deal of stability when it is plantar flexed. In this position the talus is not gripped so tightly by the malleoli, hence lateral instability appears; further, according to Anderson and his co-workers, anterior displacement of the talus on the leg is more extensive, and may amount to 7 mm or 8 mm on pressing the foot forward. They presented two clinical cases in which this ligament had been ruptured and the talus was subluxated anteriorly. Leonard pointed out

also that while an inversion injury with the foot at 90° to the leg is primarily to the calcaneofibular ligament, inversion in plantar flexion, which is how most injuries are produced, throws a strain upon the anterior talofibular ligament; he said that, clinically, that is the one most frequently injured. Nevin and Post agreed that the first ligament torn in inversion and plantar flexion is the anterior talofibular ligament, but found that in pure inversion sprains both the calcaneofibular and the anterior part of the posterior talofibular ligament were regularly injured. Leonard said that cutting the calcaneofibular ligament allowed abnormal movement at the subtalar joint at 90°, but that the foot was stable in plantar flexion. He and Anderson's group agreed that cutting both ligaments produced a markedly unstable foot; Anderson and his co-workers reported lateral, but no anteroposterior, instability from cutting the posterior talofibular and the calcaneofibular ligaments separately.

Leonard apparently did not find lateral instability to result from cutting either ligament alone, but said that cutting the posterior talofibular ligament allowed abnormal dorsiflexion of the ankle, and that cutting this and the calcaneofibular together increased the dorsiflexion; although the foot was still stable when plantar flexed, it was unstable "in the longitudinal axis of the talus" (presumably it could be subluxated posteriorly) when it was at 90°.

Colonna and Ralston have described in some detail anterior, posterior, medial, and lateral surgical approaches to the ankle joint.

Nerves and Vessels

The ankle joint is supplied chiefly by twigs from the deep peroneal and tibial nerves (Casagrande and co-workers), to which Gardner and Gray added the saphenous nerve. Casagrande and co-workers found branches from the deep peroneal nerve sometimes arising above its terminal branching and going to the anterior aspect of the capsule, and other branches regularly arising from the terminal lateral branch on the dorsum of the foot and running recurrently to the

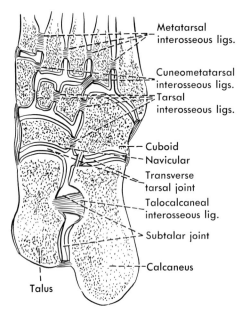

Fig. 9-66. Oblique longitudinal section through the foot. Synovial membrane is *red*.

front of the joint. Sometimes there was also a twig from the medial terminal branch of the deep peroneal. These workers described the branches of the tibial nerve, to the posterior aspect of the joint, as being typically one or two that arise from the nerve before its bifurcation into medial and lateral plantars, and course downward some 2 cm or 3 cm to reach the joint; a transverse twig from the posterior branch of the medial calcaneal nerve, supplying the posterolateral part of the capsule; and a small and inconstant branch from the lateral plantar nerve close to its origin.

The blood vessels supplying the joint are from the malleolar branches of the anterior tibial and peroneal arteries.

INTERTARSAL JOINTS

Since movements at the ankle joint are limited almost strictly to plantar flexion and dorsiflexion, it is the intertarsal joints that add to the foot the additional mobility that allows it to be inverted and adducted (supinated) or everted and abducted (pronated). Further, because the foot is so arched that the weight transmitted to it is in turn transmitted

in part to the calcaneus, in part forward to the heads of the metatarsals (these being the weight-bearing parts of the foot in contact with the ground), most of the intertarsal joints are under particular strain during weight-bearing, and must be strongly braced if the arches of the foot are not to fail. For this reason, the plantar surface of the foot is provided with particularly strong intertarsal ligaments that bind the bones together and tend to prevent collapse of the arch; in addition, the short muscles of the foot, the plantar aponeurosis, and the long tendons passing into the sole of the foot from the leg all tend to help maintain the intertarsal joints in normal positions.

Subtalar Joint

The anatomic subtalar (once talocalcaneal) joint lies between the posterior calcaneal articular surface on the body of the talus and the posterior talar articular surface of the calcaneus, but the physiologic and clinical one includes the anterior articulations, part of the talocalcaneonavicular joint, between the middle and anterior facets on the lower surface of the talar head, and the corresponding articular surfaces of the calcaneus. The capsules of these two joints are separated from each other by parts of the inferior extensor retinaculum and by ligaments (usually called the talocalcaneal interosseous ligament and so labelled in Fig. 9-66) that occupy the sinus tarsi and the tarsal canal. Smith ('58) pointed out that the interosseous ligament in the tarsal canal is separate from that in the wider sinus tarsi proper and called it the ligament of the tarsal canal; it runs obliquely downward and laterally from the sulcus tali to the sulcus calcaneus (Fig. 9-67). Smith regarded it as limiting eversion at the subtalar joint, but Cahill described it as limiting neither eversion nor inversion. Thin areolar tissue intervenes between it and the capsule of the talocalcaneal joint, and thicker tissue containing blood vessels separates it from the capsule of the talocalcaneonavicular joint. In addition to these structures in the tarsal canal, a part of the inferior extensor retinaculum (cruciate ligament) also extends into the canal from the

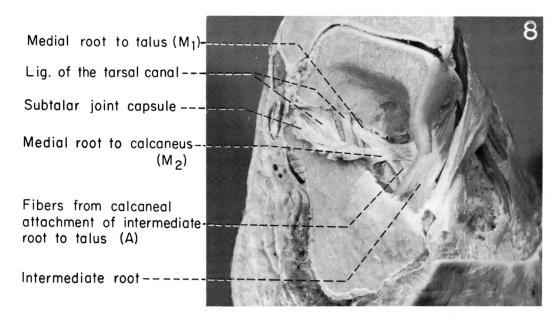

Medial root to talus (M₁)

Lig. of the tarsal canal

Subtalar joint capsule

Medial root to calcaneus (M₂)

Fibers from calcaneal attachment of intermediate root to talus (A)

Intermediate root

Fig. 9-67. A posterolateral view of a dissection of a right foot, showing the ligament of the tarsal canal (here composed of two bundles instead of the usual one) and the roots of the inferior extensor retinaculum, the lateral or superficial root of which is unlabeled. The cervical ligament (Fig. 9-68) is here hidden by the roots of the retinaculum. *A* is the oblique talocalcaneal band. (Fig. 8, Cahill DR: Anat Rec 153:1, 1965)

sinus tarsi to attach into the sulcus calcanei either anterior or posterior to the ligament of the canal and, according to Cahill, attach also to the talus.

Smith called the second part of the interosseous talocalcaneal ligament the cervical ligament. This band of fibers (Fig. 9-68) lies in the anterior part of the sinus tarsi proper (the wider lateral portion) and extends upward and medially from the calcaneus to a tubercle on the inferolateral aspect of the neck of the talus. It tends to resist inversion at the subtalar joint. This part of the sinus tarsi is also traversed by parts of the inferior extensor retinaculum, one part attaching here to the calcaneus behind the cervical ligament, and sending a strong attachment into the tarsal canal (Cahill's oblique talocalcaneal band) to attach to the talus. Lateral to the cervical ligament, some fibers of the extensor digitorum brevis arise from the calcaneus.

Talocalcaneonavicular Joint

Although this is one continuous joint cavity, its talocalcaneal part is also subtalar. This part lies anteromedial to the sinus tarsi, between articular facets on the lower surface of the head of the talus and corresponding ones on the upper surface of the sustentaculum tali, and functions with the subtalar joint. The talonavicular portion, formed between the anterior convex surface of the head of the talus and the posterior concave surface of the navicular bone, is functionally a part of the transverse tarsal joint (see below). The integrity of the talonavicular part of the joint is particularly important to the integrity of the medial side of the longitudinal arch of the foot, since the head of the talus, supported below by the calcaneus, must in turn support the navicular, which is the keystone of the arch. The capsule of this joint is thin dorsally, as are those of all the intertarsal joints, but is strengthened by fibers that extend from the neck of the talus to the dorsal surface of the navicular bone as the talonavicular ligament. Medially, this fuses with the more anterior fibers of the deltoid ligament going to the navicular bone.

Dorsolaterally a much stronger ligament,

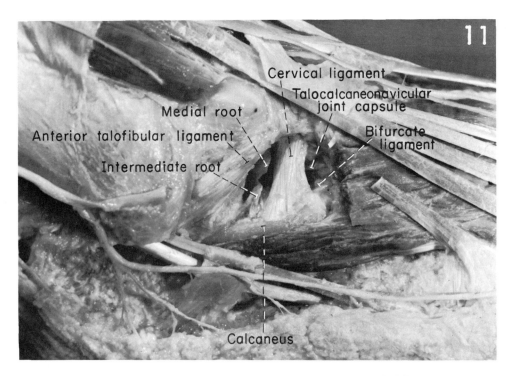

Fig. 9-68. Dissection of the right foot from the lateral side to show the cervical ligament. The remains of the medial and intermediate roots of the inferior extensor retinaculum as they pass posterior to the cervical ligament are identified. (Fig. 11, Cahill DR: Anat Rec 153:1, 1965)

the calcaneonavicular (a part of the bifurcate ligament) reinforces the capsule. The bifurcate ligament (Fig. 9-69) is attached to the upper anterior surface of the calcaneus and runs forward to divide into two parts, the calcaneonavicular ligament, attaching to the lateral surface of the navicular, and the calcaneocuboid ligament, attaching to the adjacent dorsomedial part of the cuboid. Medially, the joint capsule is strengthened by the tibionavicular part of the deltoid ligament.

The most imporant ligament of this joint is the heavy plantar calcaneonavicular ligament or "spring" ligament; it is attached posteriorly to the anterior end and the medial border of the sustentaculum tali, and passes across the ventromedial aspect of the talonavicular joint, spreading out as it does so, to attach to the entire inferior and much of the medial surface of the navicular bone. Part of the articular surface of the head of the talus rests directly on the inner surface of this ligament, which is said to be partly provided with

articular cartilage (Hardy) and partly lined by synovial membrane. It is particularly important in the support of the medial side of the arch. If it gives way it allows the head of the talus, forced downward by the weight of the body, to insinuate itself between the calcaneus and the navicular and thus both flatten the arch on its medial side and force the anterior part of the foot into abduction.

Although the "spring" ligament has been said to contain elastic tissue, Hardy found none. Downward movement of the head of the talus during eversion of the foot was found to occur as a result of rotation of the navicular bone carrying the ligament downward and laterally, not as a result of increased length of the ligament.

Calcaneocuboid Joint

The calcaneocuboid joint is a high point on the lateral side of the arch, and in consequence needs special reinforcement as does the talonavicular joint. The joint surface on the anterior end of the calcaneus is concave in one direction, convex in the other, but neither of these directions corresponds to either the mediolateral or the supero-inferior axis of the foot; the posterior articular surface of the cuboid is reciprocally shaped. The joint receives one relatively strong enforcement dorsally, the calcaneocuboid ligament, already mentioned as part of the bifurcate ligament.

The plantar surface of the joint is strengthened by two series of special fibers: Some of these belong to the plantar calcaneocuboid (short plantar) ligament, which runs from the anterior part of the lower surface of the calcaneus to the tuberosity of the cuboid; superficial to this and separated from it by loose connective tissue is the long plantar ligament (see Fig. 9-71). This, a band of considerable

strength, is the chief ligamentous support of the lateral part of the arch. It arises from much of the entire lower surface of the calcaneus in front of the tuber; its deeper fibers attach to the tuberosity of the cuboid bone (behind the groove for the peroneus longus tendon), while more superficial fibers extend across this groove, thus enclosing the peroneus longus tendon in a tunnel, to attach to the bases of the lateral three or four metatarsal bones.

Transverse Tarsal Joint

The transverse tarsal joint (Figs. 9-66 and 9-70) has two separate joint cavities, that of the calcaneocuboid and that of the anterior or talonavicular part of the talocalcaneonavicular joint. The two together form a functional

Fig. 9-69. Ligaments of the dorsum of the foot. The tarsometatarsal ligaments are *red*.

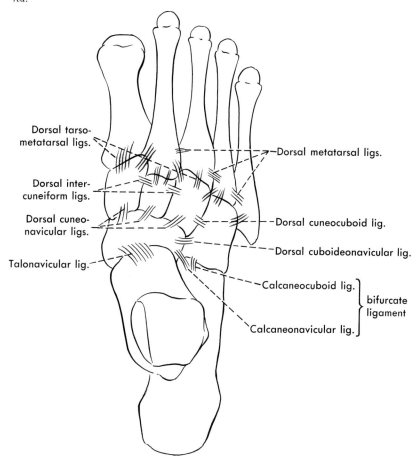

Dorsal tarso-metatarsal ligs.

Dorsal inter-cuneiform ligs.

Dorsal cuneo-navicular ligs.

Talonavicular lig.

Dorsal metatarsal ligs.

Dorsal cuneocuboid lig.

Dorsal cuboideonavicular lig.

Calcaneocuboid lig.
Calcaneonavicular lig. } bifurcate ligament

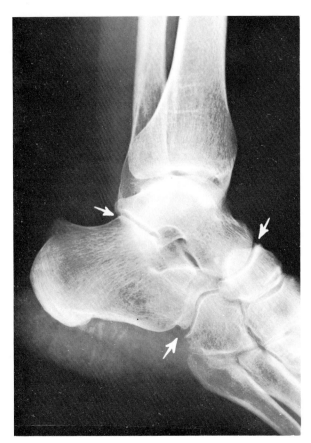

Fig. 9-70. Lateral roentgenogram of the foot. The *arrow* on the left indicates the subtalar joint, the other two *arrows* the transverse tarsal joint.

joint that permits movement of the anterior part of the foot on the posterior part. The slight amount of dorsiflexion and plantar flexion possible at this joint add to the much greater mobility in this respect obtainable at the ankle joint; also, the oblique situation of the joint adds during plantar flexion to the inversion obtainable at the subtalar joint, and during dorsiflexion to the eversion obtainable at this joint. The obliquity also accounts for the fact that eversion (pronation) and inversion (supination) are accompanied by abduction and adduction, respectively. Hicks ('53), for instance, showed that the typical movement of the fore part of the foot is a combined dorsiflexion, pronation or eversion, and abduction, or a combined plantar flexion, supination or inversion, and adduction.

The ligaments connected with the transverse tarsal joint are primarily those that have already been described as uniting the talus and the calcaneus to each other and to the navicular and cuboid bones. Neither of the two synovial cavities forming the transversal tarsal joint is continued anteriorly, since the contiguous surfaces of the navicular and cuboid bones are united by an interosseous cuboideonavicular ligament. There are also dorsal and plantar cuboideonavicular ligaments, in the positions indicated by their names.

Other Intertarsal Joints

The remaining intertarsal joints are between the navicular and the cuneiform bones, the cuboid and the cuneiforms, and between the cuneiforms. These may all form one continuous joint cavity, called the cuneonavicular, or the cuneocuboid may be separate.

The cuneonavicular part of the joint lies between the convex anterior surface of the navicular and the concave posterior surfaces of the three cuneiform bones. The capsule of this joint is strengthened dorsally and medially by rather inconspicuous dorsal cuneonavicular ligaments that pass from the navicular to the three cuneiforms; its plantar surface is supported by stronger plantar cuneonavicular ligaments, which are fused with the deep surface of the anterior continuation of the tendon of the tibialis posterior.

The cuneocuboid joint contains a relatively small cavity, which usually opens posteriorly into the cuneonavicular joint. There are dorsal and plantar cuneocuboid ligaments, and a strong interosseous one that fills up most of the anterior part of the joint.

The intercuneiform joints are extensions forward, between the three cuneiforms, from the cuneonavicular joint. The dorsal intercuneiform ligaments are weak transverse bands; the plantar intercuneiform ligaments are much stronger, and fuse with the strong interosseous intercuneiform ligaments that unite the adjacent nonarticular surfaces of the two bones. If both of these interosseous ligaments are complete, they separate the cuneonavicular joint from communication with the tarsometatarsal joints. Commonly, however, the interosseous ligament between the medial and intermediate cuneiforms does not extend dorsally to reach the dorsal intercuneiform ligament, and the cavity of the joint is therefore extended forward to communicate with the intermediate of the three tarsometatarsal joints.

Nerve and Blood Supply

The dorsal aspects of the intertarsal joints are supplied chiefly by branches from both the superficial and deep peroneal nerves (Gardner and Gray). The lateral branch of the deep peroneal nerve is largely articular and spreads widely over the dorsum of the foot. The medial branch of the nerve also gives off articular branches that go to the more medial joints, and the saphenous nerve is said (Winckler, '37) to help supply the joints on the medial border of the foot. A branch from the superficial peroneal which may accompany the peroneus longus and brevis to the lateral side of the foot (Winckler's accessory deep peroneal nerve) also supplies some of the more lateral joints when it is present (Winckler, '34; Gardner and Gray). The lateral and medial plantar nerves give twigs to the plantar aspects of the joints.

The blood supply is from branches of all the larger vessels of the foot: the dorsalis pedis and its branches, the medial and lateral plantars, and the plantar arch.

TARSOMETATARSAL AND INTERMETATARSAL JOINTS

The tarsometatarsal joints lie between the bases of the metatarsals and the front ends of the three cuneiforms and of the cuboid bone. In the articulation between the four tarsals and the five metatarsals, three synovial cavities are formed (Fig. 9-66): The medial one is between the medial cuneiform and the first metatarsal; it is separated from the part of the joint between the second cuneiform and second metatarsal by an interosseous ligament that links the medial cuneiform to the second metatarsal. The intermediate cavity lies between the medial and intermediate cuneiforms and the second and third metatarsals; it usually forms one continuous cavity with the intercuneiform and cuneonavicular articulations, and forward offshoots of this cavity also form the intermetatarsal joints between the second and third and the third and fourth metatarsals. The lateral tarsometatarsal joint is formed by the adjacent surfaces of the cuboid and of the bases of the fourth and fifth metatarsals; it is separated from the intermediate compartment by an interosseous ligament that connects the anterolateral corner of the lateral cuneiform to the base of the fourth metatarsal. A forward projection of this cavity forms the intermetatarsal joint between the fourth and fifth metatarsals.

The ligaments of the tarsometatarsal joints are dorsal, plantar, and interosseous tarsometatarsal ligaments. The dorsal ones usually unite the first metatarsal to the medial cuneiform, the second to all three cuneiforms, the third to the lateral cuneiform, the fourth to the lateral cuneiform and the cuboid, and the fifth to the cuboid (Fig. 9-69). The plantar tarsometatarsal ligaments are less regularly placed but consist likewise of longitudinally and obliquely running fibers. The interosseous cuneometatarsal ligaments are regularly at least two in number: the medial one unites the medial cuneiform to the second metatarsal, thus separating the medial tarsometatarsal joint from the intermediate one, and the lateral unites the lateral cuneiform to the fourth metatarsal, thus separating the intermediate joint from the lateral one. There may be a ligament between the anteromedial corner of the third cuneiform and the lateral aspect of the base of the second metatarsal, but it is incomplete if present, and does not divide the intermediate part of the joint cavity into two.

These joints and the following have the same nerve and blood supply as the intertarsal joints.

Intermetatarsal Joints

As already noted, the synovial cavities of the intermetatarsal joints are continuous proximally with those of the tarsometatarsal joints. The four lateral metatarsals are connected to each other by dorsal metatarsal ligaments, which pass transversely between the dorsal surfaces of the bases of the bones, and by plantar metatarsal ligaments that have a similar arrangement on the plantar surface. They are also connected by strong interosseous ligaments that bind together the adjacent surfaces of the bases and limit the forward extent of the synovial cavities. There is no joint cavity between the second and first metatarsals, and these two are bound together by an interosseous ligament only. It is said, however, that a bursa sometimes develops in this location in the midst of the interosseous fibers and communicates with the medial tarsometatarsal joint.

The interosseous ligaments between the bases of the four lateral metatarsals are particularly strong and firmly bind these together; that between the first and second metatarsals is rather weak.

Only slight gliding movements are allowed at the distal intertarsal, the tarsometatarsal, and the intermetatarsal joints, and the forepart of the foot moves as a unit at the transverse tarsal joint.

In addition to their articulations at their bases, the metatarsals are also united at their heads by a strong *deep transverse metatarsal ligament* (Fig. 9-71). This runs transversely between the heads of the adjacent metatarsals, its anterior surface blending with the plantar ligaments at the metatarsophalangeal articulations. As is true of the corresponding ligament in the hand, this ligament separates the lumbrical tendons, which run below it, from those of the interossei, which pass above it.

THE FOOT AS A WHOLE

The foot is arched longitudinally and, especially in the region of the bases of the metatarsals and the distal row of tarsals, arched transversely, and these arches have been described in various ways. The medial side (pars medialis) of the longitudinal arch is higher than the lateral side, and is usually described as passing from the tuber calcanei forward and upward through the sustentaculum tali, the head of the talus, the navicular, and the three cuneiforms to the heads of the three medial metatarsals; similarly, the lateral part of the arch, starting posteriorly at the tuber calcanei, passes through the cuboid and the two lateral metatarsals to their heads.

The transverse arch is more difficult to define. The truly transverse portion is in the region of the tarsometatarsal joints, where the distal row of tarsals and the bases of the metatarsals form a distinct arch with the concavity directly downward and medially. Posterior to this, the concavity of the plantar surface resembles a segment of a dome, so that when the two feet are held together a half dome is formed. Anteriorly, it flattens rapidly so that the inferior surfaces of the heads of the metatarsals are all in the same plane, where they share in weight bearing.

Distribution of Weight in the Foot

In quiet standing, approximately half of the weight on the limb is said to be borne by the calcaneus, half by the heads of all five of the metatarsals, and the weight on the forepart of

Fig. 9-71. Ligaments of the plantar surface of the foot. Tarsometatarsal ligaments, and those intertarsal ligaments that lie deep to the long plantar ligament, are *red*.

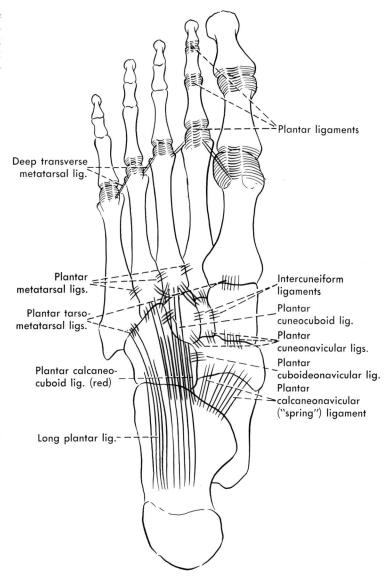

Plantar ligaments

Deep transverse metatarsal lig.

Plantar metatarsal ligs.

Intercuneiform ligaments

Plantar tarso-metatarsal ligs.

Plantar cuneocuboid lig.

Plantar cuneonavicular ligs.

Plantar cuboideonavicular lig.

Plantar calcaneo-cuboid lig. (red)

Plantar calcaneonavicular ("spring") ligament

Long plantar lig.

the foot is proportioned among the heads of the metatarsals in approximately the proportions of 2:1:1:1:1, counting the metatarsals from one to five: that is, the first metatarsal bears twice the weight of any of the others, and one third of the total weight upon the forepart of the foot (Morton and others).

Stott and co-workers did not agree with this. They found the load fairly evenly distributed across the forefoot in quiet standing, and while the maximal load on the forefoot during walking was sometimes on the first metatarsal, it was sometimes greater on the second or third. They also reported that while the load on the midfoot in seven normal persons was less than 3% of the body weight in four, and always less than 15%, it was 20.5% and 26% in persons with flatfoot; and they said that, in general, the higher the load on the midfoot, the higher the load on the lateral side of the forefoot.

The usually described distribution of weight seems to be brought about by muscular contraction and is very easily modified by slight changes in the position of the foot and the direction in which it is loaded. For instance, with all muscles relaxed and the knee loaded, 80% of the weight is borne by the calcaneus and only 20% by the metatarsals (Jones, '41), and in normal standing it is the contraction of the triceps surae that brings about the equal distribution of weight between the posterior and anterior ends of the longitudinal arch. The distribution of weight among the metatarsals is also dependent in part upon muscle contraction, for in the loaded foot of a cadaver the first metatarsal bears as much weight as the other four combined, instead of half as much. The very slight and variable activity of the muscles of the leg other than the triceps surae during quiet standing must surely produce variations in the load on various parts of the foot and probably helps to relieve the fatigue of standing.

Manter ('46) attempted to determine the distribution of weight among the tarsals, and concluded from his results that the weight is not simply parceled among the various bones, but rather the entire foot is involved in weight-bearing, and the amount of compression between any two adjacent articular surfaces is a result of the height of the arch at that point; he said that his results did not justify a separation of the longitudinal arch into medial and lateral parts, for the distribution of stresses in the arch follows the mechanical principle that stress should be greater the higher the part.

In walking, where the weight is shifted progressively from the heel to the ball of the foot, the line along which the weight falls is a straight one, which tends to be parallel with the line of progression (Elftman). Thus if one toes slightly out, weight is first borne primarily by the lateral side of the heel, and then shifted forward in a straight line through the head of the first metatarsal; if, however, one toes slightly in, the maximal weight is first borne by the inner side of the heel, and is then shifted forward toward the lateral side of the foot.

Support of the Arch

The stress on the arch is proportional to the weight borne by the ball of the foot, which in quiet standing is 25% of the weight of the body; however, when the weight is shifted onto the ball of one foot, as in taking a step, the stress on the arch is the total weight of the body, therefore four times as great as in standing; moreover, according to Jones ('41), stresses of eight or more times the normal stress may be imposed upon the arch in vigorous exercise. There were once many proponents of the concept that the intrinsic structure of the arch is not adapted for stress, and that the long muscles actively support it, relieving the plantar ligaments of this duty until the muscles fail; then, according to this concept, undue strain is thrown upon the ligaments, which give beneath the unaccustomed strain, producing flatfoot.

In contrast to this concept, Jones pointed out that the tibialis anterior cannot lend much support to the arch during plantar flexion, when support is most needed, and calculated that if the other leg muscles that are supposed to lend support (the peroneus longus, the posterior tibial, the flexor digitorum longus, and the flexor hallucis longus) were solely responsible, those small muscles would have to develop a combined tension of two and a half times that of the large triceps surae, an obvious impossibility. Both Jones and Manter ('46) also showed that the heavy plantar ligaments, including the plantar aponeurosis, are strong enough to support the statically loaded arch. More recently, numerous electromyographic studies have clearly shown that during quiet standing there is no activity in the muscles of the leg that can be associated with support of the arch. Basmajian and Stecko further showed that a load of 200 lb on one knee did not cause significant activity; with 400 lb there was variable activity, especially when the foot was dorsiflexed, plantar flexed, or inverted or everted, but they reported the average activity as "less than slight." In this study they included also plantar muscles of the foot, with the same results; and Mann and Inman, studying the muscles of the foot, found no activity during quiet standing other than short bursts of activity apparently associated with postural adjustments. Thus in the normal foot the bony and ligamentous structures are the primary support of the arch and require no muscular aid until the stress is increased by shifts in the direction of load or by walking. (Basmajian has quoted preliminary evidence that this may not be true in the flat foot.)

The short muscles in the sole of the foot do become very active upon rising on the toes and in walking, acting largely as a unit (Mann and Inman), and are apparently both stabilizing the transverse tarsal joint and adding to the flexor thrust. Muscles of the leg do also, and since some of them act across the transverse tarsal joint must surely aid in this action. In view of their relative inefficiency in supporting the arches, however, it is reasonable to assume that much of their contraction is to bring about the proper distribution of weight throughout the foot. That this is important is shown by the well-known fact that muscle imbalance can markedly distort a foot, and even slight changes in muscle imbalance will shift the distribution of weight in the foot, thereby throwing more strain than usual on certain joints and ligaments: thus weakness of the tibial muscles or relative overcontraction of the peronei will, because of the tendency to pronation so produced, throw more stress than usual on the medial side of the arch, and thus on the supporting ligaments. In this instance, the long muscles can therefore be said to contribute toward flattening of the arch, even though they play no important part in directly upholding the arch.

Interestingly, Wright and Rennels have shown that the plantar aponeurosis (and therefore presumably all the histologically similar plantar ligaments) does elongate slightly, with stiffening resistance, under load, and recover its original length when the load is removed. This slight elasticity would seem particularly suited to support of the foot. Finally, Hicks ('54) pointed out that the plantar aponeurosis probably contributes much to the increase in curvature of the arch during plantar flexion: since standing upon the toes produces dorsiflexion of them, this draws the attachments of the aponeurosis to the plantar ligaments and proximal phalanges forward, thus tightening the aponeurotic bowstring between the phalanges and the calcaneus.

Flatfoot

This is the most common abnormality of the foot. When the lateral border of the arch is low, the base of the fifth metatarsal may participate in weight-bearing (Elftman), but in typical flatfoot the weight is distributed over the whole plantar surface of the foot instead of being concentrated on the calcaneus and the heads of the metatarsals. The changes that bring this about are primarily along the medial side of the foot, and involve stretching of the talonavicular and cuneonavicular joints and an inward rolling (pronation) of the calcaneus (its lateral side being turned up rather than out; in terms of the lower surface of the calcaneus, this movement is referred to as eversion) so that the navicular and the head of the talus rest, through the soft tissues beneath them, upon the ground (e.g., Humphry). This depression of the medial side of the arch of course lengthens this side of the foot, with the result that the forepart of the foot is then forced into abduction.

It is perhaps debatable whether the primary abnormal movement is of the talus and navicular, as Humphry implied, or of the calcaneus, as Milch ('42) and Schoolfield apparently assumed, but the two movements normally go on together. The subtalar joint is so inclined, medially and downward, as to favor displacement of the head of the talus, carrying the navicular with it, if the supporting ligaments (primarily the interosseous talocalcaneal and the plantar calcaneonavicular or spring ligament) begin to give way. Further, as Schoolfield pointed out, the weight transmitted to the calcaneus by the talus is more concentrated on the medial than the lateral side of the calcaneus, and thus has a tendency to force this bone into pronation, a tendency resisted by the deltoid ligament. Obviously, medial and downward slipping of the talus on the calcaneus increases the tendency of the calcaneus to pronate, and pronation of the calcaneus increases the tendency of the talus to slip downward and forward. Milch and Schoolfield both argued, therefore, that flatfoot usually results from an inability of the deltoid ligament to prevent pronation; Milch reported success in treating the condition by reinforcing the deltoid ligament with fascial strips, Schoolfield by shortening the deltoid ligament.

Hallock said that in the usual relaxed flatfoot a shortened triceps surae is often a contributing factor to the excess strain on the plantar ligaments, and that sometimes a variation in the conformity of the subtalar joint contributes. He discussed various surgical procedures, including tendon lengthening, tendon transplants, repair of ligaments, and arthrodeses, which in varying combinations have been successfully used to correct flatfoot.

Schwartz and Heath, and Diveley have discussed nonoperative measures that can be used to relieve pain arising as a result of flatfoot and certain other disorders of the foot.

The rare condition referred to as *congenital flatfoot* or *congenital vertical talus* is characterized by dislocation of the navicular onto the upper surface of the head or neck of a talus that is in equinus. Eyre-Brook and co-workers and Wainwright have all described their experiences in correcting a number of instances of this condition. Patterson and co-workers studied anatomically two feet so affected and concluded that the relatively slight changes in the morphology of the bones could not account for the deformity, but were secondary to displacement produced by abnormally tight tendons.

Movements

Movements of the entire foot, occurring at the ankle joint, are those of plantar flexion and dorsiflexion; the talus is gripped so tightly between tibia and fibula that essentially no movements of inversion or eversion can occur here. In dorsiflexion, with the broad part of the trochlea tali forced backward, the joint is especially tight; in plantar flexion, with a narrower part of the trochlea engaged, a small amount of lateral movement can be shown. With the foot at a 90° angle to the leg, about 5° of lateral movement is normal (Leonard).

The muscles of the leg both move the foot and control movement of the leg on the weight-bearing foot. Thus, in quiet standing all the muscles of the leg except the triceps surae, which prevents the weight of the body

from producing dorsiflexion at the ankle, may be inactive, but when the weight is shifted by swaying forward or backward (Basmajian) or by standing on a slope (O'Connell) the appropriate muscles become active to prevent falling in the direction of shift. Thus in leaning forward or facing downhill the triceps becomes more active, and the peroneus longus may also become active, while in leaning backward or facing uphill it is the tibialis anterior that is the chief antigravity muscle. Similarly, in walking the tibialis anterior is involved in pulling the body forward over the fixed foot by helping to dorsiflex the leg on the foot.

The muscles producing plantar flexion at the ankle are primarily the soleus and the gastrocnemius. Both by their bulk and their attachment to the posterior end of the calcaneus (the back arm of the lever of the foot, the fulcrum of which is at the ankle joint) these muscles alone can function efficiently in plantar flexion. Even they are at a disadvantage, because of the shortness of their lever arm; Jones' ('41) calculations indicate that they must exert a pull of twice the force that is to be exerted as pressure on the plantar surface of the foot. After section of the tendo calcaneus the other plantar flexors sometimes can, sometimes cannot, produce sufficient plantar flexion to raise the individual on the balls of his feet; however, their possible contribution to plantar flexion at the talocrural joint is limited, for the mechanical disadvantage of being so close to the joint, plus the small bulk of these muscles, indicates that they can contribute only some 5% (Jones, '41) to 17% (Sutherland) to the maximal strength of plantar flexion at the ankle; moreover, their chief contribution to flexion seems to be at the transverse tarsal joint. Dorsiflexion can be brought about by all the muscles passing in front of the ankle joint: the strongest of these is the tibialis anterior; its tendency to invert the foot, a tendency shared by the other, less efficient, medially attached dorsiflexor, the extensor hallucis longus, is normally counterbalanced by the everting action of the extensor digitorum longus and peroneus tertius.

Movements at the intertarsal joints are for the most part very small gliding ones, and except for those at the subtalar and transverse tarsal joints they can be disregarded. Movements at both the subtalar and transverse tarsal joints are complex, because the planes of these joints are not in the transverse, horizontal, or vertical planes of the foot. Manter ('41) described the movements as being screwlike; Shephard described the joints as being oblique hinge joints. The two joints typically move together, and in the free foot their combined action results in a complex movement of eversion, dorsiflexion, and abduction of the foot (pronation) or inversion, plantar flexion, and adduction of the foot (supination). The muscles responsible for these actions include practically all the long muscles that reach the foot, but especially, of course, the peronei laterally and the two tibial muscles medially. With the weight on the forepart of the foot the tricips surae supinates at the subtalar joint and increases the stability of the transverse tarsal joint, and the short muscles of the foot that cross the joint then further resist dorsiflexion at this joint as they add to the plantar thrust (Mann and Inman).

METATARSOPHALANGEAL AND INTERPHALANGEAL JOINTS

These joints closely resemble those of the hand and merit little description. The metatarsophalangeal joint of each toe is formed by the concavity of the base of the proximal phalanx applied to the head of the metatarsal. The fibrous capsule, lined with synovial membrane, is strengthened dorsally by the expansion of the extensor tendon, on each side by a fan-shaped collateral ligament, and transformed on the plantar surface into the heavy plantar ligament (Fig. 9-71) attached both to the bones and to the deep transverse metatarsal ligament, and grooved on its plantar aspect for the flexor tendons. The two sesamoids of the big toe lie in the plantar ligament on this toe.

The interphalangeal joints are essentially similar to each other; the distal surface of the proximal bone is pulleylike, the proximal surface of the distal bone has a double concavity to fit the pulley. As in the case of the metatarsophalangeal joints, the extensor tendon completes the capsule dorsally, the collateral ligaments strengthen it on its sides, and the plantar ligaments represent the plantar part of it.

Movements at the metatarsophalangeal joint are flexion and extension and abduction and adduction; the latter movements are much more limited in the toes than they are in the fingers. Flexion is brought about by the long flexors, by the short flexors of the big and little toes, and by the lumbrical and perhaps the interosseous muscles; extension is brought about by the long and short extensors of the toes but especially, according to Duchenne, by the short extensors. The long and short extensors, and especially the latter, also extend the interphalangeal joints, in so far as they can be extended; the interossei contribute little or nothing to this movement in the foot, since they insert primarily into the bone and not into the long extensor tendons. The lumbricals, which do insert into the extensor tendon, may perhaps aid in extension of the phalanges.

The short flexors of the first and fifth toes act only on the proximal phalanges; the flexor digitorum brevis and the flexor digitorum longus act primarily on the middle and distal phalanges, respectively. Abduction is brought about by the abductor hallucis (when it can abduct; its chief action often seems to be flexion), the abductor digiti minimi, and the dorsal interossei, adduction by the adductor hallucis and the plantar interossei. Extension of the toes is initiated through the deep peroneal nerve; flexion, abduction, and adduction through the tibial.

ANOMALIES OF THE FOOT

Most anomalies of the foot are similar to those of the hand; they include polydactylism, cleftfoot (Fig. 9-72), tarsal fusions, anomalous tarsal bones, and the like.

A supernumerary toe may be associated with a supernumerary metatarsal (e.g., Horwitz, '40). Hammertoe, usually congenital, is

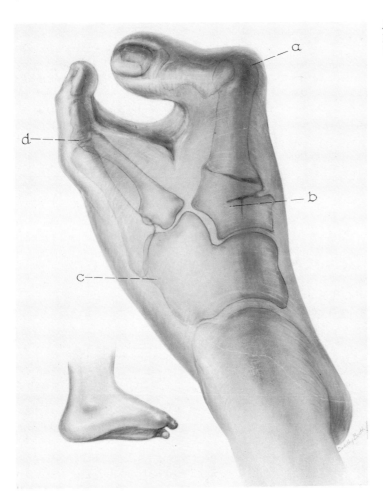

Fig. 9-72. Cleftfoot. (Meyerding HW, Upshaw JE: Am J Surg 74:889, 1947)

a fixed flexion of the toe at the proximal interphalangeal joint. According to McBride it usually affects the second toe and the ligaments of the joint are found at operation to be rigidly contracted.

Clubfoot

A deformity seen only rarely in the hand, and then usually in association with deformity of the radius, but of greater importance in the foot, is clubbing. Clubfoot (talipes) is any deformity in which the foot is twisted out of its natural shape or position, and the term even includes flatfoot (talipes planus). However, the term "clubfoot" is more commonly applied to a marked congenital abnormality, the most common deformity being talipes equinovarus (clubfoot in which the foot is plantar flexed, inverted, and adducted).

Talipes that is not congenital is very obviously brought about by muscle imbalance and is simply an exaggeration, with fixation of the foot in that position, of a normal movement. Duchenne noted, many years ago, that it is better to have all muscles of the leg paralyzed rather than to have only some of them, since in the latter case the normal muscles will, if their pull is not counterbalanced by braces, tendon transfers, or other modern treatment, gradually distort the foot in the direction into which they normally move it. Thus talipes equinus is simply a fixed plantar flexion of the foot, brought about by shortness of the triceps surae or its relative overcontraction against weakened dorsiflexors; talipes calcaneus is just the opposite, arising from contraction of the dorsiflexors of the foot against a weakened or lengthened triceps surae. In the same way, varsus (inversion and adduction) or valgus (eversion and abduction) deformity naturally results if either the invertors or the evertors of the foot exert a pull out of proportion to the antagonistic muscles; the amount of adduction or abduction of the foot associated with inversion and eversion will vary primarily according to the balance between the muscles inserting anteriorly into the foot, which have the greatest tendency to adduct or abduct the forefoot, and those inserting more posteriorly, although in a weight-bearing foot the weight alone will force the forepart of the foot into abduction if the hindpart is pronated.

Caldwell has discussed particularly various types of arthrodesis as treatment for painful and deformed feet, and Peabody, Mayer, and Reidy and his colleagues have all discussed especially tendon transpositions, which are frequently combined with arthrodesis.

In contrast to the general acceptance of the principle that muscle imbalance is the deforming force in paralytic clubfoot and may be largely responsible for recurrence of a corrected clubfoot, there has been little agreement on the origin of congenital talipes. Abnormal pressure in utero, intrinsic maldevelopment, malnutrition, muscular contracture of unknown origin, and local arrest of development have all been said to be the cause of clubfoot. The chief argument has been whether the skeletal defect is primary or secondary to some maldevelopment of soft tissues, particularly of the muscles.

According to Settle, a majority of the feet that have been dissected have had at least the talus deformed; in all 16 that he dissected this was true, the neck being particularly deformed and the articular surface for the navicular facing medially and downward instead of forward. The subtalar surfaces were also much deformed, and both the talus and navicular, and to a minor extent the calcaneus, were smaller than normal. Since the soft tissues showed no sign of being contracted or stretched, and no abnormal insertions of the muscles were found, he suggested that the condition results from a primary developmental deformity of the hind part of the foot. Irani and Sherman reported almost identical findings in 14 clubfeet they dissected and attributed it to defective cartilage formation in the sixth week of fetal life.

In contrast, Stewart said that the muscle attachments in clubfoot always diverge from the normal pattern, and suggested that differences in the abnormal muscle attachments from one foot to another account for the varying degree of severity and the difference in types of such feet. Both Flinchum and Wiley failed to find abnormal insertions that would account for clubfoot, and Wiley found no constant change in the histology of the muscles, while Flinchum described clubbing as being apparently a result of muscle imbalance brought about by inadequate length of certain muscle fibers.

Bechtol and Mossman also reported histological abnormalities of the muscles in two cases of clubfoot; in one, some of the muscles were definitely retarded in their development and the deformity was toward the more mature muscles, so that the better development of these muscles appeared to be the cause of the deformity. In the second case, all the posterior muscles of the calf presented abnormal muscle fibers that were short and apparently dying instead of growing in length with the skeleton and the other muscle fibers, and the presence of these shorter fibers appeared to be the cause of the deformity.

Wynne-Davies suggested that the genetic factor in clubfoot is probably one leading to faulty formation of connective tissue so that the foot is at some state of development easily distorted by intrauterine pressure. However, Isaacs and co-workers studied biopsies of muscles from clubfeet and found histochemical anomalies in the majority; further, all those studied by electron microscopy showed abnormal muscle fibers. They concluded that a neurogenic factor must be the cause. Ippli- tot and Ponseti, rejecting this concept, found

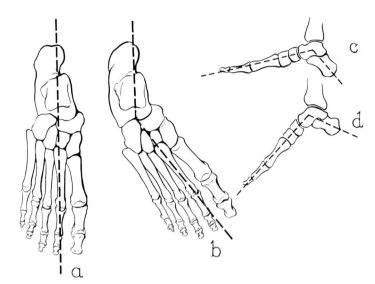

Fig. 9-73. Diagram of the adduction, *b*, and the equinus deformity *d*, in clubfoot, compared with a normal foot, *a* and *c*. (Macey HB: S Clin North America [Aug] 1937, p 1231)

marked fibrosis in muscles, tendons, fascia, and ligaments of the five clubfeet of aborted fetuses that they examined, and concluded that retraction of the fibrotic soft tissues is probably an important causal factor.

The usual congenital clubfoot, talipes equinovarus, is actually a combination of three deformities: there is an equinus or flexion deformity, typically both at the ankle and at the transverse tarsal joint; an inversion or varus deformity, for the most part centered at the subtalar joint; and an adduction deformity, taking place mostly at the transverse tarsal joint (Fig. 9-73). Kite pointed out that these deformities can occur separately, and reported 300 cases of congenital metatarsus varus, in which the deformity is one of adduction and inversion of the forepart of the foot, with no abnormality at the subtalar or ankle joints. He described this condition as being frequently overlooked at birth, but becoming more marked during the first few months of life, and expressed the opinion that while prenatal muscle imbalance is responsible for the condition, it is accentuated by the overpull of the anterior tibial muscle during the first few months of life.

Garceau and Palmer, and Critchley and Taylor, among others, regarded muscle imbalance as an important factor in the recurrence and maintenance of the inversion-adduction component of clubfoot; they reported favorable results, in cases resistant to other treatment, from transplanting the tibialis anterior tendon to the lateral border of the foot. Singer regarded anterior tibial transplant as often giving unsatisfactory results and, with other workers (*e.g.,* Fried; Gartland) who regarded abnormality of the tibialis posterior as a common cause of recurrence, advocated transferring the attachment of that muscle to an anterolateral position on the dorsum of the foot. Wagner and Butterfield said that from 90% to 95% of clubfeet respond to early adequate conservative measures; Blockey and Smith, however, described relapse as occurring in almost two thirds of previously corrected clubfeet, about half within the course of 2 or 3 years.

Tarsal Fusions

Fusion of tarsal bones associated with defects of the tibia and fibula have already been discussed. Apparently, fusion among the tarsals is rare otherwise. According to O'Rahilly, a bony bridge sometimes unites the body of the talus to the sustentaculum tali, or the anterior end of the calcaneus to the navicular. Schreiber reported 5 cases of rare congenital talonavicular fusion; Sloane described a similar fusion in one foot, while in the other foot these two bones, the calcaneus, and the cuboid were fused into one mass, the two lateral cuneiforms were fused in another, and the fourth metatarsal and toe were absent. M'Connell described a case of fusion of the navicular and cuneiform bones, as did Lusby more than 50 years later, and del Sel and Grand described one of cubonavicular synostosis.

Accessory Tarsals

O'Rahilly reported that some 30 accessory tarsal bones, including sesamoid bones associated with the tarsals, have been described; Figure 9-74 shows most of these. Trolle has studied in some detail the development of the accessory bones of the foot.

Many of the accessory bones of the foot occur very rarely; Dwight ('07) indicated that

the os trigonum, the intermetatarsal between the first and second metatarsals, the os tibiale externum, and the peroneal sesamoid are the most common accessory bones of the foot; the os vesalianum (separate tuberosity of the fifth metatarsal) is often listed as being common, but Dwight regarded it as "excessively rare." Cravener and MacElroy said that an accessory navicular bone has been reported to occur in about 14% of feet. Dwight ('11), and a number of workers since then, have reported isolated cases of the occurrence of unusual tarsal bones; Hirschtik could find no previous account of the one he described. McDougall doubted that the os trigonum is actually an accessory tarsal; he reported that

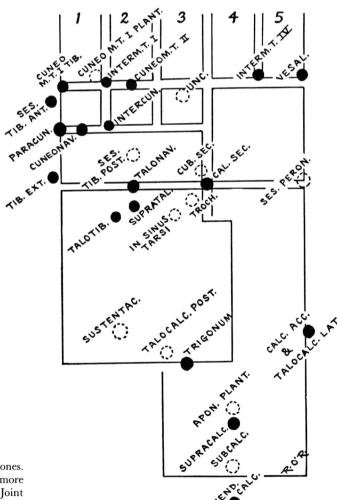

Fig. 9-74. Diagram of accessory tarsal bones. Those indicated in outline are placed more plantarward. (O'Rahilly R: J Bone Joint Surg 35-A:626, 1953)

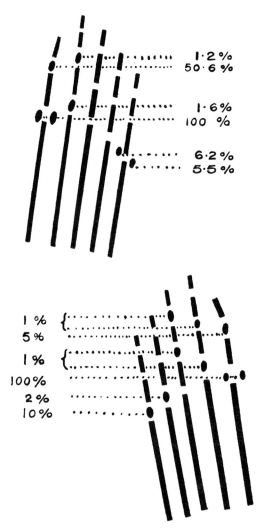

1·2 %
50·6 %

1·6%
100 %

6·2%
5·5 %

1 %
5 %
1 %
100%
2 %
10%

Fig. 9-75. Incidence of digital sesamoids as reported in two studies: Top, by dissection, diagrammed as on a left foot, and bottom, by roentgenograms, diagrammed as on a right foot. (Bizarro AH: J Anat 55:256, 1921)

both tubercles of the talus develop from separate centers of ossification and adduced evidence that the os trigonum, in the position of the lateral tubercle, is actually a lateral tubercle that has undergone epiphyseal separation or fracture.

Digital Sesamoids

Accessory sesamoids on the toes are not particularly uncommon; Figure 9-75 shows the location and percentages of the sesamoids found in two series, in one of which they were sought by dissection and maceration, and in the other by roentgenogram. Lapidus reported a case in which sesamoids were associated with all the metatarsal heads of both feet. Inge and Ferguson noted that a bipartite sesamoid associated with the big toe may be found in more than 10% of feet, and should not be mistaken for a fractured one, which is rare.

DISLOCATIONS AND FRACTURES

Dislocations and fractures about the ankle and in the foot can be particularly complicated because of the articular surfaces and ligaments that may be involved. Sprain of the ankle, the most common sprain of any joint, involves tearing the ligaments to a variable degree, and may result in very severe disability. Mayer and Pohlidal discussed various types and combinations of tearing of ligaments and fracture of the tibia and fibula that may result in loosening the grip of the malleoli on the talus. The effect of disruption of the medial and lateral ligaments of the ankle on the stability of this joint has also been briefly discussed previously.

Dislocation at the ankle joint is usually accompanied by fracture, but occasionally there is no associated fracture; according to Fonda the case he reported in 1952 was the twentieth on record. Subtalar dislocation is also rare; in this the talus retains its normal position at the ankle but the remainder of the foot is displaced. This may be inward, outward, backward, or forward.

Fracture at the ankle is usually caused by indirect violence, stumbling and twisting the ankle; according to Dickson only about 1% of fractures are due to direct violence, such as a blow. External rotation (abduction) is said to account for a majority (61%) of fractures at the ankle. The fracture may involve the tibia or the fibula alone or both together, and vary from a small chip fracture to one produced by an upward thrust on the tibia in which there is injury to the posterior lip of the tibia and a spiral fracture of the body of the bone. In children, the tibial fracture may be largely

Fig. 9-76. Three types of fracture of the epiphyses at the ankle. (Johnson EW Jr, Fahl JC: Am J Surg 93:778, 1957)

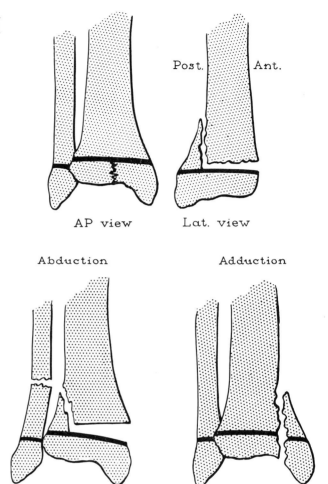

Post.   Ant.

AP view   Lat. view

Abduction   Adduction

AP view   AP view

along the epiphyseal line. The type of fracture depends largely on the mechanism of injury (Fig. 9-76).

Fracture of the calcaneus frequently involves more than one joint, and nonsurgical treatment then rarely is adequate, according to Johnson and Peterson; in their study, the calcaneocuboid joint was involved alone in 20% of cases, and it and the subtalar were involved together in 33% more. In line with this high rate of involvement of the calcaneocuboid joint, they found that triple arthrodesis gave better results than did subtalar arthrodesis. The most common fractures of the talus are those of the neck and of the posterior pro-

cess; Coltart has discussed especially fractures and dislocations of this bone, and Hawkins described a number of cases of persistent pain over the lateral part of the ankle resulting from isolated fracture of the lateral process of the talus. Isolated fracture of other tarsal bones is rare. Fractures of the navicular, cuboid, and cuneiforms are, according to Henderson, usually produced by a fall in plantar flexion that twists the foot externally or internally, or by a direct crush injury. The talus is usually injured in a fall from a height, with the knees stiff, this also being the frequent cause of calcaneal fracture.

"March" or "marching" fracture (Fig. 9-

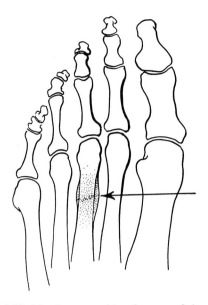

Fig. 9-77. March or marching fracture of the third metatarsal; the usual large callus, which has been mistaken for a tumor, is indicated by stipple. (Meyerding HW, Pollock GA: Surg Gynecol Obstet 67:234,1938 [by permission of Surgery, Gynecology and Obstetrics])

77) is a fracture of one of the metatarsals (commonly the second or third; see Henderson, Rider) in approximately the middle of the body, usually without any history of trauma. This appears to be produced by prolonged strain. The commonest metatarsal fracture as a result of sudden violence is that of the fifth, close to its base, produced by turning the ankle so as to throw much of the weight suddenly onto this bone (Morrison). Fractures of the phalanges are usually the result of direct trauma, as from dropping a heavy object upon the toes or stubbing the toe.

DORSUM OF FOOT

The superficial fascia of the dorsum of the foot is thin; it contains the prominent dorsal veins and the branches of the cutaneous nerves to the dorsum of the foot, already described.

The deep fascia, continuous above with the fascia of the front of the leg and strengthened

at the ankle by the thickening fibers that form the inferior extensor retinaculum, is divided by the extensor tendons on the foot into two layers, one superficial to the tendons and one deep to them. The two layers come together on the borders of the foot, where they are also attached to the first and fifth metatarsals before rounding the border to become continuous with the deep plantar fascia; deep to the fascia on the dorsum there is a potential space between it and the dorsal interosseous fascia.

MUSCULATURE

Unlike the dorsum of the hand, which presents a short extensor of the fingers only as an occasional variation, the dorsum of the foot regularly bears a short extensor of the toes (Fig. 9-78). In contrast to the expected arrangement shown, for instance, by the extensor digitorum longus, which sends tendons to the four lateral toes, the short extensor of the dorsum of the foot sends tendons to the four medial toes. The entire muscle has been called the extensor digitorum brevis, but the most medial belly of the muscle is the largest, and since its tendon goes exclusively to the big toe it is called the extensor hallucis brevis.

The *extensor hallucis brevis* arises with the extensor digitorum brevis from the anterolateral part of the upper surface of the calcaneus and from the deep surface of the inferior extensor retinaculum; its muscular belly soon separates from the most medial digitation of the extensor digitorum brevis and passes medially and distally to insert on the base of the proximal phalanx of the big toe. It may be joined by a slip of insertion from the extensor hallucis longus here; Hallisy found a medial slip of the extensor hallucis longus inserting into the proximal phalanx in 70 of 290 feet.

The *extensor digitorum brevis* may be slightly separated from the extensor hallucis brevis at its origin or the two may blend; in any case, the extensor digitorum brevis tends to form a flattened muscle on the lateral aspect of the dorsum of the foot that divides into three muscular digitations at about the point at which it passes under the peroneus tertius tendon. These digitations give rise to three tendons that attach to the long extensor tendons of the second, third, and fourth toes over the metatarsophalangeal joints; each aponeurosis formed by the long and short extensor tendons then behaves as do the dorsal aponeuroses of the fingers, expanding over each joint to form the dorsal capsule, dividing over the proximal phalanx into a middle band and two lateral bands, and inserting into the middle phalanx by the middle band, into the terminal phalanx by the two lateral bands.

The extensor digitorum brevis and extensor hallucis are supplied by the deep peroneal nerve, which gives off its last muscular branch deep to them. However, Winckler ('34) described a branch (accessory deep peroneal) of the superficial peroneal nerve as sometimes supplying the short extensors, and Lambert ('69) has found this supply in 22% of limbs. The extensor hallucis is an extensor of the proximal phalanx of the big toe only; the extensor digitorum is theoretically an extensor of the middle and distal phalanges of the second, third, and fourth toes, but actually has little effect upon these, which tend to be plantar flexed by the flexor tendons; thus the extensor brevis acts primarily on the metatarsophalangeal joints. Duchenne regarded it as the primary extensor of the toes, saying that the real function of the extensor digitorum longus is not to extend the toes, although of course it does this, but to aid in dorsiflexion at the ankle.

Lovell and Tanner said that when the tendons of the extensor digitorum brevis are well developed each one has its own synovial tendon sheath, that on the tendon to the big toe being the longest; if these tendon sheaths exist, they are commonly overlooked.

NERVES AND VESSELS

The cutaneous nerves and the superficial veins of the dorsum have already been described.

Deep Peroneal Nerve

The deep peroneal nerve emerges on the dorsum of the foot from behind the inferior extensor retinaculum, where it lies between the

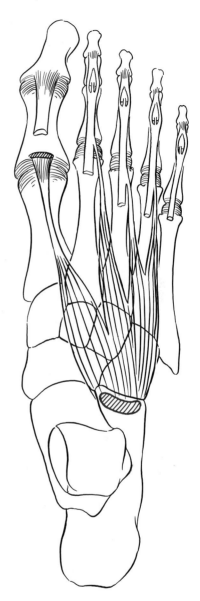

Fig. 9-78. Origin (*red*) and insertion of the extensors hallucis and digitorum brevis.

tendons of the extensor hallucis longus and extensor digitorum longus, and usually lateral to the dorsalis pedis artery (Fig. 9-79). It then divides into two branches; the medial one continues the course of the main nerve, gives off a few articular twigs, and passes toward the interspace between the first and second

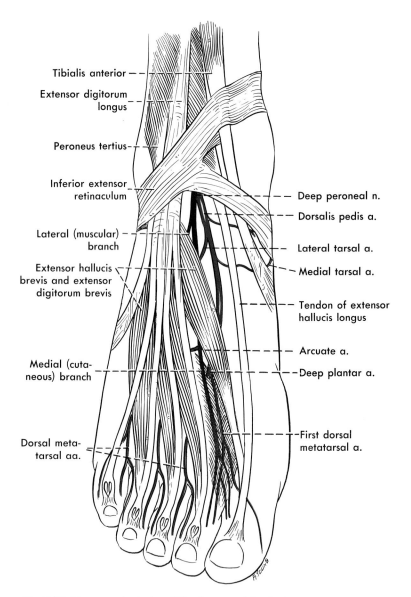

Tibialis anterior

Extensor digitorum longus

Peroneus tertius

Inferior extensor retinaculum

Lateral (muscular) branch

Extensor hallucis brevis and extensor digitorum brevis

Medial (cutaneous) branch

Dorsal metatarsal aa.

Deep peroneal n.

Dorsalis pedis a.

Lateral tarsal a.

Medial tarsal a.

Tendon of extensor hallucis longus

Arcuate a.

Deep plantar a.

First dorsal metatarsal a.

Fig. 9-79. Nerves and arteries of the dorsum of the foot.

toes, to divide into proper digital branches for the adjacent sides of these toes. The lateral branch is muscular and articular; it passes beneath the extensor digitorum brevis and gives one or more branches into the deep surface of this muscle (including the extensor hallucis brevis) and several branches that spread out over the foot to supply the intertarsal, the tarsometatarsal, and the metatarsophalangeal joints. Winckler sometimes also traced

branches from this nerve to some of the dorsal interosseous muscles, but expressed the belief that they consist of afferent fibers.

Dorsalis Pedis Artery

The dorsalis pedis artery is typically the continuation of the anterior tibial artery; it is usually defined as beginning at the level of the ankle joint, and commonly continues almost straight downward and then distally on the dorsum of the foot to end over the base of the second metatarsal by dividing into its terminal branches, the laterally directed arcuate artery and the deep plantar artery (Fig. 9-80). The latter passes between the heads of origin of the first dorsal interosseous, in the interspace between the first and second metatarsals, to join the deep plantar arch.

Huber found great variation in the dorsalis pedis artery and all its branches, so much so that the "most constant" pattern, considered in minute detail, that he found among 200 feet was present in only 11 of these. However, he also showed that most of the variations encountered are brought about simply by enlargement of certain branches of the fundamental network on the dorsum of the foot and

diminution or disappearance of certain other connecting branches (Fig. 9-81). Aside from its method of branching, the dorsalis pedis artery was of usual origin, size, and course in 147 out of 200 feet; among the remainder the artery was, in 12% of the 200 feet, so reduced in size that it was essentially absent and in another 0.5% almost as small. In 9% it deviated either laterally or medially from its usual position; in 3% it arose from the perforating branch of the peroneal artery rather than from the anterior tibial; and in 1.5% it arose from an anterior tibial that deviated toward its lower end to occupy the position between tibia and fibula usually occupied by the perforating branch of the peroneal. The usual branches of the dorsalis pedis artery in

Fig. 9-80. Two patterns of the dorsalis pedis artery and its branches. On the viewer's right, the artery is shown as a continuation of the perforating branch of the peroneal artery.

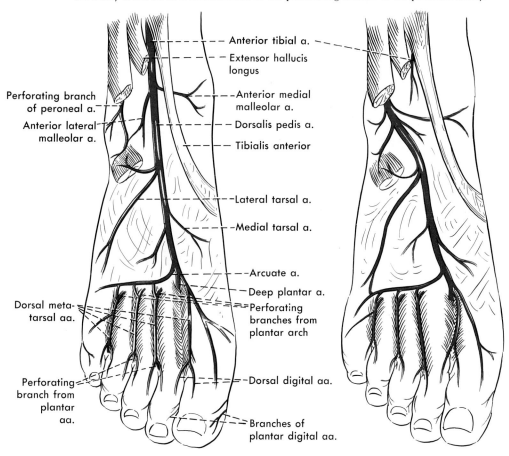

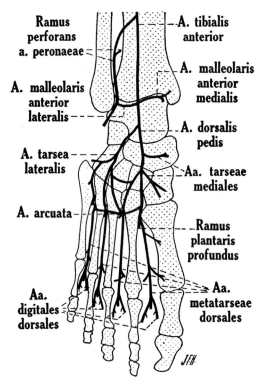

Ramus
perforans
a. peronaeae

A. malleolaris
anterior
lateralis

A. tarsea
lateralis

A. arcuata

Aa.
digitales
dorsales

A. tibialis
anterior

A. malleolaris
anterior
medialis

A. dorsalis
pedis

Aa. tarseae
mediales

Ramus
plantaris
profundus

Aa.
metatarseae
dorsales

Fig. 9-81. The basic pattern of the arterial network of the dorsum of the foot. (Fig. 1e, Huber JF: Anat Rec 80:373, 1941)

addition to its terminal ones (and the malleolar, which Huber found arising more frequently from the dorsalis pedis than from the anterior tibial) are lateral and medial tarsal arteries.

Huber found the *lateral tarsal artery* most commonly (58%) arising at about the level of junction of the neck and head of the talus. He could not identify one in three instances, in three it was extremely small, and in one it appeared to be a continuation of the perforating branch of the peroneal artery. The lateral tarsal artery typically supplies small communicating branches to the arcuate artery, and when the arcuate artery is reduced or missing these may give rise to the more lateral dorsal metatarsal arteries.

The *medial tarsal arteries* varied in number and in level of origin, but the most constant pattern was two chief ones; sometimes they arose by a common stem, and sometimes

there was only one, or there were several tiny branches representing them.

The *arcuate artery* typically arises at the base of the second metatarsal and passes laterally over the third and fourth metatarsals, giving rise to three dorsal metatarsal arteries that run forward on the second, third, and fourth dorsal interossei to supply the dorsal digital arteries to the adjacent sides of the four lateral toes. Huber found an arcuate artery in only 54% of feet; in 12% it was lacking as a result of the essential absence of the dorsalis pedis, and in 34% it was lacking even though the dorsalis pedis was otherwise essentially normal. When it was present, it fairly often arose above its usual point of origin, and it varied in how far it extended laterally, sometimes giving origin only to the second and third, or even only the second, dorsal metatarsal artery. When the arcuate artery was absent the metatarsal arteries arose from the lateral tarsal artery, from the posterior perforating branches normally given off to the metatarsals from the plantar arch, or from both sources.

The first *dorsal metatarsal artery* typically arises from the dorsalis pedis and passes distally over the surface of the first dorsal interosseous muscle; Huber found it sometimes small, sometimes arising from the plantar arch, and sometimes essentially missing. Before they divide into digital branches, the dorsal metatarsal arteries are typically connected to the plantar ones by one or two arteries that pass between the metatarsal bones just proximal to the deep transverse metatarsal ligament. The first three dorsal metatarsal arteries, in about 10% to 15% of instances, arose from the plantar vessels by means of these communications; the fourth dorsal metatarsal and the dorsal digital arteries for the adjacent sides of the fourth and fifth toes had this origin in more than 50% of cases. In between 20% and 30% of cases the dorsal metatarsal artery continued plantarward after giving off the dorsal digital vessels, running close to the web of the toes, to join the plantar metatarsal arteries distal to the deep transverse metatarsal ligament.

SOLE OF THE FOOT

The tela subcutanea of the sole of the foot is especially heavy, for it must serve as a padding between the bones and the skin. In it are embedded the fine network of superficial plantar veins and the terminal branches of the cutaneous nerves. It is continuous above with the superficial fascia or tela subcutanea of the leg and continuous around the borders of the foot with the thin superficial fascia of the dorsum, and it continues onto the toes where it contains the plantar digital nerves and vessels. Between the distal parts and heads of the metatarsals this tissue blends with the deeper tissue of the sole of the foot, surrounding the plantar metatarsal arteries and the common digital nerves, and through this also is continuous with the superficial fascia of the dorsum of the foot.

The deep fascia of the sole closely resembles the deep palmar fascia, in that it presents a strong central plantar aponeurosis, while the medial and lateral parts, covering the intrinsic muscles of the big and little toes, are thinner. From the edges of the plantar aponeurosis medial and lateral intermuscular septa pass toward the first and fifth metatarsals, or the tarsals proximal to them, but their arrangement, and the partitioning of the foot that they produce, is more complicated than in the hand. Kamel and Sakla described the medial intermuscular septum as passing deep to the abductor hallucis to attain an attachment to the first metatarsal so that the muscle lies in a medial compartment of its own. The lateral intermuscular septum passes to the fifth metatarsal, or to the cuboid and calcaneus, and segregates the muscles of the little toe into a lateral compartment. The remaining muscles, including the short flexor and adductor of the big toe, thus lie in a central compartment (Fig. 9-82).

Kamel and Sakla described further subdivision in the central compartment as resulting from a Y-shaped septum that originates from the fifth metatarsal and extends medially, with one limb passing just below (superficial to) the interossei to reach the first metatarsal

and enclose them in a deep compartment, the other passing just deep to the flexor digitorum brevis, longus, and associated lumbricals, and then turning down to attach to the origin of the medial intermuscular septum from the plantar aponeurosis. They apparently did not regard the thin fascia between the flexor digitorum brevis and the quadratus plantae and flexor digitorum longus as important, for they described these muscles as lying together in a superficial compartment, with the adductor hallucis, flexor hallucis brevis, and tendon of the flexor hallucis longus then lying in another, or middle, compartment. They said that dye injected into the superficial and middle compartments spread up the leg, but that which was injected into the others did not.

Martin found the medial intermuscular septum more complicated than this. He described it as forming not a complete septum throughout the length of the foot, but as subdividing, as it leaves the plantar aponeurosis, into three bands, each of which in turn divides into a medial and a lateral division. The medial division of the proximal band attaches to the main tendon of the tibialis posterior, the lateral division to the medial border of the cuboid; the medial division of the middle band attaches to the medial cuneiform, the lateral to the medial border of the cuboid or to that and the tissue covering the peroneus longus tendon; and the medial division of the distal band blends with fascia on the lower surface of the flexor hallucis brevis, the lateral division with either that or fascia on the lower surface of the oblique head of the adductor hallucis. These two parts of the medial septum correspond to Kamel and Sakla's description of the medial intermuscular septum and the transverse septum that turns plantarward to join the medial intermuscular septum.

Plantar Aponeurosis

As already mentioned, the central part of the deep fascia of the sole of the foot is strongly aponeurotic in character, and largely resembles the palmar aponeurosis in its form and

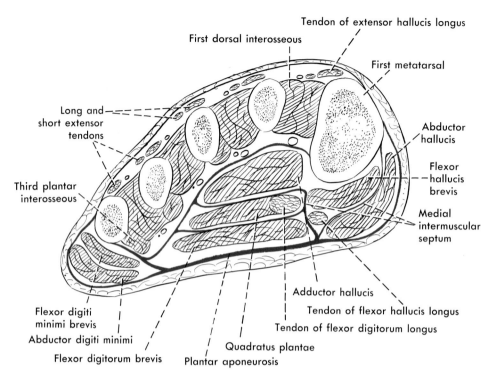

Fig. 9-82. Diagrammatic cross section through the foot to show the fascial compartments of the sole. The plantar aponeurosis and the associated fascia are in *red*.

attachments (Fig. 9-83). It differs from this in that its proximal or posterior end is attached entirely to bone, for the tendon of the plantaris, which should be continuous with the plantar aponeurosis, attaches to the posterior aspect of the calcaneus, and the flexor retinulum lies above the inferior surface of this bone. Thus the plantar aponeurosis takes origin from the medial process of the tuber calcanei and extends forward as a very strong but relatively narrow band that divides at about the middle of the foot into digitations for the five toes. As the longitudinally running bands for the toes diverge from each other, deeper-lying transverse fibers, constituting the transverse fasciculi, appear between them. Distal to the transverse fasciculi the superficial connective tissue of the sole becomes continuous with the connective tissue above the plantar aponeurosis; the digital nerves and vessels are contained in it, continuing forward between the digitations of the aponeurosis to the toes. Across the ball of the foot the digitations are united by a poorly de-

veloped band of transverse fibers, the superficial transverse metatarsal ligament, that crosses their superficial surfaces. As the digitations reach the toes they may send reinforcements to the plantar surface of the fibrous tendon sheaths on the digits (special strengthenings of the deep fascia of the digits), but largely split to pass around the sides of the flexor tendons, strengthening the sheaths laterally, and attach to the plantar ligaments and the proximal phalanges. In consequence of its attachments, a lesion of the plantar aponeurosis similar to that causing Dupuytren's contracture of the hand can cause Dupuytren's contracture of the foot; however, it is uncommon (Stoyle).

Although the fascia of the sole is generally very much thinner over the muscles of the big and little toes, it is particularly thickened, and somewhat resembles the plantar aponeurosis, where it stretches between the lateral process of the tuber calcanei and the base of the fifth metatarsal bone, and covers the abductor digiti minimi. Wright and Rennels described the plantar fascia as showing a limited elasticity, capable of recovering from a stretch of some 3.5% to 4.5% under loads of up to 2,-500 pounds, but stretching only about 1.68% under a 200-pound load.

MUSCLES AND TENDONS

The muscles and tendons of the foot can be most conveniently considered in terms of the layers that they form in the foot. From this aspect, there are four layers of muscles in the foot: the most superficial layer consists of the abductor hallucis, the flexor digitorum brevis, and the abductor digiti minimi, named from the medial side outward; the next layer, contributed to also by the long flexor tendons, is composed of the quadratus plantae and lumbrical muscles; the third layer is formed by the flexor hallucis brevis, the adductor hal-

Fig. 9-83. The plantar aponeurosis and the cutaneous nerves and vessels of the sole.

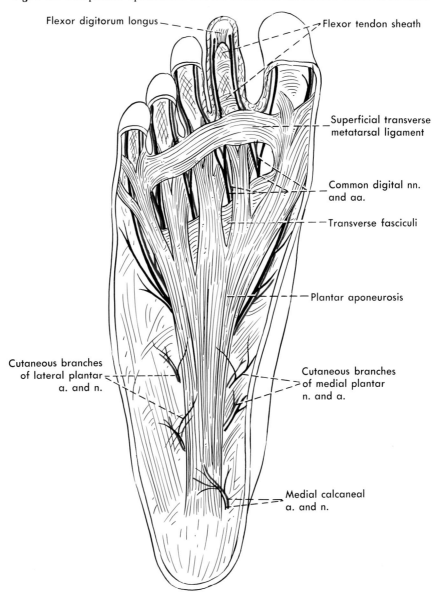

Flexor digitorum longus

Flexor tendon sheath

Superficial transverse metatarsal ligament

Common digital nn. and aa.

Transverse fasciculi

Plantar aponeurosis

Cutaneous branches of lateral plantar a. and n.

Cutaneous branches of medial plantar n. and a.

Medial calcaneal a. and n.

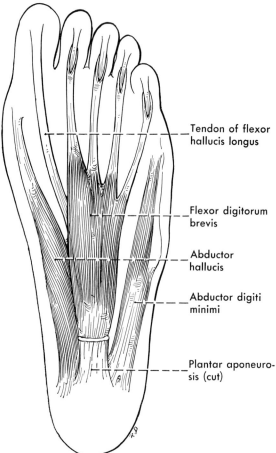

Fig. 9-84. Superficial layer of plantar muscles.

Tendon of flexor hallucis longus

Flexor digitorum brevis

Abductor hallucis

Abductor digiti minimi

Plantar aponeurosis (cut)

lucis, and the flexor digiti minimi brevis; and the last layer is composed of the interosseous muscles.

Superficial Layer

All three of these muscles (Figs. 9-84 and 9-85) arise primarily from the calcaneus, but have also attachments to adjacent fascial layers.

The *abductor hallucis,* along the medial border of the foot, arises from the medial process of the tuber calcanei and from the lower border of the flexor retinaculum. The fibers end in a flattened tendon that is inserted mostly into the medial side of the plantar surface of the base of the proximal phalanx of the big toe (less frequently onto the medial side of the base), blending here with the insertion of the medial head of the flexor hallucis brevis and

sharing with it some attachment to the medial sesamoid of the big toe. The medial and lateral plantar nerves and vessels, emerging from under cover of the flexor retinaculum, pass deep to (above) the abductor hallucis in their courses onto the plantar surface of the foot.

The *flexor digitorum brevis,* the equivalent of the flexor digitorum superficialis (sublimis) of the forearm and hand, is situated in the middle compartment, immediately above the plantar aponeurosis. It arises from the proximal part of this aponeurosis and from the medial process of the tuber calcanei, from the two intermuscular septa that separate it from the abductor hallucis and the abductor digiti minimi, and fairly frequently by a small deep slip from the tendon of the flexor digitorum longus. It gives rise to four small tendons that pass distally toward the four lateral toes (or three tendons, omitting the fifth toe) immediately superficial to the long flexor tendons and enter the digital tendon sheaths with these tendons; within these sheaths they behave exactly as do the superficial flexor tendons on the fingers, dividing into two slips to allow the long flexor tendons to pass through, and inserting on the middle phalanges of the toes. The medial and lateral plantar nerves and vessels pass from under cover of the abductor hallucis into the central compartment of the foot, and the lateral ones then course laterally and distally immediately above the flexor digitorum brevis, between it and the associated long flexor tendon and quadratus plantae.

The *abductor digiti minimi,* along the lateral border of the foot, arises from the lateral process of the tuber calcanei and (under cover of the flexor digitorum brevis) from the medial process, and from a part of the plantar surface of the calcaneus immediately adjacent to the tuber; it also arises from the intermuscular septum lying between it and the flexor digitorum brevis, and from the plantar fascia, a lateral part of which extends as a fibrous band from the calcaneus to the base of the fifth metatarsal bone. Proximal to the base of the fifth metatarsal the deep surface becomes tendinous, although the superficial surface may remain muscular for some distance beyond; the tendon passes across the smooth surface of the tuberosity of this metatarsal to continue forward on the surface of the flexor digiti minimi brevis to an insertion on the lateral side of the proximal phalanx of the little toe. Very often, some of the most lateral fibers of this muscle attach to the lateral aspect of the plantar surface of the tuberosity, and these may be sufficiently developed to constitute a separate muscle, the abductor ossis metatarsi quinti.

The actions of these muscles, except probably for the abductor hallucis, are indicated by their names: the flexor digitorum brevis flexes the middle phalanges of the four lateral toes; and the abductor digiti quinti abducts the little toe; the abductor hallucis functions regularly in plantar flexion, and while it obviously

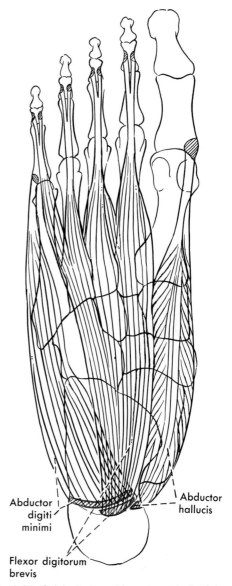

Fig. 9-85. Origin (*red*) and insertion (*black*) of the superficial plantar muscles.

can abduct in some persons, Basmajian reported that it is often so inserted that it can only flex.

The abductor hallucis and the flexor digitorum brevis are innervated by separate branches from the medial plantar nerve, which enter the deep surfaces of the muscles and bring to them fibers from the fifth lumbar and first sacral nerves (or sometimes, in addition to or in place of the first sacral, fibers

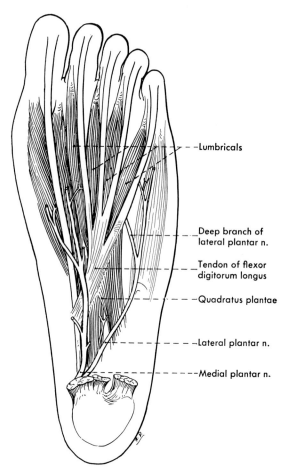

Fig. 9-86. The second layer of plantar muscles, and the long flexor tendons.

the flexor hallucis longus divides into tendons to the four lateral toes. Within the central compartment it is associated with one muscle and one set of muscles that form the second layer of muscles of the sole (Figs. 9-86 and 9-87).

The proximal muscle, the *quadratus plantae* (or *accessory flexor*), arises by two heads, from the medial and lateral sides of the plantar surface of the calcaneus; they are separated by the origin of the long plantar ligament

Fig. 9-87. Origin and insertion of the second layer of plantar muscles.

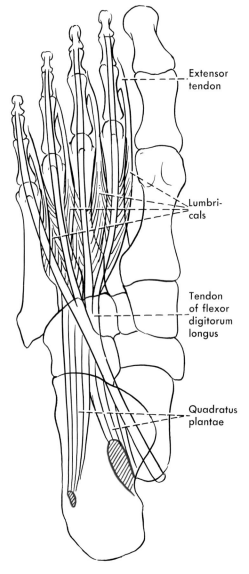

from the fourth lumbar); the abductor digiti minimi is innervated by a branch of the lateral plantar nerve, which supplies it with fibers from the first two sacral nerves.

The common variation in these three muscles is the relation of the flexor digitorum brevis to the little toe. In 33.3% of 60 feet (Nathanial) the tendon to the little toe was derived from a separate muscle slip arising from the flexor digitorum longus, and in 38.3% there was no tendon to this toe.

Second Layer
The tendon of the flexor digitorum longus passes into the central compartment of the foot deep to the abductor hallucis muscle, and after crossing superficial to the tendon of

from the calcaneus. The lateral head is largely tendinous, and takes origin also in part from the long plantar ligament; the medial head, more fleshy, arises from the medial surface of the calcaneus below the course of the flexor hallucis longus tendon along this surface, and from the medial edge of the long plantar ligament. The two heads unite into a short flattened quadrilateral belly that inserts into the tendon of the flexor digitorum longus at about its level of division into the four tendons for the toes; a part of the insertion is usually into the undivided tendon, a part of it at least into the tendon for the fifth toe, and frequently extending across this to attach also to the tendons of the fourth and third toes. Rarely, the quadratus plantae may give rise to, or contribute tendons that join, one or more of the tendons of the flexor digitorum longus (Barlow, '53). Also rarely, the medial head of the quadratus plantae may arise in the leg (a fibular origin in about 4% in one series of legs was quoted by Lewis). Lewis found the lateral head missing in 6 of 18 specimens. The lateral plantar nerve and vessels pass across the superficial surface of this muscle, between it and the flexor digitorum brevis.

The anterior muscles of the second layer are the four *lumbrical* muscles, of which the first or most medial one arises from the medial side of the tendon of the flexor digitorum longus to the second toe; the others arise from the adjacent sides of the tendons between which they lie. They normally resemble very closely the lumbrical muscles of the hand, which arise from the profundus tendons. As in the hand, each passes forward to cross superficial to (below) the deep transverse metatarsal ligament, turn dorsally around the tibial side of the digit, and insert into the extensor tendon.

The action of the quadratus plantae is to help the long flexor flex the toes; it changes the posteromedial pull of the long flexor tendons into one that produces pure flexion, and may be particularly important in flexing the toes during the part of the step in which the supporting foot is being dorsiflexed at the ankle. The lumbricals are flexors of the metatarsophalangeal joints; by virtue of their at-

tachments into the extensor aponeurosis they are also, at least theoretically, extensors of the interphalangeal joints.

The quadratus plantae is innervated by a branch from the lateral plantar nerve. The three lateral lumbricals are also innervated by the lateral plantar nerve, but the twigs to them come off the deep branch of this nerve, and therefore enter their upper surfaces. These four muscles receive fibers from the first and second sacral nerves. The first lumbrical is innervated by the medial plantar nerve (whereas the equivalent median nerve in the hand supplies the first two lumbricals). The first lumbrical of the foot is said to receive fibers from the fifth lumbar and first sacral nerves.

The tendon of the flexor hallucis longus typically gives off a slip to the deep surface of the tendon of the flexor digitorum longus to two or more toes and then runs between the medial and lateral divisions of the distal band of the medial intermuscular septum to reach the intersesamoid interval (Martin).

Third Layer

The *flexor hallucis brevis* (Figs. 9-88 and 9-89), the medial one of the three muscles forming the third layer, arises superficially from the medial intermuscular septum, in the angle between the medial and lateral divisions of the middle band (Martin) and deeply by a triangular tendinous fold attached to the medial border of the inferior aspect of the cuboid bone, from the adjacent inferior surface of the lateral cuneiform, and from the tendon of the tibialis posterior as it spreads out to its attachments in this region. The muscular belly begins only a little proximal to the tarsometatarsal joint of the big toe and lies on the plantar surface of the first metatarsal. The belly is grooved proximally, and distally divides into two parts that insert separately into the proximal phalanx of the big toe, with some attachment also to the two sesamoids; in so doing, the tendon of the medial part of the muscle blends with that of the abductor hallucis, that of the lateral part with the tendon of the adductor hallucis. The tendon of the flexor hallucis longus muscle crosses the su-

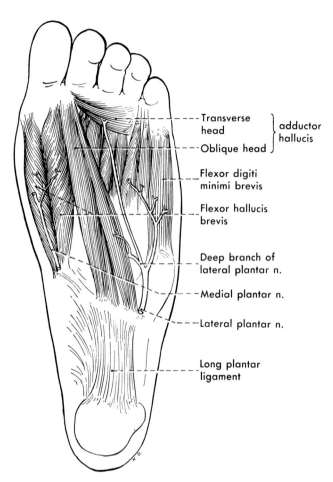

Fig. 9-88. The third layer of plantar muscles.

Transverse head ⎫
Oblique head ⎬ adductor hallucis

Flexor digiti minimi brevis

Flexor hallucis brevis

Deep branch of lateral plantar n.

Medial plantar n.

Lateral plantar n.

Long plantar ligament

perficial surface of the lateral belly of the flexor hallucis brevis obliquely to reach its position between the two sesamoids.

The *adductor hallucis* consists, as does the corresponding muscle in the palm, of an oblique and a transverse head. The oblique head is typically the larger; it arises from the bases of the plantar surfaces of the second, third, and fourth metatarsal bones and from the fibers that form the sheath of the peroneus longus tendon in its course across the plantar surface of the foot. It fills much of the concavity of the foot beneath the metatarsals from which it arises. The transverse head is smaller and more variable, but is a rather flat and slender bundle of fibers that arises from the plantar ligaments of some of the more lateral toes, and from the deep transverse metatarsal ligaments where these attach to the plantar

ligaments. The origin may or may not extend laterally as far as the ligaments associated with the little toe, and there may be no origin from the plantar ligament of the second toe. The transverse head runs almost transversely toward the medial side of the foot, deep to the flexor tendons, to join the oblique head; the common tendon of the two heads of the adductor in turn blends with the tendon of insertion of the lateral portion of the flexor hallucis brevis, shares the lateral sesamoid with this, and inserts into the lateral side of the base of the proximal phalanx of the big toe.

The lateral muscle of this group is the *flexor digiti minimi brevis.* This arises from the base of the fifth metatarsal bone and the adjacent sheath covering the peroneus longus tendon. It passes forward on the plantar surface of the fifth metatarsal, and close to its termination gives rise to a tendon that is inserted, blending with the tendon of the abductor digiti minimi, into the lateral side and flexor surface of the proximal phalanx of the little toe. This muscle is partly covered superficially by the abductor digiti minimi; sometimes some of its fibers, like the tendinous fibers of the lumbricals it resembles, can be traced into the extensor aponeurosis of the little toe; sometimes, also, some of its fibers insert into the distal part of the metatarsal bone, and if sufficiently distinct these may be regarded as forming a separate muscle, the opponens digiti minimi.

Variations of these muscles, in addition to that of the flexor digiti minimi brevis just mentioned, are for the most part minor ones of origin. Occasionally, either the flexor hallucis brevis or the adductor hallucis, or the two combined, may send a slip of insertion to the second toe. Occasionally, also, some of the fibers of the oblique head of the adductor or the medial part of the abductor hallucis insert into the first metatarsal bone, to form a rudimentary opponens hallucis.

The flexor hallucis brevis is innervated by a branch from the medial plantar nerve, usually bringing into it fibers from the fifth lumbar and first sacral nerves. The two heads of the adductor hallucis are supplied by branches, arising separately or together, from the deep branch of the lateral plantar nerve; these are said to contain fibers from the first two sacral nerves. The flexor digiti minimi brevis also receives fibers of the first two sacral nerves, but the branch coming into it is derived from the superficial, not the deep, branch of the lateral plantar nerve.

Fourth Layer

The fourth or deepest layer of muscles of the sole of the foot is formed by the interossei (Figs. 9-90 and 9-91), which are like those of the hand in that they are divided into two sets, a plantar consisting of three muscles and a dorsal consisting of four, and in that the plantar ones adduct the digits to which they are attached, the dorsal ones abduct. Although they are similar in number to the interossei of the hand, the interosseous muscles of the foot differ from these in one important

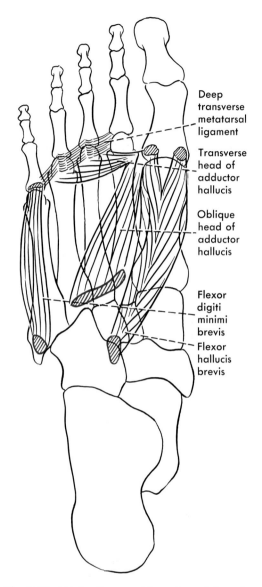

Fig. 9-89. Origin (bony origin *red*) and insertion (*black*) of the third layer of plantar muscles.

respect, namely, that they are arranged around the second digit in the foot, which represents the midline here, while those of the hand are arranged around the third or middle digit. The interosseous muscles fill the spaces between the metatarsals and project plantarward beneath the plantar surfaces of these bones.

Each plantar interosseous arises from a single metatarsal; each dorsal interosseous arises

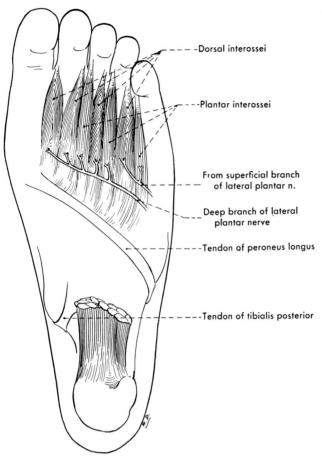

Fig. 9-90. The interossei, the deepest or fourth layer of plantar muscles.

-Dorsal interossei

-Plantar interossei

From superficial branch of lateral plantar n.

Deep branch of lateral plantar nerve

Tendon of peroneus longus

Tendon of tibialis posterior

from both the metatarsals between which it lies. The tendons of all the interossei pass distally above the deep transverse metatarsal ligaments and go to an insertion largely or entirely into the base of the proximal phalanx of the toe with which they are associated. Sarrafian and Topouzian found also some attachment to the plantar aspect of the capsule of the metatarsophalangeal joint; other fibers attach to what they called the extensor sling, apparently the equivalent of the extensor hood in the hand, and occasionally some fibers continue to the extensor aponeurosis. Through the extensor sling, they said, the interossei are weak extensors of the toes.

The three *plantar interossei* lie in the lateral three intermetatarsal spaces where they partly cover, from the plantar surface, the dorsal interosseous muscles that also lie in

these spaces. They arise from the medial sides of the third, fourth, and fifth metatarsal bones respectively, and pass across the medial sides of the metatarsophalangeal joints to be inserted into the bases of the proximal phalanges of the three lateral toes. These muscles adduct the three lateral toes toward the midline (the second toe) and the adductor hallucis substitutes for a fourth one by adducting the first toe toward this line. They are also flexors of the metatarsophalangeal joints, and if they have any insertion into the "extensor sling" or the extensor aponeurosis they presumably can help to extend the interphalangeal joints.

The four *dorsal interossei* occupy all four of the intermetatarsal spaces, the lateral three being partly covered by the plantar interossei. At their proximal ends their two heads of origin are separated by a slight gap, through which arteries pass from the dorsum to the sole of the foot, or vice versa: the deep plantar branch of the dorsalis pedis artery typically passes between the heads of the first dorsal interosseous muscle, and the proximal perforating or communicating arteries (between the plantar arch and the dorsal metatarsal arteries) pass between the heads of the second, third, and fourth dorsal interossei.

The first dorsal interosseous may arise from the adjacent surfaces of the first and second metatarsal bones, but in more than half of feet the origin from the first metatarsal is replaced by one from the peroneus longus tendon (Harbeson; Manter, '45). The second typically arises from the adjacent sides of the second and third metatarsals, the third from the third and fourth metatarsals, and the fourth from the fourth and fifth metatarsals. Each passes dorsal to the deep transverse metatarsal ligament and across the metatarsophalangeal joint to insert primarily into the base of a proximal phalanx: the first and second insert on the medial and lateral sides, respectively, of the second toe, the third and fourth on the lateral sides of the third and fourth toes. These four muscles can therefore abduct the second toe, the midline one, in either direction, and abduct the third and fourth toes from the second toe; working with the abductor hallucis and the abductor digiti minimi, the dorsal interossei therefore spread the toes. They also flex the toes at the metatarsophalangeal joints and become active with the other plantar flexors in the foot as weight is shifted forward in standing on the toes or in walking (Mann and Inman).

Variations of the interosseous muscles in the foot are apparently not at all uncommon. Manter ('45) noted that any of the dorsal interossei, but more commonly the first and the fourth, may arise by single heads of origin; that an interosseous may be double or the dorsal and plantar interossei may fuse; and that the tendon of an interosseous sometimes bifurcates to attain attachment on the adjacent surfaces of the two toes between which it lies. He said that the muscles of the second

Fig. 9-91. Origins (*red*) and insertions (*black*) of the plantar interossei, *a*, and the dorsal interossei, *b*.

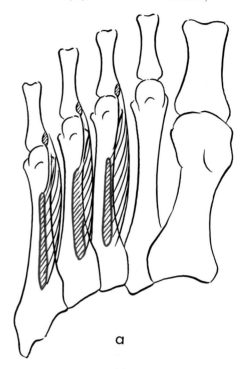

a

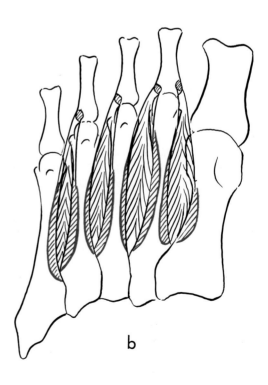

b

intermetatarsal space varied most in the extent of their origins and that the second dorsal interosseous, also one of the muscles in this space, had a bifid tendon more frequently than did the other interossei.

All the interossei are innervated by the lateral plantar nerve, which brings to them fibers from the first two sacral nerves. In this respect, therefore, they resemble the interossei of the hand, which are normally supplied by the ulnar nerve. There is, however, a minor difference: while the interossei of the hand are all supplied by the deep branch of the ulnar nerve, the two lateral interossei of the foot—the third plantar and the fourth dorsal—may be supplied by the superficial branch of the lateral plantar nerve, so that only those in the three medial intermetatarsal spaces are supplied by the deep branch.

NERVES AND VESSELS

Just before they enter the foot, the tibial nerve and the posterior tibial vessels lie side by side, the nerve usually lateral to the artery, posteromedial to the ankle. Here they lie in a compartment deep to the flexor retinaculum (laciniate ligament) between the compartments that house the tendons for the flexor digitorum longus, anteromedially, and the flexor hallucis longus, posterolaterally. Here also both the nerve and artery divide into medial and lateral plantar branches, which round the medial border of the foot and as they reach the plantar surface come to lie directly deep to the abductor hallucis, a part of which arises from the lower edge of the flexor retinaculum. Deep to the flexor retinaculum, the posterior tibial nerve is sometimes subject to compression in what has been referred to as the tarsal tunnel syndrome (Lam).

In the foot, the medial plantar nerve and vessels accompany each other, and the lateral plantar nerve and vessels accompany each other; further, the medial plantar nerve is distributed almost exactly as is the median nerve in the hand, and the lateral plantar nerve almost exactly as is the ulnar nerve in the hand. The resemblance between the arteries of the foot and hand is not quite so close, for the arteries do not form, as in the hand, a superficial arch. However, the arterial systems of the foot and hand are modifications of the same fundamental plan: the lateral plantar artery corresponds to the ulnar in the hand, the medial plantar to the superficial palmar branch of the radial; and the deep plantar branch of the dorsalis pedis, contributing to the plantar arch, corresponds to the deep palmar branch of the radial artery.

Nerves

The medial and lateral plantar nerves enter the foot side by side, in company with the arteries; as they pass beyond the flexor retinaculum they lie under cover of the abductor hallucis, and continue into the central compartment of the foot where they separate from each other. The medial plantar nerve continues forward under cover of the medial edge of the short flexor; the lateral plantar nerve runs obliquely toward the lateral side of the foot, passing between the flexor digitorum brevis and the quadratus plantae muscles, and then running forward somewhat under cover of the lateral edge of the short flexor.

The *medial plantar nerve* after separating from the lateral runs forward to appear at the medial edge of the flexor digitorum brevis (Fig. 9-92). Here, just deep to the plantar aponeurosis, it divides into its digital branches. Usually the first to arise is the proper digital branch for the medial side of the big toe, with the main stem then subsequently dividing, in a variable pattern, into three common digital nerves destined for the adjacent sides of the four medial digits.

In its course the medial plantar nerve and its digital branches give off cutaneous twigs that pierce the fascia of the sole to be distributed to the skin of the medial side of the foot. While it is in the back part of the foot, and before it has divided into digital branches, the medial plantar nerve gives off two *muscular branches* (Fig. 9-93), one of which enters the deep surface of the abductor hallucis; the other runs laterally to reach the deep surface of the flexor digitorum brevis (the equivalent, it will be recalled, of the flexor digitorum su-

KNEE, LEG, ANKLE, AND FOOT

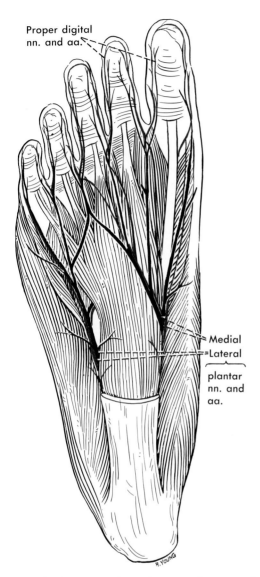

Proper digital
nn. and aa.

Medial
=Lateral
plantar
nn. and
aa.

Fig. 9-92. Plantar arteries and nerves after removal of the plantar aponeurosis.

The proper digital nerve to the big toe runs distally and slightly medially from its origin, passing obliquely through the medial intermuscular septum to lie at first on the superficial surface of the flexor digitorum brevis, supply this, and then emerge along the medial side of the big toe. The three common digital nerves run forward across the flexor digitorum brevis toward the interspaces between the toes they are to supply, lying immediately deep to the plantar aponeurosis. In

Fig. 9-93. Muscular distribution of the medial plantar nerve.

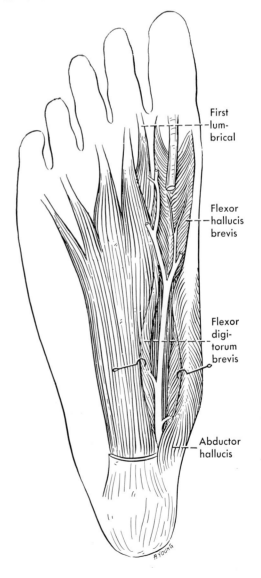

First
lumbrical

Flexor
hallucis
brevis

Flexor
digitorum
brevis

Abductor
hallucis

perficialis or sublimis in the forearm, supplied there by the median nerve). The other two muscular branches of the medial plantar nerve arise from digital ones: the flexor hallucis brevis receives a branch from the proper digital branch to the medial side of the big toe, while this lies on the superficial surface of the muscle; the first lumbrical receives a branch, on its superficial surface, from the common digital nerve to the adjacent sides of the big and second toes.

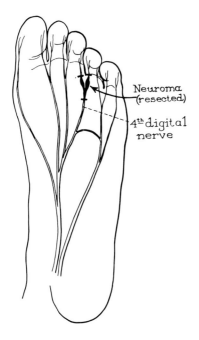

Fig. 9-94. Neuroma on the digital nerves to the third interspace. (Courtesy of Dr. H.H. Young)

the anterior part of the foot they appear, with the very tiny corresponding arteries, between the digitations that the plantar aponeurosis sends to the toes, and in this position each splits into two proper digital nerves. These supply skin of the lateral side of the big toe, both sides of the second and third toes, and the medial side of the fourth toe: thus the medial plantar nerve supplies, as does the median nerve in the hand, the skin of three and one-half digits. On the toes, the proper digital nerves supply not only the plantar surface but also skin of the distal part of the dorsum of the toes.

It may be recalled that the superficial branch of the ulnar nerve or its lateral digital branch frequently gives off a communication, of varying size, to the adjacent digital branch of the median nerve; similarly, in the foot the digital branch of the lateral plantar nerve to the adjacent surfaces of the fourth and fifth toes frequently gives a communication to the digital branch of the medial plantar nerve that supplies the adjacent sides of the third and fourth toes. This nerve, more commonly than other digital ones, may present a painful

neurofibroma (Fig. 9-94), the syndrome of which is known as *Morton's toe.*

Betts, in 1940, apparently first identified the lesion as a neurofibroma; he suggested that it occurs particularly on the third common digital nerve because that nerve's double origin somewhat anchors it and exposes it more to trauma, especially from stretch across the deep transverse metatarsal ligament. Bickel and Dockerty pointed out that a neuroma may be on this nerve when it is not formed by anastomosis between the medial and lateral plantars, that nerves to other interspaces than this may sometimes show a neurofibroma, and that even in the case of this nerve the site of the lesion is not constant: they said it sometimes lies on the branch of the medial plantar before it bifurcates, sometimes lies at the bifurcation, and sometimes is on one of the proper digital branches. They expressed the belief that trauma resulting from weight-bearing in small or ill-fitting shoes may be responsible for the lesion. Hauser ('71) agreed that it is a traumatic or inflammatory lesion, or both. Among 100 patients, most of them females, the lesion was between the second and third toes in 52, between the third and fourth in only 44, and between the fourth and fifth in four.

Nissen suggested that the cause of damage is not direct injury to the nerve, but injury to the accompanying digital artery. He said the digital artery showed changes ranging from thickening of its wall and slight adhesion to the deep transverse metatarsal ligament up to degeneration in which the vessel was represented by a fibrous strand. Fett and Pool studied the associated artery in four cases of neuroma that they removed, and said that while degenerative changes were present in one, and a second was thrombotic, the other two were entirely normal. It is apparent that the direct cause of the neuroma is not known. McKeever and others agree that the best approach for removal of such a neuroma is a dorsal one.

The *lateral plantar nerve* has a more complicated distribution than does the medial, for although its digital branches are fewer it supplies a majority of the muscles of the foot.

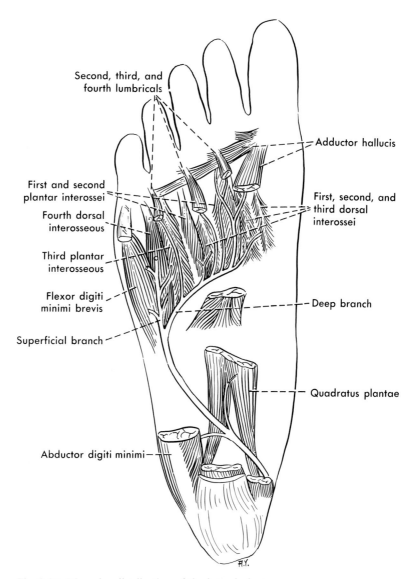

Fig. 9-95. Muscular distribution of the lateral plantar nerve.

It leaves the medial plantar nerve to run obliquely across the foot, passing from under cover of the abductor hallucis to run between the flexor digitorum brevis and the quadratus plantae, emerge between the flexor digitorum brevis and the abductor digiti minimi, and divide here into a superficial and a deep branch (Fig. 9-95).

The superficial branch continues forward on the surface of the flexor digiti minimi brevis and divides into two digital branches, a proper digital one for the lateral side of the little toe and a common digital one for the adjacent sides of the fourth and fifth toes. The latter appears, as do the common digital branches of the medial plantar, in the looser connective tissue between the digitations of the plantar aponeurosis, and divides into proper digital branches that are distributed to the adjacent sides of the two toes. As already noted, this branch frequently gives a communication to the most lateral digital branch of the medial plantar, the communication necessarily crossing the superficial surface of the flexor digitorum brevis muscle or the flexor tendons. The deep branch turns medially and more deeply, curving medially across the foot above the flexor tendons and against the plantar surface of the interossei, in company with the plantar arch.

In its course the lateral plantar nerve and its digital branches give off plantar cutaneous branches that pierce the deep fascia of the foot and supply the lateral part of the sole. While it lies between the flexor digitorum and the quadratus plantae, it gives off its muscular branch into the latter muscle, and usually at about the same level a more laterally directed branch to the abductor digiti minimi. Its other muscular branches are derived from both the superficial and the deep branches (Fig. 9-95): the superficial branch supplies the flexor digiti minimi brevis and, often, the two most lateral interossei, the third plantar and fourth dorsal; the deep branch, in its course across the foot, then supplies all the remaining interossei, gives twigs into the deep surfaces of the three lateral lumbricals, and ends in the adductor hallucis. The difference between the lateral plantar and ulnar nerves is therefore only a very slight one, in that the superficial branch of the lateral plantar nerve frequently supplies some muscles corresponding to those innervated normally by the deep branch of the ulnar nerve.

Blood Vessels

The very small arteries of the sole of the foot are not accompanied by corresponding veins, for the superficial drainage is into the still more superficial network of veins in the subcutaneous tissue. The deeper arteries are, however, accompanied by veins. The deep veins of the foot begin in the plexuses on the plantar surface of the toes, as the plantar digital veins; these in turn unite to form four plantar metatarsal veins that run backward in the metatarsal spaces, with metatarsal arteries, and unite to form a plantar venous arch that accompanies the plantar arterial arch. Medial and lateral plantar veins, arising from the ends of the arch, accompany the medial and lateral plantar arteries and unite deep to the flexor retinaculum to form the posterior tibial veins. The deep plantar venous arch and the medial and lateral plantar veins are typically doubled.

The arteries of the foot, the medial and lateral plantars (Figs. 9-96 and 9-97), are unequal in size. The *medial plantar artery* is rather small: it passes forward in the central compartment with the medial plantar nerve and gives off one or more deep branches to the muscles and other structures of the foot. Its superficial branch continues forward and divides into tiny digital branches that accompany the digital branches of the nerve across the lower surface of the flexor digitorum brevis and join the metatarsal arteries, derived from the plantar arch, of the first to third spaces. Its most medial branch usually contributes to, and may form, the branch along the medial side of the big toe.

The *lateral plantar artery*, much larger than the medial, passes obliquely laterally and forward across the foot in company with the lateral plantar nerve, and thus between the flexor digitorum brevis superficially and the quadratus plantae deeply. It emerges with the nerve to lie briefly at the lateral edge of the flexor digitorum brevis, and then turns deeply and medially, across the plantar surface of the interossei, to help form the plantar arch. In its course it gives twigs into the muscles it crosses and also branches that pass between the flexor digitorum brevis and the abductor digiti minimi to supply skin of the lateral part of the sole of the foot. Just before it turns deeply, it gives off a digital artery for the lateral side of the little toe; this may or may not send a branch of communication to the fourth plantar metatarsal artery.

The *plantar arch* curves across the foot, lying on the bases of the metatarsals and the proximal ends of the interossei and therefore deep to the long and short flexor tendons and the oblique head of the adductor hallucis. It is completed laterally by the deep branch of the lateral plantar artery, and medially, in the interspace between the bases of the first and second metatarsals, by the deep plantar branch of the dorsalis pedis. On the basis of whether the smallest diameter of the arch was lateral or medial, Vann said the deep plantar

Fig. 9-96. Diagram of the plantar arteries and the plantar arch.

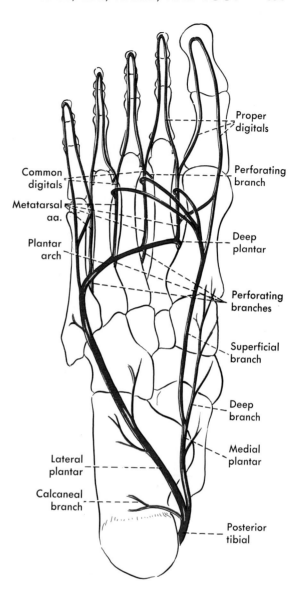

branch of the dorsalis pedis was the chief contributor to the arch in almost 81% of 361 feet, the deep branch of the lateral plantar in about 15%, and in the remainder the site of smallest diameter was doubtful. A not uncommon variation is to find the dorsalis pedis artery much reduced on the dorsum of the foot and therefore contributing little or nothing to the plantar arch. Even when a posterior tibial artery proper is not present in the leg the lateral plantar artery is present, for it then comes from the lower end of the peroneal artery.

The chief branches of the plantar arch, in addition to muscular and articular twigs, are the plantar metatarsal arteries. These are the chief supply of the adjacent sides of the five digits, since the corresponding superficial arteries are typically tiny. The four plantar metatarsal arteries run forward on the surfaces of the interossei; each divides at the bases of the digits into proper digital arteries that are distributed to the adjacent sides of the two toes between which the vessel lies. Shortly before they divide, each typically gives off a perforating branch (anterior per-

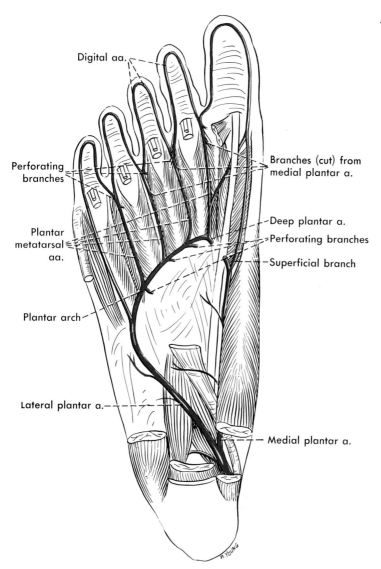

Fig. 9-97. The plantar arch.

Digital aa.

Perforating
branches

Plantar
metatarsal
aa.

Plantar arch

Lateral plantar a.

Branches (cut) from
medial plantar a.

Deep plantar a.

Perforating branches

Superficial branch

Medial plantar a.

forating branch) which passes between the heads of the metatarsal bones, proximal to the deep transverse metatarsal ligament, to join the corresponding dorsal metatarsal artery. Farther distally, in the web, and distal to the deep transverse metatarsal ligament, there may be another communication with the dorsal metatarsal artery; according to Huber, this is a termination of the dorsal metatarsal artery after it has given off its dorsal digital branches.

The other important set of branches from the plantar arch are perforating branches (posterior perforating branches) that arise from the arch as it lies on the proximal parts of the origins of the interossei. There are three perforating branches, each one of which passes dorsally in the interspace between the two heads of the dorsal interosseous muscle associated with that intermetatarsal space, to anastomose with the proximal end of the dorsal metatarsal artery. Sometimes these perforating branches form the chief source of blood to the dorsal metatarsal arteries.

References

ABBOTT LC, CARPENTER WF: Surgical approaches to the knee joint. J Bone Joint Surg 27:277, 1945

ABBOTT LC, GILL GG: Surgical approaches to the epiphysial cartilages of the knee and ankle joints. Arch Surg 46:591, 1943

ANDERSON KJ, LECOCQ JF, LECOCQ EA: Recurrent anterior subluxation of the ankle joint: A report of two cases and an experimental study. J Bone Joint Surg 34-A:853, 1952

BARLOW TE: The deep flexors of the foot. J Anat 87:308, 1953

BARNETT CH: Locking at the knee joint. J Anat 87:91, 1953

BARNETT CH, NAPIER JR: The axis of rotation at the ankle joint in man. Its influence upon the form of the talus and the mobility of the fibula. J Anat 86:1, 1952

BARNETT CH, RICHARDSON AT: The postural function of the popliteus muscle. Ann Physical Med 1:177, 1953

BARRY D, BOTHROYD JS: Tensor fasciae suralis. J Anat 58:382, 1924

BASMAJIAN JV: Muscles Alive. Their Functions Revealed by Electromyography, ed. 3. Baltimore, Williams & Wilkins, 1974

BASMAJIAN JV, LOVEJOY JF JR: Functions of the popliteus muscle in man. A multifactorial electromyographic study. J Bone Joint Surg 53-A:557, 1971

BASMAJIAN JV, STECKO G: The role of muscles in arch support of the foot. An electromyographic study. J Bone Joint Surg 45-A:1184, 1963

BECHTOL CO, MOSSMAN HW: Clubfoot: An embryological study of associated muscle abnormalities. J Bone Joint Surg 32-A:827, 1950

BERSON BL, HERMANN G: Torn discoid menisci of the knee in adults. Four case reports. J Bone Joint Surg 61-A:303, 1979

BETTS LO: Morton's metatarsalgia: neuritis of the fourth digital nerve. M J Australia 1:514, 1940

BICKEL WH: Personal communication to the author, 1968

BICKEL WH, DOCKERTY MB: Plantar neuromas, Morton's toe. Surg Gynecol Obstet 84:111, 1947

BICKEL WH, MOE JH: Translocation of the peroneus longus tendon for paralytic calcaneus deformity of the foot. Surg Gynecol Obstet 78:627, 1944

BLOCKEY NJ, SMITH MGH: The treatment of congenital club foot. J Bone Joint Surg 48-B:660, 1966

BOSWORTH DM: Transplantation of the semitendinosus for repair of laceration of medial collateral ligament of the knee. J Bone Joint Surg 34-A:196, 1952

BOYD AM, RATCLIFFE AH, JEPSON RP, JAMES GWH: Intermittent claudication: A clinical study. J Bone Joint Surg 31-B:325, 1949

BRADLEY EL III: The anterior tibial compartment syndrome. Surg Gynecol Obstet 136:289, 1973

BRANTIGAN OC, VOSHELL AF: The mechanics of the ligaments and menisci of the knee joint. J Bone Joint Surg 23:44, 1941

BRANTIGAN OC, VOSHELL AF: The tibial collateral ligament: Its function, its bursae, and its relation to the medial meniscus. J Bone Joint Surg 25:121, 1943

BRATTSTRÖM H: Shape of the intercondylar groove normally and in recurrent dislocation of patella. A clinical and x-ray anatomical investigation. Acta orthoped scand suppl 68, 1964

BURKARD O: Zwei seltene Anomalien an der Musculatur der tiefen Lage der Hinterseite des Unterschenkels. Arch f Anat u Physiol (Anat Abt) 1902, p 344

BURMAN MS, SUTRO CJ: A study of the degenerative changes of the menisci of the knee joint, and the clinical significance thereof. J Bone Joint Surg 15:835, 1933

BUTLER DL, NOYES FR, GROOD ES: Ligamentous restraints to anterior–posterior drawer in the human knee. A biomechanical study. J Bone Joint Surg 62-A:259, 1980

CAHILL DR: The anatomy and function of the contents of the human tarsal sinus and canal. Anat Rec 153:1, 1965

CALDWELL GA: Arthrodeses of the feet. Am Acad Orth Surgeons Instructional Course Lectures 6:174, 1949

CASAGRANDE PA, AUSTIN BP, INDECK W: Denervation of the ankle joint. J Bone Joint Surg 33-A:723, 1951

CAVE EF: Combined anterior–posterior approach to the knee joint. J Bone Joint Surg 17:427, 1935

CAVE EF, ROWE CR: The patella: Its importance in derangement of the knee. J Bone Joint Surg 32-A:542, 1950

CHANDLER SB: The semilunar cartilages of the knee joint. Anat Rec 103:531, 1949 (abstr)

CHILDRESS HM: Popliteal cysts associated with undiagnosed posterior lesions of the medial meniscus. The significance of age in diagnosis and treatment. J Bone Joint Surg 52-A:1487, 1970

CHO KD: Reconstruction of the anterior cruciata ligament by semitendinosus tendon. J Bone Joint Surg 57-A:608, 1975

COLONNA PC, RALSTON EL: Operative approaches to the ankle joint. Am J Surg 82:44, 1951

COLTART WD: "Aviator's astragalus." J Bone Joint Surg 34-B:545, 1952

CRAVENER EK, MACELROY DG: Supernumerary tarsal scaphoids. Surg Gynecol Obstet 71:218, 1940

CRITCHLEY JE, TAYLOR RG: Transfer of the tibialis anterior tendon for relapsed clubfoot. J Bone Joint Surg 34-B:49, 1952

CUMMINS EJ, ANSON BJ, CARR BW, WRIGHT RR, HAUSER EDW: The structure of the calcaneal tendon (of Achilles) in relation to orthopedic surgery: With additional observations on the plantaris muscle. Surg Gynecol Obstet 83:107, 1946

DANKMEIJER J: Congenital absence of the tibia. Anat Rec 62:179, 1935

DASELER ED, ANSON BJ: The plantaris muscle: An anatomical study of 750 specimens. J Bone Joint Surg 25:822, 1943

DE PALMA AF: Bursitis under fibular collateral ligament. Ann Surg 127:564, 1948

DE PALMA AF: Diseases of the knee: Management in Medicine and Surgery. Philadelphia, JB Lippincott, 1954

DEVAS MB: Shin splints, or stress fractures of the metacarpal bone in horses, and shin soreness, or stress fractures of the tibia, in man. J Bone Joint Surg 49-B:310, 1967

DICKINSON PH, COUTTS MG, WOODWARD EP, HANDLER D: Tendo Achillis bursitis. Report of twenty-one cases. J Bone Joint Surg 48-A:77, 1966

DICKSON FD: Fractures of the ankle. Am J Surg 38:709, 1937

DIVELEY RL: Functional disorders of the foot. Am Acad Orth Surgeons Instructional Course Lectures 6:152, 1949

DRIVER JR, DENISON AB: The morphology of the long accessorius muscle. Anat Rec 8:341, 1914

DUCHENNE GB: Physiology of Motion Demonstrated by Means of Electrical Stimulation and Clinical Observation and Applied to the Study of Paralysis and Deformities. Translated and edited by EB Kaplan. Philadelphia, JB Lippincott, 1949

DWIGHT T: A Clinical Atlas: Variations of the Bones of the Hands and Feet. Philadelphia, JB Lippincott, 1907

DWIGHT T: Free cuboides secundarium on both feet, with some further remarks on Pfitzner's theory. Anat Anz 39:410, 1911

ELFTMAN H: A cinematic study of the distribution of pressure in the human foot. Anat Rec 59:481, 1934

EYRE-BROOK AL: Congenital vertical talus. J Bone Joint Surg 49-B:618, 1967

FENTON RL: Surgical repair of a torn tibial collateral ligament of the knee by means of the semitendinosus tendon (Bosworth procedure). Report of twenty-eight cases. J Bone Joint Surg 39-A:304, 1957

FERGUSON LK, THOMPSON WD: Internal derangements of the knee joint: Analysis of 100 cases with follow-up study. Ann Surg 112:454, 1940

FETT HC, POOL CC: Plantar interdigital neuroma or Morton's toe. Am J Surg 78:522, 1949

FLINCHUM D: Pathological anatomy in talipes equinovarus. J Bone Joint Surg 35-A:111, 1953

FONDA MP: Dislocation of the tibiotalar joint without fracture: An unusual ski injury. J Bone Joint Surg 34-A:662, 1952

FRIED A: Recurrent congenital club-foot. The role of the M. tibialis posterior in etiology and treatment. J Bone Joint Surg 41-A:243, 1959

FULLERTON A: The surgical anatomy of the synovial membrane of the knee-joint. Brit J Surg 4:191, 1916

GARCEAU GJ, PALMER RM: Transfer of the anterior tibial tendon for recurrent club foot. A long-term follow-up study. J Bone Joint Surg 49-A:207, 1967

GARDEN RS: Ossification in the anterior cruciate ligament. J Bone Joint Surg 21:1027, 1939

GARDNER E: The innervation of the knee joint. Anat Rec 101:109, 1948

GARDNER E, GRAY D: The innervation of the joints of the foot. Anat Rec 161:141, 1968

GARTLAND JJ: Posterior tibial transplant in the surgical treatment of recurrent club foot. A preliminary report. J Bone Joint Surg 46-A:1217, 1964

GETZEN LC, CARR JE III: Etiology of anterior tibial compartment syndrome. Surg Gynecol Obstet 125:347, 1967

GHORMLEY RK: Late joint changes as a result of internal derangements of the knee. Am J Surg 76:496, 1948

GHORMLEY RK, SPEAR IM: Anomalies of the posterior tibial tendon: A cause of persistent pain about the ankle. AMA Arch Surg 66:512, 1953

GRAY JE: Congenital absence of the tibia. Anat Rec 101:265, 1948

GRISTINA AG, WILSON PD: Popliteal cysts in adults and children. A review of 90 cases. Arch Surg 88:357, 1964

GRODINSKY M: A study of the tendon sheaths of the foot and their relation to infection. Surg Gynecol Obstet 51:460, 1930

HAINES RW, MCDOUGALL A: The anatomy of hallux valgus. J Bone Joint Surg 36-B:272, 1954

HALL JE, SALTER RB, BHALLA SK: Congenital short tendo calcaneus. J Bone Joint Surg 49-B:695, 1967

HALLISY JE: The muscular variations in the human foot. A quantitative study. General results of the study: I. Muscles of the inner border of the foot and the dorsum of the great toe. Am J Anat 45:411, 1930

HALLOCK H: The surgical treatment of common mechanical and functional disabilities of the feet. Am Acad Orth Surgeons Instructional Course Lectures 6:160, 1949

HARBESON AE: Further studies on the origin of the first dorsal interosseous muscle of the foot from the tendon of the peronaeus longus. J Anat 72:463, 1938

HARDY RH: Observations on the structure and properties of the plantar calcaneo-navicular ligament in man. J Anat 85:135, 1951

HARMON PH, FAHEY JJ: The syndrome of congenital absence of the fibula: Report of 3 cases with special reference to pathogenesis and treatment. Surg Gynecol Obstet 64:876, 1937

HARRIS RI, BEATH T: The short first metatarsal: Its incidence and clinical significance. J Bone Joint Surg 31-A:553, 1949

HARROLD AJ: Congenital vertical talus in infancy. J Bone Joint Surg 49-B:634, 1967

HAUSER EDW: Extra-articular repair for ruptured collateral and cruciate ligaments. Surg Gynecol Obstet 84:339, 1947

HAUSER EDW: Interdigital neuroma of the foot. Surg Gynecol Obstet 133:265, 1971

HAWKINS LG: Fracture of the lateral process of the talus. A review of thirteen cases. J Bone Joint Surg 47-A:1170, 1965

HAXTON H: The function of the patella and the effects of its excision. Surg Gynecol Obstet 80:389, 1945

HECKER P: Study on the peroneus of the tarsus: Preliminary notes. Anat Rec 26:79, 1923

HELLER L, LANGMAN J: The menisco-femoral ligaments of the human knee. J Bone Joint Surg 46-B:307, 1964

HENDERSON MS: Fractures of the bones of the foot—except the os calcis. Surg Gynecol Obstet 64:454, 1937

HEYMAN CH, HERNDON CH: Congenital posterior angulation of the tibia. J Bone Joint Surg 31-A:571, 1949

HEYMAN CH, HERNDON CH, HEIPLE KG: Congenital posterior angulation of the tibia with talipes calcaneus. A long-term report of eleven patients. J Bone Joint Surg 41-A:476, 1959

HEYWOOD AWB: Recurrent dislocation of the patella. A study of its pathology and treatment in 106 knees. J Bone Joint Surg 43-B:508, 1961

HICKS JH: The mechanics of the foot. I. The joints. J Anat 87:345, 1953

HICKS JH: The mechanics of the foot. II. The plantar aponeurosis and the arch. J Anat 88:25, 1954

HIRSCHTICK AB: An anomalous tarsal bone. J Bone Joint Surg 33-A:907, 1951

HORWITZ MT: An investigation of the surgical anatomy of the ligaments of the knee joint. Surg Gynecol Obstet 67:287, 1938a

HORWITZ MT: Normal anatomy and variations of the peripheral nerves of the leg and foot: Application in operations for vascular diseases; study of one hundred specimens. Arch Surg 36:626, 1938b

HORWITZ T: Supernumerary metatarsal bone and toe: Case report. Am J Surg 50:578, 1940

HORWITZ T, LAMBERT RG: Patellectomy in the military service: A report of 19 cases. Surg Gynecol Obstet 82:423, 1946

HUBER JF: The arterial network supplying the dorsum of the foot. Anat Rec 80:373, 1941

HUELKE DF: The origin of the peroneal communicating nerve in adult man. Anat Rec 132:81, 1958

HUGHES ESR, SUNDERLAND S: The tibial tuberosity and the insertion of the ligamentum patellae. Anat Rec 96:439, 1946

HUMPHRY PROF: Note on the dissection of flatfoot. J Anat Physiol 25:102, 1890

HUTTER CG JR, SCOTT W: Tibial torsion. J Bone Joint Surg 31-A:511, 1949

INGE GAL, FERGUSON AB: Surgery of the sesamoid bones of the great toe: An anatomic and clinical study, with a report of forty-one cases. Arch Surg 27:466, 1933

IPPLITOT E, PONSETI IV: Congenital club foot in the human fetus. J Bone Joint Surg 62-A:8, 1980

IRANI RN, SHERMAN MS: The pathological anatomy of club foot. J Bone Joint Surg 45-A:45, 1963

ISAACS H, HANDELSMAN JE, BADENHORST M, PICKERING A: The muscles in club foot—A histological, histochemical and electron microscopic study. J Bone Joint Surg 59-B:465, 1977

JOHNSON EW JR, PETERSON HA: Fractures of the os calcis. Arch Surg 92:848, 1966

JONES FW: Voluntary muscular movements in cases of nerve lesions. J Anat 54:41, 1919

JONES KG: Reconstruction of the anterior cruciate ligament: A technique using the central one-third of the patellar ligament. J Bone Joint Surg 45-A:925, 1963

JONES RE, SMITH EC, REISCH JS: Effects of medial meniscectomy in patients older than forty years. J Bone Joint Surg 60-A:783, 1978

JONES RL: The human foot: An experimental study of its mechanics, and the role of its muscles and ligaments in the support of the arch. Am J Anat 68:1, 1941

KAMEL R, SAKLA FB: Anatomical compartments of the sole of the human foot. Anat Rec 140:57, 1960

KAPLAN EB: Discoid lateral meniscus of the knee joint. Nature, mechanism, and operative treatment. J Bone Joint Surg 39-A:77, 1957a

KAPLAN EB: Surgical approach to the lateral (peroneal) side of the knee joint. Surg Gynecol Obstet 104:346, 1957b

KAPLAN EB: The fabellofibular and short lateral ligaments of the knee joint. J Bone Joint Surg 43-A:169, 1961

KAPLAN EB: Some aspects of functional anatomy of the human knee joint. Clin Orthopaedics 23:18, 1962

KATZ MP, GROGONO BJS, SOPER KC: The etiology and treatment of congenital dislocation of the knee. J Bone Joint Surg 49-B:112, 1967

KAUFER H: Mechanical function of the patella. J Bone Joint Surg 53-A:1551, 1971

KENNEDY JC, FOWLER PJ: Medial and anterior instability of knee. An anatomical and clinical study using stress machines. J Bone Joint Surg 53-A:1257, 1971

KETTELKAMP DB, JACOB AW: Tibiofemoral contact areas—Determination and implications. J Bone Joint Surg 54-A:349, 1972

KEY JA: Partial rupture of the tendon of the posterior tibial muscle. J Bone Joint Surg 35-A:1006, 1953

KING D: Regeneration of the semilunar cartilage. Surg Gynecol Obstet 62:167, 1936a

KING D: The function of semilunar cartilages. J Bone Joint Surg 18:1069, 1936b

KITE JH: Congenital metatarsus varus: Report of 300 cases. J Bone Joint Surg 32-A:500, 1950

KOSINSKI C: The course, mutual relatiòns and distribution of the cutaneous nerves of the metazonal region of leg and foot. J Anat 60:274, 1926

LAM SJS: Tarsal tunnel syndrome. J Bone Joint Surg 49-B:87, 1967

LAMB DW: Deposition of calcium salts in the medial ligament of the knee. J Bone Joint Surg 34-B:233, 1952

LAMBERT EH: The accessory deep peroneal nerve. A common variation in innervation of extensor digitorum brevis. Neurology 19:1169, 1969

LAMBERT KL: The weight-bearing function of the fibula. A strain gauge study. J Bone Joint Surg 53-A:507, 1971

LAPIDUS PW: Sesamoids beneath all the metatarsal heads of both feet: Report of a case. J Bone Joint Surg 22:1059, 1940

LAST RJ: Some anatomical details of the knee joint. J Bone Joint Surg 30-B:683, 1948

LAST RJ: The popliteus muscle and the lateral meniscus: With a note on the attachment of the medial meniscus. J Bone Joint Surg 32-B:93, 1950

LEA RB, SMITH L: Nonsurgical treatment of tendo Achillis ruptures. J Bone Joint Surg 54-A:1398, 1972

LEACH RE, HAMMOND G, STRYKER WS: Anterior tibial compartment syndrome: Acute and chronic. J Bone Joint Surg 49-A:451, 1967

LEONARD MH: Injuries of the lateral ligaments of the ankle: A clinical and experimental study. J Bone Joint Surg 31-A:373, 1949

LEWIS OJ: The comparative morphology of M. flexor accessorius and the associated long flexor tendons. J Anat 96:321, 1962

LIPSCOMB PR, HENDERSON MS: Internal derangements of the knee. JAMA 135:827, 1947

LOVEJOY JF JR, HARDEN TP: Popliteus muscle in man. Anat Rec 169:727, 1971

LOVELL AGH, TANNER HH: Synovial membranes, with special reference to those related to the tendons of the foot and ankle. J Anat Physiol 42:415, 1908

LUSBY HLJ: Naviculo-cuneiform synostosis. J Bone Joint Surg 41-B:150, 1959

LUTFI AM: Morphological changes in the articular cartilage after meniscectomy. An experimental study in the monkey. J Bone Joint Surg 57-B:525, 1975

MANGIERI JV: Peroneal nerve injury from an enlarged fabella. A case report. J Bone Joint Surg 55-A:395, 1973

MANN R, INMAN VT: Phasic activity of intrinsic muscles of the foot. J Bone Joint Surg 46-A:469, 1964

MANN RA, HAGY JL: The popliteus muscle. J Bone Joint Surg 59-A:924, 1977

MANTER JT: Movements of the subtalar and transverse tarsal joints. Anat Rec 80:397, 1941

MANTER JT: Variations of the interosseous muscles of the foot. Anat Rec 93:117, 1945

MANTER JT: Distribution of compression forces in the joints of the human foot. Anat Rec 96:313, 1946

MAQUET PG, VAN DE BERG AJ, SIMONET JC: Femorotibial weight-bearing areas. Experimental determination. J Bone Joint Surg 57-A:766, 1975

MARTIN BF: Observations on the muscles and tendons of the medial aspect of the sole of the foot. J Anat 98:437, 1964

MATHUR PD, MCDONALD JR, GHORMLEY RK: A study of the tensile strength of the menisci of the knee. J Bone Joint Surg 31-A:650, 1949

MATSEN FA, CLAWSON DK: The deep posterior compartmental syndrome of the leg. J Bone Joint Surg 57-A:34, 1975

MAYER L: Tendon transplantations on the lower extremity. Am Acad Orth Surgeons Instructional Course Lectures 6:189, 1949

MAYER V, POHLIDAL S: Ankle mortise injuries. Surg Gynecol Obstet 96:99, 1953

MCBRIDE ED: Hammer toe. Am J Surg 44:319, 1939

MCDOUGALL A: The os trigonum. J Bone Joint Surg 37-B:257, 1955

MCGINTY JB, GEUSS LF, MARVIN RA: Partial or complete meniscectomy. A comparative study. J Bone Joint Surg 59-A:763, 1977

MCKEEVER DC: Surgical approach for neuroma of plantar digital nerve (Morton's metatarsalgia). J Bone Joint Surg 34-A:490, 1952

MCMURRAY TP: Internal derangements of the knee joint. Ann Roy Coll Surgeons England 3:210, 1948

M'CONNELL AA: A case of fusion of the semilunar and cuneiform bones. J Anat Physiol 41:302, 1907

MENSOR MC, ORDWAY GL: Traumatic subcutaneous rupture of the tibialis anterior tendon. J Bone Joint Surg 35-A:675, 1953

MEYERS MH: Isolated avulsion of the tibial attachment of the posterior cruciate ligament of the knee. J Bone Joint Surg 57-A:669, 1975

MICHELE AA: Relaxed anterior cruciate ligament. Surgery 27:588, 1950

MICHELE AA, KRUEGER FJ: Patella fractures: A method of wiring. Surgery 24:100, 1948

MILCH H: Fascial reconstruction of the tibial collateral ligament. Surgery 10:811, 1941

MILCH H: Reinforcement of the deltoid ligament for pronated flat foot: Inversion fasciodesis of os calcis. Surg Gynecol Obstet 74:876, 1942

MORRISON GM: Fractures of the bones of the feet. Am J Surg 38:721, 1937

MORTON DJ: The Human Foot: Its Evolution, Physiology and Functional Disorders. New York, Columbia University Press, 1935

MUCKART RD: Compression of the common peroneal nerve by intramuscular ganglion from the superior tibiofibular joint. J Bone Joint Surg 58-B:241, 1976

MURPHY MP, GUTEN GN, SEPIC SB, GARDNER GM, BALDWIN JM: Function of the triceps surae during gait. Compensatory mechanism for unilateral loss. J Bone Joint Surg 60-A:473, 1978

NATHANIAL D: A note on the variation of the flexor digitorum brevis. J Anat Soc India 3:103, 1954

NELSON GE JR, KELLY PJ, PETERSON LFA, JANES JM: Blood supply of the human tibia. J Bone Joint Surg 42-A:625, 1960

NEVIN JE, POST RH: Lateral ligament ankle sprains. A clinical-pathological correlation based on the production and examination of controlled sprains in preamputated extremities, J Trauma 4:292, 1964

NISSEN KI: Plantar digital neuritis: Morton's metatarsalgia. J Bone Joint Surg 30-B:84, 1948

NORLEY T, BICKEL WH: Calcification of the bursae of the knee. J Bone Joint Surg 31-A:417, 1949

NUTT JJ: Effect on the knee of a shortened gastrocnemius muscle. Am J Surg 17:113, 1932

O'CONNELL AL: Electromyographic study of certain leg muscles during movements of the free foot and during standing. Am J Physical Med 37:289, 1958

O'CONNOR BL, MCCONNAUGHEY JS: The structure and innervation of cat knee menisci, and their relation to a "sensory hypothesis" of meniscal function. Am J Anat 153:431, 1978

O'DONOGHUE DH: An analysis of end results of surgical treatment of major injuries to the ligaments of the knee. J Bone Joint Surg 37-A:1, 1955

O'DONOGHUE DH: A method of replacement of the anterior cruciate ligament of the knee. Report of twenty cases. J Bone Joint Surg 45-A:905, 1963

OETTEKING B: Anomalous patellae. Anat Rec 23:269, 1922

O'RAHILLY R: Morphological patterns in limb deficiencies and duplications. Am J Anat 89:135, 1951

OXFORD TM: Pellegrini-Stieda's disease. Am J Surg 27:543, 1935

PARSONS FG: Note on abnormal muscle in popliteal space. J Anat 54:170, 1920

PARSONS FG, KEITH A: Seventh report of the committee of collective investigation of the Anatomical Society of Great Britain and Ireland, for the year 1896–97. J Anat Physiol 32:164, 1897

PARSONS FG, ROBINSON A: Eighth report of the committee of collective investigation of the Anatomical Society of Great Britain and Ireland, for the year 1897–98. J Anat Physiol 33:189, 1898

PASCHALL J JR, GHORMLEY RK, DOCKERTY MB: Old united and ununited fractures of the patella. Surgery 26:777, 1949

PATTERSON WR, FITZ DA, SMITH WS: The pathological anatomy of congenital convex pes valgus: Post mortem study of a newborn infant with bilateral involvement. J Bone Joint Surg 50-A:458, 1968

PEABODY CW: Tendon transposition in the paralytic foot. Am Acad Orth Surgeons Instructional Course Lectures 6:178, 1949

PERLOW S, HALPERN SS: Surgical relief of pain due to circulatory disturbances of the feet: Report of a new method. Am J Surg 45:104, 1939

PICK JW, STACK JK, ANSON BJ: Measurements on the human femur: II. Lengths, diameters and angles (Concl'd). Quart Bull Northwestern Univ M School 17:121, 1943

PIERSON HH: Seven arterial anomalies of the human leg and foot. Anat Rec 30:139, 1925

PINDER IM: Treatment of popliteal cyst in the rheumatoid knee. J Bone Joint Surg 55-B:119, 1973

QUIGLEY TB: The treatment of avulsion of the colateal ligaments of the knee. Am J Surg 78:574, 1949

RAUSCHNING W: Popliteal cysts and their relation to the gastrocnemico-semimembranosus bursa. Studies on the surgical and functional anatomy. Acta orthop scandinav suppl 179, 1979

REIDY JA, BRODERICK TF JR, BARR JS: Tendon transplantations in the lower extremity: A review of end results in poliomyelitis. I. Tendon transplantations about the foot and ankle. J Bone Joint Surg 34-A:900, 1952

RIDER DL: Fractures of the metacarpals, metatarsals and phalanges. Am J Surg 38:549, 1937

ROBERTS PW: An unusual congenital anomaly of the bones of the leg. J Bone Joint Surg 12:414, 1930

ROSS JA, TOUGH ICK, ENGLISH TA: Congenital discoid cartilage, with an embryological note. J Bone Joint Surg 40-B:262, 1958

ROSS RF: A quantitative study of rotation of the knee joint in man. Anat Rec 52:209, 1932

SACHS AE, RUBINSTEIN L: Bilateral popliteal hernia: Case report. Surgery 20:385, 1946

SARRAFIAN SK, TOPOUZIAN LK: Anatomy and physiology of the extensor apparatus of the toes. J Bone Joint Surg 51-A:669, 1969

SCAPINELLI R: Blood supply of the human patella: Its relation to ischaemic necrosis after fracture. J Bone Joint Surg 49-A:563, 1967

SCHOOLFIELD BL: Operative treatment of flat-foot. Surg Gynecol Obstet 94:136, 1952

SCHREIBER RR: Talonavicular synostosis. J Bone Joint Surg 45-A:170, 1963

SCHWARTZ RP, HEATH AL: Conservative treatment of functional disorders of the feet in the adolescent and adult. J Bone Joint Surg 31-A:501, 1949

SCOTT JC: Fractures of the patella. J Bone Joint Surg 31-B:76, 1949

DEL SEL JM, GRAND NE: Cubo-navicular synostosis. A rare tarsal anomaly. J Bone Joint Surg 41-B:149, 1959

SENIOR HD: An interpretation of the recorded arterial anomalies of the human leg and foot. J Anat 53:130, 1919

SENIOR HD: Abnormal branching of the human popliteal artery. Am J Anat 44:111, 1929

SETTLE GW: The anatomy of congenital talipes equinovarus: Sixteen dissected specimens. J Bone Joint Surg 45-A:1341, 1963

SHAW JA, MURRAY DG: The longitudinal axis of the knee joint and the role of the cruciate ligament in controlling transverse rotation. J Bone Joint Surg 56-A:1603, 1974

SHEPARD E: Tarsal movements. J Bone Joint Surg 33-B:258, 1951

SILK FF, WAINWRIGHT D: The recognition and treatment of congenital flat foot in infancy. J Bone Joint Surg 49-B:628, 1967

SIMON HE, SACCHET HA: Muscle hernias of the leg: Review of literature and a report of twelve cases. Am J Surg 67:87, 1945

SIMON SR, MANN RA, HAGY JL, LARSEN LJ: Role of the posterior calf muscles in normal gait. J Bone Joint Surg 60-A:465, 1978

SINGER M: Tibialis posterior transfer in congenital club foot. J Bone Joint Surg 43-B:717, 1961

SINGH I: Variations in the metatarsal bones. J Anat 94:345, 1960

SLOANE MWM: A case of anomalous skeletal development in the foot. Anat Rec 96:23, 1946

SLOCUM DB, LARSON RL: Rotatory instability of the knee. Its pathogenesis and a clinical test to demonstrate its presence. J Bone Joint Surg 50-A:211, 1968a

SLOCUM DB, LARSON RL: Pes anserinus transplantation. A surgical procedure for control of rotatory instability of the knee. J Bone Joint Surg 50-A:226, 1968b

SMITH FB, BLAIR HC: Tibial collateral ligament strain due to occult derangements of the medial meniscus: Confirmed by operation in thirty cases. J Bone Joint Surg 36-A:88, 1954

SMITH JW: Observations on the postural mechanism of the human knee joint. J Anat 90:236, 1956

SMITH JW: The ligamentous structures in the canalis and sinus tarsi. J Anat 92:616, 1958

SNEATH RS: The insertion of the biceps femoris. J Anat 89:550, 1955

STACK RE, BIANCO AJ JR, MACCARTY CS: Compression of the common peroneal nerve by ganglion cysts. Report of nine cases. J Bone Joint Surg 47-A:773, 1966

STARR DE: Repair of old ligamentous injuries of the knee. Clin Orthopaedics 23:162, 1962

STEINDLER A: Mechanics of Normal and Pathological Locomotion in Man. Springfield, Ill., Thomas, 1935

STEVENSON PH: On an unusual anomaly of the peroneus tertius in a Chinese. Anat Rec 22:81, 1921

STEWART SF: Club-foot: Its incidence, cause, and treatment. An anatomical-physiological study. J Bone Joint Surg 33-A:577, 1951

STOTT JRR, HUTTON WC, STOKES IAF: Forces under the foot. J Bone Joint Surg 55-B:335, 1973

STOYLE TF: Dupuytren's contracture in the foot. Report of a case. J Bone Joint Surg 46-B:218, 1964

SUNDERLAND S, HUGHES ESR: Metrical and non-metrical features of the muscular branches of the sciatic nerve and its medial and lateral popliteal divisions. J Comp Neurol 85:205, 1946

SUTHERLAND DH: An electromyographic study of the plantar flexors of the ankle in normal walking on the level. J Bone Joint Surg 48-A:66, 1966

SUTHERLAND DH, COOPER L, DANIEL D: The role of the ankle plantar flexors in normal walking. J Bone Joint Surg 62-A:354, 1980

SUTRO CJ, POMERANZ MM, SIMON SM: Fabella (sesamoid in the lateral head of the gastrocnemius). Arch Surg 30:777, 1935

SUTTON FS, THOMPSON CH, LIPKE J, KETTELKAMP DB: The effect of patellectomy on knee function. J Bone Joint Surg 58-A:537, 1976

TANASESCO J GH: Lymphatiques de l'articulation du genou. Anat Anz 39:490, 1911

TAYLOR H: Cysts of the fibrocartilages of the knee joint. J Bone Joint Surg 17:588, 1935

THOMPSON FR, BOSWORTH DM: Recurrent dislocation of the patella. Am J Surg 73:335, 1947

TREANOR WJ: Personal communication to the author, 1978

TROLLE D: Accessory Bones of the Human Foot: A Radiological, Histoembryological, Comparative-anatomical, and Genetic Study. Translated from the Danish by Elisabeth Aagesen. Copenhagen, Einar Munksgaard, 1948

TROTTER M: The level of termination of the popliteal artery in the white and the Negro. Am J Phys Anthropol 27:109, 1940

VANN HM: A note on the formation of the plantar arterial arch of the human foot. Anat Rec 85:269, 1943

VOLK H, SMITH FM: "Bucket-handle" tear of the medial meniscus in a five-year-old boy. J Bone Joint Surg 35-A:234, 1953

WAGNER LC, BUTTERFIELD WL: Surgical release of contracted tissues for resistant congenital clubfeet. Am J Surg 84:82, 1952

WANG C-C, WALKER PS: Rotatory laxity of the human knee joint. J Bone Joint Surg 56-A:161, 1974

WANG JB, RUBIN RM, MARSHALL JL: A mechanism of isolated anterior cruciate ligament rupture. Case report. J Bone Joint Surg 57-A:411, 1975

WEAVER JB: Ossification of the internal semilunar cartilage. J Bone Joint Surg 17:195, 1935

WEEKS C: The surgical importance of occasional communications between the synovial sacs of knee and proximal tibiofibular joints. Am J Surg 8:798, 1930

WELNERT CR JR, MCMASTER JH, FERGUSON RT: Dynamic function of the human fibula. Am J Anat 138:145, 1973

WETZLER SH, ELCONIN DV: Calcification of the tibial

collateral ligament (Pellegrini-Stieda's disease). Am J Surg 27:245, 1935

WHITE JW: Torsion of Achilles tendon: Its surgical significance. Arch Surg 46:784, 1943

WILEY AM: Club foot. An anatomical and experimental study of muscle growth. J Bone Joint Surg 41-B:821, 1959

WILLARD DE F, NICHOLSON JT: Cyst of the semilunar cartilage. Am Surg 112:305, 1940

WILLIAMS AF: The formation of the popliteal vein. Surg Gynecol Obstet 97:769, 1953

WINCKLER G: Le nerf péronier accessoire profond. Etude d'anatomie comparée. Arch d'anat d'histol et d' embryol 18:181, 1934

WINCKLER G: Contribution à l'étude de l'innervation de la face dorsale des articulations du pied chez l'homme. Arch d'anat d'histol et d'embryol 23:39, 1937

WOLFE RD, COLOFF B: Popliteal cysts. An arthographic study and review of the literature. J Bone Joint Surg 54-A:1057, 1972

WOOD WL, ZLOTSKY N, WESTIN GW: Congenital absence of the fibula. Treatment by Syme amputation—indications and technique. J Bone Joint Surg 47-A:1159, 1965

WRIGHT DG, RENNELS DC: A study of the elastic properties of plantar fascia. J Bone Joint Surg 46-A:482, 1964

WYNNE-DAVIES R: Family studies and the cause of congenital club foot. Talipes equinovarus, talipes calcaneo-valgus and metatarsus varus. J Bone Joint Surg 46-B:445, 1964

INDEX

Fracture, of carpals (*continued*)
  of femur, 641; **641–645**
  of foot or ankle, 828; **829, 830**
  of forearm, 384; **385, 386**
  of hand, 454
  of humerus, 346, 375; **347**
  of leg, 772; **773, 829**
  marching, 829; **832**
  of menisci of knee, 766; **766**
  of patella, 746
  of pelvis, 623; **626**
  of scapula, 267
  of tarsals, 829
  of vertebral column, 130; **130–132**
Frohse, arcade of, 429
Funiculi, of spinal cord, 185
Fusion of carpal bones, 440
  spinal, disability resulting from, 115
  of tarsal bones, 771, 826

Ganglion(a), of spinal nerves, 188
  synovial, 21
Gangrene, following arterial ligation, 44; **45**
Genu recurvatum, 744
Girdle, pelvic, 619; **621, 627**
  shoulder, 259
Grafts, arterial, 46
  bone, 11
  cartilage, 12
  nerve, 63
  tendon, 38
Groove, intertubercular, variations, 318, 341; **318, 344**
Guyon's canal, 389, 472, 473

Hallux valgus, 808
Hand, *see also* Palm, 437
  anomalies, 460; **464–466**
  anomalous innervation of muscles, 522; 525*t*–526*t*
  arteries, 480, 511, 555; **481, 511, 556**
  bones and joints, 437, 454; **439–447, 449, 455–458, 460, 461**
  dorsum, 529; **521, 530, 532–536, 538, 556**
  extensor tendons in, 536, 551; **535, 536, 538, 540, 541**
  fascia, 466, 475, 500, 529, 536; **476–478, 534**
  fascial spaces, 500; **501**
  flexor tendons in, 489; **490, 492–494, 496, 497**
  infections of, 466, 504
  muscles, 498, 505, 514, 520; **506–509, 511, 513, 515–517, 519, 521**
  nerves, 468, 486, 526, 529, 550; **481, 511, 530, 532**
  palm, 466
  superficial veins and lymphatics, 467; **468, 469, 533, 534**
Hemimelia, 212

Hemipelvectomy, 625
Hemivertebrae, 118
Hernia, popliteal, 751
Hiatus, saphenous, 676; **676**
Hip, bones and joint, 619, 630, 646; **646, 648, 649**
  fracture, 623, 641; **626, 641–643**
  movements at, 656
  ossification at, 566; **566, 567**
  stability of, 652
Hood, extensor, 539; **461, 540, 541**
Hormones, and growth of bone, 8
Humerus, 341; **343–346**
  absence, 346
  blood supply, 347
  fracture, 346; **347**
  movements, 323
  ossification, 207, 209; **208–210**
  variations, 344
Humerus varus, 346
Humphry's ligament, 756
Hunter's canal, 686

Incisions, across joints, 67; **68**
Infections, of hand, 466, 491, 495
  of palmar spaces, 504
Ilium, 622; **621, 622**
Innervation, *see also* Nerve(s) *and* listings under various parts
  of blood vessels, 40; **35, 41**
  of bone, 5, 6
  cutaneous, 49, 67
   of arm, 349; **352**
   of back, 77; **79**
   of buttock, 661; **680**
   of foot, 739, 846; **739–741**
   of forearm, 388, 389
   of hand, 468; **389, 481, 530**
   of leg, 739; **739, 741**
   of lower limb, 596; **588, 589, 594–596**
   of pectoral region, 279
   of thigh, 679; **679, 680**
   of upper limb, 244; **247**
  of intervertebral disk, 111; **112**
  of joints, *see also* accounts of specific joints, 16; **17**
  of muscle, 26; **27**
  of muscle, *see also* accounts of individual muscles and nerves
   of arm, 351; **353, 359, 360**
   of back, 165
   of buttock, 671; **572*t*, 669, 673**
   of foot, 831, 846, 849; **572*t*, 847, 849**
   of forearm, 391; **393, 395, 397, 418, 421**
   of hand, 486, 512; **487, 513**
    anomalous, 522; **525–526*t***
   of leg, 788, 793; 512*t*, **777, 781, 793, 797**
   of lower limb, 599; 572*t*
   of pectoral region, 283; **299**
   of shoulder, 332